PRETREATMENT FACILITY INSPECTION

Third Edition

A Field Study Training Program

prepared by

California State University, Sacramento
College of Engineering and Computer Science
Office of Water Programs

in cooperation with the

D1568375

Industrial and Hazardous Waste Committee
California Water Environment Association

★★★★★★★★★★★★★★★★★★★★★★★★★★★★★★★★★★
Kenneth D. Kerri, Project Director
★★★★★★★★★★★★★★★★★★★★★★★★★★★★★★★★★★

for the

U.S. Environmental Protection Agency
Office of Water Enforcement and Permits
Assistance ID No. CT-901 589-01-0

2002

NOTICE

This manual is revised and updated before each printing based on comments from persons using the manual.

FIRST EDITION

    First Printing, 1988   5,000

SECOND EDITION

    First Printing, 1991   5,000

THIRD EDITION

    First Printing, 1996   4,000
    Second Printing, 2002   3,000

OPERATOR TRAINING MANUALS

OPERATOR TRAINING MANUALS AND VIDEOS IN THIS SERIES are available from the Office of Water Programs, California State University, Sacramento, 6000 J Street, Sacramento, CA 95819-6025, phone: (916) 278-6142, e-mail: wateroffice@csus.edu or FAX: (916) 278-5959.

1. *PRETREATMENT FACILITY INSPECTION,**

2. *OPERATION OF WASTEWATER TREATMENT PLANTS*, 2 Volumes,

3. *ADVANCED WASTE TREATMENT,*

4. *INDUSTRIAL WASTE TREATMENT*, 2 Volumes,

5. *TREATMENT OF METAL WASTESTREAMS,*

6. *OPERATION AND MAINTENANCE OF WASTEWATER COLLECTION SYSTEMS*, 2 Volumes,**

7. *SMALL WASTEWATER SYSTEM OPERATION AND MAINTENANCE*, 2 Volumes,

8. *WATER TREATMENT PLANT OPERATION*, 2 Volumes,

9. *SMALL WATER SYSTEM OPERATION AND MAINTENANCE,*

10. *WATER DISTRIBUTION SYSTEM OPERATION AND MAINTENANCE,* and

11. *UTILITY MANAGEMENT.*

* *PRETREATMENT FACILITY INSPECTION TRAINING VIDEOS.* This set of five training videos provides an excellent introduction to the knowledge, skills, and abilities needed by new pretreatment facility inspectors; current inspectors also will find helpful ideas to improve their performance. Each 30-minute video portrays real-world experiences and features inspectors of industrial pretreatment faclities performing actual tasks. The videos can be used alone or to enhance the training provided by the *PRETREATMENT FACILITY INSPECTION* training manual.

- MEETING THE GOAL TOGETHER — The pretreatment facility, the inspection program, responsibilities of inspectors and administrators, protection of the environment, and the importance of ethical performance.

- TAKING A CLOSER LOOK — How to conduct an inspection, scheduling, entering an industry for an inspection, level of inspection, after the walk-through, and report writing.

- STARTING AT THE SOURCE — How to inspect a metal finishing industry, on-site industrial inspections, pollution prevention, and data management.

- TAKING UP A COLLECTION — Reasons for sampling, preparation for sampling, proper sample collection and handling, transporting samples, and chain of custody.

- GOING WITH THE FLOW — Sampling and flow monitoring, instrumentation, and automatic samplers.

** Other materials and training aids developed by the Office of Water Programs to assist operators in improving collection system operation and maintenance and overall performance of their systems include:

1. *COLLECTION SYSTEMS: METHODS FOR EVALUATING AND IMPROVING PERFORMANCE.* This handbook presents detailed benchmarking procedures and worksheets for using performance indicators to evaluate the adequacy and effectiveness of existing O & M programs. It also describes how to identify problems and suggests many methods for improving the performance of a collection system.

2. *OPERATION AND MAINTENANCE TRAINING VIDEOS.* This series of six 30-minute videos demonstrates the equipment and procedures collection system crews use to safely and effectively operate and maintain their collection systems. These videos complement and reinforce the information presented in Volumes I and II of *OPERATION AND MAINTENANCE OF WASTEWATER COLLECTION SYSTEMS.*

PREFACE

The purposes of this pretreatment facility inspection home-study training program are to:

1. Develop new qualified pretreatment facility inspectors,

2. Expand the abilities of existing inspectors, permitting better service to both their employers, industry and the public, and

3. Prepare inspectors for civil service examinations.

To provide you with the knowledge and skills needed to inspect pretreatment facilities as efficiently and effectively as possible, experienced pretreatment facility inspectors, design engineers and administrators prepared the material in each chapter of this manual.

Pretreatment facilities vary from city to city and from region to region. The material contained in this program is presented to provide you with an understanding of the basic inspection aspects of pretreatment facilities. This information will help you inspect pretreatment facilities in a safe and efficient manner.

Pretreatment facility inspection is a rapidly advancing field. To keep pace with scientific and technological advances, the material in this manual must be periodically revised and updated. THIS MEANS THAT YOU, THE INSPECTOR, MUST RECOGNIZE THE NEED TO BE AWARE OF NEW ADVANCES AND THE NEED FOR CONTINUOUS TRAINING BEYOND THIS PROGRAM.

The Project Director is indebted to the many inspectors, engineers, administrators and other persons who contributed to this manual. Every effort was made to acknowledge material from the many available references in the pretreatment inspection field. Robert (Pete) Eagen served as the EPA Project Officer for the Office of Water Enforcement and Permits, and was a valuable resource, consultant and adviser. Keith Silva, EPA Region IX, provided encouragement and many helpful comments. Special thanks are due to Jay Kremer and Bill Garrett for their many contributions throughout this entire effort. Bill Garrett provided many photographs of inspectors at work in the field. Larry Hannah served as Educational Consultant. Illustrations were drawn by Martin Garrity. Special thanks are well deserved by the Program Administrator, Gay Kornweibel, who typed, administered the field test, managed the office, administered the budget, and did everything else that had to be done to complete this project successfully.

1988

KENNETH D. KERRI
PROJECT DIRECTOR

PREFACE TO SECOND EDITION

The Project Director wishes to extend a special word of thanks to Mr. Robert C. Steidel, City of Hopewell, Virginia, and Ms. Peg Hannah for their substantial contributions to the revisions in the second edition of this manual. Mr. Bill Garrett provided us with many excellent photographs.

1991

KENNETH D. KERRI
PROJECT DIRECTOR

PREFACE TO THIRD EDITION

Once again the Project Director wishes to extend his sincere thanks to Mr. Robert C. Steidel, City of Hopewell, Virginia, and Ms. Peg Hannah, CSUS Office of Water Programs, for their substantial contributions to the revision of this third edition.

1996

KENNETH D. KERRI
PROJECT DIRECTOR

OBJECTIVES OF THIS MANUAL

This manual is prepared to help pretreatment inspectors perform their jobs safely and fairly. Pretreatment inspectors have the responsibility of protecting municipal wastewater collection systems, treatment plants and the environment from the damage that may occur when industries that use hazardous or toxic materials discharge them into a wastewater collection system. This protection is achieved by regulating industrial or nondomestic dischargers who could potentially discharge toxic wastes or unusually strong conventional wastes into the collection system. These wastes could interfere with or pass through the treatment works. Pretreatment inspectors also prevent illegal discharges to storm drains which could harm the aquatic environment and contaminate public water supplies. There are five major problems that can be controlled through the proper implementation of a federal pretreatment program and a local source control program by trained and qualified pretreatment facility inspectors:

1. Interference with publicly owned treatment works (POTW) operations,
2. Pass-through of pollutants,
3. Contamination of municipal sludge,
4. Exposure of workers to chemical hazards, and
5. Contamination of public water supplies by illegal discharges into storm sewers.

Persons successfully completing this training program should have the necessary knowledge and skills to safely sample, monitor and inspect pretreatment facilities. They will be aware of their responsibilities to themselves, their employer, the public and the industries operating pretreatment facilities.

SCOPE OF THIS MANUAL

Inspectors with the responsibility of inspecting pretreatment facilities will find this manual very helpful. This manual contains information on:

1. The responsibilities of pretreatment facility inspectors,
2. The importance of pretreatment inspections,
3. The purposes of pretreatment inspection programs,
4. How to inspect a pretreatment facility,
5. The development of regulations,
6. Support items needed by pretreatment inspectors,
7. Safety in pretreatment inspection work,
8. Sampling procedures for wastewater,
9. Wastewater flow monitoring techniques,
10. Types of industrial wastes,
11. Pretreatment technology (source control),
12. Emergency response procedures,
13. The legal authority for and conceptual basis of regulations, and
14. The administration of pretreatment programs.

This manual describes situations typically encountered by pretreatment facility inspectors in most areas. By studying this manual, you will obtain the knowledge and skills you need to inspect pretreatment facilities and to analyze and solve pretreatment facility inspection problems when they occur. Pretreatment facility inspection procedures and techniques may vary with the age of the pretreatment facility, the extent and effectiveness of the facility's operation and maintenance program, and local conditions. *YOU WILL HAVE TO ADAPT THE INFORMATION AND PROCEDURES IN THIS MANUAL TO YOUR PARTICULAR SITUATION.*

Technology is advancing very rapidly in the field of pretreatment facility inspection. To keep pace with scientific advances, the material in this program must be periodically revisedand updated. This means that you, the pretreatment facility inspector, must be aware of new advances and recognize the need for continuous personal training reaching beyond this program. *TRAINING OPPORTUNITIES EXIST IN YOUR DAILY WORK EXPERIENCE, FROM YOUR ASSOCIATES, AND FROM ATTENDING MEETINGS, WORKSHOPS, CONFERENCES AND CLASSES.*

USES OF THIS MANUAL

This manual was developed to serve the needs of inspectors in several different situations. The format selected enables the manual to be used as a home-study or self-paced instruction course for inspectors unable to attend formal classes either due to shift work, personal reasons or the unavailability of suitable classes. This home-study training program uses the concept of self-paced instruction where you are your own instructor and work at your own speed. In order to certify that you have successfully completed this program, an objective test is included at the end of each chapter.

Also, this manual can serve effectively as a textbook in the classroom. Many colleges and universities have used this manual as a text in formal classes (often taught by operators or inspectors). In areas where colleges are not available or are unable to offer classes in pretreatment facility inspection, inspectors and utility agencies can join together to offer their own courses using the manual.

Cities or utility agencies can use the manual in several types of on-the-job training programs depending on the needs and situation of the agency. For example, in one type of program, a manual is purchased for each inspector. A senior inspector or a group of inspectors are designated as instructors. These inspectors help answer questions when the persons in the training program have questions or need assistance. The instructors grade the objective tests at the end of each chapter, record scores and notify California State University, Sacramento, of the scores when a person successfully completes this program. This approach eliminates any waiting while papers are being graded and returned by CSUS.

This manual was prepared to help people inspect pretreatment facilities. Please feel free to use the manual in the manner which best fits your training needs and the needs of other inspectors. We will be happy to work with you to assist you in developing your training program. Please feel free to contact

Project Director
Office of Water Programs
California State University, Sacramento
6000 J Street
Sacramento, California 95819-6025

Phone: (916) 278-6142

INSTRUCTIONS TO PARTICIPANTS
IN HOME-STUDY COURSE

Procedures for reading the lessons and answering the questions are contained in this section.

To progress steadily through this program, you should establish a regular study schedule. For example, many inspectors in the past have set aside two hours during two evenings a week for study.

The study material is contained in eleven chapters. Some chapters are longer and more difficult than others. For this reason, a few of the chapters are divided into two or more lessons. The time required to complete a lesson will depend on your background and experience. Some people might require an hour to complete a lesson and some might require three hours; but that is perfectly all right. *THE IMPORTANT THING IS THAT YOU UNDERSTAND THE MATERIAL IN THE LESSON!*

Each lesson is arranged for you to read a short section, write the answers to the questions at the end of the section, and check your answers against suggested answers. Then *YOU* decide if you understand the material sufficiently to continue or whether you should read the section again. You will find that this procedure is slower than reading a normal textbook, but you will remember much more when you have finished the lesson. Some discussion and review questions are provided following each lesson or at the end of the chapter. These questions review the important points you have covered in the lesson or chapter.

At the end of each chapter you will find an "objective test." Each test contains true or false, multiple-choice, fill-in-the-blank, or best answer types of questions. The purposes of this exam are to review the chapter and to give you experience in taking different types of exams. Mark your answers on the special answer sheet provided for each chapter. *MAIL TO THE PROGRAM DIRECTOR ONLY YOUR ANSWERS TO OBJECTIVE TESTS USING THE ANSWER SHEETS PROVIDED.* You do not have to mail any other test material.

In many of the chapters you will find one or more appendices following the objective test. The appendices contain materials such as sample wastewater discharge permit forms, an example of a typical Baseline Monitoring Report (BMR) form, and Environmental Protection Agency (EPA) guidance documents on various topics.

These materials have been included in response to many requests from pretreatment program managers and administrators for the latest EPA forms and regulations affecting pretreatment facility inspectors. *THE MATERIAL IN THE APPENDICES IS NOT REQUIRED READING TO PASS THIS COURSE, AND THE OBJECTIVE TESTS DO NOT COVER INFORMATION PRESENTED IN THE APPENDICES.*

The answer sheets you mail to the Program Director will be computer-graded in the following manner: true-false questions are considered as having only one possible answer so they have a value of 1 point. Multiple-choice questions, however, are worth 5 points since there may be more than one correct answer and all correct answers must be filled in for full credit. Best answer questions are worth 4 points. For the multiple-choice and best answer questions, the computer deducts a point for each correct answer *NOT* filled in and each *INCORRECT* response filled in. For example, if 1, 3 and 5 are the correct responses and you fill in 1 and 5, a point is deducted for failing to fill in 3. If you fill in 1, 3, 4 and 5, a point is deducted because 4 is filled in and should not be.

After you have completed the last objective test, you will find a final examination. This exam is provided for you to review how well you remember the material. You may wish to review the entire manual before you take the final exam. Some of the questions are essay-type questions which are used by some agencies for higher-level civil service examinations. After you have completed the final examination, grade your own paper and determine the areas in which you might need additional review before your next examination.

YOU ARE YOUR OWN TEACHER IN THIS PROGRAM. You could merely look up the suggested answers from the answer sheet or copy them from someone else, but you would not understand the material. Consequently, you would not be able to apply the material to the inspection of pretreatment facilities or recall it during an examination for certification or a civil service position.

YOU WILL GET OUT OF THIS PROGRAM WHAT YOU PUT INTO IT.

SUMMARY OF PROCEDURES

A. INSPECTOR (YOU)

1. Read what you are expected to learn in each chapter (the chapter objectives).

2. Read sections in the lesson.

3. Write your answers to questions at the end of each section in your notebook. You should write the answers to the questions just as you would if these were questions on a test.

4. Check your answers with the suggested answers.

5. Decide whether to reread the section or to continue with the next section.

6. Write your answers to the discussion and review questions at the end of each lesson in your notebook.

7. Mark your answers to the objective test on the answer sheet provided by the Project Director or by your instructor.

8. Mail material to the Project Director. (Send only your completed answer sheet.)

> Project Director
> Office of Water Programs
> California State University, Sacramento
> 6000 J Street
> Sacramento, California 95819-6025

B. PROJECT DIRECTOR

1. Mails answer sheet for each chapter to inspector.

2. Corrects tests, answers any questions, and returns results to inspector including explanations and solutions for missed questions.

C. ORDER OF WORKING LESSONS

To complete this program you will have to work all of the chapters. You may proceed in numerical order, or you may wish to work some chapters sooner. If your job stresses the collection of samples and the measurement of wastewater flows, you may decide to work on these chapters first.

SAFETY IS A VERY IMPORTANT TOPIC. Everyone inspecting pretreatment facilities must always be safety conscious. Inspectors daily encounter situations and equipment that can cause a serious disabling injury or illness if the inspector is not aware of the potential danger and does not exercise adequate precautions. For these reasons, you may decide to work on Chapter 5, "Safety in Pretreatment Inspection and Sampling Work," early in your studies. In each chapter, *SAFE PROCEDURES ARE ALWAYS STRESSED.*

D. COMPLETION

Following successful completion of this program, a Certificate of Completion will be sent to you. If you wish, the Certificate can be sent to your supervisor, the mayor of your town, or any other official you think appropriate. Some inspectors have been presented their Certificate at a City Council meeting, got their picture in the newspaper and received a pay raise. When you successfully complete the course, the Project Director will ask you how you would like your name to appear on your Certificate and whether you would like the Certificate sent directly to you or to your supervisor or other municipal official.

TECHNICAL CONSULTANTS

John Brady
Consultant
Red Bluff, California

Larry S. Hannah
Cal State Univ., Sacramento
Sacramento, California

Don Menno
Buffalo Sewer Authority
Buffalo, New York

Larry Bristow
Sacramento County
Sacramento, California

Steve Medbery
City & County of San Francisco
San Francisco, California

Bob Steidel
Hopewell Regional Wastewater
  Treatment Facility
Hopewell, Virginia

PROJECT REVIEWERS

Susan Adams
Texas Water Commission, District 14
510 South Congress, Suite 306
Austin, TX 78701

Fred D. Binkley
Pretreatment Coordinator
City and County of Denver
5100 Marion Street
Denver, CO 80216

Brian Carlisle
Drainage Engineer
Private Bag
Hamilton
New Zealand

Wen-Shi Cheung
City of Bakersfield
4101 Truxton Avenue
Bakersfield, CA 93309

Jack Ensminger, Coordinator
Environmental Affairs
City of Baton Rouge/EBRP
6144 Westridge Street
Baton Rouge, LA 70817

Julio S. Guerra
Industrial Waste Inspector
City of Merced
P.O. Box 2068
Merced, CA 95344

Leo Hermes
Manager of Industrial Wastes
Metropolitan Waste Control Commission 350 Metro Square Building
St. Paul, MN 55101

D.E. Hopkins
Water Quality Institute
Sumter Area Technical College
506 North Guignard Drive
Sumter, SC 29150

K.R. Jensen, Inspector
Industrial Waste Pretreatment
City of South Salt Lake
195 West Oakland Avenue
South Salt Lake, UT 84145

Sandra Kapstain
Sanitary District of Elgin
Raymond Street & Purify Drive
P.O. Box 92, Elgin, IL 60121-0092

Marv Lambert
Manager, Treatment Operations
Columbus City Utilities
123 Washington St., P.O. Box 170
Columbus, IN 47202-0170

Betty Meyer
Technical Services Division
Encina WPC Facility
6200 Avenida Encinas
Carlsbad, CA 92008-0171

Stephen E. Moehlmann
Executive Director, ABC
P.O. Box 786
Ames, IA 50010-0786

Larry Moon, Leader
Operations Certification Group
Ohio EPA
361 East Broad Street
Columbus, OH 43215

Al Pagorski, General Manager
Sanitary District of Elgin
Raymond Street & Purify Drive
P.O. Box 92
Elgin, IL 60121-0092

Vallana Piccolo
Industrial Waste Investigator
Municipality of Metropolitan
Seattle (METRO)
322 West Ewing Street
Seattle, WA 98119

Lee Powers
Industrial Waste Engineer
Metro St. Louis Sewer District
10 East Grand Avenue
St. Louis, MO 63147

Gary Pugh, Superintendent
Quality Control
Columbus City Utilities
123 Washington Street, PO Box 170
Columbus, IN 47202-0170

Bernard A. Rains, Director
Dept. of Environmental Compliance
Metro St. Louis Sewer District
10 East Grand Avenue
St. Louis, MO 63147

Clarence Scherer
Water and Environmental Technology
1012 Melody Lane
Amarillo, TX 79108

Paul Shaffer
56 North State Street
Orem, UT 84057

Joe H. Shockcor, PE
Consulting Engineer
Pomfret Road, Harding Place
P.O. Box 567
Woodstock, VT 05091

Keith Silva, Environmental Engineer
EPA Region IX
Water Management Division
215 Fremont Street
San Francisco, CA 94105

Raymond E. Stillwell
General Supervisor
Water Pollution Control Division
1001 North East Poplar
Topeka, KS 66616

Rich Thomasson, Head
Systems Maintenance Division
Washington Sub. Sanitary Comm.
4017 Hamilton Street
Hyattsville, MD 20781

Robert Townsend, PE
Senior Sanitary Engineer
Bureau of Wastewater Facilities
50 Wolf Road, Room 320
Albany, NY 12233-0001

Rubben Valenzuela, Supervisor
Industrial Waste Pretreatment
195 West Oakland Avenue
South Salt Lake, UT 84115

PRETREATMENT FACILITY INSPECTION*
COURSE OUTLINE

* Other similar courses in our training program that may be of interest to you are on treatment of metal wastestreams and industrial waste treatment. The contents of those training manuals are listed on the following two pages.

INDUSTRIAL WASTE TREATMENT, VOLUME I
COURSE OUTLINE

INDUSTRIAL WASTE TREATMENT, VOLUME II
COURSE OUTLINE

TREATMENT OF METAL WASTESTREAMS
COURSE OUTLINE

CHAPTER 1

THE PRETREATMENT FACILITY INSPECTOR

by

Patrick S. Kwok

TABLE OF CONTENTS

Chapter 1.   THE PRETREATMENT FACILITY INSPECTOR

Chapter 1.   THE PRETREATMENT FACILITY INSPECTOR

At the beginning of each chapter in this pretreatment facility inspector training manual you will find a list of *OBJECTIVES*. The purpose of this list is to stress those topics in the chapter that are most important. Contained in the list will be items you need to know and skills you must develop to inspect pretreatment facilities properly and safely.

Following completion of Chapter 1, you should be able to:

1. Describe a pretreatment facility,

2. Describe the purpose of a pretreatment facility inspection program,

3. Explain the importance of the work performed by pretreatment facility inspectors,

4. List the major duties performed by inspectors,

5. Explain how to become a pretreatment facility inspector, and

6. Find sources of further training information on how to do the jobs performed by inspectors.

PRETREATMENT WORDS
Chapter 1.   THE PRETREATMENT FACILITY INSPECTOR

At the beginning of each chapter in this manual there is a section which defines many of the words used by the pretreatment inspection profession. Reading this section now and understanding these words will help you better understand the material in the chapter. You may also wish to refer to these definitions when you encounter these words in the chapter. After some of the words you will find in parentheses a guide to pronouncing (pro-NOUN-sing) the words. The capitalized letters should be accented or spoken with slightly more emphasis than the other groups of letters. All of the words defined at the beginning of each chapter are listed at the end of this manual in the Appendix, "Pretreatment Inspection Words (Glossary)."

40 CFR 403   40 CFR 403

EPA's General Pretreatment Regulations appear in the **C**ode of **F**ederal **R**egulations under 40 CFR 403. 40 refers to the numerical heading for the environmental regulations portion of the **C**ode of **F**ederal **R**egulations (CFR). 403 refers to the section which contains the General Pretreatment Regulations. Other sections include 413, Electroplating Categorical Regulations.

ACEOPS   ACEOPS

See **A**LLIANCE OF **CE**RTIFIED **OP**ERATOR**S**, LAB ANALYSTS, INSPECTORS, AND SPECIALISTS (ACEOPS).

ACUTE HEALTH EFFECT   ACUTE HEALTH EFFECT

An adverse effect on a human or animal body, with symptoms developing rapidly.

ALLIANCE OF CERTIFIED OPERATORS,   ALLIANCE OF CERTIFIED OPERATORS,
  LAB ANALYSTS, INSPECTORS,   LAB ANALYSTS, INSPECTORS,
  AND SPECIALISTS (ACEOPS)   AND SPECIALISTS (ACEOPS)

A professional organization for operators, lab analysts, inspectors, and specialists dedicated to improving professionalism; expanding training, certification, and job opportunities; increasing information exchange; and advocating the importance of certified operators, lab analysts, inspectors, and specialists. For information on membership, contact ACEOPS, 1810 Bel Air Drive, Ames, IA 50010, phone (515) 663-4128 or e-mail: ACEOPS@aol.com.

ANAEROBIC (AN-air-O-bick)   ANAEROBIC

A condition in which atmospheric or dissolved molecular oxygen is *NOT* present in the aquatic (water) environment.

BOD (pronounce as separate letters)   BOD

See **B**iochemical **O**xygen **D**emand.

BIOCHEMICAL OXYGEN DEMAND (BOD)   BIOCHEMICAL OXYGEN DEMAND (BOD)

The rate at which organisms use the oxygen in water or wastewater while stabilizing decomposable organic matter under aerobic conditions. In decomposition, organic matter serves as food for the bacteria and energy results from its oxidation. BOD measurements are used as a measure of the organic strength of wastes in water.

BIOLOGICAL PROCESS   BIOLOGICAL PROCESS

A waste treatment process by which bacteria and other microorganisms break down complex organic materials into simple, nontoxic, more stable substances.

BIOSOLIDS   BIOSOLIDS

A primarily organic solid product, produced by wastewater treatment processes, that can be beneficially recycled. The word biosolids is replacing the word sludge.

CATEGORICAL STANDARDS   CATEGORICAL STANDARDS

Industrial waste discharge standards developed by EPA that are applied to the effluent from any industry in any category anywhere in the United States that discharges to a Publicly Owned Treatment Works (POTW). These are standards based on the technology available to treat the wastestreams from the processes of the specific industrial category and normally are measured at the point of discharge from the regulated process. The standards are listed in the Code of Federal Regulations.

CERTIFICATION EXAMINATION CERTIFICATION EXAMINATION

An examination administered by a state agency or professional association that pretreatment facility inspectors take to indicate a level of professional competence. Today pretreatment facility inspector certification exams may be voluntary and administered by professional associations. In some states inspectors are recognized as environmental compliance inspectors. Current trends indicate that more employers will encourage inspector certification in the future.

CHAIN OF CUSTODY CHAIN OF CUSTODY

A record of each person involved in the handling and possession of a sample from the person who collected the sample to the person who analyzed the sample in the laboratory and to the person who witnessed disposal of the sample.

CHEMICAL PROCESS CHEMICAL PROCESS

A waste treatment process involving the addition of chemicals to achieve a desired level of treatment. Any given process solution in metal finishing is also called a "chemical process."

CHRONIC HEALTH EFFECT CHRONIC HEALTH EFFECT

An adverse effect on a human or animal body with symptoms that develop slowly over a long period of time or that recur frequently.

CODE OF FEDERAL REGULATIONS (CFR) CODE OF FEDERAL REGULATIONS (CFR)

A publication of the United States Government which contains all of the proposed and finalized federal regulations, including environmental regulations.

CONSERVATIVE POLLUTANT CONSERVATIVE POLLUTANT

A pollutant found in wastewater that is not changed while passing through the treatment processes in a conventional wastewater treatment plant. This type of pollutant may be removed by the treatment processes and retained in the plant's sludges or it may leave in the plant effluent. Heavy metals such as cadmium and lead are conservative pollutants.

CONTAMINATION CONTAMINATION

The introduction into water of microorganisms, chemicals, toxic substances, wastes, or wastewater in a concentration that makes the water unfit for its next intended use.

CONVENTIONAL POLLUTANTS CONVENTIONAL POLLUTANTS

Those pollutants which are usually found in domestic, commercial or industrial wastes such as suspended solids, biochemical oxygen demand, pathogenic (disease-causing) organisms, adverse pH levels, and oil and grease.

EPA EPA

United States **E**nvironmental **P**rotection **A**gency. A regulatory agency established by the U.S. Congress to administer the nation's environmental laws. Also called the U.S. EPA.

EFFLUENT EFFLUENT

Wastewater or other liquid—raw (untreated), partially or completely treated—flowing *FROM* a reservoir, basin, treatment process, or treatment plant.

HARMFUL PHYSICAL AGENT HARMFUL PHYSICAL AGENT

Any chemical substance, biological agent (bacteria, virus or fungus) or physical stress (noise, heat, cold, vibration, repetitive motion, ionizing and non-ionizing radiation, hypo- or hyperbaric pressure) which:

(A) Is regulated by any state or federal law or rule due to a hazard to health;

(B) Is listed in the latest printed edition of the National Institute of Occupational Safety and Health (NIOSH) Registry of Toxic Effects of Chemical Substances (RTECS);

(C) Has yielded positive evidence of an acute or chronic health hazard in human, animal or other biological testing conducted by, or known to, the employer; or

(D) Is described by a Material Safety Data Sheet (MSDS) available to the employer which indicates that the material may pose a hazard to human health.

Also called a TOXIC SUBSTANCE. Also see ACUTE HEALTH EFFECT and CHRONIC HEALTH EFFECT. (Definition from California General Industry Safety Orders, Section 3204.)

HAZARDOUS WASTE   HAZARDOUS WASTE

A waste, or combination of wastes, which because of its quantity, concentration, or physical, chemical, or infectious characteristics may:

1. Cause, or significantly contribute to, an increase in mortality or an increase in serious irreversible, or incapacitating reversible illness; or

2. Pose a substantial present or potential hazard to human health or the environment when improperly treated, stored, transported, or disposed of or otherwise managed; and

3. Normally not be discharged into a sanitary sewer; subject to regulated disposal.

(Resource Conservation and Recovery Act (RCRA) definition.)

INFLUENT   INFLUENT

Wastewater or other liquid—raw (untreated) or partially treated—flowing *INTO* a reservoir, basin, treatment process, or treatment plant.

INSPECTOR   INSPECTOR

An Industrial Pretreatment Inspector is a person who conducts inspections of industrial pretreatment facilities to ensure protection of the environment and compliance with regulations adopted by the POTW in addition to the General and Categorical Pretreatment Regulations. Also called an Industrial Pretreatment (Waste) Inspector and a Pretreatment Inspector.

NIOSH (NYE-osh)   NIOSH

The **N**ational **I**nstitute of **O**ccupational **S**afety and **H**ealth is an organization that tests and approves safety equipment for particular applications. NIOSH is the primary federal agency engaged in research in the national effort to eliminate on-the-job hazards to the health and safety of working people. The NIOSH Publications Catalog, Sixth Edition, NIOSH Pub. No. 84-118, lists the NIOSH publications concerning industrial hygiene and occupational health. To obtain a copy of the catalog, write to National Technical Information Service (NTIS), 5285 Port Royal Road, Springfield, VA 22161. NTIS Stock No. PB-86-116-787, price, $103.50, plus $5.00 shipping and handling per order.

NPDES PERMIT   NPDES PERMIT

National **P**ollutant **D**ischarge **E**limination **S**ystem permit is the regulatory agency document issued by either a federal or state agency which is designed to control all discharges of pollutants from point sources and storm water runoff into U.S. waterways. NPDES permits regulate discharges into navigable waters from all point sources of pollution, including industries, municipal wastewater treatment plants, sanitary landfills, large agricultural feedlots and return irrigation flows.

O & M MANUAL   O & M MANUAL

Operation and **M**aintenance Manual. A manual that describes detailed procedures for operators to follow to operate and maintain a specific wastewater treatment or pretreatment plant and the equipment of that plant.

OSHA (O-shuh)   OSHA

The Williams-Steiger **O**ccupational **S**afety and **H**ealth **A**ct of 1970 (OSHA) is a federal law designed to protect the health and safety of industrial workers and also the operators and inspectors of pretreatment facilities. OSHA regulations require employers to obtain and make available to workers the Material Safety Data Sheets (MSDSs) for chemicals used at industrial facilities and treatment plants. OSHA also refers to the federal and state agencies which administer the OSHA regulations.

POTW POTW

Publicly **O**wned **T**reatment **W**orks. A treatment works which is owned by a state, municipality, city, town, special sewer district or other publicly owned and financed entity as opposed to a privately (industrial) owned treatment facility. This definition includes any devices and systems used in the storage, treatment, recycling and reclamation of municipal sewage or industrial wastes of a liquid nature. It also includes sewers, pipes and other conveyances only if they convey wastewater to a POTW treatment plant. The term also means the municipality (public entity) which has jurisdiction over the indirect discharges to and the discharges from such a treatment works. (Adapted from EPA's General Pretreatment Regulations, 40 CFR 403.3.)

pH (pronounce as separate letters) pH

pH is an expression of the intensity of the basic or acidic condition of a liquid. Mathematically, pH is the logarithm (base 10) of the reciprocal of the hydrogen ion activity.

$$pH = Log \frac{1}{[H^+]} \quad or = -Log\ [H^+]$$

The pH may range from 0 to 14, where 0 is most acidic, 14 most basic, and 7 neutral.

PHYSICAL WASTE TREATMENT PROCESS PHYSICAL WASTE TREATMENT PROCESS

Physical wastewater treatment processes include use of racks, screens, comminutors, clarifiers (sedimentation and flotation), and filtration. Chemical or biological reactions are important treatment processes, but *NOT* part of a physical treatment process.

POLLUTANT POLLUTANT

Any substance which causes an impairment (reduction) of water quality to a degree that has an adverse effect on any beneficial use of the water.

POLLUTANT POLLUTANT

Dredged spoil, solid waste, incinerator residue, sewage, garbage, sewage sludge, munitions, chemical wastes, biological materials, radioactive materials, heat, wrecked or discarded equipment, rock, sand, cellar dirt and industrial, municipal and agricultural waste discharged into water and onto the ground where subsurface waters can become contaminated by leaching. (Definition from Section 502(6) of the Clean Water Act.)

POLLUTION POLLUTION

The impairment (reduction) of water quality by agricultural, domestic, or industrial wastes (including thermal and radioactive wastes) to a degree that the natural water quality is changed to hinder any beneficial use of the water or render it offensive to the senses of sight, taste or smell or when sufficient amounts of wastes create or pose a potential threat to human health or the environment.

PRETREATMENT FACILITY PRETREATMENT FACILITY

Industrial wastewater treatment plant consisting of one or more treatment devices designed to remove sufficient pollutants from wastewaters to allow an industry to comply with effluent limits established by the US EPA General and Categorical Pretreatment Regulations or locally derived prohibited discharge requirements and local effluent limits. Compliance with effluent limits allows for a legal discharge to a POTW.

RCRA (RICK-ruh) RCRA

The Federal **R**esource **C**onservation and **R**ecovery **A**ct (10/21/76), Public Law (PL) 94-580, provides technical and financial assistance for the development of plans and facilities for recovery of energy and resources from discharged materials and for the safe disposal of discarded materials and hazardous wastes. This Act introduces the philosophy of the "cradle to grave" control of hazardous wastes. RCRA regulations can be found in the Code of Federal Regulations (40 CFR) Parts 260-268, 270 and 271.

RECEIVING WATER RECEIVING WATER

A stream, river, lake, ocean or other surface or groundwaters into which treated or untreated wastewater is discharged.

RECYCLE RECYCLE

The use of water or wastewater within (internally) a facility before it is discharged to a treatment system. Also see REUSE.

REPRESENTATIVE SAMPLE REPRESENTATIVE SAMPLE

A sample portion of material or wastestream that is as nearly identical in content and consistency as possible to that in the larger body of material or wastestream being sampled.

REUSE REUSE

The use of water or wastewater after it has been discharged and then withdrawn by another user. Also see RECYCLE.

SANITARY SEWER   SANITARY SEWER

A pipe or conduit (sewer) intended to carry wastewater or waterborne wastes from homes, businesses, and industries to the treatment works. Storm water runoff or unpolluted water should be collected and transported in a separate system of pipes or conduits (storm sewers) to natural watercourses.

SLUDGE (sluj)   SLUDGE

(1) The settleable solids separated from liquids during processing.

(2) The deposits of foreign materials on the bottoms of streams or other bodies of water.

SOLVENT MANAGEMENT PLAN   SOLVENT MANAGEMENT PLAN

A strategy for keeping track of all solvents delivered to a site, their storage, use and disposal. This includes keeping spent solvents segregated from other process wastewaters to maximize the value of the recoverable solvents, to avoid contamination of other segregated wastes, and to prevent the discharge of toxic organics to any wastewater collection system or the environment. The plan should describe measures to control spills and leaks and to ensure that there is no deliberate dumping of solvents. Also known as a Toxic Organic Management Plan (TOMP).

STANDARD METHODS   STANDARD METHODS

STANDARD METHODS FOR THE EXAMINATION OF WATER AND WASTEWATER, 20th Edition. A joint publication of the American Public Health Association (APHA), American Water Works Association (AWWA), and the Water Environment Federation (WEF) which outlines the accepted laboratory procedures used to analyze the impurities in water and wastewater. Available from Water Environment Federation, Publications Order Department, 601 Wythe Street, Alexandria, VA 22314-1994. Order No. S82010. Price to members, $164.25; nonmembers, $209.25; price includes cost of shipping and handling.

STORM SEWER   STORM SEWER

A separate pipe, conduit or open channel (sewer) that carries runoff from storms, surface drainage, and street wash, but does not include domestic and industrial wastes. Storm sewers are often the recipients of hazardous or toxic substances due to the illegal dumping of hazardous wastes or spills created by accidents involving vehicles and trains transporting these substances. Also see SANITARY SEWER.

TECHNOLOGY-BASED STANDARDS   TECHNOLOGY-BASED STANDARDS

Refers to discharge limits that a specific industry could economically achieve if they were to use the best pretreatment technology currently available, such as neutralization, precipitation, clarification, and filtration.

TOXIC   TOXIC

A substance which is poisonous to a living organism. Toxic substances may be classified in terms of their physiological action, such as irritants, asphyxiants, systemic poisons, and anesthetics and narcotics. Irritants are corrosive substances which attack the mucous membrane surfaces of the body. Asphyxiants interfere with breathing. Systemic poisons are hazardous substances which injure or destroy internal organs of the body. The anesthetics and narcotics are hazardous substances which depress the central nervous system and lead to unconsciousness.

TOXIC POLLUTANT   TOXIC POLLUTANT

Those pollutants or combinations of pollutants, including disease-causing agents, which after discharge and upon exposure, ingestion, inhalation or assimilation into any organism, either directly from the environment or indirectly by ingestion through food chains, will, on the basis of information available to the Administrator of EPA, cause death, disease, behavioral abnormalities, cancer, genetic mutations, physiological malfunctions (including malfunctions in reproduction) or physical deformations in such organisms or their offspring. (Definition from Section 502(13) of the Clean Water Act.)

TOXIC SUBSTANCE   TOXIC SUBSTANCE

Any chemical substance, biological agent (bacteria, virus or fungus), or physical stress (noise, heat, cold, vibration, repetitive motion, ionizing and non-ionizing radiation, hypo- or hyperbaric pressure) which:

(A) Is regulated by any state or federal law or rule due to a hazard to health;

(B) Is listed in the latest printed edition of the National Institute for Occupational Safety and Health (NIOSH) Registry of Toxic Effects of Chemical Substances (RTECS);

(C) Has yielded positive evidence of an acute or chronic health hazard in human, animal, or other biological testing conducted by, or known to, the employer; or

(D) Is described by a Material Safety Data Sheet (MSDS) available to the employer which indicates that the material may pose a hazard to human health.

Also called a HARMFUL PHYSICAL AGENT. Also see ACUTE HEALTH EFFECT and CHRONIC HEALTH EFFECT. (Definition from California General Industry Safety Orders, Section 3204.)

CHAPTER 1.   THE PRETREATMENT FACILITY INSPECTOR

Chapter 1 is prepared especially for new pretreatment facility inspectors and people interested in becoming inspectors. If you are an experienced inspector, you may find some new viewpoints in this chapter. First, we'll take a look at what a pretreatment facility is and does, and talk about why pretreatment programs are needed and what they accomplish. Later in the chapter you will find a detailed description of what pretreatment inspectors do and how you can become a pretreatment inspector.

1.0   THE PRETREATMENT FACILITY INSPECTION PROGRAM

1.00   What Is a Pretreatment Facility?

An industrial pretreatment facility consists of wastewater treatment processes designed to remove pollutants from wastestreams prior to discharge to the local sewer system. Pretreatment of wastes is the method by which an industry can comply with local waste discharge ordinances and also federal and state regulations. The sources, amounts and types of wastes generated at an industrial manufacturing or processing site depend on the age of the facility, raw materials used, production processes and ability to recover and recycle wastes. Some industries attempt to segregate the different wastestreams for treatment by controlling them at the source while others gather all wastewaters together for treatment at one central location. Physical, chemical and sometimes biological treatment processes are used to separate or remove pollutants from wastestreams. These treatment processes are closely controlled by operators to produce acceptable discharges.

To ensure that industrial discharges meet acceptable standards, pretreatment facility inspectors must inspect each treatment process or facility at any industrial site that discharges potentially toxic or hazardous wastewaters into sanitary sewers. The inspector accomplishes this by collecting *REPRESENTATIVE SAMPLES*[1] from the *EFFLUENTS*[2] of these treatment processes. The samples are then analyzed to ensure compliance with the pretreatment regulations and/or permit limits. Results from analyses of samples may be used to bill industries for wastewater collection and treatment service charges.

1.01   General Pretreatment Regulations

The first federal pretreatment regulations were established in 1972 under Section 307 (b) of the Clean Water Act (PL 92-500). This regulation set forth pretreatment standards mainly for the protection of publicly owned treatment works (*POTWs*).[3] In 1978 the U.S. Environmental Protection Agency developed the General Pretreatment Regulations which established mechanisms and procedures for use by state and local pretreatment programs. Most of these regulations controlled only adverse pH levels and *CONVENTIONAL POLLUTANTS*[4] such as oil and grease, biochemical oxygen demand (BOD), and suspended solids.

In 1976 the Natural Resources Defense Council sued the U.S. Environmental Protection Agency (EPA). As a result of this suit, EPA shifted its pretreatment focus from the control of conventional pollutants to *TOXIC POLLUTANTS*.[5] EPA then

---

[1] *Representative Sample.   A sample portion of material or wastestream that is as nearly identical in content and consistency as possible to that in the larger body of material or wastestream being sampled.*

[2] *Effluent.   Wastewater or other liquid—raw (untreated), partially or completely treated—flowing FROM a reservoir, basin, treatment process, or treatment plant.*

[3] *POTW.   **P**ublicly **O**wned **T**reatment **W**orks. A treatment works which is owned by a state, municipality, city, town, special sewer district or other publicly owned and financed entity as opposed to a privately (industrial) owned treatment facility. This definition includes any devices and systems used in the storage, treatment, recycling and reclamation of municipal sewage or industrial wastes of a liquid nature. It also includes sewers, pipes and other conveyances only if they convey wastewater to a POTW treatment plant. The term also means the municipality (public entity) which has jurisdiction over the indirect discharges to and the discharges from such a treatment works. (Adapted from EPA's General Pretreatment Regulations, 40 CFR 403.3.)*

[4] *Conventional Pollutants.   Those pollutants which are usually found in domestic, commercial or industrial wastes such as suspended solids, biochemical oxygen demand, pathogenic (disease-causing) organisms, adverse pH levels, and oil and grease.*

[5] *Toxic Pollutant.   Those pollutants or combinations of pollutants, including disease-causing agents, which after discharge and upon exposure, ingestion, inhalation or assimilation into any organism, either directly from the environment or indirectly by ingestion through food chains, will, on the basis of information available to the Administrator of EPA, cause death, disease, behavioral abnormalities, cancer, genetic mutations, physiological malfunctions (including malfunctions in reproduction) or physical deformations in such organisms or their offspring. (Definition from Section 502(13) of the Clean Water Act.)*

developed *CATEGORICAL STANDARDS*[6] which are *TECHNOLOGY BASED*[7] and limit the discharge of toxic pollutants to POTWs.

The objectives of the General Pretreatment Regulations (40 CFR 403.2) are to: (1) prevent the introduction into POTWs of pollutants which will interfere with the operation of the POTW, including interference with its use or disposal of municipal sludge; (2) prevent the introduction into POTWs of pollutants which pass through the treatment works into receiving waters or which might otherwise be incompatible with the treatment works; (3) improve opportunities to reclaim and recycle municipal and industrial wastewaters and sludges; and (4) reduce the health and environmental risk of pollution caused by the discharge of toxic pollutants to POTWs. One of the overall purposes of a pretreatment inspection program is to monitor industries' pretreatment facility operations and their self-monitoring programs.

The General Pretreatment Regulations have been amended several times[8] since this manual was first printed in 1988. Each time the regulations are amended, significant changes are enacted which affect the duties and responsibilities of the pretreatment inspector. It is important that you keep current on the regulations by reading the *FEDERAL REGISTER*, the Code of Federal Regulations (CFR 40), or by subscribing to a service that summarizes the EPA regulations.

1.02 Pretreatment Inspection

The General Pretreatment Regulations require the entity which is responsible for the operation of a POTW to obtain the proper legal authority and then to carry out all inspection, surveillance and monitoring procedures necessary to determine compliance or noncompliance with applicable pretreatment standards and requirements. The General Pretreatment Regulations require enforcement of this legal authority by making the approved program a part of the POTW's *NPDES PERMIT*.[9] When inspectors evaluate an industry's pretreatment program, information supplied by the industry may be considered but cannot be the only source of information used to determine compliance or noncompliance. On-site inspections and other methods are also required. Inspectors with proper credentials are authorized by local sewer-use ordinances and pretreatment regulations to enter any premises of an industrial discharger for the purposes of inspection, sampling, and examination of records required to be kept by the POTW, and to set up devices necessary to conduct sampling, monitoring and metering operations.

A good inspection program is vital to the success of the pretreatment program. The potential of an unannounced inspection at any time serves as an effective deterrent to noncompliant dischargers. Benefits from an inspection program include helping to maintain a good rapport with industry and creating both a cooperative and a regulatory presence with the operators and managers of industrial pretreatment facilities.

Table 1.1 lists the purposes of a pretreatment program and Table 1.2 lists the purposes of an inspection program. As a pretreatment inspector, you will be involved in both programs.

TABLE 1.1   PURPOSES OF A PRETREATMENT PROGRAM

- Regulate the disposal of industrial wastewater into the sanitary wastewater collection system.

- Protect the physical structures and the safety of operation and maintenance personnel of the wastewater system (collection and treatment).

- Protect the health and safety of the public and the environment.

- Achieve compliance with pretreatment regulations as required under the Federal General Pretreatment Regulations and Categorical Standards and local source control ordinances.

- Prevent illegal discharge of industrial pollutants into storm sewers.

---

[6] *Categorical Standards. Industrial waste discharge standards developed by EPA that are applied to the effluent from any industry in any category anywhere in the United States that discharges to a Publicly Owned Treatment Works (POTW). These are standards based on the technology available to treat the wastestreams from the processes of the specific industrial category and normally are measured at the point of discharge from the regulated process. The standards are listed in the Code of Federal Regulations.*

[7] *Technology-based Standards. Refers to discharge limits that a specific industry could economically achieve if they were to use the best pretreatment technology currently available, such as neutralization, precipitation, clarification, and filtration.*

[8] *There have been a series of minor amendments to 40 CFR 403 as a result of clarification and the final sewage sludge regulations, 40 CFR 503.*

[9] *NPDES Permit. National Pollutant Discharge Elimination System permit is the regulatory agency document issued by either a federal or state agency which is designed to control all discharges of pollutants from point sources and storm water runoff into U.S. waterways. NPDES permits regulate discharges into navigable waters from all point sources of pollution, including industries, municipal wastewater treatment plants, sanitary landfills, large agricultural feedlots and return irrigation flows.*

TABLE 1.2 PURPOSES OF A PRETREATMENT INSPECTION PROGRAM

- Provide a mechanism to establish or verify service charges to industrial sewer users.

- Provide a mechanism to check the completeness and accuracy of industrial users' performance and compliance records.

- Provide a mechanism to verify the accuracy of self-monitoring reports and compliance with conditions of a wastewater discharge permit or contract.

- Provide a mechanism to verify the completeness and accuracy of information contained in the initial monitoring reports and permit applications.

- Provide a mechanism to ensure there is no bypass of sampling locations or pretreatment facilities.

- Verify sampling locations to ensure samples collected are representative of what is being discharged and determine if dilution streams are present.

- Provide a mechanism to assess compliance with a hazardous materials storage ordinance or RCRA[10] requirements and the effectiveness of the spill prevention plan.

- Provide a mechanism to evaluate the effectiveness of an industry's operations and maintenance manual and the availability of qualified personnel necessary to adequately operate and maintain the pretreatment facilities.

- Provide a mechanism to verify that wastewaters forbidden from being discharged into sanitary sewers are not being "bled in" with acceptable discharges.

- Explain to industries the requirements to comply with regulations.

- Provide a mechanism to establish and maintain communication with regulatory agencies and industrial representatives.

- Provide for the beneficial use of POTW sludge.

- Maintain and provide an accurate and concise written record on each industrial discharger for the POTW's file.

QUESTIONS

Below are some questions for you to answer. You should have a notebook in which you can write the answers to the questions. By writing down the answers to the questions, you are helping yourself learn and retain the information. After you have answered all the questions, compare your answers with those given in the Suggested Answers section on page 23. Reread any sections you do not understand and then proceed to the next section. You are your own teacher in this training program, and *YOU* should decide when you understand the material and are ready to continue with new material.

1.0A   The sources, amounts and types of wastes generated at an industrial manufacturing site depend on what factors?

1.0B   What do the letters POTW mean?

1.0C   As a result of a 1976 Natural Resources Defense Council suit brought against EPA, EPA shifted its pretreatment focus from conventional pollutants to what type of pollutants?

1.0D   What are categorical standards?

1.0E   What is the value of a potential unannounced inspection?

1.1 PURPOSE OF INDUSTRIAL WASTE PRETREATMENT PROGRAMS

In years past, the primary function of an industrial waste regulatory program was to ensure that the POTW agency's capital facilities, such as sewers and treatment plants, were not damaged by industrial wastes. The role of industrial waste pretreatment programs has expanded in recent years, however, to include missions beyond the simple protection of facilities. Today the main purpose of an industrial waste pretreatment program includes protecting the environment, the POTW agency's facilities and personnel, and the local community from adverse effects due to industrial waste discharges.

1.10 Protection of Capital Facilities

The protection of capital facilities is still an important goal of industrial waste pretreatment programs. Industrial waste can damage a POTW agency's collection system facilities by clogging the sewers and by causing corrosion of the sewers. Discharge of flammable wastes can cause explosions and other harm in the collection system. Discharge of excessive quantities of wastes to overloaded sewers can cause overflows of wastewater and endanger public health and receiving waters. Sewers can be misused; examples of misuse include the discharge of storm drainage water to sanitary sewers, or the discharge of sanitary or industrial wastes to storm sewers. Treatment plant facilities of an agency can also be exposed to corrosive materials, explosions, treatment process upsets, over-capacity effects caused by slug loadings, and misuse of the treatment processes and facilities through the discharge of untreatable industrial wastes.

Excessively low or high pH wastes can cause corrosion of the structures and upsets of the biological treatment processes of a treatment plant. Flammable wastes, such as gasoline, benzene or other chemicals, can cause explosions at treatment plants or in pumping stations as well as in sewers. Toxic wastes, such as those containing heavy metals or toxic organics, can cause an upset of the activated sludge process or other biological treatment processes. The overloading of a treatment plant by solids and biochemical oxygen demand during canning season in a semi-rural area can create excessive use of treatment plant and sewer facilities if there is not adequate treatment capacity or process flexibility. This situation may require that industrial wastes be discharged to the sewers during times of low flows in the sewers and at the treatment plant in order to equalize the distribution of organic and hydraulic loadings at the treatment plant.

The equipment and facilities of a POTW represent an enormous investment of public funds that must be protected from abuse. In a broad sense, then, all the regulations that spell out how industries may dispose of wastes are designed, at least in part, to protect the public's investment in their wastewater collection and treatment systems.

1.11 Protection of Agency Personnel

The protection of POTW agency personnel (including collection system personnel) is becoming more and more important. As more is being learned about the adverse effects of industrial chemicals, POTW employees are becoming increasingly concerned about adverse effects upon their health. The POTW agency must adequately regulate industrial wastes so that treatment plant and sewer maintenance personnel cannot, for example, be exposed to possible dangers of liver cancer from excessive discharges of vinyl chloride or be subject to other illnesses that might result from excessive exposure to toxic industrial wastes. Corrosion of sewers and structures may also be dangerous to agency personnel since severe structural problems can make it hazardous for people to work in treatment plant and sewer facilities. Also corrosion damage to facilities can require significant capital outlay for repair of the facilities. Discharges of some wastes, such as cyanide wastes from metal platers, may create toxic atmospheric conditions in sewer systems which could be fatal to collection system operators. Even the excessive discharge of nontoxic food processing wastes can lead to *ANAEROBIC*[11] conditions in sewers which may generate toxic hydrogen sulfide gas.

In addition to protecting operators, the industrial waste pretreatment program should also ensure that industrial waste inspectors are properly protected during their inspections of industrial plants. Inspectors should be provided safe places to collect wastewater samples and to review whether the industrial company is adequately meeting its discharge requirements.

1.12 Protection of the Community

An increasingly important aspect of the industrial waste pretreatment program is the protection of the local community. Citizens are becoming more aware of hazardous and obnoxious problems originating from industrial wastes. Of increasing concern is the unknown hazard which may result from unseen and unknown industrial wastes. For example, accidental spills of gasoline can create potentially explosive conditions throughout a community.

In a recent problem at a hazardous waste landfill in Southern California, citizens became quite concerned when they suspected that vinyl chloride gases might be entering their homes through plumbing vents. The liquids that leached out of the soil at the nearby landfill were being disposed of through the community sewer system. Residents of the area suspected that vinyl chloride gases (part of the leachate) were rising through the sewer plumbing and vents and contaminating their homes. While the low concentrations of vinyl chloride gas existing in the landfill leachate may not have been a realistic hazard, it created such a community concern that the discharge of the leachate to the sewer system was essentially prohibited.

Citizens have the right to expect safe and reasonably pleasant service by a collection system. This service would provide rigorously for community safety and prevent any obnoxious problems from regulated wastewaters. Many industrial wastewaters contain significant quantities of odorous wastes such as hydrogen sulfide and other noxious materials. The proper regulation of such materials to ensure odor control and safety is becoming increasingly important as we try to maintain positive public relations with the community we serve.

1.13 Protection of the Environment

The protection of the environment has become a prime role of the industrial waste pretreatment program since the Federal Water Pollution Control Act of 1972. The General Pretreatment Regulations in 40 CFR 403.8 require the POTW to develop a pretreatment program. Also, local conditions may cause a POTW to develop a locally controlled and regulated pretreatment program.

The purpose of the EPA's industrial categorical regulations is essentially to prevent pollutants generated at industrial sites from passing through the POTW's collection and treatment systems into the environment either through the sludges produced by the plant or in the plant effluent. The program establishes rigorous, technology-based standards (discharge limits) for each specific type of industrial waste discharger. The local POTW agency is given the primary responsibility under EPA regulations to implement the industrial categorical program, enforce compliance by industries and thus safeguard receiving water quality against degradation from industrial wastewaters, and to prevent the production of toxic sludges due to industrial wastes.

The disposal of the solid sludge residue from the treatment processes must also be regulated to prevent environmental contamination. Some substances, such as cadmium and lead, are not changed while passing through the treatment processes. These *CONSERVATIVE POLLUTANTS*[12] are, to a large extent, removed by the POTW treatment processes and retained in the wastewater treatment plant's sludges. Disposal of these toxic sludge wastes in a way that does not endanger the environment has become a serious and costly problem for POTW agencies.

---

[11] *Anaerobic (AN-air-O-bick).   A condition in which atmospheric or dissolved molecular oxygen is NOT present in the aquatic (water) environment.*

[12] *Conservative Pollutant.   A pollutant found in wastewater that is not changed while passing through the treatment processes in a conventional wastewater treatment plant. This type of pollutant may be removed by the treatment processes and retained in the plant's sludges or it may leave in the plant effluent. Heavy metals such as cadmium and lead are conservative pollutants.*

1.14 Compliance Assistance and Enforcement of Regulations

Enforcement of federal, state and local waste discharge regulations is a major objective of industrial waste pretreatment programs. The best way to accomplish this objective is through frequent compliance assistance to each industrial company. By helping an industry comply with the regulatory goals, you may reduce or eliminate major enforcement problems. The EPA program requires that POTW agencies notify and help industrial companies to interpret and implement the federal regulations. Many companies with limited technical resources might not be knowledgeable about federal and local industrial waste regulations, especially in the early stages of a new program. Local POTW agency personnel are frequently the most knowledgeable people available and can explain the need for and the proper methods of implementing the regulations. Thus, the offering of POTW agency assistance, as time permits, to industrial companies to smooth their way toward compliance is recommended as the best and most effective method to obtain compliance.

Private consultants are available to industrial companies and should be the technical help used by such companies if the problem is even slightly complex. Whenever possible, however, POTW agency personnel should review the proposals made by consultants. The POTW agency may even review and make recommendations regarding the consultant's work prior to official submittals for approval; this is frequently helpful in accomplishing the regulatory tasks in a timely manner. However, a fine line must be drawn between the POTW assisting an industry and the agency becoming a consultant. The POTW should never place itself in the position of recommending a solution, having it fail to achieve compliance and then having the industry respond that "we did as you recommended."

Again, it is highly recommended that your agency assume a cooperative attitude with industrial companies. It is much easier to accomplish your regulatory goals through cooperation with industry personnel than it is through enforcement actions. The hiring of qualified and capable industrial waste personnel by the POTW agency is a worthwhile investment for the community. Well qualified industrial waste personnel can provide valuable compliance assistance to industry and measurably increase the overall compliance for the community.

A far less desirable alternative, of course, is to spend a significant part of your resources enforcing compliance schedules and levying and collecting fines for violations that occur. When violations do occur, enforcement actions must be consistent with the Enforcement Response Plan which the 1990 amendments to the General Pretreatment Regulations require each POTW to develop.

1.15 Pollution Prevention — Right From the Start

1.150 The Pollution Prevention Strategy

Pollution prevention is a very important part of a pretreatment inspector's job. Pollution prevention is a preventive approach to the protection of human health and the environment. The focus of pollution prevention is on minimizing the amount of waste generated as opposed to treating and recycling wastes after generation. The Pollution Prevention Act of 1990 establishes the Pollution Prevention Hierarchy (Strategy) as a national policy, declaring that:

- Pollution should be prevented or reduced at the source whenever feasible;

- Pollution that cannot be prevented should be *RECYCLED*[13] in an environmentally safe manner whenever feasible;

- Pollution that cannot be prevented or recycled should be treated in an environmentally safe manner whenever feasible; and

- Disposal or other release into the environment should be used only as a last resort and should be conducted in an environmentally safe manner.

Pollution prevention generally encompasses the following areas of endeavor:

- Process modification, material substitution and product reformulation;

- Improved process operation and maintenance;

- Good operating practices (good housekeeping); and

- Material recycle, *REUSE*[14] and recovery for in-process use.

Waste minimization is a related term primarily used in connection with hazardous wastes. Waste minimization is the reduction of hazardous waste that is generated or subsequently treated, stored or disposed of. It includes any activity that results in the reduction of the total volume or quantity of hazardous waste, or the reduction of toxicity of the hazardous waste. The goal is to minimize present and future threats to human health and the environment. Under waste minimization, activities are distinguished between source reduction and recycling. Source reduction is the reduction or elimination of the generation of waste at the source, usually within a production process. Recycling is the reuse of materials in the original process or in another process, or the reclamation of materials from a waste for resource recovery or by-product production.

1.151 Switching From Treatment to Prevention

Industrial waste treatment and pretreatment practices focus on the reduction of pollutant concentrations in wastewater discharged from an industrial facility to minimize its impact on the wastewater treatment plant or the environment. Treatment is basically a remedial process that satisfies health and safety concerns. However, since many waste treatment processes

---

[13] *Recycle. The use of water or wastewater within (internally) a facility before it is discharged to a treatment system. Also see REUSE.*
[14] *Reuse. The use of water or wastewater after it has been discharged and then withdrawn by another user. Also see RECYCLE.*

create residual sludge that needs to be landfilled, treatment of industrial wastewater may simply be shifting pollutants from one medium of the environment to another with no net reduction. Moreover, additional reduction of toxic pollutants discharged to the wastewater collection system may be required in the future to meet more stringent regulations for sludge reuse and disposal, reclaimed water recharge to groundwater, and health risk reductions from volatile organic emissions at municipal wastewater treatment plants.

Pollution prevention is a complementary tool for reducing the discharge of pollutants to a wastewater collection system or the environment. It is not new, and was applied intensively in the 1970s to meet local discharge limits without the installation of pretreatment equipment. The 1980s saw a focus on pretreatment because of the EPA Categorical Pretreatment regulations. However, in the 1990s, pollution prevention is once again the preferred choice for source control and waste management. It has the potential of minimizing or even eliminating a particular pollutant from the wastestream. Good operating practices could minimize the volume of waste entering the treatment system. Material substitution could eliminate a pollutant altogether, and a process change could eliminate a whole wastestream.

Pollution prevention programs are now being implemented in many types of industrial facilities, regardless of whether they generate only conventional pollutants or toxic/hazardous wastes as well. Food processors, paper manufacturers and textile dyers have reduced wastewater flow and minimized the amounts of conventional pollutants like suspended solids and chemical oxygen demand (COD) discharged to the sewer. The incentive is for water conservation and cost savings in sewer service charges. Facilities undergoing expansion have an additional incentive to institute pollution prevention. Expanded industrial capacity often creates a need for greater wastewater treatment plant capacity. Construction of the added treatment capacity will usually be financed (at least in part) by higher waste load charges. If the expanding industrial facility is able to increase its size or production level while reducing the waste load it generates by implementing a pollution prevention program, then its waste treatment costs could decline and additional treatment plant capacity may not be needed.

1.16 Collection of Adequate Revenue

Another important goal of many industrial waste programs is the collection of required or appropriate revenue for the program from industrial companies. Such revenue should at least reimburse the costs of an industrial waste pretreatment inspection program and any extra processing costs incurred at the POTW required for treatment of the industrial wastes. As part of the EPA's industrial waste regulatory program, companies must now pay their fair share of the costs of a POTW's wastewater treatment. This fair share has been defined in various EPA regulations, but basically it requires industrial companies to pay their pro rata (proportional) share of the annual operation and maintenance costs based upon wastewater

strength and volume. Industrial companies should also pay their proportional share of the capital recovery (including construction) costs of the POTW agency through sewer connection fees. Many industrial waste programs devote significant personnel and resources to the determination and collection of adequate revenue from industrial companies each year.

QUESTIONS

Write your answers in a notebook and then compare your answers with those on page 23.

1.1A What are the main purposes of an industrial waste pretreatment program?

1.1B What are the potential adverse effects of industrial chemicals on POTW agency employees?

1.1C Why must the solid sludge residue from wastewater treatment processes be properly disposed of?

1.1D How can a POTW agency help industrial companies?

1.2 NEED FOR PRETREATMENT FACILITY INSPECTORS

People need and want a pleasant and healthy environment that is free from *POLLUTANTS*[15] and *TOXIC SUBSTANCES*[16] today and in the future. Pollutants can enter the environment from many different sources. Solid, liquid or gaseous wastes containing toxic substances can be disposed of on land, released into the atmosphere as emissions from stacks or exhausts, discharged into sanitary sewers and ultimately surface waters, or discharged directly to surface waters. The original sources of these wastes might be from industrial processes or waste disposal techniques, residential and commercial dwellings, surface runoff from streets and highways, irrigation return flows, chemical spills, or illegal dumpings. As discussed earlier in this chapter, the discharge of hazardous wastes to sanitary sewers can create corrosive, explosive, flammable and toxic conditions in wastewater collection systems and treatment plants. These conditions can severely damage the wastewater collection and treatment facilities and can harm the operators of the facilities.

In 1981 a series of violent explosions occurred in a major section of the Louisville, Kentucky, wastewater collection system. These explosions resulted in severe damage to the sewer system, left gaping holes in city streets and damaged buildings over a wide area (Figure 1.1). The explosion was caused by the discharge into the sewers of a highly volatile industrial solvent by a soybean extraction plant. After entering the sewer system the solvent vaporized, creating an explosive atmosphere which was subsequently ignited.

Even before implementation of the U.S. Environmental Protection Agency's General Pretreatment Regulations and industrial waste source control regulations, most industries were subject to local ordinances. The effective treatment of domestic wastewaters, for which the wastewater treatment plants

---

[15] *Pollutant. Any substance which causes an impairment (reduction) of water quality to a degree that has an adverse effect on any beneficial use of the water.*

[16] *Toxic Substance. Any chemical substance, biological agent (bacteria, virus or fungus), or physical stress (noise, heat, cold, vibration, repetitive motion, ionizing and nonionizing radiation, hypo- or hyperbaric pressure) which: (A) Is regulated by any state or federal law or rule due to a hazard to health; (B) Is listed in the latest printed edition of the National Institute for Occupational Safety and Health (NIOSH) Registry of Toxic Effects of Chemical Substances (RTECS); (C) Has yielded positive evidence of an acute or chronic health hazard in human, animal, or other biological testing conducted by, or known to, the employer; or (D) Is described by a Material Safety Data Sheet (MSDS) available to the employer which indicates that the material may pose a hazard to human health. Also called a harmful physical agent. (Definition from California General Industry Safety Orders, Section 3204.)*

Fig. 1.1   Results of an explosion in a sewer
(Copyright © 1981. The Courier-Journal. Reprinted with permission.)

(POTWs) were primarily built, was seriously hindered by discharges of toxic substances to the sewer system. This sometimes resulted in an even greater discharge of pollutants to the environment. Toxic substances and partially treated wastewater sometimes reached and polluted the receiving waters, killing fish and other aquatic organisms and generally making the water unsafe for a drinking water supply and most other beneficial uses. Treatment plant sludges that were applied to land, incinerated, or landfilled contained toxic pollutants in amounts and concentrations which caused environmental damage.

Following the enactment of EPA's General Pretreatment Regulations, state regulations and local source control ordinances, municipalities and local sanitation districts employed more pretreatment facility inspectors to enforce the regulations and ordinances that applied to industrial dischargers. By limiting the types and amounts of substances that could be discharged into sanitary sewers, these regulations and ordinances shifted some responsibility for protecting the environment to the industries creating the pollution. Effective enforcement helps protect wastewater collection systems and treatment facilities from corrosive chemicals; it protects biological wastewater treatment processes from upset by toxic substances and excessive oxygen demand; it protects the quality of sludge and ash produced at the treatment plant (POTW); it protects aquatic life in receiving waters; it preserves groundwater purity; and effective enforcement of discharge regulations protects the operators of collection and treatment facilities from harmful substances. Pretreatment regulations have resulted in treatment plant sludges (*BIOSOLIDS*[17]) being used in an environmentally beneficial method, for example, as a soil amendment in land applications.

To achieve these goals of protecting the public and the environment, it has become necessary to enact and enforce far-reaching legislation such as the Clean Water Act and the Water Quality Act. And it is here, in the gap between requirements of the law and compliance by industries, that the pretreatment inspector performs his or her job. Clearly, an inspector must wear many hats. A pretreatment facility inspector conducts inspections of industrial manufacturing and pretreatment facilities to help ensure protection of the environment and compliance with pretreatment regulations. The inspector may also be a pretreatment *PROGRAM* inspector and be responsible for ensuring compliance with the program rules and regulations. Where pretreatment facilities are not installed, the inspector also visits those industries that generate

wastewater, store raw and waste residuals, and discharge wastewater to a POTW sewer. An inspector is a regulator who must be capable of explaining and enforcing the ordinance requirements and federal and state regulations to industry within the scope of the law.

If an industry violates waste discharge regulations, there could be court action to order the industry to meet the effluent limits by pretreatment, a significant fine or the industry's operation could be curtailed or shut down. The inspector's role is to work closely with the POTW's legal staff to prevent illegal dumping of industrial wastes and to gather evidence for the prosecution of illegal discharges. To do this, an inspector must be knowledgeable in many areas. For example, the inspector must fully understand the regulations that apply to the particular industry, how to gather samples that will be legally admissible as evidence of violations, and what enforcement actions may be taken in accordance with the POTW's Enforcement Response Plan.

The skill of the pretreatment inspector in walking the fine line between knowledgeable, cooperative advisor and relentless enforcer of the law will often determine whether compliance is achieved quickly and in a cost-effective manner or through expensive court battles and further restrictive legislation. The public needs skilled and dedicated inspectors who can meet the challenges of this demanding but satisfying profession.

QUESTIONS

Write your answers in a notebook and then compare your answers with those on page 23.

1.2A   List the sources from which pollutants can enter the environment.

1.2B   How did the enactment of the EPA's General Pretreatment Regulations shift responsibility for protecting the environment to industries?

1.3   HOW TO BE AN INSPECTOR

1.30   Who Does a Pretreatment Inspector Work For?

Pretreatment inspectors work for a local control authority or wastewater treatment plant (POTW) that is operated by a department of public works or water pollution control agency, for a municipality, a local sanitation district or a specific public authority. Inspectors work for a public agency that relies on inspectors to enforce industrial pretreatment regulations. They are the "watchdogs" for public agencies that must protect the public and the environment from unacceptable wastewater discharges.

1.31   Duties and Responsibilities of an Inspector
(Figures 1.2 and 1.3)

The primary role of a pretreatment facility inspector is to enforce the regulations intended to protect publicly owned treatment works (POTWs) and the environment. A pretreatment facility inspector conducts inspections of industrial manufacturing and pretreatment facilities to ensure protection of the wastewater collection system, treatment facilities, personnel and the environment. This inspection includes an investi-

---

[17] *Biosolids.   A primarily organic solid product, produced by wastewater treatment processes, that can be beneficially recycled. The word biosolids is replacing the word sludge.*

Fig. 1.2   "Plate-out" unit being examined by inspector (in hard hat) and industrial contact.
Unit is on production area floor and receiving concentrated first rinse from plating operation.
Metal ions are electroplated onto bars to remove them from solution.

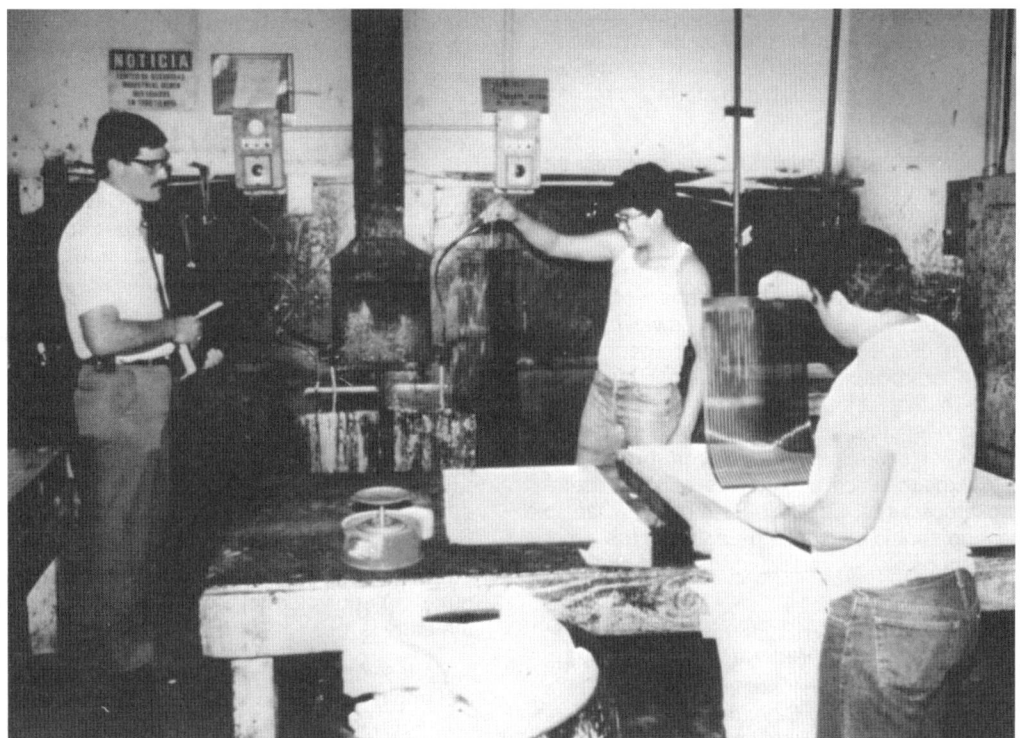

Fig. 1.3   Inspector checking production area for spill containment, proper plumbing of
wastewaters to treatment system, existence of bypasses, one-pass cooling water, uses of
chemicals, and processes not listed on discharge permit.

gation of the pretreatment facilities, raw and spent material storage areas, and also areas of the industrial facility where wastewaters (including sanitary) are generated and discharged to the sanitary sewer but no treatment exists. The inspector determines whether a pretreatment facility is properly operated and maintained and ensures that toxic pollutants in concentrations above permitted amounts do not enter the sanitary sewer system. Pretreatment inspectors also coordinate compliance efforts with industries. Protection is achieved when dischargers comply with the regulations governing the disposal of pollutants and toxic substances.

Some people view inspectors as "ambassadors" and others consider inspectors "sewer cops." Inspectors are regulators who must enforce the rules and regulations consistently, fairly, honestly and in accordance with the law. Industry relies on inspectors to be fair and to require the same level of regulatory compliance from all dischargers.

The actual job specifications for a pretreatment inspector may include: inspect and investigate industrial waste control, pretreatment and treatment facilities; investigate industrial and commercial firms to determine the nature of business, the wastewater generating processes, and the magnitude, type and strength of wastes produced; collect samples and measure flows from industrial wastewater discharges; analyze the adequacy of the wastewater pretreatment processes, determine compliance with legal requirements, discuss violations with the firm's management, and seek corrections; investigate POTW plant upsets and sewer problems such as odors and overflows to determine if the problems are of an industrial origin; conduct field tests to determine sources of offending discharges; make recommendations and prepare evidence for legal prosecution of violations; prepare enforcement actions; maintain records; and write reports.

The industrial waste inspector may also be required to discuss with and instruct companies in the preparation of discharge permit applications and to explain regulations. Additional duties not directly related to POTW operation, such as inspection and investigation of discharges to storm drains and channels, are also sometimes included.

If an industrial chemical spill occurs, an inspector may be required to decide whether the liquid spilled can be disposed of through a sanitary sewer or a storm sewer or temporarily contained by sandbags or other means prior to proper cleanup and disposal as a hazardous waste. In some states an inspector may have the authority to approve the disposal of a spill into a sanitary sewer, but the use of a storm sewer may require approval from a different pollution control agency. Extreme care must be exercised by a pretreatment inspector when evaluating alternatives and selecting the most appropriate alternative. Industries, state officials, and local fire departments rely on an inspector's expertise to interpret rules and regulations and to explain which alternatives will be acceptable spill cleanup procedures. The inspector is expected to know how to and who should evacuate the public in the event of a spill, who is responsible for controlling the spill and who can safely clean it up.

Inspectors also have an obligation to interpret and explain to industrial corporate management, city councils, and sanitation district boards of directors WHY compliance with EPA's General Pretreatment Regulations, Categorical Pretreatment Standards, and local source control ordinances is vital to the preservation of water resources and beneficial to the environment. Industry, the public, and government officials must understand the benefits of compliance in terms of protecting facilities, operators, the environment and the industry's public image, and realize that the consequences of noncompliance

could be fines, imprisonment and permit revocation which could cause a business to close.

Pretreatment inspectors are finding their jobs becoming more involved with criminal prosecution of dischargers for pretreatment violations. These activities require inspectors to develop sources of information regarding criminal violations, conduct an effective and thorough pretreatment search and investigation for evidence, ensure the collection of reliable samples and accurate lab analyses, and present the evidence in a convincing manner. An example of effective work by a pretreatment inspector is a case where company representatives were caught illegally tampering with sampling equipment. Representatives of the company were videotaped placing the suction line for the POTW's automatic composite sampler in a bucket of wastewater that complied with applicable pretreatment standards while the company was discharging acidic wastewaters which violated applicable pretreatment standards. As a result of criminal prosecution (which relied on the testimony and procedures of the pretreatment inspector for the POTW) in this case, the defendants were found guilty and were sent to jail.

1.32 Educational Requirements for Pretreatment Inspectors

Inspector positions at the entry level are generally considered technical positions with specific educational requirements. As a minimum, completion of two years of college-level biology, chemistry, environmental science or engineering is normally required. Experience as a full-time, paid laboratory technician in physical or chemical testing, wastewater treatment plant operator, sub-professional engineering position or law enforcement may be allowed as a substitute for all or part of the education requirements. Work experience is often accepted as a substitute for formal education when the inspector position is available as a promotional position from within the organization. Smaller organizations, where the inspector is required to perform a wider variety of duties, may require a bachelor's degree in biology, chemistry, environmental science or engineering at the entry level. Many organizations now require the equivalent of a BS degree for advancement beyond the entry level.

Professional qualification and registration as a civil or chemical engineer may be required for positions where approval of pretreatment systems and/or the supervision of sub-professional positions is required. This may include supervising industrial waste inspector positions.

An organization may also recognize voluntary certification programs such as those offered by the Member Associations of the Water Environment Federation in salary considerations.

1.33   Basic Competencies of Pretreatment Inspectors

The knowledge and skills required to perform the duties of a pretreatment inspector include a knowledge of the regulations, local ordinances, the industrial processes where wastewater streams are generated, the waste pretreatment technology used by industries, wastestream sampling techniques, sampling preservation procedures, and approved procedures for analyzing and testing samples of industrial wastewater.[18] Inspectors must be able to apply the pretreatment rules and regulations and local ordinances. They must also be able to inform industrial representatives which rules and regulations apply to their facilities and whether or not they are in compliance with all of their POTW permit requirements. If an industry is not in compliance, an inspector must undertake compliance enforcement actions specified in the Enforcement Response Plan and may be required to testify in court regarding the violation of an ordinance or regulation.

While inspecting an industrial facility, inspectors must be able to determine which chemical and industrial processes are being used, what types of pretreatment systems are in use, and what types and amounts of wastewaters are being generated. They should be able to determine whether or not the pretreatment facilities are being properly operated and maintained and whether all the process control instrumentation is functioning properly. Inspectors should know the correct monitoring points and sampling techniques so that accurate and representative samples are obtained. The samples taken must reflect what is being discharged into the sanitary sewer system and must be handled and stored so as to ensure their admissibility in court proceedings. A good knowledge of the approved procedures for the analysis and testing of industrial wastewaters is critical to obtain accurate information to determine whether discharges comply with the regulations. Inspectors must also know what types of on-site testing kits are available and their levels of accuracy in order to make on-site analyses to obtain a rough estimate of waste characteristics and strength being discharged.

Inspectors should be able to determine whether chemical storage facilities conform with industrial and hazardous waste regulations and whether appropriate approved SOLVENT MANAGEMENT PLANS[19] are being implemented by industries. Ultimately inspectors must be able to assess the overall management and performance of industrial facilities.

Inspectors are entrusted with the authority to inspect and cite industry for violations of industrial waste discharge permit requirements and violations of applicable regulations through enforcement procedures established by the POTW. As the representative of a POTW, inspectors should establish and maintain good communication between their agency and industrial representatives. They must have the ability to deal tactfully and effectively with their employers, industrial representatives, other regulatory officials, politicians, and members of the public.

Inspectors have an ethical responsibility to keep industrial trade secrets confidential. For example, if a particular firm has solved a difficult problem for an entire industry, the inspector could tell other companies that the "ABC Co." has found a solution to the problem, but the inspector should add that the detailed approach can't be explained because it might be confidential information.

Inspectors must be very careful when they hear "trade gossip." During the inspection of one firm, adverse comments may be made about a competitor. The comment may be deliberately misleading and primarily intended to cause problems for the competitor. A careful inspector will weigh the circumstances and investigate further before acting on such gossip.

The effectiveness of inspectors in carrying out their assignments depends on their knowledge as well as their communication skills. Inspectors must establish their own credibility with industrial representatives and let them know that they are competent and intend to enforce the regulations. Good inspectors must develop all of these characteristics in order to be effective. Table 1.3 summarizes the basic competencies of an effective pretreatment inspector and lists some additional knowledge and abilities needed by a pretreatment coordinator or program manager.

1.34   Staffing Needs and Future Job Opportunities

Time and again, public opinion polls have shown that the public is becoming seriously concerned about "clean air" and "clean water" and that they expect the environment to be protected. In response to this growing public pressure, many legislative bodies now consider control of wastewater dis-

---

[18] Approved test procedures for the collection, preservation, and analysis of samples are contained in 40 CFR Part 136, "Guidelines Establishing Test Procedures for the Analysis of Pollutants," available in CODE OF FEDERAL REGULATIONS, Protection of the Environment, 40, Parts 136-149, 2001. This publication is available from the U.S. Government Printing Office, Superintendent of Documents, PO Box 371954, Pittsburgh, PA 15250-7954. Order No. 869-044-00152-7. Price, $55.00. Part 136 lists approved test procedures of the U.S. Environmental Protection Agency, American Society for Testing and Materials (ASTM), U.S. Geological Survey (USGS), and STANDARD METHODS.* (NOTE: Some test procedures published in STANDARD METHODS have not been approved by EPA for use in meeting regulatory requirements.)*

* STANDARD METHODS.   STANDARD METHODS FOR THE EXAMINATION OF WATER AND WASTEWATER, 20th Edition. A joint publication of the American Public Health Association (APHA), American Water Works Association (AWWA), and the Water Environment Federation (WEF) which outlines the accepted laboratory procedures used to analyze the impurities in water and wastewater. Available from Water Environment Federation (WEF), Publications Order Department, 601 Wythe Street, Alexandria, VA 22314-1994. Order No. S82010. Price to members, $164.25; nonmembers, $209.25; price includes cost of shipping and handling.

[19] Solvent Management Plan.   A strategy for keeping track of all solvents delivered to a site, their storage, use and disposal. This includes keeping spent solvents segregated from other process wastewaters to maximize the value of the recoverable solvents, to avoid contamination of other segregated wastes, and to prevent the discharge of toxic organics to any wastewater collection system or the environment. The plan should describe measures to control spills and leaks and to ensure that there is no deliberate dumping of solvents. Also known as a Toxic Organic Management Plan (TOMP).

TABLE 1.3   BASIC COMPETENCIES
OF PRETREATMENT INSPECTORS AND COORDINATORS

Pretreatment inspectors should be able to demonstrate the following:

- Knowledge of regulations and responsibilities of other agencies
- Knowledge of toxic constituents in industrial waste discharges
- Knowledge of industrial processes and where wastestreams are generated
- Knowledge of spill prevention and control procedures
- Knowledge of wastewater treatment technology and process troubleshooting
- Knowledge of wastewater sampling methods
- Knowledge of approved procedures for testing of industrial pollutants, sampling techniques and control instrumentation
- Knowledge of flow measuring techniques
- Ability to inspect industrial waste treatment facilities and to verify conformance with specifications
- Ability to evaluate and select monitoring locations
- Ability to deal tactfully and effectively with the representatives of industrial and commercial firms, other public and regulatory agencies, politicians, the news media and the public
- Ability to maintain accurate records and write clear and concise reports
- Ability to read and interpret mechanical construction drawings and pipeline schematics
- Ability to keep information and trade secrets confidential
- Knowledge of and ability to identify safety hazards associated with pretreatment control functions
- Ability to understand other viewpoints and work with industries and other regulatory agencies
- Ability to prepare and maintain proper files and documentation on work performed
- Knowledge of and ability to practice professional ethics

In addition to the knowledge and abilities needed by pretreatment inspectors, pretreatment program managers or coordinators should be able to demonstrate the following:

- Knowledge of industrial user personnel
- Knowledge of collection system
- Ability to collect, preserve and deliver representative samples to the laboratory
- Ability to coordinate work with laboratory personnel
- Ability to use a computer
- Ability to properly collect, organize, enter and analyze data
- Ability to write permits
- Ability to conduct industrial surveys
- Ability to prepare notices of violation

charges polluting public waters and contamination of groundwater by toxic spills a top priority on their agendas. The amended Clean Water Act provides additional mechanisms and regulations for cleaning up the environment and pretreatment regulations now include very strict, specific requirements for industrial wastewater dischargers.

In an era demanding the control of industrial wastewater discharges, hazardous wastes, toxic chemical spills and leaking underground chemical storage and fuel tanks, the need for trained inspectors is increasing rapidly. This trend is expected to continue in the future. It is no longer sufficient to pretreat wastes and thereby reduce the hazards to POTWs and the environment. Current trends show an increasing regulatory focus on eliminating some hazards entirely through waste minimization programs, product reformulation, and process modifications. The industrial waste source control section at a wastewater treatment plant must be staffed with qualified pre-

treatment inspectors to meet the demands created by the regulations.

1.35   What Does It Take To Be a Pretreatment Facility Inspector?

When a person enters the pretreatment facility inspection field, a full commitment to this profession is required. You must be willing to learn a technical profession as well as develop the ability to communicate effectively with all types of people. As an inspector, you must be willing to adapt to new equipment, techniques and technologies used in industrial processes and must be constantly expanding your own horizons. Besides having the technical aptitude for this field, you also must feel comfortable working in a regulatory role.

Inspectors must be highly motivated and have the desire to advance up the career ladder of their organization. A college degree in engineering or the sciences is desirable. However, regardless of your previous education, this is a new and highly scientific field. You must be ready to broaden your knowledge and qualifications through continuing education and pass voluntary *CERTIFICATION EXAMINATIONS*[20] if available. By beginning your learning today, you can become a better inspector tomorrow and serve the public more effectively. By protecting the environment today, you create a better life for your family and your community for the future.

1.36 Computer Skills

Gaining computer skills and staying current in this field will make you a much better and more efficient pretreatment inspector. Inspectors are taking laptop computers and other types of miniature computers into the field on inspections to immediately record observations. This data can easily be transferred to the pretreatment program's database for access and use by everyone. Inspectors need to know how to use software database systems, word processing, local area networks, and on-line bulletin boards. The ability to use bulletin board systems (BBS) as a source of information and networking among pretreatment inspectors and managers at the federal, state and local levels is very important. This is the very real future for pretreatment inspectors if they have a telephone, computer, modem, and the correct dial-in numbers.

1.37 Training Yourself To Meet the Needs

This training course is not the only training available to help you improve your abilities. Some state water pollution control associations, vocational schools, community colleges, and universities offer training courses on both a short- and long-term basis. Many state, local and private agencies conduct training programs and informative seminars. Most state water pollution control agencies have a pretreatment section which can be very helpful in providing training programs or directing you to good programs. Some libraries can provide you with useful journals and reference books on pretreatment inspection.

This training course has been carefully prepared to help you increase your knowledge and develop the skills needed to inspect pretreatment facilities or coordinate a pretreatment program. You will be able to proceed at your own pace; you will have the opportunity to learn a little or a lot about each topic. The training manual has been prepared this way to meet the various needs of pretreatment facility inspectors, depending

on the numbers and types of pretreatment facilities that you must inspect.

To study for voluntary certification and civil service exams, you may have to learn most of the material in this manual. You will never know everything about pretreatment facility inspection and the pretreatment processes. However, when you complete this course you will be able to answer some very important questions about how, why, and when certain things happen in pretreatment facilities.

A good inspector must be capable of fairly and effectively inspecting pretreatment facilities. The remainder of this manual is developed to prepare you with the knowledge and skills to perform your duties. Chapter 4 explains how to approach and inspect a typical industry. Specific details on how to inspect various types of industries are provided in Chapter 10, "Industrial Inspection Procedures." This material will help you develop your own procedures for the industries that you will be inspecting.

1.4 BASIC REFERENCES

Throughout this manual we will be referring to CFR 40. This refers to the *CODE OF FEDERAL REGULATIONS*, Protection of the Environment, Title 40, Parts 87 to 149. Available from the U.S. Government Printing Office, Superintendent of Documents, PO Box 371954, Pittsburgh, PA 15250-7954. Parts 87-99 (Order No. 869-044-00150-1), price, $54.00; Parts 100-135 (Order No. 869-044-00151-9), price, $38.00; Parts 136-149 (Order No. 869-044-00152-7), price, $55.00.

Another helpful reference is EPA's *INDUSTRIAL USER INSPECTION AND SAMPLING MANUAL FOR POTWs*, April 1994. Office of Water (4202), U.S. Environmental Protection Agency, Washington, DC 20460. Obtain from National Technical Information Service (NTIS), 5285 Port Royal Road, Springfield, VA 22161. Order No. PB94-170271. EPA No. 831-B-94-001. Price, $67.50, plus $5.00 shipping and handling per order.

Additional important references can be obtained by contacting the US EPA Water Resources Center or the National Technical Information Service (NTIS).

● US EPA Water Resources Center
  (202) 260-7786

  Please provide the EPA Identification Number when ordering through the Water Resources Center.

---

[20] *Certification Examination. An examination administered by a state agency or professional association that pretreatment facility inspectors take to indicate a level of professional competence. Today pretreatment facility inspector certification exams may be voluntary and administered by professional associations. In some states inspectors are recognized as environmental compliance inspectors. Current trends indicate that more employers will encourage inspector certification in the future.*

- The National Technical Information Service (NTIS)
5285 Port Royal Road
Springfield, VA 22161
(800) 553-6847

    Please provide the NTIS number when ordering through NTIS.

1.5   ACKNOWLEDGMENT

Portions of the material in this chapter were written by Jay Kremer and Carl Sjoberg.

QUESTIONS

Write your answers in a notebook and then compare your answers with those on page 23.

1.3A   What does a pretreatment facility inspector do if a chemical spill occurs?

1.3B   Why should inspectors know proper sampling techniques?

1.3C   What is a representative sample?

1.3D   Inspectors must have the ability to deal tactfully and effectively with what groups of people?

1.3E   What have public opinion polls revealed about public expectations regarding the environment?

1.3F   What issues have become top priorities of many legislative bodies?

1.3G   What is a certification examination?

DISCUSSION AND REVIEW QUESTIONS

Chapter 1.   THE PRETREATMENT FACILITY INSPECTOR

DO NOT USE SPECIAL ANSWER SHEET. Please write your answers in your notebook before working the Objective Test that is located on page 26. The purpose of these questions is to indicate to you how well you understand the material in this lesson. If you have the opportunity, you might wish to discuss some of these questions with other pretreatment facility inspectors to find out how they conduct their inspections, what problems they encounter, and how they respond to these problems.

1. What is a pretreatment facility?

2. What are the objectives of categorical standards?

3. In general terms, what does a pretreatment facility inspector do?

4. Pretreatment facility inspectors must have what general types of knowledge and skills?

5. The effectiveness of inspectors in carrying out their assignments depends on what factors?

6. Who does a pretreatment facility inspector work for?

7. Improper discharges of industrial wastes can cause what types of problems in a POTW's wastewater treatment facilities?

8. What do citizens expect of a pretreatment inspector?

9. What are some of the job duties of a pretreatment facility inspector?

10. What are the responsibilities of a pretreatment inspector?

SUGGESTED ANSWERS

Chapter 1.   THE PRETREATMENT FACILITY INSPECTOR

Answers to questions on page 11.

1.0A   The sources, amounts and types of wastes generated at an industrial manufacturing site depend on:

1. Age of the facility,
2. Raw materials used,
3. Production processes, and
4. Ability to recover and recycle wastes.

1.0B   **P**ublicly **O**wned **T**reatment **W**orks.

1.0C   As a result of a 1976 Natural Resources Defense Council suit against EPA, EPA shifted its pretreatment focus to toxic pollutants.

1.0D   Categorical standards are industrial waste discharge standards that are applied to the effluent from any industry that discharges to a POTW. These are standards based on the technology available to treat the wastestreams from the processes of the specific industry.

1.0E   The value of potential unannounced inspections is that they serve as an effective deterrent to noncompliant or illegal dischargers.

Answers to questions on page 14.

1.1A   The main purposes of an industrial waste pretreatment program are to protect the environment, the POTW agency's facilities and personnel and the local community from adverse effects due to industrial waste discharges.

1.1B   Potential adverse effects of industrial chemicals on POTW agency employees include exposure to possible dangers of liver cancer from excessive discharges of vinyl chloride or other illnesses resulting from excessive exposure to toxic industrial wastes. Corrosion of sewers and structures can also be dangerous to agency personnel as severe structural problems can make it hazardous for people to work in treatment plant and sewer facilities. Discharges of cyanide wastes may create toxic atmospheric conditions, excessive discharges of food processing waste can lead to anaerobic conditions which produce toxic hydrogen sulfide gas and sample collection locations can be potentially hazardous.

1.1C   The solid sludge residue from wastewater treatment processes must be properly disposed of so that toxic effects due to pollutants in the sludge do not occur.

1.1D   A POTW agency can help industrial companies by explaining and interpreting the federal pretreatment regulations. Agency personnel can explain the need for and the proper methods of implementing the regulations.

Answers to questions on page 16.

1.2A   Pollutants can enter the environment from

1. Disposal of toxic substances on land,
2. Release of emissions from stacks or exhausts into the atmosphere,
3. Discharge of wastes into sewers and ultimately surface waters, and/or groundwater, and
4. Chemical spills and/or illegal dumpings.

1.2B   The EPA's General Pretreatment Regulations shifted the responsibility for protecting the environment by restricting the types and amounts of toxic substances that could be discharged into sanitary sewers.

Answers to questions on page 22.

1.3A   If a chemical spill occurs, a pretreatment facility inspector may be required to decide whether or not the liquid spilled can be disposed of through a sanitary sewer, a storm sewer, or must be contained, pretreated, or disposed of by other means. (The inspector should also notify the appropriate POTW and other regulatory officials.)

1.3B   Inspectors should know the proper sampling techniques so that accurate, representative, and legally admissible samples are obtained that reflect what is being discharged into the sanitary sewer system.

1.3C   A representative sample is a portion of material or wastestream that is as nearly identical in content and consistency as possible to that in the larger body of material or wastestream being sampled.

1.3D   Inspectors must have the ability to deal tactfully and effectively with their employers, industrial representatives, other regulatory officials and members of the public.

1.3E   Public opinion polls have revealed that the public expects the environment to be protected. The public is becoming more concerned about the need for "clean air" and "clean water."

1.3F   Control of wastewater discharges polluting public waters and contamination of groundwater by toxic spills have become issues of top priority in many legislative bodies.

1.3G   A certification examination is an examination administered by a state agency or professional association that pretreatment facility inspectors take to indicate a level of professional competence.

DIRECTIONS FOR WORKING OBJECTIVE TEST

1. You have been provided with a special answer sheet for each chapter. Be sure you follow the special directions provided with the answer sheets. If you lose an answer sheet or have any problems, please notify the Program Director.

2. Mark your answers on the answer sheet with a dark lead pencil. Do not use ink. Questions 1 through 11 are true or false questions. If a question is true, then mark Column 1, and if false mark Column 2. The correct answer to Question 3 is true; therefore, place a mark in Column 1. Questions 12 through 19 require you to select only the best or closest answer. The best answer for question 15 is answer 2; therefore, place a mark in Column 2. Questions 20 through 28 are multiple-choice questions which may have more than one correct answer. Question 21 has five correct answers so you should place a mark in Columns 1, 2, 3, 4, and 5 on the answer sheet.

Please mark your answers in your workbook for your record because answer sheets will not be returned to you.

3. Mail your answer sheet to the Program Director immediately after you have completed the test.

4. Answer sheets may be folded (but not into more than 3 equal parts) and mailed in a 4 x 9½-inch standard white envelope to:

Program Director
Office of Water Programs
California State University, Sacramento
6000 J Street
Sacramento, CA 95819-6025

Printed in U.S.A.   Trans-Optic® by NCS MM103984-3   321   ED11

PRETREATMENT FACILITY INSPECTION
3rd Edition
IMPORTANT

PLEASE READ INSTRUCTIONS ON REVERSE SIDE BEFORE COMPLETING THIS FORM.

Name: OPERATOR,   JOE   B.
Last   First   MI

Address: 711 MAIN ST.

INDUSTRIAL CITY, CA   98765
City   State   Zip Code

Phone: (707) 123-4567

○ If your address has changed, ☞ please fill in this circle with a Number 2 pencil.

Mail to: Professor Kenneth Kerri
Office of Water Programs
California State University, Sacramento
6000 J Street
Sacramento, California 95819-6025

For help in preparing this box, see example on reverse side. ☞

SOCIAL SECURITY NUMBER | CHAPTER NUMBER
1 2 3 4 5 6 7 8 9 | 1

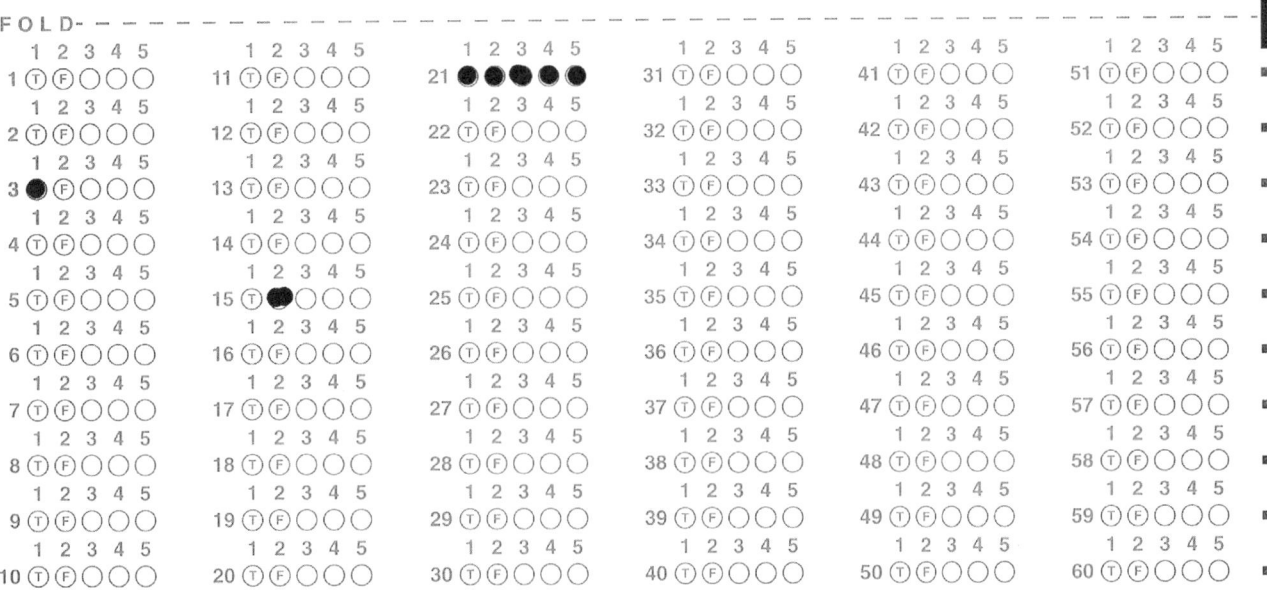

FOLD- -

IMPORTANT DIRECTIONS FOR MARKING ANSWERS

Use black lead pencil only (#2 or softer).
Make heavy black marks that fill circle completely.
Erase clearly any answer you wish to change.
Make no stray marks on this answer sheet.

1. MULTIPLE CHOICE QUESTIONS: Fill in the correct answers. If 2 and 3 are correct for question 1, mark:

   1 2 3 4 5
   1 Ⓣ ● ● ○ ○

2. TRUE — FALSE QUESTIONS: If true, fill the circle in column 1; if false, fill in column 2. If question 3 is true, mark:

   1   3 4 5
   3 Ⓕ ○ ○ ○

SAMPLE DO NOT USE

☞   URGENT!   ☜

ANSWER SHEET WILL BE GRADED INCORRECTLY IF:

▶ THIS FORM IS USED FOR ANY COURSE OTHER THAN *Pretreatment Facility Inspection, 3rd Edition*

▶ A NUMBER 2 PENCIL IS NOT USED (*Do not use ink*)

▶ CIRCLES FOR CHAPTER NUMBER & SOCIAL SECURITY NUMBER ARE NOT FILLED IN ON THE BOX ABOVE.

BEFORE completing this form, please read ALL instructions, including those printed on the REVERSE SIDE.

FOLD- -

	1 2 3 4 5		1 2 3 4 5		1 2 3 4 5		1 2 3 4 5		1 2 3 4 5		1 2 3 4 5
1	Ⓣ Ⓕ ○ ○ ○	11	Ⓣ Ⓕ ○ ○ ○	21	● ● ● ● ●	31	Ⓣ Ⓕ ○ ○ ○	41	Ⓣ Ⓕ ○ ○ ○	51	Ⓣ Ⓕ ○ ○ ○
2	Ⓣ Ⓕ ○ ○ ○	12	Ⓣ Ⓕ ○ ○ ○	22	Ⓣ Ⓕ ○ ○ ○	32	Ⓣ Ⓕ ○ ○ ○	42	Ⓣ Ⓕ ○ ○ ○	52	Ⓣ Ⓕ ○ ○ ○
3	● Ⓕ ○ ○ ○	13	Ⓣ Ⓕ ○ ○ ○	23	Ⓣ Ⓕ ○ ○ ○	33	Ⓣ Ⓕ ○ ○ ○	43	Ⓣ Ⓕ ○ ○ ○	53	Ⓣ Ⓕ ○ ○ ○
4	Ⓣ Ⓕ ○ ○ ○	14	Ⓣ Ⓕ ○ ○ ○	24	Ⓣ Ⓕ ○ ○ ○	34	Ⓣ Ⓕ ○ ○ ○	44	Ⓣ Ⓕ ○ ○ ○	54	Ⓣ Ⓕ ○ ○ ○
5	Ⓣ Ⓕ ○ ○ ○	15	Ⓣ ● ○ ○ ○	25	Ⓣ Ⓕ ○ ○ ○	35	Ⓣ Ⓕ ○ ○ ○	45	Ⓣ Ⓕ ○ ○ ○	55	Ⓣ Ⓕ ○ ○ ○
6	Ⓣ Ⓕ ○ ○ ○	16	Ⓣ Ⓕ ○ ○ ○	26	Ⓣ Ⓕ ○ ○ ○	36	Ⓣ Ⓕ ○ ○ ○	46	Ⓣ Ⓕ ○ ○ ○	56	Ⓣ Ⓕ ○ ○ ○
7	Ⓣ Ⓕ ○ ○ ○	17	Ⓣ Ⓕ ○ ○ ○	27	Ⓣ Ⓕ ○ ○ ○	37	Ⓣ Ⓕ ○ ○ ○	47	Ⓣ Ⓕ ○ ○ ○	57	Ⓣ Ⓕ ○ ○ ○
8	Ⓣ Ⓕ ○ ○ ○	18	Ⓣ Ⓕ ○ ○ ○	28	Ⓣ Ⓕ ○ ○ ○	38	Ⓣ Ⓕ ○ ○ ○	48	Ⓣ Ⓕ ○ ○ ○	58	Ⓣ Ⓕ ○ ○ ○
9	Ⓣ Ⓕ ○ ○ ○	19	Ⓣ Ⓕ ○ ○ ○	29	Ⓣ Ⓕ ○ ○ ○	39	Ⓣ Ⓕ ○ ○ ○	49	Ⓣ Ⓕ ○ ○ ○	59	Ⓣ Ⓕ ○ ○ ○
10	Ⓣ Ⓕ ○ ○ ○	20	Ⓣ Ⓕ ○ ○ ○	30	Ⓣ Ⓕ ○ ○ ○	40	Ⓣ Ⓕ ○ ○ ○	50	Ⓣ Ⓕ ○ ○ ○	60	Ⓣ Ⓕ ○ ○ ○

OBJECTIVE TEST

Chapter 1.   THE PRETREATMENT FACILITY INSPECTOR

Please write your name and mark the correct answers on an answer sheet as directed on page 24. There may be more than one correct answer to each multiple-choice question.

True-False

1. An industrial pretreatment facility consists of wastewater treatment processes designed to remove pollutants from wastestreams prior to discharge to streams and rivers.
   1. True
   - 2. False

2. The potential of an unannounced inspection at any time serves as an effective deterrent to noncompliant dischargers.
   - 1. True
   2. False

3. Corrosion damage to facilities can require significant capital outlay for repair of the facilities.
   - 1. True
   2. False

4. The EPA's pretreatment program requires that POTW agencies notify and help industrial companies to interpret and implement the federal regulations.
   - 1. True
   2. False

5. It is much easier to accomplish regulatory goals with industry through enforcement actions than it is through cooperation with industrial personnel.
   1. True
   - 2. False

6. Pretreatment regulations have resulted in treatment plant sludges (biosolids) being used as a soil amendment in land applications.
   - 1. True
   2. False

7. Pretreatment facility inspectors work for state and federal regulatory/enforcement agencies.
   1. True
   - 2. False

8. Pretreatment inspectors may be involved in industrial chemical spills.
   - 1. True
   2. False

9. Approved analytical tests are listed in *STANDARD METHODS*.
   1. True
   - 2. False

10. Inspectors have an ethical responsibility to keep industrial trade secrets confidential.
   - 1. True
   2. False

11. A good inspector must be capable of fairly and effectively inspecting pretreatment facilities.
   - 1. True
   2. False

Best Answer (Select only the closest or best answer.)

12. A representative sample is
   1. Collected during minimum flows.
   2. Obtained when industrial representatives are present.
   - 3. Similar to the larger body of wastestream being sampled.
   4. Transported to a laboratory for analysis.

13. Pretreatment inspectors with proper credentials are authorized by _____ to enter any premises of an industrial discharger for the purposes of inspection, sampling, examination of records required to be kept by the POTW, and to set up devices necessary to conduct sampling, monitoring and metering operations.
   1. Authority of POTW superintendent
   2. Law enforcement regulations
   - 3. Local sewer-use ordinances
   4. POTW NPDES permit

14. What type of waste discharge into a sewer may create a toxic atmosphere?
   1. Biochemical oxygen demand (BOD) waste from canneries
   - 2. Cyanide waste from metal platers
   3. Human waste from residences
   4. Suspended solids (SS) waste from pulp mills

15. What is the main focus of pollution prevention?
   1. Disposing of the waste generated
   - 2. Minimizing the amount of waste generated
   3. Recycling the waste generated
   4. Treating the waste generated

16. What is waste minimization?
   1. Detoxification of hazardous waste
   - 2. Reduction of hazardous waste
   3. Storage of hazardous waste
   4. Treatment of hazardous waste

17. What is the primary role of the pretreatment facility inspector?

    — 1. Enforce regulations intended to protect POTWs and the environment
       2. Ensure proper disposal of industrial sludges
       3. Prevent sewer overflows
       4. Require industry to develop a pollution prevention program

18. What do current trends indicate regarding future job opportunities for pretreatment facility inspectors?

       1. Decreasing
    — 2. Increasing
       3. Stable
       4. Unknown

19. Why do pretreatment inspectors need computer skills?

       1. Analyze laboratory results
    — 2. Become more efficient
       3. Complete forms
       4. Type reports

Multiple Choice (Select all correct answers.)

20. Conventional pollutants include

    — 1. Biochemical oxygen demand (BOD)
       2. Cyanide
       3. Hexavalent chromium
    — 4. Oil and grease
    — 5. Suspended solids (SS)

21. The objectives of the General Pretreatment Regulations are to

    — 1. Improve opportunities to reclaim and recycle sludge.
    — 2. Prevent interference with operation of POTW.
    — 3. Prevent interference with use or disposal of municipal sludge.
    — 4. Prevent introduction of pollutants which pass through treatment works.
    — 5. Reduce the health and environmental risks from pollution.

22. What is the purpose of an industrial waste pretreatment program?

    — 1. Protection of agency personnel
    — 2. Protection of capital facilities
    — 3. Protection of the community
    — 4. Protection of the environment
    — 5. Provide pollution prevention guidance to industry

23. What damages could be caused to a POTW agency's collection system by industrial wastes?

    — 1. Clogging the sewers
    — 2. Corrosion of sewers
       3. Excessive inflow and infiltration (I/I)
    — 4. Explosions from flammable wastes
       5. Lift station failure due to lack of maintenance

24. Which areas of endeavor are covered by pollution prevention?

    — 1. Good operating practices
    — 2. Material substitution
    — 3. Process modification
    — 4. Product reformulation
       5. Treatment of wastes generated

25. What does a pretreatment facility inspector inspect?

    — 1. Industrial manufacturing facilities
       2. Offices of receptionists at industrial facilities
       3. POTW facilities
    — 4. Pretreatment facilities
    — 5. Raw and spent material storage facilities

26. What are the roles of pretreatment inspectors when involved with criminal prosecution of pretreatment violations?

    — 1. Conduct investigations for evidence
    — 2. Develop sources of information regarding violations
    — 3. Ensure collection of reliable samples
    — 4. Present evidence in a convincing manner
       5. Prosecute violators in court

27. Who must pretreatment inspectors have the ability to deal with tactfully and effectively?

    — 1. Industrial representatives
    — 2. Members of the public
    — 3. Other regulatory officials
    — 4. Politicians
    — 5. Their employers

28. What activities relating to computers should pretreatment inspectors be able to perform?

    — 1. Do word processing
    — 2. Use local area networks
    — 3. Use on-line bulletin boards
    — 4. Use software database systems
       5. Wire interconnections to equipment

End of Objective Test

CHAPTER 2

PRETREATMENT PROGRAM ADMINISTRATION

by

Jay Kremer

TABLE OF CONTENTS

Chapter 2.   PRETREATMENT PROGRAM ADMINISTRATION

OBJECTIVES

Chapter 2.   PRETREATMENT PROGRAM ADMINISTRATION

Following completion of Chapter 2, you should be able to:

1. Describe the main administrative aspects of a pretreatment program,

2. Determine the sources of funding for a pretreatment program,

3. Issue and review industrial waste discharge permits,

4. Evaluate the applications of a database management system for your industrial waste program,

5. Establish a public relations program, and

6. Perform your duties in an ethical manner.

PRETREATMENT WORDS

Chapter 2. PRETREATMENT PROGRAM ADMINISTRATION

40 CFR 403 40 CFR 403

EPA's General Pretreatment Regulations appear in the **C**ode of **F**ederal **R**egulations under 40 CFR 403. 40 refers to the numerical heading for the environmental regulations portion of the **C**ode of **F**ederal **R**egulations (CFR). 403 refers to the section which contains the General Pretreatment Regulations. Other sections include 413, Electroplating Categorical Regulations.

BASELINE MONITORING REPORT (BMR) BASELINE MONITORING REPORT (BMR)

All industrial users subject to Categorical Pretreatment Standards must submit a baseline monitoring report (BMR) to the Control Authority (POTW, state or EPA). The purpose of the BMR is to provide information to the Control Authority to document the industrial user's current compliance status with a Categorical Pretreatment Standard. The BMR contains information on (1) name and address of facility, including names of operator(s) and owner(s), (2) list of all environmental control permits, (3) brief description of the nature, average production rate and SIC (or NAICS) code for each of the operations conducted, (4) flow measurement information for regulated process streams discharged to the municipal system, (5) identification of the pretreatment standards applicable to each regulated process and results of measurements of pollutant concentrations and/or mass, (6) statements of certification concerning compliance or noncompliance with the pretreatment standards, and (7) if not in compliance, a compliance schedule must be submitted with the BMR that describes the actions the industrial user will take and a timetable for completing these actions to achieve compliance with the standards.

CERCLA (SIRK-la) CERCLA

Comprehensive **E**nvironmental **R**esponse, **C**ompensation, and **L**iability **A**ct of 1980. This Act was enacted primarily to correct past mistakes in industrial waste management. The focus of the Act is to locate hazardous waste disposal sites which are creating problems through pollution of the environment and, by proper funding and implementation of study and corrective activities, eliminate the problem from these sites. Current users of CERCLA-identified substances must report releases of these substances to the environment when they take place (not just historic ones). This Act is also called the Superfund Act. Also see SARA.

CENTRALIZED WASTE TREATMENT (CWT) FACILITY CENTRALIZED WASTE TREATMENT (CWT) FACILITY

A facility designed to properly handle treatment of specific hazardous wastes from industries with similar wastestreams. The wastewaters containing the hazardous substances are transported to the facility for proper storage, treatment and disposal. Different facilities treat different types of hazardous wastes.

CONFINED SPACE CONFINED SPACE

Confined space means a space that:

A. Is large enough and so configured that an employee can bodily enter and perform assigned work; and

B. Has limited or restricted means for entry or exit (for example, tanks, vessels, silos, storage bins, hoppers, vaults, and pits are spaces that may have limited means of entry); and

C. Is not designed for continuous employee occupancy.

(Definition from the Code of Federal Regulations (CFR) Title 29 Part 1910.146.)

CONSERVATIVE POLLUTANT CONSERVATIVE POLLUTANT

A pollutant found in wastewater that is not changed while passing through the treatment processes in a conventional wastewater treatment plant. This type of pollutant may be removed by the treatment processes and retained in the plant's sludges or it may leave in the plant effluent. Heavy metals such as cadmium and lead are conservative pollutants.

CONVENTIONAL TREATMENT CONVENTIONAL TREATMENT

(1) The common treatment processes such as preliminary treatment, sedimentation, flotation, trickling filter, rotating biological contactor, activated sludge and chlorination wastewater treatment processes used by POTWs.

(2) The hydroxide precipitation of metals processes used by pretreatment facilities.

MERCAPTANS (mer-CAP-tans)   MERCAPTANS

Compounds containing sulfur which have an extremely offensive skunk-like odor; also sometimes described as smelling like garlic or onions.

NAICS   NAICS

North **A**merican **I**ndustry **C**lassification **S**ystem. A code number system used to identify various types of industries. This code system replaces the SIC (Standard Industrial Classification) code system used prior to 1997. Use of these code numbers is often mandatory. Some companies have several processes which will cause them to fit into two or more classifications. The code numbers are published by the Superintendent of Documents, U.S. Government Printing Office, PO Box 371954, Pittsburgh, PA 15250-7954. Stock No. 041-001-00509-9; price, $31.50. There is no charge for shipping and handling.

NPDES PERMIT   NPDES PERMIT

National **P**ollutant **D**ischarge **E**limination **S**ystem permit is the regulatory agency document issued by either a federal or state agency which is designed to control all discharges of pollutants from point sources and storm water runoff into U.S. waterways. NPDES permits regulate discharges into navigable waters from all point sources of pollution, including industries, municipal wastewater treatment plants, sanitary landfills, large agricultural feedlots and return irrigation flows.

POTW   POTW

Publicly **O**wned **T**reatment **W**orks. A treatment works which is owned by a state, municipality, city, town, special sewer district or other publicly owned and financed entity as opposed to a privately (industrial) owned treatment facility. This definition includes any devices and systems used in the storage, treatment, recycling and reclamation of municipal sewage or industrial wastes of a liquid nature. It also includes sewers, pipes and other conveyances only if they carry wastewater to a POTW treatment plant. The term also means the municipality (public entity) which has jurisdiction over the indirect discharges to and the discharges from such a treatment works. (Definition adapted from EPA's General Pretreatment Regulations, 40 CFR 403.3.)

RCRA (RICK-ruh)   RCRA

The Federal **R**esource **C**onservation and **R**ecovery **A**ct (10/21/76), Public Law (PL) 94-580, provides technical and financial assistance for the development of plans and facilities for recovery of energy and resources from discharged materials and for the safe disposal of discarded materials and hazardous wastes. This Act introduces the philosophy of the "cradle to grave" control of hazardous wastes. RCRA regulations can be found in Title 40 of the Code of Federal Regulations (40 CRF) Parts 260-268, 270 and 271.

SARA   SARA

Superfund **A**mendments and **R**eauthorization **A**ct of 1986. The Comprehensive Environmental Response, Compensation, and Liability Act (CERCLA), commonly known as Superfund, was enacted in 1980. The Superfund Amendments increase Superfund revenues to $8.5 billion and strengthen the EPA's authority to conduct short-term (removal), long-term (remedial) and enforcement actions. The Amendments also strengthen state involvements in the cleanup process and the Agency's commitments to research and development, training, health assessments, and public participation. A number of new statutory authorities, such as Community Right-to-Know, are also established. Also see CERCLA.

SIC CODE   SIC CODE

Standard **I**ndustrial **C**lassification Code. A code number system used to identify various types of industries. In 1997, the United States and Canada replaced the SIC code system with the North American Industry Classification System (NAICS); Mexico adopted the NAICS in 1998. See NAICS.

SIGNIFICANT INDUSTRIAL USER (SIU)   SIGNIFICANT INDUSTRIAL USER (SIU)

A Significant Industrial User (SIU) includes:

1. All categorical industrial users, and

2. Any noncategorical industrial user that

   a. Discharges 25,000 gallons per day or more of process wastewater ("process wastewater" excludes sanitary, noncontact cooling and boiler blowdown wastewaters), or

   b. Contributes a process wastestream which makes up five percent or more of the average dry weather hydraulic or organic (BOD, TSS) capacity of a treatment plant, or

   c. Has a reasonable potential, in the opinion of the Control or Approval Authority, to adversely affect the POTW treatment plant (inhibition, pass-through of pollutants, sludge contamination, or endangerment of POTW workers).

CHAPTER 2. PRETREATMENT PROGRAM ADMINISTRATION

(Lesson 1 of 3 Lessons)

2.0 PLANNING A PRETREATMENT PROGRAM

2.00 Size of the Agency

Federal regulations prescribe that if a POTW agency has wastewater flows over five MGD or significant industrial dischargers in its service area, the agency must have a federally approved industrial waste pretreatment program. At current count, there are approximately 1,650 POTW agencies within the United States operating such programs. Most POTW agencies, unless extremely small or having no industry, will require some form of industrial waste pretreatment program. The size of the agency will influence all aspects of planning and implementing an industrial waste pretreatment program.

Two factors that determine the staffing needs of an industrial waste program are the size of the agency and the amount of industrial wastewater flow. Metropolitan Chicago, New York or Los Angeles, having a wide variety of industries with numerous companies and many large wastewater dischargers, will require a sophisticated industrial waste program and a large staff. A small city, having only incidental industry, will require a much different program perhaps using only one person or a portion of one person's time to implement the program. Even in small POTW agencies, significant industrial waste problems can occur unless an adequate pretreatment program is in place. A smaller POTW agency may, in fact, be more susceptible to major industrial waste-caused problems than the large POTW agency.

2.01 Size of Wastewater Treatment Plant

The size of the wastewater treatment plant as compared to its industrial wastewater loads and characteristics is an important aspect to consider in planning an adequate pretreatment program. In general, the larger the wastewater treatment plant, the less impact one company is likely to have on the treatment system. In a small wastewater treatment plant, even a small metal plater can have a disastrous impact on the plant

processes if a spill of toxic waste occurs. In a large treatment plant, however, major industrial spills may be virtually undetectable at the treatment plant. Of particular concern is the small treatment plant with a relatively large company discharging to it. In such a situation, minor upsets and problems at the company have the potential to cause disastrous upsets or other problems at the wastewater treatment plant.

2.02 Industrial Development

The type of industrial development in the POTW agency service area is of great significance to the pretreatment program. It is not uncommon to find regional concentrations of certain types of industries. In the Southern California area, for example, there are numerous large petroleum refineries that developed when oil was discovered in Southern California. In portions of Texas and the surrounding area there are large concentrations of chemical plants, also due to the ready availability of basic petroleum resources in previous years. In some parts of Central California, large agricultural loads are placed on wastewater treatment plants during harvest season. Toxic wastes from metal plating and machine shops can be found in some areas such as Rockford, Illinois, which has a large component of metal-working industries. The industrial development in the service area will determine the characteristics of the industrial wastewater and the collection and treatment system problems which may be encountered. A review of the industries located in the service area is essential to determine the proper aim of the industrial waste pretreatment program.

2.03 Technological Capabilities of the POTW Agency

In the staffing of an industrial waste pretreatment program, the technological needs of the POTW agency must be considered. The managers or coordinators of industrial waste pretreatment programs frequently have experience in either engineering or laboratory chemistry. Industrial waste pretreatment programs require an in-depth understanding of industrial manufacturing processes, pretreatment technology, water and wastewater chemistry and how to deal with people. Generally speaking, a well-qualified technical person within the POTW agency should be assigned the responsibility to implement this program. In a small POTW agency, the only person available on staff may well be a laboratory chemist or a treatment plant operator with a strong chemistry background. This person is likely to be the only one with the technological background to adequately implement the program. At a large POTW agency there should be a significant number of employees with sufficient technical knowledge, chemical training and public relations skills to manage the pretreatment program. Currently in the United States, large agencies often have an engineer leading the industrial waste program, whereas smaller agencies tend to have science personnel such as laboratory chemists involved with industrial waste regulations.

It is preferable to have a POTW agency employee working as the lead person for the industrial waste program. If this is not possible, it may be necessary to hire a consultant to handle the technically complex details of the industrial waste regulatory program.

2.04 Enforcement Posture

When planning an industrial waste pretreatment program, it is important to consider the enforcement posture desired by the community. Two basic types of enforcement posture exist. One enforcement position is that any violation of an industrial waste effluent requirement is a crime and the company should be prosecuted for such violations. This posture will result in a great deal of legal action and even minor violations may involve lawyers' time. The legal actions can be taken through either civil or criminal court procedures.

Another basic posture is the use of administrative enforcement procedures to identify a problem and encourage industry in its efforts to make the necessary corrections. Typically, administrative actions involve issuance of various levels of notices: a warning notice or violation notice, a compliance order, an administrative order with penalty assessment, or a cease and desist order. In most cases, the industry is given some time to correct the violation, unless the violation is having an adverse impact on human health or the environment. If the problem is not corrected after a certain time limit, increasingly severe administrative actions may be taken. Assuming that the problem is not remedied, the administrative action may eventually lead to legal actions such as misdemeanor prosecutions by the local city attorney or district attorney. Administrative fines are common and may also be appropriate for those companies not complying after a reasonable time.

The use of the rigorous legal enforcement posture may be appropriate in cities or areas where industry is not a major POTW contributor. Suburban communities frequently adopt a rather rigorous attitude toward the small component of industry in their community requiring them to "toe the line" as far as effluent violations are concerned. This procedure frequently leads to confrontations between the industry and the community and the money spent in confrontational legal actions is money not spent on correction of a problem. It is suggested that unless there is a significant desire to discourage industry in a community, the better approach is the administrative enforcement posture.

Some agencies use administrative actions to encourage companies to come into compliance. Typically, such actions involve issuance of three levels of notices: a warning notice, a violation notice and a final notice of violation. Usually 30 days are allowed between the issuance of the warning and violation notices and between the issuance of the violation and

final notice of violation. To implement administrative actions such as these, the company should be informed about possible methods of correcting the violation and the required results. At the final notice of violation stage, an enforcement compliance meeting is sometimes held to discuss the resolution of the problem with the industry. At this meeting, the agency discusses the problem in detail with the company. If possible, helpful suggestions are offered to assist the company in correction of the problem. If no resolution of the problem is obtained after three administrative actions, the agency will levy fines and/or compliance schedules or refer the problem to the district attorney for criminal misdemeanor prosecution. Any POTW enforcement actions must comply with the agency's Enforcement Response Plan (ERP). Pretreatment facility inspectors must be experts in the application of their POTW Enforcement Response Plan when performing their duties.

In the implementation of these two enforcement postures, a lower level of technical expertise at the POTW may be sufficient to operate under the strict legal enforcement procedures if balanced by a higher level of legal expertise to handle the continuing legal issues that this method generates. In general, the progressive administrative enforcement procedures and the use of a higher level of technical expertise at the POTW are recommended as more effective.

QUESTIONS

Write your answers in a notebook and then compare your answers with those on page 82.

2.0A   What factors influence the type of industrial waste inspection program needed and its staffing requirements?

2.0B   Why is the size of a wastewater treatment plant an important aspect to consider in planning an adequate pretreatment inspection program?

2.0C   What type of background or experience is helpful to a pretreatment program inspector?

2.0D   What are the two basic enforcement postures for an industrial waste pretreatment inspection program?

2.05 Comparison of Programs

Industrial waste programs take many forms throughout the country. One basic difference among the programs is the type of personnel involved. In some agencies the emphasis is on a large number of industrial waste inspectors who concentrate on the detection of violations already occurring at industrial companies. These inspectors tend to have relatively low levels of technical expertise. Another industrial waste program format concentrates on a higher level of technical expertise. In such a program, engineers or chemists attempt to solve overall industrial waste problems rather than concentrate only on on-site inspections of companies. In this second type of program, there is also a need for door-to-door inspections, but the entire program is not based upon such inspections; instead, detailed discharge permit reviews provide a basis for requiring companies to install adequate industrial waste pretreatment measures before wastewater can be discharged to the sewer. In such a program, a violation does not have to be detected to generate an improvement in industrial waste treatment measures.

Other program formats focus entirely on monitoring the wastewater coming out of each company. Under such a program, the only concern is the quality of the wastewater exiting the industrial plant. A high level of technical expertise is not required. Such a program may depend upon rigorous legal enforcement since the opportunity to work cooperatively with the company to resolve technical problems may not exist. In small POTW agencies or areas where there is no significant amount of industry, salaries may not be adequate to attract people with a significant level of industrial waste program expertise, but technical expertise may be available part-time through the use of consultants. The efforts at small agencies may more appropriately be concentrated on measuring the effluent from industrial companies and determining that compliance is consistently being obtained.

For the larger POTW agencies or where significant levels of industrial development exist, it is advantageous to develop personnel who can work intelligently and cooperatively with the major industrial dischargers. For example, at some agencies personnel are assigned to specific industrial disciplines such as the petroleum production and refining industry. One person coordinates the regulation of approximately 20 major refineries and about 50 major oil field production companies. By understanding the technology involved in petroleum production and refining, the agency is able to more effectively and economically regulate industrial waste problems in this industry. A POTW agency having a large component of industrial wastewater from a specific industry may find it advantageous to develop expertise in the technical processes of this industry. The correction of industrial waste problems can be obtained more readily when the POTW agency speaks in knowledgeable terms with industrial personnel.

2.06   Resources Required

In any industrial waste program, the resources required are basically personnel and equipment. In the more complex programs, such as those at a large metropolitan POTW agency, three levels of personnel are commonly employed: high-level technical experts, mid-level technical personnel and technologist-type workers. For example, at some agencies, engineering personnel deal with the complex technology of major groups of industrial dischargers. These highly trained specialists are usually graduate-level engineers with a masters

degree or a doctorate in environmental engineering. In addition, high-level technical personnel are required in the agencies' laboratories to perform the complex analyses of the waste water required to detect industrial pollutants. These personnel may have bachelors or advanced degrees in chemistry.

Staff members with a middle level of technical expertise are employed by some agencies to inspect industrial companies. The inspection personnel generally have college degrees in the sciences, most commonly in biology or chemistry. These personnel are well qualified and develop a good understanding of the basic principles governing proper industrial waste regulation. There is currently no difficulty obtaining well-qualified inspectors with a strong biological science or chemistry background. Salary levels for industrial waste inspection work are competitive with salaries paid for biology or chemistry majors in industry.

A third level of technical expertise can be described as technologist-type workers. This level of expertise is required for setting wastewater samplers at industrial companies. The personnel doing this work generally have one to three years of college and a good practical understanding of flow measurement and chemical processes. These personnel routinely monitor industrial companies and verify that flow measurement devices and other instrumentation, such as combustible gas meters, work properly (Figure 2.1).

At agencies with less complex programs, the three levels of technical expertise may not be required. Agencies with only a small industrial waste component may find it adequate to employ only one person to do the wastewater inspection and monitoring as well as the laboratory analyses. Other agencies employ consultants to handle most of the technical aspects of the industrial waste program and use only technologist labor for periodic on-site review of the companies. Data technicians and computer programmers are often necessary in larger programs.

In addition to the on-staff, high-level technical personnel in an industrial waste program, it is frequently necessary to have legal advice. In some agencies this legal advice is provided on staff but in other agencies consulting attorneys are used. Currently, there is a shortage of attorneys who are adequately familiar with the details of EPA regulations governing industrial waste pretreatment programs.

Figure 2.2 is an organizational chart of the Industrial Waste Section at the Sanitation Districts of Los Angeles County showing how a large POTW agency is organized. The Industrial Waste Section is divided into four basic elements. The Permit Review Subsection reviews and approves industrial wastewater discharge permits prior to issuing a permit to allow an industrial waste discharge to the sewer. In general, college graduates well schooled in industrial processes and pretreatment technologies should be employed since the permit is the permission granted to an industrial user to discharge its wastewaters into the wastewater collection system.

In the Field Engineering and Inspection Subsection, LA employs some engineers to deal with ongoing technical problems such as tracing of treatment plant upsets and sewer problems reportedly caused by industrial waste discharges. An engineer also supervises the industrial waste enforcement program. The Industrial Waste Inspection Staff, as mentioned above, is composed primarily of college graduates with training in the sciences, mostly biology and chemistry. In general, college graduates who are knowledgeable in industrial processes should be employed to provide these services to the POTW. Good in-house training for all inspectors is also essential if a pretreatment program is to be successful.

Fig. 2.1   Effluent meters at an oil production facility that discharges brine water to the sewer. Flowmeter and explosimeter (LEL) are shown with continuous strip chart recorders. Light on turret is emergency warning light that flashes when explosivity reaches an unacceptable limit (10% of LEL).

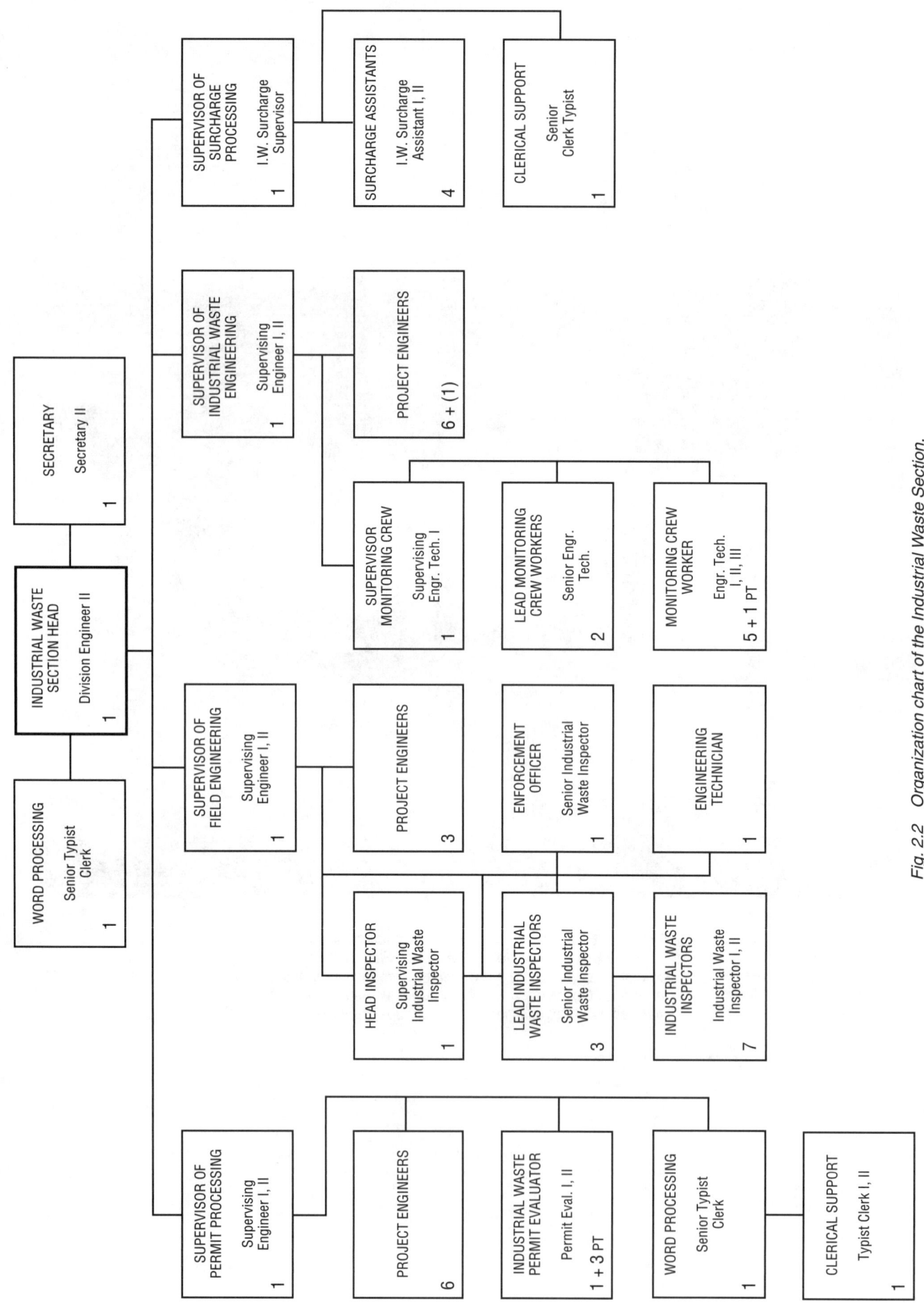

Fig. 2.2   Organization chart of the Industrial Waste Section, Sanitation Districts of Los Angeles County

In the Industrial Waste Engineering and Monitoring Subsection, graduate-level engineers are employed to develop expertise in specific fields of the larger industrial wastewater discharge groups. For example, one engineer is assigned to the metal finishing and similar inorganic chemical pollutant sources, such as battery manufacturers and reclaimers. Another engineer specializes in the petroleum production and refining industry. It is important to develop specialized experts in every pretreatment program.

The Industrial Waste Monitoring Crews are made up of personnel who have completed two to four years of college. These personnel set automatic sampling equipment at industrial companies to obtain samples for analysis at the Sanitation Districts' laboratories (Figure 2.3).

In the Surcharge and Connection Fee Subsection, personnel are employed to collect revenue from industrial dischargers. Most POTWs derive revenue from two sources: annual surcharge fees and one-time industrial sewer connection fees. The personnel in this Subsection are business-oriented employees who, in general, have completed two to four years of college training. Their function is to audit periodically the payments made by industrial companies to ensure that correct amounts of revenue are being received.

In addition to the personnel resources, which are generally by far the most costly, equipment resources are required for all industrial waste regulatory programs. These equipment resources generally include items such as cars, trucks, wastewater samplers, atmospheric monitors and grab sampling equipment. Shown in Table 2.1 is the equipment used by a Sanitation Districts' industrial waste inspector to perform daily duties. Shown in Table 2.2 is a list of equipment routinely used by an industrial waste monitoring crew in the Sanitation Districts.

2.07 Obtaining Approval of Industrial Waste Program

In developing a new industrial waste program, approvals must be obtained from several groups. Initially these approvals must be obtained from the management of the POTW agency. The program must also be discussed with local industries so that they are in general concurrence. A formal approval of the program must finally be obtained from the state (if it grants NPDES permits) and the federal EPA. Currently, most existing POTW agencies with any significant industrial waste dischargers are required by EPA to establish an industrial waste pretreatment program.

In obtaining approval of the industrial waste program from the management of the POTW agency, be aware that the reasons for industrial waste regulation may not be clearly understood by management. The main function of a POTW agency is to convey and treat wastewater. In this regard, the industrial

waste program is a secondary activity which will allow the POTW agency to perform more effectively. Since industrial waste regulation is not the prime role of the agency, it may, from time to time, suffer from inadequate attention or inadequate funding. One of the roles of industrial waste personnel must be to remind management periodically of the importance of industrial waste regulation.

Regulated industries must also understand the POTW's perspective on the need for industrial waste regulation. It is to the advantage of industry that a rigorous and evenly applied industrial waste program be established. Under current state and federal regulations, industries may face major problems if industrial wastes are not properly controlled. Lawsuits and other legal actions against companies violating environmental regulations can create severe financial burdens for these companies. Additionally, companies within an area are commonly very competitive with each other and want to be assured that all companies meet the same environmental standards and face the same costs for pollution control regulation. No one can justify allowing some companies to inadequately treat their wastewater and thereby enjoy a significant monetary advantage as compared to other companies in the same area that comply with environmental regulations.

When the pretreatment program is formulated and approved by the local POTW agency and industry, it must obtain the approval of the state (if the state grants NPDES permits) and the federal EPA. This approval process is described in 40 CFR 403. In addition to the initial pretreatment program approval process, EPA and many of the states conduct periodic audits of local pretreatment programs. The EPA has developed a detailed checklist of the items required in a local pretreatment program. (See Appendix A, "Control Authority Pretreatment Audit Checklist and Instructions.") POTW agencies must be alert to changing regulations and modify their industrial waste pretreatment programs to maintain approval by EPA.

2.08 Need for Assistance From Outside Agency

In many agencies, assistance from outside the agency will be necessary to implement or to carry forward an industrial waste program. It is preferable that at least the head of the industrial waste program be employed by the POTW agency. The use of consultants for laboratory analyses, legal prosecutions, wastewater sampling and other procedures *MAY* be feasible if properly handled. It may also be possible to collect industrial waste revenue by contracting with agencies outside the POTW. At smaller agencies, it is frequently necessary to obtain the required expertise in industrial waste by employing consulting engineers or chemists. Consultants may also be

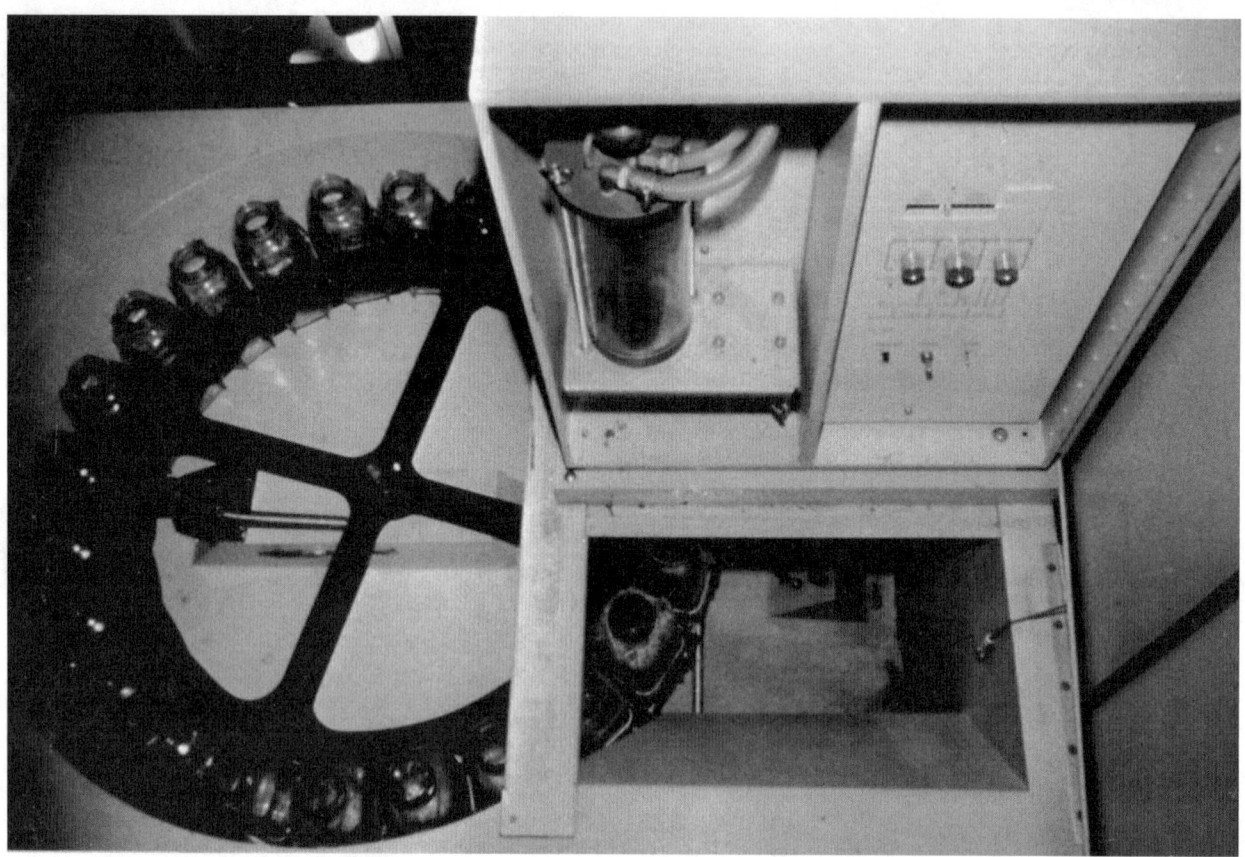

Fig. 2.3   Continuous sampler being placed in sewer lift station. This installation allows inspector to review influent at that particular junction in the sewer. Information obtained here is used to investigate long-term intermittent treatment plant upsets apparently originating from a high-risk upstream area. Each discrete sample bottle should contain the appropriate preservative. Note that the sample bottles are open to the atmosphere in the lift station and are susceptible to contamination by deposition from the atmosphere which has proven to be a significant source of error when sampling for metals.

TABLE 2.1 EQUIPMENT LIST FOR SANITATION DISTRICTS'
INDUSTRIAL WASTE INSPECTOR

Description	Number Carried	Replacement Cost	
		Each ($)	Total ($)
Sample Bottles, Vials	10	4.28	42.80
Sample Bucket and Rope	1	25.00	25.00
Camera and 2 Packs of Film	1	53.50	53.50
Mirrors, Flashlight	2	10.50	21.00
Citation Book for Warning Notices	1	—	—
Vacuum Pump	1	40.00	40.00
Package of pH Paper	1	22.47	22.47
Dye Tablets	50	0.107	5.35
Cyanide Kit	1	45.00	45.00
Chromium Kit	1	27.00	27.00
Copper Kit	1	30.00	30.00
Hydrogen Sulfide Kit (Hach)	1	20.00	20.00
Sulfide Kit (Pomeroy)	1	60.00	60.00
Thermometer	1	13.00	13.00
Explosimeter	1	600.00	600.00
Thomas Guide Map Book With Sewer Overlays	1	60.00	60.00
Consolidated Sewer Maintenance Maps	—	—	—
Permit Files and Plans	—	—	—
Laptop Computer/Data Logger	1	3,000.00	3,000.00
Cellular Phone	1	50.00	50.00
Video Camera	1	800.00	800.00
Industrial Records on Diskette	24	0.50	12.00
Microfiche Viewer	1	35.00	35.00
Business Card File	1	—	—
File Boxes	3	25.00	75.00
Identification Card	1	—	—
Oil Company Credit Card	3	—	—
Telephone Credit Card	1	—	—
CSUS and EPA Training Manuals and Forms Notebook	1	40.00	40.00
DOT Emergency Response Book	1	5.00	5.00
NIOSH Guide to Hazardous Chemicals	1	8.00	8.00
SIC Manual	1	35.00	35.00
Amber Safety Light for Vehicle	1	160.00	160.00
Manhole Lifting Equipment	1	20.00 - 100.00	40.00
Traffic Cones, Flag	6	6.00	36.00
Sample Ice Chest	1	25.00	25.00
Blue Ice	10	4.00	40.00
Hard Hat, Goggles	1	20.00	20.00
Rain Boots, Rain Suit	1	45.00	45.00
Safety Glasses	1	45.00	45.00
Misc. Hand Tools and Safety Equipment	—	—	230.00

**TABLE 2.1 EQUIPMENT LIST FOR SANITATION DISTRICTS'
INDUSTRIAL WASTE INSPECTOR** (continued)

Description	Number Carried	Replacement Cost Each ($)	Total ($)
First Aid Kit	1	35.00	35.00
Explosion Proof Flashlight	1	15.00	15.00
Aluminum Clipboard and Log Book	1	35.00	35.00
CSD Jacket and Cap	1	25.00	25.00
Drager Atmospheric Toxin Detector	1	200.00	200.00
Drager Tubes	10	3.50	35.00
5 lb. Fire Extinguisher	1	40.50	40.50
Lead Tag Seal Crimper	1	40.00	40.00
Lead Tag Seals	20	1.50	30.00
Vehicle (mid-size station wagon or minivan)	1	13,500.00	13,500.00
Total			19,721.62

**TABLE 2.2 EQUIPMENT LIST FOR SANITATION
DISTRICTS' MONITORING CREW**

Description	Number	Total Replacement Cost ($)
Automatic Samplers with Batteries	49	83,055
Spare Batteries for Automatic Samplers and Flowmeters	100	20,500
Miscellaneous Hand Tools and Safety Equipment	—	4,335
Flowmeters and Recorders	3	7,500
12-Volt D.C. Ventilation Blower With Sock	2	1,365
Tripod and Winch	1	3,300
Explosimeter-O_2-H_2S Detector	1	2,205
pH Meters with Batteries, Electrodes	10	9,600
Traffic Control Equipment	—	750
Vehicle (S-10 Pickup Truck With Camper Shell)	2	18,000
Vehicle (¾-Ton Pickup Truck With Camper Shell)	6	90,000
Vehicle (¾-Ton Pickup Truck With Utility Bed)	4	64,000
Total		304,610

very helpful in program development and in developing or revising local industrial waste discharge limits. At some agencies, all technical work is handled internally by agency staff and consulting attorneys are used to prosecute both criminal and civil actions. In general, all criminal actions must be prosecuted by city attorneys or by county district attorneys. Civil actions can be prosecuted by either the agency's in-house attorney or by consulting attorneys.

2.09 Legal Authority for Pretreatment Program

A POTW agency must have the legal authority to control industrial discharges to the POTW. Refer to Section 3.30, "Wastewater Ordinances," page 225, for a discussion of the legal authority needed.

QUESTIONS

Write your answers in a notebook and then compare your answers with those on page 82.

2.0E   Small pretreatment facility inspection agencies should concentrate their efforts on what kinds of activities?

2.0F   List the major subsections of a typical Industrial Waste Section at a large POTW.

2.0G   In developing a new industrial waste pretreatment inspection program, who must approve the program?

2.1 FUNDING OF INDUSTRIAL WASTE PRETREATMENT PROGRAMS

2.10 Sources of Funding

Industrial waste pretreatment programs generally derive funding from two sources: (1) directly from industries, and (2) from other sources within the community. Since the mid-1970s, the federal government has furnished grant funds to requesting POTW agencies for construction of wastewater treatment facilities. These funds indirectly benefit local industries since they share proportionally with the remainder of the community in the reduced local cost of wastewater treatment facilities. In the Federal Construction Grant Regulations, 40 CFR 35, EPA specifies that industries must pay their fair share of the costs of operating the POTW system (collection and treatment) as well as their proportional costs of the POTW's industrial waste pretreatment program. Companies should also pay their appropriate share of the capital costs of the collection and treatment facilities. A federal program of "Industrial Cost Recovery" of the industrial share of Federal Construction Grant Funds has, however, been voided.

A reference which may be useful in the development or revision of a revenue program is the Water Environment Federation's Manual of Practice SP-3, *FINANCING AND CHARGES FOR WASTEWATER SYSTEMS.*[1]

In many areas, it is believed that local industry is a positive benefit to the community. The community may wish to provide various incentives to encourage such industrial location and the creation of local employment. Incentives sometimes take the form of lowered charges to industrial companies for wastewater treatment and disposal. There are, however, certain minimum charge requirements for industry specified in the EPA Construction Grant Regulations. A positive benefit which communities can provide to industrial companies is the assurance that adequate wastewater treatment services are available. A major factor in the location of industry in certain areas is that their wastewater disposal needs will be adequately handled. Companies having large volumes of wastewater or wastewater with high strengths will usually locate only in those areas where wastewater treatment services are available. It is recommended that communities seeking to attract industries to their area plan ahead to have excess wastewater treatment capacity (funded by the community, not EPA) available to service these companies.

2.11 Charges to Industry

2.110 Property Taxes

Charges made directly to industry for a wastewater treatment program are generally of three basic types: (1) ad valorem taxes (according to value), (2) water bill charges, and (3) charges proportional to wastewater discharge flows and strength. *Ad valorem* taxes, such as property taxes, are proportional to the assessed valuation of the industrial company. Property tax charges were and are quite common in communities across the country. Prior to the 1960s and 1970s, many communities obtained revenue from industrial companies only through property taxes. Since property taxes and similar *ad valorem* charges are not based on the quantity and quality of the wastewater or the services provided to industry, such taxes may not appropriately charge those companies creating large demands on the collection system. For example, carpet dyeing companies or slaughterhouses may not have a large property tax bill but may create a significant burden on the wastewater collection and treatment systems. On the other hand, a large hotel or office building may pay excessive amounts of property taxes for its proportional use of the collection and treatment systems. The ability to collect adequate revenue through the use of *ad valorem* taxes has been severely restricted in some states due to maximum limits imposed on such revenue sources.

2.111 Water Bill Charges

Where the POTW agency controls the water supply agency, it is easy to place a sewer service charge (not a tax) on the water bill. Many POTWs, such as sanitation agencies, may not be able to use water bills for sewer service charges. For industrial companies, such charges may be proportional to wastewater flow (if proportional to the volume of water supplied) but not proportional to wastewater strength. In such a charge structure there should be some accounting for differences between companies and the proportion of water supply

[1] FINANCING AND CHARGES FOR WASTEWATER SYSTEMS (MOP SP-3). Obtain from Water Environment Federation (WEF), Publications Order Department, 601 Wythe Street, Alexandria, VA 22314-1994. Order No. MSP3WW. Price to members, $34.75; nonmembers, $44.75; price includes cost of shipping and handling.

discharged to the sewer. For example, a carpet dyer may discharge 85 percent of purchased water to the sewer but a soft drink manufacturer may discharge only 50 percent of its purchased water to the sewer. In some facilities "deduct meters" are installed to measure the flow not reaching the collection system. Some POTW agencies assess an add-on charge that collects additional money for high-strength wastes. The City of Los Angeles imposes a basic water bill charge to industry plus an additional charge (surcharge) that reflects the strength of the wastewater; both of these charges are collected through the water bill.

2.112 Charges Proportional to Wastewater Discharge

When EPA regulations were established after the Federal Water Pollution Control Act of 1972, industrial companies began paying for wastewater treatment services in proportion to their use of these services. The EPA regulations require that charges paid by companies be proportional to the volume of the wastewater and to the strength of the wastewater as measured by indicators such as biochemical oxygen demand (BOD) and/or pollutant concentrations (such as chemical oxygen demand (COD), suspended solids (SS), and ammonia nitrogen (NH_3–N)). Such charges, if properly calculated, can reasonably reflect the costs of providing wastewater treatment services. However, when calculated in this manner, the charges do not necessarily reflect the costs of an industrial waste control program.

At one agency, a large portion of the time of industrial waste inspectors, permit evaluation engineers, expert technical engineers and other Industrial Waste Section personnel is spent on control of dischargers of toxic pollutants. The industry responsible for the largest proportion of the agency's Industrial Waste Section time is the metal finishing industry. The large number of such companies, in comparison with other industries, and the high toxicity of their wastes make it necessary to devote a significant amount of time to control of these companies. Since the wastes from metal finishing companies are usually low in BOD, COD and SS, no significant surcharges are assessed to these companies under the EPA's wastewater quantity and quality charge program. An appropriate revenue structure would charge such companies additional fees which are proportional to the demands they place on the resources of the inspection, collection and treatment (POTW) agency. An inspection charge could be included in an annual permit renewal fee.

At one agency, all companies discharging between one and six million gallons per year are required to pay a flat rate charge based upon the volume of the wastewater discharge. To calculate the volume of the discharge, the volume measured at the influent water meter is reduced by the amount of water consumed in the company's product, water consumptively used within the plant and water used for irrigation purposes, as measured by appropriately located deduct water meters. Figure 2.4 is an example of how one company calculates its net wastewater discharge volume using water bill amounts as the starting point.

Inspectors should realize that water meter flow volumes may be in error. Comparisons of influent water meter volumes with volumes recorded on effluent wastewater meters have frequently indicated that the influent water flowmeter is in error, and subsequent calibration of the water meter has almost always verified that the water meter was reading low. Problems of this type are especially common with large influent water meters because water companies find it difficult to properly repair or calibrate such meters. The accuracy of small water meters is usually much better since they can be readily repaired, calibrated and replaced.

Some agencies require companies discharging over 50,000 gallons per day to install a wastewater flowmeter which will continuously indicate, record and totalize the wastewater flow to the collection system. Based upon the wastewater flow measurement and periodic sampling of the company for COD and SS, a strength-related total charge can be developed. The strength charge is usually based on the total mass of COD (or BOD) and SS discharged to the collection system. To offset the property taxes an industry pays for basic services, many agencies give companies credit for a basic sanitary-equivalent wastewater. The allowance of sanitary-equivalent wastewater may not be workable in states that have severely restricted the use of property taxes to pay for wastewater collection and treatment services. A recommended alternative means to obtain revenue from industry for municipal wastewater and industrial waste purposes is through user charges and similar charges proportional to the use of the collection and treatment systems.

Some cities have developed an elaborate system of categories for industries. After sampling the strength of the various industrial wastewaters, categories of industries are defined for purposes of assessing charges. On the average, the wastewater strength is assumed to be constant for companies within the industrial category and charges are based upon this average wastewater strength. Such charge structures are accurate on the average, but may significantly overstate or understate an individual company's use of the collection and treatment systems. An appeal procedure is available to companies that believe they are being overcharged.

Specific user fees are frequently applied. Some agencies charge companies for inspections, either those made by the agency or deemed necessary on an annual basis. For example, metal plating companies may be required to pay for eight inspections per year whereas small meat packing plants might pay for only two industrial waste inspections per year. Some agencies charge a fee to process an industrial wastewater discharge permit. A charge may also be appropriate for efforts made in sampling and analyzing a company's wastewater.

Legal costs may also be recovered by the agency. In legal actions with an industrial company, it frequently occurs that the court will require the industry to pay the costs incurred by the POTW agency to bring the company into court. In flagrant violations, courts sometimes impose penalty charges as a punitive measure to deter other companies from violating POTW agency regulations. Revenue from such enforcement actions is sometimes used to partially support pretreatment programs.

adjusted metered water supply table

Fiscal Year 199__/__
(Attach Additional Sheets If Necessary)

Company Name _____

Account N⁰ [   ]   ## SIC N⁰ [   ]

Sanitation Districts' Permit N⁰ [   ]

1. Metered Water Supply from Purveyor (Water Company). (Complete Table 2A and attach copies of water bills.) + [   ] mil. gal./yr.

2. Water Supply from Company Well. (Attach meter or watermaster data). + [   ] mil. gal./yr.

3. Water Received in Raw Materials. (Explain in attachments). + [   ] mil. gal./yr.

4. Wastewater Discharged to Stormwater Drainage System. (Explain how determined in attachments). Enter your NPDES Permit # _____ . − [   ] mil. gal./yr.

5. Water Lost Through Evaporation. (Explain how determined in attachments). (May include irrigation losses.) − [   ] mil. gal./yr.

6. Water Lost in Products. (Explain how determined in attachments). − [   ] mil. gal./yr.

7. Sanitary Flow Deduction − [   ] mil. gal./yr.

8. Water Gained by Other Means. (Explain how determined in attachments). + [   ] mil. gal./yr.

9. TOTAL WASTEWATER DISCHARGED TO PUBLIC SEWER. [   ] mil. gal./yr.

WATER LOSS CALCULATIONS:
IRRIGATION LOSSES

Square Footage of Land Irrigated	×	18.7*	÷	1,000,000	=	Mil. Gal. Per Year
	×	18.7	÷	1,000,000	=	

*18.7 = Gallons per square foot of irrigated area per year

COOLING TOWER

Tonnage	×	Hours of Operation Per Year	×	% Load	×	1.38*	÷	1,000,000	=	Mil. Gal. Per Year
	×		×	0._____	×	1.38	÷	1,000,000	=	
	×		×	0._____	×	1.38	÷	1,000,000	=	

*1.38 = Gallons evaporated per hour per ton.

BOILER

Horsepower	×	Hours of Operation Per Year	×	% Load	×	% Steam Loss To Atmosphere	×	3.82**	÷	1,000,000	=	Mil. Gal. Per Year
	×		×	0._____	×	0._____	×	3.82	÷	1,000,000	=	
	×		×	0._____	×	0._____	×	3.82	÷	1,000,000	=	

**3.82 = Gallons evaporated per hour per horsepower. Include gas consumption records.

SANITARY FLOW

Employees	×	Discharge Days Per Year	×	Gallons per Employee per Day	÷	1,000,000	=	Mil. Gal. Per Year
	×		×	15	÷	1,000,000	=	

Conversion Factors:
1 cubic foot = 7.48 gallons 1 acre foot = 325,900 gallons
1 gallon of water = 8.34 pounds of weight

Fig. 2.4 Adjusted metered water supply table

Some agencies use two basic charge structures for industries: (1) the annual or monthly (periodic) surcharge, and (2) the industrial sewer connection fee. A periodic surcharge is assessed based on industrial wastewater flow volume and the strength of the wastewater discharge. This charge structure allows the POTW to recover from each company the approximate cost of treating its wastewater discharge during the billing period.

Surcharge programs of this sort often rely on self-monitoring by the companies whereby they periodically report their wastewater discharge to the POTW and pay the appropriate fee. Random sampling by the POTW verifies the information submitted. Periodic audits by the POTW agency will review company information and, if the data seem questionable, the POTW may require more information or closer monitoring by the agency's staff.

Another potential source of revenue is the industrial waste sewer connection fee. This is a "one-time" charge and applicable to all new companies connecting to the collection system. The company pays this charge to expand the capital facilities, sewers and treatment plants which are necessary to handle the increased industrial wastewater discharge. The industrial sewer connection fee may be quite substantial for companies having high-volume and high-strength wastes. For large secondary-fiber paper mills, for example, the connection fee can sometimes exceed one million dollars. To help pay for the collection system, a company may also have to pay a connection fee based on the length of sewer going past the company's property (front footage).

When existing companies radically expand their wastewater discharge, usually they are also assessed additional connection fees. The baseline capacity entitlement for each industry is based either on their past surcharge payments in a certain time period or their initial purchase of capacity upon obtaining a new permit. If the industry exceeds this capacity entitlement by more than 25 percent, it must pay a new connection fee for the entire volume beyond the baseline capacity.

Obviously, a community must have available capacity ready to accommodate companies and will generally not have the time to build this capacity once a company applies for a discharge permit. Therefore, the money collected from a company for connection fees actually is a reimbursement to the POTW for collection and treatment capacity already constructed.

In summary, there are several methods by which POTWs can recover the costs of operating and administering a pretreatment program. Contact your EPA regional office to determine which methods are preferred or required in your EPA region and to request examples of ordinances or procedures of POTWs in the area using those methods. Also see Appendix II, "Pretreatment Arithmetic," at the end of this manual. Section L, "Sewer-Use Fees," contains examples of the arithmetic involved in calculating sewer-use fees.

QUESTIONS

Write your answers in a notebook and then compare your answers with those on page 82.

2.1A   List the sources of funding for an industrial waste pretreatment inspection program.

2.1B   List the basic types of charges made directly to companies for a wastewater treatment program.

2.1C   What specific user fees may be applied or charged to a company?

Please answer the discussion and review questions next.

End of Lesson 1 of 3 Lessons on PRETREATMENT PROGRAM ADMINISTRATION

DISCUSSION AND REVIEW QUESTIONS
Chapter 2.   PRETREATMENT PROGRAM ADMINISTRATION

(Lesson 1 of 3 Lessons)

At the end of each lesson in this chapter you will find some discussion and review questions. The purpose of these questions is to indicate to you how well you understand the material in the lesson. Write the answers to these questions in your notebook before continuing.

1. Why is it important that inspectors and other personnel in larger POTW agencies be technically well trained?

2. Why do industrial waste pretreatment inspection programs sometimes suffer from inadequate attention or inadequate funding?

3. What is a major concern of industry with respect to industrial waste pretreatment inspection programs?

4. What costs should companies pay with regard to waste waters discharged to a POTW?

5. What is a company's primary concern regarding waste waters when it considers locating in a community?

6. Why is EPA's quantity and quality charge program not appropriate for the metal finishing industry?

CHAPTER 2.   PRETREATMENT PROGRAM ADMINISTRATION

(Lesson 2 of 3 Lessons)

2.2   ELEMENTS OF A PRETREATMENT PROGRAM

Certain elements are common to most industrial waste pretreatment programs. The following paragraphs describe some of the basic elements believed necessary in a well-established program. Some agencies may not require all of the following elements in great depth but these elements are, in general, considered appropriate under the EPA pretreatment regulations. The following description of pretreatment program elements is not intended to be a complete listing of all the requirements.[2] Rather, it will give inspectors a general idea of the structure of a pretreatment program and an introduction to the inspector's role in various facets of the program. Chapter 3 contains more detailed descriptions of each of the required pretreatment program elements.

2.20   Permission to Discharge to the Sewer System

POTW agencies must be able to restrict the discharge of industrial wastes to the collection system. Realistically, unrestricted discharge of industrial wastes can only lead to treatment plant upsets and severe problems in the sewers, in the community, and in the receiving waters of the treatment plant. An industrial wastewater discharge permit program or similar control mechanism should be used to regulate the entrance of industrial wastewater to the collection system. All companies wishing to discharge to the collection system must obtain an industrial wastewater discharge permit.[3] This permit gathers information about the company such as location, type of industry, number of employees, flow volume, characteristics of wastewater discharges, and toxic wastes stored on site. The permit form is essentially a contractual agreement between the company and the agency: the agency agrees to provide wastewater treatment services if the industrial company complies with the requirements of the agency's *Wastewater Ordinance* and other industrial waste regulations.

Some POTW agencies have not previously used a formal permit procedure, but many are currently implementing permit procedures at least for the larger and more important companies. The EPA pretreatment regulations require that information on industrial wastewater discharges to the collection system be obtained and readily available to the POTW. The regulations further require that POTWs with approved pretreatment programs control significant industrial users (basically categorical industrial users or those with flows greater than 25,000 GPD of process wastewater) through permits or individual control mechanisms issued to each industrial user. The permit must include the following items:

1. A stated duration (not longer than five years),

2. A statement to the POTW of nontransferability or prior notification of impending transfer of ownership,

3. Effluent limitations based on applicable pretreatment standards, self-monitoring, sampling, reporting, notification, and recordkeeping requirements, and

4. A statement of applicable civil and criminal penalties for violation of pretreatment standards.

2.21   Inspection of Companies

2.210   Staffing Requirements

In all industrial waste programs it is necessary to inspect the industries to determine whether they are in compliance with industrial waste regulations. Adequately trained technical personnel must be available to enter the company's facilities and inspect the processes being used. Without reasonably accurate knowledge of what is happening within the facility, it is difficult to estimate what effect the company may have on the collection and treatment systems without full-time wastewater monitoring at the company. Knowledgeable inspectors will be able to determine, based upon their review of many types of similar companies, whether a reasonable effort is being made to meet effluent limitations. The inspection of industries is part diplomacy, part art and part science. It is definitely a discipline in which experience and education are vital to the adequate performance of an inspector's duties.

Inspection teams can be assigned responsibility for a relatively large geographical area. These inspection teams can be headed by a senior industrial waste inspector who supervises three journey-level inspectors. The team members all become generally familiar with the team area, but each of the four inspectors is responsible for control of a specific geographical area. The senior inspector monitors the daily activities of the subordinate inspectors and schedules their time. The advantage of the team concept is that more than one inspector can be called upon to handle emergency problems and several inspectors will be knowledgeable about the general team area.

The assignment of inspectors to certain geographical areas is an effective use of personnel resources. Inspectors should understand that they are responsible for the assigned geographic area and have the duty and authority to enforce wastewater discharge regulations from this area so that no

---

[2] See Appendix A, "Control Authority Pretreatment Audit Checklist and Instructions," at the end of this chapter for a more complete listing of pretreatment program elements. The audit checklist is used by the Control Authority (EPA or state) when approving or reviewing the approval of a POTW's pretreatment program.

[3] See Appendix B, "Industrial Wastewater Discharge Permit Application Form," and Appendix C, "Wastewater Discharge Permit Application for Existing Industrial Users, City of Hopewell, Virginia."

problems are created in the collection system. Although it is not a perfect analogy, the industrial waste inspector is somewhat like a "cop on the beat" who is responsible for a specific geographical area. Assignment to one area may require the inspector to develop knowledge of a wide variety of industrial plants but it also provides field personnel with clearly defined responsibilities.

Complex, long-term problems should be referred to engineering personnel so that individuals with specific expertise in certain types of industries can spend extended amounts of time working to resolve a major problem. This time for working on a single problem is, in general, not available to the field inspection personnel. The use of inspectors to solve problems taking more than a few hours to a few days removes them from their primary function of the continuing review of a large geographical area.

Field inspection personnel should, if possible, be rotated to new areas at least every two years. This allows them to see many geographical areas and develop wide experience over different areas and with different types of industries. Job rotation helps to develop well-trained and competent inspectors. Job rotation also helps to prevent stagnation of the inspector in a certain area and undue familiarity between the inspector and the companies in the area. Where inspectors spend long periods in one area, long-term contact between company personnel and inspectors may lead to special treatment of companies due to such relationships.

Many POTW agencies use one inspector for conducting inspections and/or collecting samples. Two inspectors may be used if witnesses are needed in case of legal action, or if the company has a history of significant noncompliance and one inspector is needed to look for violations during a tour while the other inspector works with the company contact. Two or more people are required for hazardous sampling situations such as in traffic or *CONFINED SPACES.*[4]

2.211   Frequency of Inspections

The frequency of inspections of pretreatment facilities varies from agency to agency and inspections are to some extent beyond the control of the individual pretreatment facility inspector or industrial waste program leader.

Recent amendments to the General Pretreatment Regulations require POTWs to inspect and sample all Significant Industrial Users (SIUs) at least once a year. The inspection and sampling may both be accomplished during a single visit. A Significant Industrial User is generally defined to be all EPA categorical companies plus those companies discharging over 25,000 gallons per day, plus other companies deemed by the agency to require special attention.

In a guidance manual entitled *PRETREATMENT COMPLIANCE MONITORING AND ENFORCEMENT GUIDANCE,* EPA recommends that the control agency prioritize their inspection frequencies based upon one or more of the following criteria:

1. Volume of industrial user wastewater discharge,

2. Type and concentration of pollutants in the wastewater discharge,

3. Problems in the POTW known or suspected to have been caused by the industrial user,

4. Past compliance history of the industrial user,

5. Daily or seasonal production or discharge variations at the industrial user, and

6. Resources of labor and equipment available to the Control Authority.

In some EPA regions it is recommended, informally, that at least one thorough inspection of a Significant Industrial User be made each year, plus a separate sampling visit where a 24-hour composite sample is obtained. During the inspection visit, a grab sample can be obtained to indicate any significant difference from the 24-hour composite sample results.

In many POTWs, the practical criteria for determining frequency of inspections may be:

1. The problems the company has created for the POTW,

2. Violations previously detected at the company, and

3. Amount of resources available for inspections.

It is recommended that a POTW establish some reasonable guidelines for frequency of inspections. Such criteria may be examined critically during the periodic EPA review of the agency's pretreatment program.

2.22   Monitoring of Industrial Wastewater Discharges

Although companies may be required to self-monitor their wastewater discharges, the POTW agency must also monitor discharges to verify that the industries' information is correct. Large agencies may require an elaborate self-monitoring program. In some instances it is suspected and observed that the self-monitoring results by some companies are better in quality than the analyses obtained by the agency. A self-monitoring report form is shown in Figure 2.5.

In addition to verifying self-monitoring information furnished by the industrial dischargers, POTW monitoring is also required to locate the source of treatment plant upsets and sewer system problems caused by industrial waste discharges. Monitoring may also be needed to verify or determine

---

[4]   *Confined Space.   Confined space means a space that: A. Is large enough and so configured that an employee can bodily enter and perform assigned work; and B. Has limited or restricted means for entry or exit (for example, tanks, vessels, silos, storage bins, hoppers, vaults, and pits are spaces that may have limited means of entry); and C. Is not designed for continuous employee occupancy. (Definition from the Code of Federal Regulations (CFR) Title 29 Part 1910.146.)*

LACSD USE ONLY		
LAB REPORT Y☐ N☐	SIGNATURE Y☐ N☐	
DATE RECEIVED	INITIALS	

SANITATION DISTRICTS OF LOS ANGELES COUNTY
Charles W. Carry, Chief Engineer & General Manager
For information please call (310) 699-7411 ext. 2900

**INDUSTRIAL WASTEWATER
SELF MONITORING REPORT**

Report due no later than **07/15/96**

PAGE 01 OF 02

PERMIT NO.
000182
SURCHARGE ACCOUNT NO.
1385321

1. Name of Company Having Wastewater Discharge
WESCAL INDUSTRIES   D

2. Has the Ownership or Occupancy Changed Since the Last Report? ☐ Yes ☐ No

3. Address of Wastewater Discharge
18033 S SANTA FE AVE RANCHO DOMINGUEZ CA 90221

4. Name of Industrial Wastewater Contact MR. JERRY CHODERA

5. Phone No. (213) 774-8500

6. Mailing Address (If Different From Above)
18033 S SANTA FE AVE RANCHO DOMINGUEZ CA

7. Sic No.(s) 3496, 3471

8. Reporting Period From: **04/01/96** To: **06/30/96**

9. (Print) Name of Company Collecting Wastewater Sample | **10. (Print) Sample Date** | **11. (Print) Sample Location(s)**

12. Daily Wastewater Discharge For Reporting Period (Gal.)
Average: _____
Maximum: _____

13. Method For Determining Wastewater Flow For Sampling Day (Z01, Z02)
☐ Direct Measurement
☐ Adjusted Metered Water Supply
☐ No Discharge During Reporting Period

14. Type of Composite Sample
☐ Time Composite
☐ Flow Proportioned Composite

15. NOTE:

CODE	PARAMETER (1)	SAMPLING METHOD	TEST RESULTS (2)	LAB ID CODE (3)
Z01	WASTEWATER FLOW, TOTAL(GPD)			
Z02	WASTEWATER FLOW,PEAK(GPM)			
101	PH (UNITS)	GRAB		
151	SUSPENDED SOLIDS (MG/L)	COMPOSITE		
206	TOTAL CYANIDE (MG/L)	GRAB		
252	SOLUBLE SULFIDE (MG/L)	GRAB		
403	TOTAL COD (MG/L)	COMPOSITE		
601	METHYLENE CHLORIDE (UG/L)	GRAB		
602	CHLOROFORM (UG/L)	GRAB		
603	1,1,1-TRICHLOROETHANE (UG/L)	GRAB		
604	CARBON TETRACHLORIDE (UG/L)	GRAB		
605	1,1-DICHLOROETHENE (UG/L)	GRAB		
606	TRICHLOROETHYLENE (UG/L)	GRAB		
607	TETRACHLOROETHYLENE (UG/L)	GRAB		
608	BROMODICHLOROMETHANE (UG/L)	GRAB		
609	DIBROMOCHLOROMETHANE (UG/L)	GRAB		
610	BROMOFORM (UG/L)	GRAB		
611	CHLOROBENZENE (UG/L)	GRAB		
612	VINYL CHLORIDE (UG/L)	GRAB		
613	O-DICHLOROBENZENE(UG/L)	GRAB		
614	M-DICHLOROBENZENE(UG/L)	GRAB		
615	P-DICHLOROBENZENE(UG/L)	GRAB		
616	1,1-DICHLOROETHANE (UG/L)	GRAB		
618	1,1,2-TRICHLOROETHANE (UG/L)	GRAB		
619	1,2-DICHLOROETHANE (UG/L)	GRAB		

(1) Report the test results from the most recent sample collected within the reporting period and include all laboratory test sheets with the self-monitoring report form.
(2) Test results are valid only if the correct sampling method is observed and the laboratory analysis is performed by a State or Sanitation Districts' approved laboratory.
(3) Indicate the appropriate laboratory certification I.D. Code for each testing parameter.

CERTIFICATION BY PERMITEE

I certify under penalty of law that this document and all attachments were prepared under my direction or supervision in accordance with a system designed to assure that qualified personnel properly gather and evaluate the information submitted. Based on my inquiry of the person or persons who manage the system, or those persons directly responsible for gathering the information, the information submitted is, to the best of my knowledge and belief, true, accurate, and complete. I am aware that there are significant penalties for submitting false information, including the possibility of fine and imprisonment for knowing violations.

Signature of responsible company official: _____ Date: _____

Print name of official: _____ Title: _____

Please submit this report to: Sanitation Districts of Los Angeles County Industrial Waste Section P.O. Box 4998 Whittier, CA 90607

Fig. 2.5   Self-monitoring report form

LACSD USE ONLY	
LAB REPORT Y ☐ N ☐	SIGNATURE Y ☐ N ☐
DATE RECEIVED	INITIALS

SANITATION DISTRICTS OF LOS ANGELES COUNTY
Charles W. Carry, Chief Engineer & General Manager
For information please call (310) 699-7411 ext. 2900

**INDUSTRIAL WASTEWATER
SELF MONITORING REPORT**

Report due no later than **07/15/96**

PAGE 02 OF 02

PERMIT NO.
000182

SURCHARGE ACCOUNT NO.
1385321

1. Name of Company Having Wastewater Discharge
WESCAL INDUSTRIES   D

2. Has the Ownership or Occupancy Changed Since the Last Report?
☐ Yes   ☐ No

3. Address of Wastewater Discharge
18033 S SANTA FE AVE RANCHO DOMINGUEZ CA 90221

4. Name of Industrial Wastewater Contact
MR. JERRY CHODERA

5. Phone No.
(213) 774-8500

6. Mailing Address (If Different From Above)
18033 S SANTA FE AVE RANCHO DOMINGUEZ CA

7. Sic No.(s)
3496, 3471

8. Reporting Period
From: **04/01/96** To: **06/30/96**

9. (Print) Name of Company Collecting Wastewater Sample

10. (Print) Sample Date

11. (Print) Sample Location(s)

12. Daily Wastewater Discharge For Reporting Period (Gal.)
Average: _____
Maximum: _____

13. Method For Determining Wastewater Flow For Sampling Day (Z01, Z02)
☐ Direct Measurement
☐ Adjusted Metered Water Supply
☐ No Discharge During Reporting Period

14. Type of Composite Sample
☐ Time Composite
☐ Flow Proportioned Composite

15. NOTE:

CODE	PARAMETER (1)	SAMPLING METHOD	TEST RESULTS (2)	LAB ID CODE (3)
620	BENZENE (UG/L)	GRAB		
621	TOLUENE (UG/L)	GRAB		
624	ETHYLBENZENE (UG/L)	GRAB		
645	TRANS-1,2-DICHLOROETHYLENE (UG/L)	GRAB		
646	BROMOMETHANE (UG/L)	GRAB		
647	CHLOROETHANE (UG/L)	GRAB		
648	2-CHLOROETHYL VINYL ETHER (UG/L)	GRAB		
649	CHLOROMETHANE (UG/L)	GRAB		
650	1,2-DICHLOROPROPANE (UG/L)	GRAB		
651	CIS-1,3-DICHLOROPROPENE (UG/L)	GRAB		
652	TRANS-1,3-DICHLOROPROPENE (UG/L)	GRAB		
653	1,1,2,2-TETRACHLOROETHANE (UG/L)	GRAB		
708	CADMIUM, TOTAL (MG/L)	COMPOSITE		
709	CHROMIUM, TOTAL (MG/L)	COMPOSITE		
712	COPPER, TOTAL (MG/L)	COMPOSITE		
714	LEAD, TOTAL (MG/L)	COMPOSITE		
718	NICKEL, TOTAL (MG/L)	COMPOSITE		
722	SILVER, TOTAL (MG/L)	COMPOSITE		
724	ZINC, TOTAL (MG/L)	COMPOSITE		

(1)   Report the test results from the most recent sample collected within the reporting period and include all laboratory test sheets with the self-monitoring report form.
(2)   Test results are valid only if the correct sampling method is observed and the laboratory analysis is performed by a State or Sanitation Districts' approved laboratory.
(3)   Indicate the appropriate laboratory certification I.D. Code for each testing parameter.

CERTIFICATION BY PERMITEE

I certify under penalty of law that this document and all attachments were prepared under my direction or supervision in accordance with a system designed to assure that qualified personnel properly gather and evaluate the information submitted. Based on my inquiry of the person or persons who manage the system, or those persons directly responsible for gathering the information, the information submitted is, to the best of my knowledge and belief, true, accurate, and complete. I am aware that there are significant penalties for submitting false information, including the possibility of fine and imprisonment for knowing violations.

Signature of responsible company official: _____ Date: _____

Print name of official: _____ Title: _____

Please submit this report to: Sanitation Districts of Los Angeles County   Industrial Waste Section   P.O. Box 4998   Whittier, CA 90607

Fig. 2.5   Self-monitoring report form (continued)

strength data for industrial sewer service charges. Monitoring of companies can be accomplished through "grab" samples taken of the industrial effluent in a short time period. Composite samples taken over a 24-hour period are more expensive to obtain but will yield a better representation of the total burden of the wastewater on the collection and treatment system.

2.23 Technical Expertise

Where any significant concentration of industry discharges to a collection system, it is desirable for the agency to have technical expertise available to coordinate the requirements of the POTW agency with company personnel. While company personnel may be well meaning, their primary function is to produce the product of the company. The company personnel do not necessarily need or want to understand the needs of the POTW agency. If POTW agency personnel are also as narrow in their technical view, it may be difficult for the POTW agency and company technical personnel to adequately communicate. *THE DEVELOPMENT OF POTW AGENCY PERSONNEL WHO UNDERSTAND INDUSTRIAL PROCESSES WILL LEAD TO A BETTER SOLUTION TO PROBLEMS.* At

small POTW agencies the allocation of personnel to develop such expertise may be difficult. If there is a significant concentration of a certain type of industry, a source of technical information for these industries is the EPA development documents used to establish industrial categorical effluent limits. These development documents furnish significant details about the industrial processes and the origin of the industrial wastewater.

At some agencies, the Industrial Waste Section includes a subsection for industrial waste engineering where project engineers are assigned specific areas of industrial waste technology as work concentration areas. As shown in Table 2.3, engineers concentrate on the industrial waste problems from specific SIC^5 number groupings of industries. The engineers develop in-depth knowledge about the industrial processes that create pollutants and the pretreatment processes which can be used to control the discharge of these pollutants. Such expertise is needed to resolve the multiple industrial waste problems found in a large metropolitan area. Again, smaller POTW agencies may not need this depth of technical knowledge but a basic level of such knowledge is desirable to adequately handle major industrial waste problems.

TABLE 2.3 AREAS OF SPECIALTY OF INDUSTRIAL WASTE PROJECT ENGINEERS

Project Engineer	Specialty Area
1	metal finishing companies, leather and leather products companies, glass manufacturing companies, clay and ceramic manufacturing companies, companies having radioactive wastes, companies having PCB wastes, cooperage houses, steam electric power generating facilities, transportation equipment and services, motor freight companies, automobile dealers, automobile repair shops, car washes
2	inorganic chemicals manufacturing companies, organic chemicals manufacturing companies, rubber and miscellaneous plastics manufacturing companies, photofinishing companies, soaps and detergents manufacturing companies, disinfecting and exterminating services, machinery and electrical manufacturing companies, measurement and control instruments manufacturing companies
3	agricultural production, pet hospitals, food products companies, grocery and food stores, eating places, hospitals, medical and dental laboratories, outpatient care facilities, commercial testing laboratories
4	primary and secondary metals manufacturing companies, textile mill companies, lumber and wood product companies, paper manufacturing companies, service stations, companies providing related services to the petroleum industry (oil well service companies), calculations of additional POTW chlorine costs due to thiosulfate discharges from refineries
5	engineer responsible for ensuring technical accuracy of field measurement studies (flow and quality surveys) and investigations/evaluations of new equipment, coordination of split sampling program with dischargers, review of industrial waste permit plans and field installations for adequacy of sampling locations, coordinating repairs to sampler pacing sockets, establishing sampling schedules
6	engineer responsible for field verification of flowmeters including investigation of their installation and operation, coordination of surveillance sampling program, maintenance of flowmeter field logbook

---

[5] *SIC Code.* **S**tandard **I**ndustrial **C**lassification Code. *A code number system used to identify various types of industries. In 1997, the United States and Canada replaced the SIC code system with the North American Industry Classification System (NAICS*); Mexico adopted the NAICS in 1998.*

* *NAICS.* **N**orth **A**merican **I**ndustry **C**lassification **S**ystem. *A code number system used to identify various types of industries. This code system replaces the SIC (Standard Industrial Classification) code system used prior to 1997. Use of these code numbers is often mandatory. Some companies have several processes which will cause them to fit into two or more classifications. The code numbers are published by the Superintendent of Documents, U.S. Government Printing Office, PO Box 371954, Pittsburgh, PA 15250-7954. Stock No. 041-001-00509-9; price, $31.50. There is no charge for shipping and handling.*

2.24   Revenue Collection

The collection of revenue from industrial companies, which is required under the EPA Pretreatment and Construction Grant Regulations, serves basically two functions: (1) to recover the costs of providing wastewater collection, treatment and disposal, and (2) to recover the costs of the POTW agency's industrial waste section functions, including pretreatment inspection. POTW agencies typically have already developed programs to recover the basic costs of providing wastewater collection and treatment services. The methods of collecting appropriate revenues from companies that do not discharge large volumes or strengths of wastewater but still place significant demands on the industrial waste function of the POTW agency are less well established. User fees to industrial groups such as metal finishers may be appropriate in this regard. Charges for industrial wastewater discharge permits, industrial waste inspections and industrial waste monitoring may place a financial burden on such companies but, as discussed earlier, EPA requires POTWs to charge dischargers their fair share of the cost of providing services.

2.25   Compliance Program/Enforcement Program

All POTW agencies with a pretreatment program are required by EPA pretreatment program guidelines to have a compliance function which encourages noncompliant companies to meet agency requirements. There are numerous compliance procedures which can be used by POTW agencies but the use of criminal or civil legal actions is perhaps the most forceful procedure outside of suspension or revocation of the permission to discharge to the collection system.

POTWs must have an adequate legal basis to require compliance and enforce the pretreatment regulations. Any violation of pretreatment requirements (effluent discharge limits, sampling, analysis, reporting and meeting compliance schedules, and regulatory deadlines) is an instance of noncompliance for which the industrial user (IU) is liable for enforcement, including penalties. If a company frequently is in violation or develops a pattern of violations, the industry may be considered in Significant Noncompliance (SNC).[6] When this occurs, the POTW reports the noncompliance to its Approval Authority, notifies the public of the noncompliance, and either initiates appropriate enforcement action or documents its reasons for not taking enforcement action.

POTWs should develop an Enforcement Response Plan which outlines the types of violations, the circumstances, and the range of appropriate responses. When determining the level of enforcement response, the technical and legal staff should consider the degree of variance from the pretreatment standards or legal requirements, the duration of noncompliance, previous enforcement actions taken against the IU, and the deterrent effect of the response on a similar facility in the regulated community. Equally important are considerations of fairness, equity, consistency, and the integrity of the pretreatment program. While the POTW may exercise a degree of flexibility in its response and may take into account any or all of the factors mentioned above, a written guide to appropriate enforcement responses clarifies the ground rules for pretreatment inspectors and for industrial users.

A key element in all enforcement responses is the timeliness with which they are initiated and bring about compliance. Since there are many types of violations, applicable legal enforcement procedures, and differences in resources available to POTWs, time frames will vary within which to initiate and complete a given response. POTWs should establish specific time frames to initiate various types of enforcement actions and follow-up activities. Figure 2.6 shows the sequence of steps taken by the Sanitation Districts of Los Angeles County to enforce compliance with industrial waste regulations.

Industrial user violations of monitoring, reporting, and treatment requirements may range from relatively minor violations (reports submitted a week late) to major violations. Each instance of noncompliance is a violation and sound enforcement policy is to review each violation and respond appropriately. Selection of the appropriate enforcement response will relate to whether the violation is major or minor, the POTW's general enforcement posture, and the types of specific circumstances described above. In most cases, a telephone call or a notice of violation from the POTW to the industry requesting an explanation will bring the problem to the attention of the industry's management. Frequently, such a notification is sufficient to correct the problem.

Industry noncompliance might cause interference with treatment plant performance or pass-through of pollutants. Violations of this type should be handled with formal enforcement action and penalties to ensure that adequate treatment and compliance is achieved promptly. In some cases, a court injunction to stop discharging temporarily will also be appropriate.

The development and implementation of an Enforcement Response Plan is an important aspect of any pretreatment program, and one that should not be ignored. In 1989 EPA and the Department of Justice filed suit against 61 POTWs for failure to properly implement and enforce their pretreatment programs. This suit makes it clear that EPA considers the enforcement of POTW regulations a crucial element in the success of a pretreatment program. Chapter 3 of this manual contains more detailed descriptions of the important elements of an enforcement response plan which includes an example of an enforcement response guide.

---

[6] *Significant Noncompliance (SNC) is defined in "Pretreatment Words" at the beginning of Chapter 3 and also in Section 3.32, PROCEDURES, item 7, and Section 3.40.*

CSD ENFORCEMENT PROCEDURES
NUMERICAL VIOLATIONS

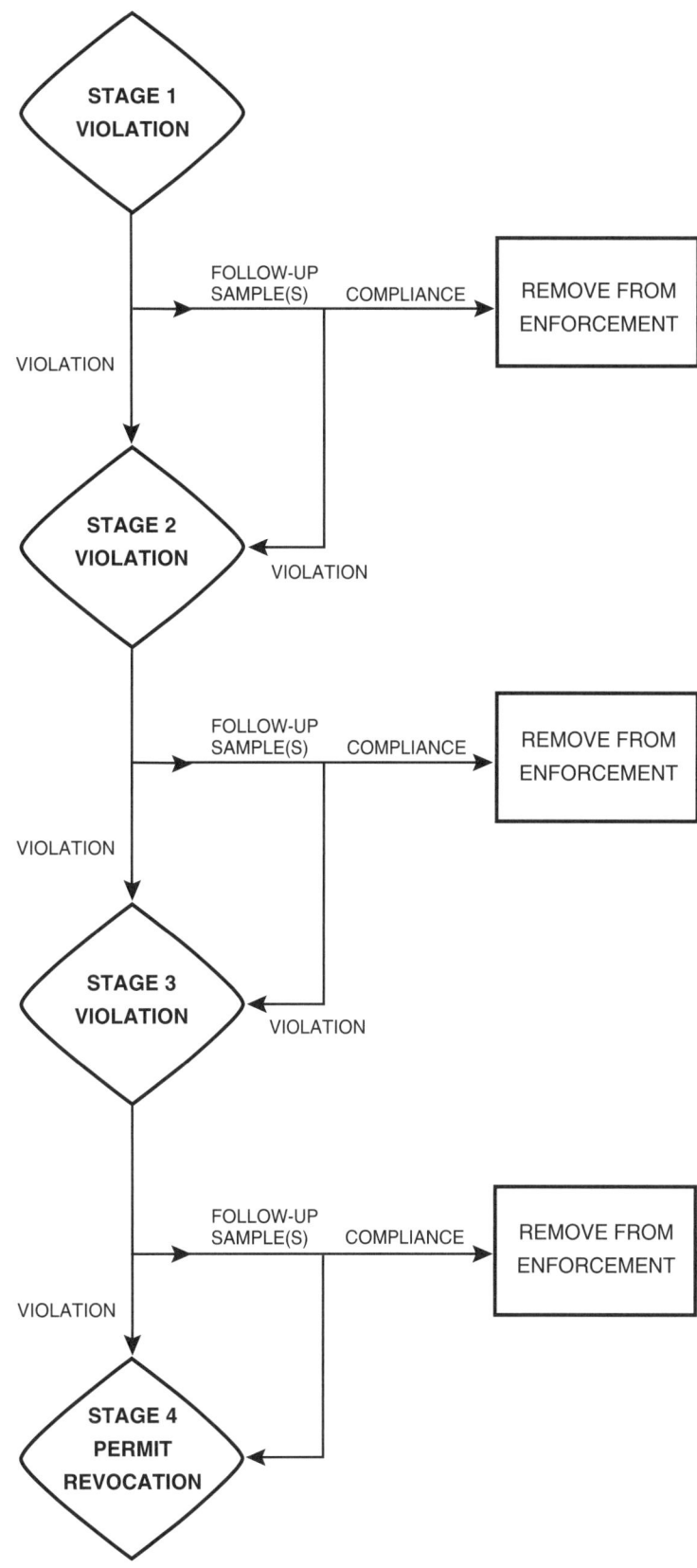

1. Issue violation notice
2. Send follow-up letter requiring written response
3. Districts obtain sample(s)

1. Issue violation notice
2. Send follow-up letter requiring written response
3. Conduct compliance meeting
4. Issue administrative complaint
5. Proposal for correction, interim compliance plan, and compliance schedule
6. Increase self-monitoring frequency
7. Districts obtain sample(s) during compliance period

1. Issue violation notice
2. Send follow-up letter requiring written response
3. Issue administrative complaint with escalated level of liquidated damages
4. Increase self-monitoring frequency
5. Districts obtain sample(s) during demonstration period
6. District Attorney referral
7. Civil action

Fig. 2.6   Industrial waste program enforcement procedures
(Source: Sanitation Districts of Los Angeles County)

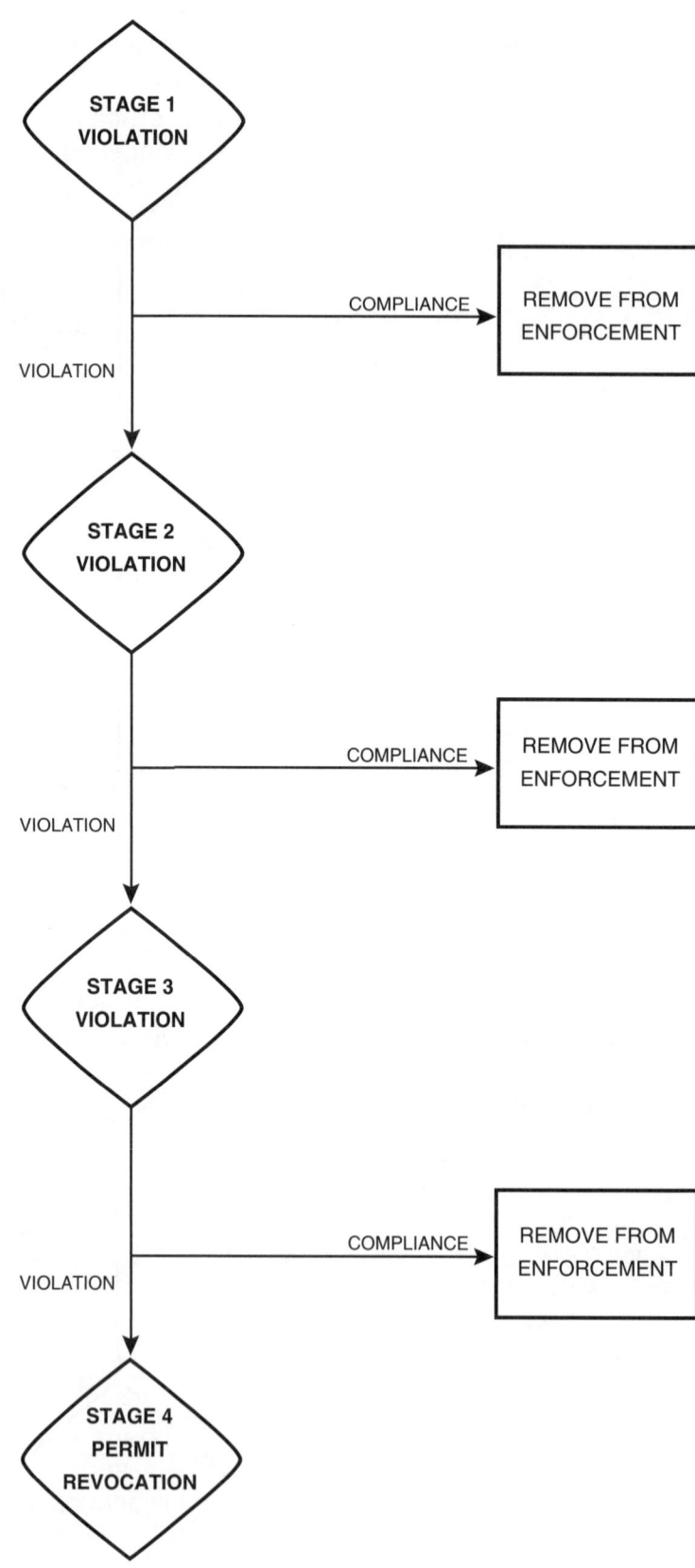

CSD ENFORCEMENT PROCEDURES
TYPE A NON-NUMERICAL VIOLATIONS

STAGE 1 VIOLATION

1. Issue violation notice
2. Send follow-up letter requiring written response

COMPLIANCE → REMOVE FROM ENFORCEMENT

VIOLATION

STAGE 2 VIOLATION

1. Issue violation notice
2. Send follow-up letter requiring written response
3. Conduct compliance meeting
4. Issue administrative complaint
5. Submit permit, information, or install within appropriate time limit
6. Reinspect

COMPLIANCE → REMOVE FROM ENFORCEMENT

VIOLATION

STAGE 3 VIOLATION

1. Issue violation notice
2. Send follow-up letter requiring written response
3. Issue administrative complaint with escalated level of liquidated damages
4. Proposal for correction and compliance schedule
5. Reinspect at end of compliance period (for non-installation)
6. No permit — revoke temporary permit
7. Non-submittal — civil action/ District Attorney referral
8. Non-installation — civil action/ District Attorney referral

COMPLIANCE → REMOVE FROM ENFORCEMENT

VIOLATION

STAGE 4 PERMIT REVOCATION

Fig. 2.6　Industrial waste program enforcement procedures (continued)
(Source: Sanitation Districts of Los Angeles County)

CSD ENFORCEMENT PROCEDURES
TYPE B NON-NUMERICAL VIOLATIONS

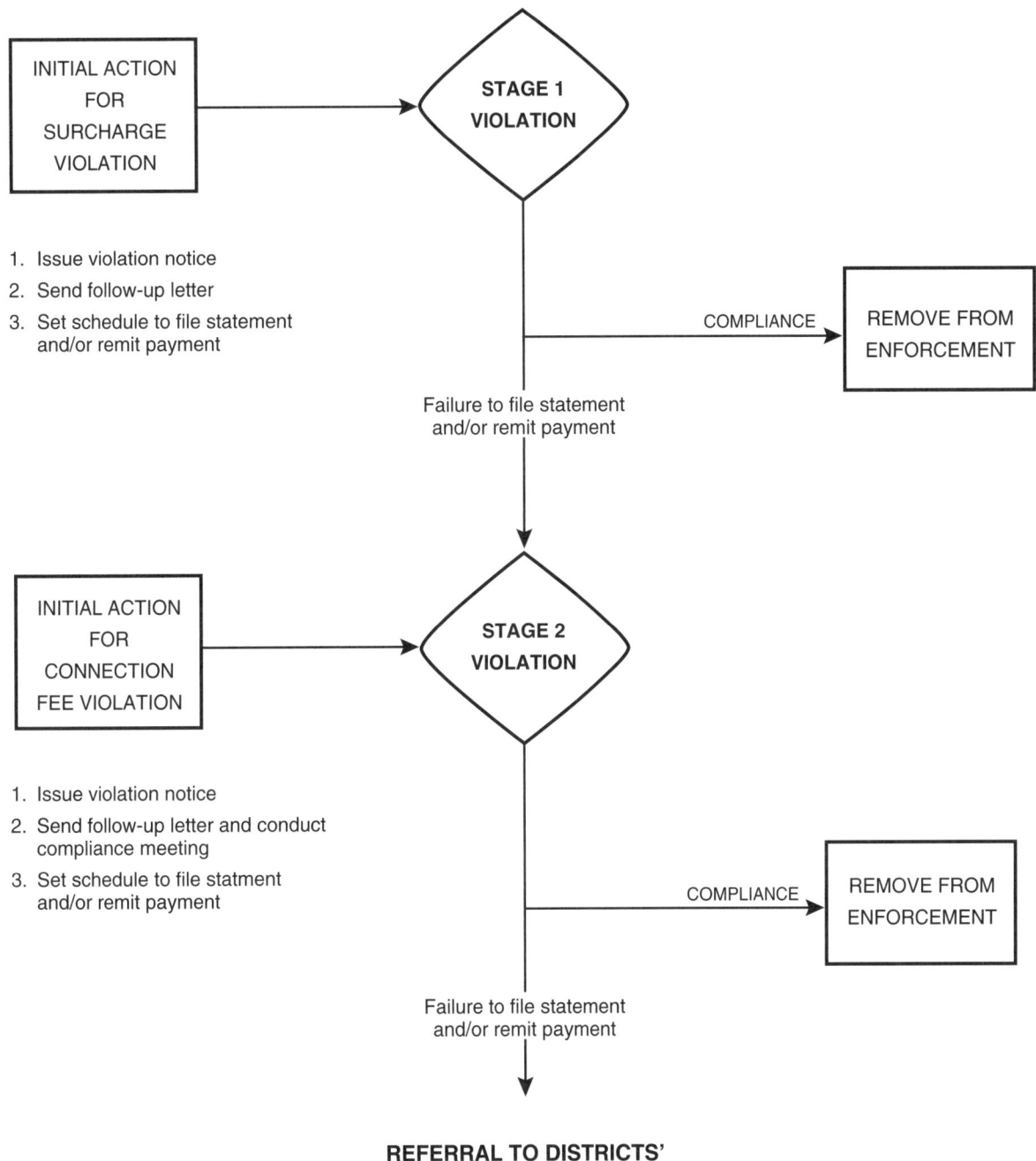

INITIAL ACTION FOR SURCHARGE VIOLATION

STAGE 1 VIOLATION

1. Issue violation notice
2. Send follow-up letter
3. Set schedule to file statement and/or remit payment

COMPLIANCE

REMOVE FROM ENFORCEMENT

Failure to file statement and/or remit payment

INITIAL ACTION FOR CONNECTION FEE VIOLATION

STAGE 2 VIOLATION

1. Issue violation notice
2. Send follow-up letter and conduct compliance meeting
3. Set schedule to file statment and/or remit payment

COMPLIANCE

REMOVE FROM ENFORCEMENT

Failure to file statement and/or remit payment

REFERRAL TO DISTRICTS' LEGAL COUNSEL

Fig. 2.6   Industrial waste program enforcement procedures (continued)
(Source: Sanitation Districts of Los Angeles County)

2.26   Laboratory Analyses

Under the EPA's pretreatment program, and in most industrial waste programs, a significant amount of chemical analysis work is necessary to determine what industrial dischargers are placing in the collection system. While there are some field test kits that inspectors can use to examine a wastewater discharge rapidly at the industrial site, these test kits are, in general, less accurate and less comprehensive than laboratory wastewater analyses (Figure 2.7). Test kit results are usually not approved for compliance action.

The complexity of laboratory analyses has increased greatly in recent years. In order to adequately analyze the full range of industrial pollutants, highly qualified technical personnel and extremely expensive laboratory instrumentation are necessary. A fully qualified wastewater analysis laboratory, staffed by POTW agency personnel, will be possible only at the largest of agencies. Most medium to small POTW agencies must necessarily contract with an outside laboratory to perform some or all of the industrial wastewater analyses. The use of outside personnel does not relieve the POTW agency of the responsibility to determine or verify what is in the industrial wastewater discharged to their POTW system. The quality control of contract laboratories should be checked regularly. Occasionally incorrect procedures are used or samples are mixed up. The POTW should regularly split samples for a comparison of results among several labs and should check for laboratory certification of these labs by state and federal labs.

2.27   Legal Assistance

In every POTW agency there will come the time when legal assistance is necessary to resolve an industrial waste problem. In large POTW agencies lawyers are frequently retained on staff to help in the resolution of such difficulties. In smaller POTW agencies and by preference for some larger agencies, consulting lawyers are employed to help in the resolution of problems. Where criminal prosecution is an enforcement option available under law to the POTW agency, the lawyer prosecuting the agency's case will necessarily be a publicly employed attorney. In the State of California, all felony violations are prosecuted by the county district attorney, while misdemeanor violations can be prosecuted by a city attorney, if available, or by the county district attorney. In civil actions involving civil torts or damages caused to the POTW agency, the agency must employ its own private attorney. Capable legal assistance will frequently more than pay for the cost of a qualified attorney.

At many agencies industrial waste personnel work closely with legal counsel to resolve industrial waste problems. These problems generally occur in two areas: (1) resolution of unpaid surcharge and connection fee bills, and (2) resolution of long-term or serious violations of effluent discharge requirements. The agency's private consulting attorney generally works on the resolution of unpaid surcharge and connection fee bills as

well as in preparation of contracts for compliance with industrial companies. The county district attorney and some city attorneys work with agencies in resolution of misdemeanor and felony violations and other environmental pollution control laws.

2.28   Computer Database for Industrial Waste Information

The EPA's pretreatment program involves such a multitude of effluent limits and categories and subcategories of companies that some means of adequately controlling and summarizing all of the various requirements is necessary. In addition, POTWs must exercise proper control of industrial wastewater discharge permits, agency and industrial self-monitoring data, and inspection results. In most POTW industrial waste programs, a computer-based data management system can greatly simplify the storage and use of such large quantities of information. Section 2.4, "Database for Pretreatment Program," more fully describes the need for a computer-based data management system and how to set up and use a system of this type.

2.29   Pollution Prevention

Pollution prevention is a very important element of a pretreatment program. With the enactment of the Pollution Prevention Act of 1990, Congress formally established pollution prevention as a national objective, placing it ahead of waste recycling, treatment, and disposal in the hierarchy of environmental management methods. A preventive approach to environmental protection can lead to improvements in environmental quality and economic efficiency by reducing harmful pollutants at the source through cost-effective changes in production, operation, and raw material use.

EPA defines pollution prevention as waste reduction prior to recycling, treatment, or disposal. Recycling conducted within a process, such as closed loop rinse water recycling, is also considered pollution prevention. Waste recycling, which takes place outside the process, is not considered pollution prevention, although waste recycling (when conducted in an environmentally safe manner) achieves the same goal as pollution prevention by reducing the need for treatment and disposal.

Basic components of a pollution prevention program include identifying pollutants of concern as well as industrial, commercial and domestic users of concern. The next step is to prioritize users of concern and then apply pollution prevention resources. Pollution prevention can be promoted among regulated and unregulated sewer users by inspections and by encouraging pollution prevention through regulatory activities, community education, and community outreach programs.

Pollution prevention programs need to convince industrial and commercial facilities that they can benefit and profit from pollution prevention. In many cases, pollution prevention might be the least expensive means of reducing unacceptable toxic discharges. Pretreatment inspectors can point out the benefits of pollution prevention to their sewer users. Through pollution prevention, companies can:

* Reduce waste monitoring, treatment, and disposal costs,

* Reduce raw material use, feed stock purchases, and manufacturing costs,

* Reduce operation and maintenance costs,

* Increase productivity and reduce off-specification products,

* Reduce regulatory compliance costs,

NOTE: When actually working in a laboratory, wear proper safety gear which includes glasses with sideshields, laboratory coat and gloves.

Fig. 2.7 Small lab and recordkeeping work area

- Reduce hazards to employees through exposure to chemicals,

- Reduce costs of environmental impairment insurance,

- Improve public image and employee morale, and

- Reduce potential liability associated with toxic waste.

QUESTIONS

Write your answers in a notebook and then compare your answers with those on page 82.

2.2A   List the basic elements of a pretreatment inspection program.

2.2B   Why should field inspectors be rotated to new geographical areas at least every two years, if possible?

2.2C   Why must the POTW agency collect revenue from industrial companies?

2.2D   How do most field test kits compare with laboratory wastewater analyses?

2.3   INDUSTRIAL WASTEWATER DISCHARGE PERMITS

2.30   Need for Permit — Benefits of a Permit System

Any POTW agency must have adequate information on the volume and quantity of industrial waste discharged to its system. Many of the serious problems encountered in proper operation of a POTW agency will originate in such industrial wastewater discharges. Serious problems such as fires and explosions in the collection system, toxic upsets of biological processes and serious structural corrosion problems in sewers and at treatment plants are most likely caused by industrial waste dischargers. A POTW agency which does not have a relatively complete body of information on industrial waste dischargers may be helpless to remedy serious operational problems in its collection and treatment facilities.

EPA requires all POTWs with approved pretreatment programs to implement a permit system (or equivalent control mechanism) to regulate the discharges of all Significant Industrial Users (SIUs). An industrial wastewater discharge permit system requiring companies to submit information and obtain a permit from the POTW agency will give the agency the information it needs to control the industrial discharges within its service area. Withholding permission to discharge to the collection system until adequate information on the company's operations and wastewater quantity and quality is obtained,

and until an adequate pretreatment system is developed, will greatly reduce the number of serious industrial waste problems and will facilitate correction of any problems that do occur.

Usually a POTW will not allow an industrial user to discharge into the sanitary sewer until adequate information on the wastewater quantity and quality is obtained. However, if the industry is new, accurate operational data may not be available. It may be necessary to issue a short-term permit or "start-up" permit to the industrial user after conventional pretreatment is installed to allow data to be collected and a final evaluation made by the POTW as to whether or not applicable pretreatment standards can be complied with.

2.31   Information Required

In the development of an industrial waste permit program, the information required will vary from POTW agency to POTW agency. It is recommended that at least the following information be obtained:

1. Exact location of industrial plant and its service connection to the POTW sewer,

2. Collection system and treatment plant into which the industrial plant discharges,

3. The general type of industrial wastewater as defined by the Federal Standard Industrial Classification (SIC) Code system or NAICS code system,

4. The quantity of wastewater discharged, in terms of both daily and hourly flows,

5. The chemical quality of the wastewater, especially regarding toxic materials,

6. The wastewater discharge pattern (slug dumps, constant discharge rate, seasonal peaks and yearly discharge quantities),

7. Information on hazardous or toxic materials stored on-site that may reach the sewer in the event of leakage or tank rupture,

8. The means for disposal of rainwater other than to the sanitary sewer,

9. The name of person in the company who is responsible for wastewater discharge problems, and

10. The pretreatment facilities needed to treat the company's wastewater.

For smaller POTW agencies serving fewer industrial companies, some of the information (such as 1, 2 and 3 above) may be readily available. For larger metropolitan areas, the information specified above may be very difficult or expensive to obtain. For POTW agencies which do not have the information described above, and where the information must be rapidly obtained, it may be appropriate to perform an industrial waste survey as recommended by EPA in their guidance documents. Since much of the information changes frequently, any short-term industrial waste survey will obtain only a momentary picture of the information. It is believed that an on-going industrial waste discharge permit program is a much more effective way to continuously obtain and update this information. The terms of the permit can require industries to notify the POTW of major changes in their discharge volumes or characteristics; this enables the POTW to update its records as changes occur.

The amount of information required may be adjusted to reflect the size and importance of the company's industrial waste discharge. For small companies, only a minimum of information (see Figure 2.8) may be required if the wastewater discharge is insignificant. For large companies, with a significant wastewater discharge quantity and large amounts of toxic materials present in the wastewater, a very detailed permit review with an in-depth information submittal may be necessary. At some agencies, three levels of industrial waste discharge permits are issued. The most complex permit information requirements are for companies where wastewater flow is over 50,000 GPD. For companies between 10,000 GPD and 50,000 GPD, or having significant amounts of toxic waste in their effluent, a moderate permit requirement is specified. For smaller companies and companies where no significant toxic discharges are present, a very simple permit procedure is used.

2.32 Permit Review Process (Figure 2.9)

The permit review process is an important part of an industrial waste program. The cost of reviewing permits is a substantial expenditure for the POTW agency, but these costs often can be recovered through a permit review charge. The use of an appropriate level of personnel in the permit review is also important. Permit reviews should be performed only by persons who are familiar with the type of pretreatment required by the company submitting the permit application. For permit renewal applications, reviewers should also be familiar with the history of the company's compliance with local, state and federal discharge requirements. The inspectors should conduct the initial permit review and make recommendations for review by as many people as are required by the local bureaucracy.

The level of information required for a permit review differs significantly between the different levels of permits. The lower level of permits requires very little in the way of permit information. For large facilities such as a petroleum refinery, paper mill or food processing plant such as a brewery, significant information is required, including full blueprints of the company property and buildings.

In the permit review process, particularly careful attention is given to the use of toxic or hazardous materials in the industrial process and to the methods for disposal of such materials. Many POTW agencies have a spill containment program which requires that companies having tanks of toxic materials contain or berm the tanks so that a tank leak or rupture will not cause a spill of such materials to the sewer. Flammable materials, cyanides, acids, heavy metals in solution and toxic solvents must be located in separate spill containment areas large enough that the rupture of the largest single tank will be contained within the bermed or diked area. For example, at metal plating companies any tank containing over ten pounds of a heavy metal or cyanide in solution must be located within a spill containment area. Cyanide tanks must be contained in a spill containment area separate from any acid tanks because mixing of wastes from cyanide and acids can generate toxic hydrogen cyanide gas. The spill containment provisions of a permit-based program will significantly reduce treatment plant upset problems due to spills of metal plating and other toxic wastes.

2.33 Conditions Established in Permit Review Process

In the permit review process, unique conditions are established for each industrial company. The final permit specifies the exact conditions which are applicable to the specific permittee. Most agencies have overall industrial effluent discharge limits which are appropriate for all companies within an agency's boundaries and which meet at least the minimum federal requirements. (See Section 3.101, "Prohibited Discharge Standards," and Table 3.1, "Comparison of Virginia Prohibited Discharge Limits," page 209.)

The spill containment program described in Section 2.32 is applied where appropriate. In addition, an agency may require all companies with over 100,000 GPD of wastewater flow and with potential flammable waste problems, such as petroleum refineries and similar plants, to install and properly operate a combustible gas detection meter. This program may be expanded to include more companies having lower volumes of wastewater discharge. The combustible gas detection program requires companies to install an automatically controlled valve which will shut off the company's wastewater discharge if the explosive gas detection meter detects a gas concentration greater than 10 percent of the lower explosive limit (LEL). By use of these meters, an agency may prevent explosions in the sewers caused by flammable industrial waste.

In the permit process, self-monitoring characterization of the wastewater is required for certain industrial companies. For wastewater discharges over 10,000 GPD and having significant quantities of toxic waste, companies may be required to monitor their wastewater discharge four times per year. For companies having small wastewater discharges and nontoxic wastes, no self-monitoring or self-monitoring at frequencies less than four times per year may be required. POTW agencies realize that some of the reports submitted to them may not reflect the worst conditions found by the company but the self-monitoring program at least requires that a responsible individual at the company periodically review the quality of the company's wastewater and sign the analysis statement sent to the inspection agency.

In some agencies' permit and surcharge programs, all companies having wastewater flows over 50,000 GPD are required to provide a full-time indicating, recording and totalizing flowmeter. These flowmeters are not required to be of any specific type but are required to be well engineered, properly designed, and calibrated at regular frequencies.

PERMIT FOR INDUSTRIAL WASTEWATER DISCHARGE
COUNTY SANITATION DISTRICTS OF LOS ANGELES COUNTY
1955 Workman Mill Road / Whittier, CA
Mailing Address: P.O. Box 4998 / Whittier, California 90607-4998
Charles W. Carry, Chief Engineer and General Manager
(310) 699-7411

PERMIT NO. _____

01 CHECK ONE: New Sewer Connection ☐   Existing Sewer Connection ☐

02 Applicant _____
(Legal Company Name)

03 Check one and fill in appropriate information

☐ Corporation   Name _____

Year Incorporated_____ State of Incorporation _____ ID#_____

☐ Partnership   Name _____ Partners _____

☐ Sole Proprietor Name _____ Business Names _____

04 Company Address _____
(Street)   (City)   (State)   (Zip)

05 Mailing Address _____
(Street)   (City)   (State)   (Zip)

06 Point of Discharge _____

07 Number of years applicant has been in business at present location _____ _____
(yrs)   (months)

08 Name of Property Owner_____
Address of Property Owner _____
(Street)   (City)   (Zip)   (Telephone Number)

09 Assessors Map Book No. _____ Page No. _____ Parcel No. _____

10 Type of Industry _____ _____
(General Description)   (Federal SIC No.)

11 Number of Employees (Full Time) _____ (Part Time) _____

12 Raw Materials Used _____
(General Description — Add Additional Sheets as Needed)

_____
(Daily Amount Used)

13 Products Produced _____
(General Description — Add Additional Sheets as Needed)

_____
(Daily Amount Produced)

14 Wastewater Producing Operations _____

_____
(Full Description — Add Additional Sheets as Needed)

15 Time of Discharge _____ AM/PM to _____ AM/PM, Shifts per Day _____, Days per Week M T W Th F Sa Su
(Circle AM or PM)   (Circle Days)

16 Wastewater Flow Rate _____ Gallons per Day _____ Gallons per Minute
(Average)   (Peak)

17 Constituents of Wastewater Discharge _____

_____
(General Description — Attach Chemical Analysis Results to the Application)

18 Person in company responsible for industrial wastewater discharge

_____ _____ _____
(Name)   (Position)   (Telephone Number)

I affirm that all information furnished is true and correct and that the applicant will comply with the conditions stated on the back of this permit form.

Date _____ , 19_____

19 Signature for Applicant _____ _____
(Company Administrative Official)   (Name)   (Position)

20 Approved/Reviewed by City or County Official   Approved by Sanitation Districts of Los Angeles County

Date _____   Date _____

For L.A. County Dept. of Public Works...☐   Expiration Date _____

City of _____   Charles W. Carry, Chief Engineer & General Manager

Name _____   By _____ _____

Position _____   Position _____

Fig. 2.8 Connection permit for industrial wastewater discharge

PERMIT SUBMITTAL EVALUATION AND APPROVAL PROCESS

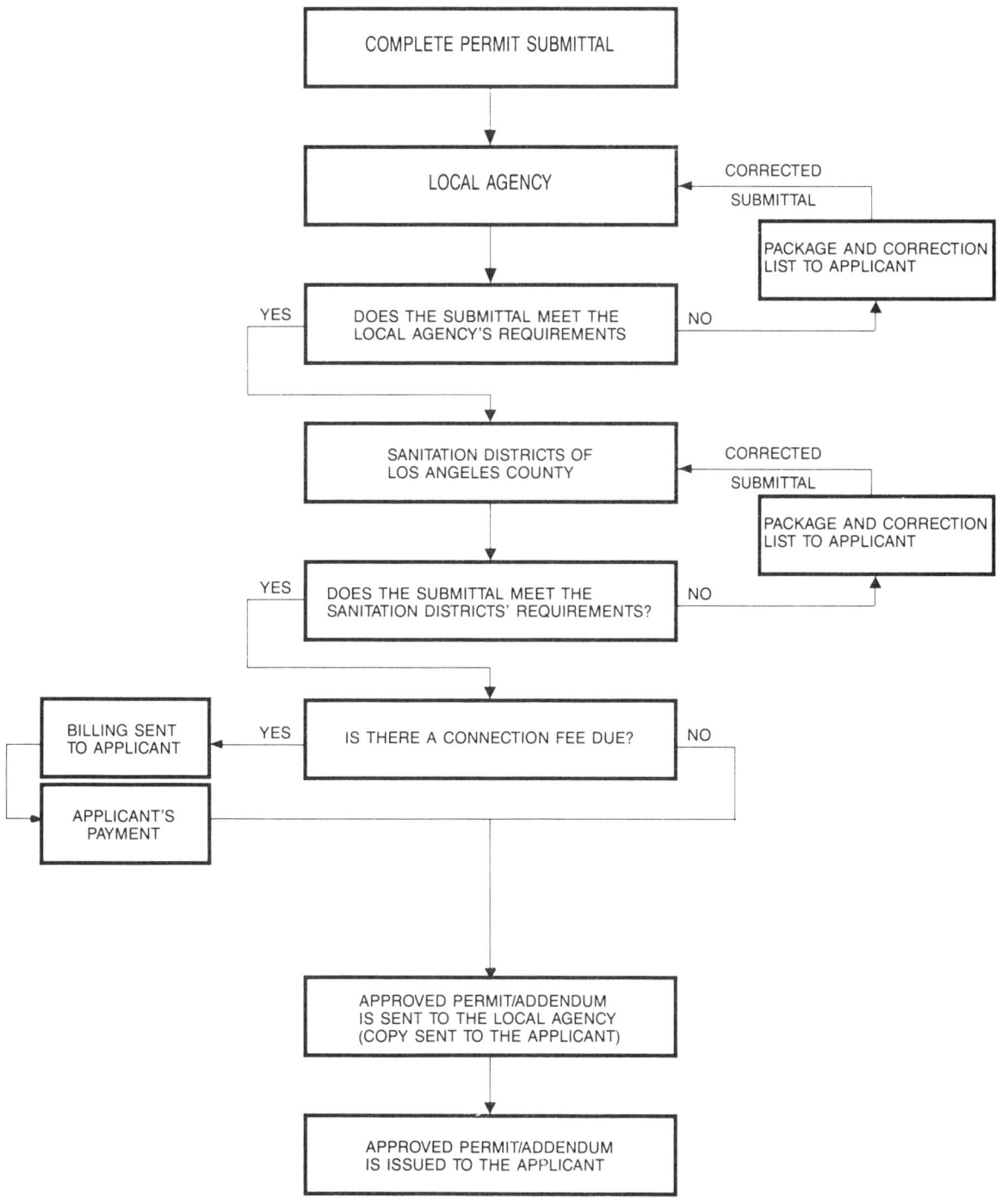

Fig. 2.9   Overview of permit review process

Small companies such as restaurants, laundromats, small photoprocessors, small dry cleaners and most small service businesses, including gas stations and auto repair shops, are usually not included in the industrial waste permit program. Such companies are also not reviewed in any great detail by industrial waste inspectors. However, some POTWs have an aggressive small-quantity-generator control program which does cover small dischargers.

Under the EPA's hazardous waste regulations, some small companies such as dry cleaners, service stations and small auto repair shops may be generators of hazardous waste. The perchloroethylene typically used at dry cleaners, as well as waste oils and radiator fluids found at gas stations and auto repair shops, may be hazardous wastes under EPA regulations. It appears that some of these smaller companies may be disposing of hazardous wastes to the sewer system where no, or inadequate, controls exist, rather than properly disposing of these wastes. In many POTW agencies, the permit and review process may have to include large numbers of such small companies if EPA believes a significant hazardous waste problem exists at these companies.

2.34 Cooperation With Other Governmental Entities

The industrial wastewater discharge permit review should be understood to be only part of the normal building and other permit processes. In a city, the industrial waste review may be a regular part of the review for a building permit or a sewer connection permit. At some agencies the industrial wastewater discharge permit is a cooperatively issued permit, issued jointly by the city and the collection system agency. As shown on Figure 2.8, the agency's permit must be signed by a city representative as well as the agency to be a valid permit. Either the agency or the city can suspend or revoke the permit to remedy problems within its service area. Both the local city and the agency determine in their review what procedures and facilities are needed to prevent problems.

Where the POTW agency does not have full control of the building permit review process, the agency should cooperate with the building permit review agency to ensure that a proper industrial waste permit review is performed. Such an industrial waste permit review will benefit both the POTW agency and the community in which the industrial facility is located. Citizens' concerns about the location of the industrial plant or the proper disposal of hazardous waste can be voiced through this permit review process.

2.35 Recordkeeping

In the permit review process, the installation of adequate pretreatment facilities should always be a primary concern. A secondary concern, however, is obtaining and handling the large amounts of company information required to complete the permit review. The proper filing of this information and the ability to retrieve parts of it when needed are significant concerns. A computer system for keeping track of this information may be appropriate for agencies that issue a large number of permits.

Some agencies have developed a computer-based data management system for industrial waste permit information. For each industrial waste permit issued, information is submitted for input to the computer system. Whenever possible all forms used by pretreatment inspectors should be designed for computer database entry. Figure 2.10 shows a typical computer input form. The computer data management system allows specified printouts of portions of this data. For example, a printout of all companies having SIC Code 3479 (NAICS 332813) may be requested; or all companies having SIC Code xxxx and located in a certain city or in a certain district could be requested; or, companies with SIC Code xxxx and having wastewater discharge volumes greater than a certain amount could be requested. The proper recording and access to industrial waste discharge permit information can be of tremendous value in tracing the origin of industrial waste discharge problems.

As the volume of paper for an industrial waste permit program increases, it is recommended that the use of microfiche or similar file reproduction systems be considered. At some agencies all permit documents, including blueprints, are periodically microfilmed and placed in microfiche files. This will greatly reduce the volume of traditional file folder files. Another advantage of a microfiche system is that file information becomes portable. By using a field viewing system, an inspector can review the permit file for a company just prior to entering the company for inspection or to locate, from the permit plans or documents, industrial waste facility components which may be difficult to locate in the field. Data entry, storage and retrieval technology is a rapidly advancing field. Pretreatment program records can be stored using microfilm, computer disks and compact disks.

2.36 Legal Considerations

The industrial wastewater discharge permit is a legally binding agreement between the POTW agency and the industrial wastewater discharger. The POTW agency agrees to provide wastewater collection, treatment and disposal for the company's wastewater if the company will comply with certain conditions. The major conditions to be met by the company should be specified on the permit document that is signed by a responsible representative of the company. The back of the permit document is an appropriate place for locating such requirements. Figure 2.11 is an example of an applicant's permit document agreement. In signing the permit, the company agrees to comply with these permit conditions.

The permit may also be a useful means to regulate companies outside the POTW's service area. Although the POTW may have some difficulty in adequately enforcing effluent regulations outside its political jurisdiction, the permit document is essentially a contractual agreement by the company to comply with the POTW's wastewater discharge requirements. Failure to comply with such requirements will at least give the POTW the right to take legal action against the company to revoke its permission to discharge to the POTW's collection system.

QUESTIONS

Write your answers in a notebook and then compare your answers with those on pages 82 and 83.

2.3A   What serious problems can result from industrial wastewater discharges into collection systems?

2.3B   List the information required for an industrial waste permit.

2.3C   What are the major concerns with regard to a recordkeeping system?

Please answer the discussion and review questions next.

```
           INDUSTRIAL WASTE PERMIT EVALUATION REPORT - C#IWP810
==============================================================================
To:  Data Control Group (Data Processing Section)    Date: _____
From: _____      Dept./Section:  I.W.      Ext: _____
Request Description:                         *************************
        Report IWP810 - Permit File Update  * JOB I.D.:  C#IWP810 *
When needed: One Week  |_|                   *************************
==============================================================================
| 1)New record - N          | 2)I.W. Permit No.         | 3)Entry |
|    Re-entry - R           |   (Insert leading zeros)  |    No.  |
|    Change Information - C |_| |_|_|_|_|_|_|_|          |  |_|_|  |
------------------------------------------------------------------------------
01| 4)Evaluation Engineer (Last name)
  |   |_|_|_|_|_|_|_|_|_|_|_|_|_|_|
  | 5)Name of Firm
  | |_|_|_|_|_|_|_|_|_|_|_|_|_|_|_|_|_|_|_|_|_|_|_|_|_|_|_|_|_|_|_|_|_|_|_|
  |
  | |_|_|_|_|_|_|_|_|_|_|_|_|_|_|_|_|_|_|_|_|_|_|_|_|_|_|_|_|_|_|_|_|_|
------------------------------------------------------------------------------
02| 6)Street No. 6a)Unit  7)Direction  8)Name                 9)Type
  | |_|_|_|_|_|  |_|_|_|  |_|  |_|_|_|_|_|_|_|_|_|_|_|_|_|_|   |_|
  |      10)City                              11)Zip Code
  | |_|_|_|_|_|_|_|_|_|_|_|_|_|_|_|_|_|_|_|_|   |_|_|_|_|_|
------------------------------------------------------------------------------
03| 12)Thomas Map: Page - Area   13)Account No.    14)District No.
  |               |_|_|  |_|_|   |_|_|_|_|_|_|        |_|_|
  | 15)Date Received(Mo,Day,Yr) 16)Action Date(Mo,Day,Yr) 17)Status
  |    |_|_| |_|_| |_|_|           |_|_| |_|_| |_|_|         |_|_|_|
------------------------------------------------------------------------------
04| 18)Working Hours(0000 - 2359)
  |    Start
  | |_|_|_|_||_|_|_|_||_|_|_|_||_|_|_|_||_|_|_|_||_|_|_|_||_|_|_|_|
  |    Stop
  | |_|_|_|_||_|_|_|_||_|_|_|_||_|_|_|_||_|_|_|_||_|_|_|_||_|_|_|_|
  |   Monday  Tuesday Wednesday Thursday  Friday  Saturday  Sunday
------------------------------------------------------------------------------
05| 19)Situs Contact, Title
  | |_|_|_|_|_|_|_|_|_|_|_|_|_|_|_|_|_|_|_|_|_|_|_|_|_|_|_|_|_|
  | 20)Telephone No.       21)Days/Wk  22)Shifts    23)SIC No.
  | |_|_|_|_|_|_|_|_|_|_|     |_|         |_|      |_|_|_|_||_|_|_|_|
  | 24)Review Engineer(Last name)      25)New Industry Code
  | |_|_|_|_|_|_|_|_|_|_|                  |_|
------------------------------------------------------------------------------
06| 26)No. Employees  27)Avg.flow(GPD)  28)Peak flow(GPM) 29)Type flow
  | |_|_|_|_|_|        |_|_|_|_|_|        |_|_|_|_|_|       (C,D,I) |_|
  | 30)Waste Character  31)WRP  32)Manhole(Dist.,Map)   33)EPA Code
  | |_|_|_|             |_|_|   |_|_|-|_|_|_|_|          |_|_||_|_||_|_|
  | 34)Pretreatment Facilities                          34A)Agency
  |    Type:  |_|_|_|_| |_|_|_|_| |_|_|_|_| |_|_|_|_|       |_|_|
  |
  |    Value: $|_|_|_| $|_|_|_| $|_|_|_| $|_|_|_|
------------------------------------------------------------------------------
07| 35)Raw Materials / Products
  |    R/P          Name              Quantity         Units
  |    |_|  |_|_|_|_|_|_|_|_|_|_|_|_|   |_|_|_|_|_|   |_|_|_|_|_|_|_|_|
  |
  |    |_|  |_|_|_|_|_|_|_|_|_|_|_|_|   |_|_|_|_|_|   |_|_|_|_|_|_|_|_|
  |
08|    |_|  |_|_|_|_|_|_|_|_|_|_|_|_|   |_|_|_|_|_|   |_|_|_|_|_|_|_|_|
  |
  |    |_|  |_|_|_|_|_|_|_|_|_|_|_|_|   |_|_|_|_|_|   |_|_|_|_|_|_|_|_|
------------------------------------------------------------------------------
```

*Fig. 2.10   Industrial waste permit evaluation report form
designed for input to computer database*

INDUSTRIAL WASTE PERMIT EVALUATION REPORT

===

09 | 36) Parameter
No. |__|__|__| |__|__|__| |__|__|__| |__|__|__| |__|__|__|

Conc. |__|__|__| |__|__|__| |__|__|__| |__|__|__| |__|__|__|

10 | 37) Permit Conditions

11 | 38) Waste Characteristics Comment

12 | 39) General Commment

13 | General Comment (Continued)

14 | 40) Additional Permit Requirements

15 | Additional Permit Requirements (Continued)

16 | Additional Permit Requirements (Continued)

17 | Additional Permit Requirements (Continued)

18 | Additional Permit Requirements (Continued)

19 | Additional Permit Requirements (Continued)

Fig. 2.10 Industrial waste permit evaluation report form designed for input to computer database (continued)

APPLICANT FOR PERMIT MUST READ THIS MATERIAL

IN CONSIDERATION OF THE GRANTING OF THIS PERMIT, the applicant agrees:

1. To furnish any additional information on industrial wastewater discharges as required by the Districts,

2. To accept and abide by all provisions of ordinances, policies and guidelines of the Districts,

3. To operate and maintain any required industrial wastewater treatment devices in a satisfactory approved manner,

4. To cooperate at all times with Districts' personnel, or their representatives, in the inspection, sampling and study of industrial wastewater facilities and discharges,

5. To immediately notify the Districts at (310) 699-7411 during normal working hours or at (310) 437-6520 or 437-1881 after 4:00 p.m. or on weekends in the event of any accident, negligence or other occurrence that causes the discharge to the sewer of any material whose nature and quantity might be reasonably judged to constitute a hazard to the public health, environment, Districts' personnel or wastewater treatment facilities,

6. To pay to the Districts annually the required surcharge or user charge fee for industrial wastewater treatment,

7. To submit, as required by the Districts, accurate data on industrial wastewater discharge flows and wastewater constituents,

8. To operate only *one* industrial wastewater discharge point to the sewerage system under the authority granted by this permit,

9. To submit additional pages as required to furnish the necessary information if there is inadequate room on the reverse side of this permit form to complete submittal of requested data,

10. To apply for a revised Districts' Industrial Wastewater Discharge Permit if any change in industrial processes, production, method of wastewater treatment or operations creates a significant change in industrial wastewater quantity or quality, or, if the quantity or quality of wastewater discharged changes by more than 25%, or, changes above any other threshold level specified in industrial waste permit requirements,

11. To provide immediate access to authorized personnel of the Districts to any facility directly or indirectly connected to the Districts' sewerage system under emergency conditions and at all other reasonable times,

12. To apply for a renewed Districts' Industrial Wastewater Discharge Permit before the five (5) year expiration date,

13. To comply with all requirements and conditions of approval specified in the permit Requirement List within the specified deadlines. Failure to do so invalidates the permit approval.

Fig. 2.11   Applicant's permit document agreement

End of Lesson 2 of 3 Lessons on PRETREATMENT PROGRAM ADMINISTRATION

DISCUSSION AND REVIEW QUESTIONS

Chapter 2.   PRETREATMENT PROGRAM ADMINISTRATION

(Lesson 2 of 3 Lessons)

Write the answers to these questions in your notebook before continuing. The question numbering continues from Lesson 1.

7. What problems can be created by unrestricted discharge of industrial wastes?

8. Why must inspectors have an adequate level of technical expertise when they enter a company and inspect the processes?

9. Why is monitoring by the POTW agency essential?

10. What important items should be given particularly careful attention during a permit review process?

11. How can citizens' concerns about the location of an industrial plant or the proper disposal of hazardous waste be voiced?

CHAPTER 2.   PRETREATMENT PROGRAM ADMINISTRATION

(Lesson 3 of 3 Lessons)

2.4   DATABASE FOR PRETREATMENT PROGRAM

2.40   Need for Computer Database System

As noted previously, there is a significant amount of data to be recorded and accessed in any reasonably large industrial waste program. The main sources of data are the following:

1. Industrial waste permit information,

2. Records of site inspections of industrial plants,

3. Industrial self-monitoring data,

4. Monitoring data by the POTW,

5. Surcharge and revenue data for the industrial companies,

6. Compliance records for EPA and local effluent limits,

7. Data on toxic pollutant concentrations in the POTW treatment plant influent, effluent and sludge, and

8. Correspondence and other written memoranda.

The large amount of data on industrial companies which is required under the EPA programs needs to be readily available for use by the POTW agency. In addition, the EPA and the various state governments will require periodic POTW reports of selected portions of the industrial waste information. In all but the smallest industrial waste programs, a computer-based data management system appears appropriate.

Any computer database management system should be carefully designed. The design must include all information which is required to be stored and the required output should be carefully specified. The value of information stored is certainly proportional to the ease with which this information can be recovered in a significant manner. When first installing a computer system, it is extremely difficult to predict what outputs will be required from the system in the future. As technology, regulations, and needs change, the output required from an industrial waste database may change significantly from year to year. Any system developed should have the flexibility to accommodate changed requirements for the data output.

Modern database management systems will allow more flexible output to be accommodated. Although mainframe systems may be more appropriate for large municipal agencies having very large masses of data, the rapid advances in personal computers, workstations and networks may make them suitable for many agencies. The client/server computing technologies now available make such systems a good choice for even the larger agencies.

2.41   Hardware Available

Personal computers are readily available and could be selected by an operator or inspector familiar with the record-keeping needs of the POTW's industrial waste program. At the time of the writing of this manual, the most widely used type of personal computer is the Intel-based family of computers formerly known as IBM PCs or PC-compatibles. Today these systems use the Pentium processor, the fifth generation of the original PC processor. These machines are extremely fast and powerful and offer an excellent price-to-performance ratio. Typically, these systems run a version of the Microsoft Windows or Windows-NT operating systems. Software for many purposes is available from local dealers. Another readily available personal computer is the Apple Macintosh. This system is also an excellent choice, although it generally costs more and offers fewer selections for off-the-shelf software packages.

It is recommended that any industrial waste program considering the purchase of a personal computer seriously analyze a number of factors such as expandability, service, relative cost, reliability, and the probability that the system will meet the needs of the program now and in the future. The trend in software development is toward producing larger, more complex software packages which can make computer hardware obsolete in a matter of a few years. It is prudent, therefore, to consider upgrades in the planning of any computer system and to expect to expand or replace computer hardware every three to five years.

A fully adequate personal computer system can be purchased for about $800 to $3,000. The price depends on several factors including processor speed, number and size of hard disks (permanent data storage), and other peripheral (add-on) equipment such as CD-ROM (laser) disk drives, extra memory, and printers. Computer vendors today sell a number of pre-configured systems that are designed to run most software well. These systems typically include a fast central processor, 128 to 512 megabytes of RAM (random access memory), a 20-gigabyte (20,000 megabytes) hard drive, color monitor, CD-ROM and floppy drives, mouse and keyboard. Such systems can often be purchased for less than $1,000. They are usually upgradeable should the need for additional disk drives or memory arise.

Printed output for reporting, correspondence, and graphics requires the addition of one or more printers to the system. For general use, there are two basic types of printers: ink jet and laser. An ink jet printer is generally a good choice where high-quality letter (correspondence) quality is not a requirement. Various brands and qualities of ink jet printers are readily available for prices ranging from $50 to $500. "Near letter quality" output and graphics are standard features. At the upper end of the price range, ink jet printers are generally more durable and faster and may offer special features such as network support and capacity to handle large paper sizes. Color printing is usually standard except at the very bottom of the price range. Laser printers, while costing more than ink jet printers (about $500 to $5,000), can print letter quality output as well as perform special tasks such as printing photographs, letterheads, complex graphical presentations with shading, and color.

Relatively large amounts of data can be stored on personal computer hard drives. Currently hard disk drives can hold up to 80 gigabytes of data. At the time of this writing, the cost of such data storage capacity is about $2.00 per gigabyte.

The equipment you will need and the cost of the system may also be influenced by the size of your collection system, the number of industrial users, the amount and types of data you intend to process, and the number of ways you want the computer(s) to help manage the system. A typical personal computer system supporting a database may have the following components:

1. The personal computer itself, including at least one hard drive (20-80 gigabyte), a floppy drive for portable data storage, a CD-ROM reader, and at least 128 megabytes of RAM (random access memory);

2. A color monitor;

3. A keyboard and a mouse;

4. A printer; and

5. Other peripheral equipment, as required, such as a modem, which allows connections to other computer systems and the Internet over telephone lines.

2.42 Software Available

As previously mentioned, most software today is written for the Intel-based computer systems (IBM-compatibles). Many large companies have developed their software to run on the Apple Macintosh as well. Such companies include Microsoft and Adobe, among others.

Typically, general-purpose software can serve most of the needs of the computer user. Such general-purpose software includes:

- Word processor,

- Database management system,

- Spreadsheet, or calculation, system,

- Graphics presentation,

- Computer utilities, such as backup programs and anti-virus programs,

- Accounting system, and

- Communication system.

For each type of software listed above, various packages from different manufacturers are readily available and most are relatively easy to use. It is a good idea to use popular software packages, since the most commonly used software tends to cost less, run better, and is better supported by the

company that developed it. Keep in mind that software is constantly being improved and upgraded. Custom-written software and less mainstream software tends to cost more. In addition, it may be difficult to get help when bugs occur or when upgrades and training are needed.

Among the many off-the-shelf software packages available, the following ones are examples of a few that are widely distributed:

- For word processing, Microsoft Word, Corel WordPerfect, and Open Office are perhaps the most widely used programs.

- In the database arena, Microsoft Access, Paradox and Fox-BASE are among the most popular.

- Microsoft Excel, Lotus 123, Quattro Pro, and Open Office are among the most popular spreadsheets.

- Microsoft Powerpoint, Corel Presentations, and Open Office are common graphics presentation packages.

- Personal accounting systems such as Quicken are very popular. Professional accounting packages tend to be more expensive. Often companies pay to have custom accounting software written to meet the particular needs of the company.

- Communications software and utilities are often included with the computer system at the time of purchase.

For the typical industrial waste program, it is highly recommended that a minimum of word processing, database management and spreadsheet software be purchased. The word processing software can be used to write letters, memos and reports dealing with industrial waste subjects. The database management systems are needed to record, organize and selectively print portions of the vast amount of data accumulated in an industrial waste program. The spreadsheet programs can handle repetitive calculations where minor changes in the calculations can result in significant additional work to obtain a final answer. Formulas can be built into a spreadsheet program so that changing one number will automatically cause the system to run through the sequence of calculations. Such systems are very useful in developing "what if" solutions using changes in significant variables and determining the results of such changes.

Computer bulletin board systems (BBS) and Internet access are other tools available to help pretreatment inspectors do a better job. Bulletin board systems provide the opportunity for inspectors to exchange information with other inspectors, coordinators and regulators. Inspectors can send messages to other inspectors and search pertinent literature. To access computer bulletin boards, inspectors need a personal computer or mainframe terminal, a modem and any necessary cables and telephone jacks to connect the modem to the computer or terminal and to the telephone system, and a communications software program. Table 2.4 lists five computer bulletin board systems and the telephone access numbers.

In addition to these bulletin boards, on-line services like Compuserve, America Online, and Prodigy offer a vast array of information in specialized fields and provide general-purpose electronic services such as e-mail. Many companies use these on-line services to provide the most up-to-date information to their customers.

At the time of this writing, the global computer network known as the Internet has established itself as a prominent and accessible source of information. Emerging technologies promise to enhance this accessibility. Today, a computer user

TABLE 2.4   COMPUTER BULLETIN BOARD SYSTEMS (BBS)

Point Source Information	EPA Office of Water-Related Information Including Pretreatment	www.epa.gov/owm
Wastewater Treatment Information	EPA Wastewater Treatment Information	www.nsfc.wvu.edu
Cleanup Information	Hazardous Waste Removal	www.clu-in.org
Technology Transfer	EPA	www.epa.gov/ttn
Wastewater Issues Discussion Group	Office of Water Programs California State University, Sacramento	www.owp.csus.edu/lists/

with only a modem and a network browser like Netscape, Internet Explorer, or Opera (free or extremely inexpensive programs) can gain access to the entire Internet by paying a small monthly fee to an Internet provider.

2.43   Advantages of Computer-Based Systems

Some of the advantages of a computer-based system are already apparent; it can handle vast quantities of data and make some sense out of this information. The computer-based systems, if properly programmed, quickly produce printouts of selected data such as all companies in a certain industry group in a certain area that have had cadmium violations within the last six months. The systems may be very useful in solving industrial waste-caused problems. Another advantage of computer-based systems is that a very large amount of data can be stored in a small space on computer hard drives and disks. A typical typewritten page will require about 2,000 bytes of memory and the equivalent of 20,000 typewritten pages can be stored on a forty megabyte hard disk. The reproduction of these pages can be done readily by recalling the information from the hard disk storage.

Complex calculations which require significant amounts of engineering time can be readily performed by a computer after the initial programming is completed. The on-line examination of large amounts of computer data records can be quite useful in making decisions about such things as whether a company has adequately paid its surcharge. Examining the surcharge records over the last several years by sequentially flashing them on a computer screen can allow a quick comparison with current surcharge payments.

2.44   Preparation of Reports and Industrial Waste Documents

In all industrial waste work, it is frequently necessary to prepare a report for review by the POTW agency management, or for review by the EPA and state supervisory agencies. It may also be desirable to report to the industry the results of analyses of samples taken by the POTW. The data presented are repetitive and may require only minor changes from time to time in order to prepare the finished report. Computer-based word processing and data management systems are particularly helpful in this regard since only the relatively minor changes need to be made; the complete report does not have to be prepared from scratch.

2.45   Field Records for Industrial Waste Personnel

In addition to submitting industrial waste data to be entered in the computer system, all field personnel should keep a record of their work in the form of a field diary noting the events that happen each day. Such records are frequently very helpful in determining when an inspector first noticed a problem or what was done to encourage correction of the problem. When taking both grab and composite samples, it is absolutely essential to record where the sample was taken, the type of sampling equipment, whether a glass or a plastic jar was used and other details of the sampling effort. Proper documentation of all samples collected must be maintained to meet legal requirements. At some agencies a portion of the field notebook records prepared by the industrial waste inspector is input to the agency's computer system. Figure 2.12 shows the computer input form for inspection data.

2.46   Laboratory Records

Any industrial waste program will generate a large number of laboratory analyses; a proper means of recording and retrieving analytical records should be provided. It is suggested that such records be maintained on a computer system using some identifying code for the industrial location from which the samples are taken. At some agencies all laboratory analyses prepared by the agency's laboratory are input to a laboratory database management system. This system allows the printout of the information or the on-line recall of the analysis information using the identification number of the industrial waste permit for each company. The laboratory analysis records can be called up or printed for any given time period (Figure 2.13). In compliance actions and surcharge reviews, it is helpful to be able to look at the laboratory data over a significant time period to determine whether the company is making an ear-

COMPUTER SERVICES SOURCE DOCUMENT

INDUSTRIAL WASTE INSPECTION REPORT - C#IWI810

To: Data Control (Computer Services)   Date: _____
From: Bill Garrett   Dept/Section: TS/IW   Ext: 2907
User Form: IWI-SRC4.GED
Job: C#IWI810   Description: Report IWI810 - Inspection File Update.

*Optional Entry Info (All Blanks Must Be Filled For Reports That Generate New Account(s)

New Record - N   * Batch   Account   I.W. Permit No.   Date   Visit
Change Info - C   Code   No.   (Insert Leading Zeros) (Mo,Day,Yr) No.

|N|

1 Name of Firm

2 Street No.   Unit   Direction   Name

Type   City   Situs Zip Code

3 Person to Contact   *Telephone No.   District No.   *SIC No.

*Thomas Map: *Inspection   Inspector   Inspection Time
Page - Area   Area   (Last Name)   Hr   Min

4 *No. Employees Shifts *Days/Wk *Water Use (100CF/M) Flow(GPM) Type Flow (C,D,I)

Waste Charact.   Violations   Corrections   Special Message

5 No Longer Used

6 General Comments

7

8

9 Waste Characteristics Comment or General Comment:

Fig. 2.12 Computer input form for industrial waste inspection report

LABORATORY DATA SYSTEM - Industrial Waste Data : Permit No. 05529   03/27/96   Page 1

S A M P L E   T E S T   R E S U L T S

DATE	LOG NO.	(101) PH PH	(151) SUSPENDED SOLIDS MG/L	(206) TOTAL CYANIDE MG/L CN	(403) TOTAL COD MG/L O	(708) CADMIUM MG/L CD	(709) TOTAL CHROMIUM MG/L CR	(712) COPPER MG/L CU	(714) LEAD MG/L PB
02/01/94	SJ71443 O-GR	9.00							
02/01/94	SJ71548 P-CO	8.12 A	22	0.067	33	<0.02 B	0.68 B	0.70 B	<0.2 B
04/06/94	SJ74605 O-GR	8.38		0.123		<0.02 B	0.39 B	0.64 B	<0.2 B
05/09/94	SJ76026 O-GR			0.012		<0.02	0.28	0.47	<0.2
05/23/94	SJ76739 S-CO	7.49	162			<0.02	0.18 B	1.22 B	<0.2
05/24/94	SJ76832 S-CO	7.44	86			<0.02 B	0.05 B	0.05 B	<0.2 B
06/01/94	SJ77091 S-CO	7.53	170			<0.02	<0.05	<0.05	<0.2
06/02/94	SJ77148 S-CO	7.30	144			<0.02	<0.05	<0.05	<0.2
06/03/94	SJ77216 S-CO	7.77	84			<0.02	<0.05	<0.05	<0.2
06/16/94	SJ77849 P-CO	7.98 A	19		77 C	<0.02	0.29	0.89	<0.2
10/26/94	SJ83553 P-GR	8.05 A	17	0.322 B		<0.02	0.48	0.67	<0.2
10/26/94	SJ83652 P-CO	9.07 A			70	<0.02	0.32	0.51	<0.2
11/29/94	SJ85777 O-GR			0.027					
01/24/95	SJ90984 P-GR			0.048					
01/24/95	SJ91033 P-CO	7.86	9		60				
03/06/95	SJ92940 P-CO	8.65 A	12		68 C	<0.02	0.24	0.45	<0.3
03/29/95	SJ94024 P-CO	7.93	22		100 C	<0.02 B	0.58 B	0.86 B	<0.2 B
04/27/95	SJ95444 P-CO	7.86	7		54	<0.01	0.34	0.46	0.24
06/07/95	SJ97142 O-GR			0.033		<0.02	0.21	0.54	<0.2
08/16/95	SJ00292 P-CO	8.36	8		46 C	<0.02	0.25	0.62	<0.2
09/11/95	SJ01265 P-CO		31		70 C	<0.02	0.93	1.68	<0.2
11/13/95	SJ04107 P-CO		11		59 C	<0.02	0.29	0.89	<0.2
11/11/96	SJ10531 P-CO		145		174	<0.02	5.14	5.78	0.4
01/12/96	SJ10532 P-GR			0.018					
02/15/96	SJ12228 P-CO		264		554	<0.02	11.3	10.0	0.5

Footnote(s): A-AVERAGE OF DUPS, B-DUPLICATE SPIKE, C-DUP & SPIKE

Fig. 2.13 Laboratory data output

LABORATORY DATA SYSTEM - Industrial Waste Data : Permit No. 05529   03/27/96   Page 2

* * * * * * * * * * * * * * * * S A M P L E   T E S T   R E S U L T S * * * * * * * * * * * * * * * *

		(718) NICKEL MG/L NI	(722) SILVER MG/L AG	(724) ZINC MG/L ZN
DATE	LOG NO.	MG/L NI	MG/L AG	MG/L ZN
02/01/94	SJ71443 O-GR	0.88		0.12
02/01/94	SJ71548 P-CO	0.92 B	< 0.2 B	0.17 B
04/06/94	SJ74605 O-GR	0.87 B		0.08 B
05/09/94	SJ76026 O-GR	0.91 B		< 0.05 B
05/23/94	SJ76739 S-CO	< 0.08	< 0.2	0.08
05/24/94	SJ76832 S-CO	< 0.08 B	< 0.2 B	0.15 B
06/01/94	SJ77091 S-CO	< 0.08	< 0.2	0.12
06/02/94	SJ77148 S-CO	< 0.08	< 0.2	0.11
06/03/94	SJ77216 S-CO	< 0.08	< 0.2	0.05
06/16/94	SJ77849 P-CO	1.17	< 0.2	0.17
10/26/94	SJ83553 P-GR	1.08	< 0.2	0.28
10/26/94	SJ83652 P-CO	0.6		< 0.05
11/29/94	SJ85777 O-GR			
01/24/95	SJ90984 P-GR			
01/24/95	SJ91033 P-CO			
03/06/95	SJ92940 P-CO	0.4	< 0.2 B	0.17
03/29/95	SJ94024 P-CO	1.2 B	< 0.2	0.18 B
04/27/95	SJ95444 P-CO	0.55	< 0.04	0.08
06/07/95	SJ97142 O-GR	0.6		< 0.05
08/16/95	SJ00292 P-CO	1.0	< 0.2	0.12
09/11/95	SJ01265 P-CO	1.6	< 0.2	0.32
11/13/95	SJ04107 P-CO	0.9	< 0.2	0.07
01/11/96	SJ10531 P-CO	5.2	< 0.2	1.56
01/12/96	SJ10532 P-GR			
02/15/96	SJ12228 P-CO	8.6	< 0.2	2.53

Footnote(s) : A-AVERAGE OF DUPS, B-DUPLICATE SPIKE, C-DUP & SPIKE

Fig. 2.13 Laboratory data output (continued)

nest effort to comply with effluent limitations or whether the current surcharge analyses are typical or changing from those obtained in the past.

QUESTIONS

Write your answers in a notebook and then compare your answers with those on page 83.

2.4A   List the main sources of data from an industrial waste pretreatment program.

2.4B   What capabilities must be included in the design of a computer database management system?

2.4C   List the basic types of general-purpose software that should be considered for use with a computer system.

2.5   INDUSTRIAL WASTE MONITORING

2.50   Industrial Self-Monitoring

As mentioned earlier, certain industrial companies are required, under EPA programs and under the program of many agencies, to self-monitor their wastewater. The self-monitoring report form commonly used by industries is shown in Figure 2.5 (pages 51 and 52). This self-monitoring report requires the industry to obtain a laboratory analysis of its wastewater and requires a responsible person in the company to observe the quality of the wastewater effluent, sign the self-monitoring form and forward it to the inspection agency. Industrial self-monitoring should comply with the following conditions:

1. Be at a frequency determined by the POTW and specified in a permit or some other mechanism,

2. Be at a location or locations and by the method (grab, composite, time or flow proportional) specified by the POTW, and

3. Include pollutants specified by the POTW as a minimum.

If the industrial user (IU) self-monitors for a pollutant more frequently than required by the POTW, the IU must report all the data it collects. However, it is widely understood that many companies, upon review of an initial sampling result by management personnel, will not submit that result to the agency. If a violation is observed, the company may discard that sample result, clean up its industrial process so that the wastewater

complies with the agency's regulations, and then obtain another discharge sample. While this may not comply with the letter of the self-monitoring requirements, it is certainly complying with the spirit of the pollution control aspects of the EPA program as well as any local POTW agency program. If a company, through a self-monitoring report, will undertake measures to improve the quality of its wastewaters so that it will not be discovered and penalized by the POTW agency, this fulfills the function of the industrial waste program without involvement of personnel from the POTW agency. However, the POTW should inform every company that it expects *ALL* data to be reported as well as corrective action taken whenever noncompliance occurs.

Realistically, some companies will, from time to time, report that their wastewater is in compliance when it is not and the company has not undertaken measures to clean up its wastewater. The requirement that a responsible company official sign the self-monitoring document under penalty of perjury should prevent widespread use of such procedures. Companies should be aware that falsification of data is a criminal offense and the POTW has the ability to prosecute guilty companies.

A significant problem that can occur during industrial self-monitoring is failure to collect representative samples and to properly preserve them. The POTW must provide industries with specific written guidelines to minimize this risk. This can be achieved by recommending compliance with published EPA Guidelines in 40 CFR Part 136[7] and *STANDARD METHODS.*[8]

2.51   Monitoring by POTW Personnel

Monitoring by POTW personnel is required under the EPA pretreatment program to verify that companies are in compliance, independent of any self-monitoring information submitted by the company. A monitoring program by the POTW agency is designed to assist and protect the company desiring to comply with environmental regulations. Such a company may be at a significant competitive disadvantage with another less ethical company if a rigorous review by the POTW is not practiced. Monitoring by the POTW encourages all companies to be more straightforward in their self-monitoring reports. Although monitoring by the POTW may in some instances be deemed unnecessary when the same information is being furnished by the company, good reasons exist for this duplication of effort. If differences exist between the POTW's monitoring and the monitoring furnished by the company, an in-depth review of the analytical data is in order.

Sampling and analysis for IU Baseline Monitoring Reports (BMRs), 90-Day Compliance Reports and Periodic Reports on Compliance may be performed by the POTW instead of the IU. This is allowed in the General Pretreatment Regulations in section 403.12(g). The original purpose for allowing POTWs to collect the data was to ensure that samples were collected properly, analysis was conducted within holding times and correct procedures were followed. Where the POTW performs the required sampling, the IU is not required to submit the compliance certification required on these reports. This compliance determination will be made by the POTW and communicated

---

[7] *40 CFR Part 136, "Guidelines Establishing Test Procedures for the Analysis of Pollutants Under the Clean Water Act." Also see Table II, "Required Containers, Preservation Techniques, and Holding Times." Available from the U.S. Government Printing Office, Superintendent of Documents, PO Box 371954, Pittsburgh, PA 15250-7954. Parts 136-149, 2001. Order No. 869-044-00152-7. Price, $55.00.*

[8] *STANDARD METHODS FOR THE EXAMINATION OF WATER AND WASTEWATER, 20th Edition. Obtain from Water Environment Federation (WEF), Publications Order Department, 601 Wythe Street, Alexandria, VA 22314-1994. Order No. S82010. Price to members, $164.25; nonmembers, $209.25; price includes cost of shipping and handling.*

to the IU. In addition, where the POTW itself collects all the information required in one of these reports, including flow data, the IU will not be required to submit the report.

2.52   Review of Analytical Data

The self-monitoring information submitted by companies may become quite voluminous. One agency could monitor as many as 1,300 companies and review their self-monitoring reports. The companies may submit from one to twelve (monthly) reports per year and the total number of reports probably is in the range of 3,000 reports per year. It is a major undertaking for the POTW to verify that the companies are submitting accurate self-monitoring reports and to verify compliance or noncompliance with the companies' effluent requirements. The reports must also be checked to see that all of the pertinent pollutants are analyzed and that a responsible individual in the company has properly signed the self-monitoring form.

Self-monitoring data could be reviewed manually by an inspector or a properly trained typist-clerk. If inadequacies are found, a form letter indicating the inadequacy is sent to the company generating the self-monitoring report. Following up to ensure that these deficiencies are properly remedied is a significant task. Although computer systems are available to determine whether companies have submitted or have not submitted their reports, no system currently exists to follow up on the individual laboratory results. Some agencies have developed and implemented a fully computerized system whereby the information from these self-monitoring reports can be input to the computer and, based on these results, various computer-printed letters and reminder forms can be automatically sent to the affected companies. It is recommended that any POTW agencies having large numbers of companies on self-monitoring programs consider the possibility of computerizing this data. Smaller POTWs having only a small number of self-monitoring companies may not find report reviews a burdensome task.

2.53   Comparison of Analytical Data

In the EPA industrial categorical pretreatment program, there are currently 26 basic categories of industries. In a few of the categories, the numerical effluent values for the entire category are basically the same. In others, such as the metal finishing category, there are several subcategories requiring special treatment and different effluent numbers. In many of the EPA categories that have mass-based standards,[9] the numbers for each individual company will be unique. These numbers may, in fact, change from year to year as the production, wastewater flow and level of pollutant discharges are changed by the company. To determine whether companies are in compliance with federal regulations will require a significant amount of review. This is especially so when a unique number is applicable to the individual company.

At some agencies all effluent requirements are converted to concentration limits[9] as this is the only practical means of day-to-day regulation of a company's wastewater discharge. Even on this basis, a concentration value for a pollutant obtained through laboratory analysis must be related to the company's effluent limits. Considering that companies may have from eight to 20 or more different pollutant effluent limitations and that the very large agencies may have close to 1,000 categorical companies, a large amount of data review is required. Typical medium-sized POTWs may have fewer than 20 categorical companies.

A computer system is capable of comparing laboratory data with effluent limitations established for the EPA categorical companies as well as comparing noncategorical industry data with the local agency's limits. If the companies are found to exceed their effluent limitations, the computer will print out a violation notice. The computer can also be programmed to produce a historical record of violations by companies to enable the POTW to determine whether the company is attempting to improve its effluent discharge record or is progressively becoming a worse violator of effluent requirements.

POTWs having any significant number of categorical dischargers should seriously consider developing a computer system which will compare those dischargers' effluent requirements with laboratory data obtained by the POTW or submitted by the industrial company. Again, smaller POTWs with small numbers of categorical dischargers may not find such a computer-based system necessary. Medium to large POTWs may find a computer-based system quite helpful in proper review and comparison of analytical data with a company's effluent limits.

2.54   Statistical Evaluation of Data

There are many occasions when a POTW needs to do a statistical evaluation of data, for example, in comparing the results of one laboratory with the results of a second laboratory. It is also frequently necessary to compare the results of the POTW's sampling and analysis to those submitted by the company. For surcharge purposes, some agencies compare the agency's sampling results with the industry-supplied sampling results. Statistical analyses such as these are more quickly and accurately accomplished by a computer than by labor-intensive manual calculations.

With many agencies the POTW's values always take precedence over a company's values. This requires that the POTW's lab have a good quality assurance/quality control (QA/QC) program. Samples can be split and results compared. However, if an industry only samples on good days, their numbers will always be better.

See Appendix II, "Pretreatment Arithmetic," Section W, "Average Limitations," at the end of this manual for additional statistical information. See problem 57 for calculations to determine if a significant industrial user (SIU) is in significant noncompliance (SNC).

QUESTIONS

Write your answers in a notebook and then compare your answers with those on page 83.

2.5A   What can be done to prevent a company from reporting that their wastewater is in compliance when it actually is not?

2.5B   What is a monitoring program by the POTW agency designed to accomplish?

---

[9] *Mass-based limits and concentration-based limits are two different types of effluent limitations; both are explained later in Chapter 3.*

2.6 PUBLIC RELATIONS

2.60 Relations With Industrial Personnel

Relations with industrial personnel can differ greatly from company to company and from individual to individual. Typically the most difficult companies to deal with are those owned by a single proprietor. This person has built a business from a very low level to a successful business and may resist government regulation. Such sole proprietor or family businesses may not see the wisdom of readily complying with government regulations and may react negatively to any government regulation. Many such companies, however, are managed by very competent people. If they can be convinced that it is reasonable and in their interests to comply with environmental regulations, they are usually quite adept at developing systems to comply with the regulations.

Generally speaking, it is easiest to deal with large companies. Personnel in these companies, especially environmental officers or those in a similar administrative function, see compliance with environmental regulations as part of doing business in modern society. Even though they may not philosophically agree with such regulations, they generally do not significantly resist or set out to block enforcement efforts. Whether they will comply as readily as the sole proprietorships is open to question but generally the personal relationships with such people are much more easily handled.

The most frustrating problem to deal with are companies that appear very cooperative but do not make measurable progress toward compliance with the regulations. It is difficult to find fault with the company that agrees with all proposals made by the POTW agency and, at least verbally, will agree to comply with such requirements. When such companies do not comply within a reasonable time period, the POTW agency should take positive steps to require compliance, through punitive measures if necessary.

2.61 Relations With Environmentalists

In dealing with environmentalists, describe the significant efforts and progress being made by your organization in reduction of pollutant discharges. Some environmentalists tend to be somewhat idealistic and not appreciative of the complex problems involved in complying with environmental regulations. An effort to inform representatives of environmental groups of the major efforts being made by a POTW agency to comply with and to enforce environmental regulations will generally reassure them that an adequate effort is underway.

The environmental movement may, to some extent, act as a counterbalancing force to a strong industrial lobbying group. The industrial waste effort at a POTW agency generally has to mediate between the POTW's responsibility to meet EPA and local regulations and the wish not to put companies out of business. To some extent the agency owes allegiance to both the industrial groups and the environmental groups within the community and should try to maintain a satisfactory contact with both groups.

POTWs must realize that federal laws are written to allow for citizens' suits and that such lawsuits are often filed by environmental organizations when the organization feels that regulatory agencies (EPA, states, and POTWs) are not enforcing prescribed laws. Therefore, failure to enforce rules and regulations can result in costly and time-consuming efforts by a POTW as well as an industry when more pressing violations and problems are occurring.

2.62 Relations With Political Leaders

The political leaders of a community are usually elected because they have successfully appealed to the wishes of the people within the community. They are closely attuned to the desires of the community for improvements in the lifestyle of the community. However, the political leaders may not be very knowledgeable about the local and national regulations affecting industrial waste control procedures. It is the duty of the industrial waste inspector or program manager to help inform political leaders about the requirements of environmental regulations and what is being done in their community to comply with these regulations.

Frequently, companies that are adversely affected by environmental regulations appeal to a political figure to help them find a solution to their problem. The politician is to some extent duty-bound to try to assist a company. Here again, the POTW should try to inform the politician of the reasons for any enforcement actions. Sometimes an offer of additional time to reach compliance may be extended to the company as evidence that the POTW agency is willing to work with the company to resolve the problems. Generally, after some reasonable additional time is given and some attempt is made to assist the company to resolve the problem, a political representative is usually willing to assist in forceful action against the company if the company refuses to comply. However, the POTW agency must stress to all parties involved that all companies must be treated equally and fairly, even with regard to any time extensions.

2.63 Relations With News Media

Working with the press and media can be intimidating. Sometimes it appears as if the only things that can happen are all bad. Water pollution control stories seem inevitably to carry a "toxic" or "pollutant" angle and knee-jerk defensiveness can become an instinctive reaction after even one bad encounter. However, poor working relationships with the press and media can create an information flow environment that is far worse. A reporter or journalist automatically sees non-cooperation and hesitancy as suspicious behavior indicating secrets, and secrets mean stories. Statements like "no comment" and an apparent reluctance to communicate with media representatives only spur reporters to take a closer look.

Reducing the description of projects, programs and goals to a simple statement is an art. But, good media and press relations are not based on just this skill. Try to develop a friendly working relationship with the reporters who are usually assigned to your organization's activities. Don't wait for emergencies. Also, it is always a good idea for your organization to appoint a spokesperson to keep the information source centralized.

A reporter usually tries to report both sides of an issue. The final article or news piece will carry the opposing viewpoint, no matter how wrong you may think it is. You still must spend the time educating the reporter on your "50 percent" of the article's content; otherwise, your side will not be heard at all.

Reporters are often rushing and are always on deadline. The more you can organize your material and prepare it for them, the more the story will contain representations of your concerns. The mode of conduct that works best with reporters is "courtesy, cooperation and candor." When first contacted by a reporter, courteously find out what the reporter is interested in and promise a call back from a representative of your organization. Cooperate and follow through. Make sure that the information is not too technical or too complex for the reporter to interpret. Reporters don't know your business, only their own. They need your help if they are even to come close to getting the story "right" as you see it. Candor is the real key. The truth is simply told and never requires alibis. Reporters are trained to seek the truth and their suspicions are easily aroused by fabrication. If they're good reporters, they'll find fabrications.

There are several good reasons why stories don't always look like you thought they would: a reporter's story is usually edited by another person; reporters do not write their own headlines; and, reporters may have received incorrect information from other sources.

Respect reporters' deadlines and they'll usually show respect for your concerns. Never hesitate to tell a reporter "I don't know" but always follow it with, "I'll find out for you." Don't ever just hope that reporters won't call again or will just go away. They won't.

In short, make sure that the right people on your staff are communicating with the media with courtesy, cooperation and candor. After all, once you've lost credibility with the press and media, you've lost it for a decade or more. Reputations of sources (you) travel fast among reporters. Always deal with the news media in a professional manner, keeping to the facts in a field situation. All questions concerning POTW policy which are beyond the responsibility of the pretreatment inspector should be referred to the POTW agency spokesperson.

Because the pretreatment inspector is often at the scene during an incident, media contact may be unavoidable. Inspectors should follow POTW agency guidelines, if available, regarding contact with the media. Always be careful not to make any statements that could even remotely be interpreted as a cause or blame for an incident.

Most POTW organizations have a policy about the dissemination of information to the media and specify which persons are authorized to make statements on behalf of the organization. Such policies are not designed to restrict media access but rather to ensure that accurate information is presented and that all levels of management are aware of incidents that may affect the organization. A specific person in management may be designated as the media contact or the organization may have a public affairs officer who handles such tasks.

The industrial waste inspector's presence on-scene may result in what could appear to be a casual contact with media personnel and the inspector may be approached for an "off-the-record" comment on a situation. Unless the inspector has been given prior authorization to consent to a media interview, organization policy should be strictly followed.

When so authorized or when the situation is such that media contact is unavoidable, the inspector should keep in mind that only known facts within the inspector's expertise should be discussed. Speculation on an incident can lead to much embarrassment or worse. In any event, all media contacts should be reported to the inspector's immediate supervisor as soon as possible.

2.64 Relations With the Public

There will be times when the pretreatment inspector will receive inquiries from the public. For example, when the POTW publishes the list of industrial users in significant noncompliance, citizens may call with questions about the impact of the noncompliance on the community. Other calls may involve "tips" concerning suspected illegal activities or questions from students undertaking science fair projects. Inspectors need to use just as much diplomacy and professional demeanor with the public as they do with industrial users.

2.65 Relations With Attorneys

Attorneys can be very helpful in avoiding costly and time-consuming litigation. Attorneys, either on staff or retainer, can assist in ordinance development and interpretation, advising on policy such as chain of custody and enforcement management, and provide another view on administrative matters in a cooperative working environment.

POTW agencies should not be too quick to use lawsuits by attorneys to settle problems. Attorneys are only required where there is a significant problem that cannot be resolved by other means. In some instances, attorneys compound rather than resolve the situation. When both the POTW agency and the company retain legal counsel to resolve a question, the question may get buried by details and other aspects of the problem about which neither of the main parties has any significant knowledge. A significant amount of effort may be spent resolving fine points of the law rather than resolving the issue in dispute. Significant effort is also required to obtain EPA assistance (if needed) and to schedule a court date.

2.66   Relations With Other POTW Agencies

The POTW agency should maintain close relationships with other POTW agencies within the area. Such relationships are helpful in developing similar programs so that a radical difference between neighboring agencies will not influence migration of one or more industries to a certain POTW system. Maintaining reasonably compatible industrial waste programs helps maintain a relatively stable industrial community in an area.

Close relationships with other POTW agencies in the area may help to solve difficult industrial waste problems. One agency may have encountered and solved a problem similar to that now causing the other agency great difficulty. Periodic meetings among the various agencies may be good forums for discussing difficult problems and the solutions which the agencies may have found for those problems.

Another advantage of close contact among POTW agencies is that problem companies that have encountered difficulty in one POTW agency may move to a neighboring agency. A warning of the difficulties one agency has had with certain company personnel may be appropriate if that company intends moving to a nearby POTW. Such a warning may allow the other POTW to more properly and more quickly resolve the problems if and when they occur.

Close relationships among agencies can be fostered by active participation in professional organizations such as the Water Environment Federation and its member associations.

2.67   Goal-Oriented Relationships

In relations with people impacting the POTW agency's industrial waste program, it is recommended that industrial waste personnel seek to develop goal-oriented relationships. For example, the reason for employing an attorney is to resolve a specific industrial waste problem; in relations with attorneys, all actions should be directed toward a resolution to the problem. Similarly, relationships with environmentalists focus on explaining to them the actions taken by the POTW agency to reduce or eliminate excessive pollutant discharges to the environment. Relations with news media personnel should be aimed at presenting a good public image for the POTW agency and in avoiding adverse publicity for the agency or agency personnel. By actively focusing on the goal you wish to achieve in a business relationship, you increase the probability of achieving it.

2.68   Avoidance of Philosophical Activism

Personnel employed by an industrial waste program in a POTW agency come into the program with their own sets of beliefs about pollution and the environment. The two philosophical extremes are represented by some environmentalists who would like to see all industrial development in the area close its doors and return the area to a forested wilderness and, at the other extreme, by people who would allow industry to discharge wastes unrestricted by the POTW agency's industrial waste program. Such philosophical positions have to be tempered to accommodate the needs of the industrial waste program. It is suggested that all incoming employees should be cautioned that the industrial waste program is not an area in which personal philosophies may guide an employee's activities. Even though an employee may be a member of the Sierra Club or the Friends of the Earth, that person must operate within the confines of the industrial

waste program as defined by the POTW agency. The EPA and local agency regulations are aimed at reducing pollution; they are not aimed at putting companies out of business. Industries may only be required to comply with the level of industrial waste control necessary to meet EPA and local agency regulations.

QUESTIONS

Write your answers in a notebook and then compare your answers with those on page 83.

2.6A   How should inspectors deal with environmentalists?

2.6B   What kinds of information should inspectors provide political leaders?

2.6C   How should inspectors deal with news media personnel?

2.7   HAZARDOUS WASTE RULES AND REGULATIONS

2.70   Federal Hazardous Waste Regulations

There have been two main federal regulatory efforts in the control of hazardous wastes. These are the Resource Conservation and Recovery Act (RCRA, pronounced RICK-ruh) and the Comprehensive Environmental Response, Compensation, and Liability Act (CERCLA, pronounced SIRK-la) of 1980.

The RCRA Act was passed by Congress in 1976. This Act established regulations to ensure that hazardous wastes were properly transported to hazardous waste landfills and that hazardous waste landfills were properly controlled so that no major sources of pollution entered the environment from these landfills. Other items were controlled under the RCRA Act but the primary focus was on ensuring that hazardous wastes were properly disposed of at treatment and disposal facilities. The RCRA Act also regulates treatment facilities which treat hazardous wastes to render them nonhazardous so that the resulting nontoxic materials can be properly disposed of in the environment.

CERCLA is the other significant piece of legislation regulating hazardous wastes. This Act is also called the Superfund Act. CERCLA was enacted primarily to correct past mistakes in industrial waste management. The focus of the Act is to locate hazardous waste disposal sites which are creating problems through pollution of the environment and, by proper funding and implementation of study and corrective activities, eliminate the problem from these sites.

The Superfund law contains severe liability provisions. Any company that disposes of waste at a site found to have problems under the Superfund law can be held liable as a "responsible party" in the cleanup of the Superfund site. Liability is joint and several, which means that all parties can jointly be held responsible for the problem or that any one party individually, if that party has enough resources, can be held responsible for all of the problems created at the site. This liability may mean that waste generators who, in good faith, disposed of their waste years ago at a legally permitted site can still be held responsible for cleaning up severe pollution problems at the site. This long-term liability for hazardous wastes is commonly known as the "cradle to grave" philosophy.

Both the RCRA and the Superfund laws make it much less advantageous to dispose of wastes on land, in ponds or with land farming methods. Since the generator of these wastes can ultimately be held responsible for any problems which arise in the future, the laws are encouraging companies to treat their wastes on site to render them nontoxic. The disposal of the clarified effluent to the sewer is then permissible and residual solids can be disposed of in more highly protected toxic waste residue repositories.

2.71 Effects on Pretreatment Programs

One of the effects of the CERCLA and RCRA laws on industrial waste pretreatment programs is that more industrial companies will be treating their wastes on site and disposing of the residual products from these wastes at the POTW. The POTW must therefore be more concerned about the proper treatment of hazardous wastes and toxic materials because other avenues for disposal elsewhere in the environment no longer exist. The level of treatment technology used by companies will continue to increase as the total destruction of toxic materials at a company site is required because the transport and disposal of such toxic materials at other community locations is prohibited.

Another effect on pretreatment programs is that certain companies which are producers of hazardous wastes under the RCRA laws may not presently be considered serious potential dischargers of toxic pollutants under the local POTW pretreatment programs or the Clean Water Act regulations. Examples of such companies are: (1) common clothes cleaners where significant quantities of perchloroethylene and other toxic solvents may be produced, (2) automobile radiator repair shops where significant quantities of heavy metals and other toxic materials may be produced, (3) gas stations and automobile repair shops where quantities of polluted waste oils may be produced, and (4) solvent reclamation locations which have little sewer discharge but significant quantities of waste solvent sludges for disposal. These companies and others are having significant problems with disposal of their residual waste materials under the RCRA and Superfund laws. As has happened in many localities where these industries are not considered major wastewater dischargers, the POTW may not have proper control measures to regulate such companies. There is a tendency for these companies to consider illegal disposal of problem wastes to the wastewater collection system rather than face the rather strenuous problems of legally disposing of their hazardous wastes under the RCRA law.

A significant concern exists in Congress and in environmental groups that such wastes may not be properly controlled under either the RCRA or Clean Water Acts. The Domestic Sewage Study was conducted in 1986 by EPA for Congress, which was at that time considering the application of the RCRA law to wastewaters discharged to POTWs. The conclusion of this study was that the EPA did not wish to implement the RCRA laws for discharges to wastewater collection systems. The study recommended that the pretreatment regulations be used to properly control such wastes through expansion of the pretreatment categories to cover any problem waste groups. This is a logical conclusion since the RCRA law was written primarily to control wastes being transported over land, land disposal of wastes, and waste disposal at treatment sites away from the site where the wastes were generated. The Clean Water Act regulations apply most appropriately under the general and categorical industrial pretreatment programs to wastes generated at an industrial site and disposed of into the sewer from that industrial site. When residue products (sludges) are created which require disposal from the industrial site, these residue products will be covered under the RCRA Act regulations.

Pretreatment inspectors should be aware that current and future environmental regulations will require that they make more decisions about what hazardous or regulated substances can be discharged to the POTW for treatment. The duty of the pretreatment inspector is to prevent IU discharges of toxic materials in toxic amounts. However, the expertise of the pretreatment inspector is necessary when materials which can be categorized as hazardous or regulated under RCRA or CERCLA are to be considered for discharge to the POTW. If the POTW has the ability to "detoxify" the waste through rearrangement of its structure or complete destruction in biological treatment or solids treatment and disposal (that is, incineration, stabilization or heat treatment), then this option of treatment and disposal must be considered as a good environmental and human health and safety option.

The POTW should use its ultimate capacity for treatment of wastewaters to provide maximum benefit to society and the environment. The POTW operating personnel rely on the pretreatment inspector to make a *DETERMINATION* of what waste materials or substances are and are not acceptable for treatment at the POTW processes and at what concentration. It then becomes the *RESPONSIBILITY* of the IU to take the information and guidance provided by the pretreatment inspector and make the decision whether or not the IU's wastewater can be discharged in compliance with applicable pretreatment

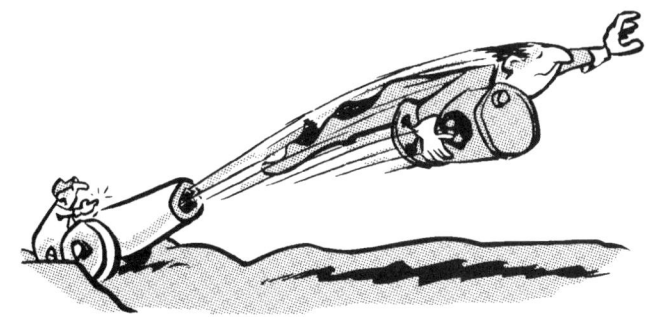

standards. Applicable pretreatment standards, which are designed to protect the POTW, POTW workers, the public and the environment, must be complied with. This is a very important and far-reaching issue for the pretreatment inspector in the 1990s.

2.72   Cleanup of Superfund Sites

As the CERCLA and SARA (amendments to CERCLA) laws take effect and Superfund sites are being cleaned up, the waste products from the cleanup of the sites will be disposed of through POTW wastewater collection systems. Most of the wastes from these sites will require some form of treatment to clean up the materials. This cleanup treatment may, in turn, generate a wastestream that includes a large volume of liquids. Proper cleanup of these liquids can ultimately allow the disposal of the material to a POTW.

Under the Superfund law, any agency accepting wastes from a Superfund site could potentially be held liable for long-term damages due to such wastes. Regulatory officials are currently discussing how a POTW and its management can be protected from the liability for later damages from Superfund wastes if such wastes are accepted by the POTW in a Superfund cleanup situation. POTWs should strive to assist the community in cleanup of Superfund sites but should also try to avoid any continuing liability problems.

2.73   Centralized Waste Treatment (CWT) Facilities[10] for Hazardous Wastes

As the sites for disposal of hazardous wastes on land become fewer and fewer, the need to properly treat these wastes will become greater. In several regions, centralized waste treatment facilities have been established. These facilities, at present, aim at certain segments of the waste treatment market. For example, one plant might specialize in the treatment of wastewaters contaminated with significant quantities of oils. These wastes originate from machine shops, airplane toilet cleanings, oily leachates from landfills, and miscellaneous oily and oil-solvent laden wastewaters. The waste cleanup stream at a centralized treatment facility could include the following elements: (1) truck unloading station, (2) coarse oil removal using a corrugated plate interceptor, (3) influent storage tanks, (4) oil-water emulsion breaking using acid, alum and polymer coagulants, (5) dissolved air flotation, (6) oil recovery from flotation scum, (7) pressure sand filtration, (8) paper cartridge filtration, (9) air stripping of volatile organic compounds, (10) incinerator burning of volatile organic off-gases, (11) effluent chlorination, (12) plate and frame filtration for sludge concentration, (13) effluent storage tanks, and (14) oil recycling storage tank facilities.

As the options for disposal elsewhere in the environment become more and more limited, the development of facilities such as these will become a prominent part of the industrial waste pretreatment picture. Pretreatment inspectors need to be familiar with all types of disposal options available to companies within their service areas.

When this manual was published EPA had proposed categorical standards for centralized waste treatment facilities. The centralized waste treatment industry is very aware of potential categorical standards and is striving to do a better job.

Because of changes in the centralized waste treatment industry, EPA may withdraw the proposed rule. EPA could simply rely on the POTW's local limits, the proper use of the combined wastestream formula, and the knowledge of the pretreatment inspector/permit writer to control these sources. Pretreatment inspectors must stay informed regarding developments in this field.

2.8   ETHICAL CONSIDERATIONS

2.80   Ethics — A Part of the Job

In any industrial waste program, ethics should be considered a significant underpinning of the program. In all industrial waste programs, the requirements placed on industry by pretreatment facility inspectors will create difficulties for the industry. There will always be attempts by industrial personnel to somehow get special treatment from the pretreatment facility inspectors and other industrial waste section personnel. Such attempts will sometimes take the form of browbeating or antagonistically discouraging the industrial waste personnel, attempts to use political influence or to gain political favors, or they may come in the form of offers of special favors to pretreatment facility inspectors. Eventually, most pretreatment inspectors will have to decide how to respond when someone seeks preferential treatment.

2.81   The Subject of Ethics

It is strongly recommended that the subject of ethics and personal honesty be directly discussed with all new agency employees. All incoming personnel must be advised that no special favors are given to industrial companies in exchange for any special consideration to the employee. Methods of persuasion can sometimes be very subtle and can involve such things as similar national backgrounds, similar religious backgrounds and other similarities of tastes or wishes. Make it clear to all personnel that all industries must be dealt with on a fair and equitable basis without any consideration or favoritism for any reason. It is recommended that personnel be counseled forcefully to avoid even the appearance of favoritism and to report to their supervisor or even to the head of the agency, when appropriate, any attempts to obtain such favoritism by industrial personnel.

---

[10] Centralized Waste Treatment (CWT) Facility.   A facility designed to properly handle treatment of specific hazardous wastes from industries with similar wastestreams. The wastewaters containing the hazardous substances are transported to the facility for proper storage, treatment and disposal. Different facilities treat different types of hazardous wastes.

Some of the problems of special treatment may not be readily apparent and may not even be intentional. Special treatment can entail such subtle situations as counseling all industrial companies to use a certain laboratory to obtain their wastewater analyses. The directing of all industrial companies to one engineering consultant for development of their pretreatment programs is, again, not ethically appropriate. Even the use of only one brand of equipment for a vital part of an industrial waste pretreatment program can be interpreted as showing undue favoritism to one company.

The only solution to the problem of ethics is to address the issue with the pretreatment inspectors. Tell the inspectors that no special favors are to be given and that the giving of special favors may result in dismissal from the agency and legal prosecution if the unethical behavior is severe. Make it known to all pretreatment inspectors that there is no tolerance of unethical behavior. Almost everyone desires to be an honest and straightforward person. Most unethical behavior stems from a lack of knowledge and a belief that such behavior is permitted.

If it is understood that such behavior is not permitted, it will not exist.

QUESTIONS

Write your answers in a notebook and then compare your answers with those on page 83.

2.7A List the two main federal regulatory efforts to control hazardous wastes.

2.7B What will be the impact on pretreatment programs when industries treat their wastes on site?

2.7C What is a Centralized Waste Treatment (CWT) Facility?

2.8A Why are ethics important?

Please answer the discussion and review questions before continuing with the Objective Test.

End of Lesson 3 of 3 Lessons on PRETREATMENT PROGRAM ADMINISTRATION

DISCUSSION AND REVIEW QUESTIONS

Chapter 2. PRETREATMENT PROGRAM ADMINISTRATION

(Lesson 3 of 3 Lessons)

Write the answers to these questions in your notebook before continuing with the Objective Test on page 84. The question numbering continues from Lesson 2.

12. The value of information stored in a computer database system depends on what factors?

13. What items should a POTW agency consider when purchasing a computer system?

14. What are the advantages of a computer-based data management system?

15. In general, what are the most difficult types of industrial companies for the inspectors to deal with?

16. What is the solution to the problem of unethical behavior by an inspector?

SUGGESTED ANSWERS

Chapter 2.   PRETREATMENT PROGRAM ADMINISTRATION

ANSWERS TO QUESTIONS IN LESSON 1

Answers to questions on page 37.

2.0A   The type and staffing of an industrial waste program are influenced in a major way by the size of the agency and the amount of industrial wastewater flow. The variety of industries and their locations with respect to each other are also important considerations.

2.0B   The size of a wastewater treatment plant is important because in a small wastewater treatment plant with low flows, even a small metal plater can have a disastrous impact on the plant processes if a spill of toxic waste occurs.

2.0C   A background in engineering or chemistry, and experience in treatment plant operations, and good public relations skills are helpful for pretreatment program inspectors.

2.0D   The two basic types of enforcement postures for an industrial waste pretreatment inspection program are: (1) any violation of an industrial waste effluent requirement is considered a crime and the company should be prosecuted for such violations, and (2) use of administrative enforcement procedures designed to identify a problem and encourage a company in its efforts to make the necessary corrections.

Answers to questions on page 45.

2.0E   The efforts at small agencies may be concentrated in measuring the effluent from industrial companies and determining that compliance is consistently being obtained.

2.0F   The major subsections of a typical large Industrial Waste Section are:

1. Permit Review,
2. Field Engineering and Inspection,
3. Industrial Waste Engineering and Monitoring, and
4. Surcharge and Connection Fee.

2.0G   Approval of a new industrial waste pretreatment inspection program must be obtained from management of the POTW agency, the state (if it grants NPDES permits), and the EPA.

Answers to questions on page 48.

2.1A   Sources of funding for an industrial waste pretreatment program are generally two: (1) directly from companies, and (2) from other sources within the community.

2.1B   The basic types of charges made directly to industry for a wastewater treatment program include: (1) property taxes, (2) water bill charges, and/or (3) charges that are proportional to wastewater discharge flow and strength.

2.1C   Specific user fees that may be applied or charged to a company include an inspection fee, a fee for processing an industrial wastewater discharge permit, and fees for sampling and analyzing a company's wastewater. Other charges include the quantity and quality surcharge, a "one-time" connection fee charge and a front footage charge.

ANSWERS TO QUESTIONS IN LESSON 2

Answers to questions on page 60.

2.2A   The basic elements of a pretreatment inspection program include:

1. Permission to discharge to the sewer system,
2. Inspection of companies,
3. Monitoring of industrial wastewater discharges,
4. Technical expertise,
5. Revenue collection,
6. Compliance/enforcement program,
7. Laboratory analyses,
8. Legal assistance,
9. Database for industrial waste information, and
10. Pollution prevention program.

2.2B   Field inspectors should be rotated at least every two years, if possible, to allow an inspector to see many geographical areas and develop wide experience over different areas and with different types of industries.

2.2C   The collection of revenue from industrial companies is required under the EPA Pretreatment and Construction Grant Regulations. This collection of revenue is basically for two functions: (1) to recover the costs of providing wastewater collection, treatment, and disposal, and (2) to recover the costs of the POTW agency's industrial waste section functions.

2.2D   Most field test kits are less accurate and less comprehensive than laboratory wastewater analyses.

Answers to questions on page 64.

2.3A   Serious problems resulting from industrial wastewater discharges include fires and explosions in the collection system, toxic upsets of biological processes and serious structural corrosion problems in sewers and at treatment plants.

2.3B   Information required for an industrial waste permit includes location, collection system and treatment plant receiving industrial discharge, type of wastewater, quantity, chemical quality, flow discharge pattern, hazardous or toxic materials stored on site, means for disposal of rainwater, name of responsible company officer, and pretreatment facilities needed for the company wastewater.

2.3C Major concerns with regard to a recordkeeping system include the proper filing of permit review information and the ability to retrieve parts of this information as needed.

ANSWERS TO QUESTIONS IN LESSON 3

Answers to questions on page 74.

2.4A The main sources of data from an industrial waste pretreatment program include:

1. Industrial waste permit information,
2. Records of site inspections of industrial plants,
3. Industrial self-monitoring data,
4. Monitoring data by the POTW,
5. Surcharge and revenue data for the industrial companies,
6. Compliance records for EPA and local effluent limits,
7. Data on toxic pollutant concentrations in the POTW treatment plant influent, effluent and sludge, and
8. Correspondence and other written memoranda.

2.4B Computer database management systems must accommodate all information which is required to be stored and must have the ability to produce the required output.

2.4C The basic types of general-purpose software that should be considered for use with a computer system include:

1. Word processor,
2. Database management system,
3. Spreadsheet, or calculation system,
4. Graphics presentation,
5. Computer utilities,
6. Accounting system, and
7. Communication system.

Answers to questions on page 75.

2.5A To prevent the widespread practice of companies reporting that their wastewater is in compliance when it actually is not, require that a company official sign the self-monitoring document under penalty of perjury.

2.5B A monitoring program by the POTW agency is designed to assist and protect the company desiring to comply with environmental regulations. Such a company may be at a significant competitive disadvantage with another less ethical industry if a rigorous review by the POTW is not practiced.

Answers to questions on page 78.

2.6A When dealing with environmentalists, try to describe the significant efforts and progress being made by the organization in reduction of pollutant discharges.

2.6B Inspectors should inform political leaders of the requirements of the environmental regulations and what is being done to comply with these regulations.

2.6C When dealing with news media personnel, inspectors should conform to agency policy regarding the release of information. If authorized to make public statements, inspectors should comment only on factual matters within their area of expertise.

Answers to questions on page 81.

2.7A The two main federal regulatory efforts to control hazardous wastes are the Resource Conservation and Recovery Act (RCRA) and the Comprehensive Environmental Response, Compensation, and Liability Act (CERCLA).

2.7B One of the impacts on pretreatment programs when industrial companies treat their wastes on site is that the residual products from these waste treatment programs will be coming to the POTW. Another effect on pretreatment programs is that certain industries which are producers of hazardous wastes under the RCRA laws may not be considered serious polluters under the local POTW pretreatment programs or the Clean Water Act regulations.

2.7C A Centralized Waste Treatment Facility (CWT) is a facility designed to properly handle treatment of specific hazardous wastes from industries with similar wastestreams. The wastewaters containing the hazardous substances are transported to the facility for proper storage, treatment and disposal. Different facilities treat different types of hazardous wastes.

2.8A Ethics are very important because all industries must be treated fairly and equitably without preferential or special treatment to any company.

OBJECTIVE TEST

Chapter 2. PRETREATMENT PROGRAM ADMINISTRATION

Please write your name and mark the correct answers on the answer sheet, as directed at the end of Chapter 1. There may be more than one correct answer to each multiple-choice question.

True-False

1. A major industrial spill will have a disastrous impact on a large wastewater treatment plant.
   1. True
   — 2. False

2. POTW enforcement actions must comply with the agency's Enforcement Response Plan (ERP).
   — 1. True
   2. False

3. Industrial wastewater treatment charges based on flows and strength of wastewater always reflect the costs of an industrial waste control program.
   1. True
   — 2. False

4. The development of POTW agency personnel who understand industrial processes will lead to a better solution to problems.
   — 1. True
   2. False

5. An industrial waste permit should include the means for disposal of runoff rainwater to the sanitary sewer.
   1. True
   — 2. False

6. Proper documentation of all samples collected must be maintained to meet legal requirements.
   — 1. True
   2. False

7. Falsification of data is a civil offense.
   1. True
   — 2. False

8. Mass-based limits and concentration-based limits are the same.
   1. True
   — 2. False

9. Working with the press and media can be intimidating.
   — 1. True
   2. False

10. EPA and POTW regulations are aimed at reducing pollution.
    — 1. True
    2. False

Best Answer (Select only the closest or best answer.)

11. What do industrial wastewater discharge permits do? Grant permission to an industrial user to discharge its wastewater into the
    1. Centralized waste treatment facility.
    2. Environment.
    3. Pretreatment facility.
    — 4. Wastewater collection system.

12. What positive benefit can communities provide industrial companies with regard to industrial wastewater?
    — 1. Adequate treatment services are available
    2. Categorical standards will not be strictly enforced
    3. Costs of industrial treatment will subsidize municipal treatment
    4. Inspection visits will be less than once a year

13. What is the most forceful compliance procedure available to a POTW?
    1. Enforcing a compliance schedule
    2. Issuing a notice of violation
    3. Leveling of a fine
    — 4. Revoking permission to discharge to a sewer

14. What is the best method to determine what industrial dischargers are placing in the collection system?
    1. Field test kits
    — 2. Laboratory chemical analyses
    3. Results of industrial self-monitoring
    4. Visual observations

15. What does a spill containment program require?
    1. Channels carrying leaks to the storm sewer
    2. Drains within a berm connected directly to the pretreatment plant
    3. Sumps within the berm for pumping leaks to the sanitary sewer
    — 4. Tanks located within a berm that will prevent leaks from reaching a sewer

16. What is the purpose of a computer modem?
    — 1. Allows connection to other computers
    2. Moves information within a database
    3. Provides fast, letter-quality printing
    4. Works with special graphics packages

17. What group acts as a counterbalancing force to the environmental movement?
    — 1. Industrial lobbying group
    2. News media
    3. POTW staff
    4. Political leaders

18. What is the mode of conduct that works best with reporters?

    − 1. Courtesy, cooperation and candor
    2. Follow your own policy
    3. No comment
    4. Technical details

19. What should be the goal of relations with news media personnel?

    1. Meeting deadlines
    − 2. Presenting a good public image
    3. Referring legal questions to attorneys
    4. Speaking only when authorized

20. What should a pretreatment facility inspector do when industrial personnel attempt to obtain favoritism?

    1. Accept offer
    2. Ignore attempt
    − 3. Report attempt to supervisor
    4. Try to avoid attempt

Multiple Choice (Select all correct answers.)

21. What factors determine the staffing needs of an industrial waste program?

    − 1. Amount of industrial waste flow
    2. Capacity of POTW
    3. Enforcement posture of POTW
    4. Magnitude of regulations
    − 5. Size of agency

22. What basic types of charges are made directly to industry for a wastewater treatment program?

    − 1. Charges proportional to wastewater discharge
    2. Income charges
    3. Sales charges
    − 4. Sewer-use fees
    − 5. Water bill charges

23. What are the basic elements of a pretreatment program?

    − 1. Frequency of inspection
    − 2. Inspection of companies
    − 3. Monitoring of discharges
    − 4. Permission to discharge
    − 5. Pollution prevention

24. Why should field inspection personnel be rotated to new areas periodically?

    1. To allow special treatment of companies
    − 2. To avoid stagnation of the inspector
    − 3. To develop wide experience
    4. To reduce travel times and transportation costs
    − 5. To see new geographical areas

25. What are the main functions of a POTW collecting revenue from industrial companies? To recover the costs of

    1. Industrial self-monitoring program.
    − 2. Industrial waste section functions.
    − 3. Pretreatment inspection.
    − 4. Providing wastewater collection, treatment and disposal.
    5. Training industrial wastewater pretreatment plant operators.

26. What factors should be considered when determining the level of enforcement response?

    − 1. Consistency
    − 2. Duration of noncompliance
    − 3. Equity
    − 4. Fairness
    − 5. Integrity of program

27. How can pollution prevention activities benefit companies?

    − 1. Improve public image and employee morale
    2. Increase operation and maintenance costs
    3. Increase regulatory compliance costs
    4. Increase waste monitoring, treatment and disposal costs
    5. Reduce productivity

28. What serious problems encountered in the proper operation of a POTW agency can originate from industrial wastewater discharges?

    − 1. Fires and explosions in collection systems
    2. Improper design of facilities
    3. Lack of adequate funding
    − 4. Serious structural corrosion problems in sewers
    − 5. Toxic upsets of biological processes

29. What factors should be considered when purchasing a personal computer?

    − 1. Expandability
    2. Life of hardware greater than five years
    − 3. Relative costs
    − 4. Service
    5. Space requirements

30. What is the purpose of database management systems?

    1. To develop "what if" solutions
    2. To handle repetitive calculations
    − 3. To organize data
    − 4. To record data
    − 5. To selectively print data

End of Objective Test

APPENDICES

CHAPTER 2. PRETREATMENT PROGRAM ADMINISTRATION

A
P
P
E
N
D
I
C
E
S

APPENDIX A

CONTROL AUTHORITY PRETREATMENT AUDIT CHECKLIST AND INSTRUCTIONS

TABLE OF CONTENTS

APPENDIX A

PART I: INTRODUCTION

OVERVIEW

The pretreatment program represents a unique partnership in the regulatory community, a partnership between Federal, State, and local regulatory agencies. The Approval Authority [the U.S. Environmental Protection Agency (EPA) or the authorized State] is responsible for ensuring that local program implementation is consistent with all applicable Federal requirements and is effective in achieving the National Pretreatment Program's goals. To carry out this responsibility, the Approval Authority determines local program compliance and effectiveness, and takes corrective actions [e.g., changes to the National Pollutant Discharge Elimination System (NPDES) permit, enforcement] where needed to bring these about. The Approval Authority currently uses three oversight mechanisms to make these determinations: (1) the program audit; (2) the Pretreatment Compliance Inspection (PCI); and (3) the Control Authority's (CA's) annual pretreatment program performance report.

The audit, which for most programs takes place once every 5 years, is the most comprehensive of the three mechanisms and provides the opportunity to evaluate all aspects of the CA's program. It also provides the opportunity to help the CA build its local program implementation capability. Since the audit was developed in 1986, its purpose was to assess the program's compliance with the regulatory requirements as they were expressed in the NPDES permit. The audit also identified areas of the CA's program that needed to be modified to bring the program into compliance with the regulations.

In recent years, both the pretreatment program itself and the tracking of program implementation compliance have undergone major changes. Revisions to the General Pretreatment Regulations [40 Code of Federal Regulations (CFR) Part 403] in response to the Pretreatment Implementation Review Task Force (PIRT) recommendations in January 1989 and the Domestic Sewage Study (DSS) findings in July 1990 resulted in numerous additions to local program requirements. This has necessitated a revision of the audit checklist. The attached checklist replaces the checklist developed in 1986. This checklist covers all the evaluated components of the previous checklist, but goes beyond and looks at the program's impact in terms of environmental effectiveness and pollution prevention.

The audit serves several important functions such as identifying needed changes to the NPDES permit and initiating enforcement action against CA noncompliance. Using this checklist, the auditor can also examine the effectiveness of the program by evaluating environmental indicators and investigating the CA's use of pollution prevention techniques to enhance the impact of the program. The new checklist is also geared toward identifying program areas where recommendations may be made to increase the effectiveness of the CA's program.

PURPOSE

The principal reason for conducting an audit is to assess the CA's program as a whole by reviewing all components and determining the program's overall effectiveness and compliance. This is done by examining the discreet portions of the whole program [e.g., legal authority, Industrial User (IU) control mechanisms, compliance monitoring, and enforcement], and making an assessment based upon how the discreet portions interact to form the whole. The specific objectives to be accomplished by conducting an audit are: determining the CA's compliance status with requirements of its NPDES permit, approved program, and Federal regulations; evaluating the adequacy and effectiveness of the program in achieving compliance and environmental goals of the program; determining whether any modifications have been made to the program; and verifying important elements of the CA's program performance reports.

EXPERIENCE NECESSARY TO CONDUCT AN AUDIT

Because the new audit checklist looks at the entire program in extensive detail and examines areas that were previously looked at only on a case-by-case basis, this checklist assumes a high level of pretreatment program expertise on the part of the auditor. He/she must be very familiar with the goals of the pretreatment program, the General Pretreatment Regulations (40 CFR Part 403), categorical standards, and EPA/State policy and guidance. He/she should also have participated in audits conducted by a senior lead auditor.

The auditor must be familiar enough with all aspects of a local pretreatment program to conduct an audit that will collect the data necessary to make a meaningful evaluation of the CA's compliance status and the effectiveness of the program in achieving its goals. At a minimum, he/she should be able to:

- Identify the category to which an industry belongs and to develop appropriate permit limits based on the process wastewater discharged. To do this, the auditor must be knowledgeable about the National categorical pretreatment standards.

- Evaluate the adequacy of the control mechanisms issued by the CA to their Significant Industrial Users (SIUs). The auditor must be able to determine whether the control mechanism meets the minimum regulatory requirements and whether it is effective in controlling the discharge of the SIU.

- Evaluate the CA's legal authority for its compliance with regulatory requirements and the ability of the CA to enforce its program throughout its service area. The auditor must have an understanding of the authorities provided to the CA by its Sewer Use Ordinance (SUO), including available remedies and procedures for taking action for IU noncompliance. He/she must also be familiar with issues related to implementing and enforcing a local program across jurisdictional boundaries and approaches to resolving such issues.

- Understand compliance monitoring requirements. The auditor's knowledge must include appropriate sampling techniques, EPA approved methods, and proper Quality Assurance/Quality Control (QA/QC) and chain-of-custody procedures so that data may be admissible as evidence in enforcement proceedings.

- Conduct a comprehensive pretreatment inspection at IU facilities and be familiar with hazardous waste requirements and spill prevention and control.

- Evaluate the CA's enforcement responses. To do so, the auditor must be knowledgeable of the various types of possible enforcement actions which are available to the Publicly Owned Treatment Works (POTW) as well as EPA/State policies and guidance on enforcement.

- Assist the CA to determine what pollution prevention techniques may enhance the local program. This requires the auditor to be knowledgeable about current efforts and policies regarding pollution prevention.

- Evaluate the environmental effectiveness of the program by examining data collected over the years by the CA concerning pollutant loadings, discharges, and other indicators.

PROCEDURES FOR CONDUCTING AN AUDIT

The audit requires extensive preparation, detailed data collection when onsite, and timely follow-up. In brief, the major steps for conducting an audit are:

- Office preparation prior to going onsite

  - Review NPDES permit file, enforcement file, and pretreatment program file

  - Review such documents as a manufacturers' guide, Resource Conservation and Recovery Act (RCRA) permit list for the municipalities involved, Toxic Release Inventory System (TRIS) data, etc., to be familiar with all industries that may contribute to the POTW

  - Notify the CA of the upcoming visit (if appropriate)

- Onsite visit

  - Entry (present credentials)

  - Review SIU files

  - Inspect selected SIUs

  - Interview program staff

  - Review POTW records and files

  - Conduct closing conference

- Follow-up

  - Prepare and distribute report

  - Enter Water Enforcement National Data Base (WENDB) data elements

  - Determine Reportable Noncompliance (RNC)/Significant Noncompliance (SNC) and enter data

  - Modify NPDES permit (if appropriate)

  - Refer for enforcement (if appropriate)

Preparation

The amount of data to be collected and evaluated during an audit is considerable and time is limited. Thus, preparation for the audit is crucial to the well-focused collection of meaningful data. The pretreatment program profile data sheets and status update sheets attached to the checklist will help the auditor compile very general program information before he/she goes onsite. The auditor should spend time obtaining information about the industrial contribution to the POTW by reviewing the CA's Industrial Waste Survey (IWS) as well as the manufacturers' guides for the municipalities covered by the local program. The auditor should also review TRIS and RCRA permitting data. After becoming familiar with the industrial picture, the auditor may want to review development documents to familiarize himself/herself with the primary industries discharging to the POTW. The auditor should also become familiar with issues affecting the POTW such as being listed on 304(1) or being involved in a Technical Review Evaluation (TRE).

File Review

Once onsite, the auditor should go through standard NPDES inspection entry procedures then explain to CA personnel what the audit will entail. Once the initial entry procedures are complete, the auditor should select IU files and conduct the file review. Files may be chosen in many ways; however, use of the scheme shown in Figure 1 is strongly recommended as best providing a reasonable representation of SIUs regulated under local program. The auditor should bear in mind that the above recommendations are for review of SIU files. This does not imply that non-SIU files ought not also be reviewed. The auditor will need to exercise his/her best professional judgment to determine the number of both SIU and non-SIU files to review. He/she should allocate 2 to 3 hours per file for a detailed review.

A
P
P
E
N
D
I
X

A

FIGURE 1.   RECOMMENDED NUMBER OF SIU FILES TO REVIEW

Total No. of SIUs	Minimum No. of Files to Review
≤ 10	5
11 – 20	5
21 – 30	8
31 – 50	10
51 – 100	15
101 – 200	20
201 – 300	25
301 – 1,000	30
1,001 – 1,500	50

The auditor should select files that demonstrate a representative cross section of the CA's IUs. He/she should evaluate both categorical and significant noncategorical IUs and give particular attention to files of SIUs newly added to the program and those with compliance issues (e.g., in SNC, having received escalated enforcement action). Special attention should also be given to CIUs without pretreatment, but reported to be in compliance with categorical standards. The auditor should also choose files based on: CIUs with complicated processes [i.e., production based, Combined Wastestream Formula (CWF) — Flow Weighted Averaging (FWA) issues, etc.]. Finally, he/she should review some files that were not reviewed during previous audits or inspections.

IU Site Visits

After the file review, the auditor should conduct as many IU site visits as possible. IU site visits are often essential to verify information found in the files. They are also helpful in making the IUs aware of the importance EPA places on the local programs. Again, the number and types of IUs to be visited should be representative of the program's industrial make-up and based on the time needed for each visit. The auditor should compile the results of the file review and site visits prior to conducting the interview portion of the audit.

Interview

During the interview portion, the auditor should talk with as many CA personnel as necessary to obtain an accurate picture of how the local program is implemented. Although the pretreatment coordinator may be familiar with proper monitoring procedures, he/she may not be completely familiar with how the program's monitoring is _actually_ being conducted, particularly in large programs. Information on what is happening in the field should be obtained from field personnel. Also, in multijurisdictional situations, it may be necessary to speak with representatives of the contributing jurisdictions to learn how the program is _actually_ being implemented in those service areas. The auditor should take detailed notes to document each interview. Also, whenever possible, he/she should collect supporting documentation to corroborate answers given by the interviewees. For instance, if a CA staff person states that a total of 26 inspections were conducted in the last calendar year, the auditor should request a copy of the CA's log or its equivalent to verify this information.

Closing Conference

After the file review, IU site visits, interviews, and other evaluations are complete, the auditor should compile all the data obtained to prepare for the closing conference. At the closing conference, the auditor should verbally present his/her findings to that point to the CA. He/she should make it clear that these findings are preliminary and subject to change once the data collected have been more thoroughly reviewed.

Follow-up

Audit follow-up will center on preparing the report and identifying the action necessary to ensure that appropriate changes to the POTW's program occur. Follow-up action may include revisions to the NPDES permit, formal enforcement action, or other action. The auditor should analyze his/her data as quickly as possible and draft the report so that it can be transmitted to the CA in a timely manner. The auditor should also enter the WENDB and RNC data in the data base. In addition, he/she should complete the appropriate NPDES Compliance Inspection Report Forms and update the Status Update and Program Profiles. The auditor should handle NPDES permit modifications and enforcement activities in accordance with EPA Regional/State policy.

As mentioned earlier, the audit requires balancing many different data gathering techniques. By balancing these techniques properly, the auditor will obtain the best result and a comprehensive look at the CA's program. The file review and IU site visits are areas that pose the greatest resource burden to the Approval Authority. EPA recommends looking at as many files and visiting as many IUs as possible with balance in mind. For example, reviewing 25 files and visiting 2 IUs does not provide the balance that would be achieved by reviewing 15 files and visiting 10 IUs at a medium sized POTW with 100 SIUs. Although this latter effort requires a greater resource commitment, it provides much more meaningful data.

CHECKLIST STRUCTURE

The audit checklist is divided into the following three sections. Regulatory citations are provided for all required program items. Items on the checklist that do not have a corresponding regulatory citation are not required, but are recommended because they would enhance the effectiveness of the program. Comment space is also provided for each item to enable adequate documentation of the findings.

Section I: File Review — evaluates the CA's performance by reviewing the IU records which the CA maintains. Unlike information obtained in interviews, a review of the CA's files provides proof that the CA is either implementing or not implementing its program. If relevant information is not found in the files, the auditor should note this problem as one of the audit findings. The File Review is suggested to be conducted first because it enables the auditor to identify issues that can be discussed during the interview, and either resolved during the closing conference or established as a finding for the report. The file review also provides a basis on which to select IUs for site visits.

Section II: Interview — is intended to evaluate the portions of program implementation that could not be evaluated adequately by looking at the IU files. This section also complements the information gained during the file review and IU site visits. For example, the file review looks at the quality of permits issued while the interview investigates the adequacy of the issuance process.

Section III: Findings — enables the auditor to organize all issues that will need to be addressed in the subsequent report. This section is organized to correspond to the subsections in Sections I and II. The areas of concern to consider are listed with corresponding regulatory and checklist question citations. This was done to assist the auditor in compiling all findings for each one.

There are five attachments to the checklist: the Pretreatment Program Status Update and Pretreatment Program Profile to be completed before the audit and updated subsequent to the audit; the IU Site Visit Data Sheet to be used at the auditor's option when conducting IU site visits during the audit; and the WENDB Data Entry Worksheet and the RNC Worksheet to be completed as part of the audit follow-up to provide input into the data base. When completed with thoroughness, the body of the checklist and its attachments will provide the auditor with the documentation needed to draft the audit report, initiate any corrective and enforcement actions needed, and enter WENDB and RNC data into the data base.

RESOURCES FOR CONDUCTING AUDITS

The resources necessary to conduct audits will vary greatly from program to program. Some variables contributing to different resource needs include: size of the POTW; number and size of SIUs; and number of jurisdictions involved. These variables will impact preparation time, time onsite, report preparation, and follow-up. The average resources needed to conduct an audit of a small local program would be 1-2 people onsite for 2 days with 2-4 hours needed for preparation and 8-16 hours needed to write the report. For a medium-size program, the Approval Authority should allow 2 people onsite for 2-3 days, 4-8 hours for preparation, and 16-24 hours to write the report. Finally, a large program is likely to require 3-5 people onsite for 3-5 days, 16-24 hours of preparation, and 24-40 hours to write the report. The Approval Authority should be aware that these are broad estimates and are provided only as a general basis for decisions regarding scheduling and staffing.

COMPUTER FILE INFORMATION

For the user's convenience, this "Control Authority Pretreatment Audit Checklist and Instructions" package includes a 5 ¼" High Density diskette containing the audit checklist form in seven files corresponding to the cover page, Sections I, II, and III, and Attachments A, B, and C. To use the checklist computer file, you must have WordPerfect Version 5.1 (with manual) and a Laserjet series III printer (with manual). Use on another printer may cause format problems. The following instructions will facilitate the use of the Pretreatment Audit Checklist computer file.

Entering Data in the Computer File

The Checklist was designed using the WordPerfect 5.1 TABLES function which uses "cells" and "columns" to create multi-structure tables. The questions on the enclosed hard copy of the checklist correspond to the cells containing text in the computer file (i.e., the "question cells"). These cells are locked to facilitate moving through the checklist; the cursor will skip over them. (*Note: to delete these cells, see Unlocking Cells, below.*) The blank spaces on the hard copy correspond to the empty cells (i.e., the "answer cells") in the file. These cells are not locked and are formatted so that the information typed into them will print in boldface italic type.

To ensure that the information in the IU Identification section prints properly, the cursor must be placed between the "[BOLD][ITALIC][italic][bold]" codes before typing the text. Working with the "REVEAL CODES" (F11 or ALT F3) displayed will ensure a consistent product. Also, each unlocked cell contains a specific number of HARD RETURNS (HRts) that ensure the table will remain intact on one page. When entering the text, the user is advised to cursor to each line. Where text automatically wraps, the same number of HRts should be deleted as the SOFT RETURNS (SRts) that were entered by the wraps. The user is cautioned that entering more SRts than there are HRts available may cause an inappropriate page break. If this occurs, it is recommended that a new table be created using the TABLE function (ALT F7) to move what is needed to the next page.

Once the text has been entered, you can use the "SAVE AS" function (F10) to give the files new names, thus keeping the original blank checklist file intact. In addition, entering data in the computer file necessitates use of several complex functions. These are described below.

Unlocking Cells

It is recognized that standardized forms such as the "Pretreatment Audit Checklist" cannot meet everyone's needs on every occasion. From time to time, it may be desirable to enter data into a locked cell. The cursor must be within the table frame and you must access the TABLE function using "ALT F7" to make any changes to a cell. Once in the TABLE function, the cursor can be moved to the locked cell and you can use the editing menu to make necessary changes. To unlock the cell, choose the following selections from the menu: 2 Format; 1 Cell; 5 Lock; 2 Off. After exiting the TABLE function, text may be entered in the cell. If you want the text to be consistent with that entered in the formatted cells, use "CTRL F8" and "F6" to italicize and bold text. The user is cautioned that entering text into unlocked cells may cause format and pagination changes that will need to be corrected by deleting space elsewhere or creating additional cells using the TABLE function.

Creating Check Marks

The diskette also includes two macros: "ALT C" to create a check mark (✓) as used in the "Yes"/"No" questions and in the file review; and "ALT X" to create check marks in boxes (✓) as used on the cover page and in the IU Identification section. When using "ALT X" you must delete the code for the empty box, place the cursor where the box code was, and type ALT X.

Entering Vertical Names

The first page of Section I: IU File Evaluation uses vertical boxes to contain the file name. To enter the IUs' names in these boxes, cursor to the first box and use the TEXT BOX (ALT F9) function choosing the following selections from the menu: 3 Text Box; 2 Edit; type Table Box number; Enter; 9 Edit; type name; ALT F9 (Graphics); 2 90%; F7 (Exit). The name will not appear in the box on the screen. To verify that the name was correctly entered, use View Document function (Shift F7, 6). Follow the same procedure for each box. You are allowed 24 character spaces (one line) for the IU's name. Do not try to make the box larger; instead, abbreviate the name to make it fit.

PART II: CONTROL AUTHORITY PRETREATMENT AUDIT CHECKLIST

AUDIT CHECKLIST CONTENTS

Cover Page and Acronym List

Section I   IU File Evaluation

Section II   Data Review/Interview/IU Site Visit(s)

Section III   Findings

☐ Attachment A   Pretreatment Program Status Update

☐ Attachment B   Pretreatment Program Profile

Attachment C   Worksheets

    ☐ IU Site Visit Data Sheet

    ☐ WENDB Data Entry Worksheet

    ☐ RNC Worksheet

Attachment D   Supporting Documentation

Control Authority (CA) name and address	Date(s) of audit

AUDITOR(S)

Name	Title/Affiliation	Telephone Number

CA REPRESENTATIVE(S)

Name	Title/Affiliation	Telephone Number
	*	

* Identified program contact

ACRONYM LIST

Acronym	Term
AO	Administrative Order
BMP	Best Management Practices
BMR	Baseline Monitoring Report
CA	Control Authority
CERCLA	Comprehensive Environmental Response, Compensation, and Liability Act
CFR	Code of Federal Regulations
CIU	Categorical Industrial User
CSO	Combined Sewer Overflow
CWA	Clean Water Act
CWF	Combined Wastestream Formula
DMR	Discharge Monitoring Report
DSS	Domestic Sewage Study
EP	Extraction Procedure
EPA	U.S. Environmental Protection Agency
ERP	Enforcement Response Plan
FDF	Fundamentally Different Factors
FTE	Full-Time Equivalent
FWA	Flow-Weighted Average
gpd	gallons per day
IU	Industrial User
IWS	Industrial Waste Survey
MGD	Million Gallons Per Day
MSW	Municipal Solid Waste
N/A	Not Applicable
ND	Not Determined
NOV	Notice of Violation
NPDES	National Pollutant Discharge Elimination System
O&G	Oil and Grease
PCI	Pretreatment Compliance Inspection
PCS	Permit Compliance System
PIRT	Pretreatment Implementation Review Task Force
POTW	Publicly Owned Treatment Works
QA/QC	Quality Assurance/Quality Control
RCRA	Resource Conservation and Recovery Act
RNC	Reportable Noncompliance
SIU	Significant Industrial User
SNC	Significant Noncompliance
SUO	Sewer Use Ordinance
TCLP	Toxicity Characteristic Leachate Procedure
TOMP	Toxic Organic Management Plan
TRC	Technical Review Criteria
TRE	Technical Review Evaluation
TRIS	Toxics Release Inventory System
TSDF	Treatment, Storage, and Disposal Facilty
TTO	Total Toxic Organics
UST	Underground Storage Tank
WENDB	Water Enforcement National Data Base

GENERAL INSTRUCTIONS

1. As noted in the Introduction, the auditor should review a representative number of SIU files. Section I of this checklist provides space to document five IU files. This should not be construed to mean that five is an adequate representation of files to review. The auditor should make as many copies of Section I as needed to document a representative number of files according to the discussion in the Introduction.

2. The auditor should ensure that he/she follows up on any and all violations noted in the previous inspection and annual report during the course of the audit.

3. Throughout the course of the evaluation, the auditor should look for areas in which the CA should improve the effectiveness and quality of its program.

4. Audit findings should clearly distinguish between violations, deficiencies, and effectiveness issues.

SECTION I: IU FILE EVALUATION

INSTRUCTIONS: Select a representative number of SIU files to review. Provide relevant details on each file reviewed. Comment on all problems identified and any other areas of interest. Where possible, all CIUs (and SIUs) added since the last PCI or audit should be evaluated. Make copies of this section to review additional files as necessary.

IU IDENTIFICATION

FILE _____ Industry name and address | Type of industry

☐ CIU 40 CFR _____ , _____ , _____

Category(ies) _____

Average total flow (gpd) | Average process flow (gpd)

☐ Other SIU   ☐ Non SIU | Industry visited during audit   Yes ☐   No ☐

Comments

FILE _____ Industry name and address | Type of industry

☐ CIU 40 CFR _____ , _____ , _____

Category(ies) _____

Average total flow (gpd) | Average process flow (gpd)

☐ Other SIU   ☐ Non SIU | Industry visited during audit   Yes ☐   No ☐

Comments

SECTION I: IU FILE EVALUATION (continued)

IU IDENTIFICATION (continued)

FILE _____ Industry name and address | Type of industry

☐ CIU 40 CFR _____ , _____ , _____

Category(ies) _____

☐ Other SIU   ☐ Non SIU

Average total flow (gpd) | Average process flow (gpd)

Industry visited during audit   Yes ☐   No ☐

Comments

FILE _____ Industry name and address | Type of industry

☐ CIU 40 CFR _____ , _____ , _____

Category(ies) _____

☐ Other SIU   ☐ Non SIU

Average total flow (gpd) | Average process flow (gpd)

Industry visited during audit   Yes ☐   No ☐

Comments

FILE _____ Industry name and address | Type of industry

☐ CIU 40 CFR _____ , _____ , _____

Category(ies) _____

☐ Other SIU   ☐ Non SIU

Average total flow (gpd) | Average process flow (gpd)

Industry visited during audit   Yes ☐   No ☐

Comments

General Comments

SECTION I: IU FILE EVALUATION (continued)

Industry Name

INSTRUCTIONS: Evaluate the contents of selected IU files; emphasis should be placed on SIU files. Use N/A (Not Applicable) where necessary. Use ND (Not Determined) where there is insufficient information to evaluate/determine implementation status. Comments should be provided in the comment area at the bottom of the page for all violations, deficiencies, and/or other problems as well as for any areas of concern or interest noted. Enter comment number in box and in the comment area at the bottom of the page, followed by the comment. Comments should delineate the extent of the violation, deficiency, and/or problem. Attach relevant copies of IU file information for documentation. Where no comment is needed, enter ✓ (check) to indicate area was reviewed. The evaluation should emphasize any areas where improvements in quality and effectiveness can be made.

File —	File —	File —	File —	File —	**IU FILE REVIEW**	**Reg. Cite**
					A. ISSUANCE OF IU CONTROL MECHANISM	
					1. Control mechanism application form	
					2. Fact sheet	
					3. Issuance or reissuance of control mechanism	403.8(f)(1)(iii)
					4. Control mechanism contents	403.8(f)(1)(iii)
					a. Statement of duration (≤ 5 years)	403.8(f)(1)(iii)(A)
					b. Statement of nontransferability w/o prior notification/approval	403.8(f)(1)(iii)(B)

Comments

					c. Applicable effluent limits	403.8(f)(1)(iii)(C)
					● Application of applicable categorical standards	403.8(f)(1)(ii)
					— Classification by category/subcategory	
					— Classification as new/existing source	
					— Application of limits for all categorical pollutants	
					— Application of TTO or TOMP alternative	
					— Calculation and application of production-based standards	403.6(c)
					— Calculation and application of CWF or FWA	403.6(d)&(e)
					— Application of variance to categorical standards	403.7
					● Application of applicable local limits	
					● Application of most stringent limit	403.8(f)(1)(ii)

Comments

SECTION I: IU FILE EVALUATION (continued)

File —	File —	File —	File —	File —	IU FILE REVIEW	Reg. Cite
					A. ISSUANCE OF IU CONTROL MECHANISM (continued)	
					d. IU self-monitoring requirements	403.8(f)(1)(iii)(D)
					• Sampling (pollutants, frequency, locations, types)	
					• Reporting requirements (e.g., periodic, resampling)	
					• Notification requirements (e.g., slug, spill, changed discharge, 24-hour notice of violation)	
					• Record keeping requirements	403.12(o)
					e. Statement of applicable civil and criminal penalties	403.8(f)(1)(iii)(E)
					f. Compliance schedules/progress reports (if applicable)	403.8(f)(1)(iii)(D)
					g. Slug discharge control plan requirement (if applicable)	403.8(f)(2)(v)

Comments

File —	File —	File —	File —	File —	IU FILE REVIEW	Reg. Cite
					B. CA COMPLIANCE MONITORING	
					1. Inspection	
					a. Inspection (at least once a year)	403.8(f)(2)(v)
					b. Inspection at frequency specified in approved program	403.8(c)
					c. Documentation of inspection activities	403.8(f)(2)(vi)
					d. Evaluation of need for slug discharge control plan (reevaluation of existing plan)	403.8(f)(2)(v)
					2. Sampling	
					a. Sampling (at least once a year)	403.8(f)(2)(v)
					b. Sampling at frequency specified in approved program	403.8(c)
					c. Documentation of sampling activities (chain-of-custody; QA/QC)	403.8(f)(2)(vi)
					d. Analysis for all regulated parameters	403.12(g)(1)
					e. Appropriate analytical methods (40 CFR Part 136)	403.8(f)(2)(vi)

Comments

File	File	File	File	File	IU FILE REVIEW	Reg. Cite
—	—	—	—	—		
					C. CA ENFORCEMENT ACTIVITIES	
					1. Identification of violations	403.8(f)(2)(vi)
					a. Discharge violations	
					• IU self-monitoring	
					• CA compliance monitoring	
					b. Monitoring/reporting violations	
					• IU self-monitoring	
					— Reporting (e.g., frequency, content)	
					— Sampling (e.g., frequency, pollutants)	
					— Record keeping	
					• Notification (e.g., slug, spill, changed discharge, 24-hour notice of violation)	
					• Slug control plan	
					• Compliance schedule/reports	
					c. Compliance schedule violations	
					• Start-up/final compliance	
					• Interim dates	

Comments

SECTION I: IU FILE EVALUATION (continued)

File	File	File	File	File	IU FILE REVIEW	Reg. Cite
—	—	—	—	—		
					C. CA ENFORCEMENT ACTIVITIES (continued)	
					2. Calculation of SNC	403.8(f)(2)(vii)
					a. Chronic	
					b. TRC	
					c. Pass through/interference	
					d. Spill/slug load	
					e. Reporting	
					f. Compliance schedule	
					g. Other violations (specify)	
					3. Response to violation	
					4. Adherence to approved ERP	403.8(f)(5)
					5. Return to compliance	
					a. Within 90 days	
					b. Within time specified	
					c. Through compliance schedule	
					6. Escalation of enforcement	403.8(f)(5)
					7. Publication for SNC	403.8(f)(2)(vii)
					D. OTHER	

Comments

SECTION I COMPLETED BY:	DATE:
TITLE:	TELEPHONE:

SECTION II: DATA REVIEW/INTERVIEW/IU SITE VISIT

INSTRUCTIONS: Complete this section based on CA activities to implement its pretreatment program. Answers to these questions may be obtained from a combination of sources including discussions with CA personnel, review of general and specific IU files, IU site visits, review of POTW treatment plants, among others. Attach documentation where appropriate. Specific data may be required in some cases.

● Write ND (Not Determined) beside the questions or items that were not evaluated during the audit; indicate the reason(s) why these were not addressed (e.g., lack of time, appropriate CA personnel were not available to answer)

● Use N/A (Not Applicable) where appropriate.

A. CA PRETREATMENT PROGRAM MODIFICATION [403.18]

1. a. Has the CA made any substantial changes to the pretreatment program that were not reported to the Approval Authority (e.g., legal authority, less stringent local limits, multijurisdictional situation)?

Yes	No

   If yes, discuss.

   b. Is the CA in the process of making any substantial modifications to any pretreatment program component (including legal authority, less stringent local limits, DSS requirements, multijurisdictional situation, etc.)?

Yes	No

   If yes, describe.

SECTION II: DATA REVIEW/INTERVIEW/IU SITE VISIT (continued)

B. LEGAL AUTHORITY [403.8(f)(1)]

	Yes	No
1. Are there any contributing jurisdictions discharging wastewater to the POTW?		

If yes, explain how the legal authority addresses the contributing jurisdictions.

	Yes	No
2. a. Has the CA updated its legal authority (e.g., SUO) to reflect changes in the General Pretreatment Regulations?		
b. Did all contributing jurisdictions update their SUOs in a consistent manner?		

Explain.

	Yes	No
3. Does the CA experience difficulty in implementing its legal authority [i.e., SUO, interjurisdictional agreement (e.g., permit challenged, entry refused, penalty appealed)]?		

If yes, explain.

C. IU CHARACTERIZATION [403.8(f)(2)(i)&(ii)]

1. How does the CA define SIU? (Is it the same in contributing jurisdictions?)

2. How are SIUs identified and categorized (including those in contributing jurisdictions)?

   Discuss any problems.

3. a. How and when does the CA update its IWS to identify new IUs (including those in contributing jurisdictions)?

   b. How and when does the CA identify changes in wastewater discharges at existing IUs (including contributing jurisdictions)?

4. How many IUs are currently identified by the CA in each of the following groups?

   a. ☐ SIUs (as defined by the CA) [WENDB-SIUs]

   ☐ CIUs [WENDB-CIUs]

   ☐ Noncategorical SIUs

   b. ☐ Other regulated noncategorical IUs (specify)

   c. ☐ TOTAL

D.   CONTROL MECHANISM EVALUATION [403.8(f)(1)(iii)]

---

1. a. How many and what percent of the total SIUs are <u>not</u> covered by an existing, unexpired permit, or other individual control mechanism? [WENDB-NOCM] [RNC-II]

		%

   b. How many control mechanisms were not issued within 180 days of the expiration date of the previous control mechanism? [RNC-II]

   If any, explain.

2. a. Do any UST, CERCLA, RCRA corrective action sites and/or other contaminated ground water sites discharge wastewater to the CA?

   b. How are control mechanisms (specifically limits) developed for these facilities?

   Discuss:

3. a. Does the CA accept any waste by truck, rail, or dedicated pipe?

   b. Is any of the waste hazardous as defined by RCRA?

Yes	No

   If a. or b. above is yes, explain.

   c. Describe the CA's program to control hauled wastes including a designated discharge point (e.g., number of points, control/security, procedures). [403.5(b)(8)]

E. APPLICATION OF PRETREATMENT STANDARDS AND REQUIREMENTS

1. What limits (categorical, local, other) does the CA apply to wastes that are hauled to the POTW (directly to the treatment plant or within the collection system, including contributing jurisdictions)? [403.1(b)(1)]

2. How does the CA keep abreast of current regulations to ensure proper implementation of standards? [403.8(f)(2)(iii)]

3. Local limits evaluation: [403.8(f)(4); 122.21(j)]

    a. For what pollutants have local limits been set?

    b. How were these pollutants decided upon?

    c. What was the most prevalent/most stringent criteria for the limits?

    d. Which allocation method(s) were used?

    e. Has the CA identified any pollutants of concern beyond those in its local limits?

    If yes, how has this been addressed?

Yes	No

4. What problems, if any, were encountered during local limits development and/or implementation?

SECTION II: DATA REVIEW/INTERVIEW/IU SITE VISIT (continued)

F. COMPLIANCE MONITORING

---

1. a. How does the CA determine adequate IU monitoring (sampling, inspecting, and reporting) frequencies?
    [403.8(f)(2)(ii)&(v)]

   b. Is the frequency established above more, less, or the same as required?

      Explain any difference.

---

2. In the past 12 months, how many, and what percentage of, SIUs were: [403.8(f)(2)(v)] [RNC-II]
   (Define the 12 month period _____ to _____.)

   a. Not sampled or not inspected at least once [WENDB-NOIN]

   b. Not sampled at least once

   c. Not inspected at least once (all parameters)?

		%
		%
		%

   If any, explain. Indicate how percentage was determined (e.g. actual, estimated).

---

3. Indicate the number and percent of SIUs that were identified as being in SNC* with the following requirements from the CA's last pretreatment program performance report. [WENDB] [RNC-II]

   SNC Evaluation Period

		%	Applicable pretreatment standards and reporting requirements
		%	Self-monitoring requirements
		%	Pretreatment compliance schedules

*SNC defined by:

POTW	
EPA	

4. What does the CA's basic inspection include? (Process areas, pretreatment facilities, chemical and hazardous waste storage areas, chemical spill prevention areas, hazardous waste handling procedures, sampling procedures, laboratory procedures, and monitoring records.) [403.8(f)(2)(v)&(vi)]

F. COMPLIANCE MONITORING (continued)

5. Who performs CA's compliance monitoring analysis?

	Performed by: CA/Contract Laboratory Name
Metals	
Cyanide	
Organics	
Other (specify)	

6. What QA/QC techniques does the CA use for sampling and analysis (e.g., splits, blanks, spikes), including verification of contract laboratory procedures and appropriate analytical methods? [403.8(f)(2)(vi)]

7. Discuss any problems encountered in identification of sample location, collection, and analysis.

8. Did any IUs notify the CA of a hazardous waste discharge? [403.12(j)&(p)]

Yes	No

   If yes, summarize.

9. a. How and when does the CA evaluate/reevaluate SIUs for the need for a slug control plan? [403.8(f)(2)(v)]

   b. How many SIUs were not evaluated for the need to develop slug discharge control plans in the last 2 years?

APPENDIX A

SECTION II: DATA REVIEW/INTERVIEW/IU SITE VISIT (continued)

G. ENFORCEMENT

1. What is the CA's definition of SNC? [403.8(f)(2)(vii)]

2. ERP implementation: [403.8(f)(5)]

   a. Status

   b. Problems with implementation

   c. Is the ERP effective and does it lead to compliance in a timely manner? Provide examples if any are available.

3. a. Does the CA use compliance schedules? [403.8(f)(1)(iv)(A)]

   b. If yes, are they appropriate? Provide examples.

Yes	No

4. Did the CA publish all SIUs in SNC in the largest daily newspaper in the previous year? [403.8(f)(2)(vii)]

Yes	No

   If yes, attach a copy.

   If no, explain.

5. How many SIUs are in SNC with self-monitoring requirements and were not inspected and/or sampled (in the four most recent full quarters)? [WENDB]

[ ]

6. a. Has the CA experienced any problems since the last inspection (interference, pass through, collection system problems, illicit dumping of hauled wastes, or worker health and safety problems) caused by industrial discharges?

Unk	Yes	No

   b. If yes, describe and explain the CA's enforcement action against the IUs causing or contributing to problems. [RNC-I]

H. DATA MANAGEMENT/PUBLIC PARTICIPATION

1. How is confidential information handled by the CA? [403.14]

2. How are requests by the public to review files handled?

3. Describe whether the CA's data management system is effective in supporting pretreatment implementation and enforcement activities.

4. How does the CA ensure public participation during revisions to the SUO and/or local limits? [403.5(c)(3)]

5. Explain any public or community issues impacting the CA's pretreatment program.

6. How long are records maintained? [403.12(o)]

I. RESOURCES [403.8(f)(3)]

1. Estimate the number of personnel (in FTEs) available for implementing the program. [Consider: legal assistance, permitting, IU inspections, sample collection, sample analysis, data analysis, review and response, enforcement, and administration (including record keeping and data management)].

FTEs

2. Does the CA have adequate access to monitoring equipment? (Consider: sampling, flow measurement, safety, transportation, and analytical equipment.)

Yes	No

   If no, explain.

3. a. Estimate the annual operating budget for the CA's program.

$

   b. Is funding expected to: stay the same, increase, decrease (note time frame; e.g., following year, next 3 years, etc.)?

      Discuss any changes in funding.

4. Discuss any problems in program implementation which appear to be related to inadequate resources.

5. a. How does the CA ensure personnel are qualified and up-to-date with current program requirements?

   b. Does the CA have adequate reference material to implement its program?

Yes	No

J.   ENVIRONMENTAL EFFECTIVENESS/POLLUTION PREVENTION

1. a. How many times were the following monitored by the CA during the past year?

	Influent	Effluent	Sludge	Ambient (Receiving Water)
Metals				
Priority pollutants				
Biomonitoring				
TCLP				
EP toxicity				
Other (specify)				

b. Is this frequency less than, equal to, or more than that required by the NPDES permit?

Less	Equal	More

Explain any differences.

2. a. Has the CA evaluated historical and current data to determine the effectiveness of pretreatment controls on:

	Yes	No
Improvements in POTW operations?		
Loadings to and from the POTW?		
NPDES permit compliance?		
Sludge quality?		

- Improvements in POTW operations?
- Loadings to and from the POTW?
- NPDES permit compliance?
- Sludge quality?

b. Has the CA documented these findings?

c. If they have been documented, what form does the documentation take?

Explain. (Attach a copy of the documentation, if appropriate.)

APPENDIX A

SECTION II: DATA REVIEW/INTERVIEW/IU SITE VISIT (continued)

J. **ENVIRONMENTAL EFFECTIVENESS/POLLUTION PREVENTION** (continued)

3. If the CA has historical data compiled concerning influent, effluent, and sludge sampling for the POTW, what trends have been seen? (Increases in pollutant loadings over the years? Decreases? No change?)

   Discuss on pollutant-by-pollutant basis.

	Yes	No
4. Has the CA investigated the sources contributing to current pollutant loadings to the POTW (i.e., the relative contributions of toxics from industrial, commercial, and domestic sources)?		

   If yes, what was found?

	Yes	No
5. a. Has the CA attempted to implement any kind of public education program?		
b. Are there any plans to initiate such a program to educate users about pollution prevention?		

   Explain.

6. What efforts have been taken to incorporate pollution prevention into the CA's pretreatment program (e.g., waste minimization at IUs, household hazardous waste programs)?

J. ENVIRONMENTAL EFFECTIVENESS/POLLUTION PREVENTION (continued)

	Yes	No
7. Does the CA have any documentation concerning successful pollution prevention programs being implemented by IUs (e.g., case studies, sampling data demonstrating pollutant reductions)?		

Explain.

K. ADDITIONAL EVALUATIONS/INFORMATION

SECTION I COMPLETED BY:	DATE:
TITLE:	TELEPHONE:

SECTION III: FINDINGS

INSTRUCTIONS: Based on information and data evaluated, summarize the findings of the audit for each program element shown below. Identify all problems or deficiencies based on the evaluation of program components. Clearly distinguish between deficiencies, violations, and effectiveness issues. This is to ensure that the final report will clearly identify required actions versus recommended actions and program modifications.

Description	Regulatory Citation	Checklist Question(s)
A. CA PRETREATMENT PROGRAM MODIFICATION		
• Status of program modifications	403.18	II.A.1

B. LEGAL AUTHORITY		
• Minimum legal authority requirements	403.8(f)(1)	II.B.2&3
• Adequate multijurisdictional agreements	403.8(f)(1)	II.B.1&3

SECTION III: FINDINGS (continued)

Description	Regulatory Citation	Checklist Question(s)
C. IU CHARACTERIZATION		
• Application of "significant industrial user" definition	403.3(t)(1)	II.C.1; Attach B.E.2
• Identify and categorize IUs	403.8(f)(2)(i)&(ii)	I.A.4.c; II.C.2&3
D. CONTROL MECHANISM		
• Issuance of individual control mechanisms to all SIUs	403.8(f)(1)(iii)	II.D.1
• Adequate control mechanisms	403.8(f)(1)(iii)	I.A.4
• Adequate control of trucked, railed, and dedicated pipe wastes	403.5(b)(8)	II.D.2&3, E.1

APPENDIX A

SECTION III: FINDINGS (continued)

Description	Regulatory Citation	Checklist Question(s)
E. APPLICATION OF PRETREATMENT STANDARDS AND REQUIREMENTS		
• Appropriately categorize, notify, and apply all applicable pretreatment standards	403.8(f)(1)(ii)&(iii); 403.5	I.A.
• Basis and adequacy of local limits	403.8(f)(4); 122.21(j)	II.E.3&4
F. COMPLIANCE MONITORING		
• Adequate sampling and inspection frequency	Approved program 403.8(f)(2)(ii)&(v)	I.B.1.a&b, 2.a&b; II.F.1&2
• Adequate inspections	403.8(f)(2)(v)&(vi)	I.B.2.c; II.F.3
• Adequate sampling protocols and analysis	403.8(f)(2)(vi)	I.B.1.c,d&e; II.F.4,5&6
• Adequate IU self-monitoring	403.8(f)(2)(iv)	I.A.4.d, C.1.b; II.F.6, G.5
• Notification of changed and hazardous waste discharges	403.12(j)&(p)	I.C.1.b; II.F.7

SECTION III: FINDINGS (continued)

Description	Regulatory Citation	Checklist Question(s)
F. COMPLIANCE MONITORING (continued)		
• Evaluate the need for SIUs to develop slug discharge control plans	403.8(f)(2)(v)	I.B.2.d; II.F.8
• Monitor to demonstrate continued compliance and resampling after violations(s)	403.12(g)(1)&(2); 403.8(f)(2)(vi)	I.A.4.d; C.1.b
G. ENFORCEMENT		
• Appropriate application of "significant noncompliance" definition	403.8(f)(2)(vii)	I.C.2; II.G.1; Attach B.I.1
• Develop and implement an ERP	403.8(f)(5)	I.C.3; II.G.2
• Annually publish a list of IUs in SNC	403.8(f)(2)(vii)	I.C.6; II.G.4
• Effective enforcement	403.8(f)(1)(iv)(A)	I.C.1.c, 4&5 II.G.2.c&d,5&6

Description	Regulatory Citation	Checklist Question(s)

H.   DATA MANAGEMENT/PUBLIC PARTICIPATION

Description	Regulatory Citation	Checklist Question(s)
● Effective data management/public participation	403.5(c)(3); 403.12(o); 403.14	II.H

I.   RESOURCES

Description	Regulatory Citation	Checklist Question(s)
● Adequate resources	403.8(f)(3)	II.I

SECTION III: FINDINGS (continued)

Description	Regulatory Citation	Checklist Question(s)
J. ENVIRONMENTAL EFFECTIVENESS/POLLUTION PREVENTION		
• Understanding of pollutants from all sources		II.J.1&3
• Documentation of environmental improvements/effectiveness		II.J.2
• Integration of pollution prevention		II.J.6

A
P
P
E
N
D
I
X

A

K. ADDITIONAL EVALUATIONS/INFORMATION

SECTION III COMPLETED BY:	DATE:
TITLE:	TELEPHONE:

ATTACHMENT A: PRETREATMENT PROGRAM STATUS UPDATE

INSTRUCTIONS: This attachment is intended to serve as an update of program status. It should be updated prior to each audit based on information obtained from the most recent PCI and/or audit and the last pretreatment program performance report.

A. CA INFORMATION

1. CA name

2. a. Pretreatment contact	b. Mailing address
c. Title	d. Telephone number

3. Date of last CA report to Approval Authority

	Yes	No
4. Is the CA currently operating under any pretreatment-related consent decree, Administrative Order, compliance schedule, or other enforcement action?		

5. Effluent and sludge quality

   a. List the NPDES effluent and sludge limits violated and the suspected cause(s).

Parameters Violated	Cause(s)

	Yes	No
b. Has the treatment plant sludge violated limits based on the following tests?		
• EP toxicity		
• TCLP		

B. PRETREATMENT PROGRAM STATUS

1. Indicate components that were identified as deficient.

	Last PCI Date:	Last Audit Date:	Program Report Date:
a. Program modification			
b. Legal authority			
c. Local limits			
d. IU characterization			
e. Control mechanism			
f. Application of pretreatment standards			
g. Compliance monitoring			
h. Enforcement program			
i. Data management			
j. Program resources			
k. Other (specify)			

ATTACHMENT A: PRETREATMENT PROGRAM STATUS UPDATE (continued)

B. PRETREATMENT PROGRAM STATUS (continued)

2. Is the CA presently in RNC for any of these violations?	Data Source	Yes	No
a. Failure to enforce against pass-through and/or interference [RNC-I] [SNC]			
b. Failure to submit required reports within 30 days [RNC-I] [SNC]			
c. Failure to meet compliance schedule milestones within 90 days [RNC-I] [SNC]			
d. Failure to issue/reissue control mechanisms to 90 percent of SIUs within 6 months [RNC-II]			
e. Failure to inspect or sample 80 percent of SIUs within the last 12 months [RNC-II]			
f. Failure to enforce standards and reporting requirements [RNC-II]			
g. Other (specify) [RNC-II]			

3. List SIUs in SNC identified in the last pretreatment program performance report, PCI, or audit (whichever is most recent).

Name of SIU in SNC	Reason for SNC	Source (PCI, Annual Report)

4. Indicate the number and percent of SIUs that were identified as being in SNC* with the following requirements from the CA's last pretreatment program performance report. If the CA's report does not provide this information, obtain the information for the most recent four full quarters during the audit.

SNC evaluation period [   ]

	%	Applicable pretreatment standards and reporting requirements
	%	Self-monitoring requirements
	%	Pretreatment compliance schedules

*SNC defined by:

| POTW | |
| EPA | |

5. Describe any problems the CA has experienced in implementing or enforcing its pretreatment program.

| ATTACHMENT A COMPLETED BY: | DATE: |
| TITLE: | TELEPHONE: |

APPENDIX A

ATTACHMENT B: PRETREATMENT PROGRAM PROFILE

INSTRUCTIONS: This attachment is intended to serve as a summary of program information. This background information should be obtained from the original, approved pretreatment program submission and modifications and the NPDES permit. The profile should be updated, as appropriate, in response to approved modifications and revised NPDES permit requirements.

A. CA INFORMATION

1. CA name

2. Original pretreatment program submission approval date

3. Required frequency of reporting to Approval Authority

4. Specify the following CA information.

Treatment Plant Name	NPDES Permit Number	Effective Date	Expiration Date

5. Does the CA hold a sludge permit or has the NPDES permit been modified to include sludge use and disposal requirements?

Yes	No

If yes, provide the following information.

POTW Name	Issuing Authority	Issuance Date	Expiration Date	Regulated Pollutants

B. PRETREATMENT PROGRAM MODIFICATIONS

1. When was the CA's NPDES permit first modified to require pretreatment implementation? [WENDB-PTIM]

2. Identify any substantial modifications the CA made in its pretreatment program since the approved pretreatment program submission. [403.18]

Date Approved	Name of Modification	Date Incorporated in NPDES Permit

ATTACHMENT B: PRETREATMENT PROGRAM PROFILE (continued)

C. TREATMENT PLANT INFORMATION

INSTRUCTIONS: Complete this section for each treatment plant operated under an NPDES permit issued to the CA.

1. Treatment plant name	2. Location address

3. a. NPDES permit number	b. Expiration date	4. Treatment plant wastewater flows
		Design [ ] MGD Actual [ ] MGD

5. Sewer System	a. Separate %	b. Combined %	c. Number of CSOs

6. a. Industrial contribution (MGD)	b. Number of SIUs discharging to plant	c. Percent industrial flow to plant
		%

7. Level of treatment	Type of Process(es)
a. Primary	
b. Secondary	
c. Tertiary	

8. Indicate required monitoring frequencies for pollutants identified in NPDES permit.

	Influent (Times/Year)	Effluent (Times/Year)	Sludge (Times/Year)	Receiving Stream (Times/Year)
a. Metals				
b. Organics				
c. Toxicity testing				
d. EP toxicity				
e. TCLP				

9. Effluent Discharge

a. Receiving water name	b. Receiving water classification	c. Receiving water use

d. If effluent is discharged to any location other than the receiving water, indicate where.

10. 301(h) waiver (ocean discharge)

	Yes	No
a. Applied for		
b. Granted		

c. Date of application [ ]

d. Date approved or denied [ ]

APPENDIX A

ATTACHMENT B: PRETREATMENT PROGRAM PROFILE (continued)

C. TREATMENT PLANT INFORMATION (continued)

11. Did the CA submit results of whole effluent biological toxicity testing as part of its NPDES permit application(s)? [122.21(j)(1) and (2)]

	N/A	Yes	No
11.			
a. If yes, did the CA use EPA-approved methods? [122.21(j)(3)]			
b. Has there been a pattern of toxicity demonstrated?			

12. Indicate methods of sludge disposal.

Quantity of sludge
- a. Land application _____ dry tons/year
- b. Incineration _____ dry tons/year
- c. Monofill _____ dry tons/year
- d. MSW landfill _____ dry tons/year

Quantity of sludge
- e. Public distribution _____ dry tons/year
- f. Lagoon storage _____ dry tons/year
- g. Other (specify) _____ dry tons/year

D. LEGAL AUTHORITY

1. a. Indicate where the authority to implement and enforce pretreatment standards and requirements is contained (cite legal authority).

b. Date enacted/adopted	c. Date of most recent revisions

2. Does the CA's legal authority enable it to do the following? [403.8(f)(1)(i-vii)]

	Yes	No
a. Deny or condition pollutant dischargers [403.8(f)(1)(i)]		
b. Require compliance with standards [403.8(f)(1)(ii)]		
c. Control discharges through permit or similar means [403.8(f)(1)(iii)]		
d. Require compliance schedules and IU reports [403.8(f)(1)(iv)]		
e. Carry out inspection and monitoring activities [403.8(f)(1)(v)]		
f. Obtain remedies for noncompliance [403.8(f)(1)(vi)]		
g. Comply with confidentiality requirements [403.8(f)(1)(vii)]		

3. a. How many contributing jurisdictions are there? _____

List the names of all contributing jurisdictions and the number of SIUs in those jurisdictions.

Jurisdiction Name	Number of CIUs	Number of Other SIUs

D. LEGAL AUTHORITY (continued)

	Yes	No
3. b. Has the CA negotiated all legal agreements necessary to ensure that pretreatment standards will be enforced in contributing jurisdictions?		

If yes, describe the legal agreements (e.g., intergovernmental contract, agreement, IU contracts, etc.)

4. If relying on contributing jurisdictions, indicate which activities those jurisdictions perform.

a. IWS update

b. Permit issuance

c. Inspection and sampling

d. Enforcement

e. Notification of IUs

f. Receipt and review of IU reports

g. Analysis of samples

h. Other (specify)

E. IU CHARACTERIZATION

	Yes	No
1. a. Does the CA have procedures to update its IWS to identify new IUs or changes in wastewater discharges at existing IUs? [403.8(f)(2)(i)]		

b. Indicate which methods are to be used to update the IWS.

- Review of newspaper/phone book
- Review of water billing records
- Review of plumbing/building permits
- Onsite inspections
- Permit application requirements
- Citizens' involvement
- Other (specify)

c. How often is the IWS to be updated?

	Yes	No
2. Is the CA's definition of "significant industrial user" consistent within the language in the Federal regulations? [403.3(t)(1)]		

If no, provide the CA's definition of "significant industrial user."

ATTACHMENT B: PRETREATMENT PROGRAM PROFILE (continued)

F. CONTROL MECHANISM

1. a. Identify the CA's approved control mechanism (e.g., permit, etc.).	
b. What is the maximum term of the control mechanism?	

2. Does the approved control mechanism include the following? [403.8(f)(1)(iii)]	Yes	No
a. Statement of duration		
b. Statement of nontransferability		
c. Effluent limits		
d. Self-monitoring requirements		
• Identification of pollutants to be monitored		
• Sampling location		
• Sample type		
• Sampling frequency		
• Reporting requirements		
• Notification requirements		
• Record keeping requirements		
e. Statement of applicable civil and criminal penalties		
f. Applicable compliance schedule		

	N/A	Yes	No
3. Does the CA have a control mechanism for regulating IU whose wastes are trucked to the treatment plant?			
4. Does the program identify designated discharge point(s) for trucked or hauled wastes? [403.5(b)(8)]			

If yes, describe the discharge point(s) (including security procedures).

G. APPLICATION OF STANDARDS

	Yes	No
1. Does the CA have procedures to notify all IUs of applicable pretreatment standards and any applicable requirements under the CWA and RCRA? [403.8(f)(2)(iii)]		

	N/A	Yes	No
2. If there is more than one treatment plant, were local limits established specifically for each plant?			

ATTACHMENT B: PRETREATMENT PROGRAM PROFILE (continued)

G. APPLICATION OF STANDARDS (continued)

3. Has the CA <u>technically evaluated</u> the need for local limits for all pollutants listed below?
[WENDB-EVLL] [403.5(c)(1); 403.8(f)(4)]

Partial Technical Evaluation (not all 10 pollutants evaluated)?

	Headworks Analysis Completed?		Technically Evaluated?		Local Limits Adopted?		Local Limit (Numeric)
	Yes	No	Yes	No	Yes	No	
a. Arsenic (As)							
b. Cadmium (Cd)							
c. Chromium (Cr)							
d. Copper (Cu)							
e. Cyanide (CN)							
f. Lead (Pb)							
g. Mercury (Hg)							
h. Nickel (Ni)							
i. Silver (Ag)							
j. Zinc (Zn)							
k. Other (specify)							

H. COMPLIANCE MONITORING

1. Indicate compliance monitoring and inspection frequency requirements.

Program Aspect	Approved Program Requirement	NPDES Permit Requirement	State Requirement	Minimum Federal Requirement
a. Inspections				
• CIUs				1/year
• Other SIUs				1/year
b. Sampling by POTW				
• CIUs				1/year
• Other SIUs				1/year
c. Self-monitoring				
• CIUs				2/year
• Other SIUs				2/year
d. Reporting by IU				
• CIUs				2/year
• Other SIUs				2/year

APPENDIX A

ATTACHMENT B: PRETREATMENT PROGRAM PROFILE (continued)

I.   ENFORCEMENT

	Yes	No
1. Does the CA's program define "significant noncompliance"?		
If yes, is the CA's definition of "significant noncompliance" consistent with EPA's? [403.8(f)(2)(vii)]		

If no, provide the CA's definition of "significant noncompliance."

	Yes	No
2. Does the CA have an approved, written ERP? [403.8(f)(5)]		

3. Indicate the compliance/enforcement options that are available to the POTW in the event of IU noncompliance. [403.8(f)(1)(vi)]

a. Notice or letter of violation		f. Administrative Order		
b. Compliance schedule		g. Revocation of permit		
c. Injunctive relief		h. Fines (maximum amount)		
d. Imprisonment		• Civil	$_____/day/violation	
e. Termination of service		• Criminal	$_____/day/violation	
		• Administrative	$_____/day/violation	

J.   DATA MANAGEMENT/PUBLIC PARTICIPATION

	Yes	No
1. Does the approved program describe how the POTW will manage its files and data?		

Are files/records [ ] computerized? [ ] hard copy? [ ] both?

	Yes	No
2. Are program records available to the public?		
3. Does the POTW have provisions to address claims of confidentiality? [403.8(f)(2)(vii)]		

ATTACHMENT B: PRETREATMENT PROGRAM PROFILE (continued)

K. RESOURCES

1. What are the resource allocations for the following pretreatment program components:

	FTEs
a. Legal assistance	
b. Permitting	
c. Inspections	
d. Sample collection	
e. Sample analysis	
f. Data analysis, review, and response	
g. Enforcement	
h. Administration?	
TOTAL	

2. Identify the sources of funding for the pretreatment program. [403.8(f)(3)]

a. POTW general operating fund ▢   d. Monitoring charges ▢

b. IU permit fees ▢   e. Other (specify) ▢

c. Industry surcharges ▢

L. ADDITIONAL INFORMATION

ATTACHMENT B COMPLETED BY:	DATE:
TITLE:	TELEPHONE:

APPENDIX A

ATTACHMENT C: WORKSHEET — IU SITE VISIT DATA SHEET

I. IU SITE VISIT REPORT FORM

INSTRUCTIONS: Record observations made during the IU site visit. Provide as much detail as possible.

Name and address of industry

Date of visit	Time of visit

Name(s) of inspector(s)

Provide name(s) and title(s) of industry representative(s).

Name	Title

Classification assigned by CA:

Provide the following documentation:

1. Describe the products manufactured or the services provided by the IU.

2. Verify CA's classification or discuss any errors.

3. Describe any significant changes in processes or flow.

4. Identify the raw materials and processes used. (Include discussion of where wastewater is produced and discharged and attach a step-by-step diagram if possible.)

5. Describe the sample location and any differences in CA and IU locations.

6. Describe the treatment system which is in place.

7. Identify the chemicals that are maintained onsite and how they are stored. (Attach list of chemicals, if available.) Discuss the adequacy of spill prevention.

8. Discuss whether hazardous wastes are stored or discharged and any related problems.

Notes:

IU SITE VISIT REPORT FORM COMPLETED BY: TITLE:	DATE: TELEPHONE:

ATTACHMENT C: WORKSHEET — WENDB DATA ENTRY WORKSHEET

II.   WENDB DATA ENTRY WORKSHEET

INSTRUCTIONS: Enter the data provided by the specific checklist questions that are referenced.

CA name

NPDES number

Date of audit

	PCS Code	Checklist Reference	Data
• Number of SIUs*	SIUs	II.C.4.a	
• Number of CIUs	CIUs	II.C.4.a	
— Number of SIUs without control mechanism	NOCM	II.D.1.a	
— Number of SIUs not inspected or sampled	NOIN	II.F.2.a	
— Number of SIUs in SNC** with standards or reporting	PSNC	Attach A.B.4	
— Number of SIUs in SNC with self-monitoring	MSNC	Attach A.B.4	
— Number of SIUs in SNC with self-monitoring and not inspected or sampled	SNIN	II.G.5	

* The number of SIUs entered into PCS is based on the CA's definition of "significant industrial user."

** As defined in 40 CFR 403.8(f)(2)(vii).

WENDB DATA ENTRY WORKSHEET COMPLETED BY: TITLE:	DATE: TELEPHONE:

ATTACHMENT C: WORKSHEET — RNC WORKSHEET

III. RNC WORKSHEET

INSTRUCTIONS: Place a check in the appropriate box on the left if the CA is found to be in RNC or SNC.

CA name

NPDES number

Date of audit

		Level	Checklist Reference
	Failure to enforce against pass-through and/or interference	I	II.G.6
	Failure to submit required POTW reports within 30 days	I	Attach A.B.2.b
	Failure to meet compliance schedule milestone date within 90 days	I	Attach A.B.2.c
	Failure to issue/reissue control mechanisms to 90% of SIUs within 6 months	II	II.D.1.b
	Failure to inspect or sample 80% of SIUs within the last 12 months	II	II.F.2.a
	Failure to enforce pretreatment standards and reporting requirements (more than 15% of SIUs in SNC)	II	I.C.1; II.G.2.
	Other (specify)	II	

SNC

	CA in SNC for violation of any Level I criterion
	CA in SNC for violation of two or more Level II criterion

For more information on RNC, please refer to EPA's 1990 Guidance for Reporting and Evaluating POTW Noncompliance with Pretreatment Implementation Requirements.

RNC WORKSHEET COMPLETED BY:	DATE:
TITLE:	TELEPHONE:

PART III: AUDIT CHECKLIST INSTRUCTIONS

SECTION I: IU FILE EVALUATION

Each of the major program components in Section I of the Checklist is listed below along with an explanation (generally an explanation of the regulatory requirement). Guidance is provided on how the auditor can evaluate the CA's (or IU's) compliance with the program requirement and on what constitutes a deficiency. Much of the information needed to do necessary evaluations will probably be in the CA's files on the individual IUs. The auditor should begin by finding out how the CA organizes their files. Some CAs have individual files for each IU and all information pertaining to that IU is in the file. Other CAs may have files segregated by subject so that all permits are in one file while all monitoring data are in another file and all correspondence in another and so on. Once the auditor has determined the file organization, he or she can move on to doing the evaluation.

Section I requires the auditor to review certain components of the CA's IU files. After reviewing each component, the auditor must determine if what he or she found was adequate or appropriate. Once this determination has been made, the auditor should decide if the information learned is worthy of comment or explanation. If comment or explanation is necessary, the auditor should put a number in the square corresponding to the component being evaluated, and the same number in the comment area followed by the explanation of what was found. It is recommended that numbering begin anew on each page.

To facilitate completion of this section, elements of each program area are listed for consideration. The regulatory citations are provided where there are specific <u>requirements</u> for that element. The auditor should be aware that not all questions on the checklist reflect regulatory requirements. Some of the questions are included to allow the auditor to better evaluate program effectiveness. This fact should be taken into consideration when developing required versus recommended actions to be taken by the CA.

IU Identification

> PURPOSE: This section is designed to provide a brief profile of the IU. This information should summarize industrial categorization, discharge characterization, and comment on compliance history and other issues of note. The auditor should briefly look through the file and fill out the information requested. Some information will be filled out at the start of the file review (e.g., name, address, etc.). Some information (e.g., category, flow, compliance status, etc.) will be obtained as the review proceeds. The auditor should enter additional information about the industry obtained from the interview with CA staff or site visit to the IU.

IU File Review

> The auditor should review each point covered in the file review to determine if there is anything worth noting to question the CA about during the interview. For instance, something the CA is doing that is out of the ordinary either positive or negative.

A.   Issuance of IU Control Mechanism

Note: This section takes a <u>comprehensive</u> look at the CA's control mechanism. The auditor should evaluate the adequacy and effectiveness of the control mechanism used. Comments should reflect an evaluation of the control mechanism for both presence and the adequacy of all control mechanism components. For each area examined in this section of the file review, the auditor should determine whether the CA met the regulatory requirement and also if the CA is effective in controlling the IUs. If the auditor determines there is a problem or deficiency (e.g., control mechanisms are not issued/reissued in a timely manner, do not contain all the elements required by the regulation, contain incorrect limits, etc.), he/she should comment on it in the area provided and explain it in the report to be attached.

A.1.   Control mechanism application form

> PURPOSE: The CA should require certain baseline data from the IU in order to write an appropriate control mechanism. Although there are several ways these data may be obtained, it is strongly recommended that the CA utilize an application form (there is no regulatory requirement). For CIUs, the Baseline Monitoring Report (BMR) may serve as an application and may then be updated for permit reissuance purposes. For each point covered or issue addressed in the file review, the auditor should also review each point to determine if there is anything worth noting to question the CA about during the interview. For instance, something the CA is doing that is out of the ordinary either positive or negative.

> FACTORS TO CONSIDER:

- If the application is being used as a BMR, it must contain all the 40 CFR 403.12(b) required elements.

- To be useful, the application should at least include IU identification, address, phone, responsible officer, a clear description of processes, the flow from each, as well as a description of any pretreatment system in place or proposed.

- Where applications are incomplete, there should be evidence that the CA followed up by requiring the applicant to submit missing data or, at least, that the CA obtained the missing data on its own.

- Where there is evidence that the data contained in the application are inaccurate, there should be evidence that the CA took an enforcement action.

A.2.   Fact sheet

PURPOSE: Individual control mechanisms issued to SIUs must contain specific conditions applicable to the IU. A fact sheet is recommended to provide data concerning decisions made in developing the control mechanism (there is no regulatory requirement).

FACTORS TO CONSIDER:

- The fact sheet should explain the basis of every IU-specific standard or requirement contained in the control mechanism, including:

  — The basis for determining that the IU is subject to a particular category and sub-category, if applicable.

  — The basis for the permit limits applied (i.e., local limits versus categorical standards, production-based limits, CWF/FWA, and mass- versus concentration-based limits).

  — The rationale behind the pollutants specified for self-monitoring.

  — Documentation for the need for any slug control plan, Best Management Practices (BMPs), and compliance schedule requirements. It should include the circumstances identified which necessitated these requirements.

A.3.   Issuance or reissuance of control mechanism

PURPOSE: The CA is required to control IU discharges to the POTW. Under 40 CFR (403.8)(f)(1)(iii), all SIU discharges are required to be controlled under individual control mechanisms (i.e., permit, order, or similar means).

FACTORS TO CONSIDER:

- If the auditor cannot locate a control mechanism or if the control mechanism is not current or valid, a deficiency should be noted. If the control mechanism has to be signed by the CA and is not signed, it may not be valid.

- The auditor should check an expired control mechanism to see if it has been or will be reissued within 180 days from the expiration of the last control mechanism.

A.4.   Control mechanism contents

PURPOSE: Individual control mechanisms issued to SIUs must contain the minimum conditions listed in 40 CFR 403.8(f)(1)(iii). The required elements to consider are elaborated upon below in A.4.a-g.

FACTORS TO CONSIDER:

- Each condition contained in the control mechanism must also be evaluated for appropriateness and accuracy. For instance, if production-based categorical standards are applied, the auditor must determine whether the IU was correctly categorized and whether the discharge limit contained in the control mechanism was correctly calculated. An explanation of each control mechanism condition is presented below.

A.4.a.   Statement of duration (≤ 5 years)

PURPOSE: The auditor should review the control mechanism to determine that the duration is not for more than 5 years.

A.4.b.   Statement of nontransferability w/o prior notification/approval

PURPOSE: The control mechanism is not allowed to be transferred without, at a minimum, prior notification to the CA and provision of a copy of the existing control mechanism to the new owner or operator.

A.4.c.   Applicable effluent limits

PURPOSE: The control mechanism must contain effluent limits based on applicable general pretreatment standards in 40 CFR 403.5, categorical pretreatment standards, local limits, and State and local law. The auditor should determine that the limits in the control mechanism are correct.

FACTORS TO CONSIDER:

- Application of applicable categorical standards includes the following:

  — Classification by category/subcategory.
  — Classification as new/existing source.
  — Application of limits for all categorical pollutants.
  — Application of Total Toxic Organics (TTO) or Toxic Organic Management Plan (TOMP) alternative.
  — Calculation and application of production-based standards.
  — Calculation and application of CWF or FWA.
  — Application of variance to categorical standards, including Fundamentally Different Factors (FDF) variances and net/gross adjustments.

- Application of applicable local limits.

- Application of most stringent limit.

A.4.d. IU self-monitoring requirements

PURPOSE: All SIUs are required to submit a report at least semiannually. For all CIUs, the semiannual report must include results of monitoring for all pollutants regulated under the applicable categorical standard and any additional applicable local limits. These requirements can be modified if the CA assumes responsibility for the sampling. The auditor should review the self-monitoring requirements contained in the control mechanism to determine whether they will be effective in identifying noncompliance considering the type and size of the facility, variability in sampling results, the IU's compliance history, etc. **(Note: the CIU is required to report on all regulated pollutants at least semiannually whether or not the requirement is included in the control mechanism.)**

FACTORS TO CONSIDER:

- Sampling:

  - Pollutants — All pollutants regulated under an applicable categorical standard must be sampled and analyzed at least semiannually.
  - Frequency — Although all SIUs are required to self-monitor for all regulated pollutants at least semiannually, these two monitoring events may not be sufficient to provide the CA with a true picture of ongoing compliance, but it is the minimum frequency.
  - Location(s) — Should be clearly identified.
  - Types of samples (e.g., 24-hour composite, grab) — To be taken for each parameter. The auditor should be aware that all pretreatment compliance monitoring must be done in accordance with the procedures specified under 40 CFR Part 136. Further, 24-hour composite samples (or their equivalent) must be used to determine compliance with categorical pretreatment standards except for the following parameters which require the use of grab samples: pH, heat, oil and grease, volatile organics, and phenols.

- Reporting requirements (e.g., periodic, resampling).

- Notification requirements (e.g., slug, spill, changed discharge, 24-hour notice of violation).

- Record keeping requirements — All SIUs are required to retain effluent self-monitoring data and other related documentation for a period of at least 3 years, throughout the course of any ongoing litigation related to the IU, and for the period of time specified by the CA.

A.4.e. Statement of applicable civil and criminal penalties

PURPOSE: All SIU control mechanisms are required to specify the penalties applicable for violation of control mechanism conditions. These penalties must include civil and/or criminal penalties in an amount up to at least $1,000 per day per violation.

FACTORS TO CONSIDER:

- The CA may also apply administrative penalties for control mechanism violations and is encouraged to do so. However, administrative penalties do not satisfy this regulatory requirement.

- If penalties are not stated, the control mechanism should cite the specific ordinance provision which establishes the penalties.

A.4.f. Compliance schedules/progress reports (if applicable)

PURPOSE: The CA must require compliance schedules where a CIU is not in compliance with a newly promulgated categorical standard. This schedule must have a final compliance date which is no later than the compliance deadline specified by the standard. The schedule must also include milestone dates and a requirement for progress reports to be submitted for each milestone [see requirement under 40 CFR 403.12(b)(7) and (c)].

FACTORS TO CONSIDER:

- Compliance schedules for compliance with a categorical standard deadline which has already passed should not be contained in the control mechanism, but in an enforcement order.

- Compliance schedules are also strongly recommended for use where any IU is out of compliance with any pretreatment standard or requirement. These schedules are also best placed in an enforcement order.

- Compliance schedules used for attaining compliance with a revised local limit by the limit's effective date should be treated similarly to those prepared for compliance with a categorical compliance date.

A.4.g. Slug discharge control plan requirement (if applicable)

PURPOSE: Where IU slug discharge control plans are required to prevent slug loadings to the POTW, they must contain the elements specified under 40 CFR 403.8(f)(2)(v): (1) A description of discharge practices, including nonroutine batch discharges; (2) a description of stored chemicals; (3) procedures for immediately notifying the POTW of slug discharges, including any discharge which would violate a prohibition under 40 CFR 403.5(b), with procedures for follow-up written notification within 5 days; and (4) if necessary, procedures to prevent adverse impact from accidental spills, including inspection and maintenance of storage areas, handling and transfer of materials, loading and unloading operations, control of plant site run-off, worker training, building of containment structures or equipment, measures for containing toxic organic pollutants (including solvents), and/or measures and equipment necessary for emergency response.

FACTORS TO CONSIDER:

- SIU control mechanisms must contain the requirement to immediately notify the CA of any slug discharge. However, it is recommended that the CA incorporate the entire slug discharge control plan into the control mechanism, making compliance with the plan a condition of discharge.

- Any plan which is less inclusive or less stringent than that required under 40 CFR 403.8(f)(2)(v) should be recorded as a deficiency.

B.   **CA Compliance Monitoring**

Note: The CA is required to do sampling and inspection of IUs to verify compliance independent of information supplied by the IU. If the CA has not undertaken any surveillance activity or no documentation exists, if documentation is insufficient, or if the CA has not sampled for all regulated parameters, the auditor should note these problems.

B.1.   **Inspection**

B.1.a. **Inspection (at least once a year)**

B.1.b. **Inspection at frequency specified in approved program**

B.1.c. **Documentation of inspection activities**

B.1.d. **Evaluation of need for slug discharge control plan (reevaluation of existing plan)**

PURPOSE: The CA is required to inspect all IUs to determine compliance with pretreatment standards and requirements independent of data submitted by the IU.

FACTORS TO CONSIDER:

- Inspection at least once a year or as specified in the approved program.

- Although the CA is required to inspect the IU once a year, or more frequently if required by the approved program, the auditor should assess the adequacy of this frequency based on the IU's compliance history, IU-specific requirements, process changes, etc.

- Documentation of inspection activities should be clear and cover every aspect of the inspection. Some CAs may use activity logs to demonstrate an inspection took place, however, the log alone will not fulfill the requirement for sufficient care to produce evidence admissible in enforcement cases [40 CFR 403.8(f)(2)(vi)].

- Evaluation of need for slug discharge control plan (reevaluation of existing plan) — The CA is required to evaluate each IU's need for a slug discharge control plan at least once every 2 years. *Note: This may also be called an accidental spill prevention plan. However, to fulfill the regulatory requirement, the plan must also address any potential nonaccidental slug discharges.*

B.2.   **Sampling**

B.2.a. **Sampling (at least once a year)**

B.2.b. **Sampling at frequency specified in approved program**

B.2.c. **Documentation of sampling activities (chain-of-custody; QA/QC)**

B.2.d. **Analysis for all regulated parameters**

B.2.e. **Appropriate analytical methods (40 CFR Part 136)**

PURPOSE: The CA is required to sample each SIU discharge point to verify compliance independent of self-monitoring data supplied by the IU. The auditor should determine that the CA has sampled the IU by reviewing sampling records, lab reports, chain-of-custody forms, etc. The auditor should examine all CA compliance sampling data in the IU's file.

FACTORS TO CONSIDER:

- Sampling frequency — At least once a year or at the frequency specified in the approved program.

- Documentation of sampling activities should include QA/QC analytical results and chain-of-custody [sample date and time; location; flow, where applicable; sampling method/type; sampler's name; sample preservation techniques; sample characteristics; dates of analyses; name of analyst; analytical technique/method (40 CFR Part 136); and analytical results].

- Sampling results should include analyses for all regulated parameters.

- Appropriate analytical methods (40 CFR Part 136) — The SIU is required to use the methods defined under 40 CFR Part 136 when collecting and analyzing all samples obtained to determine compliance with pretreatment standards. Since the CA's compliance monitoring serves to verify compliance with the same standards and to check the validity of self-monitoring data, the CA's monitoring should also be conducted in accordance with 40 CFR Part 136. While specific test procedures included in Standard Methods for the Examination of Water and Wastewater are approved under 40 CFR Part 136 for many parameters, not all the test procedures in "Standard Methods" are approved.

C.   **CA Enforcement Activities**

Note: This section serves several purposes. In this section, the auditor will determine the compliance status of the selected IUs and the corresponding response of the CA. If the IU is in noncompliance and the CA fails to identify the noncompliance, the auditor should note this on the checklist and explain the situation in the comment section. The auditor should also determine if the IU is in SNC, whether the enforcement taken by the CA was effective and followed the approved Enforcement Response Plan (ERP). If the auditor finds any problems, he/she should note these and explain the situation in the report.

C.1.   **Identification of violations**

C.1.a.   **Discharge violations**

C.1.b.   **Monitoring/reporting violations**

C.1.c.   **Compliance schedule violations**

PURPOSE: The CA is required to identify and investigate all instances of noncompliance with pretreatment standards and requirements. The auditor should verify the CA has identified all violations.

FACTORS TO CONSIDER:

● The CA must identify any and all IU noncompliance. It is recommended that the CA use a tracking system to:

— Obtain and compare sampling data with applicable limits and identify and investigate any violations. The investigation should include requiring the IU to explain the violation.

— Receive IU reports and determine their timeliness, completeness, and accuracy.

— Determine appropriate progress with compliance schedules.

● The CA must obtain enough IU discharge data to determine compliance on an ongoing basis. If the IU has a history of noncompliance and/or variability in discharge constituents and characteristics, the CA will need more frequent sampling data to determine the pattern and causes of noncompliance.

● If the IU has a history of noncompliance, has not submitted any required self-monitoring reports, or discharges pollutants for which the POTW has NPDES violations, these facts should be noted.

● The auditor should attempt to determine whether the monitoring frequency and the reports for the particular IU is sufficient to provide a true picture of compliance.

● IU self-monitoring — As discussed above, all SIUs are required to report at least twice a year, and more frequently if required by the CA.

● Where CA compliance monitoring data show instances of noncompliance, the auditor should find Notices of Violation (NOVs) provided to the IU for each instance as well as other appropriate follow-up.

● Violations of monitoring and reporting requirements must be addressed by the CA's enforcement program. IU reporting includes all notices required to be submitted by the IU [i.e., notice of a slug discharge (including accidental spills), prior notice of a changed discharge, and 24-hour notice of violation identified in self-monitoring data].

● The CA should respond to any failure by the IU to comply with compliance schedule requirements.

C.2.   **Calculation of SNC**

C.2.a.   **Chronic**

C.2.b.   **TRC**

C.2.c.   **Pass through/interference**

C.2.d.   **Spill/slug load**

C.2.e.   **Reporting**

C.2.f.   **Compliance schedule**

C.2.g.   **Other violations (specify)**

PURPOSE: The CA is required to calculate SNC in order to determine which industries to publish at least annually in the largest local daily newspaper. The CA must also report a summary of IU compliance status in its pretreatment program performance reports to the State or EPA. The auditor should evaluate the file to determine if the CA correctly calculated SNC. This can be done by reviewing violations and performing SNC calculations. *(Note: If the auditor is unfamiliar with the definition of SNC, he/she should refer to the definition in the General Pretreatment Regulations and EPA policy.)*

FACTORS TO CONSIDER:

- CAs should be evaluating SNC based on the procedures set forth in the regulations and EPA's September 9, 1991, memorandum on the <u>Application and Use of the Regulatory Definition of Significant Noncompliance for Industrial Users</u>.

- Evidence of SNC evaluation should be found and evaluated. This information may be in the CA's enforcement file, the pretreatment program performance report submitted to EPA or the State, as well as in the CA and IU sampling reports or included in the data management system. The auditor should look for any SNC violations as described below and determine whether the CA has correctly determined SNC.

C.3.   Response to violation

PURPOSE: The CA is expected to respond to every violation in an appropriate manner and consistent with its approved ERP.

FACTORS TO CONSIDER:

- If the CA has an approved ERP, did it respond to each violation as specified in the ERP?

- Effective enforcement requires a timely response by the CA to all violations. The auditor should investigate the cause of any instances where response did not occur in a timely manner.

C.4.   Adherence to approved ERP

PURPOSE: Where the CA has an approved ERP, it is required to implement that plan in all its enforcement proceedings.

FACTORS TO CONSIDER:

- Implementation of the approved ERP involves timely and appropriate enforcement and escalation of enforcement actions where violations persist. The CA should have noted and responded to any instance of noncompliance with local limits and/or categorical pretreatment standards. At a minimum, for minor violations the CA should have notified the IU of the violation through a phone call, meeting, or NOV. Instances of noncompliance with any pretreatment requirement should also have resulted in a response by the CA.

- Where the CA's actions conformed to the ERP but were not effective (i.e., they did not result in a final resolution within a reasonable length of time), the auditor should document the situation and consider whether the ERP requires modification.

C.5.   Return to compliance

C.5.a. Within 90 days

C.5.b. Within time specified

C.5.c. Through compliance schedule

PURPOSE: There are a number of criteria by which to determine effective enforcement. A return to compliance within 90 days of the initial violation is the primary goal, but even effective enforcement may take longer.

FACTORS TO CONSIDER:

- One criteria for successful enforcement is returning the IU to compliance within 90 days.

- The IU should be returned to compliance within the time specified by the CA. If the IU must come into compliance with a categorical pretreatment standard deadline or a deadline for compliance with a modified local limit, the CA should take appropriate actions (usually issuance of a compliance schedule) to ensure that the IU will meet that deadline.

- Violation of a compliance schedule deadline may indicate lack of effective enforcement. If the deadline has built-in milestone dates, the CA has the opportunity to take actions whenever the IU falls behind in its progress toward compliance. Effective action should result in achievement of compliance by the schedule's deadline.

C.6.   Escalation of enforcement

PURPOSE: The CA is expected to escalate enforcement for persistent violations.

FACTORS TO CONSIDER:

- The CA is expected to bring noncompliant users back into compliance by timely and appropriate enforcement. This requires escalation of enforcement activity for persistent violations per the CA's ERP. The auditor should look for patterns of increasingly severe enforcement actions [e.g., NOVs followed by Administrative Orders (AOs)] where the past enforcement actions have not resulted in the IU achieving consistent compliance. The auditor should evaluate dates of the enforcement actions and IU responses (provide examples).

- Where self-monitoring data show instances of noncompliance, the auditor should look for and note follow-up by the CA to any violations and determine the appropriateness of actions taken.

C.7.   Publication for SNC

PURPOSE: The CA is required to annually publish, in the area's largest daily newspaper, a list of IUs found to be in SNC. The auditor should verify that IUs in SNC, if any, were properly published.

FACTORS TO CONSIDER:

- The IU file or a central enforcement file should contain a copy or clipping of the latest notice placed in the local newspaper. The CA may keep this public notice in a separate file.

- If an IU has been in SNC at any time during the year to which the publication pertains, then the IU must be included in the published list. Even those IUs that returned to compliance and are in compliance at the time of publication must be included in the published list. IUs that are on compliance schedules (but have had or continue to have SNC violations of standards or requirements) must also be published.

- The auditor should randomly check IUs in SNC against the published list and determine whether the CA published and reported on all these IUs.

- Publication may take the form of a legal notice; however, it may be more effective in the form of an article or advertisement.

D.   Other

PURPOSE: The auditor should use this section to document any initiatives, unusual situations, or other issues of note or concern identified in the file review and not covered under the sections above.

SECTION II: DATA REVIEW/INTERVIEW/IU SITE VISIT

Each of the questions in Section II of the Checklist is listed below along with an explanation of the purpose or intent of the question. Brief guidance is provided on how the auditor can evaluate the CA's efforts. More detailed guidance on the technical aspects of each question may be found in the appendices. This section is primarily designed to be interactive between the auditor and the CA personnel. However, the information collected should not be solely from the answers provided by the CA personnel. Where possible, all answers provided by the CA should be supported by other data (monitoring reports, correspondence, etc.). The auditor should use this section to complement the data gathered through the file review and to further evaluate the effectiveness of the CA's implementation of the pretreatment program.

To facilitate completion of this section, elements of each program area are listed for consideration. The regulatory citations are provided where there are specific requirements for that element. The auditor should be aware that not all questions on the checklist reflect regulatory requirements. Some of the questions are included to allow the auditor to better evaluate program effectiveness. This fact should be taken into consideration when developing required versus recommended actions to be taken by the CA.

A.   CA Pretreatment Program Modification [403.18]

Note: The auditor should attempt to determine if any modifications have taken place without approval by the Approval Authority. He/she should also determine if any modifications are planned in the near future or are currently being worked on.

A.1.a.   Has the CA made any substantial changes to the pretreatment program that were not reported to the Approval Authority (e.g., legal authority, less stringent local limits, multijurisdictional situation)? If yes, discuss.

A.1.b.   Is the CA in the process of making any substantial modifications to any pretreatment program component (including legal authority, less stringent local limits, DSS requirements, multijurisdictional situation, etc.)? If yes, discuss.

PURPOSE: The CA is required to notify the Approval Authority of any substantial modifications it intends to make in its pretreatment program. Substantial modifications should not be made without approval by the Approval Authority.

FACTORS TO CONSIDER:

- When investigating this area, the auditor should keep in mind that program modifications are likely to be made in any of the following areas:

  — Contributing jurisdiction added.

  — Legal authority — SUO and interjurisdicational agreements.

  — Local limits — reevaluation and modification, addition or deletion of parameters.

  — Definition of SIU and/or changes in criteria for IUs to be included in the pretreatment program.

  — Control mechanisms (including IU contracts) — type (order vs. permit, etc.), content, format, or standard conditions.

  — Inspection and sampling (including self-monitoring) frequencies and/or priorities.

  — Resources committed to the program — equipment, personnel, funding.

B.   Legal Authority [403.8(f)(1)]

Note: This section is designed to investigate whether the CA has adequate legal authority to implement their program. The auditor should review the CA's legal authority/ordinance to make sure it is current with the new regulations, and to determine that the CA has adequate authority to cover any extrajurisdictional situation that may exist. The auditor should note any problems and explain them in the spaces provided on the checklist.

B.1.   Are there any contributing jurisdictions discharging wastewater to the POTW? If yes, explain how the legal authority addresses the contributing jurisdictions.

PURPOSE: The CA is responsible for the implementation and enforcement of its pretreatment program for all IUs (i.e., existing and future IUs) throughout its service area, regardless of jurisdictional boundaries. The CA should have a mechanism(s) to ensure implementation and enforcement in its contributing jurisdictions.

FACTORS TO CONSIDER:

- The CA may be relying on its SUO to regulate IUs in contributing municipalities, but may not have adequate authority to do so under State law.

- The CA may be relying on existing interjurisdictional agreements that were entered into for the purpose of guaranteeing treatment capacity and providing for payment thereof. Such agreements seldom address the needs of pretreatment program implementation. At a minimum, the agreement should require the contributing municipality to adopt and maintain a SUO which is at least as stringent and inclusive (including local limits) as the CA's SUO. Ideally, the agreement (or a supplement to the agreement) should provide for every program implementation activity.

- The CA may have no means of obtaining an adequate agreement with a contributing municipality (i.e., the CA may be required to continue providing service to the municipality) and may not have entered into a contract with extrajurisdictional IUs.

- The CA may not have entered into an agreement (or may have an inadequate agreement) with contributing municipalities which do not currently have IUs located within their boundaries. Even if zoning in such cases allows only for commercial and/or residential premises, the CA should have an agreement which requires notification to and approval by the CA should any IU request to connect to the system since zoning laws are subject to change.

B.2.a. Has the CA updated its legal authority (e.g., SUO) to reflect changes in the General Pretreatment Regulations?

B.2.b. Did all contributing jurisdictions update their SUOs in a consistent manner?

PURPOSE: The CA is required to amend its legal authority, as necessary, to be consistent with all revisions of the General Pretreatment Regulations. The amendment would be a substantial program modification and must be approved by the Approval Authority. The auditor should verify the status of the CA's legal authority.

FACTORS TO CONSIDER:

- CA may have modified its SUO without submitting proposed changes to the Approval Authority or may have enacted modifications without approval. If so, this should be noted along with the date modifications were enacted and citations of the modified provisions.

- CA may have submitted proposed changes, but has not yet received approval. The date of submission should be noted.

- The SUO may have been modified to be consistent with PIRT, but not yet modified to be consistent with DSS.

- Additional modifications (not required by PIRT or DSS) may have been made or proposed. If so, cite and explain those modifications and the reasons for them.

B.3.   Does the CA experience difficulty in implementing its legal authority [e.g., SUO, interjurisdictional agreement (e.g., permit challenged, entry refused, penalty appealed)]? If yes, explain.

PURPOSE: The CA should be able to ensure the successful implementation of its SUO provisions through the service area.

FACTORS TO CONSIDER:

- CA's SUO authorities may have been challenged as being inconsistent with State statutes or as being unconstitutional. State statutes may not provide adequate authority for the CA to take effective enforcement action. SUO may contain language that is open to interpretation.

- In general, the CA's SUO applies only to IUs within its jurisdictional boundaries. However, a few States provide authority to public utilities to regulate all users throughout their service area. In such cases, the SUO may apply to all users of the POTW.

- CA may not have an agreement with all contributing municipalities or it may have an inadequate existing agreement which cannot be modified without the mutual consent of both parties.

- Interjurisdictional agreements may not be specific enough to ensure that the contributing municipality takes adequate enforcement when required.

- Interjurisdictional agreements may not provide the CA with authority to take direct action against a violating IU where the contributing jurisdiction has failed to do so. Where this is the case, it may be that State law does not allow for such authority. Further, this authority generally does not exist in interstate situations unless special legislation has been enacted.

C.   IU Characterization [403.8(f)(2)(i)&(ii)]

Note: This section is to be used to evaluate how the CA identifies and characterizes their IUs. The auditor should determine whether the CA has any problems identifying IUs, differentiating between SIUs and non-SIUs, and further, differentiating between CIUs and significant non-CIUs. Any problems should be recorded.

C.1.   How does the CA define SIU? (Is it the same in contributing jurisdictions?)

PURPOSE: In accordance with 40 CFR 403.8(f)(1)(iii), the CA is required to issue individual control mechanisms to all its SIUs as defined under 40 CFR 403.3(t). The CA must apply equivalent or more encompassing criteria to determine which IUs must obtain individual control mechanisms. The auditor should determine what definition the CA is applying to their SIUs and whether or not the definition is equivalent or more stringent than the Federal definition.

FACTORS TO CONSIDER:

- EPA adopted its definition of SIU on July 24, 1990. Many CAs are still applying an earlier definition found in EPA's model SUO. The CA's definition of SIU may contain the 50,000 gallons per day (gpd) flow criteria of the earlier EPA definition rather than the more inclusive 25,000 gpd flow criteria of the current definition.

- Frequently, the CA's definition of SIU includes any IU which has in its discharge toxic pollutants as defined under Section 307 of the Clean Water Act (CWA). This provision is not a substitute for specifying all IUs subject to National categorical pretreatment standards since not all categorical standards regulate toxic pollutants.

- The CA's definition of SIU may include any IU whose discharge constitutes 5 percent or more of the POTW's flow. However, this is not necessarily as inclusive as the regulatory criteria of any IU whose process wastewater constitutes 5 percent or more of the POTW's hydraulic or organic capacity.

- EPA's definition includes any IU which the CA determines has a reasonable potential to adversely affect the POTW or cause a violation of applicable standards or requirements. If the CA's definition contains a criteria which includes any IU which the Director has found to have an impact on the POTW, this criteria is not as inclusive as the Federal definition.

C.2.   How are SIUs identified and categorized (including those in contributing jurisdictions)? Discuss any problems.

PURPOSE: Proper identification and categorization of SIUs is essential to the application of appropriate pretreatment standards and requirements. The CA should have procedures for determining which IUs are significant, which of those are subject to categorical standards, and the appropriate category/subcategory to apply to each SIU.

FACTORS TO CONSIDER:

- The CA may be including IUs in its program based on determinations made prior to the adoption of a Federal definition for SIU. They should have reevaluated their IWS to determine if there are any existing SIUs who were not previously included in the program.

- The CA should have procedures to determine which SIUs are subject to categorical pretreatment standards and the applicable category(ies) for those which are. These procedures should include permit application/BMR review, onsite inspection, and comparison to categorical pretreatment standard regulations, guidance documents, and/or development documents.

C.3.a.   How and when does the CA update its IWS to identify new IUs (including those in contributing jursidictions)?

PURPOSE: The CA needs to be able to identify new IUs that move into the CA's service area. The CA is also required to update its IWS at least annually [40 CFR 403.12(i)]. Generally, a system for continuous update is the most effective.

FACTORS TO CONSIDER:

- The CA should be relying on numerous sources to identify new users. Reliance on one municipal department (e.g., building permits) to identify these users is likely to result in the CA overlooking some new IUs such as those located in existing facilities. At a minimum, it is recommended that the CA verify its IWS by comparing it to another source such as water billings records at least annually.

- CAs also frequently experience difficulty in identifying new users locating in contributing municipalities. If the CA relies on that municipality to notify it of new IUs, the CA should have procedures to verify this information at least annually.

C.3.b.   How and when does the CA identify changes in wastewater discharges at existing IUs (including contributing jurisdictions)?

PURPOSE: Identification of changed discharges from existing IUs is part of the CA's IWS update and is required to be done at least annually. Again, continuous update procedures are the most effective.

FACTORS TO CONSIDER:

- Existing IUs are required to notify the CA of any changes in their facilities or process which might result in the discharge of new or substantially increased pollutants. The CA should ensure that all IUs (including those in contributing jurisdictions) are aware of this requirement.

- The CA should have procedures to review existing IUs not currently included in the program. The CA should verify current conditions at those facilities having the greatest potential for changes which may result in a change of status. Water billing records provide data for IUs that suddenly change volume of water used which is a strong indicator of a change in processes being operated.

- The CA may only update its IWS for IUs in its program when their control mechanisms are due for reissuance. If this is the case, update for existing IUs may not be occurring annually and/or may be reliant upon permit application data rather than onsite inspection data.

- If contributing municipalities are conducting their own inspections, the CA should have oversight procedures to ensure that those inspections are adequate to identify any facility changes which might result in the discharge of new or increased pollutants.

C.4.   How many IUs are currently identified by the CA in each of the following groups?

C.4.a.   SIUs (as defined by the CA); CIUs; Noncategorical SIUs

PURPOSE: The CA is required to issue control mechanisms to all SIUs in its service area. It is also required to identify those SIUs which are subject to categorical pretreatment standards and their applicable category/subcategory.

FACTORS TO CONSIDER:

- The CA generally should have the numbers of CIUs and noncategorical SIUs readily available. However, in the case of a very large program, the CA may need to obtain data from its computer system to provide these numbers. Enough time should be allowed to ensure that the auditor obtains these data during the course of the audit.

- If the CA issues control mechanisms to non-SIUs, it should still be able to identify which IUs are SIUs to ensure that all applicable pretreatment standards and requirements are being applied.

C.4.b.   Other regulated noncategorical IUs (specify)

PURPOSE: The CA is not required to regulate non-SIUs; however, many choose to regulate some or all of these IUs.

FACTORS TO CONSIDER:

- Often, the CA regulates non-SIUs strictly for revenue purposes. If this is the case, the auditor should determine what pollutants are monitored and/or what other requirements are applied to these users.

- Some CAs regulate specific categories of non-SIUs such as photo finishers, dry cleaners, and transportation centers. In such cases, the auditor should ask why the CA decided to regulate those particular IUs and how.

C.4.c.   Total

PURPOSE: Although the CA is only required to issue individual control mechanisms to its SIUs, many also control mechanisms to non-SIUs. Non-SIU control mechanisms are not required to contain the elements specified under 40 CFR 403.8(f)(1)(iii), however, it is recommended that they do.

FACTORS TO CONSIDER:

- The CA may issue control mechanisms to specific categories of industries/commercial facilities because of problems experienced from such facilities (e.g., shipping depots — oil and grease). Although these control mechanisms are not required to be as comprehensive as those for SIUs, they should contain standards and/or requirements that make sense (e.g., clean traps bi-weekly).

D.   Control Mechanism Evaluation [403.8(f)(1)(iii)]

Note: This section is designed to help the auditor evaluate the CA's issuance and reissuance of control mechanisms. The auditor should determine whether the control mechanisms used are issued/reissued in a timely manner, whether the CA is controlling all sources and whether the control mechanisms are adequate and effective. Any problems should be recorded.

D.1.a.   How many and what percent of the total SIUs are **not** covered by an existing, unexpired permit or other individual control mechanism?

PURPOSE: Under 40 CFR 403.8(f)(1)(iii), the CA is required to issue individual control mechanisms to all SIUs.

FACTORS TO CONSIDER:

- The auditor should consider how many SIUs the CA reported in question C.4 and whether the number of control mechanisms reported here matches. If it does not, the auditor should determine why the discrepancy exists.

- If the CA reports any expired and not reissued or reissued late control mechanisms, the auditor should determine the reason.

D.1.b. How many control mechanisms were not issued within 180 days of the expiration date of the previous control mechanism? If any, explain.

PURPOSE: A CA is considered to be in RNC if it fails to issue, reissue, or ratify control mechanisms for at least 90 percent of its SIUs within 180 days of the expiration date of the previous control mechanism. If the CA failed to issue or reissue all control mechanisms in the appropriate time frames, the auditor should record and explain.

FACTORS TO CONSIDER:

● The CA should have procedures which ensure timely reissuance of all control mechanisms. Control mechanisms should be issued/reissued on time; if any are not, the auditor should record this and determine the reason they were not issued/reissued on time.

● The CA may grant an administrative extension of the current control mechanism. However, only those extensions provided for due cause (e.g., awaiting the approval of revised local limits) are adequate to exempt the CA from being considered in RNC. A lack of adequate CA staff and resources or simply a failure to issue/reissue permits in a timely manner are not acceptable instances for granting an extension.

D.2.a. Do any UST, CERCLA, RCRA corrective action sites and/or other contaminated ground water sites discharge wastewater to the CA?

D.2.b. How are control mechanisms (specifically limits) developed for these facilities? Discuss.

PURPOSE: Any Underground Storage Tank (UST), Comprehensive Environmental Response, Compensation, and Liability Act (CERCLA), or RCRA correction action site which requests to discharge to the CA, even though the discharge may be of short duration, should be considered an SIU. As such, each facility must be issued a control mechanism containing all required elements.

FACTORS TO CONSIDER:

● The CA's local limits should cover the pollutants of concern to be discharged by these facilities. The CA should have prepared an IU-specific permit to address such pollutants. Unfortunately, in the case of CERCLA and RCRA facilities, there may not be much literature data available regarding secondary treatment inhibition from the applicable pollutants. The CA will have to rely upon whatever data is available and best professional judgment. Where there is doubt that these sources will ensure protection of the POTW, the CA should consider requiring/conducting a bench-scale study to obtain better data.

● The CA should be aware that receipt of hazardous wastes through a dedicated pipe or via truck into the headworks of the POTW will cause the CA to be considered a Treatment Storage and Disposal Facility (TSDF) under the RCRA permit-by-rule. The CA is then subject to applicable limits.

D.3.a. Does the CA accept any waste by truck, rail, or dedicated pipe?

D.3.b. Is any of the waste hazardous as defined by RCRA? If a. or b. above is yes, explain.

D.3.c. Describe the CA's program to control hauled wastes including a designated discharge point (e.g., number of points, control/security procedures). [403.5(b)(8)]

PURPOSE: According to 40 CFR 403.1(b)(1), the General Pretreatment Regulations apply to pollutants from all nondomestic sources subject to pretreatment standards (including prohibited discharge standards, local limits, and categorical pretreatment standards) which are indirectly discharged into or transported by truck or rail or otherwise introduced into a POTW or may contaminate sewage sludge.

Under 40 CFR 403.5(b)(8), the CA is required to prohibit the discharge of trucked or hauled pollutants except at a point which the CA designates. The auditor should determine what kind of program the CA has in place for handling hauled waste and whether any of the hauled waste qualifies as hazardous waste under RCRA. The auditor should determine if there is some kind of permitting system in place and, if so, how it is implemented.

FACTORS TO CONSIDER:

● The CA should be aware that any hazardous wastes received by the POTW from these sources are not covered by the domestic sewage exclusion provision of RCRA. Therefore, a POTW receiving such waste may be considered a TSDF and subject to "permit by rule."

● Where the CA states that it accepts only sanitary or sanitary and grease trap wastes, it should be able to demonstrate that it prohibits the discharge by the sources of any other wastes. Unless it has established (in its SUO or elsewhere in its code) that it is illegal for these sources to discharge industrial waste, the CA probably will not be able to enforce against such discharges. Even where the CA has prohibited the discharge of industrial wastes by these sources, it should have sufficient oversight procedures (e.g., manifest verification, manned discharge points, random sampling) to ensure compliance.

E.   Application of Pretreatment Standards and Requirements

Note: This section is set up to complement the file review's investigation of the CA's application of pretreatment standards. The auditor should collect information on the CA's use and understanding of pretreatment standards. He/she should try to determine whether the CA understands all issues relevant to the application of these standards. The auditor should also determine how the CA developed local limits. Any problems encountered by the CA in applying pretreatment standards or developing local limits should be recorded.

E.1.   What limits (categorical, local, other) does the CA apply to wastes that are hauled to the POTW (directly to the treatment plant or within the collection system, including contributing jurisdictions)? [403.1(b)(1)]

PURPOSE: According to 40 CFR 403.1(b)(1), the General Pretreatment Regulations apply to pollutants from all nondomestic sources subject to pretreatment standards (including prohibited discharge standards, local limits, and categorical pretreatment standards) which are indirectly discharged into or transported by truck or rail or otherwise introduced into a POTW. The auditor should determine that the appropriate limits are being applied to hauled waste.

FACTORS TO CONSIDER:

● Any nondomestic wastes from these sources must, at minimum, be subject to the CA's prohibited discharge standards and local limits.

● If the discharge contains, or is likely to contain, pollutants which may interfere with or pass through the POTW, but are not currently regulated by the CA (e.g., discharges from ground water cleanup sites), it is recommended the CA determine the allowable concentrations/loadings from such pollutants and apply them in a control mechanism issued for that discharge.

E.2.   How does the CA keep abreast of current regulations to ensure proper implementation of standards? [403.8(f)(2)(iii)]

PURPOSE: It is the CA's responsibility to keep up-to-date with all applicable regulations.

FACTORS TO CONSIDER:

● It is recommended the CA have procedures to review the Federal Register or some other publications or source that provides routine updates of the Federal Register.

● CAs frequently rely on information provided by EPA or the approved State to keep up-to-date with pretreatment and applicable RCRA revisions. This may not be adequate since such updates usually occur quarterly or less frequently.

E.3.   Local limits evaluation: [403.8(f)(4); 122.21(j)]

Note: The auditor should determine what methods were used to establish the CA's local limits, how these limits are being allocated, and whether there is any indication that the limits should be reevaluated (e.g., more pollutants covered).

E.3.a.   For what pollutants have local limits been set?

PURPOSE: The CA is required to evaluate the need for new or revised local limits. This must be a technical evaluation to determine the maximum allowable POTW headworks loading for each pollutant which will ensure protection of: the treatment plant unit processes from inhibition or upset; the receiving stream from violation of any water quality standards; compliance with any effluent or sludge use and disposal requirements in the NPDES permit; and worker health and safety.

FACTORS TO CONSIDER:

● Frequently, the local limits contained in the approved program submission were developed by a consultant and the CA may not know the methods used for their development. The CA may be able to call the consultant in or to obtain the appropriate documentation. Time should be allowed, where possible.

● A technical evaluation may have been conducted, but may have been reliant mainly upon literature values due to a lack of real data. In this case, the validity of the limits may be questionable.

E.3.b.   How were these pollutants decided upon?

PURPOSE: The CA should evaluate the need for local limits for any pollutant reasonably expected to occur in the POTW.

FACTORS TO CONSIDER:

● EPA generally recommends that limits be evaluated for 10 parameters that frequently occur in POTWs receiving industrial discharges. These parameters include: arsenic, cadmium, chromium, copper, cyanide, lead, mercury, nickel, silver, and zinc.

● The CA should also evaluate other pollutants reasonably expected to occur in the POTW. The CA may identify these pollutants in several ways, including running a priority pollutant scan on the POTW influent and identifying pollutants common to the types of industries located in its service area. The CA should be able to explain the rationale for selecting the pollutants for which local limits exist.

● The CA should consider volatile pollutants likely to be found in the collection system which may not be detectable in the POTW but are necessary to protect worker health and safety.

E.3.c.   What was the most prevalent/most stringent criteria for the limits?

PURPOSE: According to 40 CFR 122.21(j), the CA must reevaluate its local limits and submit the results with each application for a NPDES permit. Under 40 CFR 403.5(c)(1), the CA developing a pretreatment program must develop and enforce local limits to prevent interference and pass through. The CA must also continue to develop these limits as necessary.

FACTORS TO CONSIDER:

• The CA should develop local limits as part of its pretreatment program submission, when applying for a new NPDES permit, and when any substantial change in loadings occur at the plant (for instance when new IUs hook onto the system).

• The CA should develop local limits for any pollutant which is known to have caused interference or pass through or worker health and safety problems, or that has a reasonable potential to cause these problems.

E.3.d.   Which allocation method(s) were used?

PURPOSE: Federal regulations require local limits to be developed on a technical basis to prevent interference and pass through. The regulations do not specify the manner in which the CA must allocate these loadings.

FACTORS TO CONSIDER:

• The regulations require that the CA have the legal authority to establish local limits. They do not require local limits to be contained in the SUO. If the CA chooses to allocate its maximum allowable headworks loadings to all IUs on a uniform concentration basis, it is recommended that these end-of-pipe discharge limits be specified in the SUO.

• The CA may choose to allocate these loadings for specific pollutants among those IUs with the potential for those pollutants in their discharge. In this case, the limits are best placed in the IU control mechanisms.

• IU-specific limits are not required to be uniform for all IUs to which they apply. However, the CA should have a defensible rationale for its allocations. Where IU-specific limits are applied, the SUO should specify the maximum allowable headworks loadings and prohibit the discharge of those pollutants at a rate which, alone or in conjunction with other discharges, causes an exceedence of those loadings.

E.3.e.   Has the CA identified any pollutants of concern beyond those in its local limits? If yes, how has this been addressed?

PURPOSE: The CA is required to continue to develop local limits, as necessary.

FACTORS TO CONSIDER:

• If the CA has experienced a pass through or interference event caused by a pollutant not included in its list of local limits, the auditor should determine what follow-up has been done to regulate that pollutant in the future.

• Where a new SIU, particularly a ground water cleanup site has come on line and has the potential to discharge pollutants that could impact the POTW but for which it does not have a local limit, the auditor should determine the CA's approaches to recycling that pollutant.

• Pollutants which are not likely to be discharged by more than one or two IUs may be more appropriately regulated on an IU-specific basis. The CA should still have a technical rationale for these limits.

E.4.   What problems, if any, were encountered during local limits development and/or implementation?

PURPOSE: Frequently, the CA encounters difficulties in evaluating its local limits.

FACTORS TO CONSIDER:

• The State may not have developed water quality standards for the receiving stream. Data may not be available for a particular unit process used at the POTW. There may not be a point at which the CA can monitor to get a good profile of domestic contributions.

F.   Compliance Monitoring

Note: This section evaluates the CA's compliance monitoring of its IUs. The monitoring should be conducted at a frequency that will produce data that is indicative of the IUs discharge, and with care (proper sampling, analysis, and record keeping) to produce data that are supportive of enforcement actions. The auditor should record any problems that are found.

F.1.a.   How does the CA determine adequate IU monitoring (sampling, inspecting, and reporting) frequency? [403.8(f)(2)(ii)&(v)]

PURPOSE: Under 40 CFR 403.8(f)(2)(v), the CA is required to inspect and sample all SIUs at least once a year. According to 40 CFR 403.12(e), CIUs are required to submit reports twice per year and 40 CFR 403.12(h) requires the same reporting from noncategorical SIUs. Further, the CA's approved program or NPDES permit may specify required sampling, inspection, self-monitoring or reporting requirements. The auditor should determine that the CA knows how to establish proper monitoring frequencies and that they are aware of their minimum requirements.

FACTORS TO CONSIDER:

- At minimum, the CA's monitoring frequencies should be consistent with the regulatory requirements.

- The CA should also consider each IU's potential for impacting the POTW and determine monitoring frequencies accordingly.

F.1.b. Is the frequency established above more, less, or the same as required? Explain any difference.

PURPOSE: The CA should have a rationale for its monitoring frequencies. The auditor should investigate any discrepancies between required and actual monitoring frequencies.

FACTORS TO CONSIDER:

- Where monitoring frequencies are not consistent with required frequencies, the CA's rationale for its monitoring frequencies should demonstrate that the monitoring is adequate to determine ongoing compliance by all regulated IUs.

F.2.   In the past 12 months, how many, and what percentage of, SIUs were: [403.8(f)(2)(v)] (Define the 12-month period.)

F.2.a. Not sampled or not inspected at least once?

F.2.b. Not sampled at least once?

F.2.c. Not sampled at least once?

If any, explain. Indicate how percentage was determined (e.g., actual, estimated).

PURPOSE: Under 40 CFR 403.8(f)(2)(v), the CA is required to inspect and sample all SIUs at least once a year. According to 40 CFR 403.12(e), CIUs are required to submit reports twice per year and 40 CFR 403.12(h) requires the same reporting from noncategorical SIUs.

FACTORS TO CONSIDER:

- If the CA fails to inspect or sample at least 80 percent of its SIUs at least once during the past 12 months, the CA is considered to be in RNC.

Note: The auditor should be aware that CAs often establish their monitoring schedules around their reporting to the Approval Authority. Therefore, they may not have completed all the required monitoring in the last 12 months, but they will complete it before they are required to submit their annual performance report to the Approval Authority.

F.3.   Indicate the number and percent of SIUs that were identified as being in SNC (as defined by the POTW or EPA) with the following requirements from the CA's last pretreatment program performance report.

PURPOSE: The auditor must determine the number and percent of SIUs in SNC for noncompliance with applicable pretreatment standards and reporting requirements, self-monitoring requirements, and pretreatment compliance schedules for input into WENDB and to determine RNC.

F.4.   What does the CA's basic inspection include? (Process areas, pretreatment facilities, chemical and hazardous waste storage areas, chemical spill-prevention areas, hazardous waste handling procedures, sampling procedures, laboratory procedures, and monitoring records.) [403.8(f)(2)(v)&(vi)]

PURPOSE: The CA is required to inspect its IUs to determine compliance with all applicable standards and requirements. The auditor should determine that the CA is aware of all areas that need to be investigated during an inspection.

FACTORS TO CONSIDER:

- The regulations do not specify required components of an IU inspection. However, to adequately determine compliance with all applicable standards and requirements, the CA should inspect all areas indicated above.

- If the CA inspects facilities more frequently than once a year, only one inspection may need to be comprehensive. Other inspections may be limited to areas of specific concern.

F.5.   Who performs CA's compliance monitoring analysis?

PURPOSE: The CA is required to conduct its compliance monitoring and analysis in a manner that will provide admissible evidence in enforcement proceedings. The auditor should verify that the analyses are performed properly by reviewing reports and through discussions with the CA.

FACTORS TO CONSIDER:

- If the CA performs all of its own analyses or if it is performed by a contract lab, the CA should have documented that adequate procedures, equipment, and qualified personnel were used to analyze for all pollutants required to be monitored under its program.

F.6.   What QA/QC techniques does the CA use for sampling and analysis (e.g., splits, blanks, spikes), including verification of contract laboratory procedures and appropriate analytical methods? [403.8(f)(2)(vi)]

PURPOSE: The CA is required to conduct its compliance monitoring and analysis in a manner that will provide admissible evidence in enforcement proceedings. The auditor should review the QA/QC chain-of-custody procedures used by the CA to determine if they are adequate.

FACTORS TO CONSIDER:

- The analytical results for spikes, splits, and blanks should be included with the analytical data. The CA's in-house lab should have written QA/QC protocols. QA/QC protocols should be provided by the contract lab.

F.7. **Discuss any problems encountered in identification of sample location, collection, and analysis.**

PURPOSE: The CA must sample its IUs to determine compliance independent of data submitted by the IU. The auditor should investigate any problems the CA has determining the compliance status of its IUs.

FACTORS TO CONSIDER:

- Frequently, the CA requires CIUs to self-monitor after pretreatment, but conducts its own monitoring at end-of-pipe to avoid having to enter the facility. All sampling should be conducted at the same sampling point.

- Both the IU and the CA should be employing 40 CFR Part 136 procedures.

- Appropriate types of samples should be taken (i.e., composite vs. grab).

F.8. **Did any IUs notify the CA of hazardous waste discharge? [403.12(j)&(p)] If yes, summarize.**

PURPOSE: The CA is required to notify all its IUs of the requirement to notify the CA, EPA, and the State of any hazardous waste in their discharges which are subject to the requirement, as specified under 40 CFR 403.12(p). The auditor should verify that the CA notified its IUs of this requirement and determine whether any IUs contacted the CA.

FACTORS TO CONSIDER:

- Many CAs have notified their permitted IUs of this requirement, but are unaware that it applies to all IUs. Unless the CA permits all IUs, it is likely that many non-SIUs have not been notified. The IUs are still required to contact the POTW, State, and EPA even if the CA did not contact the IUs.

F.9.a. **How and when does the CA evaluate/reevaluate the need for a slug control plan? [403.8(f)(2)(v)]**

PURPOSE: The CA is required to evaluate all IUs at least once every 2 years to determine the need to develop or revise a slug discharge control plan. The auditor should determine if the CA evaluated its SIUs for the need to develop a slug control plan.

FACTORS TO CONSIDER:

- Many CAs require through their SUO that all IUs submit an accidental spill prevention plan. Although this may be adequate for non-SIUs, it is not adequate for any SIU with the potential to discharge an intentional slug load (e.g., nonroutine batch discharge).

- The requirement for a slug control plan for discharges other than accidental spills may be contained in the IU permits.

- Where the CA has conducted initial inspections for slug control plans, the auditor should determine if the CA has done a follow-up inspection within 2 years to determine the need for any revisions, as required.

F.9.b. **How many SIUs were not evaluated for the need to develop slug discharge control plans in the last 2 years?**

PURPOSE: The CA is required to evaluate each SIU for the need to develop or revise a slug discharge control plan at least once every 2 years. The auditor should combine this information with the information collected in the previous question.

G. **Enforcement**

Note: This section is designed to evaluate the CA's enforcement program. The auditor should evaluate the adequacy and effectiveness of the CA's enforcement actions by examining its definition of SNC, implementation of the SNC definition, implementation of its approved ERP, problems with the POTW, and use of compliance schedules. Any problems found by the auditor should be recorded.

G.1. **What is the CA's definition of SNC? [403.8(f)(2)(vii)]**

PURPOSE: EPA has defined the term "significant noncompliance" in 40 CFR 403.8(f)(2)(vii) and requires the CA to publish all SIUs in SNC at least once per year. The auditor should determine what the CA's definition for SNC is and whether it matches the Federal definition and subsequent guidance.

FACTORS TO CONSIDER:

- EPA's current definition of SNC replaces the earlier definition of "significant violation" which formerly provided the criteria for publication.

- If the CA's NPDES permit requires adoption of DSS regulatory revision requirements, the CA must apply EPA's definition of SNC or more stringent criteria when determining which IUs must be published.

G.2. **ERP implementation: [403.8(f)(5)]**

G.2.a. **Status**

G.2.b. **Problems with implementation**

G.2.c. Is the ERP effective and does it lead to compliance in a timely manner? Provide examples if any are available.

PURPOSE: The CA is required to develop an ERP. Once approved by the Approval Authority, the ERP constitutes a modification of the approved program. As such, the CA is obligated to conduct its enforcement activities in a manner consistent with the procedures established in the ERP. The auditor should determine whether or not the CA is following their approved ERP.

Note: If the CA does not yet have an approved ERP, the auditor should use this section to evaluate and discuss the enforcement actions the CA is taking.

FACTORS TO CONSIDER:

- If the ERP has not yet been approved, the CA has no obligation to conduct its enforcement activities in accordance with the ERP procedures.

- In some cases, the ERP may not work or may be in conflict with the CA's legal authority. This does not exempt the CA from implementing its ERP. However, where such problems are identified, the CA should be required to submit a request for modification of its ERP to correct the problem.

- Even when the CA successfully implements its ERP as approved, it may run into problems. For instance, although circular enforcement may not be apparent in the ERP, certain scenarios may result in such a situation. In any such instances, the ERP should be modified.

- The ERP should result in a return to compliance by the IU within 90 days or within the time specified in a compliance schedule or order.

G.3.a. Does the CA use compliance schedules? [403.8(f)(1)(iv)(A)]

G.3.b. If yes, are they appropriate? Provide examples.

PURPOSE: The CA should establish compliance schedules for SIUs in accordance with its approved ERP. The auditor should determine if the CA uses compliance schedules; if the CA does, the auditor should determine if they are effective.

FACTORS TO CONSIDER:

- Compliance schedules should identify specific actions the SIUs are to take and establish specific dates by which these actions are to be completed.

- Where a CIU is on a compliance schedule for achieving compliance with a categorical deadline that has already passed or will pass prior to the schedule's final compliance deadline, the compliance schedule/enforcement order should clearly state the CIU is subject to enforcement for failure to comply with a Federal deadline even though the user is in compliance with the terms of the schedule.

G.4. Did the CA publish all SIUs in SNC in the largest daily newspaper in the previous year? [403.8(f)(2)(vii)] If yes, attach a copy. If no, explain.

PURPOSE: The CA is required to publish (on an annual basis) all SIUs that had been in SNC during the reporting year. The auditor should verify that the CA did publish those IUs that were in SNC during the reporting year.

FACTORS TO CONSIDER:

- Where the CA's NPDES permit requires adoption of DSS regulatory revision requirements, publication of IUs in SNC must be based on EPA's definition of SNC or on more stringent criteria. Publication is required to appear in the largest daily newspaper in the municipality.

G.5. How many SIUs are in SNC with self-monitoring requirements and were not inspected and/or sampled (in the four most recent full quarters)?

PURPOSE: Failure by the CA to inspect and/or sample any SIU which is in SNC with self-monitoring requirements must be reported in WENDB. The auditor should determine the number of SIUs in SNC with self-monitoring that were not inspected and/or sampled and record it for WENDB. Indicate if this is an actual or estimated number.

FACTORS TO CONSIDER:

- SIUs which are not complying with self-monitoring requirements have the potential to have serious discharge violations. Therefore, failure by the CA to inspect or sample these IUs may result in allowing serious violations to continue without enforcement.

G.6.a. Has the CA experienced any problems since the last inspection (interference, pass through, collection system problems, illicit dumping of hauled wastes, or worker health and safety problems) caused by industrial discharges?

G.6.b. If yes, describe the CA's enforcement action against the IUs causing or contributing to problems.

PURPOSE: The CA must investigate and take enforcement actions against IUs causing or contributing to pass through or interference. The auditor should be aware of any effluent violations at the POTW based on Discharge Monitoring Report (DMR) data that may be due to discharges from IUs. The auditor should investigate the CA's response to any problems caused by IU discharges.

FACTORS TO CONSIDER:

- Any indications of pass through or interference should result in immediate response by the CA to determine the source(s) of the violation and take appropriate enforcement actions. Where the source(s) of the violation could not be determined, the CA should have detailed documentation of the event and the reasons why the source could not be determined.

H. Data Management/Public Participation

Note: This section is designed to evaluate the adequacy and effectiveness of the CA's data management and public participation procedures. The auditor should examine the CA's procedures for dealing with confidential information, public inquiry, public notice, and confidentiality issues impacting the program. Any problems identified should be recorded.

H.1. How is confidential information handled by the CA? [403.14]

PURPOSE: Where the CA allows for confidentiality for information determined to be proprietary, it should have procedures to guarantee that confidentiality while ensuring that IU effluent data remain available to the public and that all IU data obtained through the course of program implementation remain available to EPA and/or the approved State. The auditor should determine if the CA has procedures to handle confidential information; if the CA does, the auditor should evaluate whether they are adequate.

FACTORS TO CONSIDER:

- It is recommended that confidential information be maintained in a locked file to which only one or a few people have access. All personnel with access to confidential information should be fully conversant with the CA's confidentiality procedures.

H.2. How are requests by the public to review files handled?

PURPOSE: All IU effluent data must be made available to the public. The auditor should determine the level of interest in the program and whether the CA has a mechanism in place to handle public inquiry.

FACTORS TO CONSIDER:

- Effluent data should be maintained separately or procedures should be established to ensure that the public has ready access to these data.

H.3. Describe whether the CA's data management system is effective in supporting pretreatment implementation and enforcement activities.

PURPOSE: A well organized data management system is essential to maintaining the IWS, issuance of control mechanisms, efficient compliance tracking, and timely and effective enforcement. The auditor should evaluate the CA's data management system.

FACTORS TO CONSIDER:

- An effective management system can range from a well organized filing system to a sophisticated computer data system.
- All data on each IU should be readily accessible in the IU's file.
- The data should be organized in a reasonable manner. That is, all control mechanism components should be kept together as should all CA sampling data, etc. Organizing files by subject matter and then chronologically within the subject is recommended.
- All inspections, meetings, and telephone calls should be clearly and comprehensively documented so as to provide evidence in enforcement actions.
- All chain-of-custody and QA/QC data should be complete.

H.4. How does the CA ensure public participation during revisions to the SUO and/or local limits? [403.5(c)(3)]

PURPOSE: The auditor should determine what mechanism the CA has for ensuring adequate public comment during revisions to the program.

FACTORS TO CONSIDER:

- The CA should have procedures for public notice which include the opportunity for public comment. Frequently, these procedures are specified in the municipality's code or State code.

H.5. Explain any public or community issues impacting the CA's pretreatment program.

PURPOSE: Frequently, public/community issues affect the implementation of the CA's pretreatment program. Such issues which impede effective implementation and enforcement of the local program should be discussed.

FACTORS TO CONSIDER:

- Enforcement may be difficult where a violating IU is one of the community's major sources of revenues and employment.
- CA's practicing public outreach often find it facilitates program implementation.

APPENDIX A

H.6.   How long are records maintained? [403.12(o)]

PURPOSE: SIUs are required to maintain and retain data obtained in response to program requirements for a period of at least 3 years and/or throughout the course of any ongoing litigation related to the IU. The auditor should determine that SIUs maintain files for the appropriate length of time.

FACTORS TO CONSIDER:

- The CA should review SIU records during the course of its annual comprehensive inspection. Any problems with IU record maintenance should be noted in the inspection report and should result in an enforcement response.

I.   Resources [403.8(f)(3)]

Note: This section is designed to determine whether the CA has dedicated enough resources (i.e., personnel, equipment, and funding) to implement each program activity effectively. The auditor should bear in mind that while resources for present activities may be adequate, if the CA's activities themselves are not adequate (e.g., not regulating all SIUs), the resources may be inadequate to cover the additional work necessary to correctly implement the program. The auditor should identify any existing resource problems as well as any anticipated problems.

I.1.   Estimate the number of personnel (in FTEs) available for implementing the program. [Consider: legal assistance, permitting, IU inspections, sample collection, sample analysis, data analysis, review and response, enforcement, and administration (including record keeping and data management).

PURPOSE: The CA is obligated to have at least the number of Full-Time Equivalents (FTEs) specified in the approved program available for program implementation activities. It should have increased personnel if required to adequately implement the program. The auditor should determine the number of FTEs devoted to the program and whether a lack of resources contributes to ineffective implementation.

FACTORS TO CONSIDER:

- Frequently, the CA uses the same personnel for collection system maintenance, POTW sampling, and pretreatment sampling. With this, and with all program areas, the FTEs should reflect the number actually, and consistently, available to the program.

- If the CA uses a contract lab for sampling and/or analysis, the FTEs should reflect the approximate number the contract budget would cover.

I.2.   Does the CA have adequate access to monitoring equipment? (Consider: sampling, flow measurement, safety, transportation, and analytical equipment.) If no, explain.

PURPOSE: The CA is obligated to have at least the equipment specified in the approved program available for program implementation activities. It should have additional equipment if required to adequately implement the program. The auditor should inquire about whether or not the CA has certain basic equipment necessary to run their program.

FACTORS TO CONSIDER:

- Although not specifically required, the CA should have adequate safety equipment, including equipment for safely entering a manhole, where necessary.

- If the CA uses a contract lab, the contract budget should provide for an adequate number of analyses, including additional analyses for demand sampling, that the CA is expected to require.

I.3.a.   Estimate the annual operating budget for the CA's program.

I.3.b.   Is funding expected to: stay the same, increase, decrease (note time frame; e.g., following year, next 3 years, etc.)? Discuss any changes in funding.

PURPOSE: The CA is obligated to have at least the funding specified in the approved program available for program implementation activities. It should have increased funding if required to adequately implement the program. The auditor should inquire about the annual operating budget necessary to run the program.

FACTORS TO CONSIDER:

- Frequently, funding for the pretreatment program comes from the municipality's/Department of Public Works' general fund. A review of the CA's program funding over the past several years may be necessary to determine funding adequacy. The auditor should also inquire into any anticipated funding problems. In addition, if the audit has found that the scope of any program activity is inadequate, then funding will most likely need to be increased to bring the program into compliance.

I.4.   Discuss any problems in program implementation which appear to be related to inadequate resources.

PURPOSE: The CA is obligated to have at least the funding specified in the approved program available for program implementation activities. It should have increased funding if required to adequately implement the program. The auditor should investigate whether the funding devoted to the program seems adequate, and if there are any problems related to funding, the auditor should note it in the report.

FACTORS TO CONSIDER:

- See question I.1-3 above.

I.5.a.   How does the CA ensure personnel are qualified and up-to-date with current program requirements?

PURPOSE: In order to adequately implement the pretreatment program, all program staff need to be qualified for the positions they hold and trained to perform their jobs in a manner consistent with pretreatment program requirements. The auditor should determine whether staff seem adequately trained and note any problems in the report.

FACTORS TO CONSIDER:

- Although the CA's pretreatment coordinator may be qualified and up-to-date with program requirements, it is not uncommon to find that field and lab personnel are not.

I.5.b.   Does the CA have adequate reference material to implementing its program?

PURPOSE: In order to determine correct categorization of SIUs, the CA should have ready access to the General Pretreatment Regulations, categorical pretreatment standard regulations, and EPA's categorical pretreatment standards guidance documents. The auditor should determine whether the CA seems to have adequate access to resource material or whether resource material has an impact on the implementation of the program. The auditor should review the CA's reference material to determine whether any additional materials may be needed. The auditor should plan to provide any missing materials.

FACTORS TO CONSIDER:

- The Region/State may know that particular documents have been provided to the CA. However, some mailings never quite make it to the pretreatment staff but end up in the Public Works Department, etc. Also, when staff members leave for another position, these documents frequently leave with them.

- It is not uncommon that documents were received and shelved, but that the pretreatment staff (including inspectors) may not have reviewed them. All pretreatment personnel should be familiar with guidance material.

J.   Environmental Effectiveness/Pollution Prevention

Note: This section is designed to assist the auditor in determining whether the CA's program has evaluated and documented any environmental benefits to date. Although there are no regulatory requirements directly related to achieving environmental benefits, it is EPA's stated goal for all environmental regulatory programs. The auditor should make every effort to determine if sufficient data are being collected, analyzed, and summarized to demonstrate trends (whether positive or negative) in the years since the CA's pretreatment program implementation, particularly in the years since the last audit. All findings should be documented as thoroughly as possible.

J.1.a.   How many times were the following monitored by the CA during the past year? Metals, priority pollutants, biomonitoring, TCLP, EP toxicity, other.

J.1.b.   Is this frequency less than, equal to, or more than that required by the NPDES permit? Explain any differences.

PURPOSE: The primary goal of the pretreatment program is to improve environmental quality. Environmental monitoring is essential to determine the program's effectiveness and the accomplishment of this goal. The auditor should determine whether the CA has a monitoring program in place that will assist the CA in tracking any progress or lack of progress the program is making in enhancing environmental effectiveness.

FACTORS TO CONSIDER:

- It is recommended that the CA perform monitoring of its treatment plant(s) in order to track the environmental effectiveness of the program's implementation. The frequency should be such that enough data are collected to recognize trends of increasing or decreasing loadings in the influent, effluent and sludge.

J.2.a.   Has the CA evaluated historical and current data to determine the effectiveness of the pretreatment controls on: improvements in POTW operations, loadings to and from the POTW, NPDES permit compliance, and sludge quality?

J.2.b.   Has the CA documented these findings?

J.2.c.   If they have been documented, what form does the documentation take? Explain. (Attach a copy of the documentation, if appropriate.)

PURPOSE: A successful pretreatment program is expected to result in improved POTW operations and NPDES compliance as well as in reduced pollutant loadings. The auditor should review any data the CA has available on environmental effectiveness and record any findings. If the CA has no data, the auditor should recommend the CA start collecting data.

FACTORS TO CONSIDER:

- Environmental monitoring should demonstrate a trend of decreasing concentrations of pollutants coming to the POTW and ending up in the receiving stream and sludge.

- The cost of operating and maintaining the POTW (minus cost of living increases) should decrease due to fewer system upsets and inhibitions.

- As sludge quality improves, less expensive disposal operations may become available.
- NDPES permit compliance should improve.

J.3.   **If the CA has historical data compiled concerning influent, effluent, and sludge sampling for the POTW, what trends have been seen? (Increases in pollutant loadings over the years? Decreases? No change?) Discuss on pollutant-by-pollutant basis.**

PURPOSE: It is generally anticipated that a successfully implemented local pretreatment program will result in a decrease of pollutant loadings to the POTW and a resulting decrease in loadings to the receiving waters. Where this has not happened, the auditor should attempt to determine the causative factors. These factors should be well documented.

FACTORS TO CONSIDER:

- If all IUs were in compliance with applicable pretreatment standards prior to the CA obtaining POTW monitoring data, it is likely that no change will be seen.
- If the CA's service area has recently experienced industrial growth or a change in the character of its industries, the data may show an increase in pollutant loadings even though effective program implementation is taking place.

J.4.   **Has the CA investigated the sources contributing to current pollutant loadings to the POTW (i.e., the relative contributions of toxics from industrial, commercial, and domestic sources)? If yes, what was found?**

PURPOSE: In order to effectively control toxics discharged to the POTW, the CA needs to determine the types and amount of toxics received from the above sources. The auditor should determine what the CA is doing to evaluate and keep track of pollutant loadings to the treatment plant, specifically what kind of monitoring program the CA has in place for tracking contributions to the collection system. If no system exists, the auditor should recommend the CA start one.

FACTORS TO CONSIDER:

- Along with sampling plant influent, effluent, and sludge, it is recommended that the CA monitor points within the collection system to better characterize the contributions of toxics. This will help to determine program effectiveness and to assist in developing more appropriate local limits.

J.5.a.   **Has the CA attempted to implement any kind of public education program?**

J.5.b.   **Are there any plans to initiate such a program to educate users about pollution prevention? Explain.**

PURPOSE: Practicing pollution prevention by changing the types of products used can be a painless way for the public to make a contribution to the environment. Industries often realize significant cost savings when they adopt pollution prevention measures. Adoption of pollution prevention practices on all fronts will almost certainly result in a reduced need for enforcement as well as a decreased loading of pollutants at the POTW. However, pollution prevention has not been a raging overnight success due to the lack of public awareness of the possibilities in this area. The CA is in an ideal position to foster pollution prevention and improve its image with both its IUs and the general public. Where the CA has no pollution prevention awareness program in place, the auditor should recommend one be adopted.

FACTORS TO CONSIDER:

- CAs often see pollution prevention awareness as yet another task they are being asked to take on in an already too full work load. Frequently, they are unaware of the benefits to be reaped for both the POTW and their pretreatment program, including an eventual reduction in their work load.
- Making their IUs aware of pollution prevention need not really impact the CA's work load. They might consider bringing State pollution prevention literature out with them on IU inspections. Specific questions can then be handled by State personnel.

J.6.   **What efforts have been taken to incorporate pollution prevention into the CA's pretreatment program (e.g., waste minimization at IUs, household hazardous waste programs)?**

PURPOSE: Pollution prevention is of great importance in implementing a comprehensive pretreatment program. In order to further the CA's attainment of program goals, pollution prevention initiatives and ideas should be discussed with CA personnel.

FACTORS TO CONSIDER:

- It is hoped that, at a minimum, the CA will be talking to its IUs about pollution prevention and the benefits of pollution prevention/waste minimization to the IU.
- The auditor should be aware that the States have grants to conduct pollution prevention studies at industrial facilities. He/she should inquire as to the CA's awareness of this program. Generally, the States are encouraging CAs to recommend likely candidates for a study. The study is free of charge to the industry and carries no obligation.

J.7.   **Does the CA have any documentation concerning successful pollution prevention programs being implemented by IUs (e.g., case studies, sampling data demonstrating pollutant reductions)? Explain.**

PURPOSE: The more documentation we can provide to other CAs regarding successful IU pollution prevention programs, the more willing they will be to bring the pollution prevention message to their own IUs. The auditor should obtain all available documentation. He/she should also consider contacting the IU to ask whether the IU would be willing to be named in case studies and/or to respond to questions from interested parties.

FACTORS TO CONSIDER:

- Sometimes IUs have made recent modifications to incorporate pollution prevention measures of which the CA is unaware. In the course of the IU site visit, the auditor should ask the IU whether this has been done or is being considered.

K.   Additional Evaluations/Information

FACTORS TO CONSIDER:

- The auditor should record any activities being taken by the CA, EPA, the State, environmental organizations, or the public at large that have, or may have in the future, <u>any</u> bearing on the CA's pretreatment program. Included in these considerations should be any new initiatives (e.g., regulatory, hospital waste, river, bay, geographic targeted, result oriented initiatives).

SECTION III: FINDINGS

The intent of Section III is to provide a <u>brief</u> summary of the concerns and deficiencies identified (findings) throughout the audit in each program area. It also provides the opportunity to identify inconsistencies in information collected. For instance, information obtained through the interview process is sometimes in disagreement with information obtained during the file review. For this reason, it is strongly recommended that the auditor(s) complete Section III prior to the audit's closing conference in order to raise, and hopefully resolve, such issues at that time.

To facilitate completion of this section, elements of each program area are listed for consideration. Citations to all pertinent checklist questions are provided for each element. The regulatory citations are also provided where there are specific <u>requirements</u> for that element. The auditor should be aware that not all questions on the checklist reflect regulatory requirements. Some of the questions are included to allow the auditor to better evaluate program effectiveness. This fact should be taken into consideration when developing the subsequent report which specifies the required versus recommended actions to be taken by the CA.

When documenting findings, the auditor should take care to clearly distinguish between findings of deficiencies, violations, and program effectiveness issues. He/she should also specify whether follow-up actions are required or recommended or whether program modification is needed. Thoroughness in completing Section III of the checklist will facilitate preparation of a clear and accurate final report.

Section III should provide the framework for the report to which the checklist may be attached. Since the checklist constitutes the auditor's field documentation of findings, it should contain only the audit's factual findings

APPENDIX B

INDUSTRIAL WASTEWATER DISCHARGE PERMIT APPLICATION FORM

Note: Please read all attached instructions prior to completing this application.

SECTION A — GENERAL INFORMATION

1. Facility Name: _____

   a. Operator Name: _____

   b. Is the operator identified in 1.a. the owner of the facility? Yes [ ] No [ ]

   If no, provide the name and address of the operator and submit a copy of the contract and/or other documents indicating the operator's scope of responsibility for the facility.

   _____

2. Facility Address:

   Street: _____

   City: _____ State: _____ Zip: _____

3. Business Mailing Address:

   Street or P.O. Box: _____

   City: _____ State: _____ Zip: _____

4. Designated signatory authority of the facility:

   [Attach similar information for each authorized representative]

   Name: _____

   Title: _____

   Address: _____

   City: _____ State: _____ Zip: _____

   Phone #: (   ) _____

5. Designated facility contact:

   Name: _____

   Title: _____ Phone #: (   ) _____

SECTION B — BUSINESS ACTIVITY

1. If your facility employs or will be employing processes in any of the industrial categories or business activities listed below (regardless of whether they generate wastewater, waste sludge, or hazardous wastes), place a check beside the category of business activity (check all that apply).

Industrial Categories*

[ ] Aluminum Forming

[ ] Asbestos Manufacturing

[ ] Battery Manufacturing

[ ] Can Making

[ ] Carbon Black

[ ] Coal Mining

[ ] Coil Coating

[ ] Copper Forming

[ ] Electric and Electronic Components Manufacturing

[ ] Electroplating

[ ] Feedlots

[ ] Fertilizer Manufacturing

[ ] Foundries (Metal Molding and Casting)

[ ] Glass Manufacturing

[ ] Grain Mills

[ ] Inorganic Chemicals

[ ] Iron and Steel

[ ] Leather Tanning and Finishing

[ ] Metal Finishing

[ ] Nonferrous Metals Forming

[ ] Nonferrous Metals Manufacturing

[ ] Organic Chemicals Manufacturing

[ ] Paint and Ink Formulating

[ ] Paving and Roof Manufacturing

[ ] Pesticides Manufacturing

[ ] Petroleum Refining

[ ] Pharmaceutical

[ ] Plastics and Synthetic Materials Manufacturing

[ ] Plastics Processing Manufacturing

[ ] Porcelain Enamel

[ ] Pulp, Paper, and Fiberboard Manufacturing

[ ] Rubber

[ ] Soap and Detergent Manufacturing

[ ] Steam Electric

[ ] Sugar Processing

[ ] Textile Mills

[ ] Timber Products

* A facility with processes inclusive in these business areas may be covered by Environmental Agency's (EPA) categorical pretreatment standards. These facilities are termed "categorical users."

2. Give a brief description of all operations at this facility including primary products or services (attach additional sheets if necessary):

_____

_____

3. Indicate applicable Standard Industrial Classification (SIC)(or North American Industry Classification System (NAICS) code) for all processes (If more than one applies, list in descending order of importance.):

a. _____ b. _____ c. _____

d. _____ e. _____ f. _____

4. PRODUCT VOLUME:

PRODUCT (Brand name) (levels with others) (and no u.1)	PAST CALENDAR YEAR Amounts per day (Daily Units)		ESTIMATE THIS CALENDAR YEAR Amounts per day (Daily Units)	
	Average	Maximum	Average	Maximum
_____	_____	_____	_____	_____
_____	_____	_____	_____	_____

SECTION C — WATER SUPPLY

1. Water Sources: (Check as many as are applicable.)

[ ] Private Well

[ ] Surface Water

[ ] Municipal Water Utility (Specify City): _____

[ ] Other (Specify): _____

2. Name on the water bill: _____

Name: _____

Street: _____

City: _____ State: _____ Zip: _____

3. Water service account number: _____

4. List average water service usage on premises: [New facilities may estimate]

Type	Average Water Usage (GPD)	Indicate Estimated (E) or Measured (M)
a. Contact cooling water	_____	_____
b. Non-contact cooling water	_____	_____
c. Boiler feed	_____	_____
d. Process	_____	_____
e. Sanitary	_____	_____
f. Air pollution control	_____	_____
g. Contained in product	_____	_____
h. Plant and equipment washdown	_____	_____
i. Irrigation and lawn watering	_____	_____
j. Other	_____	_____
k. Total of A-J	_____	_____

APPENDIX B

SECTION D — SEWER INFORMATION

1. a. <u>For an existing business</u>:

   Is the building presently connected to the public sanitary sewer system?

   [ ] Yes: Sanitary sewer account number _____

   [ ] No: Have you applied for a sanitary sewer hookup?   [ ] Yes   [ ] No

   b. <u>For a new business</u>:

   (i). Will you be occupying an existing vacant building (such as in an industrial park)?

   [ ] Yes   [ ] No

   (ii). Have you applied for a building permit if a new facility will be constructed?

   [ ] Yes   [ ] No

   (iii). Will you be connected to the public sanitary sewer system?

   [ ] Yes   [ ] No

2. List size, descriptive location, and flow of each facility sewer which connects to the City's sewer system. (If more than three, attach additional information on another sheet.)

Sewer Size	Descriptive Location of Sewer Connection or Discharge Point	Average Flow (GPD)
_____	_____	_____
	_____	_____
_____	_____	_____
	_____	_____
_____	_____	_____
	_____	_____

SECTION E — WASTEWATER DISCHARGE INFORMATION

1. Does (or will) this facility discharge any wastewater other than from restrooms to the City sewer?

   [ ] Yes   If the answer to this question is "yes," complete the remainder of the application.

   [ ] No   If the answer to this question is "no," skip to Section I.

2. Provide the following information on wastewater flow rate: [New facilities may estimate]

   a. Hours/Day Discharged (e.g., 8 hours/day):

   M _____ T _____ W _____ TH _____ F _____ SAT _____ SUN _____

   b. Hours of Discharge (e.g., 9 a.m. to 5 p.m.):

   M _____ T _____ W _____ TH _____ F _____ SAT _____ SUN _____

   c. Peak hourly flow rate (GPD)   _____

   d. Maximum daily flow rate (GPD)   _____

   e. Annual daily average (GPD)   _____

3. If batch discharge occurs or will occur, indicate: [New facilities may estimate]

   a. Number of batch discharges _____ per day.

   b. Average discharge per batch _____ (GPD).

   c. Time of batch discharges _____ at _____ .
                           (days of week)   (hours of day)

   d. Flow rate _____ gallons/minute.

   e. Percent of total discharge _____ .

4. Schematic Flow Diagram — For each major activity in which wastewater is or will be generated, draw a diagram of the <u>flow of materials, products, water, and wastewater</u> from the start of the activity to its completion, showing all unit processes. Indicate which processes use water and which generate wastestreams. Include the average daily volume and maximum daily volume of each wastestream [new facilities may estimate]. If estimates are used for flow data this <u>must</u> be indicated. <u>Number each unit process</u> having wastewater discharges to the community sewer. Use these numbers when showing the unit processes in the building layout in Section H. This drawing must be certified by a State Registered Professional Engineer.

   Facilities that checked activities in question 1 of Section B are considered Categorical Industrial Users and should skip to question 6.

5. For Non-Categorical Users Only: List average wastewater discharge, maximum discharge, and type of discharge (batch, continuous, or both), for each plant process. Include the reference number from the process schematic that corresponds to each process. [New facilities should provide estimates for each discharge.]

No.	Process Description	Average Flow (GPD)	Maximum Flow (GPD)	Type of Discharge (batch, continuous, none)
_____	_____	_____	_____	_____
_____	_____	_____	_____	_____
_____	_____	_____	_____	_____

ANSWER QUESTIONS 6 & 7 ONLY IF YOU ARE SUBJECT TO CATEGORICAL PRETREATMENT STANDARDS.

6. For Categorical Users: Provide the wastewater discharge flows for each of your processes or proposed processes. Include the reference number from the process schematic that corresponds to each process. [New facilities should provide estimates for each discharge.]

No.	Regulated Process	Average Flow (GPD)	Maximum Flow (GPD)	Type of Discharge (batch, continuous, none)
_____	_____	_____	_____	_____
_____	_____	_____	_____	_____
_____	_____	_____	_____	_____

No.	Unregulated Process	Average Flow (GPD)	Maximum Flow (GPD)	Type of Discharge (batch, continuous, none)
_____	_____	_____	_____	_____
_____	_____	_____	_____	_____
_____	_____	_____	_____	_____

No.	Dilutions	Average Flow (GPD)	Maximum Flow (GPD)	Type of Discharge (batch, continuous, none)
_____	_____	_____	_____	_____
_____	_____	_____	_____	_____
_____	_____	_____	_____	_____

7. For Categorical Users Subject to Total Toxic Organic (TTO) Requirements:

   Provide the following (TTO) information.

   a. Does (or will) this facility use any of the toxic organics that are listed under the TTO standard of the applicable categorical pretreatment standards published by EPA?

      [ ] Yes   [ ] No

   b. Has a baseline monitoring report (BMR) been submitted which contains TTO information?

      [ ] Yes   [ ] No

   c. Has a toxic organics management plan (TOMP) been developed?

      [ ] Yes (Please attach a copy)   [ ] No

8. Do you have, or plan to have, automatic sampling equipment or continuous wastewater flow metering equipment at this facility?

   Current: Flow Metering   [ ] Yes   [ ] No   [ ] N/A

              Sampling Equipment   [ ] Yes   [ ] No   [ ] N/A

   Planned: Flow Metering   [ ] Yes   [ ] No   [ ] N/A

              Sampling Equipment   [ ] Yes   [ ] No   [ ] N/A

   If so, please indicate the present or future location of this equipment on the sewer schematic and describe the equipment below:

   _____
   _____

9. Are any process changes or expansions planned during the next three years that could alter wastewater volumes or characteristics? Consider production processes as well as air or water pollution treatment processes that may affect the discharge.

   [ ] Yes   [ ] No (skip question 10)

10. Briefly describe these changes and their effects on the wastewater volume and characteristics: (Attach additional sheets if needed.)

   _____
   _____

11. Are any materials or water reclamation systems in use or planned?

   [ ] Yes   [ ] No (skip question 12)

12. Briefly describe recovery process, substance recovered, percent recovered, and the concentration in the spent solution. Submit a flow diagram for each process: (Attach additional sheets if needed.)

   _____
   _____

SECTION F — CHARACTERISTICS OF DISCHARGE

All current industrial users are required to submit monitoring data on all pollutants that are regulated specific to each process. Use the tables provided in this section to report the analytical results. DO NOT LEAVE BLANKS. For all other (nonregulated) pollutants, indicate whether the pollutant is known to be present (P), suspected to be present (S), or known not to be present (O), by placing the appropriate letter in the column for average reported values. Indicate on either the top of each table, or on a separate sheet, if necessary, the sample location and type of analysis used. Be sure methods conform to 40 CFR Part 136; if they do not, indicate what method was used.

New dischargers should use the table to indicate what pollutants will be present or are suspected to be present in proposed wastestreams by placing a P (expected to be present), S (may be present), or O (will not be present) under the average reported values.

Pollutant	Detection Level Used	Maximum Daily Value		Average of Analyses		Number of Analyses	Units	
		Conc.	Mass	Conc.	Mass		Conc.	Mass
Acenaphthene								
Acrolein								
Acrylonitrile								
Benzene								
Benzidine								
Carbon tetrachloride								
Chlorobenzene								
1,2,4-Trichlorobenzene								
Hexachlorobenzene								
1,2-Dichloroethane								
1,1,1-Trichloroethane								
Hexachloroethane								
1,1-Dichloroethane								
1,1,2-Trichloroethane								
1,1,2,2-Tetrachloroethane								
Chloroethane								
Bis (2-chloroethyl) ether								
17 Bis (chloro methyl) ether								
2-Chloroethyl vinyl ether								
2-Chloronaphthalene								
2,4,6-Trichlorophenol								
Parachlorometa cresol								
Chloroform								
2-Chlorophenol								
1,2-Dichlorobenzene								
1,3-Dichlorobenzene								
1,4-Dichlorobenzene								
3,3-Dichlorobenzidine								
1,1-Dichloroethylene								
1,2-Trans-dichloroethylene								
2,4-Dichlorophenol								
1,2-Dichloropropane								
1,2-Dichloropropylene								
1,3-Dichloropropylene								
2,4-Dimethylphenol								
2,4-Dinitrotoluene								
2,6-Dinitrotoluene								
1,2-Diphenylhydrazine								
Ethylbenzene								
Fluoranthene								
4-Chlorophenyl phenyl ether								

Pollutant	Detection Level Used	Maximum Daily Value		Average of Analyses		Number of Analyses	Units	
		Conc.	Mass	Conc.	Mass		Conc.	Mass
4-Bromophenyl phenyl ether								
Bis (2-chlorisopropyl) ether								
Bis (2-chloroethoxy) methane								
Methylene chloride								
Methyl chloride								
Methyl bromide								
Bromoform								
Dichlorobromomethane								
Chlorodibromomethane								
Hexachlorobutadiene								
Hexachlorocyclopentadiene								
Isophorone								
Naphthalene								
Nitrobenzene								
Nitrophenol								
2-Nitrophenol								
4-Nitrophenol								
2,4-Dinitrophenol								
4,6-Dinitro-o-cresol								
N-nitrosodimethylamine								
N-nitrosodiphenylamine								
N-nitrosodi-n-propylamine								
Pentachlorophenol								
Phenol								
Bis (2-ethylhexyl) phthalate								
Butyl benzyl phthalate								
Di-n-butyl phthalate								
Di-n-octyl phthalate								
Diethyl phthalate								
Dimethyl phthalate								
Benzo (a) anthracene								
Benzo (a) pyrene								
3,4-benzofluoranthene								
Benzo (k) fluoranthane								
Chrysene								
Acenaphthylene								
Anthracene								
Benzo (ghi) perylene								
Fluorene								
Phenanthrene								
Dibenzo(a,h) anthracene								

Pollutant	Detection Level Used	Maximum Daily Value		Average of Analyses		Number of Analyses	Units	
		Conc.	Mass	Conc.	Mass		Conc.	Mass
Indeno (1,2,3-cd) pyrene								
Pyrene								
Tetrachloroethylene								
Toluene								
Trichloroethylene								
Vinyl chloride								
Aldrin								
Dieldrin								
Chlordane								
4,4'-DDT								
4,4'-DDE								
4,4'-DDD								
Alpha-endosulfan								
Beta-endosulfan								
Endosulfan sulfate								
Endrin								
Endrin aldehyde								
Heptachlor								
Heptachlor epoxide								
Alpha-BHC								
Beta-BHC								
Gamma-BHC								
Delta-BHC								
PCB-1242								
PCB-1254								
PCB-1221								
PCB-1232								
PCB-1248								
PCB-1260								
PCB-1016								
Toxaphene								
(TCDD)								
Asbestos								
Acidity								
Alkalinity								
Bacteria								
BOD_5								
COD								
Chloride								
Chlorine								

Pollutant	Detection Level Used	Maximum Daily Value		Average of Analyses		Number of Analyses	Units	
		Conc.	Mass	Conc.	Mass		Conc.	Mass
Fluoride								
Hardness								
Magnesium								
NH$_3$–N								
Oil and Grease								
TSS								
TOC								
Kjeldahl N								
Nitrate N								
Nitrite N								
Organic N								
Orthophosphate P								
Phosphorous								
Sodium								
Specific Conductivity								
Sulfate (SO$_4$)								
Sulfide (S)								
Sulfite (SO$_3$)								
Antimony								
Arsenic								
Barium								
Beryllium								
Cadmium								
Chromium								
Copper								
Cyanide								
Lead								
Mercury								
Nickel								
Selenium								
Silver								
Thallium								
Zinc								

SECTION G — TREATMENT

1. Is any form of wastewater treatment (see list below) practiced at this facility?

   [ ] Yes   [ ] No

2. Is any form of wastewater treatment (or changes to an existing wastewater treatment) planned for this facility within the next three years?

   [ ] Yes, describe _____   [ ] No

3. Treatment devices or processes used or proposed for treating wastewater or sludge (check as many as appropriate).

   [ ] Air flotation
   [ ] Centrifuge
   [ ] Chemical precipitation
   [ ] Chlorination
   [ ] Cyclone
   [ ] Filtration
   [ ] Flow equalization
   [ ] Grease or oil separation, type: _____
   [ ] Grease trap
   [ ] Grinding filter
   [ ] Grit removal
   [ ] Ion exchange
   [ ] Neutralization, pH correction

   [ ] Ozonation
   [ ] Reverse osmosis
   [ ] Screen
   [ ] Sedimentation
   [ ] Septic tank
   [ ] Solvent separation
   [ ] Spill protection
   [ ] Sump
   [ ] Biological treatment, type: _____
   [ ] Rainwater diversion or storage
   [ ] Other chemical treatment, type: _____
   [ ] Other physical treatment, type: _____
   [ ] Other, type: _____

4. Describe the pollutant loadings, flow rates, design capacity, physical size, and operating procedures of each treatment facility checked above.

   _____
   _____

5. Attach a process flow diagram for each existing treatment system. Include process equipment, by-products, by-product disposal method, waste and by-product volumes, and design and operating conditions.

6. Describe any changes in treatment or disposal methods planned or under construction for the wastewater discharge to the sanitary sewer. Please include estimated completion dates.

   _____
   _____

7. Do you have a treatment operator?   [ ] Yes   [ ] No

   (If Yes,) Name: _____

              Title: _____

              Phone: _____

              Full Time: _____ (specify hours)

              Part Time: _____ (specify hours)

8. Do you have a manual on the correct operation of your treatment equipment?

   [ ] Yes   [ ] No

9. Do you have a written maintenance schedule for your treatment equipment?

   [ ] Yes   [ ] No

SECTION H — FACILITY OPERATIONAL CHARACTERISTICS

1. Shift Information

Work Days		[ ] Mon.	[ ] Tues.	[ ] Wed.	[ ] Thurs.	[ ] Fri.	[ ] Sat.	[ ] Sun.
Shifts per work day:		_____	_____	_____	_____	_____	_____	_____
Empl's	1st	_____	_____	_____	_____	_____	_____	_____
per	2nd	_____	_____	_____	_____	_____	_____	_____
shift	3rd	_____	_____	_____	_____	_____	_____	_____
Shift start	1st	_____	_____	_____	_____	_____	_____	_____
and end	2nd	_____	_____	_____	_____	_____	_____	_____
times	3rd	_____	_____	_____	_____	_____	_____	_____

APPENDIX B

2. Indicate whether the business activity is:

[ ] Continuous through the year, or

[ ] Seasonal — circle the months of the year during which the business activity occurs:

   J   F   M   A   M   J   J   A   S   O   N   D

Comments: _____

_____

3. Indicate whether the facility discharge is:

[ ] Continuous through the year, or

[ ] Seasonal — circle the months of the year during which the business activity occurs:

   J   F   M   A   M   J   J   A   S   O   N   D

Comments: _____

_____

4. Does operation shut down for vacation, maintenance, or other reasons?

[ ] Yes, indicate reasons and period when shutdown occurs: _____

[ ] No

5. List types and amounts (mass or volume per day) of raw materials used or planned for use (attach list if needed:

_____

_____

6. List types and quantity of chemicals used or planned for use (attach list if needed). Include copies of Manufacturer's Safety Data Sheets (if available) for all chemicals identified.

Chemical	Quantity
_____	_____
_____	_____

7. Building Layout — Draw to scale the location of each building on the premises. Show map orientation and location of all water meters, storm drains, numbered unit processes (from schematic flow diagram), public sewers, and each facility sewer line connected to the public sewers. Number each sewer and show existing and proposed sampling locations. This drawing must be certified by a State Registered Professional Engineer.

A blueprint or drawing of the facilities showing the above items may be attached in lieu of submitting a drawing on this sheet.

SECTION I — SPILL PREVENTION

1. Do you have chemical storage containers, bins, or ponds at your facility?   [ ] Yes   [ ] No

If yes, please give a description of their location, contents, size, type, and frequency and method of cleaning. Also indicate in a diagram or comment on the proximity of these containers to a sewer or storm drain. Indicate if buried metal containers have cathodic protection.

2. Do you have floor drains in your manufacturing or chemical storage area(s)? [ ] Yes [ ] No If yes, where do they discharge to?

3. If you have chemical storage containers, bins, or ponds in manufacturing area, could an accidental spill lead to discharge to: (check all that apply).

[ ] an onsite disposal system

[ ] public sanitary sewer system (e.g., through a floor drain)

[ ] storm drain

[ ] to ground

[ ] other, specify:

[ ] not applicable, no possible discharge to any of the above routes

4. Do you have an accidental spill prevention plan (ASPP) to prevent spills of chemicals or slug discharges from entering the Control Authority's collection systems?

[ ] Yes — [Please enclose a copy with the application.]

[ ] No

[ ] N/A, Not applicable since there are no floor drains and/or the facility discharges only domestic wastes.

5. Please describe below any previous spill events and remedial measures taken to prevent their reoccurrence.

_____

_____

SECTION J — NON-DISCHARGED WASTES

1. Are any waste liquids or sludges generated and not disposed of in the sanitary sewer system?

[ ] Yes, please describe below

[ ] No, skip the remainder of Section J

Waste Generated	Quantity (per year)	Disposal Method
_____	_____	_____
_____	_____	_____
_____	_____	_____

2. Indicate which wastes identified above are disposed of at an off-site treatment facility and which are disposed of on-site.

3. If any of your wastes are sent to an off-site centralized waste treatment facility, identify the waste and facility.

4. If an outside firm removes any of the above checked wastes, state the name(s) and address(es) of all waste haulers:

a. _____   b. _____

   _____   _____

   _____   _____

   Permit No. (if applicable):   Permit No. (if applicable):

   _____   _____

5. Have you been issued any Federal, State, or local environmental permits?

[ ] Yes   [ ] No

If yes, please list the permit(s): _____

SECTION K — AUTHORIZED SIGNATURES

Compliance certification:

1. Are all applicable Federal, State, or local pretreatment standards and requirements being met on a consistent basis?

Yes [ ]   No [ ]   Not yet discharging [ ]

2. If No:

a. What additional operations and maintenance procedures are being considered to bring the facility into compliance? Also, list additional treatment technology or practice being considered in order to bring the facility into compliance.

b. Provide a schedule for bringing the facility into compliance. Specify major events planned along with reasonable completion dates. Note that if the Control Authority issues a permit to the applicant, it may establish a schedule for compliance different from the one submitted by the facility.

Milestone Activity	Completion Date
_____	_____
_____	_____
_____	_____
_____	_____

Authorized Representative Statement

I certify under penalty of law that this document and all attachments were prepared under my direction or supervision in accordance with a system designed to assure that qualified personnel properly gather and evaluate the information submitted. Based on my inquiry of the person or persons who manage the system, or those persons directly responsible for gathering the information, the information submitted is, to the best of my knowledge and belief, true, accurate, and complete. I am aware that there are significant penalties for submitting false information, including the possibility of fine and imprisonment for knowing violations.

_____   _____

Name(s)   Title

_____   _____ _____

Signature   Date   Phone

INSTRUCTIONS TO FILL OUT WASTEWATER DISCHARGE PERMIT APPLICATION

All questions must be answered. DO NOT LEAVE BLANKS. If you answer "no" to question E.1., you may skip to Section I. Otherwise, if a question is not applicable, indicate so on the form. Instructions to some questions on the permit application are given below.

SECTION A — INSTRUCTIONS (GENERAL INFORMATION)

1. Enter the facility's official or legal name. Do not use a colloquial name.

    a. Operator Name: Give the name, as it is legally referred to, of the person, firm, public organization, or any other entity which operates the facility described in this application. This may or may not be the same name as the facility.

    b. Indicate whether the entity which operates the facility also owns it by marking the appropriate box:

    (i) If the response is "No," clearly indicate the operator's name and address and submit a copy of the contract and/or other documents indicating the operator's scope of responsibility for the facility.

2. Provide the physical location of the facility that is applying for a discharge permit.

3. Provide the mailing address where correspondence from the Control Authority may be sent.

4. Provide all the names of the authorized signatures for this facility for the purposes of signing all reports. The designated signatory is defined as:

    a. A responsible corporate officer, if the Industrial User submitting the reports is a corporation. For the purpose of this paragraph, a responsible corporate officer means:

    (i) a president, secretary, treasurer, or vice-president of the corporation in charge of a principal business function, or any other person who performs similar policy- or decision-making functions for the corporation, or

    (ii) the manager of one or more manufacturing, production, or operation facilities employing more than 250 persons or having gross annual sales or expenditures exceeding $25 million (in second-quarter 1980 dollars), if authority to sign documents has been assigned or delegated to the manager in accordance with corporate procedures.

    b. A general partner or proprietor if the Industrial User submitting the reports is a partnership or sole proprietorship respectively.

    c. The principal executive officer or director having responsibility for the overall operation of the discharging facility if the Industrial User submitting the reports is a Federal, State, or local governmental entity, or their agents.

    d. A duly authorized representative of the individual designated in paragraph (a), (b), or (c) of this section if:

    (i) the authorization is made in writing by the individual described in paragraph (a), (b), or (c);

    (ii) the authorization specifies either an individual or a position having responsibility for the overall operation of the facility from which the Industrial Discharge originates, such as the position of plant manager, operator of a well, or well field superintendent, or a position of equivalent responsibility, or having overall responsibility for environmental matters for the company; and

    (iii) the written authorization is submitted to the City.

    e. If an authorization under paragraph (d) of this section is no longer accurate because a different individual or position has responsibility for the overall operation of the facility, or overall responsibility for environmental matters for the company, a new authorization satisfying the requirements of paragraph (d) of this section must be submitted to the City prior to or together with any reports to be signed by an authorized representative.

5. Provide the name of a person who is thoroughly familiar with the facts reported on this form and who can be contacted by the Control Authority (e.g., the plant manager).

SECTION B — INSTRUCTIONS (BUSINESS OPERATIONS)

1. Check off all operations that occur or will occur at your facility. If you have any questions regarding how to categorize your business activity, contact the Control Authority for technical guidance.

3. For all processes found on the premises, indicate the NAICS code, as found in the most recent edition of the North American Industry Classification System manual prepared by the Executive Office of the President, Office of Management and Budget. This document is available from the Government Printing Office in Washington, D.C. Copies of the manual may also be available at some public libraries.

4. List the types of products, giving the common or brand name and the proper scientific name. Enter from your records the average and maximum amounts produced daily for each operation for the previous calendar year, and the estimated total daily production for this calendar year. Be sure to specify the daily units of production. Attach additional pages as necessary.

SECTION C — INSTRUCTIONS (WATER SUPPLY)

4. Provide daily average water usage within the facility. Contact cooling water is cooling water that during the process comes into contact with process materials, thereby becoming contaminated. Non-contact cooling water does not come into contact with

process materials. Sanitary water includes only water used in restrooms. Plant and equipment washdown includes floor washdown. If sanitary flow is not metered, provide an estimate based on 15 gallons per day (gpd) for each employee.

SECTION E — INSTRUCTIONS (WASTEWATER DISCHARGE INFORMATION)

1. If you answer "no" to this question, skip to Section I, otherwise complete the remainder of the application.

4. A schematic flow diagram is required to be completed and certified for accuracy by a State registered professional engineer. Assign a sequential reference number to each process starting with No. 1. An example of a drawing is shown below in Figure 1. To Determine your average daily volume and maximum daily volume of wastewater flow, you may have to read water meters, sewer meters, or make estimates of volumes that are not directly measurable.

FIGURE 1. SCHEMATIC FLOW DIAGRAM

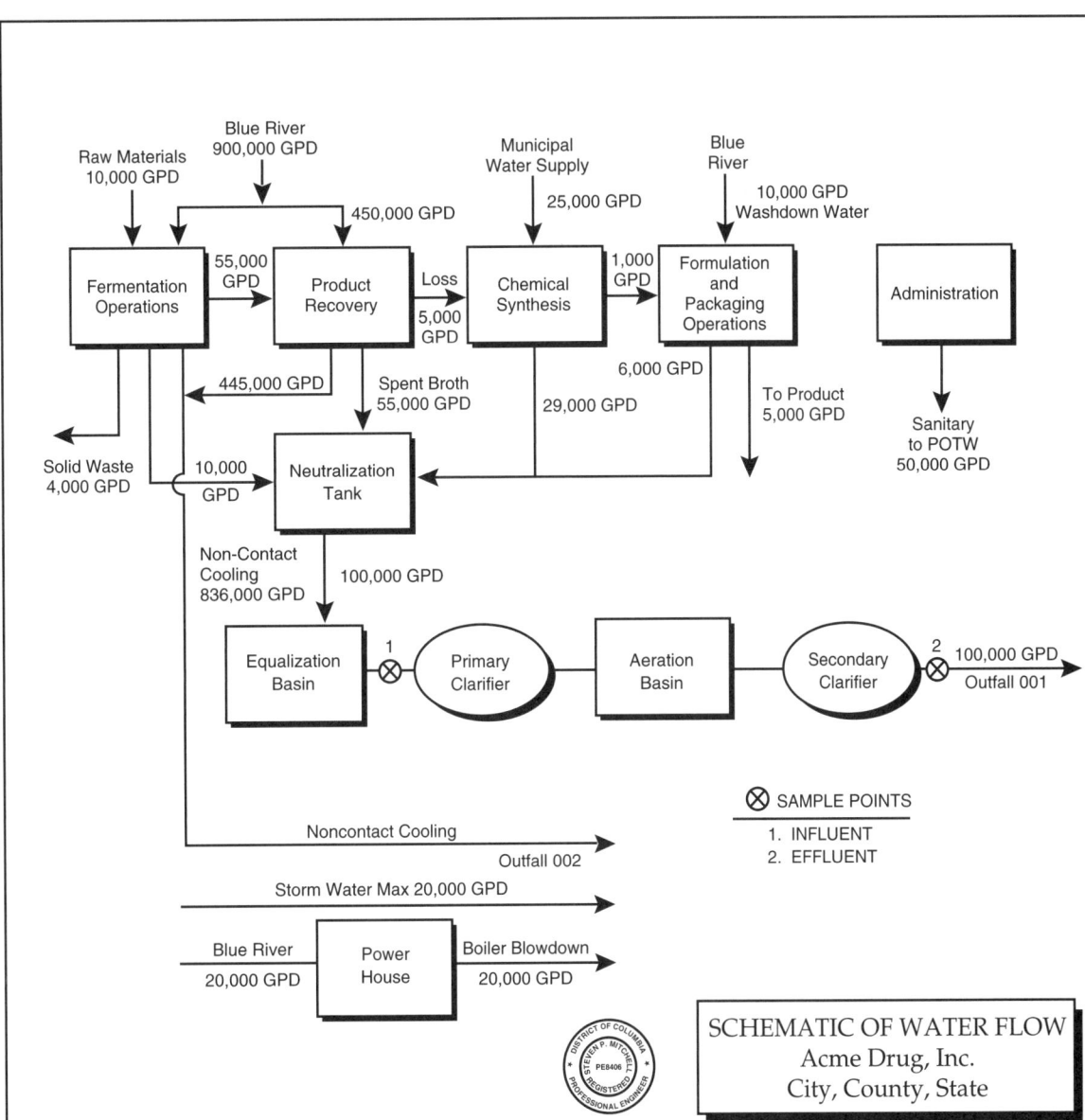

5. Non-categorical users should report average daily and maximum daily wastewater flows from each process, operation, or activity present at the facility. Categorical users should skip to question 6.

6. Categorical users should report average daily and maximum daily wastewater flows from every regulated, unregulated, and dilution process. A regulated wastestream is defined as wastewater from an Industrial process that is regulated for a particular pollutant by a categorical pretreatment standard. Unregulated wastestreams are wastestreams from an industrial process that are not regulated by a categorical pretreatment standard and are not defined as a dilution wastestream. Dilution wastestreams include sanitary wastewater, boiler blowdown, noncontact cooling water or blowdown, stormwater streams, demineralized backwash streams and process wastestreams from certain industrial subcategories exempted by EPA from catergorical pretreatment standards. [For further details see 40 CFR 403.5(e).]

7. Total Toxic Organics (TTO) means the sum of the masses or concentrations of specific toxic organic compounds found in the industrial user's process discharge. The individual organic compounds that make up the TTO value and the minimum reportable quantities differ according to the particular industrial category [see applicable categorical pretreatment standards, 40 CFR Parts 405-471].

SECTION H — INSTRUCTIONS (FACILITY OPERATIONAL CHARACTERISTICS)

2. Indicate whether the business activity is continuous throughout the year or if it is seasonal. If the activity is seasonal, circle the months of the year during which the discharge may occur. Make any comments you feel are required to describe the variation in operation of your business activity.

4. Indicate any shut downs in operation which may occur during the year and indicate the reasons for shutdown.

5. Provide a listing of all primary raw materials used (or planned) in the facility's operations. Indicate amount of raw material used in daily units.

6. Provide a listing of all chemicals used (or planned) in the facility's operations. Indicate the amount used or planned in daily units. Avoid the use of trade names of chemicals. If trade names are used, also provide chemical compounds. Provide copies of all available manufacturer's safety data sheets for all chemicals identified.

7. A building layout or plant site plan for the premises is required to be completed and certified for accuracy by a State registered professional engineer. Approved building plans may be substituted. An arrow showing North as well as the map scale must be shown. The location of each existing and proposed sampling location and facility sewer line must be clearly identified as well as all sanitary and wastewater drainage plumbing. Number each unit process discharging wastewater to the public sewer. Use the same numbering system shown in Figure 1, the schematic flow diagram. An example of the drawing required is shown below.

FIGURE 2.   BUILDING LAYOUT

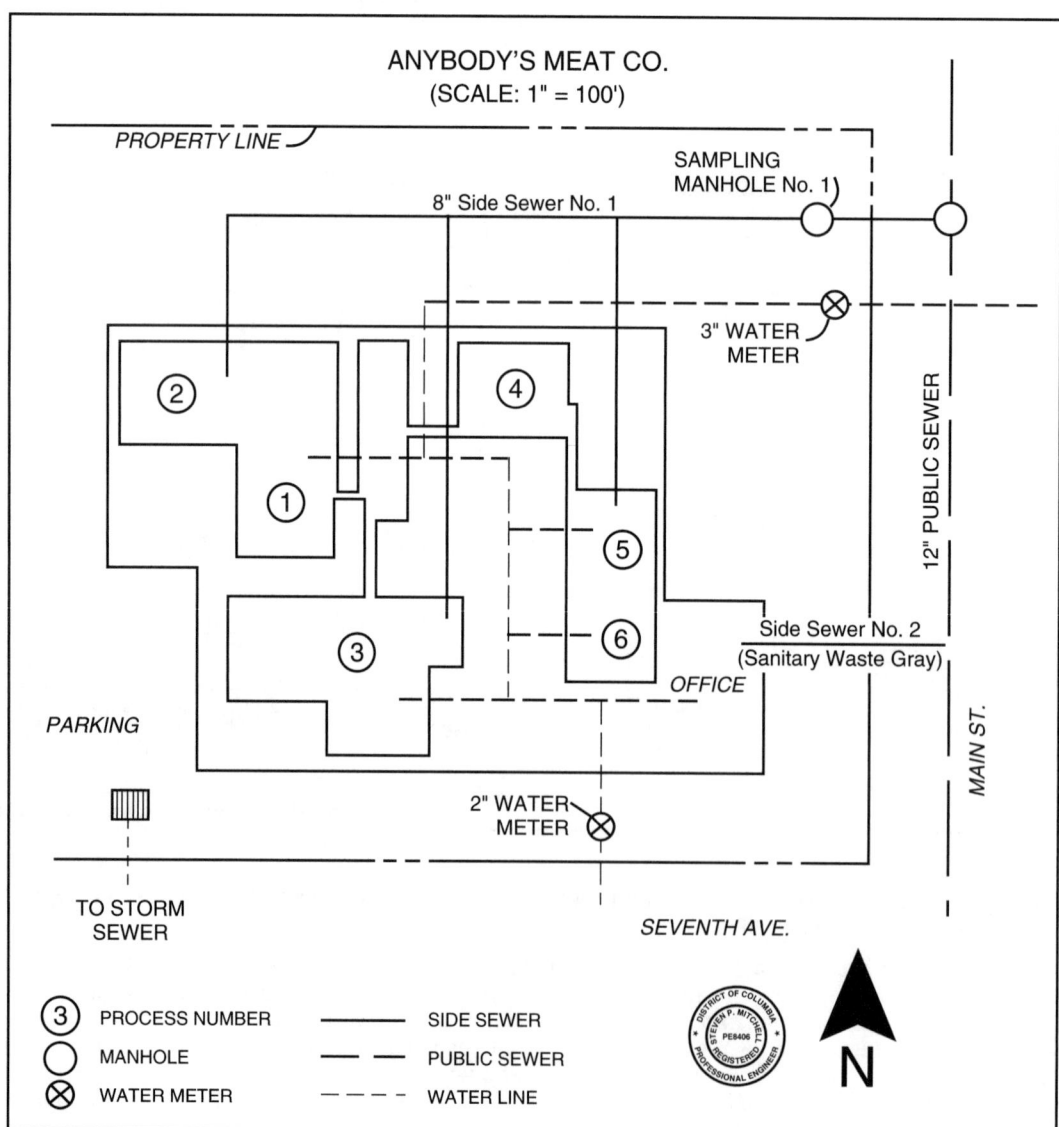

SECTION I — INSTRUCTIONS (SPILL PREVENTION)

5. Describe how the spill occurred, what was spilled, when it happened, where it occurred, how much was spilled, and whether or not the spill reached the sewer. Also explain what measures have been taken to prevent a reoccurrence or what measures have been taken to limit damage if another spill occurs.

SECTION J — INSTRUCTIONS (NON-DISCHARGED WASTES)

1. For wastes not discharged to the Control Authority's sewer, indicate types of waste generated, amount generated, the way in which the waste is disposed (e.g., incinerated, hauled, etc.), and the location of disposal.

2. Onsite disposal system could be a septic system, lagoon, holding pond (evaporative-type), etc.

3. Types of permits could be: air, hazardous waste underground injection, solid waste, NPDES (for discharges to surface water), etc.

SECTION K — INSTRUCTIONS (AUTHORIZED SIGNATURE)

See instructions for question 4 in Section A, for a definition of an authorized representative.

APPENDIX B

APPENDIX C

WASTEWATER DISCHARGE PERMIT APPLICATION FOR EXISTING INDUSTRIAL USERS, CITY OF HOPEWELL, VIRGINIA

Table of Contents

For further information or answers to questions please call Robert C. Steidel, Environmental Manager, 804-541-2210 (FAX 804-748-8284).

Schedule for Reissuance

January 3, 1995	Permit applications mailed to existing industrial users.
January 18, 1995	An informal meeting will be held at the HRWTF conference room beginning at 1:00 PM to discuss the reapplication form and the reapplication process.
April 3, 1995	Permit application due date.
October 2, 1995	Draft permits with new local limits issued for 45 day comment period.
December 5, 1995	Final industrial wastewater discharge permit issued effective January 1, 1996.

Type or print the name of the facility <u>on each page</u> of the permit application.

Section I. General Information

Facility Name	
Authorized Representative[1]	
Authorized Representative Title	
Authorized Representative Telephone Number	
Mailing Address (Street)	
Mailing Address (P.O. Box)	
City/County	
State	
Zip Code	
Facility Location (Street address or other specific identifier)	
City/County	
State	
Zip Code	
Additional Contact Persons	
Name	
Title	
Telephone Number	
FAX Telephone Number	
Name	
Title	
Telephone Number	
FAX Telephone Number	
Name	
Title	
Telephone Number	
FAX Telephone Number	
Please provide a brief description of the operations at the facility	

Operator Name	
Operator Status (Public, Private, Federal, State, other)	
Operator Mailing Address (Street)	
Operator Mailing Address (P.O. Box)	
City/County	
State	
Zip Code	
Is the Operator the Owner of the Facility?	_____Yes_____No
If the answer to the previous question is "No", please provide the following:	
Owner Mailing Address (Street)	
Owner Mailing Address (P.O. Box)	
City/County	
State	
Zip Code	
Person Completing this Application	
Title	
Telephone Number	
Average Annual Number of Employees	
Average Annual Days per Week of Operation	
Discharge Occurs	_____AM to _____PM
Days of the Week Discharge Occurs	__Su__M__T__W__Th__F__Sa
Is the Facility Subject to Annual Shutdown?	_____Yes_____No
If "Yes", please provide a brief description and include the annual quarter (if any) during which shut-down is planned.	
Is the Facility Subject to Seasonal Variation?	_____Yes_____No
If "Yes", please provide a brief description.	
Are any process changes or expansions planned during the next 5 years?	_____Yes_____No
	If yes, please attach a separate sheet to this form describing the nature of planned changes or expansions.
[1] Attach written delegation when required by applicable pretreatment standards.	

Section II. Environmental Permits Held by the Facility (please list).

Permitting Agency	Permit Type	Identifying Number

Section III. Standard Industrial Classification (SIC) Code Number(s) for the Facility
(please list all manufacturing process and the 4 digit SIC code number applicable to each)

Manufacturing Process	SIC Code Number (4 digit)

Section IV. Raw Materials and Process Additives Used at the Facility
(please list, attaching additional sheets as necessary):

Raw Materials	Raw Materials	Process Additives	Process Additives

A P P E N D I X C

Section V. Manufacturing Process Production Information
(please provide the following for each manufacturing process listed in Section III above)

Manuacturing Process	SIC Code Number	Average Rate of Production	Reporting Basis from below[1]	Units from below[2]	Production Rates are (check one) Verifiable Estimates	

[1]   a. = Day,   b. = Month,   c. = Year

[2]   A. = Pounds, B. = Tons, C. = Barrels, D. = Bushels, E. = Square Feet, F. = Gallons, G = Pieces or Units, H. = Kilograms, I. = Liters, J. = Other (please specify:_____)

Section VI. Applicable Pretreatment Standards
(please identify the pretreatment standards applicable to this facility from the following)

Chapter 31, City of Hopewell Code (sewer use ordinance)	
General Pretreatment Regulations, 40 CFR 403	
Pulp, Paper and Paperboard Manufacturing Category, 40 CFR 430	
Organic Chemicals and Plastics and Synthetic Fibers Category, 40 CFR 414	
Fertilizer Manufacturing Category, 40 CFR 418	
Inorganic Chemicals Manufacturing Category, 40 CFR 415	
Steam Electric Category, 40 CFR 423	

Please circle any of the following categorical standards that are applicable to the processes at this facility

Aluminum Forming	Electrical and Electronic Components	Leather Tanning and Finishing	Petroleum Refining	Plastics Molding and Forming
Asbestos Manufacturing	Electroplating	Meat Processing	Pharmaceuticals	Textiles Mills
Battery Manufacturing	Feedlots	Metal Finishing	Phosphate Manufacturing	
Builder's Paper	Ferroalloy Manufacturing	Metal Molding and Casting	Porcelain Enameling	
Carbon Black	Fruits and Vegetables Processing Manufacturing	Nonferrous Metals Forming	Rubber Processing	
Cement Manufacturing	Glass Manufacturing	Nonferrous Metals Manufacturing	Seafood Processing	
Coil Coating	Grain Mills Manufacturing	Paint Formulating	Soaps and Detergent Manufacturing	
Copper Forming	Ink Formulating	Paving and Roofing (Tars and Asphalt)	Sugar Processing	
Dairy Products Processing	Iron and Steel Manufacturing	Pesticides	Timber Products Manufacturing	

Section VII. HRWTF Agreement Contract Limitations and Requests

Parameter	Agreement Conditions	Existing Conditions	Requested Conditions
Average Monthly Flow (mgd)			
Maximum Daily Flow (mgd)			
Average Monthly BOD Loading (ppd)			
Average Monthly BOD Concentration (mg/L)			
Maximum Daily BOD Loading (ppd)			
Maximum Daily BOD Concentration (mg/L)			
Average Monthly TSS Loading (ppd)			
Average Monthly TSS Concentration (mg/L)			
Maximum Daily TSS Loading (ppd)			
Maximum Daily TSS Concentration (mg/L)			
Maximum Daily Temperature (Degrees C)			

From Hopewell Regional Wastewater Treatment Facility Agreement, July 1, 1975, Table II, or Agreement for Wastewater Service Between City of Hopewell and County of Prince George, April 5, 1979, or Fort Lee and the City of Hopewell UTILITY SERVICES CONTRACT (DABT59-79-C-0001), November 28, 1978.

Section VIII. Wastewater Volume Information, Water Balance

Water Use Category	Volume Discharged to HRWTF		
	mgd	gpd	gpm
Sanitary (toilets, locker rooms, food service)			
Process (rinsing, production, CIP, contact cooling)			
Cooling (noncontact)			
Boiler (blowdown)			
Wet Scrubbers (air pollution control, process)			
Stormwater (contaminated by process contact)			
Stormwater (not contaminated by process contact)			
Other (please describe:)			
Total Volume Discharged to HRWTF			

APPENDIX C

If your facility does not have periodic measurements of stormwater volume to the Regional Plant, rates of flow can be estimated by multiplying the annual amount of rainfall by the land area of the facility and then multiplying that figure by the runoff coefficient. The runoff coefficient represents that fraction of rainfall that does not seep into the ground but runs off as stormwater. The runoff coefficient is directly related to how the land in the drainage area is used. See the table below:

Description	Runoff Coefficient	Description	Runoff Coefficient	Description	Runoff Coefficient
Business, Downtown	0.7-0.95	Streets, Asphaltic	0.7-0.95	Sandy Soil, Avg 2-7%	0.1-0.15
Business, Neighborhood	0.5-0.7	Steets, Concrete	0.08-0.95	Sandy Soil, Steep, 7%	0.15-0.2
Industrial, Light	0.05-0.08	Streets, Brick	0.7-0.85	Heavy Soil, Flat, 2%	0.13-0.17
Industrial, Heavy	0.06-0.9	Drives and Walks	0.7-0.85	Heavy Soil, Avg 2-7%	0.18-0.22
Railroad Yard	0.2-0.4	Roofs	0.75-0.95	Heavy Soil, Steep, 7%	0.25-0.35
Unimproved Area	0.1-0.3	Sandy Soil, Flat, 2%	0.05-0.1		

Section IX. Production Wastewater Generation Volume in Gallons Per Day (GPD)

Manufacturing Process	SIC Code	Daily Average Flow	Maximum Daily Flow	Continuous or Intermittant Flow

Section X. Batch Discharges

(If batch discharges from the facility occur and will NOT BE EQUALIZED IN EXISTING TREATMENT UNITS (lagoons, basins or ponds), please complete the following table)

Batch Description	Annual Avg No of Batch Discharges per Day	Avg Volume of Batch Discharge (gallons)	Day of the Week Batch Discharge Takes Place	Time of day Batch Discharge Occurs	Flowrate of Batch Discharge in Gallons per Minute
Totals:					
% of Total Flow:					

Section XI. Pretreatment Devices or Processes
(Check as many as appropriate and in use at the facility for treating wastewater or sludge)

Air floatation	
Centrifuge	
Chemical precipitation	
Chlorination	
Cyclone	
Filtration	
Flow equalization	
Grease or oil separation, type _____	
Grease trap	
Grit Removal	
Ion Exchange	
Neutralization, pH correction	
Ozonation	
Reverse Osmosis	
Screen	
Sedimentation	
Septic tank	
Solvent separation	
Spill prevention	
Sump	
Biological treatment, type _____	
Rainwater diversion or storage _____	
Other chemical treatment, type _____	
Other physical treatment, type _____	
Other, type _____	
No pretreatment provided	

Section XII. Residual Disposal Information, Discharge to POTW

(please complete this table from Part III of the most recent EPA FORM R, Toxic Chemical Release Inventory Reporting Form submitted for this facility for chemicals discharged to POTW)

Chemical	CAS Number	Total Release (lb/yr)			Estimate
		0	1-499	500 - 999	

Section XIII. Other Wastes

Are any liquid wastes or sludges form this facility disposed of by means other than discharge to the sewer system? _____Yes _____No. If "No", skip forward to "**Section XIV. Slug Control Plan**". If "Yes" complete the following table.

These wastes may best be described as:

Check all applicable	Description	Est Gal or Lbs / Yr	Storage		Disposal	
			On - Site	Off - Site	On - Site	Off - Site
	Acids and Alkalies					
	Heavy Metal Sludges					
	Inks/Dyes					
	Oil and/or Grease					
	Organic Compounds					
	Paints					
	Pesticides					
	Plating Wastes					
	Pretreatment Sludges					
	Sovents/Thinners					
	Other Hazardous Wastes (specify) _____ _____					
	Other Wastes (specify) _____ _____					

Briefly describe any method(s) of storage or disposal checked above:

Section XIV. Slug Control Plan

Does the facility have a written plan that contains, as a minimum, a description of discharge practices, including non-routine batch dishcarges; description of stored chemicals, procedures for immediately notifying the POTW of slug discharges including any discharge that would violate a prohibition under 40 CFR 403.5(b) with procedures for follow-up written notification within five days, and procdures to prevent adverse impact from accidental spills, including inspection and maintenance of storage areas, handling and transfer of materials, loading and unloading operations, control of plant site run-off, worker training, building of containment structures or equipment, measures for containing toxic organic pollutants (including solvents), and/or measures and equipment for emergency response? _____ Yes _____ No

If "Yes", please provide the following regarding the document(s) that contain this information.

Document Title	Last Updated	Responsible for Maintaining the Document as Current (list job title only)

Section XV. Hazardous Waste Disposal to the Process (Sanitary) Sewer System within the Hopewell Regional Wastewater Treatment Facility

(As required by the EPA in 40 CFR 403.12(p), the following information must be completed by all Significant Industrial Users to report a substance discharged to the process (sanitary) sewer which, if otherwise disposed of, would be a hazardous waste under 40 CFR 261 (RCRA)).

Column 1	Column 2	Column 3	Column 4	Column 5	Column 6	Column 7	Column 8
Hazardous Waste Name from 40 CFR 261	Does this facility discharge more than 15 kg (6.8 lbs) but less than 100 kg (45.3 lbs) of the hazardous waste listed in column 1 per calendar month, none of which is acutely hazardous waste, to the process (sanitary) sewer system? If "Yes" then complete columns 4 through 5.	Does this facility discharge more than 100 kg (45.3 lbs) of the hazardous waste listed in column 1 per calendar month or any mass of acute hazardous waste to the process (sanitary) sewer system? If yes, the complete columns 4 through 8.	EPA Hazardous Waste Number	Specify the type of discharge to the process (sanitary) sewer system as continuous, batch, or other (please explain)	List all hazardous constituents in the hazardous waste discharged to the process (sanitary) sewer system.	Estimate concentration and mass of the hazardous constituents listed in column 6 discharged to the process (sanitary) sewer system (determined on a monthly average basis).	Estimated mass of the hazardous constituent lished in column 6 to be discharged tothe process (sanitary) sewer system in the next 12 months.

Section XVI. Process Diagram and Plant Layout Diagram

Please include 8.5" x 11" copies of facility process diagrams, schematics and a plant layout diagram with this application.

Section XVII. Certification

I certify that sampling and analysis for this application is representative of normal work cycles and the expected pollutant discharges to the Hopewell Regional Wastewater Treatment Facility.

Name (type or print)   _____

Signature   _____

Title   _____

Date   _____

We have personally examined and are familiar with the information submitted in this application, and certify under penalty of law that this information was obtained in accordance with applicable requirements. Moreover, based on our inquiry of those individuals immediately responsible for obtaining the information reported herein, we believe that the submitted information is true, accurate and complete. We are aware that there are significant penalties for sumitting false information, including the possibility of fine and imprisonment.

Therefore, we certify that this facility is _____ is not _____ in consistent compliance with the applicable pretreatment standards identified in this application. (If the facility is not in compliance, attach a schedule to provide for compliance within the shortest possible time period).

Qualified Professional

Name (type or print)   _____

Signature   _____

Title   _____

Date   _____

Authorized Representative

Name (type or print)   _____

Signature   _____

Title   _____

Date   _____

Wastewater Discharge Permit Application for Existing Industrial Users, January 1995
Section XVIII, Attachment I

Required pollutant or pollutant characteristic analysis support of this permit application.
For these analysis, the following statements are applicable:

- *Report all results in micrograms per liter* unless another unit of quantification is more appropriate (e.g.: degrees F for flash point and temperature, units for pH, number per 100 ml for fecal coliform).
- Quantification level (QL) is defined as the lowest concentration used for the calibration of a measurement system when the calibration is in accordance with the procedures published for the required method.
- Units for quantification level are micrograms per liter unless otherwise specified.
- C = 24 hour composite sample unless otherwise specified.
- G = Grab sample.
- 1/8H = 1 grab sample every eight hours. Permittee shall analyze each sample individually and report the average of the three samples. Alternative laboratory compositing methods may be approved on a case by case basis.
- Dashes mean that the QL is at the discretion of the permittee.

For this permit application, all applicants shall collect 3 samples on consecutive days within one calendar month during which normal operations are taking place. Each sample shall be analyzed for the following pollutants or pollutant characteristics.

Only applicants regulated under the EPA's OCPSF Categorical Standard, 40 CFR 414, need to complete pages H and I of this section.

Summarize all data collected in support of this permit application on these pages. <u>Do not attach or substitute copies of original laboratory sheets or certification forms.</u> HRWTF personnel will conduct an on-site inspection of the facility between April 3 and October 2 1995 as part of the reapplication review process. The original laboratory sheets and certification forms will be reviewed during that on-site inspection.

Wastewater Discharge Permit Application for Existing Industrial Users, January 1995
Section XVIII, Attachment I

Pollutant or Pollutant Characteristic	Method	Quantification Level	Sample Type	Results (by individual date and avg)			
				Date:	Date:	Date:	Avg
Arsenic, Total Recoverable	EPA 206.2	10	C				
Arsenic, Dissolved	EPA 206.2	10	G				
Barium, Total Recoverable	EPA 200.7	20	C				
Barium, Dissolved	EPA 200.7	20	G				
Cadmium, Total Recoverable	EPA 213.2	1	C				
Cadmium, Dissolved	EPA 213.2	1	G				
Chromium III[1], Total Recoverable	EPA 218.2 minus 218.4	10	C				
Chromium III, Dissolved	EPA 218.2 minus 218.4	10	G				
Chromium VI, Dissolved	EPA 218.4	10	G				
Copper, Total Recoverable	EPA 220.2	10	C				
Copper, Dissolved	EPA 220.2	10	G				
Iron, Total Recoverable	EPA 236.1 or 236.2	---	C				
Iron, Dissolved	EPA 236.1 or 236.2	---	G				
Lead, Total Recoverable	EPA 239.2	5	C				
Lead, Dissolved	EPA 239.2	5	G				
Manganese, Total Recoverable	EPA 243.1	---	C				
Manganese, Dissolved	EPA 243.1	---	G				
Mercury, Total Recoverable	EPA 245.1 or 245.2	0.3	C				
Mercury, Dissolved	EPA 245.1 or 245.2	0.3	G				
Nickel, Total Recoverable	EPA 249.2	40	C				
Nickel, Dissolved	EPA 249.2	40	G				
Selenium, Total Recoverable	EPA 270.2 or 270.3	5	C				
Selenium, Dissolved	EPA 270.2 or 270.3	5	G				
Silver, Total Recoverable	EPA 272.2	2	C				
Silver, Dissolved	EPA 272.2	2	G				
Zinc, Total Recoverable	EPA 289.1	20	C				
Zinc, Dissolved	EPA 289.1	20	G				

Wastewater Discharge Permit Application for Existing Industrial Users, January 1995
Section XVIII, Attachment I

Pesticides/PCB's							
Pollutant or Pollutant Characteristic	Method	Quantification Level	Sample Type	Results (by individual date and avg)			
				Date:	Date:	Date:	Avg
Aldrin	608	0.5	1/8 H				
Chlorpyrifos	622	---	1/8 H				
Chlordane	608	0.2	1/8 H				
DDT	608	0.1	1/8 H				
Demeton	3	---	1/8 H				
2,4-dichlorphenoxy acetic acid (2,4-D)	3	---	1/8 H				
Dieldren	608	0.1	1/8 H				
Endosulfan I	608	0.1	1/8 H				
Endosulfan II	608	0.1	1/8 H				
Endosulfan Sulfate	608	0.1	1/8 H				
Endrin	608	0.1	1/8 H				
Guthion	622	---	1/8 H				
Heptachlor	608	0.05	1/8 H				
Hexachlorocyclohexane (Lindane)	608	0.05	1/8 H				
Malathion	3	---	1/8 H				
Methoxychlor	3	---	1/8 H				
Mirex	3	---	1/8 H				
Parathion	3	---	1/8 H				
PCB-1242	608	1	1/8 H				
PCB-1254	608	1	1/8 H				
PCB-1221	608	1	1/8 H				
PCB-1232	608	1	1/8 H				
PCB-1248	608	1	1/8 H				
PCB-1260	608	1	1/8 H				
PCB-1016	608	1	1/8 H				
2-(2,4,5-Trichlorophenoxy) propionic acid (Silvex)	3	---	1/8 H				
Toxaphene	608	5	1/8 H				

APPENDIX C

Wastewater Discharge Permit Application for Existing Industrial Users, January 1995
Section XVIII, Attachment I

Base Neutral Extractables							
Pollutant or Pollutant Characteristic	Method	Quantification Level	Sample Type	Results (by individual date and avg)			
				Date:	Date:	Date:	Avg
Anthracene	625	10	1/8 H				
Benzo(a)anthracene	625	10	1/8 H				
Benzo(b)fluoranthene	625	10	1/8 H				
Benzo(k)fluoranthene	625	10	1/8 H				
Benzo(a)pyrene	625	10	1/8 H				
Chrysene	625	10	1/8 H				
Dibenzo(a,h)anthracene	625	20	1/8 H				
1,2-Dichlorobenzene	625	10	1/8 H				
1,3-Dichlorobenzene	625	10	1/8 H				
1,4-Dichlorobenzene	625	10	1/8 H				
2,4-Dinitrotoluene	625	10	1/8 H				
Di-2-ethylhexyl Phthalate	625	10	1/8 H				
Fluoranthene	625	10	1/8 H				
Fluorene	625	10	1/8 H				
Isophorone	625	10	1/8 H				
Indeno(1,2,3-cd)pyrene	625	20	1/8 H				
Naphthalene	625	10	1/8 H				
Pyrene	625	10	1/8 H				

Wastewater Discharge Permit Application for Existing Industrial Users, January 1995
Section XVIII, Attachment I

Volatiles							
Pollutant or Pollutant Characteristic	Method	Quantification Level	Sample Type	Results (by individual date and avg)			
				Date:	Date:	Date:	Avg
Benzene	624	10	1/8 H				
Bromoform	624	10	1/8 H				
Carbon Tetrachloride	624	10	1/8 H				
Chlorodibromomethane	624	10	1/8 H				
Chloroform	624	10	1/8 H				
Chloromethane	624	20	1/8 H				
Dichloromethane	624	20	1/8 H				
Dichlorobromomethane	624	10	1/8 H				
1,2-Dichloroethane	624	10	1/8 H				
Ethylbenzene	624	10	1/8 H				
Monochlorobenzene	624	50	1/8 H				
Tetrachloroethylene	624	10	1/8 H				
Toluene	624	10	1/8 H				
Trichloroethylene	624	10	1/8 H				
Vinyl Chloride	624	10	1/8 H				

APPENDIX C

Wastewater Discharge Permit Application for Existing Industrial Users, January 1995
Section XVIII, Attachment I

Acids Extractables							
Pollutant or Pollutant Characteristic	Method	Quantification Level	Sample Type	Results (by individual date and avg)			
				Date:	Date:	Date:	Avg
Pentachlorophenol	625	50	1/8 H				
Phenol[2]	625	10	1/8 H				
2,4,6-Trichlorophenol	625	10	1/8 H				

Wastewater Discharge Permit Application for Existing Industrial Users, January 1995
Section XVIII, Attachment I

Pollutant or Pollutant Characteristic	Method	Quantification Level	Sample Type	Results (by individual date and avg)			
				Date:	Date:	Date:	Avg
Miscellaneous							
Total Residual Chlorine	3	100	G				
Cyanide	EPA 335.2	10	G				
Dioxin	EPA 1613	0.00001	C				
Hardness	3	---	C				
Sulfate	3	---	C				
Tributyltin	NBSR 85-3295	---	C				
Xylenes (total)	EPA SW 846 Method 8020	---	1/8 H				
Flash Point	3	---	G				
Temperature	3	---	G				
pH	3	---	G				
BOD	3	---	C				
CBOD	3	---	C				
TSS	3	---	C				
Total Phosphorus	3	---	C				
Total Nitrogen	3	---	C				
Fecal Coliform	3	---	G				
Ammonia Nitrogen	3	---	C				
Acetone	EPA 8260	5	1/8 H				
Methyl Ethyl Ketone	EPA 8260	5	1/8 H				
Ethylene Glycol	EPA 8015	1 mg/L (direct injection), 50 ug/L (purge and trap)	1/8 H				
Tertiary Butyl Alcohol	EPA 8015	1 mg/L (direct injection), 50 ug/L (purge and trap)	1/8 H				
Methanol	EPA 8015	1 mg/L (direct injection), 50 ug/L (purge and trap)	1/8 H				
Catechol (1,2-Benzenediol)	EPA 8270	10 ug/L	1/8 H				

Wastewater Discharge Permit Application for Existing Industrial Users, January 1995
Section XVIII, Attachment I

Organic Chemicals, Plastics and Synthetic Fibers Pollutants							
Pollutant or Pollutant Characteristic	Method	Quantification Level	Sample Type	Results (by individual date and avg)			
				Date:	Date:	Date:	Avg
Acenaphthene	625	---	C				
Anthracene	625	---	C				
Benzene	624	---	G				
Bis (2-ethylhexyl) phthalate	625	---	C				
Carbon Tetrachloride	625	---	G				
Chlorobenzene	624	---	G				
Chloroethane	624	---	G				
Chloroform	624	---	G				
Di-n-butyl phthalate	625	---	G				
1,2-Dichlorobenzene	624	---	G				
1,3-Dichlorobenzene	624	---	G				
1,4-Dichlorobenzene	624	---	G				
1,1-Dichloroethane	624	---	G				
1,2-Dichloroethane	624	---	G				
1,1-Dichloroethylene	624	---	G				
1,2-trans-Dichloroethylene	624	---	G				
1,2-Dichloropropane	624	---	G				
1,3-Dichloropropylene	624	---	G				
Diethyl phthalate	625	---	C				
Dimethyl phthalate	625	---	C				
4,6-Dinitro-o-cresol	625	---	C				
Ethylbenzene	624	---	G				
Fluoranthene	625	---	C				
Fluorene	625	---	C				
Hexachlorobenzene	625	---	C				
Hexachlorobutadiene	625	---	C				
Hexachloroethane	625	---	C				
Methyl Chloride	624	---	G				
Methylene Chloride	624	---	G				
Naphthalene	625	---	C				
Nitrobenzene	625	---	C				

APPENDIX C

Wastewater Discharge Permit Application for Existing Industrial Users, January 1995
Section XVIII, Attachment I

Organic Chemicals, Plastics and Synthetic Fibers Pollutants							
Pollutant or Pollutant Characteristic	Method	Quantification Level	Sample Type	Results (by individual date and avg)			
				Date:	Date:	Date:	Avg
2-Nitrophenol	625	---	C				
4-Nitrophenol	625	---	C				
Phenanthrene	625	---	C				
Pyrene	625	---	C				
Tetrachloroethylene	624	---	G				
Toluene	624	---	G				
Total Cyanide	EPA 335.2	---	C				
Total Lead	EPA 239.2	---	G				
Total Zinc	EPA 289.1	---	G				
1,2,4-Trichlorobenzene	625	---	C				
1,1,1-Trichloroethane	624	---	G				
1,1,2-Trichloroethane	624	---	G				
Trichloroethylene	624	---	G				
Vinyl Chloride	624	---	G				

[1]   If the result of the Total Chromium analysis is less than or equal to the QL of 10 micrograms per liter, the result for Chromium III can be reported as not quantifiable.

[2]   Required continuous extraction.

[3]   Any approved method presented in 40 CFR Part 136.

CHAPTER 3

DEVELOPMENT AND APPLICATION OF REGULATIONS

by

Eddie Esfandi

TABLE OF CONTENTS

Chapter 3.   DEVELOPMENT AND APPLICATION OF REGULATIONS

OBJECTIVES

Chapter 3.   DEVELOPMENT AND APPLICATION OF REGULATIONS

Following completion of Chapter 3, you should be able to:

1. Identify the legal authorities involved in a pretreatment program,

2. List the important aspects of EPA's regulations,

3. Describe the various types of EPA standards,

4. Outline the development of categorical limits,

5. Identify types of industries and facilities subject to the categorical program,

6. Explain the electroplating and metal finishing regulations and their differences,

7. Describe the contents and significance of a Wastewater Ordinance,

8. Plan or review a pretreatment program,

9. Explain the role of local authorities in the control of pollutants,

10. Establish and implement an industrial waste enforcement program, and

11. Keep current with changing regulations.

PRETREATMENT WORDS

Chapter 3.   DEVELOPMENT AND APPLICATION OF REGULATIONS

40 CFR 403   40 CFR 403

EPA's General Pretreatment Regulations appear in the Code of Federal Regulations under 40 CFR 403. 40 refers to the numerical heading for the environmental regulations portion of the Code of Federal Regulations (CFR). 403 refers to the section which contains the General Pretreatment Regulations. Other sections include 413, Electroplating Categorical Regulations.

BASELINE MONITORING REPORT (BMR)   BASELINE MONITORING REPORT (BMR)

All industrial users subject to Categorical Pretreatment Standards must submit a baseline monitoring report (BMR) to the Control Authority (POTW, state or EPA). The purpose of the BMR is to provide information to the Control Authority to document the industrial user's current compliance status with a Categorical Pretreatment Standard. The BMR contains information on (1) name and address of facility, including names of operator(s) and owner(s), (2) list of all environmental control permits, (3) brief description of the nature, average production rate and SIC (or NAICS) code for each of the operations conducted, (4) flow measurement information for regulated process streams discharged to the municipal system, (5) identification of the pretreatment standards applicable to each regulated process and results of measurements of pollutant concentrations and/or mass, (6) statements of certification concerning compliance or noncompliance with the pretreatment standards, and (7) if not in compliance, a compliance schedule must be submitted with the BMR that describes the actions the industrial user will take and a timetable for completing these actions to achieve compliance with the standards.

BEST AVAILABLE TECHNOLOGY (BAT)   BEST AVAILABLE TECHNOLOGY (BAT)

A level of technology represented by a higher level of wastewater treatment technology than required by Best Practicable Technology (BPT). BAT is based on the very best (state of the art) control and treatment measures that have been developed, or are capable of being developed, and that are economically achievable within the appropriate industrial category.

BEST PRACTICABLE TECHNOLOGY (BPT)   BEST PRACTICABLE TECHNOLOGY (BPT)

A level of technology represented by the average of the best existing wastewater treatment performance levels within the industrial category.

CAPTIVE SHOP   CAPTIVE SHOP

Those electroplating and metal finishing facilities which in a calendar year own more than 50 percent (based on area) of material undergoing metal finishing.

CATEGORICAL LIMITS   CATEGORICAL LIMITS

Industrial wastewater discharge pollutant effluent limits developed by EPA that are applied to the effluent from any industry in any category anywhere in the United States that discharges to a POTW. These are pollutant effluent limits based on the technology available to treat the wastestreams from the processes of the specific industrial category and normally are measured at the point of discharge from the regulated process. The pollutant effluent limits are listed in the Code of Federal Regulations.

CHAIN OF CUSTODY   CHAIN OF CUSTODY

A record of each person involved in the handling and possession of a sample from the person who collected the sample to the person who analyzed the sample in the laboratory and to the person who witnessed disposal of the sample.

CODE OF FEDERAL REGULATIONS (CFR)   CODE OF FEDERAL REGULATIONS (CFR)

A publication of the United States Government which contains all of the proposed and finalized federal regulations, including environmental regulations.

COMPLIANCE   COMPLIANCE

The act of meeting specified conditions or requirements.

CONVENTIONAL POLLUTANTS   CONVENTIONAL POLLUTANTS

Those pollutants which are usually found in domestic, commercial or industrial wastes such as suspended solids, biochemical oxygen demand, pathogenic (disease-causing) organisms, adverse pH levels, and oil and grease.

CRADLE TO GRAVE   CRADLE TO GRAVE

A term used to describe a hazardous waste manifest system used by regulatory agencies to track a hazardous waste from the point of generation to the hauler and then to the ultimate disposal site.

DIRECT DISCHARGER   DIRECT DISCHARGER

A point source that discharges a pollutant(s) to waters of the United States, such as streams, lakes or oceans. These sources are subject to the National Pollutant Discharge Elimination System (NPDES) program regulations.

EFFLUENT LIMITS   EFFLUENT LIMITS

Pollutant limitations developed by a POTW for industrial plants discharging to the POTW system. At a minimum, all industrial facilities are required to comply with federal prohibited discharge standards. The industries covered by federal categorical standards must also comply with the appropriate discharge limitations. The POTW may also establish local limits more stringent than or in addition to the federal standards for some or all of its industrial users.

EXISTING SOURCE   EXISTING SOURCE

An industrial discharger that was in operation at the time of promulgation of the proposed pretreatment standard for the industrial category.

HAZARDOUS WASTE   HAZARDOUS WASTE

A waste, or combination of wastes, which because of its quantity, concentration, or physical, chemical, or infectious characteristics may:

1. Cause, or significantly contribute to, an increase in mortality or an increase in serious, irreversible, or incapacitating reversible illness; or

2. Pose a substantial present or potential hazard to human health or the environment when improperly treated, stored, transported, or disposed of or otherwise managed; and

3. Normally not be discharged into a sanitary sewer; subject to regulated disposal.

(Resource Conservation and Recovery Act (RCRA) definition.)

INDIRECT DISCHARGER   INDIRECT DISCHARGER

A nondomestic discharger introducing pollutants to a POTW. These facilities are subject to the EPA pretreatment regulations.

INTEGRATED FACILITY   INTEGRATED FACILITY

Plants which, prior to treatment or discharge, combine the metal finishing wastewaters with significant quantities (more than 10 percent of total volume) of wastewaters not covered by the metal finishing category. Also refers to any facilities covered by any of the industrial pretreatment categories and various combinations of categories.

JOB SHOP   JOB SHOP

Those electroplating and metal finishing facilities which in a calendar year do not own more than 50 percent (based on area) of material undergoing metal finishing.

NAICS   NAICS

North **A**merican **I**ndustry **C**lassification **S**ystem. A code number system used to identify various types of industries. This code system replaces the SIC (Standard Industrial Classification) code system used prior to 1997. Use of these code numbers is often mandatory. Some companies have several processes which will cause them to fit into two or more classifications. The code numbers are published by the Superintendent of Documents, U.S. Government Printing Office, PO Box 371954, Pittsburgh, PA 15250-7954. Stock No. 041-001-00509-9; price, $31.50. There is no charge for shipping and handling.

NPDES PERMIT   NPDES PERMIT

National **P**ollutant **D**ischarge **E**limination **S**ystem permit is the regulatory agency document issued by either a federal or state agency which is designed to control all discharges of pollutants from point sources and storm water runoff into U.S. waterways. NPDES permits regulate discharges into navigable waters from all point sources of pollution, including industries, municipal wastewater treatment plants, sanitary landfills, large agricultural feedlots and return irrigation flows.

NEW SOURCE   NEW SOURCE

Any building, structure, facility or installation from which there is or may be a discharge of pollutants. Construction of the facility must have begun after publication of the applicable Pretreatment Standards. The building, structure, facility or installation must also be constructed at a site at which no other source is located; or, must totally replace the existing process or production equipment producing the discharge at the site; or, must be substantially independent of an existing source of discharge at the same site.

NONINTEGRATED FACILITY   NONINTEGRATED FACILITY

Plants that produce wastewaters from different pretreatment categorical processes but do not combine the wastestreams prior to pretreatment or discharge to sewers.

POTW   POTW

Publicly **O**wned **T**reatment **W**orks. A treatment works which is owned by a state, municipality, city, town, special sewer district or other publicly owned and financed entity as opposed to a privately (industrial) owned treatment facility. This definition includes any devices and systems used in the storage, treatment, recycling and reclamation of municipal sewage or industrial wastes of a liquid nature. It also includes sewers, pipes and other conveyances only if they carry wastewater to a POTW treatment plant. The term also means the municipality (public entity) which has jurisdiction over the indirect discharges to and the discharges from such a treatment works. (Definition adapted from EPA's General Pretreatment Regulations, 40 CFR 403.3.)

RCRA (RICK-ruh)   RCRA

The Federal **R**esource **C**onservation and **R**ecovery **A**ct (10/21/76), Public Law (PL) 94-580, provides technical and financial assistance for the development of plans and facilities for recovery of energy and resources from discarded materials and for the safe disposal of discarded materials and hazardous wastes. This Act introduces the philosophy of the "cradle to grave" control of hazardous wastes. RCRA regulations can be found in Title 40 of the Code of Federal Regulations (40 CFR) Parts 260-268, 270 and 271.

SIC CODE   SIC CODE

Standard **I**ndustrial **C**lassification Code. A code number system used to identify various types of industries. In 1997, the United States and Canada replaced the SIC code system with the North American Industry Classification System (NAICS); Mexico adopted the NAICS in 1998. See NAICS.

SIGNIFICANT INDUSTRIAL USER (SIU)   SIGNIFICANT INDUSTRIAL USER (SIU)

A Significant Industrial User (SIU) includes:

1. All categorical industrial users, and
2. Any noncategorical industrial user that
   a. Discharges 25,000 gallons per day or more of process wastewater ("process wastewater" excludes sanitary, noncontact cooling and boiler blowdown wastewaters), or
   b. Contributes a process wastestream which makes up five percent or more of the average dry weather hydraulic or organic (BOD, TSS) capacity of a treatment plant, or
   c. Has a reasonable potential, in the opinion of the Control or Approval Authority, to adversely affect the POTW treatment plant (inhibition, pass-through of pollutants, sludge contamination, or endangerment of POTW workers).

SIGNIFICANT NONCOMPLIANCE   SIGNIFICANT NONCOMPLIANCE

An industrial user is in significant noncompliance if its violation meets one or more of the following criteria:

(A) Chronic violation of wastewater discharge limits, defined here as those in which sixty-six percent or more of all of the measurements taken during a six-month period exceed (by any magnitude) the daily maximum limit or the average limit for the same pollutant parameter;

(B) Technical Review Criteria (TRC) violations, defined here as those in which thirty-three percent or more of all of the measurements for each pollutant parameter taken during a six-month period equal or exceed the product of the daily maximum limit or the average limit multiplied by the applicable TRC (TRC = 1.4 for BOD, TSS, fats, oil and grease, and 1.2 for all other pollutants except pH);

(C) Any other violation of a pretreatment effluent limit (daily maximum or longer-term average) that the Control Authority determines has caused, alone or in combination with other discharges, interference or pass through (including endangering the health of POTW personnel or the general public);

(D) Any discharge of a pollutant that has caused imminent endangerment to human health, welfare or to the environment or has resulted in the POTW's exercise of its emergency authority under paragraph (f)(1)(vi)(b) of this section of the regulations to halt or prevent such a discharge;

(E) Failure to meet, within 90 days after the schedule date, a compliance schedule milestone contained in a local control mechanism or enforcement order for starting construction, completing construction, or attaining final compliance;

(F) Failure to provide, within 30 days after the due date, required reports such as baseline monitoring reports, 90-day compliance reports, periodic self-monitoring reports, and reports on compliance with compliance schedules;

(G) Failure to accurately report noncompliance; or

(H) Any other violation which the Control Authority determines will adversely affect the operation or implementation of the local pretreatment program.

SOLVENT MANAGEMENT PLAN SOLVENT MANAGEMENT PLAN

A strategy for keeping track of all solvents delivered to a site, their storage, use and disposal. This includes keeping spent solvents segregated from other process wastewaters to maximize the value of the recoverable solvents, to avoid contamination of other segregated wastes, and to prevent the discharge of toxic organics to any wastewater collection system or the environment. The plan should describe measures to control spills and leaks and to ensure that there is no deliberate dumping of solvents. Also known as a TOXIC ORGANIC MANAGEMENT PLAN (TOMP).

TOXIC ORGANIC MANAGEMENT PLAN (TOMP) TOXIC ORGANIC MANAGEMENT PLAN (TOMP)

A strategy for keeping track of all solvents delivered to a site, their storage, use and disposal. This includes keeping spent solvents segregated from other process wastewaters to maximize the value of the recoverable solvents, to avoid contamination of other segregated wastes, and to prevent the discharge of toxic organics to any wastewater collection system or the environment. The plan should address the control of spills and leaks and also ensure that there is no deliberate dumping of solvents. Also known as a SOLVENT MANAGEMENT PLAN.

TOXIC POLLUTANT TOXIC POLLUTANT

Those pollutants or combinations of pollutants, including disease-causing agents, which after discharge and upon exposure, ingestion, inhalation or assimilation into any organism, either directly from the environment or indirectly by ingestion through food chains, will, on the basis of information available to the Administrator of EPA, cause death, disease, behavioral abnormalities, cancer, genetic mutations, physiological malfunctions (including malfunctions in reproduction) or physical deformations, in such organisms or their offspring. (Definition from Section 502.(13) of the Clean Water Act.)

WASTEWATER FACILITIES WASTEWATER FACILITIES

The pipes, conduits, structures, equipment, and processes required to collect, convey, and treat domestic and industrial wastes, and dispose of the effluent and sludge.

WASTEWATER ORDINANCE WASTEWATER ORDINANCE

The basic document granting authority to administer a pretreatment inspection program. This ordinance must contain certain basic elements to provide a legal framework for effective enforcement.

PRETREATMENT ABBREVIATIONS

Chapter 3.   DEVELOPMENT AND APPLICATION OF REGULATIONS

BAT   BAT
Best Available Technology Economically Achievable.

BMR   BMR
Baseline Monitoring Report.

BPT   BPT
Best Practicable Technology Economically Available.

CERCLA   CERCLA
Comprehensive Environmental Response, Compensation, and Liability Act of 1980.

CFR   CFR
Code of Federal Regulations.

EPA   EPA
Environmental Protection Agency.

ERP   ERP
Enforcement Response Plan.

FDF   FDF
Fundamentally Different Factors.

IPCBM   IPCBM
Independent Printed Circuit Board Manufacturers.

IU   IU
Industrial User.

NAICS   NAICS
North American Industry Classification System.

NPDES   NPDES
National Pollutant Discharge Elimination System.

NRDC   NRDC
Natural Resources Defense Council.

NSPS   NSPS
New Source Performance Standards.

O & M   O & M
Operation and Maintenance.

OWWM   OWWM
Office of Water and Wastewater Management.

PDL   PDL
Prohibited Discharge Limit.

PEL   PEL
Permissible Exposure Level.

POTW   POTW
Publicly Owned Treatment Works.

PSES   PSES
Pretreatment Standards for Existing Sources.

PSNS   PSNS
Pretreatment Standards for New Sources.

RCRA   RCRA
Resource Conservation and Recovery Act.

SIC   SIC
Standard Industrial Classification.

SIU   SIU
Significant Industrial User.

TLV   TLV
Threshold Limit Value.

TOMP   TOMP
Toxic Organic Management Plan.

TSCA   TSCA
Toxic Substances Control Act.

TTO   TTO
Total Toxic Organics.

TWA   TWA
Time Weighted Average.

WOW.....─〤─....WOW!

CHAPTER 3. DEVELOPMENT AND APPLICATION OF REGULATIONS

In your work as a pretreatment inspector, there will be occasions when you will be called upon to explain pollution control regulations and various aspects of the regulatory process. For example, you will frequently be the person to inform industry representatives of the meaning of regulations that apply to them. Sometimes you may be asked to make presentations to community groups or municipal officials. Or, you could be asked to help instruct co-workers. The purpose of this chapter is to explain the basis of your legal authority to inspect industrial facilities and to teach you how to interpret and apply pollution control regulations.

3.0 OVERVIEW OF POLLUTION CONTROL AUTHORITY

3.00 Legislative History of Federal Pollution Control Regulations

Even prior to the awakening of the environmental movement in the 1960s and 1970s, many local wastewater treatment agencies had established various types of industrial waste control programs. These programs were aimed at gross industrial pollutants which were found to cause operational problems in collection systems and at treatment plants of the agency. While such programs were generally beneficial at local levels, there existed no means to ensure that all communities would be protected or that all industries would be uniformly prohibited from discharging environmentally damaging wastes.

Some national efforts had been made to protect the environment prior to 1972. The following is a list of federal pollution control efforts which led to modern water pollution regulations:

- 1899 The Rivers and Harbors Act,

- 1912 The Public Health Service Act,

- 1965 The Oil Pollution Control Act,

- 1966 The Clean Water Restoration Act, and

- 1970 The Water Quality Improvements Act.

By the early 1970s, public clamor for effective, comprehensive legislation to control water pollution led Congress to enact landmark legislation in the form of the 1972 Federal Water Pollution Control Act. For the first time in the history of the United States, Congress declared with this Act that the pollution of the nation's waters by either industries or municipalities was unlawful. The Act created a system of uniform controls on discharge of all pollutants and required a federal agency, the Environmental Protection Agency, to establish effluent limitations and standards. The major goals of this legislation were:

1. To have the nation's waters clean enough for swimming by 1983, and

2. To eliminate the discharge of pollutants by 1985.

The second goal, not yet reached, is an ambitious one which may not be attainable economically in the near future. However, it remains the goal toward which the nation's efforts are directed. The Federal Water Pollution Control Act has been amended several times since its passage and is now known as the Clean Water Act. Additional legislation further reinforces the mandate to protect the nation's waters. Two significant Acts of Congress are the 1977 Clean Water Act and the 1987 Water Quality Act.

3.01 Environmental Protection Agency

3.010 Scope of Authority

Established in 1970, the Environmental Protection Agency is the federal agency responsible for the protection of the nation's water, air and land from various pollutants. The agency is charged with developing pollution control guidelines, implementing programs to achieve the goals established by Congress, and coordinating the various pollution control activities throughout the nation. In this chapter we will be discussing the EPA's role in controlling water pollution. More specifically, we will deal with the EPA water pollution control regulations that concern industrial facilities which discharge to wastewater collection systems and treatment plants (POTWs).

3.011 Organizational Structure

EPA is divided into nine staff offices, six program offices and ten regional offices. The Office of Water and Wastewater Management (OWWM) is a program office that is responsible for development of effluent limitations and standards. The Office of Water Regulations and Standards is a suboffice of the OWWM. The Industrial Technology Division of the suboffice develops the federal categorical pretreatment standards. The Industrial Technology Division was formerly known as the Effluent Guidelines Division.

3.02 Delegation of Federal Authority

The Clean Water Act gives EPA the authority to develop and implement programs to control the flow of *TOXIC POLLUTANTS*[1] into POTWs. EPA can delegate its implementation authority to state or local officials and, for reasons of effective management, has chosen to do so. Local industrial source control programs already exist in many areas. The local POTWs are familiar with their industrial dischargers, may already have developed an extensive database about dischargers, and may have ongoing wastewater permit and administration mechanisms. By delegating its regulatory authority, EPA uses existing programs rather than replacing them with a costly new bureaucracy. Local authorities are also better able to respond quickly and effectively to wastewater collection system or treatment plant emergencies; they are better able to promptly resolve problems with industrial dischargers; and greater resources are usually available to local agencies to conduct a pretreatment program.

The process by which EPA delegates its implementation authority requires states to develop a pretreatment program that meets EPA guidelines. States with approved pretreatment programs are called pretreatment-delegated states. If a state shows that their pretreatment program meets all federal requirements, EPA will give authority to the state to implement and enforce the national pretreatment program.

Once a state's pretreatment program has been approved by EPA, the state may then delegate its regulatory authority to local POTWs for day-to-day implementation. This is accomplished through a process similar to EPA's state approval process. The state may require the local POTW to submit a pretreatment plan (Wastewater Ordinance) for review and state approval. If approved, the POTW is given the legal authority to implement the regulatory activities detailed in the ordinance.

The EPA has established certain conditions state governments must meet if they are to assume regulatory control of the local POTW pretreatment programs within their state. The regulations[2] (Section 403.10) provide that the state must have the legal authority, procedures, funding and personnel to perform the items listed below, as well as some additional specified work.

1. Incorporate pretreatment program conditions in state permits issued to POTWs.

2. Thoroughly review the monitoring reports furnished by POTWs and Industrial Users (IUs) and ensure that they are meeting federal discharge standards.

3. Thoroughly inspect and monitor POTWs and Industrial Users (IUs) to determine, independent of the above information, that the POTWs and IUs are in compliance with federal discharge standards.

4. For noncompliant POTWs and IUs, seek civil and/or criminal penalties proportional to the seriousness of the violation.

5. Approve or deny approval of local POTW pretreatment programs.

6. Approve or deny requests for variances from EPA discharge standards under the Fundamentally Different Factors and Removal Credit provisions. (Both provisions are explained in a later section of this chapter.)

In addition to the legal requirements under the EPA General Pretreatment Regulations, it may be necessary for the state to enact enabling legislation which will allow the local POTW to pass ordinances to implement the EPA pretreatment program requirements at the local level.

EPA regulations also establish requirements for local pretreatment programs. All local POTWs with over five million gallons per day (MGD) of wastewater flow or a significant level of industrial wastewater dischargers are required to establish a pretreatment program in compliance with the EPA 40 CFR 403 regulations. These regulations require a local pretreatment program to have the legal authority, procedures, funding and personnel to at least do the following:

1. Deny permission or establish conditions for the discharge of pollutants from Industrial Users (IUs) to the POTW;

2. Require compliance by IUs with applicable EPA and local pretreatment standards;

3. Control through permit, order or similar means the contribution to the POTW by each IU to ensure compliance with applicable pretreatment standards and requirements. In the case of Significant Industrial Users (SIUs), this control could be achieved through permits or equivalent individual control mechanisms issued to each user. Such control mechanisms must be enforceable;

4. Require a compliance schedule from all IUs not in full compliance with the EPA pretreatment standards;

---

[1] *Toxic Pollutant.   Those pollutants or combinations of pollutants, including disease-causing agents, which after discharge and upon exposure, ingestion, inhalation or assimilation into any organism, either directly from the environment or indirectly by ingestion through food chains, will, on the basis of information available to the Administrator of EPA, cause death, disease, behavioral abnormalities, cancer, genetic mutations, physiological malfunctions (including malfunctions in reproduction) or physical deformations, in such organisms or their offspring. (Definition from Section 502.(13) of the Clean Water Act.)*

[2] *40 CFR 403.   EPA's General Pretreatment Regulations appear in the Code of Federal Regulations under 40 CFR 403. 40 refers to the numerical heading for the environmental regulations portion of the Code of Federal Regulations (CFR). 403 refers to the section which contains the General Pretreatment Regulations. Other sections include 413, Electroplating Categorical Regulations. Significant amendments to the General Pretreatment Regulations include the PIRT Amendments (FEDERAL REGISTER, October 18, 1988) and the DSS Amendments (FEDERAL REGISTER, July 24, 1990).*

5. Require an initial *BASELINE MONITORING REPORT (BMR)*[3] plus 90-day compliance reports plus periodic self-monitoring reports from IUs to ensure they are in compliance with all applicable pretreatment standards;

6. Perform inspection and monitoring activities independent of the above information which are sufficient to ensure that the IUs are complying with the pretreatment standards (the POTW must be freely able to enter and inspect industrial property in order to perform this function);

7. Obtain remedies for noncompliance by any IU with any pretreatment standard or requirement.[4] All POTWs must be able to seek injunctive relief for noncompliance by IUs with pretreatment standards and requirements. All POTWs must also have authority to seek or assess civil or criminal penalties in at least the amount of $1,000.00 a day for each violation by IUs of pretreatment standards and requirements;

8. Identify and locate IUs affected by the EPA regulations and notify them of the applicability of these regulations;

9. Investigate and remedy instances of noncompliance by IUs in the POTW service area; and

10. Obtain effective control of industrial waste discharges that endanger public health, the environment or the operation of the POTW.

Federal regulations require that the POTW have an adequate control mechanism (for example, permit system, administrative orders) to implement the above requirements. If properly used, an industrial waste discharge permit system can incorporate many of the above requirements of the EPA's pretreatment program. A permit system will allow the POTW to:

1. Obtain necessary information from an IU;

2. Conduct periodic reviews of the IU's discharge quality and quantity;

3. Incorporate specific discharge conditions in the company's permit, including effluent limits based on applicable general and categorical pretreatment regulations, local limits, and state and local law;

4. Establish the duration of the permit (to a maximum of five years) and the conditions governing transferability or nontransferability of the permit;

5. Specify requirements for self-monitoring, sampling, reporting, notification and recordkeeping, including identification by the POTW of pollutants to be monitored, sampling locations, sampling frequencies and sample types;

6. Suspend or revoke the permit to remedy noncompliance by the IU;

7. Specify and seek applicable civil and criminal penalties for violation of pretreatment standards and requirements; and

8. Impose compliance schedules to ensure compliance within federal deadlines.

One issue which is common to many POTWs is a question of adequate jurisdiction when a discharger in a different political area is violating discharge requirements for disposal of wastewater to the POTW. EPA requires the agency (the POTW) holding the NPDES discharge permit to enforce pretreatment requirements throughout the tributary service area. If industrial dischargers are located outside the political boundaries of your POTW, it is recommended that you enter into a contractual agreement with the local agency providing collection system services to the industry. The purpose of this agreement is to give you (the POTW) authority to regulate the specific industry by administrative order or through your permit system.

The delegation of authority is a matter of direct importance to you as a pretreatment program manager or inspector. Proper legal authority is an essential part of a workable pretreatment program. Without adequate authority, an inspector's position is reduced to strictly an advisory role. For effective control, authority must be backed by effective standards and the willingness to enforce them. It is essential that you, the pretreatment program manager or inspector, be fully aware of the scope of your authority and be thoroughly familiar with the provisions of regulations you are hired to enforce.

---

[3] *Baseline Monitoring Report (BMR).   All industrial users subject to Categorical Pretreatment Standards must submit a baseline monitoring report (BMR) to the Control Authority (POTW, state or EPA). The purpose of the BMR is to provide information to the Control Authority to document the industrial user's current compliance status with a Categorical Pretreatment Standard. The BMR contains information on (1) name and address of facility, including names of operator(s) and owner(s), (2) list of all environmental control permits, (3) brief description of the nature, average production rate and SIC (or NAICS) code for each of the operations conducted, (4) flow measurement information for regulated process streams discharged to the municipal system, (5) identification of the pretreatment standards applicable to each regulated process and results of measurements of pollutant concentrations and/or mass, (6) statements of certification concerning compliance or noncompliance with the pretreatment standards, and (7) if not in compliance, a compliance schedule must be submitted with the BMR that describes the actions the industrial user will take and a timetable for completing these actions to achieve compliance with the standards.*

[4] *Pretreatment requirements which will be enforced through the remedies discussed above will include, but not be limited to, the duty to allow or carry out inspections, entry, or monitoring activities; any rules, regulations, or orders issued by the POTW; any requirements set forth in individual control mechanisms issued by the POTW; or any reporting requirements imposed by the POTW or these regulations. The POTW shall have authority and procedures (after informal notice to the discharger) immediately and effectively to halt or prevent any discharge of pollutants to the POTW which reasonably appears to present an imminent endangerment to the health or welfare of persons. The POTW shall also have authority and procedures (which shall include notice to the affected industrial users and an opportunity to respond) to halt or prevent any discharge to the POTW which presents or may present an endangerment to the environment or which threatens to interfere with the operation of the POTW.*

QUESTIONS

Write your answers in a notebook and then compare your answers with those on page 240.

3.0A   What were the two major goals of the 1972 Federal Water Pollution Control Act?

3.0B   What is the basic mission of the Environmental Protection Agency?

3.0C   Under 40 CFR 403, which POTWs are required to establish a pretreatment program?

3.1   FEDERAL REGULATIONS

3.10   National Pretreatment Program

In its efforts to control the flow of pollutants into water, EPA recognizes two types of dischargers: direct and indirect.

DIRECT DISCHARGERS are facilities (such as an industry or POTW) that discharge wastewaters directly into U.S. streams, lakes or other waters, or into the oceans. These facilities are subject to the National Pollutant Discharge Elimination System (NPDES) program regulations.

INDIRECT DISCHARGERS are facilities that discharge wastewaters to a POTW and are subject to the regulations known as Pretreatment Standards for New or Existing Sources (PSNS or PSES). An example of an indirect discharger is a metal finishing company which discharges its industrial wastewaters to the local POTW as opposed to discharging directly to the environment.

Each of these two types of dischargers is further subdivided on the basis of when the plant came into existence. These subdivisions are referred to as existing or new dischargers and may each be subject to a separate set of effluent standards. Existing and New Sources are defined as follows:

EXISTING SOURCES are industrial dischargers which were in existence at the time of promulgation of the proposed applicable Categorical Standards.

NEW SOURCES are any building, structure, facility or installation from which there is or may be a discharge of pollutants. Construction of the facility must have begun after publication of the applicable Pretreatment Standards. The building, structure, facility or installation must also be constructed at a site at which no other source is located; or, must totally replace the existing process or production equipment producing the discharge at the site; or, must be substantially independent of an existing source of discharge at the same site.

3.100   *General Pretreatment Regulations*

At the heart of the National Pretreatment Program is a set of rules and standards known as the General Pretreatment Reg-

ulations. The purpose of the General Pretreatment Regulations is to implement the Federal Water Pollution Control Act Amendments of 1972 and the Clean Water Act of 1977 as they apply to industries which discharge nondomestic wastewaters to POTWs. The regulations establish the responsibilities of the federal, state and local governments, industry and the public to implement the National Pretreatment Program. These regulations set the ground rules on the type and quantity of pollutant which may be discharged to POTWs, and they apply universally to all industrial dischargers who discharge to POTWs. The complete text of the General Pretreatment Regulations is contained in the Code of Federal Regulations under 40 CFR 403.

The General Pretreatment Regulations regulate pollutants which may:

1. Pass through the POTW's treatment system, untreated or partially treated;

2. Interfere with the POTW's treatment works; and/or

3. Contaminate the POTW's sludge.

The most common of these pollutants are conventional pollutants which are usually found in domestic, commercial or industrial wastes. These are types of pollutants which a POTW's treatment system is designed to remove. The conventional pollutants are described below.

1. BOD, Biochemical Oxygen Demand, is the rate of oxygen uptake required by the microorganisms to use the organic content of the wastewater. In other words, this is a measure of the organic strength of the wastes in water. BOD content of wastewater is usually reported as mg/L (ppm) of oxygen required by the microorganisms to use the organic matter in the wastewater during five days at 20°C (68°F).

2. SS, Suspended Solids, is a measure of the quantity of suspended materials in the wastewater. POTWs are designed to remove these solids from the wastewater. Suspended solids contents of wastewaters are usually reported as mg/L (ppm).

3. Fecal coliforms are microbial organisms normally found in the digestive tract of humans and animals. The fecal coliform bacteria are used as an indicator organism to reveal the potential presence of pathogenic (disease-causing) bacteria and other harmful organisms. Fecal coliform concentrations are usually reported as coliforms/100 milliliters.

4. pH is a measure of the basic or acidic condition of the wastewater. A pH of 7.0 is called a neutral pH. pH below 5.0 is considered an acidic pH while pH above 8.0 is considered as basic pH.

5. Oil and grease are a measure of the oil and grease content of wastewater and are generally reported as mg/L (ppm).

Any pollutant which is not classified as a conventional or toxic pollutant is a nonconventional pollutant; for example, ammonia.

The General Pretreatment Regulations apply to pollutants from nondomestic sources which are directly discharged into or transported by truck or rail or otherwise introduced to POTWs. Pollutants which are transported by truck or rail into POTWs, if discharged into the headworks of the treatment plant without prior mixing with domestic wastewater (sewage), would be covered by the *RCRA*[5] regulation (if the pollutant is a listed RCRA hazardous waste). Hazardous wastes delivered to a POTW by truck, rail, or dedicated pipeline must comply with RCRA regulations and POTWs receiving such hazardous wastes are subject to regulation under a RCRA permit.

The Federal Pretreatment Regulations do not intend to affect any pretreatment requirements, including any standards or prohibitions, established by state or local laws as long as the state or local requirements are more stringent than the applicable National Pretreatment Standards or any other requirements or prohibitions under the General Pretreatment Regulations.

The General Pretreatment Regulations require states with **N**ational **P**ollutant **D**ischarge **E**limination **S**ystem (NPDES) programs to develop a pretreatment program. These regulations also require that a POTW (or combination of POTWs operated by the same authority) with a design flow greater than five million gallons per day (MGD) must establish a pretreatment program as a condition of its NPDES permit. POTWs with a design flow of less than five MGD may also be required to establish a pretreatment program if they have industrial users subject to National Pretreatment Standards or if nondomestic wastes or wastewaters cause upsets, violations of NPDES permit conditions, or sludge contamination.

QUESTIONS

Write your answers in a notebook and then compare your answers with those on pages 240 and 241.

3.1A   EPA has developed specific regulatory requirements for what two different types of industrial dischargers?

3.1B   What is the difference between "new sources" and "existing sources"?

3.1C   In general, what types of pollutants are regulated by the General Pretreatment Regulations?

3.1D   What are the most common conventional pollutants?

3.1E   Under what conditions might a POTW with a design flow of less than five MGD be required to establish a pretreatment program?

3.101   Prohibited Discharge Standards

In the 40 CFR 403 regulations (General Pretreatment Regulations), two types of regulatory programs are established: the Prohibited Discharge Standards and the Industrial Categorical Discharge Standards.

The Prohibited Discharge Standards are relatively simple and include two General Prohibitions and eight Specific Prohibitions.

A. *GENERAL PROHIBITIONS* disallow introduction into POTWs of industrial wastewaters which:

1. Pass through the POTWs untreated, and/or

2. Interfere with the operation or performance of the POTWs.

B. *SPECIFIC PROHIBITIONS* disallow the introduction into POTWs of eight specific categories of pollutants, as follows:

1. Pollutants that create a fire or explosion hazard in the POTW's sewer system or at the treatment plant including, but not limited to, wastestreams with a closed cup flashpoint of less than 140°F or 60°C using the test methods specified in the RCRA regulation (not 40 CFR Part 136!);

2. Pollutants that are corrosive, including any discharge with a pH lower than 5.0, unless the POTW is specifically designed to handle such discharges;

3. Solid or viscous (thick like syrup) pollutants in amounts that will obstruct the flow in the collection system and treatment plant, resulting in interference with operations;

4. Any pollutant discharged in quantities sufficient to interfere with POTW operations (includes BOD, but BOD is not listed specifically);

5. Discharges with temperatures above 104°F (40°C) when they reach the treatment plant, or hot enough to interfere with biological processes at the wastewater treatment plant;

6. Petroleum oil, nonbiodegradable cutting oil, or products of mineral oil origin in amounts that will cause interference or pass-through;

7. Pollutants which result in the presence of toxic gases, vapors, or fumes within the POTW in a quantity that may cause acute (sudden and severe) worker health and safety problems; and

8. Any trucked or hauled pollutants, except at discharge points designated by the POTW.

---

[5] *RCRA (RICK-ruh).   The Federal **R**esource **C**onservation and **R**ecovery **A**ct (10/21/76), Public Law (PL) 94-580, provides technical and financial assistance for the development of plans and facilities for recovery of energy and resources from discarded materials and for the safe disposal of discarded materials and hazardous wastes. This Act introduces the philosophy of the "cradle to grave" control of hazardous wastes. RCRA regulations can be found in Title 40 of the Code of Federal Regulations (40 CFR) Parts 260-268, 270 and 271.*

The local pretreatment program is required to establish specific discharge limits for industrial facilities as needed to enforce the Prohibited Discharge Standards. Table 3.1 provides a comparison of local Prohibited Discharge Limits (PDLs) for various treatment plants.

The General Pretreatment Regulations require all POTWs with flows of more than five million gallons per day (5 MGD), and smaller POTWs that serve a significant number of industrial dischargers, to develop, implement and enforce a local pretreatment program. Such pretreatment programs must, as a condition for approval, implement the General and Specific Prohibitions listed previously.

A major component of any local pretreatment program is local limits. At a minimum, all industrial facilities must comply with Federal Prohibited Discharge Standards, and categorical companies must comply with the Categorical Pretreatment Standards. In addition to these standards, however, POTWs may also establish effluent limits that are more stringent than, or in addition to, federal standards.

3.102   Categorical Pretreatment Standards

In addition to the Prohibited Discharge Standards, the EPA has established the Industrial Categorical Pretreatment Standards. These standards were established by EPA in conjunction with its program to regulate the direct discharge to the environment of wastewaters from certain categories of industrial facilities. To date, EPA has established standards for 26 industrial categories, and regulates both direct discharges and indirect discharges passing through a POTW. For many of the categories, several subcategories have been created leading to a number of effluent limitations. The local POTW is required to enforce these regulations at categorical companies within its service area. In some special situations, the state may enforce the regulations in place of the POTW. Table 3.2 lists the industrial categories regulated by the EPA.

Early regulations to implement the Clean Water Act of 1972 originally applied effluent limits only to conventional pollutants such as pH, oil and grease, BOD, and suspended solids. As a result of lawsuits brought against EPA by environmental groups, EPA agreed to undertake a major review of industrial effluent limitations. (This effort became known as the BAT (Best Available Technology Economically Achievable) review.) It was intended to develop more stringent and far-reaching controls on industrial pollution. The net effect, however, was a major shift in EPA's pollutant control strategy from the control of conventional pollutants to control of toxic pollutants.

Toxic pollutants are compounds or classes of compounds identified by EPA to be harmful to one or more forms of plant or animal life. Originally EPA identified 129 such pollutants but subsequently deleted three, leaving 126 substances classified as toxic pollutants. These priority toxic pollutants are subdivided into two categories: inorganic pollutants and organic pollutants. Table 3.3 contains a listing of both the inorganic and organic priority toxic pollutants targeted by EPA for regulation.

POTWs are not designed to remove toxic pollutants, although some toxic pollutants are incidentally removed through the POTW's treatment system. In fact, heavy metals such as zinc and copper inhibit sulfide corrosion of the sanitary sewers by combining with the dissolved sulfide in the wastewater. However, excessive concentrations of heavy metals are toxic to the bacteria which purify the wastewater in the POTWs. According to the EPA, implementation of the categorical pretreatment program will reduce introduction of heavy metals to POTWs from 56 million to 9 million pounds per year (84 percent reduction).

In establishing the pretreatment regulations for industrial categories, EPA performed detailed reviews of the wastes created at various types of industrial sites and evaluated the types of wastewater treatment in use or which could be used. Based on these studies and the projected economic impacts on the affected industries, EPA defined a reasonable expectation for the quality of the treated wastewater.

The resulting categorical standards are technology-based standards, meaning they are based upon the available treatment technologies, in contrast to receiving water standards, which are based on the tolerance to pollutants of the stream which receives the wastewater.

QUESTIONS

Write your answers in a notebook and then compare your answers with those on page 241.

3.1F   EPA's General Pretreatment Regulations (40 CFR 403) established what two types of regulatory programs?

3.1G   The General Pretreatment Regulations specifically prohibit the introduction of what eight types of pollutants into POTWs?

3.1H   What are toxic pollutants?

3.1I   What process did EPA use to establish the Industrial Categorical Pretreatment Standards?

3.1J   What is the difference between technology-based standards and receiving water standards?

3.11   EPA Regulation Development Process

3.110   EPA Procedures

The Industrial Technology Division of the EPA Office of Water Regulations and Standards, formerly the Effluent Guidelines Division, develops the categorical regulations and limits using the following process.

As previously described, EPA divides industrial dischargers into categories or groups of similar industries for regulatory purposes. Each category is further subdivided by the facility type and point of discharge. Based on point of discharge, industries are referred to as direct or indirect dischargers. In addition, each discharger type has been subdivided based on when the industrial plant came into existence, and plants are referred to as existing or new dischargers. Each subdivision may be subject to a separate set of effluent standards.

To develop categorical pretreatment standards, EPA first studies a particular category or type of discharger, conducts

TABLE 3.1 COMPARISON OF VIRGINIA PROHIBITED DISCHARGE LIMITS
Provided by Virginia Association of Pretreatment Coordinators (1991 Survey, Updated February, 1992)

	1	2	3	4	5	6	7	8	9	10	11	12	13
Copper	0.11	3.19	1.16	0.13	0.21	1.68	3.19	1	4.5	1	1	1.52	3.19
Cadmium	0.005	0.04	1.11	0.017	0.49	0.06	0.04	0.02	2	0.02	0.02	0.004	0.04
Nickel		2.02	0.99	0.33	4.98	1.87	2.02	1	8	1	2	0.57	2.02
Chromium		4.17	2	1.5	1.39	3.82	4.17	5	5	1	2	1.38	4.17
Zinc		1.53	2.54	0.72	2.41	2.99	1.53	5	4.2	1	2	3.37	1.53
Iron			1	17				5		50			
Lead	0.014	0.45	2.05	0.4	0.32	1.44	0.45	0.1	0.3	0.1	0.2	1.03	0.45
Mercury		0.04	0.005	0.013	0.02		0.04	0.005	0.01	0.02	0.005		0.04
Silver	0.001		0.1	0.05			0.19	0.5	0.6	0.1	0.1		0.19
Cyanide		1	0.25	0.025	0.21		1	0.55		0.02	1		1
FOG	100	100			100	100	300		100	100		100	
pH (units)		>5.0		6-9	6-9	5.5-9.5	5-9	5-9	>5.0	6-9	5.5-9.5	6-9	5-9
TTO							2.13		1	2.13		2.13	2.13
Phenol			0.05					0.5		0.02			
Selenium			0.02					0.02		0.02			
Arsenic			0.02					0.05		0.05			
Barium			2					5					
Boron			0.05					1		1			
Surfactant			5										
Tin			0.2					1					
COD (<15,000 GPD)			500										
COD (>15,000 GPD)			1,500										
BOD						300							
TSS						300							
Flash Point ('F)							>140						
O&G/P							100						
Chloride								250					
Chromium +6								1					
Manganese								5					
BTEX											5		
LOCAL LIMITS ARE BASED ON WHICH OF THE FOLLOWING?													
Technology-based			X		X	X			X	X	X	X	X
BPJ	X	X	X		X		X	X	X	X			
WQS-based	X	X		X	X	X			X	X		X	
Inhibition					X								
Sludge Quality													
ARE LOCAL LIMITS MASS-BASED IN THE ORDINANCE OR PERMIT?													
YES													
NO	X	X	X	X	X	X	X	X	X	X	X	X	

POTW KEY:

1	City of Martinsville	5	Frederick County Sanitation Authority
2	City of Richmond	6	Harrisonburg-Rockingham Regional Sewerage Authority
3	City of Waynesboro	7	County of Henrico
4	Upper Occoquan Sewage Authority	8	Pepper's Ferry Regional Wastewater Treatment Authority
		9	Alexandria Sanitation Authority
		10	Rivanna Water and Sewer Authority
11	City of Roanoke		
12	Chesterfield County		
13	County of Hanover		

TABLE 3.2   LIST OF INDUSTRIAL CATEGORIES REGULATED BY EPA

Industrial Category	40 CFR Part	Regulation Status	Pending Revisions [a]
Aluminum Forming	467	Final	None
Asbestos Manufacturing	427	Final	None
Battery Manufacturing	461	Final	None
Canned and Preserved Fruits and Vegetables	407	Final	None
Carbon Black	458	Final	None
Cement Processing	411	Final	None
Coal Mining	434	Final	None
Coil Coating			
Phase I	465	Final	None
Phase II (Canmaking)	465	Final	None
Copper Forming	468	Final	Proposed
Electrical/Electronic Components			
Phase I	469	Final	None
Phase II	469	Final	None
Electroplating			
PSES only	413	Final	None
Feedlots	412	Final	None
Ferroalloy Manufacturing	424	Final	None
Fertilizer (Phosphate)			
NSPS only	418		
Glass Manufacturing	426	Final	None
Grain Mills	406	Final	None
Ink Formulating	447	Final	None
Inorganic Chemicals			
Phase I	415	Final	None
Phase II	415	Final	None
Iron and Steel Manufacturing	420	Final	None
Leather Tanning and Finishing	425	Final	None
Machinery Manufacturing and Rebuilding	( )[b]	Proposed [a]	
Metal Finishing	433 and 413	Final	None
Metal Molding and Casting (Foundries)	464	Final	None
Nonferrous Metals			
Phase I	421	Final	None
Phase II	421	Final	None
Nonferrous Metals Forming	471	Final	None
Oil and Gas (Offshore)	435	Final	Proposed
Ore Mining	440	Final	None
Ore Mining (Placer Mining)	440	Final	None

[a] Per EPA Regulatory Calendar, October 29, 1990.
[b] ( ) Indicates the Category has not been assigned a Part number.

TABLE 3.2   LIST OF INDUSTRIAL CATEGORIES REGULATED BY EPA (continued)

Industrial Category	40 CFR Part	Regulation Status	Pending Revisions [a]
Organic Chemicals and Plastics and Synthetic Fibers	414	Final	Proposed
Paint Formulating	446	Final	None
Paving and Roofing Materials	443	Final	None
Pesticides	455	Final	Proposed
Petroleum Refining	419	Final	None
Pharmaceuticals	439	Final	Proposed
Plastics Molding and Forming	463	Final	None
Porcelain Enameling	466	Final	None
Pulp, Paper and Paperboard	430	Final	Proposed
Rubber Manufacturing	428	Final	None
Soap and Detergent Manufacturing	417	Final	None
Steam Electric Power Generation	423	Final	None
Sugar Processing	409	Final	None
Textile Mills	410	Final	None
Timber	429	Final	None
Waste Treatment	( ) [b]	Proposed [a]	

[a]   Per EPA Regulatory Calendar, October 29, 1990.
[b]   ( ) Indicates the Category has not been assigned a Part number.

TABLE 3.3   PRIORITY TOXIC POLLUTANTS

1. Acenaphthene
2. Acrolein
3. Acrylonitrile
4. Benzene
5. Benzidine
6. Carbon tetrachloride (tetrachloromethane)
7. Chlorobenzene
8. 1,2,4-trichlorobenzene
9. Hexachlorobenzene
10. 1,1-dichloroethane
11. 1,2-dichloroethane
12. 1,1,1-trichloroethane
13. Hexachloroethane
14. 1,1,2-trichloroethane
15. 1,1,2,2-tetrachloroethane
16. Chloroethane
17. Bis(2-chloroethyl) ether
18. 2-chloroethyl vinyl ether (mixed)
19. 2-chloronaphthalene
20. 2,4,6-trichlorophenol
21. Parachlorometa cresol
22. Chloroform (trichloromethane)
23. 2-chlorophenol
24. 1,2 dichlorobenzene
25. 1,3 dichlorobenzene
26. 1,4 dichlorobenzene
27. 3,3-dichlorobenzidine
28. 1,1-dichloroethylene
29. 1,2-trans dichloroethylene
30. 2,4-dichlorophenol
31. 1,2 dichloropropane
32. 1,2 dichloropropylene (1,3 dichloropropylene)
33. 2,4-dimethylphenol
34. 2,4-dinitrotoluene
35. 2,6-dinitrotoluene
36. 1,2-diphenylhydrazine
37. Ethylbenzene
38. Fluoranthene
39. 4-chlorophenyl phenyl ether
40. 4-bromophenyl phenyl ether
41. Bis(2-chloroisopropyl) ether
42. Bis(2-chloroethoxy) methane
43. Methylene chloride (dichloromethane)
44. Methyl chloride (chloromethane)
45. Methyl bromide (dibromomethane)
46. Bromoform (tribromomethane)
47. Dichlorobromomethane
48. Chlorodibromomethane
49. Hexachlorobutadiene

50. Hexachlorocyclopentadiene
51. Isophorone
52. Naphthalene
53. Nitrobenzene
54. 2-nitrophenol
55. 4-nitrophenol
56. 2,4-dinitrophenol
57. 4,6-dinitro-o-cresol
58. N-nitrosodimethylamine
59. N-nitrosodiphenylamine
60. N-nitrosodi-n-propylamine
61. Pentachlorophenol
62. Phenol
63. Bis(2-ethylhexyl) phthalate
64. Butyl benzyl phthalate
65. Di-n-butyl phthalate
66. Di-n-octyl phthalate
67. Diethyl phthalate
68. Dimethyl phthalate
69. 1,2-benzanthracene (benzo(a)anthracene)
70. Benzo(a)pyrene (3,4-benzopyrene)
71. 3,4-Benzofluoranthene (benzo(b)fluoranthene)
72. 11,12-benzofluoranthene (benzo(k)fluoranthene)
73. Chrysene
74. Acenaphthylene
75. Anthracene
76. 1,12-benzoperylene (benzo(ghi)perylene)
77. Fluorene
78. Penanthrene
79. 1,2,5,6-dibenzanthracene (dibenzo(a,h)anthracene)
80. Indeno (1,2,3-cd)pyrene (2,3-o-phenylene pyrene)
81. Pyrene
82. Tetrachloroethylene
83. Toluene
84. Trichloroethylene
85. Vinyl chloride (chloroethylene)
86. Aldrin
87. Dieldrin
88. Chlordane (technical mixture & metabolites)
89. 4,4-DDT
90. 4,4-DDE (p,p-DDX)
91. 4,4-DDD (p,p-TDE)
92. Alpha-endosulfan
93. Beta-endosulfan

94. Endosulfan sulfate
95. Endrin
96. Endrin aldehyde
97. Heptachlor
98. Heptachlor epoxide (BHC-hexachlorocyclohexane)
99. Alpha-BHC
100. Beta-BHC
101. Gamma-BHC (lindane)
102. Delta-BHC PCB-polychlorinated biphenyls
103. PCB-1242 (Arochlor 1242)
104. PCB-1254 (Arochlor 1254)
105. PCB-1221 (Arochlor 1221)
106. PCB-1232 (Arochlor 1232)
107. PCB-1248 (Arochlor 1248)
108. PCB-1260 (Arochlor 1260)
109. PCB-1016 (Arochlor 1016)
110. Toxaphene
111. 2,3,7,8-tetrachlorodibenzo-p-dioxin (TCDD)

Inorganic Priority Pollutants

112. Antimony (Total)
113. Arsenic
114. Asbestos
115. Beryllium
116. Cadmium
117. Chromium
118. Copper
119. Cyanide
120. Lead
121. Mercury
122. Nickel
123. Selenium
124. Silver
125. Thallium
126. Zinc

effluent sampling in various industrial sites and establishes a database. The data are evaluated by EPA to determine the type and quantity of wastewater generated by an industrial category.

Next, EPA conducts field studies at industrial sites within a category to identify the range and capabilities of various wastewater control systems. Technical and economic analyses are performed to test the effectiveness and economic feasibility of each alternative pollution control system. Based on these studies, EPA identifies two levels of treatment technology:

> *BEST PRACTICABLE TECHNOLOGY (BPT)*. BPT level of technology represents the average of the best existing wastewater treatment performance within the industrial category.

> *BEST AVAILABLE TECHNOLOGY ECONOMICALLY ACHIEVABLE (BAT)*. BAT levels of technology represent a higher level of wastewater treatment technology than those required by Best Practicable Technology. BAT is based on the very best (state of the art) control and treatment measures that have been developed or are capable of being developed within the industrial category.

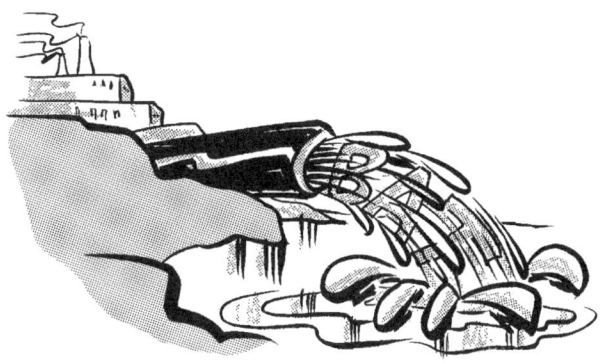

BAT technology is the basis for categorical effluent standards for direct dischargers and pretreatment standards for indirect dischargers. An additional consideration in the development of pretreatment standards, however, is the capacity of POTWs to remove various pollutants. EPA has developed data on the performance of 50 typical POTWs. It uses the information in this database to determine whether or not a typical POTW is capable of removing any of the pollutants discharged by the categorical dischargers to the same extent as BAT. If it is determined that a typical POTW is capable of removing a pollutant to the same extent as BAT, pretreatment standards for those pollutants are not promulgated for that category.

The detailed technical information compiled by EPA during these initial studies is published as "Development Documents" for the proposed regulations concerning each industrial category. These documents provide detailed technical information about the alternative technologies studies and are valuable sources of information for pretreatment inspectors.

Once the EPA has completed its preliminary studies, proposed effluent limits are published in the *FEDERAL REGISTER* as "Proposed Regulations." A brief comment period is provided to consider opinions of the parties affected by the regulations. After review of the public comments and appropriate modifications of the proposed regulations, the regulations are promulgated in final form in the *FEDERAL REGISTER*.

EPA decides on a model treatment technology for an industrial category and then issues discharge or effluent limita-

tions; these are the categorical standards for the industrial category. Usually, different standards are developed for the various types of dischargers: new and existing direct sources and new and existing indirect sources. Each of these sets of standards is described below.

EXISTING SOURCES — DIRECT DISCHARGERS. The Federal Water Pollution Control Act of 1972 required that all existing industries discharging industrial wastewater into waters of the United States were to achieve the Best Practicable Technology (BPT) economically achievable by July 1977, and to achieve the Best Available Technology (BAT) economically achievable by July 1,1983. The standards applicable to this group of dischargers are referred to as Categorical Standards.

NEW SOURCES — DIRECT DISCHARGERS. The Federal Water Pollution Control Act of 1972 required that all new industries constructed after the publication of a proposed regulation in the *FEDERAL REGISTER* which discharge industrial wastewaters into waters of the United States were to use the Best Available Technology (BAT). The regulations for this group of dischargers are called the New Source Performance Standards (NSPS).

EXISTING SOURCES — INDIRECT DISCHARGERS. The Pretreatment Standards for Existing Sources (PSES) are the category-specific regulations which apply to existing sources discharging industrial wastewater to POTWs. Table 3.4 is an example of the PSES regulations for the metal finishing category. The PSES for the metal finishing category apply to facilities which began construction before August 31, 1982 (the date for the proposed categorical regulations for the metal finishing category (40 CFR 433)).

NEW SOURCES — INDIRECT DISCHARGERS. The Pretreatment Standards for New Sources (PSNS) are the category-specific regulations which apply to new industrial sources discharging to POTWs. Table 3.5 is an example of the PSNS regulations for the metal finishing category. The PSNS for the metal finishing category apply to facilities which began construction after August 31,1982.

3.111 EPA-Regulated Categories

EPA entered into a Consent Decree in the case of Natural Resources Defense Council, Inc. (NRDC) vs. Train. This Decree required promulgation of technology-based standards for 21 industrial categories and required EPA to control toxic pollutants from these industrial categories.

In 1982 a court order modified the NRDC Settlement Agreement creating a total of 38 categories. Twelve categories were exempted under paragraph eight of the Agreement. Paragraph eight of the NRDC Agreement permits EPA to exempt any industrial category which does not warrant regulation after comprehensive evaluations. Regulations for the remaining 28 categories were promulgated by EPA.

Currently the total number of industrial categories subject to the EPA categorical program is 26. However the number of categories sometimes is cited as 29. This is because three categories, Electrical and Electronic Components Manufacturing, Inorganic Chemicals, and Nonferrous Metals Manufacturing, have Phase I and Phase II regulations. Table 3.6 is a list of these industrial categories with compliance dates for pretreatment standards for the existing sources for each category. Pretreatment standards for various industrial categories are available from EPA and state agencies. The Pretreatment Standards for Electroplating and Metal Finishing have been summarized and are included in Appendix A following the Objective Test at the end of this chapter.

3.112 Categories Exempt From EPA Categorical Regulations

Currently six industrial categories are exempted under paragraph eight of the NRDC Agreement. Following comprehensive evaluations EPA has determined that these categories (see Table 3.7) do not warrant federal regulation but they are still subject to local limits adopted by the POTWs.

3.113 Category Determination Requests

In the course of implementation of categorical pretreatment regulations, if there is a question or disagreement about applicability of a designated industrial category to an industrial discharge, the POTW or the industrial user can request a ruling by the EPA concerning the appropriateness of the industrial category. The decision of the Water Management Director of the EPA Regional Office is final.

3.114 Local Limits for Noncategorical Industries

Local authorities may establish limits for noncategorical industries when necessary. These limits also apply to categorical industries if they are more stringent than the Categorical Standards.

TABLE 3.4   PRETREATMENT STANDARDS FOR EXISTING SOURCES
FOR THE METAL FINISHING CATEGORIES

Parameter	Maximum for Any One Day, mg/L	Monthly Average Shall Not Exceed, mg/L
Cadmium (T)	0.69	0.26
Chromium (T)	2.77	1.71
Copper (T)	3.38	2.07
Lead (T)	0.69	0.43
Nickel (T)	3.98	2.38
Silver (T)	0.43	0.24
Zinc (T)	2.61	1.48
Cyanide (T) [a,b]	1.20	0.65
TTO, Interim	4.57	—
TTO, Final	2.13	—

T   Total
TTO   Total Toxic Organics

[a] Alternatively, for industrial facilities with cyanide treatment and upon agreement of the pollution control authority, a one-day maximum and a monthly average of 0.86 and 0.32 mg/L of amenable cyanide limit respectively may apply in place of the total cyanide limit specified above.
[b] Sampling for cyanide analyses to meet these standards (as for all other standards for cyanide) must be taken at the completion of the cyanide treatment processes, and prior to intermixing with any other noncyanide wastewaters in the total pretreatment system.

TABLE 3.5   PRETREATMENT STANDARDS FOR NEW SOURCES
FOR THE METAL FINISHING CATEGORIES

Parameter	Maximum for Any One Day, mg/L	Monthly Average Shall Not Exceed, mg/L
Cadmium (T)	0.11	0.07
Chromium (T)	2.77	1.71
Copper (T)	3.38	2.07
Lead (T)	0.69	0.43
Nickel (T)	3.98	2.38
Silver (T)	0.43	0.24
Zinc (T)	2.61	1.48
Cyanide (T)[a]	1.20	0.65
TTO	2.13	—

T   Total

[a] Alternatively, for industrial facilities with cyanide treatment and upon agreement of the pollution control authority, a one-day maximum and a monthly average of 0.86 and 0.32 mg/L of amenable cyanide limit respectively may apply in place of the total cyanide limit specified above.

**TABLE 3.6 INDUSTRIAL CATEGORIES SUBJECT TO THE
EPA CATEGORICAL PRETREATMENT REGULATIONS**

Industrial Category	Compliance Date
1. Aluminum Forming	10/14/86
2. Battery Manufacturing	03/09/87
3. Coil Coating	12/01/85
4. Coil Coating, Canmaking	11/17/86
5. Copper Forming	08/15/86
6. Electrical and Electronic Components Manufacturing — Phase I: TTO	07/01/84
AS	11/08/85
7. Electrical and Electronic Components Manufacturing — Phase II	07/14/86
8. Electroplating: Nonintegrated	04/27/84
Integrated	06/30/84
TTO	07/15/86
9. General Pretreatment Regulations	—
10. Inorganic Chemicals Manufacturing — Phase I	08/12/85
11. Inorganic Chemicals Manufacturing — Phase II	08/29/85
12. Iron and Steel Manufacturing	07/10/85
13. Leather Tanning and Finishing	11/25/85
14. Metal Finishing: Interim TTO	06/30/84
Iron and Steel TTO	07/10/85
Final	02/15/86
15. Metal Molding and Casting	10/31/88
16. Nonferrous Metals Forming	08/23/88
17. Nonferrous Metals Manufacturing — Phase I	03/09/87
18. Nonferrous Metals Manufacturing — Phase II	09/20/88
19. Ore Mining and Dressing (no known indirect discharger)	N/A
20. Organic Chemicals	2/90
21. Plastics Molding and Forming	*
22. Pesticides	11/18/88
23. Petroleum Refining	12/01/85
24. Pharmaceutical Manufacturing	10/27/86
25. Porcelain Enameling	11/25/85
26. Pulp, Paper and Paperboard Manufacturing	07/01/84
27. Steam Electric Power Generation	07/01/84
28. Textile Mills	*
29. Timber Products Processing	01/26/84

* No numerical pretreatment limits have been established and there is no compliance date. Industrial dischargers subject to these categories are required to comply with General Pretreatment Regulations 40 CFR Part 403 (Regulation 9 in this Table).

**TABLE 3.7 INDUSTRIAL CATEGORIES EXEMPTED
FROM THE EPA CATEGORICAL REGULATIONS**

1. Adhesives and Sealants
2. Auto and Other Laundries
3. Explosives Manufacturing
4. Gum and Wood
5. Photographic Equipment and Supplies
6. Printing and Publishing

QUESTIONS

Write your answers in a notebook and then compare your answers with those on page 241.

3.1K EPA's categorical regulations identify two levels of treatment technology. What are the two levels?

3.1L What are "Development Documents"?

3.1M What are the regulatory requirements for the control of water pollutants from indirect industrial wastewater dischargers called?

3.1N Who resolves questions or disagreements about the applicability of a designated industrial category between a POTW and an industrial user?

3.2 APPLICATION OF CATEGORICAL PRETREATMENT REGULATIONS

3.20 EPA Reporting Requirements

The General Pretreatment Regulations require all categorical dischargers to submit a number of reports to the POTW (Control Authority) authorized to implement the EPA categorical program. The reporting requirements for the categorical dischargers are as follows:

1. Baseline Monitoring Report — Compliance Schedule,

2. Final Compliance Report, 90-Day Report, and

3. Periodic Compliance Reports.

These reports must be signed by a responsible corporate officer of the industrial user (IU) and must contain the following certification statement:

"I certify under penalty of law that this document and all attachments were prepared under my direction or supervision in accordance with a system designed to assure that qualified personnel properly gather and evaluate the information submitted. Based on my inquiry of the person or persons who manage the system, or those persons directly responsible for gathering the information, the information submitted is, to the best of my knowledge and belief, true, accurate, and complete. I am aware that there are significant penalties for submitting false information, including the possibility of fine and imprisonment for knowing violations."

For purposes of signing industrial user reports, "responsible corporate officer" is defined as follows:

1. If the industrial user submitting the reports is a corporation, a responsible corporate officer means:

   a. A president, secretary, treasurer, or vice-president of the corporation in charge of a principal business function, or any other person who performs similar policy- or decision-making functions for the corporation, or

   b. The manager of one or more manufacturing, production, or operation facilities employing more than 250 persons or having gross annual sales or expenditures exceeding $25 million (in second quarter 1980 dollars), if authority to sign documents has been assigned or delegated to the manager in accordance with corporate procedures.

2. A general partner or proprietor if the industrial user is a partnership or sole proprietorship, respectively.

3. A duly authorized representative of the individual designated in paragraph 1 or 2 above if:

   a. The authorization is made in writing by the individual described in 1 or 2;

   b. The authorization specifies either an individual or a position having responsibility for the overall operation of the facility from which the industrial discharge originates, such as the position of plant manager, operator of a well, or well field superintendent, or a position of equivalent responsibility, or having overall responsibility for environmental matters for the company; and

   c. The written authorization is submitted to the Control Authority.

4. If an authorization under paragraph 3 is no longer accurate because a different individual or position has responsibility for the overall operation of the facility, or overall responsibility for environmental matters for the company, a new authorization satisfying the requirements of paragraph 3 must be submitted to the Control Authority prior to or together with any reports to be signed by an authorized representative.

These reports are subject to the provisions of 18 United States Code (U.S.C.) section 1001 relating to fraud and false statements; the provisions of sections 309(c)(4) of the Clean Water Act, as amended, governing false statements, representation or certification; and the provisions of section 309(c)(6) of the Clean Water Act regarding responsible corporate officers. Significant criminal and civil penalties await those who submit fraudulent or false statements.

For these reports, a minimum of four (4) grab samples must be used for pH, cyanide, total phenols, oil and grease, sulfide and volatile organics. For all other pollutants, a minimum of one (1) representative sample (24-hour composite sample obtained through flow proportional composite sampling techniques where feasible) must be collected for analysis. Collection and analysis of all samples must follow the procedures described in 40 CFR Part 136, Guidelines Establishing Test Procedures for the Analysis of Pollutants Under the Clean Water Act as amended.

The sampling and analysis for these reports may be performed by the Control Authority in place of the IU. Where the POTW performs the required sampling and analysis in place of the IU, the IU will not be required to submit the compliance certification stating whether applicable Pretreatment Standards or Requirements are being met on a consistent basis. The purpose behind not requiring this certification language is that the pretreatment inspector and the POTW will already know if the IU is in consistent compliance by reviewing the data the POTW collects. In addition, where the POTW itself collects all the information required for the report, including flow data and production data from the IU, the IU will not be required to submit a BMR, 90-day Compliance Report or Periodic Report on Compliance.

3.200   Baseline Monitoring Reports

A Baseline Monitoring Report (BMR) is the first report an indirect discharger must file following promulgation of a categorical standard applicable to the category. The BMR is due 180 days after the effective date of the regulations. The information required in the BMR includes:

1. Identification of the indirect discharger,

2. A list of environmental control permits,

3. A description of its operations,

4. A report on flows of regulated streams,

5. The results of sampling and analyses of the industrial wastewater discharges to determine levels of regulated pollutants in those streams,

6. A certification statement by the discharger indicating compliance or noncompliance with the applicable pretreatment standards, and

7. A description of any additional steps required to achieve compliance for noncompliant dischargers.

A typical BMR for electroplating and metal finishing categories can be found in Appendix B.

3.201   Compliance Schedule

A noncompliant industrial user who is not likely to achieve compliance with the applicable categorical pretreatment standards with minor operation and maintenance (O & M) modifications is required to submit a compliance schedule to the control authority as a part of BMR submission. This schedule must represent the shortest schedule of the industrial user's actions to achieve compliance with the applicable pretreatment standards. The final completion date on the compliance schedule must not be later than the applicable categorical compliance date specified in the regulations. If the date specified in the regulations has already passed or is impossible to meet, the final completion date must be as soon as possible. The schedule must contain specific incremental progress dates not exceeding nine months for completion of any increment. In actual practice the POTW and the industrial user commonly work together to establish an acceptable compliance schedule.

3.202   Final Compliance Report

All industrial facilities subject to the pretreatment regulations are required to submit a final compliance report. The final compliance report must be filed within 90 days after the compliance date of the regulations for all existing industrial dis-

chargers. The new categorical dischargers must file a final compliance report as soon as operations begin. A final report must contain the following information:

1. The sampling results for regulated pollutants in the industrial wastewater discharges,

2. Average and maximum daily industrial wastewater flows,

3. A statement of compliance, and

4. For noncompliant dischargers, a statement as to whether additional operation and maintenance modifications and/or pretreatment equipment is necessary to achieve compliance.

3.203   Periodic Compliance Reports

The General Pretreatment Regulations require all categorical dischargers to submit a minimum of two periodic self-monitoring reports on a semiannual basis to the POTW. Some POTWs require more frequent submission of self-monitoring reports, for example, quarterly. A periodic compliance report must contain the following information:

1. Type of facility,

2. Type of discharge,

3. Type and concentration of pollutants in the discharge, and

4. A certification statement concerning accuracy of the submitted information.

A typical periodic compliance report entitled "Critical Parameter Report Form" is included in Appendix C.

3.204   Slug Loading Reporting

All categorical and noncategorical IUs must notify the POTW immediately of all discharges that could cause problems to the POTW, including any slug loadings which would create a fire or explosion hazard; corrosive structural damage to the sewer lines or the treatment plant; solid or viscous pollutants which would cause obstruction of the flow; any pollutant, including oxygen-demanding pollutants, released in a discharge at a flow rate and/or pollutant concentration which will cause interference; heat which will inhibit biological activity; petroleum oil, nonbiodegradable cutting oil, or products of mineral oil origin in amounts that will cause interference or pass-through; or pollutants which result in the presence of toxic gases, vapors, or fumes within the POTW in a quantity that may cause acute worker health and safety problems.

3.205   Resampling to Confirm Violations by the Industrial User (IU)

If sampling performed by an IU indicates a violation, the user must notify the POTW within 24 hours of becoming aware of the violation. The IU also must repeat the sampling and analysis and submit the results of the repeat analysis to the POTW within 30 days after becoming aware of the violation, except the IU is not required to resample if:

a. The POTW performs sampling at the IU at least once per month, or

b. The POTW performs sampling at the IU between the time when the IU performs its initial sampling and the time when the IU receives the results of this sampling.

3.206   Notification of Changed Discharge

All IUs must notify the POTW in advance of any substantial change in the volume or character of pollutants in their discharge, including the listed or characteristic hazardous wastes for which the IU has submitted initial notification and those described in Section 3.207 of this manual.

3.207 Hazardous Waste Disposal Reporting

All Significant Industrial Users (SIUs) are required to notify the POTW, the EPA Regional Waste Management Division Director, and the state hazardous waste authorities in writing of any discharge into the POTW of a substance which, if otherwise disposed of, would be a hazardous waste under the Resource Conservation and Recovery Act (RCRA) and specifically contained in EPA's regulations in 40 CFR Part 261 (RCRA regulations). Information in this report must include:

- SIU identifying information,

- The name of the hazardous waste as set forth in 40 CFR Part 261,

- The EPA hazardous waste number, and

- The type of discharge (continuous, batch or other).

If the SIU discharges more than 100 kilograms of such waste per calendar month or any volume of acutely hazardous waste to the POTW, the notification should also contain the following information to the extent such information is known and readily available to the SIU:

- An identification of the hazardous constituents contained in the wastes,

- An estimation of the mass and concentration of such constituents in the wastestream discharged during that calendar month, and

- An estimation of the mass of constituents in the wastestream expected to be discharged during the following twelve months.

Any notification needs to be submitted only once for each hazardous waste discharged. However, notifications of changed discharge must be submitted as required (see previous Section 3.206). The notification requirement does not apply to pollutants already reported by the BMR, Final Compliance Report or the Periodic Compliance Reports and detailed in the POTW permit issued to the SIU.

The SIU is exempt from the requirements of this section during a calendar month in which the facility discharges no more than fifteen kilograms of hazardous wastes, unless the wastes are acute hazardous wastes as specified in the RCRA regulation. Discharge of more than fifteen kilograms of non-acute hazardous wastes in a calendar month, or any quantity of acute hazardous wastes, requires a one-time notification. Subsequent months during which the IU discharges more than these quantities of any hazardous waste do not require additional notification.

In the case of any new regulations under section 3001 of RCRA identifying additional characteristics of a hazardous waste or listing any additional substance as a hazardous waste, the IU must notify the POTW, the EPA Regional Waste Management Division Director, and state hazardous waste authorities of the discharge of such substance within 90 days of the effective date of such regulations. When submitting this information, the SIU must certify that it has a program in place to reduce the volume and toxicity of hazardous wastes generated to the degree it has determined to be economically practical.

This report can be signed by a technical representative of the SIU familiar with the source(s) of information contained within. It is not necessary to have the report signed by a responsible corporate officer, general partner, proprietor, or duly authorized representative of the corporate officer as required of other reports in this section.

3.21 Modification of Categorical Standards

General Pretreatment Regulations state the ground rules for the implementation of EPA's Categorical Pretreatment Program. Although the Categorical Pretreatment Standards apply uniformly to all affected categorical companies which discharge to POTWs, the Pretreatment Regulations specify three ways in which Categorical Pretreatment Standards could be modified:

1. Variance for fundamentally different factors (FDF),

2. Net gross adjustments, and

3. Removal credits.

3.210 Variance From Categorical Standards for Fundamentally Different Factors (FDF)

The General Pretreatment Regulations provide for modification of the Categorical Pretreatment Standards applicable to an industrial discharger if an industrial firm or an interested party can show that factors relating to the industrial user are fundamentally different from factors considered by EPA in establishing the standards. The General Pretreatment Regulations identify six factors which may be considered fundamentally different. These factors are:

1. The nature or quality of pollutants contained in the raw waste;

2. The volume of the industrial user's wastewater and quantity of effluent discharged;

3. Non-water-quality environmental impacts of control and treatment of the waste;

4. Energy requirements for the application of control technology;

5. Age, size, land availability, and configuration as relates to equipment or facilities, processes used, process modifications, and engineering aspects of application of the control technology; and

6. Cost of compliance.

The ability of EPA to grant FDF variances was challenged because of an amendment to the 1977 Act which states:

> The Administrator may not modify any requirement of this section as it applies to any specific pollutant which is on the toxic pollutant list.

In 1980, the Fourth Circuit Court of Appeals ruled that "BPT variance regulations need not exempt toxic pollutants." In 1983 the Court of Appeals for the Third Circuit reversed the above decision and ruled that "EPA could not issue FDF variances in the case of toxic pollutants." However, in February of 1985, the Supreme Court reversed the judgment of the Third Circuit Court of Appeals. EPA continues to have authority to grant FDF variances from the Pretreatment Standards.

3.211   Net Gross Calculations

A net gross credit allows the subtraction of the concentration of regulated pollutants in the intake water from the concentration level in the industrial discharger's effluent. Therefore, a net gross adjustment allows a facility to discharge a pollutant at a concentration level in excess of the applicable Federal Pretreatment Standards if it can be shown that a regulated pollutant is present in the industrial user's incoming water. To obtain a net gross adjustment, the discharger must submit a formal request to EPA that contains the following information:

1. Intake water is drawn from the same body of water that POTW discharges to;

2. The pollutants present in the incoming water will not be entirely removed by the industrial discharger's treatment system;

3. The pollutants in the intake water do not vary chemically or biologically from the pollutants limited by the applicable categorical standards; and

4. The industrial discharger does not significantly increase the concentration of pollutants in the intake water, even if the total quantity of pollutants remains the same.

3.212   Removal Credits

The Clean Water Act provides that POTWs may grant removal credits to indirect dischargers based on the degree of removal actually achieved at the POTW. If such removal credits were applied to the Categorical Pretreatment Standards, the affected industries could discharge regulated pollutants in excess of those permitted by the applicable EPA Pretreatment Standards.

A POTW with an approved pretreatment program, and approved authority to grant removal credits, has the discretion to grant removal credits to indirect industrial dischargers located within its service area.

A POTW must demonstrate consistent capability of removing 50 percent of its influent chromium through its treatment

system before it can grant removal credits equivalent to the removal rates at its system to the affected categorical dischargers. For example, a job shop electroplater located within the service area of the above POTW, where the categorical limit for chromium is 7.0 mg/L, can discharge up to 14.0 mg/L of chromium without violating the pretreatment standards.

The final sewage regulations (40 CFR Part 257, Technical Standards for the Use or Disposal of Sewage Sludge), published on February 19, 1993, contain provisions which would allow the POTW to grant removal credits. However, granting removal credits would be conditional on meeting the requirements of the sewage sludge regulations. To make granting removal credits possible under the sewage sludge regulations, 40 CFR 403.7 of the General Pretreatment Regulations was amended to make removal credits available for the following pollutants:

- Any pollutant listed in Appendix G Section I (of 40 CFR 403.7) for the use or disposal practice employed by the POTW if the requirements in 40 CFR Part 503 for that practice are met.

- Any pollutant listed in Appendix G Section II (of 40 CFR 403.7) for the use or disposal practice employed by the POTW if the concentration of the pollutant does not exceed the concentration listed in Appendix G Section II.

- Any pollutant in sewage sludge when the POTW disposes all of its sewage sludge in a municipal solid waste landfill unit that meets the criteria in 40 CFR Part 258.

In addition, Appendix G (listing the pollutants eligible for a removal credit) was added to 40 CFR Part 403.

3.22   Types of Categorical Standards

Two different types of categorical standards are used in the EPA pretreatment program: concentration-based standards and production-based standards. The production-based standards are occasionally referred to as mass-based standards.

Concentration-based standards are based on the relative strength of a pollutant in a wastestream, usually expressed as mg/L; for example, the pretreatment standard for an existing source in the metal finishing category for cadmium is 0.69 mg/L.

Production-based or mass-based standards are based on the actual mass of pollutants in a categorical wastewater stream per unit of production. Production-based standards are the most equitable type of standards. However, in practice it is difficult to develop, implement and enforce this type of standard. The production-based or mass-based standards are generally reported as pounds/square foot or milligrams/square meter of production area.

When the limits in a Categorical Pretreatment Standard are expressed only in terms of mass of pollutant per unit of production, the POTW may convert the limits to equivalent limitations expressed either as the mass of pollutant discharged per day or as the effluent concentration for purposes of calculating effluent limitations applicable to individual IUs.

A POTW calculating equivalent *MASS-PER-DAY LIMITATIONS* calculates such limitations by multiplying the limits in the Categorical Pretreatment Standard by the IU's average rate of production. This average rate of production is not based on the designed production capacity but rather on a reasonable measure of the IU's actual long-term daily production, such as the average daily production during a representative year. For new sources, actual production should be estimated using projected production.

A POTW calculating equivalent *CONCENTRATION LIMITATIONS* calculates such limitations by dividing the mass limitations derived above (see previous paragraph) by the average daily flow rate of the IU's regulated process wastewater. This average daily flow rate should be based on a reasonable measure of the IU's actual long-term average flow rate, such as the average daily flow rate during a representative year.

Equivalent limitations calculated using the procedures described in the two previous paragraphs are considered Pretreatment Standards for the purposes of Section 307(d) of the Clean Water Act and the General Pretreatment Standards. IUs will be required to comply with the equivalent limitations in place of the promulgated Categorical Standards from which the equivalent limitations were derived. Equivalent limitations are enforced under the POTW's Enforcement Response Plan (ERP) the same as local limits and Categorical Pretreatment Standards.

Many Categorical Pretreatment Standards specify one limit for calculating maximum daily discharge limitations and a second limit for calculating maximum monthly average or 4-day average limitations. Where such standards are being applied, the same production or flow figure must be used in calculating both types of equivalent limitations.

Any IU with a POTW-issued permit using equivalent mass or concentration limits calculated from a production-based standard must notify the POTW within two business days after the IU has a reasonable basis to know that the production level will significantly change within the next calendar month. Any IU not notifying the POTW of such anticipated change will be required to meet the mass or concentration limits in its permit that were based on the original estimate of the long-term average production rate.

For any IU subject to equivalent mass or concentration limits in its permit, the initial report on compliance and the periodic compliance reports must contain a reasonable measure of the IU's Pretreatment Standards expressed in terms of allowable pollutant discharge per unit of production (or other measure of operation); those reports must include the IU's actual production during the appropriate sampling or reporting period.

3.220 *Industrial Categories With Concentration-Based Standards*

EPA has promulgated concentration-based pretreatment standards for the following five categories:

1. Electrical and Electronic Component Manufacturing,

2. Leather Tanning and Finishing,

3. Metal Finishing,

4. Pharmaceutical Manufacturing, and

5. Steam Electric Power Generation.

3.221 *Industrial Categories With Mass-Based Standards*

EPA has promulgated production-based (mass-based) pretreatment standards for the following ten categories:

1. Aluminum Forming,

2. Battery Manufacturing,

3. Coil Coating,

4. Copper Forming,

5. Iron and Steel Manufacturing,

6. Metal Molding and Casting,

7. Nonferrous Metals Forming,

8. Nonferrous Metals Manufacturing,

9. Organic Chemicals, and

10. Pesticides.

3.222 *Industrial Categories With Both Concentration- and Mass-Based Standards*

EPA has promulgated both concentration- and mass-based pretreatment standards for the following six categories:

1. Electroplating,

2. Inorganic Chemicals Manufacturing,

3. Petroleum Refining,

4. Porcelain Enameling,

5. Pulp, Paper and Paperboard Manufacturing, and

6. Timber Products Processing.

3.23 Types of Wastestreams

This section defines three different types of wastestreams that are described in regulations.

Regulated streams are industrial process wastewater streams subject to a national categorical pretreatment standard; for example, wastewater produced as a result of rinsing electroplated parts in an electroplating facility.

Unregulated streams are industrial process wastewater streams not subject to a national categorical pretreatment standard; for example, wastestreams resulting as rinses from a galvanizing operation.

Dilution streams are wastewaters from nonregulated process wastewater streams that contain no pollutants subject to the national categorical pretreatment standard; for example, boiler blowdown, sanitary wastewater, or noncontact cooling water.

Except where expressly authorized to do so by an applicable Pretreatment Standard or Requirement, no IU shall ever increase the use of process water or in any other way attempt to dilute a discharge as a partial or complete substitute for adequate treatment to achieve compliance with a Pretreatment Standard or Requirement. The POTW may impose mass limitations on IUs which are using dilution to meet applicable Pretreatment Standards or Requirements, or in other cases where the imposition of mass limitations is appropriate.

3.24 Total Toxic Organics (TTO)

3.240 *Regulated Toxic Organics*

For each regulated industrial category, EPA specifies which organic compounds it will regulate. The regulated industry then must measure and report the quantities of listed pollutants in its wastestreams. Table 3.8 lists the toxic organics regulated under the electroplating and metal finishing categories. For some industrial categories, the number of toxic organics is fewer than those listed in Table 3.8. In the electroplating and metal finishing regulations, TTO is defined as the summation (total) of all quantifiable (measurable) values of components in Table 3.8 in excess of 10 micrograms per liter.

3.241 TTO Monitoring

Requiring analytical determination of all 111 organic compounds listed in Table 3.8 by all small metal finishers and electroplaters may be unreasonable and uneconomical. Therefore, in place of monitoring for TTO, the control authority may permit the industrial user to submit a *SOLVENT MANAGEMENT PLAN*[6] (also known as a Toxic Organic Management Plan (TOMP)) for the POTW's review and approval.

TABLE 3.8   LIST OF ORGANIC CHEMICALS INCLUDED IN TOTAL TOXIC ORGANICS (TTO) REGULATED UNDER ELECTROPLATING AND METAL FINISHING CATEGORIES

1. Acenaphthene
2. Acrolein
3. Acrylonitrile
4. Benzene
5. Benzidine
6. Carbon tetrachloride (tetrachloromethane)
7. Chlorobenzene
8. 1,2,4 trichlorobenzene
9. Hexachlorobenzene
10. 1,1-dichloroethane
11. 1,2-dichloroethane
12. 1,1,1-trichloroethane
13. Hexachloroethane
14. 1,1,2-trichloroethane
15. 1,1,2,2-tetrachloroethane
16. Chloroethane
17. Bis(2-chloroethyl) ether
18. 2-chloroethyl vinyl ether (mixed)
19. 2-chloronaphthalene
20. 2,4,6-trichlorophenol
21. Parachlorometa cresol
22. Chloroform (trichloromethane)
23. 2-chlorophenol
24. 1,2 dichlorobenzene
25. 1,3 dichlorobenzene
26. 1,4 dichlorobenzene
27. 3,3-dichlorobenzidine
28. 1,1-dichloroethylene
29. 1,2-transdichloroethylene
30. 2,4-dichlorophenol
31. 1,2 dichloropropane
32. 1,2 dichloropropylene (1,3 dichloropropylene)
33. 2,4-dimethylphenol
34. 2,4-dinitrotoluene
35. 2,6-dinitrotoluene
36. 1,2 diphenylhydrazine
37. Ethylbenzene
38. Fluoranthene
39. 4-chlorophenyl phenyl ether
40. 4-bromophenyl phenyl ether
41. Bis(2-chloroisopropyl) ether

42. Bis(2-chloroethoxy) methane
43. Methylene chloride (dichloromethane)
44. Methyl chloride (chloromethane)
45. Methyl bromide (dibromomethane)
46. Bromoform (tribromomethane)
47. Dichlorobromomethane
48. Chlorodibromomethane
49. Hexachlorobutadiene
50. Hexachlorocyclopentadiene
51. Isophorone
52. Naphthalene
53. Nitrobenzene
54. 2-nitrophenol
55. 4-nitrophenol
56. 2,4-dinitrophenol
57. 4,6-dinitro-o-cresol
58. N-nitrosodimethylamine
59. N-nitrosodiphenylamine
60. N-nitrosodi-n-propylamine
61. Pentachlorophenol
62. Phenol
63. Bis(2-ethylhexyl) phthalate
64. Butyl benzyl phthalate
65. Di-n-butyl phthalate
66. Di-n-octyl phthalate
67. Diethyl phthalate
68. Dimethyl phthalate
69. 1,2-benzanthracene (benzo(a)anthracene)
70. Benzo(a)pyrene (3,4-benzopyrene)
71. 3,4-Benzofluoranthene (benzo(b)fluoranthene)
72. 11,12-benzofluoranthene (benzo(k)fluoranthene)
73. Chrysene
74. Acenaphthylene
75. Anthracene

76. 1,12-benzoperylene (benzo(ghi)perylene)
77. Fluorene
78. Penanthrene
79. 1,2,5,6-dibenzanthracene (dibenzo(a,h)anthracene)
80. Indeno (1,2,3-cd)pyrene (2,3-o-phenylene pyrene)
81. Pyrene
82. Tetrachloroethylene
83. Toluene
84. Trichlorethylene
85. Vinyl chloride (chloroethylene)
86. Aldrin
87. Dieldrin
88. Chlordane (technical mixture & metabolites)
89. 4,4-DDT
90. 4,4-DDE (p,p-DDX)
91. 4,4-DDD (p,p-TDE)
92. Alpha-endosulfan
93. Beta-endosulfan
94. Endosulfan sulfate
95. Endrin
96. Endrin aldehyde
97. Heptachlor
98. Heptachlor epoxide (BHC-hexachlorocyclohexane)
99. Alpha-BHC
100. Beta-BHC
101. Gamma-BHC (lindane)
102. Delta-BHC PCB- polychlorinated biphenyls
103. PCB-1242 (Arochlor 1242)
104. PCB-1254 (Arochlor 1254)
105. PCB-1221 (Arochlor 1221)
106. PCB-1232 (Arochlor 1232)
107. PCB-1248 (Arochlor 1248)
108. PCB-1260 (Arochlor 1260)
109. PCB-1016 (Arochlor 1016)
110. Toxaphene
111. 2,3,7,8-tetrachlorodibenzo-p-dioxin (TCDD)

---

[6] *Solvent Management Plan.   A strategy for keeping track of all solvents delivered to a site, their storage, use and disposal. This includes keeping spent solvents segregated from other process wastewaters to maximize the value of the recoverable solvents, to avoid contamination of other segregated wastes, and to prevent the discharge of toxic organics to any wastewater collection system or the environment. The plan should describe measures to control spills and leaks and to ensure that there is no deliberate dumping of solvents. Also known as a Toxic Organic Management Plan (TOMP).*

To request that no monitoring be required, industrial users submit a solvent management plan to the POTW. The plan must specify (to the control authority's satisfaction) procedures for ensuring that toxic organics used do not routinely spill or leak into the wastewater and that there is no deliberate dumping of any of the solvents. The criteria for a toxic management plan include the following:

1. Identification of toxic organics used,

2. Quantity of each toxic organic used,

3. Use of each toxic organic, and

4. The method of disposal used instead of dumping, such as reclamation, contract hauling, or incineration.

In addition, the industrial users of POTWs must make the following certification as a comment on the periodic reports required by 40 CFR Section 403.12(E):

> Based on my inquiry of the person or persons directly responsible for managing compliance with the pretreatment standard for total toxic organics (TTO), I certify that, to the best of my knowledge and belief, no dumping of concentrated toxic organics into the wastewaters has occurred since filing the last discharge monitoring report. I further certify that this facility is implementing the solvent management plan submitted to the control authority.

If monitoring is necessary to measure compliance with the TTO standard, the industrial discharger needs to analyze for only those pollutants which would reasonably be expected to be present. Total toxic organics monitoring is *REQUIRED* for Baseline Monitoring Reports (BMR) and Final Compliance Reports, except where the POTW allows for oil/grease measurements to be substituted. EPA has developed a guidance manual specifically for application of the TTO standard entitled: "Guidance Manual for Implementing Total Toxic Organic (TTO) Pretreatment Standards."

QUESTIONS

Write your answers in a notebook and then compare your answers with those on page 241.

3.2A   What information is required in a Baseline Monitoring Report (BMR)?

3.2B   What is a net gross credit?

3.2C   What are the two different types of categorical standards used in the EPA pretreatment program?

3.2D   What is a regulated stream?

3.2E   In the electroplating and metal finishing regulations, what does the phrase "Total Toxic Organics" mean?

3.2F   What pollutants must a discharger analyze for in order to comply with the TTO standard?

3.25   Electroplating and Metal Finishing Regulations

The Categorical Pretreatment Standards are a series of specific industrial wastewater discharge standards developed and promulgated by EPA. Categorical standards were first developed for the direct dischargers and subsequently for indirect dischargers. Each categorical standard covers a specific industrial category; each set of standards is uniquely applicable to a particular industrial category. Within each category, these standards apply uniformly throughout the country to all industrial dischargers.

As previously mentioned, EPA initially regulated 21 industrial categories under the National Categorical Program. Two of the categories have been subdivided, creating a total of 38 categories. Twelve categories were later exempted leaving a total of 26 categories for categorical regulatory control.

Two of these categories are the electroplating and metal finishing categories. These two categories are by far the largest group of categorical dischargers. In fact, the total number of indirect electroplaters and metal finishers by far exceeds the total number of all other categorical dischargers combined.

The EPA categorical standards for the electroplating category apply to all industrial dischargers within the scope of this category. The pretreatment standards for the electroplating category are more specific and apply to all indirect dischargers covered by the electroplating category.

Because these two categories make up such a large proportion of regulated industries, the remainder of this section will discuss the application of categorical standards to the electroplating and metal finishing categories.

3.250   Regulatory Classification of Categorical Dischargers

For regulatory purposes, the EPA has divided the industrial dischargers into several different classes. Direct and indirect dischargers have already been defined. The following terms apply more specifically to the electroplating and metal finishing categories.

Job Shop

Those facilities which in a calendar year do not own more than 50 percent (area basis) of material undergoing metal finishing.

Captive Shop

Those facilities which in a calendar year own more than 50 percent (area basis) of material undergoing metal finishing.

Integrated Facility

Plants which, prior to treatment or discharge, combine the metal finishing wastewaters with significant quantities (more than 10 percent of total volume) of wastewaters not covered by the metal finishing category.

Nonintegrated Facility

Plants which have significant wastewater discharges only from operations covered by the metal finishing category.

In theory, job shops can be divided into integrated and nonintegrated; however, approximately 97 percent of all job shops are nonintegrated. Many captive shops (50 percent) are integrated facilities. Captive shops often have a complex range of operations; job shops usually perform fewer operations.

3.251 Electroplating Category

EPA divides the electroplating category into the following eight subcategories:

1. Electroplating of Common Metals,
2. Electroplating of Precious Metals,
3. Electroplating of Specialty Metals,
4. Electroless Plating,
5. Anodizing,
6. Coating,
7. Chemical Etching and Milling, and
8. Printed Circuit Board Manufacturing.

Clearly the EPA's Electroplating Regulations cover much more than just the electroplating processes. The "electroplating" title originally applied to these regulations was an unfortunate misuse of the term. In a partial attempt to correct this error, the EPA promulgated later regulations under the title of Metal Finishing Regulations. Use of the term "electroplating" has created and continues to create a great deal of confusion within the industry.

40 CFR 413 deals specifically with job shops that were in existence when the regulations were promulgated. It applies only to shops doing electroplating, metal finishing other than electroplating, and printed circuit board work if the shop discharges to a POTW.

A new job shop (construction commenced after promulgation of proposed regulations) would be covered under 40 CFR 433, as would any job shop discharging directly to the environment under an NPDES permit.

3.252 Metal Finishing Category

On July 15, 1983, the EPA promulgated categorical regulations for the Metal Finishing Category. Metal Finishing Regulations (40 CFR Part 433) apply to plants which perform one or more of the following six operations:

1. Electroplating,
2. Electroless Plating,
3. Anodizing,
4. Coating (phosphatizing, chromating, and coloring),
5. Chemical Etching and Milling, and
6. Printed Circuit Board Manufacturing.

If a facility does not perform at least one of the above six electroplating operations, it is not subject to the Metal Finishing Regulations. On the other hand, if one or more of these processes is performed at a given facility, potentially there are 40 additional processes (Table 3.9) which are subject to the Metal Finishing Regulations. Consequently the six operations mentioned above are prerequisites for coverage by the Metal Finishing Regulation.

3.253 Comparison of Electroplating and Metal Finishing Categories

Electroplating Regulations are based on BPT (Best Practicable Technology). The Metal Finishing Regulations are based on BAT (Best Available Technology). However, EPA has found that the technology for BPT for Electroplating Regulations is equivalent to the BAT for the Metal Finishing Regulations.

Job shops and independent printed circuit board manufacturers presently covered under Electroplating Regulations are exempt from Metal Finishing Regulations except for compliance with TTO. However, job shops and independent printed circuit board manufacturers are not exempt from Metal Finishing Regulations if they are discharging under an NPDES permit, or are a New Point Source discharging into a POTW.

The Electroplating Regulations limit only cadmium, lead and amenable cyanide in plants with wastewater flows of less than 10,000 gallons per day (38,000 liters per day). For plants with flows greater than or equal to 10,000 gallons per day (38,000 liters per day), Electroplating Regulations limit cadmium, copper, chromium, total cyanide, nickel, lead, silver, zinc, and total metals.

The Metal Finishing Regulations control cadmium, copper, chromium, cyanide, lead, nickel, zinc and silver for all facilities subject to the Metal Finishing Regulations, regardless of the volume of discharge. There are no effluent limits for total metals in the Metal Finishing Regulations. Besides metals and cyanide, the Metal Finishing Regulations limit TTO for all facilities subject to BOTH Electroplating and Metal Finishing Regulations.

For most regulated pollutants, the metal finishing limits are more stringent than the electroplating regulations.

Another similarity between the electroplating and the metal finishing standards is addition of Total Toxic Organics (TTO) to the metal finishing as well as electroplating facilities. TTO is a surrogate (substitute) parameter for the organic constituents of the priority pollutants (see Table 3.8, page 221).

These categorical standards are technology-based standards which means that if the right type of pretreatment is provided, these effluent limits can be achieved.

3.254 Differences Between the Electroplating and Metal Finishing Categories

The electroplating/metal finishing industry is covered by 40 CFR Part 413 (Electroplating Category), promulgated on September 7, 1979, and amended on January 18, 1981. The Metal Finishing Category is covered by 40 CFR Part 433, promulgated on July 15, 1983. All existing job shops and independent printed circuit board manufacturers remain as electroplaters (413) and all others are metal finishers (433). A facility must be either an electroplater or a metal finisher.

3.255 Determining Who Are the Categorical Companies Within a Specific Category

It is difficult to determine the exact number of electroplaters and metal finishers located within any large area served by a given POTW. Most EPA categorical regulations are based on the type of product manufactured within an industry. Both Electroplating and Metal Finishing Regulations are based on

TABLE 3.9   UNIT OPERATIONS IN THE METAL FINISHING INDUSTRY

1. Electroplating	16. Heat Treating	32. Hot Dip Coating
2. Electroless Plating	17. Thermal Cutting	33. Sputtering
3. Anodizing	18. Welding	34. Vapor Plating
4. Coating (Chromating, Phosphatizing and Coloring)	19. Brazing	35. Thermal Infusion
5. Chemical Etching and Milling	20. Soldering	36. Salt Bath Descaling
6. Printed Circuit Board Manufacturing	21. Flame Spraying	37. Solvent Degreasing
7. Cleaning	22. Sandblasting	38. Paint Stripping
8. Machining	23. Other Abrasive Jet Machining	39. Painting
9. Grinding	24. Electric Discharge Machining	40. Electrostatic Painting
10. Polishing	25. Electrochemical Machining	41. Electropainting
11. Tumbling	26. Electron Beam Machining	42. Vacuum Metalizing
12. Burnishing	27. Laser Beam Machining	43. Assembly
13. Impact Deformation	28. Plasma Arc Machining	44. Calibration
14. Pressure Deformation	29. Ultrasonic Machining	45. Testing
15. Shearing	30. Sintering	46. Mechanical Plating
	31. Laminating	

the unit operations performed at a given facility and not based on the type of products or services performed at the industry. The *SIC*[7] listing of the industrial dischargers consequently generates a larger number of potential dischargers which may be subject to categorical regulations.

In order to cover all possible sources subject to the categorical regulations for the Electroplating and Metal Finishing Categories, it should be assumed that all companies with Major Group SICs of 34 through 39 (or NAICS Groups 331 through 421) are potentially subject to these regulations. This assumption will yield a larger number of potential dischargers subject to these regulations than the actual number of dischargers. However, this may be the simplest method of identifying all dischargers subject to these regulations.

3.256   Proposed New Category

At the time this manual was being printed, EPA was planning to propose a new categorical standard: Effluent Limitations Guidelines Pretreatment Standards and New Source Performance Standards for the Metal Products and Machinery (MP&M) Category. EPA expects 10,601 facilities will be subject to this regulation (POTWs estimate there will be more than 25,000 such facilities). Most of the industrial users currently regulated as captive electroplaters or metal finishers would be covered under this new MP&M Categorical Standard. A total of 48 unit operations are proposed for regulation as well as 98 general categories of manufacturing, maintaining or rebuilding facilities. This proposed rule has the potential to significantly affect POTW pretreatment programs by expanding the number of categorical industrial users regulated. If the proposed rule is implemented, it will provide pretreatment inspectors with many additional challenges in their jobs.

QUESTIONS

Write your answers in a notebook and then compare your answers with those on pages 241 and 242.

3.2G   What two categories of industries make up the largest proportion of industries regulated by categorical standards?

3.2H   What is a job shop?

3.2I   What is an integrated facility?

3.2J   The Electroplating Regulations limit which pollutants in plants with flows of less than 10,000 gallons per day?

---

[7] *SIC Code.   Standard Industrial Classification Code. A code number system used to identify various types of industries. In 1997, the United States and Canada replaced the SIC code system with the North American Industry Classification System (NAICS*); Mexico adopted the NAICS in 1998.*

* *NAICS.   North American Industry Classification System. A code number system used to identify various types of industries. This code system replaces the SIC (Standard Industrial Classification) code system used prior to 1997. Use of these code numbers is often mandatory. Some companies have several processes which will cause them to fit into two or more classifications. The code numbers are published by the Superintendent of Documents, U.S. Government Printing Office, PO Box 371954, Pittsburgh, PA 15250-7954. Stock No. 041-001-00509-9; price, $31.50. There is no charge for shipping and handling.*

3.3 POTW PRETREATMENT PROGRAMS

Thus far in the chapter we've examined the history of pollution control regulations, the general nature of those regulations, and the manner in which legal authority to implement and enforce them can be delegated to local agencies. We'll look now at the structure and contents of a wastewater ordinance. This is the principal document through which local POTWs apply for and are given authority to regulate industrial discharges.

In your work as a pretreatment inspector or program manager you will be required to understand virtually every detail of your local wastewater ordinance. You may even participate in drafting one. The following section describes this important document in terms of what information it must include and how to go about drafting a local wastewater ordinance.

3.30 Wastewater Ordinances[8]

The basic document granting authority to administer a pretreatment inspection program is the Wastewater Ordinance or Sewer-Use Ordinance. These ordinances may vary greatly in format depending on the laws of the state within which the Publicly Owned Treatment Works (POTW) operates. However, all such ordinances must contain certain basic elements to provide a legal framework for effective enforcement. In addition, there are specific items which are unique to each inspection agency. A typical ordinance covers the following basic topics:

1. Definitions,
2. Prohibitions and limitations of wastewater discharges,
3. Control of prohibited wastes,
4. Industrial wastewater monitoring and reporting,
5. Industrial discharge permit system,
6. Enforcement procedures,
7. Penalties and costs,
8. Savings clause (if a portion of the ordinance is ruled to be invalid, the remainder of the ordinance remains valid),
9. Resolution of conflicts,
10. Effective date, and
11. Enacting clause.

Appendix D contains an excellent *EPA MODEL PRETREATMENT ORDINANCE.*

Any agency preparing a wastewater ordinance should review similar ordinances prepared by neighboring communities. By reviewing already prepared wastewater ordinances from neighboring communities as well as other communities known to have adequate industrial waste programs, you will greatly speed up the process of drafting your own POTW's ordinance.

Most ordinances regulate both domestic and industrial wastewater disposal. In some communities the industrial waste (pretreatment) inspector may be responsible for both

aspects of the ordinance. For the purpose of this discussion, we will be concerned primarily with the industrial waste control provisions. Essential elements include:

- Administration
- EPA Minimum Requirements
- General Provisions
- Specific Provisions
- Industrial Wastewater Limitations

3.300 Administration

The Wastewater Ordinance will generally be preceded by a section where specific authority for the POTW to regulate wastewater disposal is granted under state law and any regulations under which the POTW is required to operate are specified. Such sections may also make a statement as to the purpose of the organization operating the POTW and outline the intent of the ordinance.

A further statement of scope may be included defining specific limitations and defining terms to be used in interpretation of the ordinance. The scope will outline the major provisions to follow within the document, including the intent to regulate all discharges to the *WASTEWATER FACILITIES*[9] (POTW), the quantity and quality of discharges, degree of required pretreatment, fees and distribution of costs, issuance of permits, and penalties for violations of the ordinance.

The ordinance may include a policy statement as further guidance on the application of the ordinance and to set priorities for use of the wastewater collection system. Such a statement will commonly say that the primary use of the wastewater facilities is for the collection, conveyance, treatment and disposal of domestic wastewater and that industrial discharges will be subject to additional regulation to accomplish the primary objective.

Additional statements may be made limiting wastes accepted by the POTW operator to those that will not (a) damage the system, (b) create nuisances, (c) menace public health, (d) impose unreasonable costs, (e) interfere with wastewater treatment processes, including sludge disposal, (f) violate any quality requirements set by governmental regulatory agencies, or (g) be harmful to the environment.

---

[8] *INDUSTRIAL WASTE TREATMENT, Volume I. Obtain from the Office of Water Programs, California State University, Sacramento, 6000 J Street, Sacramento, CA 95819-6025. Price, $22.00. See Chapter 3, "Regulatory Requirements," Section 3.40, "Wastewater Ordinances," for additional information.*

[9] *Wastewater Facilities. The pipes, conduits, structures, equipment, and processes required to collect, convey, and treat domestic and industrial wastes, and dispose of the effluent and sludge.*

Policy statements may also encourage conservation and reuse of reclaimed wastewater and sometimes give priority consideration to those dischargers who practice water conservation. The POTW operator may also reserve the right to restrict industrial discharges during periods of upset or repair to the wastewater facilities and may specify when such discharges may occur.

3.301　EPA Minimum Requirements

Before your pretreatment program will be approved by the EPA (or by a state agency if Clean Water Act enforcement authority has been granted to the state), your wastewater ordinance must meet certain minimum standards regarding contents. It must provide pretreatment standards for industrial discharges into the system. Specific reference must be made to the incorporation of Categorical Pretreatment Standards established by EPA for specific industrial users. A POTW operator may also establish local pretreatment standards that are more stringent than, and in addition to, the EPA standards if you (the POTW) can demonstrate that such limitations are necessary to meet other state and local limitations on POTW effluent or to accomplish wastewater reclamation objectives.

The wastewater ordinance must include the details of the control mechanism you will use to regulate dischargers to the POTW, and must specify the conditions and limitations of discharge. The control mechanism may be a permit, administrative order or some other type of non-permit arrangement with dischargers. Whatever type of control mechanism you use, the ordinance must specify the records to be maintained and must outline the notification procedures to be used in the event of process changes or unauthorized discharges in excess of permit limitations. At a minimum, industrial discharges subject to Categorical Pretreatment Standards must be included. The POTW must be able to control the discharge of each industry even if that industry is located in an outlying jurisdiction.

The ordinance must grant specific inspection authority and right of access to processes generating industrial wastewaters and also pretreatment facilities. Such authority includes the ability to take samples and to examine records on the operation and maintenance of the pretreatment facility, and other waste handling or chemical records such as purchase orders or receipts and waste manifests.

The ordinance must provide for enforcement through inspections, reporting and monitoring of industrial discharges, and provide a mechanism for notification where violations are found. The ordinance must also provide penalties for violations. Such penalties may be civil or criminal and may include suspension or revocation of a permit to discharge and disconnection from the POTW system.

The POTW operator must establish a means for sharing the costs for wastewater treatment in a manner that reflects the user's fair share of the costs. General taxes such as those based on property value (ad valorem) may not be used except to retire existing bond obligations. Cost recovery may take place through permit and inspection fees, recovery of capital costs through connection charges and fees, or through charges or surcharges based on the quantity and quality of wastewater discharged into the POTW system.

3.302　General Provisions

Your wastewater ordinance must specify who is authorized to enforce the ordinance and how this authority may be delegated to subordinate employees, including the industrial waste (pretreatment) inspector. The general provisions will also specify penalties for violation of the ordinance. Such penalties may be criminal or civil and include fines, compliance orders, permit suspension or permit revocation.

Criminal penalties may be imposed for any violation of the ordinance as an infraction or misdemeanor. Specific violations such as failure to pay fees, meet permit conditions or limitations, failure to adhere to approved plans, or other acts may be named in the ordinance. To successfully prosecute a criminal violation, it is necessary to show beyond a reasonable doubt that the violation occurred. Evidence gathering is vital; be sure to use *CHAIN OF CUSTODY*[10] procedures. It may also be necessary to prove intent to avoid regulations to achieve maximum criminal penalties.

Civil remedies may be more appropriate when technical violations occur or when the primary goal is to induce the discharger to take some action. These may include improvements of pretreatment systems to meet discharge limitations and protect the POTW. Civil action in the form of injunctions may be appropriate to prevent threatened discharge of pollutants to the collection system or POTW.

An industrial waste inspector must be totally familiar with the penalty provisions of the wastewater ordinance and must know what supporting information is needed to bring action under each provision.

The EPA General Pretreatment Regulations require POTWs to verify compliance or noncompliance with the Categorical Pretreatment Standards by the affected dischargers independent of information submitted by the industrial dischargers. This implies that POTWs are required to monitor industrial wastewater dischargers. Since most pretreatment standards are based on daily maximum numbers, POTWs are required to obtain daily composite samples in order to verify compliance or noncompliance with applicable pretreatment standards. In some areas POTWs collect several samples per month or collect monthly samples instead of daily samples. If noncompliance is suspected or discovered, the frequency of sampling is increased.

The general provisions of the ordinance must include complete details of the procedures that will be followed when

---

[10] Chain of Custody.　A record of each person involved in the handling and possession of a sample from the person who collected the sample to the person who analyzed the sample in the laboratory and to the person who witnessed disposal of the sample.

violations are discovered or threatened. For example, the ordinance will provide for the issuance of notification of the discharger; it will specify the means by which notice must be served (consistent with state law); and the ordinance will even specify what information the notice must contain, such as a statement of the nature of the violation and the time limits within which a correction must take place.

The general provisions will also spell out the rights of the permittee or discharger when served by a notice. These rights may include ability to rebut or deny such charges by appeal to the administrative authority. The administrative authority, however, may be able to administratively suspend a permit to discharge until the matter is resolved. Once again, it is vital that you be thoroughly familiar with the authority specified in your local wastewater ordinance. If you are unsure of the penalty for a violation, you should contact your supervisor or the POTW's legal counsel.

As noted in the discussion on EPA requirements, fees are an important part of any pretreatment program. The manner of fee collection and the enforcement authority available to the POTW operator are normally described in the general provisions of the wastewater ordinance.

QUESTIONS

Write your answers in a notebook and then compare your answers with those on page 242.

3.3A   What is the basic document granting authority to administer a pretreatment inspection program?

3.3B   What are wastewater facilities?

3.3C   What enforcement provisions must a wastewater ordinance contain?

3.303 Specific Provisions

The sewer-use ordinance may describe in this section specific prohibitions on wastewater discharges. These provisions may prohibit the industrial user from discharging, depositing, or causing or allowing to be discharged or deposited into the wastewater collection or treatment system any wastewater containing excessive amounts or concentrations of:

1. Oil and grease,

2. Noxious materials,

3. Explosive mixtures,

4. Improperly shredded garbage,

5. Radioactive wastes,

6. Solid or viscous wastes,

7. Excessive discharge rate,

8. Toxic substances,

9. Unpolluted waters,

10. Discolored material, and

11. Corrosive wastes.

A section in the ordinance would be devoted to each of the previously listed items and describe excessive amounts or concentrations and the problems they can create.

3.304 Industrial Wastewater Limitations

Industrial wastewater limitations commonly refer to the Specific Provisions described in Section 3.303 and describe the limitations as general limitations and/or specific limitations. The general limitations prohibit the discharge or conveyance to public sewers (POTW) of any wastewater containing pollutants of such character or quantity that will:

1. Not be susceptible to treatment or will interfere with the processes or efficiency of the POTW,

2. Constitute a hazard to human or animal life, or to the stream or watercourse receiving the treatment plant effluent,

3. Violate pretreatment standards, or

4. Cause the treatment plant to violate its NPDES permit or applicable receiving water standards.

Specific limitations will list the maximum concentrations or mass limitations of pollutants allowable in industrial wastewater discharges to the POTW. This section will also contain a clause stating that dilution of any wastewater discharge for the purpose of satisfying these requirements shall be considered a violation of the ordinance.

3.31 Pretreatment Program Approval Process

When the industrial waste ordinance is drafted, it must be approved in order to go into effect. This approval has to be obtained from the management of the POTW agency, from the political body governing the POTW agency, and then from the state in which the agency is located, and ultimately from the EPA regional industrial waste pretreatment coordinator. In those cases where the state does not have an approved pretreatment program, the EPA regional industrial waste pretreatment coordinator reviews, but may not approve, the POTW's ordinance. As this process is long and may require several revisions of the proposed ordinance, it is recommended that you send a draft copy to the various approval agencies well before your final request for political enactment. In general, if a wastewater ordinance is similar to ordinances which have already been approved, no major problems with ordinance review from the state and EPA Region should be anticipated.

If major ordinance changes are anticipated, it is recommended that the POTW agency personnel contact any significant industrial groups or significantly large industrial companies to obtain their review and suggestions on the ordinance content. It may be that minor changes in the wastewater ordinance would make it significantly less burdensome to local industry and not reduce the agency's ability to control problem pollution.

The recent enactment of two amendments (PIRT and DSS) to the General Pretreatment Regulations make it necessary for POTW pretreatment program administrators to review and modify their programs to meet the requirements of these amendments. EPA and the states want the approval of modifications of some aspects of the already approved POTW pretreatment program to be subject to the same public participation as when they were initially approved. This will give the public and the IUs the opportunity to comment on the POTW pretreatment program. Therefore, some of the anticipated changes have been classified as substantial modifications to the POTW pretreatment program and the POTW must follow procedures in the General Pretreatment Regulations for submittal to the approval authority.

Substantial modifications are:

- Changes to the POTW's legal authorities;

- Changes to local limits which result in less stringent local limits;

- Changes to the POTW's permitting authority;

- Changes to the POTW's method for implementing Categorical Pretreatment Standards (for example, incorporation by reference, separate promulgation);

- A decrease in the frequency of self-monitoring or reporting required of industry;

- Changes to the POTW confidentiality procedures;

- Significant reduction in the POTW's pretreatment program resources (including personnel commitments, equipment and funding levels); and

- Changes in the POTW's sludge disposal and management practices.

Submittal approval will be according to the following steps:

- The POTW shall submit to the approval authority (EPA or state) a statement of the basis for the desired POTW pretreatment program modification, a modified program description, or such other documents the approval authority determines to be necessary under the circumstances.

- The approval authority shall approve or disapprove the modification based upon the criteria found in the General Pretreatment Regulations (40 CFR Parts 403.8(f) and 403.11(b)-(f)). A notice of the proposed POTW pretreatment program modification shall be published in the largest daily newspaper published in the municipality in which the POTW is located.

- The modification shall be incorporated into the POTW's NPDES permit after approval.

- Notice of approval shall be published in the same newspaper as the notice of the original request for approval of the modification.

For additional information see Appendix E, "Action Checklist for POTWs: Assessing the Need to Modify Pretreatment Programs in Response to DSS and PIRT."

Experience has shown that once a POTW ordinance is developed and worked with over a period of time, it is often necessary to modify it by adding or deleting sections in order to increase its usefulness as a tool for the pretreatment inspector. Pretreatment program administrators should expect to make periodic revisions as the program evolves.

3.32 Pretreatment Program Minimum Requirements

As described earlier, local pretreatment programs require legal authority, funding, staffing, an industrial dischargers database, and local effluent limitations. The implementation of a local pretreatment system involves permit processing, inspection, monitoring and enforcement, in addition to other program requirements.

The minimum requirements for a pretreatment program are as follows:

REQUIREMENTS

1. Operate the pretreatment program with legal authority enforceable in federal, state or local courts which authorizes the POTW to apply and to enforce those sections of the Clean Water Act relating to pretreatment and any regulations implementing the EPA pretreatment program. Such authority may be contained in a statute, ordinance, or joint powers agreements which the POTW is authorized to enact, enter into or implement and which is authorized by state law.

2. Deny or condition new or increased contributions of pollutants, or changes in the nature of pollutants, to the POTW by IUs where such contributions do not meet applicable pretreatment standards and requirements or where such contributions would cause the POTW to violate its NPDES permit.

3. Require compliance with applicable pretreatment standards and requirements by IUs.

4. Control through permit, order or similar means, the contribution to the POTW by each industrial user to ensure compliance with applicable pretreatment standards and requirements. In the case of Significant Industrial Users, this control shall be achieved through permits or equivalent individual control mechanisms issued to each user. Such control mechanisms must be enforceable and contain at a minimum:

- A statement of duration (in no case more than five years),

- A statement of nontransferability without, at a minimum, prior notification to the POTW and provision of a copy of the existing control mechanism to the new owner or operator,

- Effluent limits based on applicable general and specific limitations from the General Pretreatment Standards (40 CFR Part 403), Categorical Pretreatment Standards, local limits and state and local law,

- Self-monitoring, sampling, reporting, notification and recordkeeping requirements, including an identification of the pollutants to be monitored, sampling location, sampling frequency, and sample type, based on the applicable general and specific limitations from the General Pretreatment Standards (40 CFR Part 403), Categorical Pretreatment Standards, local limits and state and local law, and

- A statement of applicable civil and criminal penalties for violation of pretreatment standards and requirements, and any applicable compliance schedule. Such schedules may not extend compliance dates beyond applicable federal deadlines.

5. Require:

- The development of a compliance schedule by each IU for the installation of technology required to meet applicable pretreatment standards and requirements, and

- The submission of all notices and self-monitoring reports from IUs as are necessary to assess and assure compliance by IUs with applicable pretreatment standards including but not limited to BMRs, 90-Day Compliance Reports and Periodic Reports on Compliance.

6. Carry out all the inspection, surveillance and monitoring procedures necessary to determine, independent of information supplied by the IU, compliance or noncompliance with applicable pretreatment standards. Representatives of the POTW shall be authorized to enter any premises of any IU in which a discharge source or treatment system is located or in which records are required to be kept to ensure compliance with applicable pretreatment standards. Such authority shall be at least as extensive as the authority provided under the Clean Water Act.

7. Obtain remedies for noncompliance by any IU with any applicable pretreatment standards. All POTWs shall be able to seek injunctive relief for noncompliance by an IU with applicable pretreatment standards. All POTWs shall also have authority to seek or assess civil or criminal penalties in at least the amount of $1,000.00 a day for each violation by IUs of applicable pretreatment standards.

8. Pretreatment requirements which will be enforced through paragraph 7 above will include but not be limited to:

- The duty to allow or carry out inspections, entry, or monitoring activities,

- Any rules, regulations, or orders issued by the POTW,

- Any requirements set forth in individual control mechanisms (permits) issued by the POTW, or

- Any reporting requirements imposed by the POTW or the General Pretreatment Regulations.

The POTW shall have authority and procedures (after informal notice to the discharger) to immediately and effectively halt or prevent any discharge of pollutants to the POTW which reasonably appears to present an imminent endangerment to the health or welfare of persons. The POTW shall also have authority and procedures (which shall include notice to the affected IU and an opportunity to respond) to halt or prevent any discharge to the POTW which presents or may present an endangerment to the environment or which threatens to interfere with the operation of the POTW. (*NOTE:* The state or EPA shall have authority to seek judicial relief and may also use administrative penalty authority when the POTW has sought a monetary penalty which the state or EPA believes to be insufficient.)

9. Comply with the following confidentiality requirements:

- Any information submitted pursuant to the General Pretreatment Regulation requirements may be claimed as confidential by the submitter. Any such claim must be asserted at the time of submission in the manner prescribed on the application form or instructions, or, in the case of other submissions, by stamping the words "confidential business information" on each page containing such information. If no claim is made at the time of submission, the POTW may make the information available to the public without further notice. If a claim is asserted, the information will be treated in accordance with the procedures in 40 CFR Part 2 (Public Information).

- Information and data provided which is effluent data shall be available to the public without restriction.

- All other information which is submitted shall be available to the public at least to the extent provided by 40 CFR Part 2.302.

PROCEDURES

1. Identify and locate all possible IUs which might be subject to the POTW pretreatment program.

2. Identify the character and volume of pollutants contributed to the POTW by the IUs identified.

3. Notify IUs identified of applicable pretreatment standards and any applicable requirements under Sections 204(b) and 405 of the Clean Water Act and Subtitles C and D of the Resource Conservation and Recovery Act (RCRA). A listing of the POTW's Significant Industrial Users (SIUs) shall be prepared (as discussed on page 230) and submitted to the POTW's Approval Authority for approval. Within 30 days of that approval, the POTW shall notify each SIU of its status and of all requirements applicable to it as a result of such status.

4. Receive and analyze self-monitoring reports and other notices submitted by IUs in accordance with the self-monitoring requirements for BMRs, 90-Day Compliance Reports and Periodic Reports on Compliance.

5. Randomly sample and analyze the effluent from IUs and conduct surveillance activities in order to identify, independent of information supplied by IUs, occasional or continuing noncompliance with pretreatment standards. Inspect and sample the effluent from each SIU at least once a year. Evaluate, at least once every two years, whether each such SIU needs a plan to control slug discharges. For purposes of this plan, a slug discharge is any discharge of a nonroutine, episodic nature, including but not limited to an accidental spill or a noncustomary batch discharge. The results of such activities shall be available to the POTW's Approval Authority upon request. If the POTW decides that a slug control plan is needed, the plan shall contain, at a minimum, the following elements:

- A description of discharge practices, including nonroutine batch discharges,

- A description of stored chemicals,

- Procedures for immediately notifying the POTW of slug discharges including any discharge that would violate a specific discharge prohibition, with procedures for follow-up written notification within five days, and

- If necessary, procedures to prevent adverse impacts from accidental spills, including inspection and maintenance of storage areas, handling and transfer of materials, loading and unloading operations, control of plant site runoff, worker training, building of containment structures or equipment, measures for containing toxic organic pollutants (including solvents), and/or measures and equipment for emergency response.

6. Investigate instances of noncompliance with applicable pretreatment standards, as indicated in BMRs, 90-Day Reports on Compliance and Periodic Reports on Compliance, or indicated by analysis, inspection and surveillance activities carried out by the POTW. Sample taking and analysis and the collection of other information shall be performed with sufficient care to produce evidence admissible in enforcement proceedings or judicial actions.

7. Comply with public participation requirements in the enforcement of national pretreatment standards. These procedures shall include provisions for at least annual public notification, in the largest daily newspaper published in the municipality in which the POTW is located, of IUs which at any time during the previous twelve months were in significant noncompliance with applicable pretreatment requirements. For the purposes of this provision, an IU is in significant noncompliance if its violation meets one or more of the following criteria:

- Chronic violations of wastewater discharge limits, defined here as those in which 66 percent or more of all of the measurements taken during a 6-month period exceed (by any magnitude) the daily maximum limit or the average limit for the same pollutant parameter.

- Technical Review Criteria (TRC) violations, defined here as those in which 33 percent or more of all of the measurements for each pollutant parameter taken during a 6-month period equal or exceed the product of the daily maximum limit or the average limit multiplied by the applicable TRC. (TRC = 1.4 for BOD, TSS, fats, oil, and grease, and 1.2 for all other pollutants except pH.)

- Any other violation of a pretreatment effluent limit (daily maximum or longer term average) that the POTW determines has caused, alone or in combination with other discharges, interference or pass-through (including endangering the health of POTW personnel or the general public).

- Any discharge of a pollutant that has caused imminent endangerment to human health, welfare or to the environment or has resulted in the POTW's exercise of its emergency authority to halt or prevent such a discharge.

- Failure to meet, within 90 days after the schedule date, a compliance schedule milestone contained in a local control mechanism (permit) or enforcement order for starting construction, completing construction, or attaining final compliance.

- Failure to provide, within 30 days after the due date, required reports such as BMRs, 90-Day Compliance Reports, Periodic Reports on Compliance, and reports on compliance with compliance schedules.

- Failure to accurately report noncompliance.

- Any other violation or group of violations which the POTW determines will adversely affect the operation or implementation of the local pretreatment program.

FUNDING

The POTW shall have sufficient resources and qualified personnel to carry out the authorities and procedures described above. In some limited circumstances, funding and personnel may be delayed where:

- The POTW has adequate legal authority and procedures to carry out the pretreatment program requirements described above, and

- A limited aspect of the program does not need to be implemented immediately.

LOCAL LIMITS DEVELOPMENT

The POTW shall develop local limits (pretreatment standards developed by the POTW, in addition to EPA categorical standards and the specific prohibitions on discharge in the General Pretreatment Regulations) to protect the specific treatment processes at the treatment plant, the health and safety of the POTW workers, and the specific environmental conditions into which the POTW discharges its treated effluent, or demonstrate that they are not necessary. In addition, all POTWs with pretreatment programs must provide a written technical evaluation of the need to revise local limits (once developed) as part of their NPDES permit applications (usually every 5 years).

ENFORCEMENT RESPONSE

The POTW shall develop and implement an enforcement response plan. This plan shall contain detailed procedures indicating how the POTW will investigate and respond to instances of IU noncompliance. (See Chapter 2, Section 2.25, "Compliance Program/Enforcement Program," and Section 3.40, "Enforcement Response Plan," later in this chapter for additional information about enforcement response plans.) The plan shall, at a minimum:

- Describe how the POTW will investigate instances of noncompliance,

- Describe the types of escalating enforcement responses the POTW will take in response to all anticipated types of IU violations and the time periods within which responses will take place,

- Identify (by title) the official(s) responsible for each type of response, and

- Adequately reflect the POTW's primary responsibility to enforce all applicable pretreatment requirements and standards.

SIGNIFICANT INDUSTRIAL USER (SIU) LISTING

The POTW shall prepare a list of its industrial users meeting the definition of Significant Industrial User (SIU). The list shall identify which of the criteria in the definition of SIU is applicable to each industrial user and whether or not the industrial user has a reasonable potential for adversely affecting the POTW's operation or for violating any pretreatment standard or requirement. The POTW may at any time, at its own initiative or in response to a petition received from the IU, determine that an industrial user is not a significant industrial user. This list, and any subsequent modifications, shall be submitted to the POTW's approval authority as a nonsubstantial program modification. Discretionary designations or removals from the list by the control authority shall be deemed to be approved by the approval authority 90 days after submission of the list or modifications, unless the approval authority determines that a modification is in fact a substantial modification.

3.33 Pretreatment Program Implementation

In order for a POTW with an approved pretreatment program to implement and enforce the categorical program, it must take a number of steps.

1. Identify all industrial dischargers subject to each industrial category.

2. Summarize finalized categorical regulations for all industrial categories.

3. Educate the public about the categorical regulations through participation in professional organization meetings and other similar meetings.

4. Transmit the summaries of the various categorical regulations to affected industrial dischargers.

5. Design Baseline Monitoring Reports (BMRs) for various categories.

6. Require categorical dischargers to submit BMRs for the POTW's review and approval.

7. Evaluate and categorize the submitted BMRs.

8. Inspect and verify the status of companies claiming exemption from coverage by the categorical regulations in their BMR.

9. Incorporate general and specific categorical requirements into affected industrial dischargers' permits.

10. Initiate administrative enforcement actions against industrial dischargers not in compliance with the POTW's request for submittal of BMRs.

11. Request submittal of 90-day compliance reports for the POTW's review and approval for affected categorical dischargers.

12. Evaluate and categorize the submitted 90-day reports.

13. Initiate compliance monitoring efforts for categorical dischargers by obtaining 24-hour composite samples from affected categorical discharges.

14. Issue administrative enforcement notices to noncomplying dischargers.

15. Conduct follow-up monitoring with higher priorities for the noncomplying industrial dischargers.

16. Escalate enforcement for continuation of violations.

17. Incorporate Pretreatment Standards for New Sources (PSNS) into all new industrial wastewater discharge permits issued.

3.34 Other Local Ordinances

While the Wastewater Ordinance is the primary instrument enforced by the industrial waste pretreatment inspector, there may be other laws and ordinances which affect the disposal of industrial waste. As an inspector, you will also need to be familiar with these regulations and will need to establish good working relationships with the agencies that enforce them. The remainder of this section describes several types of local ordinances and other laws governing waste disposal that you may encounter as an industrial waste pretreatment inspector.

3.340 Local Sanitary Sewer Codes

The POTW operating agency may not actually be responsible for maintenance of local sewers or control connections to the local system. Local ordinances sometimes establish flow restrictions or allocate sewer capacity based on land use (zoning) or actual occupancy. Local sewers may have been financed by bond issues which may require payment of fees for the right to connect or discharge. Other methods of financing include industrial cost recovery and rate ordinances. The local ordinance may specify which property has the right to discharge to a specific sewer.

Local ordinances may also contain restrictions on entering manholes and require permits and inspection of any work or connection to the system. Such ordinances may require reimbursement to the local wastewater collection system agency for any damage to the system caused by an industrial discharger.

The local wastewater collection system agency may also regulate noncritical industrial wastes which may present local sewer maintenance problems but not necessarily present a problem to the wastewater treatment plant. Such wastes include those for which no EPA categorical pretreatment standard has been established. Other items of concern to the local agency include excessive grease from restaurants and food processing industries; high temperatures; low pH; high pH that may cause scaling; flammable, corrosive or toxic gases; incompatible wastes; excessive solids discharges which can cause stoppages in sewers; and dissolved and unsettleable solids.

3.341 Building Codes

These laws normally regulate all construction within a private land parcel and specify minimum standards for electrical, mechanical and plumbing systems. The Wastewater Ordinance may require the building official to determine that industrial waste discharge approval has been obtained before building permits are issued for any proposed structures. In addition, the actual construction of required facilities may still require standard building and plumbing permits.

Codes administered by the building official usually specify sizes and materials for any plumbing installed within a building and the method of connection to the sewer. Standards may also exist for fixtures such as floor drains, clarifiers, grease traps, pumps and sampling manholes. Electrical permits are generally required for any electrically operated equipment including pretreatment or monitoring systems.

The industrial waste inspector should be especially aware of building code requirements when requesting modifications to existing pretreatment systems. The inspector's job is to verify that all the necessary permits are obtained for the work. If you have any doubts about what permits are required or if you need help interpreting building codes, contact the appropriate plumbing or building inspectors. Ensuring that the completed work meets code requirements is the responsibility of the building inspector.

3.342 Underground Tank Laws

Many states have enacted laws regulating the underground storage of hazardous materials and new federal regulations are being developed by EPA. EPA issued regulations on hazardous waste tanks (40 CFR 260-265, 270 and 271) on July 14, 1986 and issued regulations on new and existing underground tanks and associated piping used to store petroleum and hazardous substances (40 CFR 280 and 281) on September 23, 1988.

The 40 CFR 280 technical standards are organized in seven subparts that progressively add requirements. The subparts deal with the following topics:

1. Program scope (applicability, definitions, the interim prohibition),

2. The standards for new and upgraded system design, construction, installation and notification,

3. General operating requirements,

4. Release detection,

5. Reporting and investigation,

6. Response and corrective action, and

7. Closure.

Although these regulations provide very specific criteria at the federal level, they are designed for implementation by state and local agencies.

Pretreatment inspectors must be aware of applicable underground tank laws in their area and be sure that the industrial facilities they inspect are in compliance or notify the appropriate regulatory agency if anything suspicious is discovered during an inspection.

3.343 Land Use Ordinances

These laws, also known as zoning ordinances, regulate a wide variety of activities on property including the type of business allowed. Many existing businesses operate as nonconforming users. That is to say, the zoning of the property they occupy has been changed at some point and, under present zoning restrictions, those businesses would not be permitted to locate there. While nonconforming users may be allowed to continue operations, zoning restrictions may affect the design and placement of any new construction, such as pretreatment facilities. You should be aware of the requirements of zoning laws, especially those regulating the storage of chemicals, setbacks that may affect placement of pretreatment facilities, restrictions on the use of outside areas as work space (pretreatment facilities may have to be enclosed), and the desig-

nation of specific areas for employee parking which may preclude use for storage of items such as sludge containers.

3.344 Hazardous Waste Laws

In the course of your job inspecting an industrial property, you may be in a position to observe violations of state or local hazardous waste laws. Usually, the disposal of hazardous waste into the sewer system in violation of pretreatment standards is also a violation of the hazardous waste laws. Some hazardous waste laws contain very severe penalties including felony provisions for knowingly causing an illegal discharge. You must be familiar enough with these laws to accurately assess what you observe. You should be aware, for example, that RCRA has provisions which allow small quantity generators (less than 100 kilograms per month) to dispose of hazardous wastes into the sanitary sewer without manifesting if the discharge is allowed by the POTW operator.

When an industrial waste violation is suspected, you will want to coordinate your activities with the health officer[11] and any other appropriate local, state and federal agencies. This will enable you to achieve the maximum protection of the public and the environment for which these laws have been designed.

QUESTIONS

Write your answers in a notebook and then compare your answers with those on page 242.

3.3D Why should POTW agency personnel contact any significant industrial groups or significantly large industrial companies to obtain their review and suggestions on the contents of or changes in an industrial waste ordinance?

3.3E List the types of local laws and ordinances that pretreatment inspectors should become familiar with.

3.3F Why is it important for inspectors to establish good working relationships with the agencies that enforce local laws and ordinances?

3.3G List the items that might be included in a local sanitary sewer code.

3.4 INDUSTRIAL WASTE COMPLIANCE PROGRAMS

Although the vast majority of industrial companies will strive to comply with reasonable effluent limitations, a small minority of companies may try to obtain a competitive advantage

---

[11] Also may be called a hazardous waste officer or surface water quality officer.

through noncompliance with such programs or through partial compliance. A well-designed industrial waste pretreatment program ensures that the vast majority of companies which are striving to comply in an ethical manner with environmental programs will be protected from the unfair competitive advantages obtained by companies that are not so ethically motivated.

The financial advantages of noncompliance with environmental programs have increased dramatically in recent years. For small metal plating companies, the cost of pollution control may be a very significant part of the total net proceeds for the company during the year. The initial cost for the purchase of pretreatment facilities, as well as the costs of operating and maintaining a treatment plant (chemicals, repair of equipment, and labor), represent a substantial capital outlay for compliant industries and a substantial capital savings for noncompliant companies.

The local pretreatment agency has a legal and ethical duty to properly implement a pretreatment program. EPA regulations allow penalties to be assessed against POTWs when they do not rigorously implement compliance actions against noncompliant companies. Failure by the POTW to properly penalize noncompliant companies can also result in the state or the federal government taking over the local POTW's pretreatment program.

3.40   Enforcement Response Plan

In the process of drafting your Wastewater Ordinance, you will be required to prepare an Enforcement Response Plan (ERP) which defines in detail the procedures the POTW will use to enforce its regulatory authority when violations occur or are suspected. As a minimum, the plan must:

1. Describe how the POTW will investigate instances of industrial user (IU) noncompliance;

2. Describe the types of escalating enforcement responses the POTW will take in response to all anticipated types of industrial user violations and the time periods within which responses will take place;

3. Identify (by title) the official(s) responsible for each type of response; and

4. Adequately reflect the POTW's primary responsibility to enforce all applicable pretreatment requirements and standards, as detailed in 40 CFR 403.8(f)(1) and (f)(2).

Any violation of pretreatment requirements (effluent discharge limits, sampling, analysis, reporting and meeting compliance schedules, and regulatory deadlines) is an instance of noncompliance for which the industrial user (IU) is liable for enforcement, including penalties. If a company frequently is in violation or develops a pattern of violations, the industry may be considered in significant noncompliance (SNC).

Major violations are those that exceed the limits frequently and/or by a large quantity; impede the determination of compliance status; or have the potential to cause or may have actually caused adverse environmental effects or health problems, or interfered with the POTW treatment capability. Any violation that meets the definition of "significant noncompliance" should be considered a major violation.

EPA has defined instances of significant noncompliance as industrial user violations which meet one or more of the following criteria:

1. Violations of wastewater discharge limits.

   a. Chronic violations. Sixty-six percent or more of the measurements exceed the same daily maximum limit or the same average limit in a 6-month period (any magnitude of exceedance).

   b. Technical Review Criteria (TRC) violations. Thirty-three percent or more of the measurements exceed the same daily maximum limit or the same average limit by more than the TRC in a 6-month period. There are two groups of TRCs:

Group I for conventional pollutants (BOD, TSS, fats, oil, and grease)	TRC = 1.4
Group II for all other pollutants	TRC = 1.2

   c. Any other violation(s) of an effluent limit (average or daily maximum) that the POTW believes has caused, alone or in combination with other discharges, interference (for example, slug loads) or pass-through or endangered the health of the wastewater treatment personnel or the public.

   d. Any discharge of a pollutant that has caused imminent endangerment to human health/welfare or to the environment and has resulted in the POTW's exercise of its emergency authority to halt or prevent such a discharge.

2. Violations of *compliance schedule milestones* contained in a local control mechanism or enforcement order for starting construction, completing construction, and attaining final compliance by 90 days or more after the schedule date.

3. Failure to provide reports for compliance schedules, self-monitoring data, or categorical standards (baseline monitoring reports, 90-day compliance reports, and periodic reports) within 30 days from the due date.

4. Failure to accurately report noncompliance.

5. Any other violation or group of violations that the POTW considers to be significant.

Turn to the end of this manual and see Appendix II, "Pretreatment Arithmetic," Problem 57, for procedures to analyze data and perform calculations to determine if a significant industrial user (SIU) is in significant noncompliance (SNC).

If an industrial user is in significant noncompliance, the POTW (1) reports this information to its approval authority (state or EPA); (2) lists the industrial user in the newspaper as being in significant noncompliance; and (3) responds to the industry with appropriate enforcement action (including penalties or contract damages), or documents in a timely manner the reasons for withholding enforcement.

Section 403.8(f)(2)(vii) of the General Pretreatment Regulations requires that the POTW publish, at least annually, in the largest daily newspaper located in the municipality serviced by the POTW, a list of industrial users who were in significant noncompliance with applicable pretreatment standards and requirements during the previous twelve months.

EPA recommends that the published list of industries who are in significant noncompliance include additional information such as duration of violation, parameters and/or reporting requirements violated, compliance action taken (if any), whether the industrial user is currently complying with a compliance schedule, and whether the industrial user has returned to compliance. The POTW may also report the type of enforcement action taken. The publishing of lists of industrial users who are in significant noncompliance subjects industrial users to public scrutiny and is an effective deterrent to continued noncompliance.

In addition to developing an Enforcement Response Plan, POTWs should develop an Enforcement Response Guide which outlines the types of violations, the circumstances, the range of appropriate responses, and the person (by title) who is responsible for making the response. Many factors need to be evaluated when determining what constitutes an appropriate response. As discussed in Chapter 2, the POTW should consider the degree of variance from the pretreatment standards, the duration of the violation, the company's history of compliance or noncompliance, and the deterrent effect of the response. Fairness, equity and consistency are also to be considered. The measure of the effectiveness of an enforcement response includes:

1. Whether the noncomplying source returns to compliance as soon as possible,

2. Whether the enforcement response establishes the appropriate deterrent effect for the particular violator and other potential violators, and

3. Whether the enforcement response promotes fairness of regulatory treatment for all comparable violators as well as for complying and noncomplying parties.

For additional information on how to develop an Enforcement Response Plan, see EPA's publication, *GUIDANCE FOR DEVELOPING CONTROL AUTHORITY ENFORCEMENT RESPONSE PLANS*, September 1989. In addition, Appendix F of this chapter contains the Enforcement Response Plan developed by the City of Hopewell, Hopewell Regional Wastewater Treatment Facility.

3.41 Administrative Compliance Programs

It seems quite obvious that it is in everyone's best interests to resolve problem situations with a minimum of coercion. A verbal warning is sometimes sufficient. There will be cases, however, when violators do not respond to verbal warnings and then you must be prepared to act more forcefully. While the POTW may ultimately have to resort to criminal or civil prosecution of a violator, most cases can be resolved without resort to such extreme action.

The following paragraphs describe an administrative compliance program that is typical of the programs used by many POTWs. Such a program exerts increasing levels of pressure on a violator while protecting both the authority of the POTW and the due process rights of the company or industry. It is important to remember that once you begin a formal process of notification, you must proceed at each step according to the guidelines of the Wastewater Ordinance. In this way, you protect the position of the POTW or regulatory agency and the rights of the alleged violator in the unlikely event the case goes to court for resolution.

There are three possible levels of response to all violations available to POTWs — no response, an informal response, or a formal response. For any violation, the POTW must review the violation and determine the appropriate response. For some violations, the response may be no action necessary at this time. The informal enforcement response can be an inspection, phone call, informal meeting, or a letter of violation to the industry. The letter of violation can be limited to a notification of the violation or can require the industry to take certain steps within certain time frames. The formal enforcement response for serious violations must be one of the following:

1. Administrative orders and compliance schedules,

2. Civil suit for injunctive relief and/or civil penalties,

3. Criminal suit,

4. Termination of service (revoke permits), and

5. Collection of contract damages.

As previously noted, many violations can be corrected with only a verbal warning from the pretreatment inspector. If you have developed good working relationships with industrial dischargers in your area, you can often resolve minor violations simply by bringing them to the attention of the appropriate company representative. It is highly recommended that you try this approach first.

If a verbal warning does not resolve the problem within a reasonable time, the next step is to begin a formal notification process. Typically, this type of administrative compliance program uses three levels of notice to encourage companies to comply with local and EPA regulations: (1) a Warning Notice, (2) a Violation Notice, and (3) a Final Notice of Violation.

A Warning Notice, such as the one shown in Figure 3.1, states in some detail the problem which requires correction and the date by which it must be corrected. The time period will usually be about 30 days unless a shorter or longer period seems appropriate.

After the Warning Notice has been signed by an authorized representative of the POTW or agency, it is sent with a cover letter to the highest local executive officer of the company. The letter is directed to the top administrative officer so there can be no misunderstanding that an enforcement problem exists within the company. For POTWs with multiple jurisdictions, a copy of this letter should also be sent to a local city official,

WARNING NOTICE OF VIOLATION

1. DISCHARGER

2. ADDRESS OF WASTEWATER DISCHARGE

3. LOCAL AGENCY

4. TIME OF VIOLATION (Date, Hour)

5. PERMIT NO.

6. SIC NO.

7. DISTRICT NO.

8. VIOLATION OF THE DISTRICTS, WASTEWATER ORDINANCE, SECTION ___ CONCERNING

9. IMPORTANT: VIOLATION MUST CEASE AND DESIST PRIOR TO: _____ (Date)

CHIEF ENGINEER AND GENERAL MANAGER

BY: _____ (Name) _____ (Date)

_____ (Title)

10. RECEIPT OF NOTICE ACKNOWLEDGED BY DISCHARGER

PRINTED NAME _____ TITLE

_____ (Signature)

Form No. 5019

Fig. 3.1   Warning notice of violation

NOTICE OF VIOLATION

1. DISCHARGER

2. ADDRESS OF WASTEWATER DISCHARGE

3. LOCAL AGENCY

4. TIME OF VIOLATION (Date, Hour)

5. PERMIT NO.

6. SIC NO.

7. DISTRICT NO.

8. VIOLATION OF THE DISTRICTS, WASTEWATER ORDINANCE, SECTION _____ CONCERNING

9. IMPORTANT: VIOLATION MUST CEASE AND DESIST PRIOR TO: _____
(Date)

10. RECEIPT OF NOTICE ACKNOWLEDGED BY DISCHARGER

CHIEF ENGINEER AND GENERAL MANAGER

BY:_____

PRINTED NAME

TITLE

(Name)

(Date)

(Signature)

(Title)

Form No. 5020

Fig. 3.2   Notice of violation

FINAL NOTICE OF VIOLATION

1. DISCHARGER

2. ADDRESS OF WASTEWATER DISCHARGE

3. LOCAL AGENCY

4. TIME OF VIOLATION (Date, Hour)

5. PERMIT NO.

6. SIC NO.

7. DISTRICT NO.

8. VIOLATION OF THE DISTRICTS, WASTEWATER ORDINANCE, SECTION _____ CONCERNING

9. IMPORTANT: VIOLATION MUST CEASE AND DESIST PRIOR TO: _____
(Date)

10. RECEIPT OF NOTICE ACKNOWLEDGED BY DISCHARGER

CHIEF ENGINEER AND GENERAL MANAGER

BY:_____

PRINTED NAME

TITLE

(Name)

(Date)

(Signature)

(Title)

Form No. 5021

Fig. 3.3   Final notice of violation

such as the director of public works, to let local officials know that a problem exists with a company within their municipal jurisdiction. Similarly, a copy should be sent to the state regulatory agency that oversees hazardous wastes and the pretreatment permit program.

If the problem for which the Warning Notice was issued is uncorrected after the time limit, then the agency's inspector or an administrative officer will issue a Violation Notice. The Violation Notice is similar to the Warning Notice but is the second level in the three-step administrative compliance program. To increase pressure on the violator, the Violation Notice implements additional procedures such as more frequent sampling of wastewater and more frequent inspections. Generally, the agency's personnel show greater concern about the seriousness of the violation. Most violations are resolved at the Warning and Violation Notice levels.

If the violation remains unresolved, a Final Notice of Violation is issued to the company. This notice and the Violation Notice are similar in format and are shown in Figures 3.2 and 3.3. Again, the details of the violation are spelled out on the notice, and for both the Violation and Final Notice of Violation, letters are sent to the highest level local company officer as well as to a local city official.

The Final Notice of Violation is the last administrative action available to the agency. At this level of notice, the agency may set up a mandatory compliance meeting where an authorized representative of the company is requested to meet with the agency's Industrial Waste Section employees to resolve the problem. Usually at such a meeting, the company representatives and the agency's engineering and enforcement personnel discuss the extent of the problem, as well as the actions necessary for correction, and a mandatory compliance schedule is established. A letter is again sent to the company, with a copy to the local city official, spelling out the items discussed at the meeting and giving a summary of the compliance schedule set for the company.

The vast majority of all violation problems are resolved through the administrative compliance program of the agency. While this program seems somewhat ponderous and requires writing a lot of form letters, it is still much less costly and easier than pursuing direct legal actions. The implementation of legal actions requires a very significant effort and a great deal of staff time is spent preparing background information,

obtaining and retaining evidence samples, furnishing witnesses, and taking and receiving depositions. Resolving the vast majority of industrial waste problems through administrative actions, even if these actions extend over some time, may be the least costly and quickest option available to most POTWs. Once legal action is undertaken, an industry could well end up spending more effort and money defending against the legal action than in solving the problem. In many cases the company needs not only to be told to comply but to be given some indication of how to comply with environmental regulations. The compliance meeting, using technically well-qualified personnel from the POTW to suggest possible solutions to the compliance problem, has been found to be quite helpful in resolving these problems. Industry must realize that it must select a solution and the solution must resolve the problem. If the solution fails, industry cannot blame the POTW for suggesting the solution.

There are many ways to administer a compliance program and the example given in this section is only one typical method. In some POTWs the inspector's supervisor issues the warning notices. Your agency may not be a typical POTW and you will have to modify these suggested procedures to meet your agency's circumstances.

3.42 Administrative Fine Penalties

A variation of the administrative compliance program in some wastewater agencies involves establishing a schedule of administrative fines. Fines are allowed by many state laws when used to penalize industries for discharging pollutants beyond their permitted limits. A fine (penalty) schedule, which must be approved by the agency's board of directors, establishes specific dollar fines for the discharge of excessive quantities of pollutants. Table 3.10 illustrates such a schedule for various pollutants.

EPA has identified four goals that should be considered in assessing penalties. These goals are:

1. Penalties should recover the economic benefit of noncompliance plus some amount for the seriousness of the violation,

2. Penalties should be large enough to deter future noncompliance,

3. Penalties should be uniform or reasonably consistent for similar instances of noncompliance, and

4. A logical basis for the calculation of penalties should exist.

EPA has issued a guidance manual[12] to assist POTWs in calculating the economic benefit an industrial user might gain by delaying installation or proper operation of pretreatment equipment and thereby violating discharge limits.

---

[12] GUIDANCE MANUAL FOR POTWs TO CALCULATE THE ECONOMIC BENEFIT OF NONCOMPLIANCE, Office of Water, U.S. Environmental Protection Agency. Obtain from National Service Center for Environmental Publications (NSCEP), PO Box 42419, Cincinnati, OH 45242-2419. EPA No. 833-B-93-007.

TABLE 3.10   FEES FOR NONCOMPLIANCE WITH PERMIT CONDITIONS AND MASS EMISSION RATES

	Dollars Per Pound Per Day in Excess of Limit			
	Routine Sampling	During and After S&E [a]	After Expiration	Batch Dumps
Arsenic	$100.00	$200.00	$300.00	$300.00
Cadmium	100.00	200.00	300.00	300.00
Chromium (Total)	100.00	200.00	300.00	300.00
Copper	100.00	200.00	300.00	300.00
Lead	80.00	160.00	240.00	240.00
Mercury	100.00	200.00	300.00	300.00
Nickel	50.00	100.00	150.00	150.00
Silver	100.00	200.00	300.00	300.00
Zinc	50.00	100.00	150.00	150.00
Cyanide (Total)	50.00	100.00	150.00	150.00
Cyanide (Free, amenable to chlorination)	100.00	200.00	300.00	300.00
PCBs, Pesticides and Total Toxic Organics	100.00	200.00	300.00	300.00
Phenols	50.00	100.00	150.00	150.00
Dissolved Sulfide	50.00	100.00	150.00	150.00

[a] S&E, Sampling and Evaluation

In a company is discharging large quantities of pollutants, fines may be a substantial motivating factor to compel compliance without resorting to severe legal actions. On the other hand, the fines could be thought of as buying pollutant discharges in some contexts. Companies that can afford it could find it less expensive to pay fines than to implement pollution control measures.

A further variation of the administrative fine procedure involves issuing violation notices and assessing administrative fines on industrial companies who violate restrictions. Upon receiving a notice of a minor violation, the company has the option of directly paying a fine to the agency or having the agency take the case to court. This procedure has worked well for some agencies over the years, although there is some question about the legality of such procedures if they are not handled properly. Appeals of fines may, in some cases, be handled at the City Council level with the assistance of an industrial user board of review consisting of industrial representatives. In some POTW areas this system is very successful in securing compliance and formal court legal action is rarely needed.

3.43   Misdemeanor and Felony Criminal Actions

The option of filing criminal actions against hard-core, noncompliant companies may not be available to all POTWs. State laws must enable the POTW to implement such criminal prosecutions. At most agencies, the law does permit the POTW to take criminal misdemeanor actions against companies who violate the POTW's industrial waste ordinance requirements.

POTWs use criminal misdemeanor actions against companies who essentially refuse to comply with the POTW's ordinance requirements, in spite of the series of in-house administrative warnings and violation notices. In years past, criminal

misdemeanor actions were not taken very seriously either by the lawyers prosecuting the cases or by the criminal justice system. The cases were heard in municipal court where the judges also dealt with felony actions involving crimes such as rape, murder, felonious assault and major theft crimes. In this context, the criminal justice system tended to believe that the pollution violations were minor white collar crimes. In spite of major actions by the agencies involving, in some cases, hundreds of hours of staff time plus thousands of dollars of wastewater analyses, the penalty administered by the court was frequently a few hundred dollars' fine plus six months' to two years' probation. Under these circumstances, going to the final step of a criminal misdemeanor prosecution was not too meaningful and was not undertaken except for the most serious violators.

Within recent years, however, attitudes about the violation of environmental pollution control laws have changed; violations are now considered a more serious problem by local law enforcement personnel. A district attorney with the City of Los Angeles instituted some major legal actions against companies found to be purposely violating environmental regulations. The courts ruled in favor of the POTW and imposed stiff penalties. With these penalties has come a new interest in applying the full penalties of the law for companies found in violation of the agency's wastewater ordinance requirements.

The Los Angeles County district attorney's office has established an Environmental Crimes Division where lawyers who specialize in interpreting environmental regulations vigorously prosecute violations of these laws. With the development of this new interest in environmental crimes by the criminal justice system, the use of misdemeanor and felony criminal actions will become increasingly common at POTWs. The EPA regulations require the POTW to pursue every legal alternative necessary to ensure compliance with permit limits. This mandate often includes criminal prosecution if a company continues to violate its discharge requirements. Although these actions may be quite effective in resolving a continuing wastewater violation problem, it is recommended that they not be taken except where a company is either a long-time violator or has demonstrated continuing disregard for the POTW ordinance requirements.

3.44   Suspension and Revocation of Wastewater Discharge Permits

The suspension or revocation of a wastewater discharge permit is an administrative action which can be taken by the POTW if authorized by the wastewater ordinance. The suspension of a permit for a defined time period is often used to encourage a company to resolve its wastewater discharge problems. The revocation of a permit is the ultimate action which can be taken to prohibit a company from discharging any more wastewater to the collection system. This is the most severe action that can be taken administratively by the POTW. In both of these actions the company should have some recourse, either administratively or legally, for actions that may not be warranted by their circumstances. Preventing a company from performing its day-to-day functions is essentially denying the company the right to exist. Such actions, therefore, should be viewed as serious, should not be undertaken lightly, and should be handled strictly in accordance with the procedures defined in the wastewater ordinance.

The suspension of a permit can be undertaken for serious violations of the wastewater ordinance involving a danger to public health or to the environment, or for creation of a significantly hazardous situation. On numerous occasions agencies have suspended a company's permit by issuing a cease and desist order. This has most commonly been done where a hazardous situation existed, such as the discharge of flammable waste to the sewer creating possible explosion hazards in the sewer. Your program should give the industry the right to appeal the permit suspension to your agency's highest authority within a specific time period. In some POTWs a permit cannot be revoked by staff, but only by the POTW's governing board of elected officials. This decision can only be appealed by industry to the circuit court.

The revocation of a discharge permit, as stated above, is the maximum administrative action which can be undertaken by the POTW. As such, a revocation action should be taken by the legislative body of the POTW. At many agencies the revocation action allows the industrial company to appear before the agency's highest authority and present information and ask questions of agency employees to determine if the revocation is appropriate. Following this review, the agency may act to revoke the industry's wastewater discharge permit if circumstances warrant the action.

3.45   US EPA and State Guidance on Enforcement of Compliance Programs

The EPA has issued several documents that provide guidance on enforcement programs. In general, these documents tend to recommend positive actions to remedy wastewater discharge problems with a minimum of coercion. Referring specifically to the enforcement of Categorical Standards, the EPA recommends that if a violation has not continued for a long time or if there is a goodwill effort to correct a violation, serious enforcement actions should not be taken. If the violation has existed for a long time or there appears to have been no goodwill attempt to remedy the violation, actions should be undertaken to control the problem.

In the implementation of penalties for noncompliance with the EPA's Industrial Categorical Regulations, EPA suggests that any economic advantage to the company for noncompliance with such regulations should be offset by penalties obtained in legal actions. The EPA recommends calculating the benefit to a company for noncompliance with an EPA limitation in the following manner. If a company was required to be in compliance with the electroplating standard in midyear 1984 and did not attain compliance until midyear 1986, the advantage gained by the company would be the interest earned on the funds needed to purchase the pretreatment equipment plus the savings obtained by not operating this equipment over the two-year period. In addition to this, EPA recommends that in flagrant cases a punitive fine should be added on top of the other fines to encourage other companies to comply with the regulations.

3.5   KEEPING CURRENT

These regulations were current when this manual was prepared. Pretreatment inspectors and program managers must be aware that regulations change when better information becomes available, waste treatment technology improves, and results of litigation are implemented. *YOU* have the responsibility to be alert for changes and to adjust your pretreatment program accordingly.

3.6   ACKNOWLEDGMENTS

Portions of the material in this chapter were prepared by Carl Sjoberg, Jay Kremer and Robert Steidel. Their contributions are greatly appreciated.

QUESTIONS

Write your answers in a notebook and then compare your answers with those on page 242.

3.4A   Under what circumstances would a discharger be considered in significant noncompliance?

3.4B   What is an administrative compliance program?

3.4C   What are the levels of notification in a typical administrative compliance program?

3.4D   How can a company be helped in a compliance meeting?

3.4E   Under what circumstances should a wastewater discharge permit be suspended?

3.5A   Why must pretreatment inspectors keep current with changes in pretreatment rules and regulations?

Please answer the discussion and review questions before continuing with the Objective Test.

DISCUSSION AND REVIEW QUESTIONS

Chapter 3.   DEVELOPMENT AND APPLICATION OF REGULATIONS

Write the answers to these questions in your notebook before continuing with the Objective Test on page 243. The purpose of these questions is to indicate to you how well you understand the material in the chapter.

1. What are the advantages of EPA's delegation of implementation authority to local agencies?

2. Why is proper legal authority an essential part of a pretreatment program?

3. What is the difference between a direct and indirect discharger?

4. The General Pretreatment Regulations were established to regulate what types of pollutants?

5. What are the eight specific categories of pollutants regulated by the Prohibited Discharge Standards?

6. According to the General Pretreatment Regulations, who must develop a local pretreatment program?

7. What are Categorical Pretreatment Standards?

8. How are categorical effluent limits developed?

9. How can Categorical Pretreatment Standards be modified?

10. How can an industrial discharger avoid monitoring for TTO?

11. Under what circumstances may a POTW operator establish local pretreatment standards that are more stringent than, and in addition to, EPA Categorical Pretreatment Standards?

12. What procedures should be followed when obtaining data for legal action?

13. What can happen if a POTW does not rigorously implement compliance actions against noncompliant companies?

14. Under what circumstances should felony criminal actions be considered?

SUGGESTED ANSWERS

Chapter 3.   DEVELOPMENT AND APPLICATION OF REGULATIONS

Answers to questions on page 206.

3.0A   The major goals of the 1972 Federal Water Pollution Control Act were:

1. To have the nation's waters clean enough for swimming by 1983, and
2. To eliminate the discharge of pollutants by 1985.

3.0B   The Environmental Protection Agency is responsible for the protection of the nation's water, air and land from various pollutants. The agency is charged with developing pollution control guidelines, implementing programs to achieve the goals established by Congress, and coordinating the various pollution control activities throughout the nation.

3.0C   Under 40 CFR 403, all POTWs with over 5 MGD of wastewater flow or a significant level of industrial dischargers are required to establish a pretreatment program.

Answers to questions on page 207.

3.1A   EPA has developed specific regulatory requirements for industrial companies that discharge directly to lakes, rivers, and oceans (surface waters) and different regulations for industries that discharge to POTWs (direct and indirect dischargers).

3.1B   An existing source is a discharger that was in existence when the proposed applicable categorical standards were promulgated.

A new source is any building, structure, facility or installation from which there is or may be a discharge of pollutants. Construction of the facility must have begun after publication of the applicable Pretreatment Standards. The building, structure, facility or installation must also be constructed at a site at which no other source is located; or, must totally replace the existing process or production equipment producing the discharge at the site; or, must be substantially independent of an existing source of discharge at the same site.

3.1C   The General Pretreatment Regulations regulate pollutants which may:

1. Pass through the POTW's treatment system, untreated or partially treated;
2. Interfere with the POTW's treatment works; and/or
3. Contaminate the POTW's sludge.

3.1D   The most common conventional pollutants are biochemical oxygen demand, suspended solids, fecal coliforms, pH, and oil and grease.

3.1E POTWs with a design flow of less than five MGD may be required to establish a pretreatment program if they have industrial users subject to National Pretreatment Standards or if nondomestic wastes or wastewaters cause upsets, violations of NPDES permit conditions, or sludge contamination.

Answers to questions on page 208.

3.1F EPA's General Pretreatment Regulations (40 CFR 403) established two types of regulatory programs, the Prohibited Discharge Standards and the Industrial Categorical Discharge Standards.

3.1G The General Pretreatment Regulations specifically disallow the introduction to POTWs of:

1. Pollutants that create a fire or explosion hazard in the POTW's sewer system or at the treatment plant;
2. Pollutants that are corrosive, including any discharge with a pH lower than 5.0, unless the POTW is specifically designed to handle such discharges;
3. Solid or viscous pollutants in amounts that will obstruct the flow in the collection system and treatment plant, resulting in interference with operations;
4. Any pollutant discharged in quantities sufficient to interfere with POTW operations (including BOD, but BOD is not listed specifically);
5. Discharges with temperatures above 104°F (40°C) when they reach the treatment plant, or hot enough to interfere with biological processes at the wastewater treatment plant;
6. Petroleum oil, nonbiodegradable cutting oil, or products of mineral oil origin in amounts that will cause interference or pass-through;
7. Pollutants which result in the presence of toxic gases, vapors, or fumes within the POTW in a quantity that may cause acute worker health and safety problems; and
8. Any trucked or hauled pollutants, except at discharge points designated by the POTW.

3.1H Toxic pollutants are compounds or classes of compounds identified by EPA to be harmful to one or more forms of plant or animal life.

3.1I EPA established the Industrial Categorical Pretreatment Standards by performing detailed reviews of the wastes created at various types of industrial sites and evaluated the types of wastewater treatment in use or which could be used. Based on these studies, the EPA defined a reasonable expectation for the quality of the treated wastewater.

3.1J Technology-based standards are based upon the available treatment technologies that could be used to remove pollutants, whereas receiving water standards are based on the tolerance to pollutants of the stream which receives the wastewater.

Answers to questions on page 216.

3.1K The levels of treatment technology identified by EPA's categorical regulations are Best Practicable Technology (BPT) and Best Available Technology (BAT).

3.1L "Development Documents" contain the technical categorical development information published for proposed regulations concerning each industrial category.

3.1M The regulatory requirements for the control of water pollutants from indirect industrial wastewater dischargers are referred to as Pretreatment Standards for Existing and New Sources (PSES) and (PSNS). (The General Pretreatment Regulations also establish several prohibited discharge criteria.)

3.1N If there is a question or disagreement about applicability of a designated industrial category to an industrial discharge, the POTW or the industrial user can request a ruling by the EPA concerning the appropriateness of the industrial category.

Answers to questions on page 222.

3.2A Information required in a Baseline Monitoring Report (BMR) includes:

1. Identification of the indirect discharger,
2. A list of environmental control permits,
3. A description of its operations,
4. A report on flows of regulated streams,
5. The results of sampling and analyses of the industrial wastewater discharges to determine levels of regulated pollutants in those streams,
6. A certification statement by the discharger indicating compliance or noncompliance with the applicable pretreatment standards, and
7. A description of any additional steps required to achieve compliance for noncompliant dischargers.

3.2B A net gross credit allows the subtraction of the concentration of regulated pollutants in the intake water from the concentration level in the industrial discharger's effluent.

3.2C The two different types of categorical standards used in the EPA pretreatment program are concentration-based standards and production-based standards (also referred to as mass-based standards).

3.2D A regulated stream is an industrial process wastestream subject to a national categorical pretreatment standard.

3.2E For each regulated industrial category, EPA specifies certain toxic organic compounds which must be measured. The amounts of all such compounds which are present in a wastestream in a concentration greater than 10 micrograms per liter are added together and the result is a measure of the "Total Toxic Organics."

3.2F If monitoring is necessary to measure compliance with the TTO standard, the industrial discharger needs to analyze for only those pollutants which would reasonably be expected to be present.

Answers to questions on page 224.

3.2G The electroplating and metal finishing categories make up by far the largest proportion of industries regulated by the categorical standards.

3.2H A job shop is an electroplating or metal finishing shop which during a calendar year does not own more than 50 percent (area basis) of the material undergoing metal finishing.

3.2I An integrated facility is a shop which, prior to treatment or discharge, combines the metal finishing wastewaters with significant quantities (more than 10 percent of total volume) of wastewaters not covered by the metal finishing category.

3.2J   The Electroplating Regulations limit only TTO, cadmium, lead and amenable cyanide in plants with wastewater flows of less than 10,000 gallons per day.

Answers to questions on page 227.

3.3A   The basic document for granting authority to administer a pretreatment inspection program is the wastewater ordinance.

3.3B   Wastewater facilities are the pipes, structures, equipment, and processes required to collect, convey, and treat domestic and industrial wastes, and dispose of effluent and sludge.

3.3C   The ordinance must provide for enforcement through inspections, reporting and monitoring of industrial discharges, and provide a mechanism for notification when violations are found. The ordinance must also provide penalties for violations.

Answers to questions on page 232.

3.3D   POTW agency personnel should communicate with industry about the contents of or changes in an industrial waste ordinance because minor changes in the ordinance could reduce the burden for a local industry to comply and not reduce the agency's ability to control problem pollution.

3.3E   The types of local laws and ordinances that pretreatment inspectors should become familiar with are:

1. Local sanitary sewer codes,
2. Building codes,
3. Underground tank laws,
4. Land use ordinances, and
5. Hazardous waste laws.

3.3F   Pretreatment inspectors must establish good working relationships with the agencies that enforce local laws because regulations often overlap with the inspector's responsibilities and because effective protection of the public and the environment requires coordinated efforts of all enforcement agencies.

3.3G   Items that might be included in a local sanitary sewer code include:

1. Flow restrictions,
2. Connection fees,
3. Which property has the right to discharge to a specific sewer,
4. Restrictions on entering manholes,
5. Requirements for permits,
6. Inspection of any work or connection to the system,
7. Reimbursement for any damage to the system caused by an industrial discharger, and
8. Restrictions on wastes not regulated by EPA standards.

Answers to questions on page 239.

3.4A   A discharger who meets any of the following criteria may be considered to be in significant noncompliance: (1) violates wastewater discharge limits, (2) violates compliance schedule milestones, (3) fails to provide required reports, (4) fails to accurately report noncompliance, and (5) does anything else the POTW considers to be a significant violation of program requirements.

3.4B   An administrative compliance program is a program which uses administrative actions to encourage companies to come into compliance with local and EPA regulations.

3.4C   The levels of notification in an administrative compliance program include a verbal warning, a Warning Notice, a Violation Notice, and a Final Notice of Violation.

3.4D   A company can be helped in a compliance meeting by using technically well-qualified personnel from the POTW to suggest possible solutions to the compliance problem.

3.4E   A wastewater discharge permit may be suspended when there is a danger to public health or to the environment, or for the creation of a significantly hazardous situation, such as the discharge of flammable waste to the sewer creating possible explosion hazards in the sewer.

3.5A   Pretreatment inspectors must keep current because regulations change when better information becomes available, waste treatment technology improves, and results of litigation are implemented.

OBJECTIVE TEST

Chapter 3.   DEVELOPMENT AND APPLICATION OF REGULATIONS

Please write your name and mark the correct answers on the answer sheet, as directed at the end of Chapter 1. There may be more than one correct answer to each multiple-choice question.

True-False

1. The Clean Water Act gives EPA the authority to develop and implement programs to control the flow of toxic pollutants into POTWs.

 — 1. True
 2. False

2. EPA gives authority to state or local officials to implement and enforce the national pretreatment program.

 1. True
 — 2. False

3. A major component of any local pretreatment program is local limits.

 — 1. True
 2. False

4. POTWs are designed to remove toxic pollutants.

 1. True
 — 2. False

5. All categorical and noncategorical IUs must notify the POTW immediately of all discharges that could cause problems to the POTW.

 — 1. True
 2. False

6. All IUs must notify POTW in advance of any substantial change in volume or character of pollutants in their industrial discharge.

 — 1. True
 2. False

7. IUs are allowed to increase the use of process water to dilute a discharge to achieve compliance with a Pretreatment Standard or Requirement.

 1. True
 — 2. False

8. A job shop is a facility which in a calendar year owns more than 50 percent (area basis) of material undergoing metal finishing.

 1. True
 — 2. False

9. Criminal penalties may be imposed for any violation of the Wastewater Ordinance as an infraction or misdemeanor.

 — 1. True
 2. False

10. Many violations can be corrected with only a verbal warning from the pretreatment inspector.

 — 1. True
 2. False

Best Answer (Select only the closest or best answer.)

11. EPA recognizes what two types of dischargers in its efforts to control the flow of pollutants into water?

 — 1. Direct and indirect
 2. Municipal and industrial
 3. Raw and treated
 4. Sanitary and storm

12. Which pollutant is a conventional pollutant?

 1. Ammonia
 2. Chromium
 3. Cyanide
 — 4. Oil and grease

13. What are receiving water standards based on?

 1. Best available technology economically achievable
 2. Categorical standards
 3. Technology-based standards
 — 4. Tolerance of stream to pollutants

14. The Clean Water Act provides that POTWs may grant removal credit to indirect dischargers based on what? Degree of removal

 — 1. Actually achieved at the POTW.
 2. By best available technology economically available.
 3. By pretreatment facility.
 4. Provided by industrial direct dischargers.

15. What two types of categorical standards are used in the EPA pretreatment program?

 — 1. Concentration-based and production-based standards
 2. Conventional and toxic standards
 3. Influent-based and effluent-based standards
 4. Priority and nontoxic standards

16. When may civil remedies in the form of an injunction be appropriate?

 1. To demand adherence to approved plans
 2. To justify meeting permit conditions
 — 3. To prevent threatened discharge of pollutants to POTW
 4. To require payment of past due fees

17. What procedure is required of a POTW for compliance with public notification requirements? Publish in largest daily newspaper in municipality a list of IUs

 1. In compliance with pretreatment requirements.
 — 2. In significant noncompliance with pretreatment requirements.
 3. That were denied permission to discharge to sewer.
 4. That were required to pay fines.

18. How could some companies look favorably on fines?

    1. Appearing in newspaper as significant noncomplier is free publicity.
    2. Company will not be able to afford pollution control equipment after paying fine.
    ─ 3. Cost of fine could be less expensive than implementing pollution control.
    4. Public will feel sorry for company.

Multiple Choice (Select all correct answers.)

19. What does a local pretreatment program need to have the legal authority, procedures, funding and personnel to do?

    ─ 1. Deny permission to discharge pollutants from industrial users (IUs) to the POTW
    ─ 2. Identify and locate IUs affected by EPA regulations
    ─ 3. Perform inspection and monitoring activities to ensure compliance with standards
    4. Put IUs out of business for noncompliance
    ─ 5. Require compliance by IUs with pretreatment standards

20. What types of pollutants or conditions are regulated by the General Pretreatment Regulations? Those that

    ─ 1. Contaminate sludge.
    ─ 2. Interfere with treatment.
    ─ 3. Pass through treatment system.
    4. Overload treatment processes.
    5. Wash out treatment reactors.

21. What categories of pollutants are regulated by the Specific Prohibitions in the Prohibited Discharge Standards? Those which

    ─ 1. Are corrosive.
    ─ 2. Are a fire or explosion hazard.
    3. Cause interference.
    4. Cause pass-through.
    ─ 5. Result in toxic gases.

22. IUs must notify the POTW of what types of slug loadings?

    ─ 1. Corrosive pollutants
    ─ 2. Fire or explosion hazards
    ─ 3. Oxygen demanding pollutants
    ─ 4. Solid or viscous pollutants
    ─ 5. Thermal pollutants

23. EPA has promulgated both concentration- and mass-based pretreatment standards for which of the following categories?

    1. Battery manufacturing
    2. Electrical and electronic manufacturing
    3. Pesticides manufacturing
    4. Pharmaceutical manufacturing
    ─ 5. Pulp, paper and paperboard manufacturing

24. What are the basic elements of a toxic management plan?

    ─ 1. Identification of toxic organics used
    2. Location of dump site
    ─ 3. Method of disposal
    ─ 4. Quantity of each toxic organic used
    ─ 5. Use of each toxic organic

25. A Wastewater Ordinance may limit what types of wastes? Those that

    ─ 1. Create a nuisance.
    ─ 2. Damage the system.
    ─ 3. Harm the environment.
    ─ 4. Impose unreasonable costs.
    ─ 5. Menace public health.

26. Which criteria are used to determine if a violation is significant noncompliance?

    1. Discharge of high level of BOD
    ─ 2. Discharge of pollutant that has caused imminent endangerment to human health
    ─ 3. Failure to accurately report noncompliance
    4. Failure to have contact person available for pretreatment inspector
    ─ 5. Failure to provide required reports within 30 days after due date

27. What factors should be considered when assessing penalties?

    1. Ability of IU to pay penalty
    ─ 2. Large enough to deter future noncompliance
    ─ 3. Penalties should be calculated on a logical basis
    ─ 4. Penalties should be uniform
    ─ 5. Recovery of economic benefit for noncompliance

28. What actions are considered by a POTW for serious violations of the Wastewater Ordinance?

    1. Administrative warnings
    ─ 2. Filing criminal actions
    3. Issuing notices of violation
    ─ 4. Revocation of wastewater discharge permit
    ─ 5. Suspension of wastewater discharge permit

END OF OBJECTIVE TEST

APPENDICES

CHAPTER 3. DEVELOPMENT AND APPLICATION OF REGULATIONS

APPENDIX A

SUMMARY OF FEDERAL PRETREATMENT STANDARDS FOR THE ELECTROPLATING AND METAL FINISHING POINT SOURCE CATEGORIES

SANITATION DISTRICTS OF LOS ANGELES COUNTY

Applicable CFR: Electroplating : 40 CFR Part 413   Published in Federal Register: Part 413: January 28, 1981
                  Metal Finishing: 40 CFR Part 433   Part 413 TTO limits: July 15, 1983
Applicable SIC: Major Groups 34 to 39   Part 433: July 15, 1983

Description of the Point Source Categories:

   The Electroplating and Metal Finishing Point Source Categories are defined as facilities that use one or more of
   the following six unit operations:

   1. Electroplating
   2. Electroless plating
   3. Anodizing
   4. Coating (phosphating, chromating and coloring)
   5. Chemical etching and milling
   6. Printed circuit board manufacturing

Facilities in the Electroplating and Metal Finishing Point Source Categories are subject to one of four sets of pretreatment
standards:

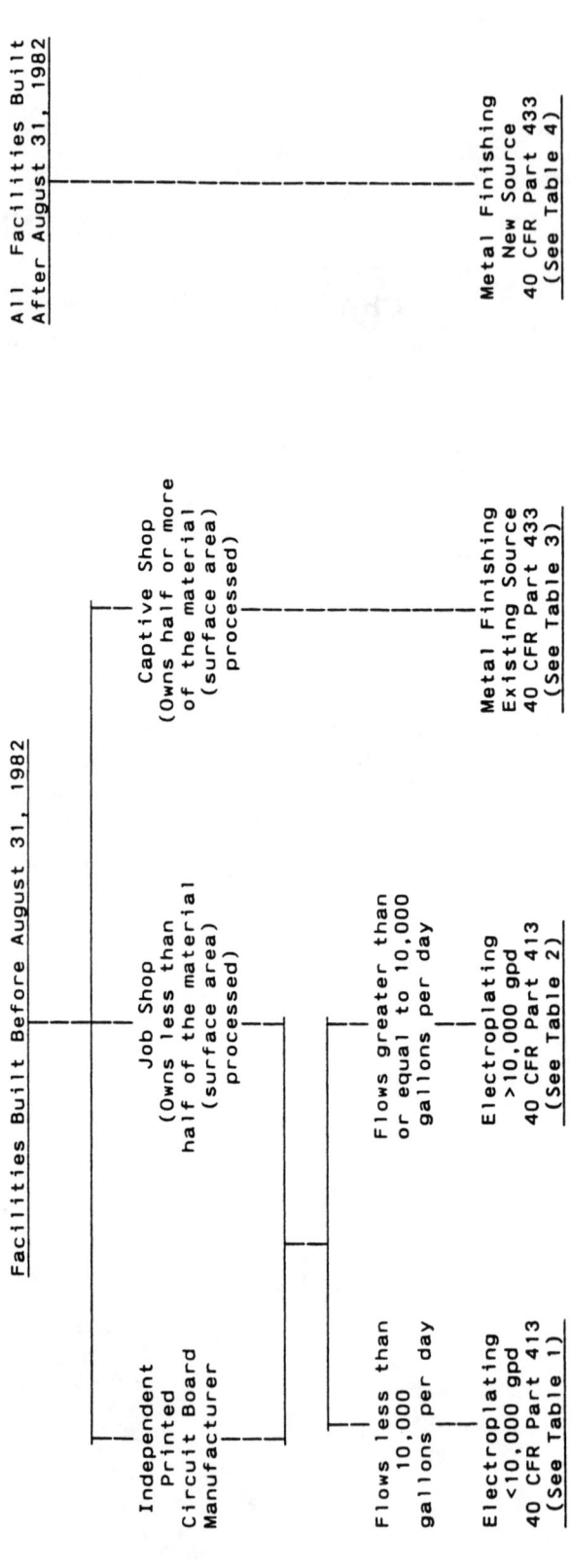

Some electroplating and metal finishing operations are regulated by EPA pretreatment standards for other point source
categories. In the point source categories listed on the next page, the electroplating and metal finishing pretreatment
standards do not apply to operations regulated by other EPA pretreatment standards.

SANITATION DISTRICTS OF LOS ANGELES COUNTY

SUMMARY OF FEDERAL PRETREATMENT STANDARDS

Electroplating and Metal Finishing Point Source Categories
Page 2 -- Exempted Operations

The Metal Finishing regulation exempts operations that are regulated under the following point source categories, except that the Metal Finishing TTO limits apply to the Iron and Steel Manufacturing Point Source Category:

1. Nonferrous Metals Smelting and Refining
2. Coil Coating
3. Porcelain Enameling
4. Battery Manufacturing
5. Iron and Steel
6. Metal Casting Foundries
7. Aluminum Forming
8. Copper Forming
9. Plastic Molding and Forming

The Electroplating regulation excludes certain operations in certain industries as follows:

1. Nonferrous Metals Smelting and Refining - Electrowinning and Electro Refining.
2. Coil Coating - Metal Surface Preparation and Conversion Coating.
3. Porcelain Enameling - Metal Surface Preparation and Immersion or Electroless Plating.
4. Battery Manufacturing - Electro Deposition of Active Electrode Materials, Electro Impregnation and Electro Forming.
5. Iron and Steel Manufacturing - Continuous Strip Electroplating.

Both the Electroplating and the Metal Finishing regulations exempt metalic platemaking and gravure cylinder preparation conducted within or for printing and publishing facilities.

SANITATION DISTRICTS OF LOS ANGELES COUNTY

SUMMARY OF FEDERAL PRETREATMENT STANDARDS

Metal Finishing Point Source Categories
Page 3 -- Regulated Unit Operations

1. Electroplating
2. Electroless Plating
3. Anodizing
4. Coating (Chromating, Phosphating, and Coloring)
5. Chemical Etching and Milling
6. Printed Circuit Board Manufacturing
7. Cleaning
8. Machining
9. Grinding
10. Polishing
11. Tumbling
12. Burnishing
13. Impact Deformation
14. Pressure Deformation
15. Shearing
16. Heat Treating
17. Thermal Cutting
18. Welding
12. Brazing
20. Soldering
21. Flame Spraying
22. Sand Blasting
23. Other Abrasive Jet Machining
24. Electric Discharge Machining
25. Electrochemical Machining
26. Electron Beam Machining
27. Laser Beam Machining
28. Plasma Arc Machining
29. Ultrasonic Machining
30. Sintering
31. Laminating
32. Hot Dip Coating
33. Sputtering
34. Vapor Plating
35. Thermal Infusion
36. Salt Bath Descaling
37. Solvent Degreasing
38. Paint Stripping
39. Painting
40. Electrostatic Painting
41. Electropainting
42. Vacuum Metalizing
43. Assembly
44. Calibration
45. Testing
46. Mechanical Plating

ELECTROPLATING AND METAL FINISHING POINT SOURCE CATEGORIES
CHEMICALS REGULATED UNDER TOTAL TOXIC ORGANICS (TTO)

001 Acenaphthene
002 Acrolein
003 Acrylonitrile
004 Benzene
005 Benzidine
006 Carbon tetrachloride (tetrachloromethane)
007 Chlorobenzene
008 1,2,4-trichlorobenzene
009 Hexachlorobenzene
010 1,2-dichloroethane
011 1,1,1-trichloroethane
012 Hexachloroethane
013 1,1-dichloroethane
014 1,1,2-trichloroethane
015 1,1,2,2-tetrachloroethane
016 Chloroethane
018 Bis(2-chloroethyl) ether
019 2-chloroethyl vinyl ether (mixed)
020 2-chloronaphthalene
021 2,4,6-trichlorophenol
022 Parachlorometa cresol
023 Chloroform (trichloromethane)
024 2-chlorophenol
025 1,2-dichlorobenzene
026 1,3-dichlorobenzene
027 1,4-dichlorobenzene
028 3,3-dichlorobenzidine
029 1,1-dichloroethylene
030 1,2-trans-dichloroethylene
031 2,4-dichlorophenol
032 1,2-dichloropropane
033 1,2-dichloropropylene (1,3-dichloropropene)
034 2,4-dimethylphenol
035 2,4-dinitrotoluene
036 2,6-dinitrotoluene
037 1,2-diphenylhydrazine
038 Ethylbenzene
039 Fluoranthene
040 4-chlorophenyl phenyl ether
041 4-bromophenyl phenyl ether
042 Bis(2-chloroisopropyl) ether
043 Bis(2-chloroethoxy) methane
044 Methylene chloride (dichloromethane)
045 Methyl chloride (dichloromethane)
046 Methyl bromide (bromomethane)
047 Bromoform (tribromomethane)
048 Dichlorobromomethane
051 Chlorodibromomethane
052 Hexachlorobutadiene
053 Hexachloromycyclopentadiene
054 Isophorone
055 Naphthalene
056 Nitrobenzene
057 2-nitrophenol
058 4-nitrophenol
059 2,4-dinitrophenol
060 4,6-dinitro-o-cresol
061 N-nitrosodimethylamine
062 N-nitrosodiphenylamine
063 N-nitrosodi-n-propylamin
064 Pentachlorophenol
065 Phenol
066 Bis(2-ethylhexyl) phthalate
067 Butyl benzyl phthalate
068 Di-N-Butyl Phthalate
069 Di-n-octyl phthalate
070 Diethyl Phthalate

071 Dimethyl phthalate
072 1,2-benzanthracene (benzo(a) anthracene
073 Benzo(a)pyrene (3,4-benzo-pyrene)
074 3,4-Benzofluoranthene (benzo(b) fluoranthene)
075 11,12-benzofluoranthene (benzo(b) fluoranthene)
076 Chrysene
077 Acenaphthylene
078 Anthracene
079 1,12-benzoperylene (benzo(ghi) perylene)
080 Fluorene
081 Phenanthrene
082 1,2,5,6-dibenzanthracene (dibenzo(,h) anthracene)
083 Indeno (,1,2,3-cd) pyrene (2,3-o-pheynylene pyrene)
084 Pyrene
085 Tetrachloroethylene
086 Toluene
087 Trichloroethylene
088 Vinyl chloride (chloroethylene)
089 Aldrin
090 Dieldrin
091 Chlordane (technical mixture and metabolites)
092 4,4-DDT
093 4,4-DDE (p,p-DDX)
094 4,4-DDD (p,p-TDE)
095 Alpha-endosulfan
096 Beta-endosulfan
097 Endosulfan sulfate
098 Endrin
099 Endrin aldehyde
100 Heptachlor
101 Heptachlor epoxide (BHC-hexachlorocyclohexane)
102 Alpha-BHC
103 Beta-BHC
104 Gamma-BHC (lindane)
105 Delta-BHC (PCB-polychlorinated biphenyls)
106 PCB-1242 (Arochlor 1242)
107 PCB-1254 (Arochlor 1254)
108 PCB-1221 (Arochlor 1221)
109 PCB-1232 (Arochlor 1232)
110 PCB-1248 (Arochlor 1248)
111 PCB-1260 (Arochlor 1260)
112 PCB-1016 (Arochlor 1016)
113 Toxaphene
129 2,3,7,8-tetrachloro-dibenzo-p-dioxin (TCDD)

TABLE 1

FEDERAL PRETREATMENT STANDARDS (40 CFR 413)

ELECTROPLATING POINT SOURCE CATEGORY

FACILITIES DISCHARGING LESS THAN 10,000 GALLONS PER DAY

Pollutant	Daily Maximum (mg/l)	4-Day Average (mg/l)
Cadmium (T)	1.2	0.7
Lead (T)	0.6	0.4
Cyanide (amenable to chlorination)	5.0	2.7
TTO[1]	4.57	----

TABLE 2

FEDERAL PRETREATMENT STANDARDS (40 CFR 413)

ELECTROPLATING POINT SOURCE CATEGORY

FACILITIES DISCHARGING 10,000 GALLONS PER DAY OR MORE

Pollutant	Daily Maximum (mg/l)	4-Day Average (mg/l)
Cadmium (T)	1.2	0.7
Chromium (T)	7.0	4.0
Copper (T)	4.5	2.7
Lead (T)	0.6	0.4
Nickel (T)	4.1	2.6
Silver (precious metals subcategory)	1.2	0.7
Zinc (T)	4.2	2.6
Cyanide (T)	1.9	1.0
TTO[1]	2.13	----
Total Metals[2]	10.5	6.8

[1]Total Toxic Organics, the sum of the toxic organic compounds listed in 40 CFR 413.02(i), found in excess of 0.01 mg/l.

[2]Total Metals is the sum of the individual concentrations of Cr (T), Cu (T), Ni (T), and Zn (T).

TABLE 3

FEDERAL PRETREATMENT STANDARDS (40 CFR 433)

METAL FINISHING POINT SOURCE CATEGORY

EXISTING SOURCES

Pollutant	Daily Maximum (mg/l)	Monthly Average (mg/l)
Cadmium (T)	0.69	0.26
Chromium (T)	2.77	1.71
Copper (T)	3.38	2.07
Lead (T)	0.69	0.43
Nickel (T)	3.98	2.38
Silver (T)	0.43	0.24
Zinc (T)	2.61	1.48
Cyanide (T)	1.20	0.65
Cyanide (amenable to chlorination)[1]	0.86	0.32
TTO[2]	2.13	----

TABLE 4

FEDERAL PRETREATMENT STANDARDS (40 CFR 433)

METAL FINISHING POINT SOURCE CATEGORY

NEW SOURCES

Pollutant	Daily Maximum (mg/l)	Monthly Average (mg/l)
Cadmium (T)	0.11	0.07
Chromium (T)	2.77	1.71
Copper (T)	3.38	2.07
Lead (T)	0.69	0.43
Nickel (T)	3.98	2.38
Silver (T)	0.43	0.24
Zinc (T)	2.61	1.48
Cyanide (T)	1.20	0.65
Cyanide (amenable to chlorination)[1]	0.86	0.32
TTO[2]	2.13	----

[1]Alternate cyanide amenable to chlorination limits for facilities with cyanide treatment, upon agreement of the pollution control authority (Sanitation Districts).

[2]Total Toxic Organics, the sum of the toxic organic compounds listed in 40 CFR 433.11(e), found in excess of 0.01 mg/l.

APPENDIX B

BASELINE MONITORING REPORT (BMR)

INDUSTRIAL WASTE SECTION
SANITATION DISTRICTS OF LOS ANGELES COUNTY
1955 WORKMAN MILL ROAD
P.O. BOX 4998
WHITTIER, CA   90607-4998
(310) 699-7411, EXT. 2900

EPA Electroplating (40 CFR 413) or Metal Finishing (40 CFR 433)
Point Source Category Regulations
(Please print or type)

I. Company Information

Name:_____ Tel: (___)_____
Situs Address:_____
_____ Zip _____
Mailing Address:_____
_____ Zip _____
Sanitation Districts' Industrial Wastewater Discharge Permit Number[1] _____
Sanitation Districts' Industrial Wastewater Discharge Permit Flow Rate _____ gal/day
Federal Standard Industrial Classification Numbers (SIC) characterizing this facility _____
Company's Industrial Waste Contact Person _____
Title:_____
Person In Charge of Local Operations:_____
Title:_____
Owner of Company (parent company or corporate entity if appropriate):_____
Address of Owner:_____

II. Category Determination

A. Unit Operations

Please check any and all of the following unit operations which are performed at your **facility**.[2] Provide a brief description of those operations in the space provided (e.g. CuCN plating, H_2SO_4 anodizing, Fe phosphatizing, $FeCl_3$ etching, etc.).

Brief Description

1 ☐ Electroplating   _____
2 ☐ Electroless Plating   _____
3 ☐ Anodizing   _____
4 ☐ Coating   _____
      (phosphatizing, chromating, etc.)

---

[1]One Baseline Monitoring Report (BMR) must be completed for **every** industrial wastewater discharge point to the sewer from your facility.

[2]If your facility has more than one sewer connection, indicate in the description area which operations discharge to which permit number.

5 ☐ Chemical Etching and Milling   _____
       (includes caustic cleaning of
       aluminum & acid cleaning of
       most metals)

6 ☐ Printed Circuit Board Mfg.   _____

7 ☐ None of the Above   _____

B. Exempt Companies

If you checked item number 7 in IIA above, your facility is not covered by the EPA Electroplating or Metal Finishing regulations. Your claim to such an exemption will be verified through inspections of your facility by Sanitation Districts' personnel.

If you believe your facility is exempt from the Electroplating and Metal Finishing regulations but may be covered by another EPA categorical regulation, please indicate that category here.

_____

As your company has denoted it is exempt by checking item 7 in IIA above, you do not need to fully complete this BMR. Please proceed to Section IX of this form, complete the certification statement and return this form to the Sanitation Districts.

C. Companies Not Exempt

If you did not check item number 7 in IIA above, please continue. Please check one of the two boxes below. If your facility had categorical installations made both before and after August 31, 1982, please include a construction timetable as a separate attachment.

1 ☐ New Source - electroplating or metal finishing operations were constructed after August 31, 1982.[3]

2 ☐ Existing Source - electroplating or metal finishing operations were constructed prior to August 31, 1982. Date facilities were constructed _____.

Please check one of the three boxes below.

3 ☐ Job Shop - Owns less than 50% (area basis) of the materials undergoing metal finishing.

4 ☐ Independent Printed Circuit Board Manufacturer - Manufactures printed circuit boards primarily for sale to other companies.

5 ☐ Captive Shop - Owns over 50% (area basis) of the materials undergoing metal finishing.

_____

[3]The term "new source" is not always easily defined. The Sanitation Districts encourage you to request our assistance in making this determination if there is any uncertainty.

A P P E N D I X B

D. Category Determination

The category your facility belongs to can be determined based on the items checked in IIC above. Please check the box below which applies to your facility.

☐   New Source Metal Finishing (40 CFR 433) if you checked box 1 and box 3, 4, or 5 in IIC.

☐   Existing Source Metal Finishing (40 CFR 433) if you checked box 2 and box 5 in IIC.

☐   Electroplating Less Than 10,000 gpd (40 CFR 413) if you checked box 2 and box 3 or 4 in IIC and your permit flowrate from page 1 is less than 10,000 gpd.[4]

☐   Electroplating Greater Than 10,000 gpd (40 CFR 413) if you checked box 2 and box 3 or 4 in IIC and your permit flowrate from page 1 is greater than 10,000 gpd.[4]

III. Flow Measurement Information - Please complete the following tables. Also, attach a schematic process flow diagram showing wastestreams, flowrates, treatment units and sampling locations.

Description of EPA Regulated Wastewater Flows[5]	Average Daily Flow (gal/day)	Maximum Daily Flow (gal/day)	Flow Description[6]		Regulated EPA Category (413, 433, etc.)	Does Wastestream Receive Pretreatment? Yes or No Describe
			E/M	I/C		

---

[4]If your permit flowrate is no longer an accurate measure of your daily discharge then your company should apply for a permit revision.

[5]Separately include all wastestreams discharging to this connection/outfall which are covered by EPA categorical regulations. Discharges regulated under categories other than 413 and 433 should also be listed here. For Metal Finishing, the six operations listed in section IIA and an additional 40 operations are regulated. For Electroplating, only the six operations listed in section IIA are regulated. Any of the additional 40 operations are considered unregulated flows under the Electroplating regulations and should be listed there.

[6]Please indicate by letter in this column whether wastewater flow value is (E) Estimated or (M) Measured, and (I) Intermittent or (C) Continuous.

Description of Unregulated Wastewater Flows[7]	Average Daily Flow (gal/day)	Maximum Daily Flow (gal/day)	Flow Description[6]			Does Wastestream Receive Pretreatment? Yes or No Describe
			E/M	I/C		

Description of Dilution Flows[8]	Average Daily Flow (gal/day)	Maximum Daily Flow (gal/day)	Flow Description[6]			Does Wastestream Receive Pretreatment? Yes or No Describe
			E/M	I/C		

A
P
P
E
N
D
I
X

B

IV. Environmental Control Permits

Please list other environmental control permits held by or for your facility.

_____
_____
_____

Does your facility have an air pollution control device which produces a discharge to the sewer?
☐ Yes ☐ No

V. Production Information

Provide a brief description of your operations from raw materials to finished product.

_____
_____
_____
_____

---

[7]Includes wastewater flows to this connection/outfall from operations not covered by EPA industrial categorical regulations and not considered dilution flows.

[8]Dilution flows include non-contact cooling water and boiler blowdown, D.I. backwash and R.O. reject water from incoming water supply treatment, wastestreams listed in Appendix D to 40 CFR 403 and sanitary wastes. Sanitary wastes should not be listed here unless they discharge through the legal sampling point.

Rates of Production - Please complete the following table, add additional pages if necessary.

	1	2	3
Name of unit production			
Unit of production			
Daily production rate			
Monthly production rate			

VI. Total Toxic Organic (TTO) Information

Dischargers subject to Metal Finishing (40 CFR 433) or Electroplating (40 CFR 413) regulations are regulated for TTO. Your company must periodically monitor its wastestream to show compliance with this limitation. EPA will allow dischargers subject to these regulations to provide the Control Authority (Districts) with a list of TTO compounds stored or used at the facility to lessen the number of parameters which must be monitored. The discharger may also submit a Toxic Organic Management Plan (TOMP) to the Districts, which if approved will allow the discharger to certify that TTO compounds have not been discharged, in lieu of TTO self-monitoring. The Sanitation Districts have guidelines for "Total Toxic Organics Monitoring Requirements and Toxic Organic Management Plans" which you should request if you do not have a copy.

Please check the following boxes which apply.

☐ TTO compounds **are** stored and/or used at this facility. A TOMP was submitted to the Districts on _____.

☐ **No** TTO compounds are stored or used at this facility. A TOMP was submitted to the Districts on _____.

☐ This company will perform self-monitoring for TTO (check one of the boxes below)

    ☐ Monitoring will be for the entire list of 111 regulated TTO compounds as shown in 40 CFR 433.11(e) or 40 CFR 413.02(i).

    ☐ Monitoring will be performed for TTO compounds expected to be present in the wastewater, as listed below. It is understood that the Districts may expand the self-monitoring TTO parameters at their discretion.

TTO compounds Stored or Used at this Facility (attach additional pages as necessary)	Is the compound listed in the adjacent column expected to be present in the wastewater? (Yes or No)
_____	_____
_____	_____
_____	_____
_____	_____
_____	_____

VII. Measurement of Pollutants

The daily maximum and average concentrations of all regulated pollutants in all regulated wastewater streams described in Section III must be provided with this BMR[9]. The wastewater must be sampled and analyzed in accordance with 40 CFR 403.12(b)(5)(iii-viii). Copies of the wastewater analysis results must be attached to this BMR when it is submitted to the Sanitation Districts. The sample results must indicate the analytical test method used for each parameter.

Representative samples will consist of a minimum of four (4) grab samples for pH, cyanide and volatile organics. All other regulated pollutants must be obtained through composite samples, flow-proportioned where feasible. Historical data may be used if it can still be considered representative of the current discharge. For new sources only, estimates of pollutant values are allowed. However, within 90 days of commencement of discharge, the new source discharger must submit a 90 day compliance report to the Sanitation Districts on an additional BMR form.

Please complete the following table describing the sampling and analytical results accompanying this BMR.

Sample Type	Sampling Date & Time	Sampling Location	Name & Address of Company Obtaining Sample	Name & Address of Laboratory Performing Analysis
Composite				
Grab #1				
Grab #2				
Grab #3				
Grab #4				

Sampling Certification

I certify that the sampling and analysis provided with this BMR is representative of normal work cycles and expected pollutant discharges to the Sanitation Districts.

Date: _____ Sign Name: _____
Print Name: _____
Job Title: _____

---

[9]Dioxin may be screened, rather than specifically analyzed, as long as the discharger attaches a statement certifying that the facility neither stores, uses or manufactures dioxin.

Using the results from the attached analyses, complete the shaded areas of the following table. Compare the sample result value to the 4-day or monthly average limit for your appropriate category to determine consistent compliance. If dilution flows from page 4 passed through the sampling point, then the limits must be adjusted.

Constituent	Sample Result	Limits - Daily Maximum / 4-Day or Monthly Average (mg/l)				Do Sample Results Show Consistent Compliance?
		413 <10,000 gpd	413 >10,000 gpd	433 Existing	433 New Source	
Cyanide[10]	mg/l	5.0 / 2.7	1.9 / 1.0	1.20 / 0.65[11]	1.20 / 0.65[11]	
Cd	mg/l	1.2 / 0.7	1.2 / 0.7	0.69 / 0.26	0.11 / 0.07	
Cr	mg/l		7.0 / 4.0	2.77 / 1.71	2.77 / 1.71	
Cu	mg/l		4.5 / 2.7	3.38 / 2.07	3.38 / 2.07	
Pb	mg/l	0.6 / 0.4	0.6 / 0.4	0.69 / 0.43	0.69 / 0.43	
Ni	mg/l		4.1 / 2.6	3.98 / 2.38	3.98 / 2.38	
Ag	mg/l		1.2 / 0.7[12]	0.43 / 0.24	0.43 / 0.24	
Zn	mg/l		4.2 / 2.6	2.61 / 1.48	2.61 / 1.48	
Total Metals[13]	mg/l		10.5 / 6.8			
Volatile Organics[14]	ug/l					
TTO[15]	ug/l	4570 ug/l	2130 ug/l	2130 ug/l	2130 ug/l	

[10]Average of all 4 CN grab sample results. Must be analyzed as CN amenable to chlorination for Electroplating Less Than 10,000 gpd. All other categories must analyze for Total CN.

[11]Under 40 CFR 433, Metal Finishers with CN wastestreams must obtain CN samples downstream of the cyanide treatment unit but prior to commingling with other wastestreams, or alternate CN limits must be calculated based on the dilution ratio of the CN wastestream flow to the effluent flow.

[12]These Electroplating >10,000 gpd silver limits only apply to Electroplating >10,000 gpd facilities which perform precious metal electroplating operations.

[13]For Electroplating > 10,000 gpd only, total metals is the sum of Cr, Cu, Ni, & Zn.

[14]Average all 4 grab sample results for each regulated volatile organic compound. Sum all average values in excess of 10 ug/l.

[15]Sum of the volatile organics value above and all other regulated toxic organic compounds detected in excess of 10 ug/l.

VIII. Statement of Compliance

An authorized representative of the company as defined in 40 CFR 403.12(k) must review the following statements of compliance which must be certified to by a qualified professional.

I hereby certify that the EPA categorical pretreatment standards which apply to this facility are being met on a consistent basis as evidenced by the attached data. ☐ *Yes* ☐ *No*

I hereby certify that dilution is not being used in lieu of treatment to meet the EPA categorical pretreatment standards. ☐ *Yes* ☐ *No*

If the answer to either of the above statements is *No*, then additional pretreatment, flow reduction or operations and maintenance measures to bring the company into compliance with the EPA categorical regulations must be proposed below. Anticipated completion dates must be provided.

1. _____
2. _____
3. _____
4. _____

A detailed compliance schedule for the above changes must be attached. This schedule shall contain increments of progress in the form of dates for the commencement and completion of major events leading to the construction and operation of additional pretreatment required for the facility to meet the EPA categorical pretreatment standards (e.g. hiring an engineer, completing preliminary plans, completing final plans, executing contract for major components, commencing construction, completing construction, etc.). A commitment to design, install or alter pretreatment or process systems to effect future compliance does not relieve your company of the requirement to immediately comply with discharge limits by whatever means necessary (cessation, impounding, hauling, etc.) until a more permanent solution is implemented.

Reviewed by: (company official's signature) _____ Date: _____
Print Name: _____ Job Title: _____

Certified by: (qualified professional's signature) _____ Date: _____
Print Name: _____
Qualifications as an Environmental Professional: _____
_____
_____
Company Name: _____
Company Address: _____

IX. Certification (by authorized company official per 40 CFR 403.12 (l))

I certify under penalty of law that this document and all attachments were prepared under my direction or supervision in accordance with a system designed to assure that qualified personnel properly gather and evaluate the information submitted. Based on my inquiry of the person or persons who manage the system, or those persons directly responsible for gathering the information, the information submitted is, to the best of my knowledge and belief, true, accurate and complete. I am aware that there are significant penalties for submitting false information, including the possibility of fine and imprisonment for knowing violations.

Date: _____
Signature of authorized company official: _____
Print name of official: _____
Title of authorized company official: _____

APPENDIX B

APPENDIX C

WASTEWATER SAMPLING AND ANALYSIS REQUIREMENTS

Wastewater contained in the regulated and nonregulated wastewater streams must be analyzed for the pollutants controlled by the EPA's Electroplating and Metal Finishing regulations. A table listing these pollutants is attached.

When doing wastewater sampling and analysis, the following precautions should be observed:

a. Flow-proportioned composite samples over the full operating day must be used where feasible. Grab samples may be permissible in special situations with Sanitation Districts' permission or where the nature of the pollutant (e.g. cyanide) requires grab sampling.

b. Three samples within a two-week period must be obtained for wastewater flow rates less than 250,000 gallons per day; six samples within a two-week period must be obtained for larger wastewater flow rates.

c. Information furnished must include the date, time and place of sampling; the methods of analysis; and a statement certifying that the samples are representative of normal work cycles and expected pollutant discharges. The location at which the wastewater sample was taken must be defined in relation to the regulated and unregulated waste streams present in the sampled wastewater.

d. If the appropriate pollutants have been included in past Critical Parameter Reports (CPRs) submitted to the Districts, these sample results may be substituted for the samples required above.

e. All laboratory analysis work should be performed by a State Certified Laboratory or a laboratory approved by the Sanitation Districts. Sampling and analysis procedures shall conform to EPA 40 CFR 136 requirements or those specified in Standard Methods for the Examination of Water and Wastewater.

f. If your company has recently monitored its wastewater for the purpose of submitting a Baseline Monitoring Report to the Sanitation Districts for the Electroplating category, this monitoring material may be resubmitted with this report if no substantial changes in effluent quality have occurred.

g. At least one analysis of the wastewater must be made to determine the presence of materials regulated as Total Toxic Organics (TTO) by EPA. Please review the listing of the compounds included under the TTO regulations as given in Appendix X. If any of these compounds are used or stored at your company, one analysis of a typical 24-hour composite wastewater sample must be obtained for such compounds (only) and the results submitted to the Sanitation Districts with this Baseline Monitoring Report.

LIST OF ORGANIC CHEMICALS INCLUDED IN TOTAL TOXIC ORGANICS (TTO) REGULATED UNDER ELECTROPLATING AND METAL FINISHING CATEGORIES

1. Acenaphthene
2. Acrolein
3. Acrylonitrile
4. Benzene
5. Benzidine
6. Carbon tetrachloride (tetrachloromethane)
7. Chlorobenzene
8. 1,2,4 trichlorobenzene
9. Hexachlorobenzene
10. 1,1-dichloroethane
11. 1,2-dichloroethane
12. 1,1,1-trichloroethane
13. Hexachloroethane
14. 1,1,2-trichloroethane
15. 1,1,2,2-tetrachloroethane
16. Chloroethane
17. Bis(2-chloroethyl) ether
18. 2-chloroethyl vinyl ether (mixed)
19. 2-chloronaphthlene
20. 2,4,6-trichlorophenol
21. Parachlorometa cresol
22. Chloroform (trichloromethane)
23. 2-chlorophenol
24. 1,2 dichlorobenzene
25. 1,3 dichlorobenzene
26. 1,4 dichlorobenzene
27. 3,3-dichlorobenzidine
28. 1,1-dichloroethylene
29. 1,2-trans-dichloroethylene
30. 2,4-dichlorophenol
31. 1,2 dichloropropane
32. 1,2 dichloropropylene (1,3 dichloropropene)
33. 2,4-dimethylphenol
34. 2,4-dinitrotoluene
35. 2,6-dinitrotoluene
36. 1,2-diphenylhydrazine
37. Ethylbenzene
38. Fluoranthene
39. 4-chlorophenyl phenyl ether
40. 4-brorophenyl phenyl ether
41. Bis(2-chloroisopropyl) ether
42. Bis(2-chloroethoxy) methane
43. Methylene chloride (dichloromethane)
44. Methyl chloride (chloromethane)
45. Methyl bromide (dibromomethane)
46. Bromoform (tribromomethane)
47. Dichlorobromomethane
48. Chlorodibromomethane
49. Hexachlorobutadiene
50. Hexachlorocyclopentadiene
51. Isophorone
52. Naphthalene
53. Nitrobenzene
54. 2-nitrophenol
55. 4-nitrophenol
56. 2,4-dinitrophenol
57. 4,6-dinitro-o-cresol
58. N-nitrosodimethylamine
59. N-nitrosodiphenylamine
60. N-nitrosodi-n-propylamine
61. Pentachlorophenol
62. Phenol
63. Bis(2-ethylhexyl) phthalate
64. Butyl benzyl phthalate
65. Di-n-butyl phthalate
66. Di-n-octyl phthalate
67. Diethyl phthalate
68. Dimethyl phthalate
69. 1,2-benzanthracene (benzo(a)anthracene)
70. 3,4-benzopyrene (benzo(a)pyrene)

APPENDIX C

LIST OF ORGANIC CHEMICALS INCLUDED IN TOTAL TOXIC ORGANICS (TTO) REGULATED UNDER ELECTROPLATING AND METAL FINISHING CATEGORIES (continued)

71. 3,4-Benzofluoranthene (benzo(b)fluroanthene)
72. 11,12-benzofluoranthene (benzo(k)fluoranthene)
73. Chrysene
74. Acenaphthylene
75. Anthracene
76. 1,12-benzoperylene (benzo(ghi)perylene)
77. Fluorene
78. Penanthrene
79. 1,2,5,6-dibenzanthracene (dibenzo(a,h)anthracene)
80. Indeno (1,2,3-cd)pyrene (2,3-o-phenylene pyrene)
81. Pyrene
82. Tetrachloroethylene
83. Toluene
84. Trichylorethylene
85. Vinyl chloride (chloroethylene)
86. Aldrin
87. Dieldrin
88. Chlordane (technical mixture & metabolites)
89. 4,4-DDT
90. 4,4-DDE (p,p-DDX)
91. 4,4-DDD (p,p-TDE)
92. Alpha-endosulfan
93. Beta-endosulfan
94. Endosulfan sulfate
95. Endrin
96. Endrin aldehyde
97. Heptachlor
98. Heptachlor epoxide (BHC-hexachlorocyclohexane)
99. Alpha-BHC
100. Beta-BHC
101. Gamma-BHC (lindane)
102. Delta-BHC PCB-polychlorinated biphenyls
103. PCB-1242 (Arochlor 1242)
104. PCB-1254 (Arochlor 1254)
105. PCB-1221 (Arochlor 1221)
106. PCB-1232 (Arochlor 1232)
107. PCB-1248 (Arochlor 1248)
108. PCB-1260 (Arochlor 1260)
109. PCB-1016 (Arochlor 1016)
110. Toxaphene
111. 2,3,7,8-tetrachlorodi-benzo-p-dioxin (TCDD)

Inorganic Priority Pollutants

112. Antimony (Total)
113. Arsenic
114. Asbestos
115. Beryllium
116. Cadmium
117. Chromium
118. Copper
119. Cyanide
120. Lead
121. Mercury
122. Nickel
123. Selenium
124. Silver
125. Thallium
126. Zinc

APPENDIX X

SANITATION DISTRICTS OF LOS ANGELES COUNTY
Charles W. Carry, Chief Engineer & General Manager
Telephone: (213) 699-7411 / From Los Angeles (213) 685-5217
INDUSTRIAL WASTEWATER
CRITICAL PARAMETER REPORT FORM

SURCHARGE ACCOUNT NO. _____

PERMIT NO. _____

_____
(Print) Name of Company Having Wastewater Discharge   SIC Number(s)

_____
(Print) Address of Wastewater Discharge

_____ _____ to _____
(Print) Sample Date   Sample Point Location   Reporting Period

DAILY WATER USE FOR REPORTING PERIOD (GAL)   AVG. _____ MAX. _____

WASTEWATER FLOW (A,B)
DETERMINED BY: ☐ DIRECT MEASUREMENT   ☐ METERED WATER SUPPLY   ☐ ADJUSTED METERED WATER SUPPLY

TYPE OF SAMPLE: ☐ GRAB   ☐ TIME COMPOSITE   ☐ FLOW PROPORTIONED COMPOSITE

CRITICAL PARAMETER VALUES

IDENT. CODE	PARAMETER 1/	2/	QUANTITY VALUES	IDENT. CODE	PARAMETER 1/	2/	QUANTITY VALUES
3 / A	WASTEWATER FLOW (Total)		gals/day	LL-370 371, 372	RADIOACTIVITY (Alpha, Beta, Gamma)		pci/l
3 / B	WASTEWATER FLOW (Peak)		gals/mi.	MM-111	TEMPERATURE		Degrees F
C-403	COD		mg/1	NN-104	COLOR		Units
D-151	SS (Suspended Solids)		mg/1	OO-253	THIOSULFATE (S)		mg/1
E-101	pH		Units	PP-703	CALCIUM		mg/1
F-155	TOTAL DISSOLVED SOLIDS		mg/1	QQ-704	MAGNESIUM		mg/1
G-201	AMMONIA (N)		mg/1	RR-719	POTASSIUM		mg/1
H-252	SULFIDE – DISSOLVED		mg/1	SS-706	BARIUM		mg/1
I-206	CYANIDE		mg/1	TT-204	NITRATE		mg/1
J-313	FLUORIDE		mg/1	UU-301	CHLORIDE		mg/1
K-707	ALUMINUM – Total		mg/1	VV-319	BROMIDE		mg/1
L-725	ANTIMONY – Total		mg/1	WW-257	SULFATE		mg/1
M-705	ARSENIC – Total		mg/1	XX-311	PHOSPHATE – ORTHO		mg/1
N-726	BERYLLIUM – Total		mg/1	620	BENZENE		µg/1
O-314	BORON – Total		mg/1	604	CARBON TETRACHLORIDE		µg/1
P-708	CADMIUM – Total		mg/1	611	CHLOROBENZENE		µg/1
Q-709	CHROMIUM – Total		mg/1	613	DICHLOROBENZENE		µg/1
R-711	COBALT – Total		mg/1	619	1, 2-DICHLOROETHANE		µg/1
S-712	COPPER – Total		mg/1	603	1, 1, 1-TRICHLOROETHANE		µg/1
T-713	IRON – Total		mg/1	657	2-CHLOROPHENOL		µg/1
U-714	LEAD – Total		mg/1	658	2, 4-DICHLOROPHENOL		µg/1
V-716	MANGANESE – Total		mg/1	663	PENTACHLOROPHENOL		µg/1
W-717	MERCURY – Total		mg/1	664	2, 4, 6-TRICHOROPHENOL		µg/1
X-732	MOLYBDENUM – Total		mg/1	602	CHLOROFORM		µg/1
Y-718	NICKEL – Total		mg/1	626	2, 4-DIMETHYLPHENOL		µg/1
Z-720	SELENIUM – Total		mg/1	624	ETHYL BENZENE		µg/1
AA-722	SILVER – Total		mg/1	601	METHYLENE CHLORIDE		µg/1
BB-723	SODIUM – Total		mg/1	607	TETRACHLOROETHYLENE		µg/1
CC-734	THALLIUM – Total		mg/1	621	TOLUENE		µg/1
DD-735	TIN – Total		mg/1	606	TRICHLOROETHYLENE		µg/1
EE-736	TITANIUM – Total		mg/1	525	HCH (Total)		µg/1
FF-724	ZINC – Total		mg/1	530	CHLORDANE (Total)		µg/1
GG-408	OIL & GREASE		mg/1	507	DDT (Total)		µg/1
HH-312	PHENOLS		mg/1	521	PCBs (Total)		µg/1
II-315	SURFACTANTS (MBAS)		mg/1	512	ALDRIN		µg/1
316	NONIONIC SURFACTANTS(NID)		mg/1	514	ENDRIN		µg/1

1/ Report all critical parameters required by the Sanitation Districts and any other critical parameter known to be present in the wastewater. Those parameters required by the Districts but known to be absent from the wastewater may be reported by placing the word absent in the appropriate space. Test procedures must be in accordance with procedures contained in the current edition of STANDARD METHODS, if applicable. Test procedures for priority organics must be run in accordance with the appropriate EPA method.

2/ If values are obtained by measurements or analyses write A in this column. Analysis values must be determined, using representative 24-hour composite samples (unless the parameter is identified by footnote 4/), by a State Certified or Districts Approved Laboratory. If values are obtained by estimate, write E in this column. Estimated values are acceptable for new plants only.

3/ Report flow rate for sampling day.

4/ Grab samples should be acquired with precautions taken to ensure that volatile constituents are preserved.

APPENDIX C

OTHER CRITICAL PARAMETERS

IDENT. CODE	PARAMETER	2/	QUANTITY VALUES	IDENT. CODE	PARAMETER	2/	QUANTITY VALUES

_____

(PRINT) Name and Address of Laboratory Preparing Analyses

_____

(PRINT) Name and Address of Company Collecting Wastewater Sample

CERTIFICATION BY PERMITTEE

I certify, under the penalty of perjury, that I have personally examined, and am familiar with, all of the information in this Critical Parameter Report. Based upon my inquiry of those persons immediately responsible for obtaining the information contained in this Report, I believe that the information is a true and correct representation of the wastewater discharged from the stated discharge point.

Signature of responsible company official: _____   Date _____

Print name of official: _____

Title of person certifying report: _____

Please submit this report to:

INDUSTRIAL WASTE SECTION
Sanitation Distrcits of Los Angeles County
P.O. Box 4998
Whittier, CA 90607

APPENDIX D

EPA MODEL PRETREATMENT ORDINANCE

ORDINANCE NO. _____

SECTION 1 — GENERAL PROVISIONS

1.1 Purpose and Policy

This ordinance sets forth uniform requirements for direct and indirect discharges of pollutants from nondomestic sources into the wastewater collection and treatment system for the (City of _____) and enables the City to comply with all applicable State and Federal laws including the Clean Water (Act 33 U.S.C. 1251 *et seq.*), and the General Pretreatment Regulations (40 CFR Part 403). The objectives of this ordinance are:

(a) To prevent the introduction of pollutants into the municipal wastewater system which will interfere with the operation of the system;

(b) To prevent the introduction of pollutants into the municipal wastewater system which will pass through the system, inadequately treated, into receiving waters or the atmosphere or otherwise be incompatible with the system;

(c) To ensure that the quality of the wastewater treatment plant sludge is maintained at a level which allows its marketability;

(d) To protect both municipal personnel who may come into contact with sewage, sludge and effluent in the course of their employment as well as protecting the general public;

(e) To preserve the hydraulic capacity of the municipal wastewater system;

(f) To improve the opportunity to recycle and reclaim wastewater and sludge from the system;

(g) To provide for equitable distribution of the cost of operation, maintenance and improvement of the municipal wastewater system; and

(h) To ensure the City complies with its NPDES permit conditions, sludge use and disposal requirements and any other Federal or State laws which the municipal wastewater system is subject to.

This Ordinance provides for the regulation of direct and indirect discharges to the municipal wastewater collection system through the issuance of permits to certain nondomestic users and through enforcement of general requirements for other users, authorizes monitoring and enforcement activities, establishes administrative review procedures, requires user reporting, and provides for the setting of fees for the equitable distribution of costs resulting from the program established herein.

This Ordinance shall apply to the City and to persons outside of the City who, by contract with the City, are included as users of the municipal wastewater system. Except as otherwise provided herein, the Superintendent of the municipal wastewater system or his designees shall administer, implement, and enforce the provisions of this ordinance. By discharging wastewater into the municipal wastewater system, industrial users located beyond the City limits agree to comply with the terms and conditions established in this Ordinance, as well as any permits or orders issued hereunder.

1.2 Administration

Except as otherwise provided herein, the Superintendent shall administer, implement and enforce the provisions of this Ordinance. Any powers granted to or duties imposed upon the Superintendent may be delegated by the Superintendent to other City personnel.

1.3 Definitions

Unless the context specifically indicates otherwise, the following terms and phrases, as used in this Ordinance, shall have the meanings hereinafter designated;

(1) *Act or "the Act."* The Federal Water Pollution Control Act, also known as the Clean Water Act, as amended, 33 U.S.C. 1251 *et seq.*

(2) *Approval Authority.* An NPDES State approved State Pretreatment Program; otherwise, the Regional Water Management Division Director of the U.S. EPA.

Note: The City should not adopt this definition verbatim. It must designate the State as the Approval Authority if the State has an EPA-approved pretreatment program. Alternatively, it must designate EPA in a non-approved State.

(3) *Authorized Representative of the Industrial User.*

1. If the industrial user is a corporation, authorized representative shall mean:

   a) the president, secretary, treasurer, or a vice-president of the corporation in charge of a principal business function, or any other person who performs similar policy or decision-making functions for the corporation, or

   b) the manager of one or more manufacturing, production, or operation facilities employing more than 250 persons or having gross annual sales or expenditures exceeding $25 million (in second-quarter 1980 dollars), if authority to sign documents has been assigned or delegated to the manager in accordance with corporate procedures.

2. If the industrial user is a partnership, association, or sole proprietorship, an authorized representative shall mean a general partner or the proprietor.

A
P
P
E
N
D
I
X

D

3. If the individual user is representing Federal, State or local governments, or an agent thereof, an authorized representative shall mean a director or highest official appointed or designated to oversee the operation and performance of the activities of the government facility.

4. The individuals described in paragraphs 1-3 above may designate another authorized representative if the authorization is in writing, the authorization specifies the individual or position responsible for the overall operation of the facility from which the discharge originates or having overall responsibility for environmental matters for the company, and the authorization is submitted to the City.

(4) *Biochemical Oxygen Demand (BOD).*   The quantity of oxygen utilized in the biochemical oxidation of organic matter under standard laboratory procedure, five (5) days at 20° Centigrade expressed in terms of weight and concentration [milligrams per liter (mg/L)].

(5) *Building Sewer.*   A sewer conveying wastewater from the premises of a user to the POTW.

(6) *Categorical Pretreatment Standard or Categorical Standard.*   Any regulation containing pollutant discharge limits promulgated by the U.S. EPA in accordance with Section 307(b) and (c) of the Act (33 U.S.C. 1317) which applies to a specific category of industrial users and which appears in 40 CFR Chapter 1, Subchapter N, Parts 405-471, incorporated herein by reference.

(7) *City.*   The City of _____ or the City Council of _____.

(8) *Color.*   The optical density at the visual wave length of maximum absorption, relative to distilled water. One hundred percent (100%) transmittance is equivalent to zero (0.0) optical density.

(9) *Composite Sample.*   The sample resulting from the combination of individual wastewater samples taken at selected intervals based on an increment of either flow or time.

(10) *Cooling Water.*   The water discharged from any use such as air conditioning, cooling or refrigeration, or to which the only pollutant added is heat.

(11) *Control Authority.*   The term "Control Authority" shall refer to the Superintendent once the City has a U.S. EPA approved pretreatment program according to the provisions of 40 CFR 403.11.

(12) *Discharge.*   The discharge or the introduction of nondomestic pollutants into the municipal wastewater system by an industrial user.

(13) *Environmental Protection Agency or U.S. EPA.*   The U.S. Environmental Protection Agency or, where appropriate, the term may also be used as a designation for the Regional Water Management Division Director or other duly authorized official of said agency.

(14) *Existing Source.*   Any source of discharge, the construction or operation of which commenced prior to the publication of proposed categorical pretreatment standards under Section 307(b) and (c) (33 U.S.C. 1317) of the Act which will be applicable to such source if the standard is thereafter promulgated in accordance with section 307 of the Act.

(15) *Grab Sample.*   A sample which is taken from a wastestream on a one-time basis without regard to the flow in the wastestream and without consideration of time.

(16) *Holding Tank Waste.*   Any waste from holding tanks such as vessels, chemical toilets, campers, trailers, septic tanks, and vacuum-pump tank trucks.

(17) *Industrial User.*   Any person who is a source of discharge.

(18) *Industrial Wastewater.*   A nondomestic wastewater originating from a nonresidential source.

(19) *Interference.*   A discharge which causes or contributes to the inhibition or disruption of the municipal wastewater system, including sewerage collection facilities, the processes or operations of the treatment plant, or the use or disposal of sewage sludge in accordance with the City's NPDES permit or any of the following regulations or permits issued thereunder (or more stringent state or local regulations): Section 405 of the Clean Water Act; the Solid Waste Disposal Act [including Title II commonly referred to as the Resource Conservation and Recovery Act (RCRA)]; any State sludge management plan prepared pursuant to Subtitle D of the SWDA; the Clean Air Act; the Toxic Substances Control Act; and the Marine Protection, Research and Sanctuaries Act.

(20) *Medical Waste.*   Isolation wastes, infectious agents, human blood and blood by-products, pathological wastes, sharps, body parts, fomites, etiologic agents, contaminated bedding, surgical wastes, potentially contaminated laboratory wastes and dialysis wastes.

(21) *Municipal Wastewater System or System.*   A "treatment works" as defined by Section 212 of the Act, (33 U.S.C. 1292) which is owned by the State or municipality. This definition includes any devices or systems used in the collection, storage, treatment, recycling and reclamation of sewage or industrial wastes and any conveyances which convey wastewater to a treatment plant. The term also means the municipal entity having responsibility for the operation and maintenance of the system.

(22) *New Source*

1. Any source of a discharge, the construction or operations of which commenced after the publication of proposed Categorical Pretreatment Standards under Section 307(c) [33 U.S.C. 1317(c)] of the Act which will be applicable to such source if the standard is thereafter promulgated in accordance with Section 307(c), provided that:

    (i)   no other source is located at that site; or

    (ii) the source completely replaces the process or production equipment of an existing source at that site; or

(iii) the new wastewater generating process of the source is substantially independent of an existing source at that site; and the construction of the source creates a new facility rather than modifying an existing source at that site.

2. For purposes of this definition, construction or operation has commenced if the owner or operator has:

(i) Begun, or caused to begin as part of a continuous on-site construction program:

(a) Any placement, assembly, or installation of facilities or equipment; or

(b) Significant site preparation work including clearing, excavation, or removal of existing buildings, structures, or facilities which is necessary for the placement, assembly, or installation of new source facilities or equipment; or

(ii) Entered into a binding contractual obligation for the purchase of facilities or equipment which are intended to be used in its operation within a reasonable time. Options to purchase or contracts which can be terminated or modified without substantial loss, and contracts for feasibility, engineering, and design studies do not constitute a contractual obligation under this definition.

(23) *Nondomestic Pollutants.* Any substances other than human excrement and household gray water (shower, dishwashing operations, etc.). Nondomestic pollutants include the characteristics of the wastewater (that is, pH, temperature, TSS, turbidity, color, BOD, COD, toxicity, odor).

(24) *Pass Through.* A discharge which exits the treatment plant effluent into waters of the U.S. in quantities or concentrations which, alone or in conjunction with an indirect discharge or discharges from other sources, is a cause of a violation of any requirements of the City's NPDES permit (including an increase in the magnitude or duration of a violation).

(25) *Person.* Any individual, partnership, copartnership, firm, company, corporation, association, joint stock company, trust, estate, governmental entity or any other legal entity, or their legal representatives, agents or assigns. This definition includes all Federal, State or local governmental entities.

(26) *pH.* A measure of the acidity or alkalinity of a substance, expressed in standard units; neutral wastewaters are numerically equal to 7 while the number increases to show increasing alkalinity and decreases to show increasing acidity.

(27) *Pollutant.* Any dredged spoil, solid waste, incinerator residue, sewage, garbage, sewage sludge, munitions, medical wastes, chemical wastes, industrial wastes, biological materials, radioactive materials, heat wrecked or discharged equipment, rock, sand, cellar dirt and agricultural wastes.

(28) *Pretreatment or Treatment.* The reduction of the amount of pollutants, the elimination of pollutants, or the alteration of the nature of pollutant properties in wastewater thereby rendering them less harmful to the municipal wastewater system prior to introducing such pollutants into the system. This reduction or alteration can be obtained by physical, chemical or biological processes, by process changes, or by other means, except by diluting the concentration of the pollutants unless allowed by an applicable pretreatment standard.

(29) *Pretreatment Standards and Requirements.* Any substantive or procedural requirement related to pretreatment, including National pretreatment categorical standards and prohibitive discharge standards imposed on an industrial user.

(30) *Prohibited Discharge Standards or Prohibited Discharges.* Absolute prohibitions against the discharge of certain defined types of industrial wastewater; these prohibitions appear in Section 2.1 of this ordinance.

(31) *Residential Users.* Persons only contributing sewage wastewater to the municipal wastewater system.

(32) *Receiving Stream or Water of the State.* All streams, lakes, ponds, marshes, watercourses, waterways, wells, springs, reservoirs, aquifers, irrigation systems, drainage systems and all other bodies or accumulations of water, surface or underground, natural or artificial, public or private, which are contained within, flow through, or border upon the State or any portion thereof.

(33) *Sewage.* Human excrement and gray water (household showers, dishwashing operations, etc.)

(34) *Significant Industrial User.* The term significant industrial user shall mean: a) industrial users subject to categorical pretreatment standards, and b) any other industrial user that: i) discharges an average of 25,000 GPD or more of process wastewater, ii) contributes a process wastestream which makes up 5 percent or more of the average dry weather hydraulic or organic capacity of the treatment plant, or, iii) is designated as significant by the City on the basis that the industrial user has a reasonable potential for causing pass through or interference.

(35) *Slug Load.* Any pollutant (including BOD) released in a discharge at a flow rate or concentration which will cause a violation of the specific discharge prohibitions in Section 2 of this Ordinance.

(36) *Standard Industrial Classification (SIC) Code or North American Industry Classification System (NAICS).* A classification pursuant to the classification manual issued by the Executive Office of the President, Office of Management and Budget.

(37) *Storm Water.* Any flow occurring during or following any form of natural precipitation and resulting therefrom, including snowmelt.

(38) *Suspended Solids.* The total suspended matter that floats on the surface of, or is suspended in, water, wastewater, or other liquid, and which is removable by laboratory filtering.

(39) *Superintendent.* The person designated by the City to supervise the operation of the municipal wastewater system and who is charged with certain duties and responsibilities by this article, or his duly authorized representative.

(40) *Toxic Pollutant.* One of 126 pollutants or combination of those pollutants listed as toxic in regulations promulgated by the Environmental Protection Agency under the provisions of Section 307 (33 U.S.C. 1317) of the Act.

(41) *Treatment Plant.*   That portion of the municipal wastewater system designed to provide treatment of sewage and industrial waste.

(42) *Treatment Plant Effluent.*   Any discharge of pollutants from the municipal wastewater system into waters of the State.

(43) *User.*   Any person who contributes, or causes or allows the contribution of sewage or industrial wastewater into the municipal wastewater system, including persons who contribute such wastes from mobile sources.

(44) *Wastewater.*   The liquid and water-carried industrial wastes, or sewage from residential dwellings, commercial buildings, industrial and manufacturing facilities, and institutions, whether treated or untreated, which is contributed to the municipal wastewater system.

This Ordinance is gender neutral and the masculine gender shall include the feminine and vice versa. Shall is mandatory; may is permissive or discretionary. The use of the singular shall be construed to include the plural and the plural shall include the singular as indicated by the context of its use.

1.3   Abbreviations

The following abbreviations shall have the designated meanings:

- *BOD*   Biochemical Oxygen Demand.
- *CFR*   Code of Federal Regulations.
- *COD*   Chemical Oxygen Demand.
- *EPA*   U.S. Environmental Protection Agency.
- *GPD*   Gallons Per Day.
- LC_{50} Lethal Concentration for Fifty Percent (50%) of the Test Organisms.
- *L*   Liter.
- *mg*   Milligrams.
- *mg/L*   Milligrams per liter.
- *NAICS*   North American Industry Classification System.
- *NPDES*   National Pollutant Discharge Elimination System.
- *O & M*   Operation and Maintenance.
- *POTW*   Publicly Owned Treatment Works.
- *RCRA*   Resource Conservation and Recovery Act.
- *SIC*   Standard Industrial Classification.
- *SWDA*   Solid Waste Disposal Act (42 U.S.C. 6901, *et seq.*).
- *TSS*   Total Suspended Solids.
- *USC*   United States Code.

SECTION 2 — GENERAL SEWER USE REQUIREMENTS

2.1   Prohibited Discharge Standards

No user shall contribute or cause to be contributed, directly or indirectly, any pollutant or wastewater which will cause interference or pass through. These general prohibitions apply to all users of the municipal wastewater system whether or not the user is subject to categorical pretreatment standards or any other National, State or local pretreatment standards or requirements. Furthermore, no user may contribute the following substances to the system:

a.   Any liquids, solids or gases which by reason of their nature or quantity are, or may be, sufficient, either alone or by interaction with other substances, to cause fire or explosion or be injurious in any other way to the municipal wastewater system. Included in this prohibition are wastestreams with a closed cup flashpoint of less than 140°F (60°C). At no time shall two successive readings on an explosion hazard meter at the point of discharge into the system or at any point in the system be more than five percent (5%) nor any single reading over ten percent (10%) of the lower explosive limit (LEL) of the meter.

b.   Solid or viscous substances in amounts which will cause interference with the flow in a sewer but in no case solids greater than one-half inch ($^1/_2$") (1.27 centimeters) in any dimension.

c.   Any fats or greases, including but not limited to petroleum oil, nonbiodegradable cutting oil, or products of mineral oil origin, in amounts that will cause interference or pass through.

d.   Any wastewater having a pH less than 5.0 or more than 9.0, or which would otherwise cause corrosive structural damage to the system, city personnel or equipment.

e.   Any wastewater containing pollutants in sufficient quantity (flow or concentration), either singly or by interaction with other pollutants, to pass through or interfere with the municipal wastewater system, any wastewater treatment or sludge process, or constitute a hazard to humans or animals.

f.   Any noxious or malodorous liquids, gases, or solids or other wastewater which, either singly or by interaction with other wastes, are sufficient to create a public nuisance or hazard to life or are sufficient to prevent entry into the sewers for maintenance and repair.

g. Any substance which may cause the treatment plant effluent or any other residues, sludges, or scums, to be unsuitable for reclamation and reuse or to interfere with the reclamation process. In no case shall a substance discharged to the system cause the City to be in noncompliance with sludge use or disposal regulations or permits issued under Section 405 of the Act, the Solid Waste Disposal Act, the Clean Air Act, the Toxic Substances Control Act, or other State requirements applicable to the sludge use and disposal practices being used by the City.

h. Any wastewater which imparts color which cannot be removed by the treatment process, such as, but not limited to, dye wastes and vegetable tanning solutions, which consequently imparts color to the treatment plants effluent thereby violating the City's NPDES permit. Color (in combination with turbidity) shall not cause the treatment plant effluent to reduce the depth of the compensation point for photosynthetic activity by more than 10 percent from the seasonably established norm for aquatic life.

i. Any wastewater having a temperature greater than 150°F (55°C), or which will inhibit biological activity in the treatment plant resulting in interference, but in no case wastewater which causes the temperature at the introduction into the treatment plant to exceed 104°F (40°C).

j. Any wastewater containing any radioactive wastes or isotopes except as specifically approved by the Superintendent in compliance with applicable State or Federal regulations.

k. Any pollutants which result in the presence of toxic gases, vapors or fumes within the system in a quantity that may cause worker health and safety problems.

l. Any trucked or hauled pollutants, except at discharge points designated by the City in accordance with Section 3.6.

m. Storm water, surface water, ground water, artesian well water, roof runoff, subsurface drainage, swimming pool drainage, condensate, deionized water, cooling water and unpolluted industrial wastewater, unless specifically authorized by the Superintendent.

n. Any industrial wastes containing floatable fats, waxes, grease or oils, or which become floatable at the wastewater temperature at the introduction to the treatment plant during the winter season; but in no case, industrial wastewater containing more than 100 mg/L of emulsified oil or grease.

o. Nonbiodegradable cutting oils, commonly called soluble oils, which form a persistent water emulsion, and nonbiodegradable complex carbon compounds.

p. Any sludges, screenings, or other residues from the pretreatment of industrial wastes.

q. Any medical wastes, except as specifically authorized by the Superintendent in a wastewater permit.

r. Any material containing ammonia, ammonia salts, or other chelating agents which will produce metallic complexes that interfere with the municipal wastewater system.

s. Any material identified as hazardous waste according to 40 CFR Part 261 except as may be specifically authorized by the Superintendent.

t. Any wastewater causing the treatment plant effluent to show a lethal concentration of fifty percent (LC_{50}) as determined by a toxicity test of ninety-six (96) hours or less, using a percentage of the discharge and aquatic test species chosen by the Superintendent.

u. Recognizable portions of the human or animal anatomy.

v. Any wastes containing detergents, surface active agents, or other substances which may cause excessive foaming in the municipal wastewater system.

Wastes prohibited by this section shall not be processed or stored in such a manner that these wastes could be discharged to the municipal wastewater system. All floor drains located in process or materials storage areas must discharge to the industrial user's pretreatment facility before connecting with the system.

2.2 Federal Categorical Pretreatment Standards

Users subject to categorical pretreatment standards are required to comply with applicable standards as set out in 40 CFR Chapter I, Subchapter N, Parts 405-471 and incorporated herein.

2.3 State Requirements

Users are required to comply with applicable State pretreatment standards and requirements set out in (State citation, if any). These standards and requirements are incorporated herein.

2.4 Specific Pollutant Limitations

The following pollutant limits are established to protect against pass through and interference. No person shall discharge wastewater containing quantities in excess of the following instantaneous maximum allowable discharge limits:

mg/L		mg/L	
_____	aldrin	_____	bis(2-ethylhexyl) phthalate
_____	arsenic	_____	BOD_5
_____	benzene	_____	cadmium
_____	benzol(a)pyrene	_____	carbon tetrachloride
_____	beryllium	_____	chlordane
_____	bis(2-chloroethyl) ether	_____	chlorinated hydrocarbons

APPENDIX D

_____ mg/L	COD		_____ mg/L	mercury
_____ mg/L	copper		_____ mg/L	molybdenum
_____ mg/L	cyanide		_____ mg/L	nickel
_____ mg/L	DDD, DDE, DDT		_____ mg/L	oil and grease
_____ mg/L	dieldrin		_____ mg/L	PCBs
_____ mg/L	dimethyl nitrosamine		_____ mg/L	phenolic compounds
_____ mg/L	fluoride		_____ mg/L	selenium
_____ mg/L	heptachlor		_____ mg/L	silver
_____ mg/L	hexachlorobenzene		_____ mg/L	total chromium
_____ mg/L	hexachlorobutadiene		_____ mg/L	total suspended solids
_____ mg/L	iron		_____ mg/L	toxaphene
_____ mg/L	lead		_____ mg/L	trichloroethylene
_____ mg/L	lindane		_____ mg/L	zinc

Concentrations apply at the point where the industrial waste is discharged to the municipal wastewater system. All concentrations for metallic substances are for "total" metal. At his discretion, the Superintendent may impose mass limitations in addition to or in place of the concentration-based limitations above. Compliance with all parameters may be determined from a single grab sample.

2.5 City's Right of Revision

The City reserves the right to establish, by ordinance or in wastewater permits, more stringent limitations or requirements on discharges to the municipal wastewater system if deemed necessary to comply with the objectives presented in Section 1.1 of this Ordinance or the general and specific prohibitions in Section 2.1 of this Ordinance.

2.6 Special Agreement

The City reserves the right to enter into special agreements with users setting out special terms under which the industrial user may discharge to the system. In no case will a special agreement waive compliance with a pretreatment standard. However, the industrial user may request a net gross adjustment to a categorical standard in accordance with 40 CFR 403.15. Industrial users may also request a variance from the categorical pretreatment standard from U.S. EPA. Such a request will be approved only if the user can prove that factors relating to its discharge are fundamentally different from the factors considered by U.S. EPA when establishing that pretreatment standard. An industrial user requesting a fundamentally different factor variance must comply with the procedural and substantive provisions in 40 CFR 403.13.

2.7 Dilution

No user shall ever increase the use of process water or in any way attempt to dilute a discharge as a partial or complete substitute for adequate treatment to achieve compliance with a discharge limitation unless expressly authorized by an applicable pretreatment standard, or in any other pollutant-specific limitation developed by the City.

SECTION 3 — PRETREATMENT OF WASTEWATER

3.1 Pretreatment Facilities

Industrial users shall provide necessary wastewater treatment as required to comply with this Ordinance and shall achieve compliance with all categorical pretreatment standards, local limits and the prohibitions set out in Section 2 above, within the time limitations specified by the Superintendent. Any facilities required to pretreat wastewater to a level acceptable to the City shall be provided, operated, and maintained at the industrial user's expense. Detailed plans showing the pretreatment facilities and operating procedures shall be submitted to the City for review, and shall be acceptable to the City before construction of the facility. The review of such plans and operating procedures will in no way relieve the user from the responsibility of modifying the facility as necessary to produce an acceptable discharge to the City under the provisions of this Ordinance.

3.2 Additional Pretreatment Measures

Whenever deemed necessary, the Superintendent may require industrial users to restrict the industrial user's discharge during peak flow periods, designate that certain wastewater be discharged only into specific sewers, relocate and/or consolidate points of discharge, separate sewage wastestreams from industrial wastestreams, and such other conditions as may be necessary to protect the municipal wastewater system and determine the industrial user's compliance with the requirements of this Ordinance.

a. Each person discharging into the municipal wastewater system greater than 100,000 gallons per day or greater than five percent (5%) of the average daily flow in the system, whichever is lesser, shall install and maintain, on his property and at his expense, a suitable storage and flow control facility to ensure equalization of flow over a twenty-four (24) hour period. The facility shall have a capacity for at least fifty percent (50%) of the daily discharge volume and shall be equipped with alarms and a rate of discharge controller, the regulation of which shall be directed by the Superintendent. A wastewater permit may be issued solely for flow equalization.

b. Grease, oil and sand interceptors shall be provided when, in the opinion of the Superintendent, they are necessary for the proper handling of wastewater containing excessive amounts of grease, flammable substances, sand, or other harmful substances; except that such interceptors shall not be required for residential users. All interception units shall be of a type and capacity approved by the Superintendent and shall be so located to be easily accessible for cleaning and inspection. Such interceptors shall be inspected, cleaned, and repaired regularly, as needed, by the owner, at his expense.

c. Industrial users with the potential to discharge flammable substances may be required to install and maintain an approved combustible gas detection meter.

3.3 Spill Prevention Plans

Industrial users shall provide protection from accidental discharge of materials which may interfere with the municipal wastewater system by developing spill prevention plans. Facilities necessary to implement these plans shall be provided and maintained at the owner's or industrial user's expense. Spill prevention plans, including the facilities and the operating procedures, shall be approved by the City before implementation of the plan.

a. Industrial users that store hazardous substances shall not contribute to the municipal wastewater system after the effective date of this Ordinance unless a spill prevention plan has been approved by the City. Approval of such plans shall not relieve the industrial user from complying with all other laws and regulations governing the use, storage, and transportation of hazardous substances.

3.4 Tenant Responsibility

Any person who shall occupy the industrial user's premises as a tenant under any rental or lease agreement shall be jointly and severally responsible for compliance with the provisions of this Ordinance in the same manner as the Owner.

3.5 Separation of Domestic and Industrial Wastestreams

All new and domestic wastewaters from rostrums, showers, and drinking fountains shall be kept separate from all industrial wastewaters until the industrial wastewaters have passed through a required pretreatment system and the industrial user's monitoring facility. When directed to do so by the Superintendent, industrial users must separate existing domestic wastestreams from industrial wastestreams.

3.6 Hauled Wastewater

Septic tank waste (septet) will be accepted into the municipal wastewater system at a designated receiving structure within the treatment plant area, and at such times as are established by the Superintendent, provided such wastes do not contain toxic or hazardous pollutants, and provided such discharge does not violate any other requirements established by the City. Permits for individual vehicles to use such facilities shall be issued by the Superintendent.

a. The discharge of industrial wastes as "industrial septet" requires prior approval and a wastewater permit from the City. The Superintendent shall have authority to prohibit the disposal of such wastes, if such disposal would interfere with the treatment plant operation. Waste haulers are subject to all other sections of this Ordinance.

b. Fees for dumping septet will be established as part of the user fee system as authorized in Section 15.

3.6 Vandalism

No person shall maliciously, willfully or negligently break, damage, destroy, uncover, deface, tamper with or prevent access to any structure, appurtenance or equipment, or other part of the municipal wastewater system. Any person found in violation of this requirement shall be subject to the sanctions set out in Section 11, below.

SECTION 4 — WASTEWATER PERMIT ELIGIBILITY

4.1 Wastewater Survey

When requested by the Superintendent all industrial users must submit information on the nature and characteristics of their wastewater by completing a wastewater survey prior to commencing their discharge. The Superintendent is authorized to prepare a form for this purpose and may periodically require industrial users to update the survey. Failure to complete this survey shall be reasonable grounds for terminating service to the industrial user and shall be considered a violation of the Ordinance.

4.2 Wastewater Permit Requirement

It shall be unlawful for significant industrial users to discharge wastewater into the City's sanitary sewer system without first obtaining a wastewater permit from the Superintendent. Any violation of the terms and conditions of a wastewater permit shall be deemed a violation of this Ordinance and subjects the industrial user to the sanctions set out in Sections 10-12. Obtaining a wastewater permit does not relieve a permittee of its obligation to obtain other permits required by Federal, State, or local law.

a. The Superintendent may require other industrial users, including liquid waste haulers, to obtain wastewater permits as necessary to carry out the purposes of this chapter.

4.3 Permitting Existing Connections

Any significant industrial user which discharges industrial waste into the municipal wastewater system prior to the effective date of this Ordinance and who wishes to continue such discharges in the future, shall, within ninety (90) days after said date, apply to the

City for a wastewater permit in accordance with Section 4.7 below, and shall not cause or allow discharges to the system to continue after one hundred eighty (180) days of the effective date of this Ordinance except in accordance with a permit issued by the Superintendent.

4.4 Permitting New Connections

Any significant industrial user proposing to begin or recommence discharging industrial wastes into the municipal wastewater system must obtain a wastewater permit prior to beginning or recommencing such discharge. An application for this permit must be filed at least ninety (90) days prior to the anticipated start up date.

4.5 Permitting Extra jurisdictional Industrial Users

Any existing significant industrial user located beyond the City limits shall submit a permit application, in accordance with Section 4.7 below, within ninety (90) days of the effective date of this Ordinance. New significant industrial users located beyond the City limits shall submit such applications to the Superintendent ninety (90) days prior to any proposed discharge into the municipal system. Upon review of such application, the Superintendent may enter into a contract with the industrial user which requires the industrial user to subject itself to and abide by this Chapter, including all permitting, compliance monitoring, reporting, and enforcement provisions herein. Alternatively, the Superintendent may enter into an agreement with the neighboring jurisdiction in which the significant industrial user is located to provide for the implementation and enforcement of the pretreatment program and requirements against said user.

4.7 Wastewater Permit Application Contents

In order to be considered for a wastewater permit, all industrial users required to have a permit must submit the following information on an application form approved by the Superintendent:

a. Name, mailing address, and location (if different from the mailing address);

b. Environmental control permits held by or for the facility;

c. Standard Industrial Classification (SIC) codes for pretreatment for the industry as a whole and any processes for which categorical pretreatment standards have been promulgated;

d. Description of activities, facilities, and plant processes on the premises, including a list of all raw materials and chemicals used at the facility which are or could accidentally or intentionally be discharged to the municipal system;

e. Number and type of employees, and hours of operation, and proposed or actual hours of operation of the pretreatment system;

f. Each product produced by type, amount, process or processes and rate of production;

g. Type and amount of raw materials processed (average and maximum per day);

h. The site plans, floor plans and mechanical and plumbing plans and details to show all sewers, floor drains, and appurtenances by size, location and elevation, and all points of discharge;

i. Time and duration of the discharge;

j. Measured average daily and maximum daily flow, in gallons per day, to the municipal system from regulated process streams and other streams as necessary to use the combined wastestream formula in 40 CFR 403.6(e);

k. Daily maximum, daily average, and monthly average wastewater flow rates, including daily, monthly, and seasonal variations, if any;

l. Wastewater constituents and characteristics, including any pollutants in the discharge which are limited by any Federal, State, or local standards, or pretreatment standards applicable to each regulated process; and nature and concentration (or mass if pretreatment standard requires) of regulated pollutants in each regulated process (daily maximum and average concentration or mass when required by a pretreatment standard). Sampling and analysis will be undertaken in accordance with 40 CFR Part 136;

m. A statement reviewed by an authorized representative of the user and certified by a qualified professional indicating whether or not the pretreatment standards are being met on a consistent basis, and if not, what additional pretreatment is necessary;

n. If additional pretreatment and/or O&M will be required to meet the standards, then the industrial user shall indicate the shortest time schedule necessary to accomplish installation or adoption of such additional treatment and/or O&M. The completion date in this schedule shall not be longer than the compliance date established for the applicable pretreatment standard. The following conditions apply to this schedule:

   (i) The schedule shall contain progress increments in the form of dates for the commencement and completion of major events leading to the construction and operation of additional pretreatment required for the user to meet the applicable pretreatment standards (such events include hiring an engineer, completing preliminary plans, completing final plans, executing contracts for major components, commencing construction, completing construction, beginning operation, and conducting routine operation). No increment referred to in (a) above shall exceed nine (9) months, nor shall the total compliance period exceed eighteen (18) months.

   (I) No later than 14 days following each date in the schedule and the final date for compliance, the user shall submit a progress report to the Superintendent including, as a minimum, whether or not it complied with the increment of progress, the reason for any delay, and if appropriate, the steps being taken by the user to return to the established schedule. In no event shall more than nine (9) months elapse between such progress reports to the Superintendent.

o.   Any other information as may be deemed by the Superintendent to be necessary to evaluate the permit application.

Incomplete or inaccurate applications will not be processed and will be returned to the industrial user for revision.

4.8   Application Signatories and Certification

All permit applications and industrial user reports must contain the following certification statement and be signed by an authorized representative of the industrial user.

"I certify under penalty of law that this document and all attachments were prepared under my direction or supervision in accordance with a system designed to assure that qualified personnel properly gather and evaluate the information submitted. Based on my inquiry of the person or persons who manage the system, or those persons directly responsible for gathering the information, the information submitted is, to the best of my knowledge and belief, true, accurate, and complete. I am aware that there are significant penalties for submitting false information, including the possibility of fine and imprisonment for knowing violations."

4.9   Wastewater Permit Decisions

The Superintendent will evaluate the data furnished by the industrial user and may require additional information. Within sixty (60) days of receipt of a complete permit application, the Superintendent will determine whether or not to issue a wastewater permit. If no determination is made within this time period, the application will be deemed denied.

SECTION 5 — WASTEWATER PERMIT ISSUANCE PROCESS

5.1   Wastewater Permit Duration

Permits shall be issued for a specified time period, not to exceed five (5) years. A permit may be issued for a period less than five (5) years, at the discretion of the Superintendent. Each permit will indicate a specific date upon which it will expire.

5.2   Wastewater Permit Contents

Wastewater permits shall include such conditions as are reasonably deemed necessary by the Superintendent to prevent pass through or interference, protect the quality of the water body receiving the treatment plant's effluent, protect worker health and safety, facilitate sludge management and disposal, protect ambient air quality, and protect against damage to the municipal wastewater system.

a.   Wastewater Permits must contain the following conditions:

   (1)   A statement that indicates permit duration, which in no event shall exceed 5 years.

   (2)   A statement that the permit is nontransferable without prior notification to and approval from the City, and provisions for furnishing the new owner or operator with a copy of the existing permit.

   (3)   Effluent limits applicable to the user based on applicable standards in Federal, State and local law.

   (4)   Self-monitoring, sampling, reporting, notification and recordkeeping requirements. These requirements shall include an identification of pollutants to be monitored, sampling location, sampling frequency, and sample type based on Federal, State and local law.

   (5)   Statement of applicable penalties for violation of pretreatment standards and requirements, and compliance schedules.

b.   Permits may contain, but need not be limited to, the following:

   (1)   Limits on the average and/or maximum rate of discharge, time of discharge, and/or requirements for flow regulation and equalization.

   (2)   Limits on the instantaneous, daily and monthly average and/or maximum concentration, mass, or other measure of identified wastewater pollutants or properties.

   (3)   Requirements for the installation of pretreatment technology or construction of appropriate containment devices, etc., designed to reduce, eliminate, or prevent the introduction of pollutants into the treatment works.

   (4)   Development and implementation of spill control plans or other special conditions including management practices necessary to adequately prevent accidental, unanticipated, or routine discharges.

   (5)   Development and implementation of waste minimization plans to reduce the amount of pollutants discharged to the municipal wastewater system.

   (6)   The unit charge or schedule of user charges and fees for the management of the wastewater discharged to the system.

   (7)   Requirements for installation and maintenance of inspection and sampling facilities and equipment.

   (8)   Specifications for monitoring programs which may include sampling locations, frequency of sampling, number, types, and standards for tests, and reporting schedules.

APPENDIX D

(9) Requirements for immediate reporting of any instance of noncompliance and for automatic resampling and reporting within thirty (30) days where self-monitoring indicates a violation(s).

(10) Compliance schedules for meeting pretreatment standards and requirements.

(11) Requirements for submission of periodic self-monitoring or special notification reports.

(12) Requirements for maintaining and retaining plant records relating to wastewater discharge as specified in Section 7.8 and affording the Superintendent, or his representatives, access thereto.

(13) Requirements for prior notification and approval by the Superintendent of any new introduction of wastewater pollutants or of any significant change in the volume or character of the wastewater prior to introduction in the system.

(14) Requirements for the prior notification and approval by the Superintendent of any change in the manufacturing and/or pretreatment process used by the permittee.

(15) Requirements for immediate notification of excessive, accidental, or slug discharges, or any discharge which could cause any problems to the system.

(16) A statement that compliance with the permit does not relieve the permittee of responsibility for compliance with all applicable Federal and State pretreatment standards, including those which become effective during the term of the permit.

(17) Other conditions as deemed appropriate by the Superintendent to ensure compliance with this Ordinance, and State and Federal laws, rules, and regulations; the term of the permit.

5.3 Wastewater Permit Appeals

Any person including the industrial user may petition to the City to reconsider the terms of the permit within ten (10) days of the notice.

a. Failure to submit a timely petition for review shall be deemed to be a waiver of the administrative appeal.

b. In its petition, the appealing party must indicate the permit provisions objected to, the reasons for this objection, and the alternative condition, if any, it seeks to place in the permit.

c. The effectiveness of the permit shall not be stayed pending the appeal.

d. If the City fails to act within fifteen (15) days, a request for reconsideration shall be deemed to be denied. Decisions not to reconsider a permit, not to issue a permit, or not to modify a permit shall be considered final administrative action for purposes of judicial review.

e. Aggrieved parties seeking judicial review of the final administrative permit decision must do so by filing a complaint with the [name of Court] for [name of County] within [insert appropriate State Statute of Limitations].

5.4 Wastewater Permit Modification

The Superintendent may modify the permit for good cause including, but not limited to, the following:

a. To incorporate any new or revised Federal, State, or local pretreatment standards or requirements.

b. To address significant alterations or additions to the industrial user's operation, processes, or wastewater volume or character since the time of permit issuance.

c. A change in the municipal wastewater system that requires either a temporary or permanent reduction or elimination of the authorized discharge.

d. Information indicating that the permitted discharge poses a threat to the City's municipal wastewater system, City personnel, or the receiving waters.

e. Violation of any terms or conditions of the wastewater permit.

f. Misrepresentation or failure to disclose fully all relevant facts in the permit application or in any required reporting.

g. Revision of or a grant of variance from categorical pretreatment standards pursuant to 40 CFR 403.13.

h. To correct typographical or other errors in the permit.

i. To reflect a transfer of the facility ownership and/or operation to a new owner/operator.

The filing of a request by the permittee for a permit modification does not stay any permit condition.

5.5 Wastewater Permit Transfer

Permits may be reassigned or transferred to a new owner and/or operator with prior approval of the Superintendent if the permittee gives at least thirty (30) days advance notice to the Superintendent. The notice must include a written certification by the new owner which:

a. States that the new owner has no immediate intent to change the facility's operations and processes.

b. Identifies the specific date on which the transfer is to occur.

c. Acknowledges full responsibility for complying with the existing permit.

Failure to provide advance notice of a transfer renders the wastewater permit voidable on the date of facility transfer.

5.6   Wastewater Permit Revocation

Wastewater permits may be revoked for the following reasons:

a.   Failure to notify the City of significant changes to the wastewater prior to the changed discharge;

b.   Falsifying self-monitoring reports;

c.   Tampering with monitoring equipment;

d.   Refusing to allow the City timely access to the facility premises and records;

e.   Failure to meet effluent limitations;

f.   Failure to pay fines;

g.   Failure to pay sewer charges;

h.   Failure to meet compliance schedules;

i.   Failure to complete a wastewater survey;

j.   Failure to provide advance notice of the transfer of a permitted facility; and

k.   Violation of any pretreatment standard or requirement or any terms of the permit or the Ordinance.

Permits shall be voidable upon nonuse, cessation of operations, or transfer of business ownership. All permits are void upon the issuance of a new wastewater permit.

5.7   Wastewater Permit Reissuance

A significant industrial user shall apply for permit reissuance by submitting a complete permit application in accordance with Section 4.7 a minimum of ninety (90) days prior to the expiration of the user's existing permit.

5.8   Municipal User Permits (optional)

In the event another municipality contributes all or a portion of its wastewater to the municipal wastewater system, the City may require this municipality to apply for and obtain a municipal user permit.

a.   A municipal user permit application shall include:

(i)   A description of the quality and volume of its wastewater at the point it enters the City's system

(ii)   An inventory of all industrial users discharging to the municipality

(iii)   Such other information as may be required by the Superintendent

b.   A municipal user permit shall contain the following conditions:

(i)   A requirement for the municipal user to adopt both a sewer use ordinance and local limits which are at least as stringent as those set out in Section 2.4

(ii)   A requirement for the municipal user to submit a revised industrial user inventory on at least an annual basis

(iii)   Requirements for the municipal user to conduct pretreatment implementation activities including industrial user permits issuance, inspection and sampling, and enforcement as needed

(iv)   A requirement for the municipal user to provide the City with access to all information that the municipal user obtains as part of its pretreatment activities

(v)   Limits on the nature, quality, and volume of the municipal user's wastewater at the point where it discharges to the municipal wastewater system

(vi)   Requirements for monitoring the municipal user's discharge

c.   Violation of the terms and conditions of the municipal user's permit subjects the municipal user to the sanctions set out in Sections 10 and 11.

SECTION 6 — REPORTING REQUIREMENTS

6.1   Baseline Monitoring Reports

Within 180 days after the effective date of a categorical pretreatment standard, or 180 days after the final administrative decision on a category determination under 40 CFR 403.6(a)(4), whichever is later, existing significant industrial users subject to such categorical pretreatment standards and currently discharging to or scheduled to discharge to the municipal system shall be required to submit to the City a report which contains the information listed in paragraph a, below. At least ninety (90) days prior to commencement of their discharge, new sources, including existing users which have changed their operation or processes so as to become new sources, shall be required to submit to the City a report which contains the information listed in paragraph a. A new source shall also be required to report the method of pretreatment it intends to use to meet applicable pretreatment standards. A new source shall also give estimates of its anticipated flow and quantity of pollutants discharged.

a. The information required by this section includes:

   (1) *Identifying Information.*   The user shall submit the name and address of the facility including the name of the operator and owner(s);

   (2) *Permits.*   The user shall submit a list of any environmental control permits held by or for the facility;

   (3) *Description of Operations.*   The user shall submit a brief description of the nature, average rate of production, and standard industrial classifications of the operation(s) carried out by such industrial user. This description should include a schematic process diagram which indicates points of discharge to the system from the regulated processes.

   (4) *Flow Measurement.*   The user shall submit information showing the measured average daily and maximum daily flow, in gallons per day, to the system from regulated process streams and other streams as necessary to allow use of the combined wastestream formula set out in 40 CFR 403.6(e).

   (5) *Measurement of Pollutants.*

       (i)   The industrial user shall identify the categorical pretreatment standards applicable to each regulated process;

       (ii) In addition, the industrial user shall submit the results of sampling and analysis identifying the nature and concentration (and/or mass, where required by the standard or City) of regulated pollutants in the discharge from each regulated process. Instantaneous, daily maximum and long term average concentrations (or mass, where required) shall be reported. The sample shall be representative of daily operations and shall be performed in accordance with procedures set out in 40 CFR Part 136.

       (iii) A minimum of four (4) grab samples must be used for pH, cyanide, total phenols, oil and grease, sulfide, and volatile organics. All other pollutants will be measured by composite samples obtained through flow proportional sampling techniques. If flow proportional composite sampling is infeasible, samples may be obtained through time proportional sampling techniques or through four (4) grab samples if the user proves such a sample will be representative of the discharge.

   (6) *Special Certification.*   A statement, reviewed by an authorized representative of the industrial user and certified to by a qualified professional, indicating whether pretreatment standards are being met on a consistent basis, and, if not, whether additional operation and maintenance (O&M) and/or additional pretreatment is required in order to meet the pretreatment standards and requirements; and

   (7) *Compliance Schedule.*   If additional pretreatment and/or O&M will be required to meet the pretreatment standards, the shortest schedule by which the industrial user will provide such additional pretreatment and/or O&M. The completion date in this schedule shall not be later than the compliance date established for the applicable pretreatment standard. A compliance schedule pursuant to this Section must meet the requirements set out in Section 4.7(n) of this Ordinance.

   (8) All baseline monitoring reports must be signed and certified in accordance with Section 4.8.

6.2   Compliance Deadline Reports

Within ninety (90) days following the date for final compliance with applicable categorical pretreatment standards, or in the case of a new source, following commencement of the introduction of wastewater into the municipal wastewater system, any industrial user subject to such pretreatment standards and requirements shall submit to the City a report containing the information described in Section 6.1(a)(5 and 6). For industrial users subject to equivalent mass or concentration limits established in accordance with the procedures in 40 CFR 403.6(c), this report shall contain a reasonable measure of the user's long term production rate. For all other industrial users subject to categorical pretreatment standards expressed in terms of allowable pollutant discharge per unit of production (or other measure of operation), this report shall include the user's actual production during the appropriate sampling period. All compliance reports must be signed and certified in accordance with Section 4.8.

6.3   Periodic Compliance Reports

Any significant industrial user subject to a pretreatment standard shall, at a frequency determined by the Superintendent but in no case less than twice per year, submit a report indicating the nature and concentration of pollutants in the discharge which are limited by such pretreatment standards and the measured or estimated average and maximum daily flows for the reporting period. All periodic compliance reports must be signed and certified in accordance with Section 4.8.

a. All wastewater samples must be representative of the industrial user's discharge. Wastewater monitoring and flow measurement facilities shall be properly operated, kept clean, and maintained in good working order at all times. The failure of an industrial user to keep its monitoring facility in good working order shall not be grounds for the industrial user to claim that sample results are unrepresentative of this discharge.

b. In the event an industrial user's monitoring results indicate a violation has occurred, the industrial user must immediately notify the Superintendent and resample its discharge. The industrial user must report the results of the repeated sampling within thirty (30) days of discovering the first violation.

6.4   Report of Changed Conditions

Each industrial user is required to notify the Superintendent of any planned significant changes to the industrial user's operations or pretreatment systems which might alter the nature, quality or volume of its wastewater.

a. The Superintendent may require the industrial user to submit such information as may be deemed necessary to evaluate the changed condition, including the submission of a wastewater permit application under Section 4.7, if necessary.

b. The Superintendent may issue a wastewater permit under Section 4.9 or modify an existing wastewater permit under Section 5.4.

c. No industrial user shall implement the planned changed condition(s) until and unless the Superintendent has responded to the industrial user's notice.

d. For purposes of this requirement, flow increases of ten percent (10%) or greater and the discharge of any previously unreported pollutant shall be deemed significant.

6.5 Reports of Potential Problems

Each industrial user shall provide protection from accidental or intentional discharges of prohibited materials or other substances regulated by this Ordinance. Facilities to prevent the discharge of prohibited materials shall be provided and maintained at the owner's or user's own cost and expense. Detailed plans showing facilities and operating procedures to provide this protection shall be submitted to the City for review and shall be approved by the City before construction of the facility. Review and approval of such plans and operating procedures shall not relieve the industrial user from the responsibility to modify the user's facility as necessary to meet the requirements of this Ordinance.

a. No industrial user which commences contribution to the system after the effective date of this Ordinance shall be permitted to introduce pollutants into the system until accidental discharge procedures have been approved by the City.

b. In the case of an accidental or other discharge which may cause potential problems for the municipal wastewater system, it is the responsibility of the user to immediately telephone and notify the City of the incident. This notification shall include the location of discharge, type of waste, concentration and volume, if known, and corrective actions taken by the user.

c. Within five (5) days following an accidental discharge, the user shall, unless waived by the Superintendent, submit a detailed written report describing the cause(s) of the discharge and the measures to be taken by the user to prevent similar future occurrences. Such notification shall not relieve the user of any expense, loss, damage, or other liability which may be incurred as a result of damage to the system, natural resources, or any other damage to persons or property; nor shall such notification relieve the user of any fines, civil penalties, or other liability which may be imposed by this Ordinance.

d. Failure to notify the City of potential problem discharges shall be deemed a separate violation of this Ordinance.

e. A notice shall be permanently posted on the user's bulletin board or other prominent place advising employees whom to call in the event of a discharge described in paragraph b, above. Employers shall ensure that all employees who may cause or suffer such a discharge to occur are advised of the emergency notification procedure.

6.6 Reports from Noncategorical Users

All industrial users not subject to categorical pretreatment standards and not required to obtain a wastewater permit shall provide appropriate reports to the City as the Superintendent may require.

6.7 Sample Collection

Except as indicated in a, below, wastewater samples collected for purposes of determining industrial user compliance with pretreatment standards and requirements must be obtained using flow proportional composite collection techniques. In the event flow proportional sampling is infeasible, the Superintendent may authorize the use of time proportional sampling.

a. Samples for oil and grease, temperature, pH, cyanide, phenols, toxicity, sulfide, and volatile organic chemicals must be obtained using grab collection techniques.

6.8 Analytical Requirements

All pollutant analyses, including sampling techniques, to be submitted as part of a permit application or report shall be performed in accordance with the techniques prescribed in 40 CFR Part 136 or, if 40 CFR Part 136 does not contain sampling or analytical techniques for the pollutant in question, in accordance with procedures approved by the EPA and the City.

6.9 Monitoring Charges

The Superintendent may recover the City's expenses incurred in collecting and analyzing samples of the industrial user's discharge by adding the costs to the industrial user's sewer charges.

6.10 Timing

Written reports will be deemed to have been transmitted at the time of deposit, postage prepaid, into a mail facility serviced by the United States Postal Service.

6.11 Recordkeeping

Industrial users shall retain, and make available for inspection and copying, all records and information required to be retained under 40 CFR 403.12(o). These records shall remain available for a period of at least three (3) years. This period shall be automatically extended for the duration of any litigation concerning compliance with this Ordinance, or where the industrial user has been specifically notified of a longer retention period by the Superintendent.

APPENDIX D

SECTION 7 — COMPLIANCE MONITORING

7.1 Inspection and Sampling

The City shall have the right to enter the facilities of any industrial user to ascertain whether the purpose of this Ordinance is being met and all requirements are being complied with. Industrial users shall allow the Superintendent or his representatives ready access to all parts of the premises for the purposes of inspection, sampling, records examination and copying, and the performance of any additional duties.

a. Where a user has security measures in force which require proper identification and clearance before entry into their premises, the industrial user shall make necessary arrangements with its security guards so that, upon presentation of suitable identification, personnel from the City, State, and U.S. EPA will be permitted to enter, without delay, for the purposes of performing their specific responsibilities.

b. The City, State, and U.S. EPA shall have the right to set up or require installation of, on the industrial user's property, such devices as are necessary to conduct sampling and/or metering of the user's operations.

c. The City may require the industrial user to install monitoring equipment, as necessary. The facility's sampling and monitoring equipment shall be maintained at all times in a safe and proper operating condition by the industrial user at the industrial user's expense. All devices used to measure wastewater flow and quality shall be calibrated periodically to ensure their accuracy.

d. Any temporary or permanent obstruction to safe and easy access to the industrial facility to be inspected and/or sampled shall be promptly removed by the industrial user at the written or verbal request of the Superintendent and shall not be replaced. The costs of clearing such access shall be borne by the industrial user.

e. Unreasonable delays in allowing City personnel access to the industrial user's premises shall be a violation of this Ordinance.

7.2 Search Warrants

If the Superintendent has been refused access to a building, structure or property or any part thereof, and if the Superintendent has probable cause to believe that there may be a violation of this Ordinance or that there is a need to inspect as part of a routine inspection program of the City designed to protect the overall public health, safety and welfare of the community, then upon application by the City Attorney, the Municipal Court Judge of the City shall issue a search and/or seizure warrant describing therein the specific location subject to the warrant. The warrant shall specify what, if anything, may be searched and/or seized on the property described. Such warrant shall be served at reasonable hours by the Superintendent in the company of a uniformed police officer of the City. In the event of an emergency affecting public health and safety, or if the industrial user consents, inspections shall be made without the issuance of a warrant.

SECTION 8 — CONFIDENTIAL INFORMATION

Information and data on an industrial user obtained from reports, questionnaires, permit applications, permits, and monitoring programs, and from City inspection and sampling activities shall be available to the public without restriction unless the industrial user specifically requests and is able to demonstrate to the satisfaction of the City that the release of such information would divulge information, processes or methods of production entitled to protection as trade secrets under applicable State law.

a. Wastewater constituents and characteristics and other "effluent data" as defined by 40 CFR 2.302 will not be recognized as confidential information and will be available to the public without restriction.

b. When requested and demonstrated by the industrial user furnishing a report that such information should be held confidential, the portions of a report which might disclose trade secrets or secret processes shall not be made available for inspection by the public but shall be made available immediately upon request to governmental agencies for uses related to this Ordinance, the National Pollutant Discharge Elimination System (NPDES) program, and in enforcement proceedings involving the person furnishing the report.

SECTION 9 — PUBLICATION OF SIGNIFICANT VIOLATORS

The City shall annually publish, in the largest daily newspaper circulated in the area where the municipal wastewater system is located, a list of the industrial users which, during the previous 12 months, were in significant noncompliance with applicable pretreatment standards and requirements. The term significant noncompliance shall mean:

a. Sixty-six percent (66%) or more of wastewater measurements taken during a 6-month period exceed the discharge limit for the same pollutant parameter by any amount;

b. Thirty-three percent (33%) or more of wastewater measurements taken during a 6-month period equals or exceeds the product of the daily maximum limit or the average limit multiplied by the applicable criteria [1.4 for BOD, TSS, fats, oils and grease, and 1.2 for all other pollutants except pH];

c. Any other discharge violation that the City believes has caused, alone or in combination with other discharges, interference or pass through (including endangering the health of City personnel or the general public);

d. Any discharge of pollutants that has caused imminent endangerment to the public or to the environment, or has resulted in the City's exercise of its emergency authority to halt or prevent such a discharge;

e. Failure to meet, within 90 days of the scheduled date, a compliance schedule milestone contained in a permit or enforcement order for starting construction, completing construction, or attaining final compliance;

f.   Failure to provide, within 30 days after the due date, any required reports, including baseline monitoring reports, 90-day compliance reports, periodic self-monitoring reports, and reports on compliance with compliance schedules;

g.   Failure to report noncompliance; or

h.   Any other violation(s) which the City has reason to believe is significant.

SECTION 10 — ADMINISTRATIVE ENFORCEMENT REMEDIES

10.1   Notification of Violation

Whenever the Superintendent finds that any industrial user has violated or is violating this Ordinance, a wastewater permit or order issued hereunder, or any other pretreatment requirement, the Superintendent or his agent may serve upon said user a written Notice of Violation. Within 10 days of the receipt of this notice, an explanation of the violation and a plan for the satisfactory correction and prevention thereof, to include specific required actions, shall be submitted to the Superintendent. Submission of this plan in no way relieves the user of liability for any violations occurring before or after receipt of the Notice of Violation. Nothing in this Section shall limit the authority of the City to take emergency action without first issuing a Notice of Violation.

10.2   Consent Orders

The Superintendent is hereby empowered to enter into Consent Orders, assurances of voluntary compliance, or other similar documents establishing an agreement with the industrial user responsible for the noncompliance. Such orders will include specific action to be taken by the industrial user to correct the noncompliance within a time period also specified by the order. Consent Orders shall have the same force and effect as administrative orders issued pursuant to Sections 10.4 and 10.5 below and shall be judicially enforceable.

10.3   Show Cause Hearing

The Superintendent may order any industrial user which causes or contributes to violation(s) of this Ordinance, wastewater permits or orders issued hereunder, or any other pretreatment requirement to appear before the Superintendent and show cause why a proposed enforcement action should not be taken. Notice shall be served on the industrial user specifying the time and place for the meeting, the proposed enforcement action, the reasons for such action, and a request that the user show cause why this proposed enforcement action should not be taken. The notice of the meeting shall be served personally or by registered or certified mail (return receipt requested) at least ten (10) days prior to the hearing. Such notice may be served on any authorized representative of the industrial user. Whether or not the industrial user appears as noticed, immediate enforcement action may be pursued following the hearing date.

10.4   Compliance Orders

When the Superintendent finds that an industrial user has violated or continues to violate the Ordinance, permits or orders issued hereunder, or any other pretreatment requirement, he may issue an order to the industrial user responsible for the discharge directing that, following a specified time period, sewer service shall be discontinued unless adequate treatment facilities, devices, or other related appurtenances are installed and properly operated. Compliance orders may also contain such other requirements as might be reasonably necessary and appropriate to address the noncompliance, including additional self-monitoring, and management practices designed to minimize the amount of pollutants discharged to the sewer. Furthermore, the Superintendent may continue to require such additional self-monitoring for at least ninety (90) days after consistent compliance has been achieved, after which time the self-monitoring conditions in the discharge permits shall control.

10.5   Cease and Desist Orders

When the Superintendent finds that an industrial user has violated or continues to violate this Ordinance, permits or orders issued hereunder, or any other pretreatment requirement, the Superintendent may issue an order to the industrial user directing it to cease and desist all such violations and directing the user to:

a.   Immediately comply with all requirements, and

b.   Take such appropriate remedial or preventive action as may be needed to properly address a continuing or threatened violation, including halting operations and/or terminating the discharge.

10.6   Administrative Fines (optional, if State law allows)

Notwithstanding any other section of this Ordinance, any user which is found to have violated any provision of this Ordinance, permits and orders issued hereunder, or any other pretreatment requirement shall be fined in an amount not to exceed [insert maximum fine allowed under State law]. Such fines shall be assessed on a per violation, per day basis. In the case of monthly or other long-term average discharge limits, fines shall be assessed for each business day during the period of violation.

a.   Assessments may be added to the user's next scheduled sewer service charge and the Superintendent shall have such other collection remedies as may be available for other service charges and fees.

b.   Unpaid charges, fines, and penalties shall, after thirty (30) calendar days, be assessed an additional penalty of twenty percent (20%) of the unpaid balance and interest shall accrue thereafter at a rate of seven percent (7%) per month. Furthermore, these unpaid charges, fines and penalties, together with interest therefrom shall constitute a lien against the individual user's property.

c.   Industrial users desiring to dispute such fines must file a written request for the Superintendent to reconsider the fine along with full payment of the fine amount within ten (10) days of being notified of the fine. Where the Superintendent believes a request has merit, he shall convene a hearing on the matter within fifteen (15) days after receiving the request from the industrial user. In the event the user's appeal is successful, the payment together with any interest accruing thereto, shall be returned to the industrial user. The City may add the costs of preparing administrative enforcement actions, such as notices and orders to assess the fine.

10.7   Emergency Suspensions

The Superintendent may suspend the wastewater permit of an industrial user, for a period not to exceed thirty (30) days, whenever such suspension is necessary in order to stop an actual or threatened discharge which reasonably appears to present or cause an imminent or substantial endangerment to the health or welfare of persons, interferes with the operation of the municipal wastewater system, or which presents or may present an endangerment to the environment.

a.   Any industrial user notified of a suspension of its wastewater permit shall immediately stop or eliminate its contribution. In the event of an industrial user's failure to immediately comply voluntarily with the suspension order, the Superintendent shall take such steps as deemed necessary, including immediate severance of the sewer connection, to prevent or minimize damage to the system, its receiving stream, or endangerment to any individuals. The Superintendent shall allow the industrial user to recommence its discharge when the user has demonstrated to the satisfaction of the City that the period of endangerment has passed, unless the termination proceedings set forth in Section 10.8 are initiated against the user.

b.   An industrial user which is responsible, in whole or in part, for any discharge presenting imminent endangerment shall submit a detailed written statement describing the causes of the harmful contribution and the measures taken to prevent any future occurrence to the Superintendent prior to the date of any show cause or termination hearing under Sections 10.3 and 10.8.

10.8   Termination of Permit

In addition to those provisions in Section 5.6 of this Ordinance, any industrial user which violates the following conditions of this Ordinance, wastewater permits, or orders issued hereunder is subject to permit termination:

a.   Violation of permit conditions,

b.   Failure to accurately report the wastewater constituents and characteristics of its discharge,

c.   Failure to report significant changes in operations or wastewater volume, constituents and characteristics prior to discharge, or

d.   Refusal of reasonable access to the user's premises for the purpose of inspection, monitoring or sampling.

Noncompliant industrial users will be notified of the proposed termination of their wastewater permit and be offered an opportunity to show cause under Section 10.3 of this Ordinance why the proposed action should not be taken.

SECTION 11 — JUDICIAL ENFORCEMENT REMEDIES

11.1   Injunctive Relief

Whenever an industrial user has violated or continues to violate the provisions of this Ordinance, permits or orders issued hereunder, or any other pretreatment requirement, the Superintendent, through the City's attorney, may petition the [insert the appropriate Court] for the issuance of a temporary or permanent injunction, as may be appropriate, which restrains or compels the specific performance of the wastewater permit, order, or other requirement imposed by this ordinance on activities of the industrial user. Such other action as may be appropriate for legal and/or equitable relief may also be sought by the City. The Court shall grant an injunction without requiring a showing of a lack of an adequate remedy at law.

11.2   Civil Penalties

Any industrial user which has violated or continues to violate this Ordinance, any order or permit hereunder, or any other pretreatment requirement shall be liable to the Superintendent for a maximum civil penalty of [insert maximum allowed under State law but not less than $1,000] per violation per day for [insert appropriate State Statute of Limitations]. In the case of a monthly or other long-term average discharge limit, penalties shall accrue for each business day during the period of the violation.

a.   The Superintendent may recover reasonable attorney's fees, court costs, and other expenses associated with the enforcement activities, including sampling and monitoring expenses, and the cost of any actual damages incurred by the City.

b.   In determining the amount of civil liability, the Court shall take into account all relevant circumstances, including, but not limited to, the extent of harm caused by the violation, the magnitude and duration, any economic benefit gained through the industrial user's violation, corrective actions by the industrial user, the compliance history of the user, and any other factor as justice requires.

c.   Where appropriate, the Superintendent may accept mitigation projects in lieu of the payment of civil penalties where the project provides a valuable service to the City and the industrial user's expense in undertaking the project is at least one hundred and fifty percent (150%) of the civil penalty.

11.3 Criminal Prosecution

a. Any industrial user who willfully or negligently violates any provisions of this Ordinance, any orders or permits issued hereunder, or any other pretreatment requirement shall, upon conviction, be guilty of a misdemeanor, punishable by a fine of not more than $1,000 per violation per day or imprisonment for not more than one year or both.

b. Any industrial user who knowingly makes any false statements, representations, or certifications in any application, record, report, plan or other documentation filed or required to be maintained pursuant to this Ordinance, or wastewater permit, or who falsifies, tampers with or knowingly renders inaccurate any monitoring device or method required under this Ordinance shall, upon conviction, be punished by a fine of not more than $1,000 per violation per day or imprisonment for not more than one year or both.

c. In the event of a second conviction, the user shall be punishable by a fine of not to exceed $3,000 per violation per day or imprisonment for not more than 3 years or both.

SECTION 12 — SUPPLEMENTAL ENFORCEMENT ACTION

12.1 Performance Bonds (Optional)

The Superintendent may decline to reissue a permit to any industrial user which has failed to comply with the provisions of this Ordinance, any orders, or a previous permit issued hereunder unless such user first files a satisfactory bond, payable to the City, in a sum not to exceed a value determined by the Superintendent to be necessary to achieve consistent compliance.

12.2 Liability Insurance (Optional)

The Superintendent may decline to reissue a permit to any industrial user which has failed to comply with the provisions of this Ordinance, any orders, or a previous permit issued hereunder, unless the industrial user first submits proof that it has obtained financial assurances sufficient to restore or repair damage to the municipal wastewater system caused by its discharge.

12.3 Water Supply Severance (Optional)

Whenever an industrial user has violated or continues to violate the provisions of this Ordinance, orders, or permits issued hereunder, water service to the industrial user may be severed and service will only recommence, at the user's expense, after it has satisfactorily demonstrated its ability to comply.

12.4 Public Nuisances (Optional)

Any violation of the prohibitions or effluent limitations of this Ordinance, permits, or orders issued hereunder is hereby declared a public nuisance and shall be corrected or abated as directed by the Superintendent or his designee. Any person(s) creating a public nuisance shall be subject to the provisions of the City Codes (insert proper citation) governing such nuisances, including reimbursing the City for any costs incurred in removing, abating or remedying said nuisance.

12.5 Informant Rewards (Optional)

The Superintendent is authorized to pay up to five hundred dollars ($500) for information leading to the discovery of noncompliance by an industrial user. In the event that the information provided results in an administrative fine or civil penalty levied against the industrial user, the Superintendent is authorized to disperse up to ten (10) percent of the collected fine or penalty to the informant. However, a single reward payment may not exceed ten thousand dollars ($10,000).

12.6 Contractor Listing (Optional)

Industrial users which have not achieved consistent compliance with applicable pretreatment standards and requirements are not eligible to receive a contractual award for the sale of goods or services to the City. Existing contracts for the sale of goods or services to the City held by an industrial user found to be in significant violation with pretreatment standards may be terminated at the discretion of the City.

SECTION 13 — AFFIRMATIVE DEFENSES TO DISCHARGE VIOLATIONS

13.1 Upset

An upset shall be an affirmative defense to an enforcement action brought against a user for violating a pretreatment standard and requirement if the following conditions are met:

a. The user can identify the cause of the upset,

b. The facility was operating in a prudent and workmanlike manner at the time of the upset and was in compliance with applicable O&M procedures,

c. The user submits, within 24 hours of becoming aware of the upset, a description of the discharge and its causes, the period of noncompliance (if not corrected, then the time the noncompliance is anticipated to end), and the steps being taken to reduce, eliminate and prevent recurrence of the noncompliance.

d. If this report is given orally, the user must also submit a written report containing such information within five (5) days unless waived by the Superintendent.

e.   Upset shall mean an exceptional incident in which there is unintentional and temporary noncompliance with categorical pretreatment standards and requirements because of factors beyond the reasonable control of the industrial user. Noncompliance caused by operational error, improperly designed pretreatment facilities, inadequate treatment facilities, lack of preventive maintenance, or careless or improper operation does not constitute an upset.

13.2   General/Specific Prohibitions

An industrial user shall have an affirmative defense to an enforcement action brought against it for noncompliance with the general and specific prohibitions in Section 2.1 of this Ordinance if it can prove that it did not know or have reason to know that its discharge would cause pass through or interference and that either: (a) a local limit exists for each pollutant discharged and the user was in compliance with each limit directly prior to and during the pass through or interference, or (b) no local limit exists, but the discharge did not change substantially in nature or constituents from the user's prior discharge when the City was regularly in compliance with its NPDES permit, and in the case of interference, in compliance with applicable sludge use or disposal requirements.

13.3   Bypass

The intentional diversion of wastestreams from any portion of an individual user's treatment facility shall be an affirmative defense to an enforcement action brought against the industrial user if the user can demonstrate that such a bypass was unavoidable to prevent loss of life, personal injury, or severe property damage. In order to be eligible for the affirmative defense, the industrial user must demonstrate that there was no feasible alternative to the bypass and submit notice of the bypass as required by 40 CFR 403.17.

SECTION 14 — SURCHARGE COSTS AND FEES

[RESERVED]

SECTION 15 — MISCELLANEOUS PROVISIONS

15.1   Pretreatment Charges and Fees

The City may adopt reasonable charges and fees for reimbursement of costs of setting up and operating the City's Pretreatment Program which may include:

a.   Fees for permit applications including the cost of processing such applications;

b.   Fees for monitoring, inspection and surveillance procedures including the cost of reviewing monitoring reports submitted by industrial users;

c.   Fees for reviewing and responding to accidental discharge procedures and construction;

d.   Fees for filing appeals;

e.   Other fees as the City may deem necessary to carry out the requirements contained herein. These fees relate solely to the matters covered by this Ordinance and are separate from all other fees, fines and penalties chargeable by the City.

15.2   Severability

If any provision of this Ordinance is invalidated by any court of competent jurisdiction, the remaining provisions shall not be affected and shall continue in full force and effect.

15.3   Conflicts

All other Ordinances and parts of other Ordinances inconsistent or conflicting with any part of this Ordinance are hereby repealed to the extent of the inconsistency or conflict.

SECTION 16 — EFFECTIVE DATE

This ordinance shall be in full force and effect immediately following its passage, approval and publication, as provided by law.

INTRODUCED the _____ day of _____, 199__.   PASSED this ____ day of ____ 199__.

FIRST READING: ____ day of ____, 199__.   AYES:

SECOND READING: ____ day of ____, 199__.   NAYS:

ABSENT:

NOT VOTING:

APPROVED by me this ____ day of ____, 199__.

PUBLISHED: _____ _____

       Publication   Date

_____   ATTEST:_____

    Mayor/Director   City Clerk

APPENDIX E

ACTION CHECKLIST FOR POTWs:
ASSESSING THE NEED TO MODIFY PRETREATMENT PROGRAMS
IN RESPONSE TO DSS (DOMESTIC SEWAGE STUDY)
AND PIRT (PRETREATMENT IMPLEMENTATION REVIEW TASK FORCE)

Attached is a checklist for POTWs to use in determining action items which may be needed to implement the DSS regulations issued on July 24, 1990 (55 FR 30082). The checklist also includes items to implement the PIRT regulations (promulgated on January 17, 1988) so that POTWs which have not yet modified their programs in accordance with those regulations may do so in conjunction with modifications to implement DSS. The items involve changes to approved pretreatment programs and to POTW's NPDES application requirements. Also attached is a brief description of procedures to be followed in amending approved pretreatment programs. It should be remembered that the regulations themselves are relatively brief, and are contained on pp. 30128-30131 of the *Federal Register* notice. The remainder of the notice comprises background and responses to public comments. Pretreatment Bulletin Issue No. 8 also describes guidance and other technical assistance available to POTWs to implement pretreatment program requirements.

We anticipate that most pretreatment POTWs will already have the legal authority to carry out most of the items on the checklist. However, some POTWs may not have the authority to issue permits or equivalent individual control mechanisms to their significant industrial users (SIUs) as required by the DSS rule, or they may not have the penalty authority required by the PIRT regulations. In those cases, the POTW must obtain such authority before these requirements can be effectively implemented. Other provisions of the DSS rule may require ordinance changes for some POTWs (that is, the changes to the specific discharge prohibitions). As a reminder, the DSS and PIRT regulations contain only the *minimum* federal requirements; POTWs may, pursuant to local authorities, establish more stringent requirements than any of the provisions contained in the general pretreatment regulations.

Many of the provisions of the DSS rule pertain to "significant industrial users." The criteria for classification have been out in guidance for several years and are now codified in the general pretreatment regulations. They are:

● Any industrial user (IU) subject to categorical standards;

● Any noncategorical IU that discharges 25,000 gallons per day or more of process wastewater to the POTW or any IU that contributes a process wastestream which makes up five percent or more of the average dry weather hydraulic or organic capacity of the POTW treatment plant;

● Any IU designated as significant by the Control Authority on the basis that the IU has a reasonable potential for adversely affecting the POTW's operation or for violating any pretreatment standard or requirement.

PROGRAM MODIFICATIONS

The PIRT rule contained requirements for approval of POTW program modifications, which are found in 40 CFR 403.18. Most POTWs will have to modify their approved programs to conform to the changes required by the PIRT and DSS regulations. There are two types of program modifications: substantial and non-substantial. Both trigger the minor NPDES permit modification requirements of 40 CFR 122.63. POTWs are encouraged to discuss all program modifications with their Approval Authority.

Substantial Program Modifications

Substantial program modifications include all those identified in 40 CFR 403.18(c)(1)-(3). These are:

● Changes to the POTW's legal authorities, control mechanisms, or confidentiality procedures;

● Less stringent local limits;

● Changes in the POTW's method of implementing categorical standards;

● Decrease in the frequency of self-monitoring or reporting required of industrial users;

● Decrease in the frequency of industrial user inspections or samplings by the POTW;

● Significant reductions in pretreatment program resources;

● Changes to sludge disposal and management practices;

● Any other program modification that causes a substantial impact on program operations, an increase in pollutant loadings at the POTW, or less stringent requirements for industrial users.

For substantial modifications, the POTW must submit to the Approval Authority a statement of the basis of the desired modification, a modified program description, or such other documents the Approval Authority requires. The Approval Authority will follow the program approval procedures, including public notice of the submitted modification(s), of 40 CFR 403.11. Substantial modifications become effective upon approval. Notice of approval must also be published in the same newspaper as the notice of request for approval.

Non-substantial Program Modifications

The POTW must notify the Approval Authority of non-substantial modifications at least 30 days before they are to be implemented by the POTW. The POTW must submit a statement similar to that required for substantial modifications. Non-substantial program modifications are deemed to be approved by the Approval Authority unless the Approval Authority determines otherwise.

ACTION CHECKLIST FOR POTWs

ACTION ITEMS	40 CFR Citation	YES	NO	PARTIAL
NPDES Application Requirements (DSS)				
Toxicity testing results. For protocols, consult the permitting authority	122.21(j)(1-3)			
Written technical evaluation of the need to revise local limits	122.21(j)(4)			
Local Limits (PIRT)				
Technically based local limits or demonstration that they are not necessary	403.8(f)(4)			
Specific Prohibitions (Ordinance)(DSS)				
Wastestreams with a closed cup flashpoint of less than 140°F	403.5(b)(1)			
Pollutants causing toxic gases, vapors, and fumes (worker health and safety)	403.5(b)(7)			
Oil (petroleum-based, biodegradable cutting oil, products of mineral oil origin) in amounts causing pass through or interference	403.5(b)(6)			
Trucked and hauled waste (designation of discharge points)	403.5(b)(8)			
Application of Pretreatment Standards (PIRT)				
Dilution is prohibited to achieve compliance with all applicable pretreatment standards and requirements	403.6(d)			
The combined wastestream formula is properly applied	403.6(e)			
Procedures for net/gross determination	403.15			
Slug Control (DSS)				
Evaluate SIUs every two years to determine need for slug control plans	403.8(f)(2)(v)			
Do plans contain the following elements?	403.8(f)(2)(v)			
Description of discharge practices and stored chemicals	403.8(f)(2)(v)(A)-(B)			
Notify the POTW of slug discharges and for followup written notification within 5 days	403.8(f)(2)(v)(C)			
If necessary, procedures to prevent adverse impact from accidental spills	403.8(f)(2)(v)(D)			
Slug Control (PIRT)				
Require IU notification for loadings that potentially would violate the prohibited discharge standards	403.12(f)			
Individual Control Mechanisms (DSS)				
Issued to all SIUs	403.8(f)(1)(iii)			
Do they contain the following elements?				
Statements of duration and non-transferability	403.8(f)(1)(iii)(A)-(B)			
Effluent limits	403.8(f)(1)(iii)(C)			
Self-monitoring, sampling, reporting, notification, and recordkeeping requirements	403.8(f)(1)(iii)(D)			
Statement of civil and criminal penalties and any compliance schedules	403.8(f)(1)(iii)(E)			

ACTION CHECKLIST FOR POTWs (continued)

ACTION ITEMS	40 CFR Citation	YES	NO	PARTIAL
IU Hazardous Waste Notification (DSS)	403.12			
Review of submitted notifications to determine which IUs to contact for more information				
SIU Self-Monitoring (PIRT)				
Sampling techniques for categoricals revised	403.12(b)(5)			
Sampling for all SIUs for periodic reports must be performed during the period covered by the report	403.12(g)(3)&(h)			
Extra sampling data from categorical SIUs must be included in periodic reports	403.12(g)(5)			
Categorical SIUs must report all violations within 24 hours and resample within 30 days	403.12(g)(2)			
90-day report requirements same as for BMR	403.12(d)			
Revised signatory requirements for categorical SIUs	430.12(l)			
SIU Self-Monitoring (DSS)				
Procedures for reviewing twice-yearly non-categorical SIU reports	403.12(h)			
Changed Discharge (PIRT and DSS)				
All users must notify POTW of changed discharge, including change in hazardous waste discharges	403.12(j)			
New Sources (PIRT)				
Revised new source definition	403.3(k)			
Require new source compliance within 90 days	403.6(b)			
BMRs submitted 90 days prior to discharge	403.12(b)			
POTW Compliance Monitoring and Enforcement (DSS)				
POTW inspection and sampling of all SIUs once a year	403.8(f)(2)(v)			
Does the POTW have an enforcement response plan that contains the following procedures	403.8(f)(5)			
Description of investigation of IU noncompliance	403.8(f)(5)(i)			
Description of escalating enforcement responses and time periods for these responses	403.8(f)(5)(ii)			
Identification of officials responsible for each type of response	403.8(f)(5)(iii)			
Reflection of POTW's primary responsibility for enforcement of pretreatment requirements	403.8(f)(5)(iv)			
Remedies (PIRT)				
Authority to assess civil or criminal penalties of at least $1,000 per day per violation	403.8(f)(1)(vi)(A)			
Annual Report (PIRT)				
Do annual reports to the Approval Authority contain the following elements?	403.12(d)			
Updated list of IUs				
Summary of IU compliance status				
Summary of POTW compliance and enforcement activities				
Other relevant information requested by the Control Authority				

APPENDIX F

ENFORCEMENT RESPONSE PLAN (ERP)

CITY OF HOPEWELL, HOPEWELL REGIONAL WASTEWATER TREATMENT FACILITY, DECEMBER 20, 1994

1 General

1.1 Definitions

Except as noted otherwise in this Environmental Response Plan (ERP), the terms and abbreviations used herein have the meaning ascribed to them in Section 31-2(b) and (c) of the Code of the City of Hopewell (City Code).

1.2 Purpose of the ERP

The purpose of the ERP is to set forth detailed procedures for the City of Hopewell (City) and the Hopewell Regional Wastewater Treatment Facility (HRWTF) to follow in investigating and responding to instances of noncompliance by industrial users of the City's Publicly Owned Treatment Works (POTW). Specifically, the ERP:

1.2.1 describes how the City and HRWTF will investigate instances of noncompliance by industrial users;

1.2.2 describes the types of escalating enforcement responses the City and HRWTF will take in response to all anticipated types of industrial user violations and the time periods within which responses will take place;

1.2.3 identifies by title the officials responsible for each type of response; and

1.2.4 reflects HRWTF's primary responsibility to enforce all applicable pretreatment standards and requirements.

1.3 Administration of the ERP

The ERP is administered by the Environmental Manager of the HRWTF (Environmental Manager) under the supervision of the Director of the HRWTF (Director). Among other duties, the Environmental Manager, in administering the ERP, shall maintain complete and accurate records of enforcement activities carried out pursuant to the ERP and render periodic reports regarding these activities to the Director at a frequency and in such detail as determined by the Director. Enforcement action in accordance with section 4 below may be necessary for noncompliance with Chapter 31 of the City Code, a permit or order issued under Chapter 31, or any other pretreatment standard or requirement. Enforcement action in accordance with section 5 below may be necessary for noncompliance with a BOD or TSS allocation limit granted under the Hopewell Regional Wastewater Treatment Facility Agreement (HRWTF Agreement).

2 Collection and Dissemination of Information

2.1 For each industrial user, the Environmental Manager shall determine what data are necessary to determine compliance with applicable pretreatment standards as well as when and how any such data can be obtained. The Environmental Manager shall specify reporting requirements for each industrial user in its permit as required by Section 31-73(a) of the City Code and shall then track the submission of such reports. If information submitted is deficient or late, the industrial user shall be notified and required to complete the submission.

2.2 All reports submitted to the Director in accordance with Chapter 31 of the City Code and all information resulting from HRWTF monitoring activities shall be retained by the Environmental Manager for at least three years as required by 40 CFR Section 403.12(o) and Section 7.9.M. of VR 680-14-01.

2.3 The Environmental Manager shall notify significant industrial users of applicable pretreatment standards and any RCRA requirements. The Environmental Manager will provide feedback to users on compliance status by reporting the results of HRWTF sampling and analysis to industrial users.

3 Sampling and Inspection of Industrial Users

3.1 The Environmental Manager shall prepare and continually update an inspection plan for field investigations, including sample collection, facility inspections and flow monitoring. Field investigations shall be used to verify compliance status, to monitor industrial user self-monitoring activities, to collect samples, to initiate emergency or remedial action, and to gather additional information. HRWTF personnel may conduct routine compliance monitoring or special monitoring in response to violations, technical problems, or support for permit modifications. Routine wastewater sampling shall be conducted for each significant industrial user. The Environmental Manager shall further develop such checklists and procedures for routine inspections as are necessary to assure that the results of each visit are documented and notify industrial users of any deficiencies found during any inspection.

3.2 The Environmental Manager shall advise Region III of the Environmental Protection Agency and the Virginia Department of Environmental Quality (Department) of its routine and special field investigation activities each year through the annual pretreatment program report. Joint investigations of industrial users with the Environmental Protection Agency or the Department may be conducted by mutual agreement.

4   Enforcement Management System (EMS) for Violation of Applicable Pretreatment Standards

The following EMS will be implemented:

4.1   Compliance Screening

Using all available information, the Environmental Manager shall conduct an initial compliance review or screening process to determine and assess compliance with schedules, reporting requirements and applicable pretreatment standards. Such screening shall be undertaken at least monthly. Reviews completed under this section are designed to identify apparent violations rather than determining an appropriate enforcement response to such apparent violation.

During the screening process, the Environmental Manager shall verify that any required reports are submitted on schedule, cover the proper time period, include all information required in the particular report and are properly signed. As part of this process, the Environmental Manager will compare the information supplied with the requirements in the industrial user's permit. Discrepancies between information provided and information required to be provided may be considered to be a violation of the EMS. To the extent possible, the industrial user will be required to correct such discrepancies immediately upon their discovery.

4.2   Enforcement Evaluation

Violations and discrepancies identified during the compliance screening process will be reviewed by the Environmental Manager to evaluate the type of enforcement response required. Any proposed enforcement action will be reviewed by the Director prior to issuance.

The Environmental Manager will set deadlines for industrial users to respond to notices of violation. If contacts and commitments are oral, they will be confirmed in writing to preserve the record.

4.3   Enforcement Response Guide (ERG)

4.3.1   Description of Terms

The terms and abbreviations used in the ERG are defined below:

Administrative Fine (AF) — Monetary penalty assessed by the Director against an industrial user in accordance with Section 31-108(h) of the City Code.

Administrative Order (AO) — Any order issued to the industrial user by the Director or City Manager in accordance with Section 31-108 of the City Code, including compliance orders, cease and desist orders, and emergency suspensions. An AO may require a corrective response plan from the industrial user. (Sample AOs are attached at Appendix A.)

Civil Action (Civ.Act.) — An action brought by the Director or the City against the industrial user seeking injunctive relief in accordance with Section 31-109 of the City Code or monetary and actual damages in accordance with Section 31-110 of the City Code.

City Attorney — The City Attorney for the City of Hopewell or another attorney in the City Attorney's office as designated by the City Attorney.

City Manager — The City Manager of the City of Hopewell or his designee.

Consent Order (CO) — An agreement between the Director and the industrial user in accordance with Section 31-108(c) of the City Code. A CO may require a corrective response plan from the industrial user. (A sample CO is attached at Appendix B.)

Criminal Prosecution (Crim.Pros.) — A criminal action taken against an industrial user in which the City seeks punitive measures against the industrial user in accordance with Section 31-111 of the City Code.

Harm — Interference or pass through or harm to the environment.

Industrial User (IU) — An industrial user as defined in Section 31-2(c) of the City Code.

Notice of Violation (NOV) — A notice issued by the Director or Environmental Manager in accordance with Section 31-108(b) of the City Code. (A sample NOV is attached at Appendix C.)

Show Cause Hearing (SC) — A formal meeting in which the industrial user is required to appear before the City Manager in accordance with Section 31-108(d) of the City Code. (A sample SC Order is attached at Appendix D.)

Termination (Term.) — Termination of wastewater service to the industrial user in accordance with Sections 31-62, 31-77, 31-108, or 31-109 of the City Code.

Warning — Any verbal communication from the Director or Environmental Manager to the industrial user (in person or by telephone) regarding possible enforcement action for potential or actual noncompliance by the industrial user. The Environmental Manager must document the warning in writing, place the documentation in the HRWTF file, and send a copy of the documentation to the industrial user.

4.3.2   Objectives of the ERG

The primary objectives of the ERG are:

1. to ensure that noncompliant industrial users return to compliance as quickly as possible;

2. to penalize industrial users for violations of pretreatment standards and requirements;

3. to deter future noncompliance by industrial users; and

4. to recover any additional expenses incurred by the Director or the City due to the noncompliance.

The Environmental Manager will periodically review the ERG to assess its effectiveness in accomplishing these objectives. When the Environmental Manager identifies improvements or amendments designed to increase the effectiveness of the ERP, he will include a discussion of such improvements or amendments in his periodic report submitted to the Director in accordance with section 1.3 above.

4.3.3   Levels of Response

The city and HRWTF have three general levels of response to noncompliance by an IU. These responses range from a verbal warning (Level I) to criminal prosecution (Level III). Each level of response has a range of responses. The official(s) responsible for initiating and implementing the enforcement actions must choose from at least one of the responses from the level of response required by the ERG. For instance, if the ERG requires a Level I response for an IU's noncompliance, the Environmental Manager must issue a warning or an NOV to the IU, or both. The range of responses increases in severity for Level II and Level III enforcement actions. However, nothing prohibits the official(s) responsible for a Level II or Level III enforcement action from implementing a less severe response (e.g., NOV) from a lower level range of responses, so long as at least one of the responses from the required range of responses is also implemented.

Level I —   Administrative enforcement actions only; initiated and implemented by the Environmental Manager.

Range of responses: Warning, NOV

Time frame for response: Within 15 business days of becoming aware of noncompliance by IU.

Level II —   Combination of administrative and civil enforcement actions; initiated by the Environmental Manager or Director and implemented by the Director or City Manager.

Range of responses: AO, AF, SC, Civ. Act.

Time frame for response: Within sixty (60) working days of the original enforcement action or the date of noncompliance.

Level III —   Combination of civil and criminal enforcement actions; initiated by the Director and implemented by the Director, City Manager, or City Attorney.

Range of responses: SC, Civ. Act., Crim. Pros., Term.

Time frame for response: As soon as practical.

In addition to any enforcement action, the Director may enter into an AO with an IU in accordance with Section 31-108 of the City Code. However, for any enforcement action implemented by the City Manager or City Attorney, any AO entered into by the Director requires the written concurrence of the City Manager.

4.3.4 Unauthorized Discharge (No Permit)		
Noncompliance	**Nature of Violation**	**Response Level**
Unpermitted discharge	IU unaware of requirement, no harm	I
	IU unaware of requirement, harm	II
	Failure to apply, even after notice by POTW	III
Nonpermitted discharge (failure to renew)	IU has not submitted application within 10 days of due date	I

4.3.5 Discharge Limit Violation		
Noncompliance	**Nature of Violation**	**Response Level**
Exceedance of applicable Pretreatment standard	Isolated, not significant	I
	Isolated, significant, no harm	II
	Isolated, with harm	II
	Recurring, no harm	II
	Recurring, harm	III

4.3.6 Other Permit Violations		
Noncompliance	**Nature of Violation**	**Response Level**
Wastestreams are diluted in lieu of treatment	Initial violation	I, II
	Recurring	III
Failure to mitigate noncompliance or halt production	No harm	II
	Harm results	III
Failure to properly operate and maintain pretreatment or hold-up facilities	No harm	I
	Harm results	II
	NPDES Permit violation results	III

APPENDIX F

4.3.7 Monitoring and Reporting Violations		
Noncompliance	**Nature of Violation**	**Response Level**
Reporting violation	Report is improperly signed or certified	I
	Report is improperly signed or certified after notification	I
	Isolated, not significant (1-5 days late)	I
	Significant (6-30 days late)	II
	Reports are frequently late or no reports at all	II
	Failure to report spill or changed condition causing harm	II
	Repeated failure to report spills causing harm or NPDES permit violation	III
	Falsification	III
Failure to monitor correctly	Failure to monitor all pollutants as required by permit	I
	Recurring failure to monitor	II
Improper sampling	Evidence of intent	III
Failure to install monitoring equipment	Delay of less than 30 days	I
	Delay of greater than 30 days	II
	Disregard of CO	III
Compliance schedules	Missed milestone by greater than 30 days, will not affect final date	I
	Missed milestone by greater than 30 days or will affect final date, good cause for delay	I
	Missed milestone by greater than 30 days or will affect final date, no good cause for delay	II
	Disregard of compliance schedule	III

4.3.8 Violations Detected During Site Visits		
Noncompliance	**Nature of Violation**	**Response Level**
Entry denied	Entry denied or consent withdrawn. Copies of records denied	III
Illegal discharge	No harm	II
	Discharge causes harm, evidence of intent or negligence, NPDES permit violation	III
Improper sampling	Unintentional sampling at incorrect location	I
	Unintentionally using incorrect sample type	I
	Unintentionally using incorrect collection technique	I
Inadequate recordkeeping	Inspector finds file incomplete to missing (no evidence of intent)	I
	Recurring	II
Failure to report	Inspector finds additional monitoring, additional files or data	I
	Recurring	II

4.3.10   Time Frames for Responses

In addition to the time frames for response listed in section 4.3.3 above, the following time frames apply:

1. All noncompliance will be identified and documented by the Environmental Manager within ten (10) business days of receiving compliance information.

2. Any noncompliance which reasonably appears to present or cause an imminent or substantial endangerment to the health or welfare of persons or which results or may result in harm will receive immediate responses.

3. Any noncompliance meeting the criteria for significant noncompliance in section 4.4 below will be addressed with an AO within thirty (30) working days of the determination of significant noncompliance (significant noncompliance will be calculated in conformance with applicable pretreatment standards).

APPENDIX F

4.4   <u>Significant Noncompliance (SNC)</u>

After completion of the compliance screening, violations will be characterized and a determination made as to whether the user is in Significant Noncompliance (SNC). Certain instances of noncompliance are not of sufficient impact to justify extensive enforcement actions. However, certain violations or patterns of violations are significant and must be identified as such. Such SNC may be on an individual or long-term basis of occurrence. Categorization of industrial users as being SNC allows HRWTF to establish priorities for enforcement action and provides a means for reporting on the significant industrial user performance summary. Instances of SNC are industrial user violations which meet one or more of the criteria:

4.4.1   Chronic violations of wastewater discharge limits, defined here as those in which sixty-six percent (66%) or more of the wastewater measurements taken during a six (6) month period exceed (by any magnitude) the daily maximum limit or the average limit for the same pollution parameter.

4.4.2   Technical Review Criteria (TRC) violations, defined here as those in which thirty-three percent (33%) or more of the wastewater measurements taken for each pollutant parameter during a six (6) month period equal or exceed the product of the daily maximum limit or the average limit multiplied by the applicable TRC. (TRC = 1.4 for BOD, TSS, fats, oil and grease, and 1.2 for all other pollutants except pH.)

4.4.3   Any other discharge violation that the Director believes has caused, alone or in combination with other discharges, interference or pass through, including endangering the health of HRWTF personnel or the general public.

4.4.4   Any discharge of a pollutant that has caused imminent endangerment to the public or to the environment, or has resulted in the Director's exercise of his emergency authority to halt or prevent such a discharge.

4.4.5   Failure to meet, within ninety (90) days of the scheduled date, a compliance schedule milestone contained in a consent order, compliance order, or wastewater discharge permit for starting construction, completing construction, or attaining final compliance, unless due to good and valid cause.

4.4.6   Failure to provide, within thirty (30) days after the due date, any reports required by Chapter 31 of the City Code or orders or permits issued under Chapter 31 of the City Code, including baseline monitoring reports, reports on compliance with categorical pretreatment standard deadlines, periodic self-monitoring reports, and reports on compliance with compliance schedules.

4.4.7   Failure to accurately report noncompliance.

4.4.8   Any other violations that the Director determines will adversely affect the operation or implementation of the local pretreatment program.

4.5   <u>Actions to be Taken When an Industrial User is in SNC</u>

4.5.1   Report such information to the Environmental Protection Agency Region III and the Department as part of the Pretreatment Annual report.

4.5.2   List the industrial user in Hopewell's newspaper in accordance with paragraph 4.6 as having significant violations as required by Section 31-107 of the City Code.

4.5.3   Address the SNC through appropriate enforcement action described in section 4.3 above.

4.6   <u>Publishing Lists of Industrial Users with Significant Violations</u>

Section 31-107 of the City Code requires the HRWTF to publish, at least annually, in the largest daily newspaper published in Hopewell, a list of industrial users which were significantly violating applicable pretreatment standards and requirements during the previous twelve (12) months. The procedures for compiling the list of such industrial users is as follows:

4.6.1   The Environmental Manager shall prepare a compliance history from HRWTF records for each individual significant industrial user.

4.6.2   The compliance history so obtained for each industrial user shall be reviewed to determine if a pattern of noncompliance exists or if the industry has been or continues to be in SNC. To the extent that an industry meets these criteria, it will be placed on the list for publication for review. This list will then be reviewed by the Director.

CHAPTER 4

INSPECTION OF A TYPICAL INDUSTRY

by

Bill Garrett

TABLE OF CONTENTS

Chapter 4. INSPECTION OF A TYPICAL INDUSTRY

OBJECTIVES

Chapter 4.   INSPECTION OF A TYPICAL INDUSTRY

Following completion of Chapter 4, you should be able to:

1. Schedule inspections of pretreatment facilities,

2. Prepare for an inspection,

3. Meet and deal with various types of industrial contacts,

4. Enter an industry for an inspection,

5. Inspect an industrial pretreatment facility, including the outfall, effluent treatment equipment and in-plant wastewater control equipment,

6. Discuss violations and complaints with industrial representatives,

7. Write reports describing results of inspections, and

8. Conduct an inspection following approved safety procedures.

PRETREATMENT WORDS

Chapter 4.   INSPECTION OF A TYPICAL INDUSTRY

BOD (pronounce as separate letters)
BOD

Biochemical **O**xygen **D**emand. The rate at which organisms use the oxygen in water or wastewater while stabilizing decomposable organic matter under aerobic conditions. In decomposition, organic matter serves as food for the bacteria and energy results from its oxidation. BOD measurements are used as a measure of the organic strength of wastes in water.

BASELINE MONITORING REPORT (BMR)
BASELINE MONITORING REPORT (BMR)

All industrial users subject to Categorical Pretreatment Standards must submit a baseline monitoring report (BMR) to the Control Authority (POTW, state or EPA). The purpose of the BMR is to provide information to the Control Authority to document the industrial user's current compliance status with a Categorical Pretreatment Standard. The BMR contains information on (1) name and address of facility, including names of operator(s) and owner(s), (2) list of all environmental control permits, (3) brief description of the nature, average production rate and SIC (or NAICS) code for each of the operations conducted, (4) flow measurement information for regulated process streams discharged to the municipal system, (5) identification of the pretreatment standards applicable to each regulated process and results of measurements of pollutant concentrations and/or mass, (6) statements of certification concerning compliance or noncompliance with the pretreatment standards, and (7) if not in compliance, a compliance schedule must be submitted with the BMR that describes the actions the industrial user will take and a timetable for completing these actions to achieve compliance with the standards.

BATCH PROCESS
BATCH PROCESS

A treatment process in which a tank or reactor is filled, the wastewater (or other solution) is treated or a chemical solution is prepared, and the tank is emptied. The tank may then be filled and the process repeated. Batch processes are also used to cleanse, stabilize or condition chemical solutions for use in industrial manufacturing and treatment processes.

BLOWDOWN
BLOWDOWN

The removal of accumulated solids in boilers to prevent plugging of boiler tubes and steam lines. In cooling towers, blowdown is used to reduce the amount of dissolved salts in the recirculated cooling water.

COD (pronounce as separate letters)
COD

Chemical **O**xygen **D**emand. A measure of the oxygen-consuming capacity of organic matter present in wastewater. COD is expressed as the amount of oxygen consumed from a chemical oxidant in mg/L during a specific test. Results are not necessarily related to the biochemical oxygen demand (BOD) because the chemical oxidant may react with substances that bacteria do not stabilize.

CEMENTATION
CEMENTATION

(1) A spontaneous electrochemical process that involves the reduction of a more electropositive (noble) species, for example, copper, silver, mercury, or cadmium, by electronegative (sacrificial) metals such as iron, zinc, or aluminum. This process is used to purify spent electrolytic solutions and for the treatment of wastewaters, leachates, and sludges bearing heavy metals. Also called ELECTROLYTIC RECOVERY.

(2) The process of heating two substances that are placed in contact with each other for the purpose of bringing about some change in one of them such as changing iron to steel by surrounding it with charcoal and then heating it.

CEMENTATION TANK
CEMENTATION TANK

A tank in which metal ions are precipitated onto scrap aluminum, steel or other metals. The collected metal can be sent to a smelter for recovery. This process does not require electric current.

CENTRALIZED WASTE TREATMENT (CWT) FACILITY
CENTRALIZED WASTE TREATMENT (CWT) FACILITY

A facility designed to properly handle treatment of specific hazardous wastes from industries with similar wastestreams. The wastewaters containing the hazardous substances are transported to the facility for proper storage, treatment and disposal. Different facilities treat different types of hazardous wastes.

CHAIN OF CUSTODY CHAIN OF CUSTODY

A record of each person involved in the handling and possession of a sample from the person who collected the sample to the person who analyzed the sample in the laboratory and to the person who witnessed disposal of the sample.

CLARIFIER CLARIFIER

Settling Tank, Sedimentation Basin. A tank or basin in which wastewater is held for a period of time during which the heavier solids settle to the bottom and the lighter materials float to the water surface.

COMPOSITE (PROPORTIONAL) SAMPLE COMPOSITE (PROPORTIONAL) SAMPLE

A composite sample is a collection of individual samples obtained at regular intervals, usually every one or two hours during a 24-hour time span. Each individual sample is combined with the others in proportion to the rate of flow when the sample was collected. Equal volume individual samples also may be collected at intervals after a specific volume of flow passes the sampling point or after equal time intervals and still be referred to as a composite sample. The resulting mixture (composite sample) forms a representative sample and is analyzed to determine the average conditions during the sampling period.

CONFINED SPACE CONFINED SPACE

Confined space means a space that:

A. Is large enough and so configured that an employee can bodily enter and perform assigned work; and

B. Has limited or restricted means for entry or exit (for example, tanks, vessels, silos, storage bins, hoppers, vaults, and pits are spaces that may have limited means of entry); and

C. Is not designed for continuous employee occupancy.

(Definition from the Code of Federal Regulations (CFR) Title 29 Part 1910.146.)

CRADLE TO GRAVE CRADLE TO GRAVE

A term used to describe a hazardous waste manifest system used by regulatory agencies to track a hazardous waste from the point of generation to the hauler and then to the ultimate disposal site.

DRAG OUT DRAG OUT

The liquid film (plating solution) that adheres to the workpieces and their fixtures as they are removed from any given process solution or their rinses. Drag out volume from a tank depends on the viscosity of the solution, the surface tension, the withdrawal time, the draining time and the shape and texture of the workpieces. The drag out liquid may drip onto the floor and cause wastestream treatment problems. Regulated substances contained in this liquid must be removed from wastestreams or neutralized prior to discharge to POTW sewers.

ELECTROLYTIC RECOVERY ELECTROLYTIC RECOVERY

A spontaneous electrochemical process that involves the reduction of a more electropositive (noble) species, for example, copper, silver, mercury, or cadmium, by electronegative (sacrificial) metals such as iron, zinc, or aluminum. This process is used for purifying spent electrolytic solutions and for the treatment of wastewaters, leachates, and sludges bearing heavy metals. Also called CEMENTATION.

GRAB SAMPLE GRAB SAMPLE

A single sample of water collected at a particular time and place which represents the composition of the water only at that time and place.

HAZARDOUS WASTE HAZARDOUS WASTE

A waste, or combination of wastes, which because of its quantity, concentration, or physical, chemical, or infectious characteristics may:

1. Cause, or significantly contribute to, an increase in mortality or an increase in serious, irreversible, or incapacitating reversible illness; or

2. Pose a substantial present or potential hazard to human health or the environment when improperly treated, stored, transported, or disposed of or otherwise managed; and

3. Normally not be discharged into a sanitary sewer; subject to regulated disposal.

(Resource Conservation and Recovery Act (RCRA) definition.)

MATERIAL SAFETY DATA SHEET (MSDS) MATERIAL SAFETY DATA SHEET (MSDS)

A document which provides pertinent information and a profile of a particular hazardous substance or mixture. An MSDS is normally developed by the manufacturer or formulator of the hazardous substance or mixture. The MSDS is required to be made available to employees and inspectors whenever there is the likelihood of the hazardous substance or mixture being introduced into the workplace. Some manufacturers are preparing MSDSs for products that are not considered to be hazardous to show that the product or substance is *NOT* hazardous.

OUTFALL

OUTFALL

(1)   The point, location or structure where wastewater or drainage discharges from a sewer, drain, or other conduit.

(2)   The conduit leading to the final disposal point or area.

OXIDATION STATE/OXIDATION NUMBER

OXIDATION STATE/OXIDATION NUMBER

In a chemical formula, a number accompanied by a polarity indication (+ or −) that together indicate the charge of a substance as well as the extent to which the substance has been oxidized or reduced in a REDOX REACTION.

Due to the loss of electrons, the charge of a substance that has been oxidized would go from negative toward or to neutral, from neutral to positive, or from positive to more positive. As an example, an oxidation number of 2+ would indicate that a substance has lost two electrons and that its charge has become positive (that it now has an excess of two protons).

Due to the gain of electrons, the charge of the substance that has been reduced would go from positive toward or to neutral, from neutral to negative, or from negative to more negative. As an example, an oxidation number of 2− would indicate that a substance has gained two electrons and that its charge has become negative (that it now has an excess of two electrons). As a substance gains electrons, its oxidation state (or the extent to which it is oxidized) lowers; that is, its oxidation state is reduced.

OXIDATION-REDUCTION POTENTIAL (ORP)

OXIDATION-REDUCTION POTENTIAL (ORP)

The electrical potential required to transfer electrons from one compound or element (the oxidant) to another compound or element (the reductant); used as a qualitative measure of the state of oxidation in wastewater treatment systems. ORP is measured in milli-volts, with negative values indicating a tendency to reduce compounds or elements and positive values indicating a tendency to oxi-dize compounds or elements.

OXIDATION-REDUCTION (REDOX) REACTION

OXIDATION-REDUCTION (REDOX) REACTION

A two-part reaction between two substances involving a transfer of electrons from one substance to the other. Oxidation is the loss of electrons by one substance, and reduction is the acceptance of electrons by the other substance. Reduction refers to the lowering of the OXIDATION STATE or OXIDATION NUMBER of the substance accepting the electrons.

In a redox reaction, the substance that gives up the electrons (that is oxidized) is called the reductant because it causes a reduction in the oxidation state or number of the substance that accepts the transferred electrons. The substance that receives the electrons (that is reduced) is called the oxidant because it causes oxidation of the other substance. Oxidation and reduction always occur simultaneously.

pH (pronounce as separate letters)

pH

pH is an expression of the intensity of the basic or acidic condition of a liquid. Mathematically, pH is the logarithm (base 10) of the reciprocal of the hydrogen ion activity.

$$pH = Log \frac{1}{[H^+]} \quad or = -Log [H^+]$$

The pH may range from 0 to 14, where 0 is most acidic, 14 most basic, and 7 neutral.

PICKLE

PICKLE

An acid or chemical solution in which metal objects or workpieces are dipped to remove oxide scale or other adhering substances.

RCRA (RICK-ruh)

RCRA

The Federal **R**esource **C**onservation and **R**ecovery **A**ct (10/21/76), Public Law (PL) 94-580, provides technical and financial assist-ance for the development of plans and facilities for recovery of energy and resources from discarded materials and for the safe dis-posal of discarded materials and hazardous wastes. This Act introduces the philosophy of the "cradle to grave" control of hazardous wastes. RCRA regulations can be found in Title 40 of the Code of Federal Regulations (40 CFR) Parts 260-268, 270 and 271.

REAGENT (re-A-gent)

REAGENT

A pure chemical substance that is used to make new products or is used in chemical tests to measure, detect, or examine other sub-stances.

REDOX (ree-DOCKS)

REDOX

Oxidation-reduction reactions in which the oxidation state of at least one reactant is raised while that of another is lowered.

REDOX REACTION

REDOX REACTION

(See OXIDATION-REDUCTION (REDOX) REACTION)

SANITARY SEWER

SANITARY SEWER

A pipe or conduit (sewer) intended to carry wastewater or waterborne wastes from homes, businesses, and industries to the treat-ment works. Storm water runoff or unpolluted water should be collected and transported in a separate system of pipes or conduits (storm sewers) to natural watercourses.

SIGNIFICANT INDUSTRIAL USER (SIU)   SIGNIFICANT INDUSTRIAL USER (SIU)

A Significant Industrial User (SIU) includes:

1. All categorical industrial users, and

2. Any noncategorical industrial user that

   a. Discharges 25,000 gallons per day or more of process wastewater ("process wastewater" excludes sanitary, noncontact cooling and boiler blowdown wastewaters), or

   b. Contributes a process wastestream which makes up five percent or more of the average dry weather hydraulic or organic (BOD, TSS) capacity of a treatment plant, or

   c. Has a reasonable potential, in the opinion of the Control or Approval Authority, to adversely affect the POTW treatment plant (inhibition, pass-through of pollutants, sludge contamination, or endangerment of POTW workers).

   SUMP

The term "sump" refers to a facility which connects an industrial discharger to a public sewer. The facility (sump) could be a sample box, a clarifier or an intercepting sewer.

SUMP CARD   SUMP CARD

A 3x5 reference card which identifies the location of a sump, lists the monitoring devices located in the sump (pH, for example), and indicates which treatment processes are connected to the sump.

CHAPTER 4.   INSPECTION OF A TYPICAL INDUSTRY

This chapter provides inspectors with an indication of the procedures to follow when inspecting any industry. The electroplating and metal finishing industry was selected as a typical industry because in many areas it is the only categorical industry requiring inspection. Because of the amount of hazardous waste produced, it is one of the most critical industries that must be inspected. Regardless of the type of industry you are inspecting, you must develop inspection procedures that are appropriate and uniform for the industry so that all industries are treated equally and fairly. More detailed inspection information regarding all types of industries is contained in later chapters and their appendices in this manual.

4.0 SAFETY

Safety cannot be overemphasized. Chapter 5 discusses the safety aspects of being an inspector in great detail. You must always remember that a safety-conscious inspector will impress industry as a true professional. Ask your industrial contact *BEFORE* you start an inspection if there are any special safety precautions or company rules and regulations that must be followed. Always follow your contact and use prescribed safety equipment. Never enter manholes or other confined spaces without testing the atmosphere for oxygen deficiency/enrichment, combustible gases, and toxic gases. *NEVER* enter a manhole or confined space when alone. *ALWAYS* have support personnel standing by to provide assistance. Use proper ventilation, wear appropriate safety equipment and protective clothing and have support personnel standing by to provide rescue assistance if needed.

4.1 SCHEDULING

4.10 Frequency of Inspection

POTWs with approved pretreatment programs must conduct at least one on-site inspection and sampling visit annually at the facilities of all Significant Industrial Users. A Significant Industrial User is generally defined to be all EPA categorical companies plus those companies discharging over 25,000 gallons per day, plus other companies deemed by the agency to require special attention.

The frequency of inspection usually depends on the POTW agency's personnel resources allotted to on-site field inspec-

tions. Some POTW agencies have a fairly rigid prescribed frequency that is tied to a fee schedule; other agencies leave day-to-day scheduling to the individual field inspectors. An analysis of self-monitoring reports may reveal the need for an inspection over and above routine visits. Chapter 2, "Pretreatment Program Administration," also discusses frequency of inspection visits in Section 2.211, "Frequency of Inspections."

Inspections should also be scheduled on the basis of the nature of the discharge. If the discharge has a history of causing problems or has a high potential for creating problems, damaging a sewer line or upsetting a wastewater treatment plant, then frequent inspections are appropriate. *UNANNOUNCED INSPECTION SCHEDULES SHOULD NEVER BECOME SO STANDARDIZED THAT DISCHARGERS CAN PREDICT AND THEREFORE PREPARE FOR AN INSPECTOR'S VISIT.*

The metal finishing industry by its nature has the potential to generate significant quantities of toxic wastes. Oil refineries, chemical manufacturing plants, centralized waste treatment facilities and some other dischargers similarly have the potential to seriously impact a collection system, wastewater treatment plant, stream or receiving water. A well-organized pretreatment program will take into account the relative impact of its individual dischargers and ensure that the potential problem areas receive the attention they warrant. Daily or seasonal production or discharge variations by the industrial users should be known by the inspector in order to anticipate increased loading at the POTW.

Records of all industrial waste-related treatment plant upsets and sewer system problems should be maintained so that patterns can be discovered that might lead to identification and resolution of recurring problems. In the early implementation of an inspection and enforcement program, a major portion of inspectors' time may be spent investigating POTW process upsets caused by slug dumping or frequent violations of industrial effluent standards. However, after the program has been operating and the permit requirements have been established, the bulk of the inspectors' time can be spent making "routine" inspections. The metal finishing industry, in particular, can cause serious POTW process upsets and sludge contamination by discharging excess heavy metals, cyanide or toxic organics.

Many treatment system upsets develop not from the toxic effects of industrial discharges, but from conventional industrial pollutants such as excess loading due to excessive biochemical oxygen demand or flow. Foaming, odors, grease and oil, explosive gases, excessive dissolved salts, excessive chlorine demand from reducing agents (such as thiosulfate), colors, slug doses of nutrients (like phosphorus or nitrogen), and a host of other problem discharges must all be kept under control. Inspectors should keep lists of potential industrial contributors (see Chapter 8, "Industrial Wastewaters," for helpful information) to each of the recurring problem discharges in their inspection area. With this type of list, an inspector will be able to identify the source of a problem quickly.

Many different circumstances may require an investigation of sewer lines. For example, sewer line investigations can be immediate emergency responses to catastrophic incidents such as explosive conditions discovered in a sewer or stoppages that cause wastewater to back up into residences or flow onto streets. Foaming in sewer lines and at treatment plants is an indication of an improper discharge. Inspectors are often called in to investigate long-term problems such as why a particular sewer line is scaling or corroding at a faster rate than expected. Inspectors should be on close working terms with sewer maintenance crews so that potential problems can be detected and corrected before sewers produce odors, explode, collapse, become plugged, or become a nuisance.

Once a problem is detected in a sewer, it is not always obvious from sewer maps alone what lines contribute to or bypass that location. The sewer system is a dynamic network of pipes. It is often altered by diversion structures, stop logs or relief lines. Changes may be made by sewer maintenance crews for seasonal flows or maintenance activities and this information may never have been relayed to other POTW units by the maintenance division. Communication and cooperation with sewer maintenance personnel are vital and may save you a considerable amount of time and effort.

4.11 Follow-Up Inspections

Initial inspections frequently generate a follow-up inspection. Minor equipment malfunctions detected in an inspection are typically discussed with the company representative and an appropriate length of time to repair the defect is granted. You should make a note in your personal calendar or a "tickle file" so that you can verify in the future that the agreed upon repairs are completed. Every POTW should have systematic procedures for tracking industrial user problems. Inspectors should maintain permanent records of all problems in office files or computers so that supervisors and inspectors can review them at any time and no one has to rely on the memory or "personal" records of an inspector.

Enforcement actions or requests to install or modify pretreatment equipment are common causes for follow-up inspections. Any operational or physical modification that could affect effluent water quality should be noted in the permanent records by the inspector.

Submittal or non-submittal of requested information frequently points out the need for a follow-up inspection. Compliance monitoring results showing violations will require an inspection. Sewer-use fee calculations may show unusual deductions for evaporative, irrigation, or product losses that must be verified by field observation. Flowmeter installations and calibrations should be inspected for compliance with agency guidelines. New or revised permit applications often raise questions that are best investigated on site.

4.12 Assignment of Personnel

The duties of a pretreatment facility inspector are highly varied. You may work by yourself or be assigned to a unit with many inspectors. Some of the larger staffs are broken into more specialized subgroups to allow shorter training periods, more efficient use of equipment and perhaps more productivity. Possible divisions within the staff might include:

1. Sampling and/or sample analysis,

2. Routine inspections,

3. Surveillance/special investigations, and

4. Emergency response/system upsets.

Perhaps the most common method for dividing the inspection staff is by geographical area. This allows an individual to become familiar with the sewers serving industries tributary to the pump stations, facilities for diverting flow from one trunk sewer to another, and treatment plants or particular high-interest spots within a region. Once assigned to an area, the inspector should develop reference lists for following up on the sewer system and treatment plant upsets most common to that locale. Reference material that an inspector would need to have on hand regarding sample summaries, permits, and past inspection results would be limited to a single region if inspectors are assigned to particular areas.

Some agencies assign inspectors to certain classes of industries such as refineries or food processors or metal finishers. This method develops expertise within that class but does not provide the personal stimulation, education and training offered by visiting a wider variety of operations. Unless similar industries are located in the same area, the inspector will not develop familiarity with the sewer networks and excessive driving time may cut productivity. In some larger cities, inspections are scheduled on the basis of postal ZIP codes.

Sometimes the seasonal nature of large dischargers may dictate reassignment of inspectors based on critical need. Routine inspections might need to be postponed to the off-season to allow concentration on industrial flows that have a strong impact on the operation of the wastewater collection and treatment systems.

Whatever the division of labor within the staff, it is advisable to rotate individuals to different assignments on a regular basis. As personnel are rotated from one industry or area to another, the need for good records of problems in a particular area or for a particular industry cannot be emphasized too much. A new inspector needs to have access to files documenting past history. Periodic rotation maximizes flexibility when any individual is unavailable due to vacation, sick leave or unplanned absence. Having several inspectors familiar with the same territory is especially valuable during crisis conditions when it comes time to attack a problem with as much expertise as you can muster. Rotation of assignments broadens the experience of the entire staff, provides "changes of scenery" that keep the job from becoming monotonous and allows each member of the staff an opportunity to handle the entire spectrum of encounters that field work has to offer.

QUESTIONS

Write your answers in a notebook and then compare your answers with those on page 343.

4.1A   What should you do *BEFORE* an inspection to protect your safety?

4.1B   What types of industries can seriously impact a wastewater treatment plant?

4.1C   List some of the reasons an inspector might have to investigate sewer lines.

4.1D   List the events that might necessitate a follow-up inspection.

4.2 ENTERING AN INDUSTRY FOR INSPECTION

4.20 Before You Go to the Door

Your agency must have the legal authority to conduct inspections of industrial dischargers. If you don't have the legal authority, you have no more right to enter an industry than the sales representatives that you'll be sitting next to in the lobby.

You must be thoroughly familiar with your legal authority and your agency's guidelines concerning the use of that authority to enter a facility for inspection.

An inspector must carry enough equipment on an inspection to gather the necessary information during the inspection. Each item on the following list could be a critical tool yet it would be difficult for one person to carry this much equipment. Reviewing past inspection reports and a diagram (Figure 4.1) of the facility you are going to inspect will help you determine which items to carry onto a particular site:

    proper personal identification

    notebook, pen and pencil

    sampling equipment (see Chapter 6)
        sample containers
        manhole opening apparatus

    field testing equipment
        pH paper

    atmospheric testing device (Figure 4.2)
        (LEL, oxygen, hydrogen sulfide)

    test kits
        cyanide
        metals (Cu, Cr, Ni, Zn)
        sulfide
        chlorine
        conductivity

    flowmeter testing devices

    camera

    personal safety equipment
        safety glasses
        hard hat
        safety toe shoes
        earplugs
        rubber gloves

Documents or reference materials frequently used should be available in your vehicle or in your field notebook:

    copy of sewer-use or wastewater ordinance

    citation forms

    local sewer map

    emergency response procedure

    list of industrial dischargers
        within your inspection area

    procedures and forms for submittal of information
        permits
        baseline monitoring reports
        facility improvement guidelines
            spill containment
            rain diversion
            copies of "standard drawings"
                clarifier
                sample box
                rain diversion system

    list of contacts for local agencies

    location of locally available services
        consultants
        testing laboratories
        waste haulers
        landfills
        centralized waste treatment facilities
        solvent recyclers

    safety guidelines (see Chapter 5)
        confined spaces
        toxic vapor exposure limits
        traffic diversion

    schematic of industrial production processes,
        plant treatment processes and
        sampling points (Figure 4.1)

The more information you have about a discharger before you enter the facility, the better prepared you'll be to perform an efficient and professional inspection. The best sources of information are this training manual, an "industrial waste survey" as described in EPA's *INDUSTRIAL USER INSPECTION AND SAMPLING MANUAL FOR POTWs*,[1] Chapter 2, "Inspecting Industrial Users," and your own agency's industrial wastewater discharge permit. From these documents you should be able to determine the following information:

    name of firm, location and phone number

    name of individual who signed the permit
        or who is in charge of overseeing
        wastewater discharge

    number of outfalls or discharges
        and location of entry to the POTW sewer

---

[1] *INDUSTRIAL USER INSPECTION AND SAMPLING MANUAL FOR POTWs*, April 1994. Office of Water (4202), U.S. Environmental Protection Agency, Washington, DC 20460. Obtain from National Technical Information Service (NTIS), 5285 Port Royal Road, Springfield, VA 22161. Order No. PB94-170271. EPA No. 831-B-94-001. Price, $67.50, plus $5.00 shipping and handling per order.

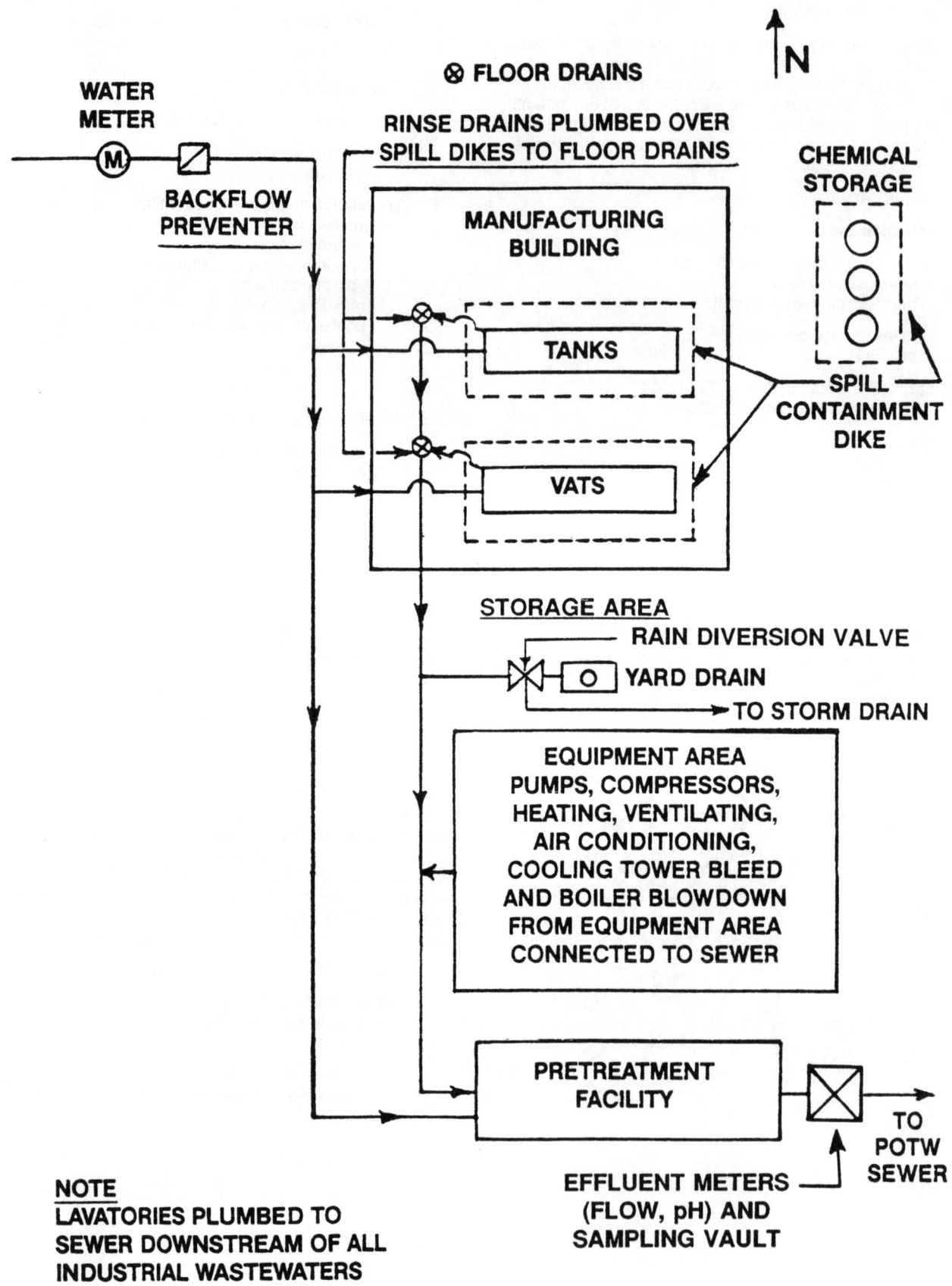

⊗ FLOOR DRAINS

RINSE DRAINS PLUMBED OVER
SPILL DIKES TO FLOOR DRAINS

WATER METER

BACKFLOW PREVENTER

CHEMICAL STORAGE

MANUFACTURING BUILDING

TANKS

VATS

SPILL CONTAINMENT DIKE

N

STORAGE AREA

RAIN DIVERSION VALVE

YARD DRAIN

TO STORM DRAIN

EQUIPMENT AREA
PUMPS, COMPRESSORS,
HEATING, VENTILATING,
AIR CONDITIONING,
COOLING TOWER BLEED
AND BOILER BLOWDOWN
FROM EQUIPMENT AREA
CONNECTED TO SEWER

PRETREATMENT FACILITY

TO POTW SEWER

EFFLUENT METERS
(FLOW, pH) AND
SAMPLING VAULT

NOTE
LAVATORIES PLUMBED TO
SEWER DOWNSTREAM OF ALL
INDUSTRIAL WASTEWATERS

Fig. 4.1   Layout of industrial site

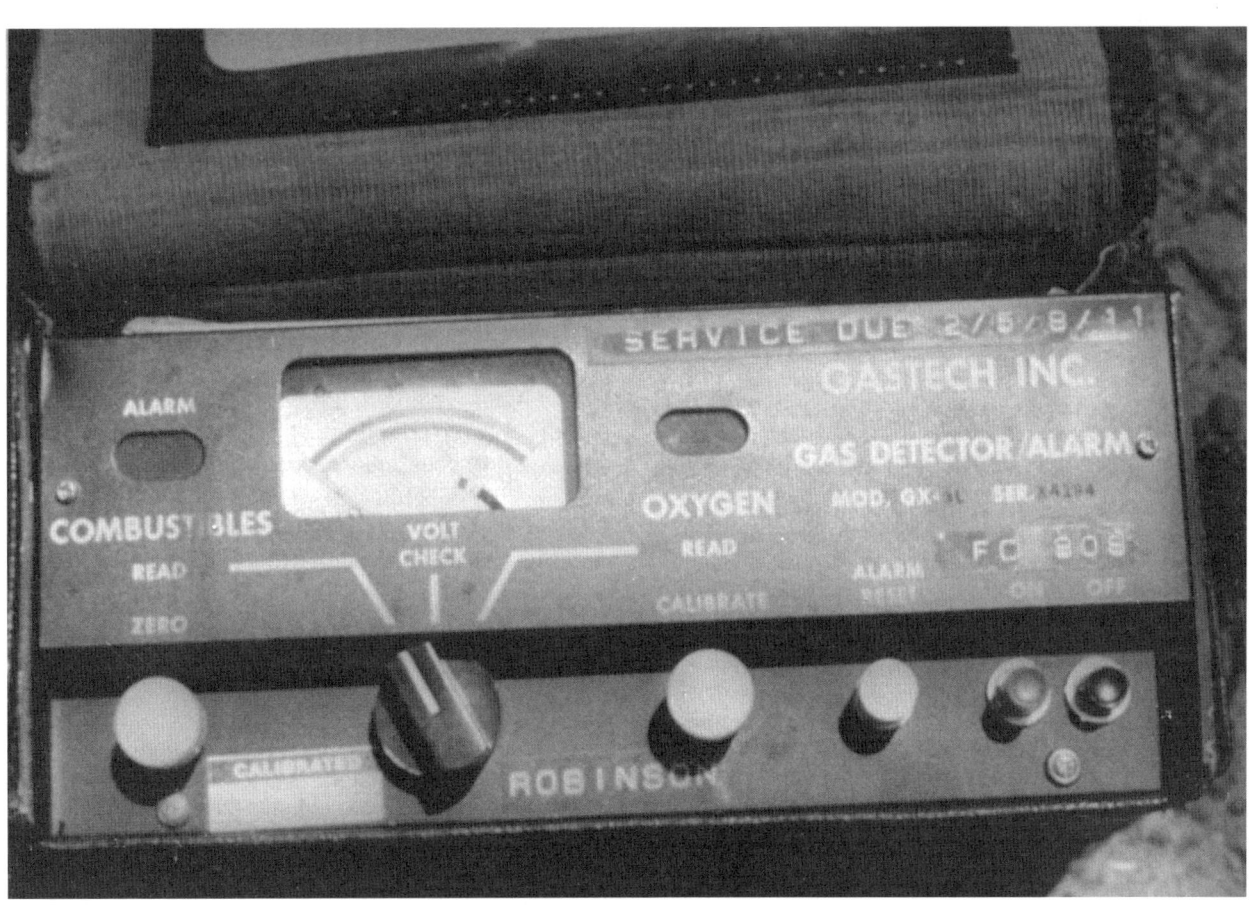

Fig. 4.2   *Closeup photo of oxygen and combustible gas meter carried by inspectors and other field personnel*

flow volume and time of discharge
    normal discharge flow based on
       historical water consumption
    water meter readings

wastewater constituents

general layout of facility

pretreatment equipment on site

production processes

production schedule
    (some companies will not release this
      information)

raw material list

location of storage of chemicals

special permit requirements
    nonstandard limits due to impact on local sewer
    numerical limit on generally uncontrolled
      substances
         total flow, peak flow or time restraint
         nutrients or COD
         particular organic compounds
         oxidizing or reducing agents

progress or status of any pertinent compliance
    schedules or enforcement actions

Beyond this basic list there are many other sources of information (permit applications, sample results, inspection reports and citations) that will yield more detailed information about the discharger.

A history of sample results will show the compliance performance as well as any trends in effluent quality. Sample summaries can come from several sources. Check the results of all grab and composite samples taken by your agency. Self-monitoring test results are valuable but tend to show a company's performance under optimum conditions. Since the discharger knows in advance when samples will be taken, the discharger has the opportunity to fine-tune the effluent treatment system, avoiding slug discharges. If surveillance samples have been taken in the sewer surrounding a company, this information may be confidential; as part of a potential legal action, it should not be discussed with the discharger without clearance from your supervisor. Be sure to follow "chain of custody" procedures with all surveillance samples.

Reviewing previous inspection reports is an excellent way to familiarize yourself with the discharger. These reports may be in the form of computer printouts, inspection files, field notebooks, *SUMP CARDS*,[2] or some other form or type of record with notes describing treatment operations. Reports should be well documented and kept on file as a permanent record of that industry's activities and compliance status. A review of the most recent inspection report will reveal whether any repairs or improvements had been requested by the last inspector. Any marginally acceptable operations or defective equipment should be noted so that progress toward compliance with agency standards can be verified.

Inspectors should be familiar with the reporting requirements for dischargers and be able to assist in explaining how and when to file necessary forms. Some of the paper flow includes the following items:

surcharge calculations and payments
    (see Chapter 2)

self-monitoring reports
    wastewater testing
    meter calibrations
       flow, atmospheric monitor, pH, *ORP*[3]

baseline monitoring report

permit revisions (see Chapter 2)
    flow change
    process or pretreatment modification

progress report due to enforcement action

POTW sampling reports

other agencies' requirements that might impact
    sewer use such as fume scrubbers or ground-
    water cleanup or hazardous waste shipments

water usage records (a substantial change can
    be due to a change of process or operation)

QUESTIONS

Write your answers in a notebook and then compare your answers with those on page 344.

4.2A   Why does an inspector need to carry equipment on an inspection?

4.2B   How can an inspector determine which items of equipment must be carried to a particular inspection site?

4.2C   Previous inspection reports may be available to inspectors in what types of form or format?

4.21 Getting in the Door (Figure 4.3)

When approaching the receptionist, security guard, office manager or whoever meets you at the entrance to a facility, it is important to maintain a businesslike demeanor and make your intention to conduct an inspection clear. Give them your business card and be prepared to show additional identification if requested to do so. *REMEMBER THAT YOU ARE THERE TO MAKE AN INSPECTION OF THE FACILITY, NOT TO SEE AN INDIVIDUAL. THE ABSENCE OF YOUR AGENCY'S NORMAL CONTACT IS NOT GROUNDS FOR POSTPONING THE INSPECTION.* If there is an issue you wish to discuss with an individual, this can be done over the telephone or by appointment, but the majority of inspections should be unannounced checks on effluent quality and treatment system performance, *NOT* visits to review paper work.

"Do you have an appointment?" Realize that one of the duties of a receptionist is to shield company employees from salespeople or distractions from their work. Restate your intention to inspect the facility for compliance with industrial wastewater discharge requirements, not to see an individual.

---

[2] *Sump Card. A 3x5 reference card which identifies the location of a sump, lists the monitoring devices located in the sump (pH, for example), and indicates which treatment processes are connected to the sump.*

[3] *ORP (Oxidation-Reduction Potential). The electrical potential required to transfer electrons from one compound or element (the oxidant) to another compound or element (the reductant); used as a qualitative measure of the state of oxidation in wastewater treatment systems. ORP is measured in millivolts, with negative values indicating a tendency to reduce compounds or elements and positive values indicating a tendency to oxidize compounds or elements.*

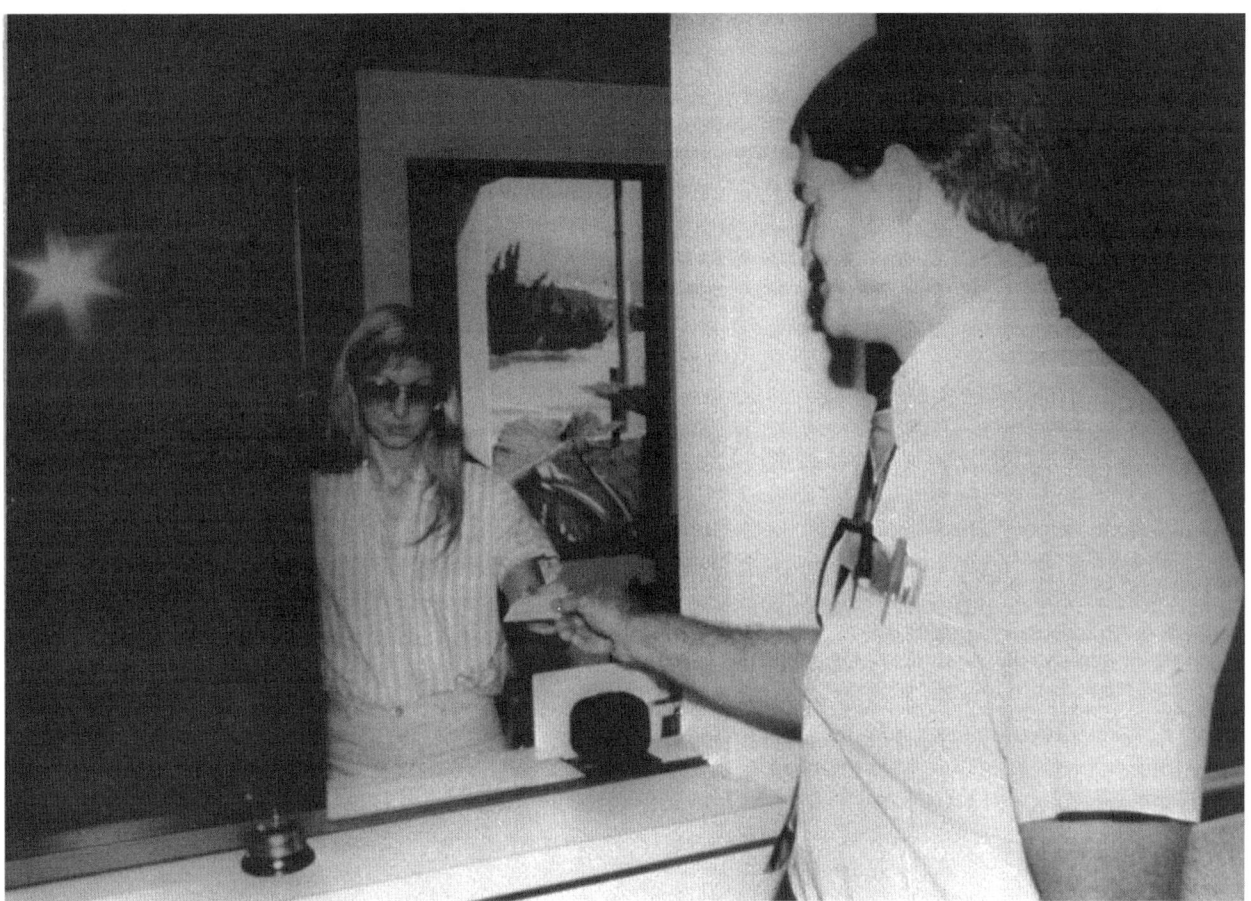

Fig. 4.3   Pretreatment facility inspector presenting business card and/or identification to industrial receptionist and announcing intention to inspect industrial facility

If appropriate, you should explain your agency's policy not to announce spot inspections. Give the name of your traditional contacts and, if they are not available, offer to wait until a substitute can be located. Don't wait longer than 15 minutes; proceed with the options below.

"There is nobody here who can authorize an inspection." Ask if the facility is open for business. If it is, then ask to see the receptionist's supervisor, personnel manager, plant manager, corporate officer or whoever is in charge. Identify yourself to any company representative you contact; explain to them that you wish to make an inspection. If there is still resistance, show them a copy of the *Sewer-Use Ordinance* or *Wastewater Ordinance*, permit or whatever authority grants you the right to inspect the facility. Do not threaten legal action but clearly state your intent to inspect. If you still cannot gain entry there are several options available:

1. Call your supervisor for direction. This is almost always a good first move. If this discharger has a history of obstructing inspections, your office may have experience handling the situation. If the company is under investigation for illegal discharge, then you may have walked in on a deliberate

dump of toxic material and immediate access would be vital to the investigation. If there is a personality clash, then bringing in a third party could soothe disharmony and reassure company representatives that they should allow the inspection.

2. Leave and immediately collect samples for analysis of suspected or possible effluent discharge violations. If access to the effluent sampling station is denied or impossible, collect samples from manholes upstream and downstream from the point of the effluent discharge to the POTW sewer.

3. Leave and try again later. This is not the recommended response, but if you are convinced that there is no illegal discharge occurring and your schedule will not allow an extended encounter, then postpone your inspection for another time. Be sure to inform your contacts that they are in violation of the provision in the ordinance, permit or whatever authority it is that grants you the right to enter the facility. Always follow up a refusal of entry as soon as possible with another inspection. Allowing individuals to illegally block your entry makes every other inspector's job more difficult.

4. Issue a citation for denial of access.[4] If the "responsible party" who signed the industrial waste (I.W.) permit is available, you should request that this person be the one to sign the citation. Advise this person that refusal to grant access may expose the company to POTW enforcement actions including civil or criminal penalties of at least $1,000 per day. If nobody will consent to sign the citation, it should be left with your contact with a notation that the contact refused to sign it. Your agency's certified, return receipt requested, follow-up letter will be addressed to the chief executive officer or the person who signed the permit.

Corrective action required as a result of the issuance of a citation for refusal to grant access could include:

a. A compliance meeting with your agency,

b. A written statement from a company official stating that access will not be denied again, and/or

c. An administrative fine or legal action (see Chapters 2 and 3).

5. Exercise your authority to enter the facility without company permission. This remedy is the most severe and should be reserved for occasions when there is reason to believe that blocking your access is an attempt to conceal an illegal activity. You should secure the highest level of approval from your agency and be sure that your agency has researched their legal authority to demand entrance for an inspection. Depending on the language of the sewer-use ordinance and discharge permit, you may or may not need to secure a search warrant before asking the local police to escort you into the facility.

Whatever course of action was followed, your next successful entry into the facility should include an attempt to normalize relations. Request that the party responsible for blocking your entrance be informed of inspectors' access rights. Request a list of backup contacts in case your normal contact is unavailable but stress that entry cannot be denied because of the absence of particular individuals.

Discuss the possibility that an off-site monitoring station may need to be installed if access continues to be hindered.

4.22 Meeting Your Contact (Figure 4.4)

Present yourself to your contact in a friendly, but an official and polite, manner. Be positive with regard to the regulations you must enforce. The discharger (your contact) should have no doubts that you believe in what you are doing and that you reflect the attitude of your POTW and its managers.

Always present your identification and clearly state why you are visiting the facility. Comply with all company regulations regarding safety, such as wearing special protective equipment. Ask your contact if any special hazards exist in the facility. Follow your contact closely; the contact is more familiar with the hazardous areas in the plant than you are. Don't touch anything unless your contact touches it first. That pallet of parts next to you may have just come out of the furnace. Watch for slippery floors and make sure your shoes have plenty of tread on them.

Companies have an affirmative duty of reasonable care to make their plant premises safe for all visitors (inspectors) and to warn them of anything that threatens an unreasonable risk to them. Inspectors should generally not sign documents that

waive rights to safe access even though such notices would probably not excuse the company from normal liability if the inspector was injured while conducting the required duties.

If you suspect that deliberate illegal discharges have been made or if there is an ongoing investigation that may lead to legal action, then this information should not be disclosed to your contact; otherwise be as open and informative as possible. If you are planning to issue a citation, it is best to do so after the tour of the facility since other violations may be discovered that should be added to the citation. Issuing a citation before the tour may generate hostility that jeopardizes the flow of information from your contact. Be sure you understand and follow your agency's policy regarding citations before issuing one.

QUESTIONS

Write your answers in a notebook and then compare your answers with those on page 344.

4.2D What is the main purpose of inspections?

4.2E What should an inspector do if a receptionist says, "There is nobody here who can authorize an inspection"?

4.2F What safety precautions should be followed when making an inspection?

4.2G At what time during an inspection should a citation be issued?

4.23 Dealing With Contacts

Do not forget that your primary objective is to inspect the facility for compliance with discharge regulations. You are not there as a consultant to solve technical difficulties or find solutions to long-standing problems for the company. If through your experience or technical expertise you can describe how similar problems have been handled successfully, you may be able to help contacts solve their problems. Be cautious about giving advice. You might like to help your local industries as much as possible, but if they attempt to correct a problem on the basis of your instructions and fail to solve the problem, they could claim the fault for a permit violation is yours since they followed your instructions.

Your contact's reaction to an inspection will be influenced by the corporate structure of the organization. You may find it difficult to get commitments to enact corrective action from a large corporation due to the many layers of responsibility that must be dealt with. On the other hand, the small, privately held firm's contact may feel so close to the operation that the contact may take it as a personal insult when told that the company is in violation. Inspecting utilities or government agencies can draw you closer into their bureaucratic mechanisms than you ever imagined possible. The important point to remember is to present your agency's position in a fair and consistent manner to all dischargers regardless of their financial strength or weakness or their corporate structure.

The personality traits encountered during inspections are diverse. Some contacts try to impress you with their spirit of social consciousness and environmental awareness even while you're writing them a violation notice for deliberately discharging toxic waste. Others express contempt for environmental regulations yet consistently exceed required perform-

[4] *Issuance of citations must be in accordance with policies described in your agency's Enforcement Response Plan. In many POTW pretreatment programs most enforcement actions, or at least the paper work and official notice, originate from the administrative office.*

*Fig. 4.4   Pretreatment facility inspector meets
industrial contact and presents business card
and/or proper identification*

ance standards. The following paragraphs describe a few examples of frequently encountered personality types and how an inspector might deal with them.

The belligerent contact tries to make the tour as unpleasant as possible, challenging the inspector's right to see various areas of the plant, berating the inspector and the inspector's employer and generally making the job as difficult as possible. The contact may be hoping that the inspection will be cut short and that the contact can intimidate the inspector from returning. The contact may be hoping that the inspector's timidity will keep the inspector from discovering the new plating line that has been added to the back end of the plant. When the contact claims that the keys that unlock the clarifier and monitoring station (Figure 4.5) are "lost," the contact is challenging the inspector to insist on access to the pretreatment equipment.

The best way to handle the belligerent contact is to keep your head and stick to business. Try not to respond to personal comments but record all abusive statements and include them in your inspection report. If you feel physically threatened, you should leave immediately and notify your supervisor. If you continue the inspection, you may wish to appeal to your contact's supervisor or another available manager to substitute someone else for the tour. Be especially careful to conduct the inspection according to your agency's guidelines. Don't overlook violations to pacify the contact or retaliate by issuing citations for minor defects that would normally receive only warning statements. Be professional. If the contact is trying to hide something, find it. If you can verify that the contact is lying to you or deliberately withholding access or needed information, then report the contact to the highest level of authority in the industrial organization.

The overly friendly "good buddy" contact may be pleasant to deal with but you must remember the same rule — follow your agency's guidelines in all matters. Remember that you are inspecting the facility for compliance with government regulations, not acting as a consultant to help the company overcome its problems.

The "in charge" contact who has the authority to undertake corrective action but has little working knowledge of the facility poses special difficulties. If you have never been through the plant, this inspection may be a case of the blind leading the blind throughout the tour. Don't be afraid to request permission to question various supervisors or maintenance personnel concerning technical aspects of the operation as you walk through the plant. Your contact will probably appreciate learning more about the operation of the plant as much as or more than you do.

It is not unusual to be given the "worker bee" contact who has a full understanding of the workings of the plant but no authority to implement any corrective action. One approach is to inspect the entire plant and then request a short meeting with the plant manager or other person of authority to outline the corrections that should be made. If the person responsible for making the necessary improvements cannot be found, then issue a citation with a follow-up letter addressed to the chief executive officer or the signer of the discharge permit.

The "public relations" contact is a person who deals with regulatory people but has neither the authority to correct violations nor any working knowledge of the plant. This contact presents a real challenge to the inspector, who must dig for information from whatever sources are available. Be aware that when you finally find someone in authority after the tour and try to explain the violations, you may have to repeat the tour with this new contact.

The "evasive" contact is one who refuses to answer questions or pretends ignorance. In this case, you will have to rely on your own investigative abilities and will probably be forced to spend a great deal of time completing the inspection. In response to this lack of cooperation, be willing to spend as much time as necessary to do a thorough investigation. Information normally acquired through conversation may need to be sought out by reviewing documents kept by the company or by government agencies. Take the time you need to do your job thoroughly.

Material Safety Data Sheets (MSDSs) are required to be available and should be reviewed if you are unfamiliar with chemicals encountered during the inspection. Liquid waste haulers' manifests are also required to be available and will show the quantity and quality of wastes hauled off site. Water bills will show consumption of incoming water. The company's Industrial Waste Discharge Permit may describe records that must be made available such as flow and pH charts, meter calibrations, discharges from particular sources such as spill containment impounds, and raw material lists. Know what records are required to be kept at the site and exercise your right to review them.

The "instructive" or "expert" contact is great for learning about processes that may not be familiar to new inspectors. There is a wealth of knowledge that can be absorbed on this job. The only caution is that as an inspector, you must not be so wide-eyed that you necessarily believe everything that the "expert" tells you. Realize that your contact is not a disinterested party, but is representing the interests of the company and may deliberately misinterpret how production or pretreatment systems operate.

Fig. 4.5   Sampling and flowmeter vault in a parking lot

NOTE: The circular cover has a hasp with a steel rod
through it which allows two different padlocks to
be attached in a way that the industry as well as
the POTW can have access to the secured vault.

The "informant" is one who comes forward with incriminating information regarding their own company, a neighboring company or a competitor. Cash rewards are offered in some regions for information that leads to conviction for violation of environmental laws. Record all details given by the informant and present this information to your supervisor. If the reported activity is outside the scope of your agency, then the information should be turned over to the appropriate authorities. Although all such reports should be investigated, the value of the information must be weighed to determine the amount of resources to spend on follow-up. Although many informants' reports are quite accurate in their details, the informant often lacks sufficient knowledge of the environmental regulations and has a wholly different perspective than the environmental agencies. Sometimes the "illegal dump" reported by an informant is the discharge of spent cleaners or quench tanks that is allowed by the discharge permit.

Always keep well-documented reports on any investigation that has the potential to lead to legal action. If the informant inquires about the results of your investigation, you should assure the informant that the "tip" was investigated but avoid discussing details or conclusions drawn from the report. Thank the informant for assistance and outline the procedure that should be followed if the informant wishes to make a formal request to view the agency's files.

4.24 Presenting Yourself

As a field inspector, you are not expected to carry blueprints of every facility to each facility. You may not even be aware of the most recent negotiations that might be going on between your office and the company being inspected. You should not be embarrassed that you are not familiar with all the various production and pretreatment system technologies. However, you should be familiar with the applicable EPA Categorical Pretreatment Standards and the General Pretreatment Regulations as they pertain to the industrial user being inspected. There are two basic guidelines that should be followed:

1. KNOW THE RULES. Know your agency's regulations and carry a copy of the *Sewer-Use Ordinance* or *Wastewater Ordinance* with you. Be familiar with current policies, guidelines and exceptions to those rules. Know what information you have the right to extract from a permittee, what constitutes a trade secret, and whether you need to apply for special clearance to gain access to any areas within the facility you're inspecting. Know the limits of your jurisdiction and when you should notify your supervisor and/or a regulatory agency if obvious violations are detected. Be familiar with the applicable EPA Categorical Pretreatment Standards and the General Pretreatment Regulations as they pertain to the industrial user being inspected.

2. FOLLOW THE RULES. Not pursuing corrective action for violations breeds disrespect for you and your organization. If you discredit your agency or your supervisors, you are encouraging disrespect toward yourself as an inspector. If you say that your superiors are fools, what does that make you? No one is going to take you more seriously than you take yourself. Be a professional at all times. Believe in what you are doing. Use whatever power or influence you have to make the organization as good as it can be. If you feel that you cannot function within the organization, then it is probably time to look for a new job.

It is a misuse of your position to push for industry performance beyond what is required by the regulations, despite your own personal environmental concerns. The notion that you are helping to save the environment while making your daily rounds makes the job rewarding but you must not lose sight of the specific goals of your program (see Chapters 1, 2, and 3). Now and then you will even encounter an industry that has discovered it can benefit (by recovery and recycling) by performing beyond the limits or intent of the law. Your job, however, is to ensure compliance with the law, nothing more and nothing less.

QUESTIONS

Write your answers in a notebook and then compare your answers with those on page 344.

4.2H   What should an inspector do when threatened physically?

4.2I   How should an inspector handle an industrial contact who has little working knowledge of the facility?

4.2J   What precautions must be exercised when dealing with an instructive contact?

4.2K   What are two basic guidelines that an inspector should consider during every inspection?

4.3 LEVELS OF INSPECTION

A "complete" inspection would involve a tour of the entire facility and include all four of the following divisions: the *OUTFALL*,[5] effluent treatment equipment, in-plant wastewater control equipment and a general tour. Depending on the size of the facility, your familiarity with the operation, the purpose of your visit and the time available, you may not be able to complete all four levels of inspection. They are presented here in order of importance.

4.30 The Outfall

Whatever your reason for visiting a company, always check the outfall. Be prepared to take a sample. It would be embarrassing to later discover that a major discharge violation occurred while you were there but you hadn't even walked out to look at the outfall. An industry may have several discharges into the sanitary sewer and each one must be inspected.

---

[5] *Outfall.   (1) The point, location or structure where wastewater or drainage discharges from a sewer, drain, or other conduit. (2) The conduit leading to the final disposal point or area.*

Be familiar with the odor of commonly encountered gases such as sulfide, ammonia, chlorine, cyanide or any gases known to be associated with the facility (see Chapter 5). Stay out of confined spaces unless you are using all prescribed safety procedures and have all the required equipment. Do not leave open manholes unattended!

The exact sample point(s) should be designated in the permit. To reduce the risk that your samples will be challenged for validity, it is a good idea to restrict your final effluent sampling to the designated sample point (see Chapter 6). Use the proper sampling technique and preserve samples properly. Contact your laboratory if you need assistance. Follow chain of custody procedures at all times. The possibility always exists that the sample results will be used in court as evidence. Offer to give a portion of samples you collect (sample split) to your contact at the facility.

Review final effluent meter printouts for evidence of recent violations. These meters might include pH, flow (see Chapter 7), temperature, Lower Explosive Limit (LEL), conductivity, oxidation-reduction potential (ORP), or some of the more advanced on-line monitoring units such as total organic carbon, solvents or heavy metals. Check automatic sampling equipment and run field tests for selected compounds. Some of this equipment may have alarm set points to notify the POTW immediately whenever an alarm condition occurs. Also check calibration records; it is a good practice to sign and date any charts that you review.

A frequently encountered dilemma is what enforcement stance to take when significant violations are documented by required effluent recording instruments. If no sample was taken during the "violation," it is difficult to determine whether a pretreatment control equipment malfunction caused the wastewater to be discharged without adequate treatment, or

if the recording equipment malfunctioned and falsely recorded a violation. In the absence of any corroborating evidence, the safe enforcement action is to proceed against the company for failure to maintain required pretreatment equipment without specifying which piece of equipment failed. Part of the company's corrective action would be to investigate the control and recording devices to determine where the failure occurred.

Field test kits (see Chapter 6) can give you an indication of effluent quality without the expense and delay of laboratory analysis but you should be aware of common interferences or other limitations on the accuracy of your test kits. The most popular field tests include pH, sulfide, temperature, conductivity, cyanide, metals (Figure 4.6) and a limited number of atmospheric tests such as hydrogen sulfide (H_2S), Lower Explosive Limit (LEL), oxygen and certain industrial solvents. Observations of the wastewater such as color, temperature and amounts of oil and settleable solids in the effluent should be recorded in your notes.

4.31 Effluent Treatment Equipment

The second level of inspection involves checking the performance of the end-of-the-line pretreatment equipment. This equipment should be described in the sewer-use permit. It may also be described in wastewater discharge permit applications, construction and operating permits, and other control mechanisms in use by the POTW to regulate the installation of pretreatment equipment. Details of any additions or modifications since the issuance of the permit need to be submitted for review by your agency.

One of the most basic pieces of equipment is the clarifier, also known as a sand and grease interceptor. Its purpose is to trap grit and settleable solids in the bottom of its chambers and to trap floatables and oil on the surface (see Figure 4.7). If the layer of oil gets too thick, it will pass to the effluent chamber. Once a clarifier's capacity to hold solids or oil is exceeded, then it must be pumped and these materials must be hauled to an acceptable landfill or dump site or disposed of in some other appropriate manner.

The generator of the material removed from the clarifier has an obligation to determine whether this material is considered a *HAZARDOUS WASTE*[6] by state and federal standards. Under the Resource Conservation and Recovery Act (*RCRA*),[7] PL 94-580, the waste must be tracked from "cradle to grave." As shown in Figure 4.22, on page 333, the manifest must be signed by the generator, the transporter, and the operator of the treatment or disposal site that is the final resting spot for the waste. A permit must be obtained by all three parties.

Detained organic material in some industrial treatment facilities may tend to generate dissolved sulfide in the water and highly poisonous hydrogen sulfide gas in the atmosphere (see

---

[6] *Hazardous Waste.   A waste, or combination of wastes, which because of its quantity, concentration, or physical, chemical, or infectious characteristics may:*
  1. *Cause, or significantly contribute to, an increase in mortality or an increase in serious, irreversible, or incapacitating reversible illness; or*
  2. *Pose a substantial present or potential hazard to human health or the environment when improperly treated, stored, transported, or disposed of or otherwise managed; and*
  3. *Normally not be discharged into a sanitary sewer; subject to regulated disposal.*
  (Resource Conservation and Recovery Act (RCRA) definition.)
[7] *RCRA (RICK-ruh).   The Federal Resource Conservation and Recovery Act (10/21/76), Public Law (PL) 94-580, provides technical and financial assistance for the development of plans and facilities for recovery of energy and resources from discarded materials and for the safe disposal of discarded materials and hazardous wastes. This Act introduces the philosophy of the "cradle to grave" control of hazardous wastes. RCRA regulations can be found in Title 40 of the Code of Federal Regulations (40 CFR) Parts 260-268, 270 and 271.*

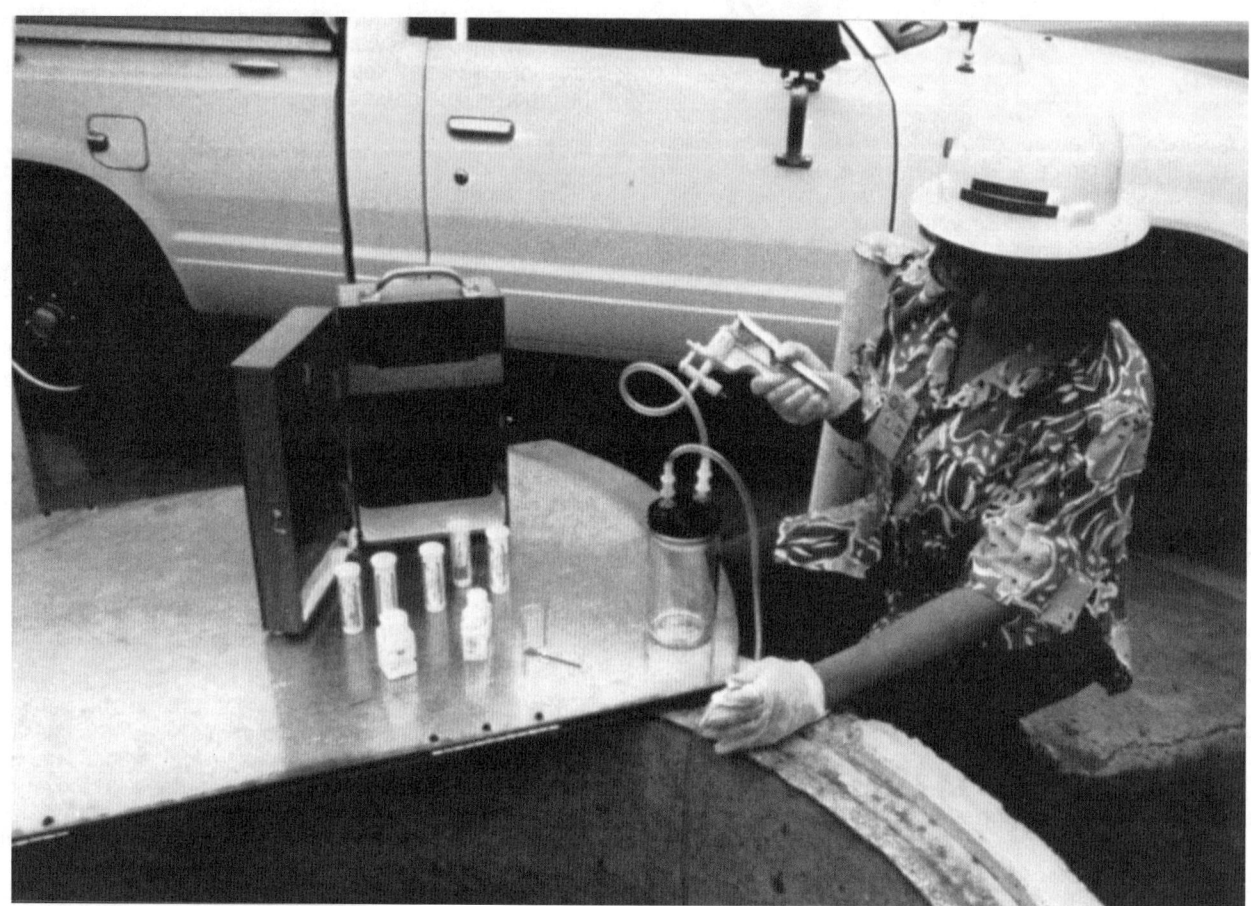

Fig. 4.6 Inspector at sampling and flowmeter vault

Inspector is drawing sample with a hand-operated
pump. Various field test kits are displayed —
pH, cyanide, chromium, nickel, zinc and copper.
The wooden box is the carrying container for
the pump, sample jars and test kits.

SAND & GREASE INTERCEPTOR

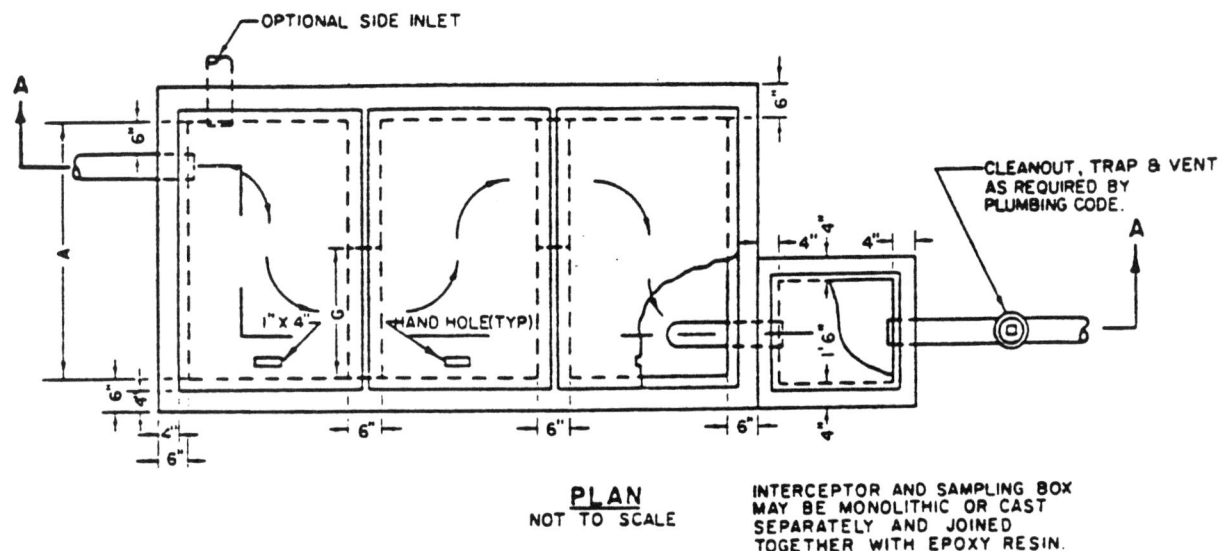

PLAN
NOT TO SCALE

INTERCEPTOR AND SAMPLING BOX
MAY BE MONOLITHIC OR CAST
SEPARATELY AND JOINED
TOGETHER WITH EPOXY RESIN.

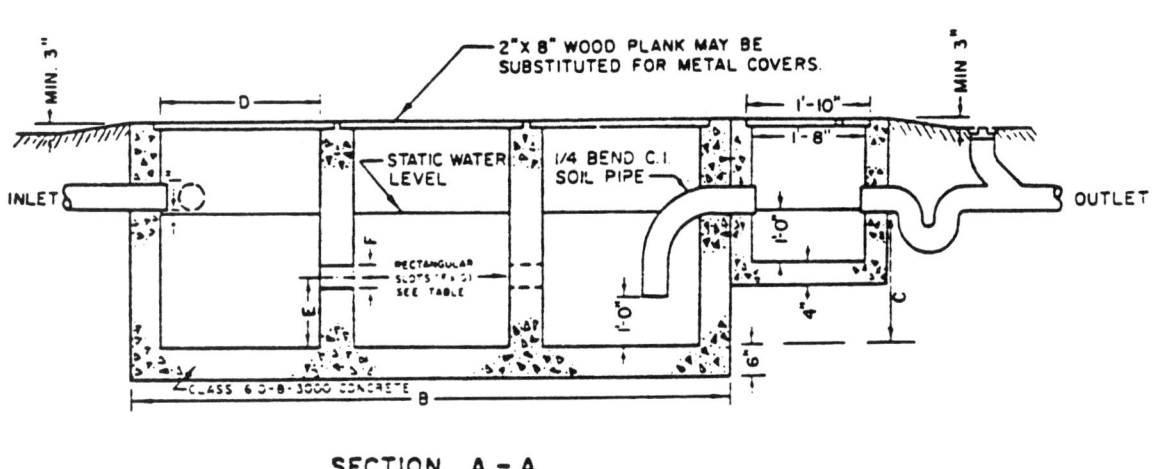

SECTION   A - A
NOT TO SCALE

NOTES:
THE APPROVAL OF THE COUNTY ENGINEER MUST BE OBTAINED BEFORE INSTALLATION
THE INTERCEPTOR TO BE CONSTRUCTED OF TYPE II PORTLAND CEMENT CONCRETE.
INTERCEPTOR EXCEEDING 6'-6" IN DEPTH MUST BE CONSTRUCTED OF REINFORCED CONCRETE.
IF INSTALLED INSIDE OF BUILDING THE TOP OF INTERCEPTOR MAY BE LEVEL WITH FLOOR
PROVIDED THAT WASTES ENTER THROUGH INLET PIPE ONLY
ALL SURFACE WATER MUST DRAIN AWAY FROM INTERCEPTOR TO EXCLUDE RAIN WATER
FROM PUBLIC SEWERS.

CAPACITY GALLONS	DIMENSIONS							COVER SIZE	METAL COVERS	PIPE SIZE
	A	B	C	D	E	F	G			
510	3'-0"	9'-6"	3'-0"	2'-6"	1'-6"	0'-4½"	1'-6"	2'-10"x3'-4"	¼" STEEL PLATE	4" MIN
866	3'-6"	10'-3"	4'-0"	2'-9"	2'-0"	0'-6"	1'-9"	3'-1" x 3'-10"	3/8" ALUMINUM PLATE	4" MIN
1260	4'-0"	12'-6"	4'-0"	3'-6"	2'-0"	0'-6"	2'-0"	3'-10"x4'-4"	3/8" ALUMINUM PLATE	4" MIN

Fig. 4.7   Sand and grease interceptor. Used by dischargers to POTW sewers with effluents containing sand and grease.

(Permission of Department of County Engineer — County of Los Angeles, California)

Chapter 5 for safety and Chapter 9 for control of sulfide). A field test for sulfide should be performed whenever the odor of rotten eggs is detected.

Open all the lids to the chambers of the clarifier for inspection. Due to the varying heights of the passages connecting the clarifier stages, the compartments will selectively trap either oil and other floatables (if the connecting passage is near the bottom) or grit and other heavy solids (if the passage is near the surface). A six- to ten-foot lightweight pole such as a long broomstick is ideal for determining the oil and solids accumulation in the clarifier. Insert the pole to the bottom of each stage of the clarifier (see Figure 4.8). Then slowly withdraw the pole while applying a steady horizontal force. You will be able to feel when the pole exits the sludge blanket on the floor of the clarifier by the abrupt lessening of resistance to the horizontal force. Sludge will be carried to the sample box and to the sewer if the sludge level is allowed to approach the height of the effluent elbow in the final stage of the clarifier.

If the oil layer on the surface of the clarifier compartments cannot be "broken" by splashing, use a "tube sampler" to determine the thickness of the floating phase (see Chapter 6).

The minimum required detention time for oil and grit removal is usually thirty minutes. EXAMPLE 1 shows how to calculate the detention time.

EXAMPLE 1

Calculate the detention time for a 1,500-gallon capacity clarifier that treats a flow of 25 gallons per minute (GPM).

Known	Unknown
Volume, gal = 1,500 gal	Detention Time, min
Flow, GPM = 25 GPM	

Calculate the detention time in minutes.

$$\text{Detention Time, min} = \frac{\text{Volume, gal}}{\text{Time, gal/min}}$$

$$= \frac{1,500 \text{ gal}}{25 \text{ gal/min}}$$

$$= 60\text{-minute detention time}$$

Some industries use a clarifier primarily to remove settleable solids from a wastestream. If a wastestream transports oil and grease, an oil/water separator is used. These devices may look similar physically, but they have special design features to serve their intended purposes.

Some permittees install holding ponds to dampen hydraulic surges, lower peak flows, hold water for off-peak discharges, provide for evaporation, cooling or settling, or to allow for *BATCH TREATMENT*[8] of certain wastestreams. Some food processors or other high-BOD generators use ponding as a method of biological treatment (see Chapter 9). The inspector's job is to verify that the system is operating as it is represented in the permit. If you are in doubt about an operation, ask if there is an operation manual for the system or if the system's procedures are included in a larger operation manual of a greater part of the industrial facility. Ask to review appropri-

ate provisions to confirm proper operation. Sewer-use fees could be dramatically underestimated if the holding ponds are misused.

If the effluent flow rate is measured and recorded on site, review the flow charts to verify that wastewater is discharged at the volume and time of day allowed by the permit. Flowmeters should conform to agency or POTW standards. They must be properly maintained and calibrated (see Chapter 7, "Wastewater Flow Monitoring").

Equipment used to remove contaminants is discussed in Chapter 9, which also contains drawings and photos of most of the equipment. Become familiar with this equipment so that you can evaluate whether it is performing satisfactorily; review maintenance and calibration records. Some of the most common pretreatment facilities encountered are:

pH control (Figure 4.9)

oil removal
  belt skimmers
  suction or surface overflow skimmers
  A.P.I. (American Petroleum Institute) separators
  air flotation

oxidation/reduction (Figure 4.10)
  cyanide
  chromium

flocculation/precipitation

lamella thickener or clarifier (Figure 4.11)

sulfide precipitation

ion exchange

reverse osmosis

ultrafiltration

neutralization

flow equalization

solids removal/dewatering
  filter press
  belt press
  centrifuge
  screens

deionization

distillation/evaporative recovery

cementation or electrolytic recovery

deplate-out units (circuit board manufacturing)

organic compound removal
  air stripping (may require atmospheric
    emission permit)
  carbon adsorption

Chapter 9, "Pretreatment Technology (Source Control)," discusses the applications of these facilities to treat industrial wastestreams before discharge to the POTW. Chapter 10, "Industrial Inspection Procedures," explains how to inspect various types of industries and what to look for when inspecting these pretreatment facilities.

---

[8] *Batch Treatment.   A treatment process in which a tank or reactor is filled, the wastewater (or other solution) is treated or a chemical solution is prepared, and the tank is emptied. The tank may then be filled and the process repeated. Batch processes are also used to cleanse, stabilize or condition chemical solutions for use in industrial manufacturing and treatment processes.*

*Fig. 4.8   Inspector with industrial contact
checking clarifier for settled solids*

Fig. 4.9 Operator standing by below-ground clarifier

White tank is ammonia injection system for pH control.
Conduit leads to first stage to detect low pH and
signal addition of ammonia. Second conduit leads
to final pH probe that sends signal to effluent
pH recording chart.

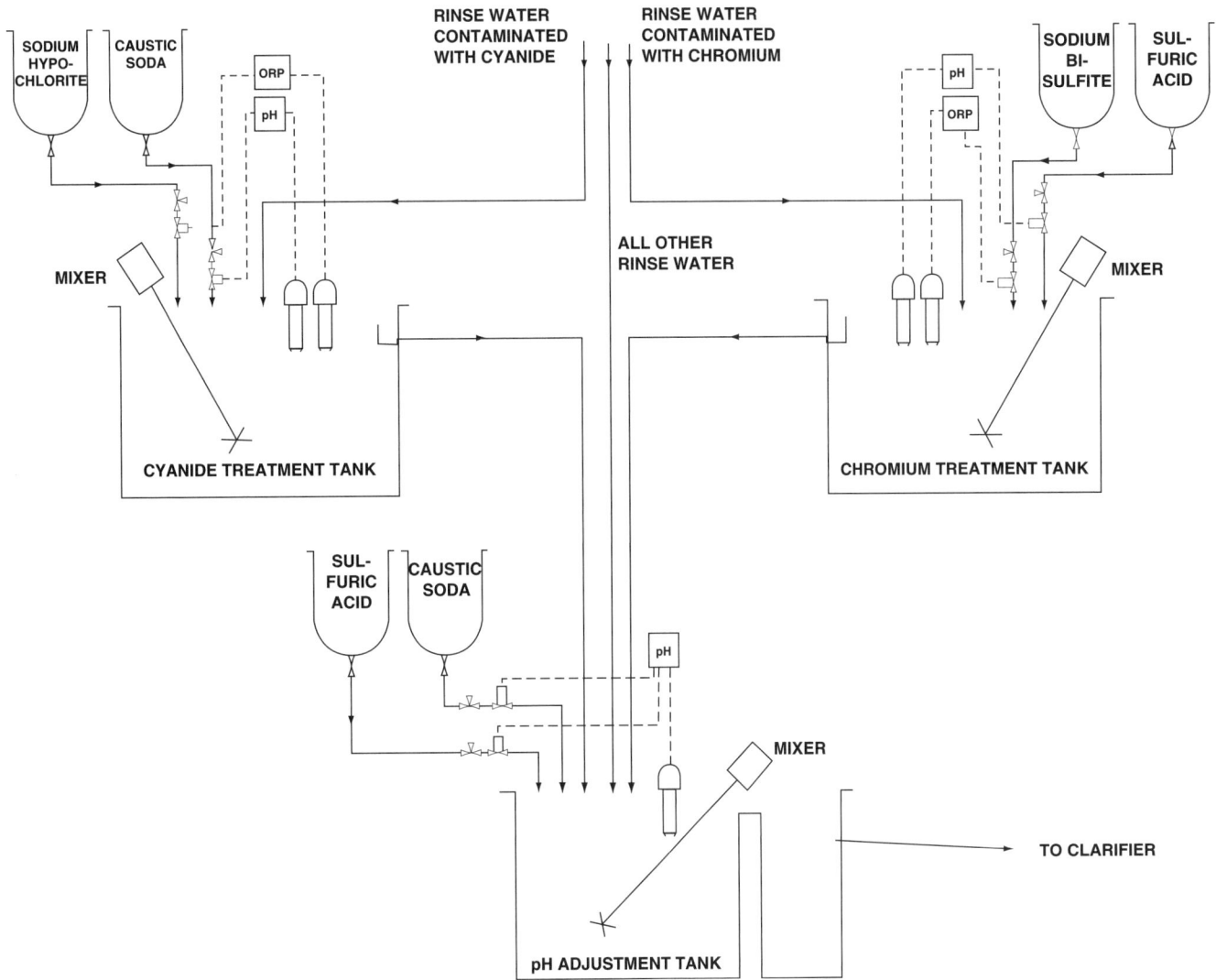

Fig. 4.10   Continuous treatment of cyanide and chromium
(Source: *UPGRADING METAL-FINISHING FACILITIES TO REDUCE POLLUTION*,
U.S. Environmental Protection Agency Technology Transfer Program)

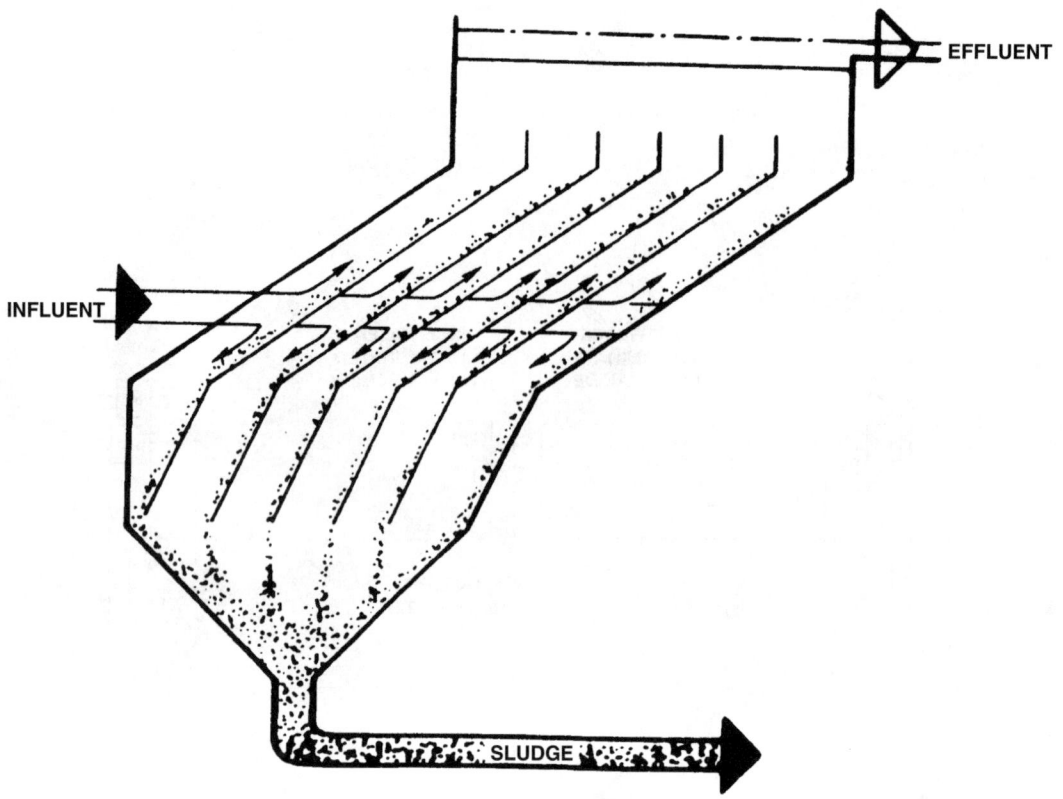

Fig. 4.11   Lamella thickener
(Courtesy of Parkson Corp.)

One of the items an inspector looks for is the existence of the potential to bypass the pretreatment system. As you walk through the pretreatment facility, try to determine by observation if flow streams remain constant through the facility. See if the flows are used in production processes or if the same flows appear to be leaving the facility. If the effluent flow appears lower than what was observed in the plant or what was entering the plant according to your observations or water meter data, then look for a bypass to somewhere.

Review the operating log and effluent test results to determine how well the system is performing. The time of day that effluent *GRAB SAMPLES*[9] are taken should be varied to account for any periodic fluctuations in wastewater quality.

Watch for additions to the system that are not shown on the permit. It has become increasingly common for companies to install groundwater cleanup systems where contaminated groundwater is pumped to a pretreatment system and then discharged to a sewer or storm drain or is reinjected. Even if the water is properly treated, the hydraulic impact on the sewer could be significant. If there is no effluent flowmeter, sewer-use fees could be greatly understated since the additional water to the sewer would not be recorded on the incoming water meter if this water meter is the only flow metering device installed.

QUESTIONS

Write your answers in a notebook and then compare your answers with those on page 344.

4.3A   What types of odors should an inspector be able to recognize?

4.3B   What types of final effluent meters are used to monitor industrial discharges?

4.3C   What is the purpose of a sand and grease interceptor?

4.3D   How can groundwater cleanup systems produce greatly understated sewer-use fees?

4.32   In-Plant Wastewater Control Equipment

Spill containment is one of the key safeguards that prevents slugs of concentrated wastes from overloading the final treatment system. Dikes and berms may be installed as permanent spill containment structures. Portable spill containers (Figure 4.12) offer flexibility in placement and are particularly useful for short-term storage of drums or other containers of dangerous materials. You may also encounter specialized containment devices such as the one shown in Figure 4.13. If your agency has written guidelines on the installation of spill con-

---

[9] *Grab Sample.   A single sample of water collected at a particular time and place which represents the composition of the water only at that time and place.*

*Fig. 4.12 Portable spill containment units protect
drums in open areas; the slots in the bottom
are for easy pick-up by forklifts.*

Fig. 4.13   Plastic spill containment barrier

tainment facilities, you should carry copies in your vehicle for distribution.

Check the structural integrity of dikes and look for poor bonding between berms and floors (Figure 4.14 on the next page). Floors in metal finishing plants should be epoxy coated for acid resistance. Within each spill containment area, check container labels (Figure 4.15) to verify that incompatible materials are not stored together. Tanks containing cyanide should *NOT* be stored in a common area with tanks containing acid. The mixing of cyanide salts with acid can generate deadly hydrogen cyanide gas.

Fig. 4.15   Inspector checks labels for incompatible substances

The facility's permit or ordinance may require the reporting of spills *BEFORE* any spilled liquids are discharged to the sewer; therefore, pumps that are activated automatically by liquid level should not be installed inside diked or bermed spill containment areas.

Any time that spills are delivered to the pretreatment system, an entry should be made in a logbook showing the date, time, operator, quantity and contents of the spill. Treatment procedures and doses of *REAGENTS*[10] should be spelled out for operators in a manual of procedures. Check for the potential to bypass the spill containment with valves, leaks or overflows. Look for floor drains that could convey spills to the POTW's collection system.

Substituting a nontoxic or nonwaste-generating process for conventional toxic waste-generating electroplating processes is the best way to control pollution. If toxic solutions must be rinsed from parts, there are several methods to reduce the volume of rinse water or to reduce the amount of plating solution dragged into the running rinse water in a plating shop. These are the kinds of improvements that should be suggested to contacts during the inspection. Realize however that to reduce the volume of rinse water without reducing the volume of drag out from the working bath that is carried into the running rinse will increase the concentration of plating solution in the final effluent. However, the addition of dilution water to meet discharge limits is illegal, and inspectors must be alert for this kind of activity.

One of the most common rinse water volume reduction techniques is to install counterflow rinsing. See Figures 4.16 and 4.17 for various rinsing configurations. Some of the other methods to reduce rinse water volume are to install conductivity- or pH-controlled rinse water activators and to use mist or spray rinses that are activated only when parts are rinsed. Air agitation in rinse tanks will increase the effective rinsing action and thus allow flow rates to be reduced.

The concentration of plating solutions in the running rinses can be reduced with the addition of spill guards between tanks, "air knives" that blow drag out back into the plating bath and static rinses that are returned to the bath. Allowing parts to momentarily "hang" over the plating bath after pulling them from the solution will allow drips to return to the bath.

Deionizers can be used to remove selected ions from rinse water. The inspector should inquire how and where deionizing resins are backwashed to strip off accumulated ions. There is always the danger that pollutants are being removed from one sewer only to be redeposited in another sewer either off site or in an area within the plant that bypasses the pretreatment system. If a commercial water conditioning service picks up the cylinders for backwashing at their own facility, they must be informed by the discharger that special precautions must be used to treat toxic metal-bearing resins.

As the price of metals increases and restrictions on the landfilling of toxic materials increase, the removal of metal from solution for reclamation becomes more commonplace. Evaporative recovery systems can be installed to return solution to the plating bath. Some facilities electroplate metals from the wastestream (Figure 4.18) or use cementation tanks (Figure 4.19, page 328) that precipitate metal ions onto scrap aluminum or steel. The collected metal can be sent to a smelter for recovery. As a product from the smelter, the discharger surrenders some of the responsibility for the material under the RCRA "cradle to grave" concept. Inspectors should regularly review current state and federal regulations to be aware of the existing situation in their area. The generator also escapes assessment for taxes levied on landfilled waste. Current state and federal regulations should be reviewed before recommending this procedure to generators of waste metals.

It is common to see preliminary chemical addition at the plating line way upstream of final effluent pretreatment equipment. This practice allows for selective treatment of a more concentrated wastestream with less interference from other streams. Sodium metabisulfite is sometimes added directly into the chromium rinse tank to reduce hexavalent chromium from the plating solution to trivalent chromium that can later be precipitated as chromium hydroxide in the general wastestream (see Chapter 9).

Metal-cyanide complexes must be broken before precipitation of the bound metal can be attempted. Poorly controlled oxidation of cyanide waste can lead to disastrous safety hazards. Improper methods during treatment can lead to evolution of deadly gases such as chlorine, hydrogen cyanide or the even more toxic waste-stream chloride. For the purposes of wastewater treatment, many electroplaters have installed a separate cyanide destruction unit process. This is a two-step process which includes first raising the pH to about 12, then adding an oxidizer such as chlorine gas or sodium hypochlorite to reduce cyanide to a harmless cyanate.

---

[10] *Reagent (re-A-gent).   A pure chemical substance that is used to make new products or is used in chemical tests to measure, detect, or examine other substances.*

*Fig. 4.14   Inspector in hard hat pointing to spill containment wall
that has been broken (portion of remaining wall is behind inspector).
Etched and stained concrete shows evidence of spill. Squeegee in
background was used to clean up spill. Industrial contact is
requested to repair spill containment wall as soon as possible.*

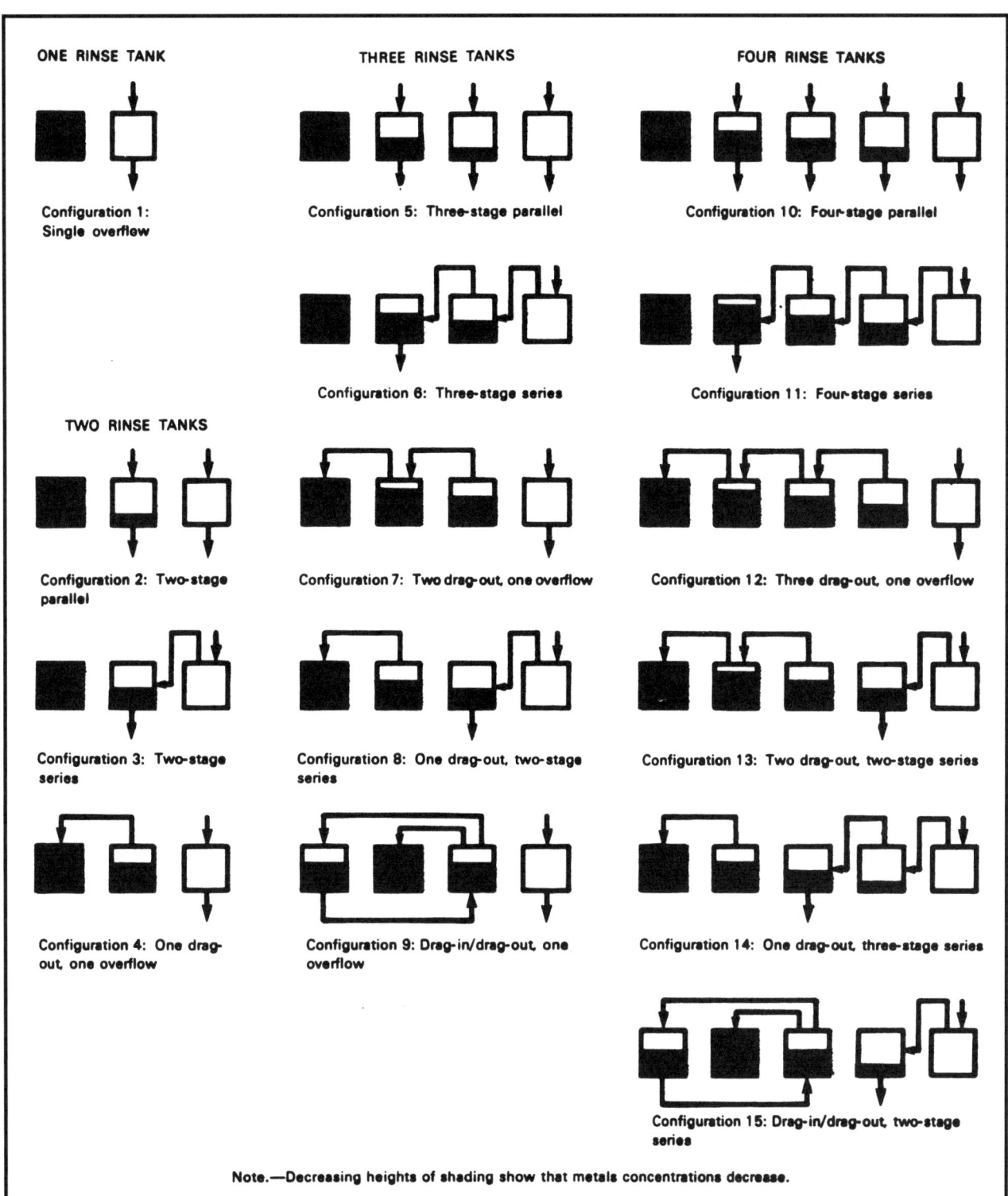

Fig. 4.16   *Various types of rinsing configurations*
(Source: *CONTROL AND TREATMENT TECHNOLOGY FOR THE
METAL FINISHING INDUSTRY, IN-PLANT CHANGES,* U.S.
Environmental Protection Agency, Washington, DC)

TOTAL RECOVERY

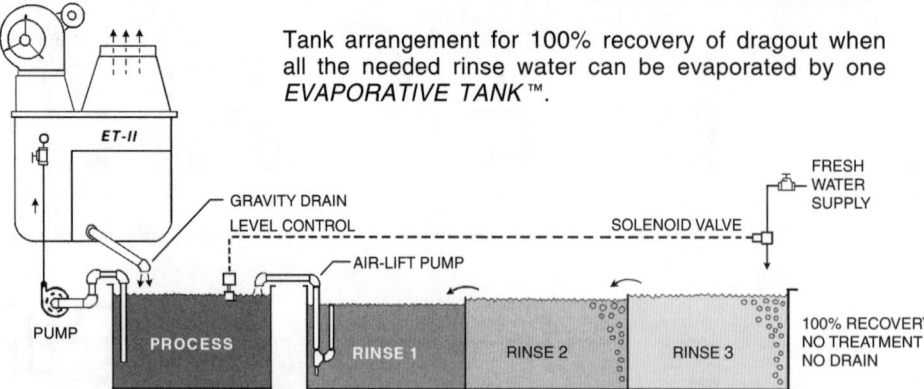

Tank arrangement for 100% recovery of dragout when all the needed rinse water can be evaporated by one *EVAPORATIVE TANK*™.

Rinse water and dragout counterflow through several rinses to the process tank. The excess water is removed from the process solution by the *EVAPORATIVE TANK*™. Heat for evaporation is taken from the process. A level control and solenoid valve can control the water flow into Rinse 3. This level control and air lift pump are available from P.P. Corp.

TOTAL RECOVERY, COLD PROCESS

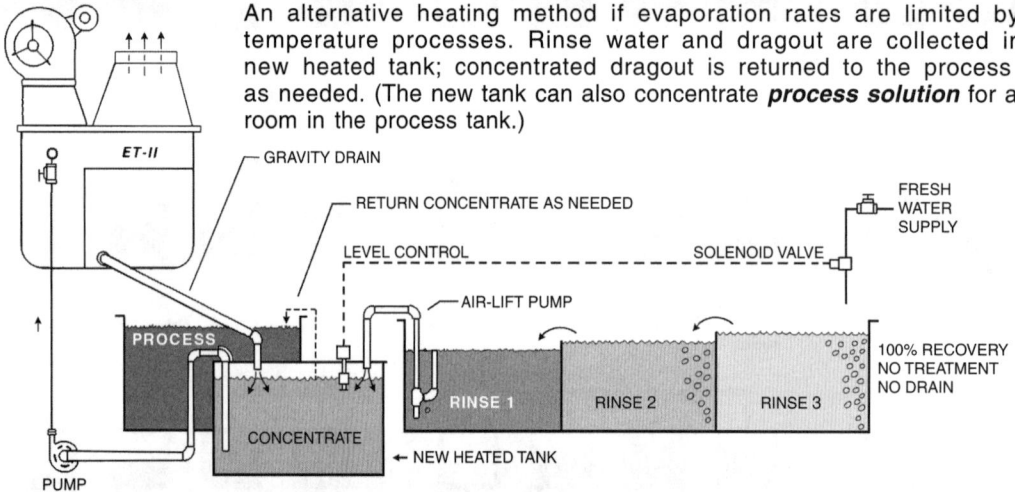

An alternative heating method if evaporation rates are limited by low temperature processes. Rinse water and dragout are collected in the new heated tank; concentrated dragout is returned to the process tank as needed. (The new tank can also concentrate **process solution** for added room in the process tank.)

NEAR TOTAL RECOVERY

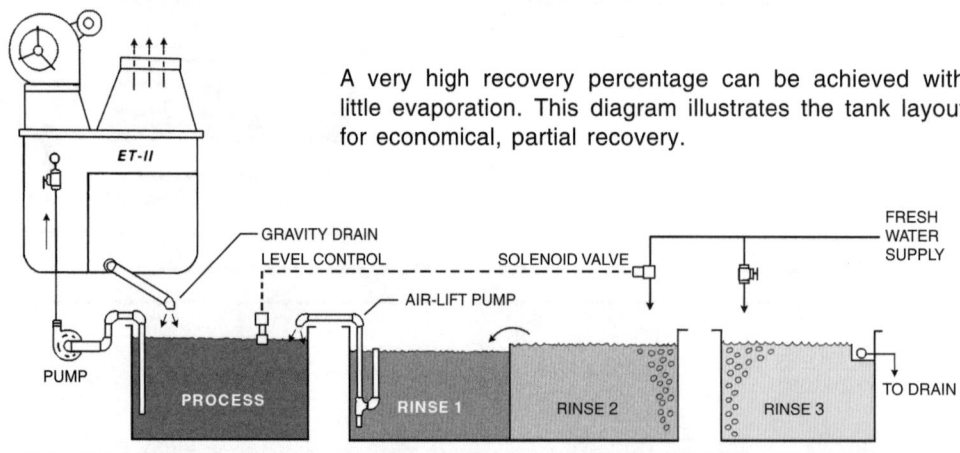

A very high recovery percentage can be achieved with little evaporation. This diagram illustrates the tank layout for economical, partial recovery.

Fig. 4.17   Rinsing system

(Permission of Poly Products Corporation)

Fig. 4.18 "Plate-out" unit being examined by inspector (in hard hat) and industrial contact. Unit is located on production area floor and receives concentrated first rinse from plating operation. Metal ions are electroplated onto bars to remove the metal ions from solution.

Proceed with caution where cyanide treatment is conducted in an enclosed area. OSHA requires that control meters be calibrated regularly and that operators be trained concerning the inherent dangers. Inspectors must verify that these requirements are being met. In addition to safety hazards that can result from poorly controlled systems, the actual oxidation of cyanide can be jeopardized. Some of the more tightly complexed metal cyanides (copper, nickel and silver) require very precise oxidizing conditions. With poorly controlled systems, these complexes probably will not be destroyed. Even with the proper pH and a positive test for excess chlorine, the cyanide might not be oxidized. When proper ORP values and pH range are maintained, the copper cyanide complex can be removed from wastewater. With poor control, inadequate treatment can be taking place and safety considerations become a serious concern.

The EPA categorical standards for metal finishers require that total toxic organics (including solvents) be controlled to a strict limit (see Chapter 3). With proper solvent recovery techniques in place, no significant amounts of solvents should reach the wastewater streams. Used solvents from vapor degreasers should be redistilled on site or captured for off-site reuse or disposal. The inspector should check the integrity of the solvent spill containment and review the hazardous waste manifests to verify that toxic waste material is being properly removed from the premises.

If rainwater is prohibited from being discharged to the sewer during a storm, then outdoor process areas must be roofed or a rain diversion system must be installed. Rain diversion hardware must be tested periodically to be sure that it will properly activate automatically when rain falls (see Figure 4.20). Typically, a small amount of rainwater is allowed to flow to the sewer (thus washing contaminated areas) before an

electrical, mechanical or air-driven valve or pump automatically diverts the rain to the storm drain or to an impound area for treatment and/or discharge after the storm is over. Impoundment capacity is usually dictated in the *Wastewater Ordinance* or the discharge permit. Before rainwater diversion equipment is installed, the sewer agency must review the proposal to be sure that the facility will not later be discharging process wastewaters with the storm water runoff to the storm drain or to the ground. Additional sewer-use fees may need to be levied if the impounded storm water is to be discharged to the sewer. The ordinance may also dictate that storm water may only be discharged during off-peak-flow hours at the POTW and only for a specified duration of time after the storm subsides.

If your *Wastewater Ordinance* prohibits the discharge of uncontaminated water, then process areas must be checked for equipment that is cooled with one-pass water. Cooling water should be recycled through an evaporative cooling tower or

Fig. 4.19   Cementation tank

"Cementation" of copper waste using scrap aluminum at a
low pH. Blue ionic copper waste is batch treated starting at the
left chamber and is allowed to sit for several days. Treated
liquid may need further final treatment to reach EPA
categorical limits. Solids are sent to reclaimer to remove
copper. No current is applied.

RAIN WATER DIVERSION SYSTEM

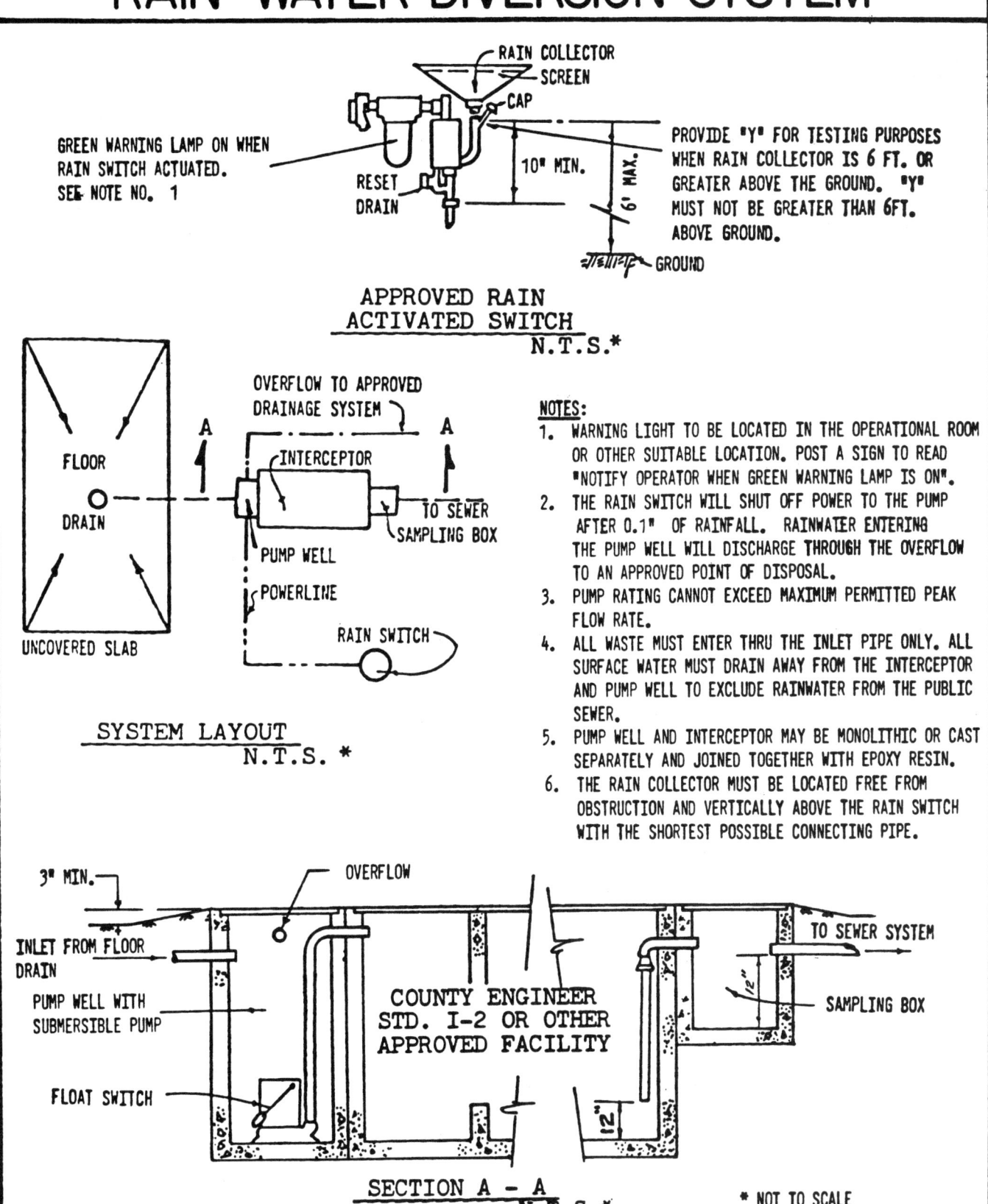

GREEN WARNING LAMP ON WHEN
RAIN SWITCH ACTUATED.
SEE NOTE NO. 1

RAIN COLLECTOR
SCREEN
CAP
RESET DRAIN
10" MIN.
6' MAX.
GROUND

PROVIDE "Y" FOR TESTING PURPOSES
WHEN RAIN COLLECTOR IS 6 FT. OR
GREATER ABOVE THE GROUND. "Y"
MUST NOT BE GREATER THAN 6FT.
ABOVE GROUND.

APPROVED RAIN
ACTIVATED SWITCH
N.T.S.*

FLOOR
DRAIN
UNCOVERED SLAB

OVERFLOW TO APPROVED
DRAINAGE SYSTEM
A   A
INTERCEPTOR
TO SEWER
SAMPLING BOX
PUMP WELL
POWERLINE
RAIN SWITCH

SYSTEM LAYOUT
N.T.S.*

NOTES:
1. WARNING LIGHT TO BE LOCATED IN THE OPERATIONAL ROOM
   OR OTHER SUITABLE LOCATION. POST A SIGN TO READ
   "NOTIFY OPERATOR WHEN GREEN WARNING LAMP IS ON".
2. THE RAIN SWITCH WILL SHUT OFF POWER TO THE PUMP
   AFTER 0.1" OF RAINFALL. RAINWATER ENTERING
   THE PUMP WELL WILL DISCHARGE THROUGH THE OVERFLOW
   TO AN APPROVED POINT OF DISPOSAL.
3. PUMP RATING CANNOT EXCEED MAXIMUM PERMITTED PEAK
   FLOW RATE.
4. ALL WASTE MUST ENTER THRU THE INLET PIPE ONLY. ALL
   SURFACE WATER MUST DRAIN AWAY FROM THE INTERCEPTOR
   AND PUMP WELL TO EXCLUDE RAINWATER FROM THE PUBLIC
   SEWER.
5. PUMP WELL AND INTERCEPTOR MAY BE MONOLITHIC OR CAST
   SEPARATELY AND JOINED TOGETHER WITH EPOXY RESIN.
6. THE RAIN COLLECTOR MUST BE LOCATED FREE FROM
   OBSTRUCTION AND VERTICALLY ABOVE THE RAIN SWITCH
   WITH THE SHORTEST POSSIBLE CONNECTING PIPE.

3" MIN.
OVERFLOW
INLET FROM FLOOR DRAIN
PUMP WELL WITH SUBMERSIBLE PUMP
FLOAT SWITCH
COUNTY ENGINEER
STD. I-2 OR OTHER
APPROVED FACILITY
TO SEWER SYSTEM
12"
SAMPLING BOX
12"

SECTION A - A
N.T.S.*

* NOT TO SCALE

Fig. 4.20   Rainwater diversion system
(Permission of Department of County Engineer —
County of Los Angeles, California)

holding pond. It can also be directed to some other operation for reuse. Cooling tower bleeds that flow to the sewer should be checked for potential overdoses of algicides, acid, chromate or other regulated compounds that might have been added to inhibit corrosion or scale in the tower.

The wastewater control techniques and equipment described in this section are the types of systems that an inspector is most likely to encounter. With regulatory emphasis shifting toward pollution prevention (as opposed to the current focus on control and treatment), many new approaches for reducing or eliminating contaminants can be expected. Pretreatment inspectors must continually seek out information about new technology and new applications and must keep abreast of changing regulations. Chapter 9, "Pretreatment Technology (Source Control)," contains additional information on equipment and methods for pollution prevention.

4.33   General Tour of Facility

After you have looked at the outfall area, the pretreatment equipment and all known wastewater generating processes, it is often worthwhile to tour the remainder of the facility. You should be looking for changes in the plant since the last permit submittal or the last time a wastewater survey was conducted. Take photographs where appropriate. Under rare circumstances an industry might refuse to allow photographs to be taken. Such a position could be justified to protect trade secrets or could be a requirement of a defense contract. Ask the company to express its concerns and justifications in writing for review by your supervisor.

A change in manufacturing process may change the parameters that should be tested at the outfall. A change in production rate may change the effluent limits if they are based on a "mass balance" equation or if the industrial user is regulated by mass-based standards found in the applicable EPA categorical point source standard. New structures may house nonpermitted wastewater generating processes.

Look for the existence of sources of wastewater not documented in the permit that could bypass the pretreatment system. Documented bypasses that carry wastewater which the permittee claims is uncontaminated by controlled substances should be field tested or periodically sampled for lab analysis and verification of its contents. Floor drains in chemical storage warehouses are typically not allowed due to the risk of

spills. If you discover any floor drains, they could be illegal sewer connections. Additional spill containment may need to be installed. Look for incompatible chemicals stored in a common area by reading the labels on the chemical drums. Storage areas for flammable chemicals should be checked for the existence of improper drains to the sewer.

Equipment installed to clean the air, the soil or the groundwater may involve a discharge to the sewer. Check air scrubbers and steam strippers for bleeds to the sewer (Figure 4.21). Well water extracted for cleanup with strippers or carbon adsorption units may be plumbed to the sewer for discharge. Also check the type of boiler and/or cooling tower maintenance chemical for regulated toxic components since most *BLOWDOWNS* [11] go to the sewer. Encourage the permittee to use the nontoxic maintenance chemicals which are now widely available.

It may become necessary to dye trace certain floor drains or undocumented connections in critical areas that you suspect might lead to the sewer (see Chapter 6 for technique). The permittee generally has an obligation to provide accurate information concerning the layout of the facility. All open floor drains in chemical processing or storage areas should be sealed or eliminated.

4.34   Summary of Inspection Items

This section lists the items or areas that should be considered during an inspection. Use this list for ideas to develop your own list of items to inspect when you actually inspect a particular industry. Chapter 10, "Industrial Inspection Procedures," contains more details for specific industries.

1. Inspect the production or manufacturing facility (all areas where wastewater is generated or produced).

    a. Review production operations and equipment.
    b. Review raw material storage.
    c. Review waste storage.
    d. Review spill containment facilities.
    e. Review air pollution devices.

2. Inspect the waste treatment facilities.

    a. Review treatment equipment and controlling devices (pH, ORP, and conductivity meters).
    b. Review waste storage.
    c. Review spill containment facilities.
    d. Review operator records on the units of operation at the facility.
    e. Check availability of procedures manual.

3. Inspect additional manufacturing areas; for example, heat treating, die casting, degreasing.

4. Inspect storage and maintenance areas.

    a. Review raw material storage (bulk storage and underground tanks).
    b. Waste 33material storage (barrel storage and underground tanks).
    c. Electroplating equipment maintenance:

        (1) Plating filters and spent cartridges.
        (2) Plating rack stripping facilities and plating barrel maintenance.

---

[11] *Blowdown.   The removal of accumulated solids in boilers to prevent plugging of boiler tubes and steam lines. In cooling towers, blowdown is used to reduce the amount of dissolved salts in the recirculated cooling water.*

*Fig. 4.21   Inspector checking fume scrubber
for pH of bleed-off that flows to sewer*

5. Inspect monitoring facilities.

   a. Collect wastewater samples.
   b. Review records of wastewater flowmeter calibration.
   c. Perform spot tests of wastestream if appropriate.

6. Records review.

   a. Review self-monitoring results of the industrial user.
   b. Review completed copies of the uniform hazardous waste manifests for all wastes hauled from the facility for disposal since last inspection.
   c. Review material safety data sheets on raw materials as necessary.

QUESTIONS

Write your answers in a notebook and then compare your answers with those on page 344.

4.3E   How can floors in metal finishing plants be made acid resistant?

4.3F   What should be an inspector's concern regarding the use of deionizers?

4.3G   Why must the oxidation of cyanide waste be carefully controlled?

4.3H   What provisions might be contained in a *Wastewater Ordinance* regarding the discharge of impounded storm water from an industrial site?

4.3I   How can an inspector determine if a floor drain leads to a sewer?

4.3J   List the items or areas that should be inspected during an inspection.

4.4   AFTER THE WALK-THROUGH

4.40   Paper Work Review

In addition to the documents or reference materials listed on page 303 of this chapter, it probably became apparent during your tour of the facility that other records need to be reviewed as well. Any sample results or other documents from your agency concerning the discharger should be presented to your contact unless court action is contemplated.

Records customarily kept in the permittee's files should be reviewed. Liquid waste haulers' manifests (Figures 4.22 and 4.23) should be checked to be sure that they are properly signed by a representative of an acceptable disposal facility and that they account for the approximate quantity of material that you estimate would be generated. Hazardous waste manifests document the shipping of hazardous waste, most of which is collected from the segregation of industrial waste or the solids gathered from the pretreatment system. Water bills will confirm that the volume of water consumption has not changed dramatically since the last permit revision. Also review any records from flowmeters used to measure wastewater discharged to the sanitary sewer.

Operations logs should be checked to see that discharges from spill containment areas and other batch discharges are being properly handled and recorded. POTWs should require industrial users to notify them of a spill. A review of the plating bath maintenance log may show excessive chemical use that would indicate the existence of a leak or spill. Calibration records for pH, ORP, temperature, explosivity, and flowmeters

will indicate the diligence exercised in maintaining the pretreatment equipment. Examine the industry's sampling records and chain of custody documentation. Review the list of chemicals stored and used at the facility and verify their locations and the availability of Material Safety Data Sheets (MSDSs).

Some states have reworded their wastewater treatment regulations with particular attention being given to industrial firms treating hazardous waste that they generate as a part of their licensed wastewater treatment system. The federal regulations (RCRA type) leave a loophole in that those so licensed do not have to provide all the training, inspection, and other activities that are required of companies that treat waste generated by others. In some states industrial firms falling into this category now have to meet the same management and operating procedural requirements as those who treat wastes generated by others.

Virtually all metal finishers and electroplaters who treat their wastewater, for either POTW or NPDES discharge, will be treating hazardous waste. Properly run wastewater treatment facilities now log these batches of hazardous wastes as separate identities. For example, alkaline cleaners are F009, acid pickles are D002, and chromate dips are D007. The treatment and disposal of these hazardous wastes should be recorded in either the wastewater treatment operations log or in a separate log devoted strictly to management of hazardous waste. In one state, these firms must also notify the local

POTW as to what hazardous wastes they are treating as a part of their wastewater treatment systems, how they are treating them, and the volumes involved. A large part of these treated hazardous wastes are being discharged to POTW wastewater collection systems. If a plater isn't treating these types of hazardous waste, then the plater might be dumping them indiscriminately and create surges (or slugs) that the normal treatment system for rinse waters will not handle. Obviously these hazardous wastes would not be dumped by a plater in that manner during effluent monitoring periods without being identified as being in violation of effluent limitations.

In some states, waste oil is considered a hazardous waste and must be disposed of under the manifest system. If the metal finisher or electroplater uses oil skimming systems to prolong the lives of process solutions, and other means to keep excessive amounts of oil out of the treated wastewater, then one would expect the manifest records to show that oil

State of California—Health and Welfare Agency
Form Approved OMB No. 2050—0039 (Expires 9-30-88)
Please print or type. *(Form designed for use on elite (12-pitch typewriter).*

Department of Health Services
Toxic Substances Control Division
Sacramento, California

UNIFORM HAZARDOUS WASTE MANIFEST	1. Generator's US EPA ID No. CAD056158573 08001	Manifest Document No.	2. Page 1 of 1	Information in the shaded areas is not required by Federal law.

3. Generator's Name and Mailing Address
ABC PLATING CO., INC
297 E WHITEWOOD AVE   90040
LOS ANGELES, CA

A. State Manifest Document Number
87388197

4. Generator's Phone (213) 512-6011

B. State Generator's ID
HAH036007444

5. Transporter 1 Company Name
URBAN DISPOSAL SERVICE
6. US EPA ID Number
CAD000389034

C. State Transporter's ID 709601
D. Transporter's Phone 213-590-8001

7. Transporter 2 Company Name
8. US EPA ID Number

E. State Transporter's ID
F. Transporter's Phone

9. Designated Facility Name and Site Address
NATIONAL RESOURCES CO
2019 W. NORMAN ST
PHOENIX, AZ 85043
10. US EPA ID Number
AZD917219131

G. State Facility's ID
H. Facility's Phone 602-233-3144

11. US DOT Description (Including Proper Shipping Name, Hazard Class, and ID Number)

		12. Containers No. / Type	13. Total Quantity	14. Unit Wt/Vol	I. Waste No.
a.	HAZARDOUS WASTE SOLID NOS ORM-E NA 9189	001 CM	110	T	State 171 / EPA/Other F006
b.					State / EPA/Other
c.					State / EPA/Other
d.					State / EPA/Other

J. Additional Descriptions for Materials Listed Above
WASTE TREATMENT SLUDGE CONTAINING HEAVY METALS

K. Handling Codes for Wastes Listed Above
a. 01   b.
c.   d.

15. Special Handling Instructions and Additional Information
WEAR SAFETY GLASSES AND GLOVES

16. **GENERATOR'S CERTIFICATION:** I hereby declare that the contents of this consignment are fully and accurately described above by proper shipping name and are classified, packed, marked, and labeled, and are in all respects in proper condition for transport by highway according to applicable international and national government regulations.

If I am a large quantity generator, I certify that I have a program in place to reduce the volume and toxicity of waste generated to the degree I have determined to be economically practicable and that I have selected the practicable method of treatment, storage, or disposal currently available to me which minimizes the present and future threat to human health and the environment; OR, if I am a small quantity generator, I have made a good faith effort to minimize my waste generation and select the best waste management method that is available to me and that I can afford.

Printed/Typed Name JOSEPH NORRIS	Signature *Joseph Norris*	Month Day Year 09 03 99

17. Transporter 1 Acknowledgement of Receipt of Materials

Printed/Typed Name GEORGE TUCKER	Signature *George Tucker*	Month Day Year 09 03 99

18. Transporter 2 Acknowledgement of Receipt of Materials

Printed/Typed Name	Signature	Month Day Year

19. Discrepancy Indication Space

20. Facility Owner or Operator Certification of receipt of hazardous materials covered by this manifest except as noted in Item 19.

Printed/Typed Name LEO TOSER	Signature *Leo Toser*	Month Day Year 09 04 99

DHS 8022 A (1/87)
EPA 8700—22
(Rev. 9-86) Previous editions are obsolete.

White: TSDF SENDS THIS COPY TO DOHS WITHIN 30 DAYS
To: P.O. Box 3000, Sacramento, CA 95812

INSTRUCTIONS ON THE BACK

Fig. 4.22   Liquid waste hauler's manifest

IN CASE OF AN EMERGENCY OR SPILL, CALL THE NATIONAL RESPONSE CENTER 1-800-424-8802; WITHIN CALIFORNIA CALL 1-800-852-7550

GENERATOR
TRANSPORTER
FACILITY

NON HAZARDOUS WASTE DATA FORM

TO BE COMPLETED BY GENERATOR

NAME __BLUE CROSS LABS__

ADDRESS __26411 GOLDEN VALLEY RD.__

EPA I.D. NO. | | | | | | | | | | | |

CITY, STATE, ZIP __SAUGUS, CA 91350__

PHONE NO. (__(805)255-0955__

CONTAINERS: No. __1__   VOLUME __4000 GAL__   WEIGHT _____

TYPE: ☒ TANK TRUCK   ☐ DUMP TRUCK   ☐ DRUMS   ☐ CARTONS   ☐ OTHER _____

WASTE DESCRIPTION __BLUE WATER__   GENERATING PROCESS __WASH DOWN__

COMPONENTS OF WASTE	PPM	%		COMPONENTS OF WASTE	PPM
1. __WATER__		__99__ %	5.		
2. __BLUE DYE__		__.5__ %	6.		
3. __PINE OIL__		__.5__ %	7.		
4.			8.		

PROPERTIES: pH _____ ☐ SOLID ☒ LIQUID ☐ SLUDGE ☐ SLURRY ☐ OTHER _____

HANDLING INSTUCTIONS: __GLOVES, GOGGLES. / COUPON# 31818.__

THE GENERATOR CERTIFIES THAT THE WASTE AS DESCRIBED IS 100% NON-HAZARDOUS.

Hernestor Leal   Ernesto Leal / 01-09-99
TYPED OR PRINTED FULL NAME & SIGNATURE   DATE

TRANSPORTER

NAME __MARTIN IND. PUMPING, INC.__

EPA I.D. NO. C A D 0 0 0 6 2 8 6 3 6

ADDRESS __P.O. BOX 1128__

SERVICE ORDER NO. __20746__

CITY, STATE, ZIP __CANYON COUNTRY, CA 91351__

PICK UP DATE __01-09-99__

PHONE NO. __(805) 251-3737__

TRUCK, UNIT, I.D. NO.

MIKE Hyden Mike Hyd   _/ 01-09-99_
TYPED OR PRINTED FULL NAME & SIGNATURE   DATE

TSD FACILITY

NAME __J.W.P.C.P.__

EPA I.D. NO. N/A | | | | | | | | | |

ADDRESS __24420 S. FIGUEROA ST.__

DISPOSAL METHOD
☐ LANDFILL   ☒ OTHER _____

CITY, STATE, ZIP __CARSON, CA 90710__

PHONE NO. __(213)775-2351__

J. Schroeder   _/ 01-10-99_
TYPED OR PRINTED FULL NAME & SIGNATURE   DATE

GEN		OLD/NEW	✓	A	TONS	
TRANS			S	B		
C/Q		RT/CD		HWDF NONE	DISCREPANCY	

Fig. 4.23 Nonhazardous waste manifest

was shipped from the plant. If the records do not show oil being shipped, then where is the oil?

4.41   Discussion of Inspection

After the inspection has been completed, discuss the findings and results with your contact. If there were no problems or serious concerns, the final discussion can be short and cordial. Thank the contact for the industry's cooperation and satisfactory performance. The following sections provide information on how to review with your industry contact violations, enforcement and complaints.

4.42   Discussion of Violations

Any notes you take describing violations or any explanations offered by your contacts concerning violations should be recorded in a manner that would be suitable for generating a report that could be presented in court. Any direct quotes should be recorded exactly as originally stated. Know the identities and titles of all your contacts.

Start the discussion with an explanation of the regulations. Present documents supporting your position, if possible. Mention other pollution control agencies' concerns and explain your obligation to report any significant violations of their standards discovered during the tour or revealed from past sampling results. Be very careful when discussing the rules and regulations of other agencies if you are not an expert in these areas because you could make incorrect demands on an industry. Do not assume that your contacts are familiar with performance or reporting requirements. Try to determine what caused the violation. Was there a known malfunction or should new pretreatment equipment be proposed?

Discuss what corrective action is required. If the violation was corrected during the inspection, is there any further action required from the permittee? Some minor corrections such as pH adjustment, plumbing repairs, sensor calibrations or rain diversion system repairs might be performed in your presence.

Discuss industrywide remedies that have been successful for solving similar problems, but make it clear that it is up to the permittee to propose the method of correction. If you carry lists of locally available services that could benefit the discharger, make these known but be careful not to show favoritism or offer endorsements.

Serious violations such as the discharge of flammables or a high volume of extremely concentrated toxic material to the sewer may require immediate major modifications that could include cessation of discharge. Most dischargers become exceptionally cooperative under emergency upset conditions but you should know what remedies you have within your power in case cooperation is not forthcoming. If you detect a major violation that could seriously impact the sewer system, you

should notify your home office immediately as well as the supervisors of the sewer maintenance crews and treatment plant operators so that diversions or other emergency action can be initiated (see Chapter 11). Do not go on to any other duties until the immediate danger has passed.

Negotiate a schedule and items for the completion of improvements to bring the facility into compliance. Also alert the facility of the reporting requirements in all compliance schedules. (In some POTW pretreatment programs, the inspector does not have the authority to negotiate a schedule for compliance with an industrial user.) If additional equipment or significant modifications are required, formal submittals will probably need to be made to your office before construction is begun. The construction schedule would be determined after review of the formal proposal. Minor improvements not requiring submittal of information should be verified by you in a follow-up inspection at a specified date.

4.43   Enforcement

All POTWs with approved pretreatment programs are required to develop an Enforcement Response Plan which describes how the POTW will investigate and respond to instances of industrial user noncompliance. As described in Chapter 3, each agency's enforcement program can be unique and will have its own schedule of escalating administrative actions. Study and understand your agency's policies regarding enforcement since different agencies have different rules, procedures and policies.

Enforcement actions can include verbal warnings, telephone calls, various levels of written notices of violations, letters explaining violations, administrative orders, compliance conferences, fines, reimbursements of agency costs and actions preliminary to the revocation of the discharge permit such as "show cause" hearings and permit suspensions. Criminal or civil action can be sought if administrative actions do not bring satisfaction. Cease discharge orders and permanent permit revocations may also be used.

If a permittee fails to correct a violation after a verbal request to comply or if the violation falls into a category that your agency defines as a citeable offense, the inspector may need to issue a written notice of violation. When issuing a citation (see Figure 4.24 for completed sample form) it is preferable to have the signer of the permit also sign the citation. If that person is unavailable, then a responsible corporate official would be second choice. The third choice would be anybody you can contact. Citations are not always written on site, but may follow shortly by registered mail with a return receipt. If nobody agrees to sign the citation, deliver it to the facility with a notation identifying the person who received it and write "refused to sign" in the space that normally would be signed. A

SANITATION DISTRICTS OF LOS ANGELES COUNTY

ATTENTION INDUSTRIAL WASTE SECTION
1955 WORKMAN MILL RD., P.O. BOX 4998, WHITTIER, CALIFORNIA 90607

NOTICE OF VIOLATION

No. V 5276

1. DISCHARGER

MURDOCK PLATING CO., INC.

2. ADDRESS OF WASTEWATER DISCHARGE

316 E. BOLSTER AVE
LOS ANGELES, CA 90040

3. LOCAL AGENCY

DEPT. OF PUBLIC WORKS

4. TIME OF VIOLATION (Date, Hour)

10-14-99   8:30

5. PERMIT NO. 14911

6. ACCT. NO. 193421

7. INSP. AREA 304

8. VIOLATION OF THE DISTRICTS, WASTEWATER ORDINANCE, SECTION 406C, 406Y **CONCERNING**

406C - DISCHARGE OF ANY WASTE HAVING A PH LOWER THAN 6.0
(FIELD PH = 4.5 BY PH PAPER METHOD 9041A)

406Y - DISCHARGE OF ANY WASTE CONTAINING DISSOLVED
SULFIDE GREATER THAN 0.1 mg/L
(FIELD DISSOLVED SULFIDE = 0.9 mg/L BY AMEROY TEST KIT)

9. IMPORTANT: VIOLATION MUST BE CORRECTED BY: IMMEDIATELY
(DATE)

10. RECEIPT OF NOTICE ACKNOWLEDGED BY DISCHARGER

MATTHEW GORDEN
PRINTED NAME

Matthew Gordon
(SIGNATURE)

V.P. - OPERATIONS
TITLE

FORM NO. 5020

CHARLES W. CARRY

CHIEF ENGINEER AND GENERAL MANAGER

BY: Maurice Nester   10-14-99
    (NAME)   (DATE)

I. W. INSPECTOR
(TITLE)

Fig. 4.24   Citation (Notice of Violation)
(Permission of Sanitation Districts of
Los Angeles County, California)

certified "return receipt requested" follow-up enforcement letter should be mailed to the permittee after each citation is issued with copies to appropriate interested parties. If at all possible make sure that the person served with the citation is the person responsible under the local ordinance and that you can testify in court that the persons who receive citations identify themselves as the responsible person.

4.44 Discussion of Complaints

If your inspection was undertaken as a result of a complaint filed by a citizen or agency, you will need to investigate the specifics of the complaint during the inspection. The source of the complaint or even its existence may be confidential. Complaints could come from in-sewer surveillance monitoring discoveries, other agencies' referrals, company informants, neighborhood reports of odors, discharges or explosions or many other sources. If the complaint came from another agency, try to read the full report before visiting the site. Always follow your agency's policy regarding release of information that you have acquired during an inspection.

4.45 Answering Questions

Finish your inspection by asking if your contact has any questions concerning the inspection. Avoid passing uncertain information; offer to research the question and call back with a definitive answer. Avoid participating in interagency squabbles. Collect information from all possible sources and verify conflicting statements by contacting the parties involved from your office.

Follow your agency's guidelines concerning approval of proposals from your contact. You can describe typical systems that have been approved but avoid making field approvals. You are not doing the discharger any favors by short-circuiting the review process and you could be contributing to the delay in finding an acceptable resolution of the problem. Request the discharger to submit plans, treatment processes, modifications and flow diagrams for approval. Notify the discharger of approval in principle in writing. Any system designed by a consultant should include a guarantee that the facility will meet design concentrations and effluent limitations.

QUESTIONS

Write your answers in a notebook and then compare your answers with those on pages 344 and 345.

4.4A List the facility's records that should be reviewed during or after an inspection.

4.4B Why should a facility's water bills be reviewed by an inspector?

4.4C How should a serious violation such as the discharge of flammables to a sewer be handled?

4.4D Who should sign a citation?

4.4E What types of complaints might cause you to undertake an inspection?

4.4F How should an inspector respond to a discharger's request for approval of pretreatment facility modifications?

4.5 REPORT WRITING

4.50 Routine Reports

A typical inspection report form that could be used to input data into a computerized summary of inspection results is shown in Figure 4.25. Space for comments is limited and there is no provision for attaching drawings, photos or supporting documents. For more detailed types of report forms, see the Appendix at the end of this chapter and also Chapter 10, "Industrial Inspection Procedures." Care must be taken to record the most critical comments. Make reference to supporting pieces of information that might be stored in other files such as sample data banks, permit files, and treatment plant upset logs.

Inspectors should retain their personal notebooks that were used in the field to record observations. These notes are valuable in reconstructing events, especially if testimony is to be given in court. Names of contacts, quotations, time of day for various events, chart readings, field test kit results and other observations should be entered in your notebook. Items that should appear in an inspection report include the following data and observation summaries:

1. Name of company official contacted,

2. Date and time of inspection,

3. Inspection of manufacturing facility,

4. Inspection of chemical storage areas,

5. Evaluation of hazardous waste generation,

6. Inspection of spill prevention and control procedures,

7. Inspection of pretreatment facilities,

8. Inspection of sampling procedures,

9. Inspection of lab procedures, and

10. Inspection of monitoring records.

Routine reports should be filed in the POTW's permanent file for the particular industry and available to all members of the POTW staff.

4.51 Special Investigation Reports

Sometimes there is too much information to fit into the standardized report form or the importance of an investigation merits a special report. Always assume that your special report will be read by someone outside your agency or by a reader who is only generally familiar with your organization. Start with a statement explaining why the report is being written. Include exact dates and times of all events. Contacts must be identified and quotations must be exact. Describe your handling of evidence and state that you followed prescribed chain of custody procedures (see Chapter 6). Give sufficient background information so that your report will "stand on its own." Refer to supporting documents if the issues are complex. Finalize your report with a recommendation for further action or a conclusion. Make sure that the report is appropriately circulated.

INDUSTRIAL WASTE INSPECTION REPORT - C#IWI810

TO: Data Control Group (Data Processing Section)   DATE: / /
FROM: George Mitchell   DEPT./SECTION: Industrial Waste   EXT: 276
REQUEST DESCRIPTION:
    Report IWI810 - Inspection File Update   * JOB I.D.:   C#IWI810 *
WHEN NEEDED: ASAP |X|

New Record - N   Account   I.W. Permit No.   Date   Visit
Change Information - C   No.   (Insert   (Mo,Day,Yr)   No.
                                       Leading Zeros)

|N|

1 Name of Firm

2 Street No. Direction   Name   Type
   City   Zip Code

3 Person to Contact   Telephone No. District No. SIC No.
   Thomas Map: Inspection   Inspector   Inspection Time
   Page - Area   Area   (Last Name)   Hr   Min

4 No. Employees Shifts Days/Wk Water Use   Flow(GPM) Type Flow
                                    (100 CF/M)   (C,D,I)
   Waste Charact.   Violations   Corrections   Special Message

5 Working Hours(0000 - 2359)
   Start
   Stop
   Monday Tuesday Wednesday Thursday Friday Saturday Sunday

6 Comments:

7 Comments:

8 Comments:

9 Waste Characteristics Comment

FORM C#IWI810

Fig. 4.25 Computer entry report form
(Permission of Sanitation Districts of
Los Angeles County, California)

Figure 4.26, "Investigation Report," and Figure 4.27, "Pollution Complaint," are examples of standardized forms used to report POTW or sewer system upsets. These forms encourage the investigator to include pertinent background information such as the source of the complaint, the location of the upset, characteristics of the upset, and what sources were investigated. When completing these reports, describe the strategy of the investigation. Include results of checks made at manholes and treatment plants. Record observations made by other agency personnel such as treatment plant operators, sewer maintenance workers and laboratory personnel. Review past POTW "upset" reports to determine if there is a correlation between the current condition and previous incidents. Once you have completed all of these activities, circulate the results of your investigation and invite comments from others involved in investigative work.

4.52 Enforcement Reports

The issuance of a citation or administrative action should always generate a report and a letter to the permittee. The standardized enforcement report shown in Figure 4.28 includes the basic elements of an enforcement action report from an inspector. The details of the violation must be described in the report. The cause of the violation, corrective action taken or proposed, the inspector's observations and recommendations as well as the contacts' reaction to the citation should be included in the report. Some of the most frequent reasons for citations to be issued are:

- Discharge of waste not allowed by the *Ordinance* (toxics),
- Discharge of waste not allowed by the permit (excessive flow or loading, oil, rain, incorrect pH, etc.),
- Violation of EPA categorical standards,
- Violations of local prohibited discharge limits,
- Failure to install and/or maintain required equipment,
- Failure to submit requested information,
- Failure to pay sewer-use fees, and
- Discharge without a valid permit.

An enforcement letter should be written after each citation is issued. The principal reasons for writing a citation follow-up letter are: (1) to be sure that the signer of the permit or the company's upper management is aware that the violation occurred, and (2) to inform the permittee how the permittee is expected to correct the violation. A copy of the citation should be enclosed. It is advisable to develop form letters and have them reviewed by the agency's legal counsel. Always send enforcement letters "return receipt requested" so you can document delivery and receipt.

Compliance meeting follow-up letters become important documents that outline what was agreed upon during the meeting. Start the letter by listing those who were in attendance and when and where the meeting was held. State exactly why the meeting was held, referring to any citations that preceded the most recent action. Briefly list the topics that were discussed during the conference. Describe the corrective action that was agreed upon and give a timetable for the

completion of corrective action. Give the permittee an idea of what the next step in enforcement activity will be if corrective action is not completed but avoid threatening statements. If this meeting was step number four in a five-step procedure to suspend the permit then make that clear. Avoid "surprises" in the letter by covering all topics thoroughly during the compliance meeting.

QUESTIONS

Write your answers in a notebook and then compare your answers with those on page 345.

4.5A What assumption should an inspector make regarding the reader of a "special report"?

4.5B How should a "special report" end?

4.5C What information should be included in a compliance meeting follow-up letter?

4.6 ADDITIONAL INFORMATION

For additional information on inspection of industries see:

1. APPENDIX at the end of this chapter, "Seattle Metro Industrial Waste Inspection Procedures," for an example of how one POTW applies the information in this chapter; and

2. Chapter 10, "Industrial Inspection Procedures," for detailed procedures on how to inspect certain industries.

4.7 ADDITIONAL READING

1. *OCCUPATIONAL SAFETY AND HEALTH GUIDANCE MANUAL FOR HAZARDOUS WASTE SITE ACTIVITIES.* Obtain from the U.S. Government Printing Office, Superintendent of Documents, PO Box 371954, Pittsburgh, PA 15250-7954. Order No. 017-033-00419-6. Price, $14.00.

2. *INDUSTRIAL USER INSPECTION AND SAMPLING MANUAL FOR POTWs,* April 1994. Office of Water (4202), U.S. Environmental Protection Agency, Washington, DC 20460. Obtain from National Technical Information Service (NTIS), 5285 Port Royal Road, Springfield, VA 22161. Order No. PB94-170271. EPA No. 831-B-94-001. Price, $67.50, plus $5.00 shipping and handling per order.

Please answer the discussion and review questions before continuing with the Objective Test.

TREATMENT PLANT/SEWER INCIDENT INVESTIGATION REPORT

I. W. Inspector(s): _____   Date: _____

Incident Reported by(Name/Title): _____   Date/Time: _____

Type of Incident: Treatment Plant I___I   Sewer I___I   No Discharge I___I

Nature of Incident: _____

Name of Treatment Plant **or** CSD Trunk Sewer and M.H. No. **or** Local Sewer: _____

Your observations and findings at the site of an incident: *(Include time of arrival at the site, duration of an incident, agencies present at the site, persons contacted, visual observations, test(s) conducted, effect on plant operations (Turbidity, aeration D.O.), did WRP divert flow? follow up investigations and recommendations, etc.)* _____

_____
_____
_____
_____
_____
_____
_____
_____
_____
_____

Source Identified: *(State if citation is issued. If the source is not positively identified, indicate the most likely source, if any, with an explanation)* _____

_____
_____
_____
_____
_____

RESULTS OF TEST KIT ANALYSIS

Sample Location	Analysis Performed	Result
_____	_____	_____
_____	_____	_____

SAMPLES SUBMITTED TO THE LAB

Sample Location	Lab Job No.	Analysis Requested
_____	_____	_____
_____	_____	_____

Fig. 4.26   Investigation report
(Permission of Sanitation Districts of Los Angeles Country)

NOTE: Back of form provides space for inspector to write a brief summary
of each company inspected and note any additional comments.

P.C.# _____
(For office use only)

CITY OF PORTLAND
Bureau of Environmental Services

POLLUTION COMPLAINT

PLEASE PROVIDE THE FOLLOWING INFORMATION AND CHECK THE CORRECT BOX

DATE: __/__/__ TIME: ___:___ a.m./p.m.

TAKEN BY: _____

ADDRESS/SITE:

_____
_____
_____

WHO? WHERE? HOW?

REPORTED BY:

☐ CALLER WOULD LIKE TO REMAIN ANONYMOUS

NAME _____
ADDRESS _____
TELEPHONE #_____

☐ TELEPHONE   ☐ LETTER
☐ IN PERSON   ☐ OTHER _____

POLLUTION TYPE:

☐ SURFACE WATER   ☐ OIL/FUEL
☐ STREET   ☐ SOLID
☐ GROUND   ☐ HAZARDOUS/TOXIC
☐ SEWER   ☐ SEWAGE
☐ OTHER _____   ☐ OTHER _____

POLLUTANT SOURCE:

☐ INDUSTRIAL
☐ COMMERCIAL
☐ RESIDENTIAL

WHO? WHAT? WHERE? WHEN? HOW?

DESCRIPTION:

_____
_____
_____
_____
_____
_____
_____
_____
_____

FORWARD TO SR/PP EMERGENCY NOTIFICATION DESK

ASSIGNED TO: _____

☐ INVESTIGATOR
☐ PERMIT MANAGER

REFERRED TO: _____

☐ DEQ   ☐ B.O.M.   ☐ U.S.C.G.   ☐ S.C.D.
☐ NUISANCE   ☐ SANITATION   ☐ OTHER _____

COMPLAINANT UPDATE:

☐ YES   ☐ NO

IF YES →

HOW?:

☐ TELEPHONE   ☐ IN PERSON   ☐ LETTER

ATTACH ADDITIONAL SHEETS AS NECESSARY

Fig. 4.27 *Pollution complaint*
(City of Portland)

I.W. Permit No. _____

Account No. _____

Memo Date _____

MEMO TO: Margaret H. Nellor
            Head, Industrial Waste Section

THROUGH: William C. Garrett
            Supervising Industrial Waste Inspector

FROM: _____

SUBJECT: _____ NOTICE, for: _____

Issued To: Mailing Address:

_____   _____

_____   _____

_____   _____

_____ Notice No. _____ was issued on _____ to subject company for violation of the

Sanitation Districts' Wastewater Ordinance, section(s): _____

_____

_____

The issuance was initiated by I.W. Inspector yes ____ no ____ or requested by _____.

The basis of the citation was _____

The notice was signed by:

(name & title) _____ , _____

of the subject company. A correction date of _____ was given on the notice. The notification

letter should be sent to the attention of:

(name & title) _____ , _____ .

(Comments - brief description of wastewater producing operations and treatment system, flow rate, cause of violation, corrective action, recommendations, observations & reaction of company to notice.)

_____

_____

_____

_____

Mark and add initial as appropriate:

cc: BG |X|, ENF |X|, CPL/Proj Eng ____ | |, JDK/Proj Eng ____, | | EE/Surch Asst ____ | |

Fig. 4.28 Standardized enforcement report
(Permission of Sanitation Districts of
Los Angeles County, California)

DISCUSSION AND REVIEW QUESTIONS

Chapter 4. INSPECTION OF A TYPICAL INDUSTRY

Write the answers to these questions in your notebook before continuing with the Objective Test on page 345. The purpose of these questions is to indicate to you how well you understand the material in the chapter.

1. How frequently should a particular industry be inspected?

2. Why are the larger staffs of industrial waste inspectors sometimes divided into specialized subgroups?

3. Why should the assignments of pretreatment inspectors be rotated on a regular basis?

4. What are the major items an inspector must consider before entering an industry to conduct an inspection?

5. How should a receptionist, security guard or office manager be approached when attempting to conduct an inspection?

6. How can an inspector help an industrial facility solve technological difficulties or find solutions to long-standing problems?

7. How can an inspector handle a belligerent industrial contact?

8. Why should an inspector always visit the outfall whenever visiting an industry?

9. What enforcement stance should be taken when significant violations are documented on required effluent recording instruments?

10. What should an inspector look for when inspecting a spill containment facility?

11. What items should be covered when discussing violations with an industrial contact?

12. Why should inspectors retain their personal notebooks that are used in the field to record observations?

SUGGESTED ANSWERS

Chapter 4. INSPECTION OF A TYPICAL INDLUSTRY

Answers to questions on page 302.

4.1A Before starting an inspection, ask your industrial contact if there are any special safety precautions or company rules and regulations that must be followed.

4.1B Industries that can seriously impact the wastewater treatment plant include metal finishing, oil refineries, chemical manufacturing plants, and centralized waste treatment facilities.

4.1C An inspector might have to investigate sewer lines as an emergency response to catastrophic incidents such as discovery of explosive conditions or a stoppage causing wastewater to back up into residences or flow onto streets. Inspectors may also be asked to investigate long-term problems such as excessive scaling or corrosion.

4.1D Events that might necessitate a follow-up inspection include:

1. Equipment malfunctions noted during an inspection,

2. Enforcement actions,

3. Requests to install or modify pretreatment equipment,

4. Submittal or non-submittal of requested information,

5. Compliance monitoring reports showing violations,

6. Unusual water losses claimed on sewer-use fee calculations, and

7. New or revised permit applications.

Answers to questions on page 306.

4.2A   An inspector needs to carry equipment on an inspection to aid in gathering the necessary information.

4.2B   An inspector can determine which items of equipment must be carried to an inspection site by reviewing past inspection reports and a diagram of the facilities.

4.2C   Previous inspection reports may be available to inspectors in the form of computer printouts, inspection files, field notebooks, sump cards, or some other form or type of record describing treatment operations. Reports should be well documented and kept on file as a permanent record of that industry's activities and compliance status.

Answers to questions on page 308.

4.2D   The main purpose of most inspections should be to make unannounced checks on effluent quality and treatment system performance.

4.2E   If a receptionist tells an inspector, "There is nobody here who can authorize an inspection," the inspector should ask if the facility is open for business. If it is, then ask to see the receptionist's supervisor, personnel manager, plant manager, corporate officer or whoever is in charge. Identify yourself to any company representatives you contact; explain to them you wish to make an inspection. If there is still resistance, show them a copy of the *Sewer-Use Ordinance*, permit or whatever authority grants you the right to inspect the facility. Do not threaten legal action but clearly state your intent to inspect.

4.2F   Comply with all company regulations regarding safety, such as wearing special protective equipment. Ask your contact if any special hazards exist in the facility. Follow your contact closely; the contact is more familiar with the hazardous areas in the plant than you are. Don't touch anything unless your contact touches it first. Watch for slippery floors and make sure your shoes have plenty of tread on them.

4.2G   Issue a citation immediately (in accordance with agency policy) if you are denied entry and this appears to be the appropriate action. If you gain entry and are planning to issue a citation, it is best to do so after the tour of the facility since other violations may be discovered that should be added to the citation. Issuing a citation before the tour may generate hostility that jeopardizes the flow of information from your contact.

Answers to questions on page 312.

4.2H   An inspector should leave immediately when threatened physically.

4.2I   If a contact has little working knowledge of the facility, the inspector should request permission to question various supervisors or maintenance personnel concerning technical aspects of the pretreatment operation.

4.2J   When dealing with an instructive contact, the inspector should not necessarily believe everything the "expert" contact tells the inspector.

4.2K   The two basic guidelines that an inspector should consider during every inspection are:

1. Know the rules, and
2. Follow the rules.

Answers to questions on page 320.

4.3A   An inspector should be able to recognize the odors of commonly encountered gases such as sulfide, ammonia, chlorine, cyanide or any gases known to be associated with a facility.

4.3B   Final effluent meters used to monitor industrial discharges include flow, pH, temperature, explosivity, conductivity, and oxidation-reduction potential (ORP).

4.3C   The purpose of a sand and grease interceptor is to trap grit and settleable solids in the bottom of its chambers and to trap floatables and oil on the surface.

4.3D   Groundwater cleanup systems could produce greatly understated sewer-use fees when the treated water is discharged into a sanitary sewer or storm sewer because this additional water is not recorded on the incoming water meter if this water meter is the only flow metering device installed.

Answers to questions on page 332.

4.3E   Floors in metal finishing plants can be made acid resistant by applying epoxy coatings.

4.3F   The inspector should inquire how and where deionizing resins are backwashed to strip off accumulated ions. There is always the danger that pollutants are being removed from one sewer only to be redeposited in another sewer either off site or in an area within the plant that bypasses the pretreatment system.

4.3G   The oxidation of cyanide waste must be carefully controlled to ensure complete oxidation and to prevent the evolution of deadly gases such as chlorine, hydrogen cyanide and cyanogen chloride.

4.3H   The *Wastewater Ordinance* might contain provisions requiring additional sewer-use fees and that storm water may only be discharged during off-peak-flow hours at the POTW and for a specified duration of time after the storm subsides.

4.3I   An inspector can tell if a floor drain leads to a POTW sewer by using a dye tracer.

4.3J   Items or areas that should be inspected during an inspection include:

1. Production or manufacturing facility,
2. Waste treatment facilities,
3. Storage and maintenance areas,
4. Monitoring facilities, and
5. Records.

Answers to questions on page 337.

4.4A   Facility records that should be reviewed during or after an inspection include:

1. Liquid waste haulers' manifests,
2. Water bills,
3. Operations logs,
4. Plating bath maintenance log,
5. Calibration records,
6. Chemicals stored and used and MSDSs,
7. Hazardous waste manifests,
8. Sampling records and chain of custody documentation, and
9. Complaints.

4.4B   An inspector should review a facility's water bills to confirm that the volume of water consumed has not changed dramatically since the last permit revision.

4.4C A serious violation such as the discharge of flammables to a sewer may require immediate modifications that could include cessation of discharge. A major violation that could seriously impact the sewer system should be reported immediately to your home office as well as the supervisors of the sewer maintenance crews and the treatment plant operators. Do not go on to any other duties until the immediate danger has passed.

4.4D A citation should be signed by the signer of the permit. If that person is unavailable, a responsible corporate official would be the second choice. The third choice would be anybody you can contact. If nobody agrees to sign the citation, write "refused to sign" in the space that normally would be signed.

4.4E An inspection could be undertaken as a result of a complaint filed by a citizen or an agency. The complaint could come from in-sewer surveillance monitoring discoveries, other agencies' referrals, company informants, neighborhood reports of odors, discharges or explosions.

4.4F An inspector should respond in writing to a discharger's request for approval of pretreatment facility modifications rather than giving verbal approval.

Answers to questions on page 339.

4.5A Always assume that your special report will be read by someone outside your agency or by a reader who is only slightly familiar with your organization.

4.5B End a "special report" with a recommendation for further action or a conclusion. Make sure that the report is appropriately circulated.

4.5C Information that should be included in a compliance meeting follow-up letter includes who was in attendance and when and where the meeting was held. State exactly why the meeting was held, referring to any citations that preceded the most recent action. List the topics that were discussed. Describe the corrective action that was agreed upon and give a timetable for the completion of corrective action.

OBJECTIVE TEST

Chapter 4.   INSPECTION OF A TYPICAL INDUSTRY

Please write your name and mark the correct answers on the answer sheet, as directed at the end of Chapter 1. There may be more than one correct answer to each multiple-choice question.

True-False

1. A safety-conscious inspector will impress industry as a true professional.

   — 1. True
   2. False

2. Periodic rotation of inspector staff maximizes flexibility.

   — 1. True
   2. False

3. Self-monitoring test results tend to show a company's performance under optimum conditions.

   — 1. True
   2. False

4. Always check the outfall when visiting/inspecting a company.

   — 1. True
   2. False

5. The addition of dilution water to meet discharge limits is legal.

   1. True
   — 2. False

6. Ionizers can be used to remove selected ions from rinse water.

   1. True
   — 2. False

7. Hazardous waste manifests document the treatment of hazardous wastes.

   1. True
   — 2. False

8. Reports should be finalized with a recommendation for further action or a conclusion.

   — 1. True
   2. False

Best Answer (Select only the closest or best answer.)

9. Why should inspectors be on close working terms with sewer maintenance crews? So that

   1. Inspectors and maintenance crews can work together.
   2. Inspectors become familiar with sewer maintenance methods.
   — 3. Potential problems can be detected and corrected.
   4. Potential sources of inflow and infiltration will be recognized by inspectors.

10. What is the purpose of an unannounced industrial inspection?

    — 1. To check on effluent quality
    2. To meet with POTW's normal contact person
    3. To review paper work
    4. To test response of industry

11. What happens when a clarifier's capacity to hold solids or oil is exceeded?

    1. Depth measurements must be collected.
    — 2. Materials must be pumped out.
    3. Materials must be washed out in effluent.
    4. Wastestream must be diverted to another clarifier.

12. What will be collected in a clarifier if the passages connecting the stages are near the bottom?

    1. Grit and other heavy solids
    2. High pH wastes
    — 3. Oil and other floatables
    4. Sludge from the wastestream

13. How can slugs of concentrated wastes be prevented from overloading the final treatment system?

    1. Chemical dosing
    2. Dilution flow
    3. Good housekeeping
    — 4. Spill containment

14. Why should container labels be checked within each spill containment area?

    1. Check to be sure sufficient chemicals are available.
    2. Confirm that storage container lids are on tight.
    3. Ensure that proper chemicals are being used.
    — 4. Verify that incompatible chemicals are not stored together.

15. Why are floor drains NOT allowed in chemical storage areas?

    1. Because they are unnecessary
    — 2. Due to risk of spills
    3. To avoid plugging drains
    4. To prevent hydraulic overloads

16. Why are compliance meeting follow-up letters important?

    1. To document who attended the meeting
    2. To notify EPA of violations
    — 3. To outline what was agreed upon during the meeting
    4. To review all past violations

Multiple Choice (Select all correct answers.)

17. What factors influence the frequency of on-site inspections?

    — 1. Analysis of self-monitoring reports
    2. Attitude of industry toward on-site inspections
    3. Location of industry
    — 4. Nature of discharge
    — 5. POTW agency's personnel resources

18. What circumstances may require an investigation of sewer lines?

    — 1. Emergency response to explosive conditions
    — 2. Foaming in a sewer line
    3. Installation of a new force main
    — 4. Scaling or corrosion in a sewer line
    — 5. Stoppages that cause flow onto streets

19. What are the options available to an inspector if a request for entry to perform an inspection is denied?

    — 1. Call supervisor for direction
    — 2. Exercise authority to enter facility without permission
    — 3. Issue a citation for denial of access
    — 4. Leave and immediately collect samples for analysis
    5. Threaten company with criminal prosecution

20. What are the levels of inspection necessary for a complete inspection?

    — 1. Effluent treatment equipment
    — 2. General tour
    3. Influent monitoring for wastewater control equipment
    — 4. In-plant wastewater control equipment
    — 5. Outfall

21. What procedures should be followed when collecting industrial effluent samples?

    1. Analyze sample when convenient
    2. Collect sample from surface of wastestream
    — 3. Follow chain of custody procedures
    — 4. Preserve samples properly
    — 5. Use proper sampling techniques

22. Final industrial effluent meters can provide continuous monitoring of what water quality indicators?

    1. BOD
    — 2. Conductivity
    — 3. LEL
    — 4. ORP
    — 5. pH

23. Why are holding ponds installed?

    — 1. Allow for batch treatment
    — 2. Dampen hydraulic surges
    — 3. Hold water for off-peak discharges
    — 4. Lower peak flows
    — 5. Provide for evaporation

24. Cooling tower bleeds that flow to the sewer should be checked for overdoses of what chemicals?

    — 1. Acids
    — 2. Algicides
    3. Caustic soda
    — 4. Chromate
    5. Sodium hypochlorite

25. What are possible types of enforcement actions?

    — 1. Administrative orders
    — 2. Cease discharge orders
    — 3. Fines
    — 4. Verbal warnings
    — 5. Written notices of violations

26. What are possible sources of complaints?

    1. Company in violation
    — 2. Company informants
    — 3. In-sewer surveillance monitoring
    — 4. Neighborhood reports of odors
    — 5. Referral from another agency

End of Objective Test

APPENDIX

CHAPTER 4.   INSPECTION OF A TYPICAL INDUSTRY

SEATTLE METRO INDUSTRIAL WASTE
INSPECTION PROCEDURES

by Vallana M. Piccolo

SEATTLE METRO INDUSTRIAL WASTE INSPECTION PROCEDURES

A. INSPECTION FIELD REPORT FORM/CHECKLIST

1) General Inspection

Two types of inspections are discussed in this appendix — the general inspection and a detailed checklist inspection of a large, complex aerospace manufacturing facility. First, refer to the Industrial Waste Inspection Field Report Form on pages 350 and 351. This form consolidates the before, during, and after inspection information. It is used for the in-field documentation but also is often used as the final file report itself (due to the limited time all inspectors have to rewrite an inspection report). This report form, and several others similar to it that Seattle Metro uses, has recently been loaded onto microcomputer floppy disks so that the inspector can easily produce a final typed inspection report for the permanent file him/herself, or simply send the field version and the disk to the secretary. Metro's inspection forms are also made up into a pad format. It's convenient to keep a pad of these forms in your field vehicle.

The Field Report Form requires standard information such as the industry's name, address, and contact person, as well as the date, time, and the inspector's name. The remaining portion addresses the background information, the data gathered during the actual inspection, and the recommended follow-up actions. These issues will be covered in more detail in the next sections.

2) Aerospace Metal Finishing Checklist

The second form at the end of this appendix (page 352) is a detailed inspection checklist which is used for the Boeing Commercial Airline Company facilities. These are large, complex, integrated aerospace manufacturing plants with hundreds of widely varying wastestreams of concern to the POTW. The checklist shows a listing of the primary data the inspector seeks and evaluates for each regulated process line at a Boeing-type facility. Aerospace manufacturing requires the extensive use of acids, alkalies, toxic heavy metals (cadmium, chromium, copper, lead, nickel, zinc), as well as cyanides and solvents to clean, anodize/electroplate, seal, and paint the aircraft components. The checklist reflects this particular type of aerospace metal finishing emphasis although it provides many useful tips for an inspection of any metal finisher. The checklist is further discussed in Section D.

B. PREPARING FOR AN INSPECTION

There are many steps the inspector (or investigator as we are called at Seattle Metro) must take before he/she even arrives at the industrial site. The purpose of this section is to briefly highlight these steps.

1) Review Files/Prepare Paper Work

It is very important to review the industrial files for the company you are about to inspect. Review the permit, the last year's compliance record (both POTW and self-monitoring), the most recent sampling result, enforcement action, compliance schedule milestones, the latest inspection report, engineering reports, water/sewer billings, and correspondence. The review will bring you up to date and fully prepare you to conduct the inspection.

From the file review you should be able to fill out the first nine items on the Inspection Field Report. Write neatly in ink as these field report originals are often subpoenaed for court actions. The inspection type and purpose should be clearly determined before you leave your base. Is it a regular unannounced permit inspection? - a scheduled meeting? - a new industry? - a complaint or spill? - or a violation follow-up visit? Similarly, be clear on the specific issues or concerns and list them briefly under "purpose" in item 6. Examples might include: solvent odor, copper spill, permit renewal, new pretreatment system, verify compliance project completed, grease buildup in city line, or anonymous tip of midnight slug discharge.

It is a good idea to also keep a bound notebook for the additional details, sketches, and other inspection or industrial contact information. All industrial inspection files or notebooks should be kept in fireproof files. It is also wise to touch base with the local city, state, or EPA office to see if they have any current problems or investigations with the industry you are going to inspect. Bring copies of the permit, the sewer ordinance, the side sewer card, the most recent sampling results, the partially filled out inspection field report, and the most recent enforcement action with you to the inspection. Another very useful tool to bring is a computerized milestone tracking printout showing all the compliance milestones, dates, violations and enforcement actions over the past year. Now your paper work is ready for the inspection. If you are responding to an emergency spill you minimize the review/paper work preparation step, grab your notebook and your safety/sampling gear (keep this in your vehicle in ready status) and head out to the site as soon as safely possible.

2) Prepare Safety/Monitoring Equipment

A standard set of safety gear and monitoring equipment should always be in good and ready condition and in your vehicle. The safety gear at Seattle Metro includes communication radios (soon to be upgraded to cellular phones), hard hat, protective coveralls, safety goggles, earplugs, protective gloves, steel toe boots, identification badges, respirators, manhole hooks, gas detection meters, and a list of emergency contact numbers.

Monitoring equipment at Metro includes gas detection meters, sampling gear for heavy metals, cyanide, oil and grease, BOD/COD/TSS, solvents, pH, and flowmeters. The sampling gear is neatly packaged in an "Emergency Response Spill Kit" which includes instructions on how exactly to take each type of sample and maintain its integrity and chain of custody. The inspector should become familiar with these procedures. A camera is often very useful during an inspection. Be familiar with its operation, make sure it has the right kind of film for the anticipated situation, and obtain any necessary photo or security clearances from the industry before the date of the inspection. Follow all safety and security procedures.

C. CONDUCTING GENERAL INSPECTIONS

Most of the hard work has already been accomplished in the file review and preparation phase. The inspection itself is primarily two steps: 1) keeping your eyes and ears open, and 2) step-by-step filling out the data on the inspection field report. First, some pointers that are easily remembered as the "4-P"s: be *PREPARED, POLITE, PROFESSIONAL*, and *PERSISTENT.*

Once on site, go in the front door and identify yourself to the receptionist (or some other employees). State who you are and that you are here to perform an inspection for pollution control. Request to see your usual contact person. Make sure you have the appropriate safety gear on and with you before proceeding any further. While most industries are familiar and cooperative with these types of inspections, the inspector occasionally must deal with very uncooperative and/or aggressive industry representatives. If such a situation occurs, show a copy of the permit, sewer ordinance, and other legal document sections that give you the authority to inspect. If the inspection is still refused, politely leave the facility and pursue the appropriate legal actions such as search warrants, court injunctions, or criminal investigation assistance from the state or federal levels.

Once inside the industrial facility, inspect the pretreatment system entirely. The types of items to check include: - is it installed according to the engineering report and compliance schedule? - is there a knowledgeable operator present? - are the pH meters operational and calibrated? - does the pH alarm work? - is the clarifier free of excess sludge? - is there general good housekeeping? - records/logs being properly maintained? - is the sample site clean and clear? - are the chemical addition systems, sludge press, solvent recovery/strippers, oil/grease separators properly working and maintained? Similarly walk through the process areas and check for spill prevention,

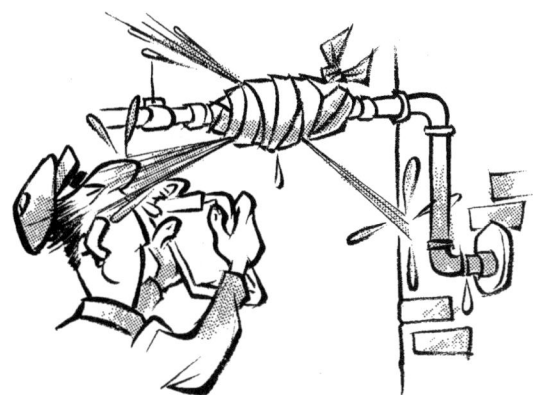

leaking tanks, floor spillage, dilution, questionable plumbing, hoses, or drains. Verify that the wastestreams are plumbed to the treatment system and that no concentrated process tanks have any plumbed connection to the sewer. Check cooling water, boiler blowdown plumbing and their maintenance chemicals (to see if toxics are used).

Next, check all the chemical and waste storage areas for spillage, illegal connections to the sewer, illegal discharges to the ground or surface waters. Verify if drums are properly stored in bermed, roofed, no-outlet areas and labeled/manifested according to state law. Note any violations that, while out of your agency's jurisdiction, should be referred for appropriate follow-up by the fire department, city, county, state, or EPA. Obviously, it helps to be familiar with the various hazardous waste and fire codes/regulations for these other agencies.

Finally, it is wise to end the inspection with a general walk-through of the entire facility both inside and outdoors. Be sure to check "back rooms" that are alleged to be just dry operations as well as storage rooms and bathrooms for signs of illegal drains or dumping. While still on site, notify the industrial representative of any/all violations and what the required corrective action is as well as what the next action is from Metro. If possible, fill out the inspection field report on site and leave a copy of it with the representative.

D. CONDUCTING A LARGE AEROSPACE PLANT INSPECTION

The same general procedures apply as previously discussed only you focus on many more details at many more wastestreams and process areas. You will see in the checklist a more detailed listing of the items to check during your inspection. The issues are easily grouped into: process tank description; accuracy/legality of the monitoring site (is it representative? does it avoid illegal dilution?); what is the level of rinse tank technology?; is excess diluting water being used?; what is the quality of the spill control plan as currently observed?; and a detailed review/inventory of the pretreatment strategy and equipment. The checklist should be used in conjunction with the inspection field report.

E. FOLLOWING UP THE INSPECTION

After you return to base, immediately notify the supervisor or check-in safety person that you have returned. Next proceed with chain of custody to deliver and log in for analysis any samples you took during the inspection. Take a shower and clean up, maintain or put away any safety gear as necessary. Notify by phone the POTW operator or any agencies of problems noted during the inspection warranting immediate attention. Notify the criminal investigation unit as appropriate. File the permit and other information which was removed for the inspection. Send any film in to be processed.

Next tackle the paper work (often the next day or two). You may determine that after final updates and review/notations that your original handwritten field report is adequate to file or to send a copy to the other agencies as is. It is best to write a cover letter to the industry with a copy of the field report (handwritten or typed) outlining the results of the inspection and any required corrective action. Copy any other appropriate agencies. If appropriate, prepare an enforcement action for any violation which was found during the inspection.

Finally, log the results of your inspection on the computer tracking system and schedule the next sampling or inspection dates as appropriate.

SEATTLE METRO INDUSTRIAL WASTE INSPECTION FIELD REPORT

1) Investigator/Date/Time: _____

2) Company Name: _____

3) Address: _____

4) Contact Person: _____

   Title/Phone: _____

5) Metro Permit: NO. _____ IW Flow _____ Category _____

   Exp. Date _____

6) Inspection/Type/Purpose: Unannounced _____ Scheduled _____

   New Company _____ Complaint _____ Spill _____

   Violation _____ Other _____

   Purpose _____

   _____

7) Last Metro Actions (Type/Date): Inspection _____

   Sample _____

   Violation _____

   Enforcement _____

8) Nature of Operation: _____

   Number of Employees: _____

9) Wastestreams to Metro: _____

   _____

   _____

10) Pretreatment System: Type _____ Flow-Thru _____

    Batch _____ Other _____

    Condition/Operation = Good _____ Fair _____ Poor _____

    Comments: _____

    _____

11) Process Areas Inspected: _____

    _____

    _____

    Condition/Operation = Good _____ Fair _____ Poor _____

    General Housekeeping = Good _____ Fair _____ Poor _____

    Condition _____

    _____

    _____

12) Metro Permit Violations: _____

    _____

    _____

    _____

13) Wastestreams to Surface/Groundwater: _____

    _____

    _____

    NPDES Permit No.: _____

14) Chemical/Waste Storage Areas: _____
_____
_____

    Condition = Good _____ Fair _____ Poor _____

    Comment _____
_____
_____

15) Hazardous Waste Drums/Labels/Manifests: OK? _____

    Problems _____
_____
_____
_____

****
****

INSPECTION RESULTS

Complied With Metro Permit?   Yes _____ No _____ N/A _____

    Comment _____
_____
_____
_____

Complied With WDOE Requirements: Yes _____ No _____ N/A _____
_____
_____
_____

Recommended Action By: Metro _____ WDOE _____ Other _____

    Priority: ASAP _____ High _____ Medium _____ Low _____

    Comments: _____
_____
_____
_____
_____

SEATTLE METRO INSPECTION CHECKLIST FOR BOEING COMMERCIAL AIRLINE COMPANY

Plant Name _____

Investigator _____

Process Line _____ Flow (GPD) _____

1) Process Tanks

   a) Type (cleaner, plating, paint booth, paint hangar, etc.)
   b) Labels
   c) Size/Gallons
   d) Maximum Concentration
   e) Condition/Leaks
   f) No. of Dead Rinses Downstream
   g) No. of Live Rinses
   h) Temperature
   i) Liners/Berms
   j) Leak Inspection Log

2) Where is the monitoring site in relation to the waste-stream?

   a) Directly downstream of metal-bearing waste-stream
   b) Downstream of any commingling with acid/caustic cleaners
   c) Downstream of hot water seal
   d) Downstream of any other metal finishing waste-streams
   e) Downstream of any regulated non-metal finishing wastes
   f) Downstream of cooling water
   g) Downstream of sanitary waste discharge

3) What type of running rinse tanks?

   a) Single
   b) Double
   c) Countercurrent
   d) Fog/Spray
   e) Baffles
   f) Other

4) Rinse tank operation/dilution issues

   a) Air mixing
   b) Shut-off devices
   c) Operator push button/foot pedals
   d) Conductivity meter/flow sticks
   e) Timer
   f) Left running when no parts/no operators in sight

5) Spill Control

   a) Berms
   b) Floor drains
   c) Chemical/waste storage
   d) Parts racked over process tank
   e) Drain back board
   f) Floor spillage contained
   g) Dead rinses/drag out tanks
   h) Emergency sewer shut-off device
   i) Emergency spill response plan/posted 24-hr phone numbers, signage
   j) Knowledgeable trained shop workers
   k) Overflow alarms

6) Pretreatment Strategy

   a) None
   b) pH control/alarms
   c) Metals precipitation/settling/clarifier
   d) Sludge press
   e) Source control
   f) Recycle/reuse/water, waste minimization
   g) Process changes to less polluting
   h) Filtration
   i) Reverse osmosis
   j) Electrolytic de-plate-out
   k) Dilution
   l) Solvent air stripping
   m) Organics — activated carbon
   n) Contract haul to hazardous waste TSD facility
   o) Batch treatment
   p) Other

CHAPTER 5

SAFETY IN PRETREATMENT INSPECTION AND SAMPLING WORK

by

Herb Schott

TABLE OF CONTENTS

Chapter 5.   SAFETY IN PRETREATMENT INSPECTION AND SAMPLING WORK

OBJECTIVES

Chapter 5.   SAFETY IN PRETREATMENT INSPECTION AND SAMPLING WORK

Following completion of Chapter 5, you should be able to:

1. Determine safety equipment and supplies needed to conduct an inspection,

2. Drive safely to and from inspection sites,

3. Identify a confined space,

4. Test the atmosphere in a confined space,

5. Enter, work in and leave a confined space safely,

6. Identify and protect yourself from hazardous materials,

7. Identify and protect yourself from physical hazards, and

8. Collect samples following safe procedures.

PRETREATMENT WORDS

Chapter 5.   SAFETY IN PRETREATMENT INSPECTION AND SAMPLING WORK

ACUTE HEALTH EFFECT

An adverse effect on a human or animal body, with symptoms developing rapidly.

ANTAGONISTIC REACTION

An interaction between two or more individual compounds that produces an injurious effect upon the body (or an organism) which is *LESS* than either of the substances alone would have produced.

AUTOIGNITION TEMPERATURE

The temperature at which a material will spontaneously ignite and sustain combustion.

CARCINOGEN (CAR-sin-o-JEN)

Any substance which tends to produce cancer in an organism.

CAUTION

This word warns against potential hazards or cautions against unsafe practices. Also see DANGER, NOTICE, and WARNING.

CHRONIC HEALTH EFFECT

An adverse effect on a human or animal body with symptoms that develop slowly over a long period of time or that recur frequently.

COMBUSTIBLE LIQUID

A liquid whose flashpoint is at or above 100°F (38°C). Flammable liquids present a greater fire or explosion hazard than combustible liquids. Also see FLAMMABLE LIQUID.

COMMUNITY RIGHT-TO-KNOW

The **S**uperfund **A**mendments and **R**eauthorization **A**ct (SARA) of 1986 provides statutory authority for communities to develop "right-to-know" laws. The Act establishes a state and local emergency planning structure, emergency notification procedures, and reporting requirements for facilities. Also see SARA and RIGHT-TO-KNOW LAWS.

COMPETENT PERSON

A competent person is defined by OSHA as a person capable of identifying existing and predictable hazards in the surroundings, or working conditions which are unsanitary, hazardous or dangerous to employees, and who has authorization to take prompt corrective measures to eliminate the hazards.

COMPOSITE (PROPORTIONAL) SAMPLE

A composite sample is a collection of individual samples obtained at regular intervals, usually every one or two hours during a 24-hour time span. Each individual sample is combined with the others in proportion to the rate of flow when the sample was collected. Equal volume individual samples also may be collected at intervals after a specific volume of flow passes the sampling point or after equal time intervals and still be referred to as a composite sample. The resulting mixture (composite sample) forms a representative sample and is analyzed to determine the average conditions during the sampling period.

CONFINED SPACE

Confined space means a space that:

A. Is large enough and so configured that an employee can bodily enter and perform assigned work; and

B. Has limited or restricted means for entry or exit (for example, tanks, vessels, silos, storage bins, hoppers, vaults, and pits are spaces that may have limited means of entry); and

C. Is not designed for continuous employee occupancy.

(Definition from the Code of Federal Regulations (CFR) Title 29 Part 1910.146.)

CONFINED SPACE, NON-PERMIT CONFINED SPACE, NON-PERMIT

A non-permit confined space is a confined space that does not contain or, with respect to atmospheric hazards, have the potential to contain any hazard capable of causing death or serious physical harm.

CONFINED SPACE, PERMIT-REQUIRED
  (PERMIT SPACE)
CONFINED SPACE, PERMIT-REQUIRED
  (PERMIT SPACE)

A confined space that has one or more of the following characteristics:

- Contains or has a potential to contain a hazardous atmosphere,
- Contains a material that has the potential for engulfing an entrant,
- Has an internal configuration such that an entrant could be trapped or asphyxiated by inwardly converging walls or by a floor which slopes downward and tapers to a smaller cross section, or
- Contains any other recognized serious safety or health hazard.

(Definition from the Code of Federal Regulations (CFR) Title 29 Part 1910.146.)

CORROSIVE MATERIAL CORROSIVE MATERIAL

A material which through its chemical action is destructively injurious to body tissues or other materials.

DANGER DANGER

The word *DANGER* is used where an immediate hazard presents a threat of death or serious injury to employees. Also see CAUTION, NOTICE, and WARNING.

DECIBEL (DES-uh-bull) DECIBEL

A unit for expressing the relative intensity of sounds on a scale from zero for the average least perceptible sound to about 130 for the average level at which sound causes pain to humans. Abbreviated dB.

FIRE POINT FIRE POINT

The lowest temperature of a liquid at which a mixture of air and vapor from the liquid will continue to burn.

FLAMMABLE LIQUID FLAMMABLE LIQUID

A liquid which by itself, or any component of it present in greater than one percent concentration, has a flashpoint below 100°F (38°C). Also see COMBUSTIBLE LIQUID.

FLASH POINT FLASH POINT

The minimum temperature of a liquid at which the liquid gives off a vapor in sufficient concentration to ignite when tested under specific conditions.

GRAB SAMPLE GRAB SAMPLE

A single sample of water collected at a particular time and place which represents the composition of the water only at that time and place.

HAZARDOUS MATERIAL
  MANAGEMENT PLAN (HMMP)
HAZARDOUS MATERIAL
  MANAGEMENT PLAN (HMMP)

A document prepared by an industry which contains copies of MSDSs (Material Safety Data Sheets) as well as additional information regarding the storage, handling and disposal of all chemicals used on site by the industry.

HAZARDOUS WASTE HAZARDOUS WASTE

A waste, or combination of wastes, which because of its quantity, concentration, or physical, chemical, or infectious characteristics may:

1. Cause, or significantly contribute to, an increase in mortality or an increase in serious, irreversible, or incapacitating reversible illness; or

2. Pose a substantial present or potential hazard to human health or the environment when improperly treated, stored, transported, or disposed of or otherwise managed; and

3. Normally not be discharged into a sanitary sewer; subject to regulated disposal.

(Resource Conservation and Recovery Act (RCRA) definition.)

IDLH IDLH

Immediately **D**angerous to **L**ife or **H**ealth. The atmospheric concentration of any toxic, corrosive, or asphyxiant substance that poses an immediate threat to life or would cause irreversible or delayed adverse health effects or would interfere with an individual's ability to escape from a dangerous atmosphere.

JSA   JSA

Job **S**afety **A**nalysis. A supervisor selects the job to be analyzed and then, with the help of the inspectors, the job is subdivided into individual steps and each step of the job is critically examined. During the examination process any potential hazards or problems associated with that step are identified by reviewing how each inspector performs the tasks of this step. The intent of this review is to see what can happen and how any potential accidents and injuries can be prevented.

LOWER EXPLOSIVE LIMIT (LEL)   LOWER EXPLOSIVE LIMIT (LEL)

The lowest concentration of gas or vapor (percent by volume in air) that explodes if an ignition source is present at ambient temperature. At temperatures above 250°F the LEL decreases because explosibility increases with higher temperature.

LOWER FLAMMABLE LIMIT (LFL)   LOWER FLAMMABLE LIMIT (LFL)

The lowest concentration of a gas or vapor (percent by volume in air) that burns if an ignition source is present.

MATERIAL SAFETY DATA SHEET (MSDS)   MATERIAL SAFETY DATA SHEET (MSDS)

A document which provides pertinent information and a profile of a particular hazardous substance or mixture. An MSDS is normally developed by the manufacturer or formulator of the hazardous substance or mixture. The MSDS is required to be made available to employees and inspectors whenever there is the likelihood of the hazardous substance or mixture being introduced into the workplace. Some manufacturers are preparing MSDSs for products that are not considered to be hazardous to show that the product or substance is *NOT* hazardous.

MUTAGENIC (MUE-ta-JEN-ick)   MUTAGENIC

Any substance which tends to cause mutations or gene changes prior to conception.

NIOSH (NYE-osh)   NIOSH

The **N**ational **I**nstitute of **O**ccupational **S**afety and **H**ealth is an organization that tests and approves safety equipment for particular applications. NIOSH is the primary federal agency engaged in research in the national effort to eliminate on-the-job hazards to the health and safety of working people. The NIOSH Publications Catalog, Sixth Edition, NIOSH Pub. No. 84-118, lists the NIOSH publications concerning industrial hygiene and occupational health. To obtain a copy of the catalog, write to National Technical Information Service (NTIS), 5285 Port Royal Road, Springfield, VA 22161. NTIS Stock No. PB-86-116-787, price, $103.50, plus $5.00 shipping and handling.

NOTICE   NOTICE

This word calls attention to information that is especially significant in understanding and operating equipment or processes safely. Also see CAUTION, DANGER, and WARNING.

OSHA (O-shuh)   OSHA

The Williams-Steiger **O**ccupational **S**afety and **H**ealth **A**ct of 1970 (OSHA) is a federal law designed to protect the health and safety of industrial workers and also the operators and inspectors of pretreatment facilities. OSHA regulations require employers to obtain and make available to workers the Material Safety Data Sheets (MSDSs) for chemicals used at industrial facilities and treatment plants. OSHA also refers to the federal and state agencies which administer the OSHA regulations.

PATHOGENIC (PATH-o-JEN-ick) ORGANISMS   PATHOGENIC ORGANISMS

Organisms, including bacteria, viruses or cysts, capable of causing diseases (giardiasis, cryptosporidiosis, typhoid, cholera, dysentery) in a host (such as a person). There are many types of organisms which do *NOT* cause disease. These organisms are considered nonpathogenic. Many beneficial bacteria are found in wastewater treatment processes actively cleaning up organic wastes.

pH (pronounce as separate letters)   pH

pH is an expression of the intensity of the basic or acidic condition of a liquid. Mathematically, pH is the logarithm (base 10) of the reciprocal of the hydrogen ion activity.

$$pH = \text{Log } \frac{1}{[H^+]} \quad \text{or} = -\text{Log } [H^+]$$

The pH may range from 0 to 14, where 0 is most acidic, 14 most basic, and 7 neutral.

RCRA (RICK-ruh)   RCRA

The Federal **R**esource **C**onservation and **R**ecovery **A**ct (10/21/76), Public Law (PL) 94-580, provides technical and financial assistance for the development of plans and facilities for recovery of energy and resources from discarded materials and for the safe disposal of discarded materials and hazardous wastes. This Act introduces the philosophy of the "cradle to grave" control of hazardous wastes. RCRA regulations can be found in Title 40 of the Code of Federal Regulations (40 CFR) Parts 260-268, 270 and 271.

RIGHT-TO-KNOW LAWS

RIGHT-TO-KNOW LAWS

Employee "Right-to-Know" legislation requires employers to inform employees (pretreatment inspectors) of the possible health effects resulting from contact with hazardous substances. At locations where this legislation is in force, employers must provide employees with information regarding any hazardous substances which they might be exposed to under normal work conditions or reasonably foreseeable emergency conditions resulting from workplace conditions. OSHA's Hazard Communication Standard (HCS) (Title 29 CFR Part 1910.1200) is the federal regulation and state statutes are called Worker Right-to-Know Laws. Also see COMMUNITY RIGHT-TO-KNOW and SARA.

SARA

SARA

Superfund **A**mendments and **R**eauthorization **A**ct of 1986. The Comprehensive Environmental Response, Compensation, and Liability Act (CERCLA), commonly known as Superfund, was enacted in 1980. The Superfund Amendments increase Superfund revenues to $8.5 billion and strengthen the EPA's authority to conduct short-term (removal), long-term (remedial) and enforcement actions. The Amendments also strengthen state involvements in the cleanup process and the Agency's commitments to research and development, training, health assessments, and public participation. A number of new statutory authorities, such as Community Right-to-Know, are also established.

SOLVENT

SOLVENT

Any substance that is used to dissolve another substance in it.

SYNERGISTIC (SIN-er-GIST-ick) REACTION

SYNERGISTIC REACTION

An interaction between two or more individual compounds which produces an injurious effect upon the body (or an organism) which is *GREATER* than either of the substances alone would have produced.

TERATOGENIC (TEAR-a-toe-JEN-ick)

TERATOGENIC

Any substance which tends to cause birth defects after conception.

THRESHOLD LIMIT VALUE (TLV)

THRESHOLD LIMIT VALUE (TLV)

The average concentration of toxic gas or any other substance to which a normal person can be exposed without injury during an average work week.

TIME WEIGHTED AVERAGE (TWA)

TIME WEIGHTED AVERAGE (TWA)

The average concentration of a pollutant based on the times and levels of concentrations of the pollutant. The time weighted average is equal to the sum of the portion of each time period (as a decimal, such as 0.25 hour) multiplied by the pollutant concentration during the time period divided by the hours in the workday (usually 8 hours). 8TWA PEL is the Time Weighted Average permissible exposure limit, in parts per million, for a normal 8-hour workday and a 40-hour workweek to which nearly all workers may be repeatedly exposed, day after day, without adverse effect.

TOXIC

TOXIC

A substance which is poisonous to a living organism. Toxic substances may be classified in terms of their physiological action, such as irritants, asphyxiants, systemic poisons, and anesthetics and narcotics. Irritants are corrosive substances which attack the mucous membrane surfaces of the body. Asphyxiants interfere with breathing. Systemic poisons are hazardous substances which injure or destroy internal organs of the body. The anesthetics and narcotics are hazardous substances which depress the central nervous system and lead to unconsciousness.

TOXIC SUBSTANCES CONTROL ACT (TSCA)

TOXIC SUBSTANCES CONTROL ACT (TSCA)

The **T**oxic **S**ubstances **C**ontrol **A**ct of 1976 gave EPA the responsibility of controlling the entry of toxic, carcinogenic or otherwise biologically active compounds into the environment. The Act placed a heavy reporting burden on industry and contained provisions allowing EPA to demand premarket testing of some chemicals.

UPPER EXPLOSIVE LIMIT (UEL)

UPPER EXPLOSIVE LIMIT (UEL)

The point at which the concentration of a gas in air becomes too great to allow an explosion upon ignition due to insufficient oxygen present.

UPPER FLAMMABLE LIMIT (UFL)

UPPER FLAMMABLE LIMIT (UFL)

The point at which the concentration of a gas in air becomes too great to sustain a flame upon ignition due to insufficient oxygen present.

WARNING

WARNING

The word *WARNING* is used to indicate a hazard level between *CAUTION* and *DANGER*. Also see CAUTION, DANGER, and NOTICE.

CHAPTER 5.   SAFETY IN PRETREATMENT INSPECTION AND SAMPLING WORK

5.0   GENERAL SAFETY CONSIDERATIONS

5.00   Responsibilities

Safety in inspection work is the concern of everyone. However, the primary responsibility for the safety training of pretreatment inspectors to ensure that they have the knowledge, skills, and equipment to perform their duties in a safe manner at all times rests with their immediate supervisor. The immediate supervisor has the responsibility to implement management's basic safety policies by scheduling regular training sessions for the inspectors, reviewing the written safety rules, reporting to management on any safety problems, initiating corrective action for any safety deficiencies or violations and disciplining the inspectors if safety rules are violated.

As management's representative in health and safety activities, the supervisor not only must develop, administer and implement the safety program for the people supervised, but is also expected to set an example to the inspectors on how to comply with all the safety requirements and procedures. In order to set a good example in the use of safe procedures, the supervisor must be thoroughly familiar with the job duties and requirements of each inspector. One way to accomplish this is by performing a Job Safety Analysis (JSA). In the JSA, the supervisor selects the job to be analyzed and then, with the help of the inspectors, the job is subdivided into individual steps and each step of the job is critically examined. Any potential hazards or problems associated with a step are identified by reviewing how each inspector performs the tasks of that step. The intent of this review is to see what can happen and how any potential accidents and injuries can be prevented. Having the inspectors participate in the JSA improves both the inspectors' and the supervisor's safety attitude and everyone gains a fuller understanding of the job and each of its components.

The knowledge gained through a Job Safety Analysis should be incorporated into a written safety program with job-specific descriptions of each inspector's duties. The safety program will also include appropriate methods to eliminate known and potential hazards and to prevent accidents from occurring. Not only will the JSA aid the supervisor in training new inspectors, but it will also be a useful guide for evaluating the inspectors' safety compliance during the performance of their regular duties. In case of an accident or an injury, the JSA will also serve as a guide in the accident investigation and preparation of the accident report by the supervisor. Appendix A contains an example of job-specific procedures that were developed in a Job Safety Analysis.

Although it is the employer's and supervisor's responsibility to provide each inspector an opportunity to work in a safe place and manner, you, the inspector, also have certain responsibilities regarding your own safety during the performance of job duties. Among these responsibilities are the obligation to participate actively in the safety program and training, to apply the safe work practices that have been learned on the job, to recognize the hazards of the job, and to take precautionary measures which will ensure both your own safety and the safety of all co-workers. You also have the responsibility and obligation to inform your supervisor of any unsafe conditions, situations and/or acts that are encountered during the performance of your duties.

Since on-the-job safety training is probably the most effective means for the exchange of information related to safe work habits, both inspectors and supervisors are encouraged to hold one-on-one discussions on safety topics. In short, safety training must become a regular part of the general on-the-job skills training that a supervisor gives to the inspectors. During the training, the supervisor must communicate effectively with the inspectors to ensure that all instructions are received and properly understood.

Another important aspect of safety training is to make inspectors aware that they must report all injuries, no matter how small, to their supervisor as soon as possible. When you are injured and do not file an accident report, you may have trouble receiving compensation if complications develop in the future. For example, if you hurt your back and complications develop three months after the accident, you may not receive any compensation for the injury unless you filed an accident report at the time of the incident.

Accident reporting rules should be strictly enforced and failure to report an injury should be considered grounds for disciplinary action. An easy way to encourage compliance with reporting requirements is to make report forms readily available so that an inspector can easily submit a signed report stating the type of injury received and the date and time that it happened. This brief injury report would at least document the incident although it does not replace a full accident investigation report which may be completed after all the facts have been obtained.

In addition to the work-related safety training supplied by your supervisor, ask your employer to provide training in first aid and cardiopulmonary resuscitation (CPR). Training to acquire and maintain these skills should be conducted by persons qualified and licensed to teach these subjects, such as your local Red Cross Chapter. Although first aid generally involves the one-time treatment of minor injuries such as scratches, cuts and small burns, knowledge of first aid and especially CPR can be a life saver in cases of emergency. Additionally, where medical personnel are not readily available — this has been interpreted as being three or four minutes away — OSHA requires employers to ensure that an adequate number of employees with a current certificate from a first aid program (American Red Cross) are available to provide emergency care for injured employees (29 CFR 1910.151).

Since the work of an industrial waste inspector involves the handling of wastewater, the health aspects of work situations should be considered. The various types of disease-producing organisms that can be encountered in the wastewater will be discussed in a later section of this chapter. However, as a general rule, the most important precautions to take are those that prevent oral or skin contact with the wastewater. Even when wearing gloves, do not eat, smoke or drink until your hands have been thoroughly washed with hot water and a disinfectant soap. Additionally, any breaks in the skin such as cuts or abrasions should be kept clean and protected to eliminate any easy entry paths for the microorganisms. Keep your clothing clean and protected by coveralls, and change clothes immediately if contamination occurs. Clothes worn on the job may be contaminated and should *NOT* be washed with household laundry.

In order to meet OSHA and state safety regulations, every organization must prepare a safety manual or written safety rules for its inspectors. However, in many instances, these rules are of a fairly general nature and apply to all inspectors and operators within a given organization. To develop an effective safety program, it is highly recommended that supervisors consult with inspectors to draft and implement specific written safety procedures that relate directly to the particular tasks being performed and to the sites being inspected. In this type of job-specific safety program, individuals working as industrial waste inspectors would use procedures (as developed through JSAs) specifically designed for testing atmospheres, opening manholes, setting samplers or any of the other job

tasks within the field. Other employees performing different tasks for the same agency would follow the procedures that relate directly to the tasks they perform.

The written safety program developed by the supervisor should also include a section which meets the *RIGHT-TO-KNOW*[1] law requirements. Part of the requirement can be met by making available copies of all the Material Safety Data Sheets (MSDSs) for the chemicals and/or chemical products that the inspectors use or encounter during their regular work period, and by training inspectors to read and interpret the contents of the MSDSs. All safety manuals and procedures should be treated as active documents that are used and updated on a regular basis or their usefulness is defeated.

The safety program should also provide for regular, documented safety meetings. At these meetings previous accidents should be discussed along with methods of preventing similar accidents in the future.

In summary, safety is everybody's business, especially the inspectors and their immediate supervisors. As an inspector you are responsible for your own safety in the field. If you encounter an unsafe condition or situation, you should remove yourself from the area and warn others. Do not return to finish an inspection until the hazardous conditions have been corrected.

QUESTIONS

Write your answers in a notebook and then compare your answers with those on page 392.

5.0A   What is the purpose of safety training for pretreatment inspectors?

5.0B   Who has the primary responsibility for the safety training of pretreatment inspectors?

5.0C   How is a Job Safety Analysis (JSA) performed?

5.0D   What precautions should an inspector take to prevent oral and skin contact with wastewater?

5.0E   How can a supervisor comply with a "right-to-know" law regarding inspectors?

5.01   Safety Equipment and Supplies Needed

In order to perform the job of an industrial waste pretreatment inspector effectively, you will need the proper equipment and supplies to safely carry out the general duties of your position. Since a significant part of an inspector's time is spent in driving to and from sampling locations and the industries being inspected, it is important that a proper vehicle be used. In general, the types of vehicles that have been found most suitable for this type of work are either vans or pickup trucks equipped with camper shells. In some instances a station wagon may also be used. However, a station wagon does not provide as much room as the other two types of vehicles and may limit the amount of equipment you can carry or make it more difficult to load or unload equipment.

---

[1] *Right-to-Know Laws. Employee "Right-to-Know" legislation requires employers to inform employees (pretreatment inspectors) of the possible health effects resulting from contact with hazardous substances. At locations where this legislation is in force, employers must provide employees with information regarding any hazardous substances which they might be exposed to under normal work conditions or reasonably foreseeable emergency conditions resulting from workplace conditions. OSHA's Hazard Communication Standard (HCS) (Title 29 CFR Part 1910.1200) is the federal regulation and state statutes are called Worker Right-to-Know Laws.*

The vehicle, regardless of which kind is chosen, should be maintained in top mechanical condition. It should be equipped with a rotating, flashing amber dome light which can be controlled from the inside of the vehicle. In addition, if much of the sampling work is done in streets with heavy traffic or if the vehicle is driven slowly at times in congested areas, it should be equipped with alternately flashing amber lights that are mounted near the top on each side of the back of the vehicle. These lights are more visible than the vehicle's regular emergency flashers and provide an extra measure of protection. Each vehicle used for inspection work should also be provided with a two-way radio for routine and emergency communications. The radio's usefulness and range can be extended by having both an inside and outside speaker as well as a "repeater" unit which allows the inspector to receive and send messages while away from the vehicle. Cellular phones also are an effective means of communication.

Safety equipment that should be routinely carried in each vehicle includes an approved first aid kit, a fire extinguisher (generally an all-purpose, ten-pound minimum, A-B-C chemical type), flares, a supply of about 15 or more traffic control cones, and spill control pillows and containers (devices to stop the flow of a spill and to remove the spill from the site). The vehicle's safety equipment should be inspected on a regular basis (quarterly) by the supervisor of the crew to whom it is assigned to ensure that everything is properly maintained. Replace all missing or used-up items immediately, especially first aid supplies. The fire extinguisher must also be given an annual maintenance check (29 CFR 1910.157) to ensure that it is fully charged and operational.

Each vehicle should also carry the inspector's personal safety gear. This includes a hard hat, safety glasses and/or goggles, a supply of different types of gloves, extra coveralls, orange traffic vest, earplugs, hand lotion or barrier cream, paper towels, rubber overshoes, safety shoes (if not already worn) and copies of the injury report form. Additionally the vehicle should carry a suitable gas detector(s) that is stored in a protective case (a padded wooden box is excellent) to prevent it from being damaged during transit, an explosion-proof flashlight or lantern, safety harness and lanyards, and respiratory protective equipment. The vehicle may also contain, if space allows, a portable blower, a tripod, and other gas detection equipment (such as gas-specific badges or indicator tubes) to test for oxygen deficiency, explosive/flammable gases and toxic gases (such as hydrogen sulfide, chlorine, cyanide, and carbon monoxide).

5.02 Planning and Preparing for an Inspection

An inspector must properly prepare for an inspection before actually performing it. The first step in planning for an inspection, after the industry has been selected, is to review

the firm's permit application and correspondence file in detail. This review serves to refresh your knowledge of the processes and operations being performed at the facility and acts as a reminder about what safety equipment you will need and the special precautions to be taken during the inspection. In many instances it is quite beneficial to prepare individual up-to-date inspection plans for each industry and keep them in the industry's file. Such a plan would list all pertinent details regarding the sampling point(s) location, safety concerns, precautions that must be taken and any other information relevant to conducting the inspection in a safe and efficient manner.

In preparing to inspect an industry, review and become familiar with any chemicals that may be used by the industry during their regular production or operations. The major source of information regarding chemical products and their hazardous nature is the MSDS for each substance (see next four pages for a copy of OSHA's Material Safety Data Sheet and typical MSDS for hydrogen sulfide). Recent changes in the regulations have made the information on the MSDS much more useful since all hazardous constituents present in one percent (10,000 mg/L) or greater concentration must now be listed. Carcinogens must be listed if concentrations are 0.1 percent or greater (29 CFR 1910.1200).

Material Safety Data Sheets must all contain certain types of information but the format in which the information is presented may vary from manufacturer to manufacturer. OSHA Form 174 (see next two pages) is a format that is frequently used but many other formats are also used. At a minimum, an MSDS must contain the following information about a chemical:

- Identity,
- Physical and chemical characteristics,
- Physical hazards,
- Health hazards,
- Primary route(s) of entry,
- Permissible exposure limit (PEL),
- Any carcinogenic factors,
- Applicable safe-handling precautions,
- Applicable control measures,
- Emergency and first aid procedures,
- Date of preparation, and
- Contact information for the preparer of the form.

Material Safety Data Sheet

May be used to comply with
OSHA's Hazard Communication Standard,
29 CFR 1910.1200 Standard must be
consulted for specific requirements.

U.S. Department of Labor

Occupational Safety and Health Administration
(Non-Mandatory Form)
Form Approved
OMB No. 1218-0072

IDENTITY *(As Used on Label and List)*

Note: Blank spaces are not permitted. If any item is not applicable, or no
information is available, the space must be marked to indicate that.

Section I

Manufacturer's Name	Emergency Telephone Number
Address *(Number, Street, City, State, and ZIP Code)*	Telephone Number for Information
	Date Prepared
	Signature of Preparer *(optional)*

Section II — Hazardous Ingredients/Identity Information

Hazardous Components (Specific Chemical Identity: Common Name(s))	OSHA PEL	ACGIH TLV	Other Limits Recommended	%(optional)

Section III — Physical/Chemical Characteristics

Boiling Point		Specific Gravity (H_2O = 1)	
Vapor Pressure (mm Hg.)		Melting Point	
Vapor Density (AIR = 1)		Evaporation Rate (Butyl Acetate = 1)	
Solubility in Water			
Appearance and Odor			

Section IV — Fire and Explosion Hazard Data

Flash Point (Method Used)	Flammable Limits	LEL	UEL
Extinguishing Media			
Special Fire Fighting Procedures			
Unusual Fire and Explosion Hazards			

(Reproduce locally)

174, Sept. 1985

Section V — Reactivity Data

Stability	Unstable		Conditions to Avoid
	Stable		

Incompatibility *(Materials to Avoid)*

Hazardous Decomposition or Byproducts

Hazardous Polymerization	May Occur		Condition to Avoid
	Will Not Occur		

Section VI — Health Hazard Data

Route(s) of Entry:　　　　Inhalation?　　　　　　Skin?　　　　　　Ingestion?

Health Hazards *(Acute and Chronic)*

Carcinogenicity　　　　NTP?　　　　　　IARC Monographs?　　　　OSHA Regulated?

Signs and Symptoms of Exposure

Medical Conditions
Generally Aggravated by Exposure

Emergency and First Aid Procedures

Section VII — Precautions for Safe Handling and Use

Steps to Be Taken in Case Material is Released or Spilled

Waste Disposal Method

Precautions to be Taken in Handling and Storing

Other Precautions

Section VIII — Control Measures

Respiratory Protection (Specify Type)

Ventilation	Local Exhaust	Special
	Mechanical *(General)*	Other

Protective Gloves	Eye Protection

Other Protective Clothing or Equipment

Work/Hygienic Practices

　　　* USGPO 1986-491-529/45775

Genium Publishing Corporation
One Genium Plaza
Schenectady, NY 12304-4690 USA
(518) 377-8854

Material Safety Data Sheets Collection:

Sheet No. 52
Hydrogen Sulfide

Issued: 7/79   Revision: B, 9/92

Section 1. Material Identification

39

Hydrogen Sulfide (H_2S) Description: Formed as a byproduct of many industrial processes (breweries, tanneries, slaughter houses), around oil wells, where petroleum products are used, in decaying organic matter, and naturally occurring in coal, natural gas, oil, volcanic gases, and sulfur springs. Derived commercially by reacting iron sulfide with dilute sulfuric or hydrochloric acid, or by reacting hydrogen with vaporized sulfur. Used in the production of various inorganic sulfides and sulfuric acid, in agriculture as a disinfectant, in the manufacture of heavy water, in precipitating sulfides of metals; as a source of hydrogen and sulfur, and as an analytical reagent.

Other Designations: CAS No. 7783-06-4, dihydrogen monosulfide, hydrosulfuric acid, sewer gas, stink damp, sulfuretted hydrogen, sulfur hydride.

Manufacturer: Contact your supplier or distributor. Consult latest *Chemical Week Buyers' Guide*[73] for a suppliers list.

Cautions: Hydrogen sulfide is a highly flammable gas and reacts vigorously with oxidizing materials. It is highly toxic and can be instantly fatal if inhaled at concentrations of 1000 ppm or greater. Be aware that the sense of smell becomes rapidly fatigued at 50 to 150 ppm, and that its strong rotten-egg odor is not noticeable even at very high concentrations.

		NFPA
R	2	
I	4	
S	3	
K	3	

NFPA diamond: H 3, F 4, R 0

HMIS
H   3
F   4
R   0
PPE*
* Sec. 8

Section 2. Ingredients and Occupational Exposure Limits

Hydrogen sulfide: 98.5% *technical*, 99.5% *purified*, and CP (*chemically pure grade*)

1991 OSHA PELs
8-hr TWA: 10 ppm (14 mg/m³)
15-min STEL: 15 ppm (21 mg/m³)

1990 IDLH Level
300 ppm

1990 NIOSH REL
10-min Ceiling: 10 ppm (15 mg/m³)

1992-93 ACGIH TLVs
TWA: 10 ppm (14 mg/m³)
STEL: 15 ppm (21 mg/m³)

1990 DFG (Germany) MAK
TWA: 10 ppm (15 mg/m³)
Category V: Substances having intense odor
Peak exposure limit 20 ppm, 10 min momentary value, 4/shift

1985-86 Toxicity Data*
Human, inhalation, LC_{Lo}: 600 ppm/30 min; toxic effects not yet reviewed
Man, inhalation, LD_{Lo}: 5700 µg/kg caused coma and pulmonary edema or congestion.
Rat, intravenous, LD_{50}: 270 µg/kg; no toxic effect noted

* See NIOSH, *RTECS* (MX1225000), for additional toxicity data.

Section 3. Physical Data

Boiling Point: -76 °F (-60 °C)
Freezing Point: -122 °F (-86 °C)
Vapor Pressure: 18.5 atm at 68 °F (20 °C)
Vapor Density (Air = 1): 1.175
pH: 4.5 (freshly prepared saturated aqueous solution)
Viscosity: 0.01166 cP at 32 °F/0 °C and 1 atm
Liquid Surface Tension (est): 30 dyne/cm at -77.8 °F/-61 °C

Molecular Weight: 34.1
Density: 1.54 g/L at 32 °F (0 °C)
Water Solubility: Soluble*; 1g/187 mL (50 °F/10 °C), 1g/242 mL (68 °F/20 °C), 1g/ 314 mL (86 °F/30 °C)
Other Solubilities: Soluble in ethyl alcohol, gasoline, kerosine, crude oil, and ethylene glycol.
Odor threshold: 0.06 to 1.0 ppm†

Appearance and Odor: Colorless gas with a rotten-egg smell.

* H_2S solutions are not stable. Absorbed oxygen causes turbidity and precipitation of sulfur. In a 50:50 mixture of water and glycerol, H_2S is stable.
† Sense of smell becomes rapidly fatigued and can not be relied upon to warn of continuous H_2S presence.

Section 4. Fire and Explosion Data

Flash Point: None reported	**Autoignition Temperature:** 500 °F (260 °C)	**LEL:** 4.3% v/v	**UEL:** 46% v/v

Extinguishing Media: Let small fires burn unless leak can be stopped immediately. For large fires, use water spray, fog, or regular foam.
Unusual Fire or Explosion Hazards: H_2S burns with a blue flame giving off sulfur dioxide. Its burning rate is 2.3 mm/min. Gas may travel to a source of ignition and flash back. **Special Fire-fighting Procedures:** Because fire may produce toxic thermal decomposition products, wear a self-contained breathing apparatus (SCBA) with a full facepiece operated in pressure-demand or positive-pressure mode. Structural firefighter's protective clothing is not effective for fires involving H_2S. If possible without risk, stop leak. Use unmanned device to cool containers until well after fire is out. Withdraw immediately if you hear a rising sound from venting safety device or notice any tank discoloration due to fire. Do not release runoff from fire control methods to sewers or waterways.

Section 5. Reactivity Data

Stability/Polymerization: H_2S is stable at room temperature in closed containers under normal storage and handling conditions. Hazardous polymerization cannot occur. **Chemical Incompatibilities:** Hydrogen sulfide attacks metals forming sulfides and is incompatible with 1,1-bis(2-azidoethoxy) ethane + ethanol, 4-bromobenzenediazonium chloride, powdered copper + oxygen, metal oxides, finely divided tungsten or copper, nitrogen trichloride, silver fulminate, rust, soda-lime, and all other oxidants. **Conditions to Avoid:** Exposure to heat and contact with incompatibles. **Hazardous Products of Decomposition:** Thermal oxidative decomposition of hydrogen sulfide can produce toxic sulfur dioxide .

Section 6. Health Hazard Data

Carcinogenicity: The IARC,[164] NTP,[169] and OSHA[164] do not list hydrogen sulfide as a carcinogen. **Summary of Risks:** H_2S combines with the alkali present in moist surface tissues to form caustic sodium sulfide, causing irritation of the eyes, nose, and throat at low levels (50 to 100 ppm). Immediate death due to respiratory paralysis occurs at levels greater than 1000 ppm. Heavy exposure has resulted in neurological problems, however recovery is usually complete. H_2S exerts most of it's toxicity on the respiratory system. It inhibits the respiratory enzyme cytochrome oxidase, by binding iron and blocking the necessary oxydo-reduction process. Electrocardiograph changes after over-exposure have suggested direct damage to the cardiac muscle, however some authorities debate this. **Medical Conditions Aggravated by Long-Term Exposure:** Eye and nervous system disorders. **Target Organs:** Eyes, respiratory system and central nervous system. **Primary Entry Routes:** Inhalation, eye and skin contact.
Acute Effects: Inhalation of low levels can cause headache, dizziness, nausea, cramps, vomiting, diarrhea, sneezing, staggering, excitability, pale

Continued on next page

No. 52   Hydrogen Sulfide   9/92

Section 6. Health Hazard Data, *continued*

complexion, dry cough, muscular weakness, and drowsiness. Prolonged exposure to 50 ppm, can cause rhinitis, bronchitis, pharyngitis, and pneumonia. High level exposure leads to pulmonary edema (after prolonged exposure to 250 ppm), asphyxia, tremors, weakness and numbing of extremeties, convulsions, unconsciousness, and death due to respiratory paralysis. Concentrations near 100 ppm may be odorless due to olfactory fatigue, thus the victim may have no warning. Lactic acidosis may be noted in survivors. The gas does not affect the skin although the liquid (compressed gas) can cause frostbite. The eyes are very susceptible to H_2S keratoconjunctivitis known as 'gas eye' by sewer and sugar workers. This injury is characterized by palpebral edema, bulbar conjunctivitis, mucous-puss secretions, and possible reduction in visible capacity.
Chronic Effects: Chronic effects are not well established. Some authorities have reported repeated exposure to cause fatigue, headache, inflammation of the conjunctiva and eyelids, digestive disturbances, weight loss, dizziness, a grayish-green gum line, and irritability. Others say these symptoms result from recurring acute exposures. There is a report of encephalopathy in a 20 month old child after low-level chronic exposure.
FIRST AID Eyes: *Do not* allow victim to rub or keep eyes tightly shut. Gently lift eyelids and flush immediately and continuously with flooding amounts of water. Treat with boric acid or isotonic physiological solutions. Serious exposures may require adrenaline drops. Olive oil drops (3 to 4) provides immediate treatment until transported to an emergency medical facility. Consult a physician immediately. **Skin:** *Quickly* remove contaminated clothing and rinse with flooding amounts of water. For frostbite, rewarm in 107.6°F (42 °C) water until skin temperature is normal. *Do not* use dry heat. **Inhalation:** Remove exposed person to fresh air and administer 100% oxygen. Give hyperbaric oxygen if possible. **Ingestion:** Unlikely since H_2S is a gas above -60 °C. **Note to Physicians:** The efficacy of nitrite therapy is unproven. Normal blood contains < 0.05 mg/L H_2S; reliable tests need to be taken within 2 hr of exposure.

Section 7. Spill, Leak, and Disposal Procedures

Spill/Leak: Immediately notify safety personnel, isolate and ventilate area, deny entry, and stay upwind. Shut off all ignition sources. Use water spray to cool, dilute, and disperse vapors. Neutralize runoff with crushed limestone, agricultural (slaked) lime, or sodium bicarbonate. If leak can't be stopped in place, remove cylinder to safe, outside area and repair or let empty. Follow applicable OSHA regulations (29 CFR 1910.120).
Ecotoxicity Values: Bluegill sunfish, TLm = 0.0448 mg/L/96 hr at 71.6 °F/22 °C; fathead minnow, TLm = 0.0071 to 0.55 mg/L/96 hr at 6 to 24 °C.
Environmental Degradation: In air, hydrogen sulfides residency (1 to 40 days) is affected by temperature, humidity, sunshine, and the presence of other pollutants. It does not undergo photolysis but is oxidated by oxygen containing radicals to sulfur dioxide and sulfates. In water, H_2S converts to elemental sulfur. In soil, due to its low boiling point, much of H_2S evaporates quickly if spilled. Although, if soil is moist or precipitation occurs at time of spill, H_2S becomes slightly mobile due to its water solubility. H_2S does not bioaccumulate but is degraded rapidly by certain soil and water bacteria. **Disposal:** Aerate or oxygenate with compressor. For in situ amelioration, carbon removes some H_2S. Anion exchanges may also be effective. A potential candidate for rotary kiln incineration (1508 to 2912 °F/820 to 1600 °C) or fluidized bed incineration (842 to 1796 °F/450 to 980 °C). Contact your supplier or a licensed contractor for detailed recommendations. Follow applicable Federal, state, and local regulations.

EPA Designations
Listed as a RCRA Hazardous Waste (40 CFR 261.33): No. U135
SARA Toxic Chemical (40 CFR 372.65): Not listed
Listed as a SARA Extremely Hazardous Substance (40 CFR 355), TPQ: 500 lb
Listed as a CERCLA Hazardous Substance* (40 CFR 302.4): Final Reportable
  Quantity (RQ), 100 lb (45.4 kg) [* per RCRA, Sec. 3001 & CWA, Sec. 311 (b)(4)]

OSHA Designations
Listed as an Air Contaminant (29 CFR 1910.1000, Table Z-1-A & Z-2)
Listed as a Process Safety Hazardous Material (29 CFR 1910.119), TQ: 1500 lb

Section 8. Special Protection Data

Goggles: Wear protective eyeglasses or chemical safety goggles, per OSHA eye- and face-protection regulations (29 CFR 1910.133). Because contact lens use in industry is controversial, establish your own policy. **Respirator:** Seek professional advice prior to respirator selection and use. Follow OSHA respirator regulations (29 CFR 1910.134) and, if necessary, wear a MSHA/NIOSH-approved respirator. For < 100 ppm, use a supplied-air respirator (SAR) or SCBA. For < 250 ppm, use a SAR operated in continuous-flow mode. For < 300 ppm, use a SAR or SCBA with a full facepiece. For emergency or nonroutine operations (cleaning spills, reactor vessels, or storage tanks), wear an SCBA. *Warning! Air-purifying respirators do not protect workers in oxygen-deficient atmospheres.* If respirators are used, OSHA requires a respiratory protection program that includes at least: a written program, medical certification, training, fit-testing, periodic environmental monitoring, maintenance, inspection, cleaning, and convenient, sanitary storage areas. **Other:** Wear chemically protective gloves, boots, aprons, and gauntlets to prevent skin contact. Polycarbonate, butyl rubber, polyvinyl chloride, and neoprene are suitable materials for PPE. **Ventilation:** Provide general & local exhaust ventilation systems to maintain airborne concentrations below the OSHA PEL (Sec. 2). Local exhaust ventilation is preferred because it prevents contaminant dispersion into the work area by controlling it at its source.[103] **Safety Stations:** Make available in the work area emergency eyewash stations, safety/quick-drench showers, and washing facilities. **Contaminated Equipment:** Separate contaminated work clothes from street clothes and launder before reuse. Clean PPE. **Comments:** Never eat, drink, or smoke in work areas. Practice good personal hygiene after using this material, especially before eating, drinking, smoking, using the toilet, or applying cosmetics.

Section 9. Special Precautions and Comments

Storage Requirements: Prevent physical damage to containers. Store in steel cylinders in a cool, dry, well-ventilated area away from incompatibles (Sec. 5). Install electrical equipment of Class 1, Group C. Outside or detached storage is preferred. **Engineering Controls:** To reduce potential health hazards, use sufficient dilution or local exhaust ventilation to control airborne contaminants and to keep levels as low as possible. Enclose processes and continuously monitor H_2S levels in the plant air. Keep pipes clear of rust as H_2S can ignite if passed through rusty pipes. Purge and determine H_2S concentration before entering a confined area that may contain H_2S. The worker entering the confined space should have a safety belt and life line and be observed by a worker from the outside. Follow applicable OSHA regulations (1910.146) for confined spaces. H_2S can be trapped in sludge in sewers or process vessels and may be released during agitation. Calcium chloride or ferrous sulfate should be added to neutralize process wash water each time H_2S formation occurs. Control H_2S emissions with a wet flare stack/scrubbing tower. **Administrative Controls:** Consider preplacement and periodic medical exams of exposed workers emphasizing the eyes, nervous and respiratory system.

Transportation Data (49 CFR 172.101)

DOT Shipping Name: Hydrogen sulfide, liquefied
DOT Hazard Class: 2.3
ID No.: UN1053
DOT Packaging Group: --
DOT Label: Poison Gas, Flammable Gas
Special Provisions (172.102): 2, B9, B14

Packaging Authorizations
  Exceptions: --
  Non-bulk Packaging: 304
  Bulk Packaging: 314, 315

Vessel Stowage Requirements
  Vessel Stowage: D
  Other: 40
Quantity Limitations
  Passenger, Aircraft, or Railcar: Forbidden
  Cargo Aircraft Only: Forbidden

MSDS Collection References: 26, 73, 89, 100, 101, 103, 124, 126, 127, 132, 136, 140, 148, 149, 153, 159, 163, 164, 168, 171, 180
Prepared by: M Gannon, BA; Industrial Hygiene Review: PA Roy, MPH, CIH; Medical Review: AC Darlington, MPH, MD

There are several ways to obtain copies of MSDSs. Each manufacturer of regulated substances is required to furnish them on request when the product is purchased. Since employers must now keep appropriate MSDSs on file for use and review by their employees, you could request copies of the MSDSs from the industry as part of their permit application. Another source is the hazardous material management plan (HMMP) that is sometimes required in a permit application submitted to the POTW for wastewater discharge. If the industry you are inspecting submitted an HMMP, ask for a copy. Generally, the HMMP will contain copies of the MSDSs as well as additional information regarding the storage, handling and disposal of any chemicals used on site by the industry. In many instances you will need to consult a combination of several sources of information to prepare for your inspection.

Always remember that even an inspector must comply with all the safety rules of the firm that is being inspected. Before starting the inspection, ask your industrial contact at the firm if there are any special precautions that must be taken or if any changes in the firm's safety policies have occurred since the last inspection. During the inspection, continually be aware of your surroundings to ensure that you do not violate any of the safety requirements of the industry. Since most inspections are performed during the industry's regular working hours, stay alert while moving through the facility and be ready to react to any kind of situation that may occur.

QUESTIONS

Write your answers in a notebook and then compare your answers with those on page 392.

5.0F   List the inspector's personal safety gear that should be carried in each inspection vehicle.

5.0G   What types of gas detection devices should be available for an inspector's use?

5.0H   What is the first step when planning for an inspection of a specific industry?

5.1 DRIVING AND TRAFFIC SAFETY REQUIREMENTS

5.10 Hazards Encountered While Driving

An inspector spends a significant part of the work day traveling to and from sampling and inspection locations. Since the actions of the other drivers on the road cannot be controlled, you must develop good driving habits and drive defensively at all times. Safe driving actually begins before you ever enter the vehicle. Make a complete check of the vehicle and its different components (lights, horn, tires) to verify that everything is operational. Check your maps to become familiar with the streets and the traffic pattern of the areas where you will be working. Choose a route that lets you avoid traffic jams and other road problems. While driving through industrial areas, pay special attention to trucks and other vehicles that can pull out of driveways or that may double park in the streets and thus create instant traffic hazards. To drive defensively, keep the following key points in mind:

1. Keep your eyes moving from side to side,

2. Aim high on curves in your steering (don't cut corners),

3. Get the total or big picture,

4. Try to make sure that other drivers can see you, and

5. Leave yourself an out or escape route.

Practice these five "seeing" habits whenever you drive. They will help significantly in avoiding collisions between two or more vehicles or a vehicle and a pedestrian.

5.11 Traffic Control in the Streets

Inspectors are often called upon to sample industries at sampling points that are located in public streets or other high-traffic (congested) areas. If at all possible, work in these areas should not be performed by one person alone. As a minimum, two people should be present when working in an area where there is oncoming traffic.

Before starting to work in the street, develop a plan to thoroughly protect the work area and to provide proper advance warning, signing and guidance to oncoming traffic. This can be accomplished by using traffic control devices such as cones, barricades, flashers, "WORK AREA" signs, flaggers and the vehicle itself to control the traffic and to provide advance warning to oncoming motorists. Position appropriate warning signs and flaggers far enough in front of the work site to give motorists time to realize that they must slow down, be alert for the activity in the area, and safely change lanes or follow a detour around the work site. The exact distances for the placement of the traffic control devices and the types of devices to be used can vary depending upon the road conditions. Traffic speed, congestion, time of day, and local regulations also influence traffic control strategies. Appendix B contains some suggested traffic control strategies and guidelines but these should be modified as needed to meet local conditions in your area.

Once the motorists have been warned, they must be safely routed around the job site. Traffic can effectively be channeled from one lane to another around the work area through proper placement of barricades and cones. If possible, place your vehicle between the sampling point and oncoming traffic so that it can act as a warning barricade (amber dome light and side lights flashing) to discourage reckless drivers from driving into the work area. While working in the street or any other high-traffic area, inspectors and all other workers should be wearing orange vests to give them higher visibility and added protection.

For additional information on how to safely route traffic around a work site, see *OPERATION AND MAINTENANCE OF WASTEWATER COLLECTION SYSTEMS*, Volume I, Chapter 4, "Safe Procedures," Section 4.3, "Routing Traffic Around the Job Site," in this series of manuals.

QUESTIONS

Write your answers in a notebook and then compare your answers with those on page 393.

5.1A   How can an inspector ensure that an assigned work vehicle is safe to operate?

5.1B   List the five "seeing" habits that should be practiced when driving defensively.

5.1C   What safety equipment should inspectors be wearing when working in a street?

5.2   CONFINED SPACE ENTRY [2]

5.20   What Is a Confined Space?

A confined space has been defined by the National Institute for Occupational Safety and Health (NIOSH) in *CRITERIA FOR A RECOMMENDED STANDARD: WORKING IN CONFINED SPACES* (December 1979) as "...a space which by design has limited openings for entry and exit, unfavorable natural ventilation which could contain or produce dangerous air contaminants, and which is not intended for continuous employee occupancy." Confined spaces are further classified by NIOSH into three categories. These are:

CONFINED SPACE, CLASS "A" — A confined space that presents a situation that is immediately dangerous to life or health (IDLH[3]). These situations include but are not limited to oxygen deficiency, explosive or flammable atmospheres and/or concentrations of toxic substances.

CONFINED SPACE, CLASS "B" — A confined space that has the potential for causing injury or illness, if preventive measures are not used, but not immediately dangerous to life and health.

CONFINED SPACE, CLASS "C" — A confined space in which the potential hazard would not require any special modification of the work procedure.

NIOSH further describes the characteristics of the three confined space classes and the considerations for entry, working in and exiting each. In general, industrial waste inspectors would not be expected to enter any Class "A" or Class "B" confined spaces. The rules and regulations regarding entering and working in a Class "C" confined space will be discussed later in this chapter.

5.21   Confined Space Atmospheric Hazards

The three most commonly encountered hazards associated with work in confined spaces are oxygen deficiency/enrichment, flammable/explosive gases, and toxic gases. It is essential that you verify the presence or absence of these hazards before entering any confined space. Standard measures of these atmospheric hazards are expressed as upper and lower explosive limits (UEL and LEL), and upper and lower flammable limits (UFL and LFL) as defined below.

UPPER EXPLOSIVE LIMIT (UEL). The point at which the concentration of a gas in air becomes too great to allow an explosion upon ignition due to insufficient oxygen present.

LOWER EXPLOSIVE LIMIT (LEL). The lowest concentration of gas or vapor (percent by volume in air) that explodes if an ignition source is present at ambient temperature. At temperatures above 250°F the LEL decreases because explosibility increases with higher temperature.

UPPER FLAMMABLE LIMIT (UFL). The point at which the concentration of gas in air becomes too great to sustain a flame upon ignition due to insufficient oxygen present.

LOWER FLAMMABLE LIMIT (LFL). The lowest concentration of a gas or vapor (percent by volume in air) that burns if an ignition source is present.

Another measure of the concentration of a gas is TLV:

THRESHOLD LIMIT VALUE (TLV). The average concentration of toxic gas or any other substance to which a normal person can be exposed without injury during an average work week.

Testing for atmospheric hazards (Figure 5.1) is usually done with a gas detection instrument which may have one or more sensors for detecting up to four different gas types. The gas detectors used to monitor each of the four gas types in the atmosphere generally use the same operating principle. The gas detection system consists of three main components: the gas sensing unit or detector which senses the gas and generates a signal indicating its presence; the circuitry for signal transmittal to a central control area; and the conversion and indicating unit which changes the signal into a readout and activates designated alarms when certain levels of the gas are detected. The power supply for the unit can either be a built-in battery pack or an external source. For example, the amount of oxygen present in the atmosphere is usually measured by a direct reading detector using an oxygen sensor cell which generates an electrical signal that is proportional to the oxygen level in the atmosphere. The signal is converted to a readout that shows the percent of oxygen present and usually activates an alarm if the oxygen concentration is less than 19.5 percent or greater than 22.5 percent.

Although many different types of gas detectors are available for testing many different types of gases, the triple detector (or tridetector) type, or combination meter as it is also known, is generally preferred for use in industrial waste inspection work.

---

[2] *For additional information on confined space entry, see OPERATION AND MAINTENANCE OF WASTEWATER COLLECTION SYSTEMS, Volume I, Chapter 4, "Safe Procedures"; Section 4.4, "Classification and Description of Manhole Hazards"; Section 4.5, "Safety Equipment and Procedures for Confined Space Entry"; Section 4.6, "Final Precautions Before Manhole Entry"; and Section 4.7, "Procedures During Manhole Entry," in this series of manuals. Also see the definition of a confined space in "Pretreatment Words" at the beginning of this chapter.*

[3] *IDLH.   Immediately Dangerous to Life or Health. The atmospheric concentration of any toxic, corrosive, or asphyxiant substance that poses an immediate threat to life or would cause irreversible or delayed adverse health effects or would interfere with an individual's ability to escape from a dangerous atmosphere.*

Fig. 5.1   Inspector checking LEL strip chart

Inspector checks strip chart for explosimeter (LEL) to determine if any explosive wastes have been recorded on the chart. The inspector looks for operators' notations such as daily maintenance checks or explanations of any unacceptable readings or malfunctions.

This type of unit has three detectors (which may be built into one case) and can monitor and sample the atmosphere simultaneously for oxygen deficiency/enrichment, flammable/explosive gases and a toxic gas, usually hydrogen sulfide (H_2S) and/or carbon monoxide (CO). This type of instrument has preset alarms for each one of the gases and usually provides the following readouts:

PROBLEM	RANGE	ALARM SET POINT
O_2 deficiency	$0 - 25\%\ O_2$	Below 19.5% and Over 22.5% O_2
Combustibles	0 – 100% LEL	10% LEL
H_2S	$0 - 50$ ppm H_2S	10 ppm
or CO	0 – 100 ppm	CO 35 ppm

For inspection work, it is recommended that a portable unit be used which has a built-in pump that draws in the sample from the immediate area or a confined space when used with the appropriate sampling line and probe. The probe should be made from a nonmetallic material so that no spark can accidentally be generated.

5.210 Toxic [4] Gases

HYDROGEN SULFIDE (H_2S) — One of the most common toxic gases encountered in the wastewater field is hydrogen sulfide (H_2S). This gas is often generated in collection lines due to bacterial action on sulfate compounds or from sulfide compounds that may be present as part of an industrial discharge. At low concentrations (less than 3 ppm), hydrogen sulfide has the odor of rotten eggs. However, at higher concentrations (20 to 30 ppm and above), it cannot be smelled since it quickly paralyzes your olfactory system (sense of

---

[4] Toxic.   A substance which is poisonous to a living organism. Toxic substances may be classified in terms of their physiological action, such as irritants, asphyxiants, systemic poisons, and anesthetics and narcotics. Irritants are corrosive substances which attack the mucous membrane surfaces of the body. Asphyxiants interfere with breathing. Systemic poisons are hazardous substances which injure or destroy internal organs of the body. The anesthetics and narcotics are hazardous substances which depress the central nervous system and lead to unconsciousness.

smell). The threshold limit value (TLV), which is defined as the average concentration of toxic gas to which a normal person can be exposed without injury during an average work week, is 10 ppm for hydrogen sulfide. At 20 to 150 ppm it can cause nervous system damage and irritation of the eyes and the respiratory system. At 500 to 1,000 ppm it causes death within minutes due to the paralysis of the respiratory center. Hydrogen sulfide can act both as an irritant and as an asphyxiant and repeated exposures at lower concentrations will result in eye damage including conjunctivitis, photophobia, tearing, pains, and blurred vision. Aside from its toxic effects, hydrogen sulfide is also a flammable gas (LEL of 4.3% and UEL of 46%) and can represent a dangerous fire hazard when exposed to heat or flame.

Although hydrogen sulfide is the most common toxic gas that can be present during sampling work, other toxic gases may also be encountered by the industrial waste inspector. These include hydrogen cyanide, chlorine and carbon monoxide. Each of these gases is discussed in the following paragraphs.

HYDROGEN CYANIDE (HCN) — This gas is an extremely toxic gas which can rapidly cause death through asphyxiation by preventing the transfer of oxygen within the bloodstream. The gas, which has a slight odor of bitter almonds, can be generated when improperly treated plating wastewaters containing cyanide salts are discharged and then allowed to react with any acidic wastewaters present in the collection system. The TLV for hydrogen cyanide is 10 ppm (the gas has an LEL of 6% and a UEL of 40%) and exposure to 100 to 200 ppm for 30 to 60 minutes is usually fatal.

CHLORINE (Cl_2) — Although chlorine is primarily used in the wastewater field as a disinfectant, it can occasionally be encountered during the sampling or inspection of an industry which uses it for cyanide destruction in its pretreatment system. Chlorine is extremely irritating to the mucous membranes of the throat and the nose and eyes since it combines with moisture to form nascent (atomic) oxygen and hydrochloric acid which will attack any tissue that they contact. Chlorine has a noticeable odor at 3.5 ppm (TLV is 1 ppm) and can cause immediate irritation to the throat at 15 ppm. Concentrations of 50 ppm are dangerous and can cause pulmonary edema (fluid in the lungs). Exposure to 1,000 ppm for even a short time can be fatal.

CARBON MONOXIDE (CO) — Carbon monoxide is a colorless and odorless gas which is primarily encountered from the exhaust of internal combustion engines. Its toxicity stems from the fact that its affinity for hemoglobin of red blood cells is over 200 times greater than oxygen. Thus exposure to CO will interfere with the transfer of oxygen to the body's tissues and may result in asphyxiation. The TLV for CO is 35 ppm with a 200 ppm concentration in air causing headaches and being

considered dangerous. Loss of consciousness occurs with exposure to higher concentrations and concentrations of 1,000 ppm may result in death.

Although many other types of toxic gases exist, most of them are not expected to be encountered during typical inspection and sampling work. An exception to this generalization is the presence of vapors from chlorinated solvents. Although the discharge of chlorinated solvents is generally prohibited or only low concentrations are allowed under local and federal regulations, accidental discharges can still occur from plating shops or electronics manufacturing facilities where these types of solvents are used for degreasing of parts or dissolving photo resists. For example, trichloroethane (TCA) can act as an irritant to the eyes and nose as well as being narcotic at higher concentrations.

Since chlorinated solvents are heavier than air, they readily displace the oxygen in any confined space and thus they can also present an asphyxiation danger. Even though many industries have switched from using the more toxic chlorinated solvents to the less toxic freons, the freons are also heavier than air and thus pose some of the same dangers as the chlorinated solvents. Fortunately, concerted efforts are being made by the industries to recycle the freons and many industries have voluntarily eliminated the use of the more toxic compounds such as trichloroethylene (TCE), chloroform, carbon tetrachloride and dichloromethane (methylene chloride).

Of the toxic gases listed, hydrogen sulfide probably presents the greatest danger because of its extreme toxicity and because of its frequent occurrence. Gas detectors are available to test for all of the gases described above. However, most combination instruments available will come equipped with either a hydrogen sulfide or carbon monoxide detector for their toxic gas monitor because of the toxic nature of these two gases and their frequent occurrence in confined spaces. Single detector units, indicator badges and indicator or detector tubes are also available for the gases listed above as well as many other gases that can be encountered in inspection or sampling work. These items should be used as supplements to the triple detector unit in monitoring the atmosphere, exposure level, and duration of any exposure to harmful constituents in the atmosphere.

5.211 Flammable/Explosive Gases

Flammable or explosive gases, mists and vapors represent the second class of hazards that can be encountered in a collection system or confined space during inspection and sampling work. The source of these materials can be leaking underground storage tanks, accidental discharges, spills, and deliberate discharges. Gasoline is one of the more common flammable substances found in a collection system because of the large number of service stations with their underground fuel storage tanks. Gasoline and other flammable liquids can travel along the bedding of pipes or in underground streams and then enter the collection system through infiltration at leaky joints. Since these materials float, they have a tendency to spread out and vaporize in the collection system. Once mixed with air in a space such as a sewer system pipe, the effect that even a small quantity of flammable liquid can have is substantial. For example, a gallon of gasoline when vaporized and mixed with air will have the same explosive force as 14 sticks of dynamite.

A flammable gas that is sometimes generated within the collection system is methane (CH_4). This gas is colorless and odorless when pure. In a sewer line, methane is often mixed with H_2S, which will give it a characteristic hydrogen sulfide gas odor. Both methane and hydrogen sulfide are generated under anaerobic (no oxygen present) conditions by bacterial action. Thus care must be taken when sampling in lines with low flows or very flat slopes since sludge pockets or layers can form which contain these gases. Disturbing the sludge pockets or layers can release sufficient quantities of these gases to present a health and fire hazard. Natural gas from leaking gas mains and home services is another flammable gas that may be encountered.

Each flammable and/or explosive gas, vapor or mist when mixed with air has its own explosive range. Within this range an ignition source can create a flame or an explosion. The gas concentration outside this range is either too low (lean) or too high (rich) to sustain combustion. The explosive range is defined as being within the lower and upper explosive limit boundaries. The point where the concentration of a gas in air is sufficient to result in a flame or explosion if an ignition source is present is the *LOWER EXPLOSIVE LIMIT (LEL)* (Figure 5.2) or *LOWER FLAMMABLE LIMIT (LFL)*. The point at which the concentration of the gas in air becomes too large and insufficient oxygen is present to have an explosion or a flame upon ignition is known as the *UPPER EXPLOSIVE LIMIT (UEL)* or *UPPER FLAMMABLE LIMIT (UFL)*.

Gas detectors are designed and calibrated to read a 0 to 100% LEL range with a 100% LEL being the point where a sufficient concentration of the gas or vapor is present to sustain and support combustion. Most gas detectors available today conform with the industrial safety standard and activate an alarm when 10% LEL is reached or exceeded (see Figure 5.3). Since each gas has its own characteristic LEL range, the gas detector must be calibrated for the specific gas that is being monitored by using either the same gas or a very similar one during the calibration procedure. For example, methane is

generally used to calibrate when checking for it and pentane is often used to calibrate when checking for gasoline vapors or similar solvents.

5.212 Oxygen-Deficient/Enriched Atmospheres

The third major type of atmospheric hazard that can be encountered during inspection work is an oxygen-deficient atmosphere. An oxygen-deficient atmosphere is one where the air with its oxygen content has been reduced by another gas or the oxygen has been removed. Normal air contains approximately 20.9 percent oxygen at sea level with the remaining constituents being primarily nitrogen (about 78.1 percent), argon (about 1 percent) and a few traces of other inert gases. The alarm point, as set on most gas detectors, is at 19.5 percent to indicate that a potentially dangerous situation exists. The effects of oxygen deprivation include hypoxia (oxygen shortage and shortness of breath) at 17 percent. Loss of consciousness occurs rapidly at 6 to 10 percent and death occurs rapidly (within minutes) at less than 6 percent.

Oxygen deficiency in an enclosed atmosphere can occur for several reasons. Some of the more common causes include bacterial action which uses up the oxygen, displacement by other gases (can be both toxic and/or inert gases such as nitrogen or carbon dioxide), oxidation of metals or other materials that depletes the oxygen level, adsorption of the oxygen onto surfaces, and combustion. Since the oxygen content of an enclosed atmosphere can change quite rapidly, the space must be continually monitored to ensure that the oxygen does not drop below an acceptable level. Also, the alarm should be activated if the oxygen level rises above 22.5 percent because high levels of oxygen increase the possibility of an explosion.

5.213 Humidity

Another atmospheric hazard in a confined space is humidity. Intense high temperature and humidity may cause suffocation. For example, a not uncommon temperature of 110°F (43°C) and 80 percent relative humidity is very dangerous. Moisture or condensation forming on manhole rungs, ladders, or concrete flooring creates slipping hazards and, in addition, moisture may react with other gases present in the manhole and cause manhole rungs and concrete to be corroded and weakened.

QUESTIONS

Write your answers in a notebook and then compare your answers with those on page 393.

5.2A   Define a confined space.

5.2B   List the three items sampled and monitored by a triple detector type or combination gas meter.

5.2C   What type of material should be used to make the probe for a gas detection device?

5.2D   How can hydrogen cyanide (HCN) be generated?

5.2E   How can chlorine gas be encountered by an inspector?

5.2F   How can an inspector encounter chlorinated solvents?

5.2G   How is an oxygen-deficient atmosphere created?

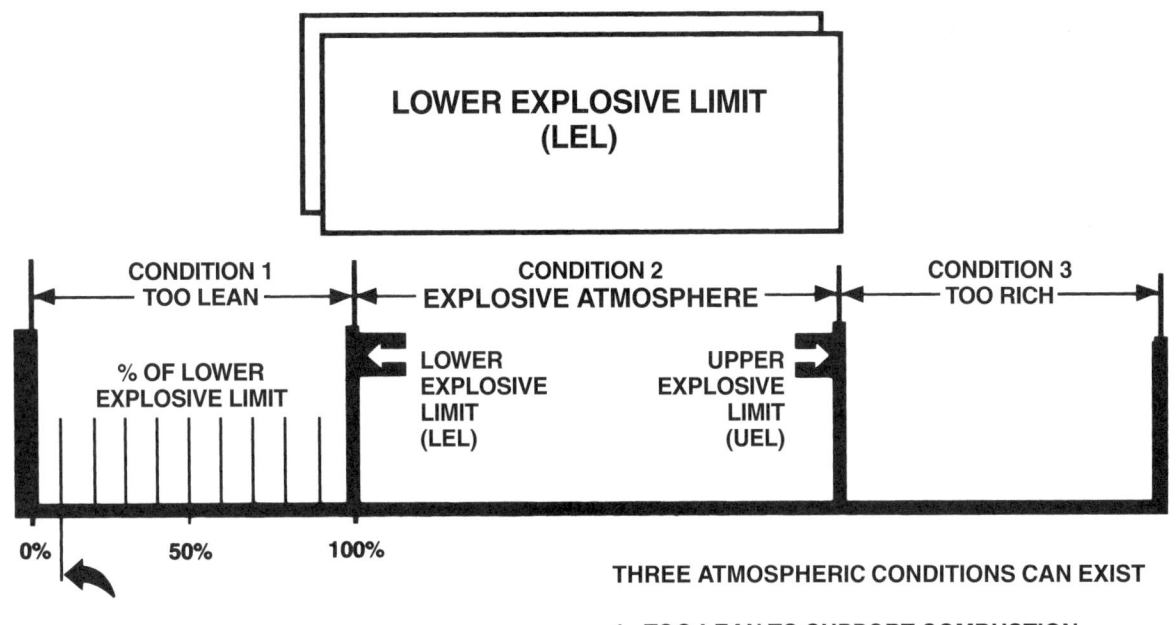

CONDITION 1
TOO LEAN

CONDITION 2
EXPLOSIVE ATMOSPHERE

CONDITION 3
TOO RICH

% OF LOWER
EXPLOSIVE LIMIT

LOWER
EXPLOSIVE
LIMIT
(LEL)

UPPER
EXPLOSIVE
LIMIT
(UEL)

0%   50%   100%

ATMOSPHERIC TESTING EQUIPMENT —
ALARM SET @ 10% LEL

THREE ATMOSPHERIC CONDITIONS CAN EXIST

1. TOO LEAN TO SUPPORT COMBUSTION
2. MIXTURE JUST RIGHT, EXPLOSION OCCURS
3. MIXTURE TOO RICH TO SUPPORT COMBUSTION

Fig. 5.2   Relationship between Lower Explosive Limit (LEL) and Upper Explosive Limit (UEL)

A sidestream of the waste is agitated to simulate turbulent conditions in the sewer. A sensor in the tank will set off an alarm if 10% of the lower explosive limit is reached.

Closeup view of explosive gas detection chamber with lid removed to show sensor. Note the turbulence of the waste that has been air-agitated to drive off explosive gases.

Fig. 5.3   Explosive gas detection chamber

5.22   Interpretation of Atmospheric Testing Results

Although the use of a gas detector is crucial for testing the safety of an atmosphere, there are certain considerations that must be kept in mind during its use. Basically, a gas detector can provide information about a potentially hazardous situation. However, it will not protect a person nor will it guarantee a safe condition since it can show a false reading. The use of a gas detector does not replace a responsible person who must make decisions regarding the potential severity of a hazard. Instead, it aids that person in making the decisions. To do so, however, *A GAS DETECTOR MUST BE PROPERLY MAINTAINED AND CALIBRATED ON A REGULAR BASIS.* All calibration data should be entered into a logbook which then becomes part of the permanent operating record for that instrument or for an agency's gas detectors (Figure 5.4). (See Appendix C for typical procedures for the use and calibration of gas detectors.) The person using the detector must be trained in its use, understand the principles involved in its operation, and be able to interpret the results and keep records of readings obtained. These readings should also be entered in the logbook or some other permanent form of recordkeeping along with the operator's initials, the monitoring location, and the date and time the readings were obtained (Figure 5.5).

If for some reason an alarm is obtained while testing the atmosphere of a manhole or another confined space which is to be entered, vent the manhole or space for at least ten minutes and then retest. If the alarm sounds again, use a portable blower with a vaporproof, enclosed motor to ventilate the manhole or space for at least five minutes. (Caution must be used in the placement of the blower so that its air intake does not pull in the exhaust of any vehicles being driven or idling nearby.) Rerun all three tests with the blower still operating. Record the results of all three tests in the logbook. If no alarm sounds during the retesting, the procedure for entering the confined space may be started *WHILE THE VENTILATING AND MONITORING OF THE CONFINED SPACE CONTINUES.*

5.23   Entering a Confined Space

The entry of confined spaces by industrial waste inspectors is generally not recommended unless absolutely necessary. Entry should never be attempted by anyone unless the person has been fully trained and meets all the requirements for entry. Since it is not feasible to discuss all the rules and requirements for entry of a confined space within the scope of the chapter, we will focus here on essential precautions and procedures that will help ensure your safety. For additional information, anyone contemplating the entry of a confined space should consult the regulations published by OSHA and all other pertinent requirements before proceeding (see Appendix D).

Several very important steps must be taken in preparation for entry into a confined space. A rescue procedure must be established by the inspector which can be implemented in case an emergency develops. The rescue procedure must have been practiced frequently enough during training sessions prior to the entry so that an efficient and calm response occurs in reaction to any emergency.

A confined space entry permit (Figure 5.6) must be issued by the responsible supervisor and/or safety department representative before the actual entry process is started. Commonly, the permit is valid for a limited time only and specifies the conditions for entry. The permit can be used by the supervisor or lead person as a checklist to ensure that all the appropriate precautions have been taken prior to starting the entry into the space. The permit forms used at some agencies are more detailed than the one shown in Figure 5.6. The permit may specify that hoisting equipment includes a tripod and anti-fall protection or that protective clothing includes hard hat, coveralls, gloves, boots and suitable eye protection. On the permit, the supervisor should specify the number of watch persons to be at the site of each confined space entry. Also under step 5, some agencies will add an oxygen content maximum of 22.5 percent.

Prior to entry into any confined space, the supervisor should assemble the crew involved and review the sequence of activities to complete the job, each person's role, safety equipment needed, and safe procedures.

The testing of the atmosphere must include tests for oxygen deficiency/enrichment, combustible gases, and toxic gases. A separate detector for measuring the carbon monoxide level is usually needed. First, check the oxygen level before any LEL measurements are made. If the oxygen level is greater or less than found under normal atmospheric conditions, adjust the LEL readings as needed. Higher than normal oxygen levels may require a lower LEL.

The equipment needed and the number of persons required for the entry to occur depend on the classification of the confined space. For a Class "C" space, the person entering the space should be outfitted with protective equipment and clothing including eye protection, a hard hat, proper foot protection, full-coverage work clothing and gloves of the appropriate material to provide hand protection. The person should also wear a full-body parachute-type harness (which will keep the body vertical when pulled up) with lanyard (made of nylon rope) attached to it by D rings (Figure 5.7). The other end of the lanyard, if possible, should be attached to a hoisting device equipped with anti-fall protection such as commercially available tripods. A second person must attend to the hoisting device at all times and a third person must be readily available to assist if needed. Communication must be maintained between the person in the space and the employee attending the hoisting device. If the confined space to be entered is a manhole, test the steps of the manhole to confirm their integrity and ability to hold the weight of the person entering the manhole.

The site supervisor who issued the entry permit determines whether or not the person who is to work within the confined space needs to wear a respirator. The recommended type of

GAS DETECTORS

R.B.N. = Replace battery nicade
R.O.2. = Replace O_2 cell
CAL. = Calibrated
B.C. = Battery charger OK
R.B.L. = Replaced battery lithium

R.M.S. = Replaced methane sensor
R.T.S. = Replaced toxic sensor
S.O. = Sent out for repair, etc.
C.O. = Checked out

Brand	Serial Number	Problems Found — What Was Done to Repair		Date & Initials
MSA	0357	CO, CAL, Batt. CK.	RAS	7-14-96
"	F-06-0734	CO, CAL, Batt. CK.	RAS	7-14-96
"	F-06-0735	CO, CAL. Batt. CK.	RAS	7-14-96
"	6192	CO, CAL, Batt. CK.	RAS	7-14-96
GASTECH	20045	CO, CAL, BC	RAS	7-14-96
"	20043	CO, CAL, BC	RAS	7-14-96
"	20040	CO, CAL, BC	RAS	7-14-96
DYN	9027	CO, CAL, BC	RAS	7-14-96
"	20061	CO, CAL, BC	RAS	7-14-96
"	20050	CO, CAL, BC	RAS	7-14-96
"	3058	CO, CAL, BC	RAS	7-14-96
"	3057	CO, CAL, BC	RAS	7-14-96
"	3056	CO, CAL, BC	RAS	7-14-96
GASTECH	20044	CO, CAL, BC	RAS	7-15-96
"	20046	CO, CAL, BC	RAS	7-15-96
"	20042	Returned from factory (Repaired & CAL by factory)	RAS	7-15-96
GASTECH	20043	CO BC Cal	KBP	8-11-96
"	20044	CO BC Cal	KBP	8-11-96
	20045	CO BC Cal	KBP	8-11-96
	20046	CO BC Cal	KBP	8-11-96
	20042	out of service — Bad Con B cell		
MSA	0357	CO BC cal	KBP	8-11-96
GASTECH	20040	CO Cal BC	KBP	8-11-96
MSA	F-06-0735	CO Cal BC	KBP	8-11-96
"	6192	CO Cal BC	KBP	8-11-96
"	F-06-0734	CO Cal BC?	KBP	8-11-96
Dynamation	3057	CO, RO2, Cal, BC	KBP	8-11-96
"	3058	CO, Cal, BC	KBP	8-11-96
"	20058	CO, RO2, Cal, BC,	KBP	8-11-96
"	3056	CO, RMS, Cal, BC,	KBP	8-11-96
"	20061	CO, Cal, BC	KBP	8-11-96
MSA	0729	CO, CAL, BC	RAS	8-14-96

Fig. 5.4   Calibration log for gas detectors

DATE	TIME	LOCATION	O_2	H_2S	L.E.L.	INSPECTOR
6/24	1500	Borden Chemical	20.6	0	6%	ZHC
6/24	1532	Borden Chemical	20.6	0	6%	ZHC
6/29	1003	Micro Automation	20.9	0	0	JR ZV
6/29	1032	LSI LOGIC	20.9	0	0	JR ZV
7/2	0935	TREND	20.9	0	0	ZV JR
7/20	1011	ZYNG	20.8	0	0	ZV JR
7/2	1024	Seagate	20.8	0	0	ZV JR
7/6	1007	Fremont Plating	20.9	0	0	ZV BX
7/6	1022	Trend Circ	20.9	0	0	ZV RM
7/6	1045	Oktel	20.9	0	0	ZV RM
7/6	1110	NUMMI	20.1	0	0	ZV RM
7/8	1010	Serra Corp	20.9	0	0	ZV JR
7/8	1030	Sumitomo	20.9	0	0	ZV JR
7/8	1045	Global Plating	20.9	0	0	ZV JR
7/8	1125	ALPHASIL	20.9	0	0	ZV JR
7/9	1015	ALPHASIL	20.9	0	0	ZV RM
7/9	1100	Serra	20.9	0	0	ZV RM
7/9	1130	Trimedia	20.5	0	0	ZV RM
7/13	0910	INLAND CONT.	20.8	0	0	ZV RM
7/13	1000	Evergreen	20.8	0	0	ZV RM
7/13	1030	H. B. Fuller	20.8	0	0	ZV RM
7/13	1430	Evergreen	20.5	0	0	ZV RM
7/14	0915	INLAND	20.5	0	12	ZV RM

Fig. 5.5   Log of gas detector readings

Confined Space Pre-Entry Checklist/Confined Space Entry Permit

Date and Time Issued: _____ Date and Time Expires: _____ Job Site/Space I.D.: _____

Job Supervisor: _____ Equipment to be worked on: _____ Work to be performed: _____

Standby personnel: _____ _____ _____

1. Atmospheric Checks: Time _____ Oxygen _____ % Toxic _____ ppm

   Explosive _____ % LEL Carbon Monoxide _____ ppm

2. Tester's signature: _____

3. Source isolation: (No Entry) N/A Yes No

   Pumps or lines blinded,
      disconnected, or blocked ( ) ( ) ( )

4. Ventilation Modification: N/A Yes No

   Mechanical ( ) ( ) ( )

   Natural ventilation only ( ) ( ) ( )

5. Atmospheric check after isolation and ventilation: Time _____

   Oxygen _____ % > 19.5% < 23.5% Toxic _____ ppm < 10 ppm H_2S

   Explosive _____ % LEL < 10% Carbon Monoxide _____ ppm < 35 ppm CO

Tester's signature: _____

6. Communication procedures: _____

7. Rescue procedures: _____

8. Entry, standby, and backup persons Yes No

   Successfully completed required training? ( ) ( )

   Is training current? ( ) ( )

9. Equipment: N/A Yes No

   Direct reading gas monitor tested ( ) ( ) ( )

   Safety harnesses and lifelines for entry and standby persons ( ) ( ) ( )

   Hoisting equipment ( ) ( ) ( )

   Powered communications ( ) ( ) ( )

   SCBAs for entry and standby persons ( ) ( ) ( )

   Protective clothing ( ) ( ) ( )

   All electric equipment listed for Class I, Division I,
      Groups A, B, C, and D, and nonsparking tools. ( ) ( ) ( )

10. Periodic atmospheric tests:

   Oxygen: ___% Time ___; ___% Time ___; ___% Time ___; ___% Time ___;

   Explosive: ___% Time ___; ___% Time ___; ___% Time ___; ___% Time ___;

   Toxic: ___ppm Time ___; ___ppm Time ___; ___ppm Time ___; ___ppm Time ___;

   Carbon Monoxide: ___ppm Time ___; ___ppm Time ___; ___ppm Time ___; ___ppm Time ___;

We have reviewed the work authorized by this permit and the information contained herein. Written instructions and safety procedures have been received and are understood. Entry cannot be approved if any brackets ( ) are marked in the "No" column. This permit is not valid unless all appropriate items are completed.

Permit Prepared By: (Supervisor) _____ Approved By: (Unit Supervisor) _____

Reviewed By: (CS Operations Personnel) _____

          (Entrant)   (Attendant)   (Entry Supervisor)

This permit to be kept at job site. Return job site copy to Safety Office following job completion.

Fig. 5.6 Confined space pre-entry checklist/permit

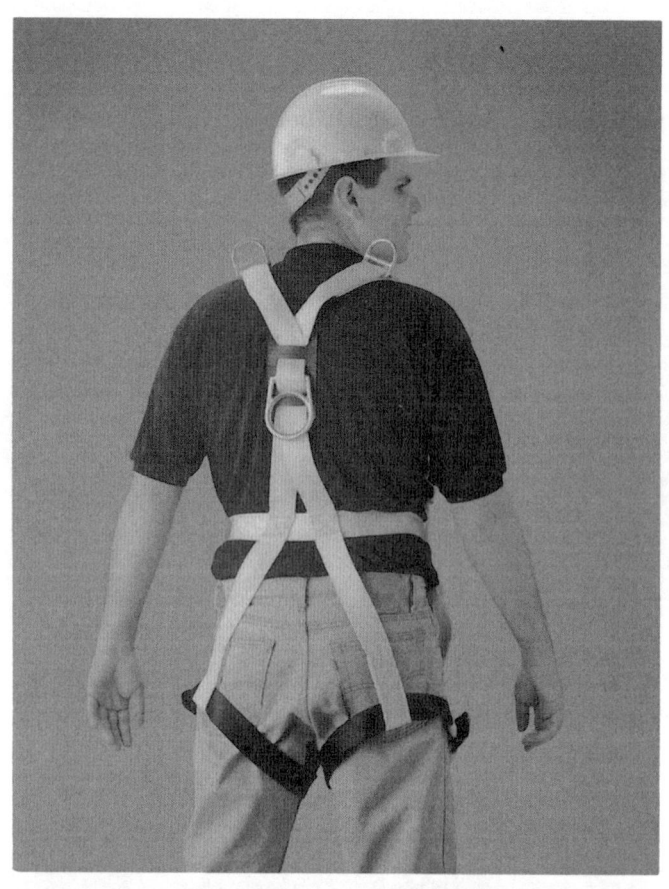

Fig. 5.7   Full-body harnesses

(Permission of Miller Equipment Division, ESB Incorporated, Franklin, PA)

respirator has its own air supply that becomes the air source for the wearer. This type of unit is often referred to as a self-contained breathing apparatus or SCBA. The volume of air that a SCBA can supply will depend on the type of unit. The length of time that the SCBA is safe to use depends on the rate of air consumption by the wearer. This rate will vary with each individual wearing the unit and the type of activity that is being performed at the time. Most SCBA units come with cylinders containing approximately a thirty- or sixty-minute supply of air, although there are some units which have a very limited supply (five minutes or less) and which are designed primarily for emergency escape purposes.

Before a person can use any respiratory protective equipment, a physician must determine that the person can physically perform the work while wearing the required respiratory equipment. This determination must have been made within the past year. The equipment cannot be used until written operating procedures for the selection, instruction and training, cleaning and sanitizing, inspection and maintenance have been established and implemented. No person should be permitted to use respiratory equipment unless they have been instructed and trained in the need, use, sanitary care and limitations of that particular piece of equipment. The use of such equipment is also prohibited if a gas-tight fit cannot be obtained due to the presence of facial or head hair or any other cause. The use of respiratory equipment is also precluded if a person is wearing contact lenses.

In summary, work within confined spaces by industrial waste pretreatment inspectors should be avoided if at all possible. If the need arises to do so, established procedures for entry, working within and exiting the confined space must be followed closely. Practice your agency's standard rescue procedure so you will be able to implement it quickly if any difficulties arise during the confined space work.

The most important things to remember about entering a confined space are to first *RECOGNIZE* what a confined space is, then to *TEST AND EVALUATE* its atmosphere prior to entry, and to continue to *MONITOR* the atmosphere while the work is being performed. Do not underestimate the seriousness of the hazards you might encounter whenever you enter or work in a confined space.

QUESTIONS

Write your answers in a notebook and then compare your answers with those on page 393.

5.2H Why is the location of a blower used to ventilate a confined space important?

5.2I What precautions must be taken *BEFORE* entering a manhole?

5.2J Why must a physician examine any person who may be required to use a respirator?

5.2K List the key points to remember about entry of a confined space.

5.3 HAZARDOUS MATERIALS

5.30 Types of Hazardous Materials Encountered During Inspections and Sampling

Chemical hazards that can be encountered while performing industrial waste inspection work include substances that are toxic or poisonous, corrosive, explosive or flammable, carcinogenic (cancer-causing), biologically active, and, in rare instances, radioactive. To minimize any potential exposure to these chemical hazards, it is important to be aware of and to understand the signs and placards that are used to designate the nature and hazard class of each of these materials.

Standards for labeling and placarding have been set by such federal agencies as the Department of Transportation (DOT), OSHA, and EPA, as well as by some individual state agencies. Much standardization has occurred in the requirements for labeling containers and the vehicles that transport hazardous materials thus making the recognition of the type of hazard relatively simple. Difficulties arise if the labels are missing, torn or illegible. In addition to the placarding, as prescribed by EPA and DOT, for hazard types such as radioactive, flammable, corrosive, or oxidizing materials, individual labels bearing the chemical or common name (not the trade name alone), plus additional information regarding the substance, must be attached in a specific manner. Some types of information that must be included on the labels are:

1. *SIGNAL OR WARNING WORDS* (OSHA recommends words like "Warning," "Caution" and "Notice" to emphasize the degree of hazard present),

2. *CAUTION OR HAZARD PHRASES* (Corrosive, Skin Irritant, Toxic),

3. *FIRST AID INFORMATION,* and

4. *STORING AND HANDLING DATA.*

Thus, a careful reading of the placards and labels on a container can give you important information regarding the hazard degree of the material and the precautions that must be taken in handling it.

As indicated earlier, Threshold Limit Values or TLVs represent conditions under which it is believed that nearly all workers may be repeatedly exposed day after day without adverse effects. TLVs are useful guidelines for the evaluation of industrial atmospheres. However, the use of TLVs to determine the safe exposure level for certain chemicals must be done with an understanding that for certain materials the TLV may actually represent a time weighted average (TWA) concentration and that the instantaneous values of the material may be higher. To calculate the approximate instantaneous values, multiply the TLV by a factor of three for materials whose TLVs are 0 to 1 ppm; multiply by 1.25 times for substances with TLVs in the range of 100 to 1,000 ppm.

Exposure limit values are often listed both in terms of TLVs and Permissible Exposure Limits (PELs). PELs represent a term used by OSHA to describe occupational health standards.

5.300 Corrosive Materials

Corrosive materials are one of the common types of chemicals that you may encounter while working as an industrial waste pretreatment inspector since they are used in many different types of industries. If a chemical, through its chemical action, is destructively injurious to body tissues or other materials, it is considered corrosive. Corrosive materials usually act on body tissue in one of the following ways:

1. Destruction of protein and/or body tissues,

2. Direct chemical attack such as dehydration, nitration (the yellowing of skin by nitric acid) or oxidation of body tissue, and

3. Disruption of cell membranes.

The most common types of corrosive chemicals are acids (both inorganic ones like hydrochloric (HCl) and sulfuric (H_2SO_4) and organic ones like acetic acid (CH_3COOH)) and caustics (also known as bases or alkalies). Although these are the most frequently encountered corrosive substances, many other materials that are not acids or bases, such as fluorine (F_2), chlorine (Cl_2) and nitric oxide (NO), are also very corrosive and hazardous.

Pure water, in which the hydrogen ion and hydroxide ion concentrations are equal to each other, is considered a neutral substance and has a pH of 7. The pH of a solution is defined as being the value of the negative log of the hydrogen ion concentration. As the hydrogen ion concentration increases, the pH drops and the solution is considered to be more acidic. If the hydrogen ion concentration decreases, the pH increases and the solution becomes more caustic or basic. In either case, increased deviation from neutrality causes the solution to become more corrosive until it reaches the point where it causes damage to tissue and becomes a hazard.

The attack on tissue by corrosive materials is usually caused by direct contact with the corrosive substance followed by absorption of the material. The most common point of exposure is on the skin, although contact of the corrosive material with the eyes is usually cause for more concern. The most serious injury generally occurs, however, when the corrosive material is ingested through either swallowing or inhalation. The seriousness of the corrosive injury will depend on the nature of the material, the length of time of the exposure and how quickly medical attention or first aid is obtained.

Exposure to corrosive materials can lead to two types of injuries. Primary injuries occur at the point of contact, are usually local in extent, and usually don't affect other bodily systems. They can be as mild as a slight reddening of the skin or as severe as the actual charring and total destruction of the tissue. Primary exposure of eye tissue to corrosive materials should always be treated as a very serious situation. Take immediate remedial action such as flushing the eyes with water and get medical attention as soon as possible. Primary exposure to the respiratory tract through inhalation usually results in serious coughing attacks and injury to the mucous membranes of the nose and throat and to the alveolar (air cell) membranes of the lung.

Secondary injuries from exposure to corrosive materials may be both local and systemic (involving the entire body) and may not be immediately apparent. In the case of inhalation of corrosives, pulmonary edema (the formation of fluid in the lungs) is usually a secondary injury that occurs shortly after the exposure has taken place.

An important consideration that should be kept in mind when dealing with corrosives is that strongly acidic substances affect body tissues in a different way than strongly basic substances. Strong acids cause an immediate reaction with the body's proteins (denaturation). This reaction causes an immediate sensation of pain which alerts the person that exposure to a corrosive chemical has taken place. On the other hand, strong bases do not cause this reaction with protein. No immediate sensation of pain is felt and the person may not even realize that an exposure to caustic material has occurred. Therefore, the period of exposure to caustics may be longer and the severity of the injury may be greater since the material had more time to be absorbed by the tissue.

5.301 Solvents and Flammable Materials

Solvents and flammable materials represent a second major category of hazardous compounds that inspectors may encounter. A solvent is any substance that is used to dissolve another substance in it. A solvent can act as a carrier to remove the dissolved substance (as is the case when freon is used to remove oils and grease from metal parts) or it can facilitate the transfer of the dissolved substance to another medium. Organic solvents, which may be flammable or toxic, or both, use the above listed properties to either dissolve the oils from the skin or, if dissolved materials are present in the solvent, they can make the skin or other tissues more permeable to the solute. In many instances, the solvent itself can be just as toxic or even more toxic than the substance dissolved in it.

Alcohols, ketones (acetone, MEK) and hydrocarbons (pentane, hexane) are all flammable solvents. So are common substances such as gasoline, kerosene and paint thinners which are used in many different businesses to clean motors and mechanical parts. Organic solvents that are nonflammable usually are halogenated (contain either chlorine, bromine, iodine, fluorine or cyanogen) materials such as the freons and the various chlorinated hydrocarbons (for example, trichloroethylene (TCE), chloroform, and carbon tetrachloride). Because of their ability to dissolve oils and greases, most organic solvents will damage the skin by dissolving its natural oils and thereby causing dry skin dermatitis or even blistering of the skin.

The inhalation of solvent vapors can also be very damaging to the respiratory system. Both hydrocarbon and chlorinated solvents, upon inhalation, can cause damage to the lungs, liver and kidneys. In many instances the inhalation or exposure to solvent fumes and vapors will cause central nervous system depression, conjunctivitis of the eyes and pulmonary edema. Because of the severe reaction that certain solvents can cause in individuals who are exposed to them, every effort should be made to minimize all contact with organic and chlorinated solvents and their vapors.

The danger of fire or explosion is also a major concern when you encounter flammable liquids. A *FLAMMABLE LIQUID* is designated as such if it, or any component of it present in greater than one percent concentration, has a flash point below 100°F (38°C). The *FLASH POINT* of a liquid is defined as the minimum temperature at which the liquid gives off a vapor in sufficient concentration to ignite when tested under specific conditions. A *COMBUSTIBLE LIQUID* is liquid whose flash point is at or above 100°F (38°C). Thus flammable liquids present a greater fire or explosion danger than combustible liquids do. The *FIRE POINT OF A LIQUID* is the lowest temperature of a liquid at which a mixture of air and vapor from the liquid will continue to burn. Both the flash point and fire point of a liquid serve as indicators of the fire hazard of that material. A third indicator for measuring the fire hazard that a substance may present is its *AUTOIGNITION TEMPERATURE,* which is the temperature at which the material will self-ignite and sustain combustion.

The relative fire hazards of various flammable liquids have been compared and several scales have been developed to indicate this grading. One of these scales classifies liquids according to the following criteria:

CLASS I — Flash point less than 100°F (38°C)

CLASS II — Flash point at or above 100 (38°C) and below 142°F (60°C)

CLASS III — Flash point at or above 140°F (60°C).

This grading is often shown on the label and placards on the containers of the material. Additionally, flammable or combustible materials should never be stored near materials that have been designated as oxidizers, such as oxygen gas, peroxides, perchlorates or similar chemicals because of the greater fire danger that this combination presents.

5.302 Poisonous and Toxic Chemicals

The third major category of hazardous materials that you might encounter during inspection work is the poisonous or toxic group of chemicals. Certain toxic gases and solvents that fall into this classification have been discussed earlier in this chapter and this section will be concerned primarily with inorganic materials, metal salts and organic compounds that are not solvents. You could encounter these chemicals in amounts from very small quantities to very large quantities contained within large storage tanks or baths.

Some inorganic materials represent a hazard due to the fact that they cause physical irritations which will result in disease or further injury over a period of time. Substances such as asbestos, found in old buildings, or silica dust, encountered in foundries or machine shops, fit into this category. Repeated exposure to these substances can cause injuries to the lungs and cause cancer many years after exposure. The frequency of exposure to these substances is relatively low so they do not represent as great a concern as the inorganic salts that are encountered much more frequently.

Inorganic salts may be toxic and/or corrosive. They are found in many different types of industries and commercial establishments. Some examples are:

1. Plating shops — cyanide and various heavy metal salts,

2. Battery manufacturing — lead salts, and

3. Tanneries — chromate and sulfide compounds.

The number of industrial establishments that use these or similar inorganic salts in production processes or generate them

as waste is very large and continually changes. Therefore, it is a good practice to verify the materials that a particular firm uses on a regular basis and to keep that information in the firm's permit file for future reference.

Non-solvent types of toxic organic compounds are also often encountered in industrial workplaces, although their use is generally regulated through a variety of laws. Chemicals such as phenols and formaldehyde are found in the plastics and adhesive industries, while materials such as phthalates can be encountered in many wastestreams due to the fact they are used in many industrial products. Even more so than the inorganic compounds, the number of organic compounds and the varying degrees of hazard that they represent is extremely large and varies from industry to industry on a case-by-case basis. The important point to remember is that both inorganic and organic chemicals are used regularly by many industries and that traces of these chemicals are often present in the industrial discharges, in the sludges from pretreatment systems, and concentrated in the process baths and raw materials of the companies.

5.303 Infectious Agents

In addition to hazardous chemicals, wastewater discharges can contain infectious agents such as viruses, bacteria, protozoa and other microorganisms. Hypodermic needles must be considered a hazardous infectious material. Certain diseases, such as typhoid fever and Asiatic cholera, both of which are caused by bacteria, are waterborne and can be spread by contaminated wastewater. Fortunately, there have been no outbreaks of cholera in the United States recently, although outbreaks still occur in other parts of the world. Diseases caused by protozoa include amoebic dysentery and other intestinal disorders. Viruses represent a third category of microorganisms found in wastewater. Although there are a wide variety of viruses in wastewater, and all human viruses should be considered as pathogenic (disease-causing), there is no evidence that contact with viruses in wastewater is causing any diseases. A possible exception to this is infectious hepatitis, which can be passed on in rare instances where exposure to raw wastewater occurs. The possibility that AIDS, which is also caused by a virus, can be contracted from exposure to raw wastewater has been discounted by researchers who have found that although the HIV virus is present in the wastes from AIDS victims, the raw wastewater environment is hostile to the virus itself and has not been identified as a mode of transmission to date.

Although tetanus is not a waterborne disease, the organism is extremely widespread in nature and can be found at treatment plants, in manholes and in the soil. Once the disease is contracted, it is very difficult to treat. Vaccination for tetanus is recommended and booster shots should be obtained every ten years. Be sure to verify the frequency of booster shots with your physician.

Since the discharge of pathogenic organisms from hospitals is generally prohibited and controlled, the major source of these agents is from everyday wastewater. The main method for avoiding the hazards associated with infectious agents in wastewater is to develop and encourage good hygiene habits such as wearing gloves at all times, washing the hands often, avoiding contact with the face and eyes, and seeking first aid for any cut or other skin injury.

In addition to the already discussed effects of hazardous materials and agents, some of these substances can also have long-term effects which are serious but which don't immediately become apparent. These are effects that occur on the cellular level. A substance can be *CARCINOGENIC* (cancer-causing), *MUTAGENIC* (causing mutations or gene changes prior to conception) and/or *TERATOGENIC* (causing birth defects after conception). Generally, both inorganic and organic compounds that have these properties are regulated by a multitude of government agencies. Known chemicals that cause any of these effects are included in OSHA's listing of regulated substances and their usage in the workplace is required to be monitored very closely. EPA, under the authority of the Toxic Substances Control Act (TSCA) of 1976, has been given the responsibility

of controlling the entry of toxic, carcinogenic or otherwise biologically active compounds into the environment. Thus the presence of toxic organics, many of which are aromatic compounds (benzene and related chemicals), pesticides, polychlorinated biphenyls or chlorinated solvents, is regulated both in the workplace and, through the total toxic organics (TTO) requirements for categorical industries, in the industrial discharge of many industries.

When dealing with the hazards of toxic substances, especially when two or more are present, the potential of a *SYNERGISTIC INTERACTION* between the toxic compounds must be considered. A synergistic interaction, or the synergism between two or more individual compounds, occurs when the individual substances interact or cause an effect upon each other which results in an injurious effect upon the body that is greater than either of the substances alone would have produced. As an example, exposure to a mixture of hydrogen sulfide (H_2S) and hydrogen cyanide (HCN) or to hydrogen sulfide after being exposed to hydrogen cyanide will produce a more severe effect than exposure to hydrogen sulfide alone at the same concentration.

QUESTIONS

Write your answers in a notebook and then compare your answers with those on page 393.

5.3A   What types of chemical hazards can be encountered while performing industrial waste inspection work?

5.3B   What information must be included on the label of a chemical?

5.3C   What is a corrosive material?

5.3D   What are the differences between your body's response to tissue exposure to an acid and a base?

5.3E   What parts of the body can be damaged by solvents?

5.3F   Flammable or combustible materials should never be stored near what types of materials?

5.31   Routes of Entry of Hazardous or Toxic Materials to the Human Body

The effects of exposure to chemicals are often characterized in terms of being acute or chronic. *ACUTELY TOXIC* chemicals are those which can cause both severe short- and long-term health effects after a single, brief exposure to them. An exposure of short duration to acutely toxic compounds can cause damage to the body's tissues, impairment of the central nervous system function, severe illness, or in some instances, death. The routes of acute exposure can be through skin and eye contact, skin absorption, inhalation or ingestion. *CHRONIC EXPOSURE* to chemicals refers to the repeated contact with a substance (large or small amount) over a long period of time which will eventually result in serious health problems. The routes of exposure are the same as those for acutely toxic substances; however, the harmful effects are not observed until quite a bit later.

5.32   Protection Against Hazardous Materials

Always wear personal protective equipment (Figure 5.8) as a guard against the effects of exposure to toxic or hazardous compounds. While working in the field collecting and handling samples, wear cotton coveralls to protect your clothing from

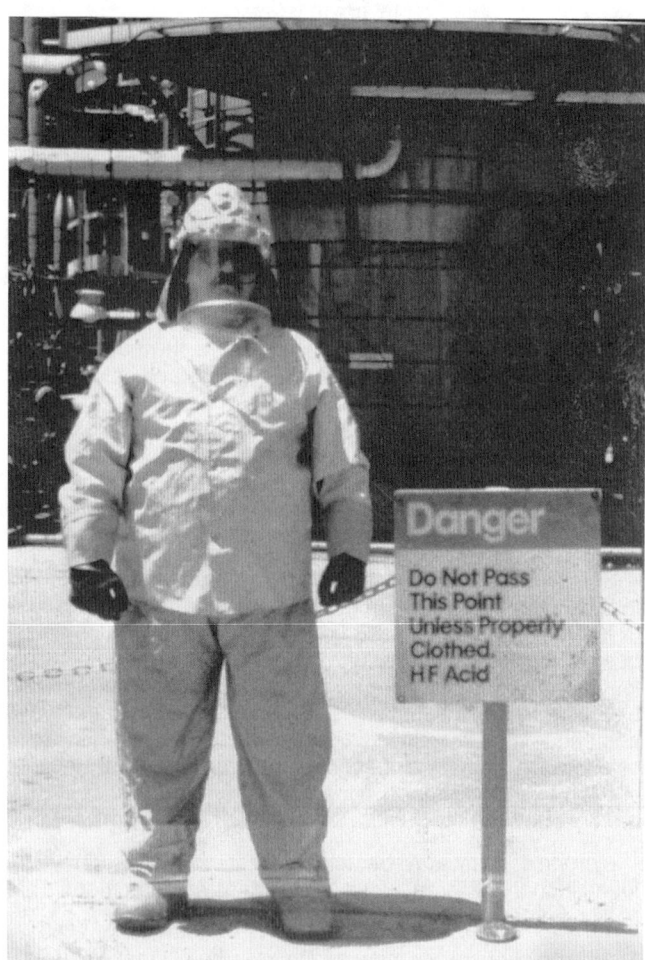

Fig. 5.8   Personal protective gear

Hydrofluoric acid (HF) alkylation plant requires that extensive protective clothing be worn by anyone who enters the area.

accidental splashes or spills of raw wastewater or chemicals. For additional protection, wear spun polyethylene-type coveralls (many of which are disposable) which are resistant to dirt, grease, corrosive and hazardous chemicals. The use of these coveralls is appropriate whenever you anticipate a heavier than usual exposure to wastewater or chemicals.

Protect your feet by wearing *SAFETY SHOES* which have steel toe, instep and heel protection. The type of soles that the shoes have is also very important. The soles should be made of a material that is high in slip resistance, does not readily abrade, and is resistant to the attack of oils, solvents and greases. Rubberized overshoes or boots worn over the safety shoes can further protect your feet from getting wet while working in puddles of wastewater or inspecting areas where spills have occurred.

Use both barrier creams (protective hand lotions) and *GLOVES* to protect your hands and prevent chemical contact with your skin. The type of gloves you need will depend to a large extent on the type of work you're doing and/or the chemicals to which you might be exposed. To protect your hands from cuts and scrapes that could result from working around rough surfaces and sharp edges or corners, a good pair of leather gloves will usually suffice. On the other hand, when working with wastewater samples that could contain potentially hazardous chemicals or infectious agents, pay close attention to the breakthrough time of the chemicals and the permeation rate (a measure of the maximum amount of chemical passing through the glove at any given time) of the gloves. A good pair of nitrile rubber gloves will provide excellent protection against acids, bases, alcohols, oils and fats, while the higher priced butyl rubber gloves have high permeation resistance to gas and water vapors and are excellent to use with dioxanes, ketones, aldehydes, alcohols and esters. Always wear appropriate hand protection to collect and handle wastewater samples. Check your gloves before each wearing to be sure there are no tears or small holes that would allow liquid to enter and thereby defeat the purpose of wearing the gloves.

Eye protection is also very important. Dust, wastewater, hazardous chemicals and radiation cause very serious eye injuries which can be avoided through the use of proper eye protection. Generally, *SAFETY GOGGLES OR GLASSES* (with side shields) made of polycarbonate material are used to provide the protection. Single-lens goggles can be worn over regular prescription safety glasses to provide additional protection around the sides of the face and eyes. Goggles and safety glasses with tinted lenses are available for use in

areas where high glare is present, while specialized goggles or face shields that can be worn over goggles are available for use in areas where welding or other radiation (UV light, lasers) is present.

The wearing of contact lenses while performing industrial waste inspection work should be strictly prohibited. If contact between a person's eyes and a hazardous substance should occur (either through direct splashing or exposure to vapor or mists) while wearing contact lenses, the severity of the effects of the exposure will be increased due to the entrapment of the hazardous substance behind the contact lenses where it cannot be readily flushed out.

RESPIRATORY PROTECTIVE EQUIPMENT is required when work is to be performed in areas where a toxic or oxygen-deficient atmosphere may be present. The equipment can be either an air-supplied unit (self-contained breathing apparatus (SCBA)) or positive pressure unit, or a negative pressure respirator (full face, half/three-quarter face or single use unit). A negative pressure respirator must not be used in an atmosphere where hazards may exceed the capability of the equipment or an oxygen-deficient condition may occur/exist. The apparatus must be fit tested for each person by using a recognized qualitative fit test procedure. One such procedure uses isoamyl acetate (banana oil); another uses an irritant smoke. Each respirator should be fit tested at least annually. If a respirator is needed for a nonemergency use, the fit testing should be done just prior to using the unit.

As previously indicated, every inspector who is to use a respirator must be instructed and trained in the need for, use, sanitary care, and limitations of the respiratory equipment that is to be used. The training should consist primarily of verbal instructions and hands-on demonstrations on the use of the respirator under simulated working conditions. Written material should only be used in a supplementary manner to support the other training. All training and instructions on the use of the respiratory equipment must be documented in writing by the inspector's supervisor and kept as part of the permanent safety record.

Respiratory safety equipment must be inspected before each use and must not be worn if a tight fit cannot be obtained. Inspectors who must wear respiratory equipment should be clean-shaven since facial hair can interfere with obtaining a good seal between the respirator's surface and the skin. Wearing of contact lenses in contaminated atmospheres with a respirator is also prohibited (29 CFR 1910.134).

Any person using a respirator must be evaluated at least annually by a physician to determine that the person is physically fit to wear the equipment. The evaluation should include the administration of a pulmonary function test to check the lung capacity of the person and should include a review of the person's medical history to verify that no conditions are present which would exclude the use of the respirator.

There are some other important considerations that should be kept in mind when using respiratory equipment. First among these is the realization that the equipment has its limitations in terms of the protection provided. For example, canister or cartridge masks are *NOT* recommended because they only provide respiratory and eye protection (full-face type) against the specific contaminant for which they were designed. Unlike a SCBA unit, they cannot be used for all materials, nor can they be used in an oxygen-deficient atmosphere. Canister-type respirators are useless where the concentration of harmful gases exceeds the absorption rate of the canister unit. Two signs of canister overloading are the detection of the odor of the gas inside the mask (assuming a good fit is present) or the canister becoming warm to the touch due to heat generation by the absorption agent (by the time this happens you may be almost dead).

The availability and use of respiratory equipment in hazardous situations can be life saving. However, the use of a respiratory unit should not be attempted or permitted until full training and instructions have been received by the user as to the type of unit to use, its limitations, and where, when and how to use it safely.

The use of the protective equipment as described above in conjunction with the practice of good sanitary habits, the receipt of proper vaccinations (tetanus and polio primarily) and the reporting and treatment of any cuts, punctures and abrasions are all excellent preventive measures against infections and infectious diseases that can be found in wastewater.

QUESTIONS

Write your answers in a notebook and then compare your answers with those on pages 393 and 394.

5.3G   What are the routes of body exposure from hazardous materials?

5.3H   Safety shoes must have what safety features?

5.3I   Why must the wearing of contact lenses while performing industrial waste inspection work be strictly prohibited?

5.3J   Where must respiratory protective equipment be worn?

5.4   PHYSICAL HAZARDS

5.40   Types of Physical Hazards Encountered During Sampling and Inspections

In addition to possible exposure to hazardous and infectious substances, industrial waste pretreatment inspection personnel are also at risk for accidents and injuries as the result of physical hazards that are encountered while inspecting and sampling industries. A number of significant accidents result from having to work in spaces where there are *OBSTRUCTIONS, PROTRUSIONS* (such as the sharp corners from supporting beams, low pipes, low ceiling), narrow aisles which limit movement, or access areas which are difficult to reach because of stored materials, drums, pallets or landscaping. In all of these situations, unless special care and precautions are taken, sampling in the area or inspection of the area can result in bumps, sprains or cuts and bruises which can be quite serious.

Another potential source of injury during inspection work is accidental contact with hot or cold objects. *HEAT HAZARDS* that can cause burns or scalds include steam and steam pipes from boilers or machinery, hot plating baths, furnaces, or heated equipment. Injuries from *COLD HAZARDS* are usually the result of coming into contact with cryogenic liquids (liquified gases such as liquid nitrogen, air or oxygen) or their containers or delivery piping. The temperature of these cryogenic liquids is many degrees below zero and contact with them can cause frostbite or more severe tissue damage. During inspections, avoid touching pipes or tanks with your bare hands.

The potential of *SHOCK HAZARDS* should be considered when working in areas where electrical wiring or electrically powered machinery is present. This includes areas where electrical panels with meters or recorders (pH, flow) that are normally looked at during an inspection are located. Since even a 110-volt circuit can kill a person, it is important not to become the ground "leg" of the circuit by standing in water, on metal, or by being in contact with metal piping while being near or touching any electrical equipment or wiring. Take care also when using extension cords to power samplers. Use only cords that have a three-pronged plug and, whenever possible, use outlets with ground fault circuit interrupters. A portable ground fault interrupter may also be used.

Many misconceptions exist about electricity. Among the more common is that low voltage (120 volts) is less dangerous than a higher voltage (240 volts, 3 phase or 480 volts, 3 phase). In fact, it is the current flow (electrons) through the body that causes the problems and injuries.

The amount of current flowing through the body depends on the voltage and resistance offered by the body. Let's take a look at the effects of current flowing through the human body. In these examples we will use milliamps. One milliamp is a thousandth (0.001) of an amp, so 0.06 amp becomes 60 milliamps and 0.83 amp becomes 830 milliamps.

- 1 milliamp or less, no sensation, not felt.

- More than 5 milliamps, painful shock. (This is 5 thousandths of an amp, 0.005 amp.)

- More than 10 milliamps, local muscle contractions sufficient to cause "freezing" of the muscles for 2.5 percent of the human population.

- More than 15 milliamps, local muscle contractions sufficient to cause "freezing" to 50 percent of the population.

- More than 30 milliamps, breathing is difficult, can cause loss of consciousness.

- 50 to 100 milliamps, possible ventricular fibrillation of the heart (uncontrolled, rapid, beating of the heart muscle).

- Over 200 milliamps, severe burns and muscular contractions, heart more apt to suffer stoppage rather than only fibrillation. (This is only 2 tenths or 200 thousandths of an amp.)

- Over a few amperes, irreparable damage to the body tissue.

The current flowing in the 120-volt circuit that lights a 7.5-watt light bulb is enough to cause severe muscular contractions of the heart. Again, the controlling factor that determines the extent of injury is the amount of resistance to the flow of current by the parts of the body through which the electricity flows.

Although not as common as some of the other hazards listed here, exposure to *RADIATION OR ULTRAVIOLET (UV)* light can occasionally be encountered during the inspection of industries where electric arc or acetylene welding (UV light) is done or where lasers are present (hospitals, electronics facilities and research laboratories).

Loud and excessive noise can present a *HEARING HAZARD* from which the ears must be protected (see Appendix E). Noise levels are measured in *DECIBELS* [5] with normal industrial background noise at about 80 decibels. Standards have been set for duration of exposure to certain decibel levels and if these levels are exceeded, then a continuing, effective hearing conservation program must be administered (Table 5.1). For inspectors this means that proper hearing protective gear must be worn when inspecting such facilities as factories, foundries, airport repair shops or any other areas where high noise levels are present.

Additional precautions that should be taken during inspection and sampling work include guarding against the hazards that can result from coming too close to moving parts of machinery and equipment (guarded and unguarded), or from slipping on floors that have water puddles, oil, hydraulic fluid, grease or other liquid leaks on them. Take extra care when climbing any ladder or steps, or when walking on planks or platforms (all should be equipped with the required guard railings) to inspect the tanks and other components of pretreatment systems. Hold on to handrails at all times and take care about the placement of your feet when climbing up and down. *BEWARE OF SLIPPERY LADDERS, STAIRS OR RAILINGS.*

TABLE 5.1 HEARING PROTECTION[a]

1. When employees are exposed to loud or extended noise, earplugs and/or protective devices shall be provided and employees shall wear them accordingly.

2. Protection against the effects of noise exposure shall be provided when sound levels exceed those shown in rule number six. These measurements shall be made on the A scale of a standard sound level meter at a slow response.

3. When employees are subjected to sound levels exceeding those listed in rule number six, feasible administrative and/or engineering controls shall be used. If such controls fail to reduce employees' exposure to within the permissible noise levels listed in this table, personal protective equipment shall be provided and used.

4. Plain cotton shall not be considered acceptable as an earplug.

5. Machinery creating excessive noise shall be equipped with mufflers.

6. Employees shall not be exposed to noise which exceeds the levels listed below for a time period exceeding those listed below.

DURATION: HOURS PER DAY		SOUND LEVEL: dBA SLOW RESPONSE
8	....	90
6	....	92
4	....	95
3	....	97
2	....	100
1½	....	102
1	....	105
½	(thirty minutes)	110
¼	(fifteen minutes or less)	115

[a] Occupational Safety and Health Standards, Table G-16, 29 CFR 1910.95; Noise Exposure.

5.41 Protective Measures Against Physical Hazards

Very similar principles apply to protecting oneself from physical hazards as apply when protecting against chemical or infectious hazards. First of all, it is very important to be able to recognize the different types of hazards that can be present, when they can occur, and what the consequences are if no protective measures are taken to guard against them.

The same protective equipment and gear used to guard against hazardous materials in many instances will also provide some protection against physical hazards. Thus safety shoes with toe, instep and heel protection will guard your feet against injuries from tripping on ground-level obstructions and protrusions, stepping on sharp objects, or from things that can be dropped on them. Oil- and grease-resistant soles aid in providing proper traction when walking on floors contaminated with spills of oil and grease or other slippery materials. Hard hats provide head protection against injuries from low obstructions, protrusions or falling objects. Safety goggles and visors protect your eyes from dust, wind, chips, and harmful rays. Properly designed and fitted earplugs or ear muffs provide

---

[5] *Decibel (DES-uh-bull). A unit for expressing the relative intensity of sounds on a scale from zero for the average least perceptible sound to about 130 for the average level at which sound causes pain to humans. Abbreviated dB.*

protection against excessive noise and sound. Leather and cotton gloves, when used in conjunction with a pair of chemically resistant ones, provide hand protection against physical hazards.

The appropriate safety gear, if worn when needed, can provide important protection. However, under no circumstances should you rely solely on the safety equipment to act as a total substitute for caution or common sense. If at all possible, hazardous conditions should be avoided or engineering controls should be introduced to reduce their severity. If hazards cannot be avoided, then only the combination of proper training, correct use of the appropriate equipment, and precaution and common sense will help prevent serious injury or even death.

5.5 SAFETY CONSIDERATIONS DURING SAMPLING

5.50 Procedure for Sampling From a Manhole

Specific precautions and procedures are to be used when setting samplers or picking up grab or composite samples at a sampling manhole. If at all possible, two persons should work as a team to do these jobs. Protective gloves, coveralls, safety shoes and eye protection should be worn during the entire sampling operation.

Before the manhole cover is moved, test the atmosphere within the manhole for the presence of any flammable/explosive gases, hydrogen sulfide (H_2S) and oxygen deficiency. Use a non-sparking (nonmetallic) probe on the gas detector's sampling hose. Insert the probe into the vent (pickup hook) hole of the lid (if available) and take readings for the three parameters (Figure 5.9). Record the readings on a designated form or in a logbook. If the readings show a safe atmosphere, slowly remove the lid by using a manhole hook or a commercially manufactured manhole cover remover. Exercise care while removing (or replacing) the lid so that it does not fall on your toes. *NEVER* use your hands to remove manhole covers. The use of manhole hooks is preferred over the use of a pick since there is generally a smaller chance of forming a spark with the manhole hook.

If a manhole has no vent holes which can be used for atmospheric testing prior to removing the lid, the lid should be partially opened using special non-sparking tools manufactured from a copper-beryllium alloy. Be very careful because a cast-iron lid and frame can spark even if using non-sparking tools. When the lid is open enough to insert an atmospheric sampling probe, proceed with the regular atmospheric testing.

Once the manhole is open, it should be allowed to air out for a minute or two. If any strange or abnormal odor is still detected after the lid has been off the manhole for a minute, allow the manhole structure to vent until the odor or vapor has dissipated.

If a high initial LEL reading is obtained, do not remove the manhole lid. Instead, attempt to locate the source of the high LEL reading so that steps can be initiated to correct a potentially dangerous situation. If an odor persists after the manhole cover has been removed and the manhole has been allowed to vent to the atmosphere for a five-minute period and if no portable blower is available to circulate new air into the manhole, replace the lid on the manhole. In either one of the above situations, notify your supervisor so that corrective action to alleviate the problem can be taken.

Always use proper lifting techniques, such as bending the knees and not the back and never twisting while you are lifting, when placing or removing the sampler. Full samplers are considerably heavier than when they are empty and therefore are definitely more awkward to lift out of the manhole in which they may be suspended or resting on the base block. When lifting or lowering samplers, take care not to let your fingers become pinched between the sampler body and the suspension harness. After the sampler has been safely placed in or removed from the manhole, replace the manhole cover. Again, take care to prevent the lid from slipping and injuring your hands or feet. This is especially true if you use your hands to handle the lid directly during its replacement.

A collapsible winch system powered by the vehicle battery and placed on the rear bumper can be used to assist in the proper lifting of automatic samplers and other heavy items. This apparatus is *NOT* rated for safety or rescue but can assist in lifting heavy loads (Figure 5.10).

Continue wearing appropriate gloves and coveralls while transferring samples and performing preliminary screening operations. Be sure that the lids of sample jars are screwed on tight and that the bottles are stored upright and secured to prevent any accidental spillage during their transportation to the laboratory. As with any other operation, good housekeeping and any other sensible precautionary measures will improve the overall safety of the sampling operations.

QUESTIONS

Write your answers in a notebook and then compare your answers with those on page 394.

5.4A   What cold hazards must inspectors try to avoid?

5.4B   How can an inspector become the ground "leg" of an electrical circuit?

5.4C   Why should safety goggles and visors be worn?

5.5A   What personal protective clothing should be worn when setting a sampler?

5.6 HAZARDS ENCOUNTERED IN THREE TYPES OF INDUSTRIES

5.60 Hazards Encountered in Metal Finishing and Plating Shops[6]

Although it is not possible to list or predict every type of chemical hazard that can be encountered in any given type of industry, including those found in metal finishing and electroplating facilities, certain hazards are encountered on a fairly frequent basis. Electroplating shops and metal plating facilities

---

[6] See TREATMENT OF METAL WASTESTREAMS, Section 7.0, "Safety. Beware of Hazardous Chemicals." Obtain from the Office of Water Programs, California State University, Sacramento, 6000 J Street, Sacramento, CA 95819-6025. Price, $11.00.

Fig. 5.9   Inspector testing atmosphere in a manhole to determine if it is safe to open manhole and sample flow

Fig. 5.10   Collapsible winch system for lifting samplers from manholes
(Courtesy of Sanitary District of Rockford)

represent an industrial category in which extensive quantities of both inorganic and organic chemicals are used on a regular basis as part of the regular production process. The largest quantities of chemicals found on site are usually the chemicals in the plating baths. The volumes of these baths are often in the hundreds or thousands of gallons and the baths contain heavy metal salt and cyanide solutions whose concentrations are often in the ounces per gallon range (over 10,000 mg/L or 1 percent solution).

Many plating baths use high voltages as part of the plating process, and many are also heated to increase the efficiency of the process. Thus, aside from being a heat hazard, the contents of the bath also become more reactive and corrosive. If the baths are hot enough, toxic and irritating fumes may also be generated which, if not properly vented and scrubbed, present an air pollution hazard to anyone exposed to them. This is particularly true in some older plating shops where the ventilation is inadequate to remove all the fumes. In these shops, unless care is taken, the time it takes to perform an inspection is often long enough to feel the effects (headache, sore throat or watery eyes) of breathing the fumes and vapors.

Many older plating shops are crowded for space and still use wet floors to collect the wastewater before it is sent on to the pretreatment system for treatment. The liquid present on the wet floor can, and usually does, contain a lot of the chemicals from the various plating baths (although usually in more dilute form). Therefore, take care to avoid contact with this material when you are inspecting plating areas.

Besides the chemical hazards present in the plating and metal finishing shop, there are also certain physical hazards. These may include wet and slippery floors, hot baths, electrical plating bars and wiring, and crowded conditions that may create obstructions. In addition to the hazards inside the shop, you may also encounter hazards in the pretreatment system of the facility. Besides the process wastewaters generated inside the shop, which can contain acids, caustics, cyanides and heavy metals, you may be exposed to the chemicals that are used in the treatment process itself. These can include chlorine or hypochlorite for the treatment of cyanide, sulfur dioxide for the reduction of hexavalent chromium, and various acids, bases and polymers to treat the heavy metals in solution. Because of the wide variety of hazards present at these types of facilities and the potential danger that they present in case of spillage or accidental discharge to the collection system, always use extreme caution when working with wastewater samples from these facilities or when inspecting them.

5.61 Hazards Encountered in the Chemical Blending/Manufacturing Industry

The chemical blending industry where cleaning agents, disinfectants, waxes and other mixtures are compounded and blended is another industrial category where a substantial number of hazardous chemicals are stored and used on site. These chemicals include ammonia, formaldehyde, phenols, chlorine or hypochlorite and various solvents and other materials. In this industry, both the raw materials and the finished products can be hazardous and care must be taken when sampling or inspecting this type of facility. For example, if the housekeeping is poor in the production area of the facility and the mixing and blending equipment has not been properly maintained, or if the tanks and blending vehicles leak, there is a danger of stepping into unknown chemicals and spills during your inspection. If, despite the precautions of wearing the appropriate protective clothing, you do make contact with an unknown liquid, immediately flush the affected area of the skin to remove the potentially hazardous material. Ask the industry representative to identify the material so that additional treatment can be sought, if necessary.

Since any and all chemicals that are used on the site can also theoretically find their way into the wastewater discharge, always wear gloves and eye protection when collecting samples. Be aware also that samples collected from separate sampling points or spills may contain incompatible pollutants, such as ammonia and hypochlorite, which should not be mixed. If any hazardous constituents are expected to be present in the wastewater samples in concentrations that could possibly be harmful, you must warn the laboratory analysts that the samples require special handling and treatment.

5.62 Hazards Encountered in the Semiconductor Industry

Although the semiconductor industry has gained a reputation for being a "clean" industry because of its clean rooms and the precise nature of its product, a surprisingly large number of hazardous chemicals are used at these facilities for the cleaning, etching and doping of the wafers. Chemicals such as freons and chlorinated solvents are used for cleaning the wafers. Acids such as hydrofluoric and fluoroboric are used in the etching process and gases and compounds such as borane (BH_3), arsine (AsH_3), hydrogen and silane as well as gallium arsenide are used to make particular types of silicon wafers. Chlorinated and aromatic solvents may also be used in the application and removal of photo resist masks.

As with the other two industries that have been discussed, precautions must be taken when sampling or inspecting the semiconductor or electronics industries. Since chemical control and cleanliness (use of clean rooms, caps and gowns) are extensively practiced in this industry, there is less chance of exposure to spilled materials due to sloppy housekeeping. The main areas of concern are the pretreatment area where wastewaters containing some of the production chemicals can be encountered and the chemical storage areas where spills can occur. Within the facility, be careful while inspecting around the various baths containing the chemicals used to etch or clean the wafers. The extensive use of the different types of gases can also present a problem if the gases are ever accidentally released or if a break in a delivery line occurs. Fortunately, most firms have excellent controls and safety measures in place to prevent such occurrences.

Of the three industrial categories discussed, the semiconductor industry probably is the one with the least potential for encountering difficulties with hazardous materials. All three types of industries, if approached with a mixture of caution and precaution, can be safely inspected and sampled. The key point, which applies in all situations, is to anticipate the unexpected and to be prepared to deal with it effectively.

QUESTIONS

Write your answers in a notebook and then compare your answers with those on page 394.

5.6A   Where are the largest quantities of chemicals found in electroplating shops and metal plating facilities?

5.6B   What should you do if an unknown liquid comes in contact with your skin during an inspection?

5.6C   For what purposes are hazardous chemicals used in semiconductor facilities?

5.7   REGULATIONS AND REFERENCES

5.70   Regulations

The standards for safety in the workplace are set by the Occupational Safety and Health Administration (OSHA), an agency of the U.S. Department of Labor. OSHA was established through the Williams-Steiger Occupational Safety and Health Act which became law on December 29, 1970, and became effective on April 28, 1971. OSHA's responsibility is to establish and enforce the rules which will ensure safe and healthful working conditions for employees (inspectors). OSHA set up the National Institute for Occupational Safety and Health (NIOSH) as a part of what was then the Department of Health, Education and Welfare. While OSHA has the major responsibility for determining safety priorities, setting standards, providing education to the employer and/or employees, maintaining a nationwide record and reporting system, and enforcement, NIOSH has a supportive role in these activities and has the responsibility for undertaking research and development programs.

In 1974 OSHA and NIOSH worked together to develop a series of complete occupational health standards for all existing substances whose permissible exposure levels were known. Initially, a series of about 400 draft technical standards were developed and adopted. Although many of these "consensus" standards concerned safety, several groups of industrial hygiene standards were also included. One of these was the American Council of Governmental Industrial Hygienist's (ACGIH's) compilation of Threshold Limit Values (TLVs). Thus, exposure limit values are often listed both in terms of TLVs and in PELs, or Permissible Exposure Limits, a term used by OSHA.

NIOSH, as part of the standard setting process, develops detailed "criteria" documents related to specific safety or chemical hazards. The documents include recommended controls and other detailed information. Additionally, OSHA has adopted many "voluntary" standards from trade associations and other governmental agencies.

OSHA regulations apply only to states that do not have their own OSHA-approved plans. Many states have their own programs similar to the national one and their safety and health rules, as contained in each state's administrative code, are generally enforced by that state's inspectors. In general, the standards (enforced through each state's plan) are as strict as those of OSHA, since most of the work done to prepare the standards is through NIOSH.

5.71   References and Information Sources

A very extensive body of literature and reference material currently exists on safety-related topics. These references include publications of a general nature covering a wide variety of topics such as industrial hygiene, first aid and general safety, as well as specialized material dealing with a single topic such as confined space entry, toxic gases or traffic control.

Copies of the actual regulations, whether OSHA or an individual state's plan, as contained in that state's administrative code, should be your primary reference in terms of what the actual requirements are and how the regulations are worded. The criteria documents prepared by NIOSH serve as reference material for individual safety or chemical hazards and contain an extensive summary of the information known about the hazard. Copies of the regulations or criteria documents can be obtained by writing or calling the appropriate state or federal agency. Many of the federal documents are available for sale through the Government Printing Office and its various regional outlets. Other important sources of information include manufacturers' literature, safety catalogs, publications such as "Industrial Hygiene News" (free to qualified subscribers from Rimbach Publishing, Incorporated, 8650 Babcock Boulevard, Pittsburgh, PA 15237-5821), and newsletters prepared by various agencies such as OSHA and NIOSH.

Government publications can serve as an important source of information since they are often available at reasonable costs or are free. To keep current with the regulations, regularly review the *FEDERAL REGISTER* (published every working day) or subscribe to a service which summarizes and interprets (on a weekly or biweekly basis) all changes related to safety and health regulations.

You can also obtain important information regarding safety-related topics by attending conferences and workshops. Often, speakers at these conferences are very knowledgeable and have practical experience in solving safety- or health-related problems, including those related to industrial waste pretreatment inspection work. Examples of helpful conferences include those presented by the various member associations of the Water Environment Federation (WEF). These conferences also give you an opportunity to meet with manufacturers' representatives, examine safety equipment first-hand, and observe or participate in safety demonstrations.

5.72   Establishing a Safety Library

The establishment and maintenance of a safety library is important and highly recommended. The library serves as a *READILY* available source of information for any unexpected or expected situations or problems that may arise. The core of the library should include some basic reference books as well as any specialized literature related to each agency's operation. As the library expands, audio and video tapes, films and slide programs should be added. (Tapes or films can also be obtained on a one-time rental basis when the cost of purchasing them is too high.)

The library should also contain copies of the training records for all employees as well as copies of accident evaluation and investigation reports. (Appendix F contains a typical accident investigation policy statement and sample reporting forms.) These items provide useful discussion topics for safety meetings and can be used to spot trends or trouble points in safety- and health-related matters. The documentation is also useful in determining what additional training might be needed by the staff.

5.73   Additional Reading

1. *SAX'S DANGEROUS PROPERTIES OF INDUSTRIAL MATERIALS*, Tenth Edition, Three-Volume Set. Obtain from John Wiley & Sons, Inc., Distribution Center, 1 Wiley Drive, Somerset, NJ 08875-1272. ISBN 0-471-37858-5. Price, $995.00, plus $5.00 shipping and handling.

2. *SAFETY TRAINING METHODS: PRACTICAL SOLUTIONS FOR THE NEXT MILLENNIUM*, Second Edition. Obtain from John Wiley & Sons, Inc., Distribution Center, 1 Wiley Drive, Somerset, NJ 08875-1272. ISBN 0-471-55230-5. Price, $120.00, plus $5.00 shipping and handling.

3. *SAFETY AND HEALTH IN WASTEWATER SYSTEMS* (MOP 1). Obtain from Water Environment Federation (WEF), Publications Order Department, 601 Wythe Street, Alexandria, VA 22314-1994. Order No. MO2001WW. Price

to members, $61.75; nonmembers, $81.75; price includes cost of shipping and handling.

4. *CRITERIA FOR A RECOMMENDED STANDARD: WORKING IN CONFINED SPACES*, NIOSH Publication No. 80-106. Obtain from National Technical Information Service (NTIS), 5285 Port Royal Road, Springfield, VA 22161. Order No. PB80-183015. Price, $34.00, plus $5.00 shipping and handling per order.

5. *A GUIDE TO SAFETY IN CONFINED SPACES* by Ted Pettit and Herb Linn. Obtain from National Institute for Occupational Safety and Health, NIOSH Publications, 4676 Columbia Parkway, Mail Stop C-13, Cincinnati, OH 45226-1998. Publication No. 87-113.

6. *GUIDELINES FOR CONTROLLING HAZARDOUS ENTRY DURING MAINTENANCE AND SERVICING*. Obtain from National Technical Information Service (NTIS), 5285 Port Royal Road, Springfield, VA 22161. Order No. PB84-199934. Price, $34.00, plus $5.00 shipping and handling per order.

A general reference (made up of thirteen individual volumes) is:

PATTY'S INDUSTRIAL HYGIENE AND TOXICOLOGY, Fifth Edition, 13-Volume Set. Obtain from John Wiley & Sons, Inc., Distribution Center, 1 Wiley Drive, Somerset, NJ 08875-1272. ISBN 0-471-31945-7. Price for all 13 volumes, $3,290.00, plus $5.00 shipping and handling. Contact publisher for prices of individual volumes.

This publication is very detailed and covers many topics related to safety, industrial hygiene, and toxicology.

QUESTIONS

Write your answers in a notebook and then compare your answers with those on page 394.

5.7A   Who sets the standards for safety in the workplace?

5.7B   What is OSHA's responsibility?

5.7C   What useful information can be gained from accident evaluation and investigation reports?

Please answer the discussion and review questions before continuing with the Objective Test.

DISCUSSION AND REVIEW QUESTIONS

Chapter 5.   SAFETY IN PRETREATMENT INSPECTION AND SAMPLING WORK

Please write your answers in a notebook before continuing with the Objective Test on page 394. The purpose of these questions is to indicate to you how well you understand the material in the chapter.

1. How can a supervisor set an example for inspectors on how to comply with all safety requirements and procedures?

2. Why should the inspector and the supervisor hold one-on-one discussions of safety topics?

3. What safety equipment should be routinely carried in an inspector's vehicle?

4. What safety precautions should an inspector take during an actual inspection of an industry?

5. What should an inspector do *BEFORE* starting work in a street?

6. Why can't hydrogen sulfide be smelled at higher concentrations?

7. What must an inspector know and be able to do to use a gas detection device?

8. How can inspectors be exposed (points of exposure) to corrosive materials?

9. How can inspectors avoid the hazards associated with infectious agents?

10. How can an inspector be protected against hazardous materials?

11. An inspector must be especially aware of the potential for electrical shock when working in what areas?

12. How can inspectors protect themselves from physical hazards?

13. What hazards may be encountered when inspecting electroplating shops and metal plating facilities?

14. Why is a safety library important?

SUGGESTED ANSWERS

Chapter 5.   SAFETY IN PRETREATMENT INSPECTION AND SAMPLING WORK

Answers to questions on page 362.

5.0A   The purpose of safety training for pretreatment facility inspectors is to ensure that they have the knowledge, skills, and equipment to perform their duties in a safe manner at all times.

5.0B   The responsibility for safety training of pretreatment inspectors rests with their immediate supervisor.

5.0C   To perform a Job Safety Analysis (JSA), the supervisor selects the job to be analyzed and then, with the help of the inspectors, the job is subdivided into individual steps and each step of the job is critically examined. During the examination process, any potential hazards or problems associated with that step are identified by reviewing how each inspector performs the tasks of this step. The intent of this review is to see what can happen and how any potential accidents and injuries can be prevented.

5.0D   Precautions an inspector should take to prevent oral and skin contact with wastewater include wearing gloves and *NEVER* eating, smoking or drinking until after having washed hands thoroughly with hot water and a disinfectant soap. Additionally, any breaks in the skin such as cuts or abrasions should be kept clean and protected to eliminate any easy entry paths for the microorganisms in wastewater.

5.0E   A supervisor can comply with a "right-to-know" law regarding inspectors by making available copies of all the Material Safety Data Sheets (MSDSs) for the chemical products that the inspectors use or encounter during their regular work period, and by training the inspectors to read and interpret MSDSs.

Answers to questions on page 368.

5.0F   Inspectors' personal safety gear that should be carried in each vehicle includes a hard hat, safety glasses and/or goggles, a supply of different types of gloves, extra coveralls, orange traffic vest, earplugs, hand lotion or barrier cream, paper towels, rubber overshoes, safety shoes (if not already worn), and copies of the injury report form.

5.0G   Gas detection devices that should be available for an inspector's use include devices to test for oxygen deficiency, explosive/flammable gases, and toxic gases (such as hydrogen sulfide, chlorine, carbon monoxide, and cyanide).

5.0H   The first step when planning for an inspection of a specific industry is to review the firm's permit application and correspondence file. This review serves to refresh the inspector's memory about the processes and operations being performed at the facility, the types of safety equipment needed, and the special precautions that must be taken during the inspection.

Answers to questions on page 369.

5.1A An inspector can ensure that a vehicle is safe to operate by making a complete check of the vehicle and its different components (lights, horn, tires) to verify that everything is operational *BEFORE* moving the vehicle.

5.1B The five "seeing" habits that must be practiced when driving defensively include:

1. Keep your eyes moving from side to side,
2. Aim high on curves in your steering (don't cut corners),
3. Get the total or big picture,
4. Try to make sure that other drivers can see you, and
5. Leave yourself an out or escape route.

5.1C When working in a street, inspectors should be wearing orange vests to give them higher visibility and added protection.

Answers to questions on page 372.

5.2A A confined space is a space which by design has limited openings for entry and exit, unfavorable natural ventilation which could contain or produce dangerous air contaminants, and which is not intended for continuous employee occupancy.

5.2B The three items sampled and monitored by a triple detector type or combination gas meter are: (1) oxygen deficiency/enrichment, (2) explosive/flammable gases, and (3) a toxic gas, usually hydrogen sulfide (H_2S) and/or carbon monoxide.

5.2C The probe of a gas detection device should be made of nonmetallic material so that no spark can be accidentally generated.

5.2D Hydrogen cyanide (HCN) can be generated when improperly treated plating wastewaters containing cyanide salts are discharged and then allowed to react with any acidic wastewaters present in the collection system.

5.2E An inspector can encounter chlorine gas during the sampling or inspection of an industry which uses chlorine for cyanide destruction in its pretreatment system.

5.2F An inspector can encounter chlorinated solvents when accidental discharges occur from plating shops or electronics manufacturing facilities where these types of solvents are used for degreasing of parts or dissolving photo resists.

5.2G An oxygen-deficient atmosphere is created through bacterial action using up the oxygen, displacement by other gases (can be toxic and/or inert gases such as nitrogen or carbon dioxide), oxidation of metals or other materials that depletes the oxygen level, adsorption of the oxygen onto surfaces, or combustion.

Answers to questions on page 379.

5.2H Caution must be used in the placement of the blower so that its air intake does not pull in the exhaust of any vehicles being driven or idling nearby.

5.2I Before the entry is started, a *RESCUE PROCEDURE* must be established which can be implemented in case an emergency develops. A proper entry permit must be issued by the responsible supervisor and/or safety department representative. Before entering a confined space, the atmosphere must be tested and all confined space entry procedures must be followed. Also, the steps of any manhole should be tested to confirm their integrity and ability to hold the weight of the person who will be using the steps to descend into the manhole.

5.2J A physician must determine whether or not a person is physically able to perform the work while wearing the required respirator.

5.2K The key points to remember about entry to a confined space are to first *RECOGNIZE* what a confined space is, then to *TEST AND EVALUATE* its atmosphere prior to entry, and to continue to *MONITOR* the atmosphere while the work is being performed.

Answers to questions on page 382.

5.3A Chemical hazards that can be encountered while performing industrial waste inspection work include substances that are toxic or poisonous, corrosive, explosive or flammable, carcinogenic (cancer-causing), biologically active, and, in rare instances, radioactive.

5.3B Information that must be included on the label of a chemical includes:

1. Chemical or common name,
2. Signal or warning words,
3. Caution or hazard phrases,
4. First aid information, and
5. Storing and handling data.

5.3C A corrosive material is one which, through its chemical action, is destructively injurious to body tissues or other materials.

5.3D Exposure to an acid results in an immediate sensation of pain which alerts the person that exposure to a corrosive chemical has taken place. Strong bases do not cause an immediate pain; therefore, the period of exposure may be longer and the injury may be more severe because the corrosive material had more time to be absorbed by the tissue.

5.3E Solvents can damage the skin, respiratory system, lungs, liver, kidneys, central nervous system and eyes.

5.3F Flammable or combustible materials should never be stored near materials that have been designated as oxidizers, such as oxygen gas, peroxides, perchlorates or similar chemicals, because of the greater fire danger that this combination presents.

Answers to questions on page 384.

5.3G The routes of body exposure from hazardous materials can be through skin and eye contact, skin absorption, inhalation or ingestion.

5.3H Safety shoes must have steel toe, instep and heel protection. The soles should be made of material that is high in slip resistance, does not readily abrade, and is resistant to the attack of oils, solvents and greases.

5.3I The wearing of contact lenses while performing industrial waste inspection work should be strictly prohibited. If contact between a person's eyes and a hazardous substance should occur (either through direct splashing or exposure to vapor or mists) while wearing contact lenses, the severity of the effects of the exposure will be increased due to the entrapment of the hazardous substance behind the contact lenses where it cannot be readily flushed out.

5.3J   Respiratory protective equipment is required when work is to be performed in areas where toxic or oxygen-deficient atmospheres may be present.

Answers to questions on page 386.

5.4A   Cold hazards that inspectors must be aware of include coming into contact with cryogenic liquids (liquified gases such as liquid nitrogen, air or oxygen) or their containers or delivery piping.

5.4B   An inspector can become the ground "leg" of an electrical circuit by standing in water, on metal, or by being in contact with metal piping while being near or touching any electrical equipment or wiring.

5.4C   Safety goggles and visors should be worn to protect your eyes from dust, wind, chips, and harmful rays.

5.5A   Protective clothing that should be worn when setting a sampler includes gloves, coveralls, safety shoes and eye protection.

Answers to questions on page 390.

5.6A   The largest quantities of chemicals in electroplating shops and metal plating facilities are usually found within the plating baths.

5.6B   If an unknown liquid comes in contact with your skin, immediately flush the affected area of skin to remove the potentially hazardous material. Ask the industry representative to identify the material so that it can be determined if additional treatment is necessary.

5.6C   Semiconductor facilities use hazardous chemicals for cleaning, etching and doping of the wafers.

Answers to questions on page 391.

5.7A   The standards for safety in the workplace are set by the Occupational Safety and Health Administration (OSHA), an agency of the U.S. Department of Labor.

5.7B   OSHA's responsibility is to establish and enforce the rules which will ensure safe and healthful working conditions for employees (inspectors). It has the major responsibility for determining safety priorities, setting standards, providing education to the employer and/or employees, maintaining a nationwide record and reporting system, and enforcement.

5.7C   Accident evaluation and investigation reports provide useful discussion topics for safety meetings and can be used to spot trends or trouble spots in safety- and health-related matters.

OBJECTIVE TEST

Chapter 5.   SAFETY IN PRETREATMENT INSPECTION AND SAMPLING WORK

Please write your name and mark the correct answers on the answer sheet, as directed at the end of Chapter 1. There may be more than one correct answer to each multiple-choice question.

True-False

1. Inspectors must report all injuries to their supervisor as soon as possible.

    �`1. True
     2. False

2. Clothes worn on the job can be washed with household laundry.

     1. True
    ➤ 2. False

3. An inspector must develop good driving habits and drive defensively at all times.

    ➤1. True
     2. False

4. One inspector can sample industries at sampling points that are located in public streets or other high-traffic (congested) areas.

     1. True
    ➤ 2. False

5. A gas detector must be properly maintained and calibrated on a regular basis.

    ➤1. True
     2. False

6. Industrial waste inspectors frequently enter confined spaces.

     1. True
    ➤ 2. False

7. Low voltage is less dangerous than high voltage.

     1. True
    ➤ 2. False

8. Plain cotton is considered acceptable as an earplug for noise protection.

     1. True
    ➤ 2. False

9. Take care to avoid contact with liquids on wet floors when inspecting plating areas.

    ➤ 1. True
     2. False

10. Always wear gloves and eye protection when collecting samples.

    ➤1. True
     2. False

Best Answer (Select only the closest or best answer.)

11. Failure to report an injury should be considered
   1. And reviewed.
   2. Failure to follow rules.
   — 3. Grounds for disciplinary action.
   4. The correct response.

12. What should you do if you encounter an unsafe condition or situation?
   1. Correct the condition or situation.
   2. Leave the area and wait for condition or situation to be corrected.
   3. Prevent others from encountering condition or situation.
   — 4. Remove yourself from the area and warn others.

13. How does an inspector determine what safety equipment should be taken on an inspection?
   1. Contact company and ask what safety equipment will be needed.
   2. Recall what safety equipment was needed for similar companies.
   — 3. Review the firm's permit application and correspondence file.
   4. Take all available safety equipment.

14. What is the *BEST* source of information regarding chemical products and their hazardous nature?
   1. EPA
   — 2. MSDS
   3. NIOSH
   4. OSHA

15. What is the most common toxic gas that is encountered during sampling work?
   1. Carbon monoxide
   2. Hydrogen cyanide
   — 3. Hydrogen sulfide
   4. Methane

16. What is the most important thing to remember about entering a confined space?
   1. Calibrate the testing equipment.
   2. Have the confined space entry permit properly completed.
   — 3. Recognize a confined space.
   4. Train the rescue crew.

17. What is the main responsibility of NIOSH?
   1. Enforcing safety standards
   2. Establishing safety rules
   3. Training pretreatment facility inspectors
   — 4. Undertaking research and development programs

18. Why is a safety library important? The safety library serves as a
   1. Location for safety meetings.
   2. Quiet place to read.
   — 3. Readily available source of safety information.
   4. Space to write reports.

Multiple Choice (Select all correct answers.)

19. What are the inspector's responsibilities regarding safety?
   — 1. Apply safe work practices learned on the job
   — 2. Inform your supervisor of unsafe conditions
   — 3. Participate actively in safety training
   — 4. Recognize hazards of the job
   — 5. Take precautionary measures which will ensure your own safety

20. What safety equipment must be in each inspector's vehicle?
   1. Acid neutralizers
   — 2. Fire extinguisher
   — 3. First aid kit
   — 4. Spill control pillows
   5. Stop signs

21. What items are an inspector's personal safety equipment?
   — 1. Earplugs
   2. Fire extinguisher
   3. First aid kit
   — 4. Gloves
   — 5. Hard hat

22. What factors influence traffic control strategies?
   1. Alertness of drivers
   — 2. Congestion
   — 3. Local regulations
   — 4. Time of day
   — 5. Traffic speed

23. What are the most commonly encountered atmospheric hazards associated with work in confined spaces?
   — 1. Flammable/explosive gases
   2. Flying insects
   3. Nitrogen gas
   — 4. Oxygen deficiency/enrichment
   — 5. Toxic gases

24. What are possible sources of flammable or explosive gases, mists and vapors in sewers?
   — 1. Accidental discharges
   — 2. Leaking underground storage tanks
   3. Refinery emissions
   — 4. Spills
   5. Storm water runoff

25. When an inspector enters a confined space, what protective clothing and equipment are needed?
   — 1. Eye protection
   — 2. Foot protection
   — 3. Full-body, parachute-type harness
   — 4. Gloves
   — 5. Hard hat

26. What items are on a confined space entry form?
   1. Estimated flow in sewers
   — 2. Explosive vapors less than 10% of LEL
   — 3. Protective equipment and rescue devices
   — 4. Space ventilation
   5. Wind direction and velocity

27. What chemical hazards can be encountered while performing industrial waste pretreatment inspection work?

    1. Carcinogens
    2. Explosive gases
    3. Noise
    4. Shocking
    5. Toxics

28. What are the major categories of hazardous materials encountered during sampling work?

    1. Corrosive materials
    2. Inorganic compounds
    3. Poisonous or toxic gases
    4. Solvents and flammable materials
    5. Synergistic reductions

29. Before a manhole cover is removed, what must the atmosphere within the manhole be tested for?

    1. Electricity
    2. Excessive noise
    3. Explosive gases
    4. Hydrogen sulfide gas
    5. Oxygen deficiency/enrichment

30. What important information is found on a Material Safety Data Sheet (MSDS)?

    1. Cost of chemical
    2. First aid measures
    3. Hazardous ingredients
    4. Personal protection
    5. Spill, leak, and disposal procedures

End of Objective Test

APPENDICES

CHAPTER 5.   SAFETY IN PRETREATMENT INSPECTION AND SAMPLING WORK

Appendices provided by
Union Sanitary District, Fremont, California

APPENDIX A

JOB SAFETY ANALYSIS

A typical JOB SAFETY ANALYSIS would be very difficult to include in this manual since it actually involves going out into the field, discussing items with the people doing the work, and possibly videotaping some of the actions or procedures to be analyzed. The remainder of Appendix A contains the "Safety Sampling Protocol" developed by the Union Sanitary District of Fremont, California, that resulted from a JOB SAFETY ANALYSIS of tasks related to setting samplers and opening manholes.

SAFETY SAMPLING PROTOCOL

1. Determine resources and staff required to do sampling, for example, one or more vehicles and one or more staff members.

2. Check to ensure that all safety equipment is on truck(s). Follow safe driving practices.

3. *SITE EVALUATION:*

   - Locate site; stop out of path of traffic and determine appropriate traffic control plan.

   - Out of traffic — Place one cone in front and one behind vehicle.

   - Put on all personal safety equipment (vest and safety bump cap).

   - Place signs and cones according to chart in the "Traffic Control for Street and Highway Participant Notebook," beginning with sign farthest from truck.

4. *OUT OF TRAFFIC FLOW:*

   - Move truck into work location; use truck as barrier between oncoming traffic and work location. If two trucks are used, place one in front and one behind work site.

   - In curbed area, use two cones, one in front and one near the back corner of vehicles.

   - In non-curbed area, use four cones, one at each corner of vehicle.

   - Keep vehicle running, emergency brake on and truck in park if vehicle has an automatic transmission. If standard transmission, shut engine off, put vehicle in gear, and lock emergency brake.

   - Perform all work within coned area.

5. *SAMPLING OF MANHOLE*

   a. Perform general site observation.

   b. Calibrate monitoring equipment away from work area; record readings.

   c. Put on any additional personal protective equipment (for example, gloves) needed.

   d. Use proper lifting techniques when setting or removing a sampler.

   e. If possible, stand upwind of manhole. Monitor vapors with manhole lid closed; record readings.

NOTE: **DO NOT REMOVE MANHOLE COVER IF ANY ALARM CONDITION (LEL, O_2 DEFICIENCY OR H_2S) EXISTS.**

- OXYGEN-DEFICIENCY ONLY:

   - Remove manhole cover; ventilate for one (1) minute; retest.

   - If alarm condition still exists, vent for ten (10) minutes and retest.

   - If condition is still present, close manhole and notify WSC Office immediately.

- LEL:

0 - 10%

   - Open manhole and vent for one minute; retest.

   - If 0 - 10% reading still present, vent for ten (10) minutes and retest.

   - Note direction of level activity. If reading remains 0 - 10%, proceed with sampling and continue monitoring. After sampling is completed, notify WSC Office and industrial user.

10 - 60%

   - Do not remove lid; continue monitoring as far away as possible from manhole. Note direction of readings.

   - Notify WSC Office.

   - Notify industrial user of problem.

60% and Above

- Do not remove lid.

- If possible, leave monitoring equipment in place and leave immediate vicinity. Remain at site.

- Notify WSC and industrial user.

- Wait at site until relieved by appropriate authority.

- H₂S:

10 - 20 ppm

- Open manhole and vent for one (1) minute; retest.

- If 10 - 20 ppm reading still present, vent for ten (10) minutes and retest.

- Note direction of level activity. If reading goes below 10 ppm, proceed with sampling and continue monitoring. If reading remains greater than 10 ppm, close lid and notify WCS Office and industrial user.

21 - 100 ppm

- Do not remove lid; continue monitoring as far away as possible from manhole. Note direction of readings.

- Notify WSC Office.

- Notify industrial user of problem.

100 ppm and Above

- Do not remove lid.

- If possible, leave monitoring equipment in place and leave immediate vicinity. Remain at site.

- Notify WSC and industrial user.

- Wait at site until relieved by appropriate authority.

Negligible O₂, LEL and H₂S Readings:

- Remove manhole cover and sample and continue monitoring of atmosphere during sampling.

- If strange odor, vapor, fumes, mist, or taste is noted, vacate immediate area around manhole and vent for up to one (1) minute.

- If unusual condition persists, replace lid.

- Notify WSC Office and industrial user of problem.

- For departure, remove vehicle, signs and cones in reverse order of setup.

- Cautiously re-enter traffic stream.

APPENDIX B

GUIDES FOR TRAFFIC CONTROL SIGNS

See the next three pages for diagrams showing the recommended placement of traffic control signs in various work situations. For additional information see *OPERATION AND MAINTENANCE OF WASTEWATER COLLECTION SYSTEMS,* Volume I, Chapter 4, "Safe Procedures," Section 4.3, "Routing Traffic Around the Job Site," in this series of training manuals.

SUGGESTED SPACING AND SIZING OF SIGNS
FOR ADVANCED WARNING RELATED TO SPEED OF TRAFFIC

Speed of Traffic	Lane Closure	Low Level Guidance Traffic Cone Spacing (Maximum)	Warning Sign to Work Site (Maximum)	Sign Legend	Sign Size
25 MPH or below	No	10'	150'	Work Area	30" x 30" or 24" x 24"
35 MPH	No	35'	250'	Work Area	30" x 30" or 24" x 24"
45 MPH	No	45'	500'	Work Area	30" x 30" or 24" x 24"
45 MPH	Yes	45'	1st sign 500'	Right or Left Lane Closed Ahead	36" x 36"
			2nd sign 150'	Work Area	30" x 30" or 24" x 24"
55 MPH	No	55'	750'	Work Area	30" x 30" or 24" x 24"
55 MPH	Yes	55'	1st sign 750'	Right or Left Lane Closed Ahead	36" x 36"
			2nd sign 150'	Work Area	30" x 30" or 24" x 24"

NOTES: 1. Flagger Ahead sign needed if flaggers are used in all cases.
2. 45 MPH or faster — high level warning sign shall be used.
3. Existing district signs may be used as determined by the lead person or supervisor.

LEGEND FOR SIGNS USED IN TRAFFIC

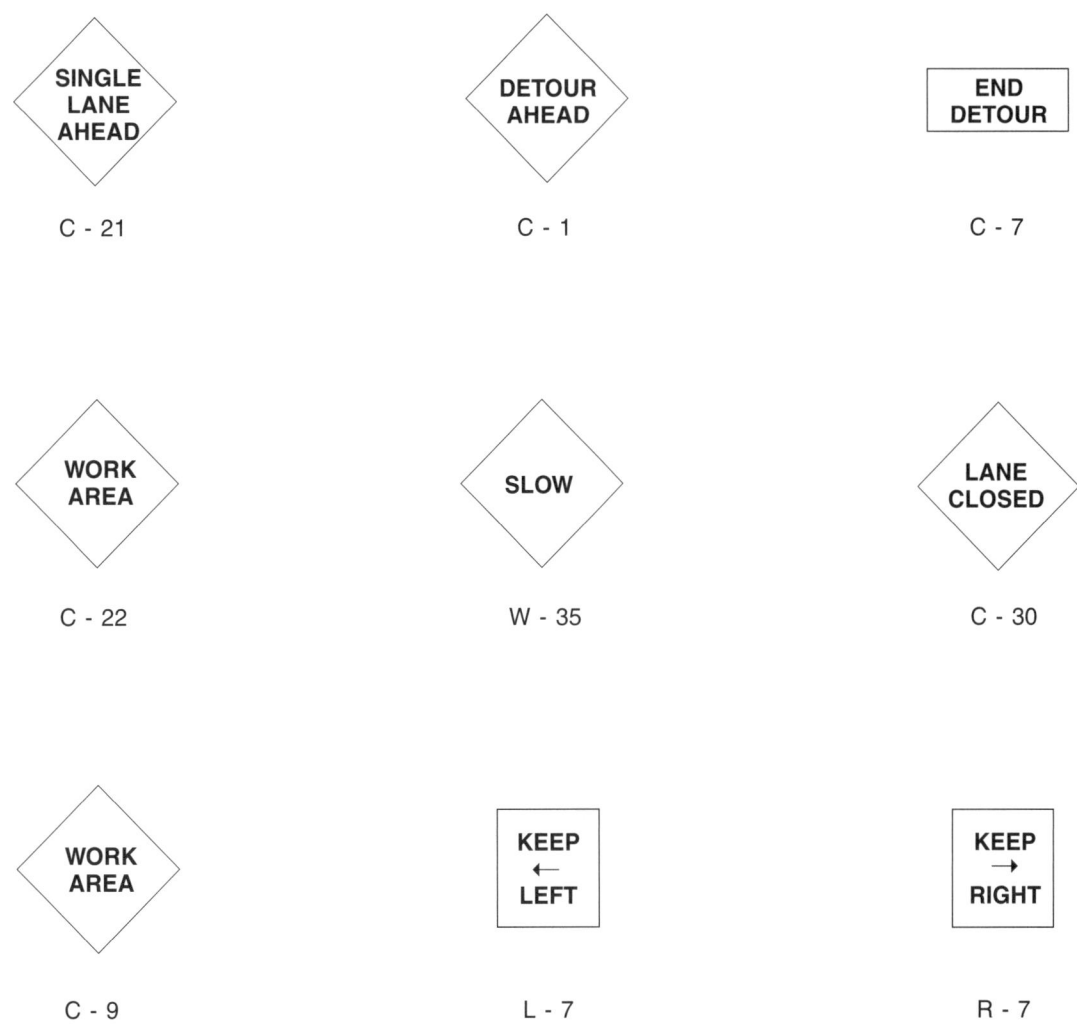

SINGLE LANE AHEAD	DETOUR AHEAD	END DETOUR
C - 21	C - 1	C - 7
WORK AREA	SLOW	LANE CLOSED
C - 22	W - 35	C - 30
WORK AREA	KEEP ← LEFT	KEEP → RIGHT
C - 9	L - 7	R - 7

NOTES: 1. Size of Sign — determined by speed of traffic.
2. Location — as specified in procedure.

TYPICAL CLOSING OF RIGHT LANE

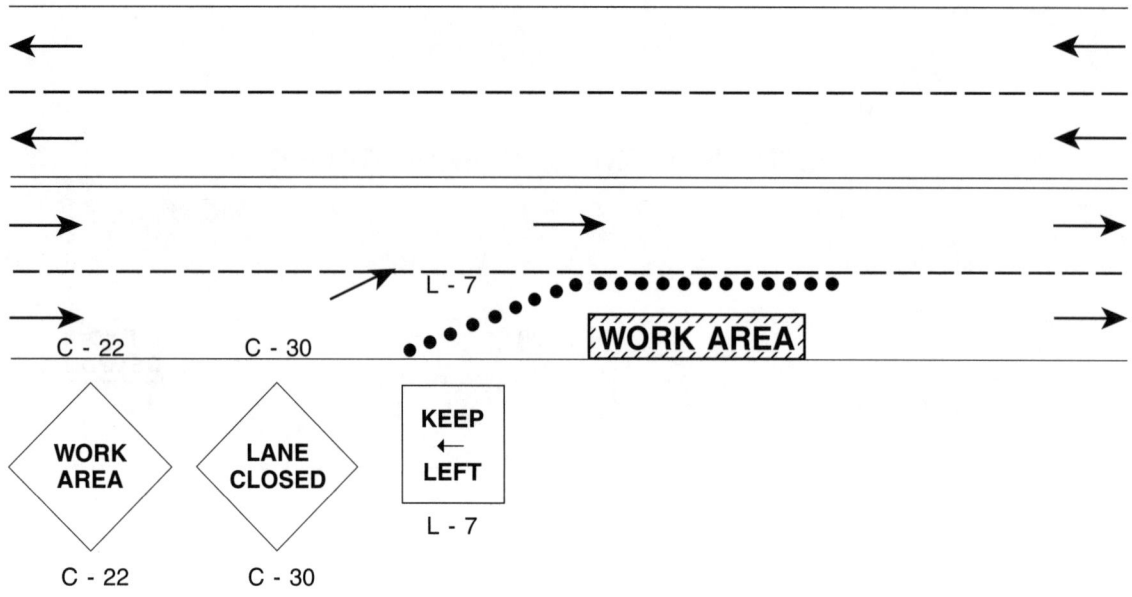

TYPICAL CLOSING OF LEFT LANE

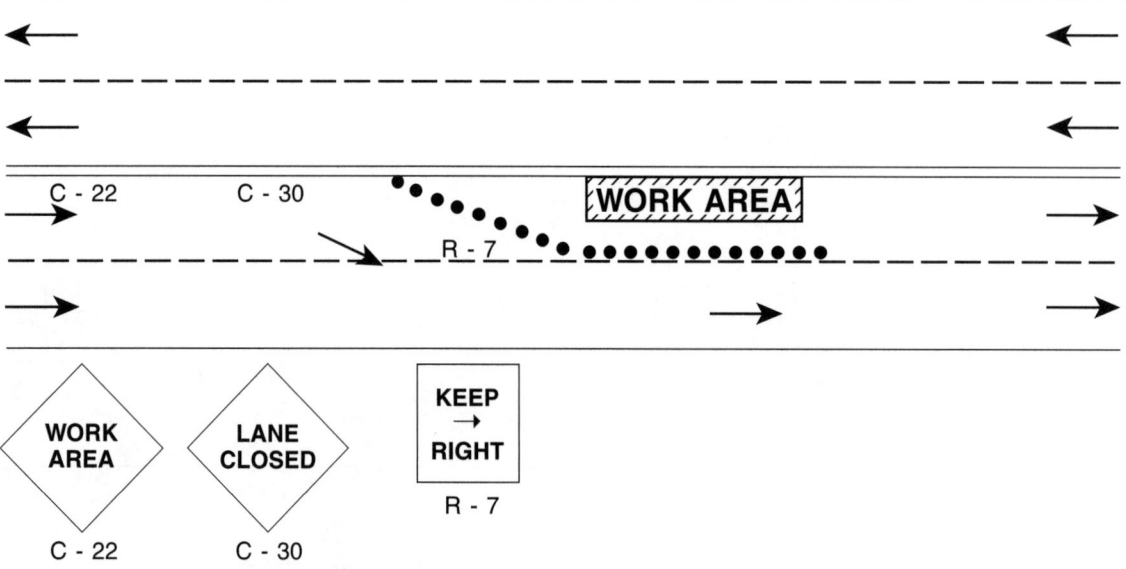

NOTES: 1. C - n = Type of Sign and Location
2. Any vehicle(s) that are part of the work area to have rotating warning light operating.

TYPICAL CLOSING OF HALF ROADWAY (Four Lanes)

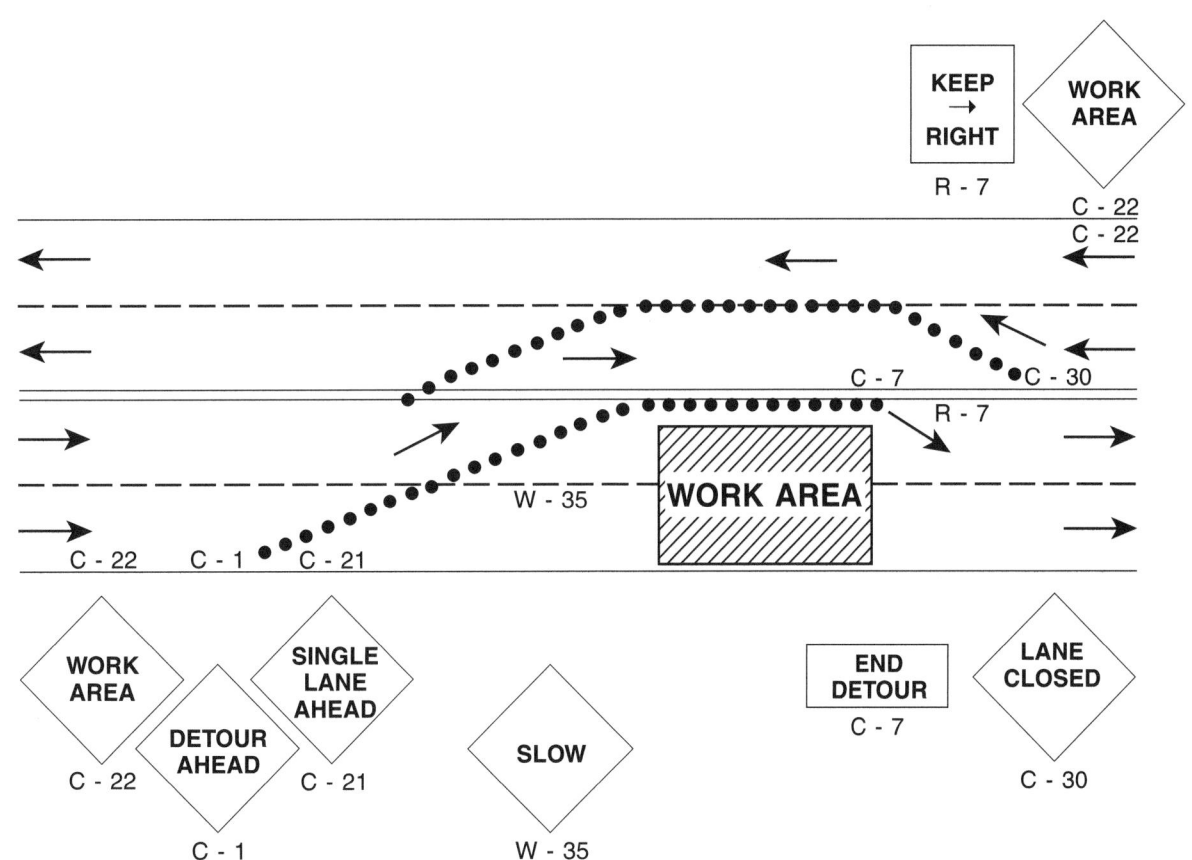

TYPICAL CLOSING OF HALF ROADWAY (Two Lanes)

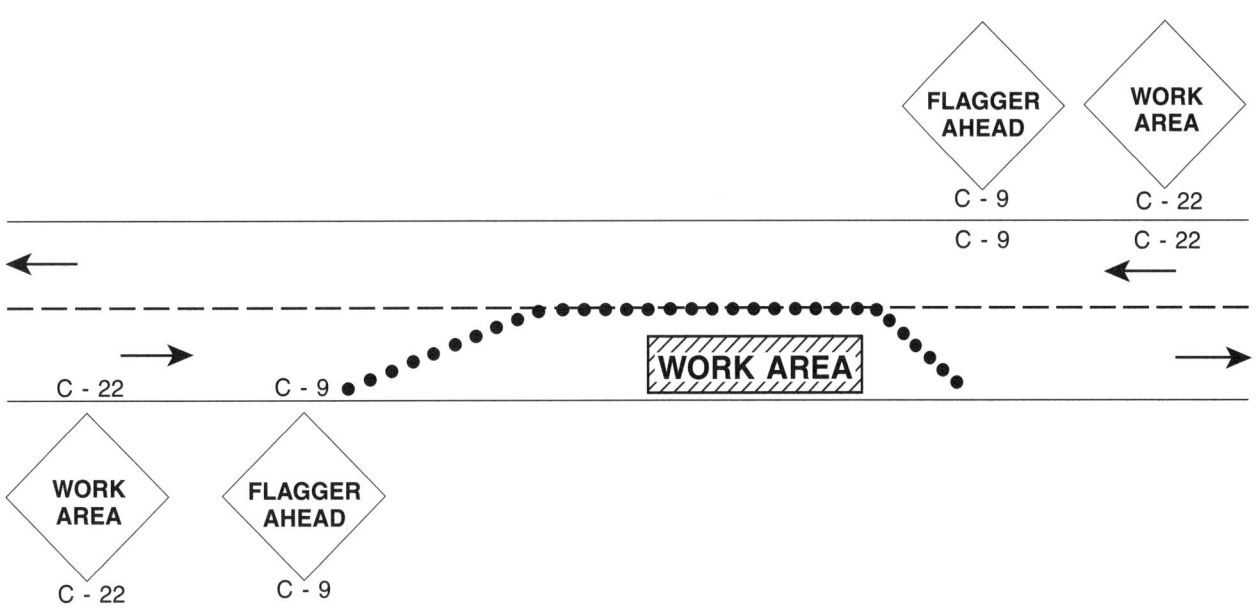

NOTES: 1. C - n = Type of Sign and Location
2. Any vehicle(s) that are part of the work area to have rotating warning light operating.

APPENDIX C

PROCEDURES FOR USE AND CALIBRATION OF GAS DETECTION DEVICES

PROCEDURE FOR USE OF GASTECH 1641

Unplug the instrument from the charger.

Check battery voltage by turning switch from OFF to V CHK. Meter should rise to near the middle or top of red band. If meter does not go into the red band *DO NOT* use the instrument. Low battery alarm is a steady tone, no lights.

Turn switch to COMB (combustibles) position and let circuits warm up. Normally, when the instrument is turned on a pulsing or steady tone will be heard as the circuits warm up.

Check that the pump is operating: FLOW indicator will be rotating.

Connect hose and probe to sample inlet. Check for leakage by placing finger over inlet. The flow indicator will slow down and/or stop if system is tight. Be sure that the probe inlet is in a gas-free area.

With the switch in the COMB (combustibles) position, lift and turn the *RED* COMB ZERO knob to move meter up and down the scale. Make alarm operate by moving meter above 10% LEL. Set meter exactly at zero and leave at zero.

Turn switch to OXY (oxygen) position, lift and turn *BLUE* OXY CAL knob to move meter up and down scale. Make alarm operate by moving meter below 19.5%. Set meter at O_2 CAL mark of 21%.

Turn switch to H_2S ZERO (hydrogen sulfide) position. Pump will stop and FLOW indicator will stop. Lift and turn *YELLOW* H_2S ZERO knob to move meter up and down scale. Make alarm operate by moving meter above alarm setting of 10 ppm. Turn meter back down. Allow instrument to stabilize for one (1) minute. Set exactly at zero using H_2S ZERO knob.

Instrument is now adjusted and ready to use.

TESTING PROCEDURE

Place the tip of the probe into a vent hole of the manhole cover (just the tip of the probe — *DO NOT PUSH PROBE INTO DUST PAN)*. Turn switch to COMB, watch meter and observe reading after 15 to 20 seconds. If no alarm lights or sounds, the manhole cover may be removed.

Lower probe and hose into manhole at various levels. *DO NOT* put end of probe into water. Test in the COMB position for 15 to 20 seconds. Switch to OXY position and test for 15 to 20 seconds. Switch to H_2S RD position and test for 15 to 20 seconds. Record results on Gas Detection Log. If any of the three alarms sounds, leave the manhole open for ten (10) minutes. Run all tests again and record results on Gas Detection Log.

If no alarm sounds on first testing, the manhole may be entered. If no alarm sounds on second testing after venting manhole for ten (10) minutes, the manhole may be entered. If alarm sounds on first and second testing, then a blower must be used to ventilate the manhole. Ventilate manhole with blower for five (5) minutes. Retest the manhole (all tests) while the blower is providing ventilation and record results on Gas Detection Log. If no alarm sounds, the manhole may be entered. A blower is to be used to ventilate the manhole continually while an employee is in the manhole.

All alarms are functional as long as the switch is in any one of the sample positions: COMB, OXY, or H_2S RD.

The instrument is to be secured at ground level with the hose and probe in the manhole and the instrument continuously testing the atmosphere. It is the responsibility of the employee at ground level, assisting the employee in the manhole, to observe the gas detector and the employee in the manhole. If an alarm sounds or a reading of the gas detector indicates a problem, the employee in the manhole is to leave the manhole.

At the end of work day (or shift), turn switch to OFF-CHG position and connect to charger.

PROCEDURE FOR USE OF GASTECH 3220 HS

Unplug the instrument from the charger.

Connect hose and probe to fitting on front of instrument. Be sure that the probe inlet is in a gas-free area.

Press POWER switch to turn instrument on. The display will indicate the results of the previous ADJUST action, in a blinking mode. At this time, the alarms will be inactive and the audible hum of the pump may be heard.

After a preset two-minute warm-up period has elapsed, the buzzer will sound once and the display will continue to blink.

When the ADJUST button is pressed and released, the instrument will automatically go through its ten (10) second adjustment sequence, after which a double tone will sound.

Then the instrument will self-adjust and the display will indicate 00% LEL, 20.9% Oxygen, and 00 ppm H_2S.

The alarm will be enabled and the instrument will now be ready for normal operation.

If the ADJUST button is not pressed, the buzzer will sound twelve (12) times at six to ten (6 – 10) second intervals. The display will continue to blink. After the twelfth time, the instrument will automatically become active but will display gas concentrations based on the previous ADJUST action in a blinking mode.

To resume normal operation, the ADJUST button must be pressed and held until the double tone is produced (approximately 15 seconds).

TESTING PROCEDURE

Place the tip of the probe into a vent hole of the manhole cover (just the tip of the probe — *DO NOT PUSH PROBE INTO DUST PAN)*. Watch meter and observe reading after 30 seconds. If no alarm sounds, the manhole cover may be removed.

Lower probe and hose into the manhole. *DO NOT* put end of probe into water. Test manhole at several depths working from top to bottom. Allow 30 seconds at each level for the instrument to test the atmosphere. Record results on Gas Detection Log. If any of the three alarms sounds, leave the manhole open for ten (10) minutes. Run all tests again and record results on Gas Detection Log.

If no alarm sounds on first testing, the manhole may be entered. If no alarm sounds on second testing after venting manhole for ten (10) minutes, the manhole may be entered. If alarm sounds on first and second testing, then a blower must be used to ventilate the manhole. Ventilate manhole with blower for five (5) minutes. Retest the manhole while the blower is providing ventilation and record results on Gas Detection Log. If no alarm sounds, the manhole may be entered. A blower is to be used to ventilate the manhole continually while an employee is in the manhole.

The instrument is to be secured at ground level with the hose and probe in the manhole and the instrument continuously testing the atmosphere. It is the responsibility of the employee at ground level, assisting the employee in the manhole, to observe the gas detector and the employee in the manhole. If an alarm sounds or a reading of the gas detector indicates a problem, the employee in the manhole is to leave the manhole.

At the end of work day (or shift), press POWER switch to turn instrument off. Connect to charger.

CALIBRATION

Before calibration of the MSA Model 360 or Model 361 can be checked, the instrument must be in operating condition. Optional calibration of equipment is made as follows:

Model 360 Calibration (Using the Model R Calibration Check Kit)

1. Attach the flow control to the 0.75% pentane/15% oxygen calibration gas tank.

2. Connect the adapter hose to the flow control.

3. Open the flow control valve.

4. Connect the adapter hose fitting to the inlet of the instrument; within 30 seconds, the LEL readout should stabilize and indicate between 47 and 55%. If the indication is not in the correct range, remove the right end of the indicator and adjust the LEL SPAN control to obtain 50%.

5. Verify the oxygen reading; it should be between 13 and 17%.

6. Disconnect the adapter hose fitting from the instrument.

7. Close the flow control valve.

8. Remove the flow control from the calibration gas tank.

9. Attach the flow control to the 300 ppm carbon monoxide calibration gas tank.

10. Open the flow control valve.

11. Connect the adapter hose fitting to the inlet of the instrument; after approximately two (2) minutes, the TOX readout should stabilize and indicate between 275 and 325 ppm. If the indication is not in the correct range, remove the right end of the indicator and adjust the TOX SPAN control to obtain 300 ppm.

12. Disconnect the adapter hose fitting from the instrument.

13. Close the flow control valve.

14. Remove the adapter hose from the flow control.

15. Remove the flow control from the calibration gas tank.

Model 361 Calibration (Using Model RP Calibration Check Kit)

(See Figure 5.11)

1. Attach the flow control to the 0.75% pentane/15% oxygen calibration gas tank.

2. Connect the adapter hose to the flow control.

3. Open the flow control valve.

4. Connect the adapter hose fitting to the inlet of the instrument; within 30 seconds, the LEL meter should stabilize and indicate between 47 and 55%. If the indication is not in the correct range, remove the right end of the indicator and adjust the LEL SPAN control to obtain 50%.

APPENDIX C

*Fig. 5.11 Portable gas indicator alarm model
(oxygen, hydrogen sulfide and combustible gas)*
(Permission of MSA)

5. Verify the oxygen reading; it should be between 13 and 17%.

6. Disconnect the adapter hose fitting from the instrument.

7. Close the flow control valve.

8. Remove the flow control from the calibration gas tank.

9. Attach the flow control to the 40 ppm hydrogen sulfide calibration gas tank.

10. Open the flow control valve.

11. Connect the adapter hose fitting to the inlet of the instrument; after approximately one (1) minute, the TOX readout should stabilize and indicate between 35 and 45 ppm. If the indication is not in the correct range, remove the right end of the indicator and adjust the TOX SPAN control to obtain 40 ppm.

12. Disconnect the adapter hose fitting from the instrument.

13. Close the flow control valve.

14. Remove the adapter hose from the flow control.

15. Remove the flow control from the calibration gas tank.

CAUTION

Calibration gas tank contents are under pressure. Do not use oil, grease or flammable solvents on the flow control or the calibration gas tank. Do not store calibration gas tank near heat or fire, or in rooms used for habitation. Do not throw in fire, incinerate or puncture. Keep out of the reach of children. It is illegal and hazardous to refill this tank. Do not attach any gas tank other than MSA calibration tanks to the flow control.

APPENDIX D

CONFINED SPACE POLICY AND PROCEDURE
(Union Sanitary District)

Policy

Union Sanitary District is committed to protecting the health and safety of its employees and to safeguard the public against harm which could result from its operations. Work in the District's confined spaces can be hazardous to workers and to the nearby public if unsafe conditions and unsafe work practices are not prevented or controlled. It is District policy, therefore, to establish, implement and maintain a systematic program to prevent and effectively control confined space hazards.

Purpose

The purpose of the program is to establish safety standards and procedures that will direct the performance of work in **all** of the District's confined spaces.

The District's objectives in directing the activities of confined space work are:

1. Prevention of injury and illness to employees and the public from confined space hazards.
2. Compliance with applicable CAL/OSHA regulations.
3. Prevention of workers compensation and legal claims from confined space accidents.
4. Maintenance of good labor and public relations through proactive identification and control of confined space hazards.

Definitions

Air Monitoring — The process of identifying the level of hazardous vapors or particulates in a sample volume of air by drawing or diffusing the air onto a collection media for purposes of analysis.

CARBON MONOXIDE ATMOSPHERE — A confined space that contains carbon monoxide toxic vapors at a concentration greater than 35 parts per million (35 ppm) of their lower toxic level.

COMBUSTIBLE ATMOSPHERE — A confined space atmosphere that contains flammable vapors or combustible particulates at a concentration greater than 10% of their lower flammability or explosion limit (LFL or LEL).

CONFINED SPACE ATTENDANT — A trained and equipped person who is stationed outside of the entrance to a confined space to assist the entrants with the entry procedures and to respond in the case of an emergency (i.e., injury, illness, fire, explosion, earthquake, etc.).

CONFINED SPACE — For purposes of this program, a confined space is any enclosed space where work may be performed that has a limited or restricted means of entry or exit due to its size, configuration or location. Some examples of confined spaces are manholes, tanks, vaults, pits, diked areas, sumps, vessels, storage bins, ducts, wells, valve boxes, digesters, aerators, etc.

A room or other such enclosed space, where entry and exit is through an unrestricted doorway located at the same elevation as the work space, is not considered a confined space. Even though the space may be hazardous due to the presence of a hazardous material or because the work performed there is dangerous, unless entry to or exit from it is restricted, it does not meet the definition of a confined space. Specific space or job safety procedures are required to protect employees who are exposed to these kinds of hazardous conditions/activities.

CONFINED SPACE ENTRY PERMIT — A District form is used to guide the supervisor or the designated employee(s) to implement all of the required Pre-entry Precautions.

The Permit must be completed and signed by the supervisor or his/her designated employee(s) prior to the first entrance into the confined space, and must be secured near the entrance to it. Every tester must sign the Permit, indicating that they have checked to see that the Pre-entry Precautions have been taken and that they are assigned to work in the confined space. The form is to be removed at the end of the confined space work and sent to the Division Manager for recordkeeping.

DESIGNATED EMPLOYEE — A trained employee who is authorized by his/her supervisor to enter a confined space and to implement the hazard control measures required by this program

HAZARD — An unsafe work practice or unsafe physical condition.

HAZARD CONTROL MEASURES — Prescribed actions implemented to effectively control inherent hazards. Examples of confined space hazard control measures would include: air monitoring; stationing an attendant at the entrance to the confined space; installing mechanical ventilation equipment; locking out and/or tagging out sources of hazardous energy, etc.

HAZARDOUS MATERIAL — A substance that is toxic, flammable, corrosive, extremely hot or cold, and/or unstable (i.e., reacts with other chemicals), or which can displace the oxygen content in a confined atmosphere.

HOT WORK — Any work activity involving an open source of ignition such as gas or electric welding or cutting.

IMMEDIATELY DANGEROUS CONFINED SPACE — A confined space that is immediately dangerous to life or health (IDLH) because of the presence of one or more of the following conditions:

1. A toxic, combustible or oxygen deficient atmosphere as detected by air monitoring.

2. A large quantity of bulk material which could engulf a person.

3. A source of fast-flowing water which could overwhelm a worker.

4. An uncontrolled form of hazardous energy (e.g., live electrical conductors; exposed operating machine parts; unshielded radio-active components, etc.).

5. Worker activity which creates one or more of the above conditions (e.g., painting, sandblasting, application of hazardous chemicals, chemical line breaking, working with live electrical conductors, hot work, etc.).

LOWER FLAMMABILITY OR EXPLOSION LIMIT (LFL, LEL) — The minimum concentration of vapors or particulates within a volume of air below which propagation of a flame will not occur in the presence of an ignition source. Below the LFL or LEL, the mixture is "too lean" to burn.

MANHOLE — A hole through which a person can enter the District's collection system. All manholes greater than 3 feet in depth, whether they are partially or completely constructed, new or existing, connected or unconnected, shall be regarded as a Permit Required Confined Space.

NON-PERMIT REQUIRED CONFINED SPACE — A confined space that does not contain, nor has recognized potential to contain, one or more of the IDLH conditions. A Confined Space Entry Permit is not required for this type of confined space because the risk from inherent hazards is low. See list of the District's Confined Spaces Locations and Classifications.

OXYGEN DEFICIENT ATMOSPHERE — A confined space atmosphere that contains less than 19.5% oxygen.

PERMIT REQUIRED CONFINED SPACE — A confined space that is not immediately dangerous to life or health, but which has a recognized potential for containing one or more of the above IDLH conditions. A Confined Space Entry Permit is required. See list of the District's Confined Spaces Locations and Classifications.

SUPERVISOR — The supervisor of an employee assigned to perform work in a confined space.

TOXIC ATMOSPHERE — A confined space atmosphere that contains 10 or more parts of Hydrogen Sulfide (H_2S) gas per million parts of air (ppm), or contains any other airborne hazardous substance at a level exceeding CAL/OSHA's 8-hour permissible exposure limit (PEL).

Procedure

Identification & Classification of Confined Spaces:

1. A District-wide team shall be established, consisting of an equal number of knowledgeable supervisors and employees to identify and classify all of the confined spaces in each division.

2. The confined spaces shall be classified in accordance with the definitions established by this program.

3. The team shall list the classified confined spaces for inclusion in this program. See the current confined space list.

4. Managers shall ensure that all of the identified confined spaces, where feasible, are visibly labeled or posted with an identifier that states: **Caution! This is a confined space. Hazard Control Measures are required. Follow the District's Confined Space Safety Procedures.**

5. Any confined space that is identified or reclassified subsequent to the effective date of this Program is to be correctly labeled or marked by division staff and the list of the District's confined spaces is to be updated and reissued.

Pre-entry Precautions: The following safety precautions shall be taken by the supervisor or his/her designated employees prior to their entrance of any of the District's confined spaces:

1. Develop a plan for safely working in the confined space. The plan should minimize the time spent in the confined space and should provide for implementation of control measures for identifiable unsafe conditions and for foreseeable unsafe work practices.

2. Review the material safety data sheets for all hazardous materials to be taken into the confined space and implement the required safeguards.

3. Extinguish all sources of ignition before approaching within 10 feet of the confined space.

4. Isolate, if possible, all hazardous substances or forms of hazardous energy connected to the confined space. This would entail separating or blocking the identified sources of hazardous substances and hazardous energy from the confined space. Pipe valves, electrical disconnect switches, and sewer plugs are examples of isolation devices.

5. Check the display of the permanently installed gas detection instrument, if one has been installed, to be certain that the level of oxygen, combustible gas, hydrogen sulfide (H_2S), and carbon monoxide (35 ppm) are within safe (no alarm) limits.

6. If a permanent gas detection instrument has not been installed for the confined space, or if it is not functioning properly, test the atmosphere inside of the entrance to the confined space with one of the District's portable gas detection meters. Open the cover or door to the confined space and insert the instrument's probe into a port or hole, if one exists. **Important:** Before inserting the probe, make sure the instrument is operating properly. Check to see that the display reads: 00% LEL, 20.9% Oxygen, 00 ppm H_2S, and carbon monoxide 0.0 ppm.

7. If there is no port or hole through which to test the atmosphere inside of the entrance to the confined space, stand upwind of the entrance and open the cover or door slightly, then insert the probe and take the reading.

8. If the gas detection meter signals an unsafe condition, leave the cover or door open with the detector probe inside, prevent unauthorized entry to and around the confined space, and notify the supervisor. **Do not enter the confined space and do not allow anyone else to enter it until the alarm condition has been eliminated.**

9. If the gas detection instrument indicates the atmosphere inside of the entrance to the confined space is below alarm levels, open the cover or door completely and insert or drop the instrument probe as far into the confined space as possible without entering it. If alarm conditions are not found, the confined space is considered safe to enter.

10. Complete and sign a Confined Space Entry Permit for all Permit Required Confined Spaces and securely attach it to, or near, the entrance of the confined space. No one is to enter a Permit Required Confined Space without having first reviewed and signed the Permit.

Entry Safeguards:

1. If no permanently installed gas detection instrument is present in the confined space, the confined space is to be entered with the probe of the portable gas meter as far in front of and below the entrant's breathing zone as possible.

2. Confined space locations where heavier than air gases may accumulate should be monitored with the portable gas detection instrument. These locations include low level gas entry points, or areas where air flow to the outside may be obstructed by equipment.

3. The gas detection instrument shall be kept running in the work area during the entire time an entrant is in a confined space to detect any harmful atmospheric conditions that may develop.

4. Confined spaces with side and top openings shall be entered from the side opening when feasible.

5. Smoking is not permitted inside of a confined space nor within a 10 foot radius around the outside of the confined space.

Hazard Control Measures for Immediately Dangerous Confined Spaces:

1. **No one is to enter a confined space in which air monitoring or observation has shown an IDLH condition as defined by this program.**

2. If an IDLH condition develops after work has begun in a confined space, the workers are to immediately evacuate the space and notify their supervisor.

3. The supervisor or his/her designated employee(s) shall secure the entrance of the confined space to prevent unauthorized entry.

4. The confined space shall not be reentered until the identified IDLH condition has been eliminated.

5. If the IDLH condition is a toxic, combustible or an oxygen-deficient atmosphere, the supervisor or his/her designated employee(s) shall open the confined space and allow it to be ventilated with outside air for a period of five minutes.

6. If the IDLH atmosphere is eliminated after five minutes of natural ventilation, as indicated by measurements taken with a gas detection instrument, work may resume in the confined space, provided the instrument is used to continuously monitor the work area.

7. If after five minutes of natural ventilation, safe levels of H_2S, combustible gases and oxygen are not achieved in the confined space, the supervisor or his/her designated employee(s) shall use mechanical ventilation equipment to purge the confined space atmosphere with outside air.

8. The mechanical ventilation equipment shall remain in operation in the confined space along with the gas detection instrument until the IDLH condition has been eliminated and work in the confined space has been completed.

9. If an IDLH condition does not clear, only trained emergency response workers may enter the confined space and only for rescue purposes.

Hazard Control Measures for Suspected IDLH Atmospheres:

1. Work may not begin or be resumed in a confined space, even if measurements with a gas detection instrument indicate that the H_2S, LEL and oxygen levels are within normal limits, if there is a **logical reason** to suspect the presence in the confined space of other hazardous atmospheric conditions which may be immediately dangerous to life or health.

    It would be logical to suspect the existence of an IDLH atmosphere, for example, from the presence of a strong or unusual odor, or from a reported or observed chemical spill upstream, upwind or within the confined source, or from the occurrence of skin or respiratory irritation while in a confined space.

2. The supervisor or his/her designated employee(s) shall keep all employees away from a confined space suspected of having an IDLH atmosphere until secondary air monitoring for suspected air contaminants can be conducted.

3. Trained environmental compliance employees shall conduct secondary air testing using a grab air sampler and appropriate collection media (i.e., Drager bellows pump and colorimetric tubes or other comparable equipment).

4. The air monitoring employee shall record the findings in writing and shall report them to appropriate supervisors, who in turn shall communicate the findings to their affected employees.

5. The supervisor shall instruct the affected workers to enter or re-enter the confined space **only** if the air monitoring findings and the readings from the gas detection instrument, together with his/her personal assessment of the suspected conditions, **confirm that a toxic, combustible, or oxygen-deficient atmosphere is not present.**

6. If an industrial spill occurs, or is suspected to have occurred, within a confined space, the appropriate Spill Incident Procedures shall be followed and a spill documentation form completed.

Hazard Control Measures for Permit Required Confined Spaces: Supervisors shall ensure, through periodic instruction and site auditing, that their designated employees implement the following Hazard Control Measures whenever they are assigned to work in a Permit Required Confined Space. They shall:

1. Implement the Pre-entry Precautions described on pages 408 and 409.

2. Complete a Confined Space Entry Permit and secure it near the entrance to the confined space prior to entering it.

3. Review and sign the Entry Permit prior to entrance.

4. Implement the Entry Safeguards described on page 409.

5. Place a rescue harness on each entrant to the confined space, attach the harness to a lanyard and connect the lanyard to a hoisting device/person located outside of the confined space.

6. Station a trained Confined Space Attendant at the entrance to the space.

7. Have at least one other employee within sight or call of the Confined Space Attendant to assist in the case of an emergency.

8. Operate a charged and calibrated gas detection instrument at the work site within the confined space for the duration of the work activity.

9. Have the following protective equipment at or near the confined space:

   a. A radio or telephone for calling 911 in an emergency.

   b. Fire extinguishers for placement inside and outside of the confined space when there is a recognized fire hazard (hot work).

   c. Mechanical ventilation equipment for use in rapid replacement of the confined space atmosphere with outside air.

   d. A first aid kit.

Control Measures for Non-Permit Required Confined Spaces: Supervisors shall ensure, through instruction and site auditing, that their designated employees implement the following Hazard Control Procedures whenever they are assigned to work in a Non-Permit Required Confined Space. They shall:

1. Implement the Pre-entry Precautions described on pages 408 and 409.

2. Execute the Entry Safeguards described on page 409.

3. Run a charged and calibrated gas detection instrument at the work site for the duration of the work activity.

4. Have a first aid kit and radio or telephone within 10 feet of the confined space.

Special Hazard Control Measures:

1. If a mechanical hoist cannot be used due to the location or configuration of a manhole or other confined space, no one shall enter it unless all of the following safeguards have been implemented:

   a. The employee is fitted into an approved parachute harness and two lanyards are attached to it.

   b. Two confined space attendants are stationed at the entrance to the confined space.

   c. The confined space attendants have determined through pre-entry lifting of the employee that they are physically able to pull him/her out of the confined space if he/she should become disabled.

2. The drivers of trucks with permanently mounted hoisting devices that will be used for work in and around manholes shall:

   a. Attach the key ring containing the truck's ignition key to the harness ring prior to attaching the lanyard to the hoisting device.

   b. Bring the truck to a complete stop at a safe distance away from traffic and the work site; put the automatic transmission selector in the park position; engage the emergency brake or hand brake; survey the site to identify any hazardous conditions; and secure the work area with the appropriate traffic control devices before allowing work to begin.

   c. Inspect, along with the confined space entrant, the entire hoisting system (i.e., hoist, tripod, winch, hoist cable, lanyard and harness) prior to use to ensure that none of the components is defective.

   d. Assist the entrant to put on the harness, connect the lanyard to the harness, and attach the vehicle keys to the harness ring.

   e. Ensure that communication is maintained with the entrant and with the third crew member throughout the work assignment.

f. Assist the entrant to exit the confined space safely when the work is complete.

g. Once the employee has safely exited the manhole, detach the lanyard from the harness; secure the hoisting device to a transport position; and remove the keys from the key ring on the harness.

Changes to Confined Space Classifications: The classification of a confined space can change, requiring the implementation of a different set of Hazard Control Measures than was anticipated. These changes can occur from an external event, such as an industrial spill of a hazardous material into the collection system, or from the performance of hazardous work within the confined space. Such work would include, but is not limited to, "hot" work, sandblasting, painting, wall repairs, application of cooling products, and use of other hazardous chemicals.

For example, if an external event or performance of hazardous work in a Non-Permit Required Confined Space creates an actual or potential IDLH condition, the space is to be reclassified accordingly and the corresponding Hazard Control Measures are to be implemented. Conversely, whenever the risk defining an Immediately Dangerous or Permit Required Confined Space is eliminated, the affected confined space would have to be reclassified in accordance with its lower risk.

The Hazard Control Measures delineated in this program are specific to the confined space classification. Any classification change is to be determined by the supervisor or his/her designated employee based upon an assessment of the risk.

The Pre-entry Precautions and Entry Safeguards are applicable to all confined space classifications.

Emergency Response and Rescue:

1. The first action to be taken by a Confined Space Attendant upon discovery of an incapacitated confined space entrant shall be to attempt to hoist/pull the victim out of the space.

2. Confined Space Attendants **shall not** enter a confined space to rescue an injured or ill employee unless **all** of the following conditions exist:

a. The victim cannot be hoisted out.

b. No alarms are sounded and displayed by the gas detection instrument.

c. The victim and the inside of the confined space can be seen.

d. The Attendant can talk to the victim and knows for certain that an IDLH condition is not present.

3. Absent all of these assurances, the Confined Space Attendant **shall not** enter a confined space for rescue purposes. Instead, the Attendant shall follow the District's emergency standard operating procedures and shall immediately telephone or radio 911 to get assistance from the emergency response team of the local fire department.

4. After notifying the response team representative of an emergency situation, the Attendant shall notify his/her supervisor of the situation and shall prepare, in accordance with his/her training, for the arrival of the emergency responders.

Equipment Calibration and Maintenance:

1. Facilities Maintenance shall assign qualified persons to regularly calibrate and/or maintain the equipment to be used in the implementation of this program. This equipment includes, but is not limited to: permanent and portable gas detection instruments, mechanical ventilation equipment, harnesses, lanyards, hoists, tripods, winches, booms, and hoist cables.

2. Each division is responsible for checking its own confined space equipment in accordance with District policy/procedure.

3. The equipment shall be calibrated and/or maintained in accordance with the specifications and recommendations of the manufacturer.

4. The permanent and portable gas detection instruments are to be checked for defects and recalibrated at least every month.

Education and Training:

1. Each division manager shall ensure that all of his/her supervisors and employees affected by this program are kept informed about its requirements and their roles and responsibilities with regard to it.

2. Supervisors shall cover the program's requirements with their new and current employees in advance of their first assignment to work in a confined space.

3. Supervisors also shall conduct annual refresher training in confined space safety for their designated employees.

4. Each division manager shall ensure that at least one drill per year, simulating an emergency situation, is conducted in the division to prepare Confined Space Attendants and their supervisors to respond effectively in an emergency. The emergency response team of the local fire department shall participate in the planning and execution of each drill.

5. Training shall be conducted to prepare Program Auditors to effectively conduct an effective program audit. The training shall cover the requirements of the program and the methods for auditing program compliance activity and documentation.

6. The Environmental Compliance Supervisor shall ensure that all of his/her employees assigned to perform secondary monitoring in accordance with the requirements of this program receive annual training covering the principles and methods of air monitoring.

Documentation and Recordkeeping:

1. Each division shall maintain a file of the confined space entry records generated by its employees. These records include:

    a. Completed Confined Space Entry Permits: The Permits must be sent at the end of each month to the division manager or his or her designated representative for review.

    b. Training documents: All training and drills conducted in support of this program shall be documented on the District's training form by the trainer or the training coordinator.

    c. Program Auditor Reports: Program Auditors must submit a copy of their completed audit report to the appropriate division and to the Safety Committee.

    d. Air Monitoring Log: The findings from secondary air monitoring performed by the Environmental Compliance Section shall be recorded on an Air Monitoring Log. The Log is to be given to the supervisor of the affected employees, who, after communicating the findings to his/her employees, is to send the Log to his/her division manager for filing.

2. The calibration and maintenance records for confined space equipment shall be kept on file in Facilities Maintenance, who shall send a copy to the division in which the equipment is kept.

3. All confined space training and entry records shall be maintained for a minimum of three years.

Program Auditing:

1. Each supervisor shall routinely conduct program compliance audits at the site of confined space work performed by their designated employees. The objective of the audits is to identify and permanently correct any misunderstanding and misapplication of the confined space procedures.

2. The Confined Space Task Force shall select one or more safety representatives from each division to serve as Program Auditors to monitor program activity (including training) and documentation. Human Resources is responsible for coordinating the implementation of this requirement.

3. The Program Auditors shall be assigned to conduct program audits in a division other than the one in which they work. A representative of the audited division shall accompany the Program Auditor as an observer.

4. Program Auditors shall become thoroughly knowledgeable of the program's requirements and of the methods for conducting a program audit.

5. A minimum of three audits per year shall be performed in each division, consisting of observation and evaluation of a representative sample of program required activity and documents.

6. Program Auditors shall submit a written report of their audit findings to the affected supervisors and managers within one week of the audit completion. A copy of the report is to be forwarded to the Safety Committee and to Human Resources Division.

7. The divisions shall correct any program deficiencies identified in the audits. When appropriate, the Safety Committee shall meet with affected persons to facilitate understanding and action on reported audit findings.

Regulations:

This program complies with the requirements of:

1. California Code of Regulations, General Industry Safety Orders, Title 8, Article 108, Confined Spaces, Sections 5156 through 5159.

2. 29 Code of Federal Regulations, Part 1910.146, proposed final rule for Permit Required Confined Spaces.

Employee Responsibility

District employees, who enter and/or monitor the District's confined spaces, are responsible for complying with the requirements of this program.

Management Responsibility

District supervisors are responsible for ensuring that compliance with the confined space safety standards is constant, consistent, and complete.

CONFINED SPACE TESTING, SAMPLING, AND ENTRY CHECKLIST
(Union Sanitary District)

These Instructions Must Be Read Prior to Completing Reverse Side of This Form.

IMPORTANT: If entry is to take place you MUST read Sections A, B, and C and fill in the Confined Space Testing, Sampling, and Entry Permit Form (see the reverse side of this checklist). If entry will NOT take place read Sections A and B, and fill in the applicable sections of the Confined Space Testing, Sampling, and Entry Permit Form on the reverse side.

Section A — Air Monitoring

- Air monitor charged and calibrated?   Yes _____ No _____
- Oxygen level between 19.5% & 20.9%   Yes _____ No _____
- Combustible gas level <10% of LEL?   Yes _____ No _____
- Hydrogen sulfide level <10 ppm?   Yes _____ No _____
- Carbon Monoxide level <25 ppm?   Yes _____ No _____
- Other air monitoring.   Yes _____ No _____ N/A _____

Section B — Pre-Entry Precautions and Entry Safeguards

- Telephone/Radio?   Yes _____ No _____
- Harness, lanyard, and hoist?   Yes _____ No _____
- Mechanical ventilation equipment?   Yes _____ No _____ N/A _____
- First aid kit?   Yes _____ No _____
- Confined space attendant?   Yes _____ No _____
- Fire extinguishers?   Yes _____ No _____
- Connected hazards isolated, if applicable?   Yes _____ No _____ N/A _____
- Ignition sources extinguished, if applicable?   Yes _____ No _____ N/A _____
- MSDS reviewed, if applicable?   Yes _____ No _____ N/A _____
- Lockout/Tagout applied, if applicable?   Yes _____ No _____ N/A _____
- Hot Work Permit, if applicable?   Yes _____ No _____ N/A _____
- Pre-entry precautions and safeguards followed?   Yes _____ No _____

Section C — Rescue Services Provided by the Local Fire Department (911)

Note: If you need rescue services, you MUST notify the Fire Department or 911 Operator that it is a **Confined Space Emergency.**

- Provide address (including X-street), and a telephone number where you can be reached in case of a disconnect
- Prepare to specify the type of call
  - Fire - Hazardous Chemicals
  - EMS - Confined Space
- Notify USD Person-in-Charge (PIC)

Union Sanitary District
Confined Space Testing, Sampling, and Entry Permit Form

Instructions on the Reverse Side of This Form Must Be Read Prior To Beginning Entry

Location	Date	Time	Gas Det. #	Inst. Read Y/N	Classification IDLH	Classification P	Classification N/P	Cont. Mon. Y/N	Record All Alarms LEL	Record All Alarms O2	Record All Alarms H2S	Record All Alarms CO	Entry Y/N	Purpose C-I-M	Attendant Signature	Emp ID # 1	Emp ID # 2	Emp ID # 3
Cellotape	1/8/96	1000	1011	Y	—			N	00	20.9	00	00	N	1	K. Signa	231		
KAO	1/8/96	1015	1011	Y	—			N	00	20.9	00	00	N	1	K. Signa	231		
Tri City Circuits	1/8/96	1025	1011	Y	—			N	00	20.9	00	00	N	—	K. Signa	231		
Resonate 001	1/8/96	1035	1011	Y	—			N	00	20.9	00	00	N	1	6-09	209		
Readrite FAC	1/8/96	1040	1011	Y	—			N	00	20.6	00	00	N	—	K. Signa	231		
HmT #1	1/9/96	0900	1012	Y	—			N	00	20.9	00	00	N	1	K. Signa	231		
HmT #2	1/9/96	0905	1012	Y	—			N	00	20.9	00	00	N	1	K. Signa	231		
LSI	1/9/96	0915	1012	Y	—			N	00	20.9	00	00	N	1	K. Signa	231		
Magnum	1/9/96	0925	1012	Y	—			N	00	20.9	00	00	N	1	K. Signa	231		
Lam R	1/9/96	0935	1012	Y	—			N	00	20.9	00	00	N	1	K. Signa	231		

If alarm conditions exists, note actions taken to rectify situation below:

Alarm Date	Location	Action Taken/Comments

Legend:

Y/N
Y = Yes
N = No

Classification
IDLH = Immediately Dangerous to Life or Health
P = Permitted
NP = Nonpermitted

Purpose
C = Construction
I = Inspection
M = Maintenace

Employee
Emp. 1 = Attendant
Emp. 2 = Entrant
Emp. 3 = 3rd Member

APPENDIX E

HEARING PROTECTION POLICY AND PROCEDURE
(Union Sanitary District)

Policy

Union Sanitary District is committed to having a safe work environment, complying with all federal, state, and District safety regulations, and creating an atmosphere that promotes safety. The Hearing Conservation Program establishes guidelines, policies, and procedures to protect District employees from hazardous noise levels that may be present in the workplace.

Purpose

To protect District employees from hazardous noise levels that may be present in the workplace by ensuring that employees receive training on the hazards of noise and the proper use, care, and limitations of hearing protectors, appropriate hearing protection, and baseline and annual audiograms.

Definitions

NOISE — Sound with potential physiological, psychological, and/or safety repercussions.

PERMISSIBLE NOISE EXPOSURE — The level of noise to which an employee can be exposed without any protective equipment or control measures is 90 dba for continuous noise and 140 dba for impulse or impact noises. A Permissible Noise Exposure Table is located on page 416.

ACTION LEVEL (AL) — The noise exposure equal to or exceeding an 8-hour, time-weighted average of 85 decibels (dba), or equivalently, a dose of 50%. State and federal regulations require the employer to administer a hearing conservation program whenever employee noise exposures exceed the AL noise standard.

DECIBEL (dba) — A unit or measurement of sound. On a decibel scale, zero is the threshold of hearing and 120 decibels (dba) is the threshold for pain. Hearing protection will be required when using any equipment or tools that produce high noise levels over 85 dba, time-weighted average.

ENGINEERING CONTROL — An attempt to decrease an employee's exposure to a hazardous noise, such as installing a muffler or insulator of some kind, or by performing maintenance on the tool or machine involved, etc.

ADMINISTRATIVE CONTROL — An attempt to decrease an employee's exposure to a hazardous noise by instituting cross-training or job rotation.

PERSONAL HEARING PROTECTOR — A device, such as earplugs or earmuffs, used to decrease the decibel level of a noise.

TIME-WEIGHTED AVERAGE — The average exposure to noise, as determined by computing the levels of dba's over a certain period of time.

Procedure

If an employee's noise exposure exceeds the permissible noise exposure or the Action Level, state and federal regulations require the District to administer a Hearing Conservation Program which must, at a minimum, contain the following elements:

1. Audiometric Testing: Annual physical exams, which at USD normally occur in October of each year. Audiometric tests will be provided at no cost to the employee.

2. Training: Supervisors will conduct training on an as-needed basis. The training will cover the effects of noise on hearing, the purpose of noise monitoring, methods of noise control, the purpose of hearing protectors, the purpose of audiometric testing, a summary of the Hearing Protection Program, and the employees' rights and responsibilities.

3. Monitoring: Noise exposure will be evaluated whenever noise levels can be reasonably expected to exceed the Action Level as determined by District management. Monitoring will occur to ensure employees comply with the Hearing Protection Program.

4. Engineering and Administrative Control: Controls will be used to reduce the noise levels to below 90 dba, whenever feasible, and exposure levels will be considered when buying new equipment.

5. Hearing Protection: Only firm rubber earplugs or open-cell, foam type (with a special covering) will be used. Boxes of earplugs and antiseptic wipes will be made available to employees. Wipes should be used often on earplugs and they should be discarded when the earplugs become worn, dirty, or are no longer protecting the ear adequately. Each division that uses hearing protection will ensure that an adequate supply of earplugs and wipes are available. Each employee can use earmuffs or earplug protection. Employees may choose either, but generally only one type should be used at a time.

6. Record Keeping: Records regarding exposure measurements, training, and compliance will be maintained in accordance with state and federal law.

7. Notifications: All employees in the Hearing Conservation Program will be notified of their exposure to levels in excess of the Action Level (AL), as well as being notified of their audiogram results. The District notifies employees of exposure through signs posted at every permanent location where the noise level is in excess of the Action Level and during training sessions.

District employees will be included in the Program if they meet one or more of the following criteria:

- Employee is exposed to continuous noise levels in excess of CAL/OSHA Action Level without regard for attenuation provided by hearing protectors.

- Employee is exposed to impulse or impact noise at or above 140 dba.

- Employee is exposed to hazardous noise levels that are highly variable and unpredictable, and management determines that the employee should be included in the Hearing Conservation Program.

The following is a partial list of noise levels present in some work areas within Union Sanitary District (from a study conducted April, 1990):

Fremont Pump Station	80-101 dba	Alvarado Treatment Plant	
Truck-mounted, gas-powered		Heating and Mixing Room	85 dba
generator	92-102 dba	Hydro Cleaning Location	85 dba
Newark Pump Station	95-105 dba	Generator Room	89 dba
Irvington Pump Station	105 dba	RBC Bldg., Blower Room	
Jackhammer in use	107-118 dba	and Air Scrubbers	96 dba
District Trucks	92-101 dba	Aeration Blower Room	106 dba

The District will take steps to control the exposure for those employees exposed to high noise levels or to any noise level in the workplace which meets or exceeds 90 dba, as estimated by the supervisor.

Employee Responsibility

Employees are responsible for following this policy as indicated and as directed by management; taking proper precautions if working in an environment where exposure to hazardous noise levels occurs and by wearing hearing protectors when required; having hearing checked annually; properly cleaning, maintaining, and storing of hearing protectors.

Management Responsibility

Management is responsible for coordinating the Hearing Protection Program as detailed above and monitoring employee compliance with this Program and with all OSHA regulations; developing and implementing training for employees in the Hearing Conservation Program who are identified as potentially exposed at or above the Action Level; ensuring that employees in the Program have their hearing checked annually.

PERMISSIBLE NOISE EXPOSURE TABLE

PERMITTED DURATION PER WORKDAY

SOUND LEVEL (dba)	HOURS-MINUTES	SOUND LEVEL (dba)	HOURS-MINUTES
90	8-0	103	1-19
91	6-58	104	1-9
92	6-4	105	1-0
93	5-17	106	0-52
94	4-36	107	0-46
95	4-0	108	0-40
96	3-29	109	0-34
97	3-2	110	0-30
98	2-38	111	0-26
99	2-18	112	0-23
100	2-0	113	0-20
101	1-44	114	0-17
102	1-31	115	0-15

When the daily noise exposure is composed of two or more periods of noise exposure of different levels, their combined effect should be considered, rather than the individual effect of each. If the sum of the following fractions: $C_1/T_1 + C_2/T_2...C_n/T_n$ exceeds unity, then, the mixed exposure should be considered to exceed the limit value. C_n indicates the total time of exposure at a specified noise level, and T_n indicates the total time of exposure permitted at that level.

APPENDIX F

ACCIDENT INVESTIGATION PROCEDURE AND FORMS
(Union Sanitary District)

Union Sanitary District's Safety Committee endorses the following accident investigation procedures. SEIU 790 and District management have reviewed these recommendations and concur with the information outlined below.

Approach

Safety is every employee's responsibility. Management is responsible for establishing and maintaining a safe work environment, which includes providing the proper tools, equipment, training, and procedures to perform work safely. Management also provides incentives and recognition for working safely. All employees — management and non-management — must act to prevent accidents and injuries as well as to detect and correct unsafe situations. The District's Safety Action Procedure emphasizes the need to take immediate individual action, if necessary, to correct an unsafe or hazardous situation.

Accident Reports

Whenever an accident occurs, the supervisor will immediately prepare the workers' compensation insurance forms. In addition, an accident report, along with recommendations, must be prepared by the supervisor and sent to the Safety Committee chairperson within four (4) working days of the accident, whenever there is job-connected time off (including job-connected medical appointments). The supervisor will use the designated accident report form. In conjunction with the supervisor's interview of the employee, the employee(s) who had the accident will also complete a written account in his/her own words. This written statement will be attached to the Accident Report and submitted to the Safety Committee Chairperson. The employee(s) may also complete his or her own Accident Report, which will be sent to the Safety Committee Chairperson, or may sign the supervisor's Accident Report. The form has a place for the employee's signature, indicating agreement with the report as completed.

NOTE: When no job-connected time off results from an accident, an Accident Report will be completed only if the supervisor, employee, safety representative, or Safety Committee Chairperson deems it necessary.

Special Investigations

A special investigation of an accident or "near miss" will be undertaken if a written request is submitted by the employee, supervisor, division manager, department director, General Manager, Safety Committee Chairperson, or any permanent employee (the request must be made through his/her safety representative). The written request must be submitted to the Safety Committee Chairperson within a reasonable time period; three (3) calendar days of the accident or near miss is the suggested period, but other time periods may be acceptable. The investigation must be concluded within fifteen (15) days from the date the request was submitted.

The employee, his/her supervisor, manager, and/or director, and the Safety Committee Chairperson will select three people to serve on the Special Accident Investigation Team. One member must be a peer of the employee who had the accident. No supervisor or manager in the division in which the accident occurred or their department director may serve on the Special Accident Investigation Team.

ACCIDENT REPORT AND SUPERVISOR'S REPORT OF EMPLOYEE INJURY

(Union Sanitary District Human Resources)

Injured Employee's Name: _____

Class Title: _____   **Division:** _____

Department: _____   **Date & Time Reported:** _____

Accident/Injury Date: _____   **Accident Weekday:** _____

Employee's Work Day: Check One: Day 1__ 2__ 3__ 4__ 5__   **Accident Time:** _____

Accident Hour: Check One: Hour 1__ 2__ 3__ 4__ 5__ 6__ 7__ 8__ 9__ 10__ 11__ 12__

Employee's Age: _____   **Accident on Overtime?** Yes   No

Years of Service with District: _____   **Months in Job:** _____

Employee's Supervisor: _____   **Title:** _____

Date Interviewed: _____   **Interviewed By:** _____

Did Employee Leave Work?   Yes _____   No _____   **Time Left:** _____

Did Employee Return to Work? Yes _____   No _____   **Time Returned:** _____

Were safety policies, procedures, practices, etc. being followed?   Yes ____   No ____

Accident Location: _____

Body Part Injured: _____

Name of Medical Facility: _____

Witnesses: _____   **Job Title:** _____

_____   _____

_____   _____

_____   _____

_____   _____

_____   _____

Injury: _____

_____

_____

Weather: Inside ___ Outside ___ Give specifics on weather including temperature, dry or wet, etc. _____

_____

Accident Report and Supervisor's Report of Employee Injury
Continued Page - 2 -

Traffic: If not a condition, write NA. If a condiction, write specifics, such as heavy, light, commercial, etc.

_____

_____

**Personal Protective
Equipment Used:**
List out, be specific

_____

_____

Equipment Used:
List out, be specific

_____

_____

_____

Description of Worksite:

_____

_____

_____

Activity:
Be specific

_____

_____

_____

_____

_____

**Describe What Happened
and Why:**
Be specific

_____

_____

_____

_____

**Prevention Findings and
Recommendations:**

_____

_____

_____

_____

**What Steps Have Been
Taken to Prevent Similar
Accidents?**

_____

_____

_____

_____

_____

Accident Report and Supervisor's Report of Employee Injury
Continued Page - 3 -

Signed: _____   **Date:** _____
                            Injured Employee

Signed: _____   **Date:** _____
                          Employee's Supervisor

Signed: _____   **Date:** _____
                         Chair, Safety Committee

Employee's Statement

**Describe
what
happened in
your own
words:**   _____

_____

_____

_____

_____

_____

_____

_____

_____

**Employee's
Comments:**   _____

_____

_____

_____

_____

_____

_____

_____

_____

_____

_____

Signed: _____   **Date:** _____
                            Injured Employee

Original: Safety Committee Chairperson
    cc: Supervisor, Division Manager, Department Director, Employee, Bragg and Associates

CHAPTER 6

SAMPLING PROCEDURES FOR WASTEWATER

by

Scott Austin

TABLE OF CONTENTS

Chapter 6.   SAMPLING PROCEDURES FOR WASTEWATER

OBJECTIVES

Chapter 6.   SAMPLING PROCEDURES FOR WASTEWATER

Following completion of Chapter 6, you should be able to:

1. Define the goal of each sampling occasion and select appropriate sampling techniques,

2. Prepare for going into the field and collecting samples at pretreatment sites and in sewers,

3. Collect, label and preserve samples,

4. Transport samples to a laboratory for analysis,

5. Analyze samples in the field using field test kits,

6. Document the sample collection and transportation procedures,

7. Install and monitor field sensors to measure constituents in industrial wastestreams,

8. Trace an illegal discharge back up a sewer to the source, and

9. Use appropriate quality assurance/quality control (QA/QC) procedures to minimize sampling errors.

PRETREATMENT WORDS

Chapter 6.   SAMPLING PROCEDURES FOR WASTEWATER

ALIQUOT (AL-li-kwot)

ALIQUOT

Portion of a sample. Often an equally divided portion of a sample.

CHAIN OF CUSTODY

CHAIN OF CUSTODY

A record of each person involved in the handling and possession of a sample from the person who collected the sample to the person who analyzed the sample in the laboratory and to the person who witnessed disposal of the sample.

CHELATING (key-LAY-ting) AGENT

CHELATING AGENT

A chemical used to prevent the precipitation of metals (such as copper).

COMPOSITE (PROPORTIONAL) SAMPLE

COMPOSITE (PROPORTIONAL) SAMPLE

A composite sample is a collection of individual samples obtained at regular intervals, usually every one or two hours during a 24-hour time span. Each individual sample is combined with the others in proportion to the rate of flow when the sample was collected. Equal volume individual samples also may be collected at intervals after a specific volume of flow passes the sampling point or after equal time intervals and still be referred to as a composite sample. The resulting mixture (composite sample) forms a representative sample and is analyzed to determine the average conditions during the sampling period.

GRAB SAMPLE

GRAB SAMPLE

A single sample of water collected at a particular time and place which represents the composition of the water only at that time and place.

PRIORITY POLLUTANTS

PRIORITY POLLUTANTS

The EPA has proposed a list of 126 priority toxic pollutants. These substances are an environmental hazard and may be present in water. Because of the known or suspected hazards of these pollutants, industrial users of the substances are subject to regulation. The toxicity to humans may be substantiated by human epidemiological studies or based on effects on laboratory animals related to carcinogenicity, mutagenicity, teratogenicity, or reproduction. Toxicity to fish and wildlife may be related to either acute or chronic effects on the organisms themselves or to humans by bioaccumulation in food fish. Persistence (including mobility and degradability) and treatability are also important factors.

REPRESENTATIVE SAMPLE

REPRESENTATIVE SAMPLE

A sample portion of material or wastestream that is as nearly identical in content and consistency as possible to that in the larger body of material or wastestream being sampled.

RETURN SLUDGE

RETURN SLUDGE

The recycled sludge in a POTW that is pumped from a secondary clarifier sludge hopper to the aeration tank.

VOLATILE (VOL-uh-tull)

VOLATILE

(1) A volatile substance is one that is capable of being evaporated or changed to a vapor at relatively low temperatures. Volatile substances also can be partially removed by air stripping.

(2) In terms of solids analysis, volatile refers to materials lost (including most organic matter) upon ignition in a muffle furnace for 60 minutes at 550°C. Natural volatile materials are chemical substances usually of animal or plant origin. Manufactured or synthetic volatile materials such as ether, acetone, and carbon tetrachloride are highly volatile and not of plant or animal origin.

CHAPTER 6.   SAMPLING PROCEDURES FOR WASTEWATER

6.0 REASONS FOR SAMPLING

NEVER "go out and take some samples." If you don't do more preparation than that statement implies, your efforts will probably be wasted. Industrial wastewater samples are never an end product; they are only a means to obtain information for some other purpose. This chapter will give you the questions to ask before you start a sampling program. It will also help to answer those questions.

6.1 PREPARATION

6.10 Why Are the Samples To Be Collected?

In your work as a pretreatment inspector you will find there are many different reasons for collecting samples. It is important that you define your purpose or goal before you begin sampling because the ultimate goal of the sampling will always affect the manner in which the sample is collected and handled.

To illustrate how your goals affect the procedures you will need to use, let's consider a hypothetical situation. The local treatment plant is not operating properly. The design of the plant and the way it is operated seem reasonable, so the quality of the incoming wastewater is suspected of causing the problems. A sampling program is proposed to protect the treatment plant from any incompatible pollutants in the wastewater.

This program will be carried out in at least four phases, each of which may involve differences in the way the samples are obtained and handled. The goal of the first phase will be to determine which pollutants are interfering with the treatment processes. Daily *COMPOSITE SAMPLE*[1] of the treatment plant influent and effluent will be collected to measure the average amount of pollutants entering and leaving the treatment plant. The amount of each pollutant detected in these samples will be compared to published tolerable concentrations[2] to determine whether any of them are likely to be toxic to the plant's biological treatment processes. Some pollutants, however, will disappear from the sample while it is being collected. Some very toxic pollutants, such as cyanide, are actually biodegradable. Other pollutants will evaporate. A variety of different sampling techniques will be necessary to find these pollutants.

Once the offending pollutants are identified, the sources of the pollutants must be located (phase two). This will require sampling in different locations than in the first phase. Different equipment and sampling procedures may even be necessary.

Identifying the problem and locating its source will not solve the problem. In the third phase, it is necessary to determine which sources are important, which sources can be controlled, and how they will be controlled. This phase may again require different sampling equipment and procedures.

Finally, the control strategy must be implemented to reduce the discharge of the problem waste material to within specified limits. The sampling and sample handling procedures must allow the laboratory to accurately measure the discharges from the regulated sources. Here again, sampling procedures will change. In many agencies the major portion of an inspector's work load involves sampling industrial discharges rather than plant wastestreams or main sewer lines. You must always design your sampling methods to meet the specific purpose you want to achieve.

The same concepts apply within an industrial setting. The equipment and procedures used to prove that the wastewater treatment system is operating properly and complying with its discharge requirements will not help to locate the source of a *CHELATING AGENT*[3] that is preventing proper precipitation of toxic metals or a toxic material that is not removed by the treatment processes.

---

[1] *Composite (Proportional) Sample.   A composite sample is a collection of individual samples obtained at regular intervals, usually every one or two hours during a 24-hour time span. Each individual sample is combined with the others in proportion to the rate of flow when the sample was collected. Equal volume individual samples also may be collected at intervals after a specific volume of flow passes the sampling point or after equal time intervals and still be referred to as a composite sample. The resulting mixture (composite sample) forms a representative sample and is analyzed to determine the average conditions during the sampling period.*
[2] *See Chapter 9, Table 9.2, "Reported Values for Biological Process Tolerance Limits of Inorganic Priority Pollutants," and Table 9.3, "Reported Values for Biological Process Tolerance Limits of Organic Priority Pollutants."*
[3] *Chelating (key-LAY-ting) Agent.   A chemical used to prevent the precipitation of metals (such as copper).*

Perhaps the most difficult type of sampling is collecting background information merely for future reference. The best assessment of future needs is only a guess. Some parameters will be needlessly monitored, and necessary parameters will be missed. Before starting any sampling program, do your best to set down your reasons and goals.

Care and accuracy are essential elements in any sampling program. The results of almost any sampling program could eventually become involved in a court action. A dispute may arise because of a difference between a company's sampling results and an agency's sampling results. Samples collected by a public agency might be evidence that a crime has been committed. Be sure to discuss and verify with your agency's attorney that your sample collection, transportation and analysis procedures will be acceptable in court. This entire chapter is devoted to procedures that, if followed and properly documented, will "stand up" in court.

6.11   How Many Samples Are To Be Collected and When Will They Be Collected?

Once the goals of a sampling program have been established, you must determine how many samples you will need to collect and when they will be collected. This will depend, to a large extent, on the purpose for the sampling.

The operator of a pretreatment system may need to collect samples more than once per hour to allow adjustments to be made when needed. A POTW may gauge the performance of the same system by collecting a single 24-hour composite sample. Both parties will be seeking the same information, whether the pretreatment system is operating within its limits, but with different goals. The POTW is monitoring compliance with regulations, but the operator is directing the functions of the system.

A single sample may be enough to determine whether a certain pollutant is present in a wastestream. A wastestream that contains traces of a regulated pollutant may need to be monitored only rarely, but a wastestream that is near its limits must be sampled more frequently.

Many industrial dischargers have the ability to store certain wastewaters for a considerable period of time, days or even weeks. It may be necessary to sample continuously for a long time to obtain realistic samples from a company that may have a significant storage time for certain wastewaters. Some facilities operate on a cyclical or seasonal basis, so they must be sampled throughout a complete cycle or season to properly describe their discharges. Samples may be collected to:

1. Check compliance with local and federal standards,

2. Verify reported or self-monitoring data,

3. Determine strength to accurately assess discharge/use fees,

4. Search for prohibited wastes, and

5. Verify industry sampling techniques and monitoring points.

Occasionally an industrial user will request that samples collected by an inspector be split (see Figure 6.1) so that the company can conduct its own tests on samples identical in content to the samples the POTW analyzes. When splitting samples, be sure to retain a large enough sample to conduct all necessary tests.

QUESTIONS

Write your answers in a notebook and then compare your answers with those on page 460.

6.0A   What is the reason for collecting an industrial wastewater sample?

6.1A   List the four phases of a sampling program to determine the source of a pollutant that is upsetting a treatment plant.

6.1B   Why might an inspector need to collect only one sample?

6.12   Who Will Analyze the Samples Once They Are Collected?

Before collecting any samples you must know who will be processing them. Many agencies and industrial facilities have in-house laboratories. This helps to ensure cooperation between the sampling personnel and the laboratory because somewhere in the organization one person is ultimately responsible for coordinating both the collection and analysis of wastewater samples. Since no third party is collecting a profit, an in-house laboratory may be less expensive than an outside laboratory if the necessary laboratory equipment is available.

When a sampling program does not generate enough samples to pay for the special equipment and trained personnel required for unusual laboratory data, it is necessary to use an outside laboratory. An outside laboratory should be well equipped, but it is likely to be less flexible than an in-house laboratory.

*Fig. 6.1   Process control lab technician, inspector and treatment
system operator splitting a sample at the clarifier*

NOTE: Personnel should be wearing personal protective gear,
including gloves and safety glasses with side shields.

Whether you are using an in-house laboratory or an outside laboratory, you must answer several questions:

1. Is the laboratory able to do the analyses that will be requested and do them accurately?

2. Are you asking for routine analyses that the laboratory is already set up to do?

3. Are you asking for special analyses that will require special attention by the laboratory? (These analyses will be expensive.)

4. Will the laboratory be willing to do the analyses the way you want them?

Many state and local agencies administer laboratory certification programs. Laboratories are usually evaluated in three areas. A certified laboratory must have adequate equipment to perform the analyses, it must have personnel with adequate training, and it must demonstrate its ability to accurately run the analyses. To evaluate accuracy, the certifying agency either splits samples with a candidate laboratory or gives it materials with known concentrations of certain materials. Usually a laboratory's certification covers a limited number of specific pollutants.

The laboratory should use approved analytical methods. The U.S. Environmental Protection Agency has developed a list of approved methods that must be used to analyze samples for monitoring NPDES compliance. This list includes approved methods developed by the U.S. Environmental Protection Agency (EPA), the American Society for Testing and Materials (ASTM), the U.S. Geological Survey (USGS), and *STANDARD METHODS.*[4] This list is found in Title 40 of the *CODE OF FEDERAL REGULATIONS,* Part 136 (40 CFR Part 136). One way to find a listing of approved analytical methods is to obtain a copy of the *CODE OF FEDERAL REGULATIONS,* Title 40, Protection of the Environment, Parts 136 to 149, which is available from the U.S. Government Printing Office, Superintendent of Documents, PO Box 371954, Pittsburgh, PA 15250-7954. Order No. 869-044-00152-7. Price, $55.00.

---

[4] *STANDARD METHODS FOR THE EXAMINATION OF WATER AND WASTEWATER. Obtain from Water Environment Federation (WEF), Publications Order Department, 601 Wythe Street, Alexandria, VA 22314-1994. Order No. S82010WW. Price to members, $164.25; nonmembers, $209.25; price includes cost of shipping and handling.*

Cooperation between the sampler and the laboratory will make or break a sampling program. You, the sampler, need results that will answer your questions. In order to give you reliable test results, the chemist depends on you to follow specific procedures. If the samples are not collected, stored and transported properly, the analyst will not be confident that the results mean anything. Remember, the chemist who analyzes your samples may be required to defend the results in court. Your laboratory can provide a large amount of information necessary to plan a sampling program. Use it.

6.13 How Are Samples Analyzed?
by Ron Myers

The analysis of wastewater samples requires the use of complex analytical procedures conducted on sophisticated laboratory instruments. These instruments are capable of measuring minute quantities of the various contaminants found in wastewater. Several of the laboratory instruments and processes commonly used in the analysis of wastewater samples are described in the following paragraphs.

ATOMIC ABSORPTION SPECTROPHOTOMETERS (AA) are used to measure the various metal and non-metal elements in water. The device uses the principle that when the atoms of an element are "excited" (raised to a higher energy level), the atoms will absorb a specific wavelength of light that is characteristic to each element. The amount of light that is absorbed is a function of the concentration of the element that is present. Since the loss of light can be measured electronically, concentrations of the element can be determined.

To accomplish this excitation of the atoms, two major techniques are available to the chemist. One technique, *FLAME AA,* involves the aspiration (drawing in by suction) of a small amount of sample into an air-acetylene or nitrous oxide-acetylene flame. Another technique, *FURNACE AA (GFAA),* involves the injection of a small amount of sample into a graphite tube that is in turn heated to a specific temperature. Generally, Furnace AA techniques are capable of detecting lower concentrations of an element than Flame AA techniques.

INDUCTIVELY COUPLED PLASMA (ICP) instruments use a principle similar to the one used in AA analysis. Atoms of an element are "excited" in a plasma torch. As the "excited" atoms relax to a lower energy level, they emit radiation. This emitted radiation can be analyzed to determine two important parameters: (1) which elements are present in the sample, and (2) the concentrations of the individual elements. Unlike AA, an ICP is capable of multi-element analyses on a single sample. The AA analyzes samples one element at a time.

ULTRAVIOLET-VISIBLE SPECTROPHOTOMETERS (UV-VIS) are laboratory instruments used to measure the results of colorimetric techniques. Many inorganic and some organic contaminants can be analyzed by determining the amount of a colored end product that is produced when the contaminant is subjected to a specific chemical reaction. The UV-VIS Spectrophotometer is capable of generating a specific wavelength of light in either the visible or ultraviolet spectrum. In addition, the instrument is capable of measuring the amount of light absorbed by a sample that is placed in the beam between the light source and a detector. From the information generated by the detector, concentrations of a substance can be determined.

GAS CHROMATOGRAPHS (CG, GLC or GSC) are instruments used to separate, detect, and quantify organic compounds. Before analyzing the organics in a wastewater sample, the compounds must be extracted from the water with a solvent. In addition, the solvent containing the extracted compounds must be cleaned and condensed. Sometimes a special step is added that chemically converts the extracted compounds into a form that can be analyzed by a GC. This process of conversion is called derivatization.

The gas chromatograph contains a heated injector port, an oven that holds the separator column, and a detector. A portion of solvent extract is injected into the heated injector port. The solvent and compounds are vaporized in the injector port. The compounds are carried onto the separator column by a continuous flow of inert gas supplied to the column. Each of the compounds contained in the solvent extract will migrate through the column at a different rate. Individual compounds will exit the column and enter the detector. As compounds are detected, an electrical signal is generated by the detector. The amplitude of the generated signal of the detector is proportional to the amount of compound present in the detector. A continuous recording of the electrical signal from the detector produces a chromatogram, which is a record of the peaks of the eluting (exiting) compounds. Identification of compounds is accomplished by comparing retention times of sample peaks to the retention times of standards. Quantification requires the measurement of peak size of sample compounds and the measurement of the peak size of standards.

A gas chromatograph can be connected to a *MASS SPECTROMETER* to produce a *GC/MS SYSTEM.* The mass spectrometer serves as a specialized detector in this system. As individual compounds pass from the GC and enter the mass spectrometer, each compound is rapidly ionized. The many ions that are produced from one compound are sorted and detected by this unit. The resulting mass spectra is a record of the number of ions sorted by their mass/charge ratio. Since the mass spectra of a compound is unique, identification of the compounds can be achieved. As compared to standard GC analysis, mass spectroscopy provides a very high degree of confidence as to the identification of organic compounds. As with standard GC analysis, GC/MS analysis provides for quantification of organic compounds.

HIGH PRESSURE LIQUID CHROMATOGRAPHS (HPLC) are similar to gas chromatographs except that the process of separation is achieved in a liquid carrier and not in a gas carrier. An HPLC contains a high-pressure pump that is capable of pumping small volumes at pressures of 150 to 5,000 psi. Since the compounds are not vaporized in HPLC, compounds that tend to be heat sensitive can be analyzed by this instrument. In addition, compounds that have high boiling temperatures are best analyzed by HPLC.

A specialized HPLC configured to analyze organic or inorganic ions is called an *ION CHROMATOGRAPH (IC).* One example of the application of this instrument is an IC configured to analyze inorganic anions (negative ions) in water. This instrument is capable of separating and detecting ions such as chloride, nitrate, nitrite, phosphate, and sulfate.

The field of laboratory instrumentation is a continually evolving area. The constant upgrading of the present instrumentation provides the chemist with the ability to lower the detection limits for various contaminants. In addition, new types of instrumentation are being developed that will aid the chemist in analyzing for more and more contaminants. Again, it is important to follow the established sampling protocols (procedures) so that the chemist and the laboratory instruments can produce reliable results.

6.14 What Kind of Sample Containers Will Be Used?

The first information that the laboratory will provide concerns the sample containers. Your sample must be the right size. For example, the oil and grease analysis usually requires a full liter of sample. You can't analyze oil and grease, suspended solids and toxic metals from a one-quart sample. On the other hand, the chemist can't shake a 55-gallon drum to obtain a well-mixed sample for analysis. Some analyses require only a few milliliters of sample. The sample should be large enough to allow all of the desired tests to be run and leave enough for retests on critical pollutants. Remember, you can't go back for more sample if the laboratory runs out. The sample volume collected for each test should be specified by the laboratory providing the analysis.

The containers must be made of a material that is compatible with the sample. Hydrofluoric acid will dissolve a glass bottle, and some industrial solvents will dissolve most plastic bottles. Usually the problems are more subtle. Some pollutants will cling to the sides of the bottle and not show up in the analysis, and other pollutants will leach out of the bottle material and contaminate the sample. Table 6.1 contains a listing of required containers, preservation techniques, and holding times for various pollutants.

The shape of the container will depend on the nature of the sample to be collected. Cup-shaped containers with tight-fitting lids work best for sludges. Sludges will not flow through the mouth of a bottle and will not pour out so the chemist must be able to reach into the container to withdraw a sample. Wide-mouth bottles work well for liquid samples because the wide mouth makes it easier to fill the bottle.

Some pollutants, especially *VOLATILE*[5] organics such as trichloroethylene, will evaporate and be lost if the samples are not kept in septum vials. A common example of a septum vial is the small bottle that a medical doctor puts a needle into when preparing an injection. The rubber cap that the needle is

pushed through is called a septum. When used for wastewater sampling, the septum allows a small volume of sample to be withdrawn without exposing the sample to the air.

Avoid using sample containers that have not first been approved by the laboratory. This approval process should be worked out with the laboratory on a long-term basis. Your samples must arrive at the laboratory in a condition that will give the laboratory confidence that the results it publishes will be representative of the materials that were first put into the containers.

QUESTIONS

Write your answers in a notebook and then compare your answers with those on page 460.

6.1C   When should a monitoring program use an outside laboratory?

6.1D   How might an inspector's sampling skills affect the results of complex analytical lab tests?

6.1E   Sample containers must be made of what kind of material?

6.1F   When collecting a sample to be analyzed for phenols, what should be the container material, preservation methods and maximum holding time (refer to Table 6.1)?

6.1G   What shape of container should be used to collect sludges?

6.15 How Should the Sample Containers Be Cleaned?

The sample containers must be clean of all traces of the pollutants to be analyzed. Otherwise, you won't know whether the pollutants found by the laboratory were present in the wastestream you sampled or were already in the container. *STANDARD METHODS* requires acid washing for many types of analyses. Acid washing should be done by the laboratory.

Another way to get clean containers is to use disposable containers. Inexpensive containers are available that are meticulously clean when received from the factory. These containers must be bought in large quantities with a single lot number so the laboratory can test blanks to confirm their cleanliness. The laboratory will carefully wash a number of new containers and analyze the wash waters for traces of pollutants. If none are found, the entire lot is assumed to be clean and can be used in the field without further cleaning.

---

[5] Volatile (VOL-uh-tull).   (1) A volatile substance is one that is capable of being evaporated or changed to a vapor at relatively low temperatures. Volatile substances also can be partially removed by air stripping. (2) In terms of solids analysis, volatile refers to materials lost (including most organic matter) upon ignition in a muffle furnace for 60 minutes at 550°C. Natural volatile materials are chemical substances usually of animal or plant origin. Manufactured or synthetic volatile materials such as ether, acetone, and carbon tetrachloride are highly volatile and not of plant or animal origin.

TABLE 6.1 REQUIRED CONTAINERS, PRESERVATION TECHNIQUES, AND HOLDING TIMES *
(Table II 40 CFR Part 136 (7-1-00 Edition))

Parameter No./Name	Container[1]	Preservation[2,3]	Max. Holding Time[4]
Table IA — Bacterial Tests:			
1-4. Coliform, fecal and total	P, G	Cool, 4°C, 0.008% $Na_2S_2O_3$[5]	6 hours.
5. Fecal streptococci	P, G	.... do	Do.
Table IB — Inorganic Tests:			
1. Acidity	P, G	Cool, 4°C	14 days.
2. Alkalinity	P, G	.... do	Do.
4. Ammonia	P, G	Cool, 4°C, H_2SO_4 to pH<2	28 days.
9. Biochemical oxygen demand	P, G	Cool, 4°C	48 hours.
10. Boron	P, PFTE, or quartz	HNO_3 to pH<2	6 months.
11. Bromide	P, G	None required	28 days.
14. Biochemical oxygen demand, carbonaceous	P, G	Cool, 4°C	48 hours.
15. Chemical oxygen demand	P, G	Cool, 4°C, H_2SO_4 to pH<2	28 days.
16. Chloride	P, G	None required	Do.
17. Chlorine, total residual	P, G	.... do	Analyze immediately.
21. Color	P, G	Cool, 4°C	48 hours.
23-24. Cyanide, total and amenable to chlorination	P, G	Cool, 4°C, NaOH to pH>12, 0.6g ascorbic acid.[5]	14 days.[6]
25. Fluoride	P	None required	28 days.
27. Hardness	P, G	HNO_3 to pH<2, H_2SO_4 to pH<2	6 months.
28. Hydrogen ion (pH)	P, G	None required	Analyze immediately.
31,43. Kjeldahl and organic nitrogen	P, G	Cool, 4°C, H_2SO_4 to pH<2	28 days.
Metals:[7]			
18. Chromium VI	P, G	Cool, 4°C	24 hours.
35. Mercury	P, G	HNO_3 to pH<2	28 days.
3, 5-8, 10, 12, 13, 19, 20, 22, 26, 29, 30, 32-34, 36, 37, 45, 47, 51, 52, 58-60, 62, 63, 70-72, 74, 75. Metals, except boron, chromium VI and mercury.	P, G	.... do	6 months.
38. Nitrate	P, G	Cool, 4°C	48 hours.
39. Nitrate-nitrite	P, G	Cool, 4°C, H_2SO_4 to pH<2	28 days.
40. Nitrite	P, G	Cool, 4°C	48 hours.
41. Oil and grease	G	Cool to 4°C, HCl or H_2SO_4 to pH<2	28 days.
42. Organic carbon	P, G	Cool to 4°C, HCl or H_2SO_4 to pH<2	28 days.
44. Orthophosphate	P, G	Filter immediately, Cool, 4°C	48 hours.
46. Oxygen, Dissolved Probe	G Bottle and top	None required	Analyze immediately.
47. Winkler	.... do	Fix on site and store in dark	8 hours.
48. Phenols	G only	Cool, 4°C, H_2SO_4 to pH<2	28 days.
49. Phosphorus (elemental)	G	Cool, 4°C	48 hours.
50. Phosphorus, total	P, G	Cool, 4°C, H_2SO_4 to pH<2	28 days.
53. Residue, total	P, G	Cool, 4°C	7 days.
54. Residue, Filterable	P, G	.... do	7 days.
55. Residue, Nonfilterable (TSS)	P, G	.... do	7 days.
56. Residue, Settleable	P, G	.... do	48 hours.
57. Residue, volatile	P, G	.... do	7 days.
61. Silica	P, PFTE, or quartz	Cool, 4°C	28 days.
64. Specific conductance	P, G	.... do	Do.
65. Sulfate	P, G	.... do	Do.
66. Sulfide	P, G	Cool, 4°C add zinc acetate plus sodium hydroxide to pH>9.	7 days.
67. Sulfite	P, G	None required	Analyze immediately.
68. Surfactants	P, G	Cool, 4°C	48 hours.
69. Temperature	P, G	None required	Analyze.
73. Turbidity	P, G	Cool, 4°C	48 hours.
Table IC — Organic Tests.[8]			
13, 18-20, 22, 24-28, 34-37, 39-43, 45-47, 56, 76, 104, 105, 108-111, 113. Purgeable Halocarbons.	G, Teflon-lined septum.	Cool, 4°C, 0.008% $Na_2S_2O_3$.[5]	14 days.
6, 57, 106. Purgeable aromatic hydrocarbons	.... do	Cool, 4°C, 0.008% $Na_2S_2O_3$,[5] HCl to pH 2.[9]	Do.
3, 4, Acrolein and acrylonitrile	.... do	Cool, 4°C, 0.008% $Na_2S_2O_3$.[5] Adjust pH to 4-5.[10]	Do.

* For sale by the U.S. Government Printing Office, Superintendent of Documents, PO Box 371954, Pittsburgh, PA 15250-7954. Price, $55.00.
** Do and do mean ditto or same as above.

TABLE 6.1   REQUIRED CONTAINERS, PRESERVATION TECHNIQUES, AND HOLDING TIMES *
(Table II 40 CFR Part 136 (7-1-00 Edition))(continued)

Parameter No./Name	Container [1]	Preservation [2,3]	Max. Holding Time [4]
Table IC — Organic Tests.[8] (continued)			
23, 30, 44, 49, 53, 77, 80, 81, 98, 100, 112. Phenols.[11] ...	G, Teflon-lined cap.	Cool, 4°C, 0.008% $Na_2S_2O_3$.[5]	7 days until extraction; 40 days after extraction.
7, 38. Benzidines [11]	.... do	.... do	7 days until extraction.[13]
14, 17, 48, 50-52. Phthalate esters [11]	.... do	Cool, 4°C	7 days until extraction; 40 days after extraction.
82-84. Nitrosamines [11, 14]	.... do	Cool, 4°C, store in dark, 0.008% $Na_2S_2O_3$.[5]	Do.
88-94. PCBs [11]	.... do	Cool, 4°C	Do.
54, 55, 75, 79. Nitroaromatics and isophorone [11]	.... do	Cool, 4°C, 0.008% $Na_2S_2O_3$.[5] Store in dark.	Do.
1, 2, 5, 8-12, 32, 33, 58, 59, 74, 78, 99, 101. Poly-nuclear aromatic hydrocarbons.[11]	.... do	.... do	Do.
15, 16, 21, 31, 87. Haloethers [11]	.... do	Cool, 4°C, 0.008% $Na_2S_2O_3$.[5]	Do.
29, 35-37, 63-65, 73, 107. Chlorinated hydrocarbons.[11] ..	.... do	Cool, 4°C	Do.
60-62, 66-72, 85, 86, 95-97, 102, 103. CDDs/CDFs.[11] aqueous: field and lab preservation	G	Cool, 0-4°C, pH<9, 0.008% $Na_2S_2O_3$.[5]	1 year.
Solids, mixed phase, and tissue: field preservation	.... do	Cool, <4°C	7 days.
Solids, mixed phase, and tissue: lab preservation	.... do	Freeze, < −10°C	1 year.
Table ID — Pesticides Tests:			
1-70. Pesticides [11]	.... do	Cool, 4°C, pH 5-9[15]	Do.
Table IE — Radiological Tests:			
1-5. Alpha, beta and radium	P, G	HNO_3 to pH<2	6 months.

[1] Polyethylene (P) or Glass (G).

[2] Sample preservation should be performed immediately upon sample collection. For composite chemical samples each aliquot should be preserved at the time of collection. When use of an automated sampler makes it impossible to preserve each aliquot, then chemical samples may be preserved by maintaining at 4°C until compositing and sample splitting is completed.

[3] When any sample is to be shipped by common carrier or sent through the United States Mails, it must comply with the Department of Transportation Hazardous Materials Regulations (49 CFR Part 172). The person offering such material for transportation is responsible for ensuring such compliance. For the preservation requirements of Table II, the Office of Hazardous Materials, Materials Transportation Bureau, Department of Transportation has determined that the Hazardous Materials Regulations do not apply to the following materials: Hydrochloric acid (HCl) in water solutions at concentrations of 0.04% by weight or less (pH about 1.96 or greater); Nitric acid (HNO_3) in water solutions at concentrations of 0.15% by weight or less (pH about 1.62 or greater); Sulfuric acid (H_2SO_4) in water solutions at concentrations of 0.35% by weight or less (pH about 1.15 or greater); and Sodium hydroxide (NaOH) in water solutions at concentrations of 0.080% by weight or less (pH about 12.30 or less).

[4] Samples should be analyzed as soon as possible after collection. The times listed are the maximum times that samples may be held before analysis and still be considered valid. Samples may be held for longer periods only if the permittee, or monitoring laboratory, has data on file to show that the specific types of samples under study are stable for the longer time, and has received a variance from the Regional Administrator under § 136.3(e). Some samples may not be stable for the maximum time period given in the table. A permittee, or monitoring laboratory, is obligated to hold the sample for a shorter time if knowledge exists to show that this is necessary to maintain sample stability. See § 136.3(e) for details. The term "analyze immediately" usually means within 15 minutes or less of sample collection.

[5] Should only be used in the presence of residual chlorine.

[6] Maximum holding time is 24 hours when sulfide is present. Optionally all samples may be tested with lead acetate paper before pH adjustments in order to determine if sulfide is present. If sulfide is present, it can be removed by the addition of cadmium nitrate powder until a negative spot test is obtained. The sample is filtered and then NaOH is added to pH 12.

[7] Samples should be filtered immediately on-site before adding preservative for dissolved metals.

[8] Guidance applies to samples to be analyzed by GC, LC, or GC/MS for specific compounds.

[9] Sample receiving no pH adjustment must be analyzed within seven days of sampling.

[10] The pH adjustment is not required if acrolein will not be measured. Samples for acrolein receiving no pH adjustment must be analyzed within 3 days of sampling.

[11] When the extractable analytes of concern fall within a single chemical category, the specified preservative and maximum holding times should be observed for optimum safeguard of sample integrity. When the analytes of concern fall within two or more chemical categories, the sample may be preserved by cooling to 4°C, reducing residual chlorine with 0.008% sodium thiosulfate, storing in the dark, and adjusting the pH to 6-9; samples preserved in this manner may be held for seven days before extraction and for forty days after extraction. Exceptions to this optional preservation and holding time procedure are noted in footnote 5 (re the requirement for thiosulfate reduction of residual chlorine), and footnotes 12, 13 (re the analysis of benzidine).

[12] If 1,2-diphenylhydrazine is likely to be present, adjust the pH of the sample to 4.0±0.2 to prevent rearrangement to benzidine.

[13] Extracts may be stored up to 7 days before analysis if storage is conducted under an inert (oxidant-free) atmosphere.

[14] For the analysis of diphenylnitrosamine, add 0.008% $Na_2S_2O_3$ and adjust pH to 7-10 with NaOH within 24 hours of sampling.

[15] The pH adjustment may be performed upon receipt at the laboratory and may be omitted if the samples are extracted within 72 hours of collection. For the analysis of aldrin, add 0.008% $Na_2S_2O_3$.

However, separate blanks must be run for each lot number, and records of bottle or container cleanliness must be maintained.

The decision to use disposable or reusable containers is strictly economic. Compare the cost of buying disposable bottles and running blanks to the cost of washing bottles and replacing broken bottles. Washing bottles will probably be attractive for smaller laboratories where the analytical work load does not use all of the available staff hours. Disposable bottles will be more attractive when it is necessary to hire personnel specifically to wash bottles.

When testing for metals, the sample containers should be cleaned using the procedures described in *SAMPLING AMBIENT WATER FOR TRACE METALS AT EPA WATER QUALITY CRITERIA LEVELS.*[6] When testing for organics, use the cleaning methods described in *COMPARISON OF VOA COMPOSITING PROCEDURES*[7] to clean the sample containers. The disinfection and cleaning of sampling equipment (automatic samplers and grab sampling equipment, tubing, buckets, dippers, graduate cylinders and the inside of coolers and other shipping containers) should be done according to procedures established by the POTW using best management practices. As previously mentioned, disposable bottles will have to be cleaned before use unless they were specifically purchased prewashed according to a QA/QC goal.

Obtaining clean glassware from the laboratory, however, is not enough to ensure that the samples transported in the glassware will be uncontaminated. It is also necessary to store and transport these samples in a manner that will avoid contamination.

6.16 Will Sample Preservation Be Necessary?

Many pollutants will change during storage and transportation. Sample preservation techniques have been developed to slow down the rate of this change. Different methods are used to preserve different pollutants. Acid will dissolve metals and keep them from adsorbing (sticking) on the sides of the storage vessel (sample container). It will prevent bacterial attack against certain pollutants. Strong bases will prevent bacterial attack against certain pollutants that might become volatile under acidic conditions. Refrigeration at 4 degrees Celsius (39 degrees Fahrenheit) is appropriate for some types of samples. Most preservation methods are only partially successful so all samples should be analyzed as soon as possible after they are collected.

The U.S. Environmental Protection Agency has published a list of preservation methods and allowable storage times for samples used in monitoring NPDES discharges. This list will be updated from time to time. It is found at 40 CFR Part 136.3(e). Table 6.1 describes the current preservation requirements for a number of pollutant analyses.

Inadequate preservation or excessive storage time might not invalidate a sample. However, a competent chemist would have to be able to justify the results using the data. Suppose XYZ Company has a 5 mg/*L* limit for pollutant A. The sample must be analyzed within 24 hours because pollutant A will slowly decay into pollutant B. A week-old sample containing 7 mg/*L* of pollutant A is proof of a violation at XYZ Company. The sample must have contained more than 7 mg/*L* of substance A when it was collected. On the other hand, a week-old sample containing 4 mg/*L* of pollutant A does not prove that XYZ Company was in compliance because some of the pollutant A may have disappeared. It may be possible to show whether this sample was or was not in compliance by measuring both pollutant A and pollutant B. Check with your chemist. (Appendix A contains a procedure for preserving samples which will be analyzed for metals.)

6.17 What Safety Precautions Are Appropriate?

Safety is discussed extensively in Chapter 5 of this manual. At this point, however, it is necessary to point out a few hazards that are of particular concern in wastewater sampling. No one can protect you from industrial hazards if you will not look for them.

Many times the location of the sampling point presents a special danger. Sampling points are frequently located in confined spaces. If there is not adequate ventilation, toxic or explosive gases could collect in the atmosphere or the air could be depleted of oxygen. *ALWAYS* test the atmosphere and follow all confined space procedures *BEFORE* entry and while working in confined spaces. The manhole covers may be too heavy or awkward for one person to lift. Discharge permits should require industrial dischargers to provide a safe location to collect representative samples of their wastewater.

---

[6] *METHOD 1669: SAMPLING AMBIENT WATER FOR TRACE METALS AT EPA WATER QUALITY CRITERIA LEVELS. Obtain from National Technical Information Service (NTIS), 5285 Port Royal Road, Springfield, VA 22161. Order No. PB96-193313. EPA No. 821-R-96-008. Price, $31.50, plus $5.00 shipping and handling per order.*

[7] *COMPARISON OF VOA COMPOSITING PROCEDURES, Office of Water, U.S. Environmental Protection Agency. Obtain from National Service Center for Environmental Publications (NSCEP), PO Box 42419, Cincinnati, OH 45242-2419. EPA No. 821-R-95-035.*

Many companies impose special safety regulations. Some companies require anyone entering their premises to wear hard-toed shoes; other companies may require eye protection or hearing protection. And, of course, it is always appropriate to have available a hard hat, boots and gloves when working near industrial facilities and industrial wastewater.

Safety regulations are designed to protect you. Know the regulations in advance and be prepared to comply with them.

QUESTIONS

Write your answers in a notebook and then compare your answers with those on page 460.

6.1H   What is the recommended procedure for cleaning sample containers for most types of sample analyses?

6.1I   How can you be assured that disposable containers are clean?

6.1J   Why is an acid used to preserve a sample that may contain metals?

6.1K   What precautions should be taken before entering a confined space to collect a sample?

6.2 SAMPLE COLLECTION

6.20 Who Will Collect the Sample?

There are actually three possible parties who may collect a sample of industrial wastewater: the industry, the control agency and an independent third party. Depending on the situation, any one of these options could be the best choice.

From a law-and-order position, monitoring by the regulatory agency is most desirable. The agency has full control over the method of collecting the sample and the analysis. If a violation is found, the agency can preserve the remaining sample as evidence. In addition, the POTW is required to verify industrial users' compliance by some means independent of self-monitoring results.

Industrial self-monitoring may be more cost effective. It is expensive for a POTW inspector to drive across town to set up a sampler and return to retrieve the sample. In addition, the laboratory work may be done at near-zero cost by the laboratory at the facility in conjunction with its routine tasks. The regulatory agency will certainly be able to function with a smaller staff if private industry is required to perform a major portion of the monitoring.

Whenever the work load does not justify a full-time laboratory, a private laboratory (consultant) should be considered. Although the consultant will charge an extra cost (profit), the specialized operation may be more efficient and less expensive overall. Industries frequently use consultants to fulfill their self-monitoring obligations. Smaller POTWs can also benefit from this arrangement. A POTW with very few industrial dischargers may be able to work with those industries to select a single consultant to do all of the monitoring. Most large POTWs require industry self-monitoring with verification sampling (for checking compliance and wastewater strength) being done by the POTW.

Each of these options has some drawbacks as well as advantages; a balanced monitoring program will probably use a combination of sampling and analysis arrangements. Industri-

al self-monitoring is only valuable where the company doing the monitoring has a sincere wish to comply with any regulations that apply. It is easy for an industry to adulterate samples or even falsify data and provide a regulatory agency with misleading information. POTWs must issue specific guidelines to ensure that samples are properly collected and preserved and thereby ensure uniformity in sampling techniques between industries. Every regulatory agency must collect, or contract to have collected, some samples of industrial wastewater independent of an industry's self-monitoring efforts.

If an industry relies on samples collected by the regulatory agency, the first indication of a problem may be a citation. Self-monitoring provides a company with information on the operation of its system and an opportunity to correct any problems before they get serious.

Federal regulations include some safeguards against companies failing to properly conduct self-monitoring. The General Pretreatment Regulations, 40 CFR Part 403, require POTWs to annually publish a list of incidents of significant noncompliance. An industry's failure to accurately report noncompliance is itself considered significant noncompliance. A recent amendment to 40 CFR Part 403 requires industrial users to report all self-monitoring data. See Chapter 3, Section 3.4, "Industrial Waste Compliance Programs," for more information on the meaning of "significant noncompliance."

6.21 How Do I Obtain a Sample That Is Representative of the Wastewater Flow?

Selecting an appropriate sampling location is essential to collecting REPRESENTATIVE SAMPLES.[8] The sampling point should be specified in the company's industrial wastewater discharge permit. It should be in a location accessible to both the agency and the industry, preferably a location which will contain all of the industrial wastewater from the facility but no sanitary wastewater. However, in some cases you may be sampling an industry for compliance with EPA Pretreatment Standards and the sampling point may be upstream at the end of the regulated process (see Figure 6.2).

A POTW should consider requiring industrial dischargers to provide access to the sampling location from outside the facility so the inspector can obtain a wastewater sample without waiting to be escorted onto the property. This benefits the company by minimizing interference with its normal operation, and it allows the inspector to collect samples before the discharger has an opportunity to adjust its pretreatment system. The sampling location should have a gate with a POTW agency lock if possible.

---

[8] Representative Sample.   A sample portion of material or wastestream that is as nearly identical in content and consistency as possible to that in the larger body of material or wastestream being sampled.

Fig. 6.2   Sampling a regulated process

Industrial users in the EPA Metal Finishing Category (40 CFR Part 433) who use cyanide must provide a sample point downstream of the cyanide treatment tank at a place before this waste is diluted by any other wastestream. The sample taken here should be a grab sample, preserved with a high pH = 12 and any chlorine present should be treated with 0.05 gm additions of ascorbic acid until KI starch paper indicates that all available chlorine has been removed.

The physical design of the sampling point must be appropriate for the type of wastewater to be sampled. In the case of an industry where extensive pretreatment is required to remove most of the pollutants in the wastewater, a sample box may be appropriate. This is a box approximately two feet square which will retain about one foot of water when there is no flow. The treated wastewater flows through this box and then directly to the sewer. A sample box is important because it retains some water that is representative of the last wastewater to leave the plant. A meaningful sample may be obtained even if the wastewater is no longer flowing. The problem with sample boxes is that they tend to collect solids. The industrial wastewater discharge permit should require the company to properly maintain its sample box, refrain from pushing accumulated solids down the drain, and specify that any material found in the sample box will be assumed to have been discharged.

Sample boxes do tend to accumulate solids, and an attempt should always be made to collect a sample that represents the wastewater entering the sewer. If you are using an automatic sampler, place the sampler tubing in the outlet pipe from the sample box. This way the sample will be collected from flowing wastewater that has already left the sample box.

Many industries, such as paper mills or food processors, do not need to remove solids in order to meet pretreatment requirements but they must be sampled to confirm that harmful pollutants are absent and to determine the strength of the wastewater. In these facilities it is necessary to find a location where the wastewater is well mixed. The tail water of a flume or a weir is often a good sampling point. Avoid closed-pipe systems where the actual wastewater cannot be observed; you can't be confident that the sample accurately represents the wastewater being discharged. A pair of 90-degree turns, not in the same plane, or an in-line mixer can induce turbulence that will provide mixing at the sampling point.

Frequently the only access to the industrial wastewater is a clarifier or a sand and grease interceptor. This equipment accumulates oil and grease on the top and sludge on the bottom; therefore, it is very difficult to get a representative sample and this should be the last choice for a sampling location. However, this equipment does contain wastewater which is similar to the wastewater being discharged and is a possible sample location if you have no better alternatives.

The sewer downstream of a clarifier will frequently have a cleanout, a capped pipe that comes off the sewer at an angle and allows cleaning equipment to be passed through the sewer. Cleanouts frequently provide access to a portion of the sewer where good samples may be obtained if the flow is visible. In one criminal investigation, hazardous waste that had been illegally poured down the discharge pipe below the sample box was recovered through a cleanout downstream from the sample box.

Many industries have manholes just outside their property. These manholes are often the best place for monitoring. Usually a weir or flume can be installed in the manhole for flow measurements and an automatic sampler for collecting samples.

Industrial wastewater can be sampled after it has entered the public sewer, but when sampling an industry in this fashion it is also desirable to collect samples at the nearest manhole upstream of the industry. Otherwise it is impossible to tell whether the pollutants came from the industry in question or an upstream source. Automatic sampling equipment installed in the public sewer lines allows an agency to monitor a facility without informing the industry, but the pollutants will have been diluted by other flows. Also, rags or debris in the sanitary wastewater will frequently foul sampling equipment.

6.22 How Will the Sample Be Obtained?

There are three basic types of samples that may be collected: grab samples, flow-weighted composites and time-weighted composites. A grab sample represents a single instant in time, the moment when the sample is collected. A composite sample is made up of a number of grab samples that are combined in a single bottle in an attempt to average the wastewater characteristics over a period of time. Composite samples may be either flow-weighted, in which case a grab sample is collected every so many gallons, or the grab sample volume is proportional to the flow rate, or time-weighted, where a grab sample is collected every so many minutes.

Grab samples represent only one point in time so a single sample will not describe the characteristics of a particular flow before or after the sample was collected. A grab sample may be useful for operating a process or for enforcing discharge standards. An operator does not always need to take the time to collect a composite sample but, instead, can collect a grab sample and know how the system is operating as soon as the analysis is completed. For an inspector the advantage of a grab sample is that it can be collected quickly, before changes can be made to improve the quality of the discharge.

Composite samples are better for characterizing a wastewater than a grab sample. A composite sample consists of a number of grab samples, each representing a single instant. For example, grab samples of 500 mL are collected every hour or for some other specific time interval during a 24-hour period and then returned to the lab. The lab, using the flow chart recorded during the sampling period, transfers a portion of each grab sample to one large sample container. The size of each grab sample transferred is proportional to the flow at the time the sample was collected (as shown below) by adding 10 mL of grab sample for every 100 gallons per day of flow.

TIME	FLOW, GPD	SAMPLE, mL
1000	600	60
1200	800	80
1400	1,000	100
1600	1,600	160
1800	2,000	200
2000	1,700	170
2200	1,200	120
2400	1,000	100
0200	800	80
0400	700	70
0600	500	50
0800	600	60

Composite Sample Size, mL = 1,250

Mixing the sample produces a single sample equal to the average of the pollutants or constituents in all of the grab samples. If enough grabs are taken, the composite sample will contain constituents approximately equal to the average of the wastewater that was discharged. See Appendix II, "Pretreatment Arithmetic," Section G, "Composite Sampling," at the end of this manual for additional information.

For the purposes of BMRs, 90-Day Compliance Reports and Periodic Reports on Compliance, a minimum of four (4) grab samples must be collected for pH, cyanide, total phenols, oil and grease, sulfide and volatile organics. For all other pollutants, 24-hour composite samples must be obtained through flow-proportional composite sampling techniques where feasible. The POTW may waive flow-proportional composite sampling for an IU that demonstrates that flow-proportional composite sampling is unfeasible. In such cases, samples may be obtained through time-proportional composite sampling techniques or through a minimum of four (4) grab samples where the IU demonstrates that this will provide a representative sample of the effluent being discharged.

The pretreatment inspector may choose to follow these guidelines in collecting samples for the POTW laboratory to analyze. However, the pretreatment inspector is not bound by these requirements in the samples that are collected. Sample collection, preservation and holding must conform with 40 CFR Part 136 standards (as amended). It is the responsibility of the pretreatment inspector to customize sampling in order to meet the purpose for which the sample is collected. In other words, are we most interested in collecting surveillance samples to determine what types of pollutants are being discharged from an IU? If so, we may want to analyze a composite sample collected from an automatic sampler and preserved with sodium hydroxide to pH greater than 12.0 to determine the cyanide concentration instead of relying on grab samples (as recommended by EPA). But, if we are concerned with obtaining a sample for enforcement of a cyanide local limit, it would be proper to collect grab samples in conformance with EPA recommendations and follow proper preservation techniques, holding time and chain of custody in order to develop data which can be used as evidence.

For the reports indicated above, the IU must take a minimum of one representative sample to compile that data necessary to determine whether or not the IU is in consistent compliance with applicable pretreatment standards. The POTW may choose to require more frequent monitoring by the IU in the wastewater discharge permit issued to the facility.

Sampling equipment must be selected based on the type of sample to be collected. Grab samples may be collected in buckets that are hung on a string and dropped into the flow. These buckets should be of a material which will not introduce pollutants of concern into the wastewater or otherwise change the sample. The bucket must be cleaned each time it is used. One way to ensure that the bucket is clean is to line it with a plastic bag. As with disposable sample containers, the laboratory must certify that the liners will not contaminate the samples.

Another way of obtaining a grab sample is with a vacuum pump and tube sampler. The sample is collected in a sealed vessel with a cap that has two places to attach plastic tubing (Figure 6.3). The tubing attached to one side of the cap is placed into the wastewater and the tubing from the other connection goes to a vacuum pump. When a vacuum is applied, the sample is drawn into the vessel. This equipment is required to obtain a sample from a cleanout and may be the best way to obtain a sample representing the discharge from a clarifier. When sampling a clarifier, weight the sample tube and lower it to the level of the outlet.

Composite samples can be made up from a series of manual grab samples, but automatic samplers, such as shown in Figures 6.4 and 6.5, are commercially available. You can adjust an automatic sampler so that a grab sample of a controlled size is collected each time a signal is received from a controller. These grab samples are placed in a single container or into separate bottles to be composited later. The illustrated equipment will even collect 24 one-hour composites, where each composite consists of several grab samples, with a single sampler collecting all samples. Time-weighted samples are obtained by using an internal clock, and flow-weighted composite samples are obtained by connecting the automatic sampler to an appropriately equipped flowmeter.

Automatic samplers can be run off batteries or plant power. For regulatory agencies, battery-powered units are probably the better choice because they can be moved from one location to another. Permanent sampling facilities will be more convenient when operated on alternating current. Automatic samplers are suitable for use with all *PRIORITY POLLUTANTS.*[9] A drawback of automatic samplers is that some pollutants, such as vinyl chloride, may be introduced by the tubing on the sampler. Sampling technology has advanced to the level where it is now possible to collect organic samples using automatic samplers.

The U.S. EPA conducted studies in 1994 which compared the results produced by VOA grab and compositing procedures.[10] In these studies, four individual grab samples of real-world

---

[9] *Priority Pollutants. The EPA has proposed a list of 126 priority toxic pollutants. These substances are an environmental hazard and may be present in water. Because of the known or suspected hazards of these pollutants, industrial users of the substances are subject to regulation. The toxicity to humans may be substantiated by human epidemiological studies or based on effects on laboratory animals related to carcinogenicity, mutagenicity, teratogenicity, or reproduction. Toxicity to fish and wildlife may be related to either acute or chronic effects on the organisms themselves or to humans by bioaccumulation in food fish. Persistence (including mobility and degradability) and treatability are also important factors.*

[10] *COMPARISON OF VOA COMPOSITING PROCEDURES, Office of Water, U.S. Environmental Protection Agency. Obtain from National Service Center for Environmental Publications (NSCEP), PO Box 42419, Cincinnati, OH 45242-2419. EPA No. 821-R-95-035.*

NOTE: Gloves should be worn when collecting samples.

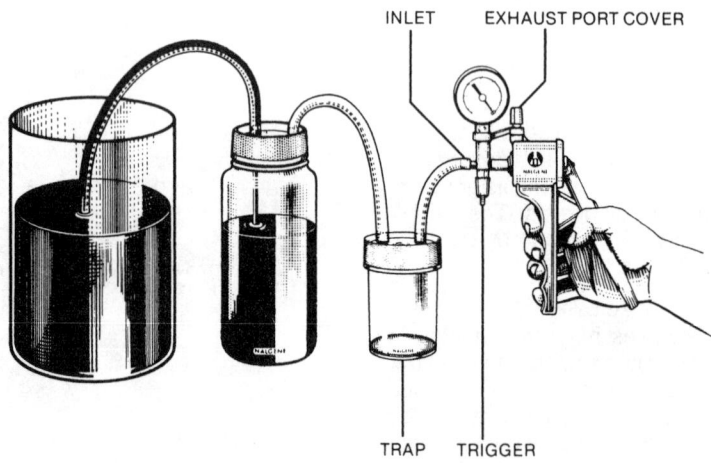

Sample is being collected in middle jar from source in left
beaker by use of a hand-operated vacuum pump.

Fig. 6.3   Obtaining a grab sample with a vacuum pump

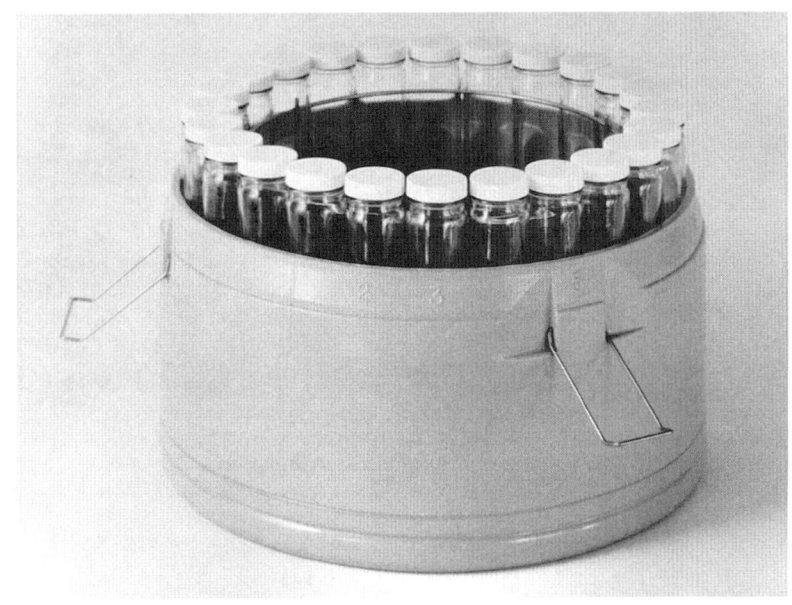

Fig. 6.4   Portable automatic sampler
(Courtesy of ISCO, Inc., Environmental Division)

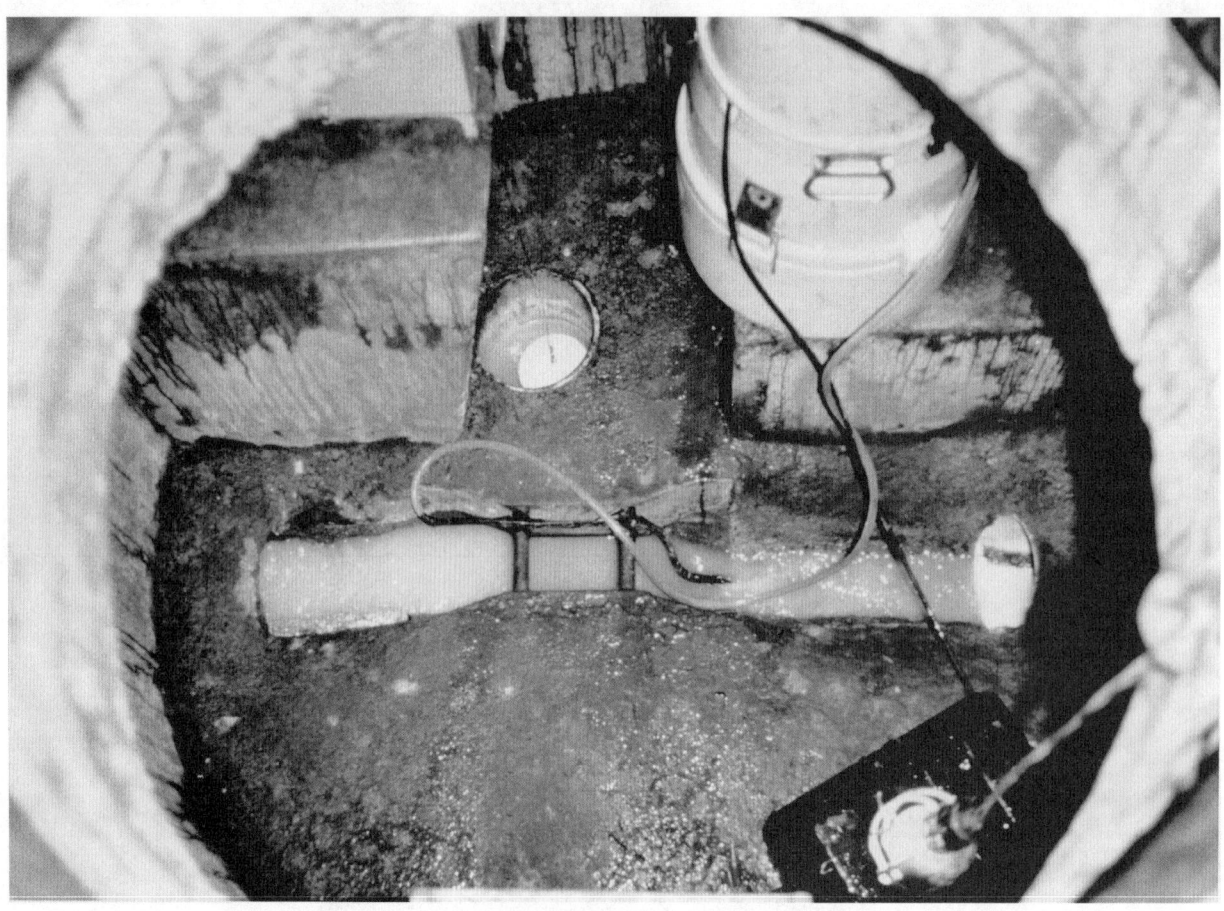

Fig. 6.5   *Portable automatic sampler in use*
(Courtesy of ISCO, Inc., Environmental Division)

effluents were collected over the course of a day. These samples were analyzed spiked or unspiked, composited, and individually by isotope dilution GC/MS, using revision C of EPA Method 1624. Grab sampling procedures used were manual and automated. Compositing procedures used were flask, purge device, and continuous. Pollutants spiked were the volatile organic GC/MS fraction of the priority pollutants plus additional compounds routinely tested for in EPA's industrial surveys.

The objective of these studies was to compare the mathematical average of the results of analyses of the manually collected individual grab samples with the results of the analyses of the composited automated grab and the flask, purge device, and continuously composited samples to determine if bias occurred in the automated grab and compositing processes. These results showed that, for the real-world samples tested, the mathematical average of the results of analysis of the individual grab samples were a few percent higher, on average, than results of the analyses of composited samples, but were a few percent lower, on average, than results of automated grab samples. The cause of these slight differences is not known. The differences are not significant for practical purposes, and the differences would likely not be discernable by non-isotope GC/MS methods.

Although there are many ingenious ways to collect composite samples, most common automatic samplers use either a peristaltic pump or a vacuum pump to draw the sample into the sample container. Each of these systems has its advantages and disadvantages. A vacuum pump, particularly in a permanent installation, may be able to draw the sample at a higher rate and allow sampling a cross section of a wastestream through several orifices. However, some such systems control the *ALIQUOT*[11] volume by filling a reservoir and then wasting the excess sample before draining the remainder to the sample container. If the wastestream contains solids that will separate quickly, this procedure will bias the sample by concentrating or discarding solids.

A peristaltic pump discharges a measured aliquot directly into the sample container so the effect of solids separation is less important. However, these systems normally sample from only one point in the wastestream.

Profile samplers are available which will either allow a grab sample to be collected at a given depth or will collect and composite samples from the top to the bottom of a wastestream. These are useful to determine the amount of sediment or scum in a clarifier.

Many pollutants can be monitored on a continuous basis. Meters are available which will measure water quality indicators such as pH, specific conductance and temperature in water, and hydrogen sulfide gas concentration and flammability in the atmosphere. These meters can also be connected to chart recorders to produce permanent records and to alarms to indicate pretreatment malfunctions. Full-time meters must be periodically checked (calibrated) against standard solutions that contain known quantities of pollutants. If you are using a meter as an enforcement tool, calibrate the meter before starting the monitoring and read the calibration standards after the study is completed. If the calibration drifts during the study, there will be two possible interpretations of the recorded meter readings. You must use the interpretation least damaging to the discharger.

6.23 What Special Precautions Are Necessary When Sampling for Total Toxic Organic (TTO) Analyses?

The term "Total Toxic Organics" (TTO) implies a universal, comprehensive list of toxic organic compounds. When this term is used in conjunction with the EPA's Categorical Pretreatment Program, however, it refers to a different list of specific chemicals for each industrial category. Since there is no universal list of TTO constituents, there is no single best procedure for obtaining TTO samples.

For sampling purposes, toxic organic compounds can be divided into two major subgroups, volatile and nonvolatile compounds. Volatile compounds (benzene, carbon tetrachloride, vinyl chloride, methylene chloride) will escape from a sample that is exposed to the atmosphere, so they must be collected in a manner that will prevent contact with air. This problem is less significant with nonvolatile compounds.

Nonvolatile toxic organics can be sampled as discussed above. However, special care must be taken to ensure that the sampling equipment is not made of materials that will contaminate the sample. Special preservation procedures, as discussed elsewhere in this chapter, may be necessary.

An automatic sampler for volatile toxic organic compounds must keep the sample isolated from the atmosphere and must not put the sample under a vacuum which will encourage volatilization. Automatic samplers are available that can collect representative composite samples of volatile toxic organic compounds.[10] Also composite samples of volatile compounds can be obtained by collecting an appropriate number of grab samples in septum vials. The laboratory can prepare a composite sample from the individual vials.

Sampling for Total Toxic Organics (TTO) requires special precautions because many of these pollutants are fairly volatile. Place the collected sample immediately under refrigeration, transport the sample to the lab and have it analyzed as soon as possible. The longer the time between sample collection and analysis, the greater the opportunity for the pollutant to volatilize and leave the sample, thus producing low results. Frequently an inspector will split a sample and give half to the industry for testing. If the industry does not submit the sample immediately to a lab and if the lab does not analyze the sample immediately, the industry could get results that indicate they are in compliance while the inspector's lab results could reveal the industry is out of compliance. Grab samples

---

[11] *Aliquot (AL-li-kwot). Portion of a sample. Often an equally divided portion of a sample.*

are recommended for TTOs rather than composites because of the volatility of many TTOs. See Appendix II, "Pretreatment Arithmetic," Section I, "Impact of Toxic Waste on Sewer System," at the end of this manual for additional information.

6.24 Can I Make Determinations in the Field Without Submitting Samples to the Laboratory?

Field testing is very useful for producing quick results which give a feel for the actual conditions when the samples are taken. Some analytic results, such as chlorine residual, pH, sulfite, and temperature, can *ONLY* be accurately obtained by testing when the sample is taken. However, most field methods are less accurate and less selective than the precise methods used by a fully equipped laboratory and, therefore, cannot be fully trusted.

Field tests are available for many pollutants. pH can be measured using portable pH meters. Hach Company makes small kits for measuring a variety of pollutants, and there is a Lamotte Kit for measuring dissolved sulfide and total sulfide. A useful type of field sampling equipment is the glass tube shown in Figure 6.6. You can collect a small amount of sample (either vapor or liquid) in the tube for analysis.

There are test strips for many pollutants. To use the test strips, dip the strip (which is chemically treated for a specific pollutant) into the wastestream. The strip will turn a specific color. Match this color with colors provided with the test strip. The matched color will indicate the concentration of the pollutant in the wastestream. Some emergency response units carry "HazCat" kits that will identify 40 different hazardous materials. (See Appendix B for an example of the procedures to use with pH test paper.)

Field tests are particularly useful when speed is more important than accuracy. Treatment plants frequently receive high-pH wastewater. This may come from a relatively innocuous process, such as cleaning food-handling equipment, or it may indicate the presence of toxic wastes. Quick tests for cyanide and, if cyanide is present, some of the common plating metals will help the treatment plant operator to react to the discharge and help the industrial waste inspector to find the source.

If you are an operator of an industrial wastewater treatment system, you may find field tests a convenient way to check the operation of the system. Be extremely cautious in doing this. One company was found guilty of discharging excessive quantities of a toxic metal even though its field tests indicated the metal was not present in its treated wastewater. The company's consultant had even confirmed the accuracy of the field test. The problem was that the company's wastewater contained a material that masked the metal from the field test but did not interfere with the rigorous laboratory procedure.

Field test kits and test strips are provided with step-by-step directions for measuring pollutants in wastestreams. To use one of these kits in the field, simply follow the directions. You must read the directions to determine the limitations and to find out if there are any pollutants which can interfere with your test results.

6.25 When To Use Grab vs. Composite Samples?

Pretreatment facility inspectors may have a difficult decision attempting to determine whether to collect a grab or composite sample. This decision is often site-specific and should be based on the characteristics of the discharge and an industry's compliance record. Composite samples may be appropriate for Categorical Standards because sampling must represent the entire flow and operation during an entire day. Grab samples may be appropriate for local limits because they often are based on instantaneous maximums. Also local limits are designed to prevent both collection system and treatment plant problems. Figure 6.7 is a sampling decision flow diagram to help inspectors decide whether to use grab or composite samples. Additional information is provided in Appendices C, D and E at the end of this chapter. Appendix C, "The Use of Grab Samples to Detect Violations of Pretreatment Standards," is an explanation of EPA's policy on grab samples vs. composite samples.

QUESTIONS

Write your answers in a notebook and then compare your answers with those on page 460.

6.2A List the three possible types of parties that may collect a sample of industrial wastewater.

6.2B Give an example of significant noncompliance with self-monitoring requirements.

6.2C What are the two types of composite samples?

6.2D What is an aliquot?

6.2E When is field testing useful?

6.3 TRANSPORTING SAMPLES

6.30 How Should the Samples Be Labeled?

At some time, your sample is likely to be placed near another sample in a similar container. In order to prevent confusion between these samples, you must attach to the sample container a label with enough information to positively identify the sample. This label should be attached before the sample is placed in any container for storage or transportation. Sample containers can sometimes be labeled before going into the field.

Labels are usually made of a tape, such as freezer tape, that can be written on. If you are using disposable containers, write the label directly on the container with an indelible marker. For large sampling studies you may want to order preprinted labels. Whichever labeling method you use, be sure that the markings will survive contact with water or other solvents that the sample could be near. Ball-point pen ink frequently runs or bleeds when it gets wet. A pencil may be a better method of marking and identifying labels.

While accurate sample labels are essential for identifying all samples, you must also record a certain amount of other relevant information. Typically, agencies require at least the ten items listed on page 444.

*Fig. 6.6   Disposable glass sampler. Sampler is designed to
provide a safe and easy method for extracting a
liquid column sample from a drum or tank.*
(Permission of Direct)

NOTE: Coveralls of appropriate material should be worn depending on the nature of the suspected pollutant in the drum.

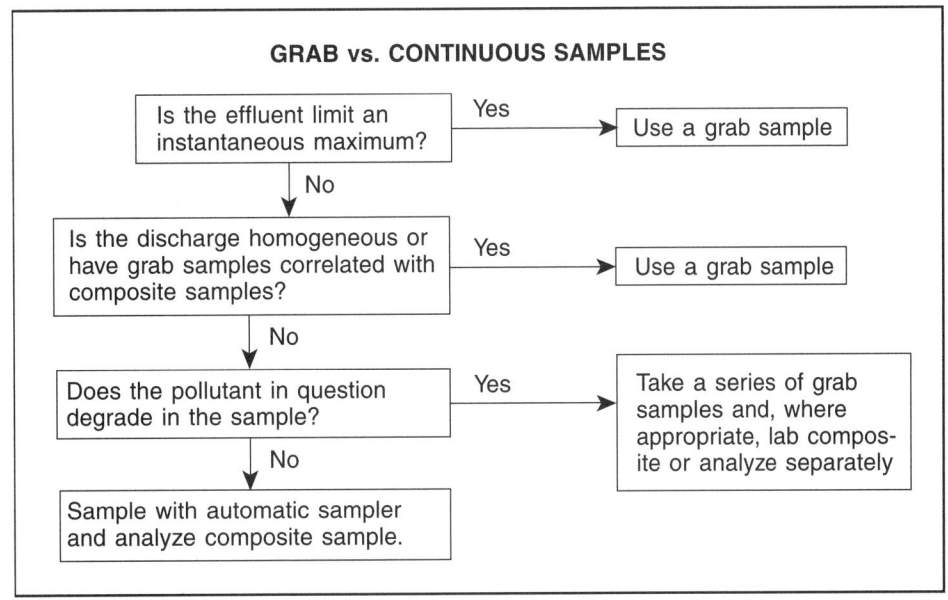

Fig. 6.7   Sampling decision flow diagram
(Provided by Keith Silva, Environmental Engineer,
Water Management Division, US EPA, Region IX)

1. Type of material sampled,

2. Type of sample (grab or composite),

3. General location where the sample was collected,

4. Specific location where sample was collected,

5. Time and date the sample was collected,

6. Who collected the sample,

7. Field observations (appearance of the wastestream, weather),

8. Laboratory analyses to be conducted,

9. Why the sample was collected, and

10. Who will receive copies of the laboratory results.

The label itself, however, may be very simple, perhaps with just a code number for identification. If the sample is collected by a treatment plant operator to check the system operation, the sample may be collected in a beaker that has the word "effluent" written on it. The operator will remember when the sample was collected, and the rest of the information is implied. Frequently the label will contain a brief description of the sample, such as items 1, 3 and 5 listed above. This identifying information will refer to a data sheet that contains the rest of the necessary information.

Figure 6.8 is an example of a data sheet you might be required to complete if you submit samples to a large laboratory. When the sample and accompanying data sheet arrive at the laboratory (Figure 6.9), a preprinted tag with an identifying number will immediately be attached to the sample container and the same identifying number will be recorded on the data sheet. This unique number will then be the major control mechanism throughout the sample processing. Although certain information from the data sheet will be entered into a computer, the technicians who work on the sample will only know the information the sampler put on the original label and that sample 352984 must be analyzed for suspended solids. The purpose of such secrecy is to protect the confidentiality of test results and to ensure unbiased analyses by the technicians. When the sample analysis has been completed, the computer will print out the necessary reports. Anyone who wishes to use the laboratory results must consult the computer reports and the sample data sheet.

6.31   How Is the Sample To Be Stored?

The purpose of storage is to maintain the integrity of the sample until an analysis can be performed. It must be protected from environments that would tend to change the nature of the sample and from people who might tamper with it. Samples sometimes contain substances that pose a danger to people who might come in contact with them. The sample storage facility must provide some degree of security, both for people and for the samples, and it must provide conditions that comply with the preservation requirements discussed earlier in this chapter.

If the same person who collected the sample will deliver it to the laboratory, they may only need a rack that will prevent breakage or spillage. If the sample is to be delivered by a courier, more secure packaging may be needed. An ice chest can be used to transport samples that must be kept cold (cool, 4°C). Consult the postal regulations before mailing any samples (also see Table 6.1, footnote 3, page 432).

For many pollutants, particularly volatile pollutants, you will want to carry a supply of field blanks while sampling. Usually an inspector's supervisor will recommend the use of a field blank if volatile pollutants are involved or questionable results have been obtained in the past. A field blank is a sample container of the same design as will be used in the sampling study that is filled in the laboratory with water that is known to be pure. This container is then carried with the sample containers used in the study to all locations in the field and stored side by side with the containers in the field. If the field blank water shows up with significant concentrations of pollutants, it may suggest that the other samples have been contaminated. If a field blank has been contaminated, the validity of test results can be questioned and the results will have little value in court. A field blank also can serve as a check on inspectors' sampling and handling procedures.

For all types of samples it is necessary to maintain a chain of custody record, as described in Section 6.41. This is a complete record of anyone who has had access to the sample at any time. Particularly in a criminal court, it may be necessary to question individually each person who handled the sample (Figure 6.10) to determine where the sample was at any given time and whether anyone might have tampered with it.

The integrity of samples can be further protected by using evidence tape. This is a special type of tape used in police work. It has an adhesive stronger than the tape. If evidence tape is placed across a cap so that the tape must be peeled back before the top can be removed, it is necessary to destroy the tape before opening the sample. Inspection of the tape will reveal if someone has tampered with the sample.

QUESTIONS

Write your answers in a notebook and then compare your answers with those on page 460.

6.3A   How much information is needed on a sample's label?

6.3B   When and why are field blanks sometimes necessary?

6.3C   What is a chain of custody record?

6.3D   When are chain of custody procedures usually appropriate?

6.4   DOCUMENTATION

6.40   What Types of Court Actions May Result?

Court actions can be classified as either civil or criminal. In a civil action, one party is accused of treating another wrongfully. The plaintiff asks the court to require the defendant to give the plaintiff something of value, usually money. Civil proceedings may be appropriate when an agency is trying to collect fees that are owed for services provided, to collect administrative fines imposed by an agency or when an agency is denying access to the sewer because of discharge or permit violations. In a criminal action, the defendant is accused of committing a crime against society. This may be a crime against an individual, such as assault, or against society at large, as in drug abuse. Discharging pollutants in excess of legal limits may be a crime against society at large or against an individual POTW.

There are two major differences between civil and criminal actions. In civil court the ruling is made based on "the preponderance of the evidence." The most likely explanation for a given set of facts is established as truth in a civil court. In a criminal court, the prosecution must prove its case "beyond a reasonable doubt." This is not "beyond a possible doubt" but it

Received in Sample Room By: _____

Time Received in Laboratory: _____

PRIORITY:
☐ Routine (0)
☐ Rush, 1 day (R)
☐ Expedite, 1 hr. (X)
☐ Upset, 1 wk. (U)

PRIORITY AUTHORIZATION _____
ext. _____

LABORATORY JOB NUMBER: _____

CHARLES W. CARRY
Chief Engineer and General Manager

SANITATION DISTRICTS OF LOS ANGELES COUNTY
P.O. Box 4998   Whittier, California 90607

For additional information regarding
this analysis call (213) 699-7411 x 259

INDUSTRIAL WASTE MONITORING RESULTS

SAMPLE IDENTIFICATION   Surcharge Account Number: _____   I.W. Permit Number: _____

Sample Source or Company Name: _____

Address: _____

Reason For Sampling: ☐ Routine   ☐ Phase I   ☐ Treatment Plant Upset   ☐ Sewer System Problem   ☐ Other: _____

Monitoring Requested By: *Kremer/* _____   Ext. _____   Sample To Be Saved After Analysis? _____

Accounting Charges: ☐ For Routine Samples Taken By Inspectors: TS14905 BI   ☐ For Routine Samples Taken By IWMC: TS14905 BM
☐ For All Special Investigation Samples: TS14905 B _____

SAMPLE COLLECTION INFORMATION

*Sample Taken From:
☐ Sample Box
☐ Clarifier
☐ Cleanout
☐ Other: _____

*Sample Method:
☐ Grab
☐ Composite-Timed With ____ minute intervals.
☐ Composite-Flow With ____ gallon intervals.

*Sample Grab or Composite Times:
At/or   ☐ A.M.
From ____   ☐ P.M.   ____ ____ 19 ____
       Hr. Min.   Month Day   Year

To ____   ☐ A.M.
          ☐ P.M.   ____ ____ 19 ____
   Hr. Min.   Month Day   Year

*Condition of Clarifier, Sample Box, Other Observations: _____

*Company Contact: Name: _____   Title: _____
*Sampled By: _____   *Submitted To Laboratory By: _____
*Total Quantity of Sample Collected: _____   *Total Quantity of Composite Prepared: _____   *Total Quantity Submitted To Lab: _____
*Sample Preservation Used: _____   *Sample Split With Company?: Y N   Name: _____
*Flow When Sampled: _____

COMPOSITE SAMPLE	GRAB SAMPLE
Period of Composite: _____ hours Totalizer Readings: Final: _____ Initial: _____ Difference: _____ Multiplier: _____ Total Flow: _____ gallons	Flow Rate At Time of Sample: _____ gpm

*Obtained From: ☐ Visual Estimate, ☐ Effluent Flow Meter, ☐ Influent Water Meter, ☐ Impossible To Tell If Flow Existed When Sampled

FIELD TEST RESULTS

*pH _____   *Temperature _____   *Color _____   *Cyanide _____
*Odor _____
*Total Sulfide _____   *Dissolved Sulfide _____   *Other _____

LABORATORY TESTS   NOTES TO ANALYSTS: _____

—— Check Tests Requested ——

CONSTITUENT	RESULTS	CODE
pH		101
SUSPENDED SOLIDS, mg/l		151
SUSPENDED SOLIDS (@ pH 7), mg/l		150
COD, mg/l 0		403
TOTAL DISSOLVED SOLIDS (TDS), mg/l		155
SULFATE, mg/l SO$_4$		257
AMMONIA NITROGEN, mg/l N		201
SOLUBLE SULFIDE, mg/l S		252
SULFITE SULFUR, mg/l S		254
THIOSULFATE SULFUR, mg/l S		253

CONSTITUENT	RESULTS	CODE
TOTAL CYANIDE, mg/l CN		206
TOTAL ARSENIC, mg/l As		705
TOTAL CADMIUM, mg/l Cd		708
TOTAL CHROMIUM, mg/l Cr		709
TOTAL COPPER, mg/l Cu		712
TOTAL LEAD, mg/l Pb		714
TOTAL MERCURY, mg/l Hg		717
TOTAL NICKEL, mg/l Ni		718
TOTAL SILVER, mg/l Ag		722
TOTAL ZINC, mg/l Zn		724
TOTAL PHENOLS, mg/l C$_6$H$_5$ OH		312
OIL AND GREASE, mg/l		408

DISTRIBUTION: Return this sheet to Industrial Waste Section Head after log-in.

*All Information to be Filled in by Field Collection Personnel.

*Fig. 6.8   Form used to record sampling information and
request laboratory analysis*

Fig. 6.9   Inspector logging sample into POTW laboratory and turning in chain of custody forms

*Fig. 6.10   Inspector and industrial contact signing sample
control sheet after splitting a sample*

does mean that if there are two reasonable explanations for a set of facts, the truth is established as being the one less damaging to the defendant. This places a much greater burden of proof on the prosecutor than in civil court. Many times evidence that is very valuable in determining a civil case will not be allowed in a criminal court.

In civil actions all parties are represented by private attorneys. A criminal case, however, is brought before the court by a public prosecutor, a representative of the district attorney, the city prosecutor, or in the case of federal crimes, the United States Attorney General. Any sample which is collected to determine compliance with applicable pretreatment standards may be used in an enforcement or judicial action. Since an inspector can never know with certainty which sample may become critical evidence in POTW enforcement actions against an IU, *ALL* sampling and analysis or other information gathering about IU activities should be conducted with sufficient care to produce evidence admissible in enforcement proceedings. The POTW's legal staff or an attorney can outline the types of evidence needed and the appropriate procedures that must be followed to ensure that the evidence will be acceptable to the court and useful to the case. If there is a reasonable doubt that someone could have tampered with a sample between the time it was collected and the time it was analyzed, there is reasonable doubt as to the accuracy of the results and the sample might not be satisfactory for use in criminal court.

In connection with criminal complaints, it is important to realize that industrial waste inspectors usually are not considered peace officers. They do not have badges, they do not carry guns, and they do not have the authority to arrest people. Industrial waste inspectors do not have to give the standard Miranda warning informing suspects of their Fifth Amendment rights unless they are acting as agents of a law enforcement agency. This can get to be a fine line, but when no peace officer is aware of an investigation, which is the case in any routine inspection or administrative enforcement action, the inspector is not an agent of a law enforcement agency.

6.41   What Is Meant by Chain of Custody?

A chain of custody record is one which shows a record of every person who has had access to a sample. Normally this means that the person collecting the sample signs a release to a person who will take custody of the sample. The person with custody normally keeps the sample under lock and key so that this person knows no other person has had access to it. The analyst must contact the custodian to obtain a portion of the sample for testing. Figure 6.11 is a sample chain of custody identification tag and Figure 6.12 is an example of a chain of custody record used in a wastewater discharge surveillance program.

Although maintaining chain of custody for all samples is a very important aspect of an industrial waste inspector's work, it is easy to become careless. Suppose you, as an inspector, collected a sample in the morning. Before submitting the sample to the laboratory, you stopped for lunch at a walk-in restaurant. If you forgot to lock your car, the defense may be able to create a reasonable doubt that someone tampered with the sample while you were eating. Evidence tape would make this more difficult. If your car was locked when you went to lunch, still locked when you got back, all the windows were up, and there was no sign of forced entry, it is not reasonable to suggest that anyone tampered with the sample. Learn the correct chain of custody procedures and practice them until they become a standard routine. (Appendix F on page 475 contains the chain of custody procedures used by the Sanitation Districts of Los Angeles County.)

6.42   What About Report Writing?

Report writing is useful in every sampling situation. A treatment system operator may only need to record the time the sample was collected and the analytical results on a table or logbook. A data sheet, such as Figure 6.8 (page 445), represents a more complicated type of report. Some sampling situations may require lengthy narrative reports describing all of the conditions at the time of sampling. Whatever the report writing requirements, if the reports are not written, it's as if the samples were never taken.

Remember that any sample may eventually, for some reason or other, wind up in court. Should a question arise concerning the sampling, the inspector's notes are needed to reconstruct the events, so they must document all the pertinent data. Tailor your recordkeeping to your needs, but be sure to collect enough information to reconstruct the sampling program. Some of the information that may be needed is listed below:

- Date and time,
- Sample location, and
- Sampling equipment.
- Observations:

    Flow rate indicated by a meter or the depth and flow velocity,

    Color,

    Odor,

    Turbidity, and

    Presence of solids.

- For automatic samplers:

    Start and stop times,

    Frequency of sampling,

    Aliquot size (the size of each individual grab sample),

    Number of bottles used, and

    Which bottles were empty after the samples were collected.

Any notes that you do keep may be used in court and your testimony will be judged according to the quality of your notes. If your notes are in a bound notebook or a notebook with sequentially numbered pages, the jury will be more likely to believe that they were produced at the time of the inspection or shortly thereafter and were not updated in preparation for the

SANITATION DISTRICTS OF LOS ANGELES COUNTY
PRETREATMENT PROGRAM — CUSTOM TAG

Sample Source: _____

This is # _____ of _____ Splits. See over for preservative.

CONTAINER LOT NUMBER:	Time	Date	Sampled By: (Signature)
Relinquished By: (Signature)	Time	Date	Sampled By: (Signature)
Relinquished By: (Signature)	Time	Date	Sampled By: (Signature)
Relinquished By: (Signature)	Time	Date	Sampled By: (Signature)
Relinquished By: (Signature)	Time	Date	Sampled By: (Signature)

LAB JOB NO.

FRACTION #_____ OF _____SPLITS

Preservative Added:		
	Preserved By: (Signature)	Time	Date
	Stored By: (Signature)	Time	Date
☐ ACID:_____	Retrieved By: (Signature)	Time	Date
☐ BASE:_____	Stored By: (Signature)	Time	Date
☐ OTHER:_____	Retrieved By: (Signature)	Time	Date
AMOUNT:_____	Stored By: (Signature)	Time	Date

Fig. 6.11 Chain of custody tag which is attached to sample container

LOGGED IN BY:_____ LABORATORY JOB NUMBER: SJS_____

CHARLES W. CARRY	SANITATION DISTRICTS OF LOS ANGELES COUNTY	For additional information regarding
Chief Engineer and General Manager	P.O. Box 4998 Whittier, California 90607	this analysis call (310) 699-7411 x 2900

SURVEILLANCE PROGRAM - CHAIN OF CUSTODY RECORD

SAMPLE IDENTIFICATION | Surcharge Account Number: | I.W. Permit Number:

Sample Source or Company Name: _____

Address: _____

Reason For Sampling: ☐ **Phase 1** ☐ Pretreatment ☐ Sewer System Problem ☐ Other: _____

Evidence Sampling Authorized By: *Martyn/Burch* _____

Accounting Charges: ☐ For Routine Samples Taken By IWMC: TS14905 BM
☐ For All Special Investigation Samples: TS14905 B_____ ☐ Other: _____

SAMPLE COLLECTION INFORMATION: Test aliquot Volume_____/_____ml. Interval____/____min. Estimated Flow Rate_____gpm.

Sample Taken From: * Sample Method: * Sample Date:

☐ d/s Manhole   ☐ Composite - Single Container   At/or   ☐ A.M.
☐ u/s Manhole   ☐ Individual Bottles _____ hr/bottle   From ____ HR. MIN.   ☐ P.M. ___ MONTH DAY __ 19 YEAR
☐ d/s Only Manhole   ☐ Individual Bottles - Composited   ☐ A.M.
☐ Other: ☐ Grab   To ____   ☐ P.M._____ 19___

Condition of Manhole, Other Observations: _____

Sampled By (Signature): _____

Total Quantity of Sample Collected:_____ Quantity submitted to lab_____ Sample Container(s) Used: ☐ C/c Other: _____

Sample Preservation Used: ☐ Ice ☐ No Cl$_2$ ☐ Sodium Thiosulfate ☐ NaOH ☐ HNO$_3$ Other: _____

Composite Sample Data: pH (905)_____ Color:_____ Other: _____

Individual Bottle Data: pH/Color/Vol.

1. ____ 6. ____ 11. ____ 17. ____ 23. ____
2. ____ 7. ____ 12. ____ 18. ____ 24. ____
3. ____ 8. ____ 13. ____ 19. ____ 25. ____
4. ____ 9. ____ 14. ____ 20. ____ 26. ____
5. ____ 10. ____ 15. ____ 21. ____ 27. ____
          16. ____ 22. ____ 28. ____

Bottle Nos. Composited: _____

Bottle Nos. Submitted to Lab: _____

LABORATORY TESTS NOTES TO ANALYSTS: _____

CONSTITUENT	CODE		CONSTITUENT	CODE
pH	101			
SUSPENDED SOLIDS, mg/l	151			
COD, mg/l O	403			
TOTAL CYANIDE, mg/l CN	206			
CYANIDE AMENABLE TO Cl$_2$	210			
TOTAL CADMIUM, mg/l Cd	708			
TOTAL CHROMIUM, mg/l Cr	709			
TOTAL COPPER, mg/l Cu	712			
TOTAL LEAD, mg/l Pb	714			
TOTAL NICKEL, mg/l Ni	718			
TOTAL SILVER, mg/l Ag	722			
TOTAL ZINC, mg/l Zn	724			

CUSTODY RECORD

Relinquished by:	Print Name	Time/Date	Received by:	Print Name
Relinquished by:	Print Name	Time/Date	Received by:	Print Name
Relinquished by:	Print Name	Time/Date	Received by:	Print Name
Relinquished by:	Print Name	Time/Date	Received by:	Print Name

Fig. 6.12 Surveillance program chain of custody record

court case. Since the court may retain the notes as evidence, it is suggested that each report start with a new page. Then notes from one case will not be confiscated along with the notes from another case.

If you are required to give testimony in the courtroom, use your notes to refresh your memory, but only to refresh your memory. If the defense attorney can get you to say that you are reading your notes instead of using them to refresh your memory, your testimony will be stricken from the record. Only your notes will be presented as evidence. The jury will then be required to decipher your notes.

QUESTIONS

Write your answers in a notebook and then compare your answers with those on pages 460 and 461.

6.4A   What is the difference between civil and criminal court actions?

6.4B   What are the two major differences between civil and criminal actions with regard to evidence and proof?

6.4C   Why should notes be kept in a bound notebook or a notebook with sequentially numbered pages?

6.5 TRACING AN ILLEGAL DISCHARGE

From time to time, an unwanted material may come into a POTW or an industrial pretreatment facility. This material might cause the facility to malfunction by damaging equipment, killing important organisms or altering chemical reactions. The EPA calls this condition "interference." If the facility is not designed to handle an unwanted or harmful material, it may go through the system without being treated. The EPA refers to this condition as "pass-through." Materials causing interference or pass-through can cause treatment facilities to violate their discharge standards.

The source of an unwanted material must be determined so that steps can be taken to prevent future discharges to the treatment system. The best way of finding the source is to have a good record of all possible sources prior to the event. For a POTW this means a permit system which records the types of pollutants discharged and the size of all connected industries. Industrial facilities may need to consult in-house

records in order to identify all possible pollutant sources affecting their own pretreatment processes.

Once an undesirable material comes into a treatment facility, it is usually too late to stop the illegal discharge. If there is a large collection system leading to the facility, the actual discharge may have stopped before the material entered the treatment plant. However, the investigation should not stop once the discharge stops. If the source of the discharge is not located, another discharge is likely. Fortunately, a thorough investigation that does not identify the source of a discharge is not necessarily a failure.

Suppose a treatment plant that serves five electroplaters suffers an upset of its biological treatment system. The treatment plant operator figures out when the material came into the plant and submits the appropriate influent sample and a sample of *RETURN SLUDGE*[12] to the laboratory. When the results come back showing that somebody dumped a lot of copper into the system, it will be too late to identify which of the five electroplaters was responsible.[13]

The industrial waste inspector should still contact each of the electroplaters and explain to them the problems that were caused by the copper discharge. Industrial dischargers must be made to understand the POTW's position. Explain to each of the dischargers who might have dumped the copper that the POTW takes seriously its responsibility to treat wastewater. Describe how a recent illegal discharge jeopardized your ability to meet those responsibilities and inform them that you are taking steps to identify the source of the illegal copper discharge. The simple fact that you personally took the time to deliver that message may be enough to prompt these five dischargers to be more careful about discharges. You still will not have identified the source of the problem, but your actions may well prevent future upsets.

Sometimes the illegal discharges will continue. In this case, since there are a limited number of sources of copper, inspect and sample all of the sources as soon as possible after each incident. One way to set up such a surveillance sampling program is to install automatic samplers (without the industries' knowledge) upstream and downstream (Figure 6.13) of each suspected discharger, as illustrated in the surveillance monitoring diagram in Figure 6.14. Program the automatic samplers (Figures 6.15 and 6.16) to collect hourly composite samples in individual bottles and try to adjust the timing of the samplers so that a volume of water is first sampled by the upstream sampler, flows past the discharger, and is then sampled again by the downstream sampler. Any increase in pollutants will then clearly have been added by the discharger.

If an ongoing situation develops where a particular type of undesirable pollutant is entering a facility frequently, but there are a large number of potential suspects, you may need to

---

[12] *Return Sludge.   The recycled sludge in a POTW that is pumped from a secondary clarifier sludge hopper to the aeration tank.*

[13] *It would not be too late if the POTW had been sampling some of the dischargers at points in the sewer system at the same time. Also the industries may have been on a self-monitoring program at the time or one of the industries may have a record of a spill.*

Fig. 6.13   Inspector setting a sampler downstream of a suspected illegal dump site
(Courtesy of ISCO, Inc., Environmental Division)

NOTE: When collecting samples proper personal safety gear should be worn, including gloves and safety glasses with side shields.

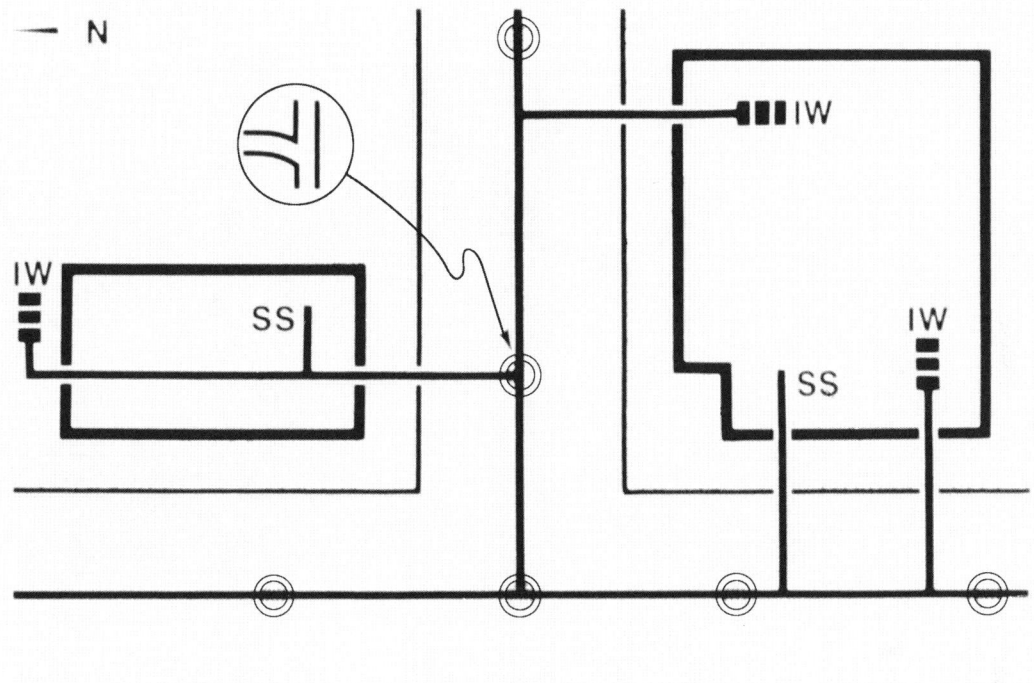

SS = SANITARY SEWER   IW = INDUSTRIAL WASTE SEWER

SURVEILLANCE MONITORING

Fig. 6.14   Surveillance monitoring diagram

Samplers must be properly located to test both upstream
and downstream of a suspected illegal dump site in order
to confirm the entry point of the waste.

Fig. 6.15 pH recorder and continuous sampler in sewer manhole to
detect illegal slug discharges over specified periods of time

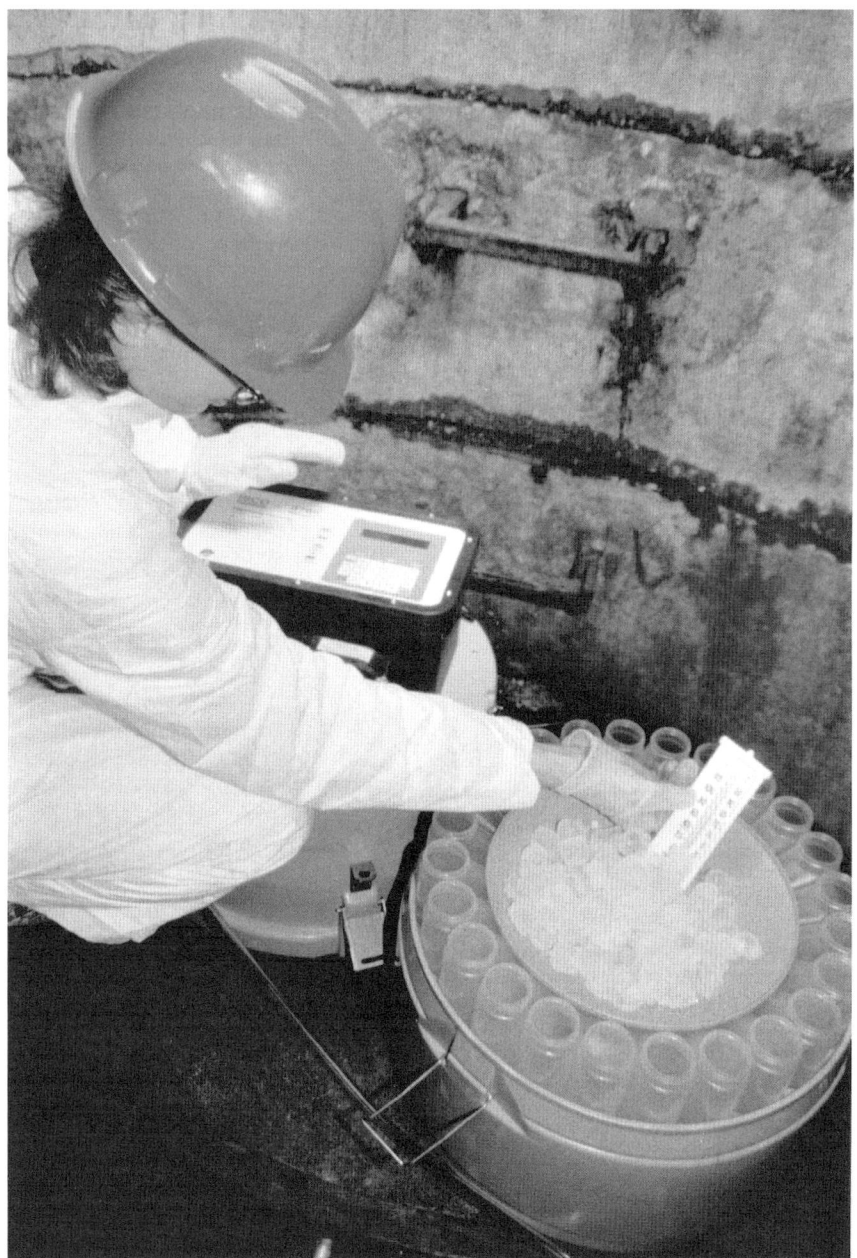

Fig. 6.16   Inspector checking sampler to determine which discrete sample bottle caught the illegal secret discharge
(Courtesy of ISCO, Inc., Environmental Division)

trace the pollutant upstream through the sewer. To do this you will need to know the locations of the major sewers and you will need a field test kit for identifying the pollutant in question or an indicator pollutant. Tracing a specific pollutant up through the wastewater collection system will require many analyses to be done quickly. The example below, which actually happened, illustrates how a field test which indicated little more than the presence or absence of an indicator material was used to track down a discharge that was incompatible with a POTW's treatment processes.

A tertiary treatment plant received a slug dose of a material which caused a high chlorine demand. The treatment plant tripled its chlorine feed rate, but it was unable to maintain a chlorine residual and attain adequate disinfection.

The POTW's records were searched for possible sources of the offending material. This treatment plant had experienced similar episodes of lesser severity in the past. They had been attributed to an oil refinery that had discharged excessive quantities of thiosulfate. The laboratory reported high levels of thiosulfate, but the industrial waste inspector who contacted the oil refinery found no indications of any unusual discharge.

This event was followed by several similar events in the following week. The thiosulfate test was therefore adapted to field kits to be used by a team of industrial waste inspectors. The adapted test did not actually measure thiosulfate concentrations but instead measured an indicator value which is chlorine demand.

Since this wastewater usually entered the plant in mid-morning, the inspectors assembled with their test kits and started testing the influent to the treatment plant at about 8:00 a.m. When the material was detected, they immediately checked the two main sewers entering the treatment plant to determine the route of the material. They then went to the major junctions of that sewer and worked their way upstream, continually finding higher concentrations of the offending material. The discharge stopped before a source was located.

The next day, the inspectors started their search at the last point where a high "thiosulfate" reading had been found. They quickly traced the high chlorine demand to a detergent manufacturer. This company was having difficulties with a certain piece of equipment and had been discharging large quantities of sulfite. The company was informed of the problems that it had created and was required to immediately pretreat its wastewater to prevent excessive discharges of sulfite.

Several lessons can be learned from this example:

1. The test kit which was intended to measure thiosulfate did not directly measure the offending pollutant. Sulfite interferes with the thiosulfate analysis. The test used by the inspectors merely indicated the presence of the problem material.

2. Locating this material depended on a large number of discharges. It was very difficult to trace it back up the sewer.

3. If there is a pattern and if the material is important, it can be traced through the sewers to its source. However, this is an expensive procedure. To trace materials upstream in the sewer requires a POTW to have the sewer maps organized in advance. The critical junction manholes in the system need to be identified so that when an unusual substance enters the plant, there is a systematic strategy already in place to guide the search and surveillance activities.

Most illegal discharges can be stopped before they happen. Through the permit process you should educate potential dischargers about the regulations that control their discharges. This alone will eliminate most potential sources of damaging discharges since most industries attempt to comply with the regulations. Periodically remind the dischargers of these regulations when you are inspecting their facilities and emphasize the importance of compliance. When necessary, insist that any potential for a damaging discharge be eliminated. Be generous, too, in acknowledging the value of compliance by dischargers who do not violate permit restrictions and those who work cooperatively to protect the POTW's treatment facilities.

Discharge permits and inspection records should provide a great deal of information concerning potential sources of materials harmful to the treatment processes. A well-prepared inspector who knows the area should be able to quickly identify the potential sources of most illegal discharges. The best way to find the source of a discharge that has already occurred is to concentrate on these sources.

QUESTIONS

Write your answers in a notebook and then compare your answers with those on page 461.

6.5A   What problems may be created by an unwanted material entering a POTW or an industrial pretreatment facility?

6.5B   How does the permit process help control illegal discharges?

6.6   QUALITY ASSURANCE/QUALITY CONTROL (QA/QC) PROCEDURES[14]

6.60   Procedures

Quality Assurance and Quality Control are tools which are necessary to maintain a specified level of quality in the measurement, documentation, and interpretation of sampling data. To produce evidence which is admissible in an enforcement action, QA and QC procedures are necessary both in the field (during sampling) and in the laboratory. The QA/QC procedures used in the field are separate from those used in the laboratory, but both are crucial for obtaining reliable data. Both laboratory and field QA/QC are discussed in this section. QA/QC procedures are used to obtain data that are both precise (degree of closeness between two or more samples) and accurate (degree of closeness between the results obtained from the sample analysis and the true value that should have been obtained). By following QA/QC procedures, the POTW's confidence in the validity of the reported analytical data is increased.

---

[14] *INDUSTRIAL USER INSPECTION AND SAMPLING MANUAL FOR POTWs, April 1994. Office of Water (4202), U.S. Environmental Protection Agency, Washington, DC 20460. Obtain from National Technical Information Service (NTIS), 5285 Port Royal Road, Springfield, VA 22161. Order No. PB94-170271. EPA No. 831-B-94-001. Price, $67.50, plus $5.00 shipping and handling per order.*

All data generated or used by the POTW must be of known, defensible, and verifiable quality. This includes data which are generated through self-monitoring at the industrial facility. Therefore, the IU should also have QA/QC procedures in place to ensure the adequacy of the data submitted as part of its periodic compliance report. (*NOTE:* All inspections, and the data obtained as a result of the inspection, have the potential to be used in an enforcement proceeding and should be treated as potential evidence to be admitted in court.)

QA consists of the program functions set up to assure the quality of measurement data while QC is the process of carrying out the procedures stated in the QA program. The QA program should be general while QC activities are specific. Also, the specific QC procedures used to assure data of good quality should be specified in the Quality Assurance and Sampling Plan developed for individual sampling events (for example, the use of duplicate samples in the field).

A QA program has two primary functions. First, it is designed to monitor and evaluate continuously the reliability (accuracy and precision) of the analytical results reported by each industrial user. This is how the quality of the data received from the IU is judged for acceptability. Second, QA should control the quality of the data to meet the program requirements. A QC program is designed to ensure the routine application of procedures necessary for the measurement process to meet prescribed standards of performance (for example, through instrument calibration and analysis of reference unknowns). A program describing the schedule for calibration is QA, while the actual calibration procedures are QC.

QA/QC functions fit into two categories, field procedures and laboratory procedures. Each of these items is discussed in greater detail in the subsections which follow.

6.61 Quality Assurance Procedures for Sampling

A QA program for sampling equipment and for field measurement procedures (for such guidelines as temperature, DO, pH, and conductivity) is necessary to ensure data of the highest quality. The inspector should recognize the importance of implementing quality assurance in sample collection to minimize such common monitoring errors as improper sampling methodology, poor sample preservation, lack of adequate mixing during compositing and testing, and excessive holding times. Again, each of these activities should be made a part of the POTW's standard operating procedures (SOP), so that all POTW sampling personnel are familiar with the proper sampling procedures. Quality assurance checks will help the inspector determine when sample collection techniques are inadequate for the intended use of the data. A field quality assurance program should contain the following elements:

- The required analytical methodology for each regulated pollutant; special sample handling procedures; and the precision, accuracy, and detection limits of all required analytical methods.

- The basis for selecting the analytical and sampling method. For example, each analytical method should consist of approved procedures. Where the method does not exist, the QA plan should state how the new method will be documented, justified, and approved for use.

- The number of analyses for QC (for example, the percentage of spikes, blanks, or duplicates), expressed as a percentage of the overall analyses (such as one duplicate sample per 10 samples) to assess data validity. Generally, the QA program should approximate 15 percent of the overall program, with 10 percent and 5 percent assigned to

laboratory QC and field QC, respectively. The QA plan should include shifting these allocations or decreasing these allocations depending on the degree of confidence established for collected data.

- Procedures to calibrate and maintain field instruments and automatic samplers.

- A performance evaluation system which allows sampling personnel to cover the following areas:

  — Qualifications of personnel for a particular sampling situation;

  — Determining the best representative sampling site;

  — Sampling techniques, including the location of sampling points within the wastestream, the choice of grab or composite samples, the type of automatic sampler, special handling procedures, sample preservation procedures, and sample identification;

  — Flow measurement, where applicable;

  — Completeness of data, data records, processing, and reporting;

  — Calibration and maintenance of field instruments and equipment;

  — Use of QC samples, such as field duplicates, or splits to assess the validity of the data; and

  — Training of all personnel involved in any function affecting data quality.

By following these QA procedures, the inspector can ensure the proper quality data from the industrial user.

6.62 Quality Control Procedures for Sampling

Sampling QC begins with calibration and preventive maintenance procedures for sampling equipment. The inspector should prepare a calibration plan and documentation record for all field sampling and analysis equipment. A complete document record should be kept in a QC logbook, including equipment specifications, calibration date, calibration expiration data, and maintenance due date. All of these activities should be reflected in the POTW's Standard Operating Procedures. The sampler should keep in mind that field analytical equipment should be recalibrated in the field prior to taking the sample. In addition to calibration procedures, the person conducting field sampling should conduct quality control checks. Quality control checks should be performed during the actual sample collection to determine the performance of sample collection techniques. In general, the most common monitoring errors are caused by improper sampling, improper preservation, inadequate mixing during compositing, and excessive holding time. The following samples should be used to check the sample collection techniques:

- Duplicate Samples (Field): Duplicate samples are collected from two sets of field equipment installed at the site, or duplicate grab samples are collected from a single piece of equipment at the site. These samples provide a precision check on sampling equipment and techniques.

- Equipment Blank: Is an aliquot of analyte-free water which is taken to and opened in the field. The contents of the blank are poured appropriately over or through the sample collection device, collected in a sample container, and returned to the laboratory as a sample to be analyzed. Equipment blanks are a check on the sampling device cleanliness.

- Field Blank: Is an aliquot of analyte-free water or solvent brought to the field in sealed containers and transported back to the laboratory with the sample containers. The purpose of the trip blank is to check on sample contamination originating from sample transport, shipping and from site conditions.

- Preservation Blank: Is an aliquot of analyte-free water (usually distilled water) to which a known quantity of preservative is added. Preservation blanks are analyzed to determine the effectiveness of the preservative, providing a check on the contamination of the chemical preservatives.

Personnel conducting sampling should be well trained in the use, cleaning, calibration, and maintenance of all instruments or samplers used. Automatic sampler tygon tubing, bottles, and the sampler itself should be cleaned prior to each sampling event. Automatic samplers should be calibrated for sample quantity, line purge, and the timing factor, if applicable. This calibration can and should be checked in the field to verify draw. The manufacturer's directions should be reviewed and followed for cleaning and calibrating all equipment.

6.63   Laboratory Quality Assurance/Quality Control

Laboratory QA/QC procedures ensure high-quality analyses through instrument calibration and the processing of control samples. The precision of laboratory findings refers to the reproducibility of test results. In a laboratory QC program, a sample is analyzed independently (more than once) using the same methods and set of conditions. Precision is estimated by comparing the measurements. Accuracy refers to the degree of difference between observed values and known or actual

values. The accuracy of a method may be determined by analyses of samples to which known amounts of a reference standard have been added.

Four specific laboratory QA procedures can be used to determine the confidence in the validity of the reported analytical data: duplicates, blanks, splits, and spiked samples.

- Duplicate Samples (Laboratory): Laboratory duplicate samples are samples which are received by the laboratory and divided (by the laboratory) into two or more portions. Each portion is separately and identically prepared and analyzed. Duplicate samples check for precision. These samples provide a check on sampling equipment and sampling techniques.

- Method Blanks: Method blanks are samples of analyte-free water that are prepared in the laboratory and analyzed by the analytical method used for field samples. The results from the analyses are used to check on the cleanliness of reagents, instruments, and the laboratory environment.

- Split Samples (Field): Field splits are collected and divided *in the field* into the necessary number of portions (for example, 2, 3, or 4) for analysis. Equally representative samples must be collected in this process, and then the samples are usually sent to different analytical laboratories. Field splits allow the comparison of analytical techniques and procedures from separate laboratories. Sampling personnel should exercise caution when splitting samples to avoid producing large differences in TSS (Total Suspended Solids). All large differences in results should be investigated and the cause identified. *NOTE:* Oil and grease samples cannot be split due to the nature of the pollutant.

- Laboratory Spiked Samples: These samples provide a proficiency check for the accuracy of analytical procedures. A known amount of a particular constituent is added to separate aliquots of the sample or to distilled/deionized water at a concentration where the accuracy of the test method is satisfactory. The amount added should be coordinated with the laboratory.

Laboratory QA/QC procedures can be quite complex. Often, the analytical procedures specify QA/QC requirements for calibration, interference checks (for ICP analyses), control samples, spiking (including the method of standard additions), blank contaminant level, and instrument tuning. Accuracy is normally determined through the analysis of blanks, standards, blank spikes, laboratory control samples, and spiked samples. Precision is determined through the comparison of duplicate results or duplicate spiked results for organic analysis. For more information on laboratory QA/QC, the POTW should contact the laboratory's Quality Assurance Management Staff or Quality Assurance Manager.

The methods used by in-house or contract laboratories to analyze industrial user samples must be methods which are EPA approved under 40 CFR 136 and thus are acceptable to a court of law as the most reliable and accurate methods of analyzing water and wastewater. Although some field test kits are useful as indicators of current conditions (and, thus, may be used for process control considerations), they are not appropriate for sampling that is conducted to verify or determine compliance. When choosing a contract lab, POTWs should obtain and review a copy of the lab's QA/QC plan.

QUESTIONS

Write your answers in a notebook and then compare your answers with those on page 461.

6.6A Why are quality assurance and quality control necessary?

6.6B What are the two functions of a quality assurance (QA) program?

6.6C What are the causes of the most common monitoring errors?

6.6D Analytical procedures often specify QA/QC requirements for what areas?

6.7 ADDITIONAL READING

INDUSTRIAL USER INSPECTION AND SAMPLING MANUAL FOR POTWs, April 1994. Office of Water (4202), U.S. Environmental Protection Agency, Washington, DC 20460. Obtain from National Technical Information Service (NTIS), 5285 Port Royal Road, Springfield, VA 22161. Order No. PB94-170271. EPA No. 831-B-94-001. Price, $67.50, plus $5.00 shipping and handling per order.

Please answer the discussion and review questions before continuing with the Objective Test.

DISCUSSION AND REVIEW QUESTIONS

Chapter 6. SAMPLING PROCEDURES FOR WASTEWATER

Write the answers to these questions in your notebook before continuing with the Objective Test on page 461. The purpose of these questions is to indicate to you how well you understand the material in the chapter.

1. Recommend a possible sequence of sampling programs to determine why a wastewater treatment plant is not performing as intended when the cause of the problem is a suspected industrial discharge.

2. How can realistic samples be obtained from a company that is trying to hide something and has the ability to store or impound certain wastestreams for a considerable period of time?

3. How does an agency setting up a sampling program decide whether to use an in-house or an outside laboratory?

4. What is a septum vial and when should it be used?

5. Why should sample containers be clean?

6. How would you determine whether to use disposable or reusable sample containers?

7. Why should a regulatory agency collect samples rather than rely on self-monitoring by dischargers?

8. Why should an industry have a self-monitoring program?

9. What factors would you consider when choosing a sampling location if you wanted a representative sample of the industrial wastewater flow?

10. Why should you avoid sampling from closed-pipe systems where the actual wastewater cannot be observed?

11. What are the advantages and limitations of using automatic sampling equipment to monitor industrial discharges after they have entered the public sewers?

12. List the information that must be recorded for each sample.

13. Why must samples be properly stored?

14. Why is an inspector not considered an agent of a law enforcement agency?

15. How can a POTW identify the source of an illegal discharge?

16. How can a suspected illegal discharger be monitored?

17. What is the purpose of a QA/QC program?

SUGGESTED ANSWERS

Chapter 6. SAMPLING PROCEDURES FOR WASTEWATER

Answers to questions on page 427.

6.0A Industrial wastewater samples are collected to obtain information for a particular purpose. Before starting any sampling program, the purposes or goals must be determined.

6.1A The four phases of a sampling program to locate a pollutant that is upsetting a treatment plant are as follows:

1. Determine which pollutants are interfering with the treatment processes,
2. Identify possible sources of offending pollutants,
3. Determine which sources are important, which sources can be controlled, and how they will be controlled, and
4. Implement a control strategy.

6.1B An inspector might need to collect only one sample to determine whether a certain pollutant is present in a wastestream.

Answers to questions on page 430.

6.1C A monitoring program should use an outside laboratory when it does not generate enough samples to pay for the special equipment and trained personnel required to produce accurate data at a reasonable cost.

6.1D Unless samples are collected, preserved and handled in accordance with established protocols (procedures), the results of laboratory tests may not accurately reflect the composition of the wastestream and test results will not be considered reliable.

6.1E Sample containers must be made of a material that is compatible with the sample.

6.1F Table 6.1 indicates that when collecting a sample to be analyzed for phenols, one should use a glass container, preserve the sample by cooling to 4°C, and add H_2SO_4 to a pH <2. The maximum holding time is 28 days.

6.1G Cup-shaped, wide-mouth containers with tight-fitting lids work best for collecting sludges.

Answers to questions on page 434.

6.1H *STANDARD METHODS* requires that for most types of sample analyses, sample containers should be cleaned by the laboratory using acid washing techniques.

6.1I Disposable containers can be considered clean after a laboratory has carefully washed a number of new containers and analyzed the wash waters for traces of pollutants. If none are found, the entire lot of containers is assumed to be clean and can be used in the field without further cleaning.

6.1J An acid is used to preserve a sample from a wastestream that may contain metals to dissolve the metals and keep them from adsorbing (sticking) on the sides of the storage vessel (sample container).

6.1K Before entering a confined space to collect a sample, *ALWAYS* test the atmosphere for toxic gases (hydrogen sulfide), explosive or flammable gases, and oxygen deficiency.

Answers to questions on page 442.

6.2A The possible types of parties who may collect a sample of industrial wastewater include the industry, the control agency and an independent third party.

6.2B An industry's failure to accurately report noncompliance is itself considered an act of significant noncompliance.

6.2C Composite samples may be either flow-weighted, in which case a grab sample is collected every so many gallons, or the grab sample volume is proportional to the flow rate, or time-weighted, where a grab sample is collected every so many minutes.

6.2D An aliquot is a portion of a sample, often an equally divided portion of a sample.

6.2E Field testing is very useful for producing quick results which give a feel for the actual conditions when the samples are taken.

Answers to questions on page 444.

6.3A Labels must have enough information to positively identify the sample.

6.3B Field blanks are often used when sampling for volatile pollutants or when past sampling produced questionable results. The field blank is carried with the sample containers used in the study to all locations in the field and stored side by side with the containers in the field. If the field blank water shows up with significant concentrations of pollutants, it may suggest that the other samples have been contaminated.

6.3C A chain of custody record is a complete written record of anyone who has had access to the sample at any time.

6.3D Chain of custody procedures are always appropriate when sampling an industrial user and they are essential when sampling results will be used in a criminal prosecution.

Answers to questions on page 451.

6.4A The difference between civil and criminal court actions is that in a civil action one party is accused of treating another wrongfully. In a criminal action, the defendant is accused of committing a crime against society.

6.4B The two major differences between civil and criminal actions with regard to evidence and proof are that in a civil court the ruling is made based on "the preponderance of the evidence." In a criminal court, the prosecution must prove its case "beyond a reasonable doubt."

6.4C Notes should be kept in a bound notebook or a notebook with sequentially numbered pages to help establish the credibility of your testimony.

Answers to questions on page 456.

6.5A Problems that may be created by an unwanted material entering a POTW or an industrial pretreatment facility include causing the facility to malfunction by damaging equipment, killing important organisms or altering chemical reactions. The material may cause the facility to violate its discharge standards.

6.5B The permit process can help control illegal discharges by educating potential dischargers of the regulations that control their discharges. This will eliminate most potential sources of damaging discharges because most industries attempt to comply with regulations once they understand what is required. The industrial waste preteatment inspector should periodically remind the dischargers of the regulations, teach them about the importance of compliance, and insist that any potential for a damaging discharge be eliminated.

Answers to questions on page 459.

6.6A Quality assurance and quality control are tools which are necessary to maintain a specified level of quality in the measurement, documentation, and interpretation of sampling data.

6.6B A quality assurance (QA) program has two primary functions. First, it is designed to *monitor and evaluate* continuously the reliability (accuracy and precision) of the analytical results reported by each industrial user. Second, QA should *control* the quality of data to meet the program requirements.

6.6C The most common monitoring errors are caused by improper sampling, improper preservation, inadequate mixing during compositing and excessive holding time.

6.6D Analytical procedures often specify QA/QC requirements for calibration, interference checks, control samples, spiking, blank contaminant level, and instrument tuning.

OBJECTIVE TEST

Chapter 6. SAMPLING PROCEDURES FOR WASTEWATER

Please write your name and mark the correct answers on the answer sheet, as directed at the end of Chapter 1. There may be more than one correct answer to each multiple-choice question.

True-False

1. The major portion of an inspector's sampling work load involves sampling industrial discharges.
   - 1. True
   - 2. False

2. Care and accuracy are essential elements in any sampling program.
   - 1. True
   - 2. False

3. Cooperation between the sampler and the laboratory will make or break a sampling program.
   - 1. True
   - 2. False

4. Sample containers must be clean of all traces of the pollutants to be analyzed.
   - 1. True
   - 2. False

5. All samples should be analyzed as soon as possible after they are collected.
   - 1. True
   - 2. False

6. Composite samples may be appropriate for local limits.
   - 1. True
   - 2. False

7. Grab samples may be appropriate for Categorical Standards.
   - 1. True
   - 2. False

8. Accurate sample labels are essential for identifying all samples.
   - 1. True
   - 2. False

9. A well-prepared inspector who knows the area should be able to quickly identify the sources of most illegal discharges.
   - 1. True
   - 2. False

10. The QA/QC procedures used in the field are the same as those used in the laboratory.
    - 1. True
    - 2. False

Best Answer (Select only the closest or best answer.)

11. What are composite samples collected to measure?
    - 1. Average amount of pollutants
    - 2. Maximum concentration of pollutants
    - 3. Pollutant discharge at specific time
    - 4. Pollutants of specific concern

12. What factor influences the decision to use disposable or reusable containers?

    1. Bulk
    2. Ease of cleaning
    — 3. Economics
    4. Weight

13. When is industrial self-monitoring valuable?

    1. When industry can perform monitoring at lower cost than POTW
    — 2. When industry has sincere wish to comply
    3. When POTW has no problems from industrial dischargers
    4. When POTW lacks necessary resources to monitor

14. When are grab samples useful?

    — 1. For enforcing discharge standards
    2. To determine average conditions
    3. To measure BOD
    4. When flow-weighted samples are needed

15. Why are some samples identified with only a number when submitted to the laboratory for analysis?

    1. To avoid improper labeling
    — 2. To ensure unbiased analyses by the technicians
    3. To facilitate computer logging and data storage
    4. To hasten the analyses and results

16. What is the main goal of sample storage procedures?

    1. To accumulate a large enough sample for analysis
    2. To allow samples to accumulate before analysis
    3. To efficiently use analysts' time
    — 4. To maintain integrity of sample until analysis

17. What is meant by chain of custody?

    1. A record of criminal action against an industrial discharger
    2. A record of each agency action leading to a notice of violation
    — 3. A record of every person who has access to a sample
    4. A record of every sample submitted to the laboratory

18. What is the function of the QA program?

    1. To assure chain of custody procedures are followed
    2. To assure collection of enough samples
    3. To assure collection of composite samples
    — 4. To assure the quality of the measurement data

Multiple Choice (Select all correct answers.)

19. Why are samples collected?

    — 1. Check compliance with local standards
    — 2. Determine strength to accurately assess discharge/use fees
    — 3. Search for prohibited wastes
    — 4. Verify industry monitoring points
    — 5. Verify self-monitoring data

20. Certified laboratories are evaluated in what areas?

    — 1. Ability to accurately run the analyses
    — 2. Adequate equipment to perform the analyses
    — 3. Personnel with adequate training
    4. Records documenting ability to perform analyses quickly
    5. Sufficient reagents available to perform analyses

21. What are gas chromatographs used to do?

    1. Detect inorganic compounds
    — 2. Detect organic compounds
    3. Quantify inorganic compounds
    — 4. Quantify organic compounds
    — 5. Separate organic compounds

22. What factors must be considered when selecting a sample container?

    — 1. Material
    2. Packing
    — 3. Shape
    — 4. Size
    5. Temperature

23. What pieces of sampling equipment must be properly cleaned?

    — 1. Automatic samplers
    — 2. Buckets
    — 3. Dippers
    — 4. Graduate cylinders
    — 5. Inside of coolers

24. What methods are used to preserve samples?

    — 1. Acids
    2. Bacteria
    — 3. Bases
    — 4. Refrigeration
    5. Storage

25. A laboratory analyst can be confident of test results provided the samples are _____ properly.

    1. Accepted
    — 2. Collected
    3. Rejected
    — 4. Stored
    — 5. Transported

26. When are disposable bottles usually a better alternative than reusable bottles?

    — 1. It is necessary to hire someone to wash bottles
    2. Smaller labs have time to wash bottles
    3. The analytical work load does not use all of the available staff hours
    4. The cost of buying disposable bottles and running blanks is more than the cost of washing and replacing broken bottles
    — 5. They are less costly than using reusable bottles

27. What are the benefits of collecting industrial wastewater samples from a sample box?

    — 1. A meaningful sample may be obtained even if the wastewater is no longer flowing
    2. Ability to collect and retain solids
    — 3. Material found in sample box can be assumed to have been discharged
    — 4. Retention of some water that is representative of the last wastewater to leave the plant
    5. Sample bottles can be safely stored in the sample box

28. Which pollutants can be monitored on a continuous basis?

    1. BOD
    — 2. Flammability
    — 3. pH
    — 4. Sulfide
    — 5. Temperature

29. Field tests are the *ONLY* accurate method for measuring which pollutants?

    — 1. Chlorine residual
       2. Cyanide
    — 3. pH
    — 4. Sulfite
    — 5. Temperature

30. When should grab samples be collected?

    — 1. If effluent limit is an instantaneous maximum
    — 2. If grab samples have been correlated with composite samples
    — 3. If the discharge is homogeneous
    — 4. If the pollutant degrades in the sample
       5. If the pollutant concentration varies with time and flow

31. What is the purpose of a field blank when collecting samples?

       1. To calibrate laboratory equipment
    — 2. To check on inspectors' sampling and handling procedures
    — 3. To document validity of field samples
       4. To ensure proper chain of custody procedures
       5. To verify accuracy of laboratory analyses

32. What information must be recorded for each sample collected?

       1. Emergency response phone number
    — 2. Field observations
       3. Name of industry contact person
    — 4. Time sample was collected
    — 5. Who collected sample

33. What information is helpful to a pretreatment inspector tracing an illegal discharge?

    — 1. Having a good record of possible sources prior to an event
    — 2. Having a permit system which records all types of pollutants discharged
    — 3. Installing automatic samplers upstream and downstream of suspected dischargers
    — 4. Knowing locations of major sewers
       5. Knowing phone numbers of regulatory officials

34. What types of errors can be minimized by a QA program?

    — 1. Improper sampling methodology
    — 2. Lack of adequate mixing during compositing
       3. Lack of adequate sample size
    — 4. Poor sample preservation
       5. Proper calibration of automatic samplers

End of Objective Test

APPENDICES

CHAPTER 6. SAMPLING PROCEDURES FOR WASTEWATER

NOTE: Appendices C, D, and E are very important EPA Memorandums that have been
reproduced for your use.

APPENDIX A

SUBJECT: Acidification of Industrial Waste Metals Samples

DATE: July 25, 1995

FROM: Paul C. Martyn
Head, Industrial Waste Section
County Sanitation Districts of
Los Angeles County

TO: All Industrial Waste Staff

Repipetor Acidification

Work stations for the acidification of industrial waste (IW) metals samples will be provided at the San Jose Creek Water Quality Lab (SJCWQL), the Los Coyotes Treatment Plant Lab and the Joint Water Pollution Control Plant Water Quality Lab Sample Receiving area. A separate, unpreserved split should be provided for pH, suspended solids, COD and any other similar inorganic method. The following procedure for metal preservation must be conducted as soon as possible, but no later than the end of the shift when the sample was collected. Under extenuating circumstances, the preservation may be conducted at the beginning of the next shift.

1. All manipulations of the sample should be performed in the hood provided for IW personnel. Safety glasses, aprons, gloves and face shields will be provided at the work stations. Safety glasses and aprons must be worn prior to sample manipulation; the other safety equipment is optional. Acid spill kits should be available in the Labs for any major spills.

2. Perform any field pH measurement prior to acidification of the sample.

3. Mix the sample thoroughly.

4. Acidification

   4.1   The stations should be equipped with a repipetor filled with 1:1 nitric acid. If the repipetor is empty, have the Lab personnel refill it for you.

   4.2   The repipetor should have an adjustment on the side of the piston that will set the amount of acid to be expressed. Set this at 5 mL per 1 liter of sample to be acidified.

   4.3   Place the mouth of the sample container directly under the outlet of the repipetor. Slowly pull up on the piston to fill it and then slowly press down to express the acid into the sample.

   4.4   Allow the sample to complete any gas generation, recap the container and mix thoroughly.

   4.5   Remove the cap and check the pH of the sample with a clean glass stirring rod and pH paper. If the sample is above a pH of 2, repeat steps 4.3 and 4.4.

5. Place an acidification sticker (available at the Lab station) on the sample container and indicate the amount and type of acid that was required.

6. Clean up any sample or acid spills in the working area. Rinse any residual acid off the outside of the sample container with a DI wash bottle.

Field Vial Acidification

Another acidification option is small screw-capped vials which are filled with 2 mL of 1:1 nitric acid. The vials can be provided upon request at the SJCWQL Stock Room.

1. All manipulations of the sample should be performed in an open area. Safety glasses, aprons, gloves, face shields and acid spill kits are recommended safety equipment. Safety glasses and aprons must be worn prior to sample manipulation; the other safety equipment is optional. All of this equipment is available at the SJCWQL Stock Room.

2. Perform any field pH measurement prior to acidification of the sample.

3. Mix the sample thoroughly.

4. Acidification

   4.1   Remove one of the small screw-capped vials that is filled with 2 mL of 1:1 nitric acid from its shipping package. NOTE: Keep any unused vials in the original shipping package during transport.

   4.2   Remove the cap from the vial and pour the contents into the sample container. Replace the cap and return the vial to the shipping case. Dispose of the empty vial(s) in the trash upon return to one of the Districts' facilities.

   4.3   Allow the sample to complete any gas generation, recap the container and mix thoroughly.

4.4   Remove the cap and check the pH of the sample with a clean glass stirring rod and pH paper. If the sample is above a pH of 2, repeat steps 4.2 and 4.3.

5. Place an acidification sticker (available from SJCWQL Sample Receiving) on the sample container and indicate the amount and type of acid that was required.

6. Clean up any sample or acid spills in the working area. Rinse any residual acid off the outside of the sample container with a DI wash bottle.

APPENDIX B

pH PAPER METHOD 9041A
(Sanitation Districts of Los Angeles County)

1.0   SCOPE AND APPLICATION

1.1   Method 9041 may be used to measure pH as an alternative to Method 9040 (except as noted in Step 1.3) or in cases where pH measurements by Method 9040 are not possible.
1.2   Method 9041 is not applicable to wastes that contain components that may mask or alter the pH paper color change.
1.3   pH paper is not considered to be as accurate a form of pH measurement as pH meters. For this reason, pH measurements taken with Method 9041 cannot be used to define a waste as corrosive or noncorrosive (see RCRA regulations 40 CFR §261.22(a)(1)).

2.0   SUMMARY OF METHOD

2.1   The approximate pH of the waste is determined with wide-range pH paper. Then a more accurate pH determination is made using "narrow-range" pH paper whose accuracy has been determined (1) using a series of buffers or (2) by comparison with a calibrated pH meter.

3.0   INTERFERENCES

3.1   Certain wastes may inhibit or mask changes in the pH paper. This interference can be determined by adding small amounts of acid or base to a small aliquot of the waste and observing whether the pH paper undergoes the appropriate changes.

CAUTION: THE ADDITION OF ACID OR BASE TO WASTES MAY RESULT IN VIOLENT REACTIONS OR THE GENERATION OF TOXIC FUMES (e.g., hydrogen cyanide). Thus, a decision to take this step requires some knowledge of the waste. See Step 7.3.3 for additional precautions.

4.0   APPARATUS AND MATERIALS

4.1   Wide-range pH paper.
4.2   Narrow-range pH paper: With a distinct color change for every 0.5 pH unit (e.g., Alkaacid Full-Range pH Kit, Fisher Scientific or equivalent). Each batch of narrow-range pH paper must be calibrated versus certified pH buffers or by comparison with a pH meter which has been calibrated with certified pH buffers. If the incremental reading of the narrow-range pH paper is within 0.5 pH units, then the agreement between the buffer or the calibrated pH meter with the paper must be within 0.5 pH units.
4.3   pH Meter (optional).

5.0   REAGENTS

5.1   Certified pH buffers: To be used for calibrating the pH paper or for calibrating the pH meter that will be used subsequently to calibrate the pH paper.
5.2   Dilute acid (e.g., 1:4 HCl).
5.3   Dilute base (e.g., 0.1 N NaOH).

6.0   SAMPLE COLLECTION, PRESERVATION, AND HANDLING

6.1   All samples must be collected using a sampling plan which addresses the considerations discussed in Chapter Nine of this manual.

7.0   PROCEDURE

7.1   A representative aliquot of the waste must be tested with wide-range pH paper to determine the approximate pH.
7.2   The appropriate narrow-range pH paper is chosen and the pH of a second aliquot of the waste is determined. This measurement should be performed in duplicate.

7.3 Identification of interference:

    7.3.1 Take a third aliquot of the waste, approximately 2 mL in volume, and add acid dropwise until a pH change is observed. Note the color change.

    7.3.2 Add base dropwise to a fourth aliquot and note the color change. (Wastes that have a buffering capacity may require additional acid or base to result in a measurable pH change.)

    7.3.3 The observation of the appropriate color change is a strong indication that no interferences have occurred.

CAUTION: ADDITION OF ACID OR BASE TO SAMPLES MAY RESULT IN VIOLENT REACTIONS OR THE GENERATION OF TOXIC FUMES. PRECAUTIONS MUST BE TAKEN. THE ANALYST SHOULD PERFORM THESE TESTS IN A WELL-VENTILATED HOOD WHEN DEALING WITH UNKNOWN SAMPLES.

8.0 QUALITY CONTROL

    8.1 All quality control data must be maintained and available for easy reference or inspection.

    8.2 All pH determinations must be performed in duplicate.

    8.3 Each batch of pH paper must be calibrated versus certified pH buffers or a pH meter which as been calibrated with certified pH buffers.

9.0 METHOD PERFORMANCE

    9.1 No data provided.

10.0 REFERENCES

    10.1 None required.

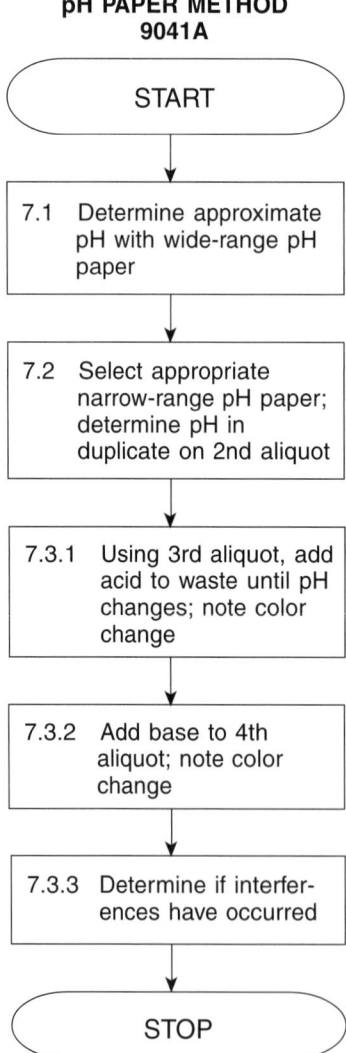

**pH PAPER METHOD
9041A**

START

7.1 Determine approximate pH with wide-range pH paper

7.2 Select appropriate narrow-range pH paper; determine pH in duplicate on 2nd aliquot

7.3.1 Using 3rd aliquot, add acid to waste until pH changes; note color change

7.3.2 Add base to 4th aliquot; note color change

7.3.3 Determine if interferences have occurred

STOP

APPENDIX C

SUBJECT: The Use of Grab Samples to Detect Violations of Pretreatment Standards

DATE: October 1, 1992

FROM: Michael B. Cook, Director
Office of Wastewater Enforcement & Compliance (WH-546)
Frederick F. Stiehl
Enforcement Counsel for Water (LE-134W)

TO: Water Management Division Directors, Regions I-X
Environmental Services
Division Directors, Regions I-X
Regional Counsels, Regions I-X

The primary purpose of this Memorandum is to provide guidance on the propriety of using single grab samples for periodic compliance monitoring to determine whether a violation of Pretreatment Standards has occurred. More specifically, the Memorandum identifies those circumstances when single grab results may be used by Control Authorities, including EPA, State or publicly owned treatment works (POTW) personnel, to determine or verify an industrial user's compliance with categorical standards and local limits. Please be aware that the concepts set out below are applicable when drafting self-monitoring requirements for industrial user permits.

REGULATORY BACKGROUND

The General Pretreatment Regulations require Control Authorities to sample all significant industrial users (SIUs) at least once per year [see 40 CFR 403.8(f)(2)(v)]. In addition, the Regulations, at 40 CFR 403.12(e), (g) and (h) require, at a minimum, that all SIUs self-monitor and report on their compliance status for each pollutant regulated by a Pretreatment Standard at least twice per year unless the Control Authority chooses to conduct all monitoring in lieu of self-monitoring by its industrial users.[1]

The Regulations, at 40 CFR 403.12(g) and (h), also specify that pollutant sampling and analysis be performed using the procedures set forth in 40 CFR Part 136. Part 136 identifies the proper laboratory procedures to be used in analyzing industrial wastewater (including the volume of wastewater necessary to perform the tests and proper techniques to preserve the sample's integrity). However, with certain exceptions, Part 136 does not specifically designate the method to be used in obtaining samples of the wastewater. Rather, section 403.12(g) and (h) require sampling to be "appropriate" to obtain "representative" data; that is, data which represent the nature and character of the discharge.

DISCUSSION OF BASIC SAMPLING TYPES

Sampling may be conducted in two basic ways. Both types of sampling provide valid, useful information about the processes and pollutants in the wastewater being sampled. The first is an "individual grab sample." An analysis of an individual grab sample provides a measurement of pollutant concentrations in the wastewater at a particular point in time. For example, a single grab sample might be used for a batch discharge which only occurs for a brief period (e.g., an hour or less). Such samples are typically collected manually but are sometimes obtained using a mechanical sampler.[2]

The second type of sample is a "composite sample." Composite samples are best conceptualized as a series of grab samples which, taken together, measure the quality of the wastewater over a specified period of time (e.g., an operating day). Monitoring data may be composited on either a flow or time basis. A flow-proportional composite is collected after the passage of a defined volume of the discharge (e.g., once every 2,000 gallons). Alternatively, a flow-proportional composite may be obtained by adjusting the size of the aliquots to correspond to the size of the flow. A time-proportional composite is collected after the passage of a defined period of time (e.g., once every two hours).

Generally, composite samples are collected using a mechanical sampler, but may also be obtained through a series of manual grab samples taken at intervals which correspond to the wastewater flow or time of the facility's operations. In some cases, composite data is obtained by combining grab samples prior to transmittal to a laboratory. At other times, the samples remain discrete and are either combined by the laboratory prior to testing or are analyzed separately (and mathematically averaged to derive a daily maximum value).[3]

---

[1] The POTW should conduct more frequent sampling and/or require more frequent self-monitoring by an industrial user if deemed necessary to assess the industry's compliance status (e.g., a daily, weekly, monthly, or quarterly frequency as appropriate).

[2] Mechanical samplers may not be used to sample for certain pollutants (e.g., those which could adhere to the sampler tubing, volatilize in the sampler, or pollutants with short holding times).

[3] Daily maximum discharge limits are controls on the average wastewater strength over the course of the operating day. They are not intended to be instantaneous limits applied at any single point during that operating day.

DETERMINING APPROPRIATE COMPLIANCE SAMPLING METHODS

EPA policy on appropriate compliance sampling types has been articulated in several pretreatment guidance manuals and regulatory preambles, and continues to be as follows:

A. Compliance With Categorical Standards

- Most effluent limits established by categorical standards are imposed on a maximum daily-average and a monthly-average basis. Generally, wastewater samples taken to determine compliance with these limits should be collected using composite methods.

- There are exceptions to the general rule. Composite samples are inappropriate for certain characteristic pollutants (i.e., pH and temperature) since the composite alters the characteristic being measured. Therefore, analysis of these pollutants should be based on individual grab samples. Alternatively, continuous monitoring devices may be used for measuring compliance with pH and temperature limits. Any exceedance recorded by a continuous monitoring device is a violation of the standard.

- Sampling wastewater from electroplating facilities regulated under 40 CFR Part 413 may be conducted using single grab samples [(assuming that the grab samples are representative of the daily discharge for a particular facility); see also preamble discussion at 44 Fed. Reg. 52609, September 7, 1979].

- A series of grab samples may be needed to obtain appropriate composite data for some parameters due to the nature of the pollutant being sampled. Examples of this situation include:

  - Sampling for parameters which may be altered in concentration by compositing or storage. These pollutants include pH-sensitive compounds (i.e., total phenols, ammonia, cyanides, sulfides);[4] and volatile organics such as purgeable halocarbons, purgeable aromatics, acrolein, and acrylonitrile.

  - Sampling for pollutants with short holding times such as hexavalent chromium and residual chlorine; and

  - Sampling for pollutants which may adhere to the sample container or tubing such as fats, oil and grease. Individual analysis for these parameters ensures that all the material in the sample is accounted for.

B. Compliance With Local Limits

- Local limits may be established on an instantaneous, daily, weekly or monthly-average basis. The sample type used to determine compliance with local limits should be linked to the duration of the pollutant limit being applied.

  - Compliance with instantaneous limits should be established using individual grab samples. Exceedances identified by composite sampling are also violations.

  - Compliance with daily, weekly or monthly average limits should be determined using composited sampling data, with the same exceptions noted in A, above.

  - Measurements of wastewater strength for non-pretreatment purposes (e.g., surcharging) may be conducted in a manner prescribed by the POTW.

GRAB SAMPLING AS A SUBSTITUTE FOR COMPOSITE SAMPLING

EPA is aware that a number of Control Authorities currently rely on a single grab sample to determine compliance, particularly at small industrial users, as a way of holding down monitoring costs. It is EPA's experience that the process activities and wastewater treatment at many industrial facilities may not be sufficiently steady-state as to allow for routine use of single grab results as a substitute for composite results. Therefore, the Agency expects composited data to be used in most cases. However, there are several circumstances when a single grab sample may be properly substituted for a single composite sample. These situations are:

- Sampling a batch or other similar short-term discharge, the duration of which only allows for a single grab sample to be taken;

- Sampling a facility where a statistical relationship can be established from previous grab and composite monitoring data obtained over the same long-term period of time;[5] and

---

[4] Certain pH-sensitive compounds can be automatically composited without losses if the collected sample is only to be analyzed for a single parameter. Additionally, a series of grab samples may be manually composited if appropriate procedures are followed.

[5] Grab sampling may provide results that are similar to composite sampling. See for example, a March 2, 1989, Office of Water Regulations and Standards(OWRS) Memorandum to Region IX describing the results of a statistical analysis of sampling data from a single industrial facility. These sampling data included both individual grab and flow-proportional, composite sampling obtained during different, non-overlapping time periods. After reviewing the data, OWRS concluded that the composite and grab sample data sets displayed similar patterns of violation for lead, copper, and total metals. In fact, the analyses did not find any statistically significant difference in the concentration values measured between the grab and composited data. Furthermore, additional statistical tests of the two data sets indicated that the means and variances for each pollutant were similar. The statistical conclusion was that the plant was judged to be out of compliance regardless of what data were analyzed.

– Where the industrial user, in its self-monitoring report, certifies that the individual grab sample is representative of its daily operation.

Except for these circumstances, Control Authorities should continue to use composite methods for their compliance sampling.

GRAB SAMPLES AS A COMPLIANCE SCREENING TOOL

Control Authorities may consider using grab samples as a compliance screening tool once a body of composite data (e.g., Control Authority and self-monitoring samples obtained over a year's time), shows consistent compliance. However, in the event single grab samples suggest noncompliance, the Control Authority and/or the industrial user should resample using composite techniques on the industrial user's effluent until consistent compliance is again demonstrated.[6]

Control Authorities may also rely on single grab samples, or a series of grab samples for identifying and tracking slug loads/spills since these "single event" violations are not tied to a discharger's performance over time.

Any time an SIU's sample (either grab or composite) shows noncompliance, the General Pretreatment Regulations, at 40 CFR 403.12(g)(2), require that the SIU notify the Control Authority within twenty-four (24) hours of becoming aware of the violation and resample within 30 days. Furthermore, EPA encourages Control Authorities to conduct or require more intensive sampling in order to thoroughly document the extent of the violation(s). Of course, the use of grab samples should be reconsidered in the event the SIU changes its process or treatment.[7]

SUMMARY

The collection and analysis of sampling data is the foundation of EPA's compliance and enforcement programs. In order for these programs to be successful, wastewater samples must be properly collected, preserved and analyzed. Although the Federal standards and self-monitoring requirements are independently enforceable, Control Authorities should specify, in individual control mechanisms for industrial users, the sampling collection techniques to be used by the industry. Generally, pretreatment sampling should be conducted using composite methods wherever possible, to determine compliance with daily, weekly or monthly average limits since this sampling technique most closely reflects the average quality of the wastewater as it is discharged to the publicly owned treatment works. Grab samples should be used to determine compliance with instantaneous limits. There are circumstances when discrete grab samples are also an appropriate, cost effective means of screening compliance with daily, weekly and monthly pretreatment standards.

In summary, there are limited situations in which single grab sample data may be used in lieu of composite data. Assuming adequate quality control measures are observed, analyses of these grab samples can indicate noncompliance with Federal, State and Local Pretreatment Standards and can form the basis of a successful enforcement action. Grab sampling can also be useful in quantifying batches, spills, and slug loads which may have an impact on the publicly owned treatment works, its receiving stream and sludge quality.

Should you have any further comments or questions regarding this matter, please have your staff contact Mark Charles of OWEC at (202) 260-8319, or David Hindin of OE at (202) 260-8547.

cc: Frank M. Covington, NEIC
       Thomas O'Farrell, OST
       Regional and State Pretreatment Coordinators
       Lead Regional Pretreatment Attorneys, Regions I-X
       Approved POTW Pretreatment Programs

---

[6] Where grab samples are used as a screening tool only (i.e., consistent compliance has been demonstrated by composite data), the results should not be used in the POTW's calculation of significant noncompliance (SNC).

[7] When POTWs choose to allow the SIU to collect single grab samples, the POTW should draft the SIU's individual control mechanism to clearly indicate that grab samples are to be obtained thereby preventing any uncertainty at a later date.

APPENDIX C

APPENDIX D

SUBJECT: Using Split Samples to Determine Industrial User Compliance

DATE: April 12, 1993

FROM: Richard G. Kozlowski, Director
              Water Enforcement Division
              U.S. Environmental Protection Agency

TO: Mr. Harold R. Otis
              Chairman, Split Sampling Task Force
              Greater Fort Wayne Chamber of Commerce

In response to your letter of January 12, 1993, and your phone conversation of February 9, 1993, with Lee Okster, I am providing a further discussion of the issues surrounding the use of split samples to determine industrial user (IU) compliance with Pretreatment Standards. In your letter and your phone conversation, you requested clarification from the Environmental Protection Agency (EPA) on three issues. First, you requested a firm definition of what constitutes "widely divergent results" when comparing split sample results. Second, when a publicly owned treatment works (POTW) splits a sample with an IU, you inquired whether a POTW must use the industrial user's data to determine compliance with pretreatment standards. Finally, you requested written authorization from the EPA to incorporate the language from our existing guidance memorandum on split samples into the Rules and Regulations of the Water Control Utility for the City of Fort Wayne.

What Are Widely Divergent Results?

As you are aware, the EPA issued a memorandum on January 21, 1992, entitled "Determining Industrial User Compliance Using Split Samples." The "widely divergent results" criterion established in this memo is to be used as an indication that a problem exists with the laboratory analysis. We did not include an indication of what constitutes "widely divergent" in our memorandum because the amount of "normal" analytical variability depends on the pollutant parameter being tested and the method being used to analyze the sample. With appropriate QA/QC, this "normal" analytical variability is small. In general, though, metals analyses have a smaller variation than organics analyses, but the magnitude of the variability depends on the pollutants being tested. Therefore, no hard and fast rules exist for determining what is widely divergent. This determination is left to the discretion of the local authority.

Must the POTW Use All Sample Results?

In the January, 1992, memorandum we state that "the POTW must use all samples which were obtained through appropriate sampling techniques and analyzed in accordance with the procedures established in 40 CFR Part 136." The memo further states "[w]hen a POTW splits a sample with an IU; the POTW must use the results from each of the split samples."

The POTW is required to sample the IU at least once per year to determine, independent of information supplied by the IU, the compliance status of that facility. If the POTW does not wish to be in a position of comparing its own data with the IU when it samples the IU's discharge, it is not required to split its samples with the IU. Furthermore, we do not recommend that the POTW use a split sample with the industry to satisfy its annual sampling requirement. The POTW should pull its own sample so that it has data which are truly independent of the IU's results.

The POTW also has the primary responsibility to ensure compliance by the IU with all applicable pretreatment standards and requirements. One way the POTW can satisfy its requirement to ensure compliance is to split a routine sample taken by the IU. If a POTW splits a routine sample taken by the IU, it must use the IU's data, in conjunction with its own, to determine the compliance status of the facility (assuming all of the data are sampled and analyzed appropriately). We encourage POTWs to split samples in this manner to verify the IU's data. In a similar fashion, if the POTW chooses to split its own sample with the IU, it must use all of the data to determine the compliance status of the facility (assuming all of the data are appropriately analyzed).

When the POTW splits a sample with an IU (whether it is a routine sample by the IU or an annual sample by the POTW) the POTW has the responsibility to determine whether the IU's results from the split sample are valid. Where an IU's results are different than the POTW's, the burden is on the IU to show that all preservation, chain-of-custody, and analytical and QA/QC methods were followed. If the IU cannot make this showing, then the analytical results from the IU should be discarded when determining the compliance status of the facility. If the IU establishes that it followed all appropriate procedures, then the POTW should review its own QA/QC program. If both the IU and POTW have followed appropriate procedures, and there is still a wide divergence, then follow-up sampling should be conducted. If follow-up sampling consistently shows IU noncompliance, or if the POTW is otherwise satisfied with the validity of its own results, it should proceed to follow its enforcement procedures.

Authorization From the EPA

In regard to your final request, the City of Fort Wayne has the authority to incorporate these procedures into its Rules and Regulations without any authorization from the EPA. As long as the City has the minimum legal authorities to implement its approved program, it has satisfied its requirements under the Federal regulations. As always, the City is encouraged to adopt the EPA's Pretreatment Guidance whenever possible.

I hope this letter responds to your questions and concerns. If you have any further questions, please feel free to call me at (202) 260-8304 or you can call Lee at (202) 260-8329.

SUBJECT: Determining Industrial User Compliance Using Split Samples

DATE: January 21, 1992

FROM: Richard G. Kozlowski, Director
              Enforcement Division

TO: Mary Jo M. Aiello, Acting Chief
              Bureau of Pretreatment and Residuals

This memo is a response to your letter of September 30, 1991, where you requested written clarification regarding the use of split samples for determining industrial user (IU) compliance under the Pretreatment Program. Specifically, you requested guidance on how to use the data from split samples for determining IU compliance in situations where split samples yield different analytical results. The fundamental question posed by your inquiry is whether all analytical results must be used when evaluating the compliance status of IUs and how to use those results for determining compliance. In situations where split samples exist and both samples were properly preserved and analyzed, POTWs should evaluate compliance with applicable Pretreatment Standards in the manner described below.

When evaluating the compliance status of an industrial user, the POTW must use all samples which were obtained through appropriate sampling techniques and analyzed in accordance with the procedures established in 40 CFR Part 136.[1] The Environmental Protection Agency (EPA) has consistently encouraged Publicly Owned Treatment Works (POTWs) to periodically split samples with industrial users as a method of verifying the quality of the monitoring data. When a POTW splits a sample with an IU, the POTW must use the results from each of the split samples.

A legitimate question arises, however, when a properly collected, preserved and analyzed split sample produces two different analytical results (e.g., one which indicates compliance and the other shows noncompliance, or where both indicate either compliance or noncompliance but the magnitudes are substantially different). In these instances, questions arise regarding the compliance status of the IU, and what should be done to reconcile the results.

There is inherent variation in all analytical measurements, and no two measurements of the same analyte (even when drawn from the same sample) will produce identical results. When a split sample is analyzed using appropriate methods, there is no technical basis for choosing one sample result over the other for determining the compliance status of a facility. Since this is the case for all split samples which have been properly analyzed, the POTW should average the results from the split and use the resulting average number when determining the compliance status of an IU. Using the average of the two sample results avoids the untenable situation of demonstrating compliance and noncompliance from the same sample.

If the split sample produces widely divergent results or results which are different over a long period of time, then the cause of the discrepancy between the analytical results should be reconciled. When this happens, the POTW should investigate Quality Assurance and Quality Control (QA/QC) procedures at each laboratory involved. For example, the POTW could submit a spiked sample (i.e., a sample of known concentration) to the laboratories involved (preferably blind) to determine which laboratory may be in error.

In situations where one or both of the analytical results is determined to be invalid, there are compliance and enforcement consequences. If one of the analytical results is determined to be invalid, the average value for that sample is also invalid. In this situation, the value for this sample should be the value of the sample which was not determined to be invalid (e.g., if the IU's results are determined to be invalid, the POTW should use its sample for assessing compliance, and vice versa). If both samples are determined to be invalid, the averaged result from that sample should be discarded and not used for compliance assessment purposes. In either case, the POTW must recalculate the compliance status of the IU using all remaining valid sample results.

In summary, whenever split samples are taken and both are properly preserved and analyzed, the POTW should average the results from each sample and use the averaged value for determining compliance and appropriate enforcement responses. Where the sample results are widely divergent, the POTW should instigate QA/QC measures at each of the analytical laboratories to determine the cause of the discrepancy. If one or both of the samples are invalid, the POTW must recalculate the compliance status of the IU using all valid results.

If you have any further questions regarding these questions, please feel free to call me at (202) 260-8304. The staff person familiar with these issues is Lee Okster. Lee can be reached at (202) 260-8329.

cc: Cynthia Dougherty
         Regional Pretreatment Coordinators
         Approved State Pretreatment Coordinators
         Bill Telliard

---

[1] See Memorandum, *"Application and Use of the Regulatory Definition of Significant Noncompliance for Industrial Users,"* U.S. EPA, September 9, 1991.

APPENDIX E

SUBJECT: Continuous pH Monitoring

DATE: May 13, 1993

FROM: Cynthia C. Dougherty, Director
Permits Division

TO: Mary Jo M. Aiello, Chief
Bureau of Pretreatment and Residuals
Wastewater Facilities Regulation Program (CN 029)
New Jersey Department of Environmental Protection and Energy

Thank you for your letter of January 25, 1993, to Jeffrey Lape of my staff regarding the New Jersey Department of Environmental Protection and Energy's (the Department) proposed policy on waivers from pH limits applicable to industrial discharges to Publicly Owned Treatment Works (POTWs). Subject to the qualifications stated below, your proposed policy is consistent with the federal regulations.

Your letter relates to the application of 40 CFR 401.17, which allows facilities that employ continuous pH monitoring to exceed certain pH limits one percent of the time. Your letter correctly notes that 40 CFR 401.17 applies only to discharges to surface waters, but inquires whether an analogous policy could be applied to discharges to POTWs.

We believe an analogous policy could be applied to discharges to POTWs subject to several restrictions. First, the federal pretreatment regulations contain a specific prohibition against discharges with a pH below 5.0, from which no waivers are allowed unless the treatment works is specifically designed to accommodate such discharges (40 CFR 403.5(b)(2)). Your letter correctly acknowledges that, except for such specifically designed treatment works, waivers below this minimum limit would not be consistent with federal regulations. Second, although federal pretreatment regulations do not include an upper pH limit applicable to all discharges, some categorical pretreatment standards do so. Waivers from the requirements of those categorical standards would not be allowed unless expressly permitted by the standards themselves.

Third, a POTW may not grant a waiver from a local limit if such waiver would cause pass through or interference. Since local limits are based on considerations at each POTW, it would not be appropriate to institute a waiver of local limits that applies statewide regardless of conditions at individual POTWs. So long as POTWs act consistently with their obligations not to allow pass through or interference, however, they might implement waivers that apply either more or less frequently than the 1% you propose. Of course, if it wishes, the State could cap all waivers at 1% and thereby be more stringent than Federal law, which requires no cap.

We note that, if a POTW wishes to provide waivers from pH limits that are technically-based and are part of the POTW's Approved Pretreatment Program, the POTW will have to modify its Approved Pretreatment Program accordingly. The Department should consider for each POTW whether the adoption of this policy is a "change to local limits, which result in less stringent local limits" and therefore requires a formal modification under 40 CFR 403.18(c)(1)(ii), or whether it constitutes a clarification of the POTW's existing local limits.

I hope that this response addresses your concerns. If you have any questions or would like to discuss this further, please call me at (202) 260-5850 or Louis Eby at (202) 260-2991.

APPENDIX F

CHAIN OF CUSTODY PROCEDURES
(Sanitation Districts of Los Angeles County)

INTRODUCTION

Custody records used at the Districts have discrete formats and contents tailored to suit the needs of different departments or agencies. All these forms, however, contain the common elements which are essential for ensuring sample integrity from collection through data reporting. This document will outline the steps which are necessary to ensure that the possession and handling of samples are traceable from the time of collection through analysis and final disposition. The process needed for generating and maintaining the required documentation, which can be used to trace the possession of a sample from the moment of collection through its introduction into evidence, will be described.

There are two central sample receiving stations at the Districts, one is located at the San Jose Creek Water Quality Laboratory and the other at the Joint Water Pollution Control Plant Water Quality Laboratory. These central receiving stations use a limited number of trained personnel for sample handling, storage and disposal. This arrangement serves to streamline and expedite the sample transfer process between the lab and sample collectors and facilitates the maintenance of sample custody and records.

The sample collector is responsible for the care and custody of the samples until they are properly dispatched or transferred to one of the receiving stations. The sample collector must ensure that each container is in his physical possession or in his view at all times, or stored in a locked area where no one can tamper with it. If evidence tape is used, the collector should place it securely on jars in such a manner that it is necessary to break the seal to open the container. Care should be taken to avoid damage to the seals when installing.

1. A sample is in custody if it is in any one of the following states:

    1.1   In a person's actual physical possession

    1.2   In view of the person after being in physical possession

    1.3   In physical possession and locked up so that no one can tamper with it (e.g., in-transit locked in vehicle)

    1.4   In a secured area restricted to authorized personnel

2. A person who has sample custody must comply with the following procedures.

    2.1   Label all samples to prevent misidentification. Gummed labels or tape is adequate and should include, at a minimum, the following information:

        2.1.1   Sample field identification number

        2.1.2   Name or initials of the collector

        2.1.3   Date and time of collection

        2.1.4   Place of collection

    NOTE: Labels should be affixed to the sample container prior to, or at the time of sampling and should be filled out <u>at the time of collection.</u>

3. All information pertinent to field sampling activities such as pH, temperature, and other measurable parameters must be recorded in a logbook. This should be a bound volume with consecutively numbered pages. At a minimum, entries in the logbook should contain the following information:

    3.1   Date and time of collection

    3.2   Purpose of sampling

    3.3   Collector's sample identification number

    3.4   Location of sampling point

    3.5   Name and address of field contact

    3.6   Producer of sample and address, if different from location

    3.7   Type of process producing sample (if known)

    3.8   Type of samples (e.g., sludge, wastewater)

    3.9   Qualitative observation of samples (e.g., color, smell)

    3.10   Number and volume of samples taken

    3.11   Description of sampling point and sampling methodology

    3.12   Sample distribution (e.g., splits, who received and how much)

Sampling conditions in the field vary widely. Rules therefore cannot be made to cover all eventualities. Sufficient information should be recorded so that anyone can reconstruct the sampling event without reliance on the collector's memory. The logbook must be stored safely. <u>Never use "white out" in logbook</u>.

4. Establishing the documentation necessary to track a sample from the time of its collection requires that a chain of custody record be maintained and accompany every sample. This sheet should contain, at a minimum, the following information:

    4.1   Sample number

    4.2   Signature of collector

    4.3   Date and time of collection

    4.4   Place and address of collection

    4.5   Sample type

    4.6   Signature of persons involved in the chain of possession

    4.7   Inclusive dates of possession.

Protect custody records from accidental spills, chemicals, etc. <u>Never use "white out" on custody documents</u>. Correct errors by drawing a single line through mistake, then write in space above or below correction line.

5. Laboratory Custody Procedures

    5.1   The laboratory must designate sample custodian(s) and alternates to act in their absence. In addition the laboratory must set aside secured storage facilities. These areas should be clean and isolated, with sufficient refrigerated space which can be locked from the outside.

    5.2   Incoming samples should be received only by the custodian who will indicate receipt by signing the chain of custody record sheet accompanying the samples and then retain the sheet as a permanent record. Couriers picking up samples shall sign jointly with the laboratory custodian.

    5.3   Immediately after receipt, the custodian should complete the log-in process and place the sample(s) in a refrigerated cooler which should remain locked, except when samples are removed or replaced by the custodians. Samples should be handled by the minimum number of persons.

    5.4   Only the custodian(s) should distribute samples to the analyst. Signatures are not required for transfers within the laboratory as long as the laboratory is maintained as a secured facility with access restricted to authorized personnel.

    5.5   The analyst is responsible for the care and custody of the samples after receipt, and should be prepared to testify that the sample was in their possession and view, or within their laboratory area at all times from the moment of receipt until returned to the custodian. Analyst should keep notes on these matters with bench sheets or in lab notebooks for all evidence samples.

    5.6   Once analysis is completed, the unused portion of the sample together with identifying labels and other documentation must be returned to the custodian. The returned, tagged samples should be retained and stored in a secured locker.

    5.7   Samples should be disposed of only after permission is received (preferably in writing) by the custodian from the laboratory director, from previously designated enforcement officials, or when it can be determined that the information is no longer required, or that the samples have deteriorated beyond use.

    These disposal procedures do not hold for tags and other lab records which are archived for an indefinite period.

6. References

    6.1   <u>Standard Methods for the Examination of Water and Wastewater</u>, 17th Ed., 1-34 (1989)

    6.2   <u>Test Methods for Evaluating Solid Wastes, SW-846</u>, Vol. II, Field Manual, Section 9.2.2.7, 3rd Ed., Sept., Nine-64 (1986)

    6.3   <u>EPA Handbook for Analytical Quality Control in Water and Wastewater Laboratories</u>, EPA-600/4-79-019, 12-4 to 12-10 (1979)

CHAPTER 7

WASTEWATER FLOW MONITORING

by

Lory Rising

Rob Wienke

Paul Martyn

TABLE OF CONTENTS
Chapter 7.   WASTEWATER FLOW MONITORING

OBJECTIVES

Chapter 7.   WASTEWATER FLOW MONITORING

Following completion of Chapter 7, you should be able to:

1. List the reasons for measuring flows,

2. Identify the conditions for measuring flows in open channels,

3. Describe the various types of open-channel flowmeters,

4. Select the appropriate instrumentation for open-channel flow measurements,

5. Determine the accuracy of open-channel flowmeters,

6. Identify the various types of closed-pipe flow metering systems,

7. Describe flow-proportioned pacing of automatic samplers, and

8. List approximate measurement methods.

PRETREATMENT WORDS

Chapter 7. WASTEWATER FLOW MONITORING

ALIQUOT (AL-li-kwot) ALIQUOT

Portion of a sample. Often an equally divided portion of a sample.

ANALOG ANALOG

The readout of an instrument by a pointer (or other indicating means) against a dial or scale. Also the continuously variable signal type sent to an analog instrument (for example, 4–20 mA).

ANALOG READOUT ANALOG READOUT

The readout of an instrument by a pointer (or other indicating means) against a dial or scale.

DIELECTRIC (DIE-ee-LECK-trick) DIELECTRIC

Does not conduct an electric current. An insulator or nonconducting substance.

DIGITAL READOUT DIGITAL READOUT

The use of numbers to indicate the value or measurement of a variable. The readout of an instrument by a direct, numerical reading of the measured value. The signal sent to such readouts is usually an ANALOG signal.

ENTRAIN ENTRAIN

To trap bubbles in water either mechanically through turbulence or chemically through a reaction.

FRICTION LOSSES FRICTION LOSSES

The head, pressure or energy (they are the same) lost by water flowing in a pipe or channel as a result of turbulence caused by the velocity of the flowing water and the roughness of the pipe, channel walls, and restrictions caused by fittings. Water flowing in a pipe loses pressure or energy as a result of friction losses. Also see HEAD LOSS.

HEAD HEAD

The vertical distance (in feet) equal to the pressure (in psi) at a specific point. The pressure head is equal to the pressure in psi times 2.31 ft/psi.

HEAD LOSS HEAD LOSS

The head, pressure or energy (they are the same) lost by water flowing in a pipe or channel as a result of turbulence caused by the velocity of the flowing water and the roughness of the pipe, channel walls, or restrictions caused by fittings. Water flowing in a pipe loses head, pressure or energy as a result of friction losses. Also see FRICTION LOSSES.

HYDRAULIC JUMP HYDRAULIC JUMP

The sudden and usually turbulent abrupt rise in water surface in an open channel when water flowing at high velocity is suddenly retarded to a slow velocity.

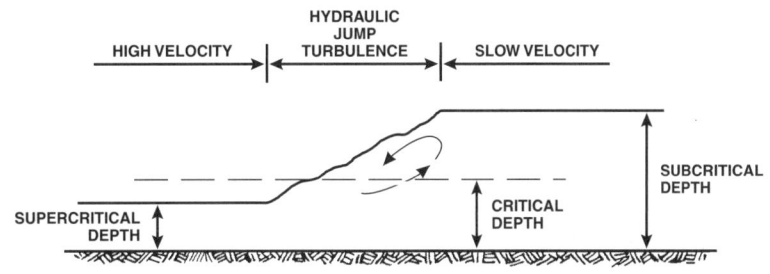

INTEGRATOR   INTEGRATOR

A device or meter that continuously measures and calculates (adds) a process rate variable in cumulative fashion; for example, total flows displayed in gallons, million gallons, cubic feet, or some other unit of volume measurement. Also called a TOTALIZER.

NAPPE (NAP)   NAPPE

The sheet or curtain of water flowing over a weir or dam. When the water freely flows over any structure, it has a well-defined upper and lower water surface.

PICKLE   PICKLE

An acid or chemical solution in which metal objects or workpieces are dipped to remove oxide scale or other adhering substances.

PITOT (pea-TOE) TUBE   PITOT TUBE

An instrument used to measure fluid (liquid or air) velocity by means of the differential pressure between the tip (dynamic) and side (static) openings.

PRIMARY ELEMENT   PRIMARY ELEMENT

The hydraulic structure used to measure flows. In open channels, weirs and flumes are primary elements or devices. Venturi meters and orifice plates are the primary elements in pipes or pressure conduits.

RECEIVER   RECEIVER

A device that indicates the result of a measurement. Most receivers use either a fixed scale and movable indicator (pointer) such as a pressure gage, or a moving chart with a movable pen like those used on a circular flow-recording chart. Also called an indicator.

RECORDER   RECORDER

A device that creates a permanent record, on a paper chart or magnetic tape, of the changes in a measured variable.

REPRESENTATIVE SAMPLE   REPRESENTATIVE SAMPLE

A sample portion of material or wastestream that is as nearly identical in content and consistency as possible to that in the larger body of material or wastestream being sampled.

SECONDARY ELEMENT   SECONDARY ELEMENT

The secondary measuring device or flowmeter used with a primary measuring device (element) to measure the rate of liquid flow. In open channels bubblers and floats are secondary elements. Differential pressure measuring devices are the secondary elements in pipes or pressure conduits. The purpose of the secondary measuring device is to (1) measure the liquid level in open channels or the differential pressure in pipes, and (2) convert this measurement into an appropriate flow rate according to the known liquid level or differential pressure and flow rate relationship of the primary measuring device. This flow rate may be integrated (added up) to obtain a totalized volume, transmitted to a recording device, and/or used to pace an automatic sampler.

SENSOR   SENSOR

A device that measures (senses) a physical condition or variable of interest. Floats and thermocouples are examples of sensors.

STILLING WELL   STILLING WELL

A well or chamber which is connected to the main flow channel by a small inlet. Waves and surges in the main flow stream will not appear in the well due to the small-diameter inlet. The liquid surface in the well will be quiet, but will follow all of the steady fluctuations of the open channel. The liquid level in the well is measured to determine the flow in the main channel.

TIMER   TIMER

A device for automatically starting or stopping a machine or other device at a given time.

TOTALIZER   TOTALIZER

A device or meter that continuously measures and calculates (adds) a process rate variable in cumulative fashion; for example, total flows displayed in gallons, million gallons, cubic feet, or some other unit of volume measurement. Also called an INTEGRATOR.

TRANSDUCER (trans-DUE-sir)   TRANSDUCER

A device that senses some varying condition measured by a primary sensor and converts it to an electrical or other signal for transmission to some other device (a receiver) for processing or decision making.

CHAPTER 7.   WASTEWATER FLOW MONITORING

7.0   NEED FOR FLOW MONITORING

Of all the things an industrial waste inspector must know or be able to do, the accurate use of appropriate flow measuring devices ranks high on the list of essential skills. Nearly every facet of your job involves flow measuring devices in one way or another. For example, to determine whether or not an industrial discharger is complying with applicable discharge limits, you must know what volume of waste is being discharged. To calculate the discharge of some pollutants, you need to know the rate or volume of flow over a period of time. To assess fair and equitable sewer-use fees, you need to know the volume of wastes each discharger generates. To effectively regulate dischargers, your disciplinary actions must be based on accurate, legally defensible measurements. And, you can only collect representative composite samples if you can accurately calculate flows in relation to time intervals. All of these functions rely heavily on the use of flow measuring devices.

Major industrial waste dischargers contributing to municipal wastewater treatment systems are often required to install effluent flow metering devices on their discharge lines to determine their wastewater contribution to the POTW system. This procedure allows for use charges to be properly assessed. In some cases, however, the POTW agency may wish to assume this responsibility in order to allow for a greater control of these flow metering devices. In any event, sewer-use charges for major contributors should accurately reflect their impact on wastewater treatment and conveyance costs. The best way to determine this impact is to directly measure each industry's discharge flows before the wastewater enters the POTW sewer.

Waste discharge requirements for most pollutants (with the most notable exceptions being pH and temperature) can either be issued on a mass or concentration basis. Concentration-based waste limits simply require that the amount of pollutants in a discharge not exceed a certain mass per unit volume, such as in a cadmium limit of 0.11 milligrams per liter (mg/L); no attempt is made to determine the particular quantity of the pollutant discharged. However, in a mass-based limit, the measured concentration of a wastewater (generally from a composite sample taken over a period of time) is applied against a corresponding measured flow rate so that a specific quantity of the pollutant in question can be limited. This procedure produces a calculated value. Without meaningful flow rate data, mass-based limits cannot be applied.

For industries with fluctuating or unusual waste discharge flow rates, composite samples taken over time using flow-proportioned techniques are most representative. In such instances, discrete time-interval samples, collected either manually or by automatic equipment, are weighted according to the prevailing flow intervals to create a single sample representative of the discharge conditions. If the flow rate during time interval A averaged twice that measured during time interval B, the amount of sample A in the flow-weighted composite would be twice that of sample B. However, the preferred way of collecting flow-weighted composite samples involves connecting an automatic wastewater sampler to a flow metering device so that the individual discrete samples in the wastewater sample are precisely and uniformly taken at predetermined, fixed flow intervals, such as one sample every 10,000 gallons. In either instance, a properly operating flowmeter is required for flow-weighted samples to be taken.

Although this chapter is primarily concerned with the procedures and concepts involved with direct effluent flow measurements, inspectors should be aware that adjusted influent flow measurements may be used in selected instances to determine discharge flow rates. In these situations, all influent sources of water, such as flow provided by the water purveyor, well water and water supplied in the industry's raw materials are considered. Evaporative losses, irrigation losses, and losses to product are subtracted from these sources to yield an estimated discharge. It should be recognized that this method is imprecise for short-term measurements due to daily variations in industrial water usage and the ability of certain dischargers to retain a significant amount of water in process tanks and storage basins. Moreover, even the long-term measurements may be inaccurate if the incoming sources of water cannot be accurately determined. Unless influent water meters have been recently calibrated (within the last four years for major meters), or newly installed, the recorded flow rates should not be accepted without question. With age, influent water meters typically underrecord actual flow rates and if such readings form the basis of the flow determination, a significant amount of the discharge will not be accounted for. However, most flowmeters are usually accurate between plus or minus one to four percent.

7.1   BASICS OF FLOW METERING

Flow can be directly measured in many different ways. The methods fall into two broad categories, open-channel flow and closed-pipe (pressure conduit) flow. In open-channel flow the water has a free surface; in closed-pipe flow the water completely fills the pipe. This chapter will focus on some commonly used systems from each of the two categories. Once you become familiar with the more common flow metering systems, it is easy to understand the operation of most other systems.

Flow is the amount of water going past a particular reference point over a certain period of time. It is measured in units of volume per unit time; typical units are cubic feet per second (CFS), gallons per minute (GPM), and million gallons per day (MGD). Flow is usually calculated by the equation:

Flow = Velocity x Cross-Sectional Area
Q = VA

For instance, Q (cubic feet per second) equals V (feet per second) times A (square feet) for the liquid area in the pipe or channel. From this basic equation you can see that the flow can be calculated by measuring both the average velocity and the area perpendicular to the moving water. Some types of flow metering systems determine the flow by directly measuring these quantities; other types of systems measure a depth or a pressure that is related to the flow.

In discussing flow metering systems, the terms "primary element" and "secondary element" are commonly used. The term "primary element" will be used to refer to the measuring structure that contains the water, and the term "secondary element" refers to the means by which measurements are taken from the primary element and converted to flow readings. The primary element may be a flume or weir (Figure 7.1). The secondary element may be a simple staff gage or automatic instrumentation which measures the depth of flow in the primary element.

In an open-channel flow metering system, the primary element may be a Palmer-Bowlus flume (Figure 7.2), a Parshall flume (Figure 7.3), or a V-notch weir and a secondary element senses the depth and converts it to flow. The instrumentation is known by the means by which the depth is sensed. Common types of instrumentation are ultrasonic transducers, floats, bubblers, capacitance probes, and pressure sensors.

The primary element of a closed-pipe flow metering system is a device installed in the pipe to affect flow conditions in such a way as to allow the flow to be measured. Meters (secondary elements) that employ a device within the pipe to create pressure conditions that are related to flow are known by the device that is installed in the pipe. These are Venturi meters, orifice plate meters, and flow nozzles and flow tubes.

The types of meters that are installed without a device in the closed-pipe system are known by the way the velocity of the flow is measured. Common meters of this type are electromagnetic meters, turbine meters, ultrasonic meters and meters that operate on the pitot (pronounced pea-TOE) tube principle.

The instrumentation of an automatic flowmeter usually contains three integrated devices to obtain data for describing the flow. A totalizer is used to continuously sum the flow (flow integrator) over an extended period; it is essentially like an odometer on a vehicle. Totalizers typically have a digital display indicating flow in tens, hundreds or thousands of gallons. A second device, the flow indicator, shows the instantaneous flow rate. These indicators may have either an *ANALOG*[1] or *DIGITAL*[2] *READOUT*. Some instrumentation may also incorporate a third device which produces a printed and graphed

display to record the instantaneous flow measurements. Graphical displays (flow charts) are desirable when measuring industrial effluent flows because they can reveal times and magnitudes of batch discharges or dumps. The instrumentation may also be equipped to pace an automatic sampler, operate an external recorder, or trigger alarms.

This chapter is organized to present the information you need to properly inspect existing flow metering systems monitoring industrial dischargers. It is not intended as a guide for selecting or designing flow metering systems since these tasks are rarely performed by pretreatment inspectors. The selection and design of flow metering systems should be performed by qualified engineers; however, an understanding of the principles on which flow metering systems are based, combined with some practical knowledge of them, will enable you to conduct thorough evaluations of the metering systems.

In many situations it is necessary to install temporary flow metering systems using portable equipment to obtain short-term flow measurements at industrial dischargers' facilities. Even in these circumstances, qualified engineers should select and design all flow metering systems.

7.2   OPEN-CHANNEL FLOW MEASUREMENT

The purpose of the open-channel primary element is to create flow conditions that produce a known relationship between flow and depth. When these conditions are created, the channel width is known and the velocity of the water does not need to be measured. Instead, the secondary element senses the depth at the established measurement point and converts it to flow based on the known relationship between depth and flow.

The accuracy of an open-channel flow metering system, properly installed and maintained, is usually less than five percent error. Instrumentation is available to detect depth changes of 0.001 foot, although some error in converting from depth to flow occurs. This error is about one to two percent of full scale. Problems with the primary element or with the instrumentation will cause inaccuracies that are much greater than the inherent inaccuracy of a properly operating flow metering system.

To verify the accuracy of an open-channel flow measurement system, check both the primary element and the instrumentation. Check the primary element by visually observing the flow through it. Certain flow characteristics (described below) will be evident when the primary element is developing the correct relationship between depth and flow. To check the instrumentation, compare its depth measurements with independent depth readings; convert depth measurements to flows and compare calculated flows with indicated and recorded flows; observe the operation of indicators and recorders; and check timing of the *TOTALIZER*[3] by comparing the rate at which it advances with your calculated flow rates.

All open-channel primary elements create readily observed flow profile characteristics by manipulating the channel slope and size. The flow is constricted and made to drop through a steep and precisely dimensioned section (the primary element) before flow through the regular channel is resumed. A

---

[1] *Analog Readout.   The readout of an instrument by a pointer (or other indicating means) against a dial or scale.*
[2] *Digital Readout.   The use of numbers to indicate the value or measurement of a variable. The readout of an instrument by a direct, numerical reading of the measured value. The signal sent to such readouts is usually an analog signal.*
[3] *Totalizer.   A device or meter that continuously measures and calculates (adds) a process rate variable in cumulative fashion; for example, total flows displayed in gallons, million gallons, cubic feet, or some other unit of volume measurement. Also called an integrator.*

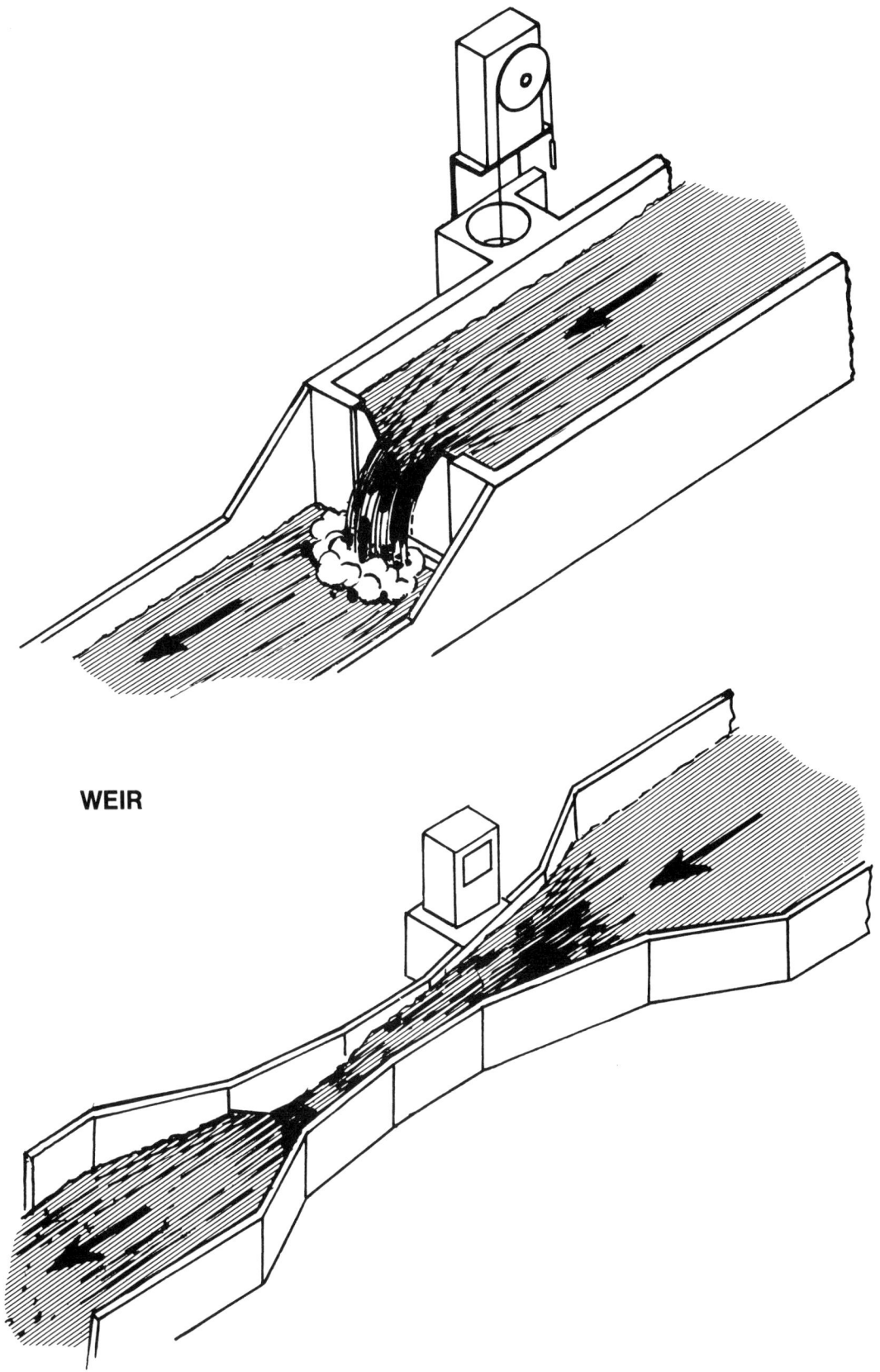

WEIR

FLUME

Fig. 7.1   Weirs and flumes are primary flow measuring devices (hydraulic structures)

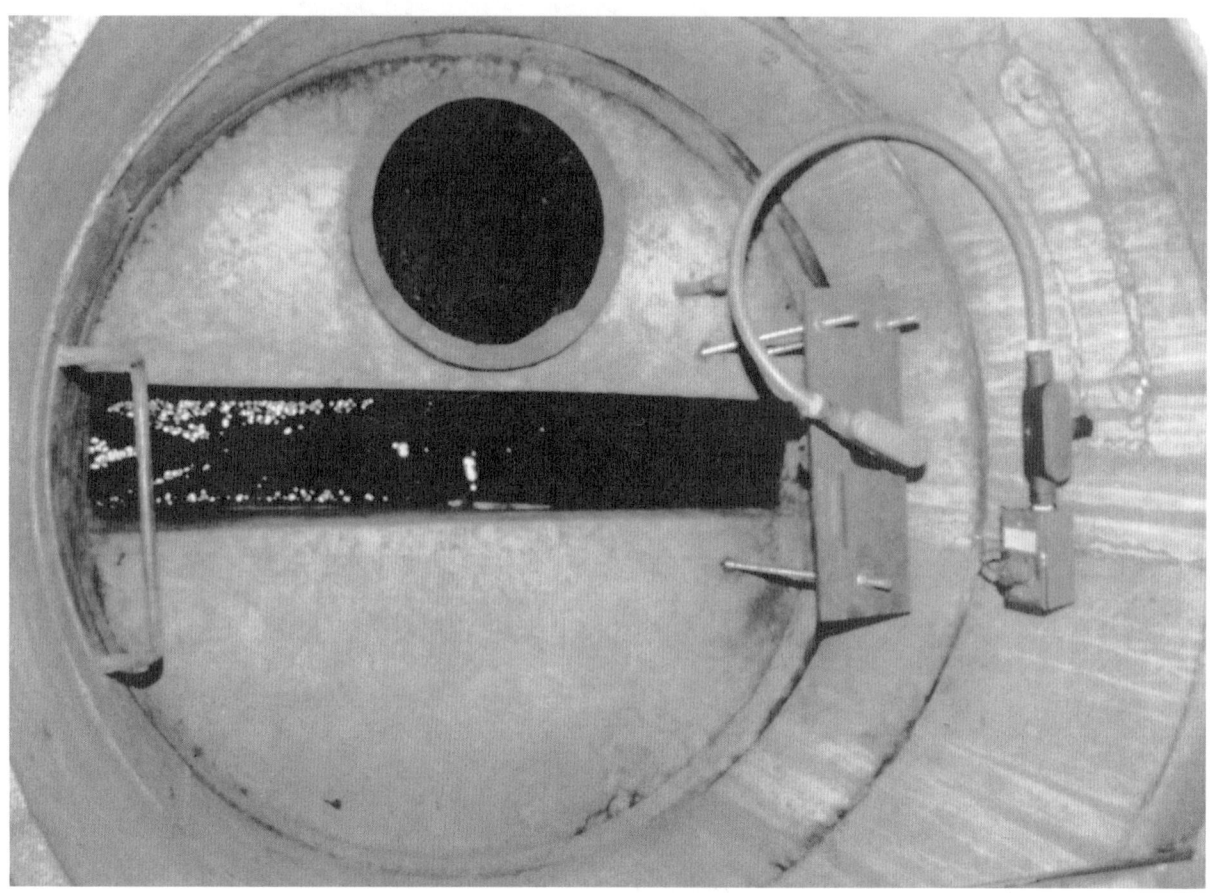

*Fig. 7.2   Palmer-Bowlus flume in manhole with ultrasonic measuring device.
The circular stilling well can be used to measure depth of the liquid by installing a
float sensor or bubbler tube. The stilling well will "dampen" flow surges.*

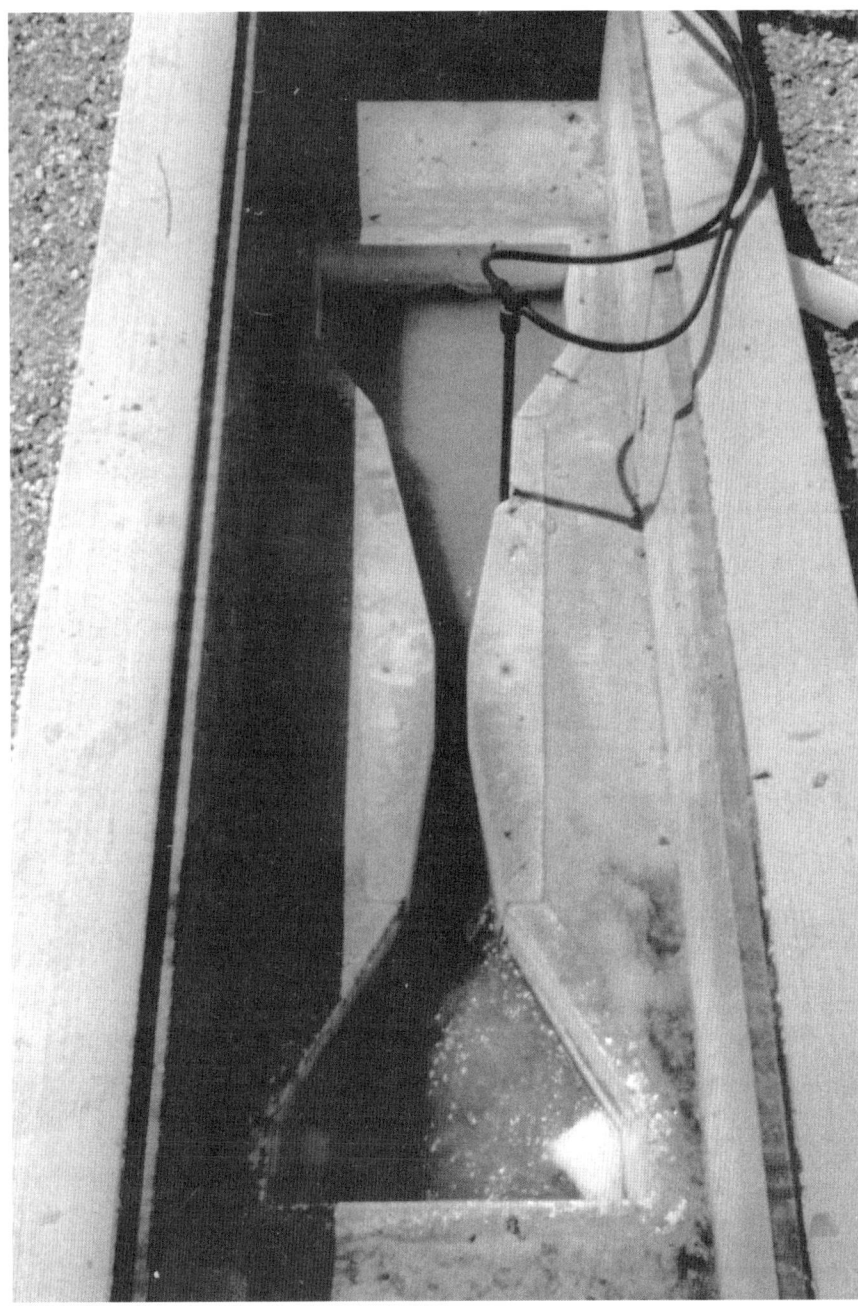

Fig. 7.3   *Small Parshall flume (3-inch throat) with bubbler in shallow installation.*
Flow moves toward bottom of picture.

known, repeatable relationship between depth and flow results in open-channel flumes.

The characteristics of a properly operating flume are distinct and easy to observe. Starting some distance upstream of the flume (primary element), the water will be relatively deep and slow moving. In comparison, as it passes through the throat of the flume it will become much shallower and faster (Figure 7.4). Downstream from the throat, the water will return to its deeper and slower moving condition. The flow is called subcritical in the upstream reach (slow and deep), supercritical where it is moving shallower and faster, and becomes subcritical again when it returns to its deeper and slower state. A hydraulic drop occurs as the flow changes from subcritical to supercritical. (The drop is more gradual in flumes than in weirs.) When the flow changes back to its original condition (from supercritical to subcritical), a hydraulic jump again occurs. The jump will usually be fairly sudden. Variations of the typical flow profile will be observed, but in all cases the approach flow must be subcritical and the change from subcritical to supercritical must be clearly evident.

A given flow can be carried in a subcritical state through a useful range of mild approach slopes and, conversely, a particular channel with a mild slope can accommodate a range of subcritical flows. Open-channel flumes and weirs (primary elements) are designed so that the change from the subcritical state to the supercritical state occurs over a wide range of flows. Because of this change, it is possible to measure the flow just by determining the depth at a specified measurement point. The flume or weir causes this change of state to occur. It also obstructs the flow by constricting it or blocking it, and usually by doing both. This causes the water to back up behind the flume or weir and build up enough energy or $HEAD^4$ to flow through it. The measuring point is located in the section where the water backs up. The depth at this point is reasonably independent of the upstream channel conditions as long as the approach flow is subcritical. The flume or weir will develop the same depth-to-flow relationship in channels of varying subcritical slopes, which enables it to be used in many different locations. The slope of the sewer line is considered during the design of the system to ensure that subcritical flow occurs.

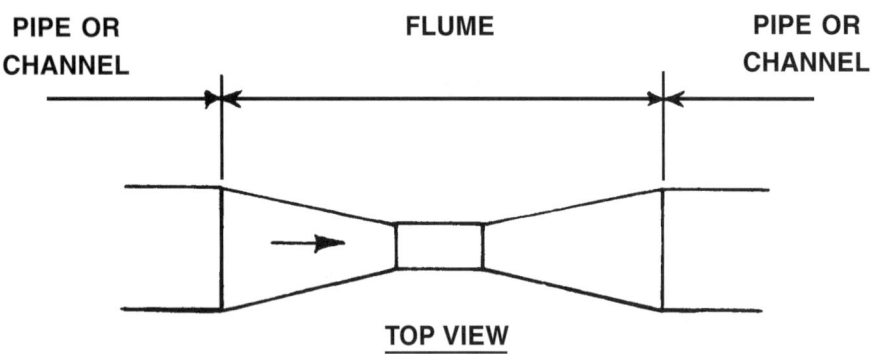

TOP VIEW

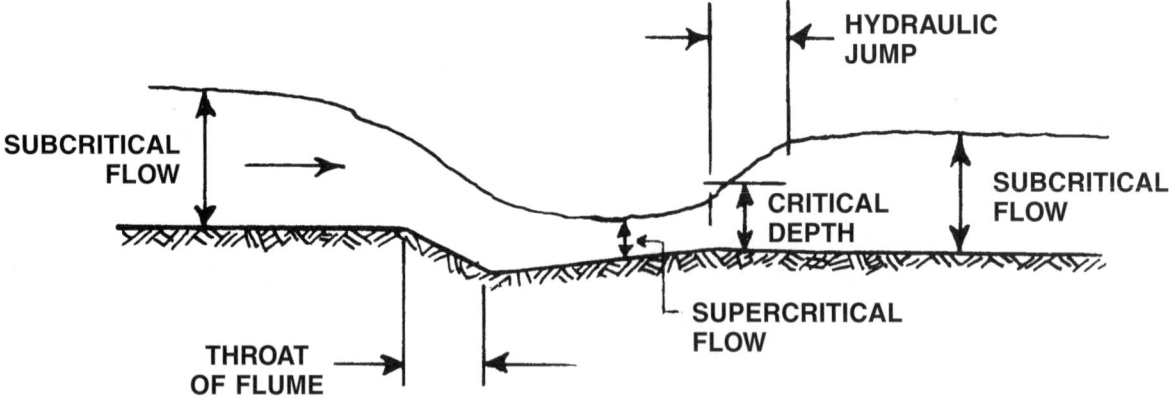

SIDE VIEW

Fig. 7.4   Flow through a flume

---

[4] Head.   The vertical distance (in feet) equal to the pressure (in psi) at a specific point. The pressure head is equal to the pressure in psi times 2.31 ft/psi.

The depth of the water in the flume or weir is measured at a particular location in the channel. The depth-to-flow relationship is only accurate at the measuring point. The depth can be measured directly from the throat or it can be measured at a stilling well (Figure 7.5). A stilling well is typically a small, circular well connected to the throat or to an upstream measuring point of the flume or weir, generally through a small-diameter pipe. The stilling well provides a calm pooling area where the depth can be more accurately measured. Because the stilling well connects at the measuring point of the flume or weir, the water level in the stilling well is the same as in the flume or weir at the measuring point. The stilling well should connect only at the measuring point for the device being used.

In some installations, another connection is made to a point downstream of the flume or weir in order to be able to flush out the stilling well. This connection must be valved closed except during cleaning of the stilling well when the flow reading from the instrument will not be used. The advantage of a stilling well is that it is not affected by wave action in a flume or weir. Also a stilling well eliminates the effects of foam in the wastewater discharge for the users of ultrasonic depth sensors and also the effects of rags for the users of sensors fouled by rags. However, frequent maintenance is often necessary to keep both the stilling well and the connection to the flume or weir clean.

A flume or weir that is properly sized, constructed, installed, and maintained will generate the correct relationship between depth and flow. Unfortunately, they are not always installed properly, are not always maintained properly, do not last forever, and occasionally the flows for which they are sized change. Changing conditions in sewer lines downstream from the flow metering system can also affect the accuracy of the flume or weir by backing up and flooding the flume or weir and producing erroneously high flow readings during flooded conditions. It is usually easy to identify a problem that would cause an incorrect relationship between depth and flow to occur. Some problems are more difficult to detect but it is often obvious when a more thorough evaluation is needed.

The flume or weir generates a predictable relationship between depth and flow over a certain range of flows. At very low flows and very high flows beyond the range for every primary element, the established relationship may not be accurate. When a system is designed, the flume or weir is sized to avoid flows in those ranges. A small amount of inaccurately measured low flow is usually not a problem. If the flow is generally within the low flow range of the device and is not likely to increase, a smaller flume liner or weir should be installed. In general, a smaller flume or weir can be inset into a larger one of the same type.

All flumes and weirs have a certain capacity which cannot be exceeded if the flow is to be measured accurately. If excessive flows occur frequently, a larger capacity flume or weir could be installed or a parallel flume or weir could be built. In these situations it is usually necessary to construct a new vault or manhole and to install larger piping or parallel piping. A facility that exceeds the flume's or weir's capacity with short-duration high peak flows may prefer to control their peak flows rather than install a new flume or weir.

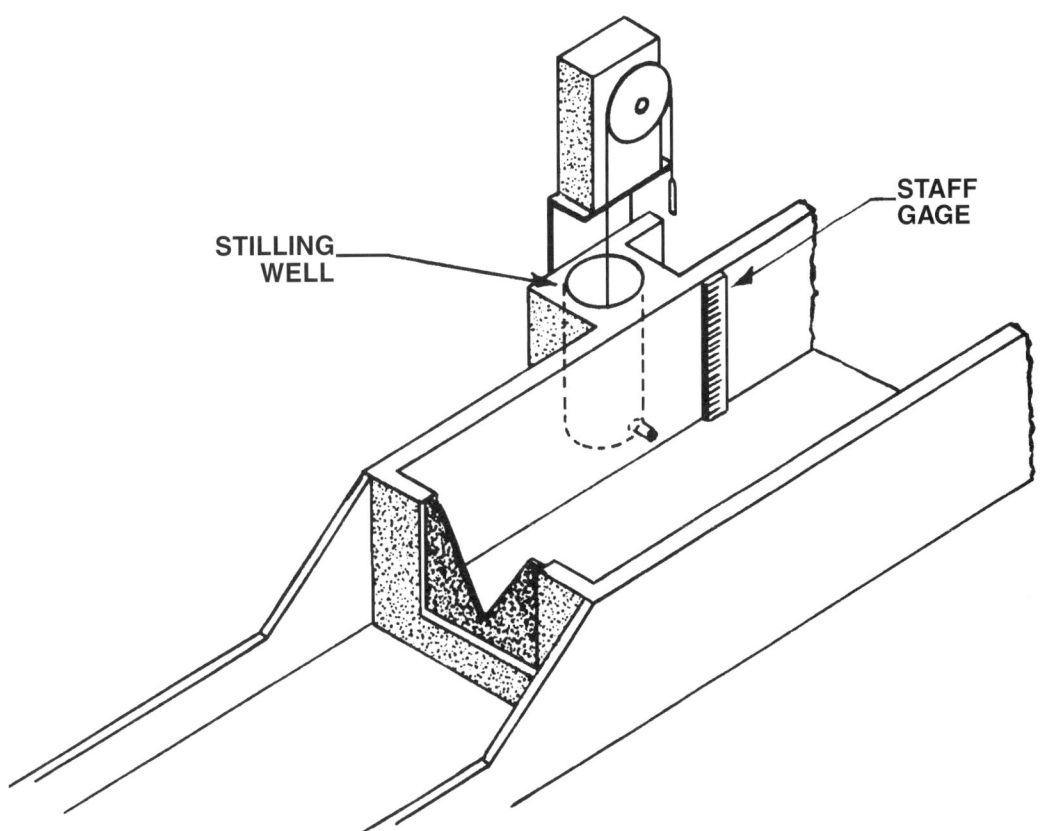

STILLING WELL

STAFF GAGE

Fig. 7.5 *Stilling well for measuring head with a weir. Also may be used to measure head with a flume.*

The flume or weir and related upstream and downstream piping must be properly installed in order for the correct depth-to-flow relationship to occur. Flumes such as the Parshall and Palmer-Bowlus types must be installed level and to specified tolerances. The slope of upstream and downstream pipes must be flatter than critical slope so flow will be subcritical. (Acceptable slopes may be difficult to find in old sewers.) Insert flumes should be securely fixed in place. When these inserts are set in grout, take care not to warp the flumes during installation. Make sure all joints are smooth so that waves will not be created. Install weirs with the weir plate in a vertical position and with the weir notch symmetrical about a vertical line. Provide proper support for the weir plate so that it does not warp.

You will need to clean flumes and weirs regularly to maintain their proper dimensions. Flumes are described as self-cleaning, but that does not mean that they do not need occasional maintenance. Use a brush, high-pressure water hose, or other similar device to remove accumulated solids. Some types of wastewater, especially oily wastewater, can rapidly coat a flume and cause significant errors to occur. Oils and greases can coat the pipe from the throat to the stilling well, thus hindering flow and creating inaccurate flow readings. Cans and rags sometimes lodge in narrow flumes. Excessive debris in the approach channel may create waves and may even cause an improper water surface profile to occur. These are often clues that the flume needs to be cleaned.

Some debris can accumulate ahead of flumes and weirs without affecting their accuracy but excessive debris in a weir increases the approach velocity of the flow, which affects the weir's accuracy. The accuracy of a weir is also sensitive to rounding of the edges of the notch. Keep the edges clean and sharp and keep the upstream face of the weir plate clean. For dependable, long-term service, each pretreatment facility should be encouraged to develop a regular schedule of cleaning and maintaining the flume or weir of their flow metering system.

QUESTIONS

Write your answers in a notebook and then compare your answers with those on page 524.

7.0A   Why must industrial waste inspectors understand wastewater flow measurement procedures?

7.0B   Where may flows be measured easily in a system?

7.1A   What are the two broad categories of methods for directly measuring flow?

7.1B   Define the terms "primary element" and "secondary element."

7.2A   What is the purpose of an open-channel flume or weir?

7.2B   How can the accuracy of an open-channel flow measurement system be checked?

7.2C   What installation precautions must be observed when placing weirs in sewers?

7.3   OPEN-CHANNEL FLOWMETERS

7.30   Palmer-Bowlus Flumes (Figures 7.6 and 7.7)

This type of flume is designed to be installed and kept level in an existing channel provided that the sewer line is on an acceptable slope and the flows do not exceed the flume's capacity. The flume is sized by the dimensions of the channel. For instance, a six-inch Palmer-Bowlus flume is used in a six-inch channel. Smaller Palmer-Bowlus flumes of the "quick-insert" type (Figure 7.8) are often used due to the ease with which their inflatable collar is inserted into the exit section of a pipe.

When installed, a Palmer-Bowlus flume is preceded by a section of straight pipe (about 25 pipe diameters long) laid on an acceptable (subcritical) slope. As shown in Figure 7.9, the point of measurement for a Palmer-Bowlus flume is located at a distance D/2 upstream from the toe of the flume, where D is the size of the flume.

The depth-to-flow relationships for Palmer-Bowlus flumes are available in tabular form. The depth, H, *IS THE VERTICAL DISTANCE BETWEEN THE FLOOR OF THE FLUME AND THE WATER SURFACE AT THE MEASURING POINT. IT IS NOT THE DISTANCE BETWEEN THE CHANNEL BOTTOM AND THE WATER SURFACE AT THE MEASURING POINT.* The distance from the channel bottom to the floor of the flume is approximately D/6. This dimension can vary considerably due to the way the flume is installed, or due to corrosion or deposition.

Subcritical flow should be observed upstream of the flume with the hydraulic drop starting to be noticeable just downstream of the measuring point (see Figure 7.4, page 488). The water should drop more noticeably with supercritical flow obvious around the downstream portion of the flume. The water surface will often show a "V" section formed by standing waves as the water enters the flume. The hydraulic jump also often has a "V" shape to it. At flumes installed in flatter sewer lines, the supercritical section tends to be less evident and to be located farther downstream than average. On steeper lines it will be more pronounced. A hydraulic jump that occurs upstream of the flume may be an indication that the upstream piping was laid at too steep a slope or that accumulated debris needs to be removed.

In some cases, the change from subcritical to supercritical flow will be evident but the hydraulic jump will not be visible. That is perfectly acceptable. The jump may occur farther downstream in the discharge pipe. A steeply sloped discharge pipe may carry the supercritical flow a considerable distance.

If the hydraulic jump seems to occur within the flume itself, or if the supercritical section does not seem to exist, the flume may be operating in a submerged condition. If the submergence is too great, the flume will no longer be accurate, as measured by a single measurement. A submerged condition can occur when the discharge pipe is not able to carry the flow. This can happen because of an improper slope of the pipe, debris in the pipe, or from flow conditions in a sewer line farther downstream that cause a backup of water into the flume. Any of these unusual conditions should be promptly investigated.

Fig. 7.6   Palmer-Bowlus flume in a sampling manhole. Flow direction is toward bottom of page.

Fig. 7.7   Palmer-Bowlus flume in specially built circular metering station. Water depth is measured with ultrasonic unit. Discharger is a galvanizer that pickles steel prior to zinc coating. Flow direction is toward bottom of page.

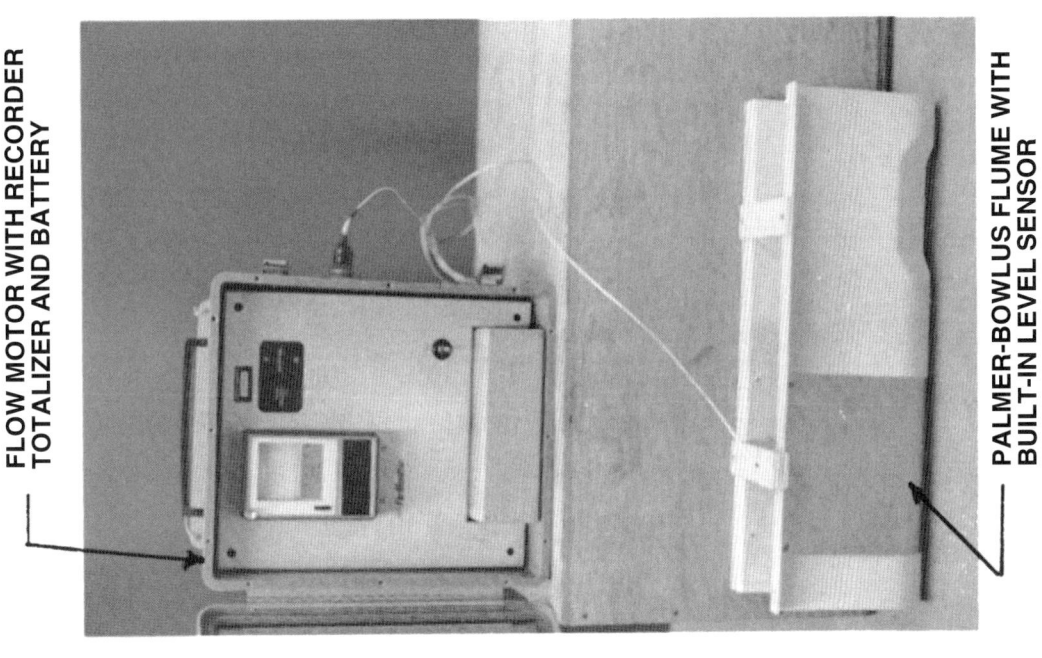

FLOW MOTOR WITH RECORDER
TOTALIZER AND BATTERY

PALMER-BOWLUS FLUME WITH
BUILT-IN LEVEL SENSOR

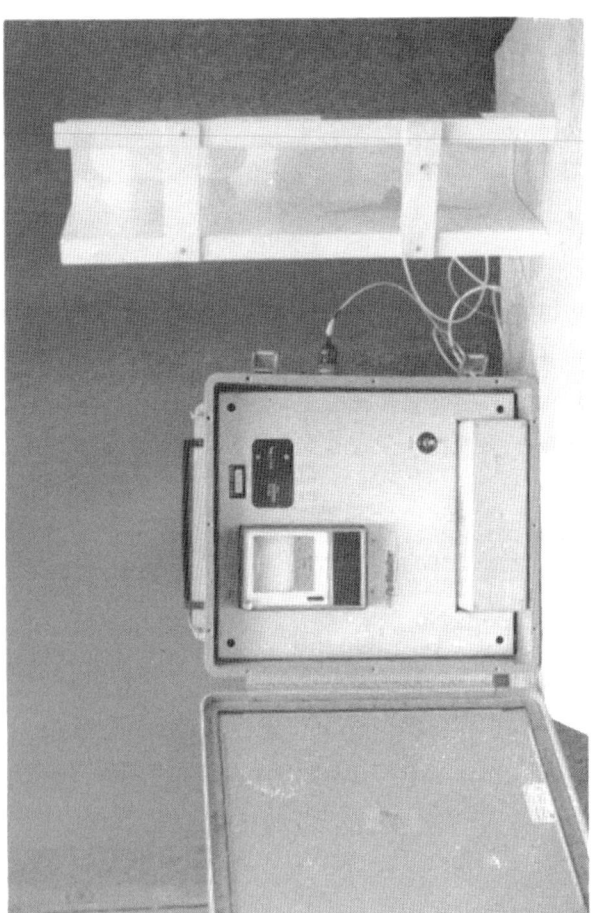

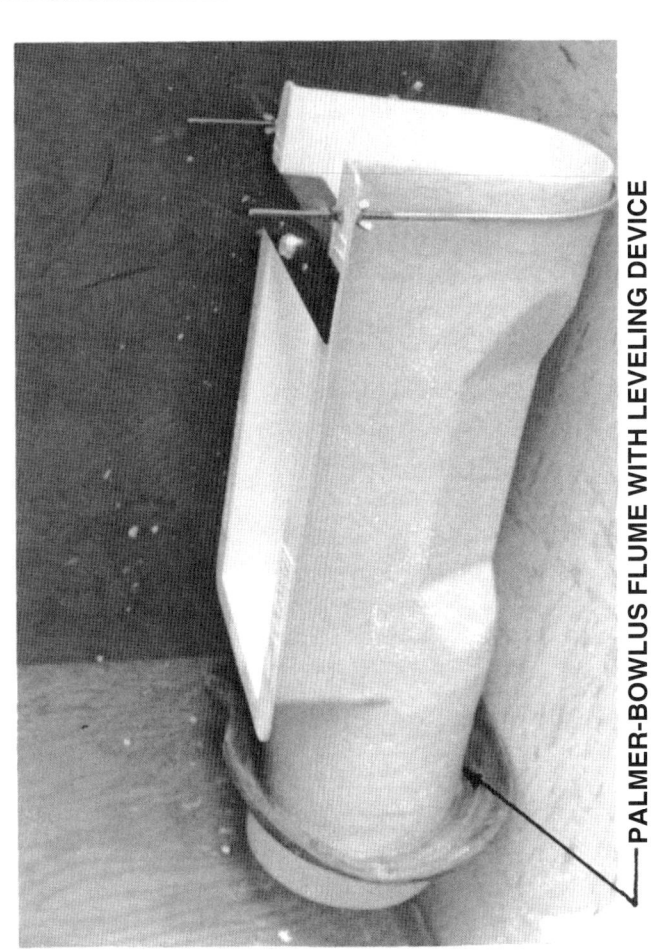

PALMER-BOWLUS FLUME WITH LEVELING DEVICE

Fig. 7.8   Portable flowmeters and recorders

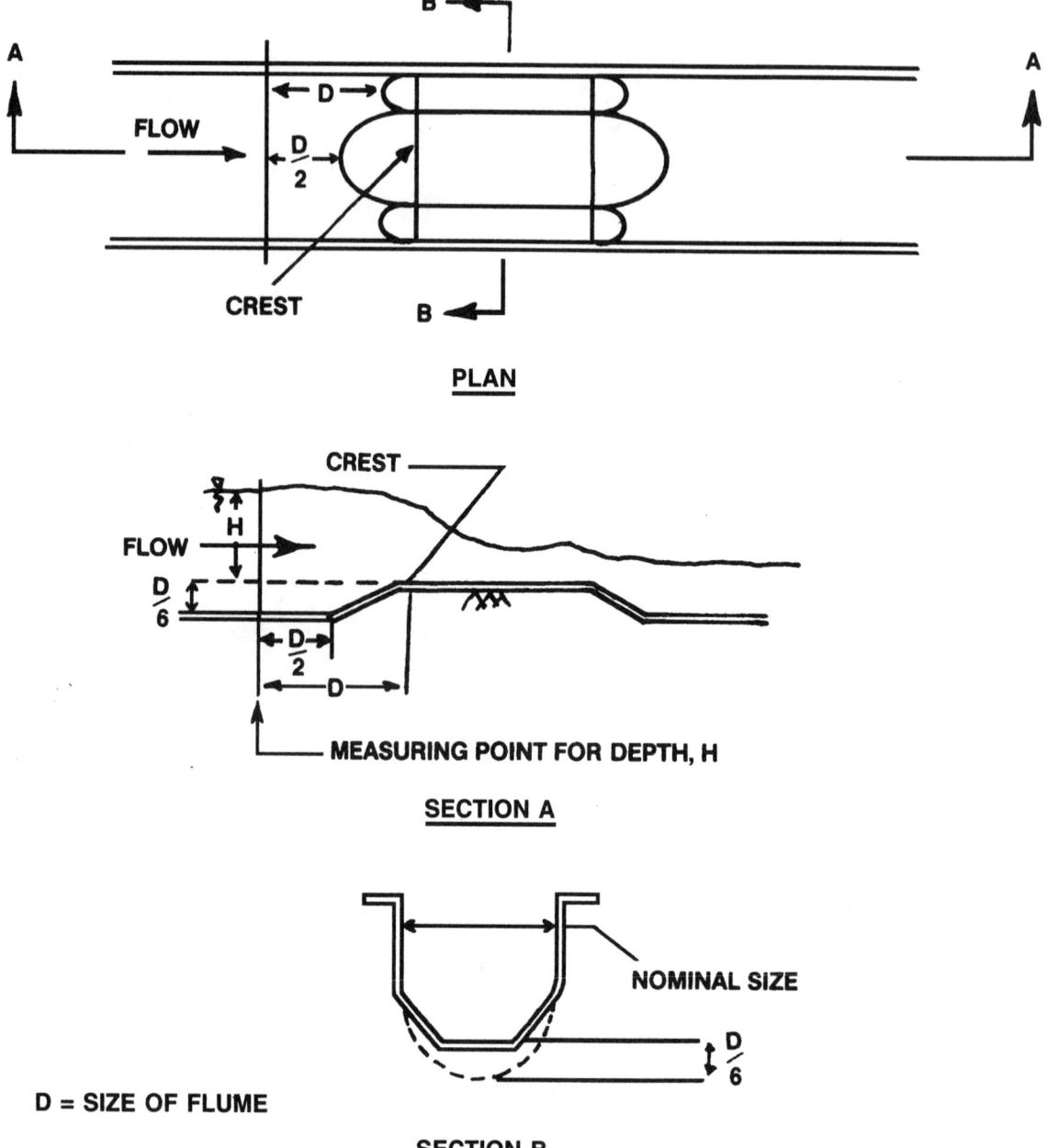

PLAN

SECTION A

D = SIZE OF FLUME

SECTION B

Fig. 7.9   Palmer-Bowlus flume

The dimensions to which a Palmer-Bowlus flume is constructed have been standardized, but in a generic sense the term Palmer-Bowlus-type flume can apply to any flume of this general shape. Be aware, however, that head-to-flow tables are not identical for different manufacturers due to slight differences in styles. For instance, another similar type of flume, the Leopold-Lagco flume, also is occasionally installed in an existing line. It has a rectangular cross section rather than a trapezoidal cross section and, consequently, produces different head-to-flow readings than a Palmer-Bowlus flume of the same nominal size.

7.31 Parshall Flumes (Figures 7.10, 7.11, and 7.12)

A Parshall flume operates on the same principles as the Palmer-Bowlus flume. The measuring point for this flume is located in the converging section at a distance of $^2/_3$ A upstream from the beginning of the throat of the flume. The distance A is the length of the converging section measured along the wall, rather than along the centerline of the flume. This point of measurement is shown in Figure 7.10.

The main advantage of a Parshall flume is that the flume will handle a wide range of flows. The flumes are available already installed in prefabricated manholes and vaults but installation in an existing sewer line may involve replacing some of the line because of the required drop in the floor of the flume.

Subcritical flow (see Figure 7.4, page 488) should occur upstream of the flume, the hydraulic drop (drop in flowing water surface) occurs in the converging section of the flume, and supercritical flow occurs in the throat of the flume. The hydraulic jump generally occurs in the throat, the diverging section, or farther downstream.

As with the Palmer-Bowlus flume, the hydraulic jump does not have to be within view. Parshall flumes are often installed to discharge to a sump or to a more steeply sloping line to prevent submergence of the flume due to water backing up in the downstream pipe.

Standard discharge formulas have been developed for the various sizes of Parshall flumes. In all of the formulas, H is the head, in feet, and is the depth of water between the floor of the channel and the water surface at the measuring point; Q is the flow, in cubic feet per second or gallons per minute.

	Q in CFS	Q in GPM
1-inch Parshall flume:	$Q = 0.338\ H^{1.55}$	$Q = 151.69\ H^{1.55}$
	$H = 2.0135\ Q^{0.6452}$	$H = 0.03916\ Q^{0.6452}$
2-inch Parshall flume:	$Q = 0.676\ H^{1.55}$	$Q = 303.39\ H^{1.55}$
	$H = 1.2874\ Q^{0.6452}$	$H = 0.02504\ Q^{0.6452}$
3-inch Parshall flume:	$Q = 0.992\ H^{1.547}$	$Q = 445.21\ H^{1.547}$
	$H = 1.0052\ Q^{0.6464}$	$H = 0.0194\ Q^{0.6464}$
6-inch Parshall flume:	$Q = 2.060\ H^{1.58}$	$Q = 924.53\ H^{1.58}$
	$H = 0.6329\ Q^{0.6329}$	$H = 0.01327\ Q^{0.6329}$
9-inch Parshall flume:	$Q = 3.070\ H^{1.53}$	$Q = 1377.8\ H^{1.53}$
	$H = 0.4804\ Q^{0.6536}$	$H = 0.008876\ Q^{0.6536}$
12-inch Parshall flume:	$Q = 4.000\ H^{1.522}$	$Q = 1795.2\ H^{1.522}$
	$H = 0.4022\ Q^{0.6570}$	$H = 0.007277\ Q^{0.6570}$

EXAMPLE 1

a. Calculate the flow in CFS through a 3-inch Parshall flume if the depth is 9 inches or 0.750 feet.

Known	Unknown
H, ft = 0.75 ft	Q, CFS

Calculate the flow in CFS.

$$Q,\ CFS = 0.992\ (H,\ ft)^{1.547}$$
$$= 0.992\ (0.75\ ft)^{1.547}$$
$$= 0.992\ (0.64)^*$$
$$= 0.635\ CFS$$

---

* 0.75 to the 1.547 power is 0.64. For example, $10^2 = 100$, $0.1^2 = 0.01$, and $25^{0.5} = 5$.

b. Calculate the depth in feet if the flow is given as 285 GPM in the same flume (1 CFS = 449 GPM).

Known	Unknown
Q, GPM = 285 GPM	H, ft

Calculate the head in feet.

$$= 0.0194\ (Q,\ GPM)^{0.6464}$$
$$= 0.0194\ (285\ GPM)^{0.6464}$$
$$= 0.0194\ (38.62)$$
$$= 0.749\ ft$$

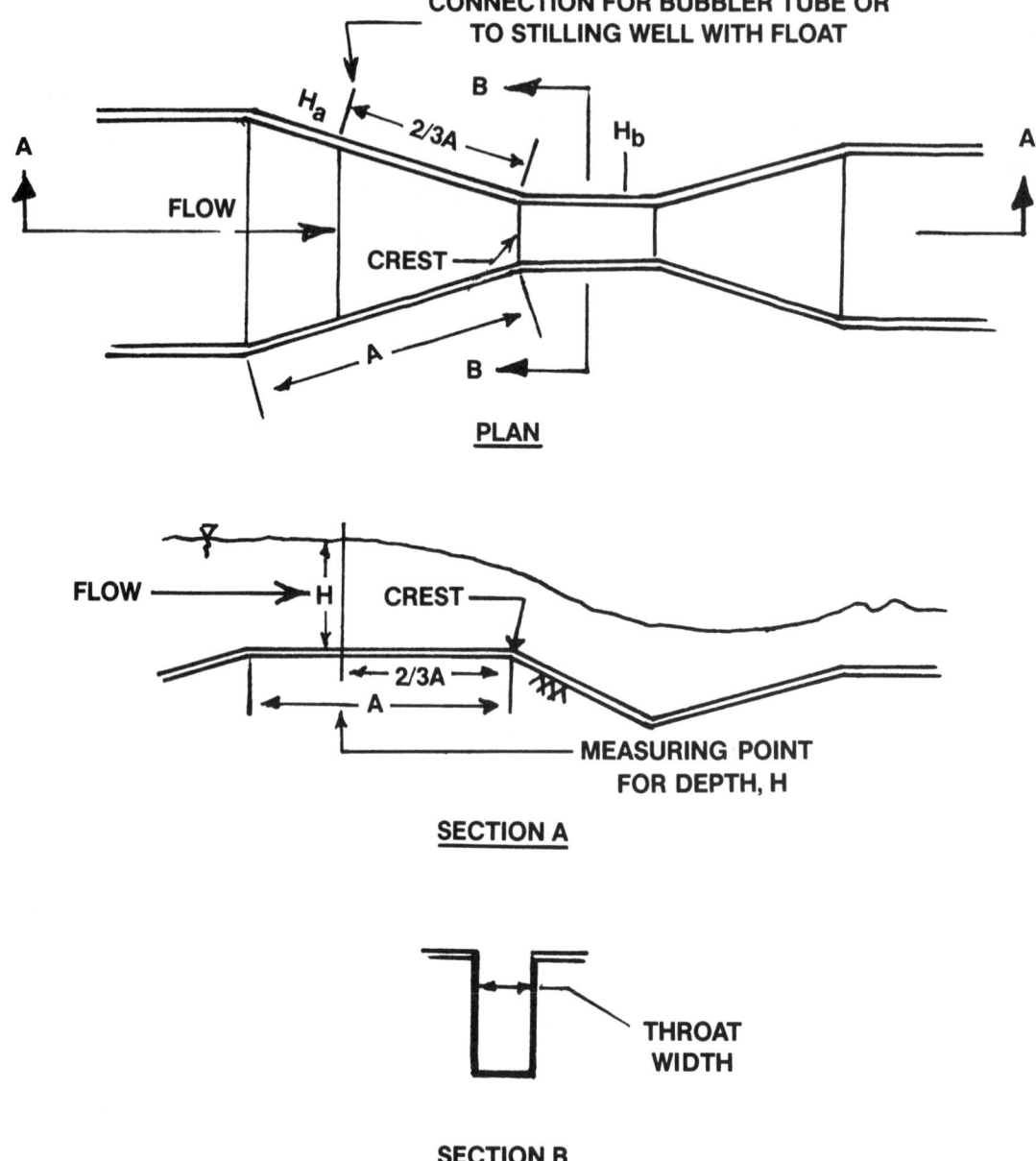

Fig. 7.10   Parshall flume

*Fig. 7.11   Large Parshall flume (12-inch throat) at paper mill with sampler in
place at top. Flow is toward top of page.*

NOTES: 1. Parshall flume on left
       2. Stilling well on right of flume
       3. Bubbler tube flow recorder on upper right

Fig. 7.12   Permanent industrial installation of a Parshall flume and stilling well. The POTW agency has temporarily installed a bubbler tube in the stilling well. This illustrates simultaneous flow measurements to verify the accuracy of the permanent installation. Flow is toward top of page.

Many flumes have a staff gage (see Figure 7.5, page 489) installed on the side of the flume for depth of flow measurements. If a staff gage is not available, measure the water depth at the appropriate location with a steel rule. The use of a wooden yardstick to measure water depth should be avoided because these devices may create a wave in the flowing water which could lead to erroneous depth measurements. Record the depth reading from the steel rule. Using the proper table or rating curve for the size of flume, use the depth of flow reading to determine the flow (see Section 7.9, "Additional Reading," Reference 2, *OPEN CHANNEL FLOW MEASUREMENT HANDBOOK,* for tables that convert depths to flows for various types and sizes of flowmeters).

Larger Parshall flumes are also available, but the sizes listed earlier in this section are the ones most commonly used to measure flows from industrial dischargers. A Parshall flume is not always installed to carry the maximum flume capacity. For instance, a flume that can accommodate a depth of three feet at the measuring point could be cut at two feet if space limitations so necessitated, although this reduces its capacity.

Parshall flumes were initially designed to be installed in irrigation systems on relatively flat surfaces and are capable of operating partially submerged. However, such operation requires additional depth measurement. Most instrumentation is not designed for that circumstance, so the flume should not be operated past a certain degree of submergence. If the hydraulic jump is located well up in the throat of the flume, further investigation is advised.

A number of other types of flumes have been developed, but generally they are used for carrying larger flows than are usually encountered at industrial facilities. These are the cutthroat flume, the San Dimas flume, and trapezoidal flumes. Many other flumes have been designed for specific applications. All of these flumes control the cross-sectional flow area and convert the depth of flow measurement to a rate of flow value.

7.32 Weirs

7.320 *Flow Conditions*

Weirs differ from flumes in that a weir is essentially a dam across the flow, as compared to a reshaping of the channel. Weirs are either broad-crested (wide in the direction of flow) or sharp-crested (Figure 7.13, page 500). The sharp-crested weir is more commonly used in measuring industrial wastewater flows than the broad-crested weir. The V-notch weir is the most common of the sharp-crested weirs because it is the most accurate flow measuring device for the small, fluctuating flows which are common for small industries.

Weirs can be installed in a variety of situations; often an existing sump will be large enough to serve as a weir box. Always provide adequate clearance below the notch for a free discharge to occur. This requirement may limit installation in existing lines if the backup of water would flood or submerge the weir.

Weirs operate on the same principles as flumes; however, they can look quite different. The approach section, which is sized so that the approach velocity is minimal, has subcritical flow. Supercritical flow occurs as the water pours through the weir notch. The flow returns to subcritical flow in the afterbay of the weir.

Under normal conditions, you will see that the flow through the notch, called the nappe (pronounced NAP) of the flow, springs away from the weir plate (Figure 7.14). This means that the weir is operating with a free discharge and that the nappe is well ventilated, or aerated; that is, air can move freely beneath the nappe. Only at low flows should the water cling to the face of the weir plate.

A weir cannot be operated in a submerged condition. The nappe of the water must fall freely into the weir afterbay. If the level in the afterbay rises too high, aeration of the nappe may cease and the measured discharge will be greater than the actual discharge. A weir should be constructed with several inches' clearance between the crest of the weir (the bottom of

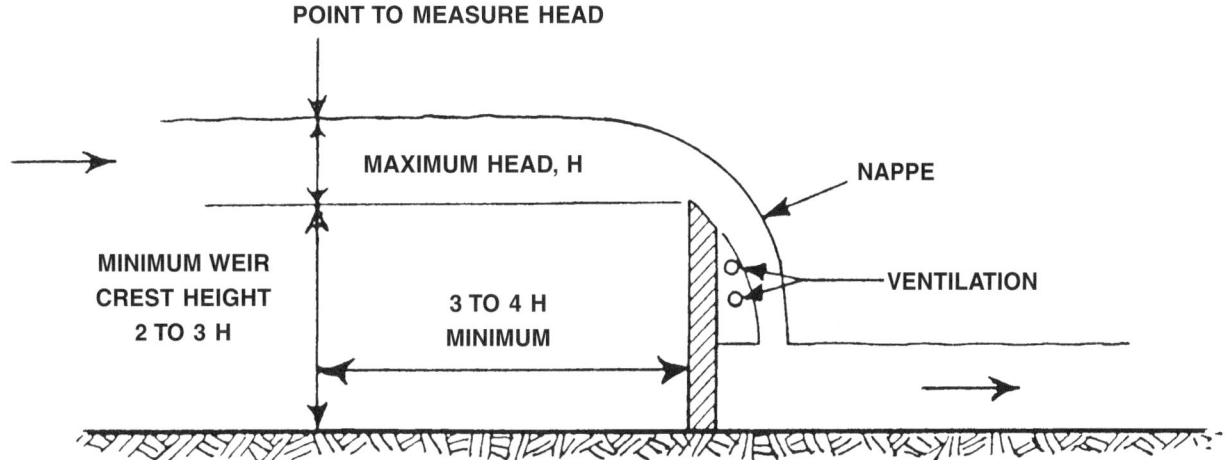

Fig. 7.14 Weir dimensions

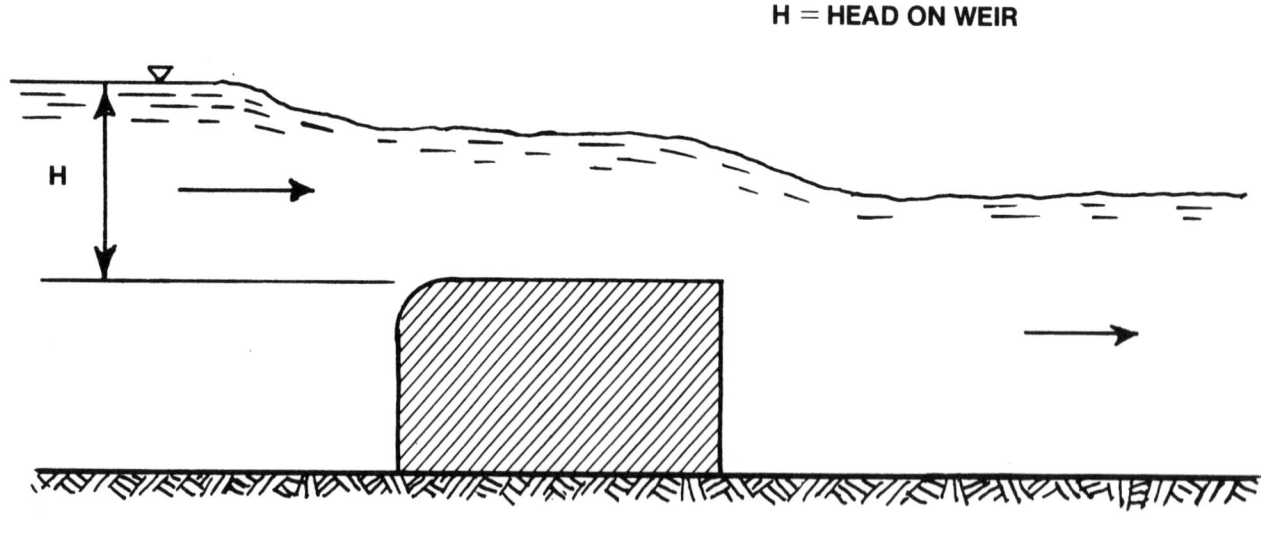

H = HEAD ON WEIR

NAPPE

SHARP-CRESTED WEIR

H = HEAD ON WEIR

BROAD-CRESTED WEIR

Fig. 7.13 Sharp-crested and broad-crested weirs

the notch) and the afterbay level (Figure 7.15). In general, a weir should be constructed with the top of the downstream pipe at least six inches below the crest of the weir. If the discharge pipe is not visible and the afterbay level is approaching the crest of the weir, it is likely that the proper depth-to-flow relationship does not exist.

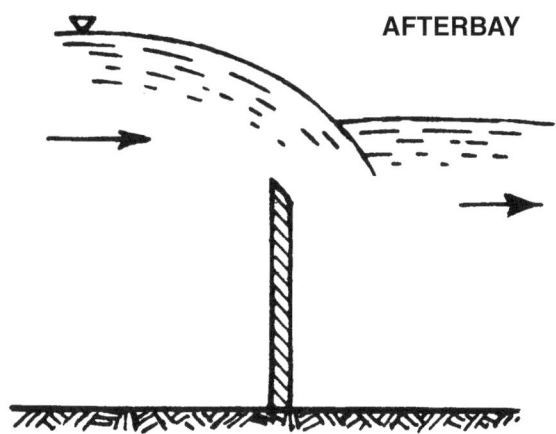

Fig. 7.15   Submerged weir

To develop the proper depth-to-flow relationship with a weir, it is generally necessary that an upstream pool be formed to dissipate the approach velocity of the flow. The dimensions (determined by qualified design engineers) of this pool are based on the maximum capacity, expressed as the depth (head) behind the weir. The absence of this pool may cause the weir to measure a lower than actual flow.

The measurement point for all types of weirs is located at a distance of about 3H to 4H upstream (or to the side) of the weir (see Figure 7.16). H is the maximum expected head on the weir. The depth of flow (head) through a weir is measured from the crest (bottom or lowest point) of the weir to the water surface at the measuring point.

7.321   V-Notch Weirs

A V-notch weir is formed by cutting a 22 1/2, 30, 45, 60 or 90 degree notch in a metal plate and fixing the plate in appropriate supports. Other materials are used for weir plates including polycarbonate (a plastic material like plexiglass). The edges of the notch must be cut and beveled to the correct dimensions. For permanent installations, the weir plate should be made of metal since the accuracy of a weir is affected by gradual rounding of the edges of the notch. The angle of the weir and the depth of the notch fix the dimensions of the upstream pool. The key elements of a V-notch weir are shown in Figure 7.16.

Two sets of equations that relate depth to flow are commonly used for V-notch weirs. These are:

V-Notch Angle (degrees)	Weir Discharge Formulas	
22½	$Q = 0.515H^{2.43}$	$Q = 0.497H^{2.5}$
30		$Q = 0.676H^{2.5}$
45	$Q = 1.045H^{2.46}$	$Q = 1.035H^{2.5}$
60	$Q = 1.446H^{2.47}$	$Q = 1.443H^{2.5}$
90	$Q = 2.490H^{2.48}$	$Q = 2.500H^{2.5}$

In all of these equations H is the head, in feet, or the depth of water between the crest (bottom of V) and the water surface, and Q is the flow in cubic feet per second.

EXAMPLE 2

Calculate the flow in CFS through a 30-degree V-notch weir if the water depth is 8 inches or 0.67 feet.

Known	Unknown
H, ft = 0.67 ft	Q, CFS

Calculate the flow in CFS.

$$Q, CFS = 0.676 (H, ft)^{2.5}$$
$$= 0.676 (0.67 ft)^{2.5}$$
$$= 0.676 (0.367)$$
$$= 0.248 CFS$$

See Section 7.9, "Additional Reading," Reference 2, *OPEN CHANNEL FLOW MEASUREMENT HANDBOOK,* for tables that convert depths to flows for various types and sizes of weirs.

The actual formula that should be used by the secondary measurement device should be determined when checking the accuracy of the system. Use the formula that is recommended by the flowmeter manufacturer. As a comparison, a 90-degree V-notch weir at a head of 3.00 feet would be discharging at 17,049 GPM (1 CFS = 449 GPM) according to the first formula, and 17,498 GPM according to the second formula. The flow of the first formula, substituted into the second formula, corresponds to a head of 2.969 feet, which is 0.031 foot or about 0.37 inch less than 3.00 feet. If the instrumentation being checked was converting depth to flow based on the second formula, and the first formula was referenced, it would be mistakenly assumed that the meter read high. The error associated with the different formulas is proportionally less for smaller weirs and at lower flows.

7.322   Rectangular Weirs

Another common type of weir is the rectangular weir. The rectangular opening may span the width of the channel in which case the weir is known as a suppressed (without end contractions) weir. Aeration of the nappe is achieved by installing vent pipes beneath the nappe (Figure 7.14). When the opening spans only a portion of the width of the channel, the weir is known as a contracted (with end contractions) weir. As with V-notch weirs, the weir pool dimensions depend on the type and capacity of the rectangular weir. The measuring point is located at about 3H to 4H upstream of the weir (Figure 7.14). The weir should be sized so that the minimum depth is about 0.2 foot and the maximum depth is about one-half the length of the crest, although greater depths can be adequately measured. Rectangular weirs will measure larger flows than V-notch weirs.

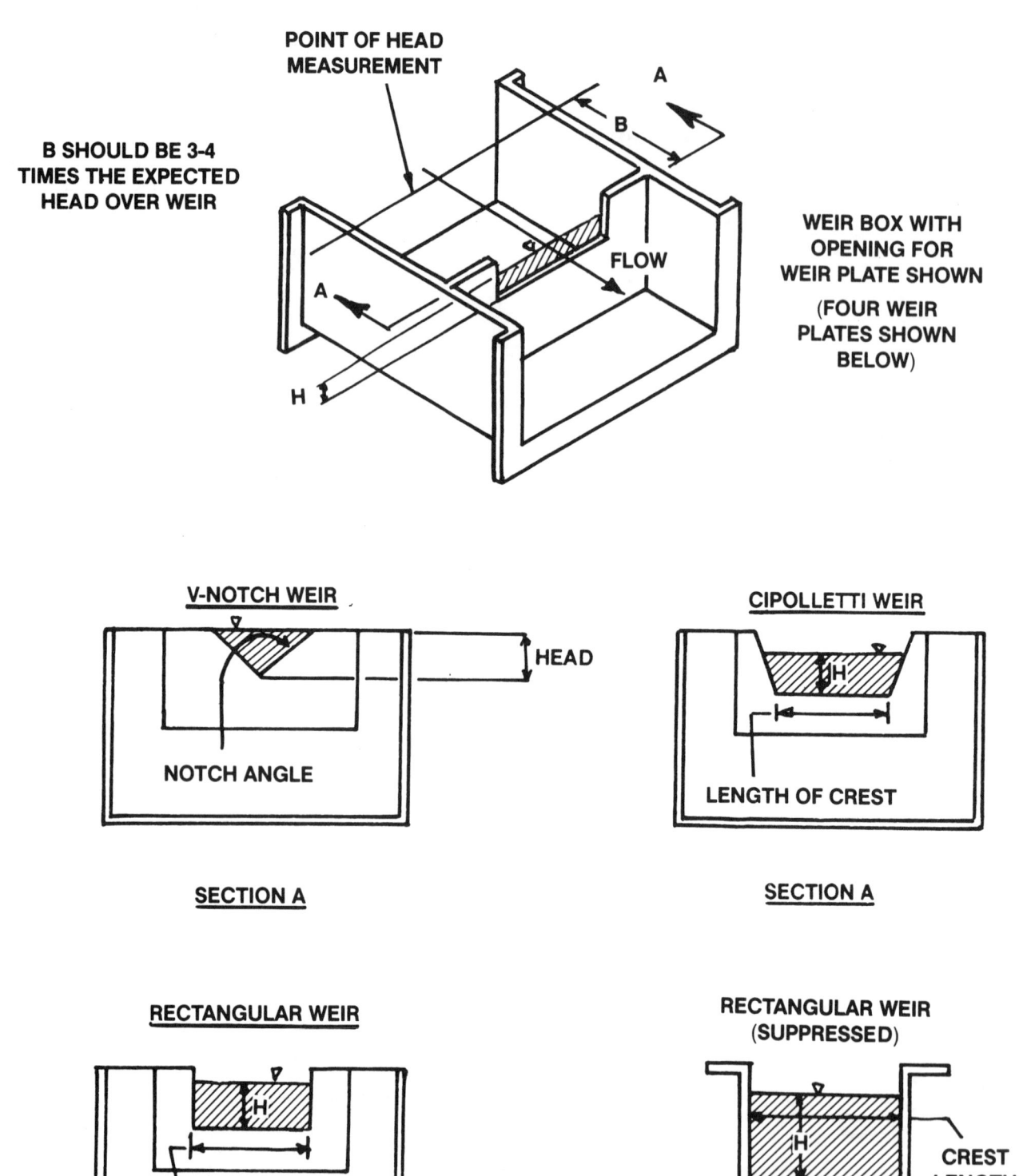

POINT OF HEAD MEASUREMENT

B SHOULD BE 3-4 TIMES THE EXPECTED HEAD OVER WEIR

WEIR BOX WITH OPENING FOR WEIR PLATE SHOWN

(FOUR WEIR PLATES SHOWN BELOW)

FLOW

V-NOTCH WEIR

HEAD

NOTCH ANGLE

SECTION A

CIPOLLETTI WEIR

H

LENGTH OF CREST

SECTION A

RECTANGULAR WEIR

H

LENGTH OF CREST

RECTANGULAR WEIR (SUPPRESSED)

H

CREST LENGTH

H = HEAD OR DEPTH BEHIND WEIR CREST
MEASURE H AT A DISTANCE B UPSTREAM IN A STILLING WELL

Fig. 7.16   Weirs

The depth-to-flow formula for suppressed rectangular weirs is usually given as:

$$Q = 3.33 \, LH^{1.5}$$

The depth-to-flow formula for contracted rectangular weirs is usually given as:

$$Q = 3.33 \, (L - 0.2H)H^{1.5}$$

In these formulas, H is the depth in feet from the crest of the weir to the water surface at the measuring point, L is the crest length in feet, and Q is the flow in cubic feet per second.

A Cipolletti weir (Figure 7.16) is quite similar to a contracted rectangular weir but has a trapezoidal-shaped opening rather than a rectangular opening. The discharge formula for this weir, with the same units as above, is usually given as:

$$Q = 3.367 \, LH^{1.5}$$

The key elements of standard weirs are shown in Figure 7.16. Several other types of sharp-crested weirs are occasionally used in flow measurement work. However, because of their unusual shapes and a resulting difficulty of construction, they are not usually selected for installation by industrial dischargers.

7.33 H-Type Flumes (Figures 7.17, 7.18, and 7.19)

H-type flumes were developed to measure runoff from agricultural watersheds and have found use in other applications. The H-flume, HS-flume, and HL-flume combine features of both weirs and flumes. Flow control is achieved at a sharp-edged opening and the flat floor allows passage of solids. These flow measurement devices are designated by the maximum depth of the flume; for instance, a 1.0-foot H-flume has a maximum head of 1.0 foot. The dimensions to which the flume is constructed, and also the point of measurement, depend on the maximum depth. The depth of flow is measured from the floor of the flume to the water surface. For the H-flume, the measurement point is located a distance of 1.05 D from the discharge tip of the flume, where D is the size of the flume (maximum head). For the HS-flume the distance is D; for the HL-flume the distance is 1.25 D.

The discharge formulas for H-type flumes are complicated, thus tables which are easy to read should be used to relate depth to flow. See Section 7.9, "Additional Reading," Reference 2, *OPEN CHANNEL FLOW MEASUREMENT HANDBOOK,* for tables that convert depths to flows for various types and sizes of weirs. The flume should discharge in a free flow condition, as with a weir, and without submergence.

H-flumes are more correctly classified as flow nozzles. Two other types of flow nozzles, the Kennison nozzle and the parabolic nozzle (Figure 7.20), are also occasionally used to measure industrial discharge flows.

Table 7.1 summarizes the significant advantages and limitations of primary flow measuring devices.

TABLE 7.1 KEY ASPECTS OF PRIMARY FLOW MEASUREMENT DEVICES

Type	Advantages	Limitations
Parshall Flumes	Basically self-cleaning. Can accommodate a wide range of flows. Generally very accurate when properly deployed.	Requires that a significant height difference exist between upstream and downstream sections. Relatively expensive to install. Smaller flumes cannot handle flows with high solids content.
Palmer-Bowlus Flumes	Self-cleaning. Inexpensive to install. Can be inserted directly into a channel. Can handle wastewaters with a high solids content.	Can be somewhat inaccurate where there is a wide range in the flow rate.
Weirs	Inexpensive to install. Good accuracy. Easy to maintain.	Not self-cleaning. Cannot accommodate wastewaters with a high solids content. Requires that a moderate height difference exist between upstream and downstream sections.
H-Flumes	Self-cleaning. Good accuracy.	May be subject to sludge buildup with low-volume, high-solids content wastewaters.

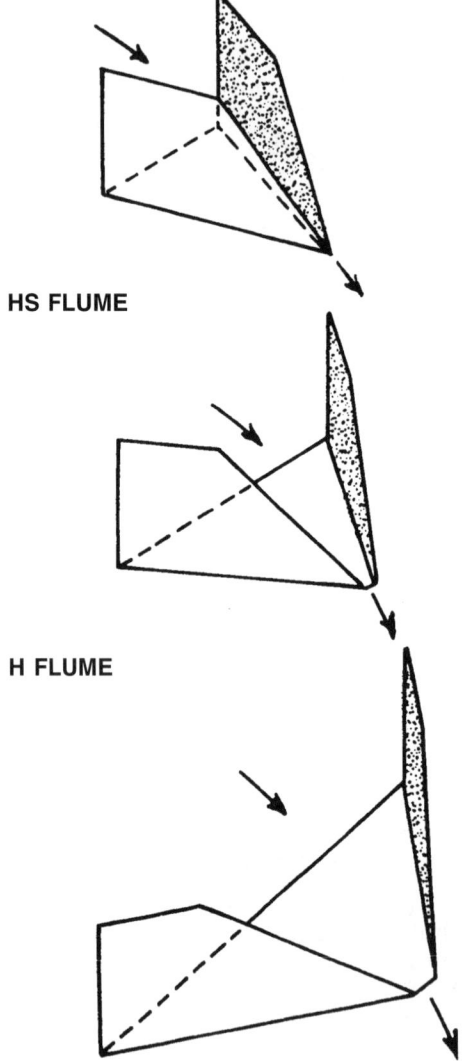

HS FLUME

H FLUME

HL FLUME

Fig. 7.17   H-type flumes

Fig. 7.18   H-flume with ultrasonic measurement of flow depth. Flow
spills into a vault connected to the sewer.

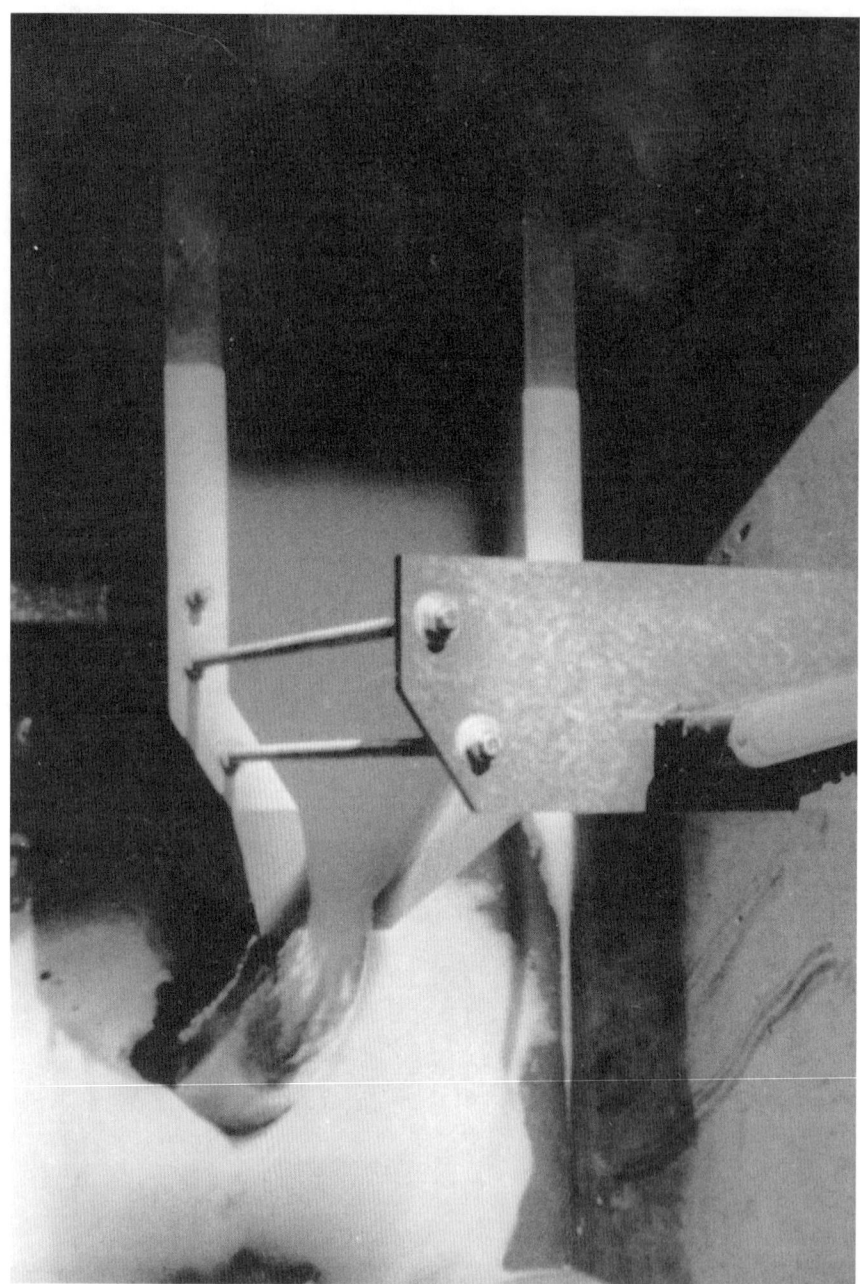

Fig. 7.19   H-flume with ultrasonic depth measurement.
Actual flow is red with heavy white foam.

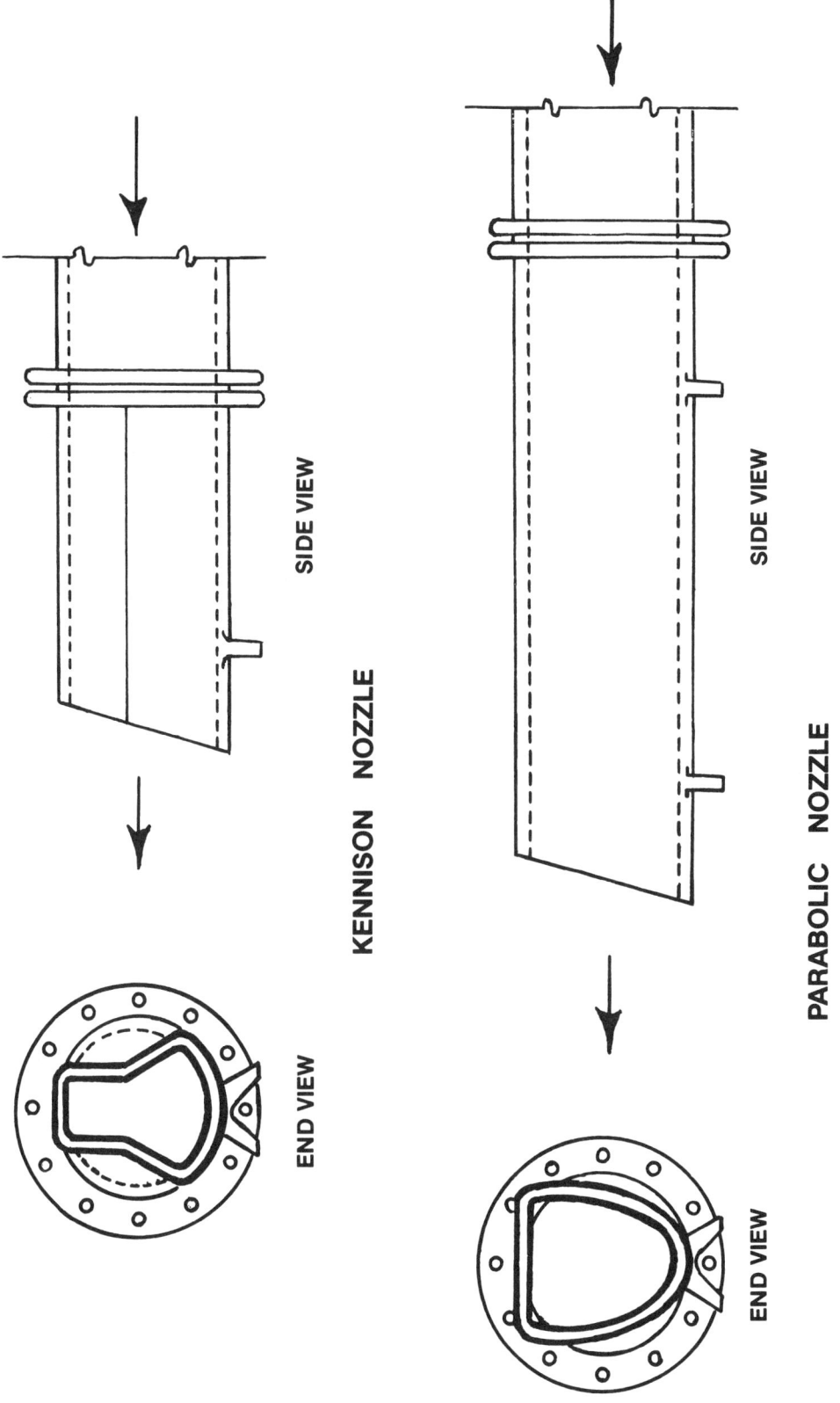

SIDE VIEW

KENNISON   NOZZLE

END VIEW

SIDE VIEW

PARABOLIC   NOZZLE

END VIEW

Fig. 7.20   Kennison nozzle and parabolic nozzle

QUESTIONS

Write your answers in a notebook and then compare your answers with those on page 524.

7.3A   How would you measure the depth to determine the flow for a Palmer-Bowlus flume?

7.3B   How would you measure the depth to determine the flow in a Parshall flume?

7.3C   How do weirs differ from flumes?

7.3D   Why should a weir *NOT* be operated in a submerged condition?

7.4 INSTRUMENTATION FOR OPEN-CHANNEL FLOW

Several different types of instruments are available for measuring open-channel flow. Generally, all of them can be installed on any type of flume or weir, at either the channel or the stilling well, although the characteristics of a particular wastewater may preclude the use of certain types of instrumentation. The function of the instrumentation is to sense the level of the water, convert the depth to flow, and to indicate, record, and totalize the flow. The instrumentation may also be used to activate an automatic sampler, and outputs are usually available for other uses.

The totalizer, indicator, and recorder should be properly labeled to prevent problems in interpreting their readings. Also the pulse output for a contact closure used in flow proportional sampling should be clearly labeled. Totalizer readings usually require that a multiplier factor be used and this factor should be clearly posted. Analog readout indicators often use a span of zero to 100 percent. The flow at 100 percent should be posted. The recorder often has the same span as the indicator, but when it differs it should be posted. The chart paper on the recorder should be regularly annotated with the time and date and the totalizer reading. Some meters are constructed without indicators and instantaneous readings of the flow must be taken directly from the recorder. The timer operation generated by the flowmeter to activate an automatic sampler should also be posted.

7.40 Float (Figure 7.21)

A float moves in conjunction with the level of water and is mounted either on a cable and pulley or on a pivoting arm. The movement of the float is translated to flow by a mechanical cam. The profile of the cam is related to the depth-to-flow relationship of the primary element. A cable and pulley float is usually installed in a stilling well and a float on a pivoting arm is installed directly in the channel.

Sources of equipment errors in a float-operated system arise from float lag; water leaking into float; temperature and humidity changes; errors in manufacturing tolerances; the buildup of grease on the float, float feed cables, and stilling well walls; and wear and tear. The effect of float lag can be minimized when taking a comparative depth measurement by taking the measurement when the flow is steady. Leaking floats can be avoided by using leakproof materials and construction. Temperature and humidity changes can cause the float cable to expand or contract, but errors due to this should be negligible. Much greater errors result from debris and grease in stilling well lines. Excessive error in manufacturing tolerances and errors due to gradual wear and tear should show up and be compensated for during calibrations of the meter. A kink in the float cable or a broken mechanical device should show as an obvious error. An accumulation of debris on the float or a foam layer on the water surface may affect the

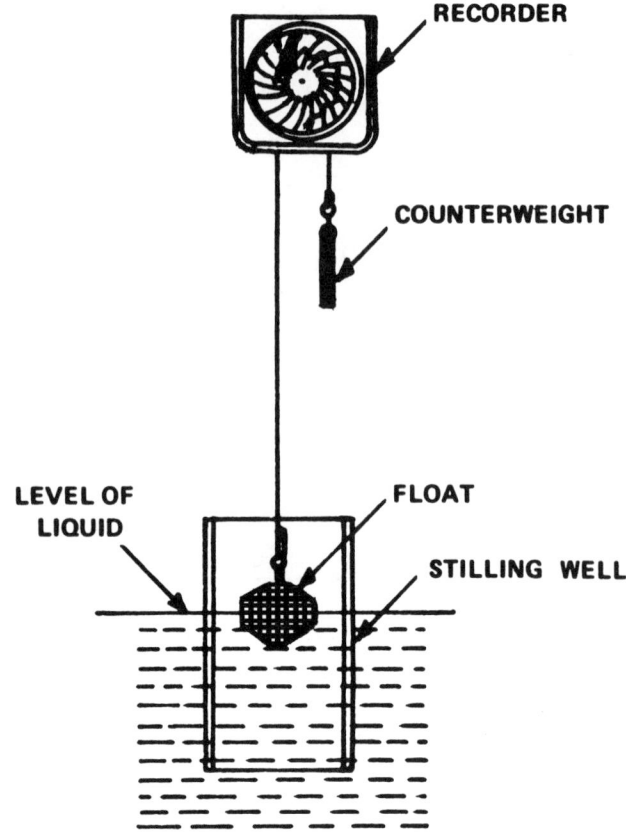

Fig. 7.21   A float is used to measure liquid level and convert the level reading to a flow measurement

accuracy of the system. As with all flow metering systems, some error will result if the meter drifts out of adjustment.

Float measuring systems should be calibrated monthly or quarterly to ensure they are being properly maintained. Maintenance includes flushing the stilling well, checking for free movement of float and cable, and proper lubrication.

7.41 Dipping Probe

With this flowmeter, a probe is lowered on a wire to the water's surface. When the tip of the probe touches the water, a circuit through the water to a ground return is completed which signals the depth of the water. The probe is raised just above the water and the process is repeated a moment later. The accuracy of the dipping probe can be affected by a foam layer on the water surface.

7.42 Bubbler (Figure 7.22)

The depth of water behind a flowmeter can be measured using a bubbler. As shown in Figure 7.22, a tube is located in the water and a constant flow of air is bubbled through the tubing. A meter measures the pressure that is necessary to maintain a constant bubble rate in an open atmosphere. The pressure corresponds to the depth of water above the end of the bubbler tube. The pressure is measured with a mechanical pressure sensor or an electrical pressure transducer to provide an electrical output proportional to depth. The distance between the end of the bubbler tubing and the lowest level of the flow must be known and compensated for to zero the meter properly.

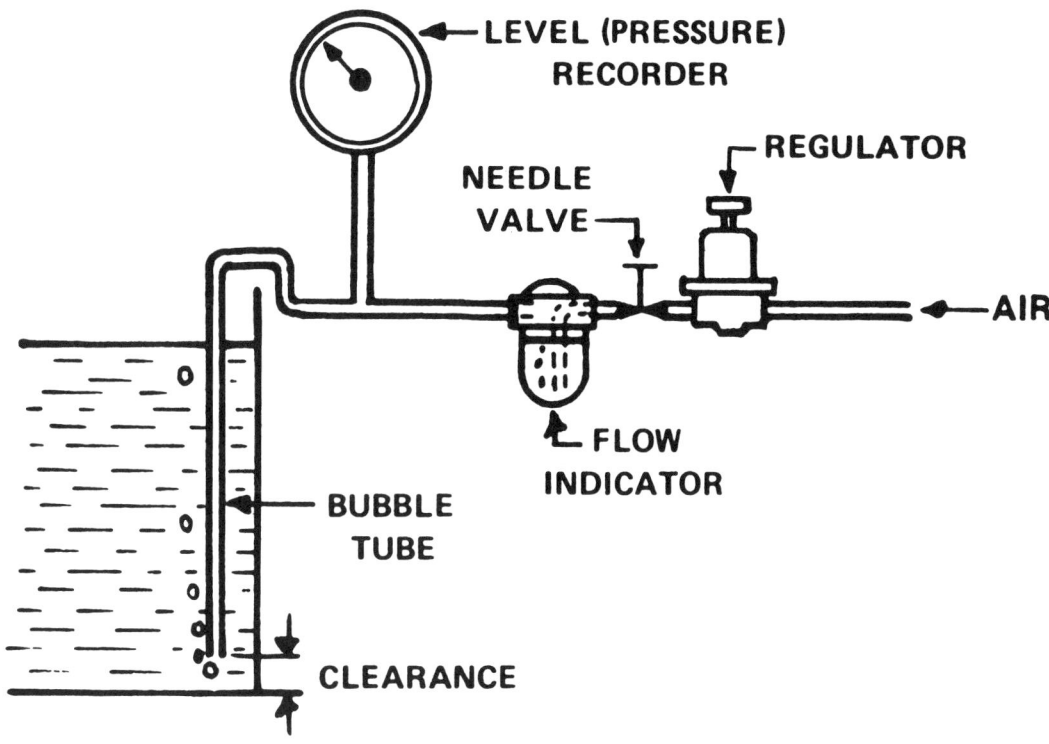

Fig. 7.22 A bubbler tube is used to measure liquid level and convert the depth reading to a flow measurement

Equipment errors arise from lag effects and the effect of temperature on the pressure sensing element, although these errors should be minor. The density of the liquid can also change and this changes the pressure sensed by the bubbler. However, this source of error should usually be minor. Significant errors can result if the end of the bubbler tubing becomes clogged from solids in the wastestream. Errors can also result if condensate forms and freezes in any dips or low points in the piping or tubing. This error invariably causes the meter to read high. This circumstance can be prevented with frequent maintenance or by installing an automatic purging system to blow out accumulated material in the tube. A split in the bubbler tubing will cause a low reading and will be indicated by an unsteady bubble rate. A kink in the bubbler tubing may also cause an inaccurate reading. Too low a bubble rate may cause the meter to fail to properly respond to changing flows; the meter should be set to the manufacturer's recommended bubble rate. Do not turn up the bubble rate to keep the bubbler tubing clean because a higher than necessary bubble rate could produce high pressure or depth readings.

7.43 Pressure Sensor

In this type of meter, a pressure transducer is placed on the bottom of the channel or stilling well. The transducer is referenced to atmospheric pressure and sends a signal in proportion to the hydraulic pressure. The hydraulic pressure depends primarily on the depth of water above the probe.

7.44 Ultrasonic Meter (Figures 7.23, 7.24, 7.25, and 7.26)

With an ultrasonic meter, the liquid level is measured by determining the time it takes an acoustic (sound) pulse generated in a transducer to reach and be reflected from the air/liquid

interface. A transmitter/receiver is housed in the transducer which is positioned over the measuring point. These meters compensate for air temperature changes. Be sure the manufacturer's recommendations are followed regarding maximum distance above the water surface and angle of transmitted signal or false readings could be produced.

Equipment errors can occur with the instrumentation that generates and receives the acoustic pulses and measures the transit time. These errors should be minor as long as the meter remains properly adjusted. The accuracy of the meter can be affected by a foam layer on the surface of the water since the sound may be reflected by the foam instead of the water. If the transducer is located over the channel, excessive wave action or solids on the surface may cause a loss of the reflected signal; if this occurs the transducer should be relocated over a stilling well. The sound wave can also be lost if the transducer is placed in a windy location, although this rarely occurs in the measurement of industrial wastewater. The meter can be affected if the face of the transducer is not kept clean. A false echo can occur if the transducer is located in a very restricted locale, although transducers mounted through a flat plate have been found to be unaffected by a false echo.

Receiver   **Transmitter**

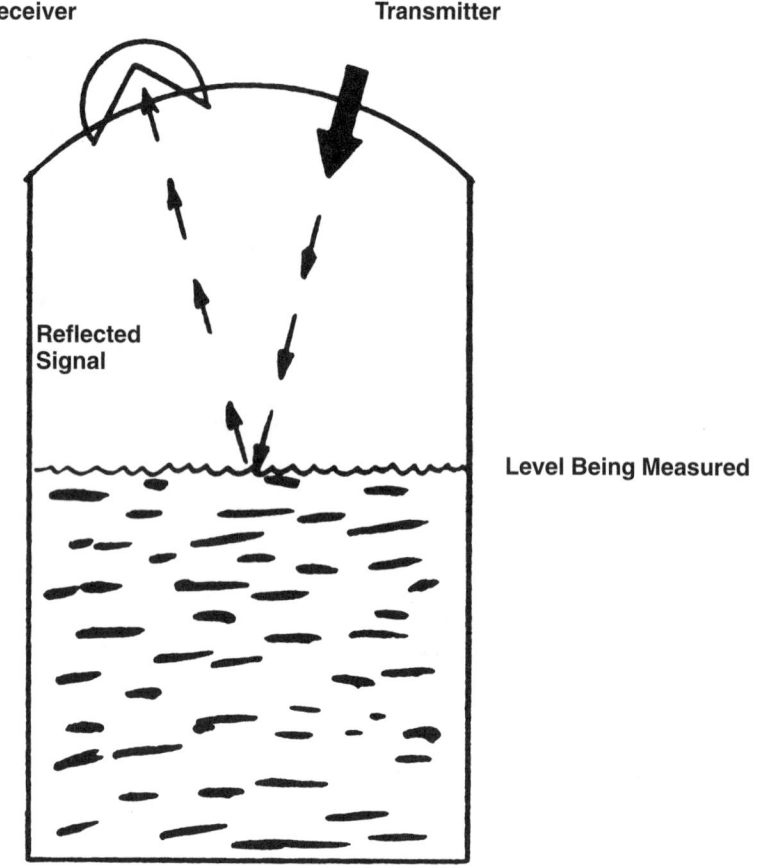

Reflected Signal

Level Being Measured

Fig. 7.23   Ultrasonic sound is used to measure liquid level and convert the level reading to a flow measurement

7.45   Capacitance Probe (Figure 7.27)

Capacitance probes determine liquid levels from a change in an electric circuit. Part of the circuit is located in the probe which is immersed in the water. This portion of the circuit consists of two conductive plates separated by a nonconducting material. The water around the probe acts as a dielectric (nonconducting) material. The capacitance of the circuit changes with the depth of water surrounding the probe. The probe should be mounted flush to the side wall of the channel or in a stilling well. Related types of flowmeters use the electrical properties of resistance or conductivity to sense the depth.

Errors with this type of system can arise if the probe is too heavily coated with solids or oil and grease; periodic maintenance will minimize this error. An improperly located probe can disturb the flow and cause an improper depth to occur in the vicinity of the probe so the probe should be located in the flow stream as recommended by the manufacturer.

7.5   METHODS OF CHECKING THE ACCURACIES OF OPEN-CHANNEL FLOWMETERS

The basic methods to check the accuracy of an open-channel flow metering system are hydraulic calibration, instrument calibration, and comparative depth measurements. Hydraulic calibration checks both the flume or weir and instrumentation as a system. Instrument calibration ensures that the instrumentation is operating correctly over the range of the system.

A comparative depth measurement is an approximate check of a meter's accuracy at a particular flow.

When a flow metering system is first installed, a hydraulic calibration is performed to verify that the system follows the standard rating curve for the system or to develop a nonstandard rating curve. Known flow volumes are introduced through the system and the accuracy of the system is compared to the calibration standard. Typically it is recommended that flow measurements be obtained for five known flows over the normal operating range of the system. The corresponding depths are also obtained. Once these five flow rate relationships are obtained they can be graphed, thereby providing the necessary rating curve. This procedure is fairly standard for all open-channel flow systems. A hydraulic calibration of the system should be performed every six months, and more often for big dischargers if appropriate. Some agencies use longer time intervals (up to once every three years) depending on their situation. Newer units are capable of simulating flows for hydraulic calibration. This is necessary to verify that the depth-to-flow relationship of the flume or weir has remained constant. An occasional system will develop a nonstandard curve over time due to wear and tear or ground settlement. The nonstandard curve can be generated by performing a hydraulic calibration. Many types of instruments will accept nonstandard curves when converting depths of water to flows.

An instrument calibration is done in order to check the performance of the instrumentation. This procedure does not

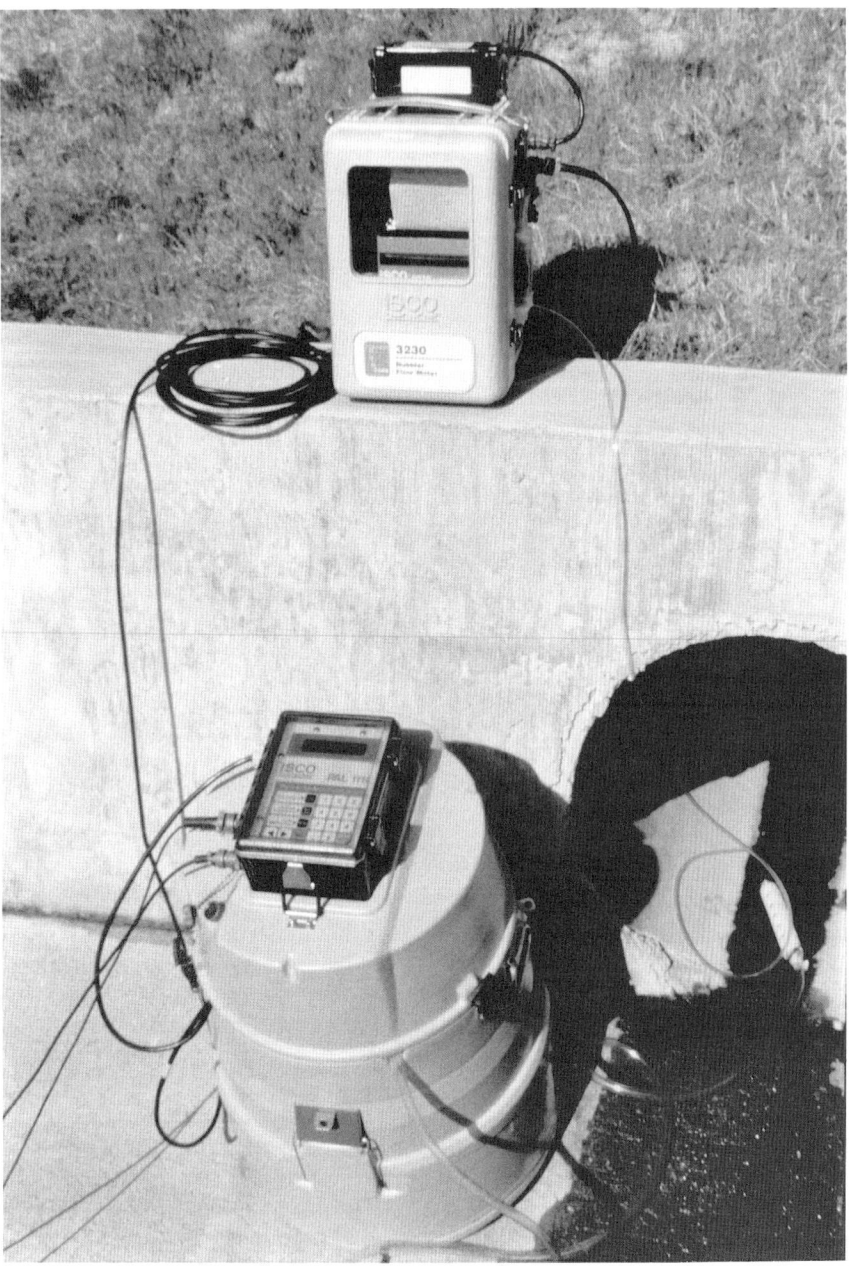

Fig. 7.24   Ultrasonic flowmeter with an automatic sampler
(Courtesy of ISCO, Inc., Environmental Division)

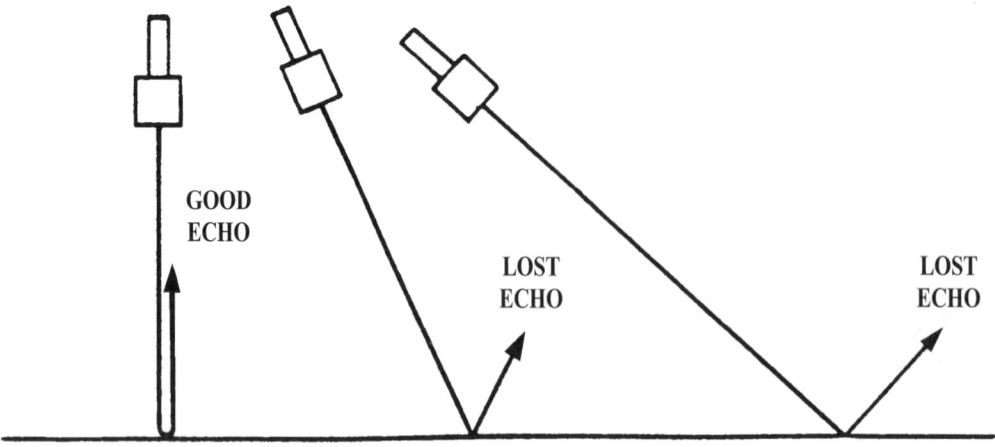

GOOD
ECHO

LOST
ECHO

LOST
ECHO

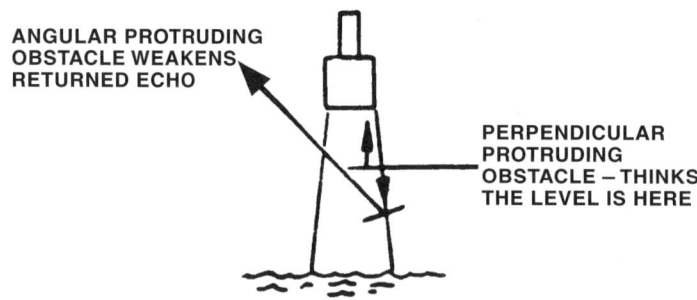

ANGULAR PROTRUDING
OBSTACLE WEAKENS
RETURNED ECHO

PERPENDICULAR
PROTRUDING
OBSTACLE – THINKS
THE LEVEL IS HERE

**NORMAL
INSTALLATION**

**ALTERNATE
INSTALLATION**

DEAD ZONE

45°

REFLECTIVE
SURFACE

MAX LEVEL

A

B

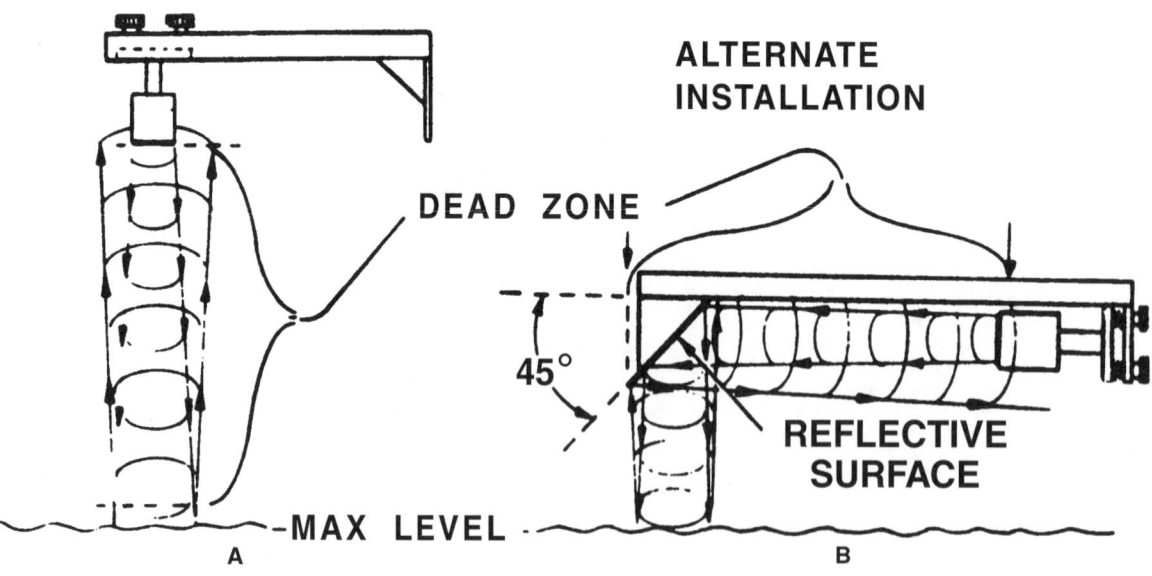

Fig. 7.25   Ultrasonic transducer installation
(Courtesy of Manning Technologies, Inc.)

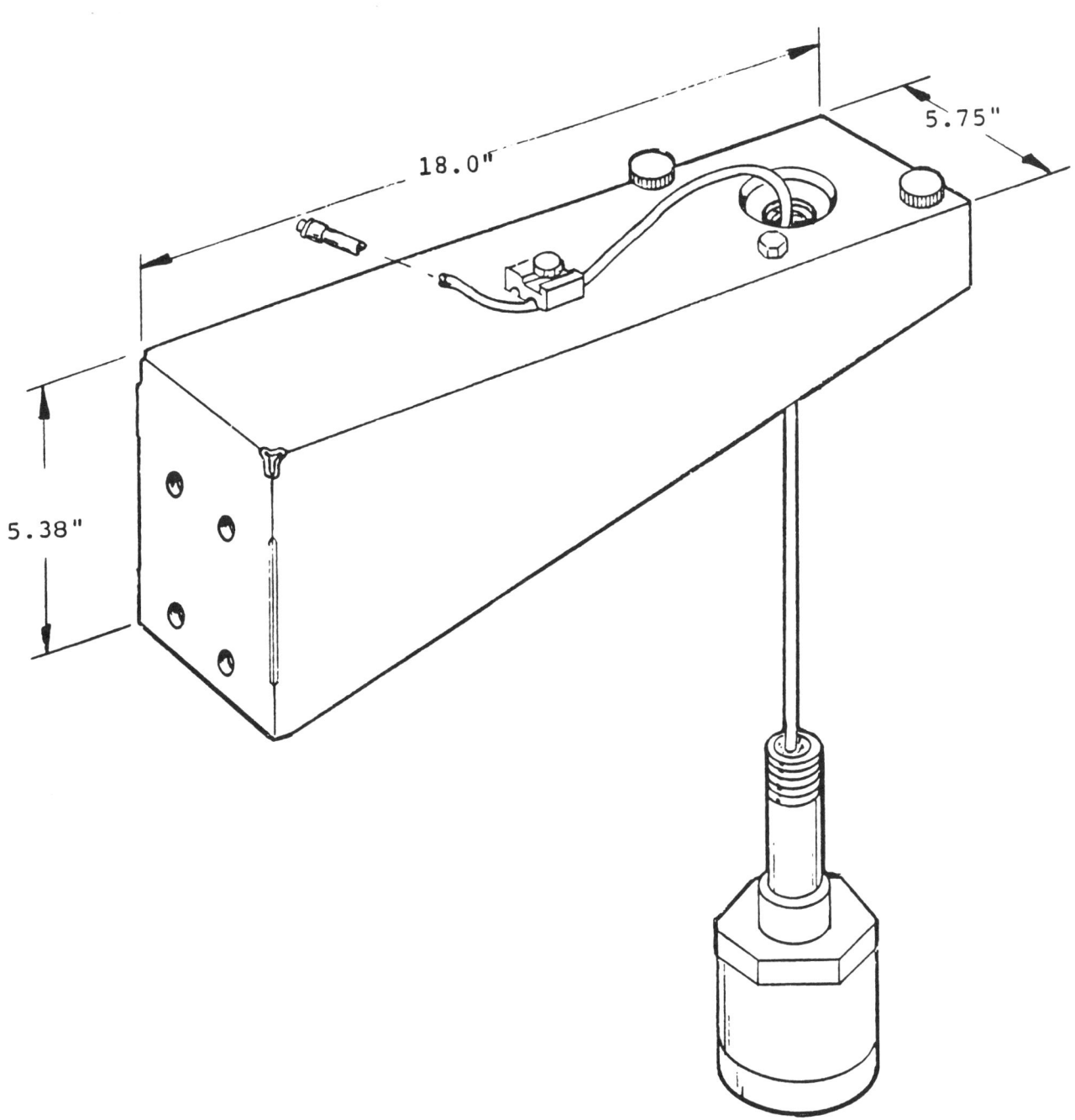

Fig. 7.26 Ultrasonic transducer and mounting bracket
(Courtesy of Manning Technologies, Inc.)

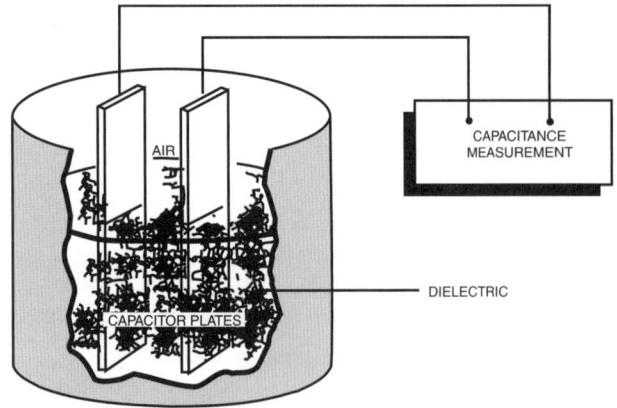

Capacitance as a function of liquid level

Fig. 7.27   Capacitance probe

require any flow through the system but instead allows for the secondary device to be removed and tested separately. Some types of instrumentation can be calibrated in place electronically.

An instrument calibration for a bubbler system, capacitance probe or pressure sensor is usually performed with a column of water into which the bubbler tube, probe or sensor is placed (Figure 7.28). The depth of the column of water (standpipe) is varied so as to simulate the flow through the primary element. The instrument should be checked at several depth levels, ranging from zero to the most common flow to the maximum head required for the system. The totalizing system of the bubbler should also be checked during this procedure to ensure its accuracy.

An instrument calibration for an ultrasonic transducer can be performed by removing the transducer from its mounting bracket and aiming it against an adjustable target plate. The meter should be checked at several distances, especially at

maximum distance (zero flow) and minimum distance (100 percent flow). Also check the totalizing system of the meter.

Hydraulic calibrations and instrument calibrations are usually performed under the supervision of a qualified engineer. An industrial waste inspector should be aware of the previously discussed calibration procedures; however, since they either require the use of extensive auxiliary equipment or a partial dismantling of the flow metering system, neither calibration procedure is likely to be performed by the inspector.

Of more importance to the inspector are the techniques needed to make comparative depth measurements to verify the accuracy of an operating flowmeter. Provided a confined space entry is not required, one person can check a flowmeter if the weir or flume and instrumentation are close together or if the flow is steady. However, simultaneous readings of depth measurements and indicated instrumentation levels are preferable and two persons are usually involved. The following are some of the required items that you should be familiar with in order to make these measurements.

It is important to check the flow rate during an average discharge period. The accuracy of a meter can be determined at any flow rate; however, observing a flow rate during an average discharge period will also allow you to observe the normal flow conditions for that particular system and to verify that the characteristics of the flow through the primary element are correct.

Some manufacturers of prefabricated open-channel elements install staff gages on the primary element at the measuring point. A staff gage is simply a ruler, and may be graduated in feet, inches, or units of flow. If an installation has such a staff gage, then a quick check of the meter's accuracy can be performed by reading the level at the staff gage and comparing it with the level indicated by the instantaneous flow rate shown on the instrumentation. It is difficult, however, to read staff gages located in deep manholes or vaults. Also, solids tend to cover the gages or obliterate the markings. It is often not possible to see exactly where the water touches the gage

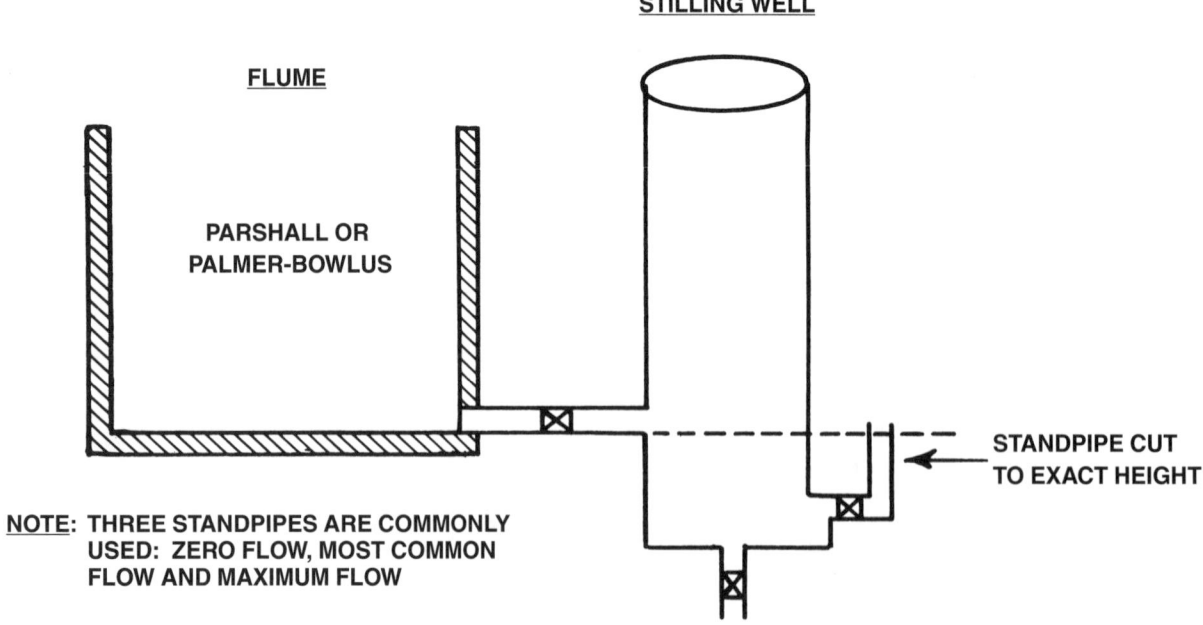

Fig. 7.28   Instrument calibration for depth measurement

and some allowance for this should be made when checking the depth of the water.

If a staff gage has not been installed on the primary device or if it cannot be easily read, then a manual measurement needs to be taken. This type of measurement can be made with a thin steel rule or a measuring tape. First, mark a reference point above the measuring point of the flume or weir and determine the distance from the zero flow level of the flume or weir to the reference point. Then, preferably during a steady flow period, measure the distance from the reference point to the surface of the water. The difference between these two distances is the depth in the depth-to-flow relationship (see Figure 7.9, page 494, and Figure 7.12, page 498).

In manually measuring the depth at the reference point, be careful not to disturb the flow while the measurement is being taken. Also, when the measurement to the water's surface is made, wave action may make it difficult to take a precise measurement. Flowmeter instrumentation is usually able to dampen out such wave action.

In order for accurate depths to be measured, care must be taken in determining the distance from the reference point to the water surface. Take the measurement to the water's surface by lowering a thin steel rule[5] into the water and noting the interference waves that form if it is not possible to clearly see the rule touch the water. When the surface of the water is touched, a drop of water will often remain on the end of the measuring stick. This drop should be knocked off before another measurement is taken or a false indication of the water surface may result.

In some situations it may not be possible to set a reference point or get close enough to the water to measure the depth with a thin steel rule. An alternative method for checking the accuracy is to use a long, thin pole that has been tapered on the lower end (with knife edges on both sides) to minimize turbulence and waves when inserting it into the flow. The wetted section of the pole can then be measured and checked against the instantaneous reading of the flowmeter. This method is not as precise as the methods previously mentioned, but provides a fairly good check of a meter's accuracy.

A number of manufacturers have developed portable equipment for checking the accuracy of an existing meter. Portable automatic flowmeters or velocity meters can be installed to provide continuous flow readings, and other types of equipment can be used to determine the instantaneous level or flow.

There are some important differences in the procedures used to take depth measurements from different types of primary elements. With Palmer-Bowlus flumes and V-notch and rectangular weirs, the measurement point is located a short distance upstream of the flume throat or weir plate. Also, as previously discussed, the desired depth is not the full depth of the water at the measurement point. To set a reference point in these cases, it is necessary to place a carpenter's level between the measurement point and the weir or flume. The reference measurement is set above the floor of the flume, or crest (notch) of the weir to the level, then the depth is determined from the distance down to the water surface at the measuring point. The depth or head is the distance from the reference point to the water surface. For weirs, the full measured depth is the head on the weir. With Parshall flumes and H-flumes, the depth of the flow is the full depth of the water at the measurement point.

If it is only possible to take a depth measurement at a Palmer-Bowlus flume by measuring the full depth of water, the assumed reference distance D/6 (see Figure 7.9, page 494) must be subtracted. However, as mentioned earlier, the actual distance from the floor of the channel to the floor of the flume may not be precisely D/6.

A reference point at a V-notch weir is usually set by placing a ruler into the notch to set the level. Due to the thickness of the ruler, it may not quite touch the very bottom of the notch and will cause a wave, thus producing an inaccurate measurement.

In many installations, it is not always possible to take the depth measurement precisely at the measuring point. With Palmer-Bowlus flumes and Parshall flumes, the measurement can be taken in the channel a short distance upstream or downstream of the measurement point. With H-flumes the measurement should be taken close to the measurement point. Since weirs form a large stilling pool, there is considerable latitude in where the measurement can be taken. In general, the depth measurement should be taken as close to the proper measuring point as conditions will allow.

A small amount of error will be associated with any method of measuring the depth of water behind a weir or flume. The magnitude of the error in the comparison check should be considered in the evaluation of a permanent flowmeter's accuracy.

---

[5] Rain gage sticks make excellent rules to measure water depths. These sticks are specially calibrated and will "wet" to the point of the water surface.

The measurement methods described in this section are not equally accurate. Errors related to the reading of a staff gage are assumed to be minor and therefore this means of determining a flow rate should be considered very accurate, provided the staff gage is properly installed and can be accurately read. Errors related to the determination of head by means of a reference point should be considered minor as long as the flow rate remains fairly constant during the check. Errors related to the use of a long tapered pole should be considered to be the greatest since the insertion of any obstruction into the flow can affect the flow conditions.

QUESTIONS

Write your answers in a notebook and then compare your answers with those on page 524.

7.4A   How is the movement of a depth-measuring float translated to flow?

7.4B   How does a bubbler measure flow depth?

7.4C   How does an ultrasonic meter measure water depth?

7.5A   What are the basic methods to check the accuracy of an open-channel flow metering system?

7.5B   How can an inspector check the accuracy of a flowmeter?

7.6   CLOSED-PIPE FLOW METERING SYSTEMS

7.60   Installation Conditions

Closed-pipe (pressure conduit) flowmeters are installed in a section of pipe that remains full under all normal discharge conditions. The pipe may flow from gravity conditions or from a pump discharge. Closed-pipe flowmeters are divided into two categories: (1) those that measure the average velocity of the flow (which is applied to the cross-sectional area of the pipe to determine the flow), and (2) those that produce a differential pressure across the meter by constricting the flow. The flow can be determined from that differential pressure.

A closed-pipe meter should be preceded and followed by five to ten pipe diameters of straight pipe to develop and maintain a satisfactory flow profile. A satisfactory profile means that the velocity is fairly uniform across the pipe. An unsatisfactory profile could occur near a bend or elbow. Manufacturers of such devices recommend that certain distances of straight pipe equal to so many pipe diameters be installed upstream and downstream of their meters.

As with open-channel meters, closed-pipe flowmeters should also be hydraulically calibrated with known flows when first installed. Instrument calibrations and hydraulic calibrations should be performed at regular intervals thereafter.

A general limitation of a closed-pipe flowmeter in the measurement of industrial wastewater is the difficulty of determining if the meter is clean. The materials present in some wastewaters can coat, clog or corrode a meter in an undesirably short period of time. This possibility should be considered in the selection of a meter. Flowmeters must be calibrated regularly (every six months) after installation.

7.61   Electromagnetic Flowmeters (Figure 7.29)

Electromagnetic flowmeters use Faraday's Law to determine flow rates. This principle states that if a conductor, in this case the water, is passed through a magnetic field, a voltage will be induced across the conductor and the voltage will be proportional to the velocity of the conductor and the strength of the magnetic field. Electromagnetic flowmeters produce the magnetic field and measure the voltage created by the movement of the water; the voltage reading is then translated to a flow measurement based on the pipe diameter. The mag meter does not have any intrusive parts, can operate over a wide range of velocities and is not sensitive to viscosity, density, turbulence, or suspended material. A minimum conductivity of the fluid is necessary; most wastewater is adequately conductive. Results can be affected by deposits of grease or oil, and some electromagnetic flowmeters are equipped with self-cleaning probes to remove these deposits from the measuring area.

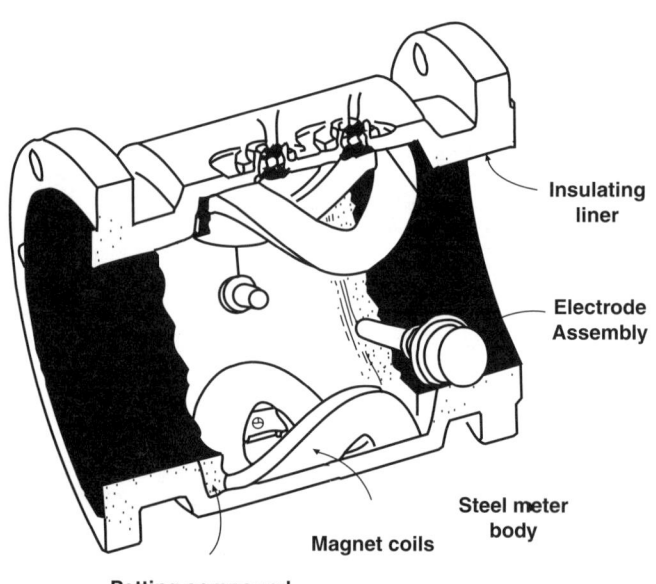

Insulating liner

Electrode Assembly

Steel meter body

Magnet coils

Potting compound

Fig. 7.29   Magnetic flowmeter
(Permission of Fischer and Porter Co.)

7.62   Turbine Meters and Propeller Meters (Figure 7.30)

Both of these meters operate on the principle that a fluid flowing past an impeller causes it to rotate at a speed proportional to the velocity of the flow. On some models the axis of the impeller is located in the direction of the flow; with others it is perpendicular to the flow. The motion of the impeller is conveyed through a mechanical device or a magnetic coupling to the registers of the meter. These meters are common in water measurement. The accuracy of the meter is affected by a poor flow profile, misalignment of the impeller, and accumulation of solids, especially oil and grease, on the impeller. Turbine and propeller meters are *NOT* used to measure flows in wastewaters that carry rags, rubber or plastic goods, and other abrasive debris or corrosive liquids.

The Doppler type of ultrasonic flowmeter makes use of the principle that a frequency shift of an ultrasonic signal occurs when the signal is reflected from a moving object; in this application, suspended solids or *ENTRAINED*[6] air bubbles in the wastewater reflect the signal. The frequency shift results in a higher returned frequency if the water is moving toward the transducer, and a lower returned frequency if the water is moving away from the transducer. The velocity of the water can be determined from the frequency shift.

Ultrasonic flowmeters are sensitive to flow profile effects. The manufacturer's recommendations for distances of upstream and downstream pipe diameters should be followed. This type of meter's accuracy is affected by pipe wall buildup and particle sound absorption. The in-line type of transducer is affected by a buildup of solids on the transducer. The clamp-on type of transducer is affected if the pipe and liner have sonic discontinuities in them or between them.

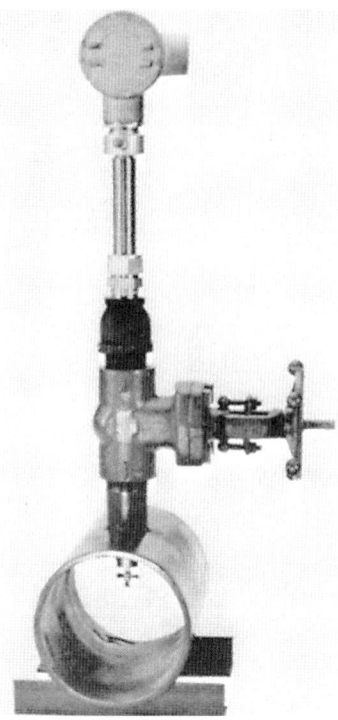

7.64 Pitot Tube Meters (Figure 7.31)

The pitot tube and similar devices measure the velocity at a single point within the pipe. With a proper length of straight pipe upstream, a pitot tube installed approximately 30 percent of the pipe radius from the inside pipe wall will give an average velocity reading. However, it may be necessary to profile the flow to find the location at which this average velocity occurs. Pitot tubes are appropriate for measuring clean water or gases rather than wastewater since they are sensitive to fouling.

7.65 Differential Pressure Systems (Figure 7.32)

An orifice plate meter consists of a thin flat plate with a precision-bored hole through it. Measuring taps are located on either side of the orifice plate to measure the pressure drop through the plate.

The Venturi meter creates a pressure differential by gradually reducing the pipeline cross section. The pressure differential is sensed from the high pressure and low pressure taps.

Flow nozzles and flow tubes use a smoothly curving inlet and a short throat section to develop the pressure differential. The flow tube has a more sharply curved inlet and a shorter throat than the flow nozzle, and the measuring taps are located closer together than with the flow nozzle.

Differential pressure meters tend to be used for measuring gases and clean water rather than wastewater. Orifice plate meters are especially sensitive to deposits on the edge of the hole in the plate. The pressure taps of each type of meter must remain clean for accurate measurements to result.

Fig. 7.30   Turbine flowmeter
(Permission of EFM)

7.63 Ultrasonic Meters

Ultrasonic flowmeters for closed-pipe flow use sound waves to measure the velocity of the water. In comparison, ultrasonic meters for open-channel flow measure distance. The velocity of the water is measured either by the transit time of the sound waves or by the "Doppler Effect." With the transit time type of meter, two transducers, each of which houses a transmitter and receiver, are located along the pipe. One transducer sends a signal in the direction of flow and the other transducer sends a signal opposite to the flow. The signal sent with the flow is received sooner than the signal sent against the flow. The difference in transit time is used to determine the velocity of the flow.

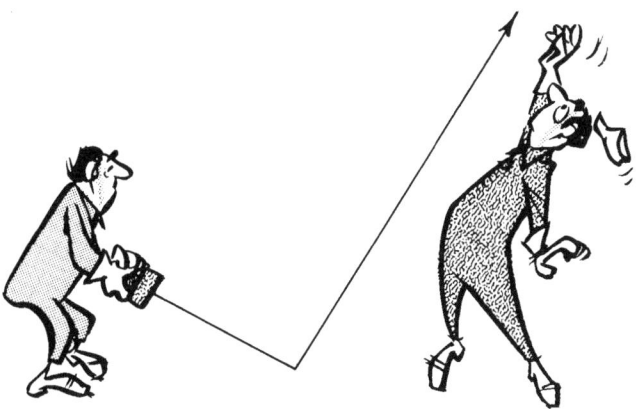

---

[6] *Entrain.   To trap bubbles in water either mechanically through turbulence or chemically through a reaction.*

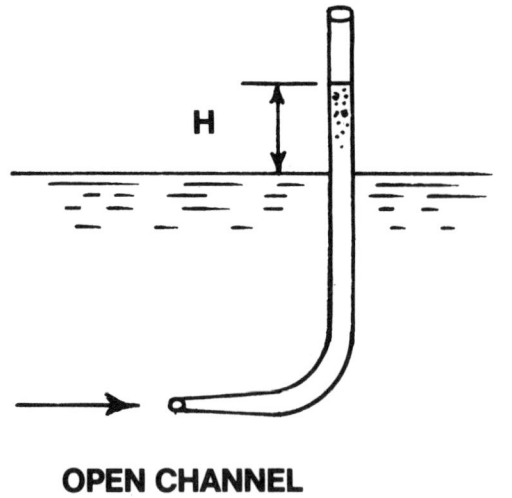
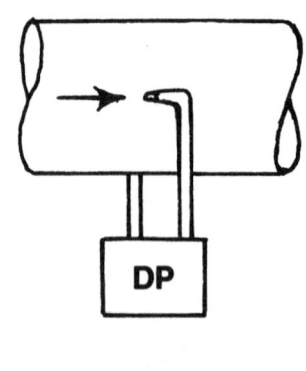

OPEN CHANNEL

PIPE

H = HEIGHT OF WATER

DP = DIFFERENTIAL PRESSURE MEASUREMENT

Fig. 7.31   Pitot tubes

7.66   Velocity Modified Flowmeters (VMFMs)

Velocity modified flowmeters are a type of hybrid between an open-channel flowmeter and a closed-channel flowmeter. This type of meter operates by sensing both depth and velocity. The velocity is determined by an electromagnetic sensor and the head is typically determined by a pressure sensor or bubbler. The sensors are secured in a ring which is inserted into the pipe. This flowmeter acts as both a primary and secondary device, therefore requiring no flume or weir. VMFMs are sized according to the pipe size into which they are inserted.

The VMFM flowmeter is especially useful for situations where the wastestream lines are occasionally submerged. The meter performs satisfactorily on clean streams and effluents, but not streams carrying solids. The depth sensor does not sense a depth greater than the pipe diameter and the meter is not affected by a slight amount of reverse flow.

The proper installation of a VMFM depends largely upon the channel or pipe into which it will be inserted. The channel should be relatively level and smooth. The flow through the VMFM should be subcritical and free of severe surface waves. Routine maintenance should be performed to keep the channel clean and free of obstructions that may affect the meter's accuracy.

7.7   FLOW-PROPORTIONED PACING OF AUTOMATIC SAMPLERS (Figure 7.33)

Most flowmeter instrumentation, for both open-channel and closed-pipe flow, includes a contact closure generating device used to pace an automatic sampler. This device generates a contact closure in proportion to flow. The sampling cycle of the sampler is triggered on each closure. The desired interval between gallons discharged is set on the flowmeter; for example, an *ALIQUOT*[7] (an individual discrete sample) could be obtained for every 1,000 gallons discharged. This method of sampling ensures that samples are taken in proportion to flow. Aliquots are not obtained when there is no flow through the sample point.

Some automatic samplers have the capability to skip contact closures. For example, with a flowmeter set at one closure per 1,000 gallons, it may be desirable to sample every 6,000 gallons. The sampler can be programmed to sample every sixth closure.

The contact closure interval to which a flowmeter and sampler are set depends on the flow at the facility and the number of aliquots that are desired during the sampling period as determined by the pretreatment inspector or the lab. It is necessary to know a facility's average daily flow and how much the flow can vary from the average. Aliquots should be obtained at least once an hour over a 24-hour period, but more frequent sampling may be advisable.

---

[7] *Aliquot (AL-li-kwot).   Portion of a sample. Often an equally divided portion of a sample.*

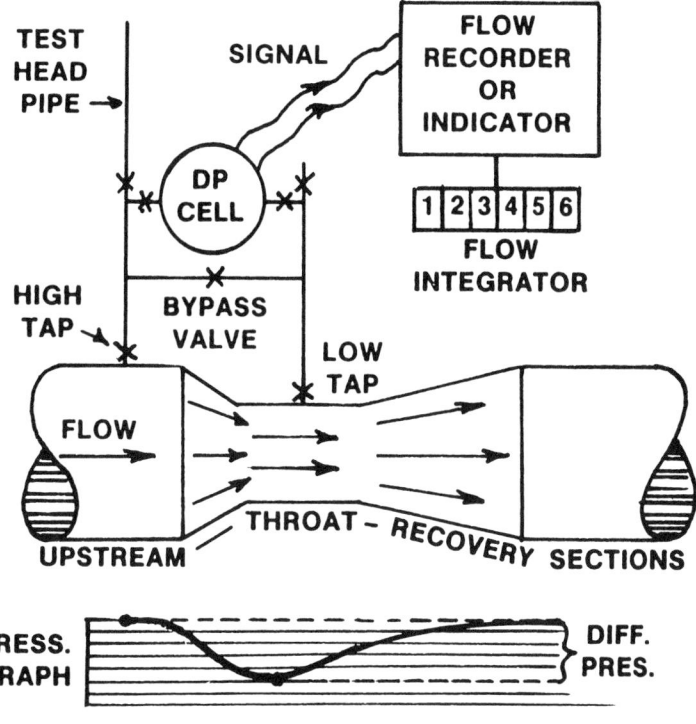

Venturi system (flow rate)

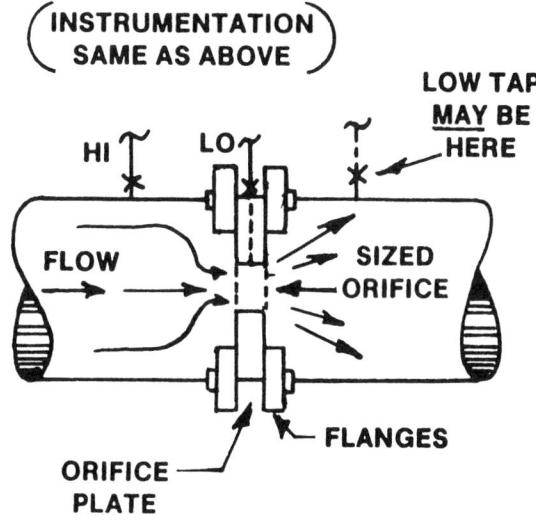

Orifice plate installation (flow rate)

Fig. 7.32   Schematic diagrams of differential pressure flow metering

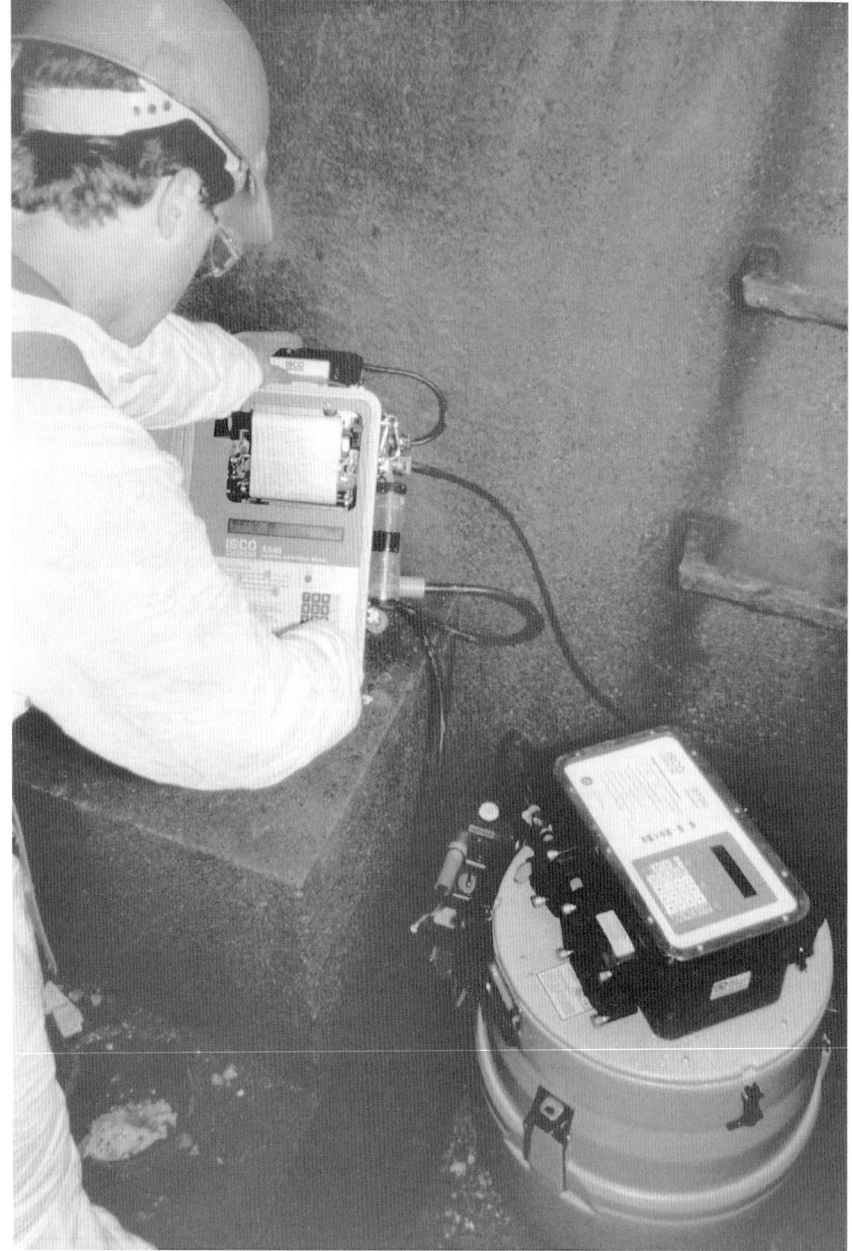

Fig. 7.33 Automatic sampling unit that can be either flow- or time-proportioned
(Courtesy of ISCO, Inc., Environmental Division)

Most automatic samplers accept an isolated (dry) contact closure from the flowmeter (the power source of the sampler provides the power for the circuit) to signal the sampler to operate. But, in order to prevent damage to the flowmeter or to the sampler, care should be taken to ensure that the flowmeter and sampler are compatible. The manufacturers of the equipment can provide the necessary information and instructions to properly interface samplers to flowmeters.

A sampler can be connected to a flowmeter in two basic ways. A permanent sampler can be directly wired to the meter, or a connector can be provided for use with portable samplers. When two samplers are to be used at the same time with the same flowmeter, the connections should be wired in parallel or through a double-pole relay. Flowmeters and samplers are often equipped with sampling connectors, but these connectors are not usually compatible between different makes of equipment. A standard connector can be specified as part of the flow monitoring system.

The flowmeter instrumentation is often installed at a considerable distance from the sample point. Sampler connectors for portable samplers should be permanently located near the sample point to keep patch cords away from pedestrian and vehicular traffic. Connectors installed in manholes or vaults should be located within easy reach so that a confined space entry is not required for sampling.

The proper operation of the flow-proportioning equipment of the flowmeter should be checked periodically. Flowmeters that generate isolated closures can be checked with an ohmmeter by placing the ohmmeter's leads on the sampler connection. The relay from the contact closure generating device is normally open, and the ohmmeter will show infinite resistance. When contact closures occur, the reading on the ohmmeter will drop to zero. Some malfunctions of the instrumentation, or improper wiring of the connector, can cause 110 V A.C. to show on the connector, which will damage a sampler. This can be checked with a voltmeter. Connectors located in manholes and vaults are vulnerable to corrosion and should be checked and cleaned regularly. Moisture in the connector can cause a short which will show on the ohmmeter.

7.8 APPROXIMATE MEASUREMENT METHODS

As previously discussed in the introduction, there are other methods for estimating discharge flow rates which do not involve direct measurement of the effluent flow. Probably the most accurate of these methods involves the use of an adjusted, metered water supply procedure in which you subtract evaporative and consumptive losses from incoming source water amounts to yield a discharge value. This calculated discharge establishes a flow rate over an extended time period; care must be taken in placing too much confidence in such flow rates over the short term, as in-plant tank holding capacities can distort the flow determination.

In using the adjusted, metered water supply method, all sources of significant discharge must first be identified. The incoming flows of most interest are metered fresh water supplied by a water purveyor or well water provided by an auxiliary well. Unless the on-line meters used with these sources are new, or have been calibrated within the last several years (preferably every four years for major compound meters), their readings should be considered suspect. Other sources of discharge should also be considered, such as the amount of wastewater received as raw material. For certain industrial dischargers this alternative source of flow can be significant (for example, brines associated with petroleum extraction op-

erations, rendering plants, and milk received for cheese production). Production reports required by governmental agencies may be of use in substantiating this source of discharge; these reports may consist of oil field extraction records, milk pooling documents or other similar filings submitted by industry. It is fairly common to measure water losses by putting meters on boiler water make-up lines, product water lines at soft drink water operations, and even some irrigation systems.

In the midwest and east where water supply usually is not as critical as in arid regions, noncontact cooling water discharged to surface waters is the most significant nonsewered loss (evaporative losses are usually greater in arid regions). The amount of noncontact cooling water lost can be determined by direct water metering or subtraction between incoming water supply consumption and measured process wastewater discharged to the sanitary sewer.

Of the nonsewered losses experienced by industrial dischargers, evaporative losses are probably the most significant in arid regions. In both cooling towers and boilers, lost or rejected heat is applied to water resulting in water being evaporated. If the amount of heat applied to the water can be determined, then the evaporative losses can be reasonably calculated after the blowdown and bleedoff have been added back in.

The amount of heat or energy required to evaporate a liquid is called the heat of vaporization. The heat of vaporization value for water at 77° Fahrenheit is 1,050 British thermal units (BTUs) per pound of water. Consequently, with one gallon of water being equal to 8.34 pounds, it would take 8,760 BTUs to evaporate this one gallon. If the amount of heat applied to the water source (either through a cooling tower, boiler or other device) can be established in terms of BTUs, then the amount of water evaporated (in gallons) can simply be determined by dividing the amount of BTUs by 8,760.

With a cooling tower, evaporation, drift and splash make up the nonsewered losses. Drift refers to the loss of small water droplets that are carried out in the exhaust gases; this loss is generally considered to be of little consequence and should not be more than 0.01 percent of the cooling tower recirculation rate. Splash loss refers to the relatively small amount of water that falls outside the cooling tower and is of little importance in the overall amount of water that is not sewered. As described above, the evaporation losses can be determined from the heat load in BTUs. The following equations may be used to determine this heat load or the BTUs applied to water losses:

$$\begin{aligned}\text{Heat Rejected,} \\ \text{BTU/yr}\end{aligned} = \text{(1 BTU/lb-°F)(Water Temperature Difference, °F)}$$

$$\text{(8.34 lbs/gal)(Recirculation Flow, GPM)}$$

$$\text{(Work Time, days/yr)(Operating Time, hr/day)(60 min/hr)}$$

where the temperature difference represents the increase across the cooling tower and the operating minutes represent the time that a heat load is being applied.

$$\begin{aligned}\text{Heat Rejected,} \\ \text{BTU/yr}\end{aligned} = \text{(Rated Tower Capacity, tons)(12,000 BTU/ton-hr)}$$

$$\text{(Work Time, days/yr)(Operating Time, hr/day)(Load Factor)}$$

where the load factor represents the ratio between the actual usage of a piece of equipment to its rated capacity (generally 0.8 for units designed for their present application) and operating hours represent the time a heat load is being applied.

Boilers are used in various types of applications at industrial facilities; the use of these boilers often determines the amount of nonsewered losses. If a boiler is used to produce steam to heat a product and no attempt is made to capture any resulting condensate, then the amount of steam produced would represent the losses. However, in most cases the condensate is either sewered or returned to the boiler to produce additional steam. These procedures can significantly reduce evaporative losses. One method of determining whether or not condensate is being discharged to a sewer (being sewered) is to add dye to the wastewater and observe the downstream sewer flows for the presence of the dye. Boiler losses due to evaporation can be determined from the following equation:

$$\begin{aligned}\text{Boiler Losses,} \\ \text{BTU/day}\end{aligned} = \text{(Rated Boiler Capacity, HP)(33,446 BTU/HP-hr)}$$

$$\text{(Operating Time, hr/day)(Load Factor)}$$

$$\frac{\text{(100\% – Condensate Returned or Sewered, \%)}}{\text{100\%}}$$

In addition to evaporative losses, there may be losses to product and landscape irrigation usage at industrial facilities. The amount of water consumed in a product has to be determined on a case-by-case basis; these losses are generally unique and can only be determined by investigating the in-plant water use. Landscape irrigation also varies; however, the amount of water applied in such instances per unit area is often fairly uniform on a regional basis. Consult local agricultural agencies to determine appropriate landscape irrigation requirements.

Long-term effluent flow rates can be reasonably estimated using evaporative loss calculations provided there are no unusual water uses present. Many dischargers within the Los Angeles County Sanitation Districts with annual effluent flow rates between one and twelve million gallons per year use this method. The following paragraphs contain examples of this method for *CALCULATING COOLING TOWER WATER LOSSES.*

A company has two cooling towers, with different types of operational data available for each of them. Cooling Tower A is known to have a recirculation rate of 400 GPM, the incoming water temperature is 85°F and the outgoing water temperature is 72°F. This cooling tower is used to cool a heat source 256 days per year, 18 hours per day. The operational data for Cool-

ing Tower B is limited to the manufacturer's literature which states that its capacity is 7 tons with a load factor of 0.8; the company uses this unit to cool a heat source 300 days per year, 8 hours per day. This cooling tower was designed for its present application. The evaporative losses associated with each of these cooling towers could be calculated in the following manner.

Cooling Tower A

Known		Unknown
Flow, GPM	= 400 GPM	Water Evaporated, gallons/year
Time, days/yr	= 256 days/yr	
Time, hr/day	= 18 hr/day	
Incoming Temp., °F	= 85°F	
Outgoing Temp., °F	= 72°F	

Cooling Tower A — Calculations

1. Calculate the heat rejected in BTUs per year by Cooling Tower A.

$$\begin{aligned}\text{Heat Rejected,} \\ \text{BTU/yr}\end{aligned} = \begin{aligned}&\text{(1 BTU/lb-°F)(Temp. In, °F – Temp. Out, °F)} \\ &\text{(8.34 lbs/gal)(Flow, GPM)(Time, days/yr)} \\ &\text{(Time, hr/day)(60 min/hr)}\end{aligned}$$

$$= \begin{aligned}&\text{(1 BTU/lb-°F)(85°F – 72°F)(8.34 lbs/gal)} \\ &\text{(400 GPM)(256 days/yr)(18 hr/day)(60 min/hr)}\end{aligned}$$

$$= 1.199 \times 10^{10} \text{ BTU/yr}$$

2. Calculate the water evaporated in gallons per year on the basis of the heat rejected from Cooling Tower A.

$$\begin{aligned}\text{Water Evaporated,} \\ \text{gallons/year}\end{aligned} = \frac{\text{Heat Rejected, BTU/yr}}{\text{8,760 BTU/gal}}$$

$$= \frac{1.199 \times 10^{10} \text{ BTU/yr}}{\text{8,760 BTU/gal}}$$

$$= 1,369,000 \text{ gallons/year}$$

Cooling Tower B

Known		Unknown
Capacity, tons	= 7 tons	Water Evaporated, gallons/year
Time, days/yr	= 300 days/yr	
Time, hr/day	= 8 hr/day	
Load Factor	= 0.8	

Cooling Tower B — Calculations

1. Calculate the heat rejected in BTUs per year by Cooling Tower B.

$$\text{Heat Rejected, BTU/yr} = \frac{\text{(Capacity, tons)(12,000 BTU/ton-hr)}}{\text{(Time, days/yr)(Time, hr/day)(Load Factor)}}$$

$$= \frac{\text{(7 tons)(12,000 BTU/ton-hr)(300 days/hr)}}{\text{(8 hr/day)(0.8)}}$$

$$= 1.6128 \times 10^8 \text{ BTU/yr}$$

2. Calculate the water evaporated in gallons per year on the basis of the heat rejected from Cooling Tower B.

$$\text{Water Evaporated, gallons/year} = \frac{\text{Heat Rejected, BTU/yr}}{8,760 \text{ BTU/gal}}$$

$$= \frac{1.6128 \times 10^8 \text{ BTU/yr}}{8,760 \text{ BTU/gal}}$$

$$= 18,411 \text{ gal/yr}$$

7.9 ADDITIONAL READING

1. *A GUIDE TO METHODS AND STANDARDS FOR THE MEASUREMENT OF WATER FLOW* by G. Kulin and P. Compton, U.S. Department of Commerce. Obtain from National Technical Information Service (NTIS), 5285 Port Royal Road, Springfield, VA 22161. Order No. COM 7510683/LL. Price, $34.00 (includes shipping), plus $5.00 handling charge per order.

2. *OPEN CHANNEL FLOW MEASUREMENT HANDBOOK* by D. Grant. Obtain from ISCO, Inc., PO Box 82531, Lincoln, NE 68501. Order No. 60-3003-041. Price, $50.00.

3. *STEVENS WATER RESOURCES DATA BOOK.* Obtain from Stevens Water Monitoring Systems, Inc., 5645 SW Western Avenue, Suite F, Beaverton, OR 97005. Order No. 31751. Price, $20.00.

7.10 ACKNOWLEDGMENT

Some of the instrumentation drawings were prepared by Leonard Ainsworth for Chapter 19, "Instrumentation," in *WATER TREATMENT PLANT OPERATION*, Volume II, in this series of training manuals.

QUESTIONS

Write your answers in a notebook and then compare your answers with those on page 524.

7.6A What are the two categories of closed-pipe flow metering systems?

7.6B How does an electromagnetic flowmeter work?

7.6C How do turbine and propeller flowmeters work?

7.7A How does flow-proportioned pacing of automatic samplers work?

Please answer the discussion and review questions before continuing with the Objective Test.

DISCUSSION AND REVIEW QUESTIONS
Chapter 7. WASTEWATER FLOW MONITORING

Write the answers to these questions in your notebook before continuing with the Objective Test on page 525. The purpose of these questions is to indicate to you how well you understand the material in the chapter.

1. What is the preferred method of collecting flow-weighted composite samples?

2. How can influent flow measurements be used to determine discharge flow rates?

3. What are the three integrated instrumentation devices in an automatic flowmeter for describing the flow?

4. What factors could cause a Palmer-Bowlus flume to be in a submerged condition?

5. Describe, in general terms, the function of instrumentation for measuring open-channel flow.

6. How can errors due to manufacturing tolerances and errors due to gradual wear and tear be detected in a float system?

7. How can you prevent the end of the bubbler tubing from becoming clogged by solids in the wastestream?

8. How is a flowmeter hydraulically calibrated?

9. Why must a closed-pipe meter be preceded and followed by a certain number of pipe diameters of straight pipe?

10. What is a general limitation of a closed-pipe flowmeter in the measurement of industrial wastewater?

11. What is the difference between closed-pipe and open-channel ultrasonic flowmeters?

12. What problems may develop with sample connectors used with flow-proportioning equipment located in manholes?

SUGGESTED ANSWERS

Chapter 7.   WASTEWATER FLOW MONITORING

Answers to questions on page 490.

7.0A   Industrial waste inspectors must understand wastewater flow measurement procedures to evaluate effluent flow metering devices at significant industrial dischargers. Accurate flow measurements are needed to determine applicable sewer-use charges, to establish the amount of pollutants discharged over a period of time, and to provide for the collection of representative composite samples of discharges of varying waste strengths.

7.0B   Effluent flows can be easily measured before discharge to a POTW sewer.

7.1A   Flow may be directly measured as either open-channel flow or closed-pipe (pressure conduit) flow.

7.1B   The term "primary element" means the measuring structure that contains the water, and the term "secondary element" refers to the means by which measurements are taken from the primary element and converted to flow readings.

7.2A   The purpose of the open-channel flume or weir is to create flow conditions in which a predictable relationship exists between flow and depth.

7.2B   The accuracy of an open-channel flow measurement system can be checked by visually inspecting the flume or weir for characteristic flow patterns. Instrumentation can be checked by comparing instrument depth measurements with independent depth measurements, observing indicators and recorders, and timing the rate of totalizer advances.

7.2C   Precautions that must be observed when installing weirs in sewers include: (1) all joints must be smooth so that waves are not created, (2) the weir plate must be vertical, (3) the weir notch must be symmetrical about a vertical line, and (4) the weir plate should be properly supported so that it does not warp.

Answers to questions on page 508.

7.3A   To measure the depth to determine the flow for a Palmer-Bowlus flume, measure the vertical distance between the floor of the flume and the water surface at the measuring point.

7.3B   To measure the depth to determine flow in a Parshall flume, measure the depth of water between the floor of the channel and the water surface at the measuring point. Using a table or rating curve for the size of flume, use the depth of flow reading to determine the flow.

7.3C   Weirs differ from flumes in that a weir is essentially a dam across the flow, as compared to a reshaping of the channel in a flume.

7.3D   A weir should *NOT* be operated in a submerged condition because the measured discharge will be greater than the actual discharge.

Answers to questions on page 516.

7.4A   The movement of a depth-measuring float is translated to flow by a mechanical cam.

7.4B   A bubbler measures flow depth by the use of a bubbler tube located in the water. A constant flow of air is bubbled through the tubing. The meter measures the pressure that is necessary to maintain a constant bubble rate. The pressure corresponds to the depth of water above the end of the bubbler tube.

7.4C   An ultrasonic meter measures water depth by determining the time it takes an acoustic pulse generated in a transducer to reach and be reflected from the air/liquid interface.

7.5A   The basic methods to check the accuracy of an open-channel flow metering system are hydraulic calibration, instrument calibration, and comparative depth measurements.

7.5B   An inspector can check the accuracy of a flowmeter by determining the depth of flow on a staff gage or by manually measuring the depth of flow.

Answers to questions on page 523.

7.6A   The two categories of closed-pipe flow metering systems are those that measure the average velocity of the flow, and those that produce a differential pressure across the meter by constricting the flow.

7.6B   Electromagnetic flowmeters use Faraday's Law to determine flow rates. This principle states that if a conductor, in this case the water, is passed through a magnetic field, a voltage will be induced across the conductor and the voltage will be proportional to the velocity of the conductor and the strength of the magnetic field.

7.6C   Turbine and propeller flowmeters operate on the principle that a fluid flowing past an impeller causes it to rotate at a speed proportional to the velocity of the flow.

7.7A   Flow-proportioned pacing of automatic samplers works on the basis of a contact closure generating device used to pace the automatic sampler. This device generates a contact closure in proportion to flow. The sampling cycle of the sampler is triggered on each closure. The desired interval between gallons discharged is set on the flowmeter and a sample is collected after every time the specified volume of water flows through the meter.

OBJECTIVE TEST

Chapter 7.   WASTEWATER FLOW MONITORING

Please write your name and mark the correct answers on the answer sheet, as directed at the end of Chapter 1. There may be more than one correct answer to each multiple-choice question.

True-False

1. The best way to determine an industrial discharger's impact on wastewater facilities is to directly measure each industry's discharge flows.

    — 1. True
    2. False

2. Concentration-based waste limits require flow measurements.

    1. True
    — 2. False

3. Flow is the amount of water going past a particular reference point over a certain period of time.

    — 1. True
    2. False

4. The depth of the water in the flume or weir can be measured at any location in the channel.

    1. True
    — 2. False

5. Flumes and weirs must be cleaned regularly.

    — 1. True
    2. False

6. A Parshall flume operates on the same principles as the Palmer-Bowlus flume.

    — 1. True
    2. False

7. The broad-crested weir is more commonly used in measuring industrial wastewater flows than the sharp-crested weir.

    1. True
    — 2. False

8. Always provide adequate clearance below a weir notch for a free discharge to occur.

    — 1. True
    2. False

9. The flow totalizer, indicator, and recorder should be properly labeled to prevent problems in interpreting their readings.

    — 1. True
    2. False

10. A small amount of error will be associated with any method of measuring the depth of water behind a weir or flume.

    — 1. True
    2. False

Best Answer (Select only the closest or best answer.)

11. What are the characteristics of a properly operating flume?

    — 1. Distinct and easy to observe
    2. Fluctuating and difficult to measure
    3. Submerged hydraulic jump
    4. Uniform and flat

12. What does a stilling well provide?

    — 1. A calm pooling area where the depth can be accurately measured
    2. A location where the totalizer is not subject to vibrations
    3. A point to measure critical flow depth
    4. A source of water for pneumatic sensing systems

13. How is the flow depth measured in a Palmer-Bowlus flume?

    1. Channel bottom to water surface at measuring point
    2. Channel bottom to water surface at throat
    — 3. Floor of flume to water surface at measuring point
    4. Floor of flume to water surface at throat

14. How can the effect of float lag be minimized?

    1. By calibrating at standard conditions
    2. By observing movement after flow changes
    3. By removing grease from the float
    — 4. By taking the measurement when the flow is steady

15. What is the purpose of a bubbler?

    1. Aerate flowmeter channel
    2. Break down foam on water surface
    — 3. Measure depth of water
    4. Mix water in stilling well

16. Why is an instrument calibrated?

    1. To calculate the flow rate
    — 2. To check the performance of the instrumentation
    3. To measure the flow
    4. To verify the maintenance program

17. Closed-pipe (pressure conduit) flowmeters are installed under what flow conditions?

    1. In section of pipe at the end of a force main
    2. In section of pipe exposed to submerged conditions
    3. In section of pipe that rarely flows full
    — 4. In section of pipe that remains full under all normal discharge conditions

18. How do electromagnetic flowmeters measure flows? By measuring the_____ created by the movement of water.

 1. Area
 2. Depth
 3. Pressure differential
 — 4. Voltage

19. Electromagnetic flowmeters are sensitive to what factor?

 1. Density
 2. Suspended matter
 3. Turbulence
 — 4. Velocity

20. Where should sampler connectors for portable samplers be located?

 1. Above ground
 2. Close to the patch cords
 3. In manholes
 — 4. Near the sample point

Multiple Choice (Select all correct answers.)

21. Why are wastewater flows measured?

 — 1. To determine applicable sewer-use charges
 — 2. To establish amount of pollutants discharged
 3. To estimate emissions of gases
 — 4. To prepare composite samples
 5. To select pipe materials

22. What are the typical units used to measure flow?

 — 1. CFS
 2. FPS
 — 3. GPM
 — 4. MGD
 5. mg/L

23. What types of instruments are used to measure flow depths?

 — 1. Bubblers
 — 2. Capacitance probes
 — 3. Floats
 — 4. Pressure sensors
 — 5. Ultrasonic transducers

24. What types of flowmeters create pressure conditions in a pipe to measure the flow?

 1. Electromagnetic meters
 — 2. Flow nozzles
 — 3. Flow tubes
 — 4. Orifice plates
 — 5. Venturi meters

25. What are the advantages of a stilling well?

 — 1. Eliminates effects of foam on water surface
 — 2. Eliminates effects of rags
 3. Frequent maintenance is necessary to keep well clean
 4. No maintenance is required to clean connection to flume
 — 5. Not affected by wave action in a flume or weir

26. When can a submerged condition occur in a flume?

 — 1. Adverse flow conditions in downstream sewer
 — 2. Debris in downstream sewer
 — 3. Downstream sewer not able to carry flow
 — 4. Improper slope of downstream sewer
 5. Supercritical sewer slope downstream of throat

27. What is the function of open-channel flow instrumentation?

 — 1. Activate automatic sampler
 — 2. Convert depth to flow
 — 3. Indicate flow
 — 4. Sense level of water
 — 5. Totalize flow

28. What are sources of equipment errors in float-operated systems?

 — 1. Buildup of grease on float
 — 2. Errors in manufacturing tolerances
 — 3. Float lag
 4. Proper lubrication
 — 5. Temperature and humidity changes

29. What are maintenance tasks for float measuring systems?

 — 1. Checking for free movement of float and cable
 — 2. Flushing the stilling well
 3. Leaking floats
 — 4. Proper lubrication
 5. Wear and tear

30. What factors can cause errors in ultrasonic flowmeter readings?

 — 1. Dirty face of transducer
 — 2. Excessive wave action on water surface
 — 3. Foam layer on water surface
 — 4. Improper generation and receipt of acoustic pulses
 5. Location of transducer over stilling well

31. How often should flow-proportioned automatic samplers be programmed to collect a sample? At least once every _____hour(s) over a 24-hour period.

 — 1. 1
 2. 2
 3. 3
 4. 4
 5. 6

32. What are the possible uses of automatic flow-proportional samplers?

 — 1. Determining strength and content of wastestream during a 24-hour period
 — 2. Detecting spills or bypasses during unusual hours
 — 3. Determining maximum and minimum discharges during a 24-hour period
 — 4. Filling bottles in samplers in proportion to flow
 — 5. Determining billing rate to discharger

END OF OBJECTIVE TEST

CHAPTER 8

INDUSTRIAL WASTEWATERS

by

Richard W. von Langen

and

Mahin Talebi

TABLE OF CONTENTS

Chapter 8. INDUSTRIAL WASTEWATERS

OBJECTIVES

Chapter 8.   INDUSTRIAL WASTEWATERS

Following completion of Chapter 8, you should be able to:

1. Describe the types of industrial wastewaters discharged to sewers,

2. Explain the difference between concentration and mass of pollutants,

3. Outline various types of manufacturing processes and the wastewaters generated from each process, and

4. Identify the sources of industrial wastewaters and describe their effects on the pretreatment system and the POTW collection, treatment, and disposal systems.

PRETREATMENT WORDS
Chapter 8.   INDUSTRIAL WASTEWATERS

ANODIZING   ANODIZING

An electrochemical process which deposits a coating of an insoluble oxide on a metal surface. Aluminum is the most frequently anodized material.

APPURTENANCE (uh-PURR-ten-nans)   APPURTENANCE

Machinery, appliances, structures and other parts of the main structure necessary to allow it to operate as intended, but not considered part of the main structure.

BIODEGRADABLE (BUY-o-dee-GRADE-able)   BIODEGRADABLE

Organic matter that can be broken down by bacteria to more stable forms which will not create a nuisance or give off foul odors is considered biodegradable.

BLOWDOWN   BLOWDOWN

The removal of accumulated solids in boilers to prevent plugging of boiler tubes and steam lines. In cooling towers, blowdown is used to reduce the amount of dissolved salts in the recirculated cooling water.

CHELATING (key-LAY-ting) AGENT   CHELATING AGENT

A chemical used to prevent the precipitation of metals (such as copper).

COMPATIBLE POLLUTANTS   COMPATIBLE POLLUTANTS

Those pollutants that are normally removed by the POTW treatment system. Biochemical oxygen demand (BOD), suspended solids (SS), and ammonia are considered compatible pollutants.

CONVENTIONAL POLLUTANTS   CONVENTIONAL POLLUTANTS

Those pollutants which are usually found in domestic, commercial or industrial wastes such as suspended solids, biochemical oxygen demand, pathogenic (disease-causing) organisms, adverse pH levels, and oil and grease.

DELETERIOUS (DELL-eh-TEAR-ee-us)   DELETERIOUS

Refers to something that can be or is hurtful, harmful or injurious to health or the environment.

DRAG OUT   DRAG OUT

The liquid film (plating solution) that adheres to the workpieces and their fixtures as they are removed from any given process solution or their rinses. Drag out volume from a tank depends on the viscosity of the solution, the surface tension, the withdrawal time, the draining time and the shape and texture of the workpieces. The drag out liquid may drip onto the floor and cause wastestream treatment problems. Regulated substances contained in this liquid must be removed from wastestreams or neutralized prior to discharge to POTW sewers.

IMMISCIBLE (im-MISS-uh-bull)   IMMISCIBLE

Not capable of being mixed.

INHIBITORY SUBSTANCES   INHIBITORY SUBSTANCES

Materials that kill or restrict the ability of organisms to treat wastes.

INTERFERENCE   INTERFERENCE

Interference refers to the harmful effects industrial compounds can have on POTW operations, such as killing or inhibiting beneficial microorganisms or causing treatment process upsets or sludge contamination.

INTERSTICES (in-TOUR-stee-sees)   INTERSTICES

Small crevices on parts where concentrated plating solutions gather during the plating process. This liquid solution becomes the drag out when the part is removed from the plating bath.

MASS EMISSION RATE   MASS EMISSION RATE

The rate of discharge of a pollutant expressed as a weight per unit time, usually as pounds or kilograms per day.

MERCAPTANS (mer-CAP-tans)   MERCAPTANS

Compounds containing sulfur which have an extremely offensive skunk-like odor; also sometimes described as smelling like garlic or onions.

MISCIBLE (MISS-uh-bull)   MISCIBLE

Capable of being mixed. A liquid, solid, or gas that can be completely dissolved in water.

NONBIODEGRADABLE (non-BUY-o-dee-GRADE-able)   NONBIODEGRADABLE

Substances that cannot readily be broken down by bacteria to simpler forms.

NONCOMPATIBLE POLLUTANTS   NONCOMPATIBLE POLLUTANTS

Those pollutants which are normally *NOT* removed by the POTW treatment system. These pollutants may be a toxic waste and may pass through the POTW untreated or interfere with the treatment system. Examples of noncompatible pollutants include heavy metals such as copper, nickel, lead, and zinc; organics such as methylene chloride, 1,1,1-trichloroethylene, methyl ethyl ketone, acetone, and gasoline; or sludges containing toxic organics or metals.

PASSIVATING   PASSIVATING

A metal plating process which forms a protective film on metals by immersion in an acid solution, usually nitric acid or nitric acid with sodium dichromate.

PICKLE   PICKLE

An acid or chemical solution in which metal objects or workpieces are dipped to remove oxide scale or other adhering substances.

REAGENT (re-A-gent)   REAGENT

A pure chemical substance that is used to make new products or is used in chemical tests to measure, detect, or examine other substances.

TOXIC   TOXIC

A substance which is poisonous to a living organism. Toxic substances may be classified in terms of their physiological action, such as irritants, asphyxiants, systemic poisons, and anesthetics and narcotics. Irritants are corrosive substances which attack the mucous membrane surfaces of the body. Asphyxiants interfere with breathing. Systemic poisons are hazardous substances which injure or destroy internal organs of the body. The anesthetics and narcotics are hazardous substances which depress the central nervous system and lead to unconsciousness.

VOLATILE (VOL-uh-tull)   VOLATILE

(1) A volatile substance is one that is capable of being evaporated or changed to a vapor at relatively low temperatures. Volatile substances also can be partially removed by air stripping.

(2) In terms of solids analysis, volatile refers to materials lost (including most organic matter) upon ignition in a muffle furnace for 60 minutes at 550°C. Natural volatile materials are chemical substances usually of animal or plant origin. Manufactured or synthetic volatile materials such as ether, acetone, and carbon tetrachloride are highly volatile and not of plant or animal origin.

CHAPTER 8.   INDUSTRIAL WASTEWATERS

8.0   IMPORTANCE OF UNDERSTANDING INDUSTRIAL WASTEWATERS

By understanding the sources and quantities of industrial wastewaters, an inspector can identify, define, and solve problems caused by industrial discharges. In addition, you will need to understand how the various industrial wastes interact with each other and with domestic wastewater to deal effectively with the impact of industrial wastewaters on your facilities. You must have a basic understanding of chemistry and chemical reactions, the manufacturing processes, biological reactions, and your own collection and treatment system. This chapter will present information about the most common types of industrial wastes, tell you how they are generated, and explain what effects industrial wastes may have on your facilities. Throughout the chapter, try to "read between the lines" to understand the methodology of investigating industrial waste sources, and look at the examples as a guide to investigative technique and thinking. By noting the methodology, you can develop a more complete understanding of industrial wastes and the mixture of wastes discharged to your POTW systems.

Wastes generated by industry may be *COMPATIBLE POLLUTANTS*[1] such as biochemical oxygen demand (BOD), suspended solids (SS), ammonia, or oil and grease, or they may be *NONCOMPATIBLE POLLUTANTS*[2] such as heavy metals (cadmium, chromium, copper) or organics (acetone, methyl ethyl ketone, toluene, 1,1,1-trichloroethylene). Other chemical constituents such as sodium, calcium, nitrate, carbonate, sulfide, and pH in excess can also affect collection and treatment facilities as well as the potential reuse of the effluent. Physical characteristics such as temperature, solubility or insolubility, and viscosity of the industrial waste can affect the chemical, mechanical, and biological activities of the conveyance and treatment systems.

Industrial wastewater discharges can have numerous adverse impacts on the wastewater collection system, treatment facilities and the environment. The discharges can corrode sewers, plug sewers and release toxic gases and obnoxious odors in the collection system. When these wastewaters reach treatment plants they can be toxic to the biological processes, they can inhibit microorganisms from performing their intended functions, and they can accumulate to undesirable levels in the microorganisms. After industrial wastes leave the treatment plant or facility in process sludges or the treated effluent, they can be toxic to microorganisms, aquatic life and people in the environment.

The frequency of generation and discharge, whether it be once per year or once per day, can have an effect on your system.

8.00   Manufacturing Processes and Wastewater Generation

Knowledge of the various manufacturing processes is important in understanding how raw materials are transformed into products, by-products and waste products, and in understanding the effects of the waste products on collection systems, the treatment plant, and the environment. Many different industries use the same basic manufacturing processes including plating, sheet forming, distilling, extraction, screening, grinding, and heating. Understanding the way these manufacturing processes produce wastes enables you to evaluate an industry's proposed methods of controlling pollution at the source by recycling, reuse, or producing a useful by-product (Figure 8.1).

---

[1] *Compatible Pollutants.   Those pollutants that are normally removed by the POTW treatment system. Biochemical oxygen demand (BOD), suspended solids (SS), and ammonia are considered compatible pollutants.*

[2] *Noncompatible Pollutants.   Those pollutants which are normally NOT removed by the POTW treatment system. These pollutants may be a toxic waste and may pass through the POTW untreated or interfere with the treatment system. Examples of noncompatible pollutants include heavy metals such as copper, nickel, lead, and zinc; organics such as methylene chloride, 1,1,1-trichloroethylene, methyl ethyl ketone, acetone, and gasoline; or sludges containing toxic organics or metals.*

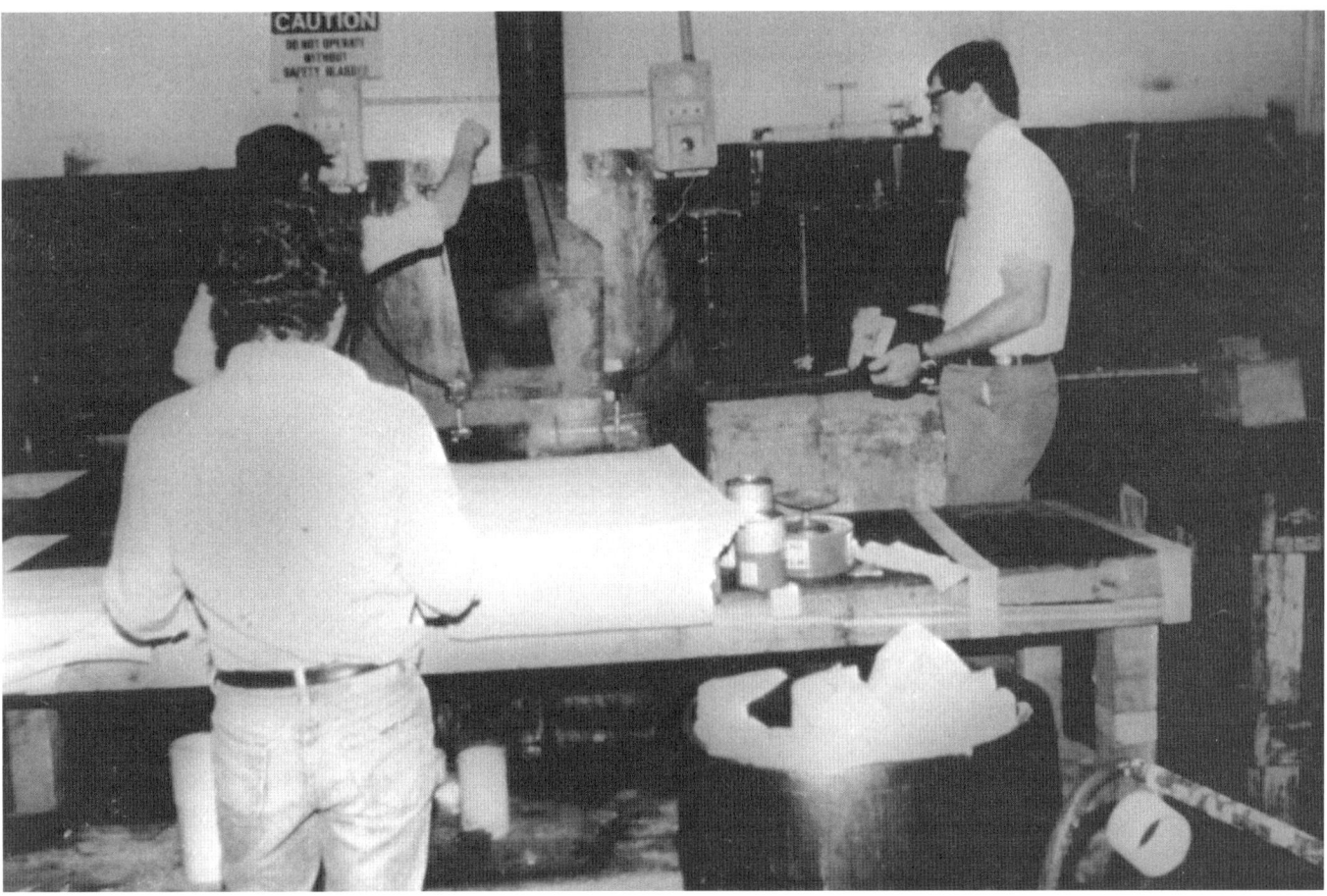

Fig. 8.1   Inspector checking production area for spill containment facilities, proper plumbing of wastewater to pretreatment system, existence of bypasses, one-pass cooling water, and use of chemicals or processes not listed on discharge permit.

For example, the metal plating process is used to plate everything from printed circuit boards to pool table legs. This manufacturing process immerses the part into a concentrated chemical solution to clean, activate, plate, or seal it. The part is then removed from the solution and transferred to another solution, usually water, to remove any excess chemical. Subsequently, the part may again be moved and dipped into another concentrated chemical solution. Let's look at some of the many different ways wastewaters can be generated and then discharged to the POTW sewer with this simple but widely used manufacturing process.

1. If the concentrated solution is *VOLATILE*,[3] reactive with the part, or may fume, dipping the part into the concentrated solution will produce an air contaminant that may have to be condensed, scrubbed, or removed. The scrubber wastewater can be discharged to the sewer.

2. If the concentrated solution tank were to burst due to an accident or tank failure, the concentrated chemical solution could go to the drain to the sanitary sewer unless there are provisions for spill containment.

3. The excess concentrated solution (*DRAG OUT*[4]) on the part may fall on the floor during the process of transferring the part from one tank to the next. This liquid may go to the drain to the sanitary sewer if not contained.

4. The drag out on the parts must be rinsed off using either still rinses or flowing rinses in order to prevent contamination of subsequent plating baths. Still rinses and the overflow from constant-flow rinses may be discharged to the POTW sanitary sewer.

5. Once the concentrated chemical solution is spent because the active chemical has been used up, it must be disposed of. A potential disposal method is to discharge this spent solution to the POTW sanitary sewer.

6. The plater was adding water to a tank and forgot to turn it off. The tank overflowed to the floor and the flow reached the POTW sanitary sewer.

7. The process of stripping a finish of unacceptable quality is a potential problem. Stripping the plated metal with either a cyanide solution or a strong acid or alkaline solution generates a quantity of very concentrated wastewaters which must be treated prior to discharge to the POTW sanitary sewers.

Wastes from all of these different sources should flow to the industry's pretreatment facility and then be discharged to the POTW's wastewater collection system. To understand the full impact of industrial wastes on your system, two other variables must also be considered: the concentration of potentially harmful constituents and the overall volume of waste discharged per unit of time. The concentration of a pollutant from an industrial source may be low, but because of a high overall volume of discharge, the total mass discharged to the sewer may impact your facilities. Conversely, even if a discharged concentration is high, no adverse effects may be noted on the POTW system if the total flow volume is low.

8.01 Effects of Industrial Wastewaters

The characteristics of industrial wastewaters, the type and size of the POTW treatment system, and the standards for sludge and wastewater disposal or reuse all have an influence on how much of an effect the industrial waste will have on the collection, treatment, disposal, and reuse systems. Waste characteristics such as temperature, pH, odor, toxicity, concentration, and flow must be evaluated to determine their acceptability. For example, in a short collection system, such as a small treatment system discharging to a trout stream, a continuous discharge of boiler *BLOWDOWN*[5] from a large power plant can be cause for concern. High temperature discharges to sewers can accelerate (1) biological degradation, (2) slime growths, (3) odor production from anaerobic decomposition, and (4) corrosion of concrete pipe and metal sewer appurtenances. The high temperature wastewater can cause a bacterial population shift in the secondary treatment causing floating sludge and reduced BOD removal efficiency. This in turn would endanger the treatment plant's ability to meet its discharge permit limits. The high temperature wastewater may also cause the plant to exceed its temperature standards to the trout stream.

On the other hand, the high temperature wastewater discharge from a power plant in a larger conveyance and treatment system located in a colder climate may, in fact, enhance the secondary treatment processes' removal efficiencies by keeping the wastewater temperature above 65°F (18°C) all year. When evaluating an industrial waste discharge, it is necessary to understand the specific characteristics of the waste and how they may affect each portion of the industry's pretreatment system and the POTW's conveyance, treatment, disposal, and reuse facilities.

QUESTIONS

Write your answers in a notebook and then compare your answers with those on page 592.

8.0A   Why is a knowledge of the various industrial manufacturing processes important for an inspector?

8.0B   What are the different reasons for immersing a part in a concentrated chemical solution in a metal plating process?

8.0C   How can a high temperature wastewater affect a POTW?

---

[3] Volatile (VOL-uh-tull).   (1) A volatile substance is one that is capable of being evaporated or changed to a vapor at relatively low temperatures. Volatile substances also can be partially removed by air stripping. (2) In terms of solids analysis, volatile refers to materials lost (including most organic matter) upon ignition in a muffle furnace for 60 minutes at 550°C. Natural volatile materials are chemical substances usually of animal or plant origin. Manufactured or synthetic volatile materials such as ether, acetone, and carbon tetrachloride are highly volatile and not of plant or animal origin.

[4] Drag Out.   The liquid film (plating solution) that adheres to the workpieces and their fixtures as they are removed from any given process solution or their rinses. Drag out volume from a tank depends on the viscosity of the solution, the surface tension, the withdrawal time, the draining time and the shape and texture of the workpieces. The drag out liquid may drip onto the floor and cause wastestream treatment problems. Regulated substances contained in this liquid must be removed from wastestreams or neutralized prior to discharge to POTW sewers.

[5] Blowdown.   The removal of accumulated solids in boilers to prevent plugging of boiler tubes and steam lines. In cooling towers, blowdown is used to reduce the amount of dissolved salts in the recirculated cooling water.

8.1 TYPES OF INDUSTRIAL WASTEWATERS

This section discusses the various types of industrial wastewaters, how they can be generally classified, the importance of knowing the frequency of generation and discharge, and the effects these discharges may have on the industry's pretreatment system. Examples of each type of industrial waste are given.

8.10 Compatible and Noncompatible Pollutants

Compatible pollutants can be defined as those pollutants that normally are removed by the POTW treatment system. Biochemical oxygen demand (BOD), suspended solids (SS), and ammonia are considered compatible pollutants. The POTW is designed to treat primarily domestic wastewater and the compatible pollutants discharged by industry.

Noncompatible pollutants are defined as those pollutants which normally are not removed by the POTW, may be considered a toxic waste, and may cause pass-through or INTERFERENCE[6] with the treatment system. Examples of noncompatible pollutants include heavy metals such as copper, nickel, lead, and zinc; organics, such as methylene chloride, 1,1,1-trichloroethylene, methyl ethyl ketone, acetone, and gasoline; and sludges containing toxic organics or metals.

Conventional pollutants can exhibit the characteristics of noncompatible pollutants and vice versa. Soluble BOD from a food industry may have some harmful effects on a POTW's secondary treatment system. The accidental discharge of ammonia by a fertilizer manufacturer may disrupt the nitrification/denitrification or stripping tower processes used by the POTW to treat ammonia. On the other hand, some of the heavy metals (usually classified as noncompatible pollutants) are used as micronutrients to aid in the production of biological mass and the reduction of BOD. Discharges of some organic chemical wastes such as acetone or isopropanol are biodegradable and, in dilute solutions, are removed by biological action in secondary treatment.

8.11 Dilute Solutions

The discharges from continuous manufacturing processes are normally dilute solutions of compatible and sometimes noncompatible pollutants. They may be discharged to the pretreatment system or directly to the POTW without any pretreatment. Manufacturing processes such as plating bath rinses, raw food cleaning, and crude oil dewatering are all examples of dilute solutions of pollutants that may be discharged directly to a POTW sanitary sewer. If a problem occurs in the manufacturing process, a probable result is that the quality of wastewater will change and it will be more laden with pollutants. Some wastewater flows from utility services, such as cooling tower and boiler blowdown, are continuous and represent the discharge of dilute solutions. (The other characteristics of utility system discharges are discussed in Section 8.15 and listed in Table 8.1 in that section.)

Another low-strength wastewater is storm water runoff from chemical handling and storage areas. Products which may have spilled on the industry's grounds are washed off during a rainstorm or during the spring thaw. The pollutant concentration is usually too dilute to require pretreatment before dis-

charge to the sewer, but exceeds the discharge standards for discharge to the surface waters. While the strength of the storm runoff may be low, the volume that must be treated in addition to normal flow to the pretreatment system or to the POTW can cause hydraulic capacity problems. Excessive flows can be diverted to storage reservoirs or basins and then gradually discharged to the pretreatment system.

A great deal of attention is presently focused on cleaning up groundwater sources that have been contaminated by leaking underground storage tanks. Cleanup projects of this nature typically involve large quantities of wastes that may contain high concentrations of solvents, fuels, heavy metals, and pesticides. Because of the public attention surrounding groundwater cleanup projects, pretreatment of the contaminated water is almost always required and the result is usually a "high quality" wastewater. In spite of effective pretreatment, return of the treated wastewaters to the environment is seldom permitted for political reasons. Instead, the POTW may be asked to receive the treated wastewater. Although the wastewater can be considered a dilute solution, it takes up hydraulic capacity in the sewer that may be needed by your system for conveyance of other contaminated flows. By working with the pollution control agencies responsible for protecting groundwater quality, it may be possible to require additional treatment, such as granular activated carbon for underground gasoline tank cleanup projects, that would allow discharge of the treated waters to surface waters rather than to the POTW.

8.12 Concentrated Solutions

Typically, concentrated solutions are batch-generated and the frequency of generation is usually not daily but weekly, monthly, annually, or even longer. These solutions are process chemicals or products that cannot be reconditioned or reused in the same manufacturing process.

Concentrated solutions such as spent plating baths, acids, alkalies, static drag out solutions, and reject product may have concentrations of pollutants hundreds or thousands of times higher than the discharge limits of the POTW or higher than can be adequately treated by the pretreatment system if discharged all at once. Because of the DELETERIOUS[7] effects of concentrated solutions, try to identify the potential sources of such discharges within your service area and ask each industry's contact person to explain specifically what treatment

---

[6] Interference.   Interference refers to the harmful effects industrial compounds can have on POTW operations, such as killing or inhibiting beneficial microorganisms or causing treatment process upsets or sludge contamination.

[7] Deleterious (DELL-eh-TEAR-ee-us).   Refers to something that can be or is hurtful, harmful or injurious to health or the environment.

and disposal methods will be used to prevent damage to your POTW facilities. Take time to examine and understand each manufacturing process so you can identify these concentrated solutions.

Remember too that you must evaluate each situation for yourself. The wastes may not be considered concentrated by the industry. For example, the ten percent sulfuric acid solution used for *PICKLING*[8] parts is considered "dilute" by comparison to the 98 percent or 50 percent stock solution that the industry uses to make up the pickling solution. When this solution is spent or can no longer be used as a pickling solution, proper treatment and disposal are required. From the manufacturer's point of view, the solution is spent and no longer concentrated. However, from a wastewater pretreatment point of view, the solution is concentrated since it contains high concentrations of acid (pH less than 1.0) and heavy metals (1,000 mg/*L*) compared to the normal pH of 1.0 to 4.0 and heavy metal concentrations of less than 100 mg/*L*.

Another source of concentrated solutions is the wastewater from equipment cleanup. While the amount of material in the process chemical bath may be considered dilute by industry standards, it forms a concentrated wastestream when discharged during the cleanup of manufacturing equipment. Cleanup wastestreams contain a high concentration of the product during the first washing of the tank, pipe, or pump. This discharge of concentrated waste is followed by successive rinses which contain less and less pollutants. If cleanup flow concentrations are not equalized, the cleanup cycle can cause problems in the pretreatment or POTW treatment system.

Spills of process chemicals to the floor, if not contained, can flow directly to the floor drain and the pretreatment or sewer system. The adverse effects on the pretreatment system and POTW are the same as those of any other concentrated solutions. This is why chemical containment areas must not have drains.

8.13 Frequency of Generation and Discharge

Important to both the operation of the industry's pretreatment system and the POTW's collection, treatment, and disposal system is the frequency of industrial waste generation and discharge. Wastewater sampling by the inspector to determine compliance is also affected by the hours of discharge. An inspector needs to understand when the waste from each process is generated and when it is discharged.

8.130 Hours of Operation Versus Discharge

Normally, the hours of operation are also the hours of discharge from an industrial facility so sampling of the discharge is necessary only during the industry's hours of operation. Because of the time it takes for the wastestream to travel through the sewer from the industry to the POTW treatment plant, the effects of the discharge may be delayed several hours but will be noticed over approximately the same number of hours of POTW plant operation as the industrial discharge. Thus, a single ten-hour shift operation starting at 0700 hours and ending at 1700 hours will only be sampled for the ten hours. The effects of the quantity and quality of wastewater discharged to the POTW conveyance system will also be felt for those ten hours only.

If the industry operates at a constant production rate, the discharge volume and chemical constituents will also be constant. For example, a coil coater that runs five million square feet per month from January through December, eight hours per day, five days per week, without any changes in manufacturing processes or chemicals, will have a constant effect on the sewer system and can be sampled in the same manner each time. However, if production increases and a second shift is added, or the work week is extended to six days per week, there is an obvious change in the hours of discharge from this facility.

An important factor affecting the actual hours of wastewater discharge to the sewer is the amount of equalization the pretreatment system provides. In order to deliver a relatively constant flow and concentration of pollutants to the industry's pretreatment system, large wastewater collection sumps, equalization tanks or storage tanks may be used. These equalization devices may also lengthen the time of discharge beyond the actual hours of operation of the manufacturing facility. While the eight-hour work shift generates 100,000 gallons of wastewater, the pretreatment system may only be designed for 150 GPM and, therefore, will require eleven hours to process the wastewater from the eight-hour production period. Since discharges from the pretreatment processes to the sewer span an eleven-hour period, sampling of the discharges will also be necessary over this longer period.

EXAMPLE 1

Known	Unknown
Wastewater Volume, gal = 100,000 gallons	Process Time, hours
Treatment Flow, GPM = 150 GPM	

Calculate the treatment process time in hours.

$$\text{Process Time, hours} = \frac{\text{Wastewater Vol, gal}}{(\text{Treatment Flow, GPM})(60 \text{ min/hr})}$$

$$= \frac{100,000 \text{ gallons}}{(150 \text{ gal/min})(60 \text{ min/hr})}$$

$$= 11.1 \text{ hours}$$

By lengthening the hours of discharge from the industry, there is an effective increase in the available hydraulic capacity of the POTW collection system because of the decreased industrial flow rates. Due to the normal diurnal (day/night) variation in domestic wastewater flows (peak flows usually occur

---

[8] *Pickle.   An acid or chemical solution in which metal objects or workpieces are dipped to remove oxide scale or other adhering substances.*

between 8:00 a.m. and 6:00 p.m.), the hydraulic capacity of a sewer may be exceeded if a large industrial flow is allowed to be discharged to the sewer during a short period of the day. Therefore, it may be necessary for the industry to discharge only at night. Sampling of this discharge would then be shifted to the nighttime hours. However, the use of flow-proportional sampling will ensure that the composite samples will be collected only during the time the facility is discharging process wastewaters.

8.131 Discharge Variations

Industries that have daily, weekly, or seasonal manufacturing cycles will show variations in wastewater generation. Business cycles for each of the various segments of the industrial community will have an effect on production, and therefore on the generation of wastewater.

The food processing industry provides a good example of daily, weekly, and seasonal variations in discharge quantity and quality. For example, an industry that processes citrus peel to make pectin is dependent on when the peel arrives at the industry's plant. This may mean anywhere from three to six days per week. As the season progresses, the type of peel changes from orange to lemon, and the sugar content changes yielding a slightly different type of wastewater. After the citrus season, the plant is completely shut down.

In certain industries, variations in the quantity of wastewater reflect the nature of the business or the business cycle of the particular business segment. In a small job shop producing printed circuit boards, it is typical to have a 30-day turnaround with sales, ordering, and development taking place during the first part of the month. Production is slow while making test boards, but once the board is developed, production proceeds at a rapid pace to produce the boards for shipment in the last week of the month.

The printed circuit board industry is subject to both downturns and upturns in the market. The major pollutant from the industry is copper and, consequently, the quantity of copper discharged to the sewer fluctuates according to market and production cycles.

Variations in the quality of industrial waste can also occur due to market forces or environmental concerns requiring a different type of product. In the metal finishing industry, for example, companies are moving from cadmium-plated metal, an environmentally more hazardous substance with more stringent discharge limitations, to zinc-plated parts. Knowledge of the industry, the manufacturing processes, and market forces are valuable tools needed by the industrial waste inspector to anticipate variations in industrial discharges.

8.14 Continuous and Intermittent Discharges

Discharges from manufacturing facilities usually reflect the type of manufacturing process used at the facility. Processes which are continuous tend to produce wastewater on a continuous basis, with relatively constant volume and quality. Batch processes, or activities that occur once per shift, per day, or per week, tend to produce an intermittent discharge. Also, as a general rule of thumb, the larger the manufacturing process, the more likelihood there is of a continuous discharge.

Examples of manufacturing processes that have continuous discharges include rinsing or cleaning of parts or food, processing of crude oil, either at the well head or refinery, air or fume scrubbing, papermaking, and leather tanning.

Intermittent discharges of wastewater are characterized by discharges of a volume of wastewater separated by a time period between discharges. These typically occur at the beginning or ending of a manufacturing process, during equipment cleanup, a spill, replacement of spent solution, or disposal of a reject product. Intermittent discharges also tend to be more concentrated and of smaller volume than the wastewater normally discharged.

For an industrial pretreatment facility, the intermittent discharges and the variations in waste generation determine the design capacity of the system. A conventional activated sludge wastewater treatment system at an adhesives, resins, and dye chemical plant may be able to handle 8,000 pounds of BOD per day, but the discharge of 100 GPM of a three percent isopropyl alcohol wastestream from a water-cooled contact condenser for one hour every 12 hours will have a significant impact on the pretreatment system. This is because thirty percent of the waste load is discharged to the system during only eight percent of the time.

EXAMPLE 2

Known		Unknown
Waste Flow, GPM	= 100 GPM	1. Waste Load, lbs BOD/day
Discharge Time, min/day	= 1 hour/12 hours	2. Waste Load, %
	or = 2 hours/24 hours	3. Discharge Time, %
	= 120 min/day	
Waste Conc, %	= 3%	
Waste Conc, mg/L	= 30,000 mg/L	
BOD, $\frac{\text{mg BOD}}{\text{mg alcohol}}$	= $\frac{0.8 \text{ mg BOD}}{\text{mg alcohol}}$	
Plant Capacity, lbs BOD/day	= 8,000 lb BOD/day	

1. Calculate the BOD waste load in pounds of BOD per day.

$$\text{Waste Load, lbs BOD/day} = \frac{(\text{Flow, gal/min})(\text{Disch Time, min/day})(\text{Waste Conc, mg/}L)}{(\text{BOD, mg BOD/mg waste})(3.785 \text{ }L\text{/gal})(1 \text{ lb/454,000 mg})}$$

$$= \frac{(100 \text{ gal/min})(120 \text{ min/day})(30,000 \text{ mg/}L)(0.8 \text{ mg BOD/mg})(3.785 \text{ }L\text{/gal})}{454,000 \text{ mg/lb}}$$

$$= 2,401 \text{ lbs BOD/day}$$

Conversion factors

$$1\% = 10,000 \text{ mg/}L$$
$$1 \text{ gal} = 3.785 \text{ liters}$$
$$1 \text{ lb} = 454 \text{ grams}$$
$$= 454,000 \text{ milligrams}$$

2. Determine the percent of waste load discharged during the two hours.

$$\text{Waste Load, \%} = \frac{(\text{Discharge Waste Load, lbs/day})(100\%)}{\text{Plant Capacity, lbs/day}}$$

$$= \frac{(2{,}401 \text{ lbs BOD/day})(100\%)}{8{,}000 \text{ lbs BOD/day}}$$

$$= 30\%$$

3. Calculate the percent of time the waste is discharged.

$$\text{Discharge Time, \%} = \frac{(\text{Discharge Time, hours/day})(100\%)}{\text{Total Time, hours/day}}$$

$$= \frac{(2 \text{ hr/day})(100\%)}{24 \text{ hr/day}}$$

$$= 8.3\%$$

8.15 Utility System Discharges

Another source of industrial wastes that may change the quantity and quality of the discharge from a facility is wastewater from the utility systems that service the manufacturing processes. These industrial wastewaters can be continuous or intermittent discharges and also follow the general rule that the larger the system the more likely the discharge will be continuous. Table 8.1 summarizes these sources and the type of discharge.

8.2 CONCENTRATION VERSUS MASS OF THE POLLUTANT

An understanding of the concentration and the mass of a pollutant in an industrial waste is needed to determine the effects on the industry's pretreatment system, the POTW collection, treatment, and disposal systems, and the sampling of the industry's discharge. The concentration of a substance in wastewater is normally expressed as milligrams per liter (mg/L) which is a measurement of the mass per unit of volume. The mass of a substance is normally expressed in pounds or kilograms and this is a weight measurement. A mass emission rate is a measurement of weight per unit time and is usually expressed as pounds or kilograms per day.

Many of the electroplating and all of the metal finishing Federal Categorical Standards are written in concentrations, whereas most of the other Categorical Standards are written as mass emission rate standards. The mass emission rate standards recognize that with more production and water, the mass of pollutant will also increase. This approach eliminates dilution of the pollutant to meet concentration limitations.

TABLE 8.1   UTILITY SYSTEM DISCHARGES OF WASTEWATER

Discharge Source	Type of Discharge C/I [a]	Remarks
Cooling Tower	C	Cooling tower blowdown may contain suspended and dissolved solids, chromium, molybdenum, and biocides. Smaller systems may discharge once per shift.
Boiler	I	Boiler blowdown contains suspended and dissolved solids, heat, and sulfite. Larger systems discharge continuously.
Air Compressor Moisture	I	During air compression moisture is condensed. No pollutants.
Cooling Water	C	Cooling water is required for larger compressors.
Inert Gas	C	Water is usually required to cool the compressor. Some moisture may also condense and contain sulfate or nitrogen oxides.
Pump, Mixer, or Equipment Seals	C	Water may be used on rotating equipment to seal or lubricate. Leaks or normal weeping will contain the product being mixed or pumped.
Demineralization System	I	Brine wastes high in total dissolved solids from softeners. Acidic and alkaline wastes from regeneration of demineralizers.
Air Scrubber	C	Scrubber blowdown contains suspended solids, the solubilized air contaminants, and treatment chemicals (typically alkaline). For smaller scrubbers or systems with high solubility of contaminants, blowdown is intermittent.

[a] C, Continuous discharge
I, Intermittent discharge

The mass emission rate of a substance can be calculated by knowing the concentration of the pollutant in the wastewater and the volume of wastewater. The mass emission rate calculation for an effluent of 10,000 gallons per day containing 4.5 milligrams per liter (mg/L) of copper is shown in *EXAMPLE 3*.

EXAMPLE 3

Known	**Unknown**
Effluent Flow, GPD = 10,000 gal/day	Mass Emission Rate, lbs/day
Waste Conc, mg/L = 4.5 mg/L	

Calculate the mass emission rate in pounds per day.

$$\text{Mass Emission Rate, lbs/day} = \frac{(\text{Flow, gal/day})(\text{Conc, mg/}L)(3.785\ L/\text{gal})}{454,000\ \text{mg/lb}}$$

$$= \frac{(10,000\ \text{gal/day})(4.5\ \text{mg/}L)(3.785\ L/\text{gal})}{454,000\ \text{mg/lb}}$$

$$= 0.38\ \text{lbs/day}$$

OR

$$\text{Mass Emission Rate, lbs/day} = (\text{Flow, MGD})(\text{Conc, mg/}L)(8.34\ \text{lbs/gal})$$

$$= (0.010\ \text{MGD})(4.5\ \text{mg/}L)(8.34\ \text{lbs/gal})$$

$$= 0.38\ \text{lbs/day}$$

NOTE: One liter of water weighs one million milligrams. Therefore, 4.5 mg/L is the same as 4.5 pounds of waste in one million pounds of water.

$$\text{Mass Emission Rate, lbs/day} = (\text{Flow, }\tfrac{\text{Mil Gal}}{\text{Day}})(\text{Conc, }\tfrac{\text{lbs}}{\text{Mil lbs}})(8.34\ \text{lbs/gal})$$

$$= \text{lbs waste/day}$$

If the industry had a permit limit of 4.12 mg/L, they would be in violation. However, if the permit limit was 0.4 lb/day, the industry would be in compliance.

The same mass of pollutant contained in different amounts of wastewater can have vastly different effects on the industry's pretreatment system. An unnoticed spill of ten pounds of copper cyanide from a process tank could increase the influent copper concentration to a pretreatment system from 10 mg/L to 500 mg/L and probably cause a pass-through of pollutants to the POTW. If the ten pounds of copper cyanide came from a new process line and was discharged on a continuous basis, the increase in influent copper concentration would only be to 15 mg/L. This slight increase in influent concentration would not be likely to cause any deterioration in the effluent quality. This example shows how the instantaneous influent mass emission rate can affect a pretreatment plant's performance.

Adverse effects on the pretreatment system can also occur from the same mass loading on the system with only small variations in concentration. A pretreatment system designed to treat 10 mg/L of copper in a flow of 100,000 gallons per day (GPD) (a mass loading of 8.34 pounds per day) could not treat 5 mg/L of copper in 200,000 GPD (still 8.34 pounds per day), because the solids separation equipment would become hydraulically overloaded (a flow of 200,000 GPD is twice a flow of 100,000 GPD).

The effects of concentration and mass on the POTW collection, treatment, and disposal systems are generally the same as described above. While the mass of the pollutant may be constant, the flow of wastewater may be much larger, causing hydraulic problems in any portion of the system and causing a pass-through of the pollutant. If the daily mass loading is the same, but the instantaneous mass emission rate is highly variable, the POTW's collection system may not equalize the slug loading of a highly concentrated solution. The result may be interference with the treatment system, causing violations of either or both effluent and sludge disposal limitations. Additional information is presented in Section 8.4, "Effects of Industrial Wastewaters on the POTW System."

QUESTIONS

Write your answers in a notebook and then compare your answers with those on page 592.

8.1A What is the difference between a compatible and a noncompatible pollutant?

8.1B List five examples of noncompatible pollutants.

8.1C Why does an inspector need to know the hours of discharge for an industry?

8.2A What are the common units of expression for the following terms: (1) concentration, (2) mass, and (3) mass emission rate?

8.2B How can high flows or concentrations cause interference or pass-through of the pollutant?

8.3 MANUFACTURING PROCESSES AND WASTEWATER GENERATION

By understanding the manufacturing processes in some of the major wastewater generating industries, the inspector can evaluate an industry's proposed solutions to solving noncompliance problems. Knowledge of source reduction and control techniques (described in Chapter 9) can also be used to reduce the industrial wastes going to the pretreatment or sewer system. And if you know what waste is generated by each industry, you can identify possible sources of industrial discharges that are having an effect on the POTW collection, treatment, or disposal systems.

This section presents information on some of the major industrial waste generators that an inspector may have to be familiar with and inspect. The information is presented in the following format:

1. General description and schematic of the manufacturing process,

2. Description of each of the manufacturing unit processes and the wastes generated, and

3. Schematic diagram of the raw materials used in each unit process and the wastes generated.

8.30   Metal Finishing Industries (Figures 8.2 and 8.3)

The largest group of industries regulated by Federal Categorical Standards is the metal finishing industry. This industry's distinguishing characteristic is the formation of a surface coating on a base material. Surface coatings are applied to provide corrosion protection, wear or erosion resistance, antifrictional characteristics, or for decorative purposes. The coating of common metals includes the processes in which a ferrous or nonferrous base material is plated with copper, nickel, chromium, brass, bronze, zinc, tin, lead, cadmium, iron, aluminum, or combinations of these metals. Precious metals plating includes the processes in which a base material is plated with gold, silver, palladium, or platinum or any of several less well known metals.

The general processes of the metal finishing industry include machining, cleaning and surface preparation, plating and coating, anodizing, and etching and chemical milling. A schematic of the general and unit processes for the metal finishers is shown in Figure 8.4.

8.300   Unit Process Description

The following descriptions provide general information for each of the above unit processes. The descriptions apply specifically to the metal finishing industry, but will help you understand similar operations in other industries.

Machining

Machining is the general process of removing stock from a workpiece with a rotating or moving cutting tool. Machining operations such as stamping, turning, milling, grinding, drilling, boring, tapping, sawing, and cut-off are included in this category.

Cleaning and Surface Preparation

In order to prepare the surface of the workpiece to form a good bond, the base metal must be properly prepared and cleaned. The cleaning process involves the removal of oil, grease, and dirt from the surface of the base material using water with or without a detergent or other dispersing agent. Electrolytic and nonelectrolytic alkaline cleaning and acid cleaning are included in this category.

1. Alkaline Cleaning — is used for removal of oily dirt or solids from workpieces. The detergent nature of the cleaning solution provides most of the cleaning action; agitation of the solution and movement of the workpiece increase cleaning effectiveness. Alkaline cleaners are classified into three types: soak, spray, and electrolyte. Soak cleaners are used on easily removed soil, spray cleaners combine the deter-

gent properties of the solution with the impact force of the spray, and electrolytic cleaning produces the cleanest surface by strong agitation of the solution during electrolysis. Also, certain dirt particles become electrically charged and are repelled from the surface. Some alkaline cleaning processes also remove oxide films.

2. Acid Cleaning — is a process in which a solution of an acid, organic acid, or an acid salt, in combination with a wetting agent or detergent, is applied to remove oil, dirt, or oxide from metal surfaces. Acid cleaning can be referred to as pickling, acid dipping, descaling, or desmutting. Heated acid solutions may be used in this process. Acid dip processes may follow alkaline cleaning prior to plating.

3. Paint Stripping — is the process of removing an organic coating from a workpiece, usually by using solvent, caustic, acid, or molten salt solutions.

4. Solvent Degreasing — is a process for removing oils and grease from the surfaces of a workpiece by the use of organic solvents, such as aliphatic petroleum, aromatics, halogenated hydrocarbons, oxygenated hydrocarbons, and combinations of these solvents. Solvent cleansing can be accomplished by applying solvents in either liquid or vapor form, but solvent vapor degreasing is normally the quicker process. Ultrasonic vibration is sometimes used in conjunction with liquid solvent degreasing processes. Emulsion cleaning is a type of solvent degreasing that uses common organic solvents in combination with an emulsifying agent.

Plating and Coating

1. Electroplating — is the production of a thin surface coating of one metal upon another by electrodeposition. Ferrous or nonferrous base materials may be coated with a variety of common metals (copper, nickel, lead, chromium, brass, bronze, zinc, tin, cadmium, iron, aluminum, or combinations thereof) or precious metals (gold, silver, platinum, osmium, iridium, palladium, rhodium, indium, ruthenium, or combinations of these metals). In electroplating, metal ions supplied by the dissolution of metal from anodes or other pieces are reduced on the workpieces (cathodes) while in either acid, alkaline, or neutral solutions.

The electroplating baths contain metal salts, alkalies, and other bath control compounds in addition to plating metals such as copper, nickel, silver, or lead. Many plating solutions contain metallic, metallo-organic, and organic additives to induce grain refining, leveling of the plating surface, and deposit brightening. Many plating operations are now using other common metal salts rather than cyanide in their plating baths. However, most precious metal baths

Fig. 8.2 Typical barrel plating production line. Floor
is diked and rinses are plumbed into standpipes.
Inspectors should be suspicious of white bucket in
lower left that could be used to transfer static drag
out or floor spills to standpipes.

Fig. 8.3   Close-up of Figure 8.2 demonstrating high volume of drag out dripping from barrel. Barrels should spend as much time as possible dripping over plating bath, NOT over floor or running rinse tank. "Spill guards" should direct drag out to plating bath.

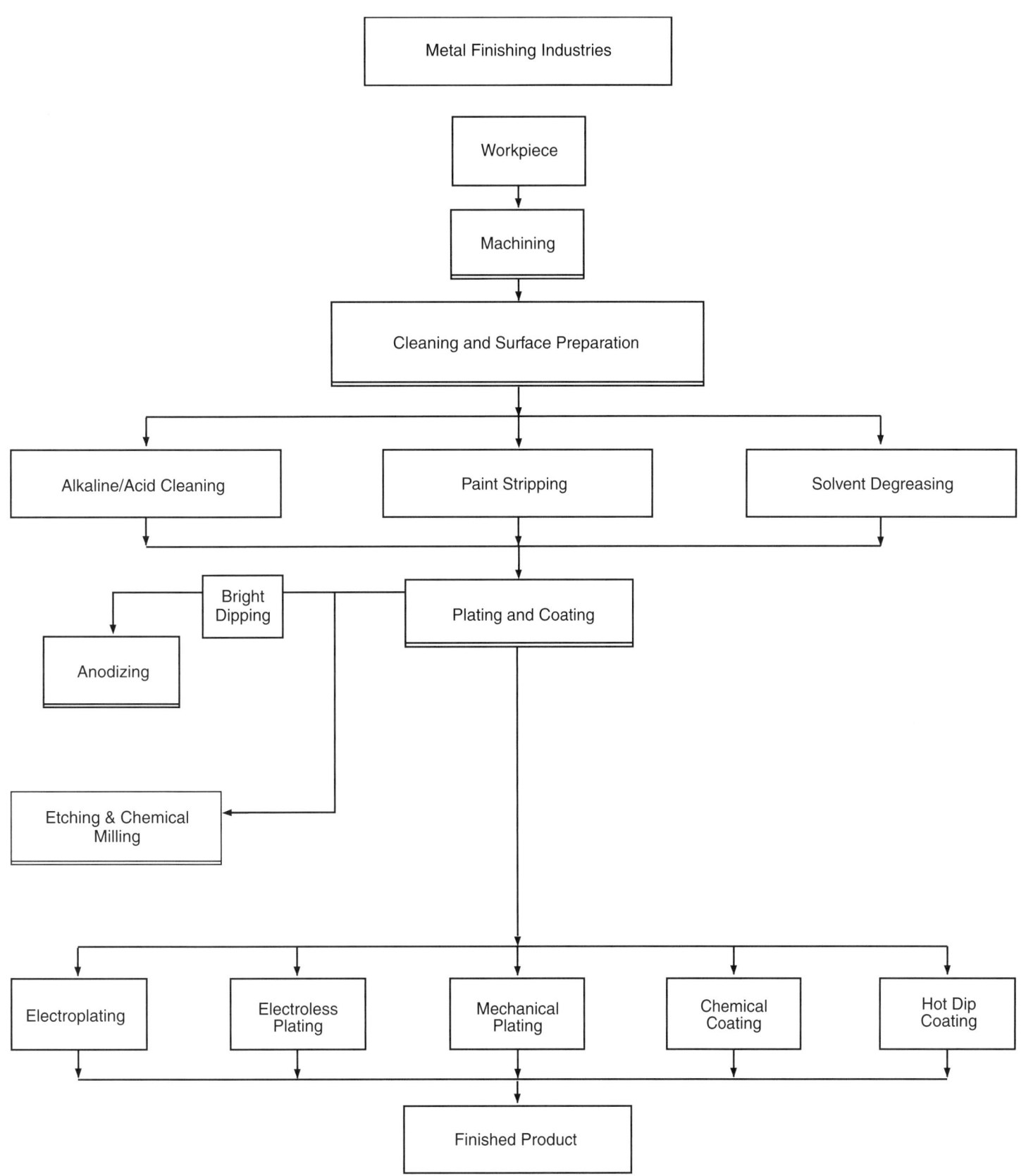

Fig. 8.4   Metal finishing industry manufacturing processes

contain cyanide salts. Cyanide is commonly used in gold plating, copper plating where a thick coat is required, and cadmium plating.

2. Electroless Plating — uses a chemical reduction/oxidation reaction to provide a uniform plating thickness on all areas of the part regardless of the shape. The most common electroless plating metals are copper and nickel. In electroless plating, the source of the metal is a salt, and a reducer such as sodium hydrophosphite or formaldehyde is used to reduce the metal ions to their base state. A complexing agent (*CHELATING AGENT*[9]), such as EDTA or Rochelle salts, holds the metal ions in solution.

3. Mechanical Plating — is the process of depositing a metal coating on a workpiece using a tumbling barrel, metal powder, and an impact medium, usually glass beads. The operation involves cleaning and rinsing processes before and after the plating operation.

4. Chemical Coatings — include chromating, phosphating, metal coloring, and passivating. In chromating, a portion of the base metal is converted to a component of the protective film formed by the coating solutions containing hexavalent chromium and active organic or inorganic compounds. Phosphate coatings are formed by the immersion of steel, iron, or zinc-plated steel in a dilute solution of phosphoric acid plus other *REAGENTS*[10] to condition the surfaces for cold-forming operations, prolong the life of organic coatings, provide good paint bonding, and improve corrosion resistance. Metal coloring chemically converts the metal surface into an oxide or similar metallic compound to produce a decorative finish. Passivating is the process of forming a protective film on metals by immersion in an acid solution, usually nitric acid or nitric acid with sodium dichromate.

5. Hot Dip Coating — is the process of coating a metallic workpiece with another metal by immersion in a molten bath to provide a protective film. The most common process of this type is galvanizing, where workpieces are coated with zinc.

Anodizing

Anodizing is an electrochemical process which converts the metal surface to a coating of an insoluble oxide. Aluminum is the most frequently anodized material. The formation of the oxide occurs when the parts are made anodic in dilute sulfuric or chromic acid solutions. The oxide layer is formed at the extreme outer surface, and as the reaction proceeds, the oxide grows into the metal. Chromic acid anodic coatings are more protective than sulfuric acid coatings and are used if a complete rinsing of the part cannot be achieved.

Etching and Chemical Milling

Etching and chemical milling are processes used to produce specific design configurations or surface appearances on parts by controlled dissolutions with chemical reagents or etchants. Chemical etching is the same process as chemical milling except the rates and depths of metal removal are usually much greater in chemical milling.

Bright Dipping

A specialized form of etching is the bright dipping process. It is used to remove oxide and tarnish from ferrous and non-ferrous materials. Bright dip solutions are mixtures of two or more acids: sulfuric, chromic, phosphoric, nitric, or hydrochloric. The process is frequently performed just prior to anodizing, and produces a bright to brilliant finished surface.

8.301   Waste/Wastewater Characteristics

The raw materials and the waste constituents most commonly found in the waste/wastewater streams generated by the metal finishing industry are presented in Figure 8.5.

Waste Types

- Heavy metals — cadmium, chromium, copper, lead, nickel, zinc, tin, aluminum, iron

- Precious metals — gold, silver, platinum, palladium, rhodium, iridium, osmium, ruthenium, indium

- Complexed metals — complexed wastes containing common and precious metals bonded with complexing solutions such as formaldehyde, hydrophosphite, ammonia solutions, EDTA, and citrate

- Acid wastewaters — typically sulfuric acid, others include nitric, hydrochloric, phosphoric, and tri-acid (sulfuric, nitric, hydrofluoric)

- Alkaline wastes — typically alkaline cleaners containing sodium hydroxide, sodium carbonate, soaps, and surfactants

- Hexavalent chromium — chromium plating solution, such as chromic acid in combination with sulfuric acid or sulfate

- Cyanide wastes — cyanide plating of copper, cadmium, zinc, brass, gold, silver; electroless plating of gold, silver; immersion of brass, silver, and tin; and cyanide stripping

- Oily wastes — free or emulsified oil and grease

- Solvent wastes — common solvents such as aliphatics, aromatics, halogenated hydrocarbons, and oxygenated hydrocarbons

---

[9] Chelating (key-LAY-ting) Agent.   A chemical used to prevent the precipitation of metals (such as copper).

[10] Reagent (re-A-gent).   A pure chemical substance that is used to make new products or is used in chemical tests to measure, detect, or examine other substances.

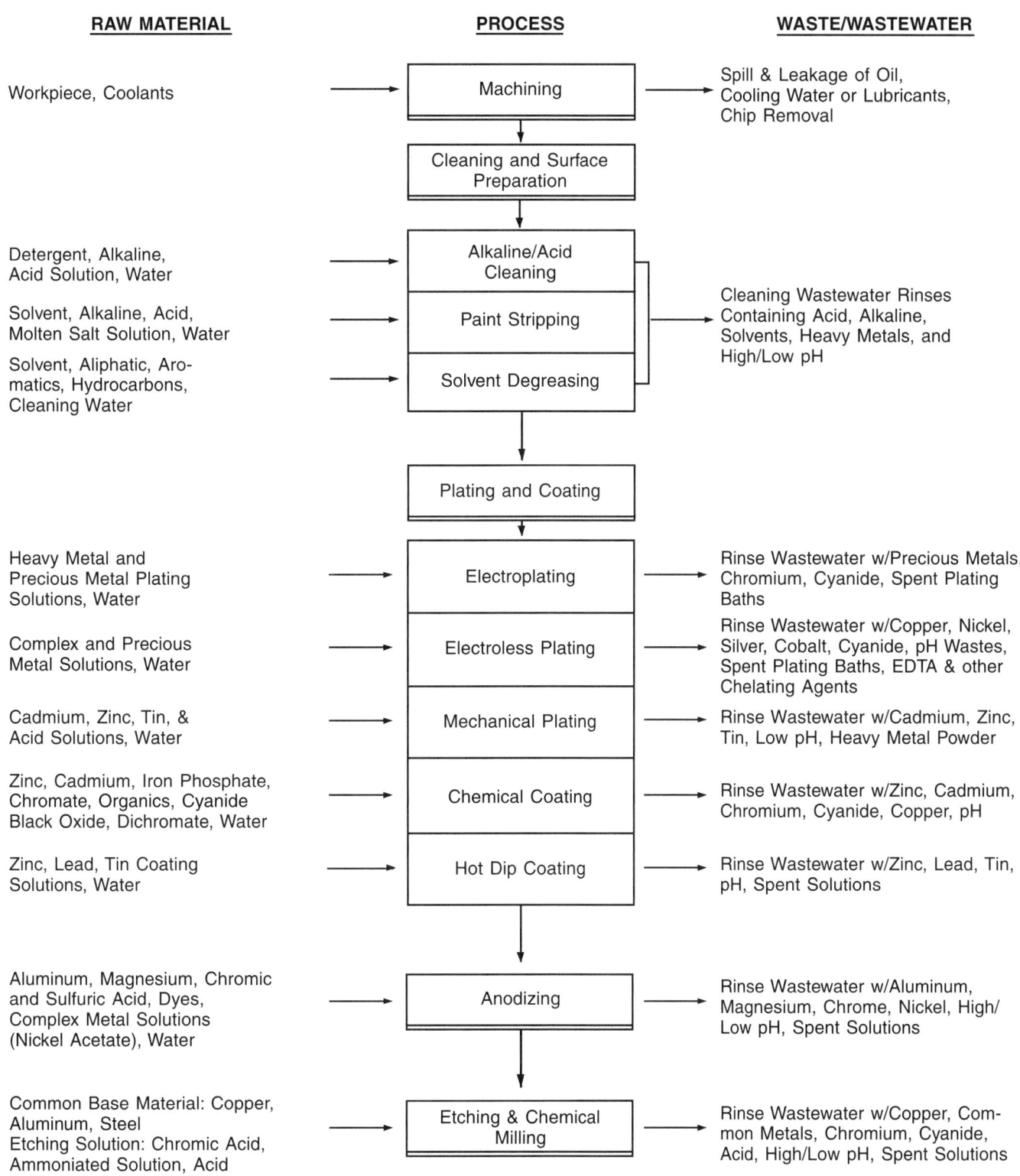

*Fig. 8.5   Waste/wastewater characteristics
for the metal finishing industry*

Water Usage/Wastewater Generation

In the metal finishing industry, the water is mostly used for rinsing workpieces, washing away spills, process fluid make-up, cooling, and washing of equipment and parts. The following paragraphs describe the high water usage processes of metal finishing manufacturers.

Rinsing — Rinse water is used to remove the film remaining on the surfaces of the workpieces after removal from the process baths. As a result, the rinse water becomes contaminated with the constituents in the preceding process solution.

Washing Away Spills — The wastewater generated from washing away spills is heavily contaminated with constituents of the process materials and dirt.

Process Fluid Makeup — Due to evaporation, drag out, or spills, process solutions (cleaning and plating solutions) are eventually used up and new process fluids have to be made up. The major component is water. Spent or contaminated process solutions are either collected for off-site disposal or conveyed to treatment facilities.

Cooling — Coolants containing petroleum or synthetic-based oils are required to lubricate and cool equipment and workpieces during many metal machining operations. The film and residues from these fluids are removed during cleaning, washing, or rinsing operations, and the resulting wastewater becomes contaminated with oil.

Washing — Washing equipment such as filters, pumps, and tanks pick up residues of concentrated process solutions, salts, or oils and may contribute a large amount of wastewater.

Deionization Bed Regeneration — Most printed circuit board manufacturers have large deionization plants on site that are regenerated with an acid solution (usually hydrochloric or sulfuric acid) and a caustic solution (usually sodium hydroxide). These regeneration streams should be thoroughly mixed and the pH adjusted to meet effluent limits prior to discharge to the sanitary sewer.

QUESTIONS

Write your answers in a notebook and then compare your answers with those on page 592.

8.3A   What metals are used by the metal finishing industry to produce a surface coating on a base material?

8.3B   List the general processes of the metal finishing industry.

8.3C   What types of chemicals are found in electroplating baths?

8.3D   List the nine common waste types found in the waste/wastewater streams generated from the metal finishing processes.

8.31   Printed Circuit Board Manufacturing

This industry is classified as a metal finishing industry and represents a significant percentage of the total industry. Printed circuit board manufacturing involves the formation of a circuit pattern of conductive metal (usually copper) on non-conductive board materials such as plastic or glass. The general processes of printed circuit board manufacturing include laminate machining, cleaning and surface preparation, electroless plating, pattern printing and masking, electroplating, and etching. The schematic of the manufacturing processes is shown in Figure 8.6.

8.310   Unit Process Description

The following is a brief description of the general processes of printed circuit board manufacturing.

Laminate Machining

Laminate machining consists of mechanical processes such as cutting to size, drilling holes, and shaping, by which the circuit boards (made of laminated materials) are prepared for the vital chemical processes. All the machining processes are dry and no liquid wastes are generated.

Cleaning and Surface Preparation

Cleaning and surface preparation is necessary to remove oil, grease, and dirt from the surface of the base material. Water, with or without detergent or other dispersing agent, is commonly used. Cleaning processes could include alkaline cleaners, acid cleaners, vapor degreasing, and scrubbing, with alkaline and acid cleaning being the most common. The alkaline cleaning process removes oil or dirt, and the acid cleaning process removes oil, dirt, and oxide from metal surfaces.

Electroless Plating

Electroless plating provides a uniform plating thickness on all areas of the part by chemical reduction/oxidation. An electroless plated surface is dense and virtually nonporous. Copper and nickel electroless plating for printed circuit boards are the most common operations. The nickel or copper salt in combination with a reducing agent is used to reduce the metal ions to their base state. A chelating agent is used to hold the nickel or copper ions in solution.

Pattern Printing and Masking

This process prints a pattern or the desired circuit configuration onto the board by using a suitable negative resist pattern for a photo resist. Ink resist, or positive resist pattern is used if plating is to be used as an etchant resist. In the case of negative resist pattern, gold or solder plating is applied to the non-photo resist areas.

Liquid photo-sensitive resists are thin coatings produced from organic solutions which, when exposed to light of the proper wavelength, are chemically changed in their solubility to certain solvents (developers). Two types are available: negative-acting and positive-acting.

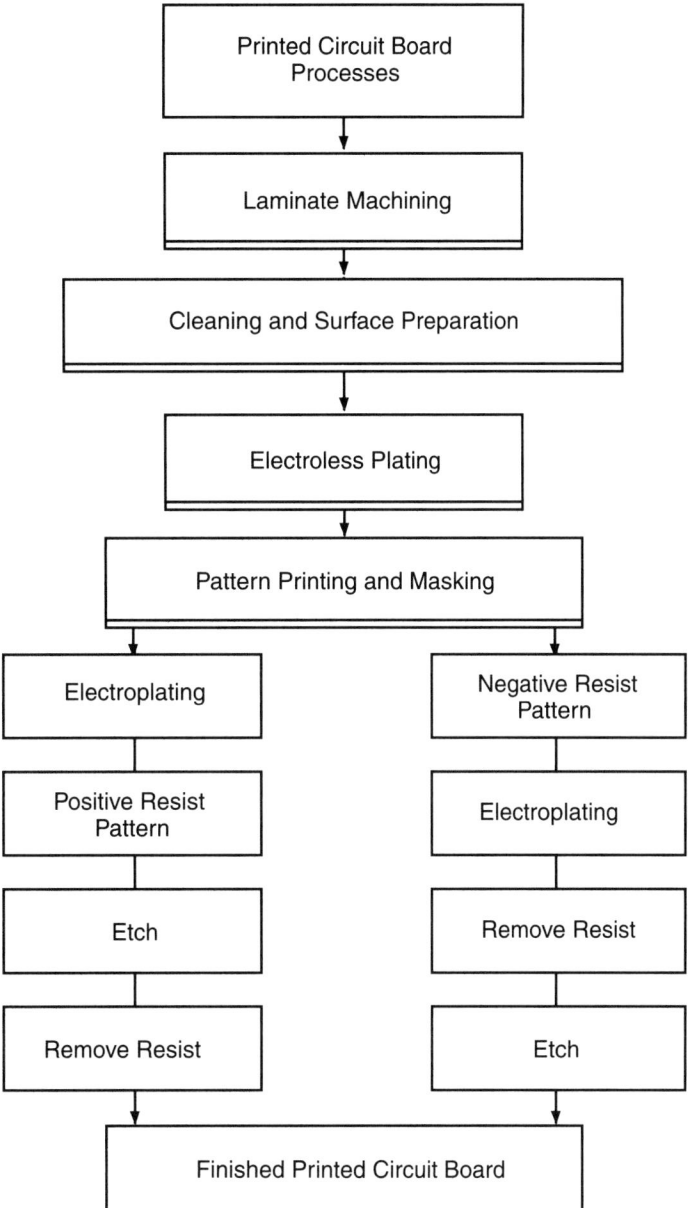

Fig. 8.6   Printed circuit board manufacturing processes

Electroplating

This process is used after the pattern printing and masking process to deposit an adherent metallic (copper, gold, or nickel) coating upon a negatively charged board by the passage of an electric current in a conducting medium. The electroplating baths contain metal salts, alkalies, and other bath-control compounds.

Etching

The last major step in chemical processing is metal removal or etching to achieve the desired circuit pattern. Typical solutions for etching are ferric chloride, nitric acid, ammonium persulfate, chromic acid, cupric chloride, hydrochloric acid, or a combination of these solutions.

8.311   *Waste/Wastewater Characteristics*

Water is used for cleaning, rinsing, spray rinsing, spill cleanup, solution replenishment, and cooling of equipment used in the manufacture of printed circuits. Rinses contain small amounts of the previous chemical bath the board was dipped in, such as cleaner solutions, copper plating bath, and resist. If the chemical bath preceding rinsing was acidic (used to activate the surface for plating), the rinse water may also contain copper and/or lead that has been solubilized.

Water used in closed-loop systems to cool or heat chemical baths or rinses may have corrosion inhibitors and/or glycols to protect and aid the heat transfer capabilities. Leaks from these systems, either directly to the sewer or into the rinse water, should be noted.

Spills to the floor may occur during the transfer of circuit boards from the chemical bath to the rinse tank. In some shops this floor wastewater is washed to a central sump and metered into the pretreatment system. This waste may contain any of the chemicals used in the facility.

Spent solutions usually make up the second largest volume of wastewaters, rinse wastewaters being first. While these solutions are no longer strong enough to be used in the manufacturing process, they still contain large amounts of alkalies, acids, and surface-active agents.

The sources of waste and wastewater production in the printed circuit board industry are similar to the metal finishing industry described in Section 8.30. Figure 8.7 shows a schematic flow diagram of waste/wastewater generation for the printed circuit board industry.

8.32   Pulp, Paper and Paperboard Industries

This industry is involved in the production of pulp, paper, and paperboard from wood and non-wood materials such as jute, hemp, rags, cotton linters, bagasse, and esparto. The pulp, paper, and paperboard industry includes three major segments: integrated mills, where pulp alone or pulp and paper or paperboard are manufactured on site; non-integrated mills, where paper or paperboard are manufactured, but pulp is not manufactured on site; and secondary fiber mills, where wastepaper is used as the primary raw material to produce paper or paperboard.

A wide variety of products including newsprint, printing and writing papers, unbleached and bleached packaging papers or paperboard, tissue papers, grease-proof papers, and special industrial papers are manufactured through the application of various process techniques. The production of pulp, paper, and paperboard involves several standard manufacturing processes and is shown schematically in Figure 8.8.

8.320   Unit Process Description

The following paragraphs are the descriptions of the pulp, paper, and paperboard manufacturing processes.

Raw Material Preparation

Depending on the form in which the raw materials arrive at the mill, log washing, bark removal, and chipping (hogging) may be used to prepare wood for pulping. These processes can require large volumes of water, but the use of dry bark removal techniques or the recycling of wash water or water used in wet barking operations significantly reduces water consumption.

Pulping

The purpose of pulping is to release the fibrous cellulose from the surrounding lignin while keeping the cellulose intact, thereby increasing the yield of useful fibers for chemical conversion or for further processing into paper or paperboard. Pulping processes vary from simple mechanical action, as in groundwood pulping, to complex chemical digesting sequences such as in the alkaline, sulfite, or semi-chemical processes.

Mechanical pulping is commonly known as groundwood. There are two basic processes: (1) stone groundwood where pulp is made by tearing fiber from the sides of short logs (called billets) using a grindstone, and (2) refiner groundwood where pulp is produced by passing wood chips through a disc refiner.

Mechanical pulps are characterized by yields of over 90 percent of the original substrate. The pulp is suitable for use in a wide variety of consumer products including newspapers, tissues, catalogs, one-time publications, and throwaway molded items.

Chemical pulping involves the use of controlled conditions of temperature and chemicals to yield a variety of pulps with unique properties. Chemical pulps are converted into paper products that have relatively higher quality standards or require special properties. There are three basic types of chemical pulping process now in common use: alkaline, sulfite, and semi-chemical.

The first alkaline process was the soda process, a forerunner of the kraft process. The kraft process is adaptable to nearly all wood species and uses a digester to cook the wood chips with caustic and sodium sulfate. The pulp is washed and screened and then sent to thickeners and filters. The cooking liquor or black liquor is concentrated by evaporation for reuse. Eighty percent of the chemical pulp made is by the kraft process.

Wood chips are fed to a direct steam-heated tank where they are digested in an aqueous (water-based) solution containing calcium bisulfite and an excess of sulfur dioxide. After digestion, the pulp is washed, screened, and thickened before bleaching. Bleaching is accomplished with chlorine dioxide and afterward a lime slurry is added to neutralize the mass. The stock is washed, thickened, filtered, and dewatered. Sulfite pulps are associated with the production of tissue and writing papers. Dissolving pulps (the highly purified chemical cellulose used in the manufacture of rayon, cellophane, and explosives) were produced solely by use of the sulfite process for many years, but because of the water pollution created, the use of this method is steadily declining.

The semi-chemical process uses the same digestion technique as described above, except sodium bisulfite buffered with sodium carbonate is used. This process is called the neutral sulfite semi-chemical (NSSC) pulping process and it has gained rapid acceptance because of its ability to use the vast quantities of inexpensive hardwoods previously considered unsuitable for producing quality pulp.

Recent advances have been made in semi-chemical pulping process technology with respect to liquor recovery systems. Three no-sulfur, semi-chemical processes have been developed: (1) the Owens-Illinois process, (2) the soda ash process, and (3) the modified soda ash process.

Secondary Fibers Pulping

Over 20 percent of the paper produced in the U.S. now comes from the repulping of recycled paper. Presently, the

RAW MATERIAL	PROCESS	WASTE/WASTEWATER

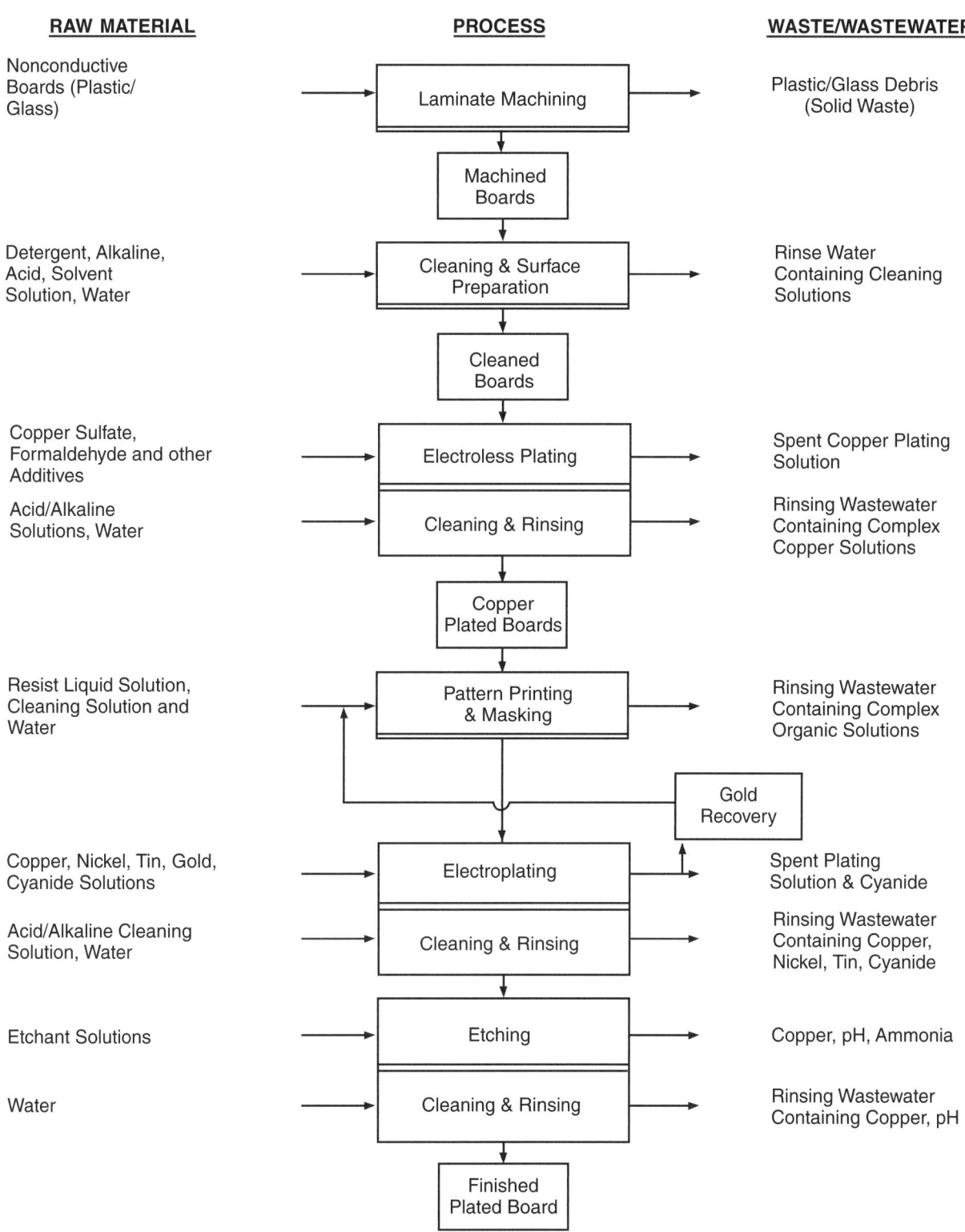

Nonconductive Boards (Plastic/Glass) → Laminate Machining → Plastic/Glass Debris (Solid Waste)

Machined Boards

Detergent, Alkaline, Acid, Solvent Solution, Water → Cleaning & Surface Preparation → Rinse Water Containing Cleaning Solutions

Cleaned Boards

Copper Sulfate, Formaldehyde and other Additives → Electroless Plating → Spent Copper Plating Solution

Acid/Alkaline Solutions, Water → Cleaning & Rinsing → Rinsing Wastewater Containing Complex Copper Solutions

Copper Plated Boards

Resist Liquid Solution, Cleaning Solution and Water → Pattern Printing & Masking → Rinsing Wastewater Containing Complex Organic Solutions

Gold Recovery

Copper, Nickel, Tin, Gold, Cyanide Solutions → Electroplating → Spent Plating Solution & Cyanide

Acid/Alkaline Cleaning Solution, Water → Cleaning & Rinsing → Rinsing Wastewater Containing Copper, Nickel, Tin, Cyanide

Etchant Solutions → Etching → Copper, pH, Ammonia

Water → Cleaning & Rinsing → Rinsing Wastewater Containing Copper, pH

Finished Plated Board

Fig. 8.7   Waste/wastewater characteristics for printed circuit board manufacturing

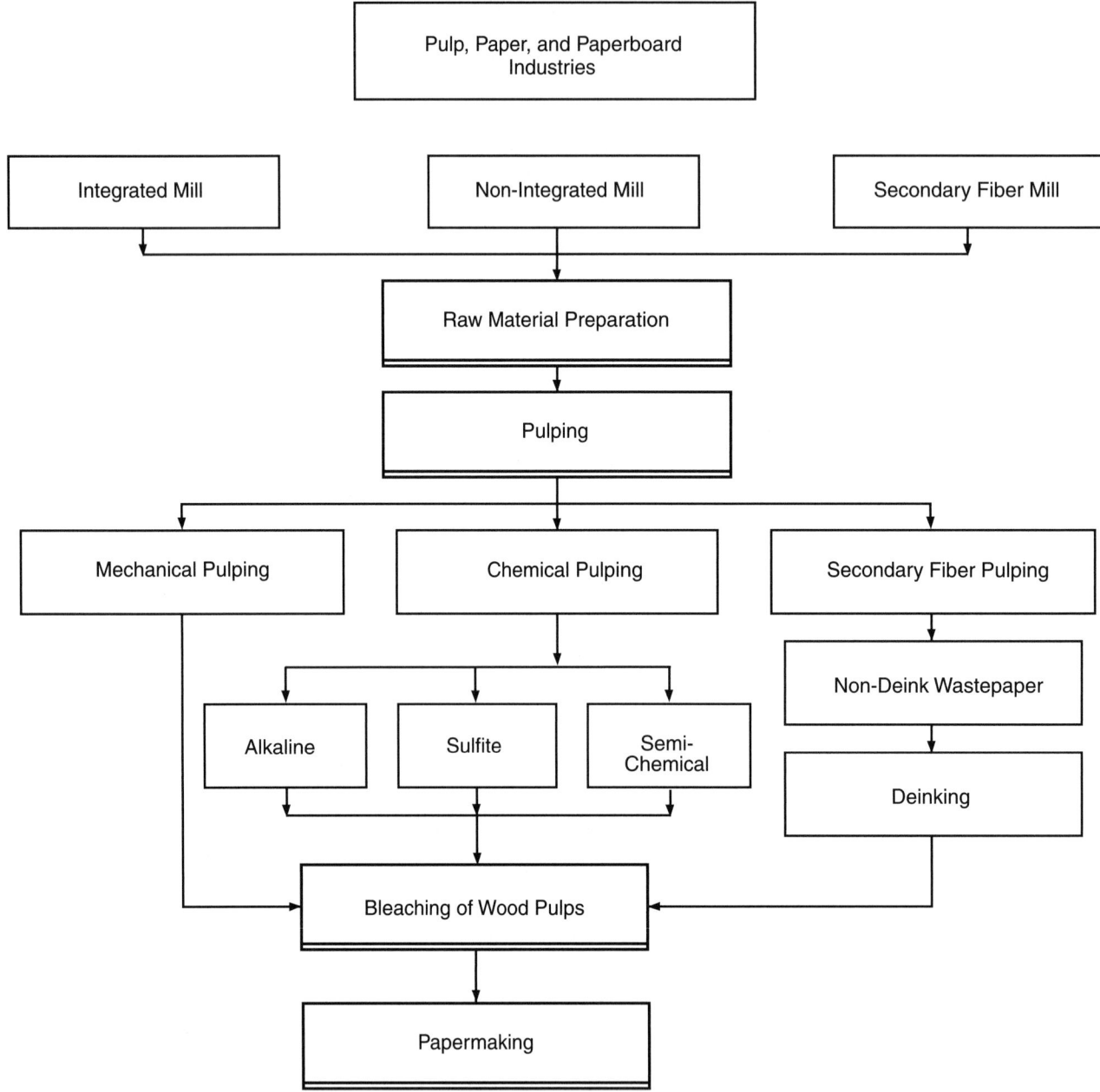

Fig. 8.8   Pulp, paper and paperboard manufacturing processes

secondary fiber field is critically dependent upon balancing available wastepaper type with the demands of the product to be manufactured. Upgrading of low quality wastepapers is difficult and costly, with inherently high discharge of both BOD and TSS to ensure adequate deinked pulp quality.

The collected material is repulped in water, cleansed of objectionable dirt and contaminants, deinked with alkali (sodium hydroxide, sodium carbonate, and $Na_2O(SiO_2)_x$, washed, cooked lightly with alkali, bleached, screened, and then handled like any other pulp.

Bleaching of Wood Pulps

After pulping, the unbleached pulp is brown or deeply colored because of the presence of lignins and resins or because of inefficient washing of the spent cooking liquor from the pulp. In order to remove these color bodies from the pulp and to produce a light colored or white product, it is necessary to bleach the pulp.

Bleaching is frequently performed in several stages using chlorine dioxide, caustic, and washing for the kraft pulping process. For sulfite pulps the bleaching sequence involves chlorination, alkaline extraction, and hypochlorite application, each followed by washing.

Mechanical pulps (groundwood) contain essentially all of the wood substrate. The use of conventional bleaching agents would require massive chemical dosages to enable brightening to levels commonly attained in the production of bleached, fully cooked kraft or sulfite pulps. Generally, mechanical pulps are used in lower quality, short-life paper products such as newsprint, telephone directories, catalogs, or disposable products. These products may be brightened slightly by the use of sodium hydrosulfite, sodium peroxide, or hydrogen peroxide. Typically, a single application in one stage is used, but two stages may be used if a higher brightness is required. Secondary fibers are often bleached to meet the requirements of specific grades.

Papermaking

Once pulps have been prepared from wood, deinked stock, or wastepaper, further mixing, blending, and addition of non-cellulosic materials, if appropriate, are necessary to prepare a suitable "furnish" for making most paper or board products. Modern stock preparation systems have preset instrumentation to control blending, addition of additives, refining, mixing, and distribution of the furnish.

The various papermaking processes are basically similar regardless of the type of pulp used or the end product manufactured. A layer of fiber is deposited from a dilute water suspension of pulp on a fine screen, called the "wire." The wire retains the fiber layer and permits water to drain through. This layer is then removed from the wire, pressed, and dried. Variations of two basic types of paper machines are commonly used. One is the cylinder machine in which the wire is mounted on cylinders which rotate in the dilute pulp furnish. The other is the Fourdrinier in which the dilute pulp furnish is deposited upon an endless wire belt. Generally, the Fourdrinier is associated with the manufacture of paper and the cylinder machine with heavier paperboard grades.

For further improvements in appearance, printability, water resistance, or texture, a coating is applied to the dry paper sheets. This may be done either on-machine or on a separate coater. Coatings may be applied by rolls, metering rods, air knives, or blades. The coating commonly is a high-density water slurry of pigments and adhesives which are blended, metered onto the fast-moving sheet, and then dried. Binders

including various starches, lattices, polyvinyl acetate, and other synthetics are used to improve the bonding of coatings to the paper. Other types of coating operations may involve the use of recoverable solvents for the application of release agents, gummed surfaces, and other films.

8.321 Waste/Wastewater Characteristics

Water is used in the following major unit operations in the manufacture of pulp, paper, and paperboard: wood preparation, pulping, bleaching, and papermaking. Each of these operations generates wastewater.

Three categories of pollutants are found in the wastewater generated from pulp and paper processes: (1) conventional pollutants, (2) toxic pollutants, and (3) nonconventional pollutants. In general the conventional pollutants, including biochemical oxygen demand (BOD) and suspended solids (SS), are the most common pollutants found in the pulp and paper mill processes.

The wastewaters from wood preparation processes are generated from three basic areas: (1) log transport, (2) log and chip washing, and (3) barking operations. Wastewater discharged during wood preparation contains large concentrations of suspended solids (bark solids) and some cellulose from the debarking operations.

In mechanical pulping, two basic processes require water: (1) the stone groundwood, and (2) the refiner groundwood processes. In stone groundwood, water serves as both a coolant and a carrier to transport pulp from the grinder. After the addition of more water to dilute the pulp slurry, it passes through coarse and fine screens to remove dirt and slivers. The typical wastewater produced by mechanical pulping contains wood fiber, suspended solids, and lignin, a slightly biodegradable organic. The wastewater generated from mechanical pulping which uses chemicals for digesting the pulp (chemi-mechanical pulping) may contain chemicals such as sodium carbonate, sodium hydroxide, and sodium sulfate.

Chemical pulping generates wastewater in three basic processes: (1) alkaline (soda), (2) sulfite, and (3) semi-chemical pulping. The wastewater sources from the above three operations are spills from the digester area, condensed digester vapors, and wastewater from wash/screening operations. The constituents and chemicals present in the chemical pulping wastewater are suspended solids, cellulose, sodium sulfite, caustic soda, sulfite salt, and ammonia.

The sources of wastewater generated from secondary fiber pulping results from deinking operations. Wastewater from this operation contains the dispersants, detergents, and solvents used to separate the ink pigment, filler, adhesives, and coating agents such as clays, calcium carbonate, and titanium dioxide.

In the bleaching process, the wastewaters are generated from bleaching, chemical preparation, and washing operations. The chemicals used in the bleaching process are sodium or zinc hydrosulfite, chlorine, calcium hypochlorite, and chlorine dioxide, and the same chemicals are found in the wastewater generated during these processes.

In papermaking processes, water is used for dilution and transportation of the pulp to the paper machine. The water is called "white water" and is drained or pressed from the paper or paperboard. The white water is sometimes reused, but is commonly directed to a biological treatment process. The wastewater sources in the papermaking operation include water losses from the preparation area and the overflow from the white water discharges.

If the paper is coated, coating solutions containing the above-mentioned chemicals are a concentrated waste with very high SS, COD, and BOD concentrations. Alkalies and carbonates are also found in some coatings.

A flow schematic of waste/wastewater generating processes of the pulp, paper, and paperboard industry is shown in Figure 8.9.

QUESTIONS

Write your answers in a notebook and then compare your answers with those on page 592.

8.3E   Describe a printed circuit board.

8.3F   List the general processes of printed circuit board manufacturing.

8.3G   List the major pulp, paper, and paperboard manufacturing processes.

8.3H   What are the three basic types of chemical pulping now in common use?

8.33   Battery Manufacturing

The battery industry makes modular electric power sources in which part or all of the fuel is contained within the unit. Batteries generate electric power directly from a chemical reaction rather than indirectly as in a heat cycle engine.

The details of battery construction vary with the type of battery. For the usual liquid electrolyte batteries, the steps include the manufacture of structural components, preparation of electrodes, and assembly into cells. Fabrication of the structural components such as cell cases or caps, terminal fittings or fixtures, electrode supports, separators, seals, and

covers are all manufacturing processes not directly involving the electrochemistry of the cell. These components may be fabricated by the battery producer or may be supplied by other manufacturers. The steps considered to be battery manufacturing operations are anode and cathode fabrication and ancillary operations. Two battery types, lead-acid and nickel-cadmium, were chosen to illustrate a range of materials, applications, and sizes. Figures 8.10 and 8.11 are simplified manufacturing process flow diagrams for these two battery types.

8.330   Unit Process Description

Battery manufacturing includes three major operations: anode production, cathode production, and cell and battery assembly.

Anodes

The active mass for anodes is usually prepared as the massive metal, finely divided metal, finely divided metal compound, or as a soluble salt of the metal which is precipitated onto a carrier or support structure. In most batteries, there is an additional support structure.

The final step in anode preparation for many types of batteries is formation, or charging, of the active mass. Formation may be carried out on individual electrodes or on pairs of electrodes in a tank of suitable electrolyte. Frequently, electrodes are formed in the cell or battery after final assembly by passing current through the electrodes to charge them.

Anodes for most lead-acid batteries and some nickel-cadmium cells are prepared from a paste of a compound of the anode metal (lead oxides or cadmium hydroxide, respectively). Additives may be mixed in, and then the paste is applied to a support structure and cured.

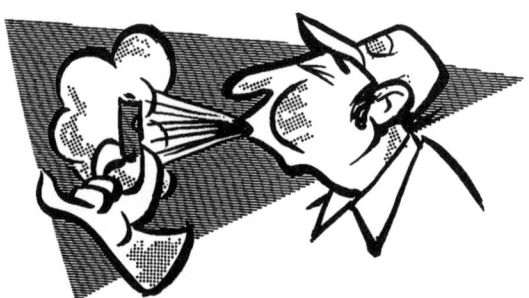

For pocket-type nickel-cadmium batteries, cadmium metal is oxidized in a high temperature air stream, then hydrated to cadmium hydroxide. Graphite, to increase conductivity, and iron oxide, to keep the cadmium in a porous state during cycling, may be mixed into the cadmium hydroxide. The cadmium for pocket-type anodes is then loaded into the pockets of the support structure made of perforated nickel or steel sheet.

Cathodes

Cathode fabrication almost always includes a rigid, current-carrying structure to support the active material. The active material may be applied to the support as a paste, deposited in a porous structure by precipitation from a solution fixed to the support as a compacted pellet, or may be dissolved in an electrolyte form which has been immobilized in a porous inert structure.

Formation of cathodes for rechargeable batteries is similar to that of anodes. Nickel cathodes may be formed outside or

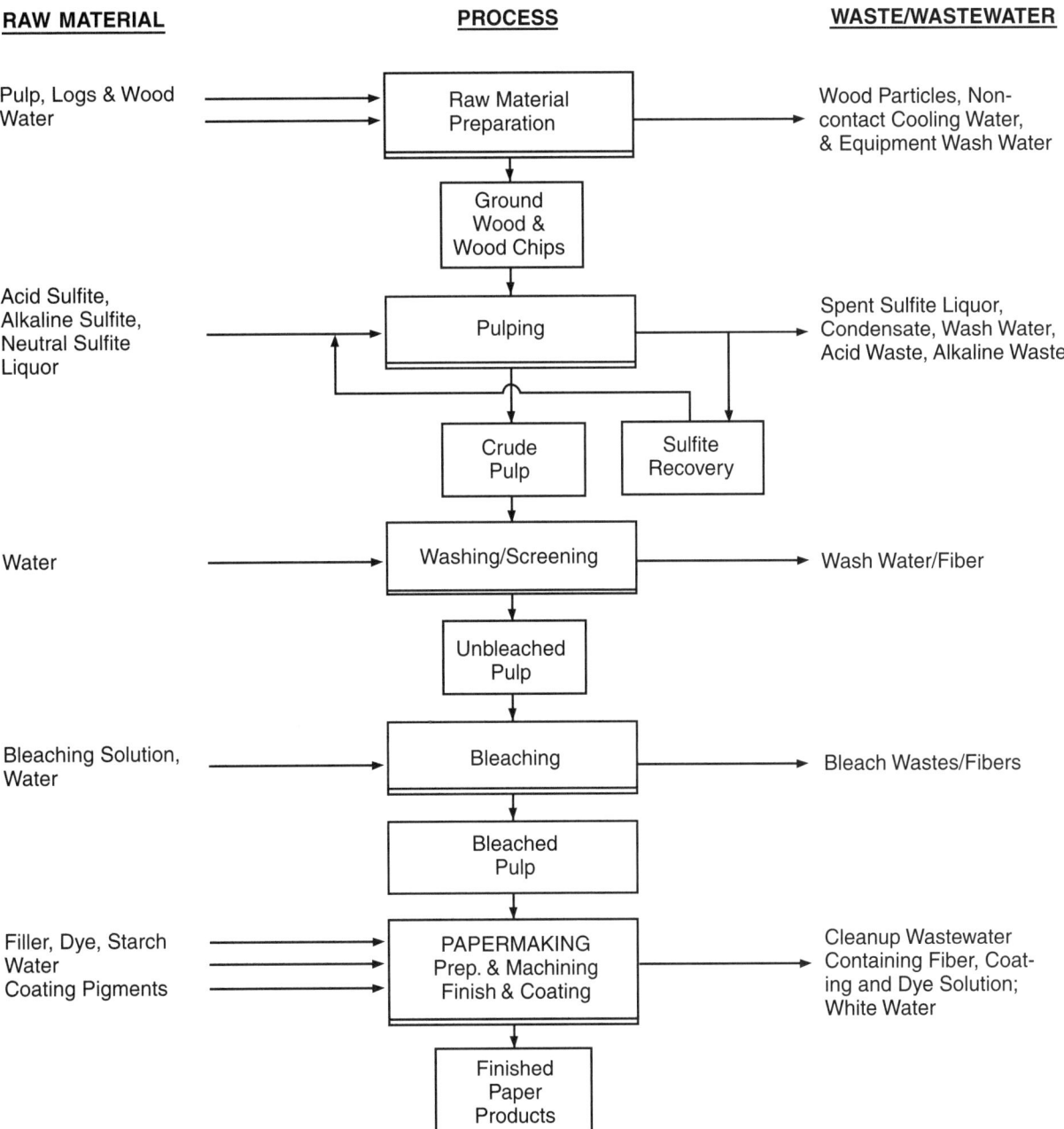

*Fig. 8.9   Waste/wastewater generating processes for
pulp, paper, and paperboard industries*

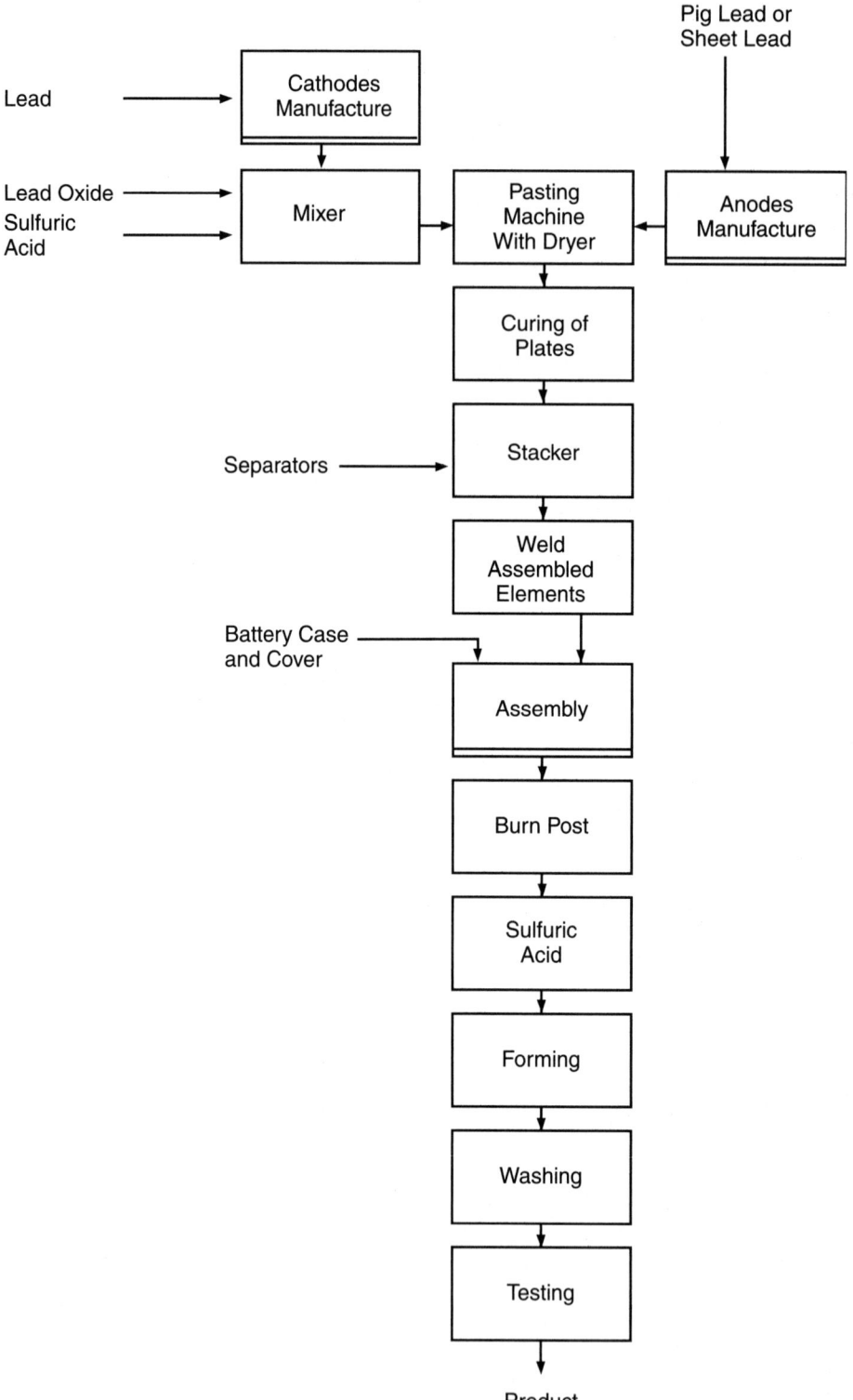

Fig. 8.10 Simplified diagram of major production operations in lead-acid battery manufacturing

(Reference: *DEVELOPMENT DOCUMENT FOR EFFLUENT LIMITATIONS GUIDELINES AND STANDARDS FOR BATTERY MANUFACTURING*, EPA, August 1984)

POSITIVE PLATE PROCESS

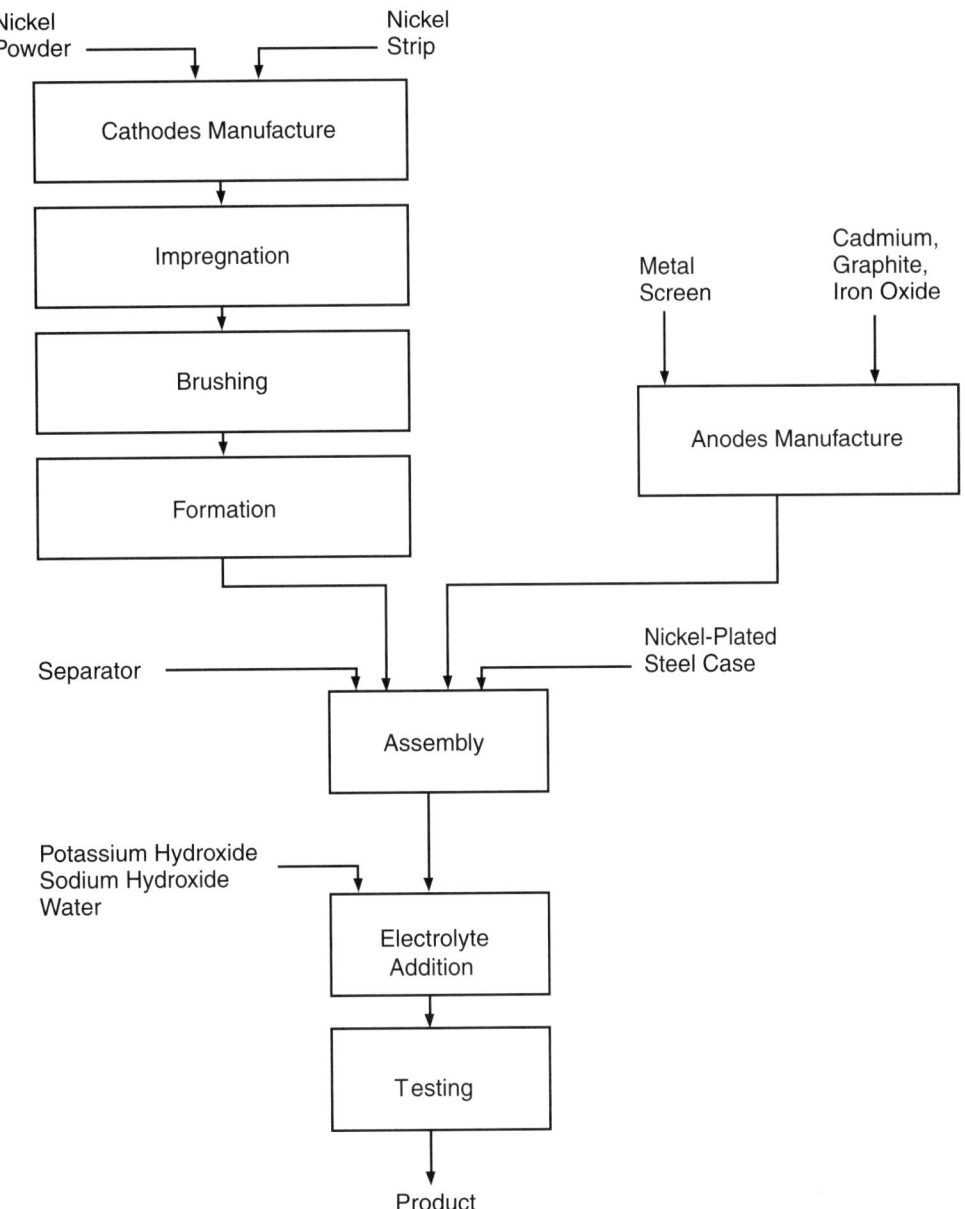

Fig. 8.11 Major production operations in
nickel-cadmium battery manufacturing

(Reference: DEVELOPMENT DOCUMENT FOR EFFLUENT LIMITATIONS GUIDELINES
AND STANDARDS FOR BATTERY MANUFACTURING, EPA, August 1984)

inside the assembled cell in the potassium hydroxide electrolyte. Lead cathodes for lead-acid batteries are handled similarly to the anodes, except they remain in lead peroxide after forming.

Preparation of the cathode-active material in the battery plant is usually restricted to the metal oxides or hydroxides. Cathode-active materials for battery types discussed here, nickel hydroxide and lead oxide, are specific to battery manufacturing and are usually produced in the battery plant. For nickel-cadmium pressed powder (pocket-electrode) cells, nickel hydroxide is produced by dissolution of nickel powder in sulfuric acid and precipitated sodium hydroxide. The resulting nickel hydroxide is centrifuged, mixed with some graphite, spray dried, compacted, and mixed with additional graphite.

Lead-acid batteries require a specific oxidation state of lead oxide (24 to 30 percent free lead) referred to by the industry as "lead oxide," which is produced by the ballmill or barton process. This lead oxide is used for both the anode and the cathode.

Ancillary Operations

Ancillary operations are all those operations unique to the battery manufacturing point source category which are not included specifically under anode or cathode fabrication. They are operations associated mainly with cell assembly and battery assembly. Also, chemical production for anode or cathode-active materials for some batteries is considered an ancillary operation.

The electrodes for rectangular nickel-cadmium batteries are placed in a stack with a layer of separator material between each electrode pair and inserted into the battery case. Almost all lead-acid batteries are assembled in a case of hard rubber or plastic with a porous separator between electrode pairs. The cells or batteries are filled with electrolyte after assembly.

Ancillary operations, besides specific chemical production, include some dry operations as well as cell washing, battery washing, and the washing of equipment and floors. Because the degree of automation varies from plant to plant for a given battery type, the specific method of carrying out the ancillary operations is not as closely identifiable with a battery type as are the anode and cathode fabrication operations.

8.331   Waste/Wastewater Characteristics

Characteristics of the wastewater generated from the battery manufacturing processes are shown in Figure 8.12. The wastewater is generated mainly from cathode production, anode production, pasting, curing, acid adding, forming, wash processes and testing (in the form of rejected batteries).

Cathode and Anode Production

Process wastewater from cathode production results from lead oxide and shell cooling processes which contain lead (up to 50 mg/L), copper, and acid. The wastewater generated from anode production contains lead particulates. The concentration of lead in wastewater generated from the process is minimal.

Pasting and Curing

One of the pasting wastewater sources is the washdown water used for cleaning of the equipment and may contain up to 850 mg/L of lead. Curing process wastewater results from steam curing of pasted plates which contain lead at very low concentrations and lead particulates.

Acid Adding and Forming

The sources of wastewater for the acid adding and forming processes are the washdown water and spills which may contain lead and acid.

Washing

The wastewater from the washing process is generated in large amounts and contains lead at low concentrations and acid.

8.34   Leather Tanning and Finishing

Leather tanning involves numerous processes to convert animal hides or skins into leather. Three major types of hides and skins are used most often to manufacture leather: cattle hides, sheepskins, and pigskins.

Animal skin is composed of outer and inner layers. The inner layer consists mainly of the protein collagen. Tanning is a reaction between collagen fibers and tannins, chromium, alum, or other tanning agents. The tanning processes stabilize and preserve the skin and make it useful.

General processes in leather tanning and finishing include three major steps: beamhouse operation, tanyard process, and retanning and wet-finishing processes. A schematic process flow diagram for leather tanning of cattle hide is shown in Figure 8.13

8.340   Unit Process Description

The following process descriptions apply to typical cattle-hide tannery manufacturing.

Beamhouse Processes

There are four subprocesses in the beamhouse processes: side and trim, soak and wash, fleshing, and unhairing. The side and trim process includes trimming and cutting the preserved hides or skin. (The hides are usually preserved with green salt or brine.) The soak and wash process restores moisture to the hides by soaking them in special processors (drums or concrete mixers with lining), and washing the hides to remove dirt and salt. The mechanical process for removing the excess flesh, fat, and muscle from the hides is called fleshing. Cold water is used to keep the fat together in chunks for later disposal. The last step in the beamhouse process, unhairing, uses three chemicals to remove the hair: calcium hydroxide, sodium sulfhydrate, and sodium sulfide. These chemicals are mixed with water and are placed in processors with hides. Higher water temperature and agitation increase the speed of the unhairing process. Shaving also is used in this process.

<u>**RAW MATERIAL**</u>   <u>**PROCESS**</u>   <u>**WASTE/WASTEWATER**</u>

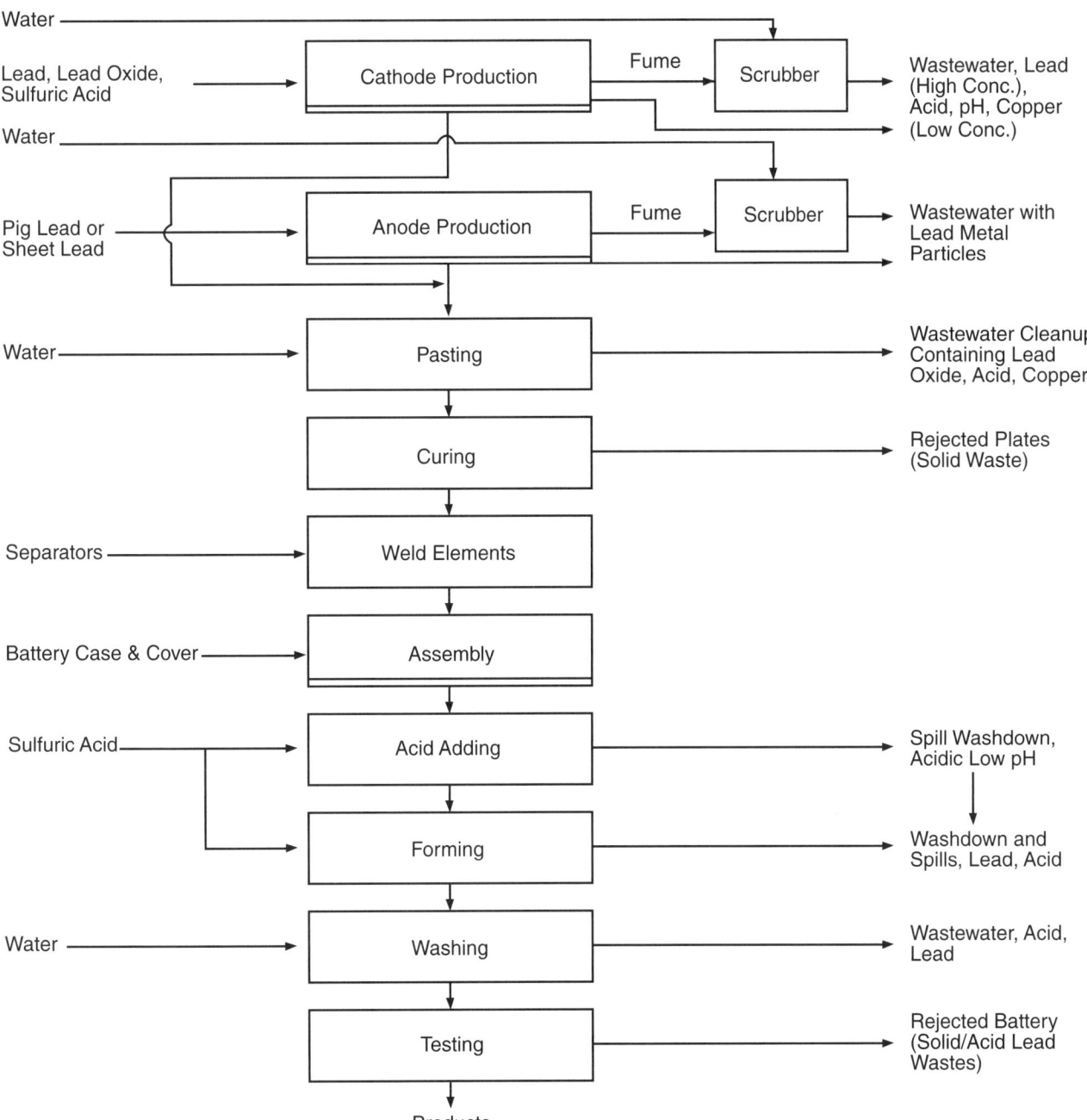

*Fig. 8.12   Waste/wastewater characteristics for
battery manufacturing*

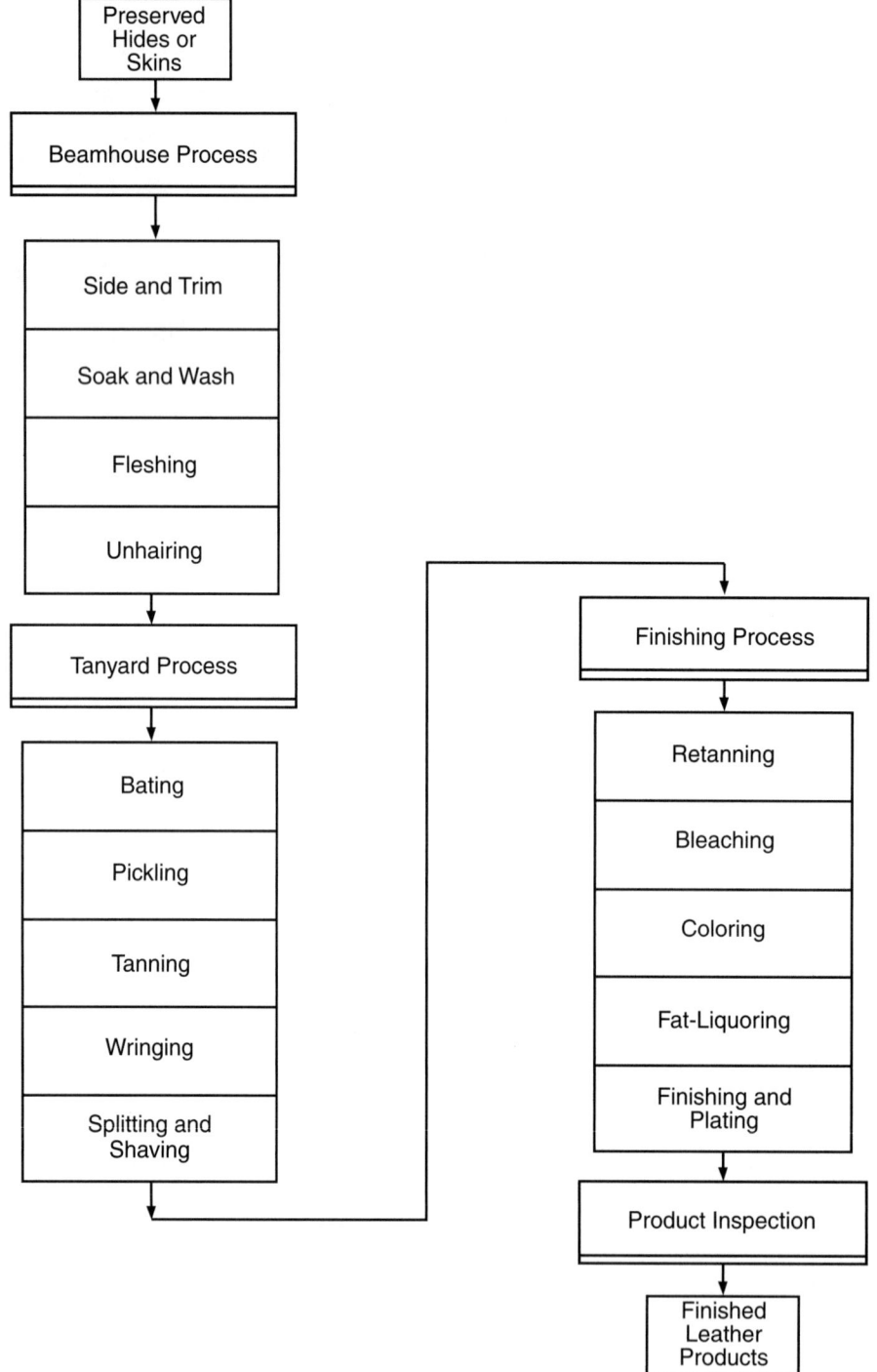

Fig. 8.13   Leather tanning and finishing process

Tanyard Processes

The tanyard processes further prepare the hides for final finishing and tanning, and include bating, pickling, tanning, wringing, and splitting.

Bating — This two-step process consists of deliming and enzyme addition. In deliming, salts of ammonium (sulfate or chloride) are used to clean the hides of any residual chemicals from previous processes. Enzyme addition facilitates separation of the protein fibers and removes the undesirable constituents.

Pickling — The pickling process prepares the hides to accept the tanning agents. This process provides an acid environment to apply chromium tanning and prevents precipitation of chromium salts. Sulfuric acid is the most common acid added to the system.

Tanning — The application of tanning agents improves the stability, heat resistance, and chemical resistance of the hides by converting the raw collagen fibers of the hide into a stable product which no longer is susceptible to decay or rotting. Chromium is the most frequently used tanning agent because it imparts better physical properties and requires a shorter processing time than other tanning agents (vegetable tanning).

Wringing — In this process, hides are fed through a machine equipped with two large rollers to wring excess moisture from them in preparation for splitting. The process yields "wet-blue stack" which is taken through the finishing process.

Splitting and Shaving — Splitting adjusts the thickness of the tanned hides and shaving removes any remaining matter from the flesh portion of the hides. The shaving machine also can adjust the thickness of the hides.

Finishing Processes

The finishing processes give the tanned hides special or desired characteristics. These processes may include retanning, bleaching, coloring, fat-liquoring, and finishing.

Retanning — The primary function of retanning is to yield certain characteristics which are lacking in the leather after the initial tanning step. Vegetable extracts and syntans (manufactured chemicals) are used in the retanning process.

Bleaching — Sodium bicarbonate and sulfuric acid are used in bleaching the hide, a process confined mainly to the sole leather industry.

Coloring — Dyes are sometimes added during the retanning process. Skin variability and color penetration are the most important factors in the coloring process.

Fat-liquoring — This process lubricates the fibers. Oils and related fatty substances replace the natural oil hides lost in the beamhouse and tanyard processes. Chemical emulsifiers are added to fat-liquor materials to permit dispersion in water.

Finishing — The finishing process is performed after the fat-liquoring process and includes settling, drying, conditioning, staking, buffing, water-base finishing, solvent-base finishing, and plating.

8.341 Waste/Wastewater Characteristics

Water usage by the leather tanning industries depends upon the type of raw materials and the processing methods used. All three basic operations (beamhouse, tanyard, and finishing processes) involve the use of some water. In general, the major sources of wastewater generated by leather tanning operations are: (1) soaking, (2) tanning and retanning with chromium, vegetable, aluminum, or other agents, (3) preparing bleach, dye, or pigment solutions, and (4) cleaning and washdown of process equipment and areas.

The beamhouse operations of trimming, soaking, fleshing and unhairing all generate wastewaters. The constituents present in wastewater produced by these processes include dirt, salt, bleach, manures, proteins, hair, and chemicals used for unhairing such as calcium hydroxide, sodium sulfhydrate, and sulfide. The quantity of manure and dirt varies with the season of the year and the origin of the hide but, in any case, the wastewater is high in BOD, suspended solids, dissolved solids, and sulfide.

The wastewater sources in the tanyard operation originate in the bating, pickling, and tanning processes. The acidic tanning solution contains high levels of chromium as the major ingredient, and zinc, lead, nickel, and copper in smaller amounts. The wastewaters from bating, pickling, and tanning processes contain dissolved hair, protein, pickling chemicals such as brine and acid solution, and tanning chemicals such as chromium at high concentrations as well as the other heavy metals mentioned previously.

The finishing operation processes of retanning, bleaching, coloring, fat-liquoring, and plating generate wastewaters containing tanning chemicals, bleaching agents, chemical dyes (chromium and organo-metallic), oil, and coating agents such as water-based organic solvents.

In conclusion, the most heavily used toxic pollutant in leather tanning is chromium, but you will also encounter other pollutants such as naphthalene, phenol, and pentachlorophenol (preservatives) and heavy metals including zinc, lead, nickel and copper (from organo-metallic dyes) in the wastewater streams. Cyanide is found in dyes.

Figure 8.14 presents a schematic diagram of raw materials used and waste/wastewater streams produced through typical leather tanning manufacturing.

8.35 Petroleum Refining Industry

The petroleum refining industry is one of the major liquid-material processing industries. It produces a large variety of fuels in gaseous and liquid forms in conjunction with many other by-products.

Crude petroleum is made up of thousands of different chemical substances including gases, liquids, and solids ranging from methane to asphalt. Most constituents are hydrocarbons, but there are significant amounts of compounds containing nitrogen, oxygen, and sulfur (up to six percent). Sulfur has always been an undesirable constituent of petroleum and efforts have been made to eliminate the sulfur from gasoline and kerosene products. At present, wherever possible, the sulfur compounds are being removed and frequently the sulfur is recovered as elemental sulfur.

Petroleum crude varies widely with each kind requiring different refining procedures. The terms "paraffin base," "asphalt," and "mixed base" are often applied to differentiate crudes on the basis of the residues produced after simple distillation. A general process flow schematic of petroleum refinery industries is presented in Figure 8.15.

8.350 Unit Process Description

The petroleum refinery processes involve the production of natural gasoline and natural gas (including liquified petroleum gas), light distillates (including motor gasolines, solvents, jet fuels, kerosene, and light heating oils), intermediate distillates (including heavy fuel oils, diesel oils, and gas oils), heavy distillates (including mineral oils, heavy oils, lubricating oils, and waxes), and residues (including lubricating oils, fuel oils, road oils, asphalts, and coke).

Crude oil is usually refined by fractional distillation to separate the various hydrocarbons prior to chemical and mechanical treatment of the fractions. The following paragraphs describe the various production processes. (Section 10.310 contains additional information about the processes involved in petroleum refining.)

Primary Distillation

The crude oil is passed through a pipe still into a fractionating tower where the lighter products, such as gasoline, kerosene, and gas oil, are removed and condensed.

Treating Process

The gasoline and kerosene from primary distillation then pass through polymerization, alkylation, and hydrogenation processes. During these processes sulfuric acid, caustic soda, and water washes are variously applied to remove impurities. The gas oil is treated and stored as light fuel oil.

Cracking and Distillation

Distillation of the remainder (light residue, asphaltic and lubricating residue) after primary distillation is then continued by vacuum distillation and catalytic cracking for further fractional distillation to separate the various hydrocarbons from crude oil.

Recovery Process

The residues from vacuum distillation are taken through solvent refining, dewaxing, decolorizing, wax refining, and coking processes. The products of these processes are lubricating oils, refined waxes and coke.

8.351 Waste/Wastewater Characteristics

Waste/wastewater streams originating from petroleum fields and refineries result from pumping, desalting, distilling, fractionation, alkylation, and polymerization processes. These wastes are of large volume, contain suspended and dissolved solids, oil, wax, sulfide, chloride, mercaptans, phenolic compounds, and sometimes large amounts of dissolved iron.

In general, the wastes from oil refineries include free and emulsified oil from leaks, spills, tank draw-off, and other sources; waste caustic, caustic sludges, and alkaline water; acid sludges and acid waters from distillate separators and tank draw-off; tank-bottom sludges; waste catalysts and filter residues; special chemicals from by-product chemical manufacturing; and cooling waters. Oil from spills and leaks can amount to as much as three percent of the total crude oil treated.

The treatment of oils with alkaline reagents to remove acidic components produces acid sludges, and the sweetening processing of oils to convert or remove mercaptans produces a series of alkaline wastes with obnoxious odors. A typical flow sheet of the petroleum refining process which produces the various wastes is shown in Figure 8.16.

QUESTIONS

Write your answers in a notebook and then compare your answers with those on page 593.

8.3I   How is electric power generated in a battery?

8.3J   List the battery manufacturing processes which generate wastewater.

8.3K   What is the purpose of the tanning segment of the tanyard processes?

8.3L   What are the major water usage processes in the leather tanning industry?

8.3M   On what basis are various types of petroleum crudes differentiated and what terms commonly describe the different types?

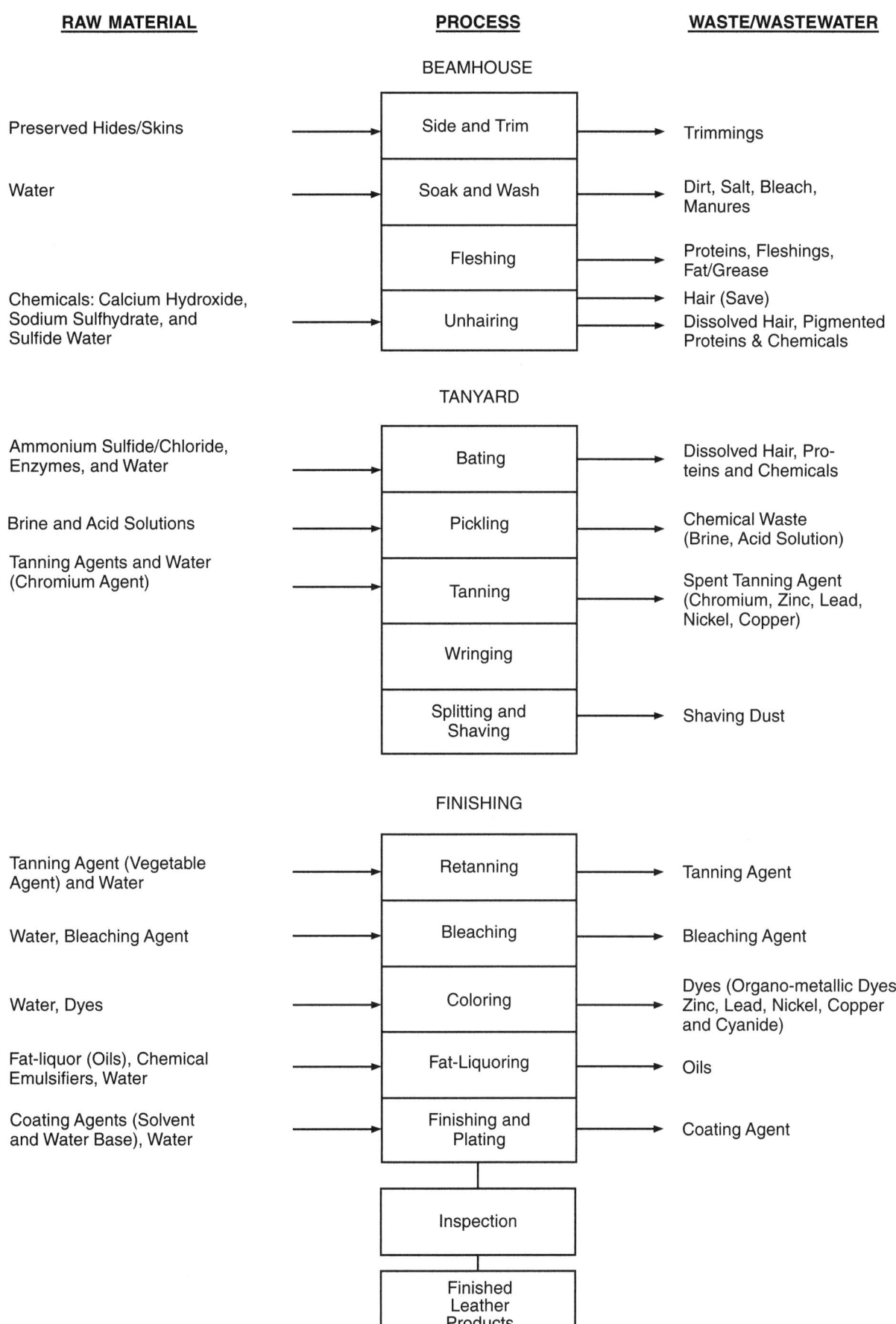

*Fig. 8.14   Leather tanning and finishing industry
waste/wastewater characteristics*

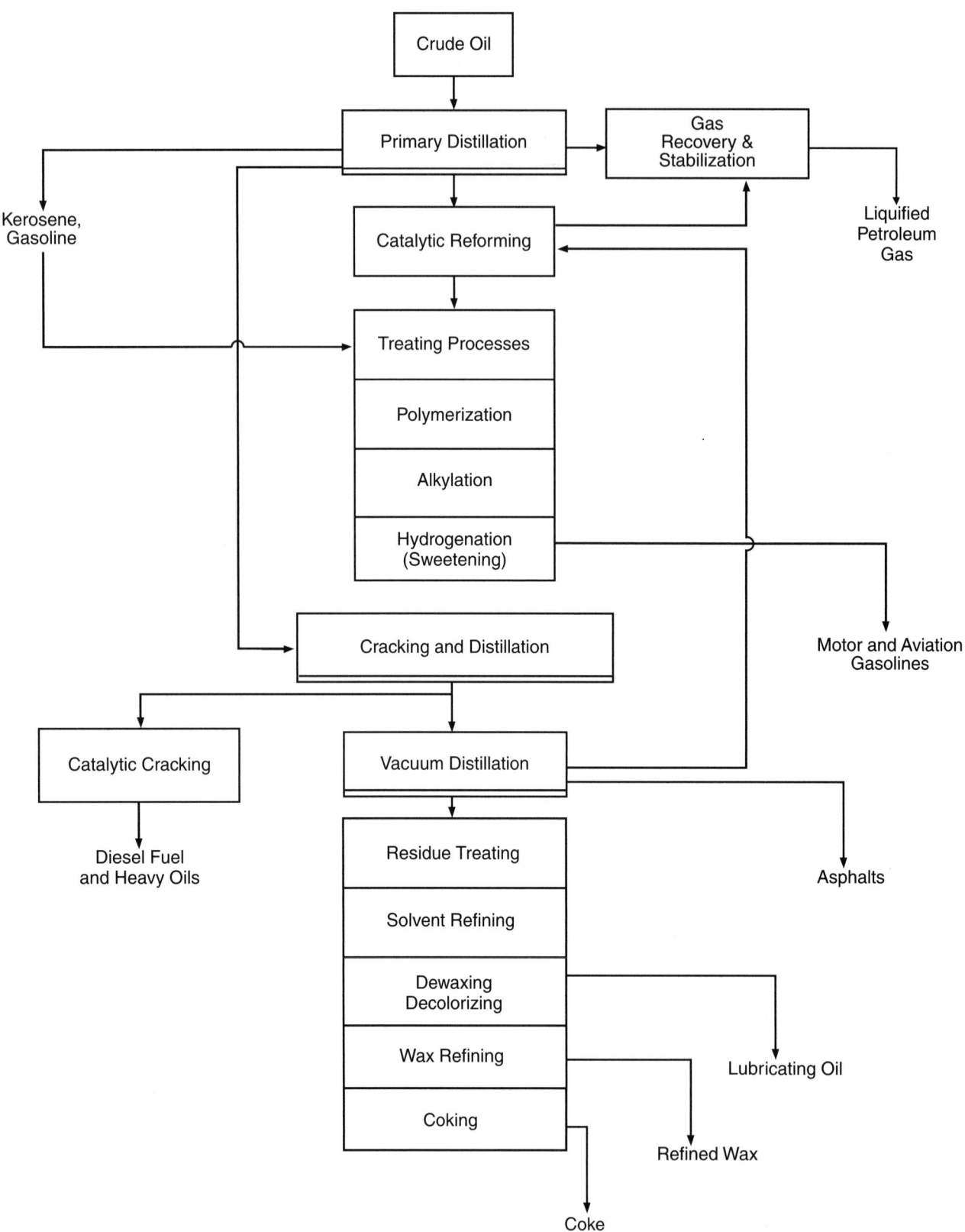

Fig. 8.15   Crude oil processing

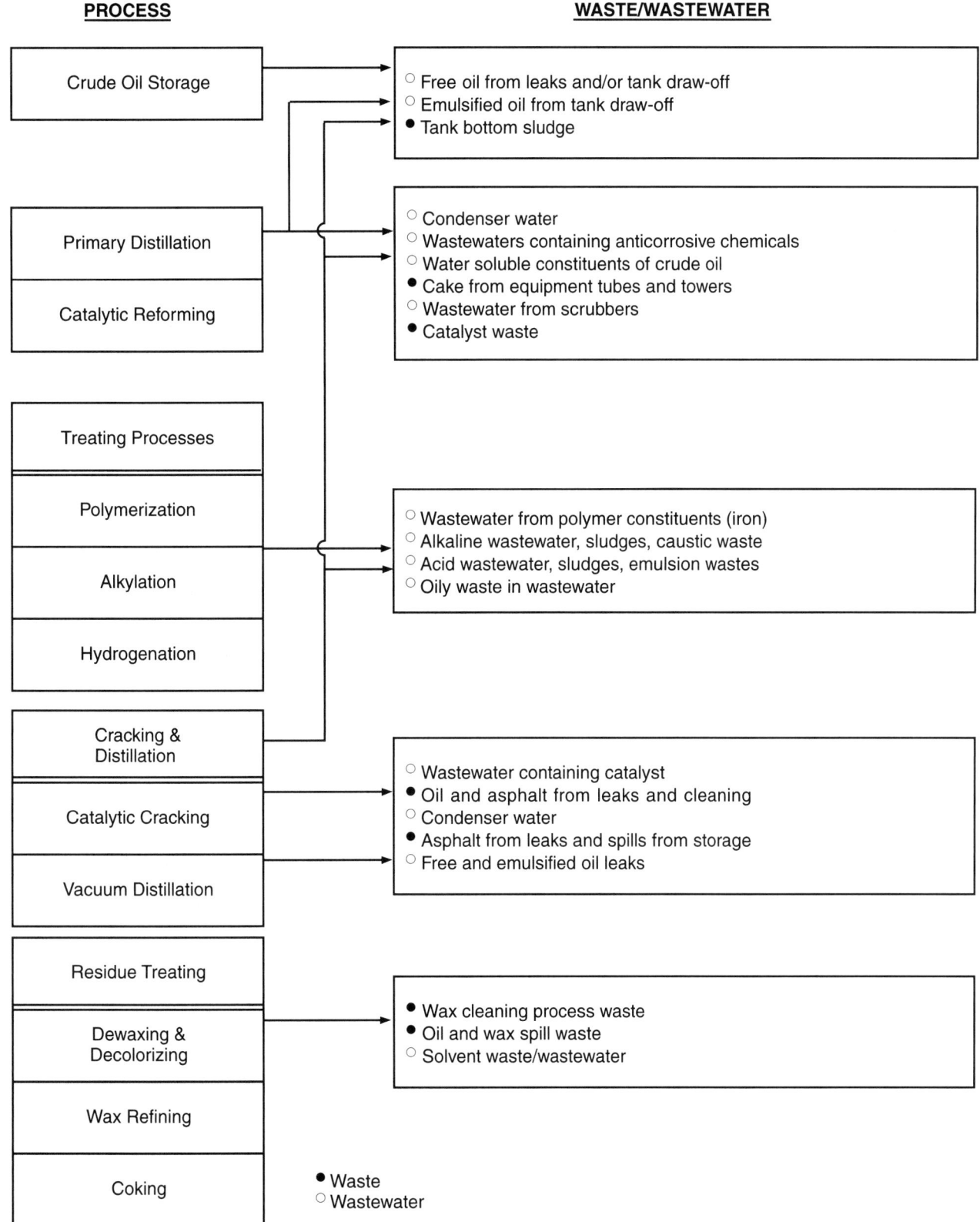

Fig. 8.16   Waste/wastewater from petroleum refining

8.36   Iron and Steel Industry

The iron and steel industry is one of the largest manufacturing operations in the United States and uses enormous volumes of water and air to produce iron and steel from raw materials to finished shapes. The manufacturing processes of the iron and steel industry are divided into twelve main processes as shown schematically in Figure 8.17. The twelve processes are listed below:

1. Cokemaking
2. Sintering
3. Ironmaking
4. Steelmaking
5. Vacuum Degassing
6. Continuous Casting
7. Hot Forming
8. Salt Bath Descaling
9. Acid Pickling
10. Cold Forming
11. Alkaline Cleaning
12. Hot Coating

8.360   Unit Process Description

Cokemaking

The production of coke is an essential part of the iron and steel industry since coke is one of the basic raw materials necessary for the reduction of iron ore in the operation of ironmaking blast furnaces. Coke is produced in by-product or beehive ovens from a carefully blended mixture of high- and low-volatility coals.

By-product cokemaking is the method used in 99 percent of the coke plants in the United States. The mixed coal is charged into the hot oven and is then heated by the external combustion of coke oven gas until all the volatile materials have been driven off and the remaining coke fuses into an incandescent mass. At the end of the coking period, the incandescent coke is pushed out of the furnace and taken directly to a quenching area where water is sprayed on the hot mass of the coke. The gas produced during the cokemaking of the coal contains valuable chemicals including ammonia, benzene, xylene, and toluene by-products. These are recovered in the by-product plant.

Beehive cokemaking is the other method in the cokemaking process and is only found in one percent of the U.S. cokemaking operations. In beehive ovens, no effort is made to recover volatile materials generated by the process so no wastewater is generated from gas cleaning as in the by-product plants.

Sintering

Sintering is the production of an agglomerate (a clump or mass of material) which is later reused as a feed material in iron and steelmaking processes. The agglomerate or "sinter" is made up of large quantities of particulate matter (fines, mill scale, flue dust). This process is used to reclaim mill scale and blast furnace fines and uses some of the fine iron ore which cannot be charged directly to the blast furnaces. These materials are mixed with fine coal in a traveling grate. This produces a sinter which can be charged directly to the blast furnace without loss of fine material.

Most of the water used in sintering operations is in the scrubbing of dusts and bases produced during the process. Quenching and cooling with large quantities of water is part of the sintering process.

Ironmaking

Ironmaking operations involve the conversion of iron-bearing materials, limestone, and coke into molten iron in a reducing atmosphere (for example, a blast furnace). A blast furnace is a tall, cylindrical furnace about 16 to 28 feet in diameter and up to 100 feet tall. The combustion gases produced in this process are a valuable heat source, but require cleaning prior to reuse. Wastewaters are generated during the scrubbing and cooling of these blast furnace off-gases.

Steelmaking

Steel is produced from iron in three types of furnaces: basic oxygen, open hearth, and electric furnaces. These furnaces receive iron produced in blast furnaces along with scrap metal and fluxing materials. Steelmaking generates large quantities of fume, smoke, and waste gases which require cleaning prior to emission to the atmosphere and wastewaters are generated during some of the gas-cleaning operations.

Oxygen furnaces convert iron to steel at much higher rates than are attainable in open hearth furnaces. In the conversion of iron to steel, the molten metal charge is reacted with oxygen, then introduced through a water-cooled lance, which burns off the impurities in a period of about 30 minutes.

The open hearth furnace produces steel (below one percent carbon) from a charge of steel scrap, iron, and slagging materials such as lime and fluorspar. As in the blast furnace, the open hearth furnace uses large quantities of water for cooling, and the gases contain high concentrations of dust.

The third process for manufacturing steel, the electric furnace process, can either produce the common grades of low-carbon steel from scrap or can be charged with alloying materials to produce specialty steels such as stainless steel or tool steel. Electric furnaces are also used for production of the ferroalloys which are the alloying element for these special steels, such as ferromanganese, ferrovanadium, and ferrochrome. In the operation of electric furnaces, fluxing materials

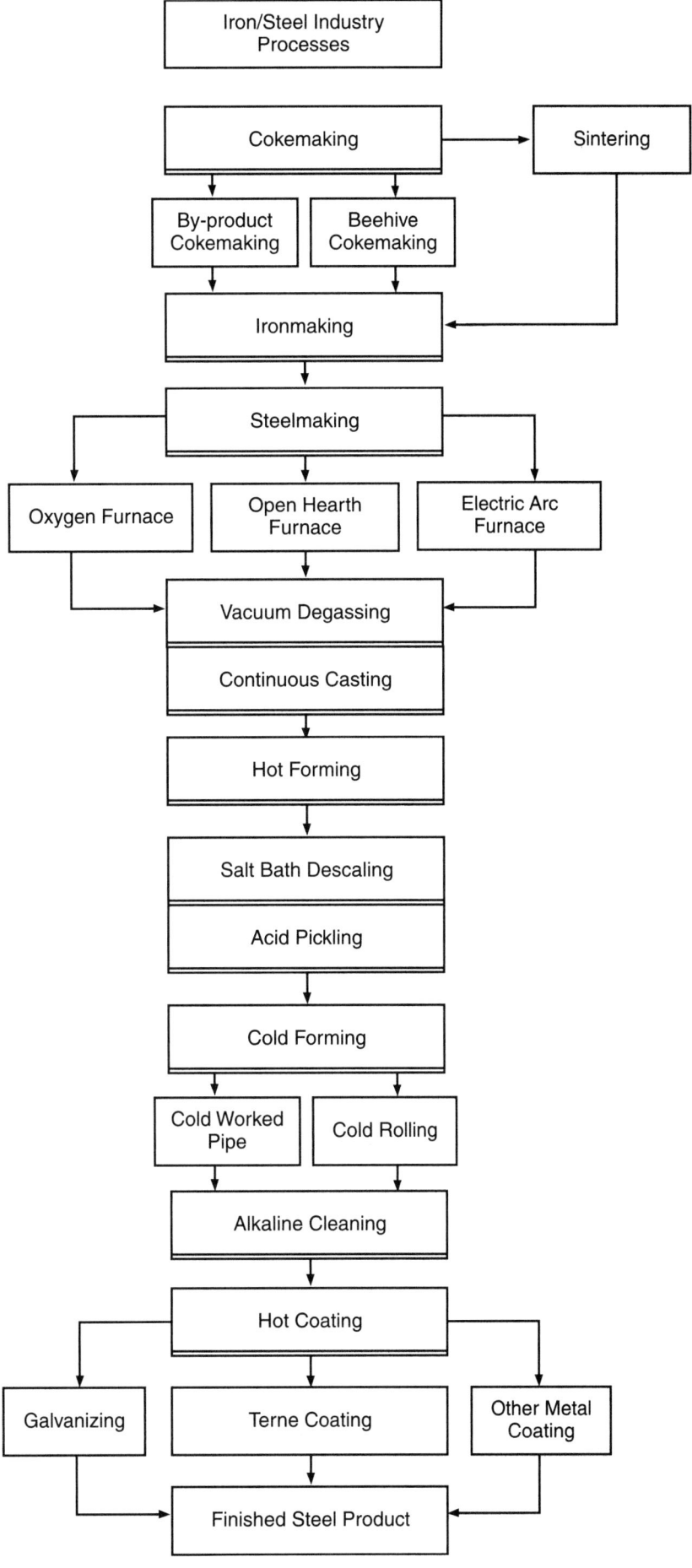

Fig. 8.17 General manufacturing process diagram for iron/steel industry

are added to the charge along with the metallic components so that a slag is produced as in the other furnace operations. The charge is reduced by an electric current.

Vacuum Degassing

Vacuum degassing is the process whereby molten steel is subjected to a vacuum in order to remove gaseous impurities. It is advantageous to remove hydrogen, nitrogen, and oxygen from the molten steel as these gases can impart undesirable qualities to certain grades of steel. In many plants, the vacuum is produced through the use of steam ejectors and barometric condensers. The particle-laden steam coming from the steam ejectors is condensed in condensers, thus producing a wastewater which requires treatment.

The industry uses various types of degassers for steels containing a variety of different components. However, it has been determined that the variations do not affect the quantity or quality of wastewaters produced in the vacuum degassing operations.

Continuous Casting

The continuous casting process is used to produce semi-finished steel directly from molten steel. The molten steel from the steelmaking operation is ladled into a receiving vessel from where it is continuously cast into water-cooled copper molds of the desired shapes. After leaving the copper mold, the semi-solidified steel is sprayed with water for further cooling and solidification. In addition to cooling, the water sprays also serve to remove scale and other impurities from the steel surface. The water that directly cools the steel and guide rollers contains particulates and roller lubricating oils.

Hot Forming

Hot forming is the steel forming process in which hot steel is transformed in size and shape through a series of forming steps to ultimately produce semi-finished and finished steel products. The steel products consist of many types of cross-sections, sizes, and lengths. Four different types of hot forming mills are used to produce the many types of hot formed products. The four types of mills (primary, section, flat, and pipe and tube) are the bases for the principal subdivisions of the hot forming process.

The hot steel is reduced in size by a number of rolling steps where contact cooling water is continuously sprayed over the rolls and hot steel product to cool the steel rolls and to flush away scale as it is broken off from the surface. Scarfing is used at some mills to remove imperfections in order to improve the quality of steel surfaces. Scarfing generates large quantities of fume, smoke, and waste gases which require scrubbing. Scrubbing of these fumes generates additional wastewater.

Salt Bath Descaling

Salt bath descaling is the operation in which specialty steel products are processed in molten salt solution for scale removal. Two types of scale removal operations are in use: oxidizing and reducing. The oxidizing process uses highly oxidizing salt baths which react far more aggressively with the scale than with base metal. This chemical action causes surface scale to crack so that subsequent pickling operations are more effective in removing the scale. Reducing baths depend upon the strong reducing properties of sodium hydride to accomplish the same purpose. During that operation, most scale-forming oxides are reduced to base metal.

Acid Pickling

Acid pickling is the process of chemically removing oxides and scale from the surface of the steel by the action of water solutions of inorganic acids. Both sulfuric acid and hydrochloric acid are used in this process. This process produces a bright sheet stripped down to bare metal suitable for finishing operations such as plating, galvanizing, and application of other coatings.

Cold Forming

The cold forming process is separated into two subprocesses: cold rolling and cold worked pipe and tube. Cold rolling is used to reduce the thickness of a steel product, produce a smooth, dense surface and develop controlled mechanical properties in the metal. An oil-water emulsion lubricant is sprayed on the material as it enters the work rolls of a cold rolling mill, and the material is usually coated with oil prior to recoiling after it has passed through the mill. The oil prevents rust while the material is further processed or formed. Oil from the oil-water emulsion lubricant is the major pollutant load in wastewaters resulting from this operation. In cold rolling three methods of oil application are used: direct application, recirculations, and combinations of the two. Because the recycle rate is dependent upon the oil application system, flow rates vary for the three systems.

In the pipe and tube subprocess of cold forming, cold, flat steel strips are formed into hollow cylindrical products with continuous flushing with water or soluble oil lubricating solutions.

Alkaline Cleaning

Alkaline cleaning baths are used to remove minerals and animal fats and oils from steel. The cleaning baths used are not very aggressive and therefore do not generate many pollutants. The alkaline cleaning solution is usually a dispersion of chemicals such as carbonates, alkaline silicates, and phosphates in water. The cleaning bath follows the rinsing operation.

Hot Coating

Hot coating processes involve the immersion of clean steel into baths of molten metal for the purpose of depositing a thick layer of the metal onto the steel surface. These metal coatings can impart such desirable qualities as corrosion resistance or a decorative appearance to the steel.

The hot coating process has been divided into three subprocesses based upon the type of coating used. Galvanizing is a zinc-coating operation. Terne coating consists of a lead and tin coating of five or six parts lead to one part tin. Other metal coatings can include aluminum, hot dipped tin, or mixtures of these and other metals.

8.361 Waste/Wastewater Characteristics

The iron and steel industry in the United States produces large amounts of wastewater. The untreated process wastewaters contain over 40,000 tons/year of toxic organic pollutants, over 120,000 tons/year of toxic inorganic pollutants, and about 14,000,000 tons/year of conventional and nonconventional pollutants. Steel industry process wastewaters are treatable by currently available treatment technologies.

By far, the largest portion of the water requirement for the steel industry process is for cooling services. The various areas requiring cooling water are as follows:

1. Coke Plant — cooling water required for gas coolers, condensers, and quenching.

2. Ironmaking — cooling water used for blast furnaces and gas cooling machines.

3. Steel Production — cooling water used for furnaces, degassing condensers, and quenching.

4. Rolling and Shaping — cooling water required for rolls, bearings, guides, saws, straighteners, and forges.

5. Heat Treating — cooling used for furnaces and quenching vats.

Another process that requires a large amount of water is the washing of blast furnace gas. The gas is usually cleaned in a knock-out drum, scrubbed with water, cooled with additional water, and finally treated through electrostatic precipitators.

The final large-scale consumption of water in the steel industry takes place in the rolling and shaping mills, where water is used for bearing cooling and for sluicing scale from rolling stands.

The raw materials and the waste constituents most commonly found in the waste/wastewater streams generated by the iron and steel industry are presented in Figure 8.18. Contaminants typically found in the iron and steel industry and the wastewater and its sources are shown below.

WATER POLLUTANTS IN A TYPICAL IRON AND STEEL INDUSTRY

Pollutant Description	Source
1. Oils:	
Rolling Oils	Rolling Mills, Cold and Hot
Lubricants	Rolling Motors, Steam Engines
Hydraulic Oils	and Gear Drives, Motive Power
Quench Oils	Pumps for Press Operations, Heat Treatment
2. Suspended Solids:	
Scale	Rolling Mills
Sand	Foundries
Burden Fines	Air Washers for Sinter Processes, Blast Furnaces
Coal and Coke	Cokemaking
3. Chemicals:	
Pickle Liquor	Acid Pickling
Acid Sludge	By-products Plants
Caustic Waste	Cleaning
Lime	Coating, Ironmaking
Cleaners	Surface Treatment
Heavy Metal	Metal Treatment and Coating
Organic (Benzene, Naphthalene)	Cokemaking

QUESTIONS

Write your answers in a notebook and then compare your answers with those on page 593.

8.3N What is the purpose of acid pickling?

8.3O What is the purpose of alkaline cleaning baths?

8.3P List the types of pollutants found in iron and steel process wastewaters.

8.37 Inorganic Chemical Industries

Inorganic chemical industries are very large and diversified and have been segmented into many different industries. Typically, inorganic chemicals are manufactured using a limited number of manufacturing processes. Different inorganic chemical producers may use the same manufacturing process, but their raw materials, process sequence, and waste/wastewater handling will vary, and produce wastes of different quality.

The inorganic chemical industries discussed in this section and listed below represent the largest types of inorganic chemical manufacturers in the United States:

Chlor-Alkali Industry	Hydrogen Cyanide Industry
Hydrofluoric Acid Industry	Sodium Dichromate Industry
Titanium Dioxide Industry	Copper Sulfate Industry
Aluminum Fluoride Industry	Nickel Sulfate Industry
Chrome Pigments Industry	

8.370 Chlor-Alkali Industry

Chlorine and its co-product caustic soda (alkali) are used in large quantities in the production of plastics, organic and inorganic chemicals, in the pulp and paper industry, in waste and wastewater treatment systems, and in a number of other industries.

The production rate in the United States is approximately nine million metric tons of chlorine per year. Over 95 percent of that amount is produced by the electrolysis of a sodium or potassium chloride solution using mercury cell and/or diaphragm cell processes.

8.3700 *GENERAL MANUFACTURING PROCESS DESCRIPTION*

Brine System

The sodium chloride solution (a brine or salt dissolved in water) is treated with sodium carbonate and sodium hydroxide to precipitate impurities such as calcium, magnesium, and iron. The precipitated hydroxides and carbonates are then settled, usually in a clarifier, and the underflow, known as brine mud, is sent to a lagoon or filtered. Brine muds from mercury

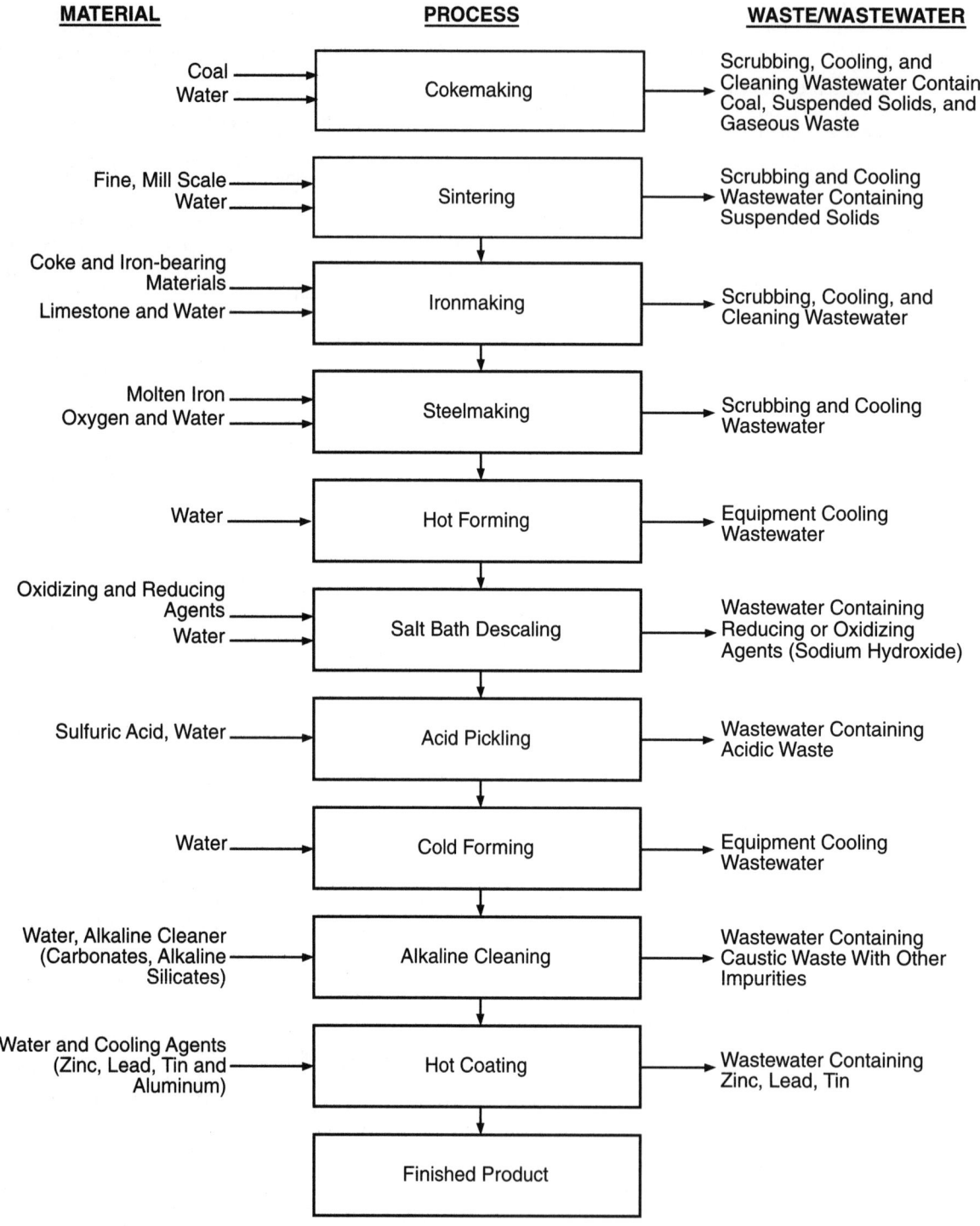

Fig. 8.18   Iron and steel industry wastewater characteristics

cell plants usually contain small amounts of mercury because the spent brine from the cells is recycled.

Mercury Cell Process

The mercury cell, in general, consists of two sections: the electrolyzer and the decomposer or denuder. Mercury flows in a thin layer at the bottom of the cell forming the cathode, and the brine flows concurrently on top of the mercury, contacting with the anode. Electric current flowing through the cell decomposes the brine, liberating chlorine gas at the anode and sodium metal at the cathode.

The material from the electrolyzer flows to a denuder (decomposer) and the spent brine is recycled to the brine purification process. Deionized water is added to the denuder and reacts with the material to form hydrogen and caustic soda.

Product Purification

Chlorine from the cell is cooled to remove water and other impurities. The condensate is usually steam stripped for chlorine recovery and returned to the brine system or discharged. After cooling, chlorine gas as a by-product is dried further by scrubbing with sulfuric acid. The diluted acid is then usually regenerated, sold, or used for pH control. When the chlorine gas is compressed and liquified, it leaves behind noncondensible gases known as tail gas. The tail gas is usually scrubbed with caustic or lime, generating a hypochlorite solution which is then decomposed, used on site, sold, or discharged with or without treatment.

The sodium hydroxide or caustic product formed at the denuder has a concentration of 50 percent sodium hydroxide. Some of the impurities present in the caustic can be removed or reduced by the addition of certain chemicals, and the caustic is then filtered. In most cases it is sent to storage or is evaporated if a more concentrated product is required.

Hydrogen gas, a by-product of the purification process, is cooled by refrigeration to remove water vapor and mercury, and can be treated further by molecular sieves or carbon. Condensate from hydrogen cooling is then discharged or recycled to the denuder after mercury recovery.

Figure 8.19 presents a general process flow diagram of chlorine production by mercury cell.

8.3701   WASTE/WASTEWATER CHARACTERISTICS

Wastewater sources for the chlor-alkali industry are mercury cell plants, cooling, scrubbing, cell washing, equipment maintenance, floor washings, and in the decomposition of sodium-mercury material in the denuder. Because most brine systems at mercury cell plants are closed systems, wastewater production in the brine system is minimal. The wastes and wastewater generated are or can be contaminated with mercury and would therefore require treatment if discharged. Waste/wastewater generation is also shown in Figure 8.19.

Brine mud produced during the purification of brine constitutes the major portion of the solid waste produced from the process. The metals commonly removed during purification are magnesium, calcium, iron, and other trace metals such as titanium, molybdenum, chromium, vanadium, and tungsten.

The major components of cell room wastewaters include leaks, spills, area washdown, and cell wash waters. The amount varies from plant to plant and depends largely on housekeeping practices. Cell room wastes contain high levels of mercury which require treatment prior to discharge.

Condensation from the cell gas is contaminated with chlorine. At some plants, the condensates are recycled to the process after chlorine recovery. Concentrated sulfuric acid is used in the dryer to remove the residual water from the chlorine gas after the first stage of cooling. In most cases, the acid is used until a constant concentration of 50 to 70 percent is reached. The spent acids can be regenerated for reuse.

The 50 percent caustic produced at the denuder is filtered to remove salt and other impurities. The filters are backwashed periodically as needed, and the backwash can be discharged to treatment or recycled to the brine system. Hydrogen produced at the denuder is cooled to remove mercury and water carried over in the gas and the condensate is either sent to treatment facilities or returned to the mercury recovery system.

8.371   Hydrofluoric Acid Industry

Hydrofluoric acid (hydrogen fluoride, HF) is produced both as solid and aqueous products. It is used in the manufacture of fluorocarbons which are used as refrigerating fluids, fire suppressants, and plastics, for pressurized packing and as dispersants in aerosol sprays. It is used in the production of aluminum, in the refining of uranium fuel, pickling of stainless steel, and for the manufacture of fluoride salts. Hydrogen fluoride is also used as an insecticide and to arrest the fermentation in brewing.

8.3710   GENERAL MANUFACTURING PROCESS DESCRIPTION

HF is the most important manufactured compound of the fluorine family in terms of the volume produced. Fluorspar (mainly CaF_2) and sulfuric acid are the raw materials used for its manufacture. Sulfuric acid having a concentration as low as 93 percent or as high as 99 percent is generally used. The chemical reaction is given as:

$$CaF_2 + H_2SO_4 + HEAT \rightarrow CaSO_4 + 2HF$$

HF generators are, in the majority of cases, externally fired rotary kilns. Acid and fluorspar are fed continuously into the rotary kiln through a screw conveyor at the forward end and gypsum is removed from the other end through an air lock. The product HF may discharge from either end.

The HF gas leaving the kiln reactor is cooled in a precooler to condense high boiling compounds. The condensables, known as drip acid, consist largely of fluorosulfonic acid (HSO_3F) and untreated sulfuric acid. The HF gas from the precooler is cooled further and condensed in a cooler/refrigeration unit. The uncondensed gas containing the HF is scrubbed with sulfuric acid and refrigerated to recover the product. The residual vent gases are scrubbed further with water to remove HF and other fluoride compounds before they are vented to the atmosphere, and the scrubbed acid liquor is returned to the kiln. Figure 8.20 is a schematic flow diagram of the manufacturing process.

The crude HF is then distilled to remove residual impurities, and the condensate, which is anhydrous HF, is stored in tanks. If aqueous HF is desired, the crude product is then diluted with water to form a 70 percent HF solution as the final product.

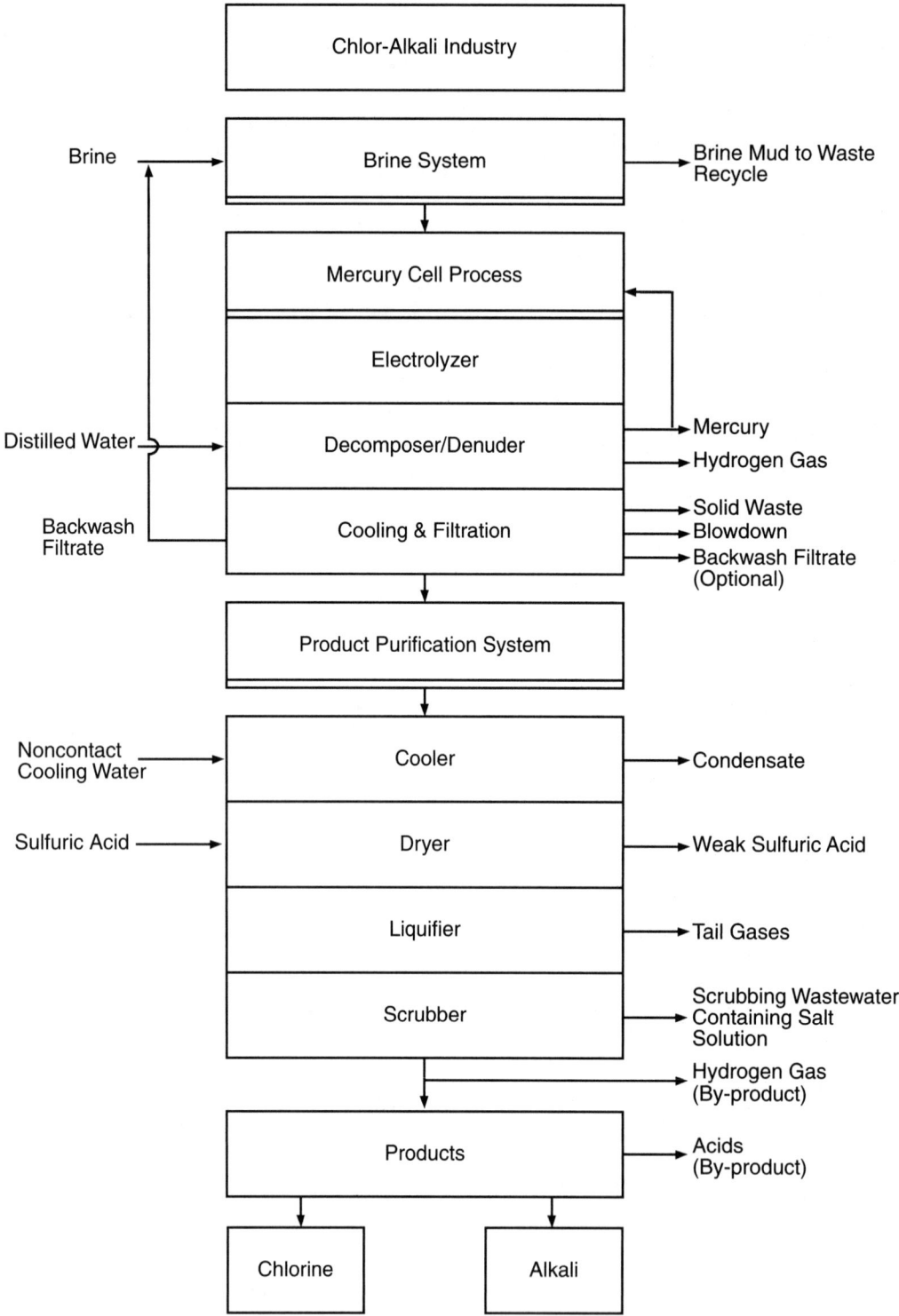

Fig. 8.19   General process diagram for production of
chlorine/caustic by mercury cells

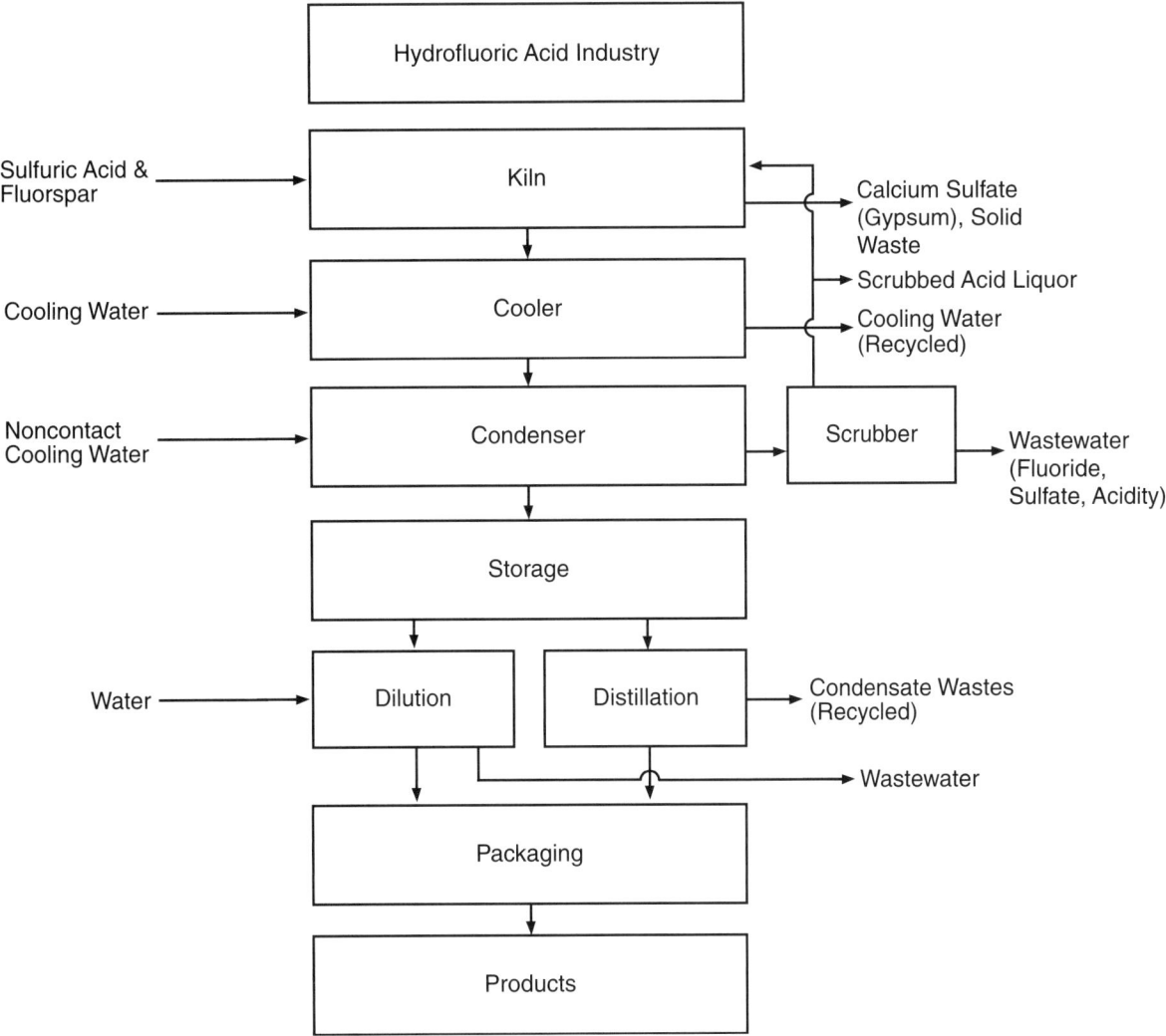

Fig. 8.20 General manufacturing process for production
of hydrofluoric acid

8.3711   WASTE/WASTEWATER CHARACTERISTICS

Waste/wastewater sources in HF production are noncontact cooling, air pollution control, product dilution, seals on pumps and kilns, and equipment area washdown. Although noncontact cooling constitutes the major wastewater production, wastes/wastewaters are also generated in other processes, such as waste from the gypsum process, scrubbing wastewater, and distillation wastewater. Figure 8.20 also indicates the waste/wastewater generating processes.

Drip acid is formed in the first stage of the cooling of the gases emitted from the kiln. Drip acid mostly contains high boiling compounds consisting of complex fluorides, especially fluorosulfonic acid, and small amounts of hydrofluoric acid, sulfuric acid, and water.

Scrubber water is another wastewater source, and in plants which practice dry disposal of gypsum, scrubber water constitutes the predominant and major source of wastewater. It contains fluoride, sulfate, and acidity. The fluoride is present as HF, silicon tetrafluoride (SiF_4), and hexafluosilicic acid (H_2SiF_6).

The distillation waste usually contains HF and water. In some cases, the vent gases from the distillation column are scrubbed before they are emitted to the atmosphere, and the resulting scrubber water requires treatment.

The other solids generated from the process and the treatment system consists of gypsum and the fluoride precipitated as calcium fluoride.

8.372   Titanium Dioxide Industry

Over fifty percent of the titanium dioxide produced is used in paints, varnishes, and lacquers. About one-third is used in the paper and plastics industries. Other uses are found in ceramics, ink, and rubber manufacturing, in curing concrete, and in coatings for welding rods.

8.3720   GENERAL MANUFACTURING PROCESS DESCRIPTION

The titanium dioxide industry has been classified further into three separate categories: sulfate process using ilmenite ore, chloride process using rutile or upgraded titanium ore, and chloride process using ilmenite ore. The section below describes the chloride process using rutile ore.

In the chloride process, the raw materials used are relatively pure materials with a high titanium and a low iron content. In this process, the ore and coke are dried and then reacted with chlorine to form titanium tetrachloride. The chemical reaction taking place in the reactor is given as:

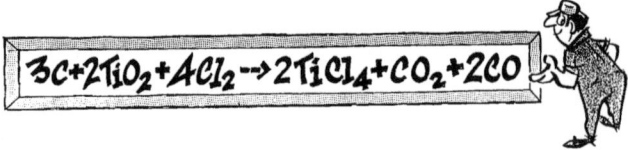

$$3C + 2TiO_2 + 4Cl_2 \rightarrow 2TiCl_4 + CO_2 + 2CO$$

The product gases leaving the reactor consist of titanium tetrachloride, unreacted chlorine, carbon dioxide, carbon monoxide and minor amounts of heavy metal chlorides. The gases are initially cooled to remove the impurities, although in some cases purification is accomplished by washing the gases. Titanium tetrachloride is liquified from the gases after the first stage of cooling by further cooling to ambient temperature. Copper, hydrogen sulfide, and, in some cases proprietary organic complexing agents, are added to purify the condensed solution.

The liquified titanium tetrachloride contains impurities such as aluminum chloride and silicon tetrachloride which are removed by distillation. The distillate is the purified titanium tetrachloride and the impurities remain as a residual which becomes waste. The tail gases from the distillation column are scrubbed with caustic soda to remove chlorine before being vented to the atmosphere. The titanium tetrachloride is vaporized and then sent to be reacted with oxygen or air in a flame at 1,500°C to produce chlorine and very fine particulate titanium dioxide. The chlorine is recovered and recycled back to the process. The titanium dioxide is then sent to the finishing operation where it is vacuum degassed and then treated with alkali, using a minimum amount of water to remove contamination. The pigment is then milled, surface treated for end-use application, dried, and packaged for sale. A generalized process flow diagram, including the wastestreams, is shown in Figure 8.21.

8.3721   WASTE/WASTEWATER CHARACTERISTICS

Wastewater sources are noncontact cooling, scrubbing of the tail gases from the purification and oxidation reactor to remove contaminants, and product finishing operations.

Wastes from cooling chlorinator gas consist of solid particles of unreacted ore, coke, iron, and small amounts of vanadium, zirconium, chromium, and other heavy metal chlorides. They are either dissolved in water and sent to the wastewater treatment facility, or disposed of in landfills as a solid waste.

Chlorinator process tail gases are scrubbed with caustic soda to remove chlorine as a hypochlorite. The main constituents of the wastewater generated are hydrogen chloride, chlorine, and titanium tetrachloride.

The liquid wastes from the finishing operation contain titanium dioxide as a suspended solid and dissolved sodium chloride formed by the neutralization of residual hydrochloric acid with caustic soda.

Distillation bottom wastes contain copper, sulfide, and organic complexing agents added during purification in addition to aluminum, silicon, and zirconium chlorides.

The uncondensed off-gases from oxidation are scrubbed with water or caustic soda to remove residual chlorine. When caustic soda is used as the scrubbing solution, the resulting solution of sodium hypochlorite is either sold, decomposed, or sent to the wastewater treatment facility. The scrubber wastestream also contains titanium dioxide particulates.

8.373   Aluminum Fluoride Industry

Aluminum fluoride is used as a raw material in the production of cryolite (sodium fluoroaluminate) which is used in the production of aluminum.

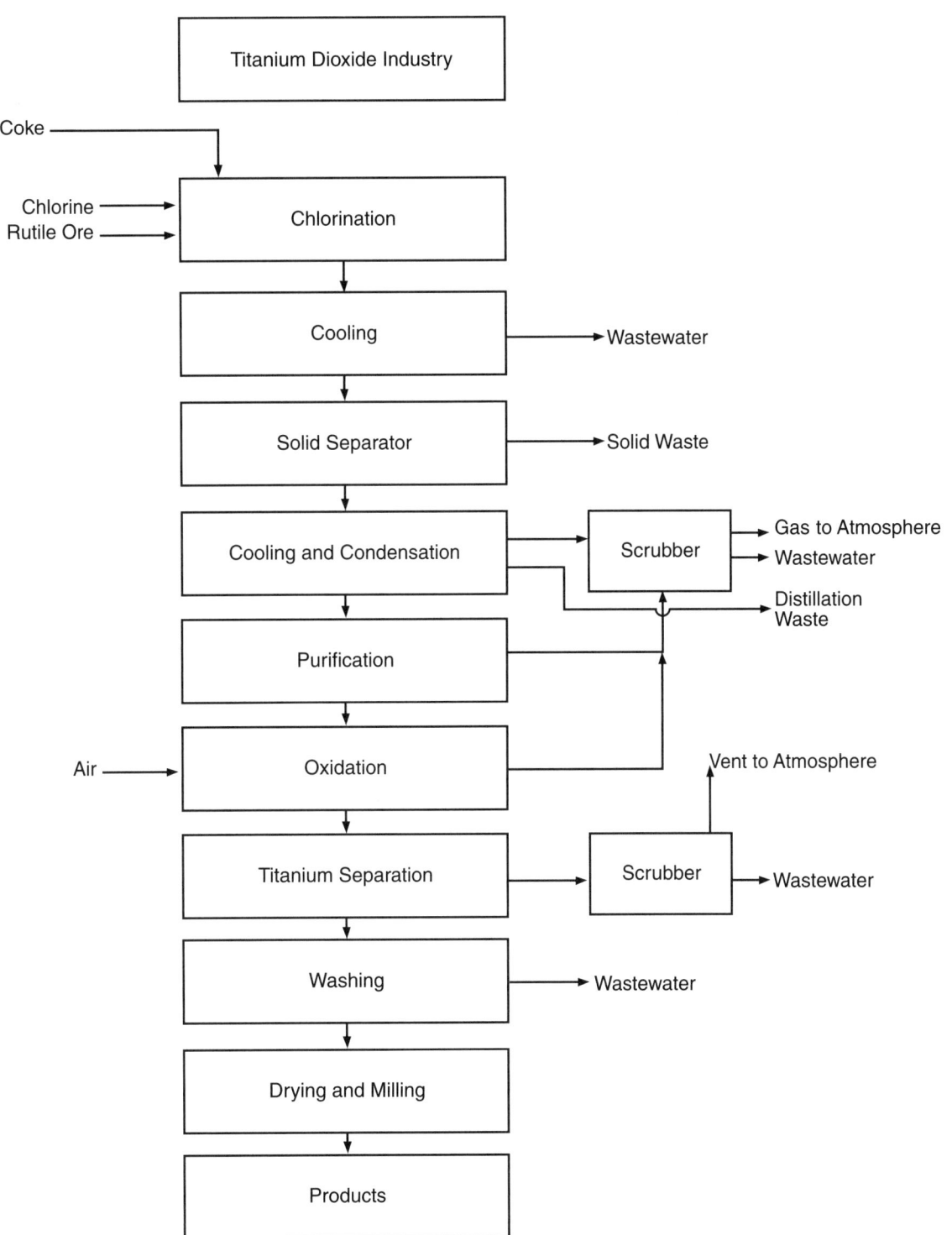

Fig. 8.21   General process diagram for production of titanium dioxide

8.3730   GENERAL MANUFACTURING PROCESS DESCRIPTION

In the dry process for the manufacture of aluminum fluoride, partially dehydrated alumina hydrate is reacted with hydrofluoric acid gas. The reaction is given as:

$$Al_2O_3 + 6HF \rightarrow 2AlF_3 + 3H_2O$$

The product, aluminum fluoride, is formed as a solid, and is cooled with noncontact cooling water before being milled and shipped. The gases from the reactor are scrubbed with water to remove unreacted hydrofluoric acid before being vented to the atmosphere. A simplified flow diagram of the process is shown in Figure 8.22.

8.3731   WASTE/WASTEWATER CHARACTERISTICS

Waste/wastewater sources of the aluminum fluoride industry are noncontact cooling of the product and scrubbing of the reacted gases before they are vented to the atmosphere. Water is also used for leak and spill cleanup and equipment washdown. Noncontact cooling water is used to cool the product coming out of the reactor; this cooling water is monitored for fluoride.

The quantity and quality of wastewater generated from floor and equipment washing operations varies and depends largely on the housekeeping practices at the individual plants.

Scrubber wastewater is the major source of wastewater requiring treatment before discharge or recycle to the scrubber. It is contaminated with hydrofluoric acid, aluminum fluoride, and aluminum oxide, and, in some cases, sulfuric acid and silicon tetrafluoride have been detected. As mentioned earlier, scrubber water constitutes the major source of wastewater in the aluminum fluoride industry.

8.374   Chrome Pigments Industry

Chrome pigments are a family of inorganic compounds primarily used as colorants in industries producing paints, ceramics, ink, paper, and cements. The various types of pigments include chrome yellow, chrome orange, molybdate chrome orange, anhydrous and hydrous chromium oxide, chrome green, and zinc yellow. Certain chromium compounds may also be used as raw materials in the manufacture of certain metals and alloys.

8.3740   GENERAL MANUFACTURING PROCESS DESCRIPTION

Anhydrous Chrome Oxide

This chromium pigment consists of two compounds: anhydrous and hydrated chrome oxide. The raw materials are blended in a mixer and then heated in an oven; the reacted material is slurried with water and filtered. The filtered solids are washed with water, dried, ground, screened, and packaged. The filtrate and the wash water are then treated with sulfuric acid to recover boric acid. The amount of anhydrous salt oxide produced by this process is approximately ten times the amount of hydrated chromic oxide produced. A general proc-

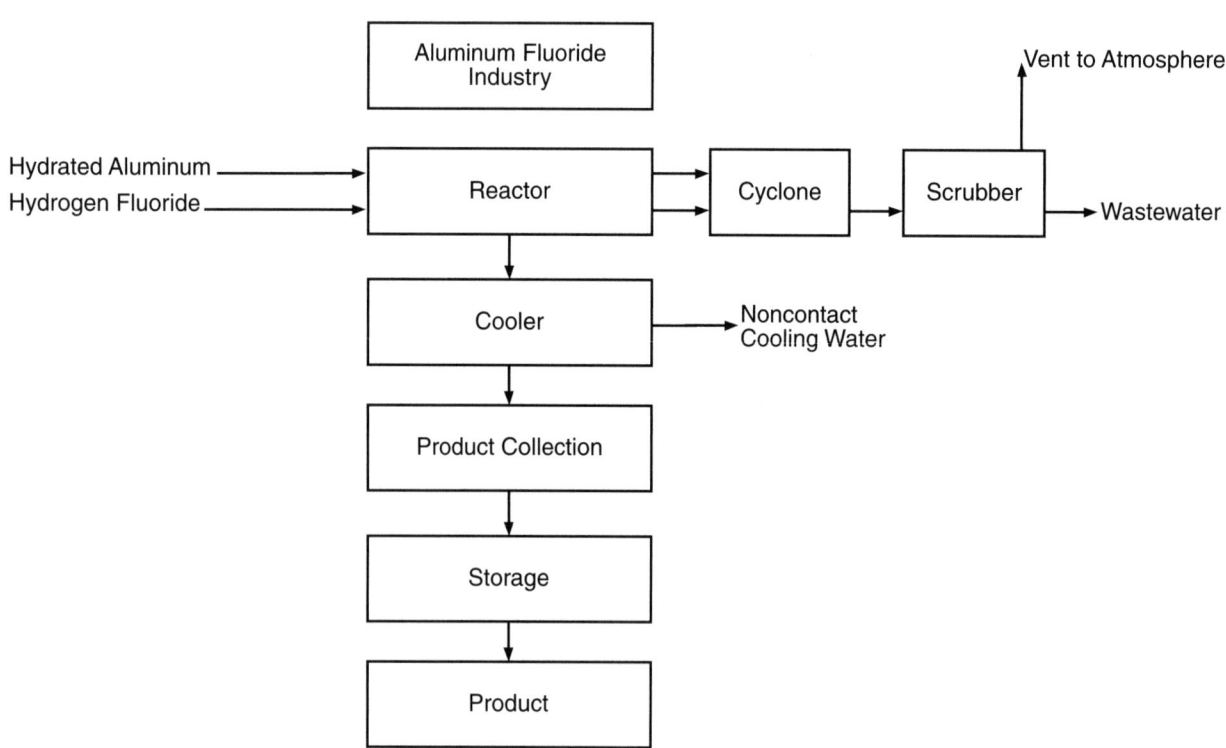

*Fig. 8.22   General manufacturing process diagram
for production of aluminum fluoride*

ess flow diagram of the preparation of anhydrous chrome oxide is shown in Figure 8.23.

Chrome Yellow and Chrome Orange

Chrome yellow is one of the more important synthetic pigments. It produces a range of hues from light greenish yellow to reddish medium yellow. Consisting mainly of lead chromate, chrome yellow and orange pigments are made by reacting dichromate, caustic soda, and lead nitrate. Lead chromate forms as a precipitate during the reaction. It is filtered and treated with chemicals for development of desired specific pigment properties, dried, milled, and packaged.

Molybdenum Orange

Molybdenum orange is made by the co-precipitation of lead chromate and lead molybdate. The resulting pigments are more brilliant than chrome oranges. The process consists of dissolving molybdic oxide in aqueous sodium hydroxide and adding sodium chromate. The solution is mixed and reacted with a solution of lead nitrate. The precipitate from the reaction is filtered, washed, dried, milled, and packaged. The filtrate is sent to the treatment facility.

Chrome Green

Chrome greens are co-precipitates of chrome yellow and iron blues. They include a wide variety of hues from very light to very dark green. Iron blues are manufactured by reaction of aqueous solutions of iron sulfate and ammonium sulfate with sodium hexacyanoferrate. Chrome green is produced by

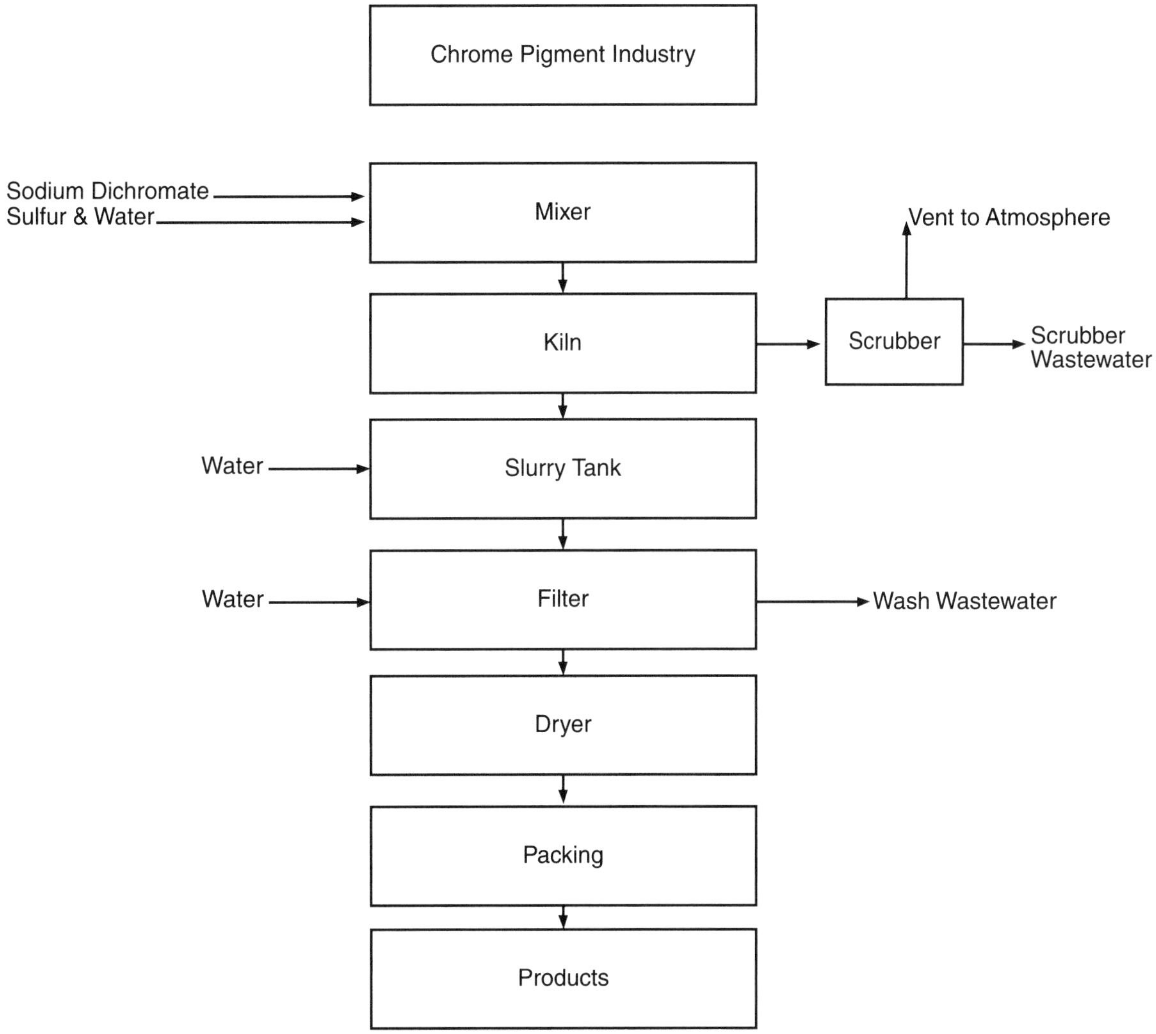

Fig. 8.23 *General manufacturing process diagram for production of anhydrous chrome oxide*

mechanically mixing chrome yellow and iron blue pigments in water.

Zinc Yellow

Zinc yellow, also called zinc chromate, is a complex compound of zinc, potassium, and chromium. It is made by the reaction of zinc oxide, hydrochloric acid, sodium dichromate, and potassium chloride. Zinc yellow is formed as a precipitate and is filtered, washed, dried, milled, and packaged for sale.

8.3741   WASTE/WASTEWATER CHARACTERISTICS

In the chrome pigments industry, wastewater sources are primarily noncontact cooling, washing the precipitated product, and boiler feed for steam generation. In some cases, water is introduced into the reactor along with the raw materials. In addition, substantial quantities of water may be used in cleaning equipment. This occurs during product changes at plants manufacturing a number of pigments. The waste constituents may include chromium, zinc, lead, and molybdenum in small quantities. (The generalized flow diagram given in Figure 8.23 applies to all chrome pigment plants.) The wastewater sources are similar for all pigment products except at chrome oxide plants, where an additional scrubber is needed. The quantity of wastewater and the pollutants vary for the different chromium pigment products since the pollutants are dependent on the raw materials used.

8.375   Hydrogen Cyanide Industry

A major portion of the hydrogen cyanide production is used in the manufacture of molding and extrusion powders and sur-face coating resins. It is also used as a fumigant for orchards and tree crops, and in electroplating.

8.3750   GENERAL MANUFACTURING PROCESS DESCRIPTION

Over 50 percent of the hydrogen cyanide manufactured is produced by the Andrussow process, while about 40 percent is a by-product from acrylonitrile manufacture. In the Andrussow process, ammonia and methane are reacted at elevated temperatures over a platinum catalyst to produce hydrogen cyanide. The reaction is given as:

$$2CH_4 + 2NH_3 + 3O_2 \rightarrow 2HCN + 6H_2O$$

The source of methane is natural gas. In addition to hydrogen cyanide, the reacted gases contain ammonia, nitrogen, carbon monoxide, carbon dioxide, hydrogen, and small amounts of oxygen. The reactor gases are cooled and then scrubbed to remove the unreacted ammonia. The recovered ammonia is recycled to the reactor.

The hydrogen cyanide is removed from the ammonia scrubber effluent gases by absorption in cold water, and the waste gases are vented to the atmosphere. The absorbed solution containing hydrogen cyanide, water, and other contaminants is distilled to produce hydrogen cyanide gas of over 99 percent purity.

Water produced during the chemical reaction forming hydrogen cyanide is purged with the distillation bottom stream and is either recycled to the absorber or discharged to the treatment facility. Figure 8.24 shows a general diagram for the manufacture of hydrogen cyanide by the Andrussow process.

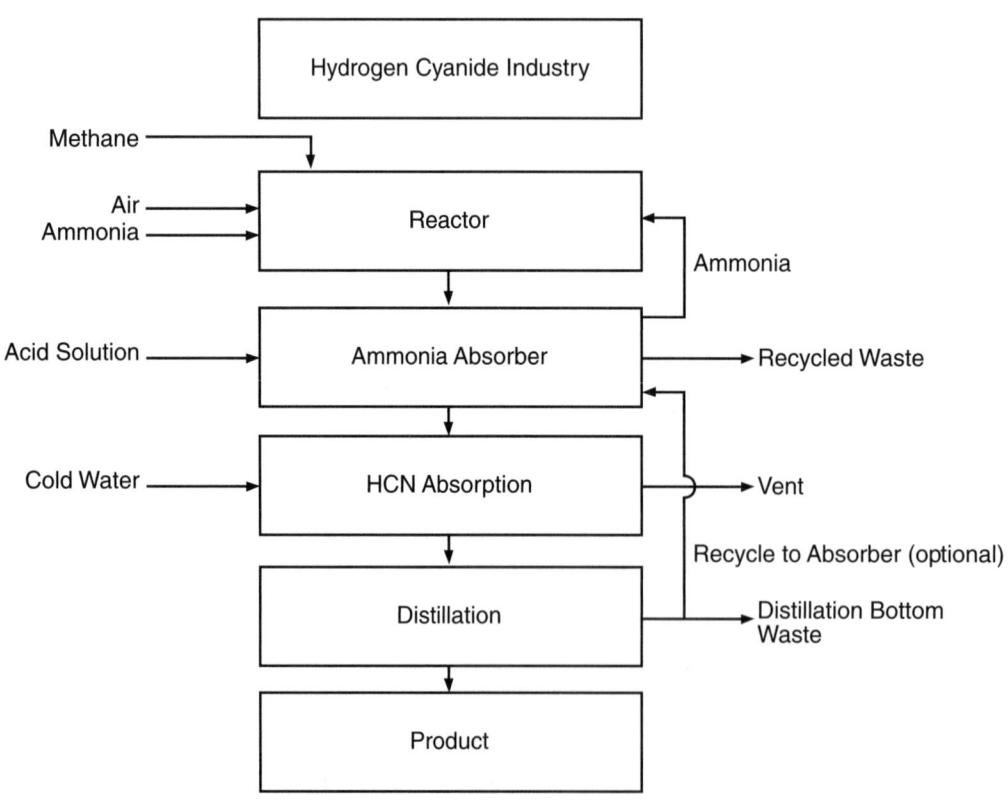

*Fig. 8.24   General manufacturing process diagram
for production of hydrogen cyanide*

8.3751 WASTE/WASTEWATER CHARACTERISTICS

Wastewater sources of the hydrogen cyanide industry are noncontact cooling in the absorber, pump seals, quenches, flare stack flushes, washdown and cleanup of tank cars, absorption of the product from reactor gases, washing equipment, cleaning up leaks and spills, and distillation bottom waste.

The wastewater from distillation contains ammonia, hydrogen cyanide, and small amounts of organic nitriles. The wastewater consists of the water produced by the reaction plus scrubber water used for the absorption of hydrogen cyanide. The absorption water is either recycled to the hydrogen cyanide absorber or discharged to the treatment facility. If the ammonia scrubber liquid is recycled, a portion of it has to be purged to control the accumulation of impurities.

Other wastewater includes leaks and spills, equipment and tank car washings, and noncontact cooling water blowdown. The noncontact cooling water may be contaminated with the product as a result of leaks. The recirculated cooling water is monitored for cyanide and the cooling tower blowdown is discharged.

8.376 Sodium Dichromate Industry

Most of the sodium dichromate produced is used in the chromic acid and pigment industries. It is used for leather tanning, metal finishing, and as a corrosion inhibitor.

8.3760 GENERAL MANUFACTURING PROCESS DESCRIPTION

The raw materials needed for the preparation of sodium dichromate are chromite ore, limestone, and soda ash. Chromite ore is a chromium iron oxide containing ferrous chromite and small amounts of aluminum, silica, and magnesia.

At the plant site, the ore is ground to a fine powder, mixed with soda ash (sodium carbonate, Na_2CO_3), and calcined (heated to remove volatile organic material) in rotary kilns at high temperatures. The reacted product is leached with hot water in a leachate tank thickened in a gravity thickener, then filtered to remove water. The thickener underflow is filtered and the filtrate is recycled to the leachate tank or thickener. The solid filter cake is dried in rotary kilns. The aluminum present in the thickened overflow is hydrolyzed and removed from the chromate solution as precipitated aluminum hydrate in slurry form. This slurry solution is centrifuged and the centrate is evaporated to give a concentrated solution of sodium chromate, which is reacted with sulfuric acid to give sodium dichromate and sodium sulfate. Sodium dichromate crystallizes out and is centrifuged; the centrate, or mother liquor, is returned to the evaporator. The sodium dichromate crystals separated in the centrifuge are dried in a rotary drum dryer and

then packaged for sale or stored for use. Figure 8.25 presents a generalized flow diagram for the production of sodium dichromate.

8.3761 WASTE/WASTEWATER CHARACTERISTICS

Wastewater sources in the sodium dichromate industry are noncontact cooling, leaching, scrubbing vent gases, and process steam for heating. Spent ore waste is the unreacted ore removed from the process as a sludge. The solids contain chromium and other impurities originally present in the ore.

The noncontact cooling water is either used on a once-through basis and discharged, or is recycled and the blowdown discharged to the treatment facility. In addition to dissolved sulfate and chloride, it may contain chromate.

The steam used for heating is recovered as condensate, while the boiler blowdown is discharged to the treatment facility. It may become contaminated with chromium escaping from the process.

The majority of aqueous streams resulting from the manufacture of sodium dichromate are recycled. Streams recycled include condensates from product evaporation and drying, product recovery filtrates, air pollution control scrubber effluents, filter wash waters, and equipment and process washdowns.

8.377 Copper Sulfate Industry

Copper sulfate is produced either as a liquid solution or dried crystals. It is used in agriculture as a pesticide, and as an additive to copper-deficient soils. It is also used in electroplating, petroleum refining, and as a preservative for wood.

8.3770 GENERAL MANUFACTURING PROCESS DESCRIPTION

Copper sulfate is produced by reacting copper with sulfuric acid, air, and water. Various forms of copper feed material are used, from pure copper to copper slag. The purity of raw materials significantly affects the quality and quantity of raw wastes generated.

Copper metal and/or copper refinery wastestreams, steam, water, sulfuric acid, and air are treated in oxidizer tanks at 100°C to produce a solution of copper sulfate. This solution is partially concentrated by evaporation.

If pure copper is used as a raw material, the resulting copper sulfate solution is pure enough to be either sold or fed to crystallizers producing copper sulfate crystals without any purification process. If impure copper feed or copper refinery waste is used, the concentrated copper sulfate solution is filtered to remove other metal impurities. Copper sulfate crystals are recovered by centrifugation, dried at approximately 110°C, screened, and then packed dry for sale. The mother liquor is recycled to the evaporator or crystallizer with some being purged to prevent impurities buildup. Figure 8.26 shows a general process flow diagram for the manufacture of copper sulfate.

8.3771 WASTE/WASTEWATER CHARACTERISTICS

Wastewater sources are copper sulfate medium and noncontact cooling water, including steam condensate. Pump seals and washdowns also generate wastewater.

Noncontact cooling water is used to cool the crystallizers and constitutes one of the main wastes. This wastestream should not be contaminated by process leaks and therefore can be discharged without treatment.

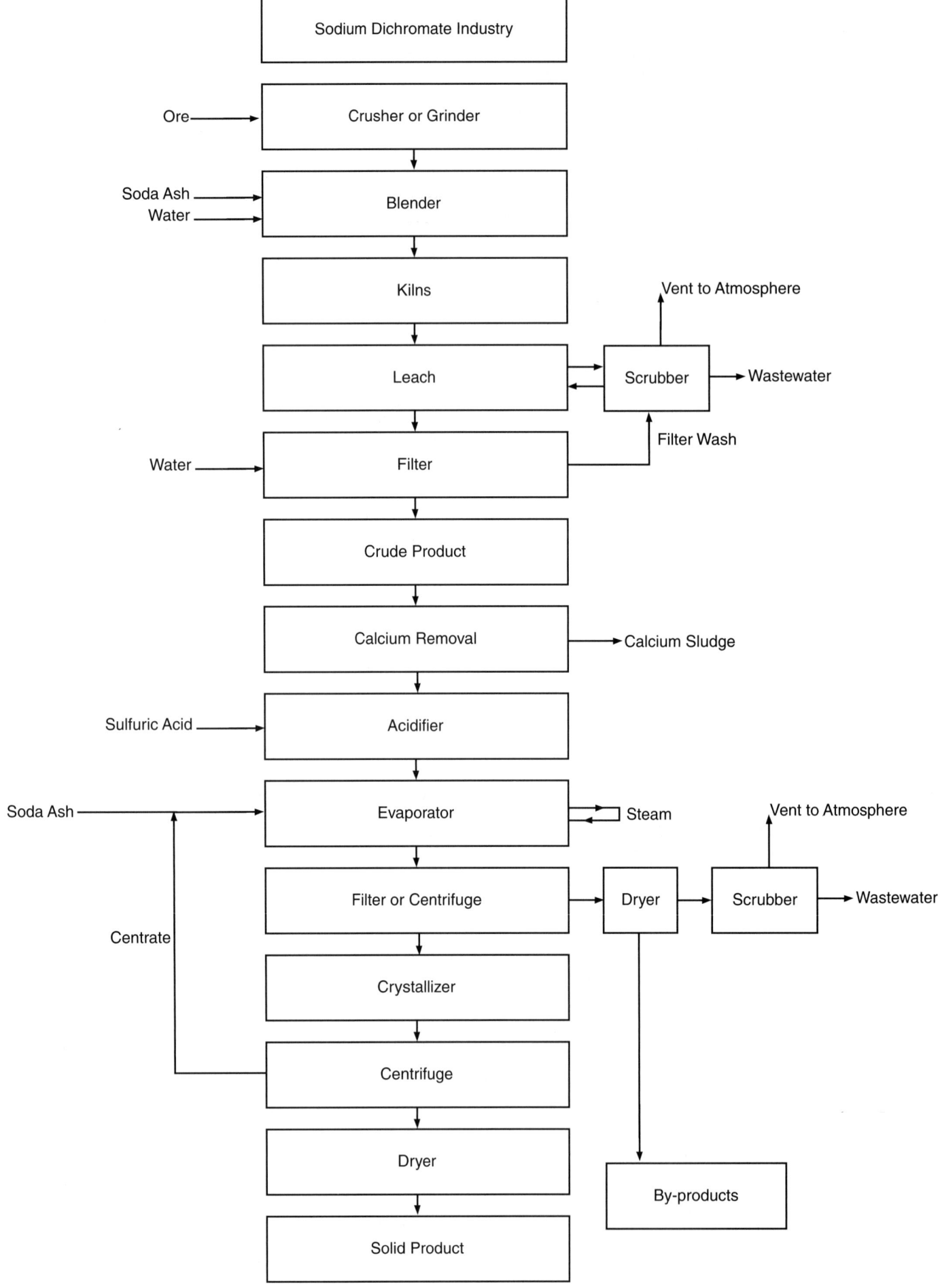

Fig. 8.25   General manufacturing process diagram for production of sodium dichromate

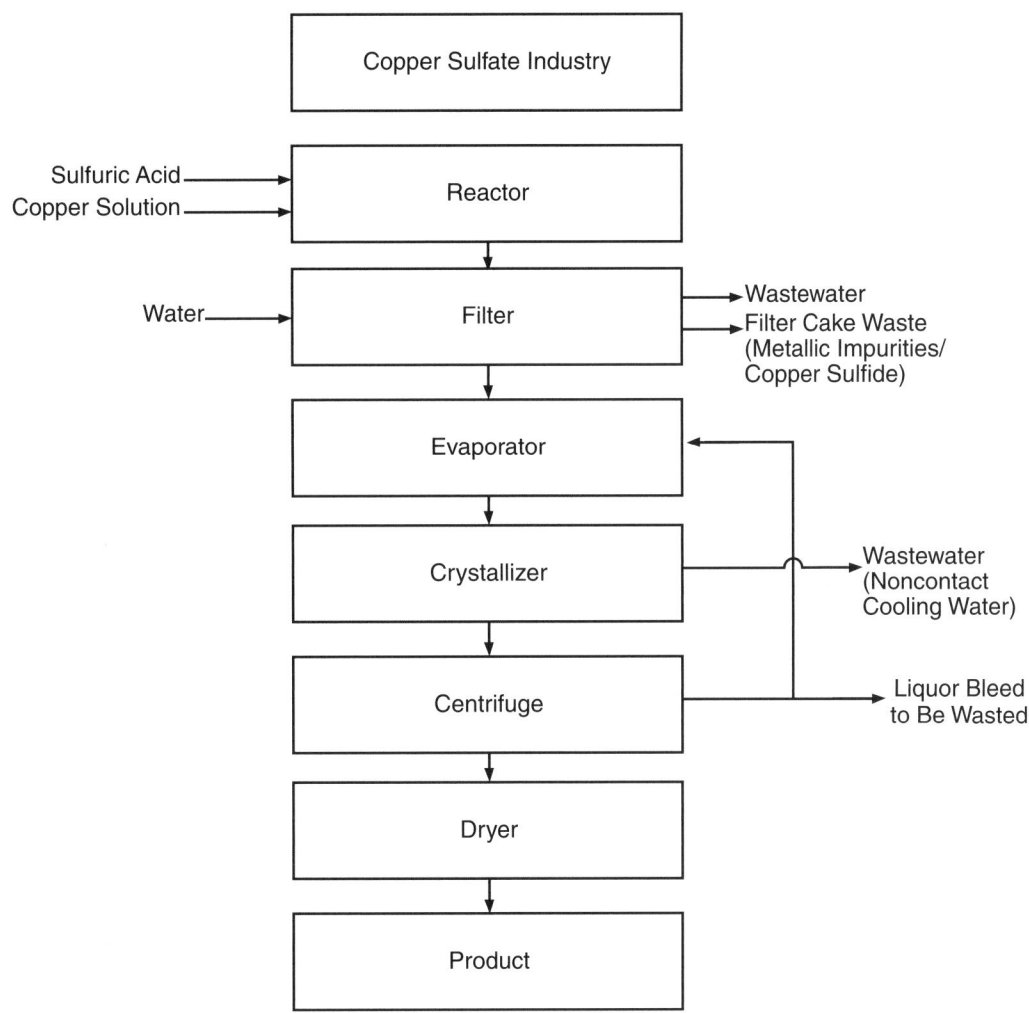

Fig. 8.26   *General manufacturing process diagram for copper sulfate*

Washdown, pump seal leaks, and spills are sources of contact wastewater. These flows, however, are relatively small and intermittent, and do not represent a major waste source. Wastewaters emanating from this source are either combined with the mother liquor, or treated and discharged.

A small portion of the mother liquor is purged periodically from the process to prevent buildup of metal impurities. The amount of purge is variable and depends on the purity of feedstock. These purges must be processed to separate metallic salts, particularly those of copper and nickel, from the impurities. Steam condensate is an additional noncontact waste that can be discharged without treatment.

In plants using impure copper or copper refinery waste as raw material, product purification by filtration generates solid waste. These filter sludges contain metallic impurities or copper sulfide and require disposal at an approved landfill.

8.378   *Nickel Sulfate Industry*

The major use of nickel sulfate occurs in the metal plating industry, but it is also used in the dyeing and printing of fabrics.

8.3780   *GENERAL MANUFACTURING PROCESS DESCRIPTION*

Nickel sulfate is produced by reacting various forms of nickel with sulfuric acid. Two different raw materials are used to produce nickel sulfate. Pure nickel or nickel oxide powder may be used as a pure material source, while spent nickel catalysts, nickel plating solutions, or residues are impure sources.

The use of impure raw materials produces a nickel sulfate solution which must be treated in a series of steps with oxidizers, lime, and sulfides to precipitate impurities. The impurities are then removed by filtration. The nickel sulfate solution can be sold or it may be crystallized, and the crystals classified, dried, and screened to produce solid nickel sulfate for sale. Figure 8.27 shows a general process flow diagram for the manufacture of nickel sulfate.

8.3781 WASTE/WASTEWATER CHARACTERISTICS

Water used in the noncontact cooling process is the main source of wastewater in the nickel sulfate industry. This stream is usually not treated before discharge. Direct process contact water is a reaction component which becomes part of the dry product.

Plants which use impure nickel raw materials generate a filter sludge which is treated as a solid waste. They also generate a small filter backwash wastestream with high impurity levels which must be treated before discharge. The filter sludges from processes using pure nickel can be recycled back to the process. Mother liquor and wastewater streams from dust control are also recycled back to the process. Washdowns, cleanups, spills, and pump leaks are periodic streams that account for the remaining wastes produced by nickel sulfate plants.

QUESTIONS

Write your answers in a notebook and then compare your answers with those on page 593.

8.3Q   List the waste/wastewater sources for the hydrofluoric acid industry.

8.3R   What are the major uses of titanium dioxide?

8.3S   What are the major sources of wastewater for the hydrogen cyanide industry?

8.4   EFFECTS OF INDUSTRIAL WASTEWATERS ON THE POTW SYSTEM

8.40   Inspector's Responsibility

The Federal Pretreatment Regulations were established to remove toxic pollutants at the source and to protect the POTW's collection, treatment, and disposal systems and the environment. Chapter 3 of this manual discusses the specific toxic and flammable constituents that industrial users are pro-

hibited from discharging to a POTW. Chapter 3 also describes how a local Industrial Waste Ordinance should specify the exact operating conditions each system must observe to prevent pass-through and interference. Setting aside questions of IU compliance or noncompliance with these regulations, this section describes the effects of industrial wastes when they do reach the POTW system.

The effects of an industrial discharge on the POTW will always depend on the characteristics and flexibility of the system, the level of skill possessed by the inspectors, laboratory analysts, and POTW operators, and the amount and type of industrial flow. Factors such as the size and length of the sewer system also influence how an industrial discharge will affect the POTW collection system. In general, the larger the system, the less effect a single industrial discharge will have on the POTW regardless of whether the industrial discharge is a slug loading or a constant discharge. Dilution and equalization of the industrial discharge occurs in the larger collection systems, thereby helping mitigate (lessen) the effect on the POTW facilities.

As the complexity of the treatment system increases from only primary treatment to tertiary treatment, the effect of an industrial discharge also increases. The higher degrees of treatment are more sensitive to upset from industrial discharges. Secondary and tertiary biological processes such as activated sludge, nitrification, denitrification, and anaerobic digestion can be upset by a toxic "overdose" of heavy metals. Tertiary physical-chemical processes such as sand filtration can be rendered useless by a pass-through of oil or a carryover of gelatinous bacteria from an upset biological process.

The effect of an industrial discharge will also increase with the increase in complexity of the equipment used in each type of treatment process. For physical-chemical processes, sedimentation is less prone to upset than filtration.

The pretreatment inspector's job is to *PREVENT* slug discharges. If a slug discharge is reported by an inspector or an industry, the POTW must be able to respond.[11] When a POTW is designed with system flexibility, the inspectors and operators may be able to lessen the effects of a slug loading that results from an accidental or illegal slug discharge. Some POTWs are equipped with equalization basins, flow control structures, chemical treatment points along the collection system, aeration basins and adequate aeration equipment, along with return sludge and effluent recycling capabilities. Facilities and equipment such as these enable the POTW to

---

[11] *See Chapter 11, "Emergency Response," Sections 11.30, "Removal of Material," 11.31, "Treatment in the Sewer," and 11.32, "POTW Process Changes."*

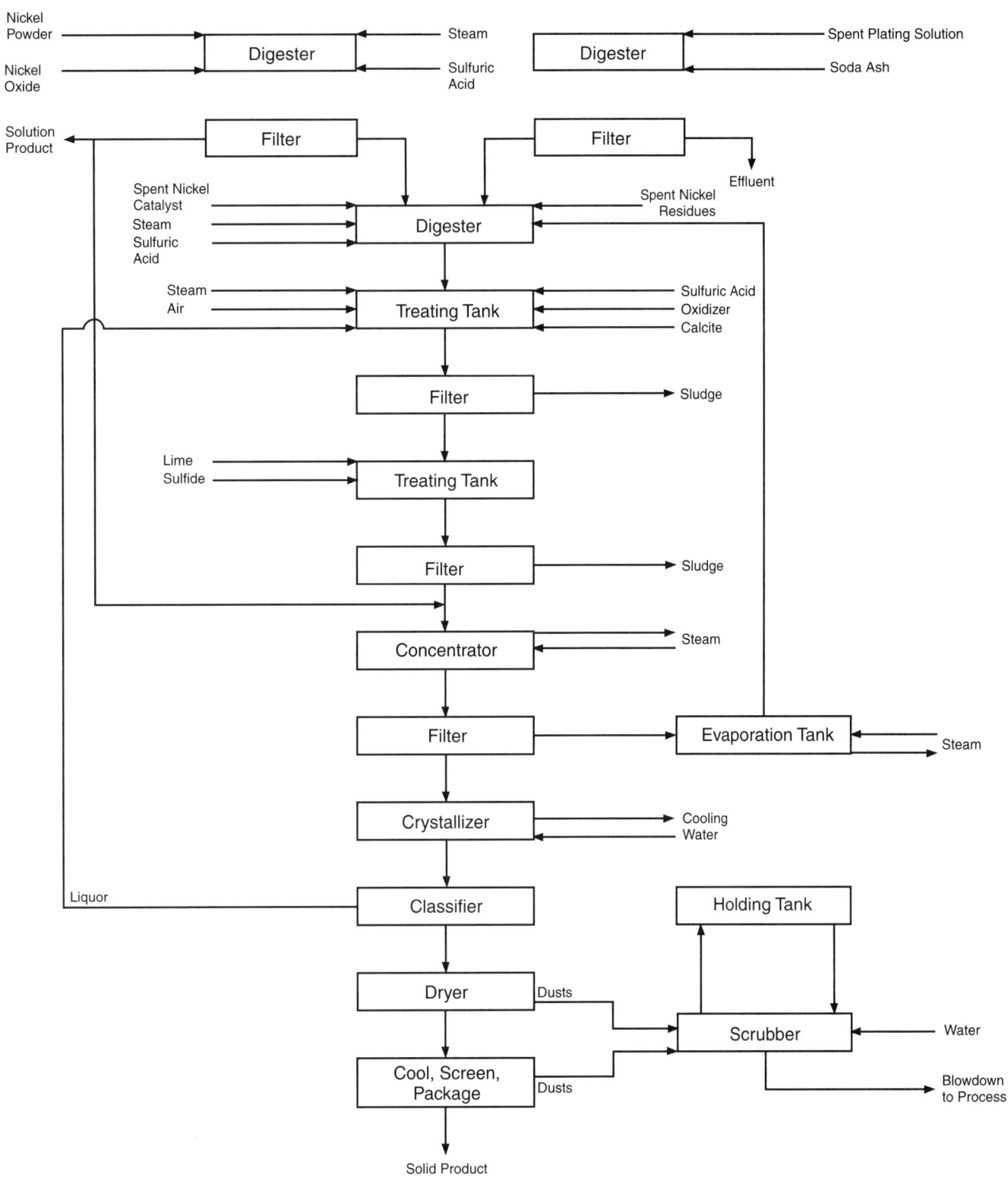

Fig. 8.27   General manufacturing process diagram for nickel sulfate

(Source: *DEVELOPMENT DOCUMENT FOR THE EFFLUENT LIMITATIONS
GUIDELINES AND STANDARDS FOR THE INORGANIC CHEMICALS
MANUFACTURING POINT SOURCE CATEGORY,* EPA, June, 1980)

modify the secondary biological treatment systems to cope with the anticipated slug load and treat the wastewater adequately to meet NPDES discharge requirements.

If the configuration of the treatment system can be easily changed, the effect of an industrial discharge may be lessened. Changing the recycle ratio on a trickling filter or altering the biomass concentration in an activated sludge system could prevent pass-through of noncompatible pollutants or air strip volatile organic compounds. Changing a two-unit process from parallel operation to series operation may help to remove high loadings of compatible pollutants.

The inspector plays an important part in identifying the type of industrial waste that is causing an unusual condition in the plant. Once the substance is identified, the operators can determine what options they might use to treat it, and the inspector can search for the industrial source and correct the problem.

The disposal of the POTW effluent and sludge are also affected by industrial discharges. The effluent discharge requirements are more stringent for water reuse than for discharge to receiving waters. The use of POTW sludge as a component in compost for resale has stricter quality requirements than those required for a sludge being landfilled. Toxic components of industrial discharges may limit the recycle and reuse options if the POTW is not properly protected from slug loadings or if contaminant concentrations reach a level that may pass through and be discharged in the effluent or sludge. When certain metals reach high enough concentrations in the sludge, then the sludge must be handled as a hazardous waste under RCRA regulations.

8.41   Effects on the Collection System

The POTW's collection system is built to handle domestic wastewater which itself contains constituents that are difficult to handle. Industrial discharges by themselves or in combination with other industrial or domestic wastewater can cause plugging, odors, erosion, corrosion, explosions, and numerous other problems. The good news, however, is that some industrial discharges contain substances that have a positive effect on the collection system, such as iron or heavy metals which are helpful in controlling sulfide corrosion (Figure 8.28). Also, cooling water discharges may produce scouring velocities in low-flow sewers.

The following sections discuss the commonly encountered effects of industrial discharges on a POTW's collection system and offer suggested solutions to these problems to give you some ideas about how they can be resolved.

8.410   Hydraulic Capacity Problems

Industries discharging large quantities of wastewater throughout the day may overload the hydraulic capacity of the local sewer or pumping station when combined with the typical diurnal (daily) fluctuation of domestic wastewater flows. The smaller the capacity of the sewer or system, and the larger the contribution by the industry, the more likely this problem is to occur. The solution may be to require the industries to store their effluent for discharge during off-peak hours.

Hydraulic capacity problems may also occur if an industry periodically discharges large quantities of wastewater during the day. The sewer or system may be able to accommodate the normal variation of flow with high flows during late morning and mid-evening hours. However, if there is a two-hour industrial discharge from cleaning a 50,000-gallon tank, the sewer and pumping station may become overloaded. Equalization of the industrial discharge may then be required to

"level out" the flow and keep the sewer from exceeding its hydraulic capacity.

The two scenarios described above can occur due to one single industry or a combination of industries on the same sewer or system. Similar industries such as food processing industries or associated industries may have similar manufacturing schedules and discharges. Thus, cleaning of tanks, reactors, or cooking pots may occur at virtually the same time. While the discharge from one industry may not cause a problem, the similar discharge schedule will combine the wastewater flows of two or more plants and cause a hydraulic overload condition. A possible solution to this problem is to have a meeting with the industries involved and discuss the problem. Explain that each of them is going to be allocated a portion of available capacity. Indicate that you will help the industries determine who can use the available capacity, when it can be used, and how much is available. Inform the industries that if a cooperative agreement can't be reached, then all will have to abide by a strict allocation of available capacity.

8.411   Plugging

Pretreatment should prevent industrial discharges containing large amounts of fibrous or stringy materials, heavy solids, adhesives, or grease that can cause plugging of sewers or pumps. Plugging may occur on the industry's site or just downstream of the discharge in the local sewer laterals. Fibrous or stringy materials are caught on rough surfaces and soon build up by entangling more solids which will eventually plug the line. These types of materials can also wind themselves around pump impellers or shafts causing the pump to fail. If problems are occurring, review permits, enforce them and require higher levels of pretreatment, if necessary.

Heavy solids such as sand, ceramic or porcelain solids, or grindings can build up in a sewer and reduce its hydraulic capacity. Solids that are not removed by pretreatment at the industrial site may be discharged during peak wastewater flows during the day and may settle in pump station wet wells or oversized sewers hundreds of yards or miles away from the actual point of discharge when the flow subsides at night. The solids then have an opportunity to compact and may not become resuspended when the flow in the sewer returns to its peak flow. This cycle of transporting the solids to a section of the collection system to settle, build up, and compact will eventually cause a restriction in the sewer.

A complete blockage may also occur if large objects are released to the sewer. Rags, tools, rejected food products, and discarded by-products may accidentally be released to the sewer due to operator carelessness or equipment malfunction. Because of their size, they can easily become wedged or entangled with other waste material and completely block the sewer.

Fig. 8.28   Sulfide-caused deterioration of sewer walls

8.412 Odors

Examples of industrial discharges that can be odorous are those from petroleum refining, petrochemical manufacturing, and food processing. Generally, the odors are produced from a compound containing sulfur, such as MERCAPTANS[12] or hydrogen sulfide. These compounds in air are detectable in the parts-per-billion range (by volume) and can cause complaints from residents and other industries. While the problem is airborne, the actual cause originates in the industrial discharge to the POTW sewer.

The first solution may be to change the manufacturing process. Sour water, which is wastewater containing high concentrations of sulfide from the petroleum refining industry, can be stripped with steam and reduced to elemental sulfur using the Klaus process. This process and other similar recovery processes have reduced the odor pollution problem while producing a saleable by-product (sulfur). Another solution may be to oxidize the offending components prior to discharge using air, hydrogen peroxide, or chlorine; or not discharge them at all.

The wastewater produced during the etherification reaction to make polyester is very odorous. Because of the quantity of organics in the wastes, it is practical to incinerate the wastes at no net fuel expense and solve the odor problem.

Another frequent odor problem is the production of hydrogen sulfide gas in sewers and pump stations. Besides an odor problem, it also presents a safety (toxic gas) problem to POTW collection system and treatment plant operators. Hydrogen sulfide gas is produced under anaerobic conditions in the sewer. Bacteria reduce the inorganic sulfate to sulfide when there is insufficient oxygen in the wastewater (less than 0.1 mg/L), thus producing hydrogen sulfide. Highly biodegradable organic discharges that must be conveyed over a long distance or are discharged to a slow-moving sewer can rapidly deplete the dissolved oxygen in the wastewater creating ideal conditions for anaerobic bacterial growth and production of hydrogen sulfide. Some suggested solutions to this problem are: require the industrial waste to be oxygenated and/or chlorinated prior to discharge; have the POTW aerate the wastewater in the collection system; periodically remove the slime layer of anaerobic growth in the system with a slug loading of alkali or chlorine; or clean the sewer with a high-velocity cleaner or a pig (a sewer cleaning device).

Industrial discharges containing high concentrations of sulfide are normally restricted. Limitations of 5.0 mg/L of total sulfide and 0.5 mg/L of dissolved sulfide are used by the County

Sanitation Districts of Orange County, California, and are typical of sulfide limits. Industrial discharges of sulfate should also be examined to determine if they are a large source of sulfur that could be reduced in the sewer environment to hydrogen sulfide.

8.413 pH Problems

The acceptable pH range for industrial discharges as regulated in many Industrial Waste or Sewer-use Ordinances is 6.0 to 9.0. In some ordinances the pH range may be widened. Remembering that a pH of 7.0 is neutral, the trend is to allow more alkaline or basic material in the discharge rather than acidic. The construction materials for sewers, pumping stations, treatment equipment, and biological processes all withstand alkaline discharges better than they withstand the discharge of corrosive acids.

Acids will corrode concrete and cast-iron sewers, concrete wet wells and tanks, the internal steel equipment in the primary and secondary clarifiers, trickling filters, aerators, and pumps. Mineral acids such as sulfuric, nitric, hydrochloric, and phosphoric acids are used extensively to clean base metals in the metal finishing industries. The fertilizer, iron and steel, mining, and petroleum industries also use vast quantities of these strong acids. Mineral acids are also used in the pretreatment systems in chromium reduction, neutralization of alkalies, and pretreatment of chelated metal plating solutions.

Discharge of acid to the sewer from a spill or due to an equipment or control instrumentation failure can cause a pH violation and damage to the collection system. During your inspections, take note of the spill containment precautions in both the manufacturing area and pretreatment area. Review the failure mode of the controls for chemical addition and ask (and get answers to) questions like: "Does the control valve for acid or alkali addition fail in the open position during a power failure?" or "Is there a high and low pH alarm on the neutralization tank?"

The organic acids such as acetic, maleic, benzoic, oxalic, and citric acids are weaker than mineral acids but, nonetheless, can have a pH of 4.0 or less. They too can corrode the sewer and are an organic load to the treatment plant. Organic acids are typically used in food processing, beverage and consumer product manufacturing, and in the manufacture of chemical intermediates.

Strong alkalies can corrode sewers and pumping stations. Aluminum is particularly affected by high pH. High pH may also precipitate metals like calcium, potentially causing a solids buildup problem in the sewer. The strong alkalies include sodium hydroxide, lime, and ammonia. These are used in the metal finishing industry to clean and chemically mill base metals. The water treatment industry uses significant quantities of lime to soften water. Pretreatment systems use these strong alkalies to neutralize industrial wastes. Because of the potential damage these substances may cause, inspectors need to review the spill containment measures of each discharger and check the failure mode of the chemical addition controls.

As previously discussed, however, the discharge of strong alkalies may actually be beneficial in removing the anaerobic slime layer from the sewer. When this is allowed, it should be done with the POTW's permission and knowledge of each dis-

---

[12] Mercaptans (mer-CAP-tans). Compounds containing sulfur which have an extremely offensive skunk-like odor; also sometimes described as smelling like garlic or onions.

charge so that the POTW influent and secondary treatment can be monitored to prevent a treatment process upset.

The discharge of out-of-pH-range wastewater will result in damage to the sewer. Over a period of time such discharges can eventually corrode the pipe completely, causing exfiltration and contamination of the groundwater. Infiltration of groundwater could occur in areas where the groundwater level is above the depth of the sewer. Industrial discharge violations of pH will also increase the maintenance requirements on pumps in the pumping stations. The damage to the pumps could eventually cause their failure, resulting in sewer backups and raw wastewater overflows.

pH problems are most severe nearest the discharge source because the wastes are not diluted by other wastewater. Frequent discharges of acidic wastes will corrode and completely etch through an industry's sewers and may result in heavy metal and/or solvent contamination of the groundwater. Use of the proper materials of construction for sewers, control of pH, and annual inspection or testing of the pipes and sewers are essential in preventing groundwater contamination from this type of source.

Too high a chlorine concentration is also corrosive to the collection system. Many platers will overchlorinate their cyanide wastewater to ensure they meet the requirements for cyanide concentrations. However, 40 to 50 mg/L excess chlorine can be corrosive to equipment and dangerous to personnel servicing a pump station.

8.414 Flammables

The discharge of flammables is potentially the most damaging industrial discharge to the collection system. Gasoline, aviation fuel, and hexane used in soybean extraction have been responsible for explosions in sewers causing losses of millions of dollars for sewers and businesses, the loss of service to hundreds of people, and the loss of life. Industries producing, distributing, and using fuels and solvents must be monitored and/or regulated to prevent discharge of these materials. Generally, these fuels and solvents are only slightly soluble in water and have a specific gravity less than water. When accidentally discharged to the sewer, they will float and accumulate in slow-moving sewers or pumping station wet wells. Any source of ignition, such as an arc from tripping a breaker or a motor or a spark created while removing a manhole cover with a pick, can cause a fire or explosion.

An additional concern with flammable wastes is the exposure of collection system crews or operators to volatile toxic substances. Immiscible (not mixable), as well as miscible (mixable), solvents, including acetone, methyl ethyl ketone, and isopropyl alcohol, when discharged to the sewer and then aerated, are volatilized and thus expose personnel to the fumes. If the concentration of fumes is high enough, an explosive atmosphere may develop, especially if the secondary treatment process is covered or uses pure oxygen. Any hydrocarbon may cause a flammable hazard in a pure oxygen activated sludge system. However, these systems are usually equipped with sensors and purge systems to prevent flammable and explosive conditions from developing.

8.415 Temperature

Heated industrial wastewaters originate from controlling manufacturing process reactions and as a by-product from utilities' production of energy. In manufacturing processes, heat is often used to increase the rate of reaction and thus creates a heated product or waste which must be cooled. Water or steam is often used directly or indirectly (by means of heat exchangers) to heat or cool the product or by-product and to transport it to the next processing step. The metal finishing industry uses steam to heat process solutions.

Accumulated solids must be removed from boilers to prevent plugging of the boiler tubes and steam lines. The discharge is called boiler blowdown. In cooling systems, single-pass cooling water and cooling tower blowdown can also contribute a heat load to the sewer. These heated industrial discharges can cause many problems in the POTW collection and treatment systems including evolution of toxic gases and odors, overheating of pump and rotating equipment bearings, shifts in the population of microorganisms used in secondary biological treatment, or even sterilization (killing of all organisms) of the wastewater.

Plastic pipe (PVC) has temperature limitations of around 104°F (40°C) and can fail if used for hot water transport. In the POTW sewer laterals, the O-rings may not be designed to withstand a constant high temperature; if they fail, exfiltration or infiltration of the collection system may occur.

QUESTIONS

Write your answers in a notebook and then compare your answers with those on page 593.

8.4A How can a tertiary sand filtration process in a wastewater treatment plant be rendered useless?

8.4B What types of heavy solids can cause plugging of sewers?

8.4C What types of industrial discharges can cause odor problems?

8.4D What industries use mineral acids and for what purposes?

8.42 Effects on the Treatment System

Industrial waste discharges damage treatment plant equipment in many of the same ways they damage the collection system. High-volume discharges can exceed the pumping capacities; plugging of mechanical equipment such as bar screens or pumps can occur from a high-solids discharge; acids and alkalies will corrode metal parts eventually causing failure; and flammables in the treatment plant are an explosive problem that can cause almost instantaneous damage. The added potential problem with industrial discharges is their effect on the treatment processes, including sedimentation and biological treatment.

8.420 Hydraulic Overload

Large flows from industrial dischargers are not usually a problem because typically even the largest industrial discharges are a small percentage of the total design flow. However, as the size of the industry's discharge increases with respect to the total POTW treatment plant capacity, and the closer the discharge is to the treatment plant, the more of an effect the discharge will have on the treatment system.

Unit processes such as neutralization, sedimentation, filtration and biological treatment operate best at a constant flow and constant loading. Large changes in the volume of flow or rapid changes in loading will decrease the efficiency of these processes. Hydraulic surges from an industry added to the normal diurnal flow can cause these rapid variations. To compensate, the treatment plant must make a series of changes in their plant operating conditions, such as changing the sludge removal rate, increasing the blower output, or increasing chlorination. The alternative is to suffer possible effluent limitation violations.

8.421 Interference

EPA defines interference as a discharge which, alone or in conjunction with discharges from other sources, inhibits or disrupts the POTW, its treatment processes or operations, its sludge processes, use or disposal, and is a cause of violation of the NPDES permit or prevents the lawful use or disposal of sludge. By working closely with the POTW treatment plant operators, the inspector can identify potential interference problems before they cause a discharge violation. Good communication between the inspector and the treatment plant operators is the most reliable way to identify changes, whether sudden or gradual, in the operation of the plant or quality of the effluent. For example, a decrease in the BOD removal rate or changes in the microbial population may be caused by treatment plant operational changes or an industrial discharge. Check first with the plant operators to see if recent changes have been made in plant processes. If not, an industrial discharge of a noncompatible pollutant such as a toxic heavy metal that inhibits growth or too much of a compatible pollutant, such as soluble BOD, could have caused the problem. If not corrected at the source (the industry), the problem could cause violation of the POTW's discharge limits. Table 8.2 illustrates other examples of how industrial discharges potentially may cause interference with treatment processes.

8.422 Influent Variability

Composition measurements of pH, temperature, and conductivity are used to detect changes in the influent. As with hydraulic surges, variability in the chemical composition of the influent wastewater can cause upsets in the treatment processes. The larger the difference between the existing influent composition and the contribution from the industrial discharge, the larger the potential for problems.

Remember that a change of one pH unit represents a 10-fold change in the concentration of acid in the influent and that secondary and anaerobic treatment are inhibited by rapid changes in environmental conditions. Operation outside of the pH range of 7.0 to 8.5 can be toxic to bacteria; however, microorganisms can become acclimated to pH levels slightly beyond this range.

Changes in conductivity normally represent increases or decreases in soluble salts, cyanide or metals. The inhibition or interference with the treatment processes is shown in Table 8.2. The normal range for conductivity depends on the water source and the amount of infiltration in the POTW influent. An average influent wastewater value of 2,500 microhms with a range of 2,000 to 3,500 microhms is typical of a large POTW. Spikes of conductivity exceeding 4,000 microhms probably indicate a dump from an industrial source. In a smaller system, spikes may be more noticeable and be a result of an industry or a home regenerating a water softener, or the discharge of a metal plating bath.

8.423 Slug Loadings

Slug loadings (also called shock loads) or batch dumps of compatible or noncompatible pollutants from industries, whether accidental or as part of normal production, may cause interference with the treatment processes or pass-through of pollutants. To assess the effect of a slug loading, you will have to consider the mass of the discharge and the resulting concentration at the treatment plant. The discharge to the sewer of concentrated solutions containing four pounds of copper may result in a five-minute concentration of 5.0 mg/L at the POTW treatment plant. While this is a significant variation from a 0.25 mg/L average influent concentration, the effect on the secondary treatment system would be negligible, and the sludge reuse potential would not appreciably suffer from a one-time occurrence. The four pounds of copper discharged would represent about a five percent increase from the normal mass of copper received.

TABLE 8.2 INTERFERENCE FROM INDUSTRIAL DISCHARGES

Source	Pollutants	Effects on Treatment System
Metal Finishing and Printed Circuit Board Manufacture	A. Heavy Metals	1. Decrease or stop biological removal rates for secondary and anaerobic treatment. 2. Prevent reuse of sludge or make it a hazardous waste.
	B. Chlorinated Solvents	1. Same effect as A. 2. Exposure of POTW workers to toxic gas.
	C. Acids	1. Destroy microbes, stopping treatment. 2. Upset anaerobic digester reducing gas production. 3. Corrode structures.
Cleaning Operations (Machinery Repair, Food Process, Clean-in-place Operations)	D. Detergents	1. Foam in secondary treatment facilities reduces settling characteristics and dewaterability.
Oil Production, Refining or Dispensing	E. Oil	1. Interferes with settling. 2. Toxic to anaerobic bacteria in large quantities reducing gas production. 3. Explosive when using a pure oxygen activated sludge system.
	F. Flammables	1. Same effects as A. 2. Explosive when it accumulates.
	G. Sulfide (Oil Production)	1. Toxic to treatment plant workers. 2. Odor complaints. 3. Increases oxygen demand and blower requirements.
	H. Salt (Oil Production)	1. Decreases oxygen transfer efficiency. 2. Inhibits biological activity.
Food Processing	I. BOD (Soluble and Insoluble)	1. Increases oxygen demand in secondary treatment. 2. May change microbiology of secondary treatment, causing secondary treatment settling problems. 3. Creates odors.
Organic Chemicals (Ketones, Alcohols)	J. Acetone, Methyl Ethyl Ketone, Isopropanol	1. If biological treatment microorganisms are acclimated, effects same as I-1. 2. If biological treatment microorganisms are not acclimated, effects same as B.
Utilities (Steam, Electricity, Cooling Towers)	K. Temperature (Warm)	1. Increases settleability, biological activity, and overall removal rates.
	L. Temperature[a] (Hot)	1. Depending on discharge point of POTW, exceeds temperature limits. 2. Changes microbiology or biological treatment efficiency. 3. Accelerates hydrogen sulfide production which causes odors and corrosion.

[a] Warm wastewaters may improve rather than interfere with treatment. Warm temperatures increase settleability, biological activity and overall removal rates.

However, if the concentration were to remain at 5.0 mg/L for a one-hour period, the biological treatment system would likely be toxicified, severely reducing or stopping biological treatment, and the sludge would be contaminated. The concentration of the slug loading as measured at the treatment plant was the same in both examples, but the second example illustrated a batch dump which was 12 times more mass than the first. It would have caused discharge violations, sludge contamination, and the biological treatment removal efficiencies would suffer until new bacteria could be cultured to return to the previous efficiency.

If slug loadings such as in the first example are allowed to continue on a daily basis, the activated sludge or trickling filter process may become acclimated and the daily discharges probably will not affect the effluent quality. The sludge, however, will now be more contaminated, possibly affecting its use as a soil amendment.

8.43   Effects on Effluent and Sludge Disposal and Reuse

Industrial discharges which, alone or in conjunction with discharges from other sources, pass through the POTW's facilities to navigable waters and cause a violation of the discharge permit are considered pass-through discharges. Pass-through of compatible and noncompatible pollutants can occur when the POTW treatment system is under stress from hydraulic or compatible waste overloads or shock loadings of toxic pollutants. When the pollutant removal efficiency decreases, the constituents from industrial discharges are found in the effluent.

Excluding slug loadings, the constituents most likely to pass through are: (1) small quantities of toxic organics that are very miscible (ketones and alcohols, if not stripped, are metabolized by secondary treatment); (2) toxic organics that are immiscible and lipophilic (having a strong affinity to fats) such as pesticides or polychlorinated biphenols; and (3) soluble heavy metals that are not used as micronutrients.

If the toxic constituents in an industrial waste are controlled on site, the level of toxics discharged to the sewer is minimal. This optimizes the recycle and reuse options of both the effluent and sludge. Effluent can be further treated for reclaimed water uses; sludge can be applied to land as fertilizer or mixed with a bulking agent and made into compost.

Industrial discharges that upset the treatment system eventually have an effect on the effluent and sludge quality. In essence, the industrial discharge has contaminated the wastewater. Instead of being a potential resource, the effluent and sludge become a liability.

8.5   CONCLUSIONS

This chapter on industrial wastes provided: (1) descriptions of various common manufacturing processes and how waste and wastewater are produced, and (2) a discussion of the effects of these wastes on the pretreatment system and the POTW collection, treatment, and disposal systems. With this information, you should be able to identify sources of industrial wastewater by following the raw materials through the manufacturing processes; you should be able to predict what wastes will be discharged; you should now know how to quantify and characterize industrial wastes to determine what effects they may have on the pretreatment and POTW systems; you should understand how industrial discharges may interact; and, finally, it is hoped that you have learned a systematic approach to investigating and mitigating problems with industrial discharges.

8.6   REFERENCES

1. *DEVELOPMENT DOCUMENT FOR EFFLUENT LIMITATION GUIDELINES, NEW SOURCE PERFORMANCE STANDARDS AND PRETREATMENT STANDARDS FOR THE PULP, PAPER, PAPERBOARD, AND BUILDERS' PAPER AND BOARD MILLS POINT SOURCE CATEGORIES* (M.M. Gorsuch, Administrator; J.D. Denit, R.W. Dellinger, and W.D. Smith, 1982. Effluent Guidelines Division, Office of Water, US EPA, Washington, DC 20460.) Obtain from National Technical Information Service (NTIS), 5285 Port Royal Road, Springfield, VA 22161. Order No. PB83-163949. EPA No. 440-1-82-025. Price, $123.00 (includes shipping), plus $5.00 handling charge per order.

   Information available on standard manufacturing processes, raw materials, water use and waste characterization, conventional pollutants, toxic and nonconventional pollutants, control and treatment technology (559 pages plus appendices).

2. *LOSS CONTROL MANAGEMENT IN THE KRAFT PULPING INDUSTRY* (G.W. Gove, J.J. McKeown, A.J. Carson, A. Hall, Grant No. R804086-01-1. Industrial Environmental Research Laboratory, Office of Research and Development, US EPA, Cincinnati, OH 45628, December 1980.) Obtain from National Technical Information Service (NTIS), 5285 Port Royal Road, Springfield, VA 22161. Order No. PB81-131971. EPA No. 600-2-80-211. Price, $54.50 (includes shipping), plus $5.00 handling charge per order.

   Presents information on a loss control strategy for the pulping, pulp washing, and chemical recovery areas of a large kraft pulpmill. Some good information on waste sources and methods of control.

3. *DEVELOPMENT DOCUMENT FOR PROPOSED EFFLUENT LIMITATION GUIDELINES, NEW SOURCE PERFORMANCE STANDARDS FOR THE METAL FINISHING POINT SOURCE CATEGORY* (Richard Kinch, August 1982, Effluent Guidelines Division, Office of Water Regulations and Standards, US EPA, Washington, DC 20460. Proposed.) Obtain from National Technical Information Service (NTIS), 5285 Port Royal Road, Springfield, VA 22161. Order No. PB83-102004. EPA No. 440-1-82-091-B. Price, $123.00 (includes shipping), plus $5.00 handling charge per order.

   Describes unit operations in the industry, water usage by operation and waste type, waste characterization of metal finishing unit operations (includes electroplating, electroless plating, anodizing, conversion coating, etching, cleaning, machining, grinding, polishing, barrel finishing, burnishing, impact formation, heat treating, thermal cutting, welding, electrical discharge, electrochemical machining, laminating, hot dip coating, salt bath descaling, solvent degreasing, paint stripping, painting, testing, mechanical plating), and waste treatment methods.

4. *DEVELOPMENT DOCUMENT FOR EFFLUENT LIMITATION GUIDELINES, NEW SOURCE PERFORMANCE STANDARDS AND PRETREATMENT STANDARDS FOR THE INORGANIC CHEMICALS MANUFACTURING POINT SOURCE CATEGORY* (Phase II) (Dr. Thomas E. Fielding, July 1984, Effluent Guidelines Division, Office of Water Regulations and Standards, US EPA, Washington, DC 20460.) Obtain from National Technical Information Service (NTIS), 5285 Port Royal Road, Springfield, VA 22161. Order No. PB85-156446. EPA No. 440-1-84-007. Price, $123.00 (includes shipping), plus $5.00 handling charge per order.

This document contains information on process wastewater sources and current treatment practices, including cadmium pigments and salts industry (water use and wastewater source characteristics, cobalt salts industry, copper salts industry, nickel salts industry, sodium chloride industry, zinc chloride industry).

5. *DEVELOPMENT DOCUMENT FOR EXISTING SOURCE PRETREATMENT STANDARDS FOR THE ELECTROPLATING POINT SOURCE CATEGORY* (J. Bill Hansen, August 1979, Effluent Guidelines Division, Office of Water and Hazardous Materials, US EPA, Washington, DC 20460.) Obtain from National Technical Information Service (NTIS), 5285 Port Royal Road, Springfield, VA 22161. Order No. PB80-196488. EPA No. 440-1-79-003. Price, $98.50 (includes shipping), plus $5.00 handling charge per order.

Information includes industry characterization and waste characterization and pretreatment technologies.

QUESTIONS

Write your answers in a notebook and then compare your answers with those on page 593.

8.4E   A hydraulic overload can cause what kinds of problems at a wastewater treatment plant?

8.4F   Define interference.

8.4G   Changes in conductivity indicate changes in what types of wastestream constituents?

Please answer the discussion and review questions before continuing with the Objective Test.

DISCUSSION AND REVIEW QUESTIONS
Chapter 8.   INDUSTRIAL WASTEWATERS

Write the answers to these questions in your notebook before continuing with the Objective Test on page 594. The purpose of these questions is to indicate to you how well you understand the material in the chapter.

1. Why does a pretreatment facility inspector need to understand the sources and quantities of industrial waste?

2. How can compatible pollutants exhibit the characteristics of noncompatible pollutants and vice versa?

3. How can spills of process chemicals reach a pretreatment system and/or POTW?

4. Identify several types or sources of intermittent discharges of wastewater.

5. Why are mass emission rate standards preferred over concentration-based standards?

6. Why is it important for an inspector to understand the manufacturing processes in the major wastewater generating industries?

7. What processes are used to clean and prepare the surface of a workpiece for a surface coating?

8. Water is used for what purposes in the metal finishing industry?

9. How is water used in the manufacture of pulp, paper, and paperboard?

10. What solid wastes are produced during the manufacturing of batteries?

11. What causes the differences in water usage among the leather tanning industries?

12. What are the products from petroleum refinery processes?

13. The effects of an industrial discharge on the POTW are dependent on what factors?

14. How can a slug loading on a POTW from an industrial discharge be mitigated?

15. How can hydraulic overloads of a POTW collection system be reduced when the overloads are caused by a specific industry?

16. How can problems caused by hydrogen sulfide from industrial discharges into a collection system be corrected?

17. What problems can be caused in sewers when an industry discharges out-of-pH-range wastewater?

18. Heated industrial discharges can cause what kinds of problems in POTW collection and treatment systems?

19. Slug loadings or batch dumps cause what kinds of problems?

20. What are pass-through discharges?

21. What can cause a pass-through discharge?

SUGGESTED ANSWERS

Chapter 8.   INDUSTRIAL WASTEWATERS

Answers to questions on page 536.

8.0A   Knowledge of the various manufacturing processes is important if the inspector is to understand how raw materials are transformed into products, by-products and waste products and the effects of the waste products on POTW collection systems and the treatment plant.

8.0B   A part could be immersed in a concentrated chemical solution in a metal plating process to clean, activate, plate, or seal the part.

8.0C   A high temperature wastewater can have both beneficial and adverse effects on a POTW. High temperature discharges to sewers can accelerate (1) biological degradation, (2) slime growths, (3) odor production from anaerobic decomposition, and (4) corrosion of concrete pipe and metal sewer appurtenances. A high temperature could cause a bacterial population shift in the secondary process causing floating sludge and reduced BOD removal efficiency. Also, the temperature discharge standards could be exceeded. In a colder climate, however, a high temperature wastewater may actually enhance the secondary processes' removal efficiencies.

Answers to questions on page 541.

8.1A   The difference between compatible and noncompatible pollutants is that compatible pollutants normally are removed by the POTW system and noncompatible pollutants normally ARE NOT removed.

8.1B   Examples of noncompatible pollutants include heavy metals such as copper, nickel, lead, and zinc; organics such as methylene chloride, 1,1,1-trichloroethylene, methyl ethyl ketone, acetone, and gasoline; and sludges containing toxic organics or metals.

8.1C   An inspector needs to know an industry's hours of discharge to set up a wastewater sampling program to determine compliance. The inspector requires an understanding of when the waste from each process is generated as well as when it is discharged.

8.2A   The common units of expression for the following terms are:

(1) Concentration is expressed as milligrams per liter (mg/L) which is a mass per unit volume measurement;

(2) Mass is expressed in pounds or kilograms; and

(3) Mass emission rate is expressed in pounds or kilograms per day and this is a mass per unit time measurement.

8.2B   If the flow of wastewater is large, the hydraulic system may become overloaded and cause pass-through of the pollutant. If a slug load of a highly concentrated solution reaches a treatment plant, the results may be interference with the treatment system and a violation of either or both effluent and sludge disposal limitations.

Answers to questions on page 548.

8.3A   The common metals used by the metal finishing industry to provide a surface coating on a base material include copper, nickel, chromium, brass, bronze, zinc, tin, lead, cadmium, iron, aluminum, or combinations thereof. In precious metals plating, a base material is plated with gold, silver, palladium, platinum, osmium, iridium, rhodium, indium, ruthenium, or combinations of these metals.

8.3B   The general processes of the metal finishing industry include machining, cleaning and surface preparation, plating and coating, anodizing, and etching and chemical milling.

8.3C   Types of chemicals found in electroplating baths include metal salts, alkalies, and other bath control compounds in addition to the plating metals. Many plating solutions contain metallic, metallo-organic, and organic additives to induce grain refining, leveling of the plating surface, and deposit brightening.

8.3D   The nine common waste types found in the waste/ wastewater streams generated from metal finishing processes include: (1) heavy metals, (2) precious metals, (3) complexed metals, (4) acid wastewaters, (5) alkaline wastes, (6) hexavalent chromium, (7) cyanide wastes, (8) oily wastes, and (9) solvent wastes.

Answers to questions on page 554.

8.3E   A printed circuit board is a board or plate of nonconductive material such as glass or plastic on which a conductive metal circuit pattern has been printed or etched.

8.3F   The general processes of printed circuit board manufacturing include laminate machining, cleaning and surface preparation, electroless plating, pattern printing and masking, electroplating, and etching.

8.3G   The major pulp, paper, and paperboard manufacturing processes include raw material preparation, pulping, secondary fibers pulping, bleaching of wood pulps and papermaking.

8.3H   There are three basic types of chemical pulping processes now in common use: alkaline, sulfite, and semichemical.

Answers to questions on page 562.

8.3I Electric power is generated directly from a chemical reaction in a battery.

8.3J Wastewater is generated mainly from the following battery manufacturing processes: cathode production, anode production, pasting, curing, acid adding, forming, wash processes, and testing.

8.3K The purpose of the tanning segment of the tanyard processes is to improve the stability, heat resistance, and chemical resistance of the hides by applying the tanning agents. Tanning agents convert the raw collagen fibers of the hide into a stable product which no longer is susceptible to decay or rotting.

8.3L The major water usage processes in the leather tanning industry are soaking and washing; tanning and retanning; preparing bleach, dye, or pigment solutions; and cleaning and washdown of the process equipment and areas.

8.3M The terms "paraffin base," "asphalt," and "mixed base" are often applied to differentiate crudes on the basis of the residues produced after simple distillation.

Answers to questions on page 569.

8.3N Acid pickling is the process of chemically removing oxides and scale from the surface of steel by the action of water solutions of inorganic acids.

8.3O Alkaline cleaning baths are used to remove minerals and animal fats and oils from steel.

8.3P Pollutants in process wastewaters from the iron and steel industry include toxic organic pollutants, toxic inorganic pollutants, conventional and nonconventional pollutants.

Answers to questions on page 582.

8.3Q The major sources of waste/wastewater for the hydrofluoric acid industry include noncontact cooling, air pollution control, product dilution, seals on pumps and kilns, and equipment area washdown.

8.3R The major uses of titanium dioxide include paints, varnishes and lacquers. Other uses include the paper and plastics industries, as well as ceramics, ink, and rubber manufacturing, in curing concrete, and in coatings for welding rods.

8.3S The major wastewater sources for the hydrogen cyanide industry are noncontact cooling in the absorber, pump seals, quenchers, flare stack flushes, washdown and cleanup of tank cars, absorption of the product from reactor gases, washing equipment, cleaning up leaks and spills, and distillation bottom waste.

Answers to questions on page 587.

8.4A A tertiary sand filtration process in a wastewater treatment plant can be rendered useless by a pass-through of oil or a carryover of gelatinous bacteria from an upset biological process.

8.4B Types of heavy solids that can cause plugging of sewers include sand, ceramic or porcelain solids, and grindings which can build up in a sewer and cause reduced hydraulic capacity. Large objects that can plug a sewer include rags, tools, rejected food products, and discarded by-products.

8.4C Examples of industrial discharges that can be odorous are those from petroleum refining, petrochemical manufacturing, and food processing.

8.4D Mineral acids such as sulfuric, nitric, hydrochloric, and phosphoric acids are used extensively to clean base metals in the metal finishing industries. The fertilizer, iron and steel, mining, and petroleum industries also use vast quantities of these strong acids. Mineral acids are also used in the pretreatment systems in chromium reduction, neutralization of alkalies, and pretreatment of chelated metal plating solutions.

Answers to questions on page 591.

8.4E A hydraulic overload can cause a decrease in the efficiency of treatments processes and possible effluent violations at the wastewater treatment plant.

8.4F EPA defines interference as a discharge which, alone or in conjunction with discharges from other sources, inhibits or disrupts the POTW, its treatment processes or operations, its sludge processes, use or disposal, and is a cause of violation of the NPDES permit or prevents the lawful use or disposal of sludge.

8.4G Changes in conductivity normally represent increases or decreases in soluble salts, cyanide or metals.

OBJECTIVE TEST

Chapter 8.   INDUSTRIAL WASTEWATERS

Please write your name and mark the correct answers on the answer sheet, as directed at the end of Chapter 1. There may be more than one correct answer to each multiple-choice question.

True-False

1. Compatible pollutants can exhibit characteristics of non-compatible pollutants and vice versa.

    — 1. True
    2. False

2. Storm water runoff from chemical handling and storage areas is considered a high-strength wastewater.

    1. True
    — 2. False

3. Chemical containment areas must have drains.

    1. True
    — 2. False

4. The major pollutant from the printed circuit board industry is cyanide.

    1. True
    — 2. False

5. Acid cleaning can be referred to as pickling.

    — 1. True
    2. False

6. Acid dip processes may follow alkaline cleaning prior to plating.

    — 1. True
    2. False

7. Steel industry process wastewaters are treated by currently available treatment technologies.

    — 1. True
    2. False

8. The pretreatment inspector's job is to prevent slug discharges.

    — 1. True
    2. False

9. The disposal of POTW sludge is affected by industrial discharges.

    — 1. True
    2. False

10. Industrial discharges that upset the POTW treatment system eventually affect sludge quality.

    — 1. True
    2. False

Best Answer (Select only the closest or best answer.)

11. Why is knowledge of the various manufacturing processes important? To understand how

    1. Companies promote environmental awareness.
    2. Industry discharges pollutants.
    3. Manufacturers produce products.
    — 4. Raw materials are transformed into waste products.

12. What are compatible pollutants?

    1. Pollutants normally found in a concentrated wastestream
    2. Pollutants normally found in a dilute wastestream
    3. Pollutants normally removed by a physical-chemical treatment system
    — 4. Pollutants normally removed by a POTW treatment system

13. Why should groundwater from leaking gasoline tank cleanup projects be required to have sufficient treatment to allow discharge to surface waters?

    1. To avoid adverse responses from the public
    2. To prevent replenishment of groundwater resources
    3. To provide additional surface water for downstream beneficial uses
    — 4. To save hydraulic capacity in the sewer for other contaminated flows

14. What does the wastewater from battery anode production contain?

    1. Acid
    2. Copper
    — 3. Lead particulates
    4. Spill washdown

15. What is the most heavily used toxic pollutant in leather tanning?

    — 1. Chromium
    2. Copper
    3. Cyanide
    4. Phenol

16. What is the major pollutant found in the wastes and wastewater generated by the chlor-alkali industry?

    1. Brine
    2. Chlorine
    — 3. Mercury
    4. Mud

17. Which acid is an organic acid?

    — 1. Acetic
    2. Nitric
    3. Phosphoric
    4. Sulfuric

18. At a large utility system, which wastewater discharge is usually an intermittent discharge?

    1. Boiler
    2. Cooling tower
    − 3. Demineralization system
    4. Pump or equipment seals

19. Where are the sources of pollutants in storm water run-off?

    1. Drag out dripping on the floor
    − 2. Products which have spilled on the grounds
    3. Rinse waters
    4. Scrubber wastewaters

20. Where are the primary sources of wastewater from the chrome pigments industry?

    1. Dryers
    2. Rinse waters
    3. Spent solutions
    − 4. Washing the precipitated product

Multiple Choice (Select all correct answers.)

21. Why does an inspector need to understand the sources and quantities of industrial wastewaters? To _____ problems caused by industrial discharges.

    1. Create
    − 2. Define
    − 3. Identify
    4. Prepare
    − 5. Solve

22. What factors must inspectors have a basic understanding of?

    − 1. Biological reactions
    − 2. Chemistry and chemical reactions
    3. Economics of industry facilities
    − 4. Manufacturing processes
    − 5. Their own collection and treatment system

23. What are some adverse impacts on wastewater collection systems from industrial wastewater discharges?

    1. Cause scouring velocities
    − 2. Corrode sewers
    − 3. Plug sewers
    − 4. Produce obnoxious odors
    − 5. Release toxic gases

24. What waste characteristics must be considered when determining the acceptability of an industrial waste discharge to a sewer?

    − 1. Concentration
    − 2. Odor
    − 3. pH
    − 4. Temperature
    − 5. Toxicity

25. Why are high-temperature discharges to sewers of concern? High-temperature discharges can accelerate

    − 1. Biological degradation.
    − 2. Corrosion of pipe.
    − 3. Odor production.
    − 4. Shift of bacterial population.
    − 5. Slime growths.

26. When do intermittent discharges occur?

    − 1. During disposal of a reject product
    − 2. During equipment cleanup
    − 3. During replacement of spent solution
    4. During rinsing or cleaning of food
    − 5. During spills

27. What are sources of dilute solutions that may be discharged to the pretreatment system or POTW?

    − 1. Plating bath rinses
    2. Reject products
    3. Spent plating baths
    4. Static drag out solutions
    − 5. Storm water runoff

28. Leaking underground tanks are often sources of high concentrations of what types of wastes?

    − 1. Fuel
    − 2. Heavy metals
    − 3. Pesticides
    4. Process rinse waters
    − 5. Solvents

29. Why is an understanding needed of the concentration and mass of a pollutant in an industrial waste? To determine the effects on

    − 1. Industry's pretreatment system.
    − 2. POTW collection system.
    − 3. POTW disposal system.
    − 4. POTW treatment system.
    − 5. Sampling of industrial discharge.

30. What are the categories of pollutants that are found in the wastewater generated from pulp and paper processes?

    − 1. Conventional pollutants
    − 2. Nonconventional pollutants
    3. Priority pollutants
    − 4. Toxic pollutants
    5. Volatile organic pollutants

31. What problems are created by the treatment of petroleum refinery wastes?

    − 1. Acid sludges
    − 2. Alkaline wastes
    3. Coliform MPN
    4. High BOD
    − 5. Obnoxious odors

32. What types of industrial discharges can cause plugging of sewers?

    − 1. Adhesives
    − 2. Fibrous materials
    − 3. Grease
    − 4. Heavy solids
    5. Soluble BOD

33. What types of discharge constituents may cause explosions in sewers?

    1. Acetic acid
    − 2. Gasoline
    − 3. Hexane
    4. Hexavalent chromium
    − 5. Methyl ethyl ketone

34. Small quantities of which pollutants are likely to pass through a POTW?

- 1. Ketones
- 2. Pesticides
- 3. Polychlorinated biphenols
- 4. Soluble heavy metals
- 5. Toxic organics

35. What type of system flexibility will allow a POTW to mitigate an industrial slug discharge?

- 1. Adequate aeration equipment
- 2. Chemical treatment points along the collection system
- 3. Effluent recycling capabilities
- 4. Equalization basins
- 5. Flow control structures

End of Objective Test

CHAPTER 9

PRETREATMENT TECHNOLOGY
(SOURCE CONTROL)

by

Michael C. Lee, Richard G. Wilson and Douglas K. Garfield

KENNEDY / JENKS / CHILTON

TABLE OF CONTENTS

Chapter 9.   PRETREATMENT TECHNOLOGY (SOURCE CONTROL)

OBJECTIVES

Chapter 9. PRETREATMENT TECHNOLOGY (SOURCE CONTROL)

Following completion of Chapter 9, you should be able to:

1. Identify sources and discharge characteristics of industrial wastestreams,

2. Establish treatment and discharge objectives,

3. Select and evaluate methods of pollution prevention and source control,

4. Identify and evaluate options for treatment of wastestreams,

5. Explain facility operational considerations, and

6. Evaluate source monitoring (compliance and operational) programs.

PRETREATMENT WORDS

Chapter 9.   PRETREATMENT TECHNOLOGY (SOURCE CONTROL)

ABSORPTION (ab-SORP-shun)   ABSORPTION

The taking in or soaking up of one substance into the body of another by molecular or chemical action (as tree roots absorb dissolved nutrients in the soil).

ADSORPTION (add-SORP-shun)   ADSORPTION

The gathering of a gas, liquid, or dissolved substance on the surface or interface zone of another material.

AEROBIC (AIR-O-bick)   AEROBIC

A condition in which atmospheric or dissolved molecular oxygen is present in the aquatic (water) environment.

ANAEROBIC (AN-air-O-bick)   ANAEROBIC

A condition in which atmospheric or dissolved molecular oxygen is *NOT* present in the aquatic (water) environment.

BUCHNER FUNNEL   BUCHNER FUNNEL

A special funnel used to separate solids (sludge) from a mixture. A perforated plate or a filter paper is placed in the lower portion of the funnel to hold the mixture (wet sludge). The funnel is placed in a filter flask and a vacuum is applied to remove the liquid (dewater the sludge).

CEMENTATION   CEMENTATION

(1)   A spontaneous electrochemical process that involves the reduction of a more electropositive (noble) species, for example, copper, silver, mercury, or cadmium, by electronegative (sacrificial) metals such as iron, zinc, or aluminum. This process is used to purify spent electrolytic solutions and for the treatment of wastewaters, leachates, and sludges bearing heavy metals. Also called ELECTROLYTIC RECOVERY.

(2)   The process of heating two substances that are placed in contact with each other for the purpose of bringing about some change in one of them such as changing iron to steel by surrounding it with charcoal and then heating it.

DIALYSIS (die-AL-uh-sis)   DIALYSIS

The selective separation of dissolved or colloidal solids on the basis of molecular size by diffusion through a semipermeable membrane.

DRAG OUT   DRAG OUT

The liquid film (plating solution) that adheres to the workpieces and their fixtures as they are removed from any given process solution or their rinses. Drag out volume from a tank depends on the viscosity of the solution, the surface tension, the withdrawal time, the draining time and the shape and texture of the workpieces. The drag out liquid may drip onto the floor and cause wastestream treatment problems. Regulated substances contained in this liquid must be removed from wastestreams or neutralized prior to discharge to POTW sewers.

ELECTRODIALYSIS   ELECTRODIALYSIS

The selective separation of dissolved solids on the basis of electrical charge, by diffusion through a semipermeable membrane across which an electrical potential is imposed.

ELECTROLYTIC RECOVERY   ELECTROLYTIC RECOVERY

A spontaneous electrochemical process that involves the reduction of a more electropositive (noble) species, for example, copper, silver, mercury, or cadmium, by electronegative (sacrificial) metals such as iron, zinc, or aluminum. This process is used for purifying spent electrolytic solutions and for the treatment of wastewaters, leachates, and sludges bearing heavy metals. Also called CEMENTATION.

ELUTRIATION (e-LOO-tree-A-shun)   ELUTRIATION

The washing of digested sludge with either fresh water, plant effluent or other wastewater. The objective is to remove (wash out) fine particulates and/or the alkalinity in sludge. This process reduces the demand for conditioning chemicals and improves settling or filtering characteristics of the solids.

EMPTY BED CONTACT TIME   EMPTY BED CONTACT TIME

The time required for the liquid in a carbon adsorption bed to pass through the carbon column assuming that all liquid passes through at the same velocity. It is equal to the volume of the empty bed divided by the flow rate.

INTERFACE   INTERFACE

The common boundary layer between two substances such as water and a solid (metal); or between two fluids such as water and a gas (air); or between a liquid (water) and another liquid (oil).

OSMOSIS (oz-MOE-sis)   OSMOSIS

The passage of a liquid from a weak solution to a more concentrated solution across a semipermeable membrane. The membrane allows the passage of the water (solvent) but not the dissolved solids (solutes). This process tends to equalize the conditions on either side of the membrane.

PERMEATE (PURR-me-ate)   PERMEATE

(1) To penetrate and pass through, as water penetrates and passes through soil and other porous materials.

(2) The liquid (demineralized water) produced from the reverse osmosis process that contains a *LOW* concentration of dissolved solids.

PICKLE   PICKLE

An acid or chemical solution in which metal objects or workpieces are dipped to remove oxide scale or other adhering substances.

POLYELECTROLYTE (POLY-ee-LECK-tro-lite)   POLYELECTROLYTE

A high-molecular-weight substance that is formed by either a natural or synthetic process. Natural polyelectrolytes may be of biological origin or derived from starch products, cellulose derivatives, and alignates. Synthetic polyelectrolytes consist of simple substances that have been made into complex, high-molecular-weight substances. Often called a POLYMER.

POLYMER (POLY-mer)   POLYMER

A long chain molecule formed by the union of many monomers (molecules of lower molecular weight). Polymers are used with other chemical coagulants to aid in binding small suspended particles to larger chemical flocs for their removal from water.

PONT-A-MOUSSON (MOO-san) PRESSURE CELL   PONT-A-MOUSSON PRESSURE CELL

A device that works like a filter press and is used to perform filter tests. A cell applies pressurized air (7 to 225 psig) with a piston (instead of using a vacuum as with a Buchner funnel) to separate solids from a mixture.

POZZOLANIC (POE-zoe-LAN-ick) PROCESS   POZZOLANIC PROCESS

The pozzolanic process mixes lime and other fine-grained siliceous materials such as fly ash, cement kiln dust, or blast furnace slag with a liquid hazardous waste to form a cement-like solid.

REAGENT (re-A-gent)   REAGENT

A pure chemical substance that is used to make new products or is used in chemical tests to measure, detect, or examine other substances.

REJECT   REJECT

The liquid produced from the reverse osmosis process that contains a *HIGH* concentration of dissolved solids.

REVERSE OSMOSIS (oz-MOE-sis)   REVERSE OSMOSIS

The application of pressure to a concentrated solution which causes the passage of a liquid from the concentrated solution to a weaker solution across a semipermeable membrane. The membrane allows the passage of the water (solvent) but not the dissolved solids (solutes). In the reverse osmosis process two liquids are produced: (1) the reject (containing high concentrations of dissolved solids), and (2) the permeate (containing low concentrations). The "clean" water is not always considered to be demineralized. Also see OSMOSIS.

SOLVENT EXTRACTION   SOLVENT EXTRACTION

The process of dissolving and separating out particular constituents of a liquid by treatment with solvents specific for those constituents. Extraction may be liquid-solid or liquid-liquid.

SUPERNATANT (sue-per-NAY-tent)   SUPERNATANT

Liquid removed from settled sludge. Supernatant commonly refers to the liquid between the sludge on the bottom and the scum on the surface of an anaerobic digester. This liquid is usually returned to the influent wet well or to the primary clarifier.

TOC (pronounce as separate letters)   TOC

Total **O**rganic **C**arbon. TOC measures the amount of organic carbon in water.

ULTRAFILTRATION   ULTRAFILTRATION

A membrane filter process used for the removal of some organic compounds in an aqueous (watery) solution.

WET OXIDATION   WET OXIDATION

A method of treating or conditioning sludge before the water is removed. Compressed air is blown into the liquid sludge. The air and sludge mixture is fed into a pressure vessel where the organic material is stabilized. The stabilized organic material and inert (inorganic) solids are then separated from the pressure vessel effluent by dewatering in lagoons or by mechanical means.

ZETA POTENTIAL   ZETA POTENTIAL

In coagulation and flocculation procedures, the difference in the electrical charge between the dense layer of ions surrounding the particle and the charge of the bulk of the suspended fluid surrounding this particle. The zeta potential is usually measured in millivolts.

CHAPTER 9.   PRETREATMENT TECHNOLOGY (SOURCE CONTROL)

As discussed in previous chapters, wastewaters from industrial sources may require pretreatment prior to discharge to wastewater collection systems leading to a publicly owned treatment works (POTW). Pretreatment standards may be imposed by the POTW through local ordinances or through administration of the EPA Federal Pretreatment Standards.

Treatment technologies that are typically used by industries to pretreat wastewater are reviewed and discussed in this chapter. From the inspector's standpoint, it is important to have a general knowledge of not only the various technologies, but also of the general types of wastes generated by each industry, the chemical and physical characteristics of the wastes and the potential toxicity of the wastes to the POTW. Potential measures that may be used by industry to reduce waste volumes and toxicity prior to "end of pipe" treatment are discussed, as are facility operational considerations which may affect pretreatment performance.

The pretreatment inspector is in an excellent position to discuss waste minimization source control (pollution prevention) issues and solutions with industry representatives. The material in this chapter could also be used by a new or planned industrial facility as a basis for developing a pretreatment program.

9.0   IDENTIFICATION OF SOURCE AND GENERAL DISCHARGE CHARACTERISTICS

9.00   Type of Industry

As described in Chapter 8, a given type of industry generally produces wastewater with certain characteristics. A listing of general industry classifications and the typical effluent characteristics of the wastewater they produce is shown in Table 9.1.

Within the broad industry classifications presented below are many subclassifications which produce specific types of wastes. At the end of this chapter in Section 9.6, several references (5, 7, 11, 13) are listed in which specific operations, their associated wastewater characteristics, and commonly used treatment processes are discussed. The EPA has published "Development Document(s) for Proposed Effluent Limitations Guidelines and New Source Performance Standards" for various industries which also contain detailed information on pollutants and treatment methods.

Another excellent reference is the US EPA's five-volume *TREATABILITY MANUAL* (Section 9.6, reference 18). The individual volumes cover such subjects as treatability data, industrial descriptions, and treatment technologies. The manual also contains detailed information on priority pollutants and treatment methods.

EPA has also developed a database containing information similar to that found in the *TREATABILITY MANUAL*. The database is called *EPA RISK REDUCTION ENGINEERING LABORATORY (RREL) TREATABILITY DATABASE*, Version 5.0 (Section 9.6, reference 16). It is available on 3.5-inch disk-ettes for use with IBM-compatible computers, and it is an excellent reference for pretreatment inspectors.

9.01   Type of Process

Within each industry there are often several different processes that can be used to produce a given product. The choice of which process to use depends upon several factors, including availability and cost of chemicals and other raw materials, cost of land, conservation measures, cost and skill level of local labor, utility costs, and pollution control costs. Two competing manufacturing processes may produce the same product, but the amount and composition of wastewater produced may differ greatly.

For example, in the paper industry several different pulping processes are used, each of which produces wastewater with different characteristics. In the steel industry, the choice of using an open-hearth, basic oxygen, or electric-arc furnace involves important water and air pollution considerations. In steel pickling, either hydrochloric or sulfuric acid can be used. Metal plating operations can produce vastly different amounts of wastewater depending on the number of rinse tanks used and the method of rinsing. In mining, metal milling and other industries, mechanical versus chemical or water-based processes can be used. The mechanical processes may produce little or no liquid wastes, but economics may dictate the use of the chemical or water-based process.

An awareness of the pollution potential from competing processes is important for anyone planning a new industrial facility or the expansion or modification of an existing one. For the waste inspector, an awareness of potential waste sources, chemical constituents in the waste, and techniques for waste reduction or treatment is equally important.

9.02   Type of Waste

As we have seen, certain types of wastes are associated with certain industries. Wastewaters with high levels of nitrate and pesticides frequently originate from agricultural operations. Polychlorinated biphenyls (PCBs) were formerly used in electric transformers, but also were used as hydraulic fluids and heat transfer fluids in other equipment. Acids, heavy metals, and cyanide are typical of metal plating wastes

TABLE 9.1   EFFLUENT CHARACTERISTICS FROM VARIOUS INDUSTRIES

Industry	Effluent Characteristics
Automotive	Oil and grease, phenols, metals, BOD, COD, acids, toxic organics
Bakery	Fats, oil and grease
Batteries	Acids, metals
Chemicals — organic	Toxic organics, BOD, COD, acids, metals
Chemicals — inorganic	Acids, metals, TDS, ammonia, phosphate
Electrical and Electronics	Metals, acids, SS, toxic organics, fluoride
Electroplating and Metal Finishing	Acids, metals, cyanide, toxic organics
Foods	BOD, COD, SS, alkalies, oil and grease
Leather Tanning and Finishing	BOD, SS, chromium, oil and grease, sulfide, fecal coliforms
Metals (primary metals smelting and refining)	Acids, metals, BOD, COD, SS, phenols, cyanide, sulfide, ammonia
Mining	Acids, metals
Paints	BOD, COD, SS, toxic organics, copper, lead, mercury
Petroleum	Oil and grease, BOD, COD, acids and alkalies, metals, sulfide, ammonia, phenols, mercaptans, heat
Pharmaceuticals	BOD, metals
Plastics and Synthetics	BOD, COD, SS, toxic organics
Power Generation (utilities)	Heat, oils, SS, TDS, metals, chlorine, PCBs
Pulp and Paper	Acids, alkalies, BOD, COD, SS, chlorine
Wood Products	SS, oil and grease, BOD, COD, wood preservatives
Rubber	BOD, SS, oil and grease
Textiles	COD, chromium, phenol, sulfide, dyes

or microelectronics industry wastes. Oil and grease, phenols, ammonia and sulfide are typical of oil refinery wastes. High BOD and SS levels often occur in food industry effluents. Knowledge of the potential sources of pollutants within your service area will help you identify the source of high levels of a certain pollutant compound present in the influent to the POTW.

The general categories of wastes (inorganic salts, metals, oil and grease, organics, pesticides, chlorinated organics) not only are characteristic of certain types of industries, but they also require different types of treatment, as will be described later in this chapter. Examples of common treatment processes are precipitation for metals; flotation for oil and grease; biological treatment for organics; and carbon *ADSORPTION*[1] for chlorinated organics. These wastewater types also have specific adverse effects on the operation of the POTW. See Table 8.2, "Interference From Industrial Discharges," page 589, for other examples of ways that wastewater can affect POTW processes.

QUESTIONS

Write your answers in a notebook and then compare your answers with those on page 665.

9.0A   What kinds of wastes could be expected in the effluent from electrical and electronics industries?

9.0B   What are the major general categories of wastes from industries?

---

[1] *Adsorption (add-SORP-shun).* The gathering of a gas, liquid, or dissolved substance on the surface or interface zone of another material.

9.1 ESTABLISHMENT OF TREATMENT AND DISCHARGE OBJECTIVES

This section provides pretreatment facility inspectors with an outline of the procedures used by engineers to design pretreatment facilities. An understanding of these concepts will help inspectors work with industries to achieve compliance with discharge limitations.

The first step is to identify and quantify the sources and discharge characteristics of wastewater from an industrial operation. The next step in designing a program to reduce wastes and treat wastewaters is to establish treatment and discharge objectives for the pretreatment facilities. The important factors that must be considered are flow, wastewater characteristics, pretreatment standards, toxicity of pollutants to POTWs, and control and monitoring guidelines.

9.10 Flow

Water requirements vary greatly between industries, but water consumption is generally determined on the basis of gallons of wastewater generated per unit of production. Since wastewater flows often vary from one plant to another within an industrial category, wastewater flow needs to be measured for actual production at a particular plant rather than relying upon industrial averages. An important first step is to identify the existing wastewater sources, associated flow, and the collection systems of the plant. In addition to existing wastewater flow, future wastewater flows should be projected based on anticipated or scheduled plant expansion, production increases, and water conservation and reuse efforts. Any increase in the future wastewater flow may be limited by the capacity of industrial waste collection systems and permitted discharge rate to the POTW sewer system.

In establishing treatment and discharge objectives, total, average, and peak flows are important. The total daily flow is typically used by the regulatory agency to identify significant dischargers and to assess the flow surcharge. The average flow during normal plant production time periods is often used to establish the size and capacity of major treatment units necessary to provide a given degree of treatment. The peak flow is used to establish the hydraulic capacity of pipelines and pumps and also flow metering devices. Performance of the treatment system during periods of peak flow should be evaluated by pretreatment inspectors to ensure compliance under these conditions. In addition, regulatory agencies often make use of the peaking factor, which is the ratio of peak flow to average flow, to assess the flow surcharge.

Water consumption needs versus water actually being used by an industrial facility should be studied critically. The literature in this field shows that in many cases water consumption can be reduced easily without harm to the product, thereby reducing the size and capacity of the needed treatment equipment.

9.11 Characteristics

Wastewater characteristics can be classified into three categories: chemical, physical, and microbial characteristics. The chemical characteristics include organics (BOD and COD), heavy metals and inorganics, cyanide, toxics, oil and grease, and pH. The physical characteristics include solids (settleable, suspended, colloidal and dissolved), odor, temperature and color. Microbial characteristics include bacteria, notably pathogens (disease-causing organisms), that must be considered.

Important wastewater characteristics which are usually treatable at the POTW are BOD, COD and suspended solids. The BOD is the rate of oxygen uptake required for the stabili-

zation of the biologically oxidizable organic matter. The COD is the measurement of the oxygen required to oxidize the total organic content. The BOD/COD ratio is sometimes used to determine the biodegradability level of wastewater. The BOD, COD and suspended solids, as well as other water quality indicators, are used by POTWs to assess surcharges on industrial discharges.

9.12 Mandated Pretreatment

Federal pretreatment standards require pretreatment of wastestreams discharged to POTWs. The standards are designed to protect the operation of POTWs and to prevent the introduction of pollutants which may be accumulated and concentrated or may pass through the POTWs. The revised EPA General Pretreatment Regulations (40 CFR Part 403) established the roles and responsibilities of federal, state and local governments and industries to implement national pretreatment standards. EPA strongly encouraged local enforcement of national pretreatment standards through development of the POTW pretreatment program.

As was discussed in Chapter 3, state and local pretreatment requirements must be established to comply with the federal standards. If more stringent than federal standards, the state or local requirements for wastewater quality take precedence. In addition, other factors which cause interference with treatment processes and exceed the plant's capacity may be considered in the state or local requirements.

9.13 Toxicity of Industrial Wastewaters to the POTW

A variety of pollutants may interfere with the operation of the POTW or may be toxic to the treatment processes including many inorganic and organic pollutants, acids, alkalies, temperature, and sulfur compounds. The degree of pretreatment required for industrial wastewater will depend on the constituents of the particular wastestream. In all cases, however, pretreatment levels must prevent upsets of POTW processes and ensure that there will be no pass-through of pollutants in the POTW effluent.

9.130 Inorganic Pollutants

Inorganic pollutants which may be toxic to biological treatment processes at POTWs at certain concentration levels and may affect the treatment processes include acids, alkalies, ammonia, alkali and alkaline earth metals including arsenic, borate (and other boron species), bromine, cadmium, chloride, chlorine, chromium, copper, cyanide, iron, lead, manganese, mercury, nickel, silver, sulfate, sulfide and zinc.

Table 9.2 presents threshold concentrations of inorganic priority pollutants reported to be inhibitory to biological treatment processes. At or above the threshold concentrations, these pollutants may inhibit or interfere with the normal functioning of biological treatment processes. In particular, copper, lead, silver, chromium, arsenic and boron have been found to be toxic to microorganisms present in biological treatment systems. In addition, free chlorine and fluorine are also toxic to the biological processes of the POTW. Wastestreams containing combinations of pollutants may exhibit lesser or greater inhibitory characteristics than indicated.

High concentrations of heavy metals in the treatment plant sludge may limit the application of sludge in land spreading, and pass-through of heavy metals may cause the POTW to violate its NPDES Permit limitations.

For sludge digestion processes, it has been reported that copper at a concentration of 100 mg/L is toxic, chromium and nickel are toxic at concentrations of 500 mg/L and sodium at high concentrations (>3,500 mg/L) is also toxic. Potassium and ammonia are considered toxic at concentrations of 4,000 mg/L. These toxic values are higher than the values in Table 9.2, which lists the *THRESHOLD OF INHIBITORY EFFECTS*.

9.131 Organic Pollutants

Many organic pollutants interfere with POTW processes, such as alcohols, phenols, cyanide, chlorinated hydrocarbons, pesticides, organic nitrogen compounds, surfactants, oil and grease, and miscellaneous organic chemicals. Chlorinated compounds such as chloroform, carbon tetrachloride and 1,1,1-trichloroethane are toxic to the anaerobic digestion process at very low concentrations (0.5 to 1 mg/L). Table 9.3 presents threshold concentrations of organic priority pollutants that inhibit biological treatment processes. As shown, phenols in small amounts are toxic to biological processes (pentachlorophenol, 0.95 mg/L, and phenol, 200 mg/L). Oil and grease in excess of 50 mg/L may interfere with biological treatment.

9.132 Effects of pH

In general, pH values less than 6 or greater than 8 hinder biological treatment processes. A change of pH may also cause changes in the solubilities of heavy metals and their concentrations. Consequently, the toxicity of soluble heavy metals to the biological processes may be intensified by an acidic or basic pH condition.

9.133 Effects of Temperature

High and low temperatures of wastewater affect the microorganism activities of conventional biological processes and can upset or reduce the treatment efficiency. Furthermore, density

changes due to temperature variations may cause interferences or upsets in the clarification processes. The increased density and viscosity of low-temperature wastewater can affect the settling and floating properties of the solids.

9.134 Sulfide Compounds

Sulfide compounds (especially hydrogen sulfide, H_2S) are known to inhibit the anaerobic digestion process under certain conditions. Naturally occurring sulfate in water supplies is also present in wastewater. Sulfate can be chemically reduced to sulfide and to hydrogen sulfide by bacteria under anaerobic conditions in the sewer system and sludge digesters. The conversion of dissolved sulfide present in wastewater to hydrogen sulfide gas can also occur under acidic (low pH) conditions. Release of hydrogen sulfide gas to the atmosphere is a potential toxic hazard to collection system and treatment plant

TABLE 9.2   REPORTED VALUES FOR BIOLOGICAL PROCESS TOLERANCE LIMITS OF INORGANIC PRIORTY POLLUTANTS

Pollutant	Threshold of Inhibitory Effect, mg/L					
	ACTIVATED SLUDGE		ANAEROBIC DIGESTION		NITRIFICATION	
	Russell[a]	EPA[b]	Russell[a]	EPA[b]	Russell[a]	EPA[b]
Arsenic	0.1	0.04-0.4	1.5			0.1-1
Cadmium	1.0	0.5-10	0.02	0.02-1	5.2	5-9
Chromium (VI)	1		5		0.25	
Chromium (III)	10		50			
Chromium (Total)		1-20		1.5-50		0.25-1
Copper	1.0	0.1-1	0.5	0.5-100	0.48	0.05-0.5
Cyanide	0.1	0.05-20	4	0.1-4	0.34	0.3-20
Lead	0.1	0.1-10		50-250	0.5	0.5-1.7
Mercury	0.1	0.1-5.0	1,365	1,400		2-12.5
Nickel	1	1-5	10	2-200	0.25	0.25-5
Silver	5	0.03-5			0.25	
Zinc	0.03	0.3-20	1.5	1-10	0.03	0.01-1

[a] Russell, L. C., Cain, C. B., and Jenkins, D. J., "Impact of Priority Pollutants on Publicly Owned Treatment Works and Processes: A Literature Review, Purdue Industrial Waste Conference," 37 (1982), 871-882. *CAUTION:* The published Purdue paper has the values in the Anaerobic Digestion and Nitrification columns interchanged and is incorrect.
[b] *GUIDANCE MANUAL FOR PREVENTING INTERFERENCE AT POTWs.* References: US EPA (1981a), Russell, et al. (1983), Geating (1981) and US EPA (1986a).

NOTE: Values reported in literature can be in error and reference to original papers is very important. Values reported in literature may refer to soluble fraction or to total amount of a pollutant.

ACKNOWLEDGMENT: Many thanks to Chris Cain for his assistance in sorting out the conflicts in the literature.

TABLE 9.3 REPORTED VALUES FOR BIOLOGICAL PROCESS TOLERANCE LIMITS OF
ORGANIC PRIORITY POLLUTANTS[a]

Pollutant	Threshold of Inhibitory Effect		
	ACTIVATED SLUDGE, mg/L[b]	ANAEROBIC DIGESTION, mg/L[b]	NITRIFI-CATION, mg/L[b]
Acenaphthene	NI[c] at 10		
Acrolein	NI at 62		
Acrylonitrile	NI at 152	5	
Benzene	125		
Benzidine	5	5	
Carbon Tetrachloride	NI at 10	2.9	
Chlorobenzene	NI at 1	0.96[d]	
1,2,4-Trichlorobenzene	NI at 6		
Hexachlorobenzene	5		
1,2-Dichloroethane	NI at 258	1	
1,1,1-Trichloroethane	NI at 10		
Hexachloroethane	NI at 10		
1,1-Dichloroethane	NI at 10		
1,1,2-Trichloroethane	NI at 5		
1,1,2,2-Tetrachloroethane	NI at 201	20	
Bis-(2-chloroethyl) ether	NI at 10		
2-Chloroethyl vinyl ether	NI at 10		
2-Chloronaphthalene	NI at 10		
2,4,6-Trichlorophenol	50		
parachlorometa cresol	NI at 10		
Chloroform	NI at 10	1	10
2-Chlorophenol	NI at 10		
1,2-Dichlorobenzene	5	0.23[d]	
1,3-Dichlorobenzene	5		
1,4-Dichlorobenzene	5	1.4[d]	
1,1-Dichlorethylene	NI at 10		
1,2-Trans-dichloroethylene	NI at 10		
2,4-Dichloropropane	NI at 75		
1,2-Dichloropropane	NI at 182		
1,3-Dichloropropylene	NI at 10		
2,4-Dimethylphenol	NI at 10		
2,4-Dinitrotoluene	5		
2,6-Dinitrotoluene	5		
1,2-Diphenylhydrazine	5		
Ethylbenzene	NI at 10		
Fluoranthene	NI at 5		
Bis-(2-chlorisopropyl) ether	NI at 10		
Methylene chloride		100	
Chloromethane	NI at 180	3.3	
Bromoform	NI at 10		
Dichlorobromomethane	NI at 10		
Trichlorofluoromethane	NI at 10	0.7	
Chlorodibromomethane	NI at 10		
Hexachlorobutadiene	NI at 10		
Hexachlorocyclopentadiene	NI at 10		
Isophorone	NI at 15.4		
Naphthalene	500		
Nitrobenzene	500		
2-Nitrophenol	NI at 10		
4-Nitrophenol	NI at 10		
2,4-Dinitrophenol	1		150
N-Nitrosodiphenylamine	NI at 10		
N-Nitroso-di-N-propylamine	NI at 10		
Pentachlorophenol	0.95	0.2	
Phenol	200		4
Bis-(2-ethylhexyl) phthalate	NI at 10		
Butyl benzyl phthalate	NI at 10		
DI-N-butyl phthalate	NI at 10		
DI-N-octyl phthalate	NI at 16.3		
Diethyl phthalate	NI at 10		

**TABLE 9.3   REPORTED VALUES FOR BIOLOGICAL PROCESS TOLERANCE LIMITS
OF ORGANIC PRIORITY POLLUTANTS[a] (continued)**

Pollutant	Threshold of Inhibitory Effect		
	ACTIVATED SLUDGE, mg/L[b]	ANAEROBIC DIGESTION, mg/L[b]	NITRIFI- CATION, mg/L[b]
Dimethyl phthalate	NI at 10		
Chrysene	NI at 5		
Acenaphthylene	NI at 10		
Anthracene	500		
Fluorene	NI at 10		
Phenanthrene	500		
Pyrene	NI at 5		
Tetrachloroethylene	NI at 10	20	
Toluene	NI at 35		
Trichloroethylene	NI at 10	20	
Aroclor-1242	NI at 1		
Aroclor-1254	NI at 1		
Aroclor-1221	NI at 1		
Aroclor-1232	NI at 10		
Aroclor-1016	NI at 1		

[a]   Source: Russell, L. L., Cain, C. B., and Jenkins, D. J., "Impact of Priority Pollutants on Publicly Owned Treatment Works and Processes: A Literature Review, Purdue Industrial Waste Conference," 37 (1982), 871-882.
[b]   Threshold of pollutant in mg/L on treatment process.
[c]   NI = no inhibition at tested concentration. No concentration is listed if reference lacked concentration data.
[d]   % wt/wt dry solids. In excess of 50 mg/L may cause interference with biological treatment.

employees and is the source of many odor complaints received by POTW facilities.

Hydrogen sulfide gas is often found in confined areas where the ventilation is poor and where wastewater flow is retarded in the treatment plant. Other likely areas include large sewers flowing slowly and nearly full, wet pits, or at the end of force mains. When atmospheric hydrogen sulfide concentrations reach 500 to 750 ppm, it can cause loss of consciousness and death can result from falling or drowning. At concentrations of 200 to 300 ppm, rapid loss of smell and breathing irritation occur and when it is strong enough in the atmosphere to cause death (1,000 to 2,000 ppm), the victim may not even smell it because hydrogen sulfide quickly deadens olfactory nerves.

9.14   Control and Monitoring Guidelines

At a minimum, effluent wastewater must be monitored to ensure compliance with local and/or EPA-mandated limits. A thorough monitoring program also enables the pretreatment operator to make process adjustments to avoid permit violations. To be most effective, the monitoring program should at least routinely analyze water quality indicators such as suspended solids, dissolved solids, BOD, COD, nitrogen, phosphorus, alkalinity, and oil and grease at the plant influent, effluent, and at certain pretreatment processes.

The monitoring of water quality indicators should be tailored to a particular industry's wastewater characteristics. For biological treatment processes, metals necessary for the growth of microorganisms, such as calcium, cobalt, copper, iron, magnesium, manganese, and zinc as well as inorganic and organic pollutants which may be inhibitory to the biological processes (as shown in Tables 9.2 and 9.3) should be analyzed. In addition, discharge parameters such as flow rates, BOD, SS, COD and phosphorus should be measured to establish the basis for large contributor and high strength surcharges by the regulatory agencies.

QUESTIONS

Write your answers in a notebook and then compare your answers with those on page 665.

9.1A What factors are involved in the establishment of treatment and discharge objectives for pretreatment facilities?

9.1B What does COD measure?

9.1C Why are sulfide compounds in industrial effluents an important concern?

9.2 SOURCE CONTROL (POLLUTION PREVENTION)

Source control is a term used to describe activities which reduce or eliminate wastewater and pollutants that would otherwise exit the manufacturing process and enter the POTW sewer. One approach to source control, pollution prevention, focuses on activities which prevent the generation of wastes. Another approach, waste minimization, seeks to reduce the volume or toxicity of wastes being generated. Waste minimization usually refers to reducing the volume or toxicity of hazardous wastes.

The beneficial effects of source control, in addition to protection of the public and the environment, are cost savings from the reduced chemical and water use, reduction of disposal costs and liability, and reduction in the size of the wastewater pretreatment system.

Pretreatment inspectors visit industrial dischargers every day and are generally familiar with the operation of a company and its management. Inspectors are in a unique position to promote awareness of pollution prevention opportunities and to inform industries of successful pollution prevention practices.

A logical first step in industrial wastewater treatment is an evaluation of source control options and methods, starting with the manufacturing process itself and proceeding to end-of-pipe treatment methods only after the waste concentrations and volumes have been minimized to the extent feasible. Various modifications to the manufacturing process or use of water and material conservation techniques can be applied to control or eliminate pollutant discharges at their source. The techniques discussed below focus primarily on the metal finishing industry but can be applied to many industries using the same or similar manufacturing processes.

9.20 Modification of Manufacturing Process

9.200 Change in Raw Materials

The chemicals used as raw materials to produce a product can sometimes be exchanged for other chemicals which have a much lower pollution potential. Many companies have been successful in removing all cyanide and chromium from cleaners, plating baths, or pickling solutions, thus eliminating these materials from wastewater discharges. There may be suitable non-chromate and non-cyanide substitutes available from metal finishing industry suppliers which can help reduce the size and complexity of wastewater treatment plant equipment. Similarly, non-chromate cooling tower treatment chemicals are available which can potentially eliminate a source of chromium in wastewaters as cooling towers are "blown down."

For a particular application, many different cleaning solvents are generally available for use. Choosing the one to use involves a number of considerations, including price, performance, flammability, vapor pressure, toxicity, threshold limit value (TLV), and water pollution potential. In some cases, water-based cleaner can be substituted for an organic solvent. Although this material substitution is encouraged by hazardous waste and industrial health and safety personnel, this practice increases the amount of wastewater generated at a facility. Therefore, the benefits of using a water-based cleaner versus a solvent-based cleaner should be weighed against the potential negative impact of increased wastewater volume if a water-based cleaner is used.

In chemical plants and in other facilities where chemical reactions can generate potentially toxic by-products, it is necessary to consider the pollution potential of not only the raw materials but also the chemical reaction products they produce.

Substitution of one raw material for another may involve some capital costs as well as additional operation and maintenance expenses. Substituted materials may be more expensive, require more employee attention to a process, and increase the need for quality control testing. Capital costs may be incurred if a powdered material is substituted for a liquid form which requires different equipment.

9.201 Modify Manufacturing Process

Process modifications in existing manufacturing facilities can be costly, require long implementation schedules, and require extensive employee training. On the other hand, advances in technology sometimes make new processes economically competitive when compared with the full costs of operating older systems. In the painting and paint and coatings removal processes, examples of process modifications include use of powder coat paints, electrostatic painting, laser paint stripping, flashlamp stripping, cryogenic coating removal, and dry ice blasting instead of aqueous alkaline stripping or use of high-pressure water jet blasting.

9.202 Modify Process Equipment

Another key to source control is matching the process equipment capacity or size with the manufacturing need. One semiconductor manufacturer reportedly reduced the volume of waste solvent from a manufacturing line simply by replacing the pump in a chemical supply line with a smaller pump. This is an example of a process equipment change which provided a degree of source control. New process equipment can sometimes be purchased or existing equipment modified to greatly reduce or even eliminate wastewater discharges. An example is the use of air-cooled equipment in place of water-cooled equipment.

9.203 Change in Operating Guidelines

In many cases operating guidelines such as chemical concentration, process temperature and dwell or residence time in process tanks can be changed with no effect on product quality. Several cases have been reported in which the concentration of chemicals in cleaning or plating solutions has been lowered, resulting in reductions in the quantity of heavy metals and other constituents in the influent to an industrial wastewater pretreatment plant. The "a little is good, more is better" philosophy which has been known to affect operating guidelines in many small shops operating under loose or nonexistent process specifications can have a needlessly detrimental effect on wastewater quality. Similarly, it may be possible to raise total dissolved solids (TDS) levels in some boilers and cooling towers to reduce blowdown and wastewater flow without significantly affecting steam or cooling water purity.

The mode of operation can have a great effect on wastewater quality and quantity. Generally speaking, rinsing operations conducted in a batch mode generate less wastes than those conducted in a continuous mode. A semiconductor manufacturer reduced the amount of hydrofluoric acid generated by making a switch from continuous to batch rinsing. However, an occasional "bad batch" as well as spillage associated with filling and emptying process vessels can account for a significant amount of wastewater loading. On the other hand, control of process variables in many chemical manufacturing operations is facilitated by using automatic control and continuous operations, rather than using manual control and batch operations. The choice between batch and continuous operations frequently involves trade-offs such as these but should at least be examined as a potential means of reducing wastewater quantities.

9.21 Water and Material Conservation, Recovery and Reuse

9.210 Housekeeping and Materials Management

Many of the measures discussed in this section have the advantage of not only reducing wastewater flows, but also reducing water and chemical costs. One of the most rewarding and least expensive means of tackling the waste treatment problem in many facilities is to undertake a thorough analysis of "housekeeping" practices and operating procedures.

Accidental spills, leaks, and drips of process solutions to the floor can contribute significantly to effluent contamination. Fill lines, hoses, and connections should be located where spills will be contained. As much as possible, material distribution should be done through pumps and pipes rather than through manual pouring of buckets or barrels. This practice reduces the chance of spills. If materials must be poured by hand, employees should be trained to clean up spills using minimal quantities of washdown water, or no washdown water if possible. Employees should be trained to close valves and place drip buckets or absorbent material in locations where drips from valves or faucets are likely to occur. As a part of routine housekeeping, repair all water, steam and condensate leaks since these also add to wastewater volumes.

Chemical storage areas should be cleaned and organized, with separate areas for compatible liquids and dry solids. Broken bags and leaking drums can be sources of sewer or storm drain contamination, so containers in poor condition should not be tolerated. Install some type of containment around these storage areas to prevent the release of spilled chemicals, and install some form of protection from rainwater intrusion. Examples of containment systems include berms, curbs, walls or trenches leading to a collection sump (Figure 9.1).

A review of purchasing procedures is advisable to avoid having chemical supplies on hand so long that the containers deteriorate. Raw materials should be ordered on an as-needed basis rather than in bulk discounted quantities that may never be used. In addition, employees should be trained to use raw materials on a first-in, first-out basis and to routinely check for expiration dates. A quality assurance program should be implemented to ensure that raw materials are not off-specification when they are received from the vendor. If possible, raw materials should be purchased in packaging that will not have to be rinsed prior to disposal or reuse.

A mechanism should be established requiring the signature of a waste disposal person on a purchase order before a new raw material can be purchased. When an industry wants to purchase a new raw material, a waste disposal person should review the material constituents and comment if those constituents or the use of the new material in a process will contribute to greater toxicity or quantity of wastewater.

Dry cleaning techniques, such as rag wipes or squeegees, should be encouraged and implemented where feasible, and equipment cleaning using aqueous (water-based) cleaners should be scheduled on an as-needed basis rather than on a set schedule that may be more frequent than necessary. Maintenance schedules should be reviewed to reduce the number of times wastes are generated by maintenance-related shutdowns.

9.211 Modification To Reduce Use

A reduction in water use and the reuse of process wastewaters are key methods of source control. A facility may be discharging to the sewer a substantial amount of clean, non-contact cooling water which can be reused, sometimes with recovery of the heat content of the water. Figure 9.2 shows a

Fig. 9.1 Inspector discusses need for spill containment for solvent and chemical storage area. Storage area has roof and diking and raised ramp to allow forklift access without permitting spills to reach yard.

Fig. 9.2 Heat recovery from wastestream
Heat from wastewater generated at a commercial laundry is
recovered in these heat exchanger tubes.

commercial laundry where heat from the wastewater is recovered in heat exchanger tubes. Water used to cool air compressors or air compressor aftercoolers could be used as boiler makeup water. Rectifier cooling water can potentially be used as rinse water. Air conditioner condensation can be used as makeup to cooling towers. In many cases the change from once-through cooling water to recirculated cooling water, using a cooling tower, may decrease water usage and wastewater volumes by up to 90 percent. Noncontact cooling water may also be discharged to surface waters or cooling/evaporation ponds in accordance with applicable NPDES Permits.

A number of devices can be used to reduce water flow, such as calibrated flow orifices, high-water level shutoff valves, timers, foot valves, and conductivity controllers. Conductivity controllers permit water additions to process and rinse tanks only when needed to maintain a desired water quality rather than maintaining a constant water makeup and overflow, such as is the case with many metal plating lines. The use of spray rinsing or countercurrent rinsing can also reduce water use and wastewater flows.

The most obvious sources of pollution in the metal finishing industry are the drag out, large-volume water rinses, and various processing baths with subsequent rinses. The amount of pollutants contributed by drag out is a function of the shape of the parts, plating procedures, and type of process solutions.

Drag Out — Many devices and procedures can be used successfully to reduce drag out. Most drag out reduction methods are inexpensive to implement and quickly pay for themselves through savings in process chemicals. The reduction of drag out will decrease the need for treatment chemicals and the volume of sludge which must be properly disposed of.

The most common method of minimizing drag out contamination is installation of a drag out tank to capture lost plating solutions and return them to the bath. For some process solutions, return of drag out may be impractical. For example, in the case of a processing bath that becomes steadily diluted in use, the return of drag out will simply increase the frequency of dumping. Other suggested methods for drag out reduction include controlling plating solutions by use of wetting agents, which allows the part to release the unneeded plating bath and reduces drag out loss by as much as 50 percent. By carefully positioning parts on the rack to drain freely, drag out can be further minimized. Also, workpiece withdrawal velocity has a marked effect on drag out volume. The faster an item is pulled out of the tank, the thicker the drag out layer will be.

One way to minimize drag out in a barrel plating operation is to rotate the barrel while it is being withdrawn from the plating solution to allow excess solution to drain back into the plating tank. This technique is enhanced further by allowing the barrel to rotate in the up position over the plating tank for a period of time prior to moving on to the next step in the plating process.

Simple drag out recovery techniques are shown in Figures 9.3 and 9.4.

Rinsing — The most effective means of reducing water use and waste treatment costs is to alter rinsing techniques. In the conventional rinsing technique, rinse waters that follow plating solutions typically contain 10 mg/L to 20 mg/L of the metal being plated. Changes can range from simple piping alterations for recycling rinse water to installation of two or three additional rinse tanks that are arranged to combine the advantages of series and recovery rinsing. The following are the suggested methods for reducing rinse water contamination.

- Rinse Water Recycling — A simple method of water conservation that is becoming more widespread involves the reuse of rinse water in two or more rinse tanks. Rinse water that is used in one process may sometimes be suitable for use in another less critical process (Figure 9.5).

- Multiple Rinse Tanks — The most common methods are parallel rinsing and series rinsing. In a parallel rinse tank arrangement, as illustrated in Figure 9.6, the rate of water flow to each tank should be the same to obtain the optimal water savings. A series or countercurrent rinse tank arrangement in which parts or workpieces move opposite the direction of flow produces a greater water savings than can be achieved in the parallel system. As shown in Figure 9.6, water flows into the rinse tank farthest away from the process tank.

- Flow Restrictors — These devices are usually installed for rinse tanks that have a constant production rate. They eliminate the option of using more rinse water.

- Process Spent Solutions — The relative volume of process solution waste is usually not large but because the chemical concentration is relatively high, the pollution effect may be considered serious. There are a number of techniques for recovering or reusing process solutions. Some examples of chemical recovery and reuse are reprocessing of oil, reclamation of oil, recycling of oil, regeneration of etchants, and drag out recovery.

- Process Modification — This can reduce the amount of water required for rinsing and/or reduce the load of certain pollutants on a waste treatment facility. For example, a rinse step might be eliminated by using an activation solution followed by a rinse. Another process modification would be to change from a high-concentration plating bath to one with a weaker concentration.

- Substitute Solutions — Substitute bath solutions and plating processes which are less toxic, such as non-chromic acid, non-cyanide plating solutions, and trivalent chromium plating, are constantly being developed. Some alternative process solutions may require unique treatment.

9.212 Physical Recovery/Separation

Wastewater treatment can obviously be avoided for chemicals that are recovered and reused (Figure 9.7). In some cases, the solution from a drag out recovery rinse or from spills can be used as a source of makeup water (Figure 9.8) for process tanks. (Drag out recovery rinse comes from an immersion tank in which production parts are rinsed following immersion in a process solution tank.) In heated process tanks, water must be added to account for evaporation. The contents of the drag out recovery rinse tank can often be used for this purpose, thereby conserving chemicals and reducing wastewater treatment requirements. Water makeup flow is typically countercurrent to the movement of parts through a plating or other chemical processing line. It may also be possible to reduce drag out by extending the dwell time of parts over the process tank, by using a misting spray rinse directly over the tank, or by using air knives to remove process solution from parts prior to entering a rinse tank.

Several of the treatment technologies discussed later in this chapter can be used for product recovery. These recovery processes are obviously easiest to justify where the chemicals

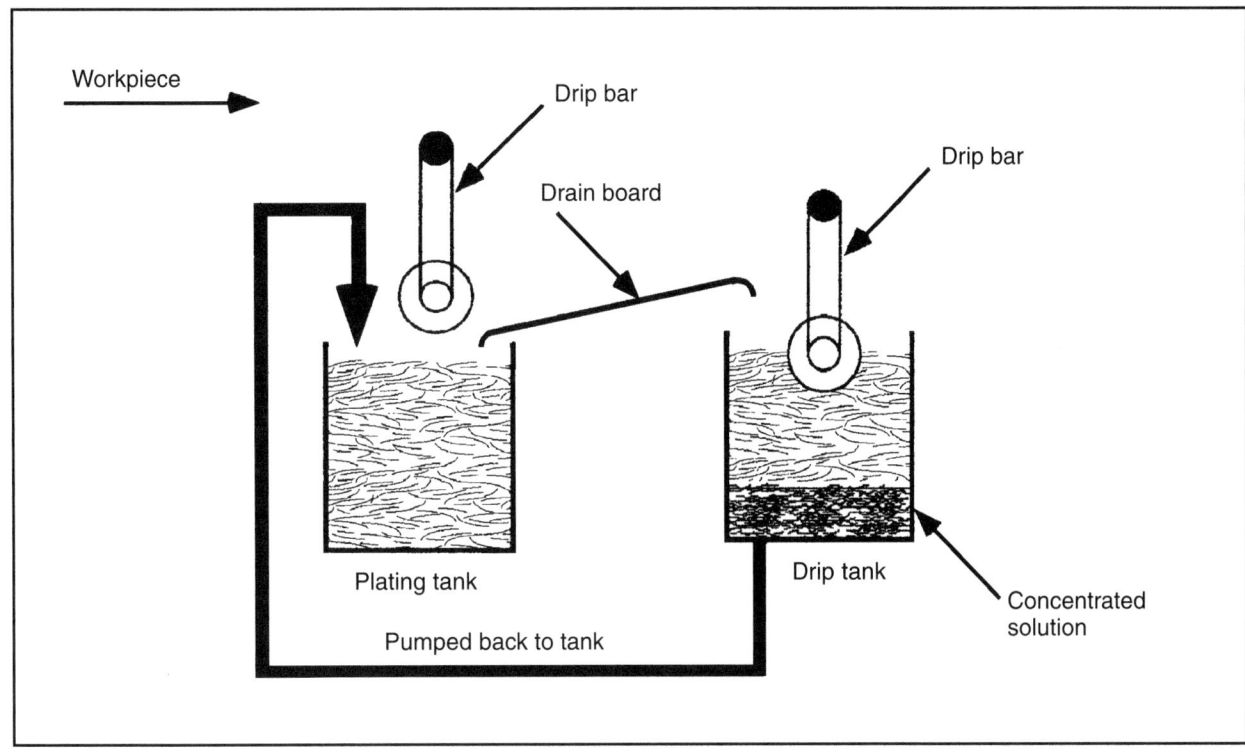

Fig. 9.3   Simple drag out recovery devices

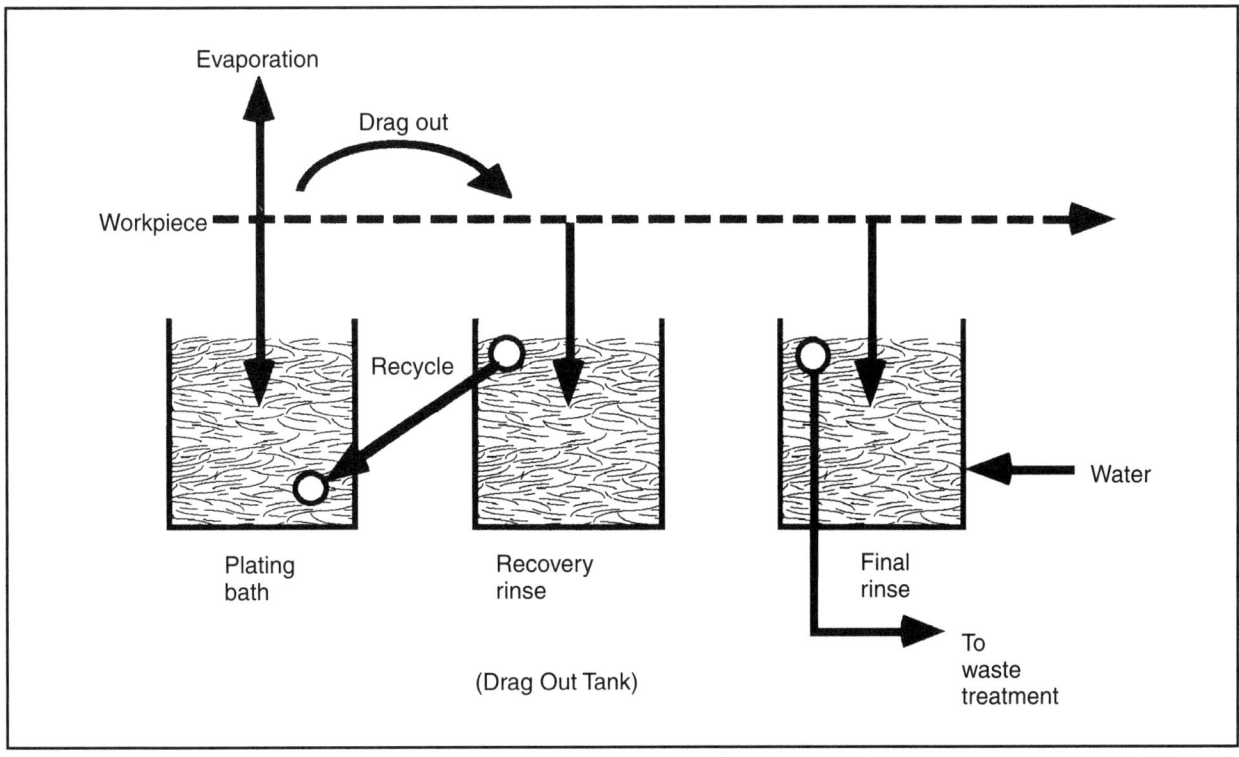

Fig. 9.4   Recovery with a drag out tank

Fig. 9.5 Concentrated ink waste recovery

In order to meet a strict color limit in wastewater, this ink manufacturer catches
concentrated wastes, labels the containers, and uses the captured rinses as
makeup for subsequent batches.

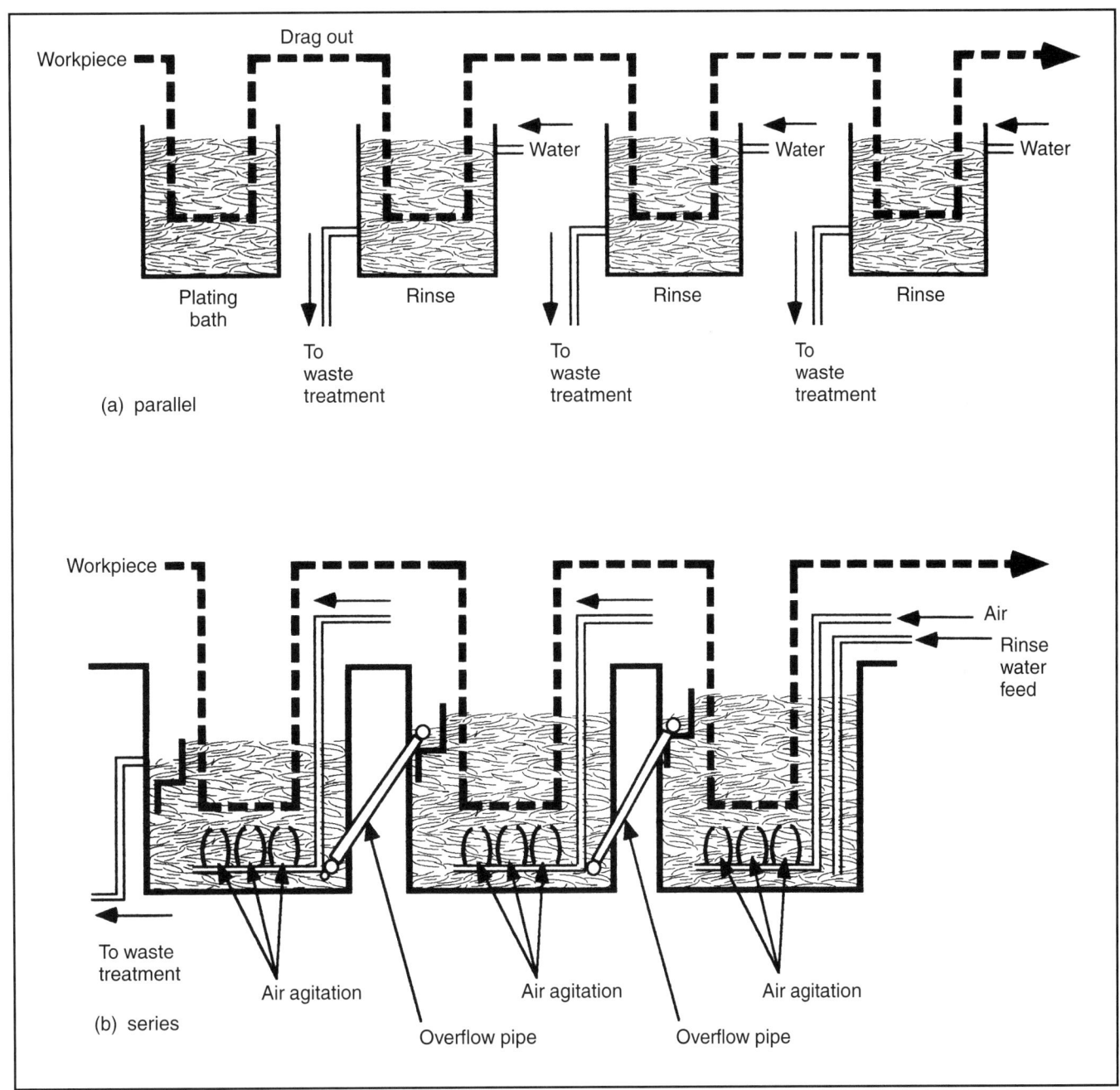

Fig. 9.6 *Three-stage rinse systems: (a) parallel and (b) series with outboard arrangement*

Fig. 9.7 Product recovery and reuse

Dry, mixed chemicals should be swept up for reuse rather than washed to the sewer.

Fig. 9.8   Product recovery and reuse

Spills and cleanup water from this bleach packaging plant are stored in the
large fiberglass tank and used as makeup for the next batch. Floor drains
have been sealed to prevent accidental loss of spilled product.

involved are valuable. Gold and other noble metals have been recovered using both ion exchange and evaporation techniques. Evaporation (Figure 9.9) has many potential applications when chemicals are simply diluted and are not mixed with other process streams. Other technologies such as *REVERSE OSMOSIS*,[2] *ELECTRODIALYSIS*[3] and *ULTRAFILTRATION*[4] are being used to an increasing extent as wastewater treatment costs rise and the various technologies are developed and refined specifically for chemical recovery applications. Another developing technology is a crystallization process for caustic recovery from aluminum finishing and chemical milling operations.

9.213   Wastewater Segregation/Treatment

Whether process wastes are intended for recovery and recycle (Figure 9.10) or for treatment, segregation (Figure 9.11) of wastestreams is often essential. The presence of a minor impurity in a waste can make recycling totally impractical. In wastewater treatment, hexavalent chromium and cyanide are most economically treated at the source, before they are mixed with other wastes. Wastes containing floating oil and grease are normally most effectively treated before mixing with other wastestreams and before other treatment operations such as neutralization and precipitation.

9.214   Waste Exchange/Management

A material might qualify as a waste in one industry and not in another. For example, an electroplater of artificial human heart valves might find a plating solution no longer within the firm's exacting tolerances and, therefore, consider it a waste. However, the same solution could very well be useful to a plater of baby shoes used as paperweights.

In some states, private organizations or state agencies act as clearinghouses for waste exchange transactions. Waste exchange organizations will also be familiar with local recycling and processing centers and can refer generators to one. Within large industrial operations it may even be possible to use an existing process to reprocess waste into a usable form.

9.215   Waste Minimization Summary

All of the techniques discussed in the preceding sections are waste minimization methods which represent true source control prior to industrial wastewater treatment. A comprehensive waste minimization effort should include a consideration of all the potential methods discussed in this section before proceeding to the design of a wastewater treatment system. Another important aspect of waste minimization is an effective personnel training program. If operators realize the importance of waste minimization and know how to minimize waste, significant reductions can be achieved in many situations.

Alison Gemmell and Philip Lo contributed to the revision of this section.

QUESTIONS

Write your answers in a notebook and then compare your answers with those on page 665.

9.2A   What is source control?

9.2B   List several important pollution source control techniques.

9.2C   What changes in operating guidelines could be considered by industry to reduce wastewater discharges?

9.2D   How can drag out from rinsing processes be reduced?

9.2E   Where are the most obvious sources of pollution in the metal finishing industry?

9.3   OPTIONS FOR TREATMENT

The treatment system must be designed to correspond to the plant's wastewater flow rates and composition and to meet the discharge or final effluent water quality requirements. For many wastes, preliminary treatability studies should be performed to establish design criteria and treatment performance. Prior to performing the treatability study, the engineer must evaluate the treatment options, select the appropriate treatment processes, and finally incorporate these processes into a flow pattern designed to achieve the treatment goal. A design engineer may provide a flexible design in anticipation of potentially more stringent future requirements.

9.30   Wastewater Treatment Technologies

The primary wastewater treatment technologies used by industry are described and discussed in the following sections. The technologies are grouped by type of treatment into five categories: physical treatment, chemical treatment, biological treatment, land treatment and thermal treatment.

---

[2] *Reverse Osmosis (oz-MOE-sis).   The application of pressure to a concentrated solution which causes the passage of a liquid from the concentrated solution to a weaker solution across a semipermeable membrane. The membrane allows the passage of the water (solvent) but not the dissolved solids (solutes). In the reverse osmosis process two liquids are produced: (1) the reject (containing high concentrations of dissolved solids), and (2) the permeate (containing low concentrations). The "clean" water is not always considered to be demineralized.*
[3] *Electrodialysis.   The selective separation of dissolved solids on the basis of electrical charge, by diffusion through a semipermeable membrane across which an electrical potential is imposed.*
[4] *Ultrafiltration.   A membrane filter process used for the removal of some organic compounds in an aqueous (watery) solution.*

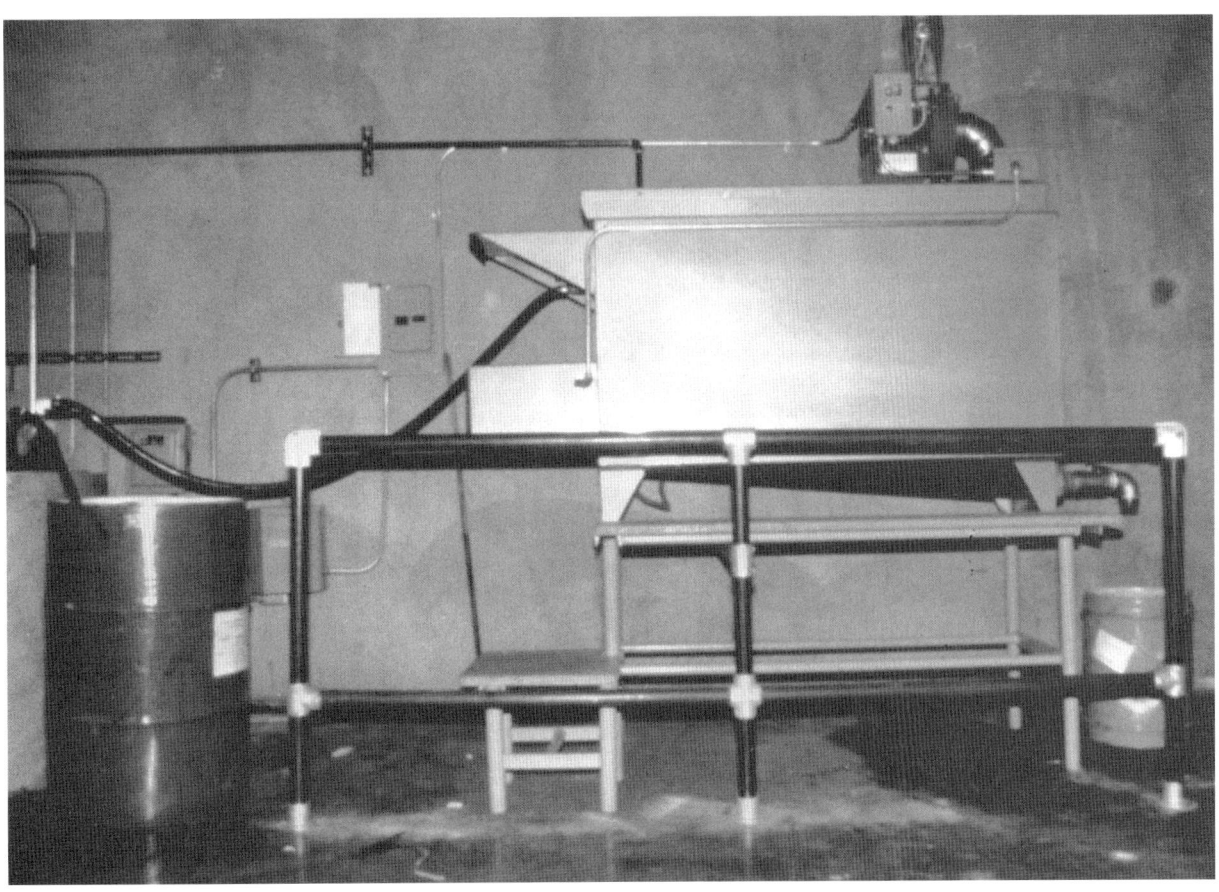

Fig. 9.9   Liquid waste evaporator

Concentrating a dilute waste on site can sometimes yield a reusable material
or greatly reduce the waste volume to be hauled off site.

Fig. 9.10   Process waste recycling

Food processing waste can often be squeegeed into containers for reuse as
animal feed rather than being water-washed to the sewer.

Fig. 9.11 Waste segregation

Parts cleaning solvents should be segregated for reclamation rather than
diluted or blended with other wastes.

9.300 Physical Treatment

Physical treatment technologies include processes which separate components of a wastestream or change the physical form of the wastewater without altering the chemical structure of the constituent materials. These processes are useful for: (1) separating hazardous materials from an otherwise nonhazardous wastestream so that they may be treated in a more concentrated form; (2) separating certain components for different treatment processes; (3) preparing a wastestream for further treatment by other processes; and (4) removing prohibited solids (such as large solids) from the wastestream. The most commonly used physical processes for wastewater treatment are equalization, screening, sedimentation, flotation, filtration, solar evaporation, evaporation, distillation, adsorption, and stripping. These processes are described and their applications are discussed below.

Equalization. In contrast to most of the other physical treatment processes, equalization is a process of combining, rather than separating, wastestreams to dampen fluctuations in flow rates and composition prior to discharge or further treatment by physical, chemical or biological methods. Equalization is normally accomplished in sumps, holding basins, ponds or tanks which act as a "wide spot in the line" to minimize hour-to-hour variations and to prevent shock loads from upsetting downstream processes. Equalization basins or tanks are usually mixed to enhance equalization and to prevent solids from settling out. Blended liquids may flow by gravity from the equalization basin or they may be pumped at a controlled flow rate, equal to the average incoming flow rate.

Many industrial wastewater discharges originate in washdowns, tank cleanings, batch operations and spills. Equalization not only dampens these flow variations, but also often permits the opportunity to neutralize wastewaters by mixing acidic and basic streams which might otherwise be discharged from the plant at different times. Concentrated wastestreams can be collected in temporary holding tanks. Equalization is then achieved by slowly blending the concentrated stream with more dilute wastewater.

Equalization basins may have to be quite large to be effective, depending on the flow rates and concentrations involved. Space and cost considerations may sometimes limit the degree to which equalization can be used by industry. Despite these factors, equalization is an often-overlooked technology which has the capability to increase the effectiveness of downstream treatment equipment and systems.

Screening. Screening is widely used to remove large particles from wastewater. It is primarily used as a pretreatment process before other solid/liquid separation processes or to protect equipment downstream from problems created by large solid particles. The three types of screens commonly used in wastewater treatment are rotating drum, vibrating, and stationary. The rotating drum screen is a revolving cylinder covered with the screen media, through which the liquid passes. The solids are scraped from the screen as it rotates so that a clean surface is always present for the incoming liquid. The vibrating screen consists of a flat screen tilted so that the solids fall off one side as the unit vibrates. The stationary screen is simply a fixed screen tilted so that the solids fall off as the liquid passes through. It has the advantage of no moving parts but is harder to keep clean than the other two types. Figures 9.12 and 9.13 depict a stationary bar screen with an automatic cleaning mechanism and a rotating drum microstrainer.

Sedimentation. Sedimentation is a process for removing suspended solids from wastestreams. This is usually accomplished by providing sufficient time and space in tanks or holding ponds for gravitational settling to occur. Continuous flow sedimentation vessels can either be circular or rectangular. They often contain a series of baffles and weirs to direct the water flow, to separate the liquids from the solids, and to collect and thicken the settled solids (sludge). Packaged inclined plate (lamella) settlers are often used by industry because the stack of parallel plates provides the same effective surface area for settling in a much smaller vessel than a traditional circular tank clarifier. Figures 9.14 and 9.15 depict typical circular and parallel plate clarifiers. With both types of clarifiers, the equipment normally contains a flocculation section near the inlet with slow mixing to minimize turbulence and enhance the formation of larger solid particles which settle faster. A means of thickening and mechanically removing the settled sludge is also often included in the clarifier.

Clarifiers are normally designed on the basis of overflow or surface loading rate, defined as liquid flow rate divided by tank surface area. Other design considerations are solids loading rate and retention time. Retention times are typically one to four hours, while overflow rates generally vary from 400 to 1,500 gal/day/sq ft. Low overflow rates increase the removal efficiency of suspended solids, but larger tanks are required. The parallel plate clarifiers and sedimentation tanks attempt to solve this problem by providing a large number of channels where turbulence and water flow rates are low and the depth particles have to settle to reach a solid surface is also low.

Sludge that settles on the bottom of clarifiers is typically only one to seven percent solids, so it is often processed further to increase solids content and thereby reduce the volume

Fig. 9.12   Bar screen with automatic cleaning mechanism
(Courtesy of Envirex, A Rexnord Company)

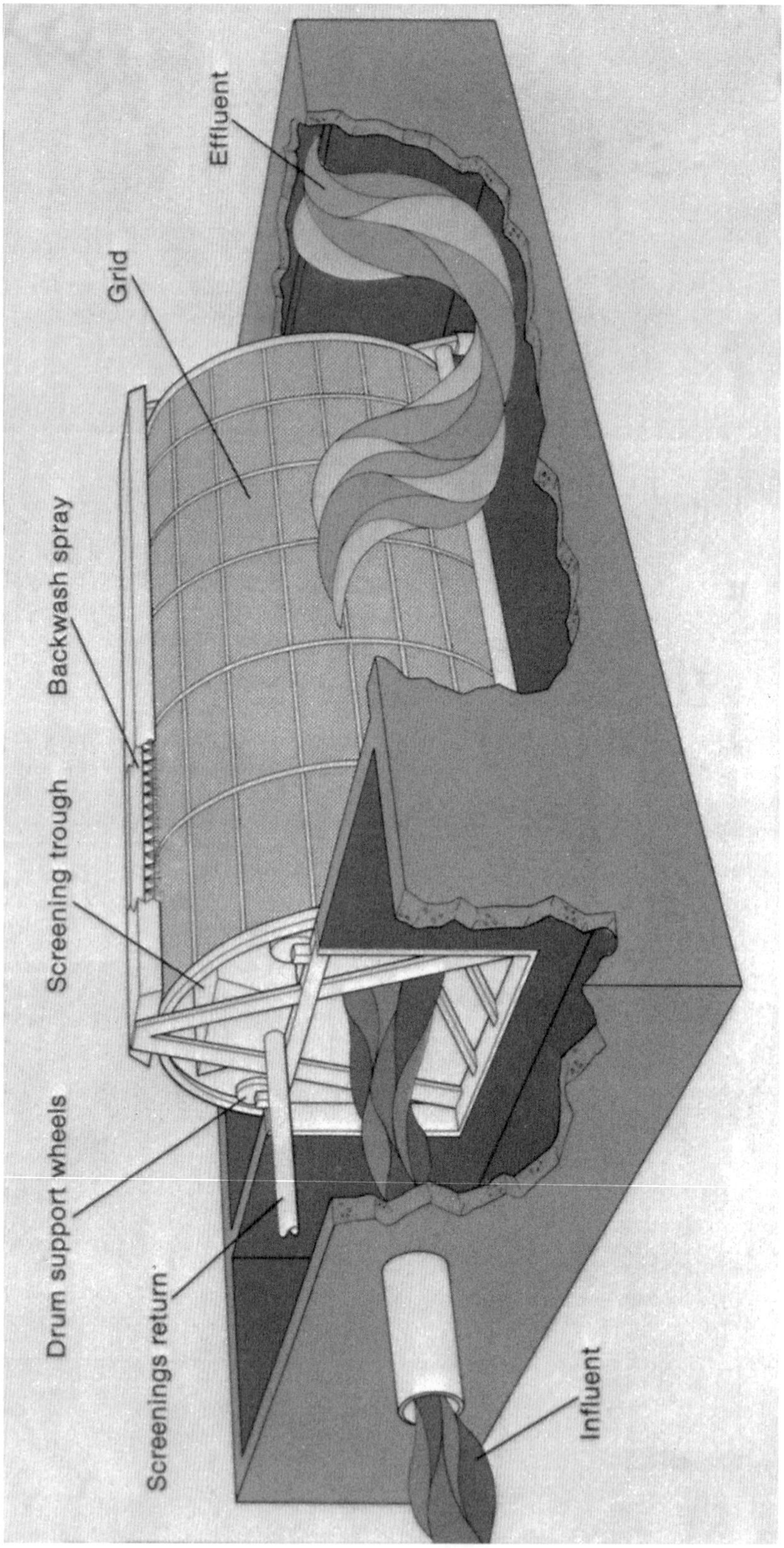

Fig. 9.13 Microstrainer used for removal of fine suspended solids from wastewater
(Courtesy of Envirex, A Rexnord Company)

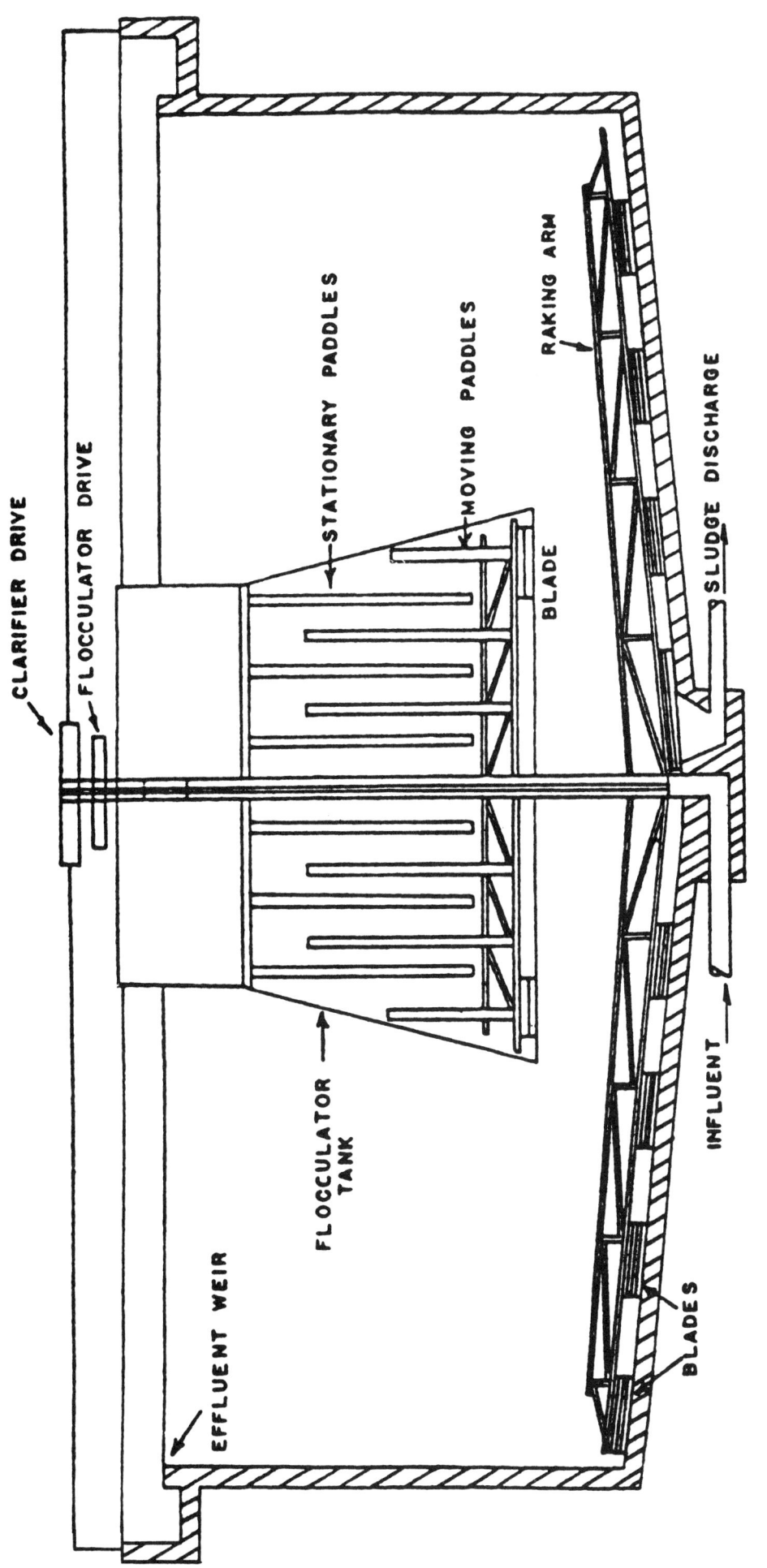

Fig. 9.14   Typical circular clarifier

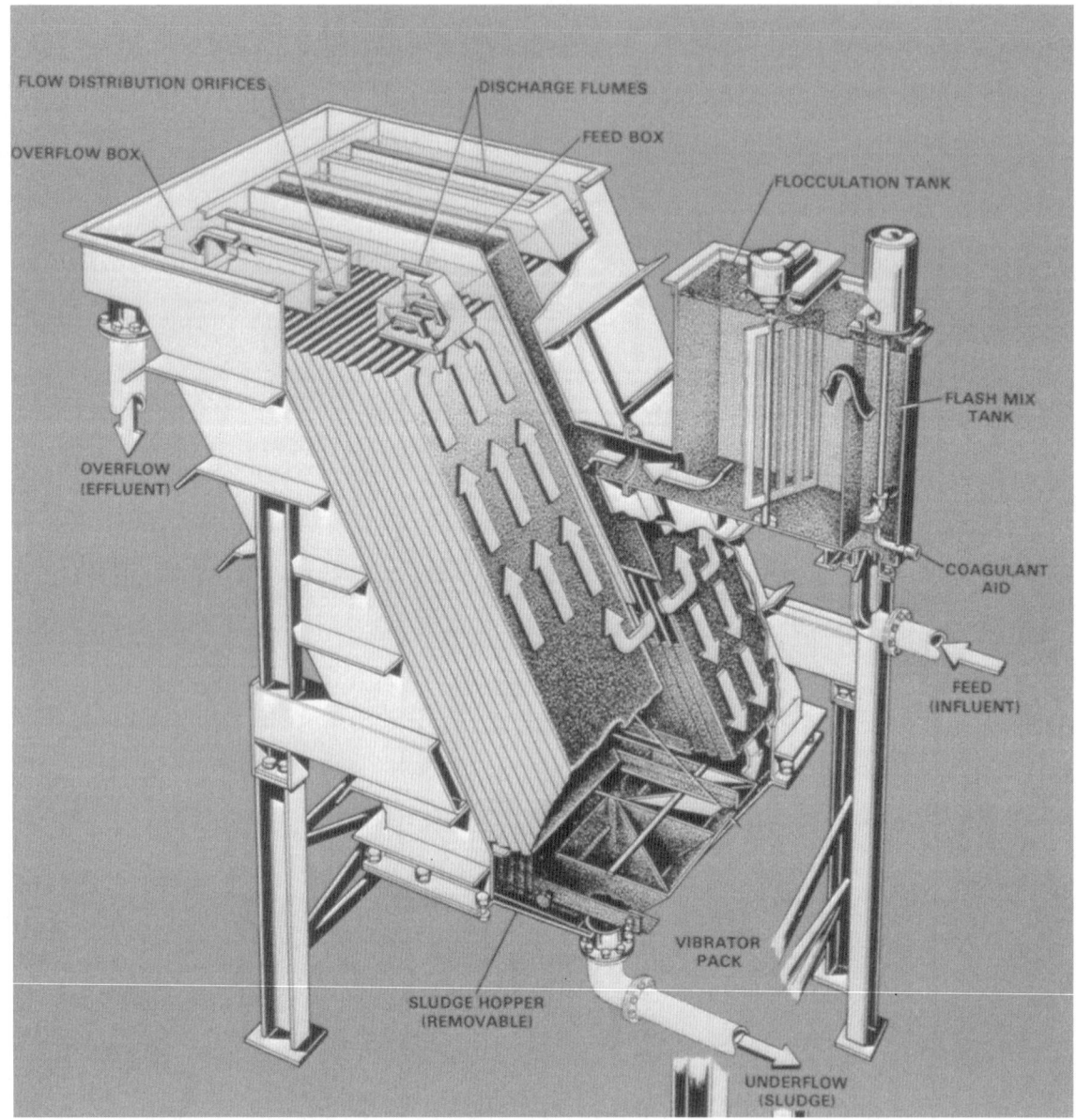

Fig. 9.15   Inclined plate (lamella) clarifier
(Courtesy of Parkson Corp.)

and cost of disposal of this material. Techniques for sludge de-watering are discussed later in this chapter.

Various chemical coagulating or flocculating agents are often added to wastewater just ahead of the clarifier to agglomerate (combine) the solid particles into larger particles that settle faster. Traditionally, lime, alum (aluminum sulfate), and ferrous or ferric chloride have been used as coagulants. More recently, a wide range of organic *POLYELECTROLYTES*[5] have been used to promote particle growth. The polyelectrolytes are classified as cationic, anionic or nonionic *POLYMERS*[6] and are generally used at dosages of 0.25 to 5 mg/L. The low dosages help minimize the increase in sludge volume caused by the addition of the coagulant itself.

Like all wastewater treatment processes, effective sedimentation requires regular attention of operating personnel to the sedimentation tanks or clarifiers. Collected solids must be regularly removed from the sedimentation tank and chemical feed solutions and feed rates must be maintained for optimum operation.

Another device used for sedimentation is the hydrocyclone, shown in Figure 9.16. Hydrocyclones are most effective for removal of large, heavy solids such as sand. They can potentially be used for recovery of valuable heavy materials such as gold, silver or mercury. Hydrocyclones are also used to remove solids from groundwater prior to process or domestic use and to continuously remove solids from recirculated cooling tower water.

The sedimentation process is widely used in the metal finishing, electroplating and printed circuit board industries to remove the solids generated by precipitation of heavy metals from wastestreams. In these industries the sedimentation process is normally used *AFTER* chemical treatment. Other descriptions in this section deal with using sedimentation as an initial physical process *PRIOR* to chemical treatment.

Flotation. Flotation is a process for removing solids or oil and grease by floating the constituents to the surface with or without the addition of air bubbles. Flotation is particularly useful for removing lighter particles not readily removed by sedimentation. For materials such as oil and grease that are less dense than water and which readily separate from waste-streams, standard separation equipment such as the API separator shown in Figure 9.17 is commonly used. Without the addition of air bubbles, both lighter-than-water materials (oil) and heavy particles (sludge) can be collected and removed by this process, so the technologies of sedimentation and flotation are combined in one vessel.

A slightly different piece of flotation equipment is the corrugated plate interceptor shown in Figure 9.18. This vessel accomplishes oil skimming and solids settling using the same parallel plate technology that is used in lamella settlers.

Dissolved air flotation (DAF) and induced air flotation (IAF) are two specific flotation processes. Both involve the addition of air and are often used when gravity separation alone is not sufficient. The air bubbles are formed by pressurizing the incoming wastewater to 30 to 70 psi in the presence of air and then releasing the air-water mixture into the flotation tank at atmospheric pressure. Figure 9.19 shows three ways the dissolved-air flotation process can be applied. As with sedimentation, flocculating agents may be used to improve the agglomerating of particles. Where oil/water emulsions are present, demulsifying agents are added to break the emulsion and allow separation to occur. The flotation process is widely used in the refining, meat packing, paint, poultry processing, paper milling and baking industries to achieve 80 to 99 percent removal of suspended and floating materials.

Filtration. Filtration is a process for separating liquids and solids using various types of filter media. The process goal of filtration may be dry solids or clarified liquid or both. Filters can be used to "polish" the effluent from the sedimentation or flotation process (clarified liquid) or to dewater sludges from the sedimentation process (dry solids). The filtering media can be sand, gravel or other particles, fabrics of woven fibers, felts and nonwoven fibers, porous or sintered (solidified) solids, or polymer membranes. Filters are also classified by direction of flow (up or down), the driving force (gravity, pressure or vacuum), the bed homogeneity (single, dual or multimedia), and the filtration depth (surface or deep bed).

Common types of filters include filter presses, pressure leaf filters, rotary drum filters, horizontal vacuum filters, cartridge filters, sand filters and bag filters. Figures 9.20, 9.21 and 9.22 show a granular media filter with automated backwash, a belt filter press and a plate filter press. Filters can be designed to operate in a batch or continuous mode and the cleaning (solids removal and backwashing) can be a manual or an automatic operation. Automatic backwashing is normally activated when a preset differential pressure across the filter is reached. There are many continuous cleaning sand filters in use by the metal finishing industry.

Many factors need to be considered in the selection and sizing of filtration equipment, including slurry character, liquid and solids throughput, performance requirements and permissible materials of construction. Important equipment-related factors are type of cycle (batch or continuous), driving force, desired separation sharpness, cake washing capability

---

[5] *Polyelectrolyte (POLY-ee-LECK-tro-lite). A high-molecular-weight substance that is formed by either a natural or synthetic process. Natural polyelectrolytes may be of biological origin or derived from starch products, cellulose derivatives, and alignates. Synthetic polyelectrolytes consist of simple substances that have been made into complex, high-molecular-weight substances. Often called a polymer.*

[6] *Polymer (POLY-mer). A long chain molecule formed by the union of many monomers (molecules of lower molecular weight). Polymers are used with other chemical coagulants to aid in binding small suspended particles to larger chemical flocs for their removal from water.*

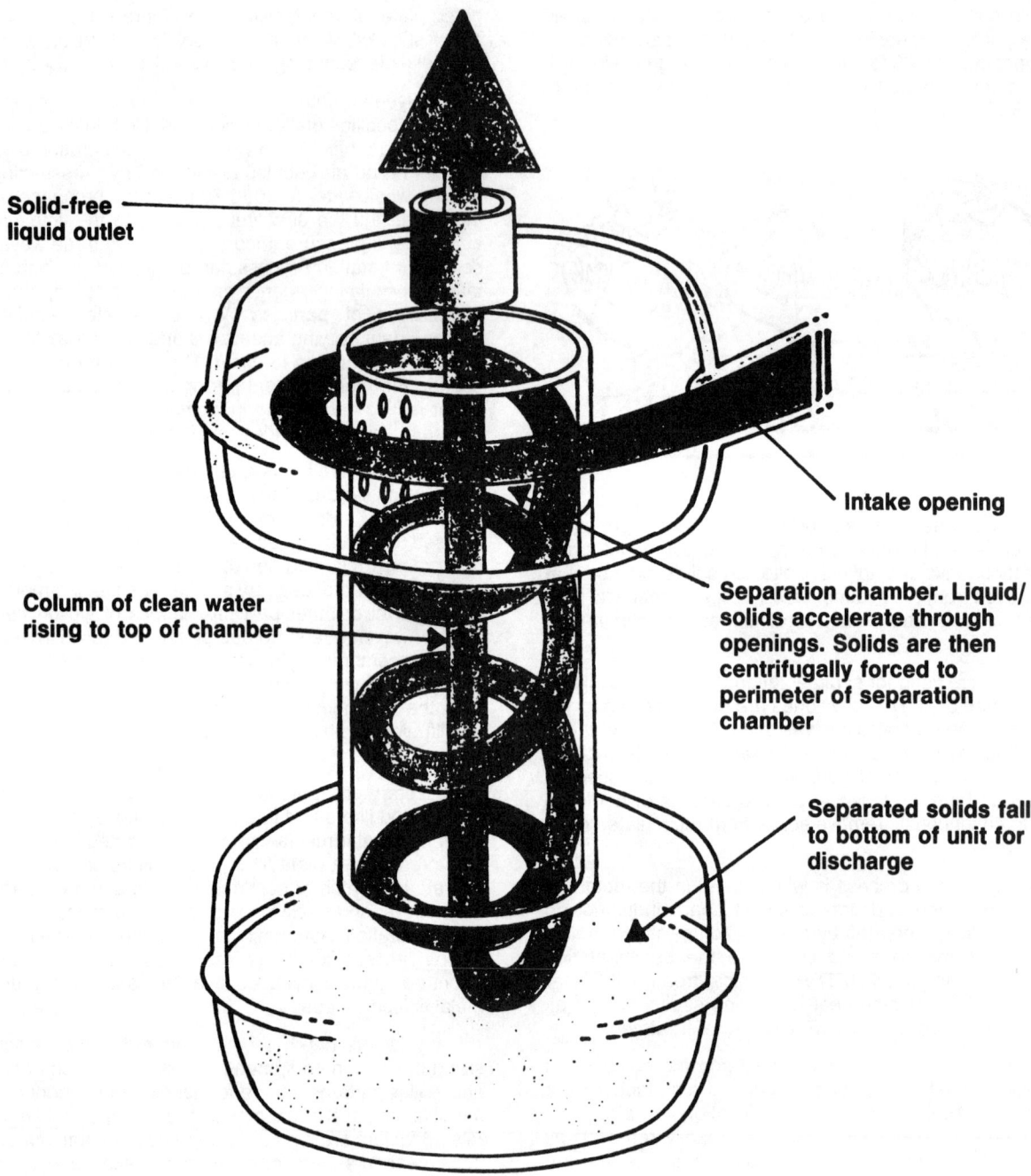

Solid-free liquid outlet

Intake opening

Column of clean water rising to top of chamber

Separation chamber. Liquid/solids accelerate through openings. Solids are then centrifugally forced to perimeter of separation chamber

Separated solids fall to bottom of unit for discharge

Fig. 9.16   Hydrocyclone
(Courtesy Claude Laval Corporation)

Fig. 9.17   API separator
(Courtesy of Envirex, a Rexnord Company)

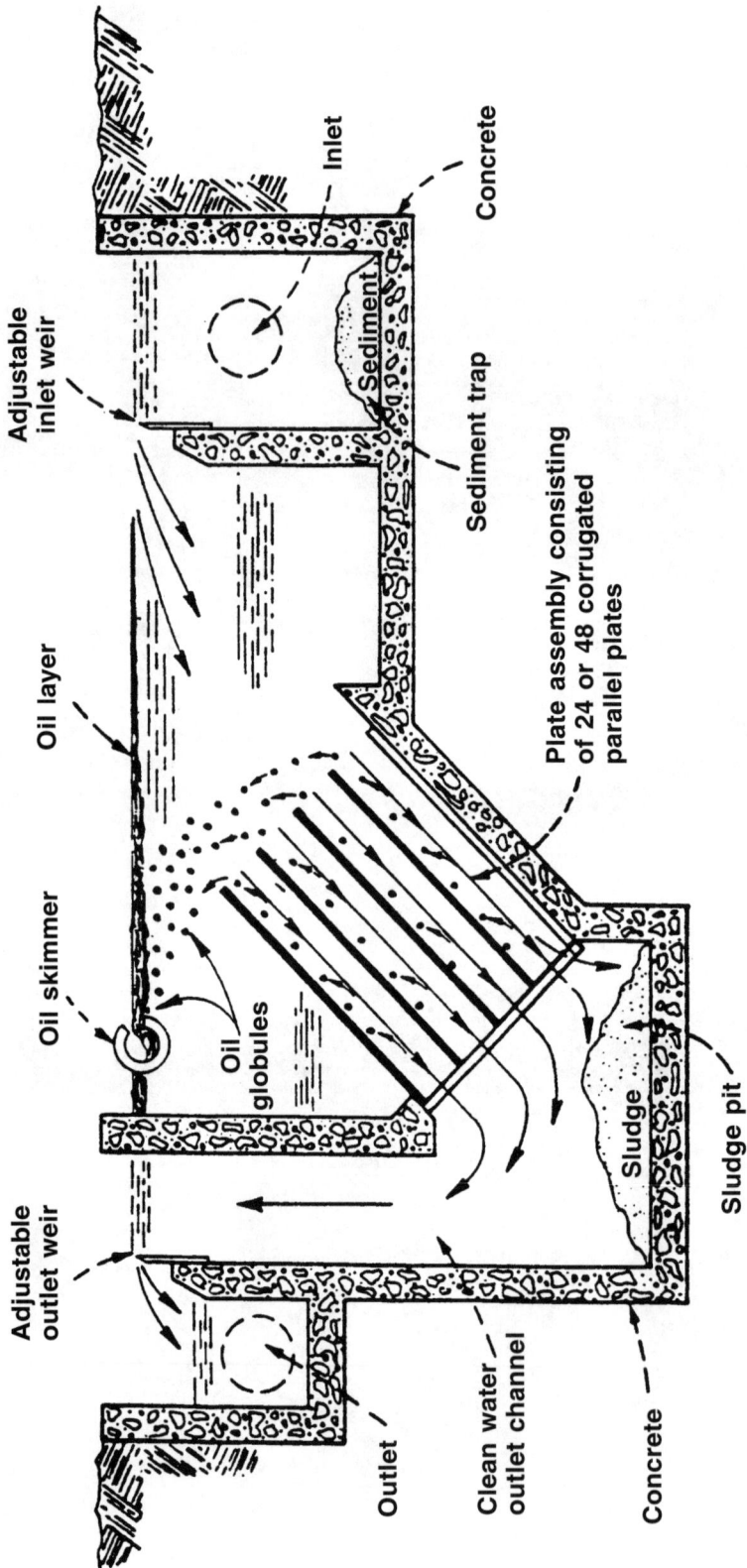

Fig. 9.18 Corrugated plate interceptor

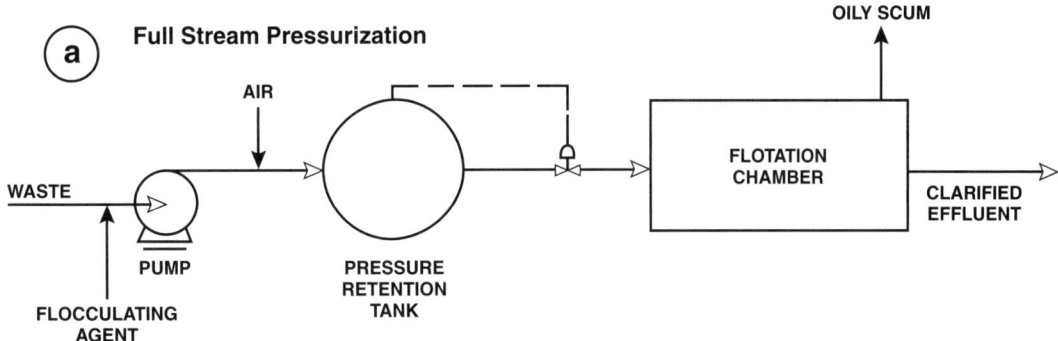

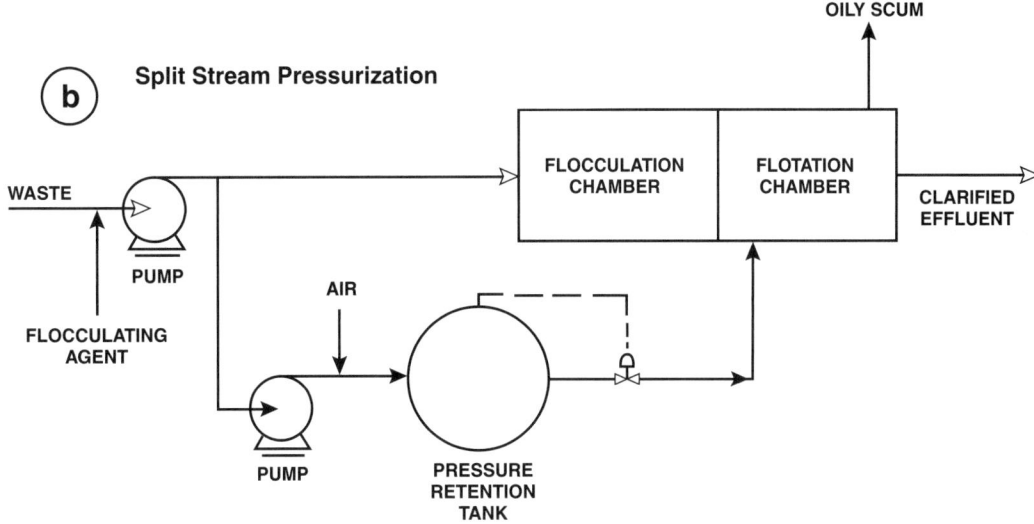

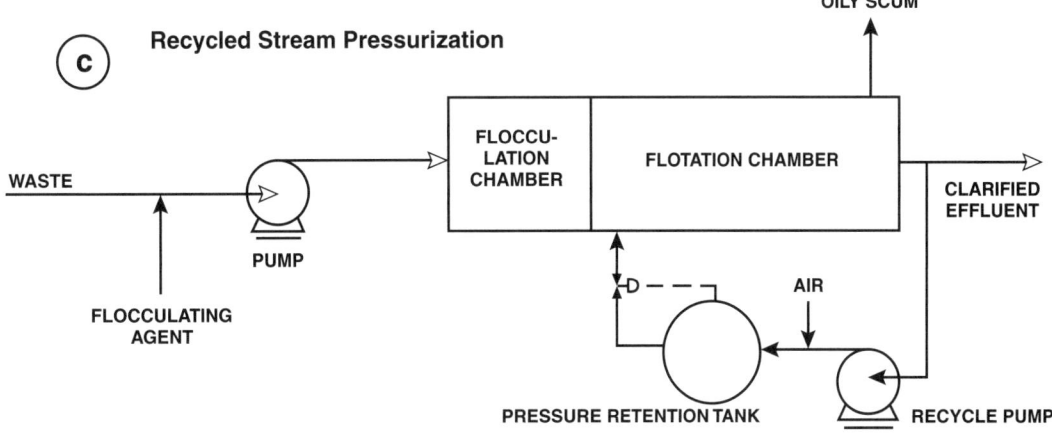

Fig. 9.19   Pressurized dissolved-air flotation process flow diagrams

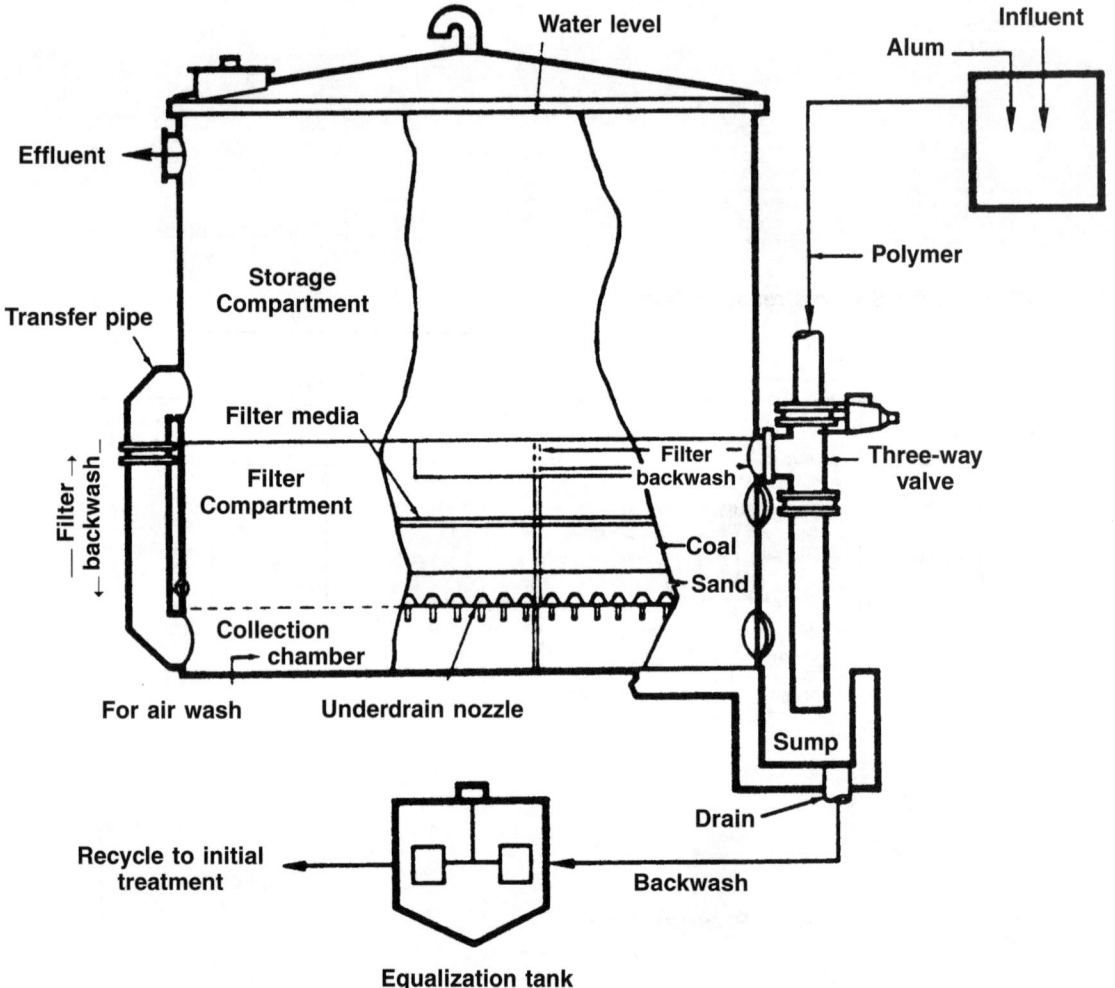

Fig. 9.20   Typical automatic, granular media filter system

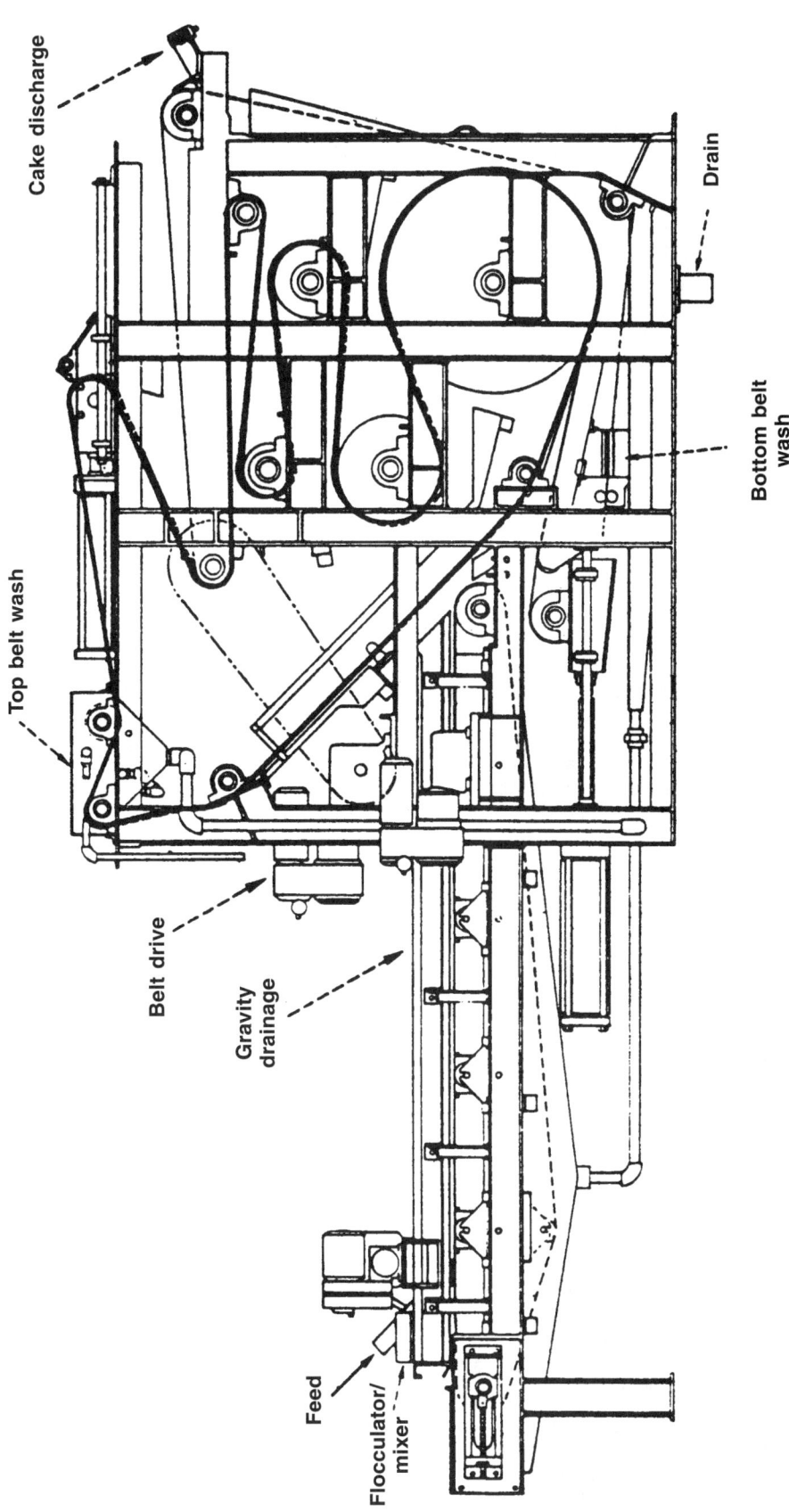

Cake discharge

Drain

Top belt wash

Bottom belt wash

Belt drive

Gravity drainage

Feed

Flocculator/ mixer

Fig. 9.21 Belt filter press

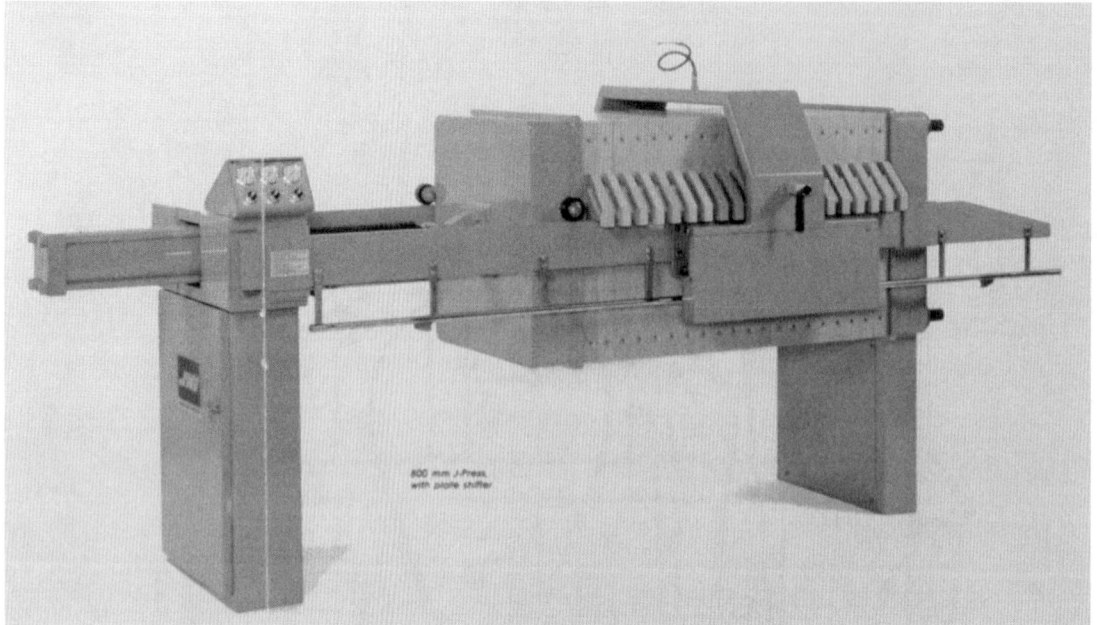

Fig. 9.22   Plate and frame filter press
(Courtesy of JWI, Inc.)

(to remove impurities from the filter cake when it is a usable material) and cost. Cost considerations should include not only installed cost, but also equipment life and costs associated with operating labor, backwash water, energy, maintenance and replacement filter media. The final selection of a filter type and media often involves a trade-off between initial capital costs and continuing operating and maintenance costs.

Solar Evaporation. Solar evaporation is a process in which wastewaters are reduced in volume through simple evaporation in uncovered ponds or surface impoundments. This treatment method is coming under increased scrutiny due to the potential for both the emission of toxic compounds to the air and discharge to land and groundwater of toxic priority pollutants. Consequently, this treatment method may have a much higher cost than in past years due to design, construction and monitoring requirements imposed by federal and state regulations. Considerable land is required for this process and the ideal climate for evaporation (hot, dry and windy) may not always be present, especially on a year-round basis. Despite these factors, it may still be possible to evaporate many aqueous wastes to dry solids at a reasonable cost using this method. The residuals remaining after evaporation may have to be treated further or disposed of directly in a properly permitted land disposal facility.

Evaporation. Non-solar evaporation is accomplished by heating a waste solution or slurry, usually with steam condensing on metal tubes inside a vessel. The waste material to be evaporated flows inside the tubes, and often a vacuum is applied to reduce the boiling point of the liquid.

Evaporation is a proven, well-developed, although expensive, industrial process. It is most widely used for separating water from inorganic solutions and slurries, but it also can be used for concentrating sludges containing organic solvents. Evaporation is currently used in the electroplating, paper, and fermentation industries to concentrate wastewater solutions

for recovery of valuable constituents. It can also be used to remove water from sludge to produce incinerable (burnable) solids and to reduce slurry volumes prior to land disposal. In the electroplating industry, evaporation can be used to concentrate and recycle rinse waters containing toxic metal cyanides or chromium salts, thus eliminating a potential disposal problem. Figure 9.23 shows a simple evaporator designed for such a use.

Distillation. Distillation is a process for separating organic liquids with different boiling points. The mixed liquid stream is exposed to increasing amounts of heat and the various components of the mixture are vaporized and then condensed and recovered. Although distillation is often used in oil refineries and chemical plant production processes, small and

Fig. 9.23   Compact evaporator designed for concentration and recovery of plating solutions from rinse water
(Courtesy of Industrial Filter and Pump Manufacturing Company)

simple batch stills are available to purify and recover a variety of organic compounds. Figure 9.24 shows a typical batch still recovery system. Packaged stills are available in a variety of sizes from several manufacturers. Halogenated hydrocarbons and other organic solvents are currently recovered from degreasing and dry cleaning operations using simple distillation. Commercial solvent recyclers use stills to recover volatile organic solvents. Vacuum distillation often allows the most cost-effective separation of volatile organic compounds from largely nonvolatile residues such as paint residue or still bottoms.

Adsorption. Adsorption is a separation process which is often used for removing low concentrations of organic contaminants (solvents, pesticides, PCBs and phenols) and some inorganic contaminants (cyanide and chromium) from wastewaters. The most common adsorbent is activated carbon, although synthetic resins and activated alumina have been used in some specific cases.

In activated carbon treatment, wastewater is typically passed through a tank or vessel filled with carbon granules. Because of its extensive pore structure, activated carbon has a large surface area per unit of mass. The organic contaminants are removed by adsorbing (clinging) to the surface of the carbon granules. When the carbon's capacity to adsorb specific components is exhausted, the carbon can be replaced or regenerated with heat or a suitable extraction solvent.

The main applications of this process are for the removal of volatile and nonvolatile organic compounds at dilute concentrations in wastestreams. Adsorption is especially applicable to the treatment of nonbiodegradable organics.

Activated carbon is available in powder and granular forms. Powdered activated carbon is usually fed into the process as a water slurry and removed downstream by sedimentation or filtration. Granular activated carbon is used as a packed media in flow-through columns or beds. Due to the relatively high cost of carbon and since spent carbon containing a variety of organic materials may be considered a hazardous waste in some states, regeneration processes to recover the organic materials and reuse the carbon may be economically justified in some cases. Figures 9.25 and 9.26 show a typical packaged carbon adsorption system and a typical arrangement for steam regeneration of the carbon and recovery of the removed organics. The water-cooled condensers shown in Figure 9.26 are used to recover and purify the solvent which has been steam-desorbed from activated carbon; the heating source (steam or electric reboiler) is associated with the solvent recovery still.

Stripping. Stripping is a process in which volatile components of an aqueous-organic mixture are removed by contact with the stripping fluid. Stripping is usually done in a countercurrent flow column containing packing materials or perforated plates. Steam stripping is most effectively used with solutions containing between one and ten percent of organics. Air stripping has been used for treatment of wastewater and contaminated groundwater. Figure 9.27 shows an air stripping column at a municipal landfill and Figure 9.28 is the conceptual diagram of an air stripper designed to remove volatile organics from groundwater. Highly volatile compounds with very low solubility in water are the most easily stripped.

In air stripping, air pollution control equipment may be required to remove the organics from the exhaust air stream. With steam stripping, the organics are usually condensed and recovered, and may be recycled after further purification steps.

Other Physical Processes. Other physical processes that have been or could be used to pretreat industrial wastewater include centrifugation for solid-liquid separation (water clarification or sludge dewatering), dialysis, electrodialysis, reverse osmosis, ultrafiltration, solvent extraction, and electrolytic recovery.

DIALYSIS[7] is a membrane separation process which uses electrolysis as the primary driving force to transport components from one side of the membrane to the other. The rayon industry uses dialysis to recover caustic soda, and the electrolytic copper refining industry uses it as a means of separating soluble impurities from the process stream. Dialysis has been used in industrial processes for over fifty years, and although caustics, acids and cyanides can be recovered from aqueous wastestreams by dialysis, actual applications in wastewater pretreatment are few.

Electrodialysis is a membrane separation process which relies upon electric forces as the primary driving force for transport and uses membranes which selectively allow passage of ionic components of a wastestream. Electrodialysis has been used successfully on acidic streams containing a single principal metal ion, such as in acid nickel-plating baths. It has also been used in the desalting of food products such as whey. Pilot work has been done on the demineralization of wastewater treatment plant effluent, acid mine drainage treatment, and the treatment of plating wastes and rinse waters. Electrodialysis has also been used to treat hydrogen fluoride and ammonium fluoride effluents from glass and quartz etching facilities.

Reverse osmosis (RO) is used for the removal or recovery of dissolved organic and inorganic materials; the process is applicable to the removal of soluble metals and most total organic carbon (*TOC*[8]) components in wastestreams. Reverse osmosis is used extensively in the production of high-purity water and other demineralization applications. Reverse osmosis separates dissolved materials in solution by pressure filtration through a semipermeable membrane at a pressure great-

---

[7] *Dialysis (die-AL-uh-sis). The selective separation of dissolved or colloidal solids on the basis of molecular size by diffusion through a semipermeable membrane.*

[8] *TOC (pronounce as separate letters).* **T**otal **O**rganic **C**arbon. *TOC measures the amount of organic carbon in water.*

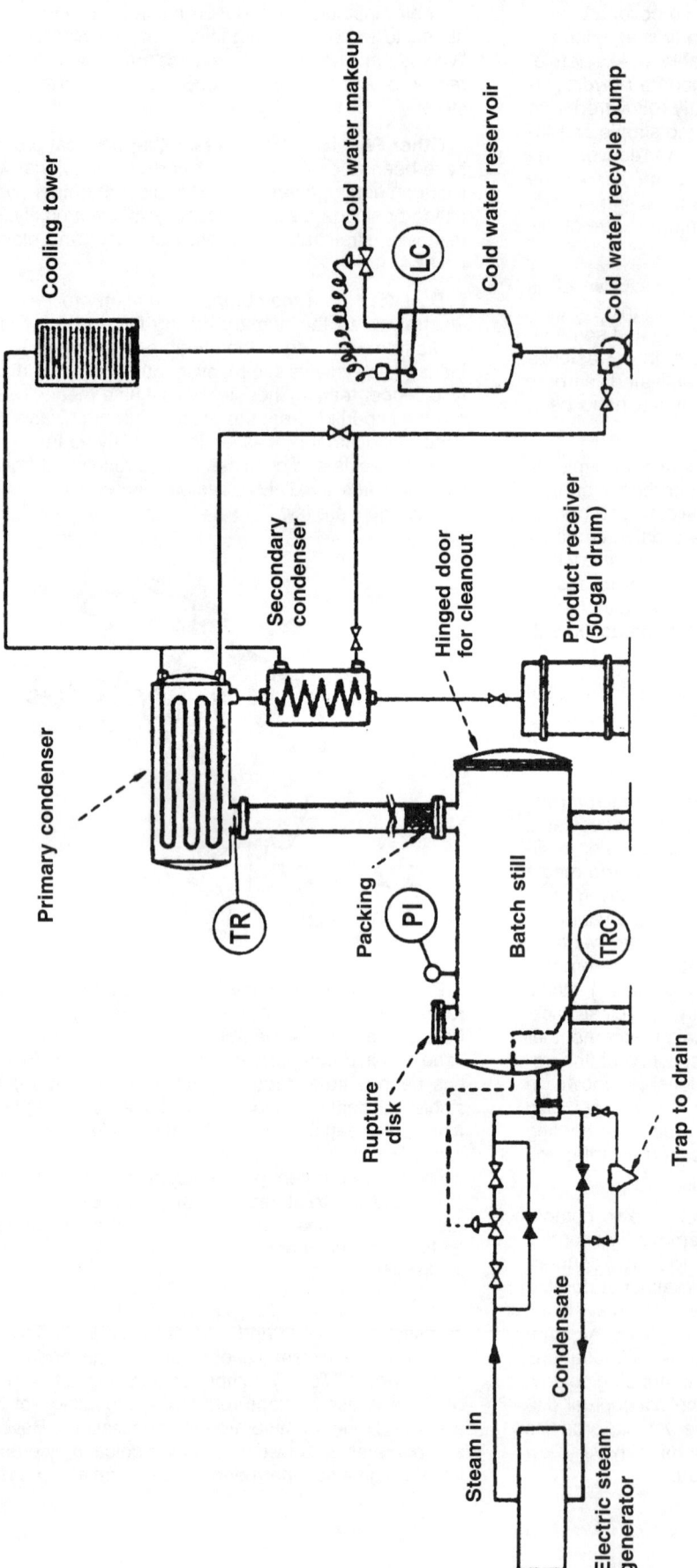

Fig. 9.24   Batch still system for recovery of waste solvents

TR — Temperature Recorder
PI — Pressure Indicator
LC — Level Control
TRC — Temperature Controller

Fig. 9.25 A packaged carbon adsorption system
(Courtesy of Calgon Carbon Corporation)

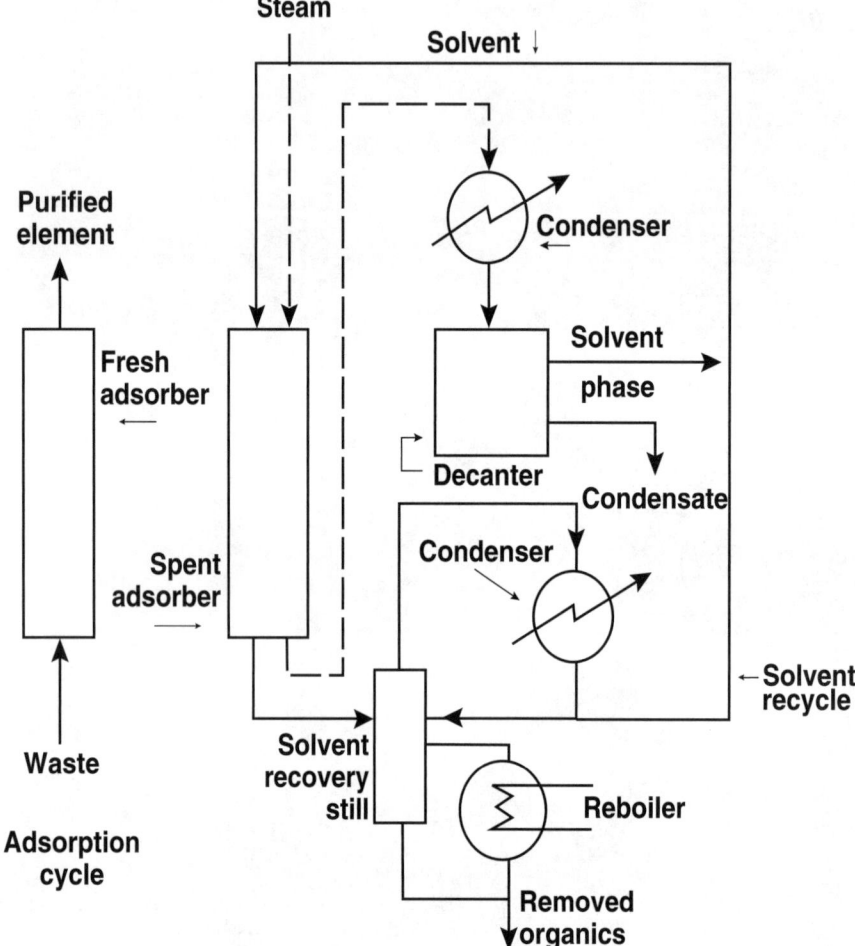

Fig. 9.26   Solvent regeneration process for activated carbon

Fig. 9.27 Air stripping column at municipal landfill

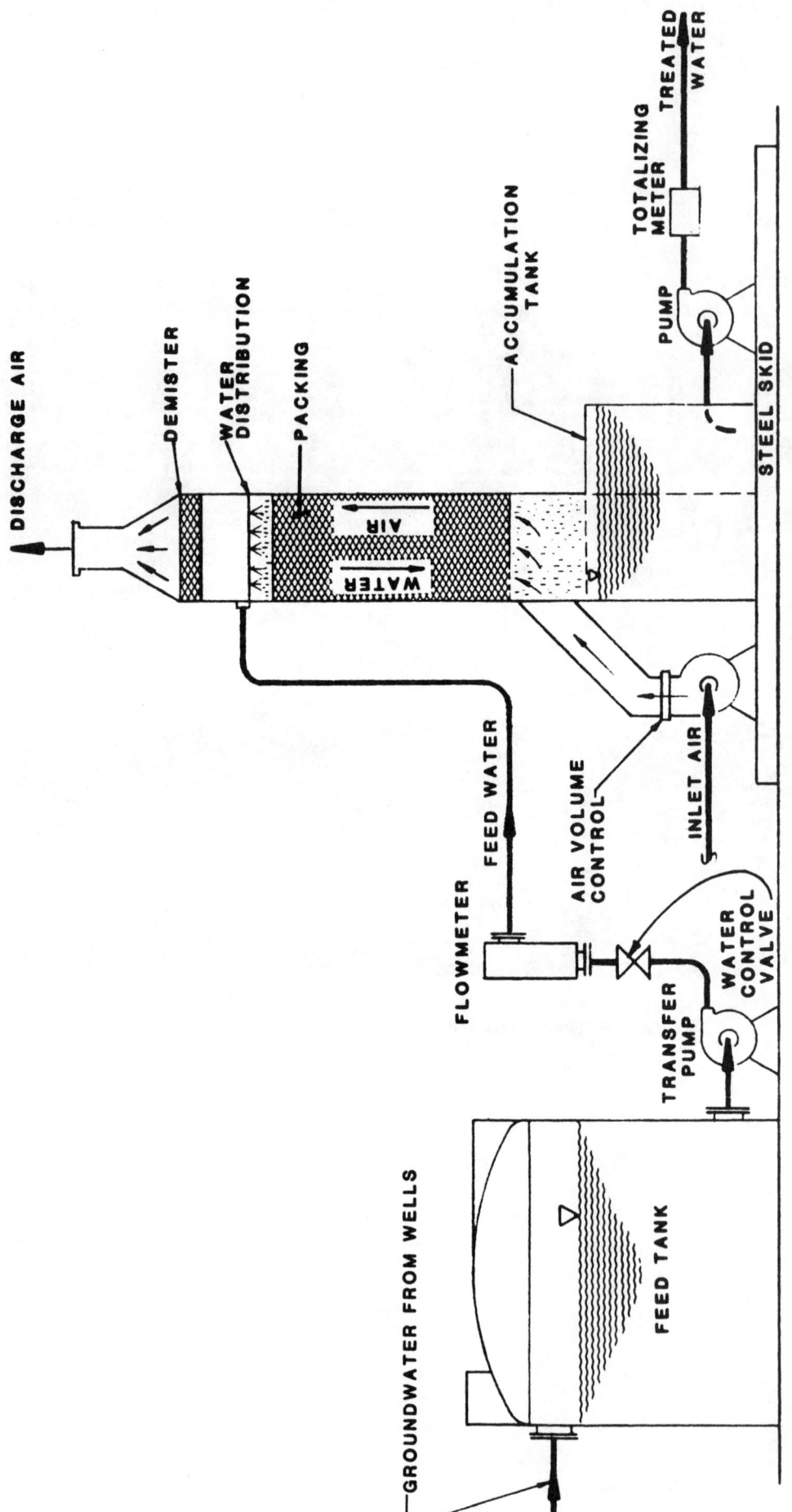

Fig. 9.28   Conceptual diagram of an air stripping groundwater treatment system
(Permission of Kennedy/Jenks/Chilton)

er than the natural osmotic pressure caused by the dissolved materials in the wastewater. Operating pressures generally vary from 200 to 1,500 psi. Commonly used membrane materials are nylon, cellulose acetate and polybenzimida-zolone (PBIL). The membrane packages can be configured as plate and frame, tubular, spiral wound and hollow fiber. RO is used in the adhesives and sealant industries to remove BOD and COD; in the electroplating industry to remove TOC, copper cyanide, zinc and nickel; in the synthetic rubber industry to remove BOD, COD, TOC, and oil and grease; and in the textile industry to remove TOC. Figure f shows an RO system installed for wastewater treatment and recycle.

Ultrafiltration is similar to reverse osmosis but allows low-molecular-weight components such as salts to pass through the membrane with the water phase. Higher-molecular-weight components, such as organic compounds, are not transported through the membrane and thus can be concentrated for recovery or disposal. Ultrafiltration systems operate at pressures ranging from 10 to 100 psi. In contrast to reverse osmosis, the membranes for ultrafiltration can be made from materials which will withstand high levels of organics and both low and high pH. Ultrafiltration is used to concentrate high-molecular-weight components of solutions for recovery or to produce a purified water stream for reuse or disposal. The main industrial applications of ultrafiltration have been treatment of paint wastes from electrocoat painting, as depicted in Figure 9.30, protein recovery from cheese and whey, the separation of metal cutting and machine oil from oil-water emulsions, and recovery of polyvinyl alcohol from textile sizing wastes.

SOLVENT EXTRACTION[9] is a process for removing organic substances from aqueous streams. In the process, a solvent that is immiscible (normally forms a separate phase) with the liquid wastestream is added to the stream where it combines with the organics. The subsequent extraction of the solvent/organics mix is often carried out in a series of mixing and settling operations. Another physical treatment process such as stripping, distillation or adsorption is then used to separate the organics from the solvent. Solvent extraction is most often encountered as a treatment method when the wastestream contains valuable organics. Two applications are the removal of up to five percent phenol from coke indus-

try wastes and the removal and recovery of toxic dyes. It is most useful for removing organics which cannot be decomposed biologically.

ELECTROLYTIC RECOVERY[10] (also known as cementation) is a process in which there is an electrochemical reduction of metal ions at the cathode where these ions are reduced to elemental metal (Figure 9.31, page 646). It is used primarily to remove metal ions from solution. Electrolytic recovery can be used to recover copper, tin, silver and other metals from metal plating, etching and printed circuit board operations.

The mining industry has been using electrolytic methods to produce metal from ores for many years and it is now increasingly being used to recover metals from wastestreams. There are two main types of electrolytic recovery systems for metals: high surface area (HSA) systems and electrowinning. In HSA systems the cathode is made from material with very high surface area such as stainless steel wool or carbon and the metal is plated onto it. HSA works on dilute solutions because of the large cathode area. A disadvantage is that metal cannot be recovered directly from the cathode and must be disposed of or stripped chemically or electrochemically to form a concentrated solution.

In electrowinning a concentrated solution is plated onto a metal cathode in a method very similar to conventional plating. The metal can then be stripped off the cathode and sold as scrap. Because a concentrated solution is required this can only be used on plating bath dumps or on dilute solutions that have been concentrated. Methods for concentrating solutions include HSA, ion exchange, and evaporation.

Ion transfer is another form of electrolytic recovery, using a porous partition through which ions can migrate freely in accordance with their electrical charge and applied potential. This process has been used to recover chrome and copper from plating bath rinse tanks.

QUESTIONS

Write your answers in a notebook and then compare your answers with those on page 665.

9.3A Define physical treatment technologies or processes.

9.3B Why would an industry use a screening process?

9.3C What are the two major purposes for using filtration processes?

9.3D What is the difference between solar and non-solar evaporation processes?

9.301 Chemical Treatment

Chemical treatment technologies treat wastewaters by altering the chemical structure of the constituents so that they can be removed from the wastestream prior to discharge, with the added result being the production of less hazardous or nonhazardous residual materials in some cases. Chemical processes are typically attractive because they produce minimal air emissions, operational requirements are typically not

---

[9] Solvent Extraction. The process of dissolving and separating out particular constituents of a liquid by treatment with solvents specific for those constituents. Extraction may be liquid-solid or liquid-liquid.

[10] Electrolytic Recovery. A spontaneous electrochemical process that involves the reduction of a more electropositive (noble) species, for example, copper, silver, mercury, or cadmium, by electronegative (sacrificial) metals such as iron, zinc, or aluminum. This process is used for purifying spent electrolytic solutions and for the treatment of wastewaters, leachates, and sludges bearing heavy metals. Also called cementation.

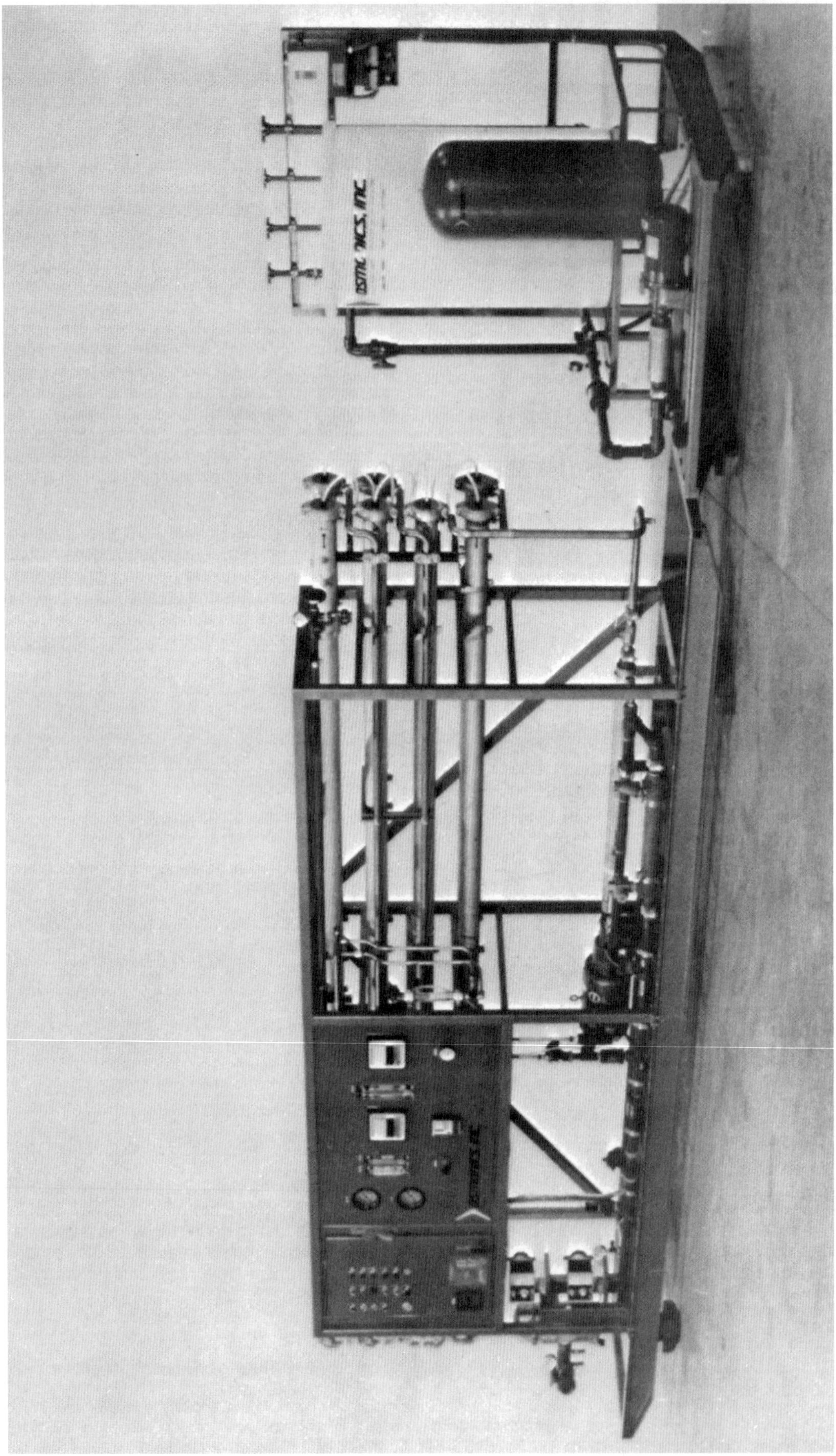

Fig. 9.29   Reverse osmosis system for recovery of plating salts from nickel plating rinse water
(Courtesy of Osmonics, Inc.)

Fig. 9.30 Ultrafiltration system
(Courtesy Koch Membrane Systems, Inc.)

highly technical, they are not labor intensive, and the processes are normally easily implemented at industrial sites. However, chemical processes often generate sludges and usually require solids handling equipment. The choice of chemicals used for neutralization, coagulation and flocculation can greatly influence the amount of sludge produced upon precipitation. Chemical treatment operations commonly used in treating wastewaters are described in the following paragraphs.

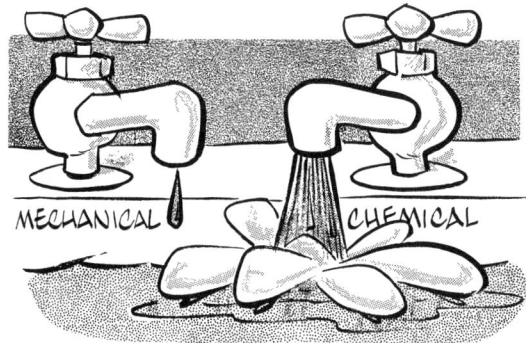

Neutralization. Neutralization is a common process used to adjust the pH of wastestreams by adding acids or bases to produce a solution which is neutral or within the acceptable range for further treatment or disposal. This is a technically and economically proven process in wastewater treatment. Sulfuric and hydrochloric acids are commonly used for neutralization of alkaline or basic streams, while sodium hydroxide (caustic soda), calcium hydroxide (hydrated lime), calcium oxide (quicklime), calcium carbonate (limestone), sodium bicarbonate (soda ash) and ammonia are used for neutralization of acids. Neutralization processes can be controlled automatically through the use of pH sensors, controllers and chemical feed pumps as shown in Figures 9.32, 9.33, and 9.34. Solids separation processes, such as sedimentation and filtration, are normally required if neutralization results in the formation of precipitates or suspended solids.

In addition to compliance with wastewater discharge limits, pH adjustment by neutralization is used for several other important reasons. First, biological treatment at a downstream POTW facility proceeds optimally at a pH near 7. Small deviations from this value may reduce biological treatment efficiency, while large differences may result in total inactivation of the bacteria. Secondly, low pH wastewaters may corrode sewers and could potentially cause the release of hydrogen sulfide gas from wastewaters containing sulfide. The adjustment and control of pH is used to achieve allowable metal concentration limits by forming metal hydroxide precipitates. The solubility of metallic ions such as lead, mercury, zinc, cadmium, copper, chromium and nickel is directly related to pH.

Examples of wastestreams which may be treated by neutralization are sulfuric or hydrochloric acid pickle liquors from steel cleaning, alkaline or acid metal plating wastes, spent acid catalysts, acid sludges, wash water from the petrochemical industry and leather tanning wastes, and regeneration wastes from deionization plants (used in many electronic component and printed circuit board plants).

Precipitation. Precipitation is a process for removing soluble compounds contained in a wastestream by forming an

Fig. 9.31   Cementation treatment (electrolytic recovery) designed for continuous flow

Low pH, copper-bearing waste is put in contact with aluminum shavings in the taller tank
on the left. Alkaline and other wastewaters are combined in the shorter tank on the right
where a pH controller raises the pH by calling for caustic solutions from the dark drum
in the middle with the chemical feeder on the top. Some flocculation and precipitation
occurs in the shorter tank on the right. The effluent pH recorder is on the wall.

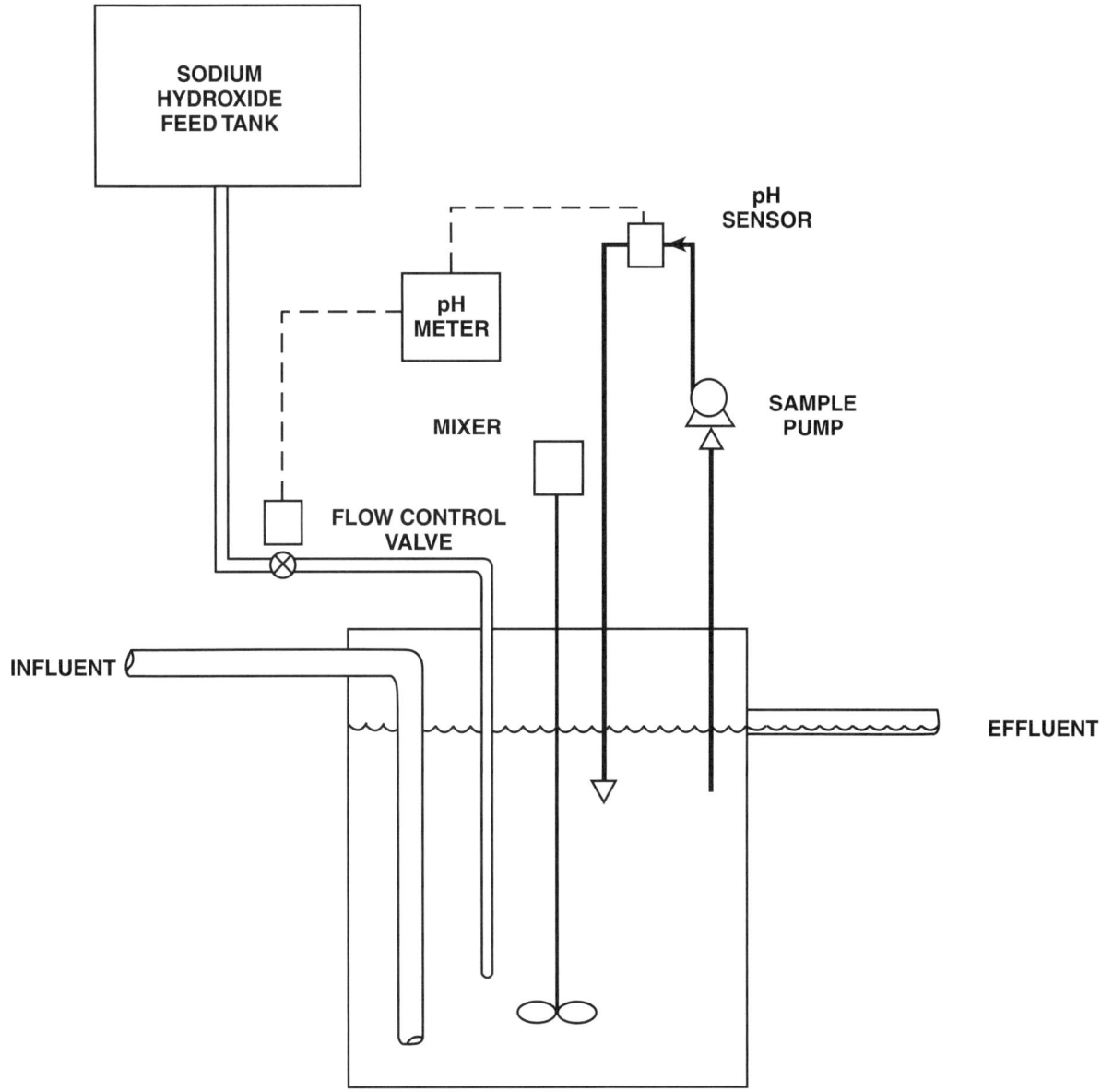

— — — Dashed line represents signal line

Fig. 9.32   Typical automated continuous-flow
acid waste neutralizing system

*Fig. 9.33   Caustic and sulfuric acid tanks at sewer neutralization
station holding reagents to control excessively high or low pH*

NOTE: The two tanks pictured above should have appropriate
hazard warning labels (in addition to the existing labels)
to comply with OSHA's Hazard Communication Standard.

Fig. 9.34   Mix motor at top keeps solution in pH neutralizing chamber mixed.
Probe at lower left senses pH and calls for injection of ammonia to raise pH.
Pipes turn white from frost as ammonia is added due to cooling of pipes by
flowing ammonia and freezing of moisture on pipes.

insoluble precipitate. A specific chemical is added to cause the formation of the insoluble precipitate. The process is specifically applicable to streams containing heavy metals. Most metals form hydroxide precipitates by the addition of lime or caustic soda. Metals can also be precipitated as sulfide or carbonate complexes. The precipitation process is well developed and usually combined with solids removal processes in wastewater treatment (Figure 9.35).

As mentioned above, for each metal there is an optimum pH which corresponds with minimum solubility. The range of pH is generally 8 to 11, as shown in Figure 9.36. In the case of a wastestream containing several different metals, pH adjustment and metals precipitation can theoretically be carried out in several steps to achieve optimum removal of each compound. In practice, however, a pH value is normally chosen which will result in the best removal of a critical metal or will minimize total metal concentrations.

Since metal sulfides generally have lower solubilities than the corresponding metal hydroxides, as shown in Figure 9.37, compounds such as ferrous sulfide or sodium sulfide can be added to wastewaters for removal of metals to very low levels. This process must be carefully controlled to avoid the generation of toxic hydrogen sulfide gas or the discharge of toxic levels of sulfide in the effluent.

Precipitation has applications in the iron, steel and copper industries for the removal of metals from pickling wastewaters; in the metal finishing industry for removing metals such as cadmium, chromium, and nickel from rinse water and spent plating bath solutions; in the electronics industry for removing copper from spent etching solutions; in the metal finishing industry for removing nickel from stripping solutions; and in the inorganic chemical industry for removing metals from a variety of wastestreams.

Ion Exchange. Ion exchange resin beads are generally composed of synthetic molecular weight polyelectrolytes that are insoluble in water. Essentially these are polymers that contain a loosely held ion in their structure that they can exchange for another ion of the same charge. Cationic resins generally exchange hydrogen ions (H^+) for other positive ions such as metal ions and anionic resins exchange hydroxyl ions (OH^-) for negative ions like carbonate. These resins can be designed to be very selective in the ions they remove so that systems can be tailored to a specific application.

Ion exchange is a process normally used to remove dissolved inorganic ions from aqueous solutions. The liquid is passed through a fixed bed of natural or synthetic resin. In ion exchange, one type of ion contained in the water is adsorbed onto an insoluble solid material and replaced by an equivalent quantity of another ion of the same charge. In essence, the resin exchanges ions with the solution resulting in the removal of selected inorganics from the solution as they attach themselves to the resin. The liquid is passed through the resin bed until the ion exchange resin can no longer effectively remove the contaminant. The resin is then regenerated by passing a concentrated brine, acid or alkaline solution through the bed. This process rinses the inorganics from the resin and returns the resin to its original effectiveness. The regeneration streams can usually be batch treated or bled into a continuous neutralization process.

A common example of ion exchange is water softening, where calcium and magnesium ions are adsorbed to the resin in exchange for sodium ions. The process is also used to produce deionized water (Figure 9.38). The ion exchange resin can either be cationic, anionic or a combination of the two. Ion exchange can be used for acid solutions containing noble metals, salt solutions, and aqueous heavy metals solutions. It can also be used in treating electroplating wastestreams containing chromium and cyanide and mixed wastestreams from metal finishing operations.

The principal disadvantages of this process are the need to treat the regenerate solutions, the need to pre-filter solutions before ion exchange, the downtime required for regeneration, upper concentration limits beyond which the process is not feasible, and cost.

Oxidation/Reduction. Chemical oxidation/reduction processes involve the exchange of electrons to convert toxic compounds into simpler, less toxic chemicals. Two of the most common applications of oxidation are conversion of cyanide to cyanate and then to nitrogen gas and carbon dioxide and conversion of sulfide to sulfate or to elemental sulfur. Oxidation of cyanide is carried out under alkaline conditions using chlorine, sodium hypochlorite or potassium permanganate. The first-stage conversion (cyanide to cyanate) is usually carried out at a pH of 10 to 11, while the second-stage conversion (cyanate to nitrogen) is normally carried out at a pH of 8.5. This operation can be automated through the use of pH and oxidation reduction potential (ORP) controllers. A pH controller is typically used to regulate the output of a chemical metering pump to control the addition of sodium hydroxide solution. Simultaneously, an ORP controller is used to regulate the output of a separate metering pump which adds sodium hypochlorite solution. The addition of sodium hypochlorite solution will raise the millivolt (mV) reading of the ORP meter which controls the sodium hypochlorite level. First-stage cyanide oxidation is normally carried out at an ORP set point of about +300 mV, while the second-stage reaction is carried out at an ORP set point of about +700 to 800 mV. Oxidation of sulfide can be accomplished with air (oxygen), chlorine, potassium permanganate or hydrogen peroxide.

The most common use of a chemical reduction process in wastewater treatment is the conversion of hexavalent chromium (chromate) to trivalent chromium. This is necessary because hexavalent chromium is soluble in water over a wide range of pH levels, while trivalent chromium that results from the reduction step can be precipitated as a hydroxide at a pH near 8 to 8.5. The reduction of hexavalent chromium to trivalent chromium is normally carried out at a pH of 2 to 3 using sulfur dioxide, sodium bisulfite or sodium metabisulfite. The reaction can be automatically controlled using pH and ORP instrumentation. Sulfuric acid is typically metered in to maintain the proper pH, while the ORP controller is used to control the addition of sodium metabisulfite solution to lower and maintain

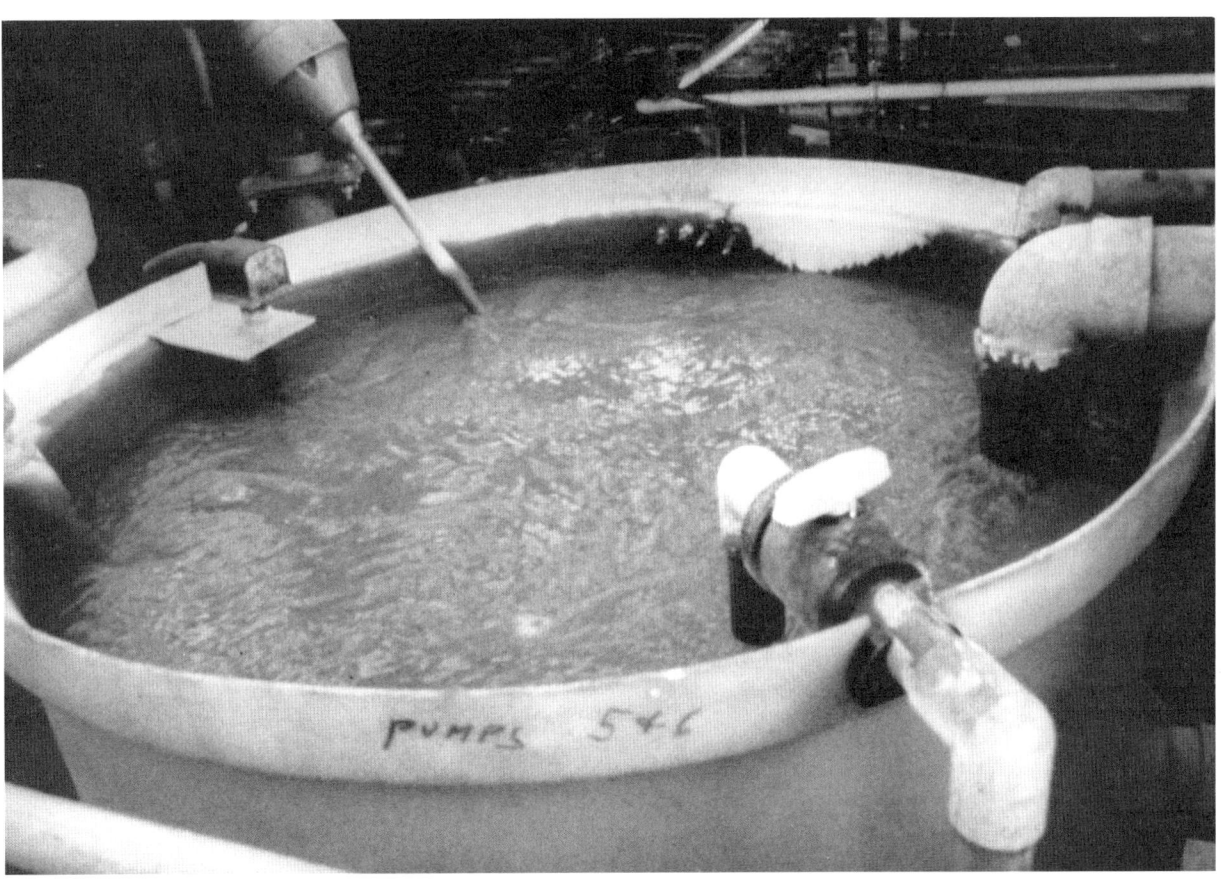

Fig. 9.35   Mixing tank for precipitation of metals
Mix motor at top agitates solution. Tiny ports in sidewall at top of tank (to the
right of mixer shaft) inject acid, base and polymer reagents.

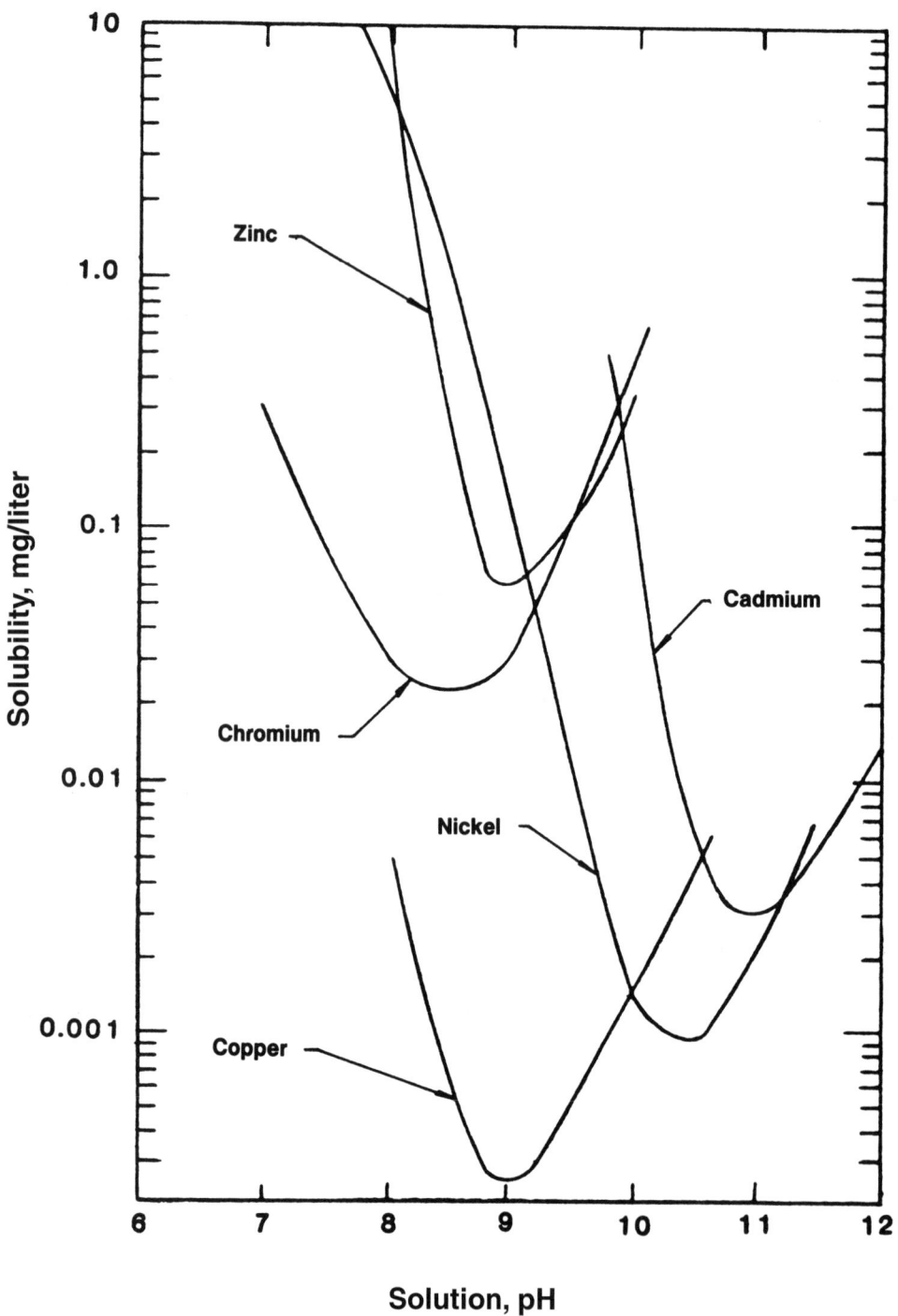

Fig. 9.36   *Effect of pH on metal hydroxide solubility*
(Source: Bennett and Philipp)

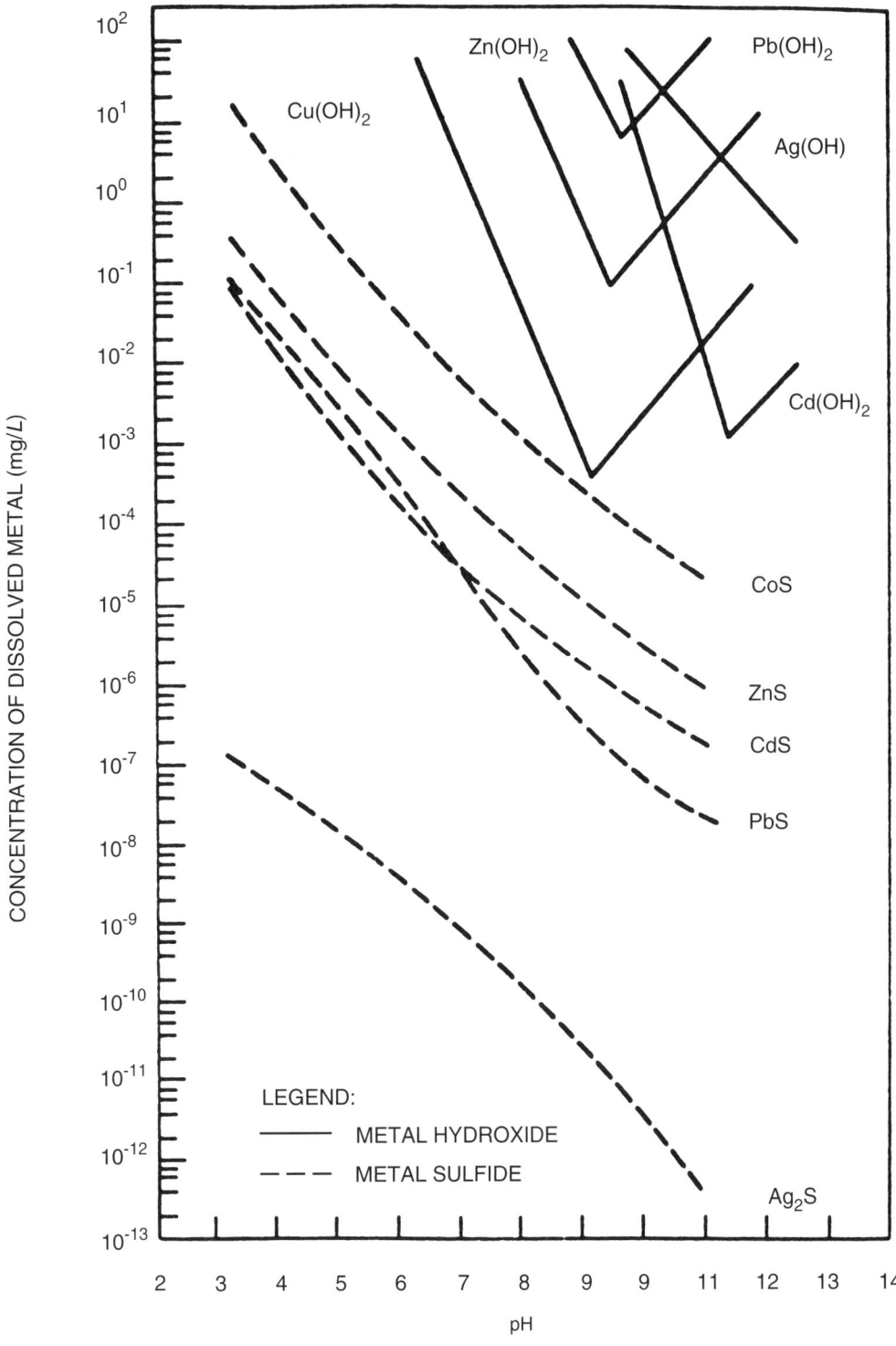

Fig. 9.37   Metal hydroxide and sulfide solubility as a function of pH
(Source: Dawson/EPA)

Fig. 9.38   The three tanks from left to right are the mixed bed (barely visible), cationic and anionic resin beds used to remove metals from solution by the deionizing process. Once the active sites on the resins are exhausted, beds are backwashed with an acid and a base. The regenerant is then collected and batch treated by adjusting pH and filtering out the metal hydroxides from solution in the box in front of the operator.

the ORP set point at about +300 mV. For both oxidation and reduction processes, a change in pH will also change the ORP value by about 60 mV per one pH unit, so it is necessary to control the pH while controlling the ORP. Trivalent chromium is then removed by alkaline precipitation at a pH between 8.0 and 8.5.

Dechlorination. Dechlorination is a specialized treatment technology used to strip chlorine atoms from highly chlorinated toxic compounds such as PCBs in order to produce a nontoxic residue. The process may also be used on chlorinated pesticides. Several companies have recently developed and marketed this technology and some offer mobile treatment units to handle such jobs as one-time treatment of electrical transformer fluids. Dechlorination is also used after cyanide oxidation by chlorine compounds to remove the residual chlorine required to obtain complete oxidation of amenable cyanides.

QUESTIONS

Write your answers in a notebook and then compare your answers with those on page 665.

9.3E   Describe the general principle on which chemical treatment technologies are based.

9.3F   Describe the neutralization process.

9.3G   Why and how is the precipitation process used to treat wastestreams?

9.3H   What is the main purpose of the ion exchange process?

9.302 Biological Treatment

Biological wastewater treatment involves processes in which living organisms are used to decompose organic wastes into (1) water and gases such as carbon dioxide and methane, (2) simpler organics such as aldehydes and organic acids, and (3) microbial matter. Typically, the microorganisms used in a biological process are already present in the raw wastewater and are recycled in the process. In some instances microorganisms developed to degrade specific compounds are introduced into a wastestream.

Except for inorganic nitrogen and sulfur compounds, biological treatment systems typically do not alter inorganics. The presence of toxic compounds in the wastestream can kill microorganisms and cause processes to not function properly (upset process). Because high concentrations of inorganics and low concentrations of toxics can severely inhibit biological decomposition activity, chemical or physical treatment may be required prior to biological treatment. Temperature extremes can also inhibit biological treatment processes by either killing the organisms used in the treatment process or lowering their activity to unacceptable levels.

Stabilization Ponds. Stabilization ponds are large shallow ponds in which wastes are allowed to decompose biologically over long periods of time. Stabilization ponds are classified as aerobic, aerobic-anaerobic (facultative) and anaerobic. Aerobic ponds are used primarily for the treatment of soluble organic wastes. The aerobic pond contains bacteria and algae in suspension that aerobically biodegrade the wastes. The aerobic conditions are maintained throughout the depth of the pond, normally by using mechanical means to increase the amount of dissolved oxygen in the pond.

The facultative ponds are the most commonly used type and have been applied to the treatment of a wide variety of industrial wastes. Stabilization of wastes is achieved by a combination of aerobic, anaerobic and facultative bacteria within three zones. The top (aerobic) zone contains aerobic bacteria and algae. The aerobic bacteria biodegrade wastes. The bottom (anaerobic) zone contains anaerobic bacteria which are actively decomposing accumulated solids. The intermediate zone is partly aerobic and anaerobic and wastes are biodegraded by facultative bacteria.

Anaerobic ponds are especially effective in the stabilization of strong organic wastes. Anaerobic conditions are maintained throughout the pond. The anaerobic ponds are usually followed by aerobic or facultative ponds.

Stabilization ponds have been widely used to provide final polishing of wastewater to meet effluent standards. Usually, only wastewaters containing less than one percent solids and biodegradable organics are suitable for stabilization by these methods.

Aerated Lagoons. An aerated lagoon is a large, shallow, aerated, earthen reservoir in which wastewater is treated on a flow-through basis. Aerated lagoons use the same biological processes as activated sludge systems (described below); however, aerated lagoons have longer retention times to decompose organic matter and there is no sludge return. Aerated lagoons are classified by the extent of mixing as aerobic or aerobic-anaerobic.

Aerated lagoons have been successfully operated for petroleum wastewaters, textile wastewaters and refinery wastewaters. The process is not considered appropriate for wastewater with a highly variable organic and metal content. It is also not recommended for high solids content wastewaters.

Activated Sludge. The activated sludge process biologically converts organic matter in a wastestream to carbon dioxide and simple organics under aerobic conditions. The conventional activated sludge process consists of an aeration tank, a clarifier and a sludge recycle system. Many modifications to the conventional activated sludge process have been adopted and standardized for different applications. The activated sludge process is considered to be a well-developed treatment technology.

Activated sludge has been reported to successfully treat refinery, petrochemical and other biodegradable organic wastewaters. Removal efficiencies as high as 95 percent have been reported for certain organic priority pollutants. The solids content of the wastestream should be less than one percent and the hazardous constituents should be primarily biologically degradable organics for this process to be effective.

Trickling Filters. A trickling filter consists of a bed of rock or synthetic media with attached bacteria. The wastes are applied at the top of the trickling filter and are allowed to trickle through the media. Microorganisms which grow on and coat the surfaces of the media decompose organic matter in the wastestream. Trickling filters are classified by hydraulic or organic loading as super-rate (roughing), high-rate or low-rate.

The high-rate trickling filter recirculates filter effluent or final effluent to achieve the same removal efficiency as the low-rate filter. The highly dependable low-rate trickling filter is capable of producing a consistent effluent quality in spite of varying influent strengths.

Trickling filters have been widely used in the United States to treat domestic wastewater since 1908. Trickling filters are also applicable to the same types of wastewaters as other biological treatment systems (wastewater containing up to one percent organic suspended matter). Trickling filters are reported to have successfully treated waste constituents such as acetaldehyde, acetic acid, acetone, acrolein, alcohols, benzene, butadiene, chlorinated hydrocarbons, cyanides, ephichlorohydrin, formaldehyde, formic acid, ketones, monoethanolamine, propylene dichloride, resins and rocket fuel.

Anaerobic Digestion. Anaerobic digestion is a biological process in which microorganisms metabolize organic matter in the absence of free or dissolved oxygen. This process provides a high degree of waste stabilization with a low production of biological sludges. The end products of this process include methane gas, carbon dioxide and microbial cell mass. Anaerobic digestion can be classified by digestion techniques as the suspended growth and attached growth (fixed film) processes.

Anaerobic digestion is traditionally used for treatment of municipal wastewater sludges. Recently, this process has gained increased attention in the industrial waste treatment field due to energy concerns and a better understanding of the process.

Anaerobic digestion processes have been used successfully to treat certain food and fermentation industrial wastes. These industries include sugar, potato, baker's yeast, breweries, distillers and dairy industries.

9.303   Land Treatment

Land treatment is a dynamic process involving the interaction of waste, soil, climate and biological activity to degrade, immobilize or deactivate the waste constituents. Different application methods are used in land treatment. In a land farming operation, wastes are applied on or into soil and regularly tilled to enhance biological degradation. For the land spreading operation, wastes are applied on top of the soil and are allowed to evaporate and degrade biologically. Suitability of land treatment is generally limited to biodegradable organic wastes, including waste biosludge, tank bottom sludges, separator sludge, emulsion solids and cooling tower sludges. Land farming is most often used for oily wastes from petroleum refinery operations. It has been practiced by many refineries since the mid-1950s.

When evaluating a land treatment system for industrial waste sludges, it is important to know whether the sludge will be inhibitory to the microorganisms present in the soil or whether these microorganisms will be able to biodegrade the waste sludge. Land treatment is not acceptable for certain industrial wastes, especially those containing volatile organics or heavy metals or for those wastes that are highly persistent and toxic. Potential problems with this technology include the concentration of trace metals in the topsoils, the production of short-term odors during application and the potential for air pollution as volatile products are produced during degradation of some wastes. Land treatment is generally not practiced as a form of pretreatment prior to discharge to the POTW.

9.304   Thermal Treatment

Thermal treatment processes consist of high-temperature (300°F to 3,000°F or 150°C to 1,650°C) technologies that reduce the volume and/or break down toxic components of wastes into simpler and less toxic forms. Thermal treatment includes processes conducted with oxygen present (incineration) or in an oxygen-starved environment (pyrolysis). Thermal treatment processes offer several advantages:

1. Significant reduction of waste volume,

2. Potential recovery of energy,

3. High destruction and/or removal efficiency, and

4. Suitability for a broad spectrum of hazardous wastes.

The disadvantages of thermal treatment processes can include:

1. The requirement of supplemental fuel,

2. Air emissions of toxic by-products,

3. Air emissions of oxides of nitrogen and particulate matter, and

4. Generation of residues and scrubber effluents.

Supplemental air pollution control and wastewater treatment equipment may be needed, and the residues remaining after incineration need to be disposed of in an appropriate land disposal facility.

Most thermal treatment processes are based on incineration. Several of the common incineration/pyrolysis processes are liquid injection systems, rotary kiln (cement kiln), fluidized-bed incineration, multiple-hearth incineration, infrared incineration and co-incineration. The most commonly used thermal treatment processes are liquid injection systems and the rotary kiln. There are also several emerging processes for treating industrial hazardous wastes including circulating bed combustion, high temperature fluid wall, electromelt, supercritical water oxidation, molten salt destruction, and plasma arc torch.

Liquid Injection. In the liquid injection system, liquid wastes and some slurried wastes are atomized and injected into an incinerator where they are burned in suspension. The incinerator consists of a refractory-lined combustion chamber and a burner. Liquid injection incinerators are the most widely used hazardous waste incinerators in the United States.

Liquid injection has been used to destroy a variety of wastes including phenols, PCBs, still and reactor bottoms, solvents, polymer wastes, herbicides and pesticides. The system is not recommended for heavy metal wastes, high-moisture content wastes, and high-inorganic content wastes.

The advantages of liquid injection include capability of incinerating a wide range of wastes and relatively low capital, operating and maintenance costs. The disadvantages include susceptibility to nozzle plugging and limited ability to handle solids.

Rotary Kiln. A rotary kiln incinerator consists of large, slightly inclined, refractory-lined steel cylinder and a secondary combustion chamber. Wastes (solid, liquid, or gaseous) are injected into the kiln at the higher end and are passed through the combustion zone as the kiln rotates. Rotary kilns (cement kilns) have been widely used in the United States. In addition, they are being used to incinerate waste solvents and still bottoms from solvent reclaiming operations and are being considered to incinerate chlorinated wastes in several states.

Rotary kilns have been used to combust a wide variety of industrial wastes including PCB-contaminated oil, obsolete munitions, obsolete chemical warfare agents, polyvinyl chloride waste, waste paint and solvents, and bottoms from solvent reclamation operations. The system is not recommended for heavy metal wastes, inorganic salts and high-organic content wastes.

The advantages of rotary kilns are the capability to burn combustible wastes in any physical form and the ability to accept wastes from a variety of feed mechanisms. The disadvantages are the high installation costs and the need for air pollution control to remove particulates.

Fluidized Bed. In fluidized-bed incinerators, wastes are injected and incinerated in a hot, suspended bed of sand at a temperature of approximately 1,500°F (815°C). Air is blown into the incinerator below the sand bed to expand and fluidize the bed. Ash residue is mostly carried out with the flue gas.

Fluidized-bed incinerators are normally used in large plants and are mostly used for wastewater treatment sludge incineration. The incinerators are normally reliable, but their complex nature requires the use of trained personnel. The exhaust gas, containing ash and water vapor, requires air pollution control equipment to remove the ash.

Emerging Technologies. Emerging technologies include circulating bed combustion, high temperature fluid wall, electromelt, supercritical water oxidation, molten salt destruction, and plasma arc torch processes.

The circulating bed combustion process operates in the same manner as fluid bed systems, except with higher velocities which circulate alumina within the system. A water wall type boiler serves as the combustor for the incineration of hazardous wastes. The system has been tested on chlorinated hydrocarbons and chlorinated solid wastes and found to be successful. This system offers on-site treatment capability.

The high temperature fluid wall process pyrolyzes waste by heat radiation in the reactor. The reactor consists of a tubular core of porous refractory material jacketed and insulated in a fluid-tight vessel. The core is heated externally by electric heating elements and radiates energy to waste materials fed into its tubular center. Bench scale tests have been demonstrated for the destruction of explosive, radioactive and nitrate wastes, phosphate sludge and oil containing PCBs. A simulated destruction of PCB-contaminated soil has been performed by one pilot system in the San Francisco area. One full-scale (50 tons per day) system has been installed in Illinois for treating PCB-contaminated soil.

In the electromelt process, molten glass acts as the heat transfer medium to decompose organic wastes in a glass furnace. The glass is heated electrically to a constant temperature of 2,200°F (1,200°C). Waste feed may be hydrocarbons, water or a mixture of both, with dirt, sludge, plating residues and still bottoms. The products are simple compounds such as carbon dioxide, water vapor and hydrogen chloride. Process equipment is available rated from 50 pounds per hour up to 50,000 pounds per hour. Two pilot units have been tested; however, a full-scale system has not been installed. It is reported that the electromelt process is applicable to most wastestreams. Destruction and/or removal efficiency (DRE) of greater than 99.978 percent has been reported.

The supercritical water oxidation process solubilizes organics and breaks down higher-molecular-weight compounds of liquid wastestreams into smaller molecules under high pressure. Subsequently, these smaller molecules are oxidized to CO_2 and H_2O when oxygen is injected into the system. Water above 705°F or 374°C (supercritical region) acts as an excellent solvent and decomposition agent. Bench scale testing has produced a DRE of greater than 99.99 percent for DDT, PCBs and hexachlorobenzene. A pilot plant that can treat up to 1,000 gallons per day of diluted waste has been tested. It has been reported that the supercritical water oxidation system will soon be available commercially. This process may cost less than other systems for treating diluted waste.

The molten salt destruction process uses an incinerator which consists of a refractory-lined vessel with a molten bed of sodium carbonate. Carbon and hydrogen in the wastes are converted to CO_2 and H_2O in the molten bed. Metals, halogens, phosphorus and arsenic are retained as inorganic salts. The molten salt incinerator has been successful in destroying chemical warfare agents, pesticides and PCBs in a pilot scale plant.

In the plasma arc torch process, either solid or liquid wastes are pyrolyzed into combustible gases by exposure to a gas which has been energized by an electrical discharge. The plasma gas is extremely destructive to the molecular makeup of the waste.

QUESTIONS

Write your answers in a notebook and then compare your answers with those on pages 665 and 666.

9.3I   What do biological wastewater treatment processes involve?

9.3J   Wastewaters from what sources can be treated by the activated sludge process?

9.3K   What wastewater constituents have been successfully treated by trickling filters?

9.3L   What do thermal treatment processes consist of?

9.3M   List the two most commonly used thermal treatment processes.

9.31   Sludge Treatment/Management

Sludges are solid or semi-solid materials or slurries that are generated in many of the previously discussed treatment processes. Sludges usually require additional treatment to remove water (thereby reducing the volume) prior to final disposal. Treatment may also be required to stabilize or detoxify the sludge. Sludge treatment and management is complicated since sludge is often composed of substances having the unfavorable characteristics of untreated wastewater and yet only a small part of the sludge is solid matter. Sludge produced from biological treatment processes is composed of organic matter which will decompose and become offensive. Sludge generated from physical-chemical treatment processes contains mostly nonbiodegradable matter.

The primary purpose of sludge treatment is to reduce the water content and stabilize the sludges. Treatment processes include (1) concentration or thickening, (2) digestion, (3) conditioning, (4) dewatering and drying, and (5) incineration, wet oxidation or stabilization and solidification.

9.310   Sludge Concentration or Thickening

The purpose of sludge concentration or thickening is to increase the solids content of waste activated sludge or mixtures of primary and waste activated sludges. Gravity and flotation thickeners are commonly used to accomplish this. In a gravity thickener, primary or waste activated sludge is fed into the thickening tank and stirred gently by sludge collection trusses or vertical pickets to allow densification. Supernatant is returned to the primary settling tank of the wastewater treatment plant, and the thickened sludge is pumped from the bottom of the tank to the sludge digesters or dewatering equipment.

Flotation thickeners usually operate at higher loading rates than gravity thickeners. Separation of solids from the liquid phase is achieved by introducing fine gas (usually air) bubbles into the liquid phase. The float (solids) is then skimmed off the top of the water surface.

9.311   Digestion

Sludge with high organic content can be stabilized by anaerobic or aerobic digestion. Anaerobic digestion of sludge has been widely used by municipal wastewater treatment plants to stabilize sludge. The anaerobic digestion process reduces organic matter to water, carbon dioxide, methane and other inert end products. It also reduces the quantity of solids that require subsequent processing and/or disposal.

Aerobic digestion is generally more suited to the treatment of excess biological sludge than to primary sludge. The aerobic digestion process has been used primarily in small plants, particularly those with extended aeration and contact stabilization. Aerobic digestion produces a biologically stable end product. The high cost of power to supply the required oxygen is the major disadvantage of the aerobic digestion process.

9.312   Conditioning

Heat treatment and the addition of chemicals are commonly used techniques for conditioning sludge to improve its dewatering characteristics. Chemicals used include ferric chloride, lime, alum and organic polymers. Chemical conditioning is used in advance of flotation, vacuum filtration, pressure filtration or centrifugation. Complete and thorough mixing of the sludge and the coagulant is essential for proper conditioning.

Heat treatment involves heating the sludge under pressure for a short period of time. Heat treatment results in coagulation of the solids, breakdown of the gel structure and a reduction of the water affinity of the sludge solids. Other conditioning processes include *ELUTRIATION*[11] and freezing.

9.313   Dewatering and Drying

Dewatering and drying processes are used to remove major portions of the water content from sludge. Dewatering sludge is commonly performed by spreading on drying beds, vacuum filtration, centrifugation and pressure filtration. Heat drying of sludge is used to remove the moisture from the wet sludge for incineration or fertilization processes.

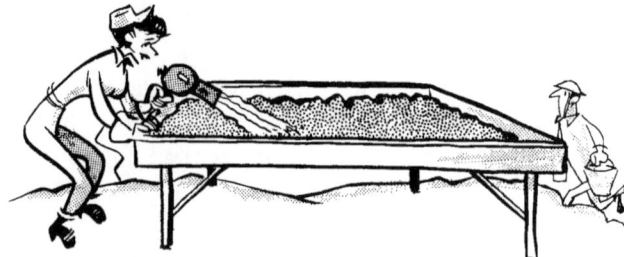

9.314   Incineration and Wet Oxidation[12]

Incineration and wet oxidation are used to reduce the organic content and the volume of sludge. The incineration process converts the sludge into an inert ash for disposal. Multiple-hearth furnaces, flash-drying systems and fluidized bed furnaces can be used to incinerate the sludge. The wet oxidation process oxidizes the raw sludge at an elevated temperature and pressure (for example, 175 to 315°C and 600 to 3,000 psig).

9.315   Stabilization and Solidification

The stabilization process chemically changes the waste characteristics of sludge whereas solidification treatment generally solidifies the waste without chemical change. However, the terms "stabilization" and "solidification" are frequently interchanged and are also referred to as "fixation."

Most of the currently available stabilization/solidification processes are suitable for treating inorganic wastes and sludges. Wastes containing more than 10 to 20 percent organic compounds are generally not amenable to this treatment method. Several suppliers of stabilization/solidification treatment processes and/or chemicals are working at solving the problem of interference from wastes containing a high percentage of organic compounds and new developments are expected in the near future.

Stabilization/solidification processes include the following: cement-based processes; *POZZOLANIC PROCESSES*[13]; thermoplastic techniques; organic polymer techniques; self-cementing; glassification/mineral synthesis; and encapsulation. The most commercially available processes are the cement-based and pozzolanic processes.

9.316   Disposal

Sludge and ash can potentially be disposed of either on land or in the sea. The most common methods of land disposal are spreading on soil, lagooning and landfill disposal. Digested sludge may be carried offshore in barges or sludge vessels and dumped or it may be pumped to deep water through an ocean outfall. Sludge can be disposed of by incineration and the resulting ash transported to a landfill. Current require-

---

[11] *Elutriation (e-LOO-tree-A-shun).   The washing of digested sludge with either fresh water, plant effluent or other wastewater. The objective is to remove (wash out) fine particulates and/or the alkalinity in sludge. This process reduces the demand for conditioning chemicals and improves settling or filtering characteristics of the solids.*

[12] *Wet Oxidation.   A method of treating or conditioning sludge before the water is removed. Compressed air is blown into the liquid sludge. The air and sludge mixture is fed into a pressure vessel where the organic material is stabilized. The stabilized organic material and inert (inorganic) solids are then separated from the pressure vessel effluent by dewatering in lagoons or by mechanical means.*

[13] *Pozzolanic (POE-zoe-LAN-ick) Process.   The pozzolanic process mixes lime and other fine-grained siliceous materials such as fly ash, cement kiln dust, or blast furnace slag with a liquid hazardous waste to form a cement-like solid.*

ments for EPA under RCRA dictate that EPA must develop alternatives to land disposal for sludges produced by the metal finishing industry.

9.32 Review of Treatability Studies

Industrial operations and processes create such a diversity of wastes that treatability studies are usually necessary to select appropriate treatment processes and to evaluate the effectiveness of the selected process. For example, organic industrial wastewaters are usually deficient in nitrogen and phosphorus that are essential for biological treatment. Bench-scale and/or pilot-scale tests will demonstrate whether or not a particular waste is treatable by the proposed processes and will provide the necessary data for conceptual design of the treatment system. In some industries, such as the metal finishing industry, virtually all of the viable treatment technologies are well defined. Treatability studies are primarily needed for new industries, new technologies and unique situations.

9.320 Physical/Chemical Treatment

Treatability studies for physical/chemical treatment processes include studies of settling, zone settling, filtration, flotation, carbon adsorption, precipitation/coagulation, ion exchange, chromium reduction, and cyanide oxidation, as well as iron removal tests.

Settling studies are conducted to evaluate the characteristics of flocculant settling of suspensions at the expected ranges of suspended solids. The settling data provide the overflow rate and retention time requirements for sizing the settling tank.

Zone settling studies are performed to determine the unit area (square feet per pounds of solids per day) or mass loading of the clarifier. By plotting the height of the sludge interface versus the time of settling, the unit area and the overflow rate can be calculated.

Filtration studies can be used to select optimum filter media and size, to determine filtration rate, length of filter run, and head loss to provide design guidelines and operating conditions of the filter system.

Flotation studies are conducted to determine the downward velocity, the quantity of solids to be floated per unit area and time requirements. The study also measures the rise of sludge interface with time at various air-to-solids ratios. Based on these data, the required surface area for flotation can be determined.

Batch equilibrium carbon adsorption isotherm studies are used to determine the effectiveness of carbon in adsorbing organic compounds. The isotherms are established by plotting the equilibrium concentration of a particular organic compound in water versus the equilibrium concentration of the compound adsorbed on the carbon. Various dosages of carbon are mixed with wastewater for a sufficient contact time to obtain data for developing isotherms. Continuous column studies are then used to develop the breakthrough curves for estimating the adsorption capacities under different conditions. These conditions include *EMPTY BED CONTACT TIME*,[14] particle size, flow rate and initial concentration for the adsorption units.

Precipitation/coagulation studies are conducted to determine the optimum pH and chemical dosage for removal of metals and some organic compounds (color). The studies are performed either with a jar test or *ZETA POTENTIAL*[15] control. The reagents most frequently used are aluminum sulfate (alum) for clarification purposes, and lime, sodium hydroxide or sodium carbonate to adjust the pH or to reduce dissolved solids. Other reagents such as activated silica and polyelectrolytes are being used as aids to facilitate flocculation and sedimentation. Lime softening tests are also conducted with a jar test apparatus.

Ion exchange studies are used to determine resin exchange capacity, regeneration efficiency and regeneration frequency of the ion exchange unit.

Iron removal tests are conducted to determine the type of oxidants and other necessary coagulants and flocculants for iron removal.

Chromium reduction studies are typically conducted to develop the relationship between chrome reduction and consumption of the reducing agent. The relationship includes the quantity of chemical required, the concentration and volumes of sludge produced, and the oxidation-reduction control point.

Cyanide oxidation treatability studies include determining the presence of interfering chelators and complex cyanides such as ferricyanide along with investigating conventional and innovative methods of cyanide treatment.

9.321 Biological Treatment

To determine whether or not an industrial wastewater is amenable to biological treatment, several laboratory or pilot studies are carried out. These include activated sludge studies, contact stabilization studies and anaerobic treatability studies. The activated sludge studies are conducted to deter-

---

[14] *Empty Bed Contact Time.* The time required for the liquid in a carbon adsorption bed to pass through the carbon column assuming that all liquid passes through at the same velocity. It is equal to the volume of the empty bed divided by the flow rate.
[15] *Zeta Potential.* In coagulation and flocculation procedures, the difference in the electrical charge between the dense layer of ions surrounding the particle and the charge of the bulk of the suspended fluid surrounding this particle. The zeta potential is usually measured in millivolts.

mine design criteria such as aeration time, settling time and sludge recycle rate for different suspended solids concentrations (mixed liquor suspended solids).

Contact stabilization studies are performed to define the required contact time and stabilization period. Design criteria such as oxygen transfer ratio, oxygen utilization rate, and sludge settling rate are also defined.

Anaerobic treatability studies are conducted to determine typical BOD and COD removal percentages, type of reactor, temperature, nutrient requirements, organics loading rates, biomass gas production and its composition and potential toxicity effects from heavy metals and other chemicals.

None of the studies described in this section are required in the metal finishing industry because the wastestreams and treatment technology are well known.

9.322   Sludge Treatment

Vacuum filterability and pressure filterability are evaluated in dewatering studies for sludge treatment. The vacuum filterability test is conducted with a *BUCHNER FUNNEL*[16] to determine the filtration characteristics of the sludges, the optimum coagulant dosage and the compressibility of sludges. The pressure filterability test uses a *PONT-A-MOUSSON PRESSURE CELL*[17] to establish the coefficient of compressibility of filter cakes and their limiting dryness.

QUESTIONS

Write your answers in a notebook and then compare your answers with those on page 666.

9.3N   Why are sludges treated?

9.3O   How can sludge with a high organic content be stabilized?

9.3P   How and why is sludge conditioned?

9.3Q   What is the purpose of batch equilibrium carbon adsorption isotherm studies?

9.4   FACILITY OPERATIONAL CONSIDERATIONS

9.40   Management Commitment and Support

There should be a commitment to compliance with pollution laws at the corporate and facility management levels. Unless this commitment exists, adequate funds for installation, operation, and maintenance of source control equipment will not be available. Furthermore, even if a proper treatment system is installed, it will not function properly unless adequate operation and maintenance personnel are provided to run and maintain the plant. Ideally, compliance with environmental laws should be a factor in evaluating managerial employee performance, in addition to production, cost control and other traditional measures.

9.41   Performance

The "bottom line" for any treatment system is its performance or ability to meet discharge standards. Since proven treatment technologies exist to allow industry to meet Federal Pretreatment Regulations and local Industrial Waste Ordinances, lack of appropriate and available technology is seldom a valid reason for noncompliance. The proper equipment and systems are not always installed, however, due to financial constraints or lack of technical understanding of wastestream characteristics, flow rates and generation patterns. The need for segregation and pretreatment of some individual wastestreams (for example, oil emulsion breaking and separation, chromium reduction, cyanide oxidation and concentrated waste batch treatment) is often overlooked. Specialized equipment or chemicals that may be necessary because of chelating agents or metal complexes may not be made available.

Poor performance is frequently traceable to operation of a system at higher than design flow rates or with solids loading or influent pH extremes beyond design specifications. Treatment systems may have been designed based on average flows (including periods of nonoperation); however, production rates may have increased since the installation of the system. In addition, operations personnel are often unaware of the effect that process start-ups, shutdowns and batch dumps can have on treatment system performance.

Poor performance may be caused by equipment which is inoperable due to lack of maintenance attention, dirty or uncalibrated sensors for process control instrumentation, lack of chemicals in makeup tanks, excessive sludge in clarifiers and for a variety of other reasons relating to inattention on the part of facility operating personnel. The inspector of industrial pretreatment facilities should be aware of the potential causes of poor performance discussed above when performing inspections. Additional factors which can relate to the ability of a facility to consistently meet discharge standards are discussed in the sections that follow.

9.42   Reliability

The reliability of a wastewater treatment system is a function of many variables related to proper design and maintenance. One of the most important is the inclusion of installed

---

[16] *Buchner Funnel.   A special funnel used to separate solids (sludge) from a mixture. A perforated plate or a filter paper is placed in the lower portion of the funnel to hold the mixture (wet sludge). The funnel is placed in a filter flask and a vacuum is applied to remove the liquid (dewater the sludge).*

[17] *Pont-a-Mousson (MOO-san) Pressure Cell.   A device that works like a filter press and is used to perform filter tests. A cell applies pressurized air (7 to 225 psig) with a piston (instead of using a vacuum as with a Buchner funnel) to separate solids from a mixture.*

spare equipment such as spare pumps, redundant level controls and alarms and redundant process sensors. Equally important is the requirement of having adequate spare parts and equipment on hand. Finally, system reliability is greatly enhanced by having a regular schedule of routine maintenance which includes greasing and oiling rotating equipment, cleaning and calibrating pH and other sensors and controllers, cleaning weirs and flumes, and monitoring the sludge levels in clarifiers and sumps.

9.43 Maintainability

An important factor related to reliability is maintainability. Major treatment system components such as pumps, mixers and sensors should be designed and installed to permit easy access for routine service (cleaning, lubrication) and for removal and replacement. Ideally, it should be possible to service equipment without emptying or entering tanks or having to shut down the treatment system.

Good system maintenance also requires adequate staffing of qualified mechanics, electricians and instrument repair personnel to keep the system running. Where this is not possible, contract maintenance services can be used.

9.44 Site Selection

The selection of an appropriate site for construction of a pretreatment facility requires careful evaluation to ensure ease of installation and safe start-up and operation. A treatment system requires a certain minimum amount of space because of the time necessary for chemical reactions to go to completion and because of the time necessary for solids settling to take place. Clarification can be accomplished in less space with an inclined plate settler and by using the proper coagulating and flocculating agents, and chemical reaction times can sometimes be minimized through the proper choice of reagent and by operating at optimum conditions of pH, temperature and pressure. Innovative design is sometimes necessary to provide adequate space for treatment at a small facility.

A complicated treatment system that works well in the laboratory or as a pilot plant may not work well as a continuous full-scale plant. The treatment system should be designed for operational simplicity, using field-proven equipment and processes whenever possible.

As with maintenance, adequate operational staffing is necessary for proper system operation. If less than full-time staffing is available, the system should be designed with a high degree of automatic control, installed spare equipment and alarms which are audible in areas that are staffed whenever the facility is in operation.

9.45 Flexibility

A properly designed treatment system will have the built-in flexibility to handle peak flows, minimum flows and varying pollutant concentrations by the proper sequencing of treatment equipment and tanks and the use of equalization vessels. However, discharge limit violations can still occur as a result of flow surges resulting from batch dumps, spills and start-up and shutdown of process equipment. These fluctuations can often be minimized by an awareness on the part of manufacturing facility personnel of the effect of their actions on effluent quality. Process tanks may be *SLOWLY* drained to the treatment plant rather than being dumped all at once. A production process can sometimes be gradually started or shut down. Alternatively, these wastewater surges can be handled by designing a treatment system that will handle the peak flows and concentrations or by providing holding tanks for concentrated solutions that can be bled into the treatment system at rates within its design capacity. Without proportional control feed systems, systems that are designed to handle *PEAK DEMAND* will lack the sensitivity required to handle normal load conditions.

9.46 Safety and OSHA Requirements

As with any other piece of equipment or system, a wastewater treatment plant should be designed to comply with applicable safety regulations. Commonly cited areas of OSHA violations include lack of guards for rotating equipment, the use of temporary or substandard wiring, the use of electrical equipment not designed for watertight or explosion-proof service, the lack of proper guardrails, failure to maintain proper access aisle widths and the lack of proper ventilation from chemical mixing/reaction tanks.

9.47 Hazardous Waste Requirements

Sludges generated by wastewater treatment systems may be classified as hazardous wastes. Waste process solutions temporarily stored prior to on-site or off-site treatment or disposal may also be classified as hazardous wastes. In either of these cases, the wastes must be handled, stored, transported and disposed of in accordance with the requirements of the Resource Conservation and Recovery Act (RCRA). Individual states may impose additional requirements.

9.48 Operating Plans, Log Sheets, and O & M Manuals

A well-designed system is of little use if no one understands how to operate and maintain it. An operation and maintenance (O & M) manual should be readily available for use by operators, maintenance personnel, management and facility inspectors. The O & M manual should contain a description of the system; flow, piping and instrumentation drawings; start-up and shutdown procedures; troubleshooting guides; and routine maintenance checks. Manufacturers' literature on system components is normally included in the manual.

The best way to document wastewater treatment system performance is with a daily operating log sheet where flows, levels, concentrations, pH readings and the operating status of pumps and mixers are recorded. Similarly, maintenance checklists and complete records on repair of equipment should be maintained. The degree of complexity of the O & M manual and the operating log sheets will obviously depend on the size and complexity of the system.

A well-prepared O & M manual will also include an explanation of the theoretical considerations used in the design and use of the systems so that the operating personnel will know *WHY* they are doing what they are doing, and not just doing something because it is in the manual. The manual should also be available for training operators. Operators will do a better job if they know why they are doing what they are required to do.

O & M manuals must be continually reviewed and updated as necessary. As time goes on, operators gradually develop more efficient methods of operating the wastewater treatment system, and these methods need to be documented. As a normal part of continuing, long-term operation, treatment units are sometimes added or taken out of service at the treatment facility. When this happens, the O & M manual needs to be revised to remain a usable tool.

9.49 Operator Training

A properly trained and motivated operator is a critical element in the success of any pretreatment facility. Inspectors should find out what training the operators receive as this will be an indication of the performance that can be expected of pretreatment operators and facilities. Operator certification is highly desirable if such a classification system is offered locally by a federal, state, local or professional association.

9.5 SOURCE MONITORING (COMPLIANCE AND OPERATIONAL)

Source monitoring is essential for demonstrating compliance/noncompliance and for achieving the best possible control at the source.

9.50 Tests and Continuous Monitoring

Some degree of source monitoring is necessary to ensure compliance with regulations. Monitoring by the local sanitation agency may take the form of periodic grab or 24-hour composite sampling. This is often conducted on a quarterly basis, with the frequency increasing if discharge limits are consistently exceeded.[18] Industrial sources are often required to conduct some type of self-monitoring as well, usually flow measurement. In some cases the flow measurement can be estimated from water meter readings on water supplied to the facility. In other cases it is necessary to directly measure water flow as a basis for computing sewer surcharge rates and the daily mass discharge of specific pollutant compounds.

A facility may also be required to periodically collect effluent samples, have the samples analyzed for specific water quality indicators and report the results to the local wastewater agency. Some facilities are required to continuously monitor and record data on water quality indicators such as pH and temperature. The wastewater agency may require some additional water quality indicators to be measured and recorded on a frequent basis. The additional tests are specific for each type of industry, such as metals or total toxic organics. (See Chapter 3, "Development and Application of Regulations," for detailed information about monitoring requirements.)

Besides required compliance monitoring, it is usually in a facility's best interest to conduct operational monitoring to optimize performance of its wastewater treatment equipment. Water quality indicators such as flow, pH, turbidity and suspended solids content can usually be checked easily without expensive instrumentation. For flumes and V-notch weirs where liquid height is proportional to flow, the flow can be easily measured. Visual checks are usually sufficient to monitor the proper operation of clarifiers and pH can easily be checked with a portable pH meter. Test kits are available to allow either the waste inspector or facility personnel to perform simple tests for a variety of specific pollutants in wastewater.

Continuous monitoring with on-line instrumentation is normally limited to flow, temperature, pH and perhaps oxidation/reduction potential (in the case of chromium and cyanide treatment systems). Operational monitoring for these and other water quality indicators is frequently done at various points before and within the wastewater treatment plant. Monitoring for proper control of wastewater treatment equipment is essential for obtaining the best possible performance from the treatment system. Effluent monitoring is usually too late for good process control.

9.51 Records and Reports

No matter what type of monitoring is done, it is important for the industrial pretreatment facility to keep complete, permanent records of monitoring results and any reports that may have been prepared and submitted to wastewater agencies. Documentation of system performance can itself be an aid in spotting trends and improving performance. The documentation can help prove a facility operator's good faith efforts in trying to comply with discharge limits, and the very existence of monitoring and reporting documentation is evidence of a certain level of concern that pollution standards be met.

If a facility is cited for violations of discharge limits, records showing compliance on days other than the official sampling

---

[18] *EPA requires that an out-of-compliance discharge must be resampled within 21 days after the date of collection of the sample that was in violation of discharge requirements.*

day or showing improvement in performance since the date of violation can be helpful.

9.6 REFERENCES

1. Beall, J.F. and McGathen, R. "Guidelines for Wastewater Treatment. Part 1. How to Minimize Wastewater." Metal Finishing, September, 1977.

2. California, State of, Department of Health Services. 1986. "Alternative Technology for Recycling and Treatment of Hazardous Wastes, A Third Biennial Report." Toxic Substances Control Division.

3. California, State of, Governor's Office of Appropriate Technology. 1981. "Alternatives to the Land Disposal of Hazardous Waste, An Assessment for California." Toxic Waste Assessment Group.

4. Clifford, D., Subramonian, S., and Sorg, T.J. "Removing Dissolved Inorganic Contaminants from Water." Environmental Science and Technology, November, 1986.

5. Dawson, G.W. and Mercer, B.W. *HAZARDOUS WASTE MANAGEMENT.* Obtain from John Wiley & Sons, Inc., Distribution Center, 1 Wiley Drive, Somerset, NJ 08875-1272. ISBN 0-471-82268-X. Price, $185.00, plus $5.00 shipping and handling.

6. Eckenfelder, W.W., Jr., et al. "Wastewater Treatment." Chemical Engineering, September 2, 1985.

7. Kemmer, F.N. *THE NALCO WATER HANDBOOK.* Obtain from McGraw-Hill, PO Box 545, Blacklick, OH 43004-0545. ISBN 0-07-045872-3. Price, $99.95, plus nine percent of order total for shipping and handling.

8. Lawler, D.F. "Removing Particles in Water and Wastewater." Environmental Science and Technology, September, 1986.

9. Lee, M.C. et al., 1985. "Alternative Hazardous Waste Treatment Technologies." Presented at 1985 HAZMAT West, Long Beach, CA. December 3-5, 1985.

10. Mackie, J.A. and Niesen, K. "Hazardous Waste Management: The Alternatives." Chemical Engineering, August 6, 1984.

11. Metcalf & Eddy, Inc. *WASTEWATER ENGINEERING: TREATMENT, DISPOSAL, AND REUSE,* Fourth Edition. Obtain from McGraw-Hill, PO Box 545, Blacklick, OH 43004-0545. ISBN 0-07-041878-0. Price, $117.19, plus nine percent of order total for shipping and handling.

12. Perry, R.H. and Green, D.W. *PERRY'S CHEMICAL ENGINEER'S HANDBOOK,* Seventh Edition. Obtain from McGraw-Hill, PO Box 545, Blacklick, OH 43004-0545. ISBN 0-07-049841-5. Price, $150.00, plus nine percent of order total for shipping and handling.

13. *PRETREATMENT OF INDUSTRIAL WASTES* (MOP FD-3). Obtain from Water Environment Federation (WEF), Publications Order Department, 601 Wythe Street, Alexandria, VA 22314-1994. Order No. MF2003WW. Price to members, $81.75; nonmembers, $101.75; price includes cost of shipping and handling.

14. Weber, W.J. and Smith, E.H. "Removing Dissolved Organic Contaminants from Water." Environmental Science and Technology, October, 1986.

15. *EPA TREATABILITY MANUALS.* Obtain from National Technical Information Service (NTIS), 5285 Port Royal Road, Springfield, VA 22161. The five volumes of the manual are sold separately. The prices are:

Volume 1	Order No. PB80-223050 (EPA No. 600-8-80-042A)	$166.00
Volume 2	Order No. PB80-223068 (EPA No. 600-8-80-042B)	$155.00
Volume 3	Order No. PB80-223076 (EPA No. 600-8-80-042C)	$133.00
Volume 4	Order No. PB80-223084 (EPA No. 600-8-80-042D)	$81.50
Volume 5	Order No. PB80-223092 (EPA No. 600-8-80-042E)	$51.00

Shipping is $5.00 per order.

16. *EPA RISK REDUCTION ENGINEERING LABORATORY (RREL) TREATABILITY DATABASE,* Version 5.0. Obtain from National Service Center for Environmental Publications (NSCEP), PO Box 42419, Cincinnati, OH 45242-2419. EPA No. 600-C-93-003A. Free.

17. *FACILITY POLLUTION PREVENTION GUIDE,* Risk Reduction Laboratory, Office of Research and Development, U.S. Environmental Protection Agency. Obtain from National Service Center for Environmental Publications (NSCEP), PO Box 42419, Cincinnati, OH 45242-2419. EPA No. 600-R-92-088. Free.

QUESTIONS

Write your answers in a notebook and then compare your answers with those on page 666.

9.4A   Why is management commitment and support important for an industrial wastewater pretreatment facility?

9.4B   A routine maintenance schedule for a wastewater treatment system should include what items?

9.4C   Why should a properly designed treatment system have built-in flexibility?

9.5A   Why is source monitoring essential?

9.5B   Why should a wastewater treatment facility conduct operational monitoring?

9.5C   Why is documentation of wastewater treatment system performance important?

Please answer the discussion and review questions before continuing with the Objective Test.

DISCUSSION AND REVIEW QUESTIONS

Chapter 9.   PRETREATMENT TECHNOLOGY (SOURCE CONTROL)

Write the answers to these questions in your notebook before continuing with the Objective Test on page 666. The purpose of these questions is to indicate to you how well you understand the material in the chapter.

1. What items does industry consider when selecting the processes to produce a given product?

2. What factors should be considered when estimating existing and expected flows from a discharger?

3. What is the basis for the establishment of the degree of pretreatment required for an industrial waste?

4. Why is source control important for an industry?

5. How can pollutant discharges be controlled at their source?

6. The amount of pollutants contributed by drag out depends on what factors?

7. Physical treatment processes are used for what purposes?

8. What is an equalization process and how is it accomplished?

9. Evaporation is used by industries for what purposes?

10. What happens to spent activated carbon?

11. Why are chemical treatment processes attractive to industry?

12. Why do some wastestreams require neutralization or pH adjustment?

13. What are the principal disadvantages of the ion exchange process?

14. What factors can inhibit or upset biological waste treatment processes?

15. Land treatment involves the interaction of what factors?

16. What are the advantages of thermal treatment processes?

17. What are the disadvantages of thermal treatment processes?

18. What are the advantages and disadvantages of rotary kilns?

19. What is the major difference between sludge produced from biological treatment processes and sludge generated from physical-chemical treatment processes?

20. Why are chromium reduction studies conducted?

21. List the important facility operational considerations for an industrial pretreatment plant.

22. What problems with the basic design could prevent an industrial wastewater treatment facility from meeting wastewater discharge standards?

SUGGESTED ANSWERS

Chapter 9.   PRETREATMENT TECHNOLOGY (SOURCE CONTROL)

Answers to questions on page 606.

9.0A   Wastes that could be expected in the effluent from electrical and electronics industries include metals, acids, SS, toxic organics and fluoride.

9.0B   The major general categories of wastes from industries include inorganic salts, metals, oil and grease, organics, pesticides and chlorinated organics.

Answers to questions on page 611.

9.1A   Important factors involved in the establishment of treatment and discharge objectives for pretreatment facilities include flow, wastewater characteristics, pretreatment standards, toxicity of pollutants to POTWs, and control and monitoring guidelines.

9.1B   COD is a measure of the oxygen required to oxidize the total organic content of the sample or wastewater.

9.1C   Sulfide compounds in industrial effluents are of concern because they inhibit the anaerobic digestion process under certain conditions; the conversion of dissolved sulfide to hydrogen sulfide gas produces a gas that is toxic to humans; and hydrogen sulfide gas is the source of many odor complaints received by POTW facilities.

Answers to questions on page 620.

9.2A   Source control is a term used to describe activities which reduce or eliminate wastewater and pollutants that would otherwise exit the manufacturing process and enter the POTW sewer.

9.2B   Important pollution source control techniques are: (1) modification of the manufacturing process through a change in raw materials, processes, equipment or operating guidelines, and (2) water and material conservation, recovery and reuse through good housekeeping and materials management, modifications to reduce use, wastewater segregation/treatment, and waste exchange/management.

9.2C   Changes in operating guidelines that could be considered to reduce wastewater discharges include chemical concentration, process temperature and dwell or residence time.

9.2D   Drag out from rinsing processes can be reduced by extending the dwell time of parts over the process tank, by using a misting spray rinse directly over the tank, or by using air knives to remove process solution from parts prior to entering a rinse tank.

9.2E   The most obvious sources of pollution in the metal finishing industry are the drag out, large volumes of rinse waters, and various processing baths with subsequent rinses.

Answers to questions on page 643.

9.3A   Physical treatment technologies or processes separate components of a wastestream or change the physical form of the wastewater without altering the chemical structure of the constituent materials.

9.3B   Screening is used to remove large particles from wastewater. It is used as a pretreatment process before other solid/liquid separation processes or to protect equipment downstream from problems created by large solid particles.

9.3C   Filtration may be used to dry solids or to clarify liquids or both. Filters can be used to "polish" the effluent from the sedimentation or flotation process (clarified liquid) or to dewater sludges from the sedimentation process (dry solids).

9.3D   Solar evaporation is a process in which wastewaters are reduced in volume through simple evaporation by sunlight in uncovered ponds or surface impoundments. Non-solar evaporation is accomplished by heating a waste solution or slurry, usually with steam condensing on metal tubes inside a vessel.

Answers to questions on page 654.

9.3E   Chemical treatment technologies treat wastewaters by altering the chemical structure of the constituents so that they can be removed from the wastestream prior to discharge, with the added result being the production of less hazardous or nonhazardous residual materials in some cases.

9.3F   Neutralization is a common process used to adjust the pH of wastestreams by adding acids or bases to produce a solution which is neutral or within the acceptable range for further treatment or disposal.

9.3G   Precipitation is used to treat wastestreams to remove soluble compounds by forming an insoluble precipitate. A specific chemical is added to cause the formation of the insoluble precipitate.

9.3H   The ion exchange process is used to remove dissolved inorganic ions from aqueous solutions.

Answers to questions on page 657.

9.3I   Biological wastewater treatment involves processes in which living organisms are used to decompose organic wastes into (1) water and gases such as carbon dioxide and methane, (2) simpler organics such as aldehydes and organic acids, and (3) microbial matter.

9.3J   Wastewaters from refineries, petrochemical facilities and other biodegradable organic wastewaters can be treated by the activated sludge process.

9.3K Trickling filters are reported to have successfully treated wastewater constituents such as acetaldehyde, acetic acid, acetone, acrolein, alcohols, benzene, butadiene, chlorinated hydrocarbons, cyanides, ephichlorohydrin, formaldehyde, formic acid, ketones, monoethanolamine, propylene dichloride, resins and rocket fuel.

9.3L Thermal treatment processes are based on incineration. They consist of high temperature (300°F to 3,000°F or 150°C to 1,650°C) technologies that reduce the volume and/or break down toxic components of wastes into simpler and less toxic forms.

9.3M The two most commonly used thermal treatment processes are liquid injection systems and the rotary kiln.

Answers to questions on page 660.

9.3N Sludges are treated to remove water, thereby reducing the volume prior to final disposal. Treatment may also be required to stabilize or detoxify the sludge.

9.3O Sludge with a high organic content can be stabilized by either anaerobic or aerobic digestion.

9.3P Sludge is conditioned by the addition of chemicals and heat treatment to improve its dewatering characteristics.

9.3Q The purpose of batch equilibrium carbon adsorption isotherm studies is to determine the effectiveness of carbon in adsorbing organic compounds.

Answers to questions on page 663.

9.4A Management commitment and support is important for an industrial wastewater pretreatment facility because unless this commitment exists, adequate funds for installation, operation, and maintenance of source control equipment will not be available. Furthermore, even if a proper treatment system is installed, it will not function properly unless adequate operational and maintenance personnel are provided to run and maintain the plant.

9.4B Items that should be included in a routine maintenance schedule for a wastewater treatment system are greasing and oiling rotating equipment, cleaning and calibrating pH and other sensors and controllers, cleaning weirs and flumes, and monitoring the sludge levels in clarifiers and sumps.

9.4C A properly designed treatment system must have built-in flexibility to handle peak flows, minimum flows and varying pollutant concentrations by the proper sequencing of equipment and tanks and the use of equalization vessels.

9.5A Source monitoring is essential for demonstrating compliance/noncompliance and for achieving the best possible control at the source.

9.5B A wastewater treatment facility should conduct operational monitoring to optimize performance of its wastewater treatment equipment.

9.5C Documentation of wastewater treatment system performance is important as an aid in spotting trends and improving performance. The documentation can help prove a facility operator's good faith efforts in trying to comply with discharge limits, and the very existence of monitoring and reporting documentation is evidence of a certain level of concern that pollution standards be met.

OBJECTIVE TEST

Chapter 9.   PRETREATMENT TECHNOLOGY (SOURCE CONTROL)

Please write your name and mark the correct answers on the answer sheet, as directed at the end of Chapter 1. There may be more than one correct answer to each multiple-choice question.

True-False

1. Federal pretreatment standards require pretreatment of wastestreams discharged to POTWs.

  - 1. True
  2. False

2. Pretreatment levels must prevent upsets of POTW processes.

  - 1. True
  2. False

3. High concentrations of heavy metals in the treatment sludge may limit the application of sludge in land spreading.

  - 1. True
  2. False

4. Pollution prevention focuses on activities which prevent the generation of wastes.

  - 1. True
  2. False

5. Waste minimization seeks to reduce the volume or toxicity of wastes being generated.

  - 1. True
  2. False

6. Material or chemical distribution should be by manual pouring of buckets or barrels rather than by pumps and pipes.

  1. True
  - 2. False

7. Hexavalent chromium and cyanide are most economically treated at the source, before mixing with other wastes.

  - 1. True
  2. False

8. Equalization basins are installed to increase the fluctuations of flows.

   1. True
   — 2. False

9. Sedimentation is widely used in the metal finishing, electroplating and printed circuit board industries to remove solids generated by precipitation of heavy metals.

   — 1. True
   2. False

10. Evaporation is a proven, well-developed, inexpensive industrial process.

    1. True
    — 2. False

11. Spent activated carbon may be considered a hazardous waste in some states.

    — 1. True
    2. False

12. Electrolytic recovery is a process in which there is an electrochemical oxidation of metal ions at the cathode where these ions are oxidized to elemental metal.

    1. True
    — 2. False

13. Anaerobic digestion is a biological process in which microorganisms metabolize organic matter in the presence of free or dissolved oxygen.

    1. True
    — 2. False

14. Sludge generated from physical-chemical treatment processes contains mostly biodegradable matter.

    1. True
    — 2. False

15. Batch equilibrium carbon adsorption isotherms are used to determine the effectiveness of carbon in adsorbing organic compounds.

    — 1. True
    2. False

Best Answer (Select only the closest or best answer.)

16. Why must pretreatment effluent wastewater be monitored?

    1. To determine surcharge fees
    — 2. To ensure compliance with mandated limits
    3. To monitor treatment process performance
    4. To provide information for process adjustment

17. How should the release of spilled chemicals in storage areas be prevented?

    — 1. Install some type of containment around the storage areas.
    2. Place chemicals in leakproof containers.
    3. Purchase chemicals that do not flow.
    4. Use floor drains to prevent release.

18. How should maintenance schedules be optimized?

    1. Cleaning up spills after they occur
    — 2. Reducing times wastes are generated by maintenance-related shutdowns
    3. Scheduling maintenance when production ceases
    4. Waiting for system failures

19. How can wastewater treatment for chemicals be avoided?

    1. By diluting chemicals
    2. By hydroxide precipitation
    3. By ion exchange
    — 4. By recovering and reusing chemicals

20. Equalization basins have the capability to

    1. Allow physical treatment of wastes.
    2. Develop biological treatment cultures.
    — 3. Increase the effectiveness of downstream treatment equipment.
    4. Provide opportunity for chemical treatment.

21. Why does industry use packaged inclined plate (lamella) settlers?

    1. Cost more than traditional circular tank clarifier
    2. Flocculation section near inlet is not needed
    — 3. Plates provide same surface area in smaller vessel than clarifier
    4. Plates provide surface area for biological treatment

22. What is the ion exchange process normally used to remove?

    — 1. Dissolved inorganic ions
    2. Dissolved organic ions
    3. Excess hydrogen ions
    4. Volatile organic ions

23. What is the most common use of the chemical reduction process in wastewater treatment?

    1. Conversion of cyanide to cyanate and then to nitrogen gas and carbon dioxide
    — 2. Conversion of hexavalent chromium to trivalent chromium
    3. Conversion of high pH to low pH
    4. Conversion of sulfide to sulfate or to elemental sulfur

24. Where are the residues remaining after incineration disposed of?

    — 1. In a land disposal facility
    2. In a sludge digester
    3. In the POTW's sewer
    4. In the receiving waters

25. What does the incineration process convert sludge into?

    1. Biosolids
    — 2. Inert ash
    3. Sludge cake
    4. Soil conditioner

Multiple Choice (Select all correct answers.)

26. What types of wastes are found in food industry effluents?

    — 1. BOD
    2. Cyanide
    3. Nitrate
    4. PCBs
    — 5. SS

27. What types of flows are important when establishing treatment and discharge objectives?

    — 1. Average flow
    2. Minimum flow
    — 3. Peak flow
    — 4. Total daily flow
    5. Weekend flow

28. What factors are used by POTWs to assess surcharges on industrial discharges?

    - 1. BOD
    - 2. COD
    - 3. Flow
    4. PCBs
    - 5. SS

29. What information should a pretreatment inspector be aware of with regard to types of industrial production processes?

    1. Availability of raw material
    - 2. Chemical constituents in the wastewater
    - 3. Potential wastewater sources
    - 4. Techniques for wastewater reduction
    5. Utility costs

30. What pollutants may interfere with the operation of the POTW or may be toxic to the treatment processes?

    - 1. Alkalies
    - 2. Inorganic pollutants
    - 3. Organic pollutants
    - 4. Sulfur compounds
    - 5. Temperature

31. What are the benefits of source control?

    - 1. Cost savings
    2. Increased size of pretreatment facility
    - 3. Protection of public
    - 4. Reduced chemical costs
    - 5. Reduced liability

32. How can manufacturing processes be modified? By change in

    - 1. Manufacturing processes.
    - 2. Operating guidelines.
    3. Plant location.
    - 4. Process equipment.
    - 5. Raw materials.

33. What are the primary wastewater treatment technologies used by industry?

    - 1. Biological
    - 2. Chemical
    3. Disinfection
    - 4. Land
    - 5. Physical

34. Why is solar evaporation coming under increased scrutiny?

    - 1. Discharge of toxic compounds to land
    - 2. Emission of toxic compounds to air
    3. Higher energy costs
    4. Reduced monitoring requirements
    - 5. Release of toxic compounds to groundwater

35. What contaminants are removed by adsorption?

    - 1. Cyanide
    - 2. PCBs
    - 3. Pesticides
    - 4. Phenols
    - 5. Solvents

36. Which of the following processes are chemical treatment processes?

    1. Filtration
    - 2. Neutralization
    - 3. Oxidation/reduction
    - 4. Precipitation
    5. Stripping

37. What are the principal disadvantages of the ion exchange process?

    - 1. Costs
    - 2. Downtime required for regeneration
    3. Lower concentration limits for effective ion exchange
    - 4. Need to pre-filter solutions before ion exchange
    - 5. Need to treat the regenerate solutions

38. What is the reliability of a wastewater treatment system a function of?

    - 1. Adequate spare parts
    - 2. Installed spare equipment
    - 3. Proper design
    - 4. Redundant level controls and process sensors
    - 5. Regular preventive maintenance schedule

39. What are the commonly cited areas of OSHA violations at wastewater treatment facilities?

    - 1. Electrical equipment not designed for explosion-proof service
    - 2. Lack of guards for rotating equipment
    - 3. Lack of proper ventilation for chemical mixing/reaction tanks
    4. Location of chlorine vents at floor level
    - 5. Use of temporary or substandard wiring

40. What concepts do pretreatment inspectors need to understand to work with industries to achieve compliance with discharge standards?

    - 1. Mandated pretreatment requirements
    2. Industrial organizational structure
    - 3. Toxicity of industrial wastewater to the POTW
    - 4. Wastewater characteristics
    - 5. Wastewater flows

End of Objective Test

APPENDIX

POLLUTION PREVENTION OPPORTUNITY CHECKLISTS FOR SELECTED INDUSTRIES

Checklist 1.   POLLUTION PREVENTION OPPORTUNITIES FOR THE CHEMICAL MANUFACTURING INDUSTRY*

The keys to pollution prevention are good operating practices and production process modifications. Wastes are usually generated from the mishandling of materials and the inadvertent production of off-spec materials.

Y/N OPPORTUNITIES	COMMENTS
I. Good Operating Practices in Material Input, Storage and Handling	
—— Inventory control	First in, first out to prevent expiration
—— Designate material storage area	Provide protection, spill containment; keep area clean and organized; give one person the responsibility to maintain the area
—— Return obsolete materials to suppliers	Suppliers are the best persons to handle them
—— Segregate wastestreams, especially nonhazardous from hazardous	Prerequisite for recovery and reuse
—— Store packages properly and shelter from weather	To prevent damage, contamination and product degradation
—— Prevent and contain spills and leaks by proper equipment maintenance and increased employee training and supervision	To prevent the generation of wastes
—— Minimize traffic through material storage area	To reduce contamination and dispersal of materials
—— Improve quality of feed by working with suppliers or installing purification equipment	Impurities in feedstream can be major contributors to waste
—— Reexamine need for each raw material	Need for a raw material that ends up as waste may be reduced or eliminated by modifying the process and control
—— Replace raw material containing hazardous ingredients with nonhazardous ones	To avoid the use of hazardous materials and the generation of hazardous wastes
—— Use off-spec material	Occasionally, a process can use off-spec material because the particular quality that makes the material off-spec is not important to the process
—— Improve product quality	Product impurities may be creating wastes at customers' plants; effect should be discussed with customers
—— Use inhibitors and continuously upgrade	Inhibitors prevent unwanted side reactions or polymer formation
—— Reformulate products from powder to pellet	To reduce dust emissions and waste generation
—— Reuse inert ingredients when flushing solids handling equipment	To minimize need for disposal
—— Change shipping containers, both for raw materials and products	To avoid disposal, change to reusable containers, totebins or bulk shipments
—— Recover product from tankcars and tank trucks	To minimize product drained from tanks going to waste
II. Production Process Modifications	
Reactors: The reactor is the heart of the process and can be a primary source for waste products. The quality of mixing is the key.	
—— Improve physical mixing in a reactor	Install baffles, a high RPM motor for the agitator, a different mixing blade design, multiple impellers, pump recirculation or an in-line static mixer
—— Distribute feeds better for better yield and conversion, both for inlet and outlet	Add feed distributor to equalize residence time through fixed bed reactor to minimize under- and overreactions that form by-products
—— Improve ways reactants are introduced into the reactor	Get closer to the ideal reactant concentrations before the feeds enter the reactor to avoid secondary reactions which form unwanted by-products in the premixing of reactants

Checklist 1. POLLUTION PREVENTION OPPORTUNITIES FOR THE CHEMICAL MANUFACTURING INDUSTRY*
(continued)

Y/N	OPPORTUNITIES	COMMENTS
____	Improve catalyst and continuously upgrade	Catalyst has a significant effect on reactor conversion and product mix; changes in the chemical makeup of a catalyst, the method by which it is prepared, or its physical characteristics can lead to substantial improvements in catalyst life and effectiveness
____	Provide separate reactor for recycle streams	The ideal reactor conditions for converting reactor streams to usable products are different from those in the primary reactor; this separation affords optimization for both streams
____	Better heating and cooling techniques for reactors	To avoid hot spots that would give unwanted by-products
____	Consider different reactor design	The classic stirred-tank batch mix reactor is not necessarily the best choice. A plug flow reactor offers the advantage that it can be staged, and each stage can be run at different conditions for optimum product mix and minimum waste generation
____	Improve control to maintain optimal conditions in reactor	To increase yield and decrease by-product; at a minimum, stabilize conditions in reactor operation frequently if advanced computer control is not available
	Heat Exchangers: Heat exchangers can be a source of waste, especially with products that are temperature-sensitive. Reducing tube-wall temperature is the key.	
____	Use lower pressure steam	To reduce tube-wall temperature
____	Desuperheat steam	To reduce tube-wall temperatures and increase the effective surface area of the exchanger because the heat transfer coefficient of condensing steam is ten times greater than that of superheated steam
____	Install a thermocompressor	To reduce tube-wall temperature by combining high and low pressure steam
____	Use staged heating	To minimize degradation, staged heating can be accomplished first using waste heat, then low pressure steam and finally, desuperheated high pressure steam
____	Use on-line cleaning techniques for exchangers	Recirculating sponge balls and reversing brushes can be used to reduce exchanger maintenance, and also to keep the tube surface clean so that lower temperature heat sources can be used
____	Use scraped-wall exchanger	To recover saleable products from viscous streams, such as monomers from polymer tar
____	Monitor exchanger fouling	Sometimes an exchanger fouls rapidly when plant operating conditions are changed too fast or when a process upset occurs; monitoring can help to reduce such fouling
____	Use noncorroding tube	Corroded tube surfaces foul more quickly than noncorroded ones
	Pumps: Preventing leaks is the key.	
____	Recover seal flushes and purges	Recycle to the process where possible
____	Use sealless pumps	Use can-type or magnetically driven sealless pumps
	Furnaces: Avoiding the hot tube-wall temperature is the key.	
____	Replace coil	Alternative designs should be investigated whenever replacement becomes necessary
____	Replace furnace with intermediate exchanger	Use a high temperature intermediate heat transfer fluid to eliminate direct heat
____	Use existing steam superheat	Sufficient superheat may be available to heat a process stream, avoiding exposure of the fluid to the hot tube-wall temperature of a furnace

Checklist 1.   POLLUTION PREVENTION OPPORTUNITIES FOR THE CHEMICAL MANUFACTURING INDUSTRY*
(continued)

Y/N	OPPORTUNITIES	COMMENTS
	Distillation Column: A distillation column typically produces waste in three ways:	
	• Allowing impurities to remain in a product from inadequate separation	
	• Forming waste within the column itself through polymerization from the high reboiler temperature in the column	
	• Losing products through venting or flaring from inadequate condensing	
____	Increase reflux ratio (if column capacity is adequate) for better separation	Increase the ratio by raising the pressure drop across the column and increasing the reboiler temperature using additional energy
____	Add section to column for better separation	The new section can have a different diameter and can use trays or high efficiency packing
____	Retray or repack column for better separation	Repack to lower pressure drop across a column and decrease the reboiler temperature; large-diameter columns have been successfully packed
____	Change feed tray for better separation	Match the feed conditions with the right feed tray in the column through valving changes
____	Insulate	Good insulation prevents heat losses and fluctuation of column conditions with weather
____	Improve feed distribution	Especially for a packed column
____	Preheat column feed	Preheating improves column efficiency and also requires lower temperatures than supplying the same heat to the reboiler; often the feed can be preheated by cross exchange with another stream
____	Remove overhead products from tray near top of column	To obtain a higher purity product if it contains a light impurity
____	Increase size of vapor line	To reduce pressure drop and decrease the reboiler temperature
____	Modify reboiler design	A falling film reboiler, a pumped recirculation reboiler, or high-flux tubes may be preferred to the conventional thermosiphon reboiler for heat-sensitive fluids
____	Reduce reboiler temperature	General temperature reduction techniques include using lower pressure steam or desuperheated steam, installing a thermocompressor and using an intermediate transfer fluid
____	Lower column pressure	To decrease reboiler temperature; the overhead temperature, however, will also be reduced which may create a condensing problem
____	Improve overhead condensers	To capture any overhead losses through retubing, condenser replacement or supplemental vent condenser addition
____	Improve column control	Similar to improving reactor control
____	Forward vapor overhead to the next column	Use a partial condenser and introduce the vapor stream to the downstream column
	Piping: A simple piping change can result in a major reduction of waste.	
____	Recover individual wastestream	Segregation is crucial for reuse
____	Avoid overheated lines	Review the amount and temperature of heat-sensitive materials in lines and in vessel tracing and jacketing
____	Avoid sending hot materials to storage	To prevent excessive venting and degradation of products
____	Eliminate leaks	To prevent waste generation

Checklist 1. POLLUTION PREVENTION OPPORTUNITIES FOR THE CHEMICAL MANUFACTURING INDUSTRY*
(continued)

Y/N OPPORTUNITIES	COMMENTS
___ Change metallurgy or use lining	Metal may cause a color problem or act as a catalyst for the formation of by-products
___ Monitor major vents and flare system and recover vented products	Storage tanks, tank cars and tank trucks are common sources of vented products; install a condenser or vent compressor for recovery
___ Consider "pipeless" batch processing	Reactants are transported in process vessels, eliminating the need for pipe cleaning and providing better prevention of contamination and greater flexibility in scheduling
Process control: Modern technology allows computer control system to respond more quickly and accurately than human beings.	
___ Improve on-line control	Good process control reduces waste by optimizing process conditions and reducing plant trips and wastes
___ Optimize daily operation	A computer can be programmed to analyze the process continually and optimize the conditions to prevent waste
___ Automate start-ups, shutdowns and product change-over	To bring the plant to stable conditions quickly to minimize the generation of off-spec wastes
___ Program plant to handle unexpected upsets and trips	To minimize downtime, spills, equipment loss and waste generation
Miscellaneous:	
___ Avoid unexpected trips and shutdowns	A good preventive maintenance program, adequate spare equipment and adequate warning system for critical equipment
___ Use wastestreams from other plants	Internal waste exchanges are feasible, but wastestreams should be adequately characterized
___ Reduce number and quantity of samples	Review sampling frequency and procedure and recycle the samples
___ Find a market for waste product	Wastes can be converted to saleable by-products with additional processing and creative salesmanship
___ Install reusable insulation	Particularly effective on equipment where the insulation is removed regularly to perform maintenance
III. Good Operation and Maintenance Practices for Equipment Cleaning and Changeover	
___ Avoid unnecessary equipment cleaning	Explore the feasibility of eliminating cleaning step between batches
___ Maximize equipment dedication	Dedicating tanks to one product will reduce clean-out and save time and labor cost for changeover
___ Recover more products	Scraping down tanks, pigging or blowing lines can recover more product and reduce wastes
___ Use less cleaner	High pressure sprays, pressurized air, steam and heated cleaning bath can reduce the amount of cleaner used and disposed of as waste
___ Reuse cleaner	Reclaim and reuse cleaner if feasible
___ Consider alternative cleaning methods and less hazardous cleaners	Mechanical cleaning such as plastic media blasting and ultrasonic cleaning, together with more biodegradable cleaner, can reduce waste volume and toxicity
___ Standardize cleaning products used in plant	To maximize recovery potential

APPENDIX

* Prepared by Philip Lo, Industrial Waste Section, County Sanitation Districts of Los Angeles County.

ACKNOWLEDGEMENT: Materials for production process modification were adapted from Ken Nelson, Dow Chemicals USA, "Use These Ideas to Cut Wastes," *HYDROCARBON PROCESSING,* March 1990.

Checklist 2. POLLUTION PREVENTION OPPORTUNITIES FOR THE CHEMICALS FORMULATING INDUSTRY*

Including Pesticides, Chemicals and Paints

PROCESS MODIFICATION	PROCESS OPERATION AND MAINTENANCE	MATERIAL RECYCLE, REUSE AND RECOVERY	GOOD HOUSEKEEPING
Use continuous processes Batch processes involve more frequent mix tank cleaning. **Use pumps and pipes** For raw material and product transfer, use closed systems as much as possible. Reduces potential for spills. **Segregate wastestreams** Increases recovery potential and treatment efficiency. Segregate water-based streams to reduce amount of solvent to wastewater treatment system. **Dedicate equipment** Reduces need for tank rinsing between batches. **Substitute nonhazardous raw materials** Avoids the production of hazardous wastes.	**Raw material purity** Use high quality raw material in batch to minimize contamination and reduce waste. **Use wiper blades on mix tanks** Physically wiping down sides of mix tanks will reduce amount of rinse water required. **Perform preventive maintenance** Routinely check for leaks in valves and fittings. Repair immediately. **Optimize inventory and production schedule using computer** Minimizes need for changeover and consolidates batch production. **RINSING** 1. **Use jet sprays with pressure booster pump** Reduces amount of required rinse water. 2. **Use water knife spray** Reduces amount of required rinse water. 3. **Use steam cleaner** Reduces waste use.	**Install drip pan for filling line** Recover and recycle product to filling reservoir. **Reuse mix tank rinse water in next process batch** **Reuse floor wash water** Treat and reuse as wash water or equipment rinse water (including empty drum rinsing). Treatment technologies include chemical precipitation, biological treatment, activated carbon adsorption, air or steam stripping, hydrolysis, chemical oxidation, and resin adsorption. **Reuse container rinse water** Treat and reuse both solvent and water rinses for containers. **Recover materials for reuse from wastestreams** Separate and concentrate materials using membrane separation and evaporative technologies. **Collect and reuse storm water** Reuse as floor wash water.	**Control inventory** Do not allow material to exceed shelf life. Use materials on a first-in, first-out basis. Do not get rid of expired products by discharging to wastewater treatment system. **Use mop floor washing** Mops and squeegees reduce amount of wash water required. **Reduce use of containers** Less water will be required for rinsing. Have suppliers deliver materials in tank trucks directly to on-site storage tanks, or use returnable tote bins. **Buy appropriate amounts** Buy materials in small quantities if only small amounts are required. Savings on large quantity purchases can be lost if unused material must be disposed of or is discharged to laboratory sink drains. **Manage laboratory samples** Do not allow concentrated lab samples to be discharged to wastewater treatment system. Recycle into process. **Cover outdoor storage** Divert clean storm water away from material storage and handling areas. **Install spill containment** Spills can be contained and managed appropriately rather than draining to wastewater treatment system and causing system upsets.

* ACKNOWLEDGEMENT: Materials for production process modification were adapted from Ken Nelson, Dow Chemicals USA, "Use These Ideas to Cut Wastes," *HYDROCARBON PROCESSING*, March 1990. Reviewed by Charlie Henderson, Rohm and Haas Southern California Incorporated.

Checklist 3. POLLUTION PREVENTION OPPORTUNITIES FOR THE DAIRY PROCESSING INDUSTRY*

Including Fluid Milk, Spread Processing, Cheese/Whey Processing, Process Cheese Manufacturing

PROCESS MODIFICATION	PROCESS OPERATION AND MAINTENANCE	MATERIAL RECYCLE, REUSE AND RECOVERY	GOOD HOUSEKEEPING
Install water desludging system on milk and whey separators instead of desludging with product Most applicable for non-fluid milk separators (check with U.S.D.A.). Reduces amount of product to drain. **Review product formulation** Look at all ingredients which make up the product. Evaluate each for impacts from delivery, handling, spillage, cleanup and ultimate disposal. Consider alternative materials which may have less environmental impact without compromising product quality. **Dedicate equipment as much as possible** Reduces the need for frequent cleanings between batches. For example, a bottling plant could have separate process and filling lines for milk system and juice system.	**Automate cheese salting process** Reduces amount of human error, spillage and overuse of salt resulting in additional salt drippings for disposal. **Review cleaning chemicals** Check to see environmental impacts of cleaning chemicals. Can alternative chemicals be used which have less environmental impact without compromising cleaning ability? For example, substitute nitric acid for phosphoric acid to reduce plant phosphorus discharge. **Integrate environmental considerations into all aspects of processing** Remember there are always two product streams, one to market, the other to "drain" or disposal; both are equally important and must be considered in making production decisions. Know your processes well; then you can find ways to reduce pollution. **Use high quality raw materials** So batches will not become contaminated and have to be managed as a waste. **Perform preventive maintenance** Routinely check for leaks in valves and fittings. Repair immediately. **Install CIP (Cleaning In Place) monitoring systems on automated CIP processes** Monitor flow, time, temperature and conductivity. Can provide information leading to reduction of water and cleaning chemicals, optimization of system.	**Install product reclaim system in milk intake (receiving station)** System involves automated CIP (Cleaning In Place) system which uses potable water for initial truck, milk line and silo rinses. **Product reclaim** "Burst" rinse truck tanks with potable water and chase milk/water slurry to collection tank. **Product reclaim** Air blow milk lines to silo and "burst" flush. Chase milk/water slurry to collection tank. **Product reclaim** In non-fluid milk applications, pump water/milk slurry (from collection tank) to milk silos (check w/U.S.D.A.). **Product reclaim** As milk silo is emptied, "burst" rinse silo with potable water and pump milk/water slurry to collection tank. In non-fluid milk applications, pump milk/water slurry from collection tank to manufacturing milk silo (check w/ U.S.D.A.). **Install membrane systems for cleaning salt brines** Reduces the number of times salt brine is discharged. The retentate is segregated for alternative disposal. **Install "product reclaim" systems on condensed whey storage and loadout system** Very similar to milk receiving product reclaim system for storage tanks, whey concentrate lines and truck tanks. **Collect separate "desludge" for alternative disposal** If handled properly this high strength waste could be used as a supplement to animal feed, for example, hog feed.	**Covered outdoor storage** Divert clean storm water away from material storage and handling areas. **Install spill containment** Spills can be contained and managed appropriately rather than draining to wastewater treatment system and causing system upsets. **Spill cleanup procedures** Establish procedures for what to do with a spill. Reduces chance of spill being discharged to wastewater treatment plant. **Provide slope for milk trucks during unloading** Trucks must be emptied as completely as possible before washing. This reduces the potential for product going to the drain. Can be done with ramps, or actually sloping the floor (1/4" per foot is desirable). **Provide high pressure nozzles on all water hoses** Allows for adequate cleaning with less water. Also automatically shuts off water stream when not in use. **Perform dry sweepings before floor washing** Picks up loose bits of cheese curd and powder so that they would not get washed down the drain. **Provide drip pans** Place in areas of leaky valves, seals, barrel or block draining areas to collect product wastes rather than having them run directly onto the floor and the floor drains.

Checklist 3.   POLLUTION PREVENTION OPPORTUNITIES FOR THE DAIRY PROCESSING INDUSTRY* (continued)

PROCESS MODIFICATION	PROCESS OPERATION AND MAINTENANCE	MATERIAL RECYCLE, REUSE AND RECOVERY	GOOD HOUSEKEEPING

	Install computer-controlled processing systems
	Reduces the potential for "bad" runs which will subsequently need to be disposed of.
	
	Adopt a definite waste prevention program and build an educational program
	Helps to ensure that all plant personnel are aware of waste prevention concerns.
	
	Instruct plant personnel completely in proper operation and handling of all dairy plant processing equipment
	Major losses are due to poorly maintained equipment and to negligence of inadequately trained and insufficiently supervised personnel.
	
	Repair or replace all worn out and obsolete equipment and/or parts of equipment including sanitary valves, fittings and pumps
	To the extent possible, drips and leaks occurring during the processing run should be collected in containers and not allowed to go down floor drains.
	
	Install suitable liquid level controls
	Install controls with automatic pump stops, alarms, and other devices at all points where overflows could occur, such as storage tanks, processing tanks, filler bowls and Cleaning In Place tanks.
	
	Use care in materials handling
	Avoid spillage of cased, canned or barreled dairy products and product ingredients.
		Reuse "cow" water (or condensate of whey) instead of potable water for the following applications (U.S.D.A. dependent):
		● Boiler Feed Waste
		● Floor Washes
		● External Truck Washes
		● CIP (Cleaning In Place)
		   Makeup Water for
		   Intermediate Rinses
		● Pump Seal Water
		
		Explore possible reverse osmosis polishing of cow water with subsequent disinfection
		Could lead to further uses of water in place of well or city water.
		
		Process salt drippings with nano-filtration membrane system
		U.S.D.A. approval needed. Can "desalt" salt drippings which can then be blended back into sweet whey.
		
		Install reclaim systems for acid/caustic
		Can use simple gravity separation techniques such as cone bottom tanks to segregate solids and decant chemicals for next day's first wash. Eliminates the need to dump all chemicals each day.
		
		Look at feasibility of dryer, scrubber water reuse
		Segregate and reuse this material rather than dumping to drain. Need to review any reuse process with U.S.D.A.
		
		Perform dry cleanup on dryers before wet wash
		Remove and collect as much dry product as possible before wet washing. Can be used as animal feed.
		
		Study the plant and develop a material balance
		Determine where losses occur and take steps to modify and replace unsatisfactory equipment. Where improper maintenance is the cause of losses, a specific maintenance program could be instigated and maintained.
			Screen removable traps in floor drains
			By providing screened traps, large particulates are able to be collected for alternative disposal or reuse as "fishbait" or animal feed. Screened traps also prevent undesirable materials from entering the process waste system, such as pump seals, bags, string and gloves.
			
			Check raw product quality, for example, antibiotics, in small quantity increments
			Sample milk on delivery and evaluate for regulated constituents before mixing with large bulk quantities to be processed. Minimizes large quantities of raw product needing disposal.
			
			Develop procedures for handling returned product
			By-product outlets minimize waste hauling.

* Prepared by Rob Lamppa, Engineering Department, Land O'Lakes, Inc., Arden Hills, MN 55126. Phone: (612) 481-2741.

Checklist 4. POLLUTION PREVENTION OPPORTUNITIES FOR THE DRY CLEANING INDUSTRY*

The keys are to ensure proper PERC (perchloroethylene) separation in the water separator and to warn against illegal disposal of still bottom residuals to the sewer. The preferable program is zero discharge to sewer, with the water and still bottom transported for off-site reclamation by a contract service.

PROCESS MODIFICATION	PROCESS OPERATION AND MAINTENANCE	MATERIAL RECYCLE, REUSE AND RECOVERY	GOOD HOUSEKEEPING
Convert to "dry-to-dry" machine Reduces solvent vapor losses.	(Mainly for minimizing air emissions)	**Contract collection of cartridge filter, separator water and still bottom for off-site reclamation** Provides for PERC reclamation and condensate treatment under proper expert supervision.	
Install solvent leak detectors Monitor for vapor losses.	**Keep lids on containers** Reduces evaporation and spills.		
Use refrigeration/condensation system Reduces vapor losses.	**Label all raw material containers** Prevents unnecessary disposal.	**Redistill still bottom with more water following boil-down** Recovers more solvent and reduces solvent content in wastestream.	
Use a reclaiming dryer Reduces solvent losses.	**Store containers shake-proof** Prevents spills in an earthquake.	**Use cartridge stripper to remove solvent from cartridge** Recovers more solvent.	
Redesign separator with baffles and decant taps Reduces PERC entrainment and affords better decanting of water.	**Replace seals regularly on dryer, deodorizer and aeration valves** Reduces emission leaks.		
Allow only batch discharge of decant water from separator after visual inspection Check for inadequate separation before sewer discharge.	**Replace door gasket on button trap**		
Steam strip cleaning filter cartridge Permits recovery of solvent.	**Replace gaskets around cleaning machine door or tighten enclosure**		
Substitute low temperature laundering for dry cleaning for applicable fabric Avoids unnecessary PERC use.	**Repair holes in air and exhaust ducts**		
	Secure hose connects and couplings		
	Clean lint screens Avoids clogging fans and condensers.		
	Open button traps and lint gaskets only long enough to clean		
	Check baffle assembly in cleaning machine bi-weekly		
	Check air relief valves for proper enclosure		
	Adjust IN and OUT condensing coil temperatures on heater to within 10°F of each other		

* Prepared by Philip Lo, Industrial Waste Section, County Sanitation Districts of Los Angeles County, Whittier, CA.

ACKNOWLEDGMENT: Some materials were adapted from the Alaskan Health Project, "Waste Reduction Tips for Dry Cleaners," 1987.

APPENDIX

Checklist 5.   POLLUTION PREVENTION OPPORTUNITIES FOR THE FLUID MILK PROCESSING INDUSTRY*

PROCESS MODIFICATION	PROCESS OPERATION AND MAINTENANCE	MATERIAL RECYCLE, REUSE AND RECOVERY	GOOD HOUSEKEEPING
Dedicate equipment as much as possible Reduces the need for frequent cleanings between batches. For example, a bottling plant could have separate process and filling lines for milk system and juice system.	**Review cleaning chemicals** Check to see environmental impacts of cleaning chemicals. Can alternative chemicals be used which have less environmental impact without compromising cleaning ability? For example, substitute nitric acid for phosphoric acid to reduce plant phosphorus discharge. **Integrate environmental considerations into all aspects of processing** Remember there are always two product streams, one to market, the other to "drain" or disposal; both are equally important and must be considered in making production decisions. Know your process well; then you can find ways to reduce pollution. **Perform preventive maintenance** Routinely check for leaks in valves and fittings. Repair immediately. **Install CIP (Cleaning In Place) monitoring systems on automated CIP processes** Monitor flow, time, temperature and conductivity. Can provide information leading to reduction of water and cleaning chemicals, optimization of system. **Install computer-controlled processing system** Reduces the potential for "bad" runs which will subsequently need to be disposed of. **Adopt a definite waste prevention program and build an educational program** Helps to ensure that all plant personnel are aware of waste prevention concerns.	**Install an air blow system in milk intake (receiving station)** System involves automated CIP (Cleaning In Place) system which uses pressurized air to chase milk to storage silos. This removes as much product as possible before cleaning. **Collect separator "desludge" for alternative disposal** If handled properly this high strength waste could be used as a supplement to animal feed, for example, hog feed. **Install reclaim systems for acid/caustic** Can use simple gravity separation techniques such as cone bottom tanks to segregate solids and decant chemicals for next day's first wash. Eliminates the need to dump all chemicals each day. **Study the plant and develop a material balance** Determine where losses occur and take steps to modify and replace unsatisfactory equipment. Where improper maintenance is the cause of losses, a specific maintenance program could be instigated and maintained.	**Install spill containment** Spills can be contained and managed appropriately rather than draining to wastewater treatment system and causing system upsets. **Spill cleanup procedures** Establish procedures for what to do with a spill. Reduces chance of spill being discharged to wastewater treatment plant. **Provide slope for milk trucks during unloading** Trucks must be emptied as completely as possible before washing. This reduces the potential for product going to the drain. Can be done with ramps or by actually sloping the floor ($^1/_4$" per foot is desirable). **Provide high pressure nozzles on all water hoses** Allows for adequate cleaning with less water. Also automatically shuts off water stream when not in use. **Provide drip pans** Place in areas of leaky valves and seals to collect product wastes rather than have these run directly on the floor and to the floor drains. **Check raw product quality, for example, antibiotics, in small quantity increments** Sample milk on delivery and evaluate for regulated constituents before mixing with large quantities to be processed. Minimizes large quantities of raw product needing disposal. **Develop procedures for handling return product** By-product outlets minimize waste hauling.

APPENDIX

Checklist 5. POLLUTION PREVENTION OPPORTUNITIES FOR THE FLUID MILK PROCESSING INDUSTRY* (continued)

PROCESS MODIFICATION	PROCESS OPERATION AND MAINTENANCE	MATERIAL RECYCLE, REUSE AND RECOVERY	GOOD HOUSEKEEPING
	Instruct plant personnel completely in proper operation and handling of all dairy plant processing equipment Major losses are due to poorly maintained equipment and to negligence of inadequately trained and insufficiently supervised personnel. **Install suitable liquid level controls** Use controls with automatic pump stops, alarms, and other devices at all points where overflows could occur, such as storage tanks, processing tanks, filler bowls and Cleaning In Place tanks. **Use care in materials handling** To avoid spillage of cased or packaged dairy products and product ingredients. **Repair or replace all worn out and obsolete equipment and/or parts of equipment including sanitary valves, fittings and pumps** To the extent possible, drips and leaks occurring during the processing run should be collected in containers and not allowed to go down floor drains.		

* Prepared by Rob Lamppa, Engineering Department, Land O'Lakes, Inc., Arden Hills, MN 55126. Phone: (612) 481-2741.

Checklist 6.   POLLUTION PREVENTION OPPORTUNITIES FOR THE MEAT PACKING (BEEF) INDUSTRY*

The keys to pollution prevention are to prevent product and contaminants from entering the wastestream and to reduce water use to a minimum.

MANAGEMENT'S ROLE	PROCESS MODIFICATION	GOOD HOUSEKEEPING	PRETREATMENT OPTIONS
Maintain motivation and support The success or failure of a program will depend on management's attitudes and actions. **Educate the employees** Provide instruction on proper water use, waste management, and cleaning procedures. Update training on a regular basis. **Appoint a water/waste supervisor** Allow the supervisor(s) to control water use. **Develop a job description for all personnel** Make waste management and waste reduction part of the job description. **Develop a preventative maintenance program** Maintain your plant at maximum efficiency to reduce waste. **Monitor cleanup** Strive for an efficient, environmentally conscious sanitation crew. **Plan a system for by-product recovery** Your wastes can become someone else's resources. **Conduct planning sessions** Get management and employees together to discuss suggestions for waste reduction.	**Allow sufficient time for the carcass to bleed after slaughter** Collect as much blood as possible at the slaughter site rather than allowing the blood to enter the packing line. **Collect the blood** Use troughs and curbs where necessary to direct the flow of blood. **Dry clean the paunch** Remove and collect paunch contents. **Transfer paunch contents using a dry system** Use a screw conveyor or an air-energized system. **Collect all trimmings** Do not let solids enter the wastestream. Keep all by-products off the floor. **Use a high-pressure, low-volume water system for all wet cleaning** Adjust the pressure as needed so minimum water pressure and water volume are used for any cleaning operation. **Install mechanical and automatic controls where necessary** In addition to employee monitoring, mechanical and automatic controls can assist in reducing waste.	**Dry clean as much as possible before washdown** Do not use the hose as a broom. When solid material is kept out of the water stream, treatment costs are less because the solids do not have to be removed from the wastewater stream. **Maintain a water record system** Monitor water use, and waste load and wastewater discharge quantities. Use only as much water as necessary. **Do not let blood coagulate** Coagulated blood requires large amounts of water for removal. **Trim all loose particles from the carcass** By removing blood clots from the neck area and loose particles from the carcass, you can prevent these by-products from entering the wastestream when the carcass is rinsed with a high-pressure washer. **Segregate and collect all by-products** This includes not only the slaughterhouse but also the holding pens and the retail cutting room. **Continuously look for areas in the plant where waste reduction can be implemented.** Observing separate processing areas helps you gain insight into how each area contributes to overall waste production. These observations provide opportunites for integrating water use reduction with solid waste reduction. **Use the correct detergent for cleaning** Using the correct detergent in the correct amount allows for cleaning with minimal rinsing.	**Flow equalization** **Screening** Static, vibrating and rotary screens can be used to capture solid particles. **Centrifuges** **Grease and suspended solids separation** Settling basins and dissolved air flotation, along with electrocoagulation and lignosulfate, can be used to separate grease and suspended solids from the wastewater. **Biological processes** Aerobic methods include basins, trickling filters, and contact stabilization. Anaerobic methods include basins and digesters.

* Prepared by John Polanski, Research Assistant, Minnesota Technical Assistance Program (MnTAP), University of Minnesota, Suite 207, 1313 Fifth Street S.E., Minneapolis, MN 55414 and Dr. Roy E. Carawan, Ph.D., Professor, Department of Food Science, North Carolina State University, Cooperative Extension Service, Box 7624, 129 Schaub Hall, Raleigh, NC 27695-7624. Printed with permission of the Minnesota Technical Assistance Program and North Carolina State University. © Copyright 1993.

Checklist 7. POLLUTION PREVENTION OPPORTUNITIES FOR THE METAL FABRICATION INDUSTRY*

PROCESS MODIFICATION	PROCESS OPERATION AND MAINTENANCE	MATERIAL RECYCLE, REUSE AND RECOVERY	GOOD HOUSEKEEPING
Change to UV-cured coatings Eliminates use of carrier solvents, maximizes paint transfer efficiency, and minimizes overspray wastes. **Change to powder coatings** Eliminates use of carrier solvents, maximizes paint transfer efficiency, and overspray powder can be collected, filtered and reused. **Change to synthetic fluids** Synthetic fluids are less susceptible to contamination, therefore have a longer useful life. **Change to gas coolant** Use a gas coolant for certain applications instead of a liquid. **Mechanize drag out** Eliminates possibility of employee using too short a drag out time, maintains product QA/QC standards if timing is set properly. **Install drip bars** Drip bars allow personnel to drain part hands-free without waiting so personnel will not use too short a drag out time. **Reduce pockets on parts** Place parts on drag out rack to minimize chances of chemical pooling in corners or in other pockets. **Use countercurrent rinses** These rinses dramatically reduce the amount of water required for rinsing and, therefore, reduce the amount of wastewater to be treated or sent for metal recovery.	**Optimize bath concentrations** Only replace plating chemical when necessary. Lengthens bath life. **Agitate rinse bath** Agitation promotes better rinsing. Agitate water or part. **Lengthen drag out time** Allows more chemical to drip back to process tank, so reduces the amount of chemical introduced in rinse water. **Establish drag out timing** Post drag out times at tanks to remind employees. **Use foot pump or photosensor to activate rinse** These items allow use of water only when processing parts. A photosensor may be used on automatic plating lines. **Use demineralized water** For mixing purposes, use of high quality water mitigates contamination problems of machining fluid. **Install bath filter** Filter can remove particulates and trace contaminant organics in the process bath, lengthens bath life. Use a filter that can be unrolled, cleaned and reused. **Extend life of fluid** Clean fluid through filtration and clarification and use of specialized biocides. **Raw material purity** Use high quality raw materials in bath so bath will not become contaminated as quickly.	**Recycle metalworking fluids** Extend usable life of fluid through filtration, skimming, dissolved air flotation, coalescing, hydrocycloning, centrifugation and pasteurization. **Reuse high performance fluids** Reuse hydraulic fluids that no longer meet spec for less stringent spec cutting oils. **Reuse secondary rinse** Reuse second rinse as primary rinse or makeup for cleaning solutions. **Reuse deionized rinse water** Depending on product, this rinse water can be reused in a plating bath as evaporated water makeup. **Ion exchange on rinse water** Ion exchange can be used to concentrate metals in rinse waters and metal can be recovered from the ion exchange acid regenerant stream. **Reverse osmosis** Concentrate drag out for reuse in plating bath; the water stream can also be reused. **Electrodialysis** Recover chromium from hard chromium plating baths and rinse waters. **Electrowinning** Recover metals from spent plating baths or ion exchange acid regenerant streams. **Evaporation** Concentrate drag out for reuse; the water condensate can also be reused.	**Seal and wiper replacement** Reduces chance of oil contamination of metalworking fluid if seal or wiper should fail. **Keep fluids from floor drains** Do not allow discharge of spills or spent fluids to sewer. Eliminate floor drains if necessary. **Install drain boards or drip guards** Boards and guards minimize spillage between tanks and are sloped away from rinse tanks so drag out fluids drain back to plating tanks.

Checklist 7.   POLLUTION PREVENTION OPPORTUNITIES FOR THE METAL FABRICATION INDUSTRY* (continued)

PROCESS MODIFICATION	PROCESS OPERATION AND MAINTENANCE	MATERIAL RECYCLE, REUSE AND RECOVERY	GOOD HOUSEKEEPING
Use spray or fog rinsing Reduces rinse water quantity required and can also be used over plating baths.		**Reuse mild acid rinse water** Use mild acid rinse water as influent to rinse following alkaline cleaning bath. Improves efficiency of rinse so less rinse water is required. **Reuse alkaline rinse water** Reuse rinse water from an alkaline cleaner operation to rinse parts from an acid cleaning operation. **Reuse spent acid/alkaline** Spent acid can be used to neutralize an alkaline wastestream. Spent alkali can be used to neutralize an acid wastestream.	

* Prepared by Alison Gemmell, Del Monte Foods, Walnut Creek, CA.

Checklist 8.   POLLUTION PREVENTION OPPORTUNITIES FOR THE METAL FINISHING INDUSTRY*

PROCESS MODIFICATION	PROCESS OPERATION AND MAINTENANCE	MATERIAL RECYCLE, REUSE AND RECOVERY	GOOD HOUSEKEEPING
Use spray or fog rinsing Reduces rinse water amount required and can also be used over plating baths. **Use countercurrent rinses** These rinses dramatically reduce the amount of water required for rinsing and therefore reduce the amount of wastewater to be treated or sent for metal recovery. **Mechanize drag out** Eliminates possibility of employee using too short a drag out time, maintains product QA/QC standards if timing is set properly. **Convert to dry floor** Reduces chances of spills reaching floor drains or causing upset in wastewater pretreatment plant.	**Increase bath temperature** Evaporates bath water so relatively clean waste rinse water can be reused as bath makeup water. Reduces solution viscosity so more chemical drains back to process tank during drag out. DO NOT USE ON CYANIDE OR HEXAVALENT CHROMIUM BATHS. **Use deionized (DI) water** Use DI water in plating baths, static rinses and, if practical, in running rinses. DI water reduces impurities in the plating bath to extend its life and minimizes the precipitation of minerals in water as sludge. **Raw material purity** Use high quality raw materials in bath so bath will not become contaminated as quickly.	**Segregate wastestreams** Increases recovery and treatment technology efficiencies. Acidic/alkaline. Chrome/non-chrome. Concentrated/dilute. Chelated/non-chelated. Cyanide/non-cyanide. **Reuse deionized rinse water** Depending on product, this rinse water can be reused in a plating bath as evaporated water makeup. **Ion exchange on rinse water** Ion exchange can be used to concentrate metals in rinse waters and metal can be recovered from the ion exchange acid regenerant stream. **Reverse osmosis** Concentrate drag out for reuse in plating bath; the water stream can also be reused.	**Control inventory** Do not allow material to exceed shelf life. Use materials on a first-in, first-out basis. **Install spill containment** Spills can be contained and managed. Reduces wastewater treatment upsets. **Spill cleanup procedures** Establish procedures for what to do with a spill. Reduces chance of spill being discharged to wastewater treatment plant or environment. **Buy appropriate amounts** Buy materials in small quantities if only small amounts are required. **Perform preventive maintenance** Routinely check for leaks in valves and fittings. Repair immediately.

Checklist 8. POLLUTION PREVENTION OPPORTUNITIES FOR THE METAL FINISHING INDUSTRY* (continued)

PROCESS MODIFICATION	PROCESS OPERATION AND MAINTENANCE	MATERIAL RECYCLE, REUSE AND RECOVERY	GOOD HOUSEKEEPING
Use different process Replace toxic cadmium plating with relatively nontoxic aluminum ion vapor deposition to achieve metal hardening properties. **Eliminate cyanide baths** Change to a noncyanide plating bath. Alternative chemistries are available with the exception of copper strike. **Install drip bars** Drip bars allow personnel to drain part hands-free without waiting so personnel will not use too short a drag out time. **Use static rinses** Static rinses usually follow the plating bath and capture the most concentrated drag out for return to the plating bath or for metal recovery. **Reduce pockets on parts** Place parts on drag out rack to minimize chance of chemical pooling in corners or in other pockets.	**Optimize bath concentrations** Only replace plating chemical when necessary. Lengthens bath life. **Reduce bath dumps** Optimize bath operation so bath dumps are infrequent. **Eliminate intermittent jobs** Stop performing small plating operations that generate intermittent wastestreams that personnel are not familiar with treating. **Agitate rinse bath** Agitation promotes better rinsing. Agitate water or part. **Lengthen drag out time** Allows more chemical to drip back to process tank, so reduces the amount of chemical introduced in rinse water. **Use foot pump or photosensor to activate rinse** These items allow use of water only when processing parts. A photosensor may be used on automatic plating lines. **Install bath filter** Filter can remove particulates and trace contaminant organics in the process bath; lengthens bath life. Use a filter that can be unrolled, cleaned and reused. **Use conductivity sensor** This sensor gives an indication of the cleanliness of the rinse water. Sensor can be designed to trigger clean rinse water flow when the tank water gets too dirty. Also allows better QA/QC. **Install flow restrictors** **Install flow control meter**	**Reuse spent acid/alkaline** Spent acid can be used to neutralize an alkaline wastestream. Spent alkali can be used to neutralize an acid wastestream. **Electrowinning** Recover metals from spent plating baths or ion exchange acid regenerant streams. **Evaporation** Concentrate drag out for reuse; the water condensate can also be reused. **Electrodialysis** Recover chromium from hard chromium plating baths and rinse waters. **Reuse mild acid rinse water** Use mild acid rinse water as influent to rinse following alkaline cleaning bath. Improves efficiency of rinse so less rinse water is required.	**Establish drag out timing** Post drag out times at tanks to remind employees. **Cover outdoor storage** Divert clean storm water away from storage areas. **Install drain boards or drip guards** Boards and guards minimize spillage between tanks and are sloped away from rinse tanks so drag out fluids drain back to plating tanks.

* Prepared by Alison Gemmell, Del Monte Foods, Walnut Creek, CA.

Checklist 9.   POLLUTION PREVENTION OPPORTUNITIES FOR THE OIL AND GAS EXTRACTION INDUSTRY*

The keys are to maximize oil separation and minimize spillage. Fail-safe devices are necessary since many of these discharge locations are unattended.

Y/N	OPPORTUNITIES	COMMENTS
___	Install free water knock-out tank with sufficient detention time and pressure relief valve	To maximize oil/water separation, especially at high flow
___	Install inlet and outlet baffles in gravity separation tank	To prevent short-circuiting
___	Install high level alarm, remote dialer and pump shutoff in separation tank	To prevent tank overflow and to alert operator at remote location
___	Install oil/water interface sensor at the bottom of the oil retention baffle of the separation tank, or install automatic oil skimming equipment	To prevent over-accumulation of oil and carryover to the water discharge
___	Install heater treater if appropriate	To enhance oil/water separation, especially for low viscosity crude oil
___	Install elevated "gooseneck" surge box at discharge end of the separation tank	To prevent accidental spill of floating oil in tank from excessive drawdown of water
___	Provide steady flow from tank to secondary oil removal units like dissolved air flotation unit or WEMCO depurator	Flotation units work best with uniform flow rate and oil and grease content of a few hundred mg/L
___	Provide buffer storage capacity between separation device and the discharge point to the sewer	To allow for additional storage safeguards for system malfunction, especially for unattended locations, until the next operator visit
___	Add polymer and adjust dosage frequently	To enhance oil and water separation
___	Cover separator and flotation units and capture volatile organics emissions through vacuum suction	Volatile organics may be removed through chilling, activated carbon or catalytic oxidation
___	Install explosimeter, if available, or oil sensor, together with auto shutoff valve for sewer discharge and flow diversion; provide temporary holding capability	To minimize oil discharge to sewer in the event of accidental spill or overloading of the flotation unit
___	Store treated water for reuse	Use treated water for reinjection
___	Discharge brine after further treatment to salt water channel, if available	High salt content in oil field brine water may affect reuse of treated municipal wastewater

* Prepared by Philip Lo, Industrial Waste Section, County Sanitation Districts of Los Angeles County, Whittier, CA.

Checklist 10. POLLUTION PREVENTION OPPORTUNITIES FOR THE PETROLEUM REFINING INDUSTRY*

The keys to pollution prevention for the petroleum refining industry are, for the short term, waste segregation, good operating practices and source control for oily wastes. For the medium term, the driving force is probably product reformulation, which has resulted in production changes in meeting limitations for airborne toxic compounds and vapor pressure in fuel products. For the longer term, the keys may be more targeted hydrocarbon rebuilding and reforming to produce the desirable fuel components, while avoiding the undesirable toxic ones. More specifically, removal of undesirable precursors in reforming, isomerization, catalytic conversions and expanded use of hydrogenation may hold the most promise.

(The following checklist presents a compilation of pollution prevention opportunities. However, since every refinery is unique, some of the opportunties may be more applicable to one refinery than to another. Please use the checklist with caution.)

Y/N	OPPORTUNITIES	COMMENTS
	I. Good Operating Practices	
___	Specify lower bottom sludge and water content for crude oil supply	To reduce wastes in storage tank and desalter through improved separation of water and bottom sludge at extraction
___	Use recycled water as makeup water for crude desalter	Recycled water quality is sufficient for desalting
___	Reroute desalter water with emulsifiers to intermediate tankage	To minimize emulsifier carryover to API separator
___	Segregate and dispose of ballast water to salt water channel, if available	To minimize brine contamination of treated water for reuse
___	Eliminate moisture contact with oxygenates (MTBE and methanol) in storage and process	To minimize water contamination of oxygenates
___	Replace desalting with an aggressive chemical treatment system for applicable situation, through crude oil dehydration in tankage with emulsion breaker, chloride reduction with caustic injection, ammonia replacement with neutralizing amine, film inhibitor feed rate optimization and anti-foulant injection to debutanizer heat exchanger (*Oil & Gas Journal*, 3/20/1989, pg. 60)	To eliminate desalter water blowdown, which could be high in benzene and emulsifiers, while maintaining corrosion protection
___	Segregate and discharge blowdown and water treatment regenerate to salt water channel or truck to ocean outlet, if available	To minimize brine contamination of treated water for reuse
___	Use corrosion-resistant lines in storage and slop oil tanks	To minimize sludge formation and need for tank cleaning
___	Install agitator in crude oil storage tanks	To minimize sludge accumulation
___	Avoid high shear pumping of oily wastes; use Archimedean screw pumps as appropriate	To minimize emulsion formation
___	Install tank cover and seal	To minimize emission loss and moisture entry
___	Upgrade to non-leaking pump seals	To eliminate leak losses
___	Replace packed pump with mechanical seal pump	To eliminate leak losses
___	Install sealless pump	To eliminate leaks and fugitive emissions, though such pumps have limited service
___	Maintain pump seals regularly	To prevent leaks
___	Recycle seal flushes and purges	To minimize wastes for treatment
___	Recycle seal and bearing cooling water stream	To minimize wastes for treatment
___	Pave process area	To minimize dirt entry to sewer
___	Install cover for sewer drain	To minimize dirt entry to sewer
___	Load and unload catalytic fines in closed system	To prevent fines from becoming wastes
___	Recover coke fines for sale with coke	To prevent solids entry to sewer
___	Reuse recycled water for washdown if quality is desirable	To minimize need for discharge
___	Integrate process units to pass processing streams from one unit to the next, if appropriate	To avoid intermediate tankage but may lose operational flexibility

Checklist 10.   POLLUTION PREVENTION OPPORTUNITIES FOR THE PETROLEUM REFINING INDUSTRY* (continued)

Y/N	OPPORTUNITIES	COMMENTS
____	Blend fuels in-line	To avoid blending tankage
____	Install closed-loop sampling system	To flush materials back to the tank or pipeline and minimize volatile compound emissions and hydrocarbon discharge to wastewater
____	Use computer software to track all hazardous materials and wastes	To better manage virgin materials and wastestreams
____	Return oily wastewater and sludge from distribution and sales terminals to refinery as permitted by federal and state recycling regulations	To afford proper handling of oily wastes
____	Segregate scrap metals for sale	To reclaim metals for reuse
____	Recondition valves and vessels for reuse	To further reduce scrap metal wastes
____	Recover and reuse sandblasting grit as blasting media or as a light aggregate in concrete production	To minimize need for grit disposal, but beware of lead and heavy metal contamination

Storm Water Management

Y/N	OPPORTUNITIES	COMMENTS
____	Selectively cover loading rack and process areas to divert rainwater	To preclude rainwater contamination
____	Segregate storm water collection system from process drainage	To prevent cross contamination of storm water
____	Impound rainwater in collection basin or tank as appropriate	To hold water pending determination of treatment needs
____	Sweep streets and provide weirs in catch basin to exclude dirt	To prevent dirt entry to storm drain
____	Keep tank farm and process area clean, including secondary containment areas	To avoid contaminating rainwater
____	Reuse rainwater after gravity recovery of oil and solids	To minimize need for discharge
____	Discharge rainwater to public storm drain system under NPDES permit	To avoid using sewer capacity
____	Dike process area that drains to storm water collection system as appropriate	To prevent contamination of storm water
____	Regularly clean out drainage system to remove accumulated dirt	To minimize contamination of storm water

Firefighting Water and Spillage Management

Y/N	OPPORTUNITIES	COMMENTS
____	Install tank overfill prevention system	To prevent spills
____	Pave areas under pipe rack	To facilitate leak detection
____	Contain spillage with diking and absorbent materials	To minimize spreading of spillage
____	Recover and reuse spillage	To minimize need for disposal
____	Impound firefighting water in rainwater basins or storage tanks as appropriate	To hold and test before discharge or reuse
____	Prevent automatic overflow of storm drain to wastewater collection system	To prevent spills and firefighting water that entered the storm drain system from overwhelming the wastewater treatment system

Groundwater and Contaminated Soil Cleanup

Y/N	OPPORTUNITIES	COMMENTS
____	Recover floatable oil for reuse	To recover oil at source and avoid entrainment in transport
____	Pretreat and reinject treated groundwater if appropriate	To eliminate need for discharge to sewer
____	Reuse hydrocarbon contaminated soil as filler in asphalt paving manufacture	To avoid need for disposal

Checklist 10.   POLLUTION PREVENTION OPPORTUNITIES FOR THE PETROLEUM REFINING INDUSTRY* (continued)

Y/N	OPPORTUNITIES	COMMENTS
——	Reuse soil with mineral contents similar to shale as raw material substitute for cement kiln; reuse in pre-heater and calciner kiln is preferred, to maximize volatile hydrocarbon destruction	To avoid need for disposal

II.   Production Process Modifications

Separation Process

Y/N	OPPORTUNITIES	COMMENTS
——	Improve separation in distillation column through various means including the following: • Increase the reflux ratio, • Add a new section to the column, • Match feed condition with the right feed tray, • Preheat column feed, • Install reusable insulation to prevent heat loss and fluctuation of column condition with weather	To increase yield and the separation of volatiles, for example, benzene
——	Lower the reboiler temperature in distillation column through various means including the following: • Retray column to lower pressure drop, • Increase size of vapor line to reduce pressure drop, • Use lower pressure steam or desuperheated steam, • Install a thermocompressor, and • Lower column pressure	To minimize degradation and waste generation from high reboiler temperature
——	Improve overhead condensers to capture overhead losses	To minimize flaring and emissions

Conversion and Upgrading Processes

Y/N	OPPORTUNITIES	COMMENTS
——	Improve conversion in reactors through various means including the following: • Distribute feeds better at inlets and outlets, • Upgrade catalysts continuously, • Provide separate reactor for recycled streams for more ideal reactor conditions, • Adjust heating and cooling to avoid hot spots, • Improve control to maintain optimum conditions in reactor, • Use inhibitors to minimize unwanted side reactions	To improve yield and conversion and minimize the formation of undesirable compounds from side reactions
——	Filter catalyst fines from decanter oil from the fluid catalytic cracking unit (FCCU)	To recover and reuse catalyst
——	Reclaim hydroprocessing catalysts for metals and alumina	To recover the metals on the catalysts like cobalt and molybdenum, as well as those removed from oil like nickel and vanadium; the alumina carrier is also recovered
——	Recycle catalyst for bauxite in cement manufacturing	To minimize need for disposal
——	Recover fluoride from spent caustics from an HF alkylation process by calcium precipitation	To produce calcium fluoride solids for use in cement industry or as fluxing agent in glass and steel industries
——	Reuse spent fluidized catalytic cracking unit (FCCU) catalysts in residue FCCU	To reuse catalysts in another FCCU where higher metal content on the catalysts can be tolerated
——	Reactivate catalysts for reuse	To reuse catalysts after the nickel and vanadium deposits are removed
——	Regenerate spent sulfuric acid by commercial reclaimer using incineration	To regenerate the acid and avoid neutralization
——	Reclaim extraction solvents like sulfolane and sulfinol	To recover solvents for reuse, with the residuals going for feed to a sulfuric acid plant because of their high BTU and sulfur contents

APPENDIX

Checklist 10.   POLLUTION PREVENTION OPPORTUNITIES FOR THE PETROLEUM REFINING INDUSTRY* (continued)

Y/N	OPPORTUNITIES	COMMENTS
	Product Treatment	
___	Minimize the amount of caustic and rinse water used for product treatment through better contacting and recycling	To minimize need for treatment of wastewater
___	Consider hydrotreating for pollutant removal	To eliminate the use of caustic and water in product treatment
___	Send spent caustics to reclaimer	To reclaim cresylic and naphthenic compounds for sale
___	Reuse spent sulfuric caustics for paper manufacturing	To reuse the caustics if the strength is high enough
___	Regenerate clay from jet fuel filtration by washing with naphtha and drying by steam heating and feeding to furnace	To recycle filter clay
	Equipment Cleaning — Heat Exchangers	
___	Use lower pressure steam	To reduce tube-wall temperature and sludge formation
___	Desuperheat steam	To reduce tube-wall temperature and increase the effective surface area of the exchanger because the heat transfer coefficient of condensing steam is ten times greater than that of superheated steam
___	Install a thermocompressor	To reduce tube-wall temperature by combining high and low pressure stream
___	Use staged heating	To minimize degradation, staged heating can be accomplished first using waste heat, then low pressure steam and finally, desuperheated high pressure steam
___	Use on-line cleaning techniques for exchangers	Recirculating sponge balls and reversing brushes can be used to reduce exchanger maintenance and also to keep the tube surface clean so that lower temperature heat sources can be used
___	Use non-corroding tube	Corroded tube surfaces foul more quickly than uncorroded ones
	Waste Gas Treatment	
___	Regenerate di-ethanol-amine (DEA) using slip stream filtration in addition to carbon filtration	To remove degradation products and prolong DEA life
___	Substitute Sulften Sulfur Recovery Process for Beavon Process	To avoid generation of spent Stretford Solution which contains vanadium
___	Regenerate activated carbon from gas scrubbing	To avoid need for disposal
	Wastewater and Sludge Treatment	
___	Add forebay skimming for API separator	To recover more hydrocarbons for recycle
___	Use floating roof on treatment tanks and drains	To minimize air emissions
___	Use pressurized air in flotation	To minimize air emissions
___	Pretreat desalter water blowdown before commingling with other oily wastes, using absorption with light oil, or stripping with steam, nitrogen, methane or vacuum	To pretreat the high concentration of benzene and, possibly, emulsifiers in the desalter water blowdown
___	Thicken sludge in sludge tank and decant supernatant	To aid in sludge dewatering
___	Treat sludge with heat and chemicals to release more oil and water	To further reduce hydrocarbon content in sludge
___	Dewater sludge to cake form	To minimize water content and remove some oil
___	Reclaim hydrocarbons in sludge by feeding it to a delayed coker which produces fuel grade coke	To dispose of solids and to reclaim hydrocarbon value
___	Use solvent extraction to remove hydrocarbons from sludge	To treat sludge for disposal and recover hydrocarbons

Checklist 10. POLLUTION PREVENTION OPPORTUNITIES FOR THE PETROLEUM REFINING INDUSTRY* (continued)

Y/N	OPPORTUNITIES	COMMENTS
___	Use high temperature sludge drying to desorb hydrocarbons	To treat sludge for disposal and recover hydrocarbons
___	Feed sludge cake to cement kiln for energy recovery	To recycle sludge for its energy value
___	Evaluate gasification of oily wastes	To convert waste to usable methane
	Utility Production — Steam, Hydrogen	
___	Use closed-loop cooling water system	To minimize water loss
___	Demineralize cooling tower feedwater	To reduce cleaning and waste generation
___	Use polymers for boiler feedwater treatment	To reduce boiler cleaning
___	Collect condensate for reuse	To avoid sewer discharge
___	Use non-chromate corrosion inhibitor	To minimize chromate emissions and also chromate treatment in blowdown
___	Reclaim hydrogen plant catalysts	To recover materials in catalysts
III.	**Product Reformulation and Material Substitution**	
___	Reformulate leaded gasoline to unleaded alternative with MTBE	To eliminate lead from gasoline and product storage tanks
___	Reduce benzene and other volatile hydrocarbons in gasoline through reblending with oxygenates like MTBE	To decrease emissions of air toxics and smog-forming volatile organics

* Prepared by Philip Lo, Industrial Waste Section, County Sanitation Districts of Los Angeles County, Whittier, CA.

APPENDIX

Checklist 11.   POLLUTION PREVENTION OPPORTUNITIES FOR THE PHOTO PROCESSING INDUSTRY*

PROCESS MODIFICATION	PROCESS OPERATION AND MAINTENANCE	MATERIAL RECYCLE, REUSE AND RECOVERY	GOOD HOUSEKEEPING
Change to silver-less film Silver does not end up in rinse waters or spent fix baths when this type of film is used. Examples include vesicular, diazo, and electrostatic films. **Add ammonium thiosulfate** Addition of this chemical to silver-contaminated bath extends the useful life of the bath. **Add acetic acid** Acetic acid added to the fix bath keeps the pH low to maximize soluble complexes, therefore, extend bath life. **Purchase new machines** Purchase new developing machines that use less rinse water (for example, countercurrent rinsing) and/or have squeegees or air blades to reduce drag out from chemical baths to rinse waters.	**Adjust replenishment** Adjust chemical replenishment rates and wash water flow rates on photoprocessor to optimize bath life and reduce wastewater quantity.	**Metal replacement canister** Install canister on rinse waters and on effluent of wastewater electrowin to recover silver from solution and reduce toxicity of wastewater. **Install electrowin** Install electrowin unit on first rinse and developer wastestreams to recover silver from solution and reduce toxicity of wastewater. **Install rinse water recycling** Reduces wastewater. **Recover fix bath silver** Install electrowin unit on fix bath of photoprocessor. Extends bath life. **Segregate wastestreams** Spent fix baths should be segregated from rinse waters and developer solutions because silver recovery is more efficient on the more concentrated spent wastestream.	**Control inventory** Do not allow material to exceed shelf life and then have to be discarded as waste. Use materials on a first-in, first-out basis. Do not discharge expired products to wastewater treatment system. **Keep lids on solutions** In storage, keep lids on bulk solutions to prevent oxidation and contamination. **Spill containment** Store chemicals and metal recovery system in double-wall tanks or in diked area to prevent accidental discharges.

* Prepared by Alison Gemmell, Del Monte Foods, Walnut Creek, CA.

Checklist 12.   POLLUTION PREVENTION OPPORTUNITIES FOR THE PRINTED CIRCUIT BOARD MANUFACTURING INDUSTRY*

PROCESS MODIFICATION	PROCESS OPERATION AND MAINTENANCE	MATERIAL RECYCLE, REUSE AND RECOVERY	GOOD HOUSEKEEPING
Mechanize drag out Eliminates possibility of employee using too short a drag out time, maintains product QA/QC standards if timing is set properly. **Reduce pockets on parts** Place parts on drag out rack to minimize chances of chemical pooling in corners or in other pockets. **Use static rinses** Static rinses usually follow the plating bath and capture the most concentrated drag out for return to the plating bath or for metal recovery.	**Lengthen drag out time** Allows more chemical to drip back to process tank, so reduces the amount of chemical introduced in rinse water. **Establish drag out timing** Post drag out times at tanks to remind employees. **Use conductivity sensor** This sensor gives an indication of the cleanliness of the rinse water. Sensor can be designed to trigger clean rinse water flow when the tank water gets too dirty. **Use foot pump or photosensor to activate rinse** These items allow use of water only when processing parts. A photosensor may be used on automatic plating machines.	**Segregate wastestreams** Increases recovery and treatment technology efficiencies. Acidic vs. alkaline. Concentrated metal (spent baths) vs. dilute metal (rinse water streams). Chelated vs. non-chelated streams. **Reuse deionized rinse water** Depending on product, this rinse water can be reused in a plating bath as evaporated water makeup. **Ion exchange on rinse water** Ion exchange can be used to concentrate metals in rinse waters and metal can be recovered from the ion exchange acid regenerant stream.	**Install drain boards or drip guards** Boards and guards minimize spillage between tanks and are sloped away from rinse tanks so drag out fluids drain back to plating tanks. **Spill cleanup procedures** Establish procedures for what to do with a spill. Reduces chance of spill being discharged to wastewater treatment plant. **Perform preventive maintenance** Routinely check for leaks in valves and fittings. Repair immediately. **Control inventory** Do not allow material to exceed shelf life. Use materials on a first-in, first-out basis.

Checklist 12. POLLUTION PREVENTION OPPORTUNITIES FOR THE PRINTED CIRCUIT BOARD MANUFACTURING INDUSTRY* (continued)

PROCESS MODIFICATION	PROCESS OPERATION AND MAINTENANCE	MATERIAL RECYCLE, REUSE AND RECOVERY	GOOD HOUSEKEEPING
Use countercurrent rinses These rinses dramatically reduce the amount of water required for rinsing and therefore reduce the amount of wastewater to be treated or sent for metal recovery.	**Agitate rinse bath** Agitation of water or part promotes better rinsing.	**Reuse spent acid/alkaline** Spent acid can be used to neutralize an alkaline wastestream. Spent alkali can be used to neutralize an acid wastestream.	**Buy appropriate amounts** Buy materials in small quantities if only small amounts are required.
Use spray or fog rinsing Reduces rinse water quantity required and can also be used over plating baths.	**Reduce bath dumps** Optimize bath operation so bath dumps are infrequent.	**Reverse osmosis** Concentrate drag out for reuse in plating bath; the water stream can also be reused.	**Cover outdoor storage** Divert clean storm water away from storage areas.
Eliminate chelated baths Change to a nonchelated plating bath to improve metal wastewater treatment. Chelated streams make it difficult to precipitate metal in wastewater treatment system.	**Install flow restrictors** Restrictors automatically reduce the amount of rinse water, so operators do not need to adjust inlet valves.	**Evaporation** Concentrate drag out for reuse; the water condensate can also be used.	**Install spill containment** Spills can be contained and managed. Reduces wastewater treatment upsets.
Buy efficient etch machine An efficient etch machine results in less copper in rinse water.	**Install drip bars** Drip bars allow personnel to drain parts hands-free without waiting, so personnel will not use too short a drag out time.	**Electrowinning** Recover metals from spent plating baths or ion exchange acid regenerant streams.	
Convert to dry floor Reduces chance of spills reaching floor drains or causing upset in wastewater pretreatment plant.	**Install bath filters** Filters can remove particulates and trace contaminant organics in the process bath. Lengthens bath life. Use filter that can be unrolled, cleaned and reused.	**Recover particulate copper** On debur operation, recover particulate copper, using centrifuge or paper filter. Reuse water.	
Electroless tanks Use continuous passivation on stainless steel components in electroless plating tanks to prevent copper plate-out. Copper plate-out needs to be stripped with nitric acid. Reduces amount of spent nitric acid that needs to be treated.	**Raw material purity** Use high quality raw materials in bath so bath will not become contaminated as quickly.	**Reuse mild acid rinse water** Use mild acid rinse water as influent to rinse following alkaline cleaning bath. Improves efficiency of rinse so less rinse water is required.	
	Increase bath temperature Evaporates bath water so relatively clean waste rinse water can be reused as bath makeup water. Reduces solution viscosity, so more chemical drains back to process tank during drag out.	**Recover copper sulfate** On microetch line, recover copper sulfate crystals directly from etch tank and reuse crystals in copper electro-plating baths.	
	Use deionized (DI) water Use DI water in plating baths, static rinses and, if practical, in running rinses. DI water reduces impurities in the plating bath to extend its life and minimizes the precipitation of minerals in water as sludge.	**Reclaim etchant** Send etchant to an off-site reclaimer instead of treating etchant in wastewater treatment system.	
	Optimize bath concentrations Only replace plating chemical when necessary. Lengthens bath life. Replace chemicals on electroless copper baths to extend bath life. Automate chemical replacement through on-line analyzers and chemical flowmeters.	**Recycle photoresist stripper** Decant spent photoresist stripper from polymer residue and recycle stripper. Do not discharge spent stripper to wastewater treatment system.	

* Prepared by Alison Gemmell, Del Monte Foods, Walnut Creek, CA.

APPENDIX

Checklist 13. POLLUTION PREVENTION OPPORTUNITIES FOR THE PRINTING INDUSTRY*

PROCESS MODIFICATION	PROCESS OPERATION AND MAINTENANCE	MATERIAL RECYCLE, REUSE AND RECOVERY	GOOD HOUSEKEEPING
Change to silver-less film Silver does not end up in rinse waters or spent fix baths when this type of film is used. Examples include vesicular, diazo, and electrostatic films. **Change to ultraviolet inks** Will not dry in ink fountain overnight. Reduces need for fountain cleaning. **Install laser platemaking** Eliminates need for photoprocessing. Expensive. **Install electronic imaging** Eliminates need for photoprocessing. Expensive. **Use flexographic process** Replaces metal etch processes for plate processing. **Purchase new machines** 1. **Purchase new developing machines** that use less rinse water (for example, countercurrent rinsing) and/or have squeegees or air blades to reduce drag out from chemical baths to rinse waters. 2. **Purchase waterless paper and film developing machines** to reduce the volume of fix waste. **Use non-dry aerosol** Special non-drying aerosol materials can be sprayed on ink fountains to keep them from drying out overnight. Fewer ink fountain cleanings required. **Add acetic acid** Acetic acid added to the fix bath keeps the pH low to maximize soluble complexes, therefore extend bath life. **Add ammonium thiosulfate** Addition of this chemical to silver-contaminated bath extends the useful life of the bath.	**Adjust replenishment** Adjust chemical replenishment rates and wash water flow rates on photoprocessor to optimize bath life and reduce wastewater quantity. **Run similar jobs at once** Minimizes need for cleaning between jobs.	**Recycle inks** Recycle inks to make black ink instead of discharging to sewer. **Segregate wastestreams** Spent fix baths should be segregated from rinse waters and developer solutions because silver recovery is more efficient on the more concentrated spent fix wastestreams. **Install rinse water recycling** Reduces wastewater. **Metal replacement canister** Install canister on rinse waters and on effluent of wastewater electrowin to recover silver from solution and reduce toxicity of wastewater. **Recover fix bath silver** Install electrowin unit on fix bath of photoprocessor. Extends bath life. **Install electrowin** Install electrowin unit on first rinse and developer wastestreams to recover silver from solution and reduce toxicity of wastewater.	**Control inventory** Do not allow material to exceed shelf life and then have to be discarded as waste. Use materials on a first-in, first-out basis. Do not discharge expired products into wastewater treatment system. **Return used ink to vendor** Purchase ink from distributors who will take back unused or spent ink so ink will not be discharged to the sewer. **Keep lids on solutions** In storage, keep lids on bulk solutions to prevent oxidation and contamination. **Minimize spills** Use dry method cleanups. **Spill containment** Store chemicals in diked area to prevent accidental discharges.

* Prepared by Alison Gemmell, Del Monte Foods, Walnut Creek, CA.

Checklist 14. POLLUTION PREVENTION OPPORTUNITIES FOR THE KRAFT SEGMENT OF THE PULP AND PAPER INDUSTRY*

The keys to pollution prevention for the kraft segment of the pulp and paper industry are to purchase preservative-free wood fiber, maintain and operate chippers to produce chips of uniform dimensions, use chlorine dioxide for pulp bleaching and reuse treated wastewaters as makeup water for the log flumes.

PROCESS MODIFICATION	PROCESS OPERATION AND MAINTENANCE	MATERIAL RECYCLE, REUSE AND RECOVERY	GOOD HOUSEKEEPING
Process pulp with oxygen in an alkaline environment After conventional or extended pulping processes but prior to bleaching with chlorine and/or chlorine derivatives.	**Operate recovery boilers within selected operating guidelines** Minimizes total reduced sulfur emissions. Reduces atmospheric emissions of total reduced sulfur. Improves boiler efficiency.	**Reuse treated wastewater as makeup water for the log flumes** Conserves water and reduces discharge loadings of total suspended solids and BOD_5.	**Install retaining walls to contain chips and fugitive dusts from the chip piles** Minimizes chip and dust losses.
Purchase wood fiber (rough wood chips, sawdust) that has not been treated with wood preservatives, particularly chlorophenols Improves product and effluent quality.	**Use relatively low charges of anthraquinone in digester** Accelerates pulping process and increases pulp yield. Reduces effluent color and BOD_5.	**Recycle scrubber water to the causticizing circuit**	**Modify barge unloading operations to use front loader instead of clam shell** Minimizes chip and dust losses.
Modify existing (continuous) digesters or install new (batch or continuous) digesters to produce brownstock pulp of lower Kappa number Accomplishes greater degree of delignification with recoverable pulping chemicals as opposed to delignification with disposable bleach plant chemicals that need to be disposed of.	**Use high shear mixers for bleach plant chemical additions** Ensures efficient chemical application and minimizes localized overchlorination which generates unwanted chlorinated compounds.	**Collect and recover steam condensates from pulp dryers and paper machines**	**Designate material storage area** Provide protection, spill containment; keep area clean and organized; give one person the responsibility to maintain the area.
Steam strip digester, evaporator and turpentine condensates for the removal of total reduced sulfur and BOD_5 (methanol, acetone) Steam-to-feed ratios of 15 to 20 percent are necessary to achieve efficient BOD_5 removal. Stripper overheads are combusted in power boilers. Stripper bottoms are reused for brownstock washing.	**Maintain and operate chippers to produce chips of uniform dimensions with emphasis on chip thickness control** Improves yield and reduces rejects. More uniform pulping occurs which results in lower bleach plant chemical consumption.	**Reuse acid stage filtrates as dilution and wash water on first bleaching stage and on log sequence bleach lines, and use filtrate from the second extraction stage in the first extraction stage**	**Store packages properly and shelter from weather** Prevents damage, contamination and product degradation.
Convert recovery boilers to a low odor design by the elimination of direct contact evaporator, modification of secondary and tertiary combustion air systems and installation of a new economizer section	**Use instrumentation to monitor and control the application of chlorine and/or chlorine dioxide in the first bleaching stage** Maintain chlorine multiple or Kappa factor within ranges where chlorinated compounds are minimized.		**Minimize traffic through material storage area** Reduces contamination and dispersal of materials.
Use dry drum debarkers instead of hydraulic debarkers	**Install a total reduced sulfur control system for collection and combustion of vent streams from brownstock washers, foam tanks, black liquor filters, oxidation tanks and storage tanks**		
Install chip screening device designed to produce chips of uniform thickness	**Install new or modify existing brownstock pulp screening and deknotting systems so that black liquid is recovered and recycled to recovery system**		
Use chlorine dioxide instead of elemental chlorine in the first bleaching stage			

* Compiled by Mischelle Mische and Philip Lo, County Sanitation Districts of Los Angeles County, Whittier, CA. Information taken from US EPA Region X's "Model Pollution Plan for the Kraft Segment of the Pulp and Paper Industry," September 1992, EPA#910/9-92-030.

APPENDIX

Checklist 15.   POLLUTION PREVENTION OPPORTUNITIES FOR THE RADIATOR REPAIR INDUSTRY*

The keys are to minimize drag in and drag out of contaminants to and from the caustic boil-out tank and test tank, thereby reducing the amount of contaminants discharged with the flush booth water. Zero discharge to the sewer is possible; however, additional treatment and specialized equipment may be required. Waste minimization techniques for zero discharge may be somewhat different.

PROCESS MODIFICATION	PROCESS OPERATION AND MAINTENANCE	MATERIAL RECYCLE, REUSE AND RECOVERY	GOOD HOUSEKEEPING
Use a low-zinc flux Reduces zinc level in discharge; however, it may not produce a satisfactory soldering job. Experiment first.	**Remove as much oil as possible from oil cooler using compressed air** Minimizes drag in to boil-out tank.	**Filter solids and reuse test tank water when bath is cloudy** Minimizes tank changing.	**Secondary containment** For test tank, boil-out tank and sludge storage area to prevent spills and leaks from contaminating discharge.
Use smaller process tanks with ultrasonic cleaning Reduces volume of waste generated.	**Remove and recycle antifreeze, if any** Minimizes drag in.	**Reuse flushing booth rinses** Use for tank makeup to minimize water use.	
Use smaller test tanks For efficient operation to minimize volume of wastewater.	**Use higher pressure and lower water flow in flushing booth** Minimizes water use and need for disposal.	**Minimize use of cleaner or flux containing metal chelating compounds** Reduces interference with waste-water treatment.	
	Provide hang bars over caustic boil-out tank and test tank Allows convenient draining.		
	Provide drain board between tanks Minimizes spillage to floor.		
	Blow out residual test tank or caustic solution to tank using compressed air Minimizes drag out to flushing booth.		
	No soldering over the test tank Provide drip pans to collect excess solder or flux to minimize zinc and lead contamination of test water.		
	Pre-rinse radiator over boil-out tank using fog spray Minimizes drag out.		
	Maintain and monitor boil-out tank Minimizes need for tank changing.		
	Upon changing of tank, remove the solids, reuse the liquid and reconstitute the bath Minimizes volume of waste upon bath changing.		

* Prepared by Philip Lo and revised by Suzanne Wienke, County Sanitation Districts of Los Angeles County, Whittier, CA.

CHAPTER 10

INDUSTRIAL INSPECTION PROCEDURES

by

Bill Garrett

TABLE OF CONTENTS

Chapter 10.   INDUSTRIAL INSPECTION PROCEDURES

OBJECTIVES

Chapter 10. INDUSTRIAL INSPECTION PROCEDURES

Following completion of Chapter 10, you should be able to:

1. Explain to POTW officials and the regulated community the intent of pretreatment facility inspections and the importance of industrial inspections and monitoring activities,

2. Perform inspection tasks in a safe manner,

3. Prepare for and conduct on-site industrial inspections, and

4. Inspect the following types of industries:

    a. Electroplating and metal finishing,
    b. Petroleum refineries,
    c. Pulp and paper industries,
    d. Chemical manufacturing,
    e. Food processing,
    f. Dairy products,
    g. Rendering,
    h. Small businesses, and
    i. Centralized waste treatment facilities.

CHAPTER 10.   INDUSTRIAL INSPECTION PROCEDURES

10.0   REVIEW OF PREVIOUS CHAPTERS

All of the chapters in this manual have been prepared to provide you with the knowledge and skills needed to inspect industrial pretreatment facilities. Chapter 4, "Inspection of a Typical Industry," introduced you to the procedures for inspecting pretreatment facilities in order for you to have an understanding of what is required and the material presented in the following chapters. Chapter 4 explained the importance of knowing and following safe procedures, scheduling of inspections, how to enter an industry to conduct an inspection, the different levels of an inspection, what to do after walking through an industrial inspection, and the report writing after an inspection is completed. The Appendix of Chapter 4 contained a summary of the "Seattle Metro Industrial Waste Inspection Procedures." Now would be a good time to review the information in Chapter 4.

Chapter 5 stressed the importance of "Safety in Pretreatment Inspection and Sampling Work." Be sure to review this chapter whenever you have any questions regarding safe procedures. Always follow the slogan, "BETTER SAFE THAN SORRY." Chapter 6 discussed "Sampling Procedures for Wastewater." Chain of custody procedures must be followed at all times so that the results of any tests will be valid in court. Chapter 7 covered "Wastewater Flow Monitoring," which is very important when determining sewer-use service charges. Flowmeters should be calibrated on a regularly scheduled basis. Chapter 8 and 9, "Industrial Wastewaters" and "Pretreatment Technology (Source Control)," provide you with information essential for inspecting industrial manufacturing sites and pretreatment facilities. Inspectors need to know how wastestreams are produced and the waste constituents in these streams, as well as the purpose of pretreatment processes and the expected results and effluent quality. Whenever you are preparing for an inspection or conducting one, review the pertinent material in this manual so you will always be properly prepared.

10.1   IMPORTANT CONSIDERATIONS

10.10   Regulatory Intent

Always keep in mind the intent or objectives of a pretreatment facility inspection. The General Pretreatment Regulations (40 CFR Part 403) require industrial inspections and monitoring to ensure compliance with all applicable pretreatment regulations. Section 403.8(f)(2)(v) states that POTWs must develop procedures to:

> Randomly sample and analyze the effluent from industrial users and conduct surveillance and inspection activities in order to identify, independent of information supplied by industrial users, occasional and continuing noncompliance with Pretreatment Standards...

Inspections and monitoring are two of the most important *ONGOING* tasks in the implementation of a local pretreatment program. Information collected during inspections and monitoring activities will be the basis for all compliance and enforcement activities taken by POTWs against industrial users in violation of pretreatment standards and requirements.

10.11   Importance of Industrial Inspections

Industrial inspections are valuable in the overall implementation of pretreatment programs because they provide:

1. A means to check the completeness and accuracy of the industrial user's performance/compliance records,

2. A basis for deciding on and conducting monitoring activities at the industry,

3. A means for communicating and developing a good working relationship with industrial users,

4. A mechanism for maintaining current data on industrial users and determining the users' compliance status,

5. A means to evaluate construction of pretreatment facilities,

6. A means to assess the adequacy of a user's self-monitoring and reporting program and the terms of the industrial discharge permit,

7. A means to assess the potential for spills, and

8. A means to evaluate the operation and maintenance activities of an industrial user's pretreatment system.

Current data on industrial dischargers is necessary in a pretreatment program to identify sources of problems and to provide a foundation for developing or amending local discharge limits for industrial users. Industrial inspections will help the POTW maintain current data on all its industrial users, and, if performed on a frequent basis, the POTW should not need to repeat the industrial waste survey.[1]

---

[1] *Industrial Waste Survey.   A survey of all companies that discharge to a POTW. The survey identifies the magnitude of the wastewater flows and pollutants in the discharge.*

Properly implemented on-site industrial inspections set the groundwork for monitoring tasks associated with the pretreatment program. Industrial inspections can furnish the POTW with information needed to plan future monitoring activities (for example, compliance/enforcement activities, operating data, sampling frequencies and locations, safety considerations, and laboratory considerations/requirements). In addition to this type of information, the inspection can provide information about needed changes in existing procedures at the industrial facility. For example, if a firm changes its processes to produce fewer pollutants, the inspector might recommend that the POTW's analyses also be cut back to match the new discharge. Similarly, a series of inspections can show a trend or a change that necessitates a modification of monitoring schedules or frequency, or even a modification of the industry's discharge permit. In this way, inspections can be a useful tool for adjusting and refining the POTW's pretreatment program to more efficiently allocate resources and conduct its own operations.

An inspector should also check the industrial user's self-monitoring procedures and equipment to ensure that data obtained by these procedures and equipment are proper and accurate. In this regard, *YOUR* inspection should provide answers to the following questions:

1. Do the user's sampling locations sample the effluent from all process and nonprocess wastewater systems?

2. Is the sampling location specified in the permit adequate for the collection of a representative sample of the wastewater?

3. Is the user's sampling technique adequate to ensure the collection of a representative sample?

4. Will the user's permit sampling and monitoring requirements yield representative samples? and

5. Are the parameters specified in the user's permit adequate to cover all pollutants of concern that may be discharged by the permittee?

Industrial inspections can also help you establish and maintain a good rapport with the industrial users. The inspection is a good time for exchanging ideas and concerns. During an industrial inspection, POTW inspectors can inform the industrial user of any new or updated pretreatment regulations or offer technical assistance on pretreatment techniques, particularly to smaller industries.

10.12 Importance of Monitoring Activities

Accurate flow measurements and representative samples of industrial users' discharges are probably the most important tools an inspector can have for enforcing local standards and other legal authority provisions. Only with these tools will you be able to determine compliance with applicable regulations and wastewater discharge requirements. POTW personnel visit industrial users to collect samples much more often than to conduct a complete industrial inspection. Because of the potential for contamination of samples and significant errors in the results, it is essential that you exercise extreme care in selecting representative sampling locations, proper equipment

and appropriate sampling and analysis protocols (procedures). Sampling considerations and references for sampling and analytical procedures are provided in Chapter 6, "Sampling Procedures for Wastewater."

10.13 Safety

Industrial inspections and sampling activities are often carried out in hazardous situations. POTW personnel must be adequately trained to identify potential hazards and take necessary precautions to avoid dangerous situations. The supervisor is responsible for ensuring the safety of subordinates and should, therefore, document that they have received proper safety equipment and training. Chapter 5, "Safety in Pretreatment Inspection and Sampling Work," discusses the necessary safety equipment and safe procedures. Some of the more common and most dangerous hazards that POTW personnel will encounter while performing industrial pretreatment inspections and sampling activities are discussed in the paragraphs below.

1. Atmospheric Hazards

Confined spaces such as manholes, metering vaults and other poorly ventilated areas may contain inadequate amounts of oxygen or collect toxic gases such as hydrogen sulfide, carbon monoxide, and hydrogen cyanide as well as explosive gases such as methane. Before entry into any confined space, always determine the adequacy of the atmosphere with an oxygen deficiency/combustion (LEL)/hydrogen sulfide gas detection meter and continuously ventilate the confined space. Follow all confined space procedures and never enter when alone.

2. Physical Hazards

Exercise care when removing manhole covers and entering manholes for sampling and observation activities. Wear protective clothing including hard hat, safety glasses, gloves, and coveralls. Whenever you must enter a manhole, follow confined space entry procedures. Wear an approved harness and safety rope, preferably attached to a portable winch. Use a canvas bucket or sling attached to a rope to move sampling equipment and tools in and out of the manhole. If there is any doubt about the soundness of the manhole steps, use a portable ladder when entering or leaving a manhole.

3. Traffic Hazards

Always use properly placed traffic cones, markers, warning signs and barricades to divert traffic around sampling locations and manholes.

QUESTIONS

Write your answers in a notebook and then compare your answers with those on page 751.

10.0A List the major topics covered in Chapter 4, "Inspection of a Typical Industry."

10.1A The General Pretreatment Regulations require that POTWs must develop procedures to accomplish what tasks?

10.1B What items should be verified when inspecting an industrial user's self-monitoring procedures and equipment?

10.1C List the most common and dangerous hazards encountered by POTW personnel while performing industrial pretreatment inspections and sampling activities.

10.2 CONDUCTING ON-SITE INDUSTRIAL INSPECTIONS

10.20 Content of an Industrial Inspection

The industrial inspection provides an excellent opportunity for POTW pretreatment facility inspectors to collect information about an industry and evaluate its compliance with pretreatment standards and requirements. Presented below is a listing of the information which should be collected and documented during an on-site inspection.

1. Industry name,
2. Site address,
3. Correspondence address,
4. Contact name, title and telephone number,
5. Year the industry was established on site,
6. Number of employees per shift,
7. Applicable NAICS (North American Industry Classification System) or SIC (Standard Industrial Classification) codes,
8. A schematic of the water flow through the industry and the location of all wastewater discharge lines that flow to the POTW system; the schematic should also include the layout of major plant features,
9. A description of each discharge (including any batch discharges) specifying the amount, chemical nature, frequency and destination of each discharge,
10. A description and process flow diagram of each major product line and process used within the plant, particularly processes which may be subject to Federal Categorical Pretreatment Standards,
11. A detailed description and appropriate sketches of existing pretreatment facilities, including operating data, if available,
12. A list of pollutants of interest at the plant. The list should be divided into two categories: (1) pollutants that come in direct contact with the water that is discharged to the POTW, and (2) pollutants that do not come in direct contact but have a potential to enter the wastewater due to spills, machinery malfunctions or other similar problems,
13. Identification of appropriate sampling locations,
14. Availability of sampling results performed by the industry,
15. Proximity of chemical storage to floor drains and whether floor drains discharge to storm or sanitary sewers,
16. A description of spill control practices the industry uses. Information about past spills, unusual discharges or temporary problems with any of the process units that may affect the wastewater discharge should be included,
17. A description of air pollution control equipment that may generate a wastestream and the discharge or disposal method and location,
18. A description of how waste residuals (solids or floatables) are handled, stored and/or disposed of,
19. A description of proposed or recent changes to the industry's processes that would affect the discharge characteristics or sampling locations,
20. A description of any operational problems or shutdowns of pretreatment facilities,
21. Identification of specific hazards and establishment of procedures to ensure safety of POTW personnel while at the industrial facility, and
22. Other information as may be necessary.

See Chapter 4 for typical forms that can be used by inspectors to record the information collected during an on-site inspection. Different POTWs will use different variations of the inspection forms depending on specific circumstances; however, the type of information collected is generally the same. POTWs with many industrial users often use a form that facilitates entering data into a computer in order to manage the information in an industrial user database.

10.21 Procedures for Conducting On-Site Industrial Inspections

Adequate preparation and training of POTW pretreatment inspectors is essential for using the industrial inspection to its fullest capabilities. In order to ensure that adequate steps are taken to prepare for the inspection and that all the necessary information is collected during the on-site visit, a checklist has been prepared as a guide for pretreatment inspectors (Table 10.1). The checklist covers three areas: preparation for the inspection, conducting the on-site inspection, and follow-up compliance activities. Once pretreatment facility inspectors have conducted several on-site industrial inspections and are familiar with the necessary steps, the checklist may only be necessary for occasional reference and training of new inspectors. See Chapter 4 for additional details.

10.22 Conducting Monitoring Activities

An effective POTW pretreatment program must include the ability to measure wastewater flows, to collect wastewater samples, and to arrive at meaningful and supportable analytical results. Such a monitoring program has two goals. The first

TABLE 10.1   INDUSTRIAL INSPECTION PROCEDURES CHECKLIST[a]

A.   PREPARATION FOR THE INSPECTION

____ 1.   Determine the need for inspection of the given industry. This exercise will involve review of the industrial user's permit and/or applicable regulations and discharge limits.

____ 2.   Review existing files for available information about the industry. Existing information such as plant layouts, process flow diagrams, compliance schedule (if applicable), and wastewater analytical data should be taken along during the inspection and verified for accuracy.

____ 3.   Review water and sewer records to determine water usage and verify connection to the sanitary sewer.

____ 4.   Review literature about unfamiliar industrial processes which may be encountered (see Chapter 8). Prepare specific questions to be asked about industrial processes to be encountered (see Appendix A at the end of this chapter).

____ 5.   If appropriate, contact the industrial user to establish a convenient date and time to perform the inspection. In most cases, no advance notice should be given.

____ 6.   Prepare sample containers and sampling equipment if monitoring activities may be performed.

____ 7.   Confirm availability of your co-inspector.[b]

B.   THE ON-SITE INSPECTION

____ 1.   Conduct a peripheral examination of the industrial user. Note the size of the industry, additional buildings, outside chemical storage, and location of the sanitary sewer.

____ 2.   Observe the physical characteristics of the wastestream flowing in the sanitary sewer from the industrial user, if access is available. Obtain samples if appropriate.

____ 3.   Establish contact with the chief executive officer, plant manager or engineer or another person in similar authority.

____ 4.   Request a pre-inspection meeting/discussion with the industry representative(s) to:

 • Provide an overview of the local pretreatment program and how it affects the industry.

 • Explain the purpose of your visit.

 • Emphasize that any process information necessary for the inspection report which the industry feels is proprietary can be handled as confidential information. However, advise the industry that effluent data is public information subject to public access through appropriate means.

 • State the POTW's intent to work cooperatively with industry to meet the goals and requirements of the pretreatment program and the National Pretreatment Policy including Categorical Pretreatment Standards.

 • Describe the information you wish to collect during the inspection. Offer the industry official an opportunity to review the inspection report form that you intend to complete.

 • Provide the industry with any written information about the pretreatment program, if available.

 • Answer any questions for the industry representative about the purpose of the visit or about the pretreatment program.

____ 5.   Obtain the basic biographical information about the industry such as industry name and address, contact name, title and phone number, number of employees, and general overview of the business.

---

[a] Source: EPA Region 10.

[b] Many POTW agencies use one inspector to conduct inspections and collect samples. Two inspectors may be used if witnesses are needed in case of legal action, or if the industry is in significant noncompliance, or if one inspector is needed to look for violations during a tour while the other inspector works with the industrial contact. Two or more people are required for hazardous sampling situations such as in traffic or confined spaces.

TABLE 10.1   INDUSTRIAL INSPECTION PROCEDURES CHECKLIST (continued)

_____ 6.   Request a complete tour of the facility and obtain all necessary information to complete the industrial inspection. If the industry manufactures a product, it may be advantageous to follow the process in sequence so that flow diagrams can be prepared.

_____ 7.   Document the exact locations of all sampling points used by the industry. This is especially important if the combined wastestream formula is used by the industry to determine discharge standards.

_____ 8.   Check for implementation of an Accidental Spill Prevention Control Plan at the industry. Comment as appropriate on the operation and effectiveness of the plan.

_____ 9.   During an inspection it should be determined if sampling inside the industry will be necessary. Sampling should be conducted at this time or scheduled for a later time with the industry. Unannounced (unscheduled) sampling may be done at any time in the sanitary sewer or at the industry.

_____ 10. Results of any sampling activities should be incorporated into the inspection report.

_____ 11. Complete the inspection report as soon as possible after the site visit to ensure its accuracy. The final report must be signed and dated upon completion by all inspectors involved. If the industry has requested that specific process information remain confidential, that information should be handled as such. Data on the effluent characteristics cannot be considered proprietary.

_____ 12. If no follow-up activities are required, the report may be appropriately filed.

C.   FOLLOW-UP COMPLIANCE ACTIVITIES

_____ 1.   When all the information has been evaluated, the final conclusion in the inspection report should indicate whether or not the industrial user is in compliance with applicable pretreatment standards and whether any further action is needed by the POTW at this time. Recommendations with regard to future monitoring may be included, where appropriate, such as:

- If the industrial user has been consistently in compliance and has had no major problems, then the monitoring frequency might be reduced or abbreviated.

- Conversely, if the monitoring visit results show problems with pretreatment facilities, chemical handling, or other violations, then the POTW may want to increase the monitoring frequency, modify the industrial discharge permit and/or request additional information from the industrial user.

_____ 2.   If the industrial inspection or sampling results identify problems or violations, the appropriate POTW staff must be notified and copies of the report must be made available to them. A POTW staff person should be assigned to follow through with the problem/violation until it is satisfactorily resolved. The POTW should:

- Notify the industrial user of the problem/violation by issuing a written notice of violation,

- Consider conducting additional sampling to verify violations,

- Establish or require the development of a compliance schedule,

- If appropriate, request that enforcement proceedings be taken against the industrial user,

- Ensure that remedial actions have been taken by the industrial user,

- Keep POTW management informed of the status of compliance/enforcement actions, and

- Submit a final report to the file once corrective actions have been completed.

_____ 3.   If any of the industrial user's processes are subject to Federal Categorical Pretreatment Standards, then the POTW must:

- Notify the industrial user of its responsibilities [403.8(f)(2)(iii)],

- Submit a category determination request to the Approval Authority [403.6(a)], if appropriate,

- Require the development of a compliance schedule for the installation of technology required to meet applicable pretreatment standards [403.8(f)(1)(iv)(A)], and

- Require the submission of all notices and reports (baseline monitoring report, self-monitoring reports) from the industrial user [403.8(f)(1)(iv)(B)].

_____ 4.   Finally, all reports, communications and data on each industrial user should be filed in such a manner that the information is readily available to the appropriate POTW staff.

is to determine the impact of industrial wastes on the POTW's collection and treatment systems including impacts on treatment plant operations, sludge management alternatives, and receiving stream quality. The second goal is to evaluate compliance by all industrial users with applicable pretreatment standards and requirements.

The POTW's monitoring program may consist of any combination of in-house sampling and analytical capabilities or contracted services. However, the POTW must ensure that the monitoring program consists of properly trained personnel, accepted sampling and analytical procedures, and accurate recordkeeping to ensure the validity of the results. Data collected from a POTW's monitoring program will likely be the basis for establishing local discharge limitations, determining compliance by industrial users, and collecting sewer-use service charges; therefore, extreme care must be taken during all phases of the monitoring program. Section 403.8(f)(2)(vi) of the Regulations states: ".... Sample taking and analysis and the collection of other information shall be performed with sufficient care to produce evidence admissible in enforcement proceedings or in judicial actions."

Sampling and analysis of an industrial user's effluent is conducted to accomplish one or more of the following objectives:

1. Verify compliance with wastewater discharge limitations,

2. Verify self-monitoring data,

3. Verify that parameters specified in the industrial user's permit are consistent with wastewater characteristics,

4. Support reissuance and revision of permits,

5. Support enforcement action, and

6. Collect sewer-use service charges to cover the industrial user's fair share of the collection and treatment costs.

See Chapter 6, "Sampling Procedures for Wastewater," and Chapter 7, "Wastewater Flow Monitoring," for additional information.

10.23 Items To Be Inspected

Chapter 4 identified the items listed below as important factors that should be checked during an on-site inspection. They are repeated here to help you develop your own checklists for industrial facilities that you are assigned to inspect.

1. Inspect the production or manufacturing facility (all areas where wastewater is generated or produced).

   a. Review production operations and equipment.

   b. Review raw material storage.

   c. Review waste storage.

   d. Review spill containment facilities.

   e. Review air pollution devices.

2. Inspect the waste treatment facilities.

   a. Review treatment equipment and controlling devices (pH, ORP, and conductivity meters).

   b. Review waste storage.

   c. Review spill containment facilities.

   d. Review pretreatment operator records concerning the maintenance of the pretreatment equipment.

3. Inspect additional manufacturing areas, for example, heat treating, die casting and degreasing.

4. Inspect storage and maintenance areas.

   a. Examine raw material storage records (bulk storage and underground tanks).

   b. Examine waste material storage areas (barrel storage and underground tanks).

   c. Review electroplating equipment maintenance:

      (1) Plating filters and spent cartridges.

      (2) Plating rack stripping facilities and plating barrel maintenance.

5. Inspect monitoring facilities.

   a. Collect wastewater samples.

   b. Review records of wastewater flowmeter calibration.

   c. Perform spot tests of wastestreams, if appropriate.

6. Records review.

   a. Review self-monitoring results of the industrial user.

   b. Review completed copies of the uniform hazardous waste manifests for all wastes hauled from the facility for disposal since last inspection.

   c. Review material safety data sheets on raw materials as necessary.

The remainder of this chapter contains detailed procedures on how to inspect specific industries. The intent of this material is to provide you with examples and ideas so you will know what to look for when conducting on-site investigations. Also see Appendix A, "Pertinent Questions for Inspections of Selected Industrial Groups," at the end of this chapter.

QUESTIONS

Write your answers in a notebook and then compare your answers with those on page 751.

10.2A   For purposes of an inspection, a list of pollutants of interest at an industrial plant should be divided into what two categories?

10.2B   What information about spills should be gathered during an inspection?

10.2C   What items should be noted during a peripheral examination of an industrial user?

10.2D   List the two goals of a monitoring program.

10.2E   List the facilities or areas that should be examined during an on-site industrial inspection.

10.3   HOW TO INSPECT SPECIFIC TYPES OF INDUSTRIES

10.30   Job Shop Electroplating Facility
by Robert C. Steidel

The inspection of an industrial facility for compliance with pretreatment standards can be approached through a method of proper background preparation, consistent inspection technique, and proper documentation and reporting of observations and results. By applying each procedure described above with the proper training and knowledge, you will be fulfilling your responsibility to protect the POTW and enforce the environmental rules and regulations established to protect the public and the environment.

10.300   Inspection Preparation

The physical areas and the processes of an industrial facility where wastewaters or other wastes are generated and where chemicals are used or stored can have an impact on the POTW and must be inspected. These areas vary according to the manufacturing processes taking place and therefore influence the sequence of the inspection and your preparations as an inspector. You must have a basic understanding of the wastewater generated at the facility and the types of manufacturing processes you will most likely encounter. EPA development documents on the industrial category and Chapter 8 are good sources of basic information if the industry is totally unfamiliar to you.

The theoretical facility you are about to audit discharges an average of 65,000 gallons per day (11,500 L/day) during two

eight-hour shifts. The facility employs 23 people and typically operates from 11:00 p.m. on Sundays to 5:00 p.m. on Fridays. Process wastewaters account for 80 percent of the total discharge to the sanitary sewer. The unit operations taking place at the facility include cyanide-cadmium and cyanide-copper electroplating, anodizing, passivating, cleaning, chromating, phosphate coating (zinc), degreasing and baking. Even if you have never been to this facility, you should know this information from the wastewater discharge permit application or baseline monitoring report (BMR) completed by the industrial user and the permit subsequently granted by the POTW. Additional information about this job shop available for your review through this source includes:

- Raw materials used,

- Source(s) and discharge flow rates to the sanitary sewer,

- Known pollutants present and their historical concentrations,

- Schematic diagram of the facility showing the locations of process equipment, sampling points, wastewater treatment, waste storage areas, and spill containment areas,

- Environmental permits held by the facility,

- The different wastestreams generated at the facility disposed of through contract waste treatment facilities,

- The applicable pretreatment standards for the facility (and all relevant information if these standards are adjusted by use of the combined wastestream formula or removal credits),

- The compliance schedule under which the facility operates (if any), and

- The name(s) of contact personnel at the industrial facility.

Transfer relevant information from the POTW file to the inspection form (such as Appendix B, "Checklist for Inspection of Significant Dischargers") you will use during the facility inspection. Make notes in the applicable section(s) of the inspection form about equipment you will want to observe or issues you will want to cover once you are at that point in the inspection. For example, if the industrial user has recently submitted a letter stating that a sand filter has been installed after the clarifier discharge, you will want to obtain pertinent information such as backwash rates, size and capacity of the sand filter along with physically verifying that the unit was installed. If you do not use a form (and it is recommended that one be provided so that inspection technique is consistent from facility to facility and among inspectors), at least prepare a written outline of the information found in the file. Examine the most recent correspondence in the file for letters of violation issued by the POTW based upon sampling results. Also review the response from the industrial facility concerning these violations. Make careful notes about treatment units or spill containment areas which should be inspected based on problems discovered during past inspections. Be sure that wastewater discharge permits are current and that the information contained in those applications is consistent with any other data you may have at your disposal. These data would include water consumption data, POTW sampling results of the wastewater discharge, and quantities of wastes or sludges generated at the facility that are hauled away under a manifest system for off-site disposal at a properly licensed contract waste treatment facility.

10.301 Conducting the Inspection

Before you enter the industrial facility, first visit the sampling points and collect wastewater samples if this is part of your responsibility. The purpose of obtaining these samples before beginning the inspection is to allow collection of an unbiased wastewater sample. Be sure to follow proper chain of custody procedures and split samples with the industrial user if this is required by the policies of the POTW. Observe the operation of any flow metering equipment that may be present at the sampling point. Note on your inspection report any signs of corrosion in the sampling manhole and document this fact with photographs. Visually observe the characteristics of the wastewater being discharged; record any unusual observations for further discussion during the inspection.

Upon entering the industrial facility, meet with the representative of the industrial user to explain the purpose and nature of the inspection; indicate whether it is an annual audit, routine enforcement inspection, or demand compliance inspection as a result of past violations. This is the only time you are expected to do the majority of the talking. During this stage, when the ground rules are being laid and the objectives of the inspection are being explained, it is reasonable for the inspector to initiate more than 50 percent of the discussion. From this point in the inspection forward, however, you must actively listen and try to establish two-way communication. This approach is essential because many times during an inspection the industrial representative will provide important information which needs to be properly documented.

This is the appropriate time to identify any safety procedures the industrial facility has established for visitors. Obtain any preliminary information the industrial representative wishes to communicate prior to entering the manufacturing facility, but do not allow too much time to elapse between arrival at the facility and actually entering the production area to begin the inspection.

The order in which you inspect the facility will be based on your own preference unless specific procedures have been established by the POTW. Your outline or inspection form can provide a structure for the inspection; many inspection forms categorize the industrial facilities into the following major areas:

- Production Area
- Chemical Storage Area
- Waste Storage Area
- Wastewater Treatment Area
- Sampling Facilities

Note that spill containment is not listed as a separate inspection group. Spill containment facilities must be provided in all areas listed above. Therefore, you will review spill containment facilities constantly throughout the inspection.

When you first enter a facility, such as a job shop electroplater, the layout of the facility may appear to be confusing. In many cases this is necessary when job shop work is being performed because multiple tasks must be accomplished in a small amount of processing area to allow the facility to be profitable. However, it is important to note immediately upon entering the facility whether or not the shop appears to be well organized and the plating equipment is well maintained. This will give you an immediate indication about how the wastewater treatment system is being maintained and operated.

If you think of the production area in terms of types of wastewater streams that are generated, you can avoid spending unnecessary time in several areas. For example, the production area can be subcategorized as sources of the following wastewater streams:

- Wet processing areas, including

    1. Wastewaters discharged to the wastewater treatment system, and

    2. Wastewaters which bypass the wastewater treatment system.

- Noncontaminated process wastewaters which include

    1. Cooling waters discharged to surface waters, and

    2. Dilution waters such as running eyewash stations or drinking fountains.

- Dry processing areas.

By dividing the facility as shown above, you can quickly examine the areas that typically do not generate wastewater of concern. For instance, the dry processing areas which house the baking oven, vapor degreaser and parts to be plated, shipped or stored within the shop can be quickly inspected to verify that this area does not have an impact on the wastewater discharge to the sanitary sewer system. However, if raw materials (especially solvents for the degreaser) are stored in these areas, be sure proper spill containment is provided.

Similarly, when inspecting noncontaminated process wastewaters such as cooling waters discharged to surface waters, you only need to verify that they do not enter into the wastewater discharge and that chemical additives such as chromates or organic phosphates do not exceed surface water discharge limits. Facilities which leave eyewash stations or water fountains continually running may be trying to dilute the pollutant concentration in the wastewater discharge. Make note of such situations in your inspection report and advise the industrial representative that a discharge of this type is a violation of the General Pretreatment Regulations. If there are safety concerns which require the eyewash station to be left running on a continuous basis, ask the industrial user to document the specific concerns in a letter and submit it to the POTW for review.

Air scrubber blowdown can be contaminated with heavy metals from mists or exhausts. This source of wastewater should be discharged into the wastewater treatment system and not discharged directly to the sanitary sewer system. Also, in a plating facility it is very important that boiler blow-down receive adequate wastewater treatment. Boiler blow-down can be a source of large concentrations of heavy metals and cyanide from leaking steam coils in plating solutions.

The wet processing area of the facility will occupy the largest percentage of your time and talents. In our hypothetical electroplating facility, wastewaters from the anodizing and cleaning operation are intentionally routed to bypass the final metals removal pretreatment system and are routed through the pH control system and then to the sanitary sewer system.

These wastewaters are generated in the last rinse of a non-counterflow rinsing system, and have been sampled and analyzed by the industrial facility. Since the pollutant concentrations and loadings were low in these wastewaters, the company was allowed to bypass the metals removal pretreatment system in order to reserve capacity in the system for more polluted wastestreams.

During your inspection, however, you must verify that there have not been any changes in the work flow at the facility which could alter the pollutant characteristics in these very dilute rinse waters. For instance, are parts which are immersed in the cyanide-cadmium plating solution allowed to pass over these rinse tanks? If so, the drag out on the parts could drip into the rinse tanks increasing the pollutant concentration as they move through the air over the top of the rinse tanks. In addition, you must check to be sure that a passivating tank has not been added using these same rinse tanks as the anodizing and cleaning operations. Such a process layout could drag in more concentrated solutions which must be rinsed and treated before being discharged to the sanitary sewer system. These more concentrated solutions may cause discharge violations.

Wastewaters which bypass the wastewater treatment system entirely must be properly protected against accidental spillage of process solutions. Inspect the trenches, sumps and piping by which wastewaters that bypass the wastewater treatment system reach the sanitary sewer outfall and periodically sample these wastestreams. Verify that the spill containment in these areas is adequate to protect against such occurrences.

It is also necessary to inspect any process equipment which is the source of process wastewaters being discharged into the company's wastewater pretreatment system. The purpose is to verify that the physical condition of the process equipment is not contributing to the pollutant loading on the wastewater pretreatment system through leaks, drips or malfunctioning controllers. As wastewaters are conveyed to the pretreatment system designed for treatment of dilute wastewaters, they must be protected against the accidental addition of concentrated process chemicals. Concentrated process chemicals are materials such as full-strength plating solutions or stripping solutions, whereas dilute wastewaters are the process rinse solutions and tank dumps for which the treatment system was designed.

The spill containment system must be inspected for integrity. The dikes, dams and trenches should show evidence of being properly maintained. To determine this, look for leaks around joints or obvious evidence that solutions within the spill containment system have not been removed on a timely basis.

Spill containment trenches that are full will be of no use if a major spill should occur.

As this is a job plating facility, do not expect the equipment and facilities to be in perfect operating condition. However, it should be apparent that some care is taken in the operation and maintenance of the facility. Be sure that filters for processing baths are located within the spill containment for the processing system. These filtering operations can be sources of large volumes of concentrated process solution which can be released into the spill containment system due to the failure of fittings or hose connections. As these filters operate under pressure, it is possible that large volumes can be spilled in short periods of time. The plating tanks themselves should have good structural integrity. Drip boards between plating tanks should be in place. The use of large quantities of oil-dry underneath a plating tank or pump is not a good sign and should be noted.

The clarity of the rinse tanks which subsequently discharge into the wastewater treatment system should *NOT* be such that you can read "head" or "tail" from a dime on the bottom of the tank. This would indicate too much water is being used for rinsing purposes and causing a hydraulic loading which could be in excess of design capacity of the treatment system. Look for rinse tanks to be properly mixed through the use of air spargers. The appropriate use of dead rinses after a plating bath as a primary rinse helps to conserve chemicals if the plating bath is operated at a temperature hot enough to lose volume through evaporation. The contents of the dead rinse can be cycled back into the plating solution as makeup waters, thus returning valuable raw materials to the process.

Plating barrels and racks should be well maintained; barrels should drain freely as they rotate out of the plating solution or rinse tanks, and barrel holes need to be cleaned to allow proper drainage. The plating racks should not have tears or holes in the protective coating covering them since this would drag out excessive volumes of plating solution. Check the condensate traps for steam coils in the plating solution to be sure that backsiphoning of process solution is not occurring. These sources can contain small volumes of very high-strength wastewaters discharged either to the sanitary sewer directly or to the wastewater pretreatment system. Observe the actions of the production personnel to see if they are using all stages of fog or counterflow rinses. Sometimes shortcuts are taken to improve output rates.

While you are still in the manufacturing area, stand and observe the actions of the industrial facility personnel as they go about their jobs (Figure 10.1). Do they appear to be competent, well trained, and conscientious in their responsibilities to produce an acceptable product? If so, it is a good bet the operation and maintenance of the wastewater treatment facility will probably reflect the same amount of pride in workmanship. If the facility representative describes to you an innova-

Fig. 10.1   Observe performance of production personnel
and appearance of manufacturing facilities

tive design he or she has made in the manufacturing process, you may draw the conclusion that the same type of innovative thinking is being applied to the wastewater pretreatment system.

The waste and raw material storage areas require close inspection. In the case of the particular job shop electroplater we are discussing, these areas are combined and located in the shipping and receiving department. In addition to wastewater treatment sludges and concentrated process solutions that cannot be treated (such as cyanide-cadmium stripping solution), the short-term inventory of process chemicals is stored here in segregated areas within a spill-contained area of the plant. You should verify that cyanides have been stored separately from acids and that both wastes and raw materials are located in areas where lift trucks or other activities taking place in the shipping/receiving area will *NOT* cause leakage of these stored materials. Review the transportation route of raw materials and wastes through the facility to verify that it has been properly designed to pose the least risk for accidental spillage of these materials to the sanitary sewer system.

The wastewater pretreatment area needs to be inspected for evidence of proper operation and maintenance. You will need access to several types of information collected by the industrial user. Review the operating parameters for the wastewater pretreatment system and examine strip chart recorders for various measurements such as flow, pH and ORP on the various treatment processes. Most important at this point in the inspection, you should document whether proper operating methods are being followed by the industrial personnel and verify the methods used by the wastewater pretreatment system operator to collect operating data. This information is vital for establishing a historical pattern of treatment whereby optimum operating conditions can be identified and maintained. Review the current O & M manual with the pretreatment system operator to discuss sections that have been updated since the last inspection. Review the type and frequency of data collected on the wastewater pretreatment system operation, particularly the daily flow readings, to verify the accuracy of the data reported in the industry's wastewater discharge permit applications and reports.

Review the plating chemical addition records for evidence of excessive raw material usage. This may indicate that leaks of process chemicals are occurring. Manifests of wastes such as sludges and wastewaters (for example, concentrated cyanide-cadmium stripping solutions which cannot be treated in the conventional alkaline chlorination treatment module at the industrial user's site) should be examined for volume and frequency of off-site disposal. Verify that properly permitted treatment, storage and disposal facilities are listed as accepting/receiving these wastes. Depending on the policy of the POTW, you may need to crosscheck these manifests with the applicable environmental agency regulating transportation, storage and disposal of wastes covered under such a manifesting system.

As you inspect the wastewater treatment system, try to form a general opinion, as you did when inspecting the manufacturing areas, about whether or not the equipment appears to be operated and maintained in a competent manner. Do not expect the job shop electroplater's treatment system to look like the photographs in most textbooks and manuals on metal finishing wastewater treatment. In many cases there will be evidence of corrosion of metal parts due to the humid and corrosive atmosphere in the wastewater treatment area. Expect to find lime or polymer spillage along with evidence of minor pipe leakage. However, proper operation should be demonstrated by the fact that mixers are in place and are op-

erating, evidence of calibration of pH and ORP meters, and a sense of pride or at least an attitude expressed by the operator to the inspector that he or she knows what is required to produce an acceptable wastewater discharge. Verify that the design and operation of the wastewater treatment system provides for redundancy and that local, off-the-shelf stock replacement parts are available for critical components such as pumps.

Review the operation of the wastewater pretreatment system with the industrial operator. Confirm that the hours of operation are the same as or longer than those of the manufacturing facility. Verify that when the manufacturing processes and wastewater pretreatment systems are shut down at the end of a day, pumps that could discharge a slug of untreated wastewater as a result of an accidental spill routed to the wastewater pretreatment system are either locked out by the operator or their power is routinely disconnected. Verify that the operator is properly collecting the operating data discussed above. Check to see that adequate stocks of treatment chemicals for use in the wastewater pretreatment facility such as lime, chlorine and caustic soda are available.

If the industrial user is required to self-monitor, verify the sampling location with the operator or other responsible industrial representative who collects the samples. Ensure that proper sampling procedures, bottles, preservatives, holding times and analytical methods are being observed. Review the industry's self-monitoring results for compliance with applicable pretreatment standards and, if possible, compare these data with the data collected by the POTW.

Observe the wastewater as it is treated in the various units of operation. Document observations such as the turbidity in the discharge from the clarifier (Figure 10.2) or the hydraulic peak loading effect of pumps discharging into the surge tank or treatment modules from collection sumps in the shop. Note the recovery time, as measured by the pH meter, when slugs of acid or alkaline wastewaters are received by the neutralization tank. Record your observations concerning the appearance of the sludge generated in the wastewater treatment system. Determine what happens to liquids and solids from filter presses (Figure 10.3). Only through training and first-hand experience will you know which types of information and how much detail has to be collected to reach reasonable conclusions about the efforts of the industrial facility to operate its wastewater pretreatment system to comply with applicable pretreatment regulations.

As you near the end of your inspection, return with the industry representative to the sampling points previously inspected and review any issues or observations documented for discussion.

At the close of the inspection, either give the industrial representative a copy of the inspection report (properly signed and dated) or indicate that a follow-up letter documenting the

*Fig. 10.2   Observe the effluent from the clarifier;
look for evidence of floc*

Fig. 10.3 Determine disposal procedures for liquids
and solids from filter presses

results will be sent within a specified time frame. Indicate whether a copy of the report or follow-up letter will detail any further actions that need to be taken by the industrial user. Be sure to properly discuss any violations you have found at the facility with the representative of the industrial user before leaving the site.

If there are supplemental observations you need to document in field notes, do so as soon as possible while the events are fresh in your mind. Take time now to document any other information you were unable to record earlier (because you were too busy actively listening). This might include the industry's plans to install heat treating at the facility, the need to add a third production shift or the long-term plan to install batch cyanide treatment for high-strength, cyanide-bearing wastestreams to be treated on site instead of hauling these wastewaters to a contract treatment facility. Use these field notes as the basis of subsequent inspections and as long-term reference information to be contained in the inspector's active file.

For additional questions to consider when inspecting electroplating and metal finishing facilities, see Appendix A, "Pertinent Questions for Inspections of Selected Industrial Groups," all SIC 347- numbers.

10.302 Conclusions

To efficiently conduct an industrial compliance inspection, take time to prepare yourself before entering the facility. Use the suggested technique of breaking down the areas to be inspected within the facility into manageable units by distinguishing between those which can be quickly inspected and those that will occupy most of your time and expertise. Use active listening skills and encourage two-way communication. Prepare complete and accurate records documenting your inspection and report your findings to both the POTW and the industrial user. Both the POTW and the industrial user will benefit from a competent and professional inspection.

QUESTIONS

Write your answers in a notebook and then compare your answers with those on page 751.

10.3A   Why should an inspector visit an industry's sampling points before visiting the industrial facility?

10.3B   When inspecting an industrial facility, spill containment facilities should be inspected in which areas?

10.3C   What should the inspector do or tell an industrial representative at the close of an inspection?

10.31 Petroleum Refineries
by Choong Hee Rhee and Bill Garrett

Petroleum refineries are extremely complex operations that are beyond the scope of this training manual to explain in detail. Only the broadest types of inspection information are contained in this section.

10.310 General Process Descriptions

In the simplest terms, a fuel refinery is a manufacturing plant for gasoline. Every operation is geared toward making more and better gasoline. The first operation is atmospheric distillation of crude oil to separate it into different boiling range fractions, including the light-ends like propane and butane, the intermediate fractions like gasoline and fuel oil, and the heavy-ends like asphalt. The residual or bottom from the atmospheric distillation then goes through vacuum distillation to extract more fuel fractions from the residual.

The market demands more gasoline and fuel oil than the heavier intermediates and residuals. Therefore, various additional refining processes were introduced to produce more gasoline from the unused intermediate fractions. Hydrocarbon cracking is the first operation, in which the heavier fractions are broken down into the lighter fractions through the use of catalyst (fluid catalytic cracking), or through a combination of hydrogen and catalyst (hydrocracking). For the heaviest residuals, thermal cracking with heat is used in the so-called coker operation.

Besides breaking down the heavier fractions to form more fuel fractions, the reverse process of hydrocarbon rebuilding is also used to make more gasoline by combining lighter fractions like butanes into heavier hydrocarbons. Alkylation is the process in which isobutane is combined with unsaturated hydrocarbons like propylene to form longer chain hydrocarbons of the fuel fraction range. This process is catalyzed by either sulfuric acid or hydrofluoric acid. Besides alkylation, polymerization is also used to link together lighter hydrocarbons to form heavier ones.

In addition to producing more gasoline, the market also demands better gasoline with higher octane numbers. In general, branched-chain hydrocarbons like isobutane, cyclic compounds like methyl cyclohexane and aromatic ring compounds like benzene have higher octane numbers. Therefore, processes were developed to rearrange the hydrocarbons to form these products. Reforming is the major process in which straight-chain hydrocarbons are converted into branched-chain isomers, cyclic compounds, and aromatic ring compounds in a catalytic process. Isomerization is also used to produce branched-chain isomers from straight-chain ones. Besides the higher octane numbers, lower sulfur content is also demanded for the fuels, and hydrotreating is used to remove sulfur through its combination with hydrogen to form hydrogen sulfide.

Besides the refining processes, there are a few processes that supply the gases used in the refining process and other processes that manage the sidestreams. The hydrogen plant makes hydrogen gas for use in the hydrotreater and the hydrocracker. The amine plant removes hydrogen sulfide from the waste gases, and the sulfur plant converts the hydrogen sulfide into elemental sulfur. Lastly, the refinery gas plant separates the different gases like butane, propane and isobutane, reusing some as feedstock and burning the rest as fuel.

10.311 Pollutants

Petroleum refineries discharge enormous quantities of wastewater compared to other industries. Virtually every refinery operation, from primary distillation through final treatment, requires a large volume of process and cooling water. Although the sheer volume of wastes from oil refineries looms as a large problem, the American Petroleum Institute (API) reports that 80 to 90 percent of the total wastewater used by the average refinery is used only for cooling purposes and is not contaminated except by leaks in the lines. As a general guideline, a topping plant discharges two to ten gallons of wastewater per barrel of crude oil processed and a topping and cracking plant generally discharges 10 to 40 gallons of wastewater per barrel of crude oil processed, depending on the types of manufacturing processes.

A combined refinery wastewater contains various fractions of dissolved and suspended mineral and organic sulfur compounds as liquids and sludges. The pollutants listed below are commonly found in the refinery wastewater.

1. Oil and grease
2. Hydrogen sulfide
3. Mercaptans (thiols)
4. Chromium
5. Zinc
6. Lead
7. Copper
8. Ammonia (N)
9. Fluoride
10. O,M,P-Creso

11. Phenol
12. Toluene
13. Benzene
14. O,M,P-Xylene
15. Alkalinity
16. pH (high side)
17. Thiosulfate
18. Combustible gas
19. Other organosulfur compounds

The oil and grease may appear in wastewater as free oil, emulsified oil, and soluble oil. The major problems in the refinery wastewater are the offensive mercaptan odors, high concentrations of corrosive hydrogen sulfide, and combustible gases.

10.312   Sources of Wastewater

Figure 10.4 shows sources of refinery wastewater generated by typical petroleum refinery operations, from primary distillation through final treatment. Some refineries segregate uncontaminated cooling waters from contaminated wastewaters and discharge them to the storm drain system under National Pollutant Discharge Elimination System (NPDES) permits. Figure 10.5 shows a highly simplified layout of the way oil moves through a refinery.

10.313   Sources of Important Pollutants

1. Hydrogen Sulfide (H_2S) and Ammonia (NH_3)

Sour water, which may contain as much as 10,000 mg/L of hydrogen sulfide and 7,000 mg/L of ammonia, is usually treated in a sour water stripper. The hydrogen sulfide and ammonia are present as ammonium bisulfide (NH_4SH) which is the salt of a weak base (NH_4OH) and a weak acid (H_2S). In solution, this salt undergoes hydrolysis to form free ammonia and free hydrogen sulfide. A well-designed sour water stripper removes greater than 99.8 percent of the free hydrogen sulfide and greater than 95 percent of the free ammonia.

2. Thiosulfate

Thiosulfate ($S_2O_3^{2-}$) mainly occurs as a result of the oxidation of sulfide. Some of the unremoved hydrogen sulfide (H_2S) is dissociated to hydrogen sulfide ion (HS^-) and hydrogen ion (H^+); the hydrogen sulfide ion (HS^-) is converted to thiosulfate.

3. Oil and Grease

As shown in Figure 10.4, sources of oil and grease in wastewaters discharged from petroleum refineries are runoff from diked and undiked tank areas, tank bottom discharges, pump and compressor cooling waters, circulating cooling water blowdowns and desalter water discharges.

4. Mercaptans

Mercaptans smell like sweet onion, skunk odor or dead animal odors. Crude oil contains small amounts of mercaptans, but most of the mercaptans are generated during the manufacturing processes. For example, mercaptans are formed when one of the hydrogen atoms in hydrogen sulfide (H_2S) is replaced by a hydrocarbon group.

5. Zinc

Some refineries are using liquid zinc sulfate for sulfide control because it is less expensive than alternative control measures. However, the use of zinc sulfate is not always recom-

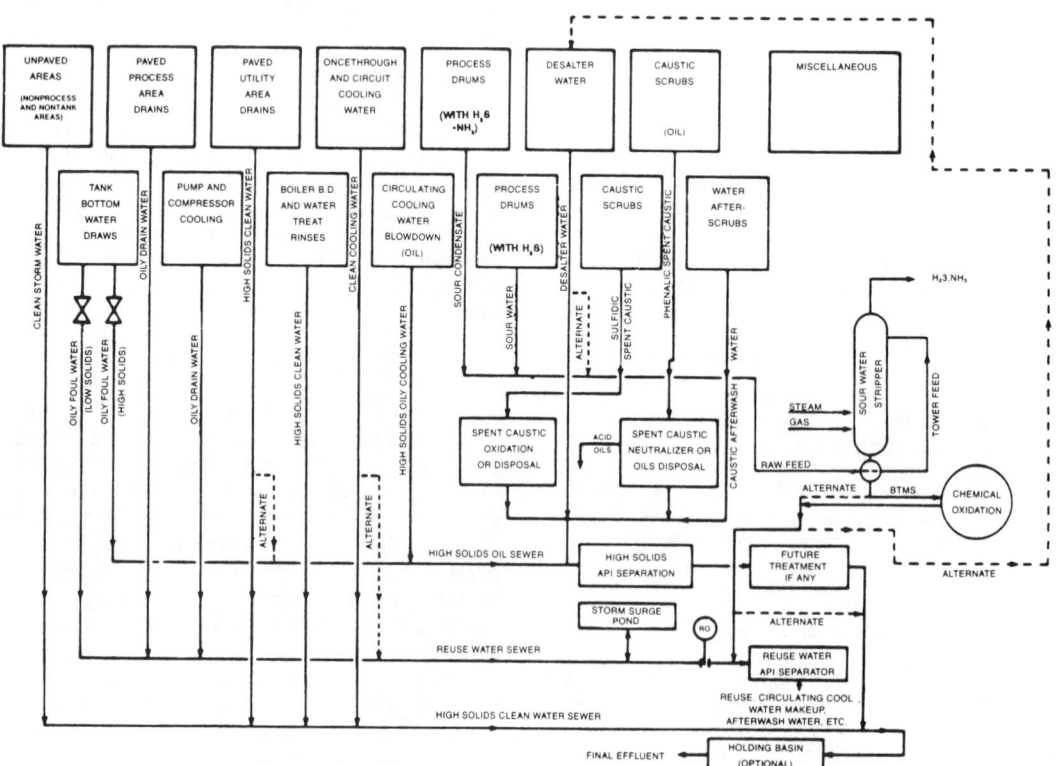

Fig. 10.4   Sources and treatment of refinery wastewaters

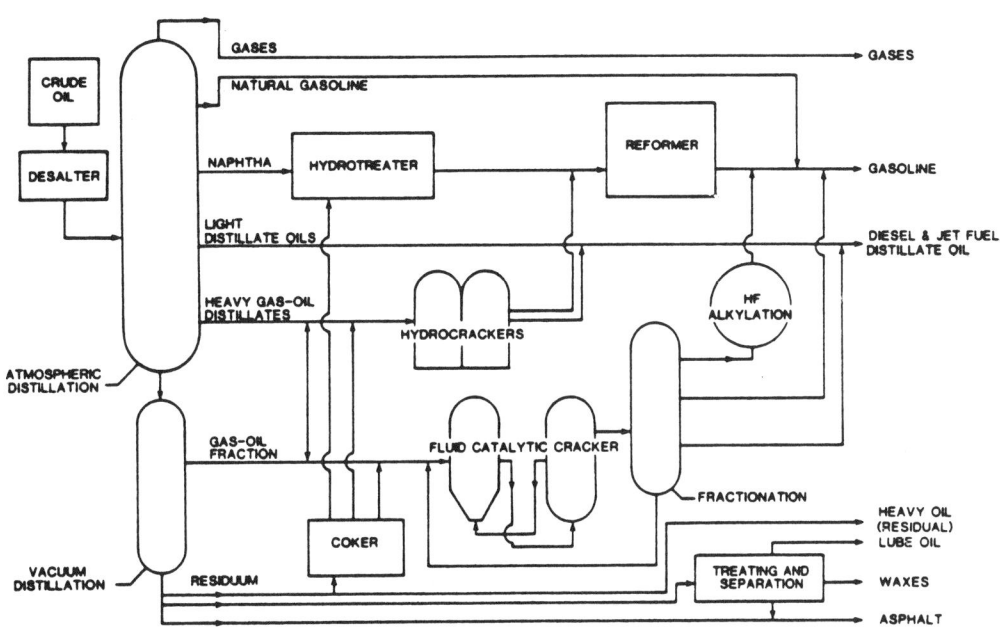

Fig. 10.5   Highly simplified petroleum refinery process flow diagram

mended since it can cause a refinery's wastewaters to exceed discharge limits for heavy metals.

6. Combustible Gas

Refinery wastewaters contain flammable materials such as methane, hydrogen sulfide, benzene, toluene and xylene. These combustible gases must be controlled by limiting the concentrations of gases. Typically, a lower explosive limit (LEL) of ten percent is established as a control limit.

10.314   *What To Inspect*

Inspection of a petroleum refinery wastewater treatment system should focus on the following major components:

1. Flow Metering System

Inspect the flow indicator, totalizer and recording chart. Check these measurements to verify that the three numerical summary values are equal or reasonably close together.

2. Sulfide Monitor

When you inspect the sulfide monitor, check the indicator and recording chart to verify that the numerical summary values of these two items are equal or reasonably close together. At this time, the two types of commercially available monitors are oxidation-reduction potential meters and colorimetric analyzers. To test for the concentration of dissolved sulfide, use a field test kit and analyze an instantaneous grab sample. If the result of the field test is different from the sulfide indicator and the recording chart, the result of the test and the accuracy of the automatic system should be discussed with the discharger. Your inspection should also include a review of the designated frequency of sulfide tests and an analysis of whether or not the use of an automatic system is appropriate.

3. Oil and Grease

An automatic on-line monitor for oil and grease in petroleum refinery wastewater is not available at this time. The only test

method permitted by EPA is the Partition-Gravimetric Method (5520B of *STANDARD METHODS FOR THE EXAMINATION OF WATER AND WASTEWATER*, 20th Edition).

4. pH Monitor

Most petroleum refineries performing cracking operations require automatic pH monitors with recording charts. Inspectors should check the pH of the wastewater by using a portable pH meter or pH indicator strips. If the result of the field test is different from the value shown on the pH recording chart, discuss the result of the field test and the accuracy of the automatic pH monitor with the discharger.

5. Combustible Gas Monitoring System

The monitor, which includes indicator and recording charts, must be capable of detecting any combustible gases contained in the wastewater. Numerical values (% LEL) of the indicator and the recording chart must be equal or similar to the field tested value. If the field test value is greatly different from the values shown on the automatic system, a recommendation on proper location of the sensing head assembly and modification of the system should be made. Insist on immediate corrective action if explosive conditions are detected during an inspection. Be prepared to initiate emergency response procedures as presented in Chapter 11.

6. Wastewater Diversion System

Any off-specification wastewater and all spills (Figure 10.6) must be diverted to a holding tank or a holding basin for subsequent pretreatment. Diversion for a high LEL can be accomplished with a gearmotor-operated valve. Diversion for other off-specification wastewaters can be done manually.

7. Wastewater Treatment System

Inspection of petroleum refinery wastewater treatment systems should also include the API (American Petroleum Institute) separator, DAF (Dissolved Air Flotation) unit, IAF (Induced Air Flotation) unit, circular or rectangular clarifier,

Fig. 10.6   Fuel loading dock at oil refinery.
Floor drains in this area should be
plumbed to the pretreatment facility.

storage basin, pH neutralization system, and the chemical oxidation system for sulfide and thiosulfate. Examine the operators' maintenance logs to determine that these systems are functioning properly.

8. Sampling Device

Many petroleum refineries are required to install a 24-hour composite sampler or hourly composite sampler to obtain and retain hourly samples each day. Samples are collected at each site where a petroleum wastestream discharges to a POTW sewer. Examination of these samples is convenient for field investigation on treatment plant upsets, sewer problems and other related problems. Be familiar with the best location to conduct off-site monitoring (Figure 10.7) so that you can quickly determine if a refinery is contributing to problems in the POTW's collection system and/or treatment facility.

10.32 Pulp and Paper Industry
by Choong Hee Rhee

Pulp and paper manufacturing processes also discharge immense quantities of wastewater compared to other types of industries. The quantity of wastewater produced by pulp and paper manufacturing processes probably is the second largest among the industrial categories (the largest is the petroleum refineries). The following types of products are produced by this industry.

PULP	PAPER
Bleached kraft pulp	Newspaper
Unbleached kraft pulp	Groundwood fine paper
Bleached sulfite pulp	Fine paper
Unbleached sulfite pulp	Tissue paper
Mechanical (ground) pulp	Currency bond
Semi-mechanical pulp	Cigarette paper
Semi-chemical pulp	Typewriter paper
Thermal-mechanical pulp	Roofing felt
Soda (alkali) pulp	Paperboard
Deinked pulp	Corrugated medium
	Plasticized paper

It is generally recognized by this industry that a pulp mill discharges approximately 5,000 to 60,000 gallons of wastewater per ton of pulp produced and 10,000 to 80,000 gallons of wastewater per ton of paper produced.

10.320 Pollutants

A combined pulp and paper mill wastewater contains extremely high concentrations of suspended solids and various organic and inorganic pollutants. The following pollutants are most likely contained in the pulp and paper mill wastewater:

1. pH (low side),
2. Hydrogen sulfide,
3. Volatile solids,
4. Suspended solids,
5. COD and BOD,
6. Oil and grease,
7. Pentachlorophenol,
8. Trichlorophenol,
9. PCBs,
10. Alkalinity, and
11. Color.

10.321 Sources of Wastewater

Figure 10.8 shows sources of a combined pulp and paper mill wastewater generated from a typical manufacturing process. The major portion of the pollutants originate in the pulping process. The wastewater discharged from the chemical pulping process contains no lignin and noncellulosic substances but contains colors. Because of its color, this wastewater is called "black liquor." On the other hand, the wastewater discharged from mechanical pulping processes contains lignin, pectin, wax, and resin.

As seen in Figure 10.8, the papermaking process adds various fillers, chemicals, sizings and dyes to improve the quality of the paper products. Beating and washing initially produce a rather strong waste, which is progressively diluted as the washing proceeds. Paper machines, either the Fourdrinier type or cylinder type, also produce an enormous quantity of wastewater. The typical color of a papermaking process is white. Because of its color, this wastewater is called "white water."

10.322 Important Wastewater Characteristics

The effluent characteristics vary somewhat depending on the type of pulping, bleaching and fiber recycling practices. Suspended solids range from 30 to 3,000 mg/L, total dissolved solids range from 1,000 to 5,000 mg/L. The five-day BOD values range from 100 to 200 mg/L and COD values range from 100 to 7,000 mg/L. One important fact to know as an inspector is that cellulose is difficult to biodegrade (could have a low BOD, but a high COD).

The wastewater discharged from a papermaking process tends to have a low pH because the optimum pH range for paper forming is between 4.0 and 4.7. Some paper mills use hydrochloric acid for cleaning their machines, further adding to low pH values.

10.323 What To Inspect

In general, pulp and paper mill wastewaters are treated using the following means:

1. Recovery of fiber by save-alls and/or a filter press,
2. Sedimentation or flotation to remove suspended matter,
3. Chemical destruction or precipitation to remove color,
4. Activated sludge (if necessary) to remove oxygen demanding matter,
5. Storage in a settling pond or equalization chamber to smooth out flows,
6. Oil and grease removal devices for the deinking process, and
7. Sulfide destruction devices.

An inspection of pulp and paper mill wastewater treatment systems should include checking the following monitoring devices:

1. Flow metering system,
2. Sulfide monitor,
3. pH monitor, and
4. Sampling device.

Refer to Section 10.314 for additional details about inspecting each of these systems.

In the foreground is the oil-refinery's off-site sampling station that allows access to wastewater recording instruments without entering the process area. The refinery is in the background past the rail cars.

Closeup view of off-site sample vault for oil refinery

Fig. 10.7   Off-site sampling stations

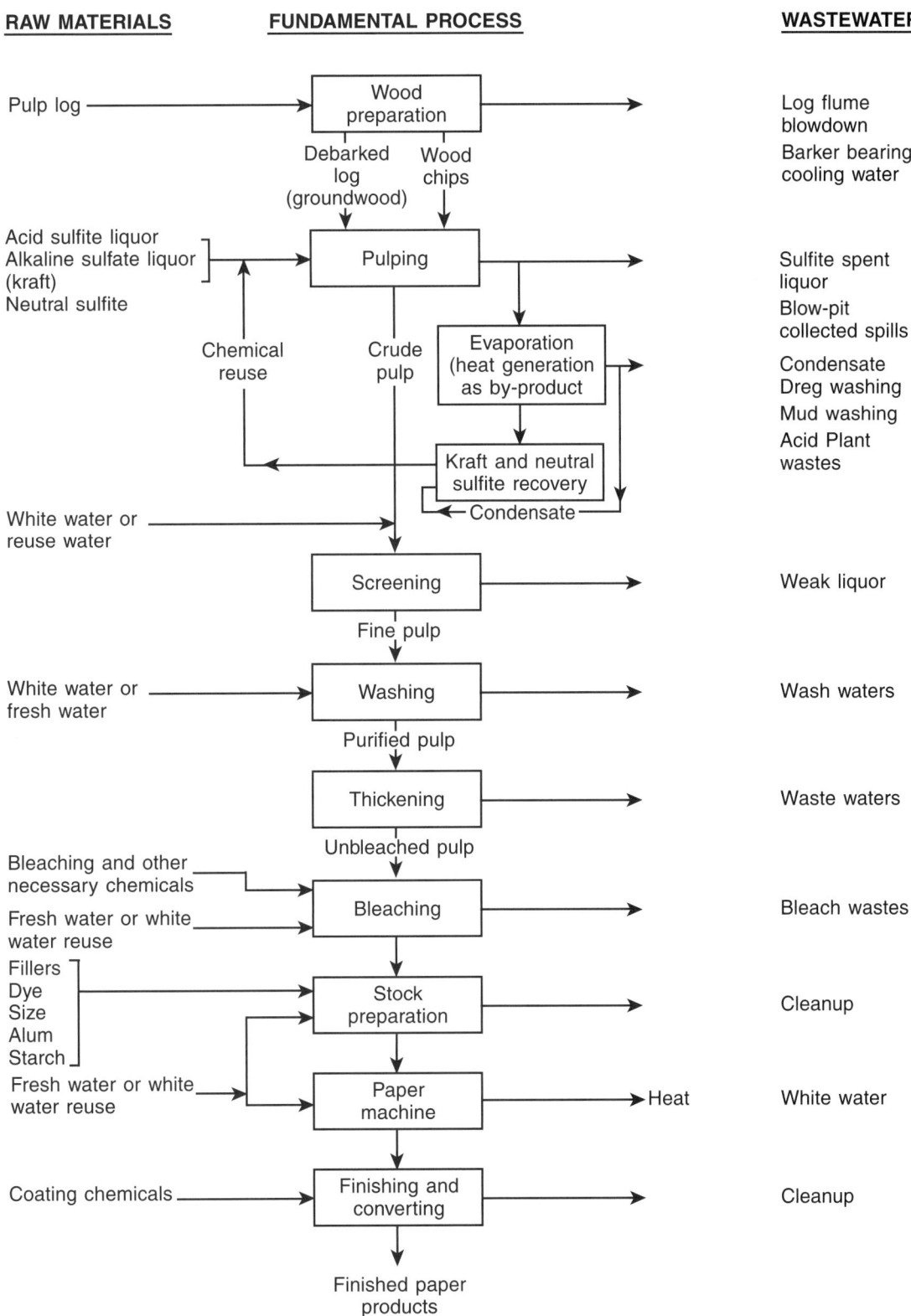

*Fig. 10.8   Highly simplified pulp and paper mill
(combined) process flow diagram*

QUESTIONS

Write your answers in a notebook and then compare your answers with those on pages 751 and 752.

10.3D   Oil and grease from refineries may appear in wastewater as what types or forms of oil?

10.3E   What are the major problems in oil refinery wastewaters?

10.3F   List the monitoring devices that should be included in an inspection of a petroleum refinery wastewater treatment system.

10.3G   List the pollutants most likely to be found in the wastewater from a pulp and paper mill.

10.33   Chemical Manufacturing
by Annie Ko

To give you a general understanding of the chemical manufacturing processes and the wastewaters they generate, the following sections describe the manufacture of calcium sulfonate, a lube oil additive which is used as a detergent in engine oils.

10.330   Outline of Manufacturing Process Steps
(Figure 10.9)

1. Sulfonation of natural oils and synthetic feeds by reaction with oleum, followed by the addition of water.

2. Separation of acid sludge. The separated acid sludge is further concentrated in a batch process and disposed of off site.

3. Neutralization of the acid feed by caustic to form sodium sulfonate. Sodium brine is generated, separated out, and sent to the wastewater treatment system.

4. Conversion of the sodium sulfonate to calcium sulfonate by reacting with calcium chloride. Calcium brine is also generated and sent to the wastewater treatment system.

5. Product sulfonate is washed with water. The calcium-containing wash water is directed to treatment.

6. Filtration to remove solids.

7. Calcium sulfonate is blended with oil to form product.

10.331   Composition of Wastewater

Elements of the various wastewater streams to be treated include sodium and calcium sulfonates, spent caustic, oils, and suspended solids.

10.332   Outline of Wastewater Pretreatment Steps
(Figure 10.10)

1. The various process water streams are pumped from interceptors throughout the plant and combined in an equalization tank which is designed to give a long residence time. Limited oil and solids removal takes place here.

2. Wastewater is then pumped through a corrugated plate interceptor (CPI) for primary removal of free floatable and settleable material. The flow is controlled to maintain *LAMINAR*[2] conditions through the plates. Separated oil is skimmed off the top of the CPI and sent to a slop oil receiver. The settled grit is removed periodically by a vacuum truck. Both oil and grit are disposed of off site.

3. Organic polymers are metered into the wastewater stream to enhance separation by coalescing (bringing together) oil droplets and agglomerating (gathering together) small solid particles.

4. Flocculation takes place in the dissolved air flotation (DAF) unit. The wastestream is saturated with air at several atmospheres in the air dissolving tank and then depressurized in the flotation tank. Flocculated oil and solid particles rise and the resulting float is skimmed off for concentration and disposal.

5. Effluent from the DAF is neutralized by the addition of sulfuric acid to adjust the pH down to 7. The purpose of this is to prevent scaling in the sewer caused by the deposition of insoluble calcium salts at high pH. Automatic pH control and recording are provided.

6. The neutralized wastewater passes through a heat exchanger and is cooled to meet the sewer discharge temperature requirement.

7. The cooled effluent goes through a final settling in an interceptor before being discharged through a sample box to the sewer.

When inspecting a chemical manufacturing facility such as the one in this example, pay particular attention to the items described in the following sections.

---

[2] *Laminar.   Laminar flow is smooth or viscous flow; not turbulent flow.*

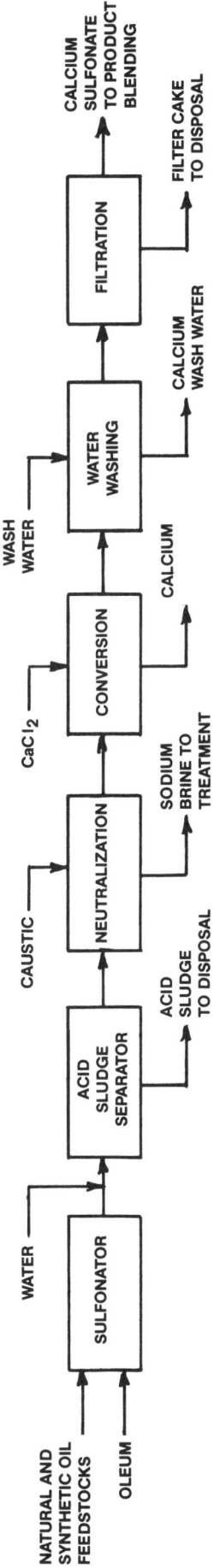

Fig. 10.9   Process schematic of organometallic chemical process

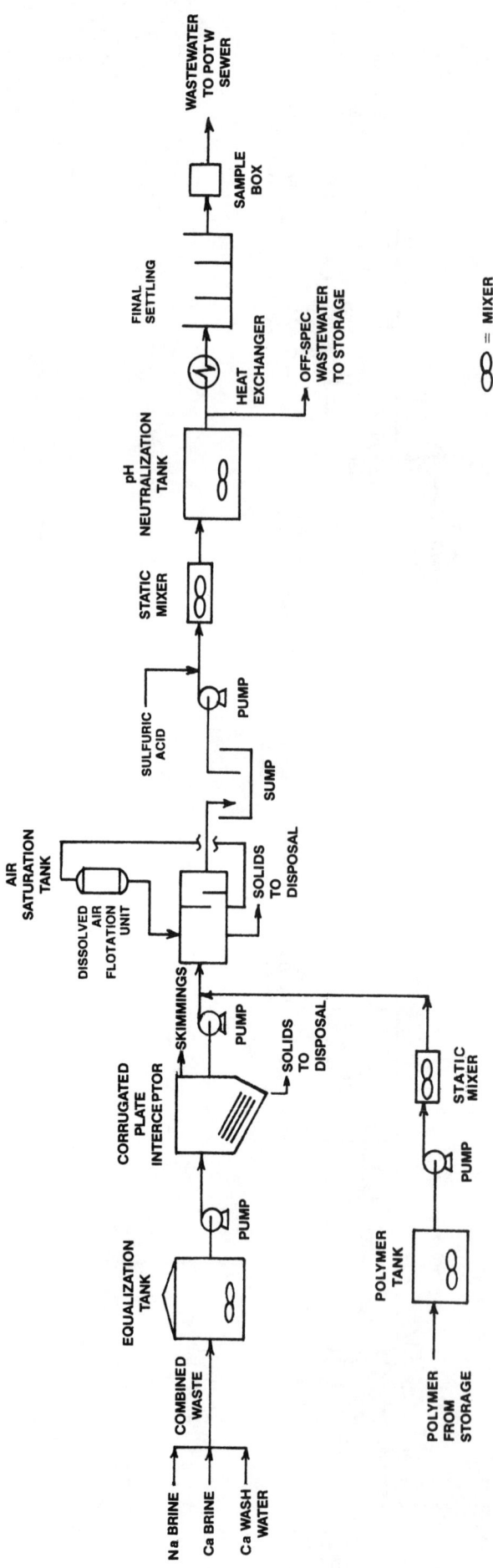

Fig. 10.10 Wastewater treatment process schematic for organometallic chemical manufacturing wastestreams

10.333 The General Manufacturing Area

1. In the chemical storage and process areas, look for possible spills and leaks from equipment. Make sure that there is adequate containment for spills to prevent hazardous chemicals from entering the sewer inadvertently.

2. Look for signs of new construction and changes in equipment which may indicate a process change; a change in wastewater quality and quantity may result. The company's industrial wastewater discharge permit may need to be amended or revised accordingly.

3. If flammable solvents are used at the facility in sufficient quantities to require the installation of a combustible gas detection system, check that the system is being calibrated regularly and maintained regularly.

4. Check hazardous waste manifests to see what, how, and where wastes are disposed of off site. Determine if these disposal methods appear satisfactory.

10.334 The Wastewater Treatment Area

1. Check wastewater at the outfall to see if it appears normal. Look out for unusual conditions in appearance and odor, and attempt to trace their source. If oil is being carried over, the CPI and DAF are probably not operating correctly.

2. Check pH of the wastewater with a test strip and compare with the pH indicator reading. Check at least the past week's recorded values on the recorder chart for compliance with the effluent standard.

3. Inspect the operation of the corrugated plate interceptor (CPI) to see if free oil or solids are being carried over the effluent weir.

4. Check the pressure gage on the air dissolving tank of the DAF to see if adequate air pressure is maintained. See if the mechanical skimmer appears to be in working order.

5. Check the indicated temperature of the effluent wastewater from the heat exchanger downstream of the neutralization system.

6. Talk with the plant operator on duty and see if the operator appears to be adequately trained in the operation of the wastewater treatment system and is familiar with emergency procedures.

7. Check laboratory records of self-monitoring of wastewater quality.

8. Obtain a sample of wastewater from the sample box for analysis of pollutants that are listed on the discharge permit.

QUESTION

Write your answer in a notebook and then compare your answer with the one on page 752.

10.3H List the pollutants found in the wastestreams from the manufacturing process for calcium sulfonate.

10.34 Food Processing and Dairy Products
by Mark Miller

10.340 Food Processing Wastestreams

The purpose of this section is to familiarize you with the processes, sources of wastewater, and methods of wastewater treatment in food processing industries and to describe

appropriate inspection procedures. The following types of food processing facilities are presented:

1. Meat processing,

2. Cheese processing, and

3. Vegetable/fruit processing.

Schematic drawings of the processes and wastewater treatment systems are shown for illustrative purposes in the appropriate paragraphs.

In studying this section, you will notice that several pretreatment systems are used by various types of food processors. The pretreatment methods discussed here are not necessarily specific for the types of industries presented. In other words, a pretreatment system that is used in the meat packing industry may also be applicable to the vegetable/fruit processing industry (with some modifications). It is hoped that by presenting a different method of pretreatment for each category of food processing that is described, you will gain a broader understanding of waste treatment methods used by these industries.

One of the primary reasons there is so much diversity in food processing pretreatment systems is that they discharge wastes to a variety of points. If the wastewater is discharged directly to a navigable waterway (that is, a stream, river, lake, or ocean), federal regulations apply and a National Pollutant Discharge Elimination System (NPDES) permit is required. Therefore, a higher degree of treatment is required. If the wastewater is discharged to a Publicly Owned Treatment Works (POTW), the degree of pretreatment may be dictated by local sewer discharge limits. Also, annual surcharge rates imposed by POTWs on industrial dischargers sometimes become so costly that elaborate pretreatment systems are economically feasible.

10.341 Meat Packing and Processing Facility

The meat packers and poultry processors are among the largest users of water in the food processing industry. The wastewater generated from these facilities exhibits the following characteristics:

1. The discharge is a higher strength (BOD) wastewater than domestic wastewater.

2. The wastewater contains appreciable concentrations of fats, oils and greases (FOG) which are generally more easily biodegraded than mineral or petroleum oils.

3. The wastewater contains proteinaceous material which deaminates (changes to ammonia) to form large concentrations of ammonia.

4. The wastewater is readily biodegradable, and usually contains sufficient concentrations of nutrients such as nitrogen and phosphorus to accomplish stabilization by conventional treatment methods.

A process schematic of a typical meat packing facility is shown in Figure 10.11. The main source of wastewater is from floor washdown and equipment cleaning operations.

The two pretreatment processes commonly used by the meat packing industry are screening followed by dissolved air flotation (DAF) (Figure 10.12). Screening removes the large particulate matter such as meat scraps and offal (waste parts) from the wastewater. This unit process may not have an appreciable effect on the removal of FOG, but in removing the coarse solids, it recovers a by-product that is used by renderers. Screening is also beneficial since it prevents clogging of

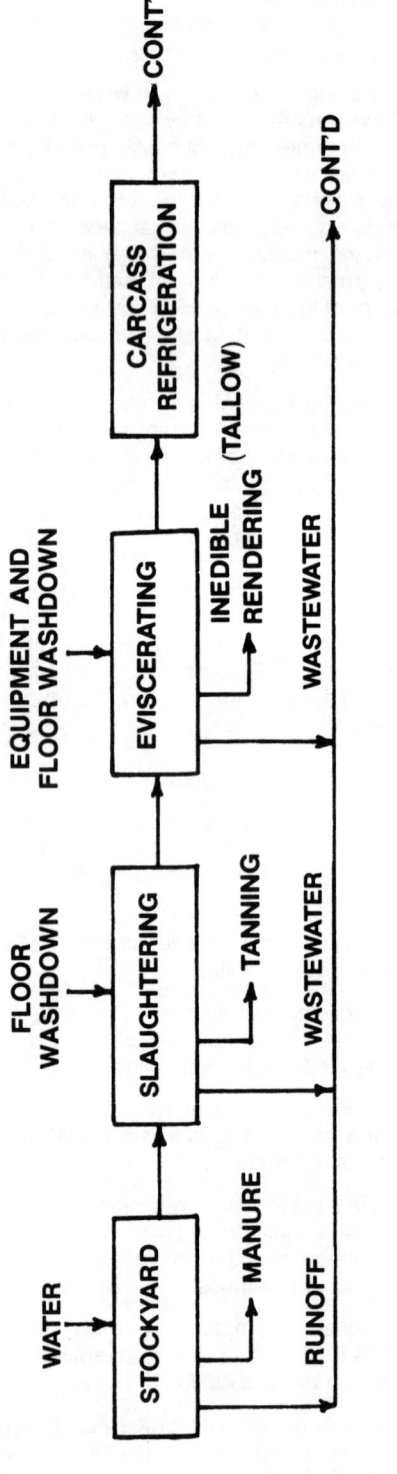
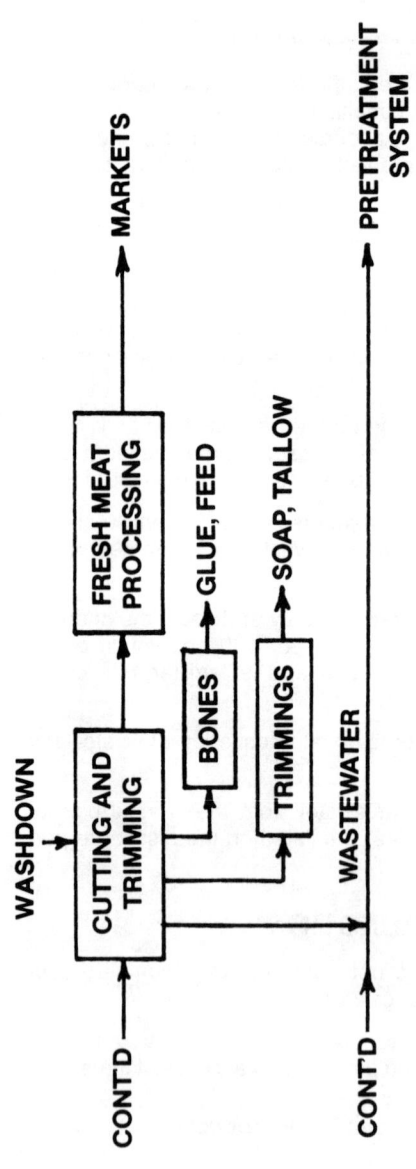

Fig. 10.11   Process schematic of a typical meat packing facility

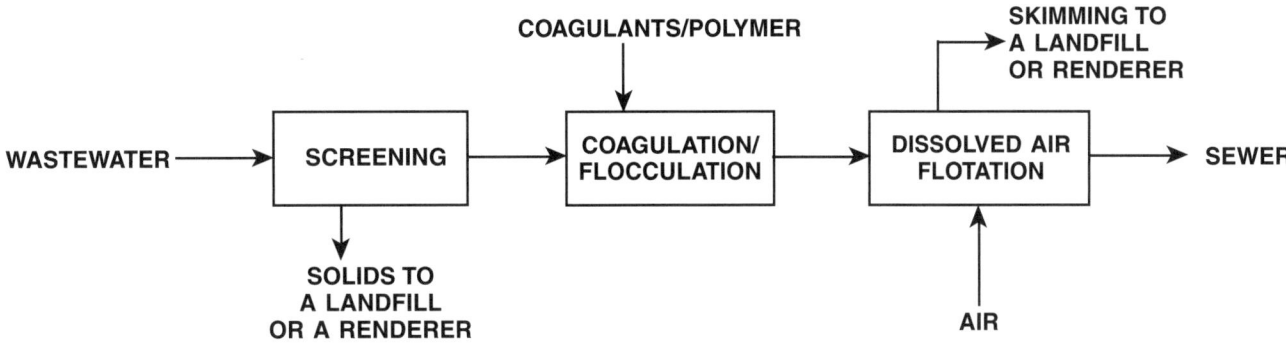

Fig. 10.12   Pretreatment processes of a typical meat packing facility

downstream pumps and pipes by reducing the solids loading on downstream treatment processes.

DAF is a process used to separate solids from wastewater thereby enhancing the removal of FOG, SS, BOD (and COD). This process is based on the principle that fine air bubbles released into the DAF unit attach themselves to solid particles and droplets of oil and rise to the surface where the particles are mechanically skimmed. In this process, solids with densities less than, equal to or slightly greater than water quickly rise to the surface with the air bubbles. These bubbles carry solid particles which form a scum blanket. The scum blanket is then mechanically skimmed from the surface.

The DAF's efficiency may be enhanced by adding chemical coagulants to agglomerate the suspended solids. Typical coagulants include alum, ferrous and ferric sulfate, and ferric chloride. Polymers are sometimes used in conjunction with the metal salts as a coagulant aid. These polymers help to remove the minute (tiny) particles that would not otherwise be removed by dissolved air flotation.

The solids removed in the screening and DAF processes should be recovered for rendering. Renderers can process these solids into soap, tallow, and animal feed. Thus, pretreating the wastewater produces a recoverable by-product readily used by renderers while at the same time reducing the organic loading on the POTW, which lowers the annual surcharge costs the industrial discharger is assessed by the POTW.

10.342 *Cheese Manufacturing Facility* (Figure 10.13)

The main by-product of cheese production is whey. Whey is an opaque, greenish-yellow fluid that remains following the separation of curd from the fluid when converting milk into cheese. Whey is produced in direct proportion to the amount of cheese produced; for every pound of milk used to make cheese, approximately 10 percent ends up as cheese curd and 90 percent as whey.

Whey is an organic substance which contains about one percent protein, five percent sugar (mainly lactose), and other nutrients. It also contains about 6.0 to 6.5 percent solids. Thus, whey is a very high strength by-product in terms of oxygen demand and solids loading, and efforts

should be made to recover it as a resource rather than dispose of it as a waste.

Although a whey recovery system may be capital intensive, it produces a marketable by-product in the form of pet foods, livestock feed and milk fat which may be used for making butter. An additional benefit of a whey recovery system is a reduction in the solids and organic loading on the POTW which translates into a savings on the surcharges for wastewater treatment and disposal.

As shown in the whey treatment schematic (Figure 10.14), the whey first passes through a fine mesh sieve which screens the curd fines from the liquid. These curd fines may be recovered for use in pet foods. The centrifugal separator is used to separate the milk fat from the whey, and the milk fat may be used for making butter. The ultrafiltration and reverse osmosis units are used to remove the protein and lactose, respectively, from whey.

Ultrafiltration and reverse osmosis are similar in principle of operation. Both use semipermeable membrane films and pressure to force water and smaller solute molecules through the membrane. The major differences between ultrafiltration and reverse osmosis are the pore size of the membrane and the pressure required to drive molecules through the pores.

The principle of osmosis and hence reverse osmosis is best illustrated by an example. Consider a vessel that is divided into two chambers. The chambers are separated by a semipermeable membrane; one chamber contains water while the other side contains a solution (water plus dissolved substances). When both sides of the vessel are under the same pressure, osmosis will occur whereby the water passes through the membrane and dilutes the solution until osmotic equilibrium is reached. The osmotic pressure is provided by the solute (dissolved substances). Larger molecules exert less osmotic pressure than smaller molecules.

Reverse osmosis occurs when a pressure is applied to the solution side of the vessel that is greater than the osmotic pressure. This drives the water from the solution side of the membrane to the water side, thus concentrating the dissolved substances in the solution side.

Not all cheese makers will have a pretreatment system as elaborate as described here. However, as the cost of discharging to POTW sewers continues to rise, it may become economically feasible for cheese manufacturers to install an ultrafiltration/reverse osmosis system.

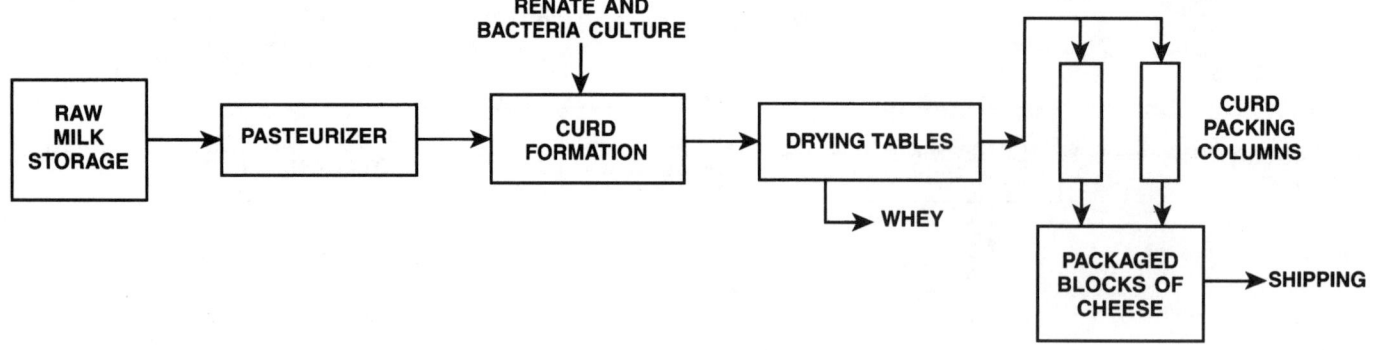

Fig. 10.13   Cheese processing schematic

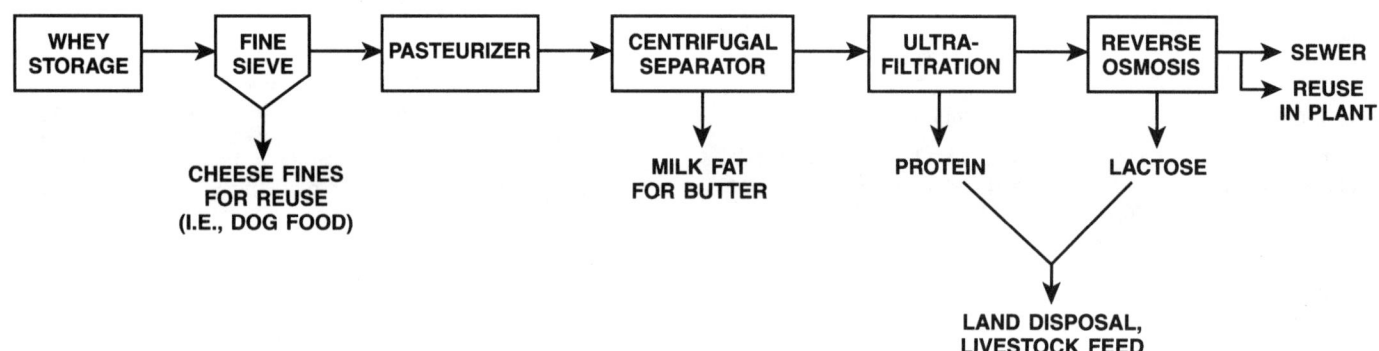

Fig. 10.14   Whey treatment schematic

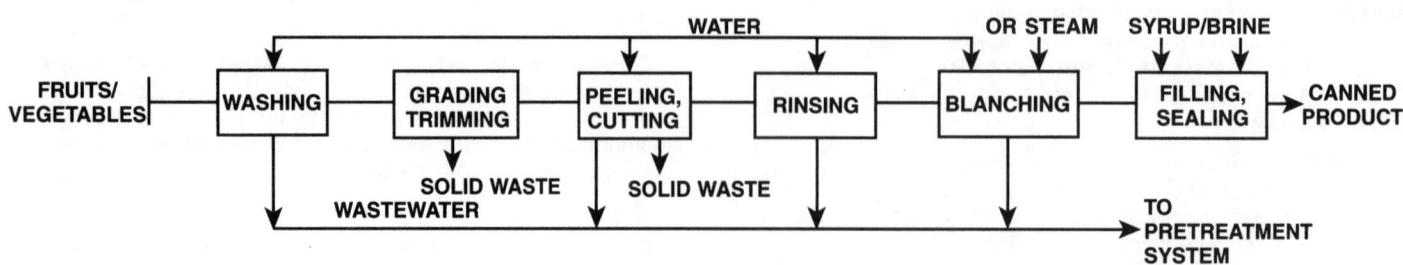

Fig. 10.15   Vegetable/fruit canning process schematic

10.343 Vegetable/Fruit Canning

Food processing plants generate wastewater in processing the fruit/vegetable and in floor and equipment cleanup operations. Water used in food processing includes:

1. Raw product washing,
2. Peeling and rinsing,
3. Blanching and cooling,
4. Heat processing (including preheating, cooking, pasteurizing),
5. Vacuum condensing,
6. Product conveying,
7. Syrup and brine preparation,
8. Can cooling, and
9. Product freezing.

A process schematic indicating water use and wastewater generation is shown in Figure 10.15 (page 726). Some of the process steps are explained in greater detail in the following paragraphs.

One of the principal uses of water in a vegetable/fruit processing facility is for washing and cleaning the raw products. The purpose of the washing step is to remove:

1. Microorganisms, particularly mold and spoilage bacteria,
2. Soil, leaves, stems and other debris,
3. Insect eggs, and
4. Insecticide residue.

The peel of fruits and vegetables is removed by several techniques which include hydraulic pressure, immersion in hot water or caustic, steam exposure, mechanical knives, abrasion, hot air blast, exposure to flame and infrared radiation.

Fruits are commonly peeled by a combination of processes which may include mechanical knives followed by immersion in a caustic solution and rinsing. Root crops are usually peeled by abrasion or immersion in a caustic solution followed by hydraulic or abrasive peel removal. Caustic soda (NaOH) will soften the outer "skin" of many fruits and vegetables enabling the peel to be removed simply by washing the treated fruit or vegetable. Caustic peeling and rinsing may be the largest source of both hydraulic and organic loading from a food processing plant.

Blanching of vegetables is accomplished by exposing the vegetables to hot water or steam. The purpose of blanching is to inactivate enzymes, preserve color, optimize the texture and/or to precook. Water requirements for blanchers are usu-

ally low, but the process extracts a considerable amount of soluble organic matter and it can be a major contributor to the total organic loading from a food processing facility.

The treatment of process wastewater is relatively simple (Figure 10.16). Screens, similar to the type used in the meat packing plants, may be used to remove the larger suspended particles such as peels, leaves and stems. A sedimentation basin may be used to settle solids such as the sand and soil from washing the raw products. To enhance solids settling, the treatment system may be expanded to include coagulation and flocculation upstream of the sedimentation basin. Coagulants may be iron salts or aluminum sulfate (alum) and polymer. The addition of coagulants will help to remove the suspended solids, including the colloidal matter.

10.344 Inspection Procedures

Before a food processing facility is inspected, it is suggested that the inspector review the facility's permit files. This will familiarize you with the company by providing necessary background information such as the type of industry, products produced, size of the facility, and type of pretreatment system. In particular, review the plans to become familiar with the plant layout and the type of pretreatment system the POTW approved. Also review any correspondence in the file to learn of any special discharge limits, such as restrictions on flow volume or strength (SS, COD), that may apply to the company or to determine if the company has experienced difficulty complying with the discharge limits. Then review the company's monitoring data to determine the strength of their discharge in terms of oxygen demand (as measured by BOD or COD) and suspended solids (SS), and also to learn of the

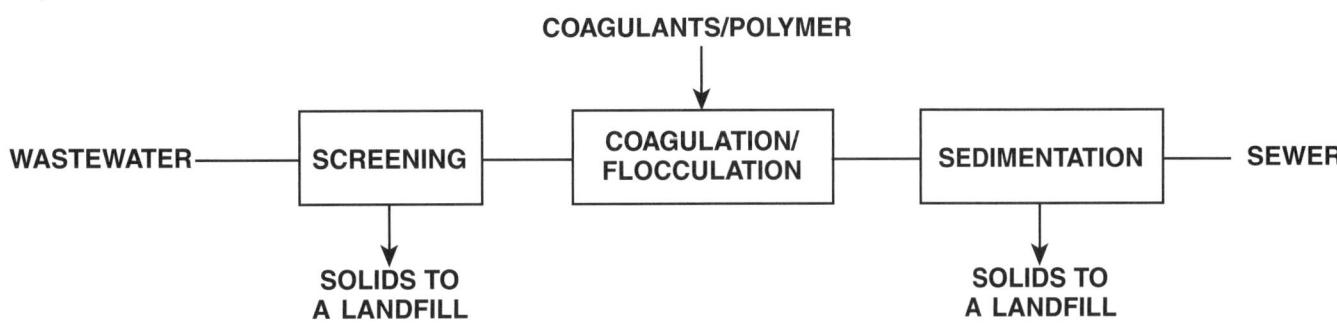

Fig. 10.16 Food processing wastewater treatment schematic

other waste constituents that are routinely monitored in their wastewater (for example, pH, oil and grease, dissolved sulfide, dissolved solids, ammonia).

Upon arriving at the facility, inspect the process areas that generate the wastewater and the industrial wastewater collection system. Ask the plant manager if there have been any modifications to the process operations that may affect the wastewater quality. For example, if a potato processor switches from prewashed potatoes to unwashed potatoes coated with soil, the solids content of the wastewater could increase greatly.

One of the primary reasons for visiting an industrial discharger is to inspect the pretreatment system and obtain a sample of the discharge. Inspect the pretreatment system to ensure that all of the equipment is operating properly. Visually compare the untreated wastewater to the final discharge to get a "feel" for the degree of SS or oil and grease removal the system is achieving. Look at the wastewater as it flows through the treatment system noting if the system is attaining good oil/water separation, good "floc" formation, and good settling characteristics. Look for "loaded" clarifiers, that is, a clarifier that contains an excessive amount of oil and grease and/or solids.

The treated effluent monitoring station should also be inspected (if applicable). The flowmeter should be inspected to verify that it is recording and totaling the wastewater flow to the sewer. Check that the flowmeter recorder has an adequate supply of paper. If the company has a pH recorder, inspect it to make sure it is operating properly by comparing the pH of the discharge as determined by pH paper to the pH as determined by the pH recorder. If the pH probe is in calibration, the difference between the pH paper and the pH recorder should not exceed 0.5 pH units. Also, review the pH chart to see if the company has had any high or low pH discharges. (Generally, POTWs require that the pH be maintained between 6.0 and 10.0.)

Lastly, obtain a sample of the discharge. Follow proper preservation techniques to ensure that sample degradation will not be significant as the sample is transported to the laboratory. Usually, keeping the sample in a sealed container in an ice chest is sufficient. The values for pH, COD, BOD and dissolved sulfide can be significantly changed if samples are not properly chilled and sealed. Label the sample jar so that it can be identified later. The company's name, date, time of sampling, and sampling location along with results of your field tests (pH, temperature, sulfide) should be included on the sample container.

10.345 *Inspecting a Dairy Products Processing Plant*
by Larry Bristow

Inspection of a dairy products processing plant does not involve critical factors encountered in industries subject to pretreatment standards or those with chemicals or other potentially harmful materials that could be discharged into the

sewers. Your main interest will probably be to see that sampling and flow measuring equipment is functioning correctly, and that no "slug" loads have been or will be discharged.

The most significant wastes from dairy products processing are whey, wash waters from products (such as buttermilk), and general cleanup, such as cleaning of tanks and processing equipment.

Considering the high strength (BOD) of dairy products wastes, and depending on the size of the POTW, some restrictions may have been placed on the processing plant, such as requiring the use of sumps or screens. Records of cleaning screens or pumping sumps should be examined. If cottage cheese is produced, check on procedures used to keep spilled material out of the POTW sewer.

In the equipment maintenance areas, check for spill protection. There should be no floor drains where solvents are used for parts cleaning. Solvents will usually be stored in a storage room approved by the fire department; if not, check the floor drains in the storage area to be sure they do not flow directly to the sewer.

Dairy products processing plants require large-capacity refrigeration systems. These will generally use ammonia as the refrigerant, due to the efficiency of ammonia systems. The large pressure vessels used in these systems represent a very severe fire hazard. For fire emergencies, "dump" systems are used to quickly dispose of the ammonia in order to avoid a catastrophic explosion. The dump systems mix water with the ammonia and discharge it to the sewer or to a holding area. These systems are usually reviewed by the POTW and fire department to verify that the required amount of water would be available during a fire (water tower or public water supply). Remember that an electrically powered well pump on the premises may fail due to power failure during a fire emergency.

If inadequate dilution water is used in this type of emergency, an explosive atmosphere could develop in the sewers. Generally, you should be able to verify that the dilution will maintain an ammonia level at about two percent or less.

In the event of this type of emergency, ammonia fumes may escape from around manhole covers and cause safety problems along the path of the sewer. Police and fire departments must be trained to handle this situation. Sewer maintenance personnel must not enter manholes in the flow path, and the POTW operators must be notified. POTW wastewater treatment processes can be severely affected by a large slug of ammonia. During your inspection, verify that adequate notification procedures have been established.

It may be desirable, depending on the size and complexity of the plant, to develop an inspection form that lists each item that should be inspected, with appropriate spaces for remarks. In this day of computers/word processors/copying machines, special forms for each industrial category or user are feasible. This can greatly simplify the inspector's job, help avoid oversights, and aid in explaining results of the inspection to the industry representative.

QUESTIONS

Write your answers in a notebook and then compare your answers with those on page 752.

10.3I   What pretreatment processes are commonly used by the meat packing industry?

10.3J   What pretreatment processes are used to remove protein and lactose from whey?

10.3K   List the items that should be checked when inspecting a dairy products processing plant.

10.35   Rendering
by Steve Medbery and Stephen Todd

10.350   Types of Waste Products

Rendering is an industrial process that takes fish or marine animals, or animal by-products and waste, and converts them to usable products such as fats, oils, tallow and other protein-containing (proteinaceous) solids. As an industry, rendering has existed for over 150 years, making it one of the country's oldest recycling industries. Originally, rendering was an adjunct to the slaughtering and packing industries and was done on the same site ("on-site" plants). Today, however, most rendering is done at off-site facilities, or by independent renderers. All of the edible rendering is done on site in captive plants since sanitation and inspection requirements are similar to those for the meat and poultry packing industry, and these are already in place at a packing/processing plant. Edible rendering, however, still amounts to less than ten percent of the industry, and is not described in this section. Descriptions in this section will apply only to rendering of meat and poultry products, and specifically exclude fish or marine animal rendering. The operations discussed are independent rendering operations rather than on-site or captive plants.

10.351   Equipment and Processes

There are two basic types of rendering systems in use today, the batch system and continuous systems (see Fig-

ures 10.17, 10.18 and 10.19). A batch system follows a repetitive cycle of charging the system with raw material, cooking under controlled conditions, and discharging the material. It is a dry rendering process, one in which no water or steam is added, and moisture is eliminated by evaporation. This procedure is used today mostly in smaller plants and in plants that handle certain types of raw material, such as blood and feathers. The advantages of batch systems are lower capital costs, ability to handle small, "unusual" materials (that is, hog hair and poultry feathers), and ease of replacement of component parts.

Continuous systems are slowly replacing batch systems, although they probably will never replace them completely. They do offer inherent advantages. Since the retention time of a continuous system is usually much less than that required for a batch system, the material has less exposure to heat. This provides improved product control and better odor and aerosol control, which, in turn, means reduced cleanup, labor, and space requirements. The greatest disadvantage is that when part of the system breaks down, the whole system is down. Therefore, a comprehensive and thorough program of regular maintenance is required for reliable operation. Most continuous systems are designed so that one person can control the entire operation from a central control panel. New continuous, "low temperature" systems have been introduced into the United States from Denmark and Norway.

10.352   Water Use and Wastewater Characteristics

Water is used in the rendering plant for the condensation of cooking vapors, plant cleanup, truck and barrel washing, odor control, and boiler makeup water.

Wastewater is generated from raw material receiving (washdown), condensing cooker vapors, hot plant cleanup, and truck and barrel washing (see Figure 10.20). The sources of the largest volume of wastewater generated are plant cleanup, raw material receiving, and truck and barrel washing. Plant cleanup alone accounts for about 30 percent of the total flow. These wastewater sources can be controlled with adequate planning of operations and maintenance by management.

Plant spills are the most common causes of untreated raw material discharged to the POTW sewer system, unless the spill is "dry cleaned"; that is, the spilled material is picked up rather than simply hosed down into the drains. Since rendering plants produce saleable greases and oils, hosing down to the sewer can be expensive and wasteful. Instead, spilled material should be contained and placed back into the system, and anything left behind should be cleaned with an absorbent material and disposed of as solid waste.

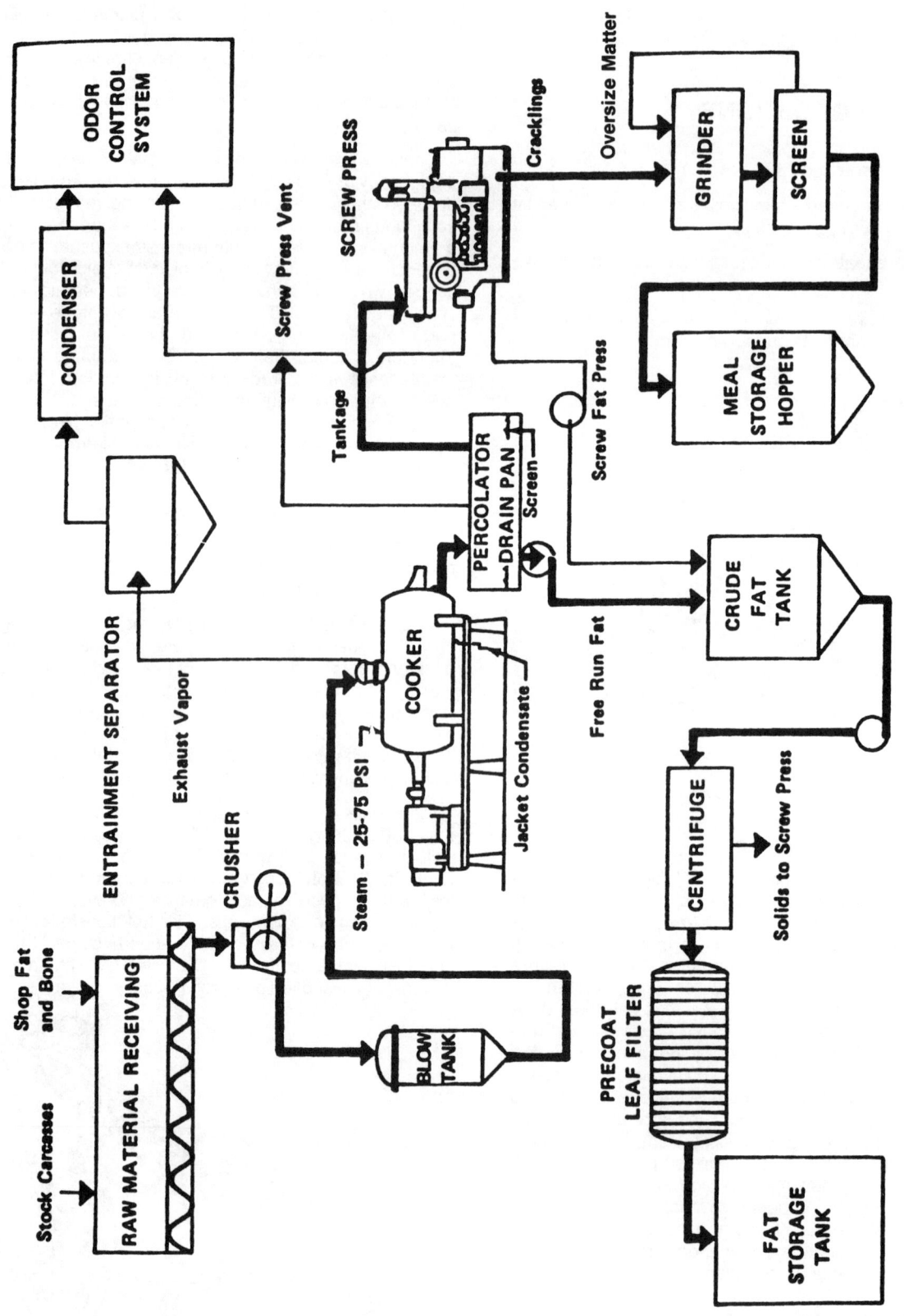

Fig. 10.17   Batch cooker rendering process

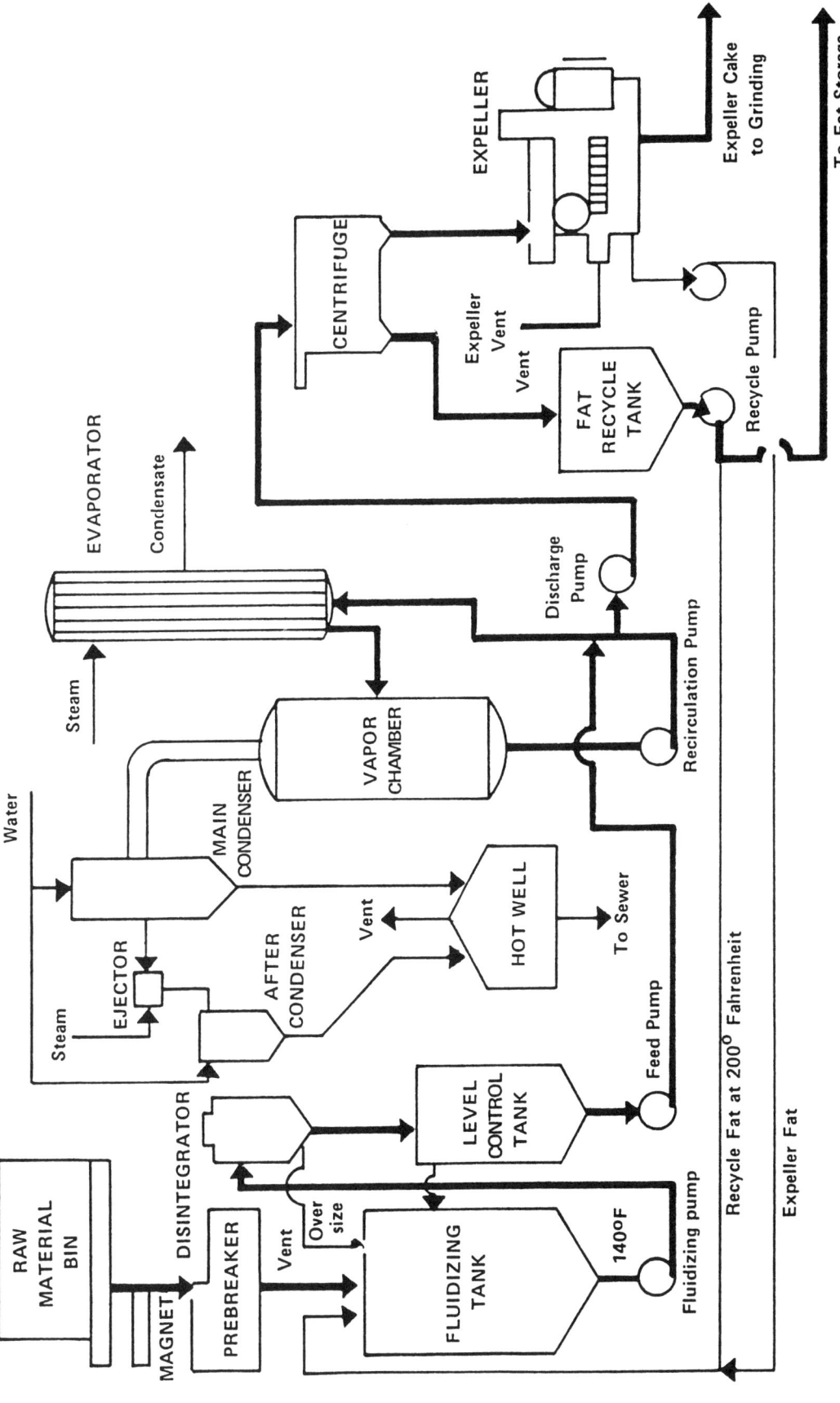

Fig. 10.18 Continuous cooker using Carrier-Greenfield process

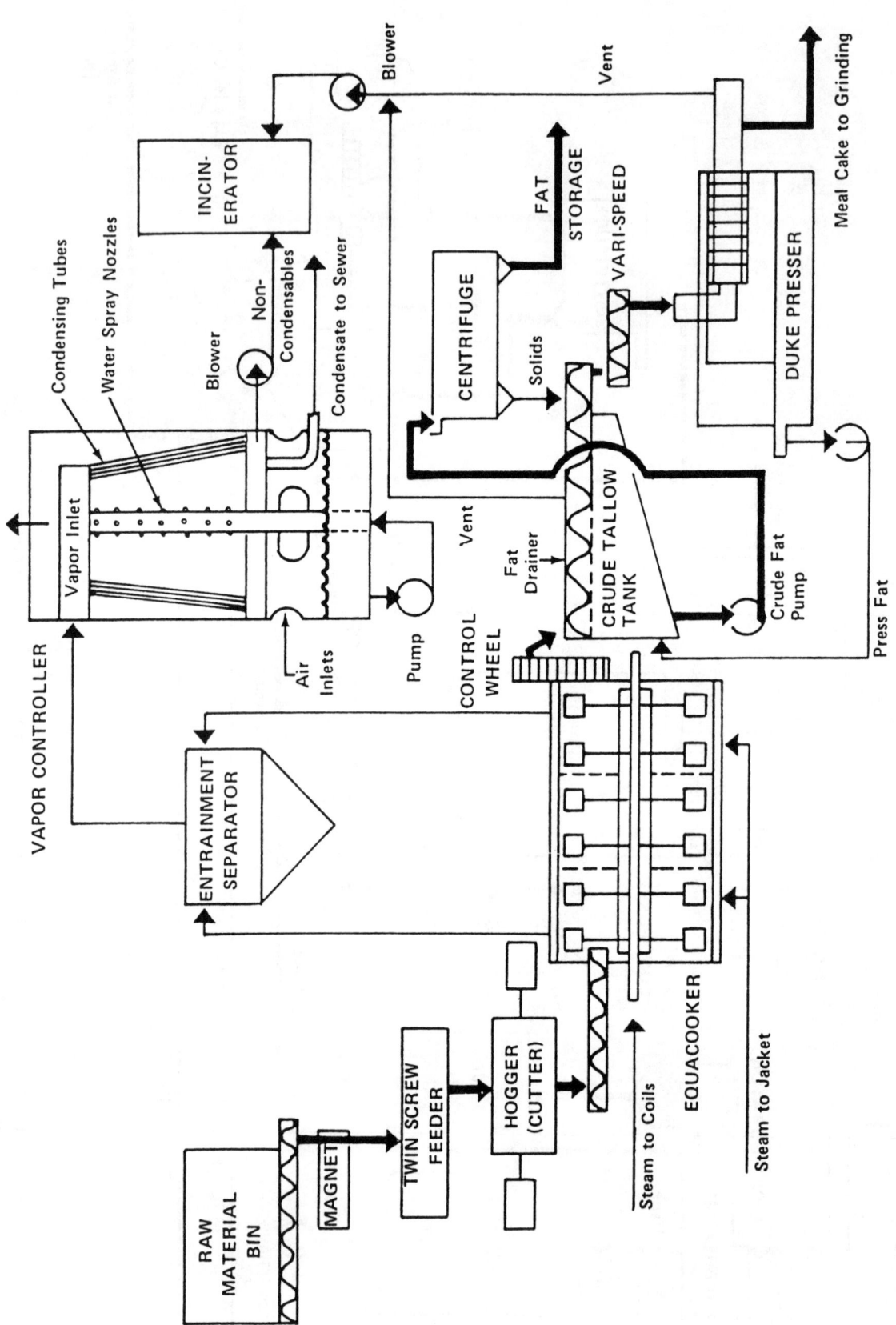

Fig. 10.19   Continuous cooker, Duke process

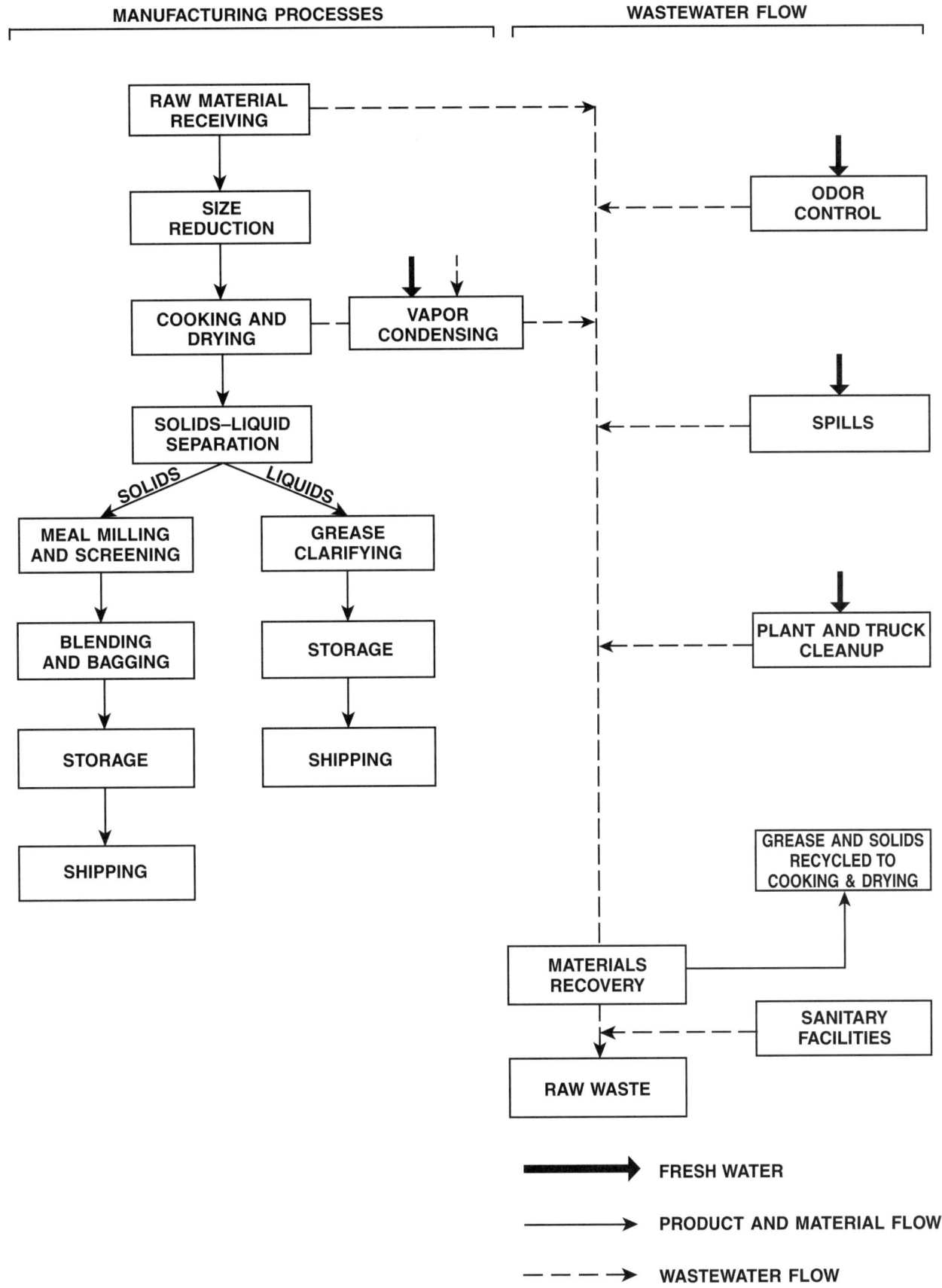

Fig. 10.20   Typical rendering process and wastewater flow arrangement

The volume of wastewater generated is primarily a function of the type of condenser used. Plants using a barometric log condenser use twice as much water as plants using either shell tube or air condenser systems. Air scrubbers for odor control are common in the rendering industry. The volume of water consumed by these units can vary greatly, depending on the type and condition of equipment, the degree of recycling, and the number of scrubbers. A typical plant, however, has two or three scrubbers which account for 5 to 10 percent of the wastewater volume.

Temperature indirectly causes the discharge of greases and fats to the sewer. Unless material recovery or pollution control equipment is operated at an optimum temperature, the separation of grease from the solution is suppressed, and significant amounts of greases and fats will be lost to the sewer system. This is expensive because of product loss and the higher maintenance costs involved in keeping the plumbing system free-flowing. This is also costly in terms of accelerated sewer-use service charges. Proportionally high water use in terms of the amount of raw material processed indicates an inefficiently operated plant or indicates serious leaks. It is not unusual for a plant to use several thousand gallons of water a day over what is required with no one aware of the problem until an audit or similar study brings it to someone's attention.

10.353 Pollutants of Concern

The pollutants of primary interest as determined by the EPA for the rendering industry are BOD, TSS, oil and grease, fecal coliforms, ammonia, phosphorus, and pH. COD can be used in place of BOD measurements if BOD determination is impractical or excessively time-consuming. For our discussion here, the emphasis will be placed on oil and grease, TSS, COD, and dissolved sulfide which is of interest in odor control. In the rendering industry, BOD can be correlated with COD, TSS and oil and grease in raw wastewater.

Other pollutants of concern from the rendering industry are total dissolved solids (TDS), total volatile solids (TVS), ammonia nitrogen, Kjeldahl nitrogen, nitrate/nitrite, phosphorus, chloride, total coliform, pH, acidity, alkalinity and temperature. Most of these parameters can be related to BOD either directly or by inference. Typical values have been established by EPA research for all parameters listed.

Typical BOD values for raw wastewater from a rendering plant range from 1,000 to 5,000 mg/L. Grease and oil are the most critical of the above parameters, and their reduction will generally result in a decrease in oxygen demand, and therefore, a reduction of BOD/COD. Spills and plant cleanup are the principal sources of grease/oil discharge to the sewer. Therefore, it should be relatively easy to control this discharge with the use of appropriate equipment that is maintained in proper working order.

10.354 Treatment Alternatives

The type of treatment needed to produce an acceptable wastewater discharge depends on several factors: type of processing equipment (batch or continuous), type of condenser (barometric log, shell and tube, or air condensers), source of pollutants, types of raw materials used in the process, and products manufactured. All systems incorporate some level of grease/oil recovery; however, discharge to the POTW sewer system without additional treatment of the wastewater could result in unacceptably high concentrations of grease/oil, total suspended solids (TSS), BOD and COD. Varying degrees of treatment are necessary, from primary to tertiary, depending on the process wastestream generated.

The most common methods used for primary treatment include static, vibrating and rotary screens for solids removal; catch basins for grease/oil and settleable solids removal; and dissolved air flotation (DAF) for the removal of fine suspended solids and additional grease/oil removal. DAF is currently the most efficient pretreatment technology available for plant removal of pollutants. It is, however, user-sensitive. Operators must know the capabilities and limitations of the equipment.

Secondary treatment processes include aerobic lagoons, anaerobic contact systems, aerated and aerobic lagoons, and activated sludge. Some tertiary systems presently in use include chemical precipitation, sand filters, microscreen/microstrainers, denitrification, ammonia stripping and ion exchange.

10.355 Inspection of Plants

The inspection process consists of preliminary work, an initial meeting at the plant, a tour of the facilities, an exit conference, and follow-up activities.

PRELIMINARY WORK

Before touring the plant, or even before the first meeting with plant personnel, it is necessary to review the discharger's file for all pertinent information. This includes the control document (permit or order), layout plan (preferably as-built drawings), schematic flow chart, and applicable correspondence. If this is an initial inspection, try to get a plan in advance of the meeting. Information to be sought includes discharger requirements and status, the type of process used (continuous or batch), the equipment manufacturer, the number of systems used (some plants use several separate, parallel systems), the type and number of condensers, modifications to the system, sewer lines, pretreatment equipment, and the age of the plant. Besides the physical layout of the plant, certain data can be inferred from information found on the plans (for example, the type of condenser will greatly affect the volume of wastewater generated, and the age of the plant is an indication that maintenance might be more critical than at a new plant). If the plan is not available beforehand (or at all), this information should be obtained at the first meeting.

It is also desirable to review the water consumption records, if available. Pay particular attention to wide variations and ask why they occurred. If not a production variation, there might be unknown leaks within the plant.

Since it is helpful to know all that one can about an industry when inspecting it, it would be an asset to an inspector to read as much as possible about the industry before arriving at the plant. This is especially true of the rendering industry where very little is known by people outside the industry, and very little published material is available. It is suggested that any inspector, whether experienced with rendering plants or not, obtain and read EPA's *DEVELOPMENT DOCUMENT* for the renderer segment of the meat products point source category (see Section 10.357, reference 1) as soon as possible in their work with the industry, preferably before they ever visit a plant. It is also an invaluable reference work for the library of any local agency that deals with the rendering industry. This EPA document also contains a glossary of specialized terms essential for discussion within the industry.

INITIAL MEETING

When you arrive at the site for your inspection, begin with an initial meeting with management to explain your purpose (entry procedures are covered in greater detail in Chapter 4, "Inspection of a Typical Industry"). The information discussed in the previous section, "Preliminary Work," should be obtained at this point if it was not previously available. Find out the type of process being used and the name of the manufacturer of the equipment. Knowing this simplifies the job of inspection considerably since there are fewer than a half dozen manufacturers[3] of equipment, the technology is well known, and their systems are essentially the same (from site to site).

The equipment of greatest concern to the industrial waste pretreatment inspector is the pretreatment/recovery systems). All rendering systems make some provision for oil/grease recovery or removal from the process stream before discharge to the wastestream. The efficiency of this system is an indication of how efficiently a plant is being run. If a plant is poorly run, pollutants, especially grease/oil, will be present in high concentrations in the wastestream. This could subject the pretreatment equipment to loadings beyond its capability, which will result in high discharges of grease/oil to the treatment facility. In a well-run plant the opposite will be true. Keep in mind that the technology is available in the industry to produce pollutant discharges well within the discharge limits of most local agencies. The industry does, however, have a very high potential for the accidental or "unintentional" high concentration discharges. The complicated nature of the equipment plumbing is a chronic concern to process management because it requires ongoing, skilled maintenance. A small valve can malfunction and disrupt the whole system. Frequently, after a problem has been solved, one hears the statement: "We found a check valve installed backwards."

In general, manufacturers of rendering equipment include material recovery systems but not pretreatment equipment. Renderers, therefore, may add on anything from very basic primary to elaborate tertiary treatment systems. All plants will have some form of primary treatment, whether it is the use of rotating or static screens, catch basins, or more sophisticated flotation systems. As was mentioned earlier, a DAF unit is probably the single most effective system available for removal of grease/oil from rendering plant wastestreams. It is also user-sensitive and prone to problems if operated by inexperienced or inadequately trained personnel. Therefore, if the discharger has a DAF unit on line, be sure to ask how long has it been in operation, which personnel operate it, and what training they have received. In general, it takes some time for plant personnel to become experienced DAF operators, and some experimentation is required to make the unit compatible with the existing rendering system. It is not uncommon for problems to develop shortly after a DAF unit is put on line. The inspector should be aware of this as a possible problem area.

Secondary treatment systems include various types of aerobic and anaerobic lagoons and activated sludge treatment processes. Lagooning is uncommon in a city environment because of escalating property values, odor problems and improving alternate technologies. In some cases, tertiary systems might be used, especially if the plant receives unusual raw materials that produce undesirable pollutants in such quantity or kind that conventional primary or secondary systems cannot adequately pretreat the resulting wastestreams. Denitrification and phosphorus removal are examples of potential alternative technologies. As an inspector, you should be familiar with all types of systems and their intended uses, including "unusual" systems.

At this point, in the first meeting, you should have a complete plan of the plant showing process and pretreatment equipment as well as the plant connections to the city's sewer system. If the plan does not include this information, do not proceed until it can be provided. Also find out if the renderer conducts any peripheral or auxiliary operations, such as truck repair. Determine where process wastewater from these operations is discharged and include these installations in your inspection. Look for and check the operation of alarms on the equipment. Have they been installed? Are they required? How do they work? Are they operational? These alarms should be tested manually during the inspection. Are they loud enough to be heard (and seen) in a noisy plant? If installed, they should be on the last stage of the material recovery system, and any ponds, tanks, or DAF units and installed between the recovery system and the final discharge. Find out before touring the plant the type of vapor control/odor control equipment in use, being aware that if a caustic scrubber is used, you might encounter hazardous conditions in the vicinity of the recirculation tank; use caution when near this equipment.

FACILITY INSPECTION

Before entering the plant, prepare yourself for what you will encounter. Rendering plants can be an assault on the senses. They are hot, noisy, and generate obnoxious odors. A fine layer of grease is found on everything, including the floor, which may also be wet and therefore slippery. In the raw material receiving area you may see anything from entrails, bones, and parts of animals to blood and "dead stock." One should be prepared mentally for such an encounter, and prepared physically by wearing proper shoes and appropriate clothing. A much better (and quicker) job of inspection will be done by someone who is prepared for what may be encountered than by someone unsuspecting of what they are about to inspect. The first impulse of most people is to get out of the place as quickly as possible, which of course devalues the inspection. If you know the physical layout beforehand, and have prepared for the plant environment, you can go directly to the equipment (recovery and pretreatment) of primary interest with only a brief inspection of the processing equipment.

When inspecting the processes from beginning to end, note the general condition of the equipment. Is it old or new? Rundown or well-maintained? Does the equipment appear to have a lot of leaks? A special concern should be the condition of the plumbing. Look for evidence of old leaks (stains, missing/mismatched plant equipment, and puddles).

---

[3] Since there are so few manufacturers of rendering equipment and machinery, it is relatively easy to get replacement parts and make system modifications.

Old and extensively repaired equipment is not necessarily a cause for concern, but it does mean the plant must have first-class maintenance people. Find out who they are and what experience they have. On the other hand, a spotless plant does not necessarily imply a well-run operation. This could mean the plant is merely being hosed down with hot water. Although somewhat common, this is not a good idea; this practice can overload the treatment system, and/or upset the "chemistry" of the rendering process. As you walk through the plant you will be able to form an impression about the discharger's attitude toward maintenance. In some plants, rags can be found holding back leaks because parts are "not available" or are "on order." A poorly maintained facility implies cost cutting, an attitude that could possibly extend to pretreatment equipment, which is non-revenue generating.

Ask the operator of the rendering recovery process how the system works to clarify any specific questions about operating details and to evaluate the knowledge of the equipment operator.

If the pretreatment system consists of screens and/or holding tanks (unlikely), note the size of the tanks so you can calculate their retention time. Ask the crew what retention time is required for the system to work, how often it is cleaned, how often grease is skimmed from the top and solids removed from the bottom, how cleaning is done (manually or automatically), and what is done with this material (it should be put back into the system). If screens are part of the system, ask how often are they cleaned, what is the procedure, and what is done with recovered materials (they should be disposed of as solid waste).

Make note of how the rendering equipment is connected into the pretreatment equipment — plumbing, valves and tanks — since this is one of the major problem areas. Very few, if any, pretreatment systems consist merely of screens and settling tanks; most will have at least a DAF (or some air flotation system), possibly followed by an activated sludge system and, if necessary, additional secondary and tertiary treatment processes. With these systems, you will want to know: How long ago was the device(s) installed? What specialized training has the crew had in its operation? Does operation of the system require a large staff or is it essentially automatic? Is a flocculating agent used? These questions could apply to any settling tank or to a DAF unit. Most do use a flocculating agent to aid in removing solids. Has the system been modified? Most DAF units now incorporate a recycling system through a pressurized tank to aid in grease separation. Make sure the system is essentially the same as the one shown on the plans and that the drawings are complete. If the plant does have problems, this is the most likely area for them to occur.

Also make sure you know the layout for the air scrubber/air condenser systems, as these will be the most likely sources of odor problems. In this regard, as you tour the plant, check to see if there are many open doors or windows. Rendering plants are usually designed to maintain a "negative pressure" inside so that any air leaving the plant passes through the odor control system. Therefore, the windows and doors should be closed (the plants are often very hot).

Check the alarm systems on the appropriate equipment as you move through the plant (recovery vessels, sumps, ponds or lagoons; anything that can overflow to the floor). Note where the discharge line from the final piece of pretreatment equipment is located, and follow it downstream to its entry point into the POTW sewer. Ideally, the local agency will have a sampling manhole provided by the discharger. If this is not available, an entry point must be found between the discharge point and entry to the sewer. Look for a sampling box near or at the discharge line near the plant. Confirm that the discharger has only one exit line.

Take a quick tour of the plant's laboratory facilities. These are usually small quality control labs without much equipment; they generally only test for what they are producing (tallow comes in several grades). Find out what tests they run. It is possible that at a later date, if problems arise, their sampling might be extended to the constituents tested for by the local agency. Reliability (of results) could be a problem here. Before you leave, take an inventory of the on-site chemicals, determine their use, and check to make sure they are stored properly. This should complete your inspection of the plant.

At the conclusion of the tour, you should have an exit conference with the discharger. At this conference you will want to clarify any questions you may have about the plant operations. You should also review the plant's paper work and management operations. Do they keep a record of spills and other problems (do they keep a daily log at all)? Do all employees know what to do if a spill occurs? Whom do they contact? What city (or local) agencies are to be notified? Do they know and understand how (and why) the alarm systems operate? Finally, you should make certain all parties have a clear understanding of what is to be done in an emergency.

10.356 Summary Closing Comments and Follow-Up Action

As you tour the outside of the plant, note how strong the odor of the plant is and, as you are approaching (or leaving), note how far away the odor persists. If you have complaints from the public, they will most certainly be in this regard. Although it should be done before the plant is ever built, determine the direction of prevailing winds in the area.

The condition of the sewers downstream from the plant is of critical importance. If they are old, undersized, or worn out, you will most assuredly have problems. If grease enters the sewers and dries (and it will), it can create virtually solid dams inside sewer structures. Are there pump stations downstream? If they do not function properly, backups into the plant can cause serious problems. Depending on the local agency's policies on these matters, it might be productive to suggest to the discharger that they have their sewer lines cleaned out periodically, as there is some evidence that a coating of grease on the interior of sewers appears as high grease/oil in the sampling procedure, and might indicate a problem where none exists. Also check the renderer's truck repair facility (if they have one). Check the discharger's vehicle washing procedure — where and when it is done, and whether the runoff goes through the pretreatment facilities (it should). Complete the inspection report and prepare any correspondence or enforcement actions as soon as possible.

10.357 Suggested Reading

1. *DEVELOPMENT DOCUMENT FOR PROPOSED EFFLUENT LIMITATIONS GUIDELINES AND NEW SOURCE PERFORMANCE STANDARDS FOR THE RENDERER SEGMENT OF THE MEAT PRODUCTS POINT SOURCE CATEGORY*, August 1974. (This document is essential.) Published by Effluent Guidelines Division, Office of Water and Hazardous Materials, U.S. Environmental Protection Agency, Washington, DC 20460. Obtain from National Technical Information Service (NTIS), 5285 Port Royal Road, Springfield, VA 22161. Order No. PB-253572. EPA No. 440-1-74-031a. Price, $51.00 (includes shipping), plus $5.00 handling charge per order.

2. "RENDER" magazine. A bimonthly trade magazine published under the auspices of the National Renderers Association (NRA). About the only publication available on the subject—current subjects and new technologies covered—invaluable. To subscribe, contact Render Magazine, 2820 Birch Avenue, Camino, CA 95709. Free to qualified parties.

QUESTIONS

Write your answers in a notebook and then compare your answers with those on page 752.

10.3L List the sources of wastewater from a rendering industry.

10.3M What questions should an inspector ask if a rendering facility is using a DAF unit?

10.3N Why should an inspector check the condition of the sewers downstream from a rendering plant?

10.36 Other Industries and Small Businesses
by Susan Adams

This section contains outlines describing: (1) processes and wastestreams associated with specific industries, (2) suggested pretreatment methods or equipment, (3) what to look for on an inspection, and (4) pollutants to test for.

10.360 Paint Formulation

SIC Code: 2851   NAICS: 32551

Processes and Wastestreams:

Manufacture of metallic paints; asphalt paints; flat wall paint — oil, primers and sealers; gloss enamels and floor enamels; house paint, exterior oil and primer; latex house paint; latex wall paint, texture coatings, and fillers; semigloss, satin or eggshell enamel; stains — penetrating or wiping, and fillers; varnish and varnish stains.

Solvent-based paints should have no wastewater discharge. Water-base or latex paint processes will have a discharge from washing equipment between batches. Wastestreams can include washdown and wastewater treatment sludges, raw materials packaging, emission control filters, spoiled batches and spill cleanup materials. All wastestreams may contain hazardous constituents.

Suggested Pretreatment:

In-plant control with reduction and reuse of wastewater.
Chemical treatment with pH adjustment, addition of polymer or coagulant, mixing, settling, discharge of supernatant, and recycling or proper disposal of solids.
No discharge of solvent-based wastes.
Containment of sludges for off-site disposal.

What to Look for on Inspection:

Spill Prevention Plan with detailed list of chemicals, quantities, storage location, and use.
Type of paints formulated (water-base or solvent-base) and ingredients used.
Location of floor drains and sinks.
Cleanup procedures.
Disposal or reuse procedures for bad batches and related recordkeeping.
On-site disposal of solids or a scavenger service used.
Check EPA RCRA registration and records for hazardous waste shipments.
Use of water reduction and recycling techniques.
Method of treatment for wastewater.
Process wastewater segregated from domestic wastewaters.
Method used to measure water and wastewater flows.
Location of sampling port for process and total flows.

Pollutants to Test for in Wastewater Discharge:

BOD*
COD*
Oil and grease
Solids (TSS)
Heavy metals: As, Cr, Cu, Cd, Ni, Zn, Pb, Hg*

Organics: chlorinated hydrocarbons, aliphatic hydrocarbons, chlorinated phenols. Specifically, benzene, ethyl-benzene, methylene chloride, ethyl acetate, butyl acetate, xylene, naphthalene, di-n-butyl phthalate, tetrachloroethylene, toluene, carbon tetrachloride, di(2-ethylhexyl) phthalate.

---

* may be found in large amounts

Discharge Limits: (established by POTW)

BOD — if exceeds 250 mg/*L*, surcharged at rate of
_____/pound/month.

COD — if exceeds 450 mg/*L*, surcharged at rate of
_____/pound/month.

TSS — if exceeds 200 mg/*L*, surcharged at rate of
_____/pound/month.

Oil and grease — not to exceed 200 mg/*L*.

Metals — maximum concentration in mg/*L* (ppm):

As	0.05	Cd	0.7
Cr	5.0	Cu	4.5
Hg	0.005	Ni	1.0
Pb	0.1	Zn	5.0

Organic pollutants: total of all organics listed in Section 307(A)(1) of the Clean Water Act not to exceed 2.0 mg/*L*.

Applicable Permit Parameters:

See permit.

10.361   *Automotive Repair*

SIC Codes: 7538, 7539   NAICS: 811111, 811118

Typical Processes:

General automotive repair
Oil change and lubrication
Transmission repair
Engine work
Brake work
Other repair work

Suggested Pretreatment:

Segregate waste oil and grease in hold/haul tanks.
Segregate all waste solvents for recycle or off-site disposal.
Use grease trap sized properly for effluent.
Use dry sweep absorbent material for oil spills.

What to Look for on Inspection:

Floor drains and where they are plumbed.
Use and storage of chemicals (oil, antifreeze, brake and transmission fluid, solvents).
Use of steam cleaning.
Size and location of tank for waste oil.
Waste solvent containment.
Grease trap for storm runoff, if needed.
Grease trap for sewer effluent.
Evidence of regular trap and tank pumping and proper disposal.
Determine water usage and wastewater flows.
Locate sampling port and collect representative samples.

Pollutants to Test for:

Oil and grease
Organic solvents
COD
Solids
pH
Sulfide
Heavy metals (spent radiator coolant)

10.362   *Car Washes*

SIC Code: 7542   NAICS: 811192

Processes and Equipment:

Service bays for washing cars or trucks.
Can be separate coin-operated facility or part of service station or delivery service.

Suggested Pretreatment:

Recycle system.
Mud box for each bay.
Grease and sand trap interceptor (minimum 50 GPM or 600 gallons for 4 bays; 10 GPM or 120 gallons for each additional bay or single bay hand or portable washer type).

What to Look for on Inspection:

Are solids and oil regularly removed from traps?
What is water consumption?
Are wash areas bermed to protect environment from run-off?
How much grease and sand is in interceptor?
How is sludge handled? and by whom? and how often?
What type of cleaners, solvents, and waxes are used?
Are commercial trucks cleaned here?
Are chemical tankers cleaned here?
What method is used to determine water and wastewater usage?
Locate sampling port and collect representative samples. Is storm water runoff diverted away from sanitary sewer and into storm sewer (unless combined sewer system)?

Pollutants to Test for:

Oil and grease
Solids
COD
BOD
Metals
Organics

10.363   *Film Processing and Photographic Developing*

SIC Code: 7384   NAICS: 812921, 812922

Processes and Equipment:

Developing and printing film, silkscreening.
Darkroom, photographic equipment, lab sinks, any lab equipment using water.

Suggested Pretreatment:

Silver recovery system (Figure 10.21).
pH adjustment.
Organics containment.
Sulfide control.

What to Look for on Inspection:

Evaluate performance of pretreatment equipment.
List of chemicals used.

Fig. 10.21   Ion exchange tanks installed to capture silver from photographic wastes

Check for use of rinse waters and floor drains.

Method of disposal for developers, fixers and bleaches. Is process wastewater segregated from domestic flow?

Check silver recovery system for regular maintenance.

Locate sampling port for process waste and for total flow and collect representative samples.

Are sulfur compounds degrading into sulfide in the clarifier due to excessive residence time or lack of routine clarifier cleaning?

Pollutants to Test for:

Organics
Ammonia
pH
Silver
COD
BOD
Sulfide

10.364   Grocery Stores

SIC Code: 5411   NAICS: 44511, 44512

Processes or Facilities:

Meat market
Deli or food service
Bakery
Produce preparation area

Suggested Pretreatment:

Properly sized grease trap (interceptor).
Garbage grinder.

What to Look for on Inspection:

Deli, bakery, and meat cutting areas connected to grease trap.

Check fixtures to determine proper size for grease trap.

Check grease trap for size, design and adequate maintenance.

Determine method for monitoring water and wastewater flows.

Locate sampling port and collect representative samples.

Check size of discharged particles. Are screens and garbage grinders properly maintained?

Are trapped organics fermenting? Check pH and sulfide levels.

Pollutants to Test for:

Oil and grease
COD
BOD
Solids
pH
Sulfide

10.365   Hospitals

SIC Codes: 8062, 8063, 8069   NAICS: 62211, 62221

Processes Include:

General medical and surgical.
Mental, psychiatric or geriatric care.
Other hospital and health care services.
Kitchen, laundry and laboratory services.

Suggested Pretreatment:

Wastes contained for off-site disposal.
Silver recovery system for darkroom.
Lint trap and waste heat recovery for laundry service.
Grease trap for kitchen service.
Incineration unit for biological waste.

What to Look for on Inspection:

Evaluate performance of pretreatment equipment.
General layout of facility including locations for food and laundry service; morgue; types of laboratories with major equipment such as x-ray and radiation.
Location of floor drains, sinks, and any machines that are plumbed to sanitary sewer.
List of chemicals, their use, storage, and methods of disposal.
Solvent handling, storage and disposal methods.
Procedures for handling hazardous and infectious wastes.
Use of cleaners and germicides.
Determine method used for water and wastewater flow monitoring.
Locate sampling port and collect representative samples.

Pollutants to Test for:

BOD
COD
pH
Silver
Solids
Oil and grease
Organics
Radioactivity

10.366   Hotels and Motels

SIC Code: 7011   NAICS: 72111, 72112, 721191, 721199

Facilities Include:

Laundry area
Kitchen facility
Swimming pool

Suggested Pretreatment:

Lint trap for laundry.
Grease trap for kitchen.

What to Look for on Inspection:

Evaluate performance of pretreatment equipment.
Check kitchen area for properly sized trap (see "Restaurants" for sizing instructions in Section 10.370).
Landscaping and swimming pool should be separately water metered.
Swimming pool drains to storm sewer through a filter.
Use of water conservation techniques.
Determine method for monitoring water and wastewater flows.
Locate sampling port and collect representative samples.

Pollutants to Test for:

Oil and grease
BOD
COD
Solids
pH
Chlorine and TDS (if swimming pool wastewater goes into sanitary sewer)

10.367   Industrial Laundries

SIC Code: 7219   NAICS: 812332

Processes Include:

Washing work uniforms, shop towels, rugs, dust mops.
Use of dyes.

Suggested Pretreatment:

Lint filters or bar screens.
Large sand and oil traps (500 gallon minimum).
Waste-heat recovery system.

What to Look for on Inspection:

Proper maintenance of lint screens.
Main trench drain that collects wastewater from all washers.
Number of washers that discharge to drain at same time.
Storage and use of solvents and shop towel dyes.
Sludge accumulation and disposal (Figure 10.22).
Are delivery trucks maintained and washed on premises?
Determine method for metering water and wastewater.
Locate sampling port and collect representative samples.
Oil and grease removal system (Figure 10.22).

Pollutants to Test for:

COD
Solids
Oil and grease
Organic solvents
Temperature

10.368   Radiator Repair

SIC Code: 7539   NAICS: 811118

Processes Include:

Radiator repair, pressure testing, checking for leaks, and soldering.

Suggested Pretreatment:

pH neutralization (Figure 10.23).
Recycle system (Figure 10.23).
Hold/haul tank.

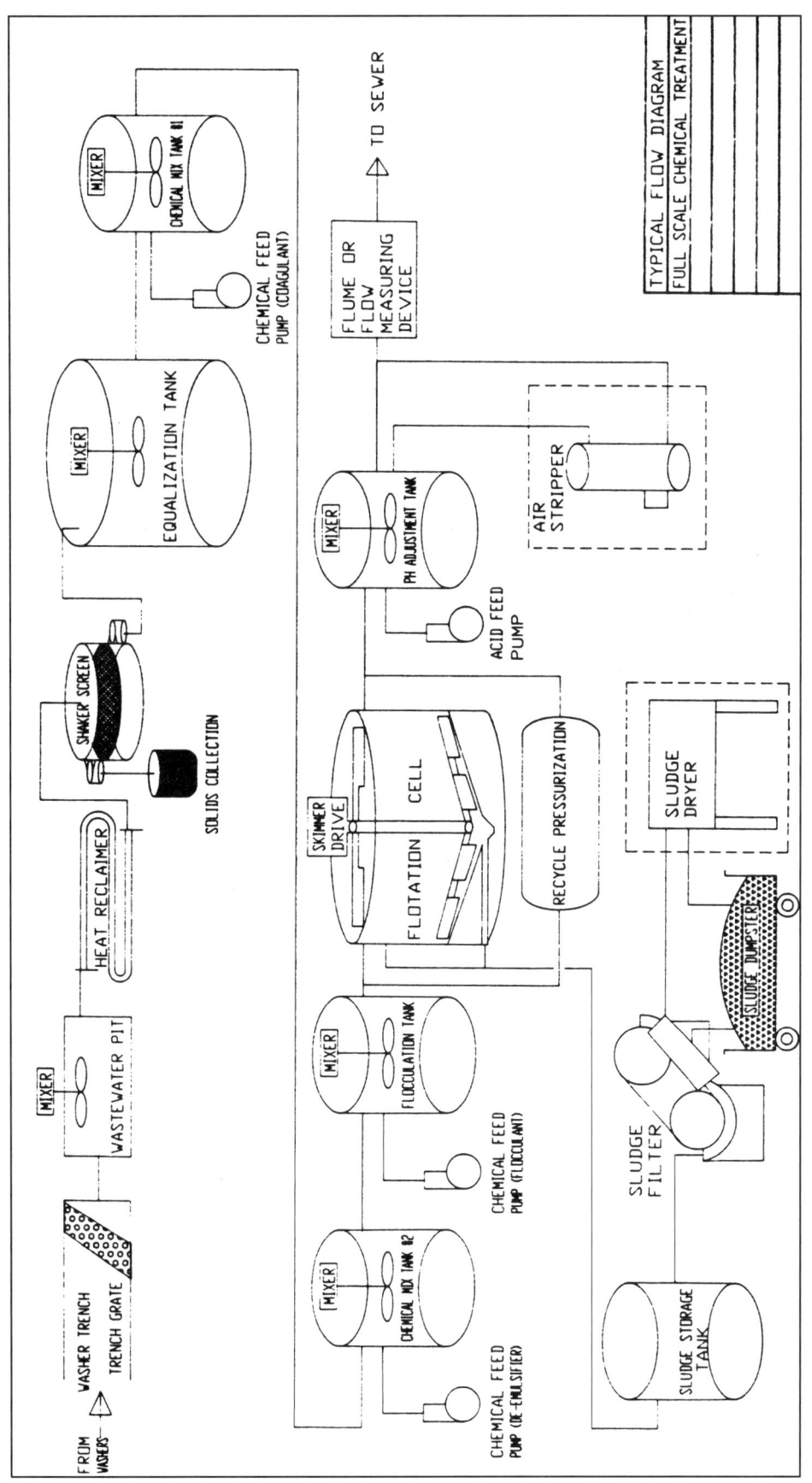

Fig. 10.22 Oily waste treatment system at an industrial laundry

Radiator shop's metal-bearing wastestreams are collected for pH adjustment and metals precipitation

Radiator coolant (ethylene glycol) is segregated for recycling

Fig. 10.23   Radiator repair shop waste handling methods

What to Look for on Inspection:

Evaluate pH control system.
Location of floor drains and cleaning tanks.
Dye test to verify destination of wastewater lines.
Use and storage of chemicals.
Check caustic cleaning and hydrotesting tank.
Methods for disposal of spent tank solutions.
Procedures for disposal of antifreeze.
Method used for measuring water and wastewater usage.
Locate sampling port and collect representative samples.

Pollutants to Test for:

Heavy metals
pH
COD
BOD
Solids
Organics

10.369 Research and Testing Laboratories

SIC Codes: 7391, 7397   NAICS: 54138

Facilities Include:

Laboratory or other physical research and development facility.
Products testing facility.
Wet chemistry and instrumental analyses facility.

Suggested Pretreatment:

Neutralization tank.
Wastes contained for off-site disposal.
Solvent recovery or off-site disposal.

What to Look for on Inspection:

Solvent storage procedures.
Use and storage of chemicals.
Basic layout of facility including analyses performed.
Procedures for disposal of wastes, especially solvents.
Check lab sinks, floor drains and safety showers for discharge to POTW sewer.
Determine method used for monitoring water and wastewater flows.
Locate sampling port and collect representative samples.

Pollutants to Test for:

Organics
pH
COD
BOD
Heavy metals

10.3610 Restaurants

SIC Code: 5812   NAICS: 7221, 7222, 7223

Processes Include:

Food preparation, service, and cleanup.

Suggested Pretreatment:

Grease, oil and sand interceptor (grease trap) with two compartments (primary having 7-minute detention time and secondary having 5-minute detention time).
Trap/interceptor sized up for garbage grinder, dishwasher, or wok stove (144-gallon capacity for each unit).
30-gallon under-sink model traps approved for restaurants without dishwashers, garbage grinders, wok stoves, or full food service.
Limit use of acids and enzymes.
Have regular schedule for maintenance of trap.
Most traps should be located outside for better odor control and maintenance.

What to Look for on Inspection:

Check all kitchen fixtures to determine proper size for grease trap.
Dye test, if necessary, to verify destination of wastewater lines.
Check trap for size, design, and adequate maintenance.
Disposal of spent cooking grease.
Check records to verify frequency for pumping and proper disposal of grease trap waste.
Check floor mat and garbage can washing areas. They should be protected from rainwater and plumbed to grease trap.
List of chemicals and cleaners used.
Determine water and wastewater usage (methods for flow monitoring).
Locate sampling port and collect representative samples.

Pollutants to Test for:

BOD
COD
pH
Oil and grease
Solids

QUESTIONS

Write your answers in a notebook and then compare your answers with those on page 752.

10.3O List the sources of wastewater in a paint formulation industry.

10.3P How can an inspector verify the destination of wastewater lines?

10.4 CENTRALIZED WASTE TREATMENT FACILITIES
by Clifford Lum

A major problem facing industry today is the proper treatment and disposal of hazardous liquid wastes. Traditionally, liquid hazardous wastes that could not be treated on site for disposal to sanitary sewers were hauled to hazardous waste landfills. As more stringent regulations are applied and as nearby hazardous waste disposal sites are closed, industries are forced to truck their wastes longer distances at extremely high costs or find alternative methods of disposal. A new type of "industry" has emerged to deal with this environmental problem. Centralized Waste Treatment Facilities (CWTFs) (Figure 10.24) have been established that are capable of treating a variety of hazardous and nonhazardous liquid wastes from multi-generator sources so that the treated liquid wastes are acceptable for disposal in sanitary sewers. These facilities represent a viable alternative to landfills for disposal of hazardous and nonhazardous liquid wastes.

In the areas within the jurisdiction of the Los Angeles County Sanitation Districts (LACSD), most of the CWTFs are designed to handle oily wastes, metal-bearing wastes, or specific nonhazardous wastes. Oily wastes may include landfill leachate, gas condensate wastes, machine coolants, and catalyst wastes. Metal-bearing wastes include spent plating baths, acidic and basic solutions, and concentrated rinse waters. Nonhazardous wastes may include food industry wastes, restaurant grease trap wastes, animal/vegetable oil wastes, and car wash clarifier wastes. The LACSD has developed a program for regulating and controlling CWTFs which involves a "three-pronged" regulatory approach to ensure compliance with effluent limitations. The three "prongs" consist of: (1) a qualification program for new generators to determine whether the CWTF can successfully treat the waste, and screening of the incoming wastes upon arrival at the CWTF; (2) overseeing the proper design and operation of the treatment system; and (3) testing the treated wastewater before discharge.

Step 1 requires new waste generators to complete a questionnaire that is evaluated by the facility laboratory. If the waste is determined to be *PROBABLY* treatable, a sample is obtained from the generator for bench-scale testing. The purpose of the tests is to determine if the waste can be successfully treated by the CWTF and to determine that organic and inorganic priority contaminants do not exceed limits for acceptance as specified in various operating permits. If testing indicates that the waste can be treated successfully, chemical

and physical analyses are performed and the results are reviewed by the facility's management. A decision is then made concerning acceptance of the generator's waste for treatment and the generator and all regulating agencies are notified of the decision.

Step 1 also involves screening (Figure 10.24) all incoming wastes from approved generators prior to acceptance and off-loading to verify compliance with the original waste profile and treatability study. The generator's file is reviewed and a representative sample is taken. The sample is then evaluated for treatability by bench-scale testing. If the test results permit load acceptance, the waste is unloaded and pumped to a receiving tank.

Step 2 involves the design and submittal of the treatment system and operational procedures by the facility's management. The submittal is evaluated by the regulating POTW to determine the adequacy of the system to treat the targeted wastes. Any deficiencies and additional requirements are determined at this point and resolved prior to permit issuance.

Step 3 requires testing of the treated water prior to discharge to the sanitary sewer system. A sample is drawn from the final holding tank while the tank is being agitated. If all parameters are within acceptable effluent limits, the batch is discharged to the sewers. In the event that the analysis shows the water is unacceptable for discharge, the batch is pumped to the emergency holding tanks for subsequent testing and pretreatment.

Fig. 10.24   Check of incoming load at Centralized Waste Treatment Facility

Once a load is determined to be acceptable, it is processed through the treatment system. Nearly all CWTFs provide initial screening and grit removal facilities prior to the incoming holding tanks. Subsequent treatment processes are dependent upon the type of waste the facility has targeted. The following are typical treatment schemes for CWTFs:

1. Oily Wastes (Figure 10.25)

   - Demulsification
   - Oil/water separation and solids removal
   - pH neutralization
   - Flocculation
   - Dissolved air flotation
   - Filtration
   - Absorption (polishing)
   - Carbon adsorption (for priority organics) (Figure 10.26)
   - Sludge dewatering
   - Vented air scrubbing and incineration

2. Metal Finishing Wastes (Figure 10.27)

   - Waste segregation
   - Cyanide destruction
   - Chromium reduction
   - pH neutralization
   - Flocculation
   - Clarification
   - Sludge conditioning and thickening
   - Sludge dewatering
   - Filtration
   - Air stripping and/or carbon adsorption

3. Nonhazardous Wastes (Figure 10.28)

   - Oil skimming and solids separation
   - Demulsification
   - Oil/water separation
   - pH neutralization
   - Flocculation
   - Clarification
   - Sludge thickening
   - Sludge dewatering

All CWTFs are required to discharge the treated wastewaters to effluent holding tanks for analysis. The batch discharge scheme provides for additional control of the treatment process. Emergency holding tanks are required for storage of any wastewater that is not in compliance with discharge limitations. In general, storage must be provided for the largest volume of tanks in the treatment chain. These wastewaters are subsequently retreated until acceptable for discharge to a sanitary sewer.

Storage must also be provided for impounded rainwater from the pretreatment areas and off-loading areas. Impounded rainwater must be tested and, if found to contain unacceptable levels of contaminants, processed through the treatment system.

An inspection of a CWTF consists primarily of inspecting hauling manifests and other recordkeeping. Proper records of loads accepted, incoming load screening procedures and analyses, load rejections, and treated load analyses should be inspected to verify operating procedures and compliance with effluent limitations. Other areas of concern are:

1. Inspection of the spill containment facilities for the off-loading, waste storage and treatment areas, and treatment chemical storage areas.

2. Inspection of the flow monitoring and pH recording instrumentation and the condition of the designated sampling point.

3. Inspection of the rainwater holding tanks or ponds and rainwater diversion facilities.

4. Observance of off-loading procedures.

5. Verification of proper maintenance of treatment equipment and adequate supply of treatment chemicals.

6. Verification of proper maintenance of emergency shutdown facilities and emergency transfer and storage tanks.

NOTE: The centralized waste treatment industry is changing and pretreatment inspectors need to keep up with changes in the industry and also changes in the industrial categorical standards when new rules are implemented. The current federal regulations concerning Centralized Waste Treatment (CWT) facilities are contained in the Code of Federal Regulations, Title 40 Part 437. Obtain from Superintendent of Documents, U.S. Government Printing Office, PO Box 371954, Pittsburgh, PA 15250-7954. Stock No. 869-044-00159-4, price, $55.00.

QUESTIONS

Write your answers in a notebook and then compare your answers with those on page 752.

10.4A What types of metal-bearing wastes can be treated at centralized waste treatment facilities?

10.4B List the types of nonhazardous wastes that may be treated at a centralized waste treatment facility.

10.4C What happens when an approved generator delivers wastewater to be treated at a centralized waste treatment facility?

10.5 ACKNOWLEDGMENTS

Portions of the material in this chapter were adapted from EPA Region X's "Pretreatment Implementation Guidance Manual" provided by Mr. Bob Robichaud. Special thanks go to all the people who contributed to the sections on how to inspect specific industries.

Please answer the discussion and review questions before continuing with the Objective Test.

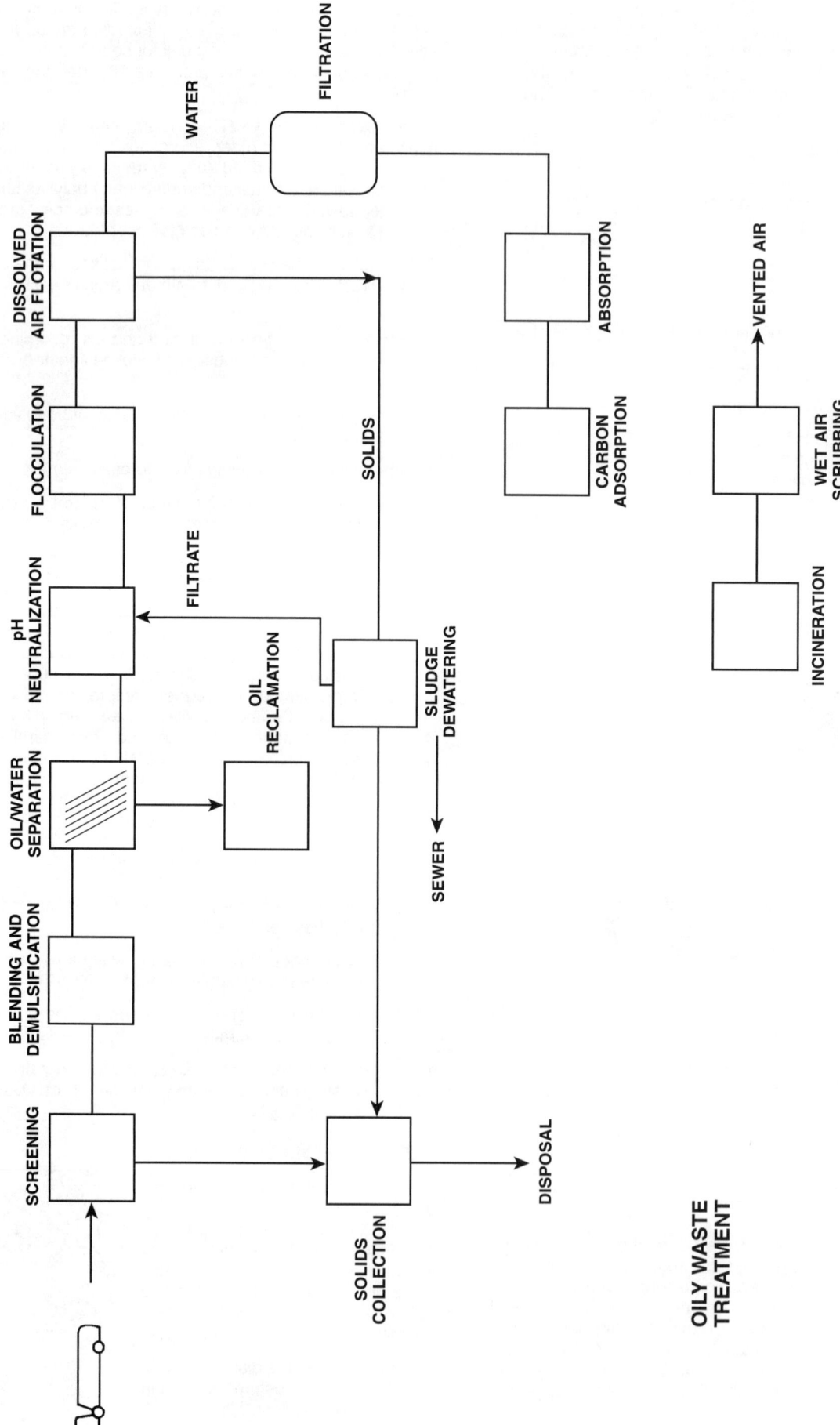

Fig. 10.25 Oily waste treatment

Fig. 10.26   Granular activated carbon is often used to remove
toxic organics or as a final polishing step to remove oil and grease
at Centralized Waste Treatment Facilities

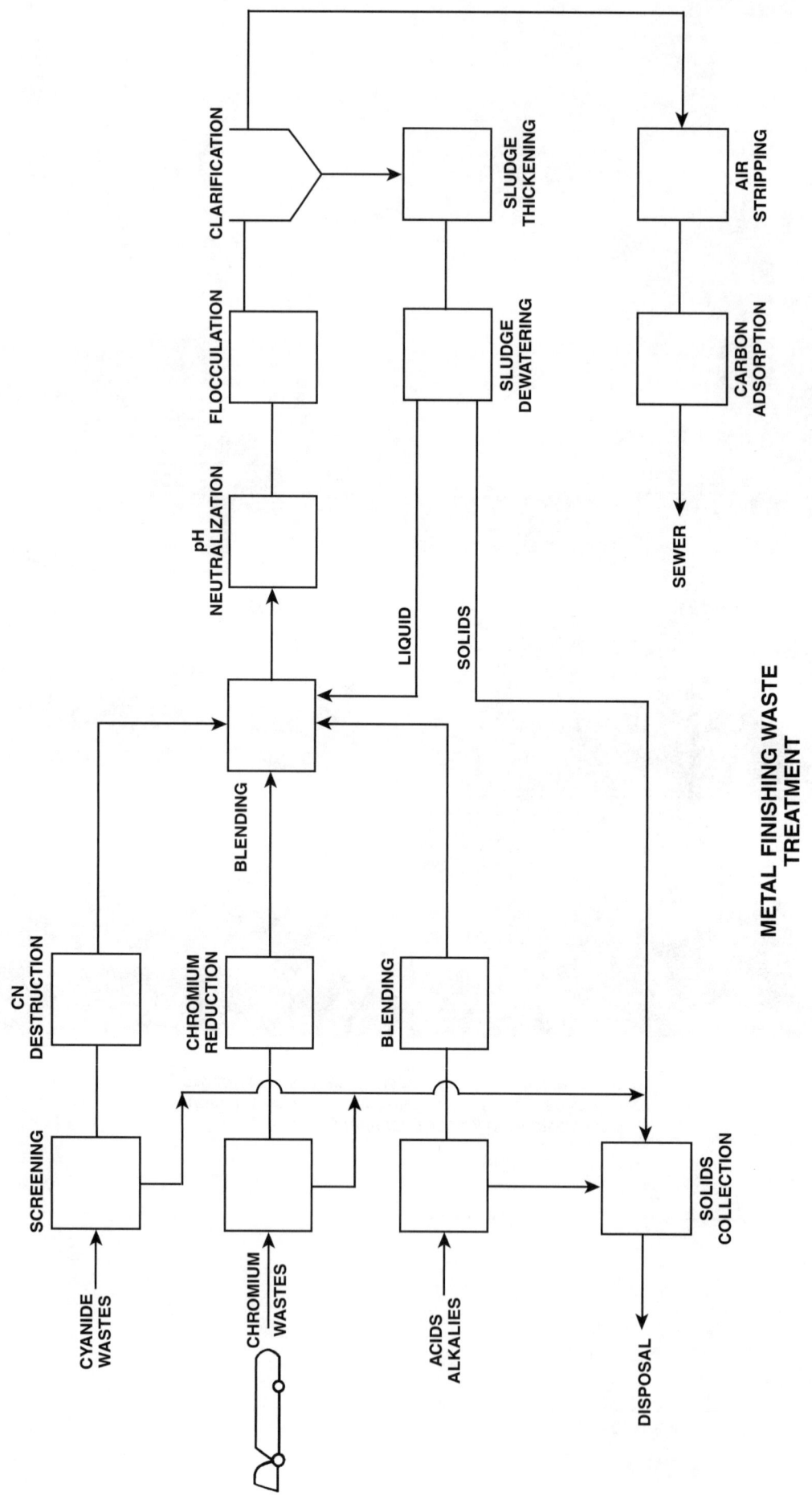

METAL FINISHING WASTE TREATMENT

Fig. 10.27   Metal finishing waste treatment

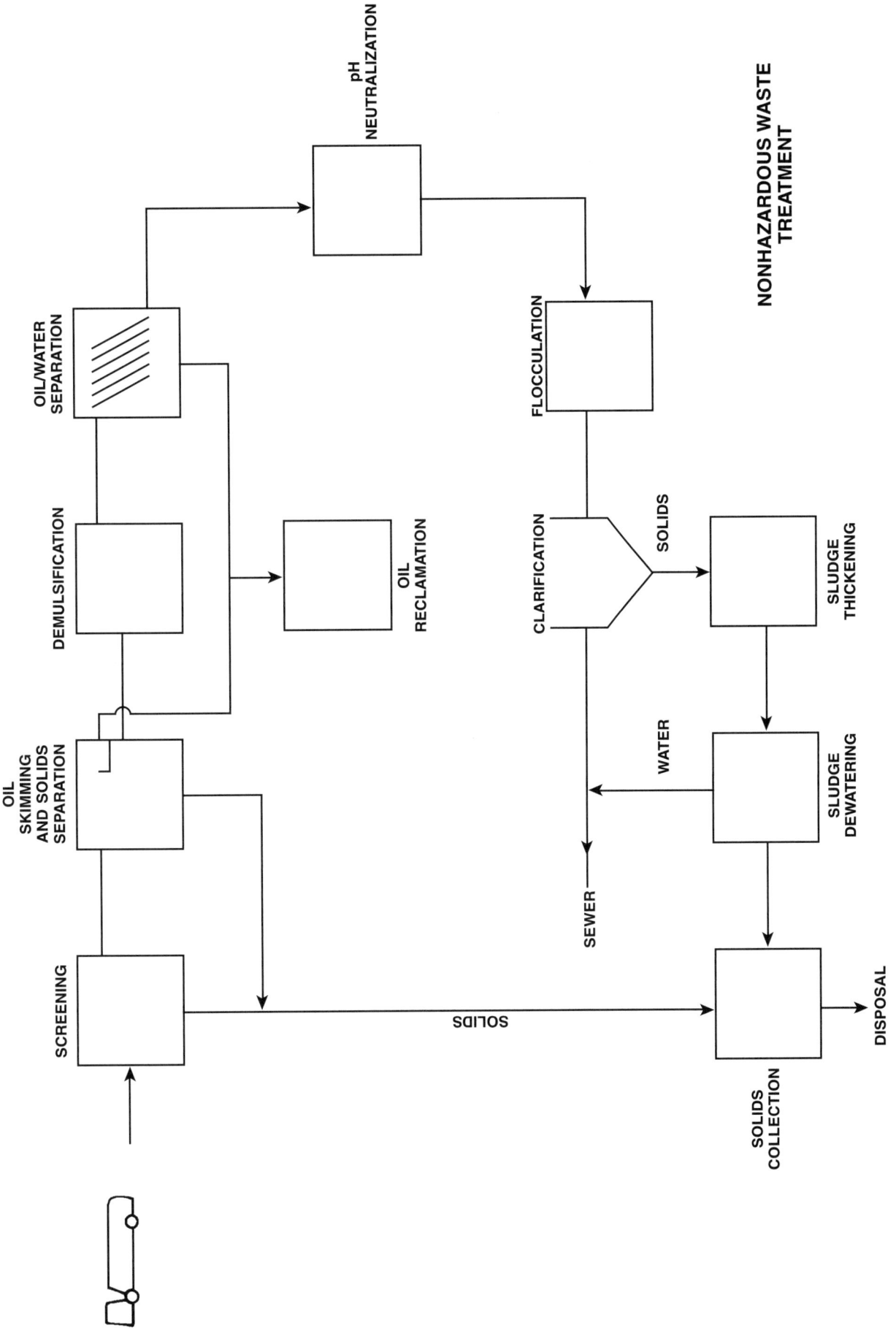

Fig. 10.28 Nonhazardous waste treatment

DISCUSSION AND REVIEW QUESTIONS

Chapter 10.   INDUSTRIAL INSPECTION PROCEDURES

Write the answers to these questions in your notebook before continuing with the Objective Test on page 753. The purpose of these questions is to indicate to you how well you understand the material in the chapter.

1. What is the main reason for conducting pretreatment facility inspections and monitoring?

2. Why is current data on industrial dischargers necessary in a pretreatment program?

3. Why should an inspector check the industrial user's self-monitoring procedures and equipment?

4. Why must accurate flow measurements and representative samples of industrial users' discharges be obtained?

5. Data collected from a POTW's monitoring program could be used for what purposes?

6. What information should be gathered during an industrial plant inspection regarding waste residuals (solids or floatables)?

7. What types of existing information regarding an industry to be inspected should be taken along during an inspection and verified for accuracy?

8. Under what circumstances might electroplating wastewaters from an anodizing and cleaning operation be intentionally allowed to bypass a wastewater pretreatment system and be discharged directly to the sanitary sewer system?

9. What are the sources of mercaptans in refinery wastewaters?

10. What should happen to off-specification wastewater at a petroleum refinery?

11. Why do wastewaters discharged from a papermaking process have a tendency toward low pH?

12. Why are organic polymers metered into the wastestreams from the manufacture of calcium sulfonate?

13. What items should be inspected in a calcium sulfonate manufacturing area?

14. In the event of an emergency in which ammonia gets dumped into a sewer, what should the inspector do or make sure gets done?

15. What items should be checked regarding alarms when inspecting an industrial facility?

16. Why should an inspector ask the operator of a rendering recovery process how it works?

17. What should an inspector look for when inspecting research and testing laboratories?

18. Why have centralized waste treatment facilities emerged as a new type of industry?

SUGGESTED ANSWERS

Chapter 10.   INDUSTRIAL INSPECTION PROCEDURES

Answers to questions on page 702.

10.0A   Major topics covered in Chapter 4, "Inspection of a Typical Industry," include the importance of knowing and following safe procedures, scheduling of inspections, how to enter an industry to conduct an inspection, the different levels of inspections, what to do after walking through an industrial inspection, and the report writing after an inspection is completed.

10.1A   The General Pretreatment Regulations require that POTWs must develop procedures to randomly sample and analyze the effluent from industrial users and conduct surveillance and inspection activities in order to identify, independent of information supplied by industrial users, occasional and continuing noncompliance with Pretreatment Standards.

10.1B   When inspecting an industrial user's self-monitoring procedures and equipment, the inspector should verify that:

1. The user's sampling locations sample the effluent from all process and nonprocess wastewater systems,
2. The sampling location specified in the permit is adequate for the collection of a representative sample of the wastewater,
3. The user's sampling technique is adequate to ensure the collection of a representative sample,
4. The user's permit sampling and monitoring requirements will yield representative samples, and
5. The parameters specified in the user's permit are adequate to cover all pollutants of concern that may be discharged by the permittee.

10.1C   The most common and dangerous hazards encountered by POTW personnel while performing industrial pretreatment inspections and sampling activities include atmospheric, physical and traffic hazards.

Answers to questions on page 706.

10.2A   For purposes of an inspection, a list of pollutants of interest at an industrial plant should be divided into two categories: (1) pollutants that come in direct contact with the water that is discharged to the POTW, and (2) pollutants that do not come in direct contact but have a potential to enter the wastewater due to spills, machinery malfunctions, or other similar problems.

10.2B   During an inspection, the information that should be gathered about spills includes a description of spill control practices the industry uses and information about past spills.

10.2C   Items that should be noted during a peripheral examination of an industrial user include size of the

industry, additional buildings, outside chemical storage, and location of the sanitary sewer.

10.2D   The first goal of a monitoring program is to determine the impact of industrial wastes on the POTW's collection and treatment systems including impacts on treatment plant operations, sludge management alternatives, and receiving stream quality. The second goal is to evaluate compliance by all industrial users with applicable pretreatment standards and requirements.

10.2E   Facilities or areas that should be examined during an on-site industrial inspection include:

1. Production or manufacturing facility,
2. Waste treatment facilities,
3. Additional manufacturing areas,
4. Storage and maintenance areas, and
5. Monitoring facilities.

Answers to questions on page 713.

10.3A   An industry's sampling points should be visited before an inspector enters an industrial facility to allow collection of an unbiased wastewater sample.

10.3B   When inspecting an industrial facility, spill containment facilities should be inspected in production areas, chemical storage areas, waste storage areas, wastewater treatment areas and sampling facilities.

10.3C   At the close of an inspection the inspector should either give the industrial representative a copy of the inspection report (properly signed and dated) or indicate that a follow-up letter will be sent within a specified time frame which will document the results of the inspection. Explain that either a copy of the report or follow-up letter will detail further actions that need to be taken by the industrial user. Be sure to properly discuss any violations you have found at the facility with the representative of the industrial user before leaving the site.

Answers to questions on page 720.

10.3D   Oil and grease from refineries may appear in wastewater as free oil, emulsified oil, and soluble oil.

10.3E   The major problems in oil refinery wastewater are the offensive mercaptan odors, the high concentrations of corrosive hydrogen sulfide, and the combustible gases.

10.3F   Monitoring devices that should be examined during the inspection of a petroleum refinery wastewater treatment system include flow metering, sulfide monitor (automatic on-line monitor not available at this time), pH, combustible gas meter and sampling devices.

10.3G   Pollutants most likely to be found in the wastewater from a pulp and paper mill include pH (acidic), hydrogen sulfide, volatile solids, suspended solids, COD and BOD, oil and grease, pentachlorophenol, trichlorophenol, PCBs, alkalinity, and color.

Answer to question on page 723.

10.3H   Pollutants found in the wastestreams from the manufacturing process for calcium sulfonate include sodium and calcium sulfonates, spent caustic, oils, and suspended solids. (The wastestreams usually also require cooling before discharge to the POTW sewer.)

Answers to questions on page 729.

10.3I   The two pretreatment processes commonly used by the meat packing industry are screening followed by dissolved air flotation (DAF).

10.3J   The ultrafiltration and reverse osmosis pretreatment processes are used to remove protein and lactose, respectively, from whey.

10.3K   Items that should be checked when inspecting a dairy products processing plant include:

1. Records of cleaning screens and pumping sumps,
2. Procedures to keep spilled cottage cheese out of the POTW sewer,
3. Spill protection in equipment maintenance areas,
4. Absence of floor drains where solvents are used for parts cleaning,
5. Solvent storage in a fire department approved storage room, and
6. Amount of water available in ammonia "dump" systems during a fire.

Answers to questions on page 737.

10.3L   Sources of wastewater from a rendering facility include washdown from raw material receiving, condensed cooker vapors, plant cleanup, and truck and barrel wash water. Plant spills are the most common sources of untreated raw material being discharged to the POTW sewer system, unless the spill is "dry cleaned."

10.3M   If an industry is using a DAF unit, the inspector should ask how long the unit has been in operation, which personnel operate it, and what training have they received.

10.3N   The inspector should check the condition of sewers downstream from a rendering plant because if they are old, undersized, or worn out, problems will occur. If grease enters the sewers and dries, it can create solid dams of grease inside the sewer.

Answers to questions on page 743.

10.3O   Sources of wastewater in a paint formulation industry include discharges from washing water-base or latex paint process equipment between batches. Wastestreams can include washdown and wastewater treatment sludges, raw materials packaging, emission control filters, spoiled batches and spill cleanup materials.

10.3P   An inspector can verify the destination of wastewater lines by dye testing.

Answers to questions on page 745.

10.4A   Types of metal-bearing wastes that may be treated at a centralized waste treatment facility include spent plating baths, acidic and basic solutions, and concentrated rinse waters.

10.4B   Nonhazardous wastes that may be treated at a centralized waste treatment facility include food industry wastes, restaurant grease trap wastes, animal/vegetable oil wastes, and car wash clarifier wastes.

10.4C   All incoming wastes from approved generators are screened prior to acceptance and off-loading to verify compliance with the original waste profile and treatability study. The screening involves the collection of a representative sample and an evaluation for treatability by bench-scale testing.

OBJECTIVE TEST

Chapter 10.   INDUSTRIAL INSPECTION PROCEDURES

Please write your name and mark the correct answers on the answer sheet, as directed at the end of Chapter 1. There may be more than one correct answer to each multiple-choice question.

True-False

1. An inspection can provide information about needed changes in existing procedures at the industrial facility.

    — 1. True
    2. False

2. Inspectors must review spill containment facilities constantly throughout the inspection.

    — 1. True
    2. False

3. Evidence of excessive raw material usage may indicate that leaks of process chemicals are occurring.

    — 1. True
    2. False

4. Petroleum refineries discharge small quantities of wastewater compared to other industries.

    1. True
    — 2. False

5. Most mercaptans produced at oil refineries are generated during the waste treatment processes.

    1. True
    — 2. False

6. At a petroleum refinery any off-specification wastewater and all spills must be diverted to a holding tank or holding basin for subsequent pretreatment.

    — 1. True
    2. False

7. Wastewaters containing cellulose could have a high BOD, but a low COD.

    1. True
    — 2. False

8. Larger molecules exert greater osmotic pressure than smaller molecules.

    1. True
    — 2. False

9. A poorly maintained facility implies cost cutting and an attitude that could possibly extend to pretreatment equipment, which is non-revenue generating.

    — 1. True
    2. False

10. At a centralized waste treatment facility, storage must be provided for impounded rainwater from the pretreatment areas and off-loading areas.

    — 1. True
    2. False

Best Answer (Select only the closest or best answer.)

11. What are the most important ongoing tasks of a local pretreatment program?

    1. Inspections and enforcement
    — 2. Inspections and monitoring
    3. Monitoring and enforcement
    4. Sampling and chain of custody

12. What should be done before entering an industrial facility for a pretreatment inspection?

    — 1. Collect wastewater samples
    2. Notify receptionist of your intent to inspect
    3. Phone industrial contact for an appointment
    4. Park in an appropriate place outside the facility

13. What should an inspector do with regard to violations discovered during an inspection?

    — 1. Discuss violations with industry before leaving site
    2. Help industry fix violations before leaving site
    3. Make a note of violations for future reference
    4. Notify enforcement authority of violations

14. Where should floor drains in the fuel loading dock area be plumbed to?

    1. POTW sewer
    — 2. Pretreatment facility
    3. Receiving water
    4. Storm drain

15. In a meat packing and processing facility, what should be done with the solids removed in the screening and DAF processes?

    1. Applied to land as a soil conditioner
    2. Processed for livestock food
    — 3. Recovered for rendering
    4. Treated in an anaerobic digester

16. In a dairy products processing plant, where should there be no floor drains?

    1. Where processing facilities are washed
    2. Where products are processed
    — 3. Where solvents are used for parts cleaning
    4. Where tanks are cleaned

17. What is the purpose of centralized waste treatment facilities?

    1. To treat acid sludge from petroleum refineries
    2. To treat fats, oils and greases from meat packing plants
    3. To treat hazardous liquid wastes
    4. To treat high-BOD wastes from food processors

18. What tasks can a POTW pretreatment inspector perform during an industrial inspection?

    1. Enter confined spaces without testing the atmosphere for hazardous conditions
    2. Follow their POTW's safety rules and regulations even if they are not as strict as the industry's safety rules and regulations
    3. Inform the industrial user of any new or updated pretreatment regulations
    4. Provide proprietary information that an industry's competition has developed to solve a pretreatment problem

19. Before entering an industrial facility for an inspection, why should you first visit the sampling points and collect wastewater samples?

    1. To allow collection of an unbiased wastewater sample
    2. To ensure following proper chain of custody procedures
    3. To minimize time required of the industrial contact
    4. To present the industrial contact with the results of the analysis of the sample

20. What types of wastes are treated by centralized waste treatment facilities?

    1. Agricultural wastes
    2. Oily wastes
    3. Spent activated carbon
    4. Waste activated sludge

Multiple Choice (Select all correct answers.)

21. What is information collected during inspections and monitoring used for?

    1. For all compliance and enforcement activities
    2. For allocation of funds to sampling and analysis
    3. For preparation and justification of budgets
    4. For receiving water surveillance activities
    5. For rotation of inspectors to areas of need

22. What are the most important tools an inspector can have for enforcing local standards and other legal authority provisions?

    1. Accurate flow measurements
    2. Cooperative legal staff
    3. Established laboratory QA/QC program
    4. Reasonable sewer-use ordinance
    5. Representative samples

23. What are the most common and most dangerous hazards that pretreatment inspectors will encounter?

    1. Atmospheric hazards
    2. Mental hazards
    3. Physical hazards
    4. Technical hazards
    5. Traffic hazards

24. What information should an inspector collect to describe an industrial discharge?

    1. Amount
    2. Chemical nature
    3. Destination
    4. Frequency
    5. Operators on duty

25. What are the major procedures inspectors follow for conducting on-site industrial inspections?

    1. Conducting on-site inspection
    2. Driving safely to inspection site
    3. Following up on compliance activities
    4. Preparing for inspection
    5. Requesting law enforcement officials to stand by if needed

26. Which of the following activities should be performed during the preparation for an inspection?

    1. Determine need for inspection of a given industry.
    2. Incorporate results of sampling activities into the inspection report.
    3. Notify the industrial user of violations.
    4. Prepare sample containers and sampling equipment if monitoring activities may be performed.
    5. Review existing files for available information about the industry.

27. Which of the following activities should be performed during an on-site inspection?

    1. Check the accidental spill prevention control plan of the industry.
    2. Conduct a peripheral examination of the industrial user.
    3. Observe the physical characteristics of the wastestream in the sewer.
    4. Require the development of a compliance schedule.
    5. Review existing files for available information about the industry.

28. Which of the following activities should be done during follow-up compliance activities?

    1. Check the accidental spill prevention control plan of the industry.
    2. Document the exact locations of all sampling points used by the industry.
    3. Increase monitoring frequency of industry.
    4. Modify the industrial discharge permit.
    5. Request enforcement proceedings be taken against industry.

29. Why is sampling and analysis of an industrial user's effluent conducted?

    1. To certify analytical laboratories
    2. To collect sewer-use service charges
    3. To prepare for an on-site inspection
    4. To verify compliance with wastewater discharge limitations
    5. To verify self-monitoring data

30. Inspection of spill containment systems includes looking for evidence of what factors?

    1. Emergency response plans
    2. Leaks around joints
    3. Proper maintenance
    4. Solutions being contained
    5. Trenches actually full of solutions

31. Oil and grease may appear in wastewater in what forms of oil?
    1. Cracked oil
    - 2. Emulsified oil
    - 3. Free oil
    4. Sequestered oil
    - 5. Soluble oil

32. What water quality indicators can be significantly changed if samples are not properly chilled and sealed?
    - 1. BOD
    - 2. COD
    - 3. Dissolved sulfide
    4. Hardness
    - 5. pH

33. What water quality indicators should be measured in the field?
    1. BOD
    2. COD
    - 3. pH
    - 4. Sulfide
    - 5. Temperature

34. Which items should an inspector review with the operator of a pretreatment facility?
    - 1. Current O & M manual
    - 2. Documentation of operating methods being followed
    - 3. Operating parameters
    - 4. Strip chart recorder information
    - 5. Type and frequency of data collected

35. What information should be collected and documented during an on-site pretreatment inspection?
    - 1. A description of each discharge
    - 2. A schematic of water flow through the industry
    - 3. Identification of appropriate sampling locations
    - 4. Number of employees per shift
    - 5. Proximity of chemical storage to floor drains

End of Objective Test

APPENDICES

CHAPTER 10. INDUSTRIAL INSPECTION PROCEDURES

APPENDIX A

PERTINENT QUESTIONS FOR INSPECTIONS
OF SELECTED INDUSTRIAL GROUPS*

* Adapted from the Pretreatment Program of the Washington Suburban Sanitary Commission, Laurel, Maryland.

APPENDIX A

PERTINENT QUESTIONS FOR SELECTED INDUSTRIES

INDUSTRY	STANDARD INDUSTRIAL CLASSIFICATION*	QUESTIONS
Auto Body Repair and Paint Shops	7532, 7536	- What is the chemical usage, destination, and storage of paint, thinner, other solvents in proximity to floor drains? - Do paint booths have a water curtain? - What are the floor washdown procedures (frequency, water usage, detergents)? - What are the paint spray gun cleaning procedures (destination)?

Auto Parts and Supplies: Wholesale and Retail	5013, 5531	- Look for floor drains first.
		  • If floor drains present, quantify storage of oils, transmission and brake fluids, and any other fluids.
		  • Batteries, battery acid.
		- Any machining or repair (see auto repair questions).

Auto Repair (Mechanical): Engine and Transmission Work	7538, 7539	- How are parts degreasers and solvents used? - What are the servicing procedures? - How are fluids such as oil, transmission fluid, brake fluid, and anti-freeze stored? - Look for drains under service bays. Where are the drains' destinations? - Are there gas/oil interceptors? - How is waste oil stored and what is its eventual destination?
Auto Wash	7542	- Is there any system for water reclamation such as a settling tank? - If a settling tank exists, how is sludge from it handled and by whom? - What types of cleaners are used? Do any specialty cleaners such as tire cleaners contain solvents? Do waxes contain solvents? - Is there an oil and grease separator on discharge lines? - Are liquids stored near floor drains? - What is water consumption?
Blueprinting and Photo-copying	7334	- Is offset printing done? - What type of blueprinting machines? With some the ammonia is totally consumed, while others will have a spent ammonia solution to dispose of. - Is there any significant amount of ammonia storage? Are there floor drains nearby?
Canned and Preserved Fruits and Vegetables	2099, 2032 2033, 2034 2035, 2037	- What detergents and techniques are used in washing the fruits and vegetables before rinsing? - Besides water used for washing, rinsing and cooling, is water also used for conveyance? - Is peeling done chemically (that is, caustic soda, surfactants to soften the cortex)? Is there any discharge? - Does the facility have a grease and solids recovery system? Any other pretreatment before discharge? - What is water consumption? How much is incorporated into product? - What percent of water use is recycled? Does this include any uncontaminated water (for refrigeration, machinery, etc.)? - What types of processing brines are used, if any? Are these sewered directly or pretreated first? - Are larger remains of processed waste fruit and vegetables ground up and sewered or used as by-products? - Are containers made? Are they washed or sterilized? - Are domestic wastewater and process wastewater segregated? - Does a representative sampling point exist?
Dairy Industry	All 202- numbers	- What products are manufactured at this plant? - Are bottles washed? Are containers washed or sterilized? Any chemicals used in the washdown? - What chemical cleaners are used for equipment washdown? (Acids such as muriatic, sulfuric, phosphoric, and acetic, surfactants, caustic soda, soda ash, and phosphates are commonly used as cleaners.)

*Please refer to pages 758 and 759 for NAICS code.

INDUSTRY	STANDARD INDUSTRIAL CLASSIFICATION	QUESTIONS
Dairy Industry (continued)	All 202- numbers	- Are acids properly neutralized before they are discharged? - Are any by-products wasted? (Buttermilk, whey, skim milk are very high in solids and BOD.) - How are spoiled materials disposed of? - What sources of uncontaminated cooling water are in the plant? Recirculated or once-through? If once-through, what is the destination? (Cooling water is used in some pasteurization processes, for condensation, refrigeration systems to cool the ammonia compressor jacket, and space air conditioning.) - What is the water consumption per day on the average? How much is incorporated into the product? - Are there any pretreatment units such as settling or grease traps or filtering devices or flocculating tanks? - What are the frequency and volume of boiler blowdown, if any? Additives? - Is there segregation of process and domestic wastewater? - Does an accessible sampling point exist?
Eating Establish- ments (Restaurants)	5812	- Is there a grease interceptor (describe, size)? How is it serviced (frequency)? How is grease disposed of? - What is done with spent cooking grease? How much is generated? - What janitorial cleaners are there (types, usage, storage)? - Do they have an automatic dishwasher (hours/day usage, water consumption, discharge water temperature)? Is it connected to grease interceptors? - How many sinks are there? How are they used? (Pots and pans or hand washing.) - What is the destination of the grill cleaning residuals?
Electric Services	4911, 4931	Steam Electric Power Generation: - Are plants coal, oil, or gas fired? - What type of boiler pretreatment is used (ion exchange, additives, etc.)? What are wastes? - How frequent is major boiler blowdown? Volumes? - Are there air pollution control devices which use water? - Are there ash handling systems which use water? - What is the source of condenser cooling water? - What is done with waste oils? Are they filtered and reused? PEPCO Substations (Potomac Electric Power Company): - Look for oil and solvent storage, proximity to floor drains. - Any contact cooling water discharge? - Possibility of leaking transformer oil? What would its destination be? What is percentage of PCB constituent of oil?
Electro- plating and Metal Finishing	All 347- numbers	- How often are cleaning solutions (both acidic and alkaline) changed and how are they disposed of? Volumes? Discharged as slugs? - Are degreasing agents used? If so, what type and how is the sludge handled? Is solvent redistilled, if used? Any cooling water used? - What types of chemicals make up plating baths? Is there cyanide? Chromium? Ammonium persulfate (etching)? - Are plating (concentrated) bath solutions ever disposed of? If so, how? - Is there any water reuse within the plant? Pretreatment? Any water-cooled machinery? Discharge to and volumes? - If masking is used, are photographic processes involved? (Circuit boards.) - Are there metal finishing operations (coloring, brightening, etc.) associated with the plating operations? - If metal coloring is present, are organic dyes used? - Is the plant manager aware of how much plating is done in terms of surface area (square feet, etc.)?

APPENDIX A

INDUSTRY	STANDARD INDUSTRIAL CLASSIFICATION	QUESTIONS
Electro-plating and Metal Finishing (continued)	All 347- numbers	- How is process wastewater from the plating room channeled to the treatment plant or sewers? Are there floor drains in the plating area? Are floors washed down regularly? - Are running rinses used? Countercurrent? Any still or dead rinses used? Discharged to where? - Any interconnections to public water supply and processing (cross connections)? Any backflow preventers? - How is sludge disposed of? If scavenger, who? - Are waste haulers' manifests being kept in accordance with state and federal laws? - How are liquids stored? Any nearby floor drains leading to sewer? - What is water consumption? Source? - Are domestic wastewater and process wastewater segregated (and cooling water, if applicable)? Do representative sampling points exist?
Fiberglass Insulation	3296	- What methods are used to bind and cool the glass after it has been drawn into fibers? What wastes are generated from this phase? Are these wastes sewered or pretreated prior to discharge? - What method is used for collecting the glass fibers (that is, wire mesh conveyors, flight conveyors, etc.)? What methods are used to clean the conveyors of any glass fibers? Is this process shut down or in service while cleaned? What type of cleaning agent is used? Is the wastewater sewered? - Are wet air scrubbers used? Is wastewater sewered or pretreated first (that is, sedimentation of particulate matter)? - How are backings applied (if applicable)? Heat? Adhesives? - Any segregation of domestic and process wastewater?
Fuel Oil Dealers	5983	- Record storage capacity (above or underground?). - Is above-ground storage diked? Is there any leakage access to storm or sanitary sewer? - Are any oils or fuels stored inside building (proximity to floor drains)? - What type of absorbent is used for spills? How much is stored (proximity to floor drains)?
Funeral Services	7261	- Embalming room chemical usage (how much formalin)? What percentage of usage is discharged to sewer? How much blood discharged/day? What is its destination? Any other chemicals? - Chemical storage (floor drains)? - Embalming table (washing and cleaning procedures) detergents and disinfectants used?
Gasoline Service Stations	5541	- Waste oil storage — is there a waste oil receptacle? • Drums or tank (proximity to floor drains)? - How are parts degreasers and solvents used? How are they serviced? - Fluids storage (transmission fluid, brake fluid, antifreeze, etc.), proximity to drains? - Look for drain trough under service bays. Is drain connected to sanitary drain? - Is there a gas/oil interceptor? (Describe.) - How is water removed from gasoline tanks?
Hospitals	8062, 8063, 8069	- What is the general layout of the facility (types of labs, x-ray equipment, morgue, laundry, food services, etc.)? - How are the chemicals used and stored (quantities and destination, proximity to floor drains)? - Are there any special procedures for handling hazardous or infectious wastes? Names of any haulers picking up such wastes? Are waste haulers' manifests being kept in accordance with state and federal laws? - What are the cleaning procedures (types and quantities of cleaners and germicides used)? - See "Radioactive Materials" for further questions.

INDUSTRY	STANDARD INDUSTRIAL CLASSIFICATION	QUESTIONS
Laundries	7211, 7213, 7214, 7215, 7216, 7219	- Is dry cleaning done? If so, what is the solvent? - Is sludge generated? Disposal? - If solvent is used, is it redistilled on site? Does this generate uncontaminated cooling water? Where is it discharged? - Do washers have lint traps, settling pits? - How many pounds of laundry are washed per day? - What is temperature of effluent? Is heat exchange system used? - Are printers' rags, shop rags, or other industrial materials cleaned? - What types of detergents and additives are used? What is pH of effluent? - Are laundry trucks maintained and washed on site? If so, how are waste oils, etc., handled? Are floor drains leading to sewer in vicinity of vehicles? - Any boiler blowdown (volumes, frequency of discharge)? Are there any additives such as chromate? Where does the discharge go? - Any loss due to evaporation? Volume estimate? - What is water consumption, source of water? - Does appropriate sampling point exist? - Is there separation of process wastewater and domestic wastewater? - Is the water reclaimed? If so, what is the volume of water and how is it used?
Leather Tanning and Finishing	3111	- What method was used to preserve the received hides? (NOTE: Hides preserved with salt will result in a high dissolved solids level in the effluent.) - What types of skins and/or hides are tanned? (NOTE: If sheepskins or goatskins are tanned, there will be a separate solvent or detergent degreasing operation.) - Is hair saved or pulped (that is, chemically dissolved)? (NOTE: In a save hair operation with good recovery of hair, the contribution to the effluent strength is substantially lower than in the pulp hair operation.) - Is deliming accomplished by treating with mild acids or by bating? What is the destination of these wastes? - What types of tannin are used? (NOTE: Chrome and vegetable tannins are the most common. A combination of tannins may also be used.) - How is chromium discharged into the sewer controlled? - Are chemicals stored near floor drains? (This is a very appropriate question to ask since many liquid chemicals are used in the leather tanning industry.) - How is sulfide controlled? - Are tannins recycled and/or chemically recovered? - Any pretreatment units? - If sludge is generated, how is it disposed of? - Any water-cooled machinery used? Discharged to? - Any boiler blowdown to sewer; frequency and volume? Any additives such as chromate? - Any segregation of domestic and process wastewater? - Does a representative sampling point exist?

Lumber and Building Materials: Retail	5211	- How are paints, thinner and other solvents, adhesives (glue), roofing materials (tar) stored?
		  • Proximity to floor drains?
		- Does paint mixing involve water (sinks)?
		  • Possibility of spillage, cleanup?
		- Cutting machinery; possibility of spillage? Destination of spills and cooling water.

Machine and Sheet Metal Shops	3599, 3444	- What type of product is manufactured? - Are cutting oils used and are they water soluble? - Are hydraulic oils used? - Would any of these oils ever be discharged to the sewer? - Are any degreasing solvents or cleaners used? What are the chemical makeup and/or brand names of the degreasers and how are they

INDUSTRY	STANDARD INDUSTRIAL CLASSIFICATION	QUESTIONS
Machine and Sheet Metal Shops (continued)	3599, 3444	used? How are the spent degreasing chemicals or sludges disposed of? Is degreasing rinse water discharged to the sewer? - Is there any water-cooled equipment such as a vapor degreaser or air compressor? Where is it discharged, frequency, volume? - Is any painting done on the premises? How are waste thinners or paints disposed of? Is a water curtain used for control of solvents entering the air and is contaminated water discharged? - Is any type of metal finishing done, such as anodizing, chromating, or application of a black oxide coating or an organic dye? What are the chemicals used, volumes consumed, and destinations of the finishing chemicals? - Are there any floor drains where any of the chemicals and oils are stored and used? - What is water consumption? - Are there any pretreatment units, traps, etc.? - Any segregation of domestic and process wastewater? - Any representative sampling point?
Metal Heat Treating Shops	3398	- Are cyanide salts used in heat treating? - What kinds of metal are heat treated? - What fluids are used for quenching metals? Are these ever changed and discharged to the sewer? - Are sludges ever removed from the quenching tank? How are the sludges disposed of? - Is any of the metal cleaned before or after heat treating? Are any degreasing solvents or cleaners used and how are they used? - Are there any water-cooled quenching baths, vapor degreasers or other equipment? Discharge to? Volume? - Are there any floor drains in the work or chemical storage areas? - Any boiler blowdown, frequency and volume to sewer? Any additives? - What is water consumption? - Are domestic and process wastewater segregated? - Any representative sampling points?
Nursing Care Facilities	8051, 8059	- Food service (see restaurant questions). - Is there any chemical usage (lab facility)? - What janitorial chemicals are there (usage, destination and storage of germicides and disinfectants)? - What is the frequency and amount of high-pressure boiler blowdown? What additives are used? To where discharged?
Organic Chemicals	2865, 2869	- Are processes batch or continuous? - If batch processes are used, how frequent is cleanup and what is done with wastes? - Are waste disposal services or scavengers used? If so, for what wastes? Are they licensed? Are required manifest records maintained? - What types of solvents are stored in bulk? - What are the sources and points of discharge for cooling waters? Are these contaminated or not contaminated? Is there an NPDES permit for discharge to surface waters? - What is the frequency and amount of high-pressure boiler blowdown? What additives are used (any chromate)? To where discharged? - Is there water in contact with catalysts used; in cleaning catalyst beds, for example? - List all products and raw materials. - Are there laboratories for research and for product testing? How are laboratory wastes disposed of? - Is water used in boiler feed or in processing pretreated? If so, how? What wastes are generated? - Are storage areas near drains leading to sewer? - Are there any chemical reaction or purification techniques, such as crystallization, filtration or centrifugation, which produce wastewater and/or sludge wastes? What is the destination of these wastestreams?

INDUSTRY	STANDARD INDUSTRIAL CLASSIFICATION	QUESTIONS
Organic Chemicals (continued)	2865, 2869	- Any pretreatment units? - Is deionized water used? How is it generated (on site)? Are columns regenerated on site? Use acids, caustics? To where discharged? - Is water tower used? Frequency, volume, discharged where? Any additives such as chromate? - Is there segregation of process and domestic wastewater?
Paint and Ink Formulation	2851, 2893, 2899, 3951, 3952, 3955	- Are oil-base or water-base inks manufactured? - What types of inks are made? - What types of paints are manufactured? Water- or solvent-base? - What are the pigments made of? Any heavy metals? - Are extenders used? - Are biocides added? Mercury? - Are solvents used? If so, what are they? - What are the resin types? - What other ingredients are used in formulating the product? - Is there any discharge to the sewer system (washdown and/or bad batches)? Are any chemicals used to clean production equipment? - Are there any floor drains in chemical and mixing areas? - Is there a scavenger service? If so, for what waste? - Is there on-site disposal of solids by burial? - Any water-cooled machinery used? Where is cooling water discharged to? - Any boiler blowdown to sewer, frequency, volume, additives? - Is process wastewater segregated from domestic? - Any representative sampling points?
Paper Mills	2611, 2621, 2631	- What are the products manufactured at the plant? - Which specific chemicals are used in the process? - Is pulp bleached? If so, what is the process and what chemicals are used? - Are any chemicals manufactured on site? (Chlorine dioxide, hypochlorite, etc.) Any discharged from these operations? - Any recovery systems (white water recycle, cooking liquor regeneration, cooling water reuse, etc.)? - Where is cooling water used in the plant (condensers, vacuum pumps, compressors)? Where is it discharged? - Describe the types of, size, filters, coatings, finishes, etc., in paper-making. - What happens to bad batches or liquids in case of equipment failure? (To the sewer or treatment plant?) - How is sulfide controlled? - How much water is consumed, on the average? Source of water? - Any boiler/water tower blowdown, frequency, volume, additives? - Any representative sampling points? - Is domestic wastewater segregated from process wastewater?
Paving and Roofing	2951, 2952, 3996	Tar and Asphalt: - Does wastewater from wet air scrubbers used on the oxidizing tower discharge directly to the sewer? Is it treated and recycled? - What method(s) are used to control the temperature of the oxidizing tower (that is, water)? Is this water discharged or recycled? - What treatment methods are used to remove suspended solids or oil from wastewater (that is, catch basins, grease traps, sedimentation, oil skimmers)? - Is water or air used to cool asphalt products? If water, is it contact or noncontact? If contact, is this water discharged directly to sewer? (NOTE: Mist spray used alone causes the largest amount of solids present in wastewater.) - Any water-cooled machinery used? Where is cooling water discharged?

INDUSTRY	STANDARD INDUSTRIAL CLASSIFICATION	QUESTIONS
Paving and Roofing (continued)	2951, 2952 3996	- Are solvents used/stored? Are floor drains nearby? - Any boiler blowdown, to sewer? Frequency, volume, additives? - Is process wastewater segregated from domestic wastewater? - Representative sampling point?
Pharmaceuticals Manufacturing	2833, 2834	- What types of processes are used to manufacture product(s)? (Fermentation, biological and natural extraction, chemical synthesis, mixing/compounding and formulation.) - If processes include fermentation and/or chemical synthesis, are these continuous or batch-type operations? - If chemical synthesis is involved, what processing steps (crystallization, distillation, filtration, centrifugation, vacuum filtration, solvent extraction, etc.) produce wastewaters? Are these wastewaters discharged to the sewer system? - What types of solvents are used, if any? How are spent solvents disposed of? How stored? Floor drains nearby? - Is raw water intake purified? If yes, by what method — ion exchange, reverse osmosis, water softening, etc.? What types and volumes of wastes are generated? Frequency of discharge? - What is done with the spent beer generated by fermentation? - Regarding equipment and floor washdown, are any chemical cleaners used? What is frequency? Volume of water used? Destination of wash water? - Is there any chance of spills or batch discharges? - Does any equipment such as condensers, compressors and vacuum pumps require the use of once-through uncontaminated cooling water? If so, where do these waters discharge? - Is there a research lab in the plant? What are the wastes generated in the facility? How controlled? - Is process wastewater segregated from domestic wastewater? - Any representative sampling point?
Photographic Processing	7384	It is important to determine what type of chemistry is used because some of the chemicals may be toxic while others may not. - What types of films are developed? Are prints made? Give an estimate of how much total processing is done per day. How many automatic processors are used and how long are they in operation per day? - What chemical brands are used: Kodak, 3M, GAP, etc.? What type of process chemistry is used: C-41, E-6, CP-30, etc.? What are the names of each chemical used in each process, what are the volumes used and which chemicals discharge to the sewer? Do any of the chemicals contain cyanide? - Is silver recovery practiced? Is bleach regeneration practiced and if so, is it done within the lab? What are the processes and wastes involved? - What is the wastewater flow from each of the photographic processing operations? Does the rinse water on the processors run continuously or does it shut off when film is not being processed? How often are the processors cleaned and the chemicals changed? - What chemicals, if any, are used to clean the processor rollers and trays? Are there any floor drains where the chemicals are mixed or stored? - Is there any pretreatment, pH control? - What is water consumption? - Is process wastewater segregated from domestic? - Is there any representative sampling point?
Plastic and Synthetic Materials Manufacturing	2821, 2823, 2824	- What is the product manufactured? - What are the raw materials used including accelerators and inhibitors? Are any known toxics (such as cyanide, cadmium or mercury) used in manufacturing the product? - What type of polymerization process is used? Does the process use a water or solvent suspension? What are the wastes generated from

INDUSTRY	STANDARD INDUSTRIAL CLASSIFICATION	QUESTIONS
Plastic and Synthetic Materials Manufacturing (continued)	2821, 2823, 2824	the process; what are the possible contaminants; how are the wastes disposed of? - Are there any product washing operations? Are reactor vessels washed down between batches? Is water or a solvent used? Would these wastes be discharged to the sewer? - Is cooling water, heating water or steam used and is it contact or non-contact? What is the destination of these streams? - Is there any boiler blowdown to the sewer; frequency, volume, additives? - Is there segregation of domestic and process wastewater? - Is there any representative sampling point?
Printing	All 27- numbers	Some of the following questions may apply while others may not; experience will be the best guide. - What kind of printing is done: offset, letterpress, silkscreen or other type of printing? - If offset printing is done, is film processing and plate developing done in the shop? - If film processing is done, is an automatic film processor used or are trays used? Does the processor's rinse water run continuously or does it shut off after processing is completed? How often are the processor's chemical tanks cleaned out and what volume is discharged to the sewer? How much developer, fixer and stop bath (if applicable) are used and are these discharged to the sewer? Is silver reclamation practiced? Is cyanide used at all for further reducing negatives? Are phototypositors used and, if so, what chemicals are discharged? - If plate developing is done, what type of plates are used? If they are aluminum plates, are they developed with a subtractive color key additive developer? What are the names of the developers and what quantities are used? Is the developer washed off the plates to the sewer or wiped off with a rag? How many plates are developed? - If paper plates are used, what type of processor is used and what are the names, volumes and destination of the chemicals used? If a silver process is used, is silver reclamation practiced? - In the press room, what type of fountain solution is used and would this ever be discharged during normal use or cleanup operations? What type of solvent is used to clean the presses and how is this applied? Would this solvent ever be discharged or does it become associated with rags? Are these rags washed on the premises or are they picked up by a commercial laundry? What is the name of the laundry? Are there any floor drains where the solvent or ink is stored? Are any of the presses water-cooled? Are there any waste oils from the presses? - If letterpress printing is done, is old lead type smelted in the shop and, if so, are the molds water-cooled? What type of solvent is used to clean the presses and type? How is solvent applied; is it ever discharged to the sewer? - If silkscreen printing is done, what kind of photo-sensitive coating and what volume is used? What kind of developer is used and is it discharged? Is a solvent or other cleaner used to clean the screen after printing? Is this discharged to the sewer? Are the screens used over again for making new stencils or are they thrown away? - If a different type of printing is done, what kind is it and what are the names and volumes of the chemicals used? Are these discharged to the sewer or collected and disposed of? Who would pick up the collected chemicals? - Is any water-cooled machinery used? Discharged to? - Is process wastewater and domestic wastewater segregated? - Are there any representative sampling points?
Retail Bakeries	5461	- What are the washdown and cleanup procedures (sequence of steps)? - What is the number of washdowns/day? - What types of detergents are used?

APPENDIX A

INDUSTRY	STANDARD INDUSTRIAL CLASSIFICATION	QUESTIONS
Retail Bakeries (continued)	5461	- What is the number of floor drains? - How are the cleaning agents stored? - How are the baking ingredients stored? - What is the amount of deep fry grease? How is it disposed of? - Are there oil and grease interceptors? (Describe)
Rubber Processing	2822, 2891, 3011, 3021, 3052, 3061, 3069	Synthetic or Natural: - What products are manufactured? - Is the rubber natural or synthetic? If synthetic rubber is used, is it polymerized on site and would it be water or solvent suspension? Is there a discharge associated with the process? - What are the ingredients of the rubber, including all additives? What kind of anti-tack agents are used? Are any known toxics used in the plant? - Are there any waste oils from rubber mixers or other processes which require disposal and, if so, how are they disposed of? - What type of forming process is used? Is cooling water contact or noncontact? Is it recirculated or discharged? If contact cooling water is used, is it treated in any way before discharge? What contaminants would be in the water? - Is any wastewater associated with the curing process (for example, steam condensate) and what would the contaminants be? - Is rubber reclaimed and if so, what type of process is used? Are any chemical agents used and how are these disposed of when spent? - Are any final coatings applied to the rubber, paint, plastics, etc.? Are any wastes or wastewater associated with the process and how would they be disposed of? - Does the plant have air pollution control equipment? - Are any liquids stored? Near floor drains leading to sewer? - Are there any water-cooled machines? Discharge to, volume? - Water consumption? - Is there any boiler blowdown to sewer (frequency, volume, additives)? - Are process and domestic wastewater segregated? - Is there any representative sampling point?

Schools	8211	- Elementary Schools.
		  • Cafeteria (see restaurants).
		  • What janitorial chemicals are used (usage, destination, storage)?
		- Junior High Schools.
		  • Cafeteria (see restaurants).
		  • What janitorial chemicals are used (usage, destination, storage)?
		  • What are the chemical disposal practices (labs)?
		  • What agents are disposed of to the sewer (amounts), paint, thinner (Art Department)?
		  • What solvents, paints and stains are used? What is their destination? How stored (Wood/Metal Shop)? Any access to floor drains?
		- High Schools — same as above.
		  • What chemicals do the Auto Mechanics and Cosmetology Departments use and store?

Scrap and Waste Materials	5093	- How are the materials processed: welding or smelting of metals? • Processing machinery cooling water (contact or noncontact), frequency of discharge? - How is the oil stored? Describe, including capacity. - Is there any other liquid storage or reclamation?

INDUSTRY	STANDARD INDUSTRIAL CLASSIFICATION	QUESTIONS
Seafood Processing Industry	2091, 2092	**GENERAL:** - What types of seafood are handled? Fish? Shellfish? Lobsters? - Is processing performed or just distribution? - Is there an ice machine, cooler and/or freezer? Are they water-cooled? If so, what is the destination of the cooling water? - Are there traps in floor drains? What is frequency of cleaning? What is destination of trapped solids? - Is washdown performed? Is soap or disinfectant used with wash water? What is the approximate volume and destination of wash water? **SPECIFIC:** **Bottom Fish** - Is descaling performed? If so, what method (manual or mechanical)? If mechanical, what is volume and destination of flush water? - Is filleting performed? Is brine solution used in conjunction with filleting process? If so, what is volume and destination of brine? - Is glazing performed? If so, is glaze tank dumped? Are there additives in glaze tank? If dumped, what is volume, frequency, and destination? - Is brine solution used? If so, is tank dumped? If so, what is volume, frequency, and destination? **Herring** - How are fish transported into plant? If pumped with water, what happens to water? - Is fluming used? If so, what happens to flume water? **Lobsters** - Is cooking performed? If so, is cook water dumped? If so, what is volume, temperature, frequency, and destination? **Shellfish** - Is prewashing performed? If so, what is volume and destination of wash water? - Is cooking performed? If so, what is volume, temperature, and destination of cooking water? - Is fluming employed? If so, what is volume and destination of flume water? - Is clam juice evaporated for broth? If so, what is volume, temperature, and destination of condenser water? - Is brine flotation used (for oysters only)? If so, is brine tank dumped? If so, what is volume, frequency, and destination? **Menhaden** (a fish) - How are fish transported into plant? If pumped with water, what is volume and destination of water? - Is cooking performed? Is there discharge from cooking operations? If so, what is volume, temperature, destination? - Is pressing performed? If so, what happens to wastewater? Is an evaporator used? If so, what is volume, temperature, destination? **Whiting** - How are fish transported into plant? If flumed, what happens to flume water? If discharged, what is volume and destination?

APPENDIX A

INDUSTRY	STANDARD INDUSTRIAL CLASSIFICATION	QUESTIONS
Seafood Processing Industry (continued)	2091, 2092	**Sardines** - Are fish preserved? If so, is brine solution ever dumped? If so, what is volume, frequency, and destination? - Are fish steamed? If so, what happens to wastewater? - Is washing performed? If so, what is volume, temperature and destination of wash water? - Are process wastewater and domestic wastewater segregated?
Soap and Detergent Manufacturing	2841	**GENERAL:** - Are only soaps manufactured, detergents, or both? Classify the plant. - Is foaming a problem in POTW sewer or treatment plant? - How is cooling water used? Discharge to? - How are liquid materials stored? Floor drains nearby leading to sewer? - Are air scrubbers used? Do these use water? Caustics? - In product purification steps, how are filter backwashes handled? - Are process and domestic wastewater segregated? - Are there any representative sampling points? **SOAP:** - What is the basic process used for manufacturing soap: Batch kettles? Fatty acid neutralization? Other? - Is process batch or continuous? If batch, what is frequency and volume of reactor clean-out? - Is waste soap from processing sewered? - Are perfumes and additives used? If so, what are they? **DETERGENT:** - What additives are used in the product? - How are spray drying towers cleaned?
Sugar Processing	2061, 2062	- Are both liquid and crystalline sugar produced? - What type of system exists in the plant for "sweet water" recovery? - Are ion exchange systems used? If so, what are the backwashing systems likely to produce as wastes? How frequent is backwash? - If charcoal filtering systems are used, does any wash water or transport water go to the sewer? - Are trucks or other heavy equipment maintained? Washed? Any floor drains leading to sewers? Any traps? - What bulk chemicals are stored and how? (Examples are acids used in liquid sugar production.) - What happens to filter sludges in the plant? What type of filter aids are used? - Is cooling water used? Discharge to? - What is the frequency of boiler blowdown and what are the additives used and volumes discharged? - From cleaning of equipment, what wastes are sewered and what wastes are recycled through the plant (examples are filters, evaporation pans, screens, etc.)? - Are process and domestic wastewater segregated? - Is there any representative sampling point?
Textile Mills	All 22- and 23-numbers	- What products are manufactured in the mill? What is the approximate production of the mill? - What types of fibers are used in the fabric? - Does the raw fiber require cleaning before spinning and weaving? - Are the fibers or fabrics scoured, mercerized, fulled, carbonized or bleached? What chemicals and rinsing operations are used and what is the destination of these wastes? - Is any kind of sizing applied and if so what kind is it?

INDUSTRY	STANDARD INDUSTRIAL CLASSIFICATION	QUESTIONS
Textile Mills (continued)	All 22- and 23-numbers	- Is desizing practiced and what are the chemicals used? Are these chemicals discharged to the sewer? - Is dye applied to the fabrics? What are the types and chemical constituents of the dyes and are the spent dye solutions and rinse waters discharged to the sewer? - Are any antistatic agents applied to synthetic fibers before spinning and weaving operations? Would these be removed from the fabric and subsequently enter the wastewater discharged to the sewer? - Are any further finishing operations practiced such as printing or application of various coatings? - What is the volume of wastewater generated by each chemical process? - Are any pretreatment processes used before discharge of wastewater to the sewer? - Is water-cooled machinery used? Discharged to? Volume? - Are any liquids stored near floor drains leading to sewer? - Is there any boiler blowdown to sewer; frequency, volume, additives? - Are process wastewater and domestic wastewater segregated? - Is there any representative sampling point?
Universities	8221	- Is a map of the campus available to inspectors? - Can a master list of chemicals used on campus be provided? Which chemicals are used most? - Is there an organized waste chemicals pick-up program? How many pick-ups/year? How many gallons picked up/year? Who is scavenger(s)? Licensed? Frequency of scavenger pick-ups? Central storage location(s) for waste chemicals that have been picked up? Are required manifest records maintained? - Are radioactive materials handled on campus? If yes, in what capacity? Are any wastes generated? If yes, how are these wastes disposed of? - Are there any photo developing facilities on campus? Any printing facilities? - Is there any prototype PC board work in the electronics labs on campus? - How are pathogenic organisms disposed of? - Are there any pretreatment facilities (marble chip acid traps, dilution pits, etc.)? - How much water is consumed/year? Has a study been done to account for all water uses (cooling water, laboratory wastewaters, cooling tower and boiler blowdowns, evaporation and drift from cooling towers, lawn irrigation, etc.)? Which, and how many of each, of the following units that usually discharge uncontaminated water to the sewer does the university have in operation at various times: stills, cold rooms, diffusion pumps, centrifuges, electron microscopes, x-ray diffraction units, electrophoresis units, air compressors, ice machines, fermentors? What is total campus population, including employees? How many reside on campus? - Are there floor drains near liquid chemical storage areas (such as Building and Grounds, chemical "supermarkets," waste chemical storage area(s), fuel oil tank(s)?
Veterinary Services	0741, 0742	- What is the chemical usage and storage; quantities, destination, floor drain proximity? • Alcohol, germicides, pesticides, cleaners, medicines? - Are there washing baths; detergents used and discharge procedures, any hair clogging problems? - What is done with excreta material (animal boarding)? - Are there any special procedures for infectious wastes?
Woodworking Shops	243X	- What is the chemical usage? Look for solvents, thinners, paints, stains, cutting oils, adhesives, etc. - How are brushes cleaned? Are any spray guns used; how are they cleaned?

A P P E N D I X A

INDUSTRY	STANDARD INDUSTRIAL CLASSIFICATION	QUESTIONS

Woodworking Shops (continued)	243X	- Chemical storage; proximity to floor drains?
		- How is cutting machinery lubricated and cooled?
		  • Are cutting oils discharged?
		  • Is there any cooling water (contact or non; recirculated or discharged)?

Radioactive Materials	No SIC	- What is the maximum quantity of each radionuclide used, stored, and discharged at the facility?
		- How are liquid and solid radioactive wastes being disposed of?
		  • Are they being hauled away? If so, what is the name of the hauler and what is the destination of the waste? Are manifests available?
		  • Are they being discharged to the sanitary sewer? If so, how often and what are the maximum concentrations in curies?
		- Is there a copy of a radioactive user license?
		- Is there a copy of any protocols for handling radioactive materials at the facility?
		- Is there a copy of any logs pertaining to radioactive discharges?

Steam Supply and Non-contact Cooling	No SIC	**STEAM SUPPLY:** - Is system high- or low-pressure steam? - What, if any, are boiler additives? Do they contain chromate? - How frequent and what is the quantity of boiler blowdown? - Is major cleaning and maintenance done? How often? - Are ion exchange systems used for boiler feed water? If so, what types of wastes are generated? **NONCONTACT COOLING WATER:** - Are cooling towers used? If so, what are the additives? - How frequently are towers blown down? Where does blowdown go? - Are closed systems ever bypassed?

APPENDIX B

CHECKLIST FOR INSPECTION OF SIGNIFICANT DISCHARGERS

(Pretreatment Committee of the Virginia Association of
Municipal Wastewater Dischargers)

I. GENERAL INFORMATION

Industry name: _____

Facility telephone number: _____

Plant manager: _____

Is the Wastewater Permit needed? _____

Permit #: _____

Permit expiration date: _____

Industry contact/title: _____

Industry contact's telephone number/fax number: _____

Site address: _____

_____

Correspondence address: _____

_____

Inspection/Type/Purpose: Unannounced _____   Scheduled _____   New Company _____

Complaint _____   Spill _____   Violation _____   Other _____

Purpose _____

_____

_____

Last POTW action (type/date):   Inspection: _____

Audit: _____

Sample: _____

Violation: _____

Enforcement: _____

Identify the action(s) implemented to address the violation: _____

_____

_____

Name of receiving POTW: _____

Nearest pump station: _____

Date of inspection: _____   Time of inspection: _____

A
P
P
E
N
D
I
X

B

Participants:

Name _____ Title _____ Phone No. _____

Name _____ Title _____ Phone No. _____

Name _____ Title _____ Phone No. _____

(1) Principal reviewer: _____

(2) Industry contact: _____

(3) POTW representative: _____

(4) Other: _____

Is this a Significant Industrial User (SIU)? If SIU, why? (Flow greater than 25,000 GPD, categorical industry, 10% POTW flow, reasonable cause to be classified as SIU.) _____

_____

_____

Is the SD subject to categorical pretreatment standards?   Yes ☐   No ☐   If yes, list standards and applicable subcategories:

_____

_____

Type of operation or products and applicable Standard Industrial Classification (SIC) code(s):

_____

Date that industry was established on site: _____

Operation Schedule: Weekday _____

                                      Saturday _____

                                      Sunday _____

Number of hours per day discharge occurs: _____

Annual shutdown planned? When and how long? _____

Percent of discharges that are batch? _____

Number of employees per shift: _____

Total daily flow of process waste: _____

Total cooling water use: _____

Total boiler water use: _____

Boiler blowdown: _____

Cooling tower blowdown: _____

Total daily flow of sanitary waste: _____

Storm water: _____

Wet scrubber (air pollution control equipment): _____

Total flow: _____

Are the sanitary and industrial wastewater streams combined?   Yes ☐   No ☐

Prior to wastewater treatment?   Yes ☐   No ☐

Prior to connecting to the POTW sanitary sewer?   Yes ☐   No ☐   N/A ☐

Has the production at the facility changed in any way since the last reporting period or inspection? If so, explain and provide the current production rate? Please note that if your facility is regulated by production-based categorical standards and your production rate has changed by twenty percent (20%) or more, then the permit limits have to be adjusted to reflect the changes.

II.   WATER AND WASTEWATER INFORMATON

Bill Metering Information:

Name	Type (POTW/Eff./ Well)	Size	Serial Number	Location

Present the most recent water/sewer bill: Month _____   CCF _____ .

Provide a summary of the water flow through the facility. Sketch or attach a schematic of all wastewater discharge lines which combine to flow to the POTW system. Superimpose this schematic on a site plan or floor plan of the facility, if possible.

Describe all the wastewater generated at the facility. Include the quantity, chemical nature, frequency, duration, and destination of each discharge.

BOILER DISCHARGES

    a. Frequency of boiler blowdown: _____

    b. Amount of water discharged during a blowdown event: _____

    c. How often are the boilers drained and cleaned? _____

       When cleaned where is the wastewater disposed? _____

    d. Possible contaminant in the boiler water discharges: _____

COOLING WATERS

    a. Identify source(s) of noncontact cooling waters and the amount of discharges: _____

    b. Is the noncontact cooling water pretreated before discharge?   Yes ☐   No ☐

       If yes, what type of pretreatment? _____

    c. Identify the sources and amount of contact cooling water: _____

    d. Is the contact cooling water pretreated before discharging? If so, what type of pretreatment? _____

    e. Possible contaminant in the contact cooling water? _____

FLOOR WASHINGS

    a. What is the frequency of floor washings? _____

b. Approximate amount of water? _____

c. How was the water use calculated? _____

d. Is the floor washing water pretreated before being discharged? If so, what is the type of pretreatment and how is the water being disposed? _____
_____

EQUIPMENT WASHING

a. List equipment, frequency of washing and volume of water used:

Equipment	Frequency	Volume Used/Wash Water

Identify the discharges and receiving stream of all wastes that are not connected to the sewerage system.

Provide a description and process flow diagram of each major product line and process used within the plant. Be sure to include any processes that may be subjected to Federal Categorical Pretreatment Standards.

Product	NAICS/ SIC Code	Avg. Rate of Production	AVERAGE FLOW AND CONVENTIONAL LOADING				
			Daily Avg. MGD	Max. Daily MGD	Flow Continuous or Batch?	Daily Avg. BOD	Daily Avg. TSS

III.   INDUSTRIAL PROCESSES AND PRETREATMENT

List raw materials used:

Describe the basic industrial process and any constituent unit operations. Include auxiliary or utility processes, such as boiler or cooling tower blowdown and heating or cooling streams which discharge to the POTW. Sketch or attach a block process flow diagram, noting which process steps generate wastewater. Please include information concerning any of the following wastes that are discharged to the POTW system. The list should include but not be limited to the following wastes: photo processing or x-ray wastes, laboratory wastes, oil/water separators, vehicle maintenance/washing areas, tanks used for chemical cleaning or etching of any metal parts or rinses from metal cleaning, and radioactive wastes, any solvents or wastes from solvent recovery systems, grease traps, or any other chemical or unusable wastes. Indicate which of these wastewater streams receive some form of pretreatment. Have shell-and-tube condensers been considered for the replacement of any contact barometric condensers?

A
P
P
E
N
D
I
X

B

Were modifications to the process or discharged wastewater incorporated since the last inspection?   Yes ☐   No ☐

If yes, explain.

_____

_____

Most recent water flow schematic was obtained on: Date: _____

Most recent process diagram was obtained on: Date: _____

List Pretreatment Permit conditions, including the location and description of the outfall, parameters monitored and limits frequency of monitoring:

List chemicals at the plant, categorized as follows: (1) chemicals that come into direct contact with the water that is discharged to the POTW, and (2) chemicals that do not come into direct contact, but have the potential to enter through spills, malfunctions, etc. Include the MSDS for these chemicals.

(1) _____   (2) _____

(1) _____   (2) _____

(1) _____   (2) _____

(1) _____   (2) _____

(1) _____   (2) _____

(1) _____   (2) _____

Have there been or will there be any changes in the industrial processes which may impact the discharge permit limitations?

Yes ☐   No ☐   If yes, explain. _____

_____

Does the facility have any air pollution control equipment which generates wastestreams?   Yes ☐   No ☐

If yes, describe the flow rate, composition, and the discharge method and location: _____

_____

Is the facility an RCRA hazardous waste generator (either through the basic process or residuals from treatment processes)?

Yes ☐   No ☐   If yes, provide EPA/Virginia Hazardous Waste I.D. number. _____

Has the POTW notified the industry of RCRA obligations?   Yes ☐   No ☐

If the industry is a hazardous waste generator, have the hazardous waste notification requirements been filled?

Yes ☐   No ☐

Describe the methods for handling, storing, and disposing of solid waste residuals. (Record name and business address of any contract haulers. Also identify the containment systems, labeling, barriers, type of waste.):

Provide a copy of the most recent manifest.

Are there any residual chemicals disposed to the process sewer?   Yes ☐   No ☐

If yes, complete the following information:

RESIDUAL DISPOSAL TO THE PROCESS SEWER		
Chemical or Chemical Category	CAS Number	Pounds/Year

Describe the pretreatment system used by the facility. If the system has multiple process steps, provide a block diagram indicating the treatment steps and their sequence. Attach copies of vendor specifications and drawings, and actual operating data, if these are available:

Pretreatment System: Type _____   Flow-through _____   Batch _____

Other _____

PRETREATMENT PROCESS

Is the pretreatment facility properly operated and maintained? (Pertinent characteristics to check might include the availability of standby power, alarm systems, operational manuals, calibration of control instrumentation, and disposal of sludges and routing of liquid return from sludge dewatering equipment.)

IV.   SAMPLING

Does the facility have a control manhole for sampling access?   Yes ☐   No ☐

If so, where is it located? (If possible, note on the wastewater discharge schematic for Section II of this checklist.)

_____

_____

Does such a control manhole provide access to a wastestream that is "end-of-pipe" for the industry before discharge to the POTW?
Yes ☐ No ☐

Is this wastestream a combined process wastestream? Yes ☐ No ☐

If yes, are the wastestreams combined prior to pretreatment? Yes ☐ No ☐ N/A ☐

If the industry has several wastestreams regulated by categorical standards, are other safe locations available that are appropriate for sampling at the end of these processes?

_____

_____

Are flowmeters and pH meters properly calibrated? Yes ☐ No ☐

Date of last calibration: _____

Person or firm doing the calibration: _____

If there is not a safe and practical alternative to sampling a combined wastestream, accurate flow rates for the regulated process streams and any dilution flows must be obtained from the industry and recorded here. Dilution flows include sanitary waste, noncontact cooling water, boiler blowdown, and other process wastestreams which are exempt from categorical pretreatment standards. (Note whether dilution flows tie into the process wastestream before or after any pretreatment.)

Automatic sampler description (flow or time proportional, pacing calculations): _____

_____

Provide the grab sampling procedure: _____

_____

_____

Provide on-line instrumentation description (pH, TOC, flowmeter, etc.): _____

_____

_____

Has the industry identified any specific hazards at the sampling location(s)? Yes ☐ No ☐

If so, what are they and have the POTW personnel been notified of such? _____

_____

_____

Does the industry perform chemical analyses required for self-monitoring "in-house"? Yes ☐ No ☐

If no, record the name and business address of any contracted private laboratory: _____

_____

_____

What QA/QC review procedures has the industry followed in determining laboratory services? _____

_____

_____

APPENDIX B

Are proper sampling procedures followed (clean containers, etc.)? Provide the information on laboratory procedures as follows:

LABORATORY PROCEDURE	
Analysis	**Reference Method**
BOD$_5$	
TSS	
pH	
Temperature	
Flowmeter Calibration	

Are sampling results performed by the facility readily available?   Yes ☐   No ☐

Are samples being properly preserved?   Yes ☐   No ☐

Is the industry doing its own pH?   Yes ☐   No ☐

Is the pH analysis performed within 15 minutes of collection?   Yes ☐   No ☐

Are chain of custody procedures being documented?   Yes ☐   No ☐

V.   SPILL PREVENTION

Describe spill or slug control methods used by the industry. Does it have a Slug Prevention Control and Countermeasures (SPCC) plan? Is there a past history of spills that were not contained? If yes, describe the event in detail including how and what was spilled and what actions followed:

If the facility has an SPCC plan the following information should be completed:

Document Title	Last Updated	Responsible	Document Satisfies Elements

Element Number	Element Description
1.	Describe discharge practices including nonroutine batch discharges.
2.	Identify procedures for promptly notifying POTW of slug discharges and complete follow-up written notification.
3.	Describe any necessary procedure to prevent accidental spills, including maintenance of storage areas, handling and transfer of materials, loading and unloading operations, and control of plant site runoff.
4.	Describe stored chemicals.
5.	Describe any necessary measures for building containment structures or equipment.
6.	Describe any necessary measures for controlling toxic organic pollutants.
7.	Describe any necessary procedures and equipment for emergency response.
8.	Describe any necessary follow-up practices to limit the damage suffered by the POTW treatment plant or environment.

Are diked chemical storage areas of sufficient size and in proper structural condition to provide for containment of their contents?
Yes ☐   No ☐

Are chemical storage areas located in close proximity to floor drains?   Yes ☐   No ☐

If so, do the floor drains discharge to the sanitary or storm sewer? _____

Are employees informed of the need to keep unauthorized chemicals out of the sanitary sewer?   Yes ☐   No ☐

If yes, by what means: _____

Are chemicals or wastewater pumps totally sealed, or are shafts sealed with packing or mechanical seals? If packing is used, where is leakage directed? _____

_____

_____

_____

VI.   OTHER INFORMATION

Does the industry have a solvent/toxic organic management plan (how solvents are used, stored and disposed of)?
Yes ☐   No ☐

How does industry report TTO?   Analysis ☐   Certification Statement ☐   N/A ☐

If the industry is subject to the electroplating, electronics or metal finishing standards, and has submitted a solvent/toxic organic management plan, has there been any change to the contents and conditions outlined by the plan?

_____

_____

_____

If the facility has any other permit with other agencies, provide the following information:

ENVIRONMENTAL CONTROL PERMITS		
Permitting Agency	Permit Type	Identifying Number

VII. RECORDS

If the industry is subject to categorical pretreatment standards, did it submit a Baseline Monitoring Report (BMR) with the required contents to the control authority?

Yes ☐ No ☐ N/A ☐ If not, briefly explain the reasons for not doing so, or list any deficiencies in the content of the BMR:

_____

_____

If categorical, has the industry submitted the 90-day compliance report, and does it submit the required semiannual self-monitoring reports?

Yes ☐ No ☐ N/A ☐ If no, briefly explain: _____

_____

If yes, do the reports address the sampling parameters required by the categorical pretreatment standards?

Yes ☐ No ☐ N/A ☐

Is the industry on a compliance schedule for the installation of any technology required to meet the applicable pretreatment standards? Yes ☐ No ☐ If so, note the progress of the industry in following this schedule: _____

_____

Is proper signatory authority being used? Yes ☐ No ☐

Authorized signatory/title: _____

Is certification statement being submitted? Yes ☐ No ☐

Are records available for at least three (3) years? Yes ☐ No ☐

<u>Does the industry submit all monitoring data performed in accordance with 40 CFR Part 136 to the POTW in its periodic compliance report?</u>

_____

_____

VIII. POLLUTION PREVENTION

Identify all the pollution prevention measures undertaken in the past twelve (12) months:

_____

_____

_____

IX. INSPECTION NOTES

CHECKLIST

☐ Production Area   ☐ Laboratory

☐ Chemical Storage Area   ☐ Machinery Repair Areas

☐ Floor Washing Procedures   ☐ Solid/Shipping Areas

☐ Boiler Room   ☐ Pretreatment Unit

☐ Cooling Tower   ☐ Records

☐ Air Pollution Control Devices   ☐ Sampling Points

List all the actions that are to be taken prior to the next inspection and their target dates for completion.

ACTION ITEMS:

_____

_____

Floor drain use: Any improperly placed floor drains? _____

_____

Located stains or other indications of chemical spills at or near floor drains, or sewer entry points? If yes, identify:

_____

_____

Permit violation noted/suspected:

_____

_____

Process area/production: Conditions/operations/general housekeeping:

_____

_____

End-of-pipe discharge to POTW:

_____

_____

On-site chemical storage condition: Any leaky containers? Identify:

_____

_____

Wastewater pretreatment: Conditions/operations:

_____

_____

Hazardous waste generation area: Drums/labels/manifests: OK?

Problems: _____

_____

Laboratory: Sample collected? Yes ☐ No ☐ If yes, was the procedure correct? _____

Was the chain of custody complete? _____

Miscellaneous: Photographs taken? Yes ☐ No ☐ If yes, were the photographs dated? _____

Were any deficiencies noted during inspection? Yes ☐ No ☐ Will the industry be notified in writing to correct the problem area? Yes ☐ No ☐

Permit change: Is a permit change warranted? Yes ☐ No ☐

Deficiency correction: _____

Date/time of reinspection: _____

Have the deficiencies been corrected? Yes ☐ No ☐

Inspector's signature _____

X. REQUIREMENTS/RECOMMENDATIONS/COMMENTS:

_____

_____

_____

_____

_____

_____

_____

_____

XI. CERTIFICATION:

I have personally examined and am familiar with the information submitted in this document and attachments. Based on my inquiry of those individuals responsible for obtaining the information reported herein, I believe that the submitted information is true, accurate and complete. I am aware that there are significant penalties for submitting false information including the possibility of fine and imprisonment.

Date: _____   Time: _____

Facility contact: _____

Title: _____

Signature: _____

City inspector: _____

Signature: _____

CHAPTER 11

EMERGENCY RESPONSE

by

Bill Garrett

TABLE OF CONTENTS

Chapter 11.   EMERGENCY RESPONSE

OBJECTIVES

Chapter 11.   EMERGENCY RESPONSE

Following completion of Chapter 11, you should be able to:

1. Plan a response procedure for an emergency,

2. Determine who should be involved in an emergency,

3. Respond to an emergency,

4. Identify the material,

5. Determine the affected area,

6. Limit the impact of the incident, and

7. Prepare enforcement proceedings.

PRETREATMENT WORDS

Chapter 11.   EMERGENCY RESPONSE

CARCINOGEN (CAR-sin-o-JEN)   CARCINOGEN

Any substance which tends to produce cancer in an organism.

CENTRALIZED WASTE TREATMENT (CWT) FACILITY   CENTRALIZED WASTE TREATMENT (CWT) FACILITY

A facility designed to properly handle treatment of specific hazardous wastes from industries with similar wastestreams. The waste-waters containing the hazardous substances are transported to the facility for proper storage, treatment and disposal. Different facilities treat different types of hazardous wastes.

MUTAGENIC (MUE-ta-JEN-ick)   MUTAGENIC

Any substance which tends to cause mutations or gene changes prior to conception.

TERATOGENIC (TEAR-a-toe-JEN-ick)   TERATOGENIC

Any substance which tends to cause birth defects after conception.

CHAPTER 11.   EMERGENCY RESPONSE

NOTICE: Procedures for responding to emergencies and the laws, rules and regulations regarding who would respond and how to respond to emergencies may vary among state and local agencies. The procedures and agencies listed in this chapter are provided to give you ideas on how you could respond. *YOU* must find out your local procedures and responsible agencies. Every POTW should have a written spill containment procedure and every inspector must be thoroughly familiar with these guidelines. If this information is not available at your facility, use this chapter as a guide and work with the appropriate agencies in your area to develop emergency response procedures for your needs.

11.0   PLANNING

11.00   What To Expect

In a serious emergency, usually created by accidents, you could encounter a life threatening situation that calls for you to coordinate your actions with a large network of other agencies, private companies and rescue teams. Inspectors could be called upon to assist emergency efforts in reacting to poisonous gases, large volumes of toxic materials or concentrated acids or bases in the sewer.

Catastrophic emergencies that pretreatment facility inspectors could encounter include railroad or trucking accidents involving toxic, flammable or explosive chemicals or strong acids or bases which are highly corrosive. Also, these types of accidents could involve unknown substances. Regardless of the type or cause of the accident, a hazardous liquid could reach a collection system or storm drain and create the potential for an emergency throughout the community, collection system, treatment plant and the environment which receives the plant's effluent and sludge. Regional disasters such as earthquakes, floods, power outages or acts of sabotage would

also qualify as occasions to implement an emergency response plan, but industrial waste inspectors might not be as closely involved.

The most likely industrial sources for catastrophic discharges for large POTWs are oil refineries, chemical plants, fuel transfer stations or any facility that handles large volumes of flammable or extremely toxic material. The most common example of a sewer condition that warrants enacting an emergency response plan is the discovery of gasoline or other flammable or explosive liquid or vapor in the sewer. The explosion in Louisville, Kentucky, in 1981 is a dramatic example of what can happen when explosive gases ignite in a public sewer (Figure 11.1). As a result of the explosion, a large portion of the sewer collapsed and streets fell into the void. Over $20 million was collected from the responsible discharger to rebuild the sewer system.

Locations of high-risk facilities should be noted on a sewer map with the route of flow to the treatment plant clearly marked. Inspections of these facilities must include a review of the industry's spill prevention and emergency response plans. An industry's spill protection facilities must be properly maintained and function as designed. This may prevent the need for an emergency response and the hazards associated with such incidents can be avoided. Maps should be updated periodically to reflect changes in the sewer system due to the installation of relief lines, flow control structures, stoplogs, cross connections or changes implemented by the discharger. The capabilities to divert flow throughout the collection system should also be noted on the map.

Infiltration of groundwater into sewers can contribute explosive or toxic liquids and gases, especially if the water table is above the top of the sewer. Locating the source of explosive or toxic substances and then planning and implementing corrective action can sometimes involve long and costly investigations (Figure 11.2).

Other essential information that should be noted regarding the collection system includes:

1. Can the system either minimize gas accumulation or enhance the opportunities for release of gases?

2. Is the system free-flowing or surcharged?

3. Are there stagnant or dead spots in the system downstream from an industry's point of discharge?

4. Where are the sections of flowing wastewater that are aerobic and anaerobic?

For small POTWs any major dischargers in the service area could be the source of a catastrophic discharge, especially an industry where large volumes of toxic materials are used. In some small communities electroplating and metal finishing industries are the only significant dischargers to the POTW. For example, in one small community a captive plating shop (electronics manufacturer) and a large job shop plater contribute about 25 percent of the flow to the POTW. In the case of a

Fig. 11.1   Results of an explosion in a sewer
(Copyright © 1981, The Courier-Journal. Reprinted with permission.)

Fig. 11.2   Sewer maintenance workers checking atmosphere and
groundwater for liquid or gaseous compounds at a test well near a sewer.
Groundwater infiltration is suspected of carrying objectionable compounds
through leaks in the joints of the sewer pipe.

major fire, these firms could create a situation where incompatible process solutions and stored chemicals intermix (Figure 11.3). The water used to put out the blaze could carry hazardous wastes to the POTW.

Another potential source for catastrophic discharges is a *CENTRALIZED WASTE TREATMENT (CWT) FACILITY*[1] that stores, transports, treats and disposes of hazardous wastes. This type of facility could specialize in metal finishing and electroplating wastes or the wastes from pesticide manufacturing firms.

If your agency is not prepared for a hazardous materials emergency when it occurs, the agency may face financial liability, public accountability, and the responsibility for injuries or even deaths.

11.01 Laws and Regulations

Below is a partial list of local, state and federal laws and regulations that could be pertinent to dealing with hazardous materials in Los Angeles County. These are listed to inform you of the types of local and state laws that could be applicable to your situation.

1. California Hazardous Waste Control Law, Health & Safety Code, Division 20, Chapter 6.5
Section 25500 - 25521 places requirements on companies dealing with hazardous materials to notify the local response agency (usually the fire department) that they use hazardous materials and to develop an emergency response plan. Also known as the "disclosure and right-to-know" law.

2. Regulations for the Management of Hazardous Materials, California Code of Regulations, Title 22, Division 4, Chapter 30, deals with hazardous waste management, and defines hazardous and extremely hazardous materials and wastes.

3. Regulations for the Transportation and Packaging of Hazardous Materials, Code of Federal Regulations, Title 49, Parts 100-199.

4. Resource Conservation and Recovery Act (RCRA), PL 94-580, Subtitle C. This is the principal federal statute dealing with hazardous waste generation, handling and disposal. Section 3002 (5) describes the "manifest" system used to keep track of wastes from "cradle to grave."

5. Hazardous Waste Regulation (US EPA) Code of Federal Regulations, Title 40

   a. 40 CFR Part 112, Oil Pollution Prevention Requirements to prevent the discharge of oil into navigable water. This regulation contains the requirements for the development of Spill Prevention Control and Countermeasure (SPCC) Plans.

   b. 40 CFR Part 125, Support K, Criteria and Standards for the NPDES requiring Best Management Practices to prevent spills by industrial users which are not otherwise regulated by categorical standards.

   c. 40 CFR Part 302, Designation, Reportable Quantities and Notification for *HAZARDOUS SUBSTANCES (NOT WASTES)* under the Comprehensive Environmental Response, Compensation, and Liability Act (CERCLA), which must be reported to the National Response Center upon release into the environment.

   d. 40 CFR Part 403, General Pretreatment Regulations for Existing and New Sources of Pollution, Sections 403.5 (a) and 403.5 (b)(1)(2)(3) and (4). All of these sections prohibit interference with the treatment plant processes by pollutants.

6. County Health Emergencies Health & Safety Code, Division 1, Part 2, Chapter 1

7. Proposition 65 This law put restrictions on the use and discharge of *CARCINOGENS*,[2] *MUTAGENS*[3] and *TERATOGENS*.[4] It also requires notification of authorities of illegal discharges of hazardous waste, under specified conditions.

8. CERCLA. Comprehensive Environmental Response, Compensation, and Liability Act of 1980. Established "superfund" to provide emergency cleanup money that is to be repaid by responsible parties.

9. TSCA. Toxic Substances Control Act of 1976. Regulates the manufacture and distribution of potentially toxic substances. Describes required testing and premanufacturing clearances.

10. OSHA. Occupational Safety and Health Act. Requires the establishment of standards for any toxic material "which must adequately assure, to the extent feasible, that no employee will suffer material impairment of health or functional capacity."

---

[1] *Centralized Waste Treatment (CWT) Facility.* A facility designed to properly handle and treat specific hazardous wastes from industries with similar wastestreams. The wastewaters containing the hazardous substances are transported to the facility for proper storage, treatment and disposal. Different facilities treat different types of hazardous wastes.
[2] *Carcinogen (CAR-sin-o-JEN).* Any substance which tends to produce cancer in an organism.
[3] *Mutagen (MUE-ta-JEN).* Any substance which tends to cause mutations or gene changes prior to conception.
[4] *Teratogen (TEAR-a-toe-JEN).* Any substance which tends to cause birth defects after conception.

Fig. 11.3   Fire at electroplating production line caused by meltdown of plastic tanks. Photo shows need for spill containment and emergency response plan wherever toxic chemicals are used or stored.

11.02 Establish a Notification List

The team that responds to an emergency may consist of members from many diverse groups. The pretreatment field inspector or whoever is first on the scene should be able to make a minimum number of calls to fully activate emergency procedures. Ideally the inspector should only have to make one call to a person in the home office who would then notify everyone else on the notification list. A list of who should be notified in case of an emergency should be posted in the home office and carried by every inspector. The concepts discussed in this section must be established as *WRITTEN* response procedures developed prior to an actual emergency to enable an efficient response. Properly developed and understood procedures will help avoid conditions of confusion, stress, and uncertainty which often accompany a response to an emergency spill, accident or other type of disaster.

In Los Angeles County the primary response core is the "Hazmat" (hazardous materials) team (Figure 11.4) for which the Fire Department, the Sheriff's Department and the County Health Department are the lead agencies. Many other agencies attend Hazmat training sessions and participate in responding to environmental disasters. In some areas, especially smaller communities, POTW personnel may be the only available water pollution control experts. If this is the situation in your area, pretreatment inspectors should become involved in, or organize, multiagency HAZMAT incident response drills. Your POTW's notification list should include the local primary response unit (probably the fire department) as well as selected additional agencies and private firms that could be of assistance.

It is also very important to notify the various departments within your own agency that would need to respond to an emergency as quickly as possible. The sewer maintenance department will probably be the most active responders at the site. Treatment plant operators must be notified so that they can prepare for any impact the emergency could have on the plant. Home telephone numbers for key personnel should be circulated to inspectors and all "front line" members of the emergency response team.

QUESTIONS

Write your answers in a notebook and then compare your answers with those on page 808.

11.0A What are the most likely sources of catastrophic discharges into sewers?

11.0B List the sources of laws and regulations that could be pertinent to dealing with hazardous materials in your area.

11.0C Where should the list of persons to be notified in case of an emergency be located?

11.1 WHO COULD BE INVOLVED

The following emergency notification list was developed for use in Los Angeles County. It is presented as an example of the kind of notification list that should be developed for every region. The members of this list who would be notified would depend on the particular emergency. Use this material as a guide for developing your own listing and to be sure all needed local, state and federal agencies are included in the spill response. Also be sure to include phone numbers on your list.

11.10 Local

1. Fire Department — always dial (911) for initial notification. The dispatcher will notify the local fire department and then the appropriate Hazmat unit as needed:

   a. Hazmat #76 — Newhall

   b. Hazmat #87 — Industry

   c. Hazmat #105 — Rancho Dominguez

2. Chief Administrative Officer, Disaster Services (frequently this person notifies everyone else on the local, state and federal lists)

3. Sheriff, Emergency Operations Bureau

4. County Health Department, Hazmat mobile van cellular phone

5. County Health Department, Radiological Department

6. Sanitation Districts

7. Department of Public Works (formerly County Engineers & Flood Control & Road Department)

   a. Inspection

   b. Flood Control

   c. Road Department

8. Agricultural Commissioner

9. Board of Supervisors

10. South Coast Air Quality Management District

11. Harbor Patrol

11.11 State

1. Health Department, Hazardous Waste Management

2. Highway Patrol, on-highway spill notification

3. Office of Emergency Services, off-highway spill notification

4. Hazard Evaluation System & Information Service (HESIS), toxic spills

5. Department of Transportation — Caltrans, Disaster Assistance

6. Regional Water Quality Control Board

7. Fish & Game

8. Oil & Gas Division

9. Air Resources Board

10. Cal OSHA

Fig. 11.4   LAPD hazardous materials specialist

11.12 Federal

1. EPA, RCRA Superfund Hotline (800/424-9346)

2. United States Coast Guard

3. National Response Center (800/424-8802)

4. Red Cross

5. TSCA Hotline (202/554-1404)

6. Nuclear Regulatory Commission

7. Department of Treasury, Alcohol, Tobacco and Firearms, explosives

11.2 INCIDENT COMMAND AND CONTROL

11.20 Initial Response

Activate the notification procedure immediately whenever any hazardous substance has been spilled, released, dumped or abandoned. If you are the first on the scene, try to evaluate the situation as quickly as possible without endangering yourself. At this stage of the emergency you must remain at a safe distance from the incident. Consider performing the following tasks:

1. Use personal protective equipment:

   a. Clothing, and

   b. Self-contained breathing apparatus (if trained and qualified).

2. Determine location of problem and affected area.

3. Estimate volume and characteristics of spilled material (liquid, gaseous, dry; explosive, toxic, asphyxiant, corrosive).

4. Estimate potential harm to:

   a. People,

   b. Property, and

   c. Environment.

5. Assess rescue needs.

6. Control public access to spill area; evacuate as needed.

7. Assess immediate remedial actions and options (diking, catch basins). Implement immediately if possible and if such actions will not threaten the safety of those involved.

8. Set up a command center with a line of communication.

9. Evaluate the situation as it impacts the operation of the treatment plant. Are collection systems, lift stations, wet wells, and the operations and maintenance people who work in the collection system and treatment plant at risk as a result of the spill?

If the primary response team is on the scene, they will have established a central command post and will be exercising their emergency response procedures. The on-scene command center must be a safe area known to all responders. Cellular telephone or "beeper" contact with your office is advisable. It is best to establish a single contact person to deal with the media so that misinformation is not released and response personnel will not be distracted.

Keep a log of events at the central command center and record your own activities. Your notes will be useful in reconstructing the events and preparing a case for court if enforcement actions are initiated.

11.21 Identify the Material

If the spill is from a known industrial facility, then a representative from that facility is the best source of information to identify the material. Survey the actual situation before initiating any sampling procedures. Many times a sample can be collected without entering a manhole by using a bottle suspended on a rope or attached to a pole. Always monitor the air quality at the opening while sampling, even if entry is not required. Always put on the appropriate safety clothing and gear for the situation. This should include a hard hat, steel-toed shoes, coveralls and safety glasses. Follow all confined space procedures. If you must enter a manhole, get two people to assist you. Have a safety line and a tripod at the surface capable of lifting you out of the manhole if necessary. Have available and calibrated an atmospheric tester that can detect oxygen deficiencies, lower explosive limit (LEL) and toxic gases (hydrogen sulfide and hydrogen cyanide). A blower for ventilation is essential. Establish an emergency line of communication in case you or your party are injured by the spill during sampling. Be familiar with the characteristics of any hazardous wastes you may encounter and know the appropriate sampling techniques.

If the material is unknown and flowing in a sewer, try to *SAFELY* draw a sample for field tests or laboratory analysis.

Test the atmosphere in the manhole for flammable/explosive gases before lifting the cover. The act of opening the manhole may produce a spark that could ignite these gases. (For additional information, see Appendix A, "How to Prevent Explosive Conditions.") Beware of poisonous hydrogen sulfide gas (rotten egg odor) that will be liberated when concentrated acid contacts domestic wastewater. Also beware of toxic hydrogen cyanide (HCN) gas (almond-like odor). The wastewater from plating shops may contain cyanide which is liberated under acidic conditions.

If a hazardous waste from a particular type of industry appears to be a problem, phone calls to industries that could be the source may be appropriate. Emphasize to each industry the need for cooperation and the importance of proper identification so effective corrective actions can be taken.

You can get information and assistance 24 hours a day concerning the identification of hazardous materials and recommended corrective action for handling the material by calling:

Industry responsible for spill (if known)
CHEMTREC (800/424-9300) or
Poison Information Center
 L.A. County Medical Association
 (or a similar association in your locality)

The location of underground utilities can be determined by phoning "Underground Service Alert" listed in the white pages of most phone books. Representatives for various utilities and fuel transporters can be asked to identify the locations of their underground facilities and to investigate possible pipeline leakages that may infiltrate wastewater collection systems.

11.22   Determine Affected Area

Use the methods discussed in Chapter 6 for tracing the material upstream through the sewer to determine the source (Figure 11.5). Explosive gases, sulfide, chlorine, ammonia, mercaptans and highly odorous compounds can be traced upstream using odor or gas detection meters. Inspectors must follow established safety procedures and precautions. Do not allow any individuals to "sniff" their way upstream toward the source because the safe limit of exposure could easily be exceeded (see Chapter 5).

Use sewer maps to help identify the travel route in the sewer. By identifying the compound and knowing the sewers tributary to the site, the number of sources to investigate should be greatly reduced. It is a good idea to mark sewer maps in advance showing tributary areas to each lift station. This procedure can save precious time in an emergency when you need to determine the route of flow quickly.

QUESTIONS

Write your answers in a notebook and then compare your answers with those on page 808.

11.1A   What major categories of persons or agencies should be included on an emergency notification list?

11.2A   How should an inspector who is first on the scene of an emergency respond?

11.2B   Why must a log of events be kept at the central command center during an emergency?

11.2C   Why should the affected area of a spill be determined?

11.3   LIMITING THE IMPACT OF THE INCIDENT

11.30   Removal of Material

It is sometimes possible for a sewer maintenance crew to divert a contaminated flow to a sewer that bypasses the treatment plant or other critical area. Stoplogs, sandbags or sewer plugs can be installed to contain the flow so that vacuum trucks can extract the material from the sewer. Care must be taken to use explosion-proof pumps when drawing liquid out of the sewer. Treatment plant wet wells or pump stations can be used as reservoirs to trap floating spills so that they can be removed. It may be advantageous to "back up" or "block off" a sewer so as to reduce headspace for volatile (toxic or explosive) gases to collect.

Establish contacts with private firms that offer emergency assistance such as explosives experts, spill cleanup crews,

Fig. 11.5   Two pretreatment facility inspectors tracing an industrial spill through a sewer. Safety gear includes emergency flashers on car, blinking light on car roof, traffic cones in street, red flag to warn and divert traffic, safety vests and hard hats. Inspector measures Lower Explosive Limit (LEL) before opening manhole. Sample is collected with bucket on rope and field tested using several kits to measure pH, cyanide, sulfide and heavy metals. Any odors detected are recorded.

waste haulers, vacuum truck operators and centralized waste treatment facilities *BEFORE* you need their services. Companies that offer large-scale emergency cleanup assistance in your area should be listed with their phone numbers and emergency contact persons.

11.31   Treatment in the Sewer

If the offending material is acid, then pH neutralization in the sewer using caustic soda, soda ash, lime or other base may be appropriate. A sample of the contaminated wastewater could be titrated with the neutralizing solution to determine the dosage and feasibility of this approach. Consultation with CHEMTREC or other available experts may determine that addition of some commonly available chemical might lessen the impact of the spill.

The introduction of foam is sometimes effective in suppressing the volatilization of some chemicals. The most common procedure used to draw offensive gases out of the sewer is to install air suction units at key manholes (Figure 11.6). This is the type of blower used by most sewer maintenance crews for routine clearing of the sewer atmosphere prior to entry for repairs or inspections of trunk sewers. The use of internal combustion engines on blowers in the presence of some volatile gases may be a source of a spark which could cause an explosion. The inspector must be aware of the hazardous nature of the spilled material and its compatibility with this method of response.

The same treatment procedures used in sanitary sewers could be applied whenever a liquid spill reaches a storm sewer. An explosion in a storm sewer could be even more structurally destructive than an explosion in a sanitary sewer due to the larger volume of headspace available to collect volatile gases. Toxic substances in flowing storm sewers can have a more serious impact on receiving waters than if they were diluted or buffered by flowing through a POTW.

Fumes extracted from sewers can be scrubbed or released to the atmosphere. Sewers can be partially blocked off to contain the spill and control the volume of the headspace to the optimum level for drawing off vapors (Figure 11.7).

11.32   POTW Process Changes

If the toxic material does reach the treatment plant, efforts must be made to limit damage to the personnel, structures, equipment and microorganisms in the plant. Written procedures should be posted in the control room so that operators can take immediate action to divert or bypass the flow to minimize impact on the plant. Diversion structures, valves and

pumps that are usually maintained in "standby" condition should be routinely exercised so that equipment and personnel will be ready for emergencies.

Once the spill enters the treatment works, experienced operators may be able to make adjustments in the treatment process such as changing the solids wasting rate or adjusting the aeration rate to lessen the impact of the toxin on the microorganisms.

Treatment plant personnel should prepare and follow a *WRITTEN* spill containment or control plan. This plan should present operating personnel with options for diversion, detention, dilution and emergency treatment. With practice, all personnel involved with containing or controlling a spill will know what actions to take during the stress of responding to an emergency. A practice or dry run of the written procedures should be undertaken every six months as a routine part of spill prevention planning.

11.4   ENFORCEMENT

One of the major differences between routine investigations discussed in Chapters 2, 4 and 6 and investigations of major spills is that there is a higher likelihood that permit revocation and/or court actions will be pursued if the source of the spill is found. Successful prosecution may bring felony rather than misdemeanor penalties for guilty dischargers. For this reason, take special precautions to protect the integrity of samples that might be introduced as evidence. Use chain of custody procedures and proper preservation techniques as described in Chapter 6, "Sampling Procedures for Wastewater."

Keep a detailed log of all events including times of events, names and methods of contacting witnesses, quotations, samples collected, field test kit results, and identities of participating workers. Request copies of reports from all agencies present. Spill or slug loads to the POTW should be documented in the POTW's pretreatment file for the industry involved. Documentation should include:

1. Type of spill or slug,

2. POTW response to notification,

3. POTW response to the discharge, and

4. Effect of spill on POTW.

*Fig. 11.6 Sewer maintenance workers setting up a blower at a manhole
to clear atmosphere caused by presence of hydrogen sulfide gas,
gasoline, or other noxious vapors.*

*Fig. 11.7   Specially fabricated device to control flow level in a sewer and
trap any floating liquid. Unit is inserted into exit pipe at bottom of a
manhole. Sluice gate can be raised with a rope from topside to control
depth of water that is backed up behind unit.*

11.5 SPILL REPORTING

In order to respond to an emergency situation of a material spilled into the sanitary sewer, the POTW must have in place a procedure by which notification of the spill can be transmitted to the treatment plant. The Appendix to this chapter contains copies of memos and letters outlining the procedures established by the Sanitary District of Rockford, Illinois, to receive such notification. Appendix B outlines the industrial spill reporting system with the dedicated spill reporting telephone number and documentation to be completed by the person who receives first notice of the occurrence. Appendix C is the letter which is sent to industrial users notifying them of the most recent method of spill reporting. Appendix D presents the written reporting form required by the District from the user after the spill has occurred. A program of this type should be established by the POTW to ensure (1) that proper notification reaches the pretreatment facility inspector; (2) that a follow-up investigation takes place after the fact; and (3) that a formalized statement is filed by the industrial user concerning the spill.

QUESTIONS

Write your answers in a notebook and then compare your answers with those on page 808.

11.3A Why may it be advantageous to "back up" or "block off" a sewer during a spill event?

11.3B How could a spilled acid be treated in a sewer?

11.3C How can the volatilization of some chemicals be suppressed in a sewer?

11.4A What is the biggest difference between standard investigations and investigations of major spills with regard to enforcement?

11.5A Why must a pretreatment facility inspector be notified of a spill?

11.6 TYPICAL RESPONSES TO EMERGENCIES
by John Brady

11.60 Notification

Typically the POTW receives notification of an emergency problem from the local area where the incident occurred. Notification could come from the industry involved, the fire department, police or sheriff's office. These incidents could be caused by chemical trucks or gasoline tankers involved in traffic accidents (Figure 11.8), gas station overflows resulting from tank truck drivers overfilling gas station tanks, and chemical spills, discharges or accidents involving industry.

As a pretreatment facility inspector, you may be notified by the POTW to report to the scene of the emergency. If you discover the emergency, immediately notify the POTW or your supervisor who will initiate notification procedures of other personnel and agencies on a notification list.

11.61 Accidental Spill

An accidental spill may be reported by industry, the police department or the fire department (Figure 11.9). These agencies should follow the procedures presented in Section 11.5, "Spill Reporting." They notify the POTW and the person receiving the notification records the essential information. See Appendix E, "Spill Report," for another example of the telephone log sheet used by the Sanitation Districts of Los Angeles County to record vital information about reported spills. After an accidental spill has been reported, the following tasks should be performed.

1. Find out from person reporting spill the critical information concerning the type of spill.

   a. Flammable (fuels)

   b. Toxic

   c. Asphyxiating

   d. Location

   e. Time occurred or discovered

If the spill does not contain flammable or asphyxiating substances, then the procedures in Section 11.62, "Unknown Material Reported in Collection System," may be followed.

2. Notify collection system supervisors and treatment plant operators of the problem, the expected collection system flow route from the area where the spill occurred to the POTW, and the expected time of arrival at lift stations and the treatment plant.

3. If the material in the spill is known, determine the best method of handling the spill. Consider:

   a. Volume to be dealt with,

   b. Possible treatment and disposal methods, and

   c. Possible dangers or impacts on POTW system, including personnel, biological processes, equipment and the environment.

4. Possible actions to be taken (as appropriate to the situation):

   a. Select an appropriate treatment process

      (a) Neutralization only,

      (b) Chemical, physical or biological treatment,

   b. Determine required level or degree of treatment,

   c. Review existing capabilities for treating the material,

   d. Remove any floatable and settleable material,

   e. Select ultimate disposal site for all removable material,

   f. Estimate effectiveness of treatment method selected,

   g. Determine what will happen to material not removed or material that flows through the treatment plant system and into the POTW effluent or sludge, and

   h. Notify other agencies involved with the hazardous material and/or those affected by contaminated effluent or sludge.

5. If material in the spill is hazardous to life and property (toxic, flammable, or asphyxiating) and an immediate threat to the POTW system and the public, the fire department will probably respond as the lead agency, with assistance from

*Fig. 11.8   Roadside emergency response. Chemical tanker overturned.
Responders should contain spilled material for removal by vacuum truck,
NOT hose spilled material into sewers, storm drains or ground surface.
Establish procedures BEFORE emergency occurs.*

Fig. 11.9   Fire truck responding to a toxic spill. The fire department should be aware of chemicals stored and used at industrial sites.

the police. The pretreatment facility inspector still has an extremely important role. The inspector must know the characteristics of the material that must be controlled, the type of equipment needed to control and correct the situation, and how to do the job. The inspector under these circumstances will be acting primarily as a liaison between the POTW agency and the other agencies involved. These agencies will rely on the inspector's knowledge of the collection system and the capabilities of the POTW agency to assist in trapping and treating the material in the system. Upon arrival at the scene of the accident, you (the inspector) should report to the person in charge (fire, police) and identify yourself as a representative of the POTW. Also inform the person in charge that the POTW's personnel are capable of:

- Following the material in the collection system,

- Supplying special equipment (gas detection devices such as Lower Explosive Limit (LEL) and oxygen deficiency),

- Providing personnel for functions such as construction of dikes and/or diversion facilities, and

- Operating safety equipment, pumps, generators or compressors.

6. Unusual situations that could occur and which pretreatment facility inspectors should be aware of and prepared to respond to include:

- Combined collection systems (sanitary sewers and storm drains) which may present special problems for containment,

- Flow routes that may pass through high-density population and/or commercial/business areas,

- Flow routes that may pass near or under other potentially dangerous sites such as refineries or fuel depots, and

- The possibility of changing flow direction by adjustment and control of gates and lift stations.

The main function of the pretreatment facility inspector in this situation is to assist the command center by forwarding information to the POTW agency and relaying information from the POTW agency to the command center. The inspector should not be responsible for evacuation of residential neighborhoods or commercial developments, routing of traffic or press releases.

11.62 Unknown Material Reported in Collection System

Unknown materials in collection systems may be reported by sewer maintenance crews, POTW operators, or by the public in the form of an odor complaint. If the unknown material is neither flammable nor asphyxiating (these materials require notification of other agencies), then it falls under the jurisdiction of the POTW agency. The sections of the POTW agency that would be involved include pretreatment, industrial waste, collection system, treatment plant and management. After an unknown material that is neither flammable nor asphyxiating has been reported, the tasks listed below should be performed.

1. Notify pretreatment inspector with investigation instructions.

2. Sample unknown material:

   a. Attempt to identify material, and

   b. Record characteristics such as odor, color and flow.

3. If material is determined to be flammable or asphyxiating, the pretreatment inspector should initiate notification of the persons on the local emergency response list (see Section 11.02, "Establish a Notification List"). This should require making only one call to the appropriate office staff person who will notify the appropriate agencies. After making this one notification call, the inspector should be free to deal with the other tasks required to cope with the emergency problem.

4. If the material is not an immediate threat to life and property, then notification could be limited to:

   a. Treatment plant,

   b. Collection system personnel,

   c. Local water agencies that use surface waters downstream from POTW's discharge (the notification should be factual and not create a situation of unnecessary concern or panic), and

   d. Other concerned or involved agencies (state pollution control agency).

5. Attempt to locate the source of the material entering the collection system.

   a. Start working upstream from the last known location of the material in the system (this may be the treatment plant or at a manhole in the collection system).

   b. Use collection system maps to determine possible flow routes and sources. Additional crews may be helpful if you need to track the flow route of the material back through the collection system.

   c. Identify which industries could be sources of the material in the system.

   d. If the material stops flowing in the system, thus preventing tracing the material to its source, then laboratory identification of the material is critical. Use laboratory identification to determine which industries could have discharged the material. Compare the results of industry self-monitoring and POTW agency monitoring programs on the day of the discharge to see if the material is reported in any of these data sources. Also look for industries that may not have been required to sample on the day that the material was discharged.

11.7 REFERENCES

The technical references listed below are provided to serve as references when preparing emergency response plans. After the chemicals used by industries have been identified, these references can help you to determine the adverse impacts that could occur if these chemicals are involved in spills or accidents and to identify *SAFE* remedial action.

HAWLEY'S CONDENSED CHEMICAL DICTIONARY, 14th Edition. Obtain from John Wiley & Sons, Inc., Distribution Center, 1 Wiley Drive, Somerset, NJ 08875-1272. ISBN 0-471-38735-5. Price, $149.00, plus $5.00 shipping and handling.

THE MERCK MANUAL OF DIAGNOSIS AND THERAPY, 17th Edition. Obtain from Merck Publishing Group, PO Box 2000, WBD-120, Rahway, NJ 07065. Price, $35.00.

EMERGENCY HANDLING OF HAZARDOUS MATERIALS IN SURFACE TRANSPORTATION—2000 EDITION. Obtain from Bureau of Explosives Publications, PO Box 1020, Sewickley, PA 15143. Price, $71.00, plus $10.50 shipping and handling.

SITTIG'S HANDBOOK OF TOXIC AND HAZARDOUS CHEMICALS AND CARCINOGENS, Fourth Edition, Three-Volume Set. Obtain from William Andrew Publishing, 13 Eaton Avenue, Norwich, NY 13815. ISBN 0-8155-1459-X. Price, $495.00.

"Preparing for Hazardous Materials Emergencies," Carl Smith, *OPFLOW*, AWWA, Volume 13, Number 3, March 1987.

QUESTIONS

Write your answers in a notebook and then compare your answers with those on page 808.

11.6A   How is a pretreatment facility inspector notified of an emergency?

11.6B   What critical information must you seek to determine about a hazardous spill?

11.6C   If spill material is known, what information is considered to determine the method of handling the spill?

Please answer the discussion and review questions before continuing with the Objective Test.

DISCUSSION AND REVIEW QUESTIONS

Chapter 11.   EMERGENCY RESPONSE

Write the answers to these questions in a notebook before continuing with the Objective Test on page 809. The purpose of these questions is to indicate to you how well you understand the material in the chapter.

1. How can sewer maps be used to assist with emergency responses?

2. When an inspector must decide who to notify of an emergency, on what basis is that decision made?

3. Why should a single contact person be designated to deal with the media during an emergency?

4. If a spill is flowing into a sewer, what should an inspector do?

5. How can spill material be removed from a sewer?

6. How can the operator of a POTW respond to a spill of a toxic material?

7. During an emergency response situation, inspectors at the site should keep a detailed log of what events?

8. Why must a POTW have a written spill notification procedure?

SUGGESTED ANSWERS

Chapter 11. EMERGENCY RESPONSE

Answers to questions on page 795.

11.0A   The most likely sources for catastrophic discharges into sewers are oil refineries, chemical plants, fuel transfer stations or any facility that handles large volumes of flammable or extremely toxic material. Groundwater infiltration can also be a source of dangerous liquids or gases which enter sewers.

11.0B   Sources of laws and regulations that could be pertinent to dealing with hazardous materials include local, state and federal sources.

11.0C   A list of who should be notified in case of an emergency should be posted in the inspector's home office and carried by every inspector.

Answers to questions on page 798.

11.1A   The major categories of persons or agencies that should be included on an emergency notification list include needed federal, state and local agencies associated with public safety and welfare.

11.2A   When an inspector arrives on the scene of an emergency, the inspector should activate the notification procedure as soon as possible. The situation should be evaluated as quickly as possible in terms of (1) use of personal protective equipment, (2) location of problem and affected area, (3) volume and characteristics of spilled material, (4) potential harm to people, property and environment, (5) rescue needs, (6) control of public access to spill area, (7) immediate remedial actions and options, (8) setting up a command center with a line of communications, and (9) impacts the situation may have on the collection system, treatment plant and personnel.

11.2B   A log of events must be kept at the central command center during an emergency to be used as a reference when reconstructing the events and preparing a case for court if enforcement actions are initiated.

11.2C   The affected area of a spill should be determined to reduce the number of possible sources you will need to investigate and thereby save precious time in an emergency.

Answers to questions on page 803.

11.3A   It may be advantageous to "back up" or "block off" a sewer so as to reduce headspace for volatile (toxic or explosive) gases to collect.

11.3B   A spilled acid could be treated in a sewer by pH neutralization using caustic soda, soda ash, lime or other base.

11.3C   The volatilization of some chemicals can be suppressed in a sewer by the introduction of foam. Gases can be removed by blowers and sewers can be "backed up" or "blocked off" to reduce headspace.

11.4A   The biggest difference with regard to enforcement actions between standard investigations and investigations of major spills is that there is a higher likelihood that permit revocation and/or court actions will be pursued if the source of the major spill is found.

11.5A   Pretreatment inspectors must be notified of spills so that follow-up action can be taken after the spill.

Answers to questions on page 807.

11.6A   A pretreatment facility inspector may be notified of an emergency by personally discovering the emergency or being notified by the POTW agency. The POTW may be notified by the police or fire departments, sewer maintenance crews, POTW operators, or by the public.

11.6B   Critical information about a hazardous spill includes whether the spill material is flammable (fuels), toxic, or asphyxiating; the location of the spill; and the time the spill occurred or was discovered.

11.6C   Information considered when determining how to handle spill material includes volume to be dealt with, possible treatment and disposal methods, and possible dangers or impacts on POTW systems, including personnel, biological processes, equipment and the environment.

OBJECTIVE TEST

Chapter 11.   EMERGENCY RESPONSE

Please write your name and mark the correct answers on the answer sheet, as directed at the end of Chapter 1. There may be more than one correct answer to each multiple-choice question.

True-False

1. The most common example of a sewer condition that warrants enacting an emergency response plan is the discovery of gasoline or other flammable or explosive liquid or vapor in the sewer.

   —1. True
   2. False

2. Locations of high-risk facilities should be noted on a sewer map with the route of flow to the treatment plant clearly marked.

   —1. True
   2. False

3. Water used to put out a fire could carry hazardous wastes to the POTW.

   —1. True
   2. False

4. If you are the first on an emergency scene, try to evaluate the situation as quickly as possible without endangering yourself.

   —1. True
   2. False

5. Cyanide is liberated under basic conditions.

   1. True
   —2. False

6. The use of internal combustion engines on sewer ventilation blowers is a safe procedure.

   1. True
   —2. False

7. An explosion in a sanitary sewer could be even more structurally destructive than an explosion in a storm sewer due to the larger volume of headspace available to collect volatile gases.

   1. True
   —2. False

8. Toxic substances in flowing storm sewers can have a more serious impact on receiving waters than if they were diluted or buffered by flowing through a POTW.

   —1. True
   2. False

9. Spills or slug loads to the POTW should be documented in the POTW's pretreatment file for the industry involved.

   —1. True
   2. False

10. The POTW must have in place a procedure by which notification of a spill can be transmitted to the treatment plant.

    —1. True
    2. False

Best Answer (Select only the closest or best answer.)

11. How should capabilities to divert flow throughout a collection system be handled?

    —1. Clearly marked on a map
    2. Diverted to low-quality receiving waters
    3. Not allowed
    4. Transferred to storage basins

12. What problem can be caused by infiltration of groundwater into sewers?

    —1. Contribution of explosive or toxic liquids and gases
    2. Contribution of increased spare capacity in treatment plant
    3. Contribution of reduced flows in sewers
    4. Contribution of sediment from washed out sands

13. How should an inspector be able to activate emergency response procedures?

    1. By calling the police and fire departments
    2. By contacting local TV and radio stations
    —3. By making one call to a person in the home office
    4. By phoning the collection system and treatment plant superintendents

14. Who should deal with the news media during an emergency?

    —1. A single contact person
    2. All persons contacted by the news media
    3. Anyone who is available
    4. The person in charge of the cleanup

15. Why should an inspector keep notes during an emergency? Notes will be useful in preparing

    1. A briefing for public officials.
    —2. A case for court if enforcement action is initiated.
    3. A meeting for news media.
    4. An informational pamphlet for the public.

16. What type of pump must be used when drawing liquid out of a sewer?

    1. Centrifugal pumps
    —2. Explosion-proof pumps
    3. Submersible pumps
    4. Trash pumps

17. How should treatment plant personnel respond to a spill arriving at the plant?

    — 1. By following a written spill control plan
    2. By monitoring the flow of the spill through the plant
    3. By preparing a written spill control plan
    4. By routing the spill through the plant

18. What is the appropriate response to chemicals spilled from an overturned tank truck?

    1. Allow chemicals to flow onto ground surface.
    — 2. Contain chemicals.
    3. Hose chemicals into sewers.
    4. Wash chemicals into storm drains.

19. What is the main function of a pretreatment facility inspector during an emergency?

    — 1. Assisting command center with communications to and from POTW
    2. Evacuating residential neighborhoods
    3. Preparing press releases
    4. Routing traffic around emergency site

20. How should an inspector respond when an unknown material that is neither flammable nor asphyxiating is found in a collection system?

    1. Assume the unknown material was an accidental spill.
    2. Disregard unknown material until next time.
    — 3. Try to trace unknown material back to industrial source.
    4. Wait for unknown material to arrive at treatment plant.

Multiple Choice (Select all correct answers.)

21. When should an emergency response procedure be activated? Whenever any hazardous substance has been

    — 1. Abandoned.
    2. Contained.
    — 3. Dumped.
    — 4. Released.
    — 5. Spilled.

22. What industrial discharges could create emergencies?

    — 1. Concentrated acids
    — 2. Concentrated bases
    — 3. Large volume of toxic materials
    4. Large volumes of water
    — 5. Poisonous gases

23. What are the most likely sources of catastrophic discharges?

    1. Canneries
    — 2. Chemical plants
    3. Dairies
    — 4. Fuel transfer stations
    — 5. Oil refineries

24. Which of the following information is critical information on sewer maps?

    — 1. Capabilities to divert flows
    — 2. Installations of relief lines
    — 3. Locations of high-risk facilities
    — 4. Routes of flows to treatment plants
    5. Sewer pipe materials

25. Which tasks must be performed by the first inspector to arrive at the scene of an emergency?

    — 1. Activate notification procedure.
    — 2. Assess rescue needs.
    — 3. Control public access to spill area.
    — 4. Determine need for personal protective equipment.
    — 5. Estimate potential harm to people.

26. How can the source of a spill be determined?

    — 1. Identifying the material
    — 2. Phoning industries and asking if spill occurred
    3. Responding to odor complaints
    — 4. Tracing material upstream
    5. Waiting for industry to notify agency

27. How can toxic or explosive gases in sewers be controlled or suppressed?

    1. Acids
    — 2. Blocking off a sewer
    3. Bases
    — 4. Blowers
    — 5. Foams

28. What are possible responses of operators of a POTW when a toxic spill reaches a treatment plant?

    — 1. Adjust aeration rate.
    — 2. Change solids wasting rate.
    — 3. Divert flow to emergency storage.
    4. Increase effluent chlorination rate.
    5. Short-circuit flow to digester.

29. What items should an inspector examine at high-risk industrial facilities with regard to emergency response?

    1. Chemical ordering plan
    — 2. Emergency response plan
    3. Future expansion plan
    4. Spare parts inventory plan
    — 5. Spill prevention plan

30. How can the impact of a hazardous material spill be minimized?

    — 1. POTW process changes
    — 2. Preventing material from entering the sewer
    — 3. Removing material from the sewer
    4. Telephoning industry and asking them to prevent spills
    — 5. Treatment in the sewer

End of Objective Test

APPENDICES

CHAPTER 11.   EMERGENCY RESPONSE

APPENDIX A

HOW TO PREVENT EXPLOSIVE CONDITIONS

NEED TO PREVENT EXPLOSIVE CONDITIONS

In 1981, a 12-foot semi-elliptical sewer in Louisville, Kentucky, exploded due to a hexane spill from a soybean processing plant (see Figure 11.1, page 791). Fifty miles of sewer was affected, with 13 miles of sewer collapsing in on itself. In 1991, a 15-foot combined sewer in Guadalajara, Mexico, exploded, killing hundreds of people and causing millions of dollars in damages. Many sewerage agencies attempt to protect themselves from these types of catastrophic events and have taken many differing actions to do so. At the heart of the problem of protecting sewer systems from fire or explosion is the accurate detection of flammable materials.

Three things must be present for a fire or explosion to occur: fuel, an oxidizer, and an ignition source. These items are often referred to as the triangle of combustion.

Fuel

For a fire or explosion to occur, a fuel must be mixed with oxygen in a particular concentration range and an ignition source of particular minimum energy must be introduced into the system. For each different fuel there is a mixture that will support combustion. These mixtures or ranges of mixtures are referred to as the lower explosion limit (LEL) and upper explosion limit (UEL).

Oxidizer

For the case of a fire or explosion in a sewer environment, air is the most likely oxidizer. Air is roughly 21 percent (by volume) oxygen and 79 percent inert gases (nitrogen, carbon dioxide, argon). In some wastewater treatment facilities pure oxygen is used for the secondary biological treatment processes so it will be useful to evaluate flammable characteristics in oxygen-enriched environments also. Although less likely to be a flammability-related issue with sewers, there are other oxidants than oxygen. A few examples include hydrogen peroxide, chlorine gas, and ozone.

Ignition Source

For the combustion of a flammable mixture to occur, a minimum activation energy is required. This energy causes the combustion reaction to proceed, thus releasing more energy and propagating the combustion. Some examples of the spontaneous ignition temperature (in air) for several flammable fuels are listed in Table A-1.

TABLE A-1   SPONTANEOUS IGNITION TEMPERATURES

Fuel	°F	°C
Acetone	1042	561
Benzene	1097	592
n-Butane	807	431
Carbon disulfide	248	120
Diethyl ether	366	186
Ethyl alcohol	738	392
n-Hexane	501	261
Methane	1170	632
n-Octane	464	240
iso-Octane	837	447

Other ignition sources include electrical sparks (switches, motors), static electricity discharges, lightning, sparks from impacts, friction, automobiles, matches, and cigarettes.

TYPES OF FIRE OR COMBUSTION

Pool Fires

Pool fires occur in open spaces where usually a liquid fuel has been spilled and has found an ignition source. The reaction rate or combustion rate is limited by the transport of oxygen (air) to the flame front. Pool fires can occur on water where a lighter than water fuel has been spilled and an ignition source has been found. Dangers include radiant heat (burns) and smoke inhalation. Smokey pool fires (crude oil) produce less radiant heat risks and have flame surface fluxes in the magnitude of 20 to 50 kW/sq m. Lighter hydrocarbon pool fires typically have flame surface fluxes of 50 to 140 kW/sq m, while LNG and LPG are the "hottest" at over 200 kW/sq m.

Jet Fires

Jet fires occur where a fuel finds an ignition source as it is being ejected from a pipe or vessel. Transport of oxygen (air) is usually enhanced, due to the geometry of the fire, resulting in a potential for a "hotter" fire. Flame surface fluxes for jet fires have been reported in the area of 200 kW/sq m.

BLEVE (Boiling Liquid Expanding Vapor Explosions)

BLEVEs are normally associated with the failure of a pressure vessel which contains flammable hydrocarbons. A propane tank that bursts as a result of a fire beneath it is a good example. Some overpressure may result from the initial pressure release from the tank, but usually the effects from a BLEVE are thermal damages from the resulting rising fireball. BLEVEs are normally of short duration, reported in the range of 15 to 45 seconds.

Explosion

An explosion or detonation occurs when the flame front speed (rate of combustion) travels faster than the speed of sound (subsonic flame front speed during normal combustion is called deflagration). This causes a shock or pressure wave much like a sonic boom. This shock wave and resulting pressure wave is referred to as "overpressure" and, depending on magnitude, can cause personal injury and structural damage. The most likely explosive incident to be encountered in a sewer environment is a confined vapor cloud explosion. Combustible mixtures contained in confined structures tend to explode when an ignition source is introduced because temperatures and pressures in the system increase very quickly, thus increasing the reaction rate which causes the temperature and pressure to increase more rapidly (a runaway reaction). Increasing the concentration of the fuel or the concentration of oxygen increases the reaction rate and decreases the required ignition energy, both contributing to the potential for explosion.

Combustion in Oxygen-Enriched Environments

In an oxygen-enriched environment, the LEL mixture levels for most constituents are roughly equivalent to that of air. Thus a LEL meter calibrated for fuel and air could be used in an oxygen and fuel environment and would provide conservative LEL indications. However, an enriched oxygen/gas mixture could support the combustion of substantially richer fuel mixtures than that of air and fuel. Methane, for example, has an UEL in air of 14 percent versus a 61 percent UEL for oxygen. Thus the specific energy (magnitude of destruction) of a rich hydrocarbon/oxygen mixture would be much higher than a rich mixture of hydrocarbon/air. Also, it is interesting to note that some compounds that are not flammable in air become flammable in oxygen-enriched environments.

EFFECTS OF FIRES AND EXPLOSIONS

Thermal Exposure Dangers

Exposure from extreme thermal radiation can cause first, second, or third degree burns when heat flux densities are exceeded for a specified length of time.

Damage from Explosions

Unconfined vapor cloud explosions can occur when large amounts of flammable materials are mixed with the right amount of air and an ignition source is found. Detonations from VCEs (vapor cloud explosions) usually generate overpressures in the range of 3 to 5 psi at the center of the explosion.

It is speculated that most of the sewer explosions to date have been unconfined vapor cloud explosions. Observations of manholes lifting six feet into the air during a sewer explosion are consistent with about 5 psi overpressure. Collapses of older unreinforced elliptical brick sewers (Kentucky) after an explosion are also consistent with about 5 psi.

FACTORS AFFECTING FLAMMABILITY

Solubility of Flammable Compounds

Many flammable organic compounds are soluble in water. It is important to note that some compounds one would consider insoluble (oil) are slightly soluble in water, often enough to make the mixture flammable. In general, solubility of semi-miscible (partially mixable) liquids increases with temperature but the solubility of gases decreases with temperature.

Vapor-Liquid Equilibria

For multi-component liquid mixtures it is useful to estimate the liquid mixture's potential to generate combustible gases. Often sample results for air or wastewater samples need to be reviewed for flammability and the relationship between the two does not appear to be well understood. There are a couple of forms of readily available thermodynamic equilibria data in the form of either Henry's law constants (H) or aqueous partition coefficients (K) at 25 degrees Centigrade. Definitions are given below:

For Henry's law: $Ya = HXa$

For aqueous partition coefficients: $K = Ya/Xa$

- Pa = Partial pressure for component A
- Pt = Pressure total
- Ya = mole fraction for component A in gas phase

$$= \frac{Pa}{Pt}$$

= % by volume for dilute mixtures at 1 atm

- Xa = mole fraction for component A in liquid phase

$$= \frac{\text{moles A}}{\text{moles total}}$$

For a given concentration of a water soluble substance in the liquid phase, the maximum amount of this substance that can be present in the gas phase will be determined by the equilibrium data. Often less than equilibria amounts of components are distributed in the vapor phase due to diffusion limitations. The headspace in a trunk sewer may be a good example of a vapor phase that is at less than its equilibrium concentration. To approach equilibrium concentrations in the vapor phase with dissolved constituents from the water phase, intimate contact between the gas and liquid phases is required. The pure oxygen biological reactors may be a good example of intimate contact of liquid and vapor phases.

Screening for Dangerous Compounds

From an explosion perspective, the compounds that present the biggest risk to the sewerage system are flammable liquids of low solubility that are lighter than water and have low boiling points or high vapor pressures (volatile). Liquid hydrocarbons are confirmed to pose a significant danger of fire or explosion in a wastewater mixture. Oxygenates, alcohols, ketones and other solvents, although flammable, tend to be less dangerous due to increased solubilities.

Order of Magnitude Discharges

It might be useful to quantify what order of magnitude of flammable material discharge could cause an explosive condition in an oxygen-enriched secondary biological reactor. For the example, the following assumptions will be made:

1. The flammable material will pass through primary settling treatment dissolved in wastewater.

2. The size of the secondary reactors totals 200 million gallons.

3. The most dangerous time for flammable discharges is at low flow or night time; assume flow is 100 MGD.

4. The feed to the secondary reactors will be dissolved and well mixed.

5. The hydraulic detention time is 1.8 hours in the reactors.

EXAMPLE 1: 1,1,3 Trimethyl Pentane (iso-octane) C-8

LEL: 0.74% v/v
Partition Coefficient K = 180,518 = Y/X
Solubility at 25°C = 2.2 mg/L
Aqueous LEL Concentration at 25°C = 0.26 mg/L
Assume 2,2,4 Trimethyl Pentane = 7.2 lbs/gal

1. Calculate pounds of water per minute entering secondary reactors.

$$\text{Mass In, lbs/min} = \frac{(\text{Flow, MGD})(8.34 \text{ lbs/gal})}{(24 \text{ hr/day})(60 \text{ min/hr})}$$

$$= \frac{(100 \text{ MGD})(8.34 \text{ lbs/gal})}{(24 \text{ hr/day})(60 \text{ min/hr})}$$

$$= 0.579 \text{ Million lbs/min}$$

2. Estimate the flow of 2,2,4 Trimethyl Pentane in pounds per minute and gallons per minute to produce a LEL concentration of 0.26 mg/L or 0.26 ppm.

$$
\begin{aligned}
\text{Quantity to Produce} \\
\text{LEL, lbs/min}
\end{aligned}
\begin{aligned}
&= (\text{Mass In, M lbs/min})(\text{LEL, ppm}) \\
&= (0.579 \text{ M lbs/min})(0.26 \text{ lb/m lbs}) \\
&= 0.15 \text{ lb/min} \\
&= \frac{0.15 \text{ lb/min}}{7.2 \text{ lbs/gal}} \\
&= 0.021 \text{ GPM}
\end{aligned}
$$

3. Estimate the slug load quantity to reach the LEL in pounds and gallons.

$$
\begin{aligned}
\text{Slug Load, lbs} &= (\text{Detention Time, hr})(60 \text{ min/hr})(\text{Quantity, lbs/min}) \\
&= (1.8 \text{ hr})(60 \text{ min/hr})(0.15 \text{ lb/min}) \\
&= 16.2 \text{ lbs} \\
&= \frac{16.2 \text{ lbs}}{7.2 \text{ lbs/gal}} \\
&= 2.25 \text{ gal}
\end{aligned}
$$

EXAMPLE 2: Cyclohexane C6

LEL: 1.09% v/v
Partition Coefficient K = 10,883 = Y/X
Solubility at 25°C = 56.1 mg/L
Aqueous LEL Concentration at 25°C = 4.70 mg/L

Estimated continuous quantity to LEL: 2.72 lbs/min Cyclohexane (0.375 GPM)

Estimated slug load quantity to LEL: 294 lbs Cyclohexane (40.5 gal)

EXAMPLE 3: Benzene C6

LEL: 1.17% v/v
Partition Coefficient K = 308 = Y/X
Solubility at 25°C = 1,780 mg/L
Aqueous LEL Concentration at 25°C = 163.9 mg/L

Estimated continuous quantity to LEL: 95 lbs/min Benzene (13.1 GPM)

Estimated slug load quantity to LEL: 10,260 lbs Benzene (1,413 gal)

EXAMPLE 4: Methyl-Tert-Butyl-Ether (MTBE)

LEL: 1.01% v/v
Partition Coefficient K = 31 = Y/X
Solubility at 25°C = 36,626 mg/L
Aqueous LEL Concentration at 25°C = 1,511 mg/L

Estimated continuous quantity to LEL: 875 lbs/min MTBE (120 GPM)

Estimated slug load quantity to LEL: 94,500 lbs MTBE (13,027 gal)

Although these examples are conservative, they provide an order of magnitude to assess risk. At discharge multiples of 3 or 5 over the conservative estimate illustrated for some of these compounds, flammability or explosion could still very easily pose a real threat to a pure oxygen secondary reactor. A slug load of 45 gallons of gasoline, for example, has enough 2,2,4 trimethyl pentane (15 percent by volume) in the gasoline blend to exceed the flammable conditions estimated in Example 1 by a factor of 3. It also illustrates that as flammable compounds increase in solubility, they decrease in flammable threat due to ease of dilution below flammable concentrations.

COMBUSTIBLE GAS SENSING

Currently, four types of combustible gas sensing technologies are commercially available: catalytic detection, infrared detection, photoionization detection (PID), and flame ionization detection (FID). Each method has advantages and disadvantages. It is important to understand the limitations of a combustible gas sensor and to understand that a reading on an instrument is merely a guide and may not accurately reflect the actual real world conditions.

A study was conducted to determine the catalytic sensor response factors for 168 chemicals. The sensor was calibrated for methane and the testing was performed by EPA in 1983. Response factors ranged from 0.8 for isobutane to 78 for dimethyl styrene for test concentrations of 10,000 ppmv (1.0% v/v). Response factor is defined as actual concentration/instrument response. A response factor of 1 for all chemicals would be ideal, however, the data indicate a significant potential for reduced instrument indication when the sensor is calibrated with methane. Some chemicals often seen in sewer systems that might be candidates for reduced responses are the following:

Chemical	Response Factor
Toluene	2.32
Dichloromethane (Methylene Chloride)	3.63
Xylene, para	5.35
Tetrachloroethylene (perc)	11.46
Freon 12	11.83
Styrene	36.83

Infrared (IR) Sensors (Figure A-1)

An infrared (IR) sensor operates by measuring a difference in absorbance of an IR beam of particular frequency and bandwidth (Beer's Law). IR sensors used for flammability detection are usually calibrated for hydrocarbons and measure an absorbance in the C-H stretch frequency (2,900 to 3,100 cm^{-1}).

Characteristics of the HNU Photoionizer and Organic Vapor Analyzer (OVA)

The HNU Photoionizer and the Century Organic Vapor Analyzer (OVA) are used in the field to detect a variety of compounds in air. The two instruments differ in their modes of operation and in the number and types of compounds they detect (Table A-2). Both instruments can be used to detect leaks of volatile substances from drums and tanks, determine the presence of volatile compounds in soil and water, make ambient air surveys, and collect continuous air monitoring data. If personnel are thoroughly trained to operate the instruments and to interpret the data, these instruments can be valuable tools for helping to decide the levels of protection to be worn, assisting in determining other safety procedures, and determining subsequent monitoring or sampling locations.

OVA (Figure A-2)

The OVA operates in two different modes. In the survey mode, it can determine the approximate total concentration of all detectable species (gases and vapors) in air. With the gas

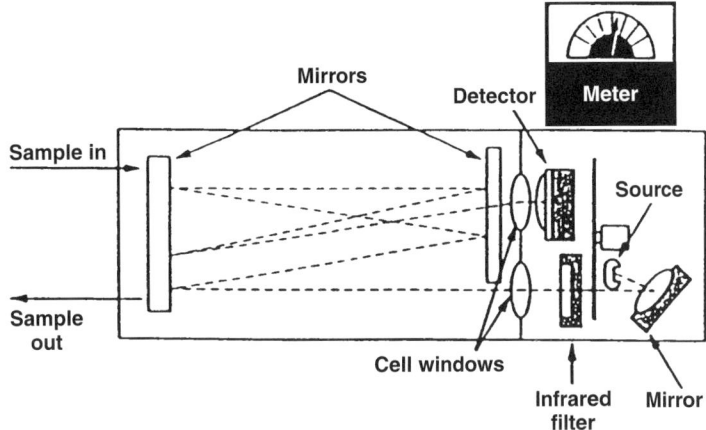

Advantages of IR sensors include:

1. No zero drift,
2. No sulfur fouling,
3. No spike fouling, and
4. Much longer service life.

Disadvantages of IR sensors include:

1. Gas specific — does not perform broad screening well,
2. Varying response factors for different gases (RF = indicated/actual),
3. Measures a surrogate (substitute) for flammability, does not directly measure the actual combustibility of the vapor environment being sampled,
4. IR measurement cell is delicate, and
5. Dirt or moisture on the measurement cell mirror will cause a false positive reading.

Fig. A-1   Infrared sensor

TABLE A-2   COMPARISON OF OVA AND HNU SENSORS

	OVA	HNU
Response	Responds to many organic gases and vapors.	Responds to many organic and some inorganic gases and vapors.
Application	In survey mode, detects total concentrations of gases and vapors. In GC mode, identifies and measures specific compounds.	In survey mode, detects total concentrations of gases and vapors. Some identification of compounds possible, if more than one probe is used.
Detector	Flame ionization detector	Photoionization detector
Limitations	Does not respond to inorganic gases and vapors. No temperature control.	Does not respond to methane. Does not detect a compound if probe has a lower energy than compound's ionization potential.
Calibration Gas	Methane	Benzene
Ease of Operation	Requires experience to interpret correctly, especially in GC mode.	Fairly easy to use and interpret.
Detection Limits	0.1 ppm (methane)	0.1 ppm (benzene)
Response Time	2 – 3 seconds (survey mode)	3 seconds for 90% of total concentration
Maintenance	Periodically clean and inspect particle filters, valve rings, and burner chamber. Check calibration and pumping system for leaks. Recharge battery after each use.	Clean UV lamp frequently. Check calibration regularly. Recharge battery after each use.
Useful Range	0 – 1,000 ppm	0 – 2,000 ppm
Service Life	8 hours; 3 hours with strip chart recorder.	10 hours; 5 hours with strip chart recorder.

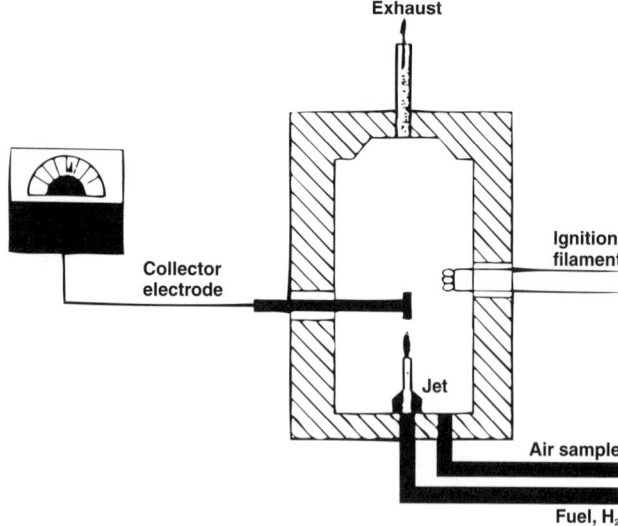

Features:

1. Responds to virtually all organic compounds with roughly the same sensitivity.
2. Linearity is good for ranges as high as 10^3.
3. Does not respond to water or carbon dioxide.
4. Has minimal effects from sample flow, pressure and temperature changes.
5. Support-fuel supply for flame detector required.

Fig. A-2   Organic vapor analyzer (OVA)

chromatograph (GC) option, individual components can be detected and measured independently, with some detection limits as low as a few parts per million (ppm).

In the GC mode, a small sample of ambient air is injected into a chromatographic column and carried through the column by a stream of hydrogen gas. Contaminants with different chemical structures are retained on the column for different lengths of time (known as retention times) and hence are detected separately by the flame ionization detector. A strip chart recorder can be used to record the retention times, which are then compared to the retention times of a standard with known chemical constituents. The sample can either be injected into the column from the air sampling hose or injected directly with a gas-tight syringe.

The OVA is internally calibrated to methane by the manufacturer. When measuring methane, it indicates the true concentration. In response to all other detectable compounds, however, the instrument reading may be higher or lower than the true concentration. Relative response ratios for substances other than methane are available. To correctly interpret the readout, it is necessary to either make calibration charts relating the instrument readings to the true concentration or to adjust the instrument so that is reads correctly. This is done by turning the ten-turn, gas-select knob, which adjusts the response of the instrument. The knob is normally set at 300 when calibrated to methane. Secondary calibration to another gas is done by sampling a known concentration of the gas and adjusting the gas-select knob until the instrument reading equals the true concentration.

The OVA has an inherent limitation in that it can detect only organic molecules. Also, it should not be used at temperatures lower than about 40 degrees Fahrenheit because gases condense in the pump and column. It has no temperature control, and since retention times vary with ambient temperatures for a given column, absolute determinations of contaminants are difficult. Despite these limitations, the GC mode can often provide tentative information on the identity of contaminants in air without relying on costly, time-consuming laboratory analysis.

HNU (Figure A-3)

The HNU portable photoionizer detects the concentration of organic gases as well as a few inorganic gases. The basis for detection is the ionization of gaseous species. The incoming gas molecules are subjected to ultraviolet (UV) radiation, which is energetic enough to ionize many gaseous compounds. Each molecule is transformed into charged ion pairs, creating a current between two electrodes. Every molecule has a characteristic ionization potential (I.P.), which is the energy required to remove an electron from the molecule, yielding a positively charged ion and the free electron.

Three probes, each containing a different UV light source, are available for use with the HNU. Energies are 9.5, 10.2, and 11.7 electron volts (eV). All three detect many aromatic and large-molecule hydrocarbons. The 10.2 eV and 11.7 eV probes, in addition, detect some smaller organic molecules and some halogenated hydrocarbons. The 10.2 eV probe is the most useful for environmental response work, as it is more durable than the 11.7 eV probe and detects more compounds than the 9.5 eV probe.

The primary HNU calibration gas is benzene. The span potentiometer knob is turned to 9.8 for benzene calibration. A knob setting of zero increases the sensitivity to benzene approximately tenfold. As with the OVA, the instrument's response can be adjusted to give more accurate readings for specific gases and eliminate the necessity for calibration charts.

While the primary use of the HNU is as a quantitative instrument, it can also be used to detect certain contaminants, or at least to narrow the range of possibilities. Noting instrument response to a contaminant source with different probes can eliminate some contaminants from consideration. For instance, a compound's ionization potential may be such that the 9.5 eV probe produces no response, but the 10.2 eV and 11.7 eV probes do elicit a response. The HNU does not detect methane.

The HNU is easier to use than the OVA. Its lower detection limit is also in the low ppm range. The response time is rapid;

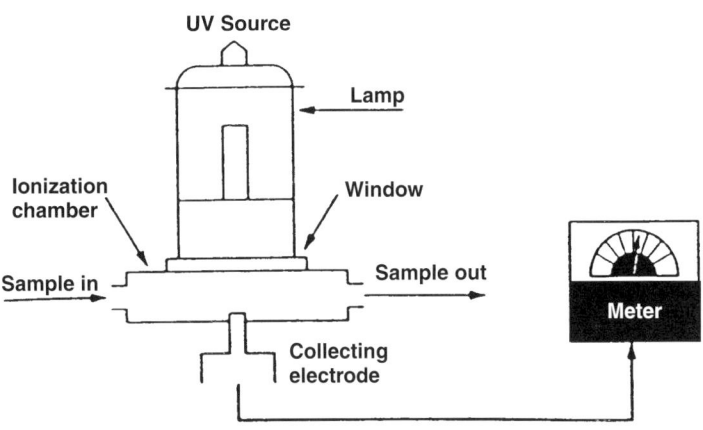

Features:

1. Sensitive, nondestructive detector with a wide dynamic range.
2. Measures many inorganic as well as many organic vapors.
3. No fuel supply required.
4. Simple to use.

Fig. A-3 HNU photoionizer

the meter needle reaches 90 percent of the indicated concentration in 3 seconds.

Both the OVA and the HNU can monitor only certain vapors and gases in air. Many nonvolatile liquids, toxic solids, particulates, and other toxic gases and vapors cannot be detected. Because the types of compounds that the HNU and OVA can potentially detect are only a fraction of the chemicals possibly present at an incident, a zero reading on either instrument does not necessarily signify the absence of air contaminants.

The instruments are generally not specific, and their response to different compounds is relative to the calibration gas. Instrument readings may be higher or lower than the true concentration. This can be an especially serious problem when monitoring for total contaminant concentrations if several different compounds are being detected at once. In addition, the response of these instruments is not linear over the entire detection range. Care must therefore be taken when interpreting the data. All identifications should be reported as tentative until they can be confirmed by more precise analysis. Concentrations should be reported in terms of the calibration gas and span potentiometer or gas-select-knob setting.

Since the OVA and HNU are small portable instruments, they cannot be expected to yield results as accurate as laboratory instruments. They were originally designed for specific industrial applications. They are relatively easy to use and interpret when detecting total concentrations of known contaminants in air, but interpretation becomes more difficult when trying to identify the components of a mixture. Neither instrument can be used as an indicator for combustible gases or oxygen deficiency.

The OVA (Model 128) is certified by Factory Mutual to be used in Class I, Division 1, Groups A, B, C, and D environments. The HNU is certified by Factory Mutual for use in Class I, Division 2, Groups A, B, C, and D.

RATIONALE FOR RELATING TOTAL ATMOSPHERIC VAPOR/GAS CONCENTRATIONS TO THE SELECTION OF THE LEVEL OF PROTECTION

The objective of using total atmospheric vapor/gas concentrations for determining the appropriate level of protection is to provide a numerical criterion for selecting Level A, B, or C protection. In situations where the presence of vapors or gases is not known, or if present, the individual components are unknown, personnel required to enter that environment must be protected. Until the constituent substances and particulates can be determined and respiratory and/or body protection related to the toxicological properties of the identified substances, total vapor/gas concentration, with judicious interpretation, can be used as a guide for selecting personal protective equipment.

Although total vapor/gas concentration measurements are useful to a qualified professional for the selection of protection equipment, caution should be exercised in interpretation. An instrument does not respond with the same sensitivity to several vapor/gas contaminants as it does to a single contaminant. Also since total vapor/gas field instruments see all contaminants in relation to a specific calibration gas, the concentration of unknown gases or vapors may be over- or underestimated.

Suspected carcinogens, particulates, highly hazardous substances, or other substances that do not elicit an instrument response may be known or believed to be present. Therefore, the protection level should not be based solely on the total vapor/gas criterion. Rather, the level should be selected case by case, with special emphasis on potential exposure and chemical and toxicological characteristics of the known or suspected material.

Factors for Consideration

In using total atmospheric vapor/gas concentrations as a guide for selecting a level of protection, a number of other factors should also be considered:

- The uses, limitations, and operating characteristics of the monitoring instruments must be recognized and understood. Instruments such as the HNU Photoionizer, Century Organic Vapor Analyzer (OVA), MIRAN Infrared Spectrophotometer, and others do not respond identically to the same concentration of a substance or respond to all substances. Therefore, experience, knowledge, and good judgment must be used to complement the data obtained with instruments.

- Other hazards may exist such as gases not detected by the HNU or OVA (for example, phosgene, cyanides, arsenic, chlorine), explosives, flammable materials, oxygen deficiency, liquid/solid particles, and liquid or solid chemicals.

- Vapors/gases with very low toxicities could be present.

- The risk to personnel entering an area must be weighed against the need for entering. Although this assessment is largely a value judgment, it requires a conscientious balancing of the variables involved and the risk to personnel against the need to enter an unknown environment.

- The knowledge that suspected carcinogens or substances extremely toxic or destructive to skin are present or suspected to be present (which may not be reflected in the total vapor/gas concentration) requires an evaluation of factors such as the potential for exposure, chemical characteristics of the material, limitations of instruments, and other considerations specific to the incident.

- What needs to be done on site must be evaluated. Based upon total atmospheric vapor concentrations, Level C protection may be judged adequate; however, tasks such as moving drums, opening containers, and bulking of materials, which increase the probability of liquid splashes or generation of vapors, gases, or particulates, may require a higher level of protection.

- Before any respiratory protective apparatus is issued, a respiratory protection program must be developed and implemented according to recognized standards (ANSI Z88.2-1980).

Level A Protection (500 to 1,000 ppm Above Background)

Level A protection provides the highest degree of respiratory tract, skin, and eye protection if the inherent limitations of the personal protective equipment are not exceeded. The range of 500 to 1,000 parts per million (ppm) total vapors/gases concentration in air was selected based on the following criteria:

- Although Level A provides protection against air concentrations greater than 1,000 ppm for most substances, an operational restriction of 1,000 ppm is established as a warning flag to:

  1. Evaluate the need to enter environments with unknown concentrations greater than 1,000 ppm,

  2. Identify the specific constituents contributing to the total concentration and their associated toxic properties,

  3. Determine more precisely the concentrations of constituents,

  4. Evaluate the calibration and/or sensitivity error associated with the instrument(s), and

  5. Evaluate instrument sensitivity to wind velocity, humidity and temperature.

- A lower limit of 500 ppm total vapors/gases in air was selected as the value to consider upgrading from Level B to Level A. This concentration was selected to fully protect the skin until the constituents can be identified and measured and substances affecting the skin excluded.

- The range of 500 to 1,000 ppm is sufficiently conservative to provide a safe margin of protection if readings are low due to instrument error, calibration, and sensitivity; if higher than anticipated concentrations occur; and if substances highly toxic to the skin are present.

With properly operating portable field equipment, ambient air concentrations approaching 500 ppm have not routinely been encountered on hazardous waste sites. High concentrations have been encountered only in closed buildings, when containers were being opened, when personnel were working in the spilled contaminants, or when organic vapors/gases were released in transportation accidents. A decision to require Level A protection should also consider the negative aspects: higher probability of accidents due to cumbersome equipment, and most importantly, the physical stress caused by heat buildup in fully encapsulating suits.

Level B Protection (5 to 500 ppm Above Background)

Level B protection is the minimum level of protection recommended for initially entering an open site where the type(s), concentration(s), and presence of airborne vapors are unknown. This level of protection provides a high degree of respiratory protection. Skin and eyes are also protected, although a small portion of the body (neck and sides of head) may be exposed. The use of a separate hood or hooded, chemical-resistant jacket would further reduce the potential for exposure to this area of the body. Level B impermeable protective clothing also increases the probability of heat stress.

A lower limit of 500 ppm total atmospheric vapor/gas concentration on portable field instruments has been selected as the upper restriction on the use of Level B. Although Level B personal protection should be adequate for most commonly encountered substances at air concentrations higher than 500 ppm, this limit has been selected as a decision point for a careful evaluation of the risks associated with higher concentrations. These factors should be considered:

1. The necessity for entering unknown concentrations higher than 500 ppm wearing Level B protection,

2. The probability that substance(s) present are severe skin hazards,

3. The work to be done and the increased probability of exposure,

4. The need for qualitative and quantitative identification of the specific components,

5. Inherent limitations of the instruments used for air monitoring, and

6. Instrument sensitivity to winds, humidity, temperature, and other factors.

Level C Protection (Background to 5 ppm Above Background)

Level C provides skin protection identical to Level B, assuming the same type of chemical protective clothing is worn, but lesser protection against inhalation hazards. A range of background to 5 ppm above ambient background concentrations of vapors/gases in the atmosphere has been established as guidance for selecting Level C protection. Concentrations in the air of unidentified vapors/gases approaching or exceeding 5 ppm would warrant upgrading respiratory protection to a self-contained breathing apparatus.

A full-face, air-purifying mask equipped with an organic vapor canister (or a combined organic vapor/particulate canister) provides protection against low concentrations of most common organic vapors/gases. There are some substances against which full-face, canister-equipped masks do not protect, or substances that have very low Threshold Limit Values or Immediately Dangerous to Life or Health concentrations. Many of the latter substances are gases or liquids in their normal state. Gases would only be found in gas cylinders, while the liquids would not ordinarily be found in standard containers or drums.

Every effort should be made to identify the individual constituents (and the presence of particulates) contributing to the total vapor readings of a few parts per million. Respiratory protective equipment can then be selected accordingly. It is exceedingly difficult, however, to provide constant, real-time identification of all components in a vapor cloud with concentrations of a few parts per million at a site where ambient concentrations are constantly changing. If highly toxic substances have been ruled out, but ambient levels of a few parts per million persist, it is unreasonable to assume only self-contained breathing apparatus should be worn. The continuous use of air-purifying masks in vapor/gas concentrations of a few parts per million gives a reasonable assurance that the respiratory tract is protected, provided that the absence of highly toxic substances has been confirmed.

Full-face, air-purifying devices provide respiratory protection against most vapors at greater than 5 ppm; however, until more definitive qualitative information is available, a concentration(s) greater than 5 ppm indicates that a higher level of respiratory protection should be used. Also, unanticipated transient excursions may increase the concentrations in the environment above the limits of air-purifying devices. The increased probability of exposure due to the work being done may require Level B protection, even though ambient levels are low.

Instrument Sensitivity

Although the measurement of total vapor/gas concentrations can be a useful adjunct to professional judgment in the selection of an appropriate level of protection, caution should be used in the interpretation of the measuring instrument's readout. The response of an instrument to a gas or vapor cloud containing two or more substances does not provide the same sensitivity as measurements involving the individual pure constituents. Hence the instrument readout may overestimate or underestimate the concentration of an unknown composite cloud. This same type of inaccuracy could also occur in measuring a single unknown substance with the instrument calibrated to a different substance. The unique characteristics of each instrument must be considered in conjunction with the other guidelines in selecting the appropriate protective equipment.

Using the total vapor/gas concentration as a criterion to determine levels of protection should provide protection against concentrations greater than the instrument's readout. However, when the upper limits of Levels C and B are approached, serious consideration should be given to selecting a higher level of protection. Cloud constituent(s) must be identified as rapidly as possible and levels of protection should be based on the toxic properties of the specific substance(s) identified.

APPENDIX B

INDUSTRIAL SPILL REPORTING SYSTEM

The material presented in Appendix B is adapted from the "Industrial Spill Reporting" system of the Sanitary District of Rockford, Illinois. Use this information to prepare or compare with your reporting system.

To: Key Personnel

From: Pretreatment Inspection Supervisor

Subject: Industrial Spill Reporting

Effective with the new phone system, a new 24-hour dedicated spill reporting telephone line has been established at the District. This is a dedicated phone number for spill reporting only. This number is in service 24 hours a day, seven days per week and is the only number industry, the fire departments, sheriff's department, IEPA, USEPA and the U.S. Coast Guard have been instructed to call in case of a spill to the sanitary sewer system.

When this phone number is called, your telephone will ring but the red LED light will flash by the PVR button. Please press this button and answer by informing the caller they have reached the Sanitary District of Rockford spill reporting number and asking the caller if they have a spill to report.

When the spill line rings, you should prepare to take the following information:

1. What was spilled into the sewer?

2. Where was it spilled into the sewer?

3. How much was spilled?

4. When was it spilled?

5. Who spilled it?

6. Name of the person reporting the spill.

7. Phone number where they can immediately be called back.

8. Who else has been contacted about the spill?

9. Have there been any injuries from the spill?

10. Is there a known fire or explosion hazard as a result of the spill?

11. Is the spill currently being contained or is an attempt being made to contain the spill?

Use a copy of the attached form to obtain the information. Have the caller stay calm. Try to get as much detailed information as you think you can, taking into account that people get excited in these situations and that the first information on a spill can be sketchy or just plain wrong. Ask the caller to spell trade names, compounds or company names you are not entirely familiar with.

The Sanitary District of Rockford's procedures also list who should be notified by phone during regular working hours and off-duty hours.

SANITARY DISTRICT OF ROCKFORD

Accidental Spill Reporting Form
To Be Completed by Treatment Plant Personnel

This accidental spill report was received on

_____ ,19 ___ at _____ a.m./p.m. by _____

What was spilled into the sewer? _____

_____

_____

Where was it spilled into the sewer? (Address & location if possible)

How much was spilled? _____

When was it spilled? _____ a.m./p.m.

Who spilled it? _____

_____

Name of person reporting the spill (print) _____

_____

Phone number where this person can immediately be called back

_____

Who else has been contacted about this spill?   IEPA _____   USEPA _____

Fire Department _____   (specify which Fire Dept.) _____

Sheriff's Dept. _____   Police Department _____   State Police _____

U.S. Coast Guard _____   Other _____

_____

Have there been any injuries as a result of this spill? _____

_____

Is there a known fire or explosion hazard as a result of the spill?

If yes, explain: _____

_____

_____

Is the spill currently being contained or is an attempt being made to contain the spill? _____ If yes, explain how:

_____

_____

_____

_____

APPENDIX C

NOTIFICATION LETTER TO INDUSTRY REGARDING SPILL REPORTING

Re: Spill Reporting

Ladies and Gentlemen:

In the past your company has received information from the Sanitary District of Rockford concerning the spill reporting requirements of SDR Ordinance 361. This program has been a great success due to your cooperation in reporting spills promptly and taking proper preventive actions to lessen the chance for accidental spillage.

Effective with this letter, the telephone number for spill reporting will change to allow for more direct communication. The new telephone number for spill reporting only is 397-9422. This number is to be used 24 hours a day, seven days a week for spill reporting purposes only. Please immediately update your internal spill reporting plan to reflect this change.

As in the past, should you have need to report a spill, please be prepared to provide the following information:

1. Your name and company,

2. The location of spill,

3. Type of waste spilled,

4. Estimated volume of the spill,

5. Is there an immediate fire or explosion hazard?

6. Phone number where the caller can be called back.

After you have reported a spill and within 15 days after the spill, you will have to submit a detailed written statement on the incident. The District will provide forms for this purpose. This statement will describe the cause of the incident and measures taken to prevent similar future occurrences.

Enclosed you will find a placard developed by the Environmental Quality Committee of the Rockford Area Chamber of Commerce as a community service project. Please use these placards as part of your spill prevention program. You may want to write the new spill reporting telephone number on the bottom of this placard before posting.

Should you have any further questions or comments, do not hesitate to contact this office.

Sincerely,

SANITARY DISTRICT OF ROCKFORD

Richard W. Eick
Plant Operations Manager

RCS:pb

cc: IEPA
       Rockford Fire Department
       Loves Park Department
       Illinois State Police
       Rockford Police Department
       Loves Park Police Department
       Time Weir, Plant Operations
       Rockford Area Chamber of Commerce

APPENDIX D

INDUSTRIAL SPILL REPORTING FORM

Dear Industrial User:

The Sanitary District's General Pretreatment Ordinance No. 361, which became effective October 1, 1982, contains provisions for reporting to the District spills of pollutants (Article IV, Section 10B and C).

B. Immediate Notification.

In the case of an accidental or deliberate discharge of compatible or incompatible pollutants which cause interference at the POTW or violate regulatory requirements of this ordinance, it shall be the responsibility of the Industrial User to immediately telephone and notify the District of the incident. The notification shall include name of caller, location and time of discharge, type of wastewater, concentration and volume.

C. Written Report.

Within fifteen (15) days following such an accidental or deliberate discharge the Industrial User shall submit to the Director a detailed written report on forms to be provided by the District, describing the cause of the discharge and the measures to be taken by the User to prevent similar future occurrences. Follow-up reports may be required by the District as needed. Such report, or reports, shall not relieve the Industrial User of any expense, loss, damage or other liability which may be incurred as a result of damage to the POTW, fish kills, or any other damage to persons or property; nor shall such report relieve the User of any fines, civil penalties, or other liability which may be imposed by this ordinance or otherwise. Failure to report accidental or deliberate discharges may, in addition to any other remedies sought by the District, result in the revocation of the discharger's wastewater discharge permit.

Please find enclosed several copies of the Accidental Discharge Reporting Form as discussed under Section 10C of the ordinance. Your cooperation concerning spills is needed and appreciated by the Sanitary District in our efforts at minimizing their impact on the treatment processes and final effluent quality.

If you have any questions, please do not hesitate to call.

Sincerely,

SANITARY DISTRICT OF ROCKFORD

Richard W. Eick
Plant Operations Manager

RWE/ra

cc: File

Enclosure

ACCIDENTAL DISCHARGE REPORTING FORM

(Sanitary District of Rockford)

This form must be completed and returned to the District Director within fifteen (15) days following the report of an accidental or deliberate discharge to the sanitary sewer. Completion of this form is a requirement of Ordinance 361 (Article IV, Section 10C) and does not relieve the User of any liabilities due to the accidental discharge. Prompt and accurate reporting does reflect that the User is attempting to address the problem.

Company Name: _____

Address: _____ Phone: _____

Person completing this form: _____

Title: _____

Time and Date accidental discharge started and stopped:

Started _____ a.m./p.m. on _____ (date) and

Stopped_____ a.m./p.m. on _____ (date).

Type of material spilled: _____

_____

_____

Volume of spill (give units): _____

Chemical analysis of a representative sample of the spilled material. Show concentration of all compounds in the spilled material. If a sample of the spilled material is not available, list all known contents present in the discharged material.

COMPOUND	CONCENTRATION (mg/L)

Location of accidental discharge:

Plant process area _____ Material storage area _____

In-plant transfer area _____ Shipping/Receiving area _____

Other (specify) _____

Is spill containment present in the area where the accidental discharge occurred?

Yes _____ No _____

Is spill containment present in other areas within the plant?

Yes _____ No _____

Describe the cause of the reported discharge:

_____

_____

_____

_____

Describe what actions were taken at the time to control the spill (e.g. sealed floor drain, use of sorbants or foams, etc):

Did the spill receive any type of treatment?

Yes _____   No _____

If yes, please describe:

_____

_____

_____

Was any part of the spill contained and prevented from discharge to the sanitary sewer?

Yes _____   No _____

If yes, please describe how that waste was disposed of.

_____

_____

_____

Describe fully what measures will be taken to prevent similar accidents in the future.

_____

_____

_____

_____

_____

Anticipated time schedule in which the above-stated measures will be completed.

_____

_____

_____

_____

_____

This accidental discharge was reported to the District on _____

_____ (date) at _____ a.m./p.m. by _____ (name),

_____ (title).

APPENDIX E

SPILL REPORT

(Sanitation Districts of Los Angeles County)

Date/Time: _____   Name of Person Taking Call: _____

QUESTIONS	ANSWERS
A. Description of Caller 1. Name of caller	Does caller request anonymity? Y / N
2. Relationship of caller to incident (employee, witness, WRP Operator, etc.)	
3. Caller's Address	
4. Caller's Phone Number	
B. Details 1. Name & address of potentially responsible party or location of incident	
2. Date, time, and duration of incident	
3. Chemical name, quantity and concentration (odor description)	

4. Additonal remarks
 	
   a. How was material lost (leaking tanks, employee error, accident, etc.)?
 	
   b. How did material enter sewer?
 	
   c. Is sample available? Where held?
 	
   d. Other remarks	

Forward this report AS SOON AS POSSIBLE to ONE of the following, from top to bottom:

   1. Bill Garrett   JAO ext. 2907

   2. ASI   JAO ext. 2924

   3. Harry Mehta   JAO ext. 3516

   4. Leon S. Directo JAO ext. 2904

Inspector: after investigation, attach report or write findings on back of this form and submit to Bill Garrett; a chronological file labeled "Referrals" will be maintained in the Supervisor's office. Draft "Response to Referral" or "Referral" letter if appropriate.

APPENDIX E

APPENDIX

PRETREATMENT FACILITY INSPECTION

A
P
P
E
N
D
I
X

APPENDIX I. FINAL EXAMINATION AND SUGGESTED ANSWERS

This final examination was prepared *TO HELP YOU* review the material in this manual. The questions are divided into four types:

1. Short Answer,

2. True-False,

3. Best Answer, and

4. Multiple Choice.

To work this examination:

1. Write the answer(s) to each question in your notebook,

2. After you have worked a group of questions (you decide how many), check your answers with the suggested answers at the end of this exam, and

3. If you missed a question and don't understand why, reread the material in this manual.

If you need practice working pretreatment arithmetic problems, please see the next section in the Appendix on "Pretreatment Arithmetic."

Since you have already completed this course, you do not have to send your answers to California State University, Sacramento.

Short Answer

1. In general terms, what does a pretreatment facility inspector do?

2. The effectiveness of inspectors in carrying out their assignments depends on what factors?

3. Improper discharges of industrial wastes can cause what types of problems in a POTW's wastewater treatment facilities?

4. What is a major concern of industry with respect to industrial waste pretreatment inspection programs?

5. What costs should companies pay with regard to wastewaters discharged to a POTW?

6. Why is monitoring by the POTW agency essential?

7. What important items should be given particularly careful attention during a permit review process?

8. What items should a POTW agency consider when purchasing a computer system?

9. Why is proper legal authority an essential part of a pretreatment program?

10. What is the difference between a direct and indirect discharger?

11. What are Categorical Pretreatment Standards?

12. How can Categorical Pretreatment Standards be modified?

13. How frequently should a particular industry be inspected?

14. What are the major items an inspector must consider before entering an industry to conduct an inspection?

15. Why should an inspector always visit the outfall whenever visiting an industry?

16. What should an inspector look for when inspecting a spill containment facility?

17. How can a supervisor set an example for inspectors on how to comply with all safety requirements and procedures?

18. What safety precautions should an inspector take during an actual inspection of an industry?

19. What should an inspector do *BEFORE* starting work in a street?

20. What must an inspector know and be able to do to use a gas detection device?

21. An inspector must be especially aware of the potential for electrical shock when working in what areas?

22. What hazards may be encountered when inspecting electroplating shops and metal plating facilities?

23. How does an agency setting up a sampling program decide whether to use an in-house or an outside laboratory?

24. Why should sample containers be clean?

25. How would you determine whether to use disposable or reusable sample containers?

26. Why should a regulatory agency collect samples rather than rely on self-monitoring by dischargers?

27. Why must samples be properly stored?

28. What is the purpose of a QA/QC program?

29. How can influent flow measurements be used to determine discharge flow rates?

30. What factors could cause a Palmer-Bowlus flume to be in a submerged condition?

31. How can you prevent the end of the bubbler tubing from becoming clogged by solids in the wastestream?

32. What is a general limitation of a closed-pipe flowmeter in the measurement of industrial wastewater?

33. Why does a pretreatment facility inspector need to understand the sources and quantities of industrial wastes?

34. Identify several types or sources of intermittent discharges of wastewater.

35. Why are mass emission rate standards preferred over concentration-based standards?

36. How is water used in the manufacture of pulp, paper, and paperboard?

37. How can a slug loading on a POTW from an industrial discharge be mitigated?

38. What are pass-through discharges?

39. What can cause a pass-through discharge?

40. What factors should be considered when estimating existing and expected flows from a discharger?

41. What is the basis for the establishment of the degree of pretreatment required for an industrial waste?

42. Why do some wastestreams require neutralization or pH adjustment?

43. What factors can inhibit or upset biological waste treatment processes?

44. What is the major difference between sludge produced from biological treatment processes and sludge generated from physical-chemical treatment processes?

45. Why is source control important for an industry?

46. What is the main reason for conducting pretreatment facility inspections and monitoring?

47. Why is current data on industrial dischargers necessary in a pretreatment program?

48. What information should be gathered during an industrial plant inspection regarding waste residuals (solids or floatables)?

49. What types of existing information regarding an industry to be inspected should be taken along during an inspection and verified for accuracy?

50. Why do wastewaters discharged from a papermaking process have a tendency toward low pH?

51. How can sewer maps be used to assist with emergency responses?

52. Why should a single contact person be designated to deal with the media during an emergency?

53. How can spill material be removed from a sewer?

54. Why must a POTW have a written spill notification procedure?

True-False

1. The potential of an unannounced inspection at any time serves as an effective deterrent to noncompliant dischargers.
   1. True
   2. False

2. The EPA's pretreatment program requires that POTW agencies notify and help industrial companies to interpret and implement the federal regulations.
   1. True
   2. False

3. It is much easier to accomplish regulatory goals with industry through enforcement actions than it is through cooperation with industrial personnel.
   1. True
   2. False

4. An industrial waste permit should include the means for disposal of runoff rainwater to the sanitary sewer.
   1. True
   2. False

5. Proper documentation of all samples collected must be maintained to meet legal requirements.
   1. True
   2. False

6. Mass-based limits and concentration-based limits are the same.
   1. True
   2. False

7. EPA gives authority to local officials to implement and enforce the national pretreatment program.
   1. True
   2. False

8. All categorical and noncategorical IUs must notify the POTW immediately of all discharges that could cause problems to the POTW.
   1. True
   2. False

9. IUs are allowed to increase the use of process water to dilute a discharge to achieve compliance with a Pretreatment Standard or Requirement.
   1. True
   2. False

10. Criminal penalties may be imposed for any violation of the Wastewater Ordinance as an infraction or misdemeanor.
    1. True
    2. False

11. One inspector can sample industries at sampling points that are located in public streets or other high-traffic (congested) areas.
    1. True
    2. False

12. Industrial waste inspectors frequently enter confined spaces.
    1. True
    2. False

13. Low voltage is less dangerous than high voltage.
    1. True
    2. False

14. The major portion of an inspector's sampling work load involves sampling industrial discharges.
    1. True
    2. False

15. All samples should be analyzed as soon as possible after they are collected.
    1. True
    2. False

16. Composite samples may be appropriate for local limits.
    1. True
    2. False

17. The best way to determine an industrial discharger's impact on wastewater facilities is to directly measure each industry's discharge flows.
    1. True
    2. False

18. Concentration-based waste limits require flow measurements.
    1. True
    2. False

19. Flow is the amount of water going past a particular reference point over a certain period of time.
    1. True
    2. False

20. The depth of the water in the flume or weir can be measured at any location in the channel.
    1. True
    2. False

21. Chemical containment areas must have drains.
    1. True
    2. False

22. The pretreatment inspector's job is to prevent slug discharges.
    1. True
    2. False

23. The disposal of POTW sludge is affected by industrial discharges.
    1. True
    2. False

24. Federal pretreatment standards require pretreatment of wastestreams discharged to POTWs.
    1. True
    2. False

25. Pollution prevention focuses on activities which prevent the generation of wastes.
    1. True
    2. False

26. Material or chemical distribution should be by manual pouring of buckets or barrels rather than by pumps and pipes.
    1. True
    2. False

27. Evaporation is a proven, well-developed, inexpensive industrial process.
    1. True
    2. False

28. Anaerobic digestion is a biological process in which microorganisms metabolize organic matter in the presence of free or dissolved oxygen.
    1. True
    2. False

29. Inspectors must review spill containment facilities constantly throughout the inspection.
    1. True
    2. False

30. Evidence of excessive raw material usage may indicate that leaks of process chemicals are occurring.
    1. True
    2. False

31. At a centralized waste treatment facility, storage must be provided for impounded rainwater from the pretreatment areas and off-loading areas.
    1. True
    2. False

32. The most common example of a sewer condition that warrants enacting an emergency response plan is the discovery of gasoline or other flammable or explosive liquid or vapor in the sewer.
    1. True
    2. False

33. The use of internal combustion engines on sewer ventilation blowers is a safe procedure.
    1. True
    2. False

34. Spills or slug loads to the POTW should be documented in the POTW's pretreatment file for the industry involved.
    1. True
    2. False

Best Answer (Select only the closest or best answer.)

1. A representative sample is
    1. Collected during minimum flows.
    2. Obtained when industrial representatives are present.
    3. Similar to the larger body of wastestream being sampled.
    4. Transported to a laboratory for analysis.

2. What is the main focus of pollution prevention?
    1. Disposing of the waste generated
    2. Minimizing the amount of waste generated
    3. Recycling the waste generated
    4. Treating the waste generated

3. What do industrial wastewater discharge permits do? Grant permission to an industrial user to discharge its wastewater into the
    1. Centralized waste treatment facility.
    2. Environment.
    3. Pretreatment facility.
    4. Wastewater collection system.

4. What is the best method to determine what industrial dischargers are placing in the collection system?
    1. Field test kits
    2. Laboratory chemical analyses
    3. Results of industrial self-monitoring
    4. Visual observations

5. What does a spill containment program require?
    1. Channels carrying leaks to the storm sewer
    2. Drains within a berm connected directly to the pretreatment plant
    3. Sumps within the berm for pumping leaks to the sanitary sewer
    4. Tanks located within a berm that will prevent leaks from reaching a sewer

6. EPA recognizes what two types of dischargers in its efforts to control the flow of pollutants into water?

   1. Direct and indirect
   2. Municipal and industrial
   3. Raw and treated
   4. Sanitary and storm

7. Which pollutant is a conventional pollutant?

   1. Ammonia
   2. Chromium
   3. Cyanide
   4. Oil and grease

8. What are receiving water standards based on?

   1. Best available technology economically achievable
   2. Categorical standards
   3. Technology-based standards
   4. Tolerance of stream to pollutants

9. Why should inspectors be on close working terms with sewer maintenance crews? So that

   1. Inspectors and maintenance crews can work together.
   2. Inspectors become familiar with sewer maintenance methods.
   3. Potential problems can be detected and corrected.
   4. Potential sources of inflow and infiltration will be recognized by inspectors.

10. What is the purpose of an unannounced industrial inspection?

   1. To check on effluent quality
   2. To meet with POTW's normal contact person
   3. To review paper work
   4. To test response of industry

11. What happens when a clarifier's capacity to hold solids or oil is exceeded?

   1. Depth measurements must be collected.
   2. Materials must be pumped out.
   3. Materials must be washed out in effluent.
   4. Wastestream must be diverted to another clarifier.

12. Why are floor drains *NOT* allowed in chemical storage areas?

   1. Because they are not necessary
   2. Due to risk of spills
   3. To avoid plugging drains
   4. To prevent hydraulic overloads

13. What should you do if you encounter an unsafe condition or situation?

   1. Correct the condition or situation.
   2. Leave the area and wait for condition or situation to be corrected.
   3. Prevent others from encountering condition or situation.
   4. Remove yourself from the area and warn others.

14. What is the most common toxic gas that is encountered during sampling work?

   1. Carbon monoxide
   2. Hydrogen cyanide
   3. Hydrogen sulfide
   4. Methane

15. What are composite samples collected to measure?

   1. Average amount of pollutants
   2. Maximum concentration of pollutants
   3. Pollutant discharge at specific time
   4. Pollutants of specific concern

16. What is the main goal of sample storage procedures?

   1. To accumulate a large enough sample for analysis
   2. To allow samples to accumulate before analysis
   3. To efficiently use analyst's time
   4. To maintain integrity of sample until analysis

17. What is meant by chain of custody?

   1. A record of criminal action against an industrial discharger
   2. A record of each agency action leading to a notice of violation
   3. A record of every person who has access to a sample
   4. A record of every sample submitted to the laboratory

18. What are the characteristics of a properly operating flume?

   1. Distinct and easy to observe
   2. Fluctuating and difficult to measure
   3. Submerged hydraulic jump
   4. Uniform and flat

19. How is the flow depth measured in a Palmer-Bowlus flume?

   1. Channel bottom to water surface at measuring point
   2. Channel bottom to water surface at throat
   3. Floor of flume to water surface at measuring point
   4. Floor of flume to water surface at throat

20. What is the purpose of a bubbler?

   1. Aerate flowmeter channel
   2. Break down foam on water surface
   3. Measure depth of water
   4. Mix water in stilling well

21. How do electromagnetic flowmeters measure flows? By measuring the _____ created by the movement of water.

   1. Area
   2. Depth
   3. Pressure differential
   4. Voltage

22. What are compatible pollutants?

   1. Pollutants normally found in a concentrated wastestream
   2. Pollutants normally found in a diluted wastestream
   3. Pollutants normally removed by a physical-chemical treatment system
   4. Pollutants normally removed by a POTW treatment system

23. What is the most heavily used toxic pollutant in leather tanning?

   1. Chromium
   2. Copper
   3. Cyanide
   4. Phenol

24. Which acid is an organic acid?

    1. Acetic
    2. Nitric
    3. Phosphoric
    4. Sulfuric

25. Where are the sources of pollutants in storm water run-off?

    1. Drag out dripping on the floor
    2. Products which have spilled on the grounds
    3. Rinse waters
    4. Scrubber wastewater

26. How should the release of spilled chemicals in storage areas be prevented?

    1. Install some type of containment around the storage areas.
    2. Place chemicals in leakproof containers.
    3. Purchase chemicals that do not flow.
    4. Use floor drains to prevent release.

27. Equalization basins have the capability to

    1. Allow physical treatment of wastes.
    2. Develop biological treatment cultures.
    3. Increase the effectiveness of downstream treatment equipment.
    4. Provide opportunity for chemical treatment.

28. Why does industry use packaged inclined plate (lamella) settlers?

    1. Cost more than traditional circular tank clarifier
    2. Flocculation section near inlet is not needed
    3. Plates provide same surface area in smaller vessel than clarifier
    4. Plates provide surface area for biological treatment

29. Where are the residues remaining after incineration disposed of?

    1. In a land disposal facility
    2. In a sludge digester
    3. In the POTW's sewer
    4. In the receiving waters

30. What should be done before entering an industrial facility for a pretreatment inspection?

    1. Collect wastewater samples
    2. Notify receptionist of your intent to inspect
    3. Phone industrial contact for an appointment
    4. Park in an appropriate place outside the facility

31. What should an inspector do with regard to violations discovered during an inspection?

    1. Discuss violations with industry before leaving site
    2. Help industry fix violations before leaving site
    3. Make a note of violations for future reference
    4. Notify enforcement authority of violations

32. Before entering an industrial facility for an inspection, why should you first visit the sampling points and collect wastewater samples?

    1. To allow collection of an unbiased wastewater sample
    2. To ensure following proper chain of custody procedures
    3. To minimize time required of the industrial contact
    4. To present the industrial contact with the results of the analysis of the sample

33. What types of wastes are treated by centralized waste treatment facilities?

    1. Agricultural wastes
    2. Oily wastes
    3. Spent activated carbon
    4. Waste activated sludge

34. How should an inspector be able to activate emergency response procedures?

    1. By calling the police and fire departments
    2. By contacting local TV and radio stations
    3. By making one call to a person in the home office
    4. By phoning the collection system and treatment plant superintendents

35. Why should an inspector keep notes during an emergency? Notes will be useful in preparing

    1. A briefing for public officials.
    2. A case for court if enforcement action is initiated.
    3. A meeting for news media.
    4. An informational pamphlet for the public.

36. What type of pump must be used when drawing liquid out of a sewer?

    1. Centrifugal pumps
    2. Explosion-proof pumps
    3. Submersible pumps
    4. Trash pumps

37. How should an inspector respond when an unknown material that is neither flammable nor asphyxiating is found in a collection system?

    1. Assume the unknown material was an accidental spill.
    2. Disregard unknown material until next time.
    3. Try to trace unknown material back to industrial source.
    4. Wait for unknown material to arrive at treatment plant.

Multiple Choice (Select all correct answers.)

1. Conventional pollutants include

    1. Biochemical oxygen demand (BOD)
    2. Cyanide
    3. Hexavalent chromium
    4. Oil and grease
    5. Suspended solids (SS)

2. The objectives of the General Pretreatment Regulations are to

    1. Improve opportunities to reclaim and recycle sludge.
    2. Prevent interference with operation of POTW.
    3. Prevent interference with use or disposal of municipal sludge.
    4. Prevent introduction of pollutants which pass through treatment works.
    5. Reduce the health and environmental risks from pollution.

3. What damages could be caused to a POTW agency's collection system by industrial wastes?

    1. Clogging the sewers
    2. Corrosion of sewer
    3. Excessive inflow and infiltration (I/I)
    4. Explosions from flammable wastes
    5. Lift station failures due to lack of maintenance

4. What factors determine the staffing needs of an industrial waste program?

   1. Amount of industrial waste flow
   2. Capacity of POTW
   3. Enforcement posture of POTW
   4. Magnitude of regulations
   5. Size of agency

5. What are the basic elements of a pretreatment program?

   1. Frequency of inspection
   2. Inspection of companies
   3. Monitoring of discharges
   4. Permission to discharge
   5. Pollution prevention

6. What factors should be considered when determining the level of enforcement response?

   1. Consistency
   2. Duration of noncompliance
   3. Equity
   4. Fairness
   5. Integrity of program

7. What types of pollutants or conditions are regulated by the General Pretreatment Regulations? Those that

   1. Contaminate sludge.
   2. Interfere with treatment.
   3. Pass through treatment system.
   4. Overload treatment processes.
   5. Wash out treat reactors.

8. What are the basic elements of a toxic management plan?

   1. Identification of toxic organics used
   2. Location of dump site
   3. Method of disposal
   4. Quantity of each toxic organic used
   5. Use of each toxic organic

9. What factors should be considered when assessing penalties?

   1. Ability of IU to pay penalty
   2. Large enough to deter future noncompliance
   3. Penalties should be calculated on a logical basis
   4. Penalties should be uniform
   5. Recovery of economic benefit for noncompliance

10. What circumstances may require an investigation of sewer lines?

   1. Emergency response to explosive conditions
   2. Foaming in a sewer line
   3. Installation of a new force main
   4. Scaling or corrosion in a sewer line
   5. Stoppages that cause flow onto streets

11. What are the options available to an inspector if a request for entry to perform an inspection is denied?

   1. Call supervisor for direction
   2. Exercise authority to enter facility without permission
   3. Issue a citation for denial of access
   4. Leave and immediately collect samples for analysis
   5. Threaten company with criminal prosecution

12. What are the levels of inspection necessary for a complete inspection?

   1. Effluent treatment equipment
   2. General tour
   3. Influent monitoring for wastewater control equipment
   4. In-plant wastewater control equipment
   5. Outfall

13. What procedures should be followed when collecting industrial effluent samples?

   1. Analyze sample when convenient
   2. Collect sample from surface of wastestream
   3. Follow chain of custody procedures
   4. Preserve samples properly
   5. Use proper sampling techniques

14. What safety equipment must be in each inspector's vehicle?

   1. Acid neutralizers
   2. Fire extinguisher
   3. First aid kit
   4. Spill control pillows
   5. Stop signs

15. What factors influence traffic control strategies?

   1. Alertness of drivers
   2. Congestion
   3. Local regulations
   4. Time of day
   5. Traffic speed

16. What are possible sources of flammable or explosive gases, mists and vapors in sewers?

   1. Accidental discharges
   2. Leaking underground storage tanks
   3. Refinery emissions
   4. Spills
   5. Storm water runoff

17. When an inspector enters a confined space, what protective clothing and equipment are needed?

   1. Eye protection
   2. Foot protection
   3. Full-body, parachute-type harness
   4. Gloves
   5. Hard hat

18. What are the major categories of hazardous materials encountered during sampling work?

   1. Corrosive materials
   2. Inorganic compounds
   3. Poisonous or toxic gases
   4. Solvents and flammable materials
   5. Synergistic reductions

19. What important information is found on a Material Safety Data Sheet (MSDS)?

   1. Cost of chemical
   2. First aid measures
   3. Hazardous ingredients
   4. Personal protection
   5. Spill, leak, and disposal procedures

20. Why are samples collected?

    1. Check compliance with local standards
    2. Determine strength to accurately assess discharge/use fees
    3. Search for prohibited wastes
    4. Verify industry monitoring points
    5. Verify self-monitoring data

21. What factors must be considered when selecting a sample container?

    1. Material
    2. Packing
    3. Shape
    4. Size
    5. Temperature

22. What methods are used to preserve samples?

    1. Acids
    2. Bacteria
    3. Bases
    4. Refrigeration
    5. Storage

23. Field tests are the *ONLY* accurate method for measuring which pollutants?

    1. Chlorine residual
    2. Cyanide
    3. pH
    4. Sulfite
    5. Temperature

24. What information must be recorded for each sample collected?

    1. Emergency response phone number
    2. Field observations
    3. Name of industry contact person
    4. Time sample was collected
    5. Who collected sample

25. What are the typical units used to measure flow?

    1. CFS
    2. FPS
    3. GPM
    4. MGD
    5. mg/L

26. What are the advantages of a stilling well?

    1. Eliminates effects of foam on water surface
    2. Eliminates effects of rags
    3. Frequent maintenance is necessary to keep well clean
    4. No maintenance is required to clean connection to flume
    5. Not affected by wave action in a flume or weir

27. What factors can cause errors in ultrasonic flowmeter readings?

    1. Dirty face of transducer
    2. Excessive wave action on water surface
    3. Foam layer on water surface
    4. Improper generation and receipt of acoustic pulses
    5. Location of transducer over stilling well

28. How often should flow-proportioned automatic samplers be programmed to collect a sample? At least once every _____ hour(s) over a 24-hour period.

    1. 1
    2. 2
    3. 3
    4. 4
    5. 6

29. Why does an inspector need to understand the sources and quantities of industrial wastewaters? To _____ problems caused by industrial discharges.

    1. Create
    2. Define
    3. Identify
    4. Prepare
    5. Solve

30. What waste characteristics must be considered when determining the acceptability of an industrial waste discharge to a sewer?

    1. Concentration
    2. Odor
    3. pH
    4. Temperature
    5. Toxicity

31. What are sources of dilute solutions that may be discharged to the pretreatment system or POTW?

    1. Plating bath rinses
    2. Reject products
    3. Spent plating baths
    4. Static drag out solutions
    5. Storm water runoff

32. What types of discharge constituents may cause explosions in sewers?

    1. Acetic acid
    2. Gasoline
    3. Hexane
    4. Hexavalent chromium
    5. Methyl ethyl ketone

33. What types of wastes are found in food industry effluents?

    1. BOD
    2. Cyanide
    3. Nitrate
    4. PCBs
    5. SS

34. What types of flows are important when establishing treatment and discharge objectives?

    1. Average flow
    2. Minimum flow
    3. Peak flow
    4. Total daily flow
    5. Weekend flow

35. What information should a pretreatment inspector be aware of with regard to types of industrial production processes?

    1. Availability of raw material
    2. Chemical constituents in the wastewater
    3. Potential wastewater sources
    4. Techniques for wastewater reduction
    5. Utility costs

36. What are the benefits of source control?

    1. Cost savings
    2. Increased size of pretreatment facility
    3. Protection of public
    4. Reduced chemical costs
    5. Reduced liability

37. What are the primary wastewater treatment technologies used by industry?

    1. Biological
    2. Chemical
    3. Disinfection
    4. Land
    5. Physical

38. Which of the following processes are chemical treatment processes?

    1. Filtration
    2. Neutralization
    3. Oxidation/reduction
    4. Precipitation
    5. Stripping

39. What are the most common and most dangerous hazards that pretreatment inspectors will encounter?

    1. Atmospheric hazards
    2. Mental hazards
    3. Physical hazards
    4. Technical hazards
    5. Traffic hazards

40. What information should an inspector collect to describe an industrial discharge?

    1. Amount
    2. Chemical nature
    3. Destination
    4. Frequency
    5. Operators on duty

41. Which of the following activities should be performed during an on-site inspection?

    1. Check the accidental spill prevention control plan of the industry.
    2. Conduct a peripheral examination of the industrial user.
    3. Observe the physical characteristics of the wastestream in the sewer.
    4. Require the development of a compliance schedule.
    5. Review existing files for available information about the industry.

42. Oil and grease may appear in wastewater in what forms of oil?

    1. Cracked oil
    2. Emulsified oil
    3. Free oil
    4. Sequestered oil
    5. Soluble oil

43. What water quality indicators can be significantly changed if samples are not properly chilled and sealed?

    1. BOD
    2. COD
    3. Dissolved sulfide
    4. Hardness
    5. pH

44. What industrial discharges could create emergencies?

    1. Concentrated acids
    2. Concentrated bases
    3. Large volumes of toxic materials
    4. Large volumes of water
    5. Poisonous gases

45. What are the most likely sources of catastrophic discharges?

    1. Canneries
    2. Chemical plants
    3. Dairies
    4. Fuel transfer stations
    5. Oil refineries

46. How can the source of a spill be determined?

    1. Identifying the material.
    2. Phoning industries and asking if spill occurred.
    3. Responding to odor complaints.
    4. Tracing material upstream.
    5. Waiting for industry to notify agency.

47. What items should an inspector examine at high-risk industrial facilities with regard to emergency response?

    1. Chemical ordering plan
    2. Emergency response plan
    3. Future expansion plan
    4. Spare parts inventory plan
    5. Spill prevention plan

SUGGESTED ANSWERS FOR FINAL EXAMINATION

Short Answer

1. In general terms, a pretreatment facility inspector conducts inspections of industrial pretreatment facilities to ensure protection of wastewater collection and treatment facilities and personnel, and to ensure protection of the environment through compliance with regulatory standards by industrial dischargers.

2. The effectiveness of inspectors in carrying out their assignments depends on their knowledge as well as their communication skills. Inspectors must establish their own credibility with industrial representatives and let them know that they are competent and mean business.

3. Improper discharges of industrial wastes can cause corrosion, explosions, treatment process upsets, over-capacity effects caused by slug loadings and misuse of the treatment processes and facilities.

4. A major concern of industry with respect to industrial waste pretreatment inspection programs is that all companies meet the same environmental standards and face the same costs for pollution control regulation.

5. Companies should pay their fair share of the costs of operating the POTW system (collection and treatment) as well as their proportional costs of the POTW's industrial waste pretreatment program. Companies should also pay their appropriate share of the capital costs of the collection and treatment facilities.

6. Monitoring by the POTW is essential to verify self-monitoring information furnished by the industrial dischargers. Agency monitoring is also required to locate the source of treatment plant upsets and sewer system problems caused by industrial waste discharges. Monitoring may also be needed to verify or determine strength data for industrial sewer service charges.

7. Important items that should be given particularly careful attention during a permit review process include use of toxic or hazardous materials in the industrial process and the methods for disposal of such materials. Spill containment procedures must also be reviewed.

8. Items a POTW agency should consider when purchasing a computer system include expandability, service, relative cost, and the ability of the system to meet the needs of the program now and in the future.

9. Proper legal authority is important because it is a requirement for EPA approval of a pretreatment program. Without adequate authority, the inspector's position is reduced to strictly an advisory role. For effective control, authority must be backed by effective standards and the willingness to enforce them.

10. The difference between a direct and indirect discharger is that direct refers to a direct discharge to surface waters such as streams, lakes or the oceans and indirect refers to a discharge to a POTW.

11. Categorical Pretreatment Standards are industry-specific, technology-based standards limiting the discharge of wastewaters from industrial facilities.

12. Categorical Pretreatment Standards can be modified by:

    1. Variance for fundamentally different factors (FDF),
    2. Net gross adjustments, and
    3. Removal credits.

13. All POTWs with approved pretreatment programs must inspect and sample Significant Industrial Users (SIUs) at least once a year. The frequency of inspection for a particular company depends on the agency's personnel resources that have been allotted to on-site inspections. Some agencies have a fairly rigid prescribed frequency that is tied to a fee schedule; other agencies leave day-to-day scheduling to the individual field inspectors. Inspection schedules should never become so standardized that dischargers can predict and therefore prepare for an inspector's visit.

14. Major items that an inspector must consider before entering an industry to conduct an inspection include:

    1. Legal authority,
    2. Equipment,
    3. Documents, or reference materials,
    4. Essential information regarding industry and discharge,
    5. Previous inspection reports and sample test results, and
    6. Reporting requirements for dischargers.

15. An inspector should always visit an outfall whenever visiting an industry because it would be embarrassing to later discover that there had been a major discharge violation during the visit and the inspector had not even walked out to look at the outfall.

16. When inspecting a spill containment facility, examine the structural integrity of dikes and look for pool bonding between berms and floors. Tanks containing cyanide should not be contained in a common area with tanks containing acid. Pumps that are activated automatically by liquid level should not be installed inside spill containment areas. Check for the potential to bypass the spill containment with valves, leaks or overflows.

17. A supervisor can set an example for inspectors on how to comply with all safety requirements and procedures by becoming thoroughly familiar with the job duties and requirements of each inspector. This is usually accomplished by performing a Job Safety Analysis (JSA).

18. Before the start of an inspection, the inspector should ask the industrial contact if there are any special safety precautions that must be taken. During the inspection the inspector must continually be aware of the surroundings and the part of the facility that is being inspected to ensure that no breach of the safety requirements or rules of the firm occurs. The inspector must maintain an alert state of mind while going through the facility and be ready to react to any situation that may occur.

19. Before starting work in a street, the inspector must establish a plan to thoroughly protect the work area and to provide proper advance warning, signing and guidance to oncoming traffic. This can be accomplished by using traffic control devices such as cones, barricades, flashers, WORK AREA signs, flaggers and the vehicle itself to control the traffic and to provide advance warning to oncoming motorists.

20. An inspector using a gas detection device must be trained in its use, understand the principles involved in its operation, and be able to interpret the results and keep records of readings obtained.

21. An inspector must be especially aware of the potential for electrical shock when working in areas where electrical wiring or electrically powered machinery is present. This includes areas where electrical panels with meters or recorders (pH, flow) that are normally looked at during an inspection are located.

22. Hazards that may be encountered when inspecting electroplating shops and metal plating facilities include heat, reactive and corrosive chemicals, toxic and irritating fumes, wet and slippery floors, electrical plating bars and wiring and crowded conditions.

23. An agency setting up a sampling program decides whether to use an in-house or an outside laboratory based on the costs of testing and the number and types of tests they will be requesting.

24. Sample containers must be clean of all traces of the pollutants to be analyzed. Otherwise, you won't know whether the pollutants found by the laboratory were present in the wastestream you sampled or were already present in the container.

25. The decision to use disposable or reusable sample containers is strictly economic. The cost of buying disposable bottles and running blanks must be compared to the cost of washing bottles and replacing broken bottles.

26. POTW pretreatment program regulations require the POTW to verify industrial users' compliance by some means independent of self-monitoring results. From a law-and-order position, monitoring by a regulatory agency is often desirable because it gives the agency full control over the method of collecting the sample and the analysis. If a violation is found, the agency can preserve the remaining sample as evidence.

27. Samples must be properly stored to maintain the integrity of the sample. The sample must be protected from environments that would tend to change the nature of the sample and from people who might tamper with it. Some samples may represent a danger to people who might come in contact with the sample. The sample storage facility must provide some degree of security, and it must provide conditions that comply with preservation requirements.

28. The purpose of a quality assurance/quality control (QA/QC) program is to maintain a specified level of quality in the measurement, documentation, and interpretation of sampling data.

29. Influent flow measurements may be used to determine discharge flow rates by considering all influent sources of water and subtracting all water losses, such as evaporative losses and water used for irrigation or water used in making products.

30. A Palmer-Bowlus flume could be in a submerged condition when a discharge pipe is not able to carry the flow, when there is an improper slope of the downstream pipe, due to debris in the pipe, or from flow conditions in a sewer line farther downstream that cause a backup of water into the flume.

31. Prevent the end of the bubbler tubing from becoming clogged with solids by frequent maintenance or by installing an automatic purging system to blow out accumulated material in the tube.

32. A general limitation of a closed-pipe flowmeter in the measurement of industrial wastewater is the difficulty of determining if the meter is clean. The materials present in some wastewaters can coat, clog or corrode a meter in an undesirably short period of time.

33. Pretreatment facility inspectors need to understand the sources and quantities of industrial wastes so they can identify, define and solve problems these wastes cause to the pretreatment system, the Publicly Owned Treatment Works (POTW) conveyance system and treatment plant, and the disposal of treated effluent and sludge.

34. Types or sources of intermittent discharges of wastewater include activities at the beginning and ending of a manufacturing process, cleanup of the equipment, a spill, replacement of spent solution and disposal of a reject product.

35. Mass emission rate standards are preferred because they recognize that with more production and water, the mass of pollutant will also increase. This approach eliminates dilution of the pollutant to meet concentration limitations.

36. Water is used in the manufacture of pulp, paper and paperboard for wood preparation, pulping, bleaching, and papermaking. Water can be used as a medium of transport, a cleaning agent and a solvent or mixer.

37. A slug loading on a POTW from an industrial discharge can be mitigated by the skill of the inspectors and POTW operators and the flexibility of the treatment processes. Equalization basins, flow control structures, chemical treatment points along the collection system, aeration basins configured to allow complete mix or step aeration, and pumping and piping to allow parallel or series operation of treatment processes all provide flexibility in the operation of the POTW treatment system.

38. Industrial discharges which, alone or in conjunction with discharges from other sources, pass through to the POTW's facilities to navigable waters and cause a violation of the discharge permit are considered pass-through discharges.

39. Pass-through of compatible and noncompatible pollutants can occur when the POTW treatment system is under stress from hydraulic or compatible waste overloads or shock loadings of toxic pollutants. When the pollutant removal efficiency decreases, the constituents from industrial discharges are found in the effluent.

40. Factors that should be considered when estimating existing and future flows from a discharger include existing wastewater sources, associated flow, and the collection systems for the plant. In addition to existing wastewater flow, future wastewater flows should be projected based on anticipated or scheduled plant expansion, production increases, and water conservation and reuse efforts.

41. The degree of pretreatment required for industrial waste is primarily established so as to prevent treatment process upsets and to avoid the pass-through of pollutants in the POTW effluent.

42. Wastestreams require neutralization or pH adjustment to comply with waste discharge limits, for optimal performance of biological treatment processes at a POTW, and to prevent corrosion of sewers and release of hydrogen sulfide gas from sulfide-containing wastewaters.

43. Biological waste treatment processes can be upset by toxic compounds in wastestreams. High concentrations of inorganics and low concentrations of toxics can severely inhibit biological decomposition activity. Temperature extremes can also inhibit biological treatment processes by killing the organisms used in the treatment process or lowering their activity to unacceptable levels.

44. The major difference between sludge produced from biological treatment processes and physical-chemical processes is that biological processes produce a sludge that is composed of organic matter which will decompose and become offensive. Sludge generated from physical-chemical treatment processes contains mostly nonbiodegradable matter.

45. Source control is important for an industry because an effective program produces cost savings from the reduced chemical and water use, reduction of disposal costs and liability, and reduction in size of the wastewater pretreatment system.

46. The main reason for conducting pretreatment facility inspections and monitoring is to ensure compliance with all applicable pretreatment regulations.

47. Current data on industrial dischargers is necessary in a pretreatment program to identify sources of problems and to provide a foundation for developing or amending local discharge limits for industrial users.

48. Information gathered during an industrial inspection regarding waste residuals (solids or floatables) should include how the waste residuals are handled, stored and/or disposed of.

49. Existing information regarding an industry to be inspected that should be taken along during an inspection and verified for accuracy include plant layouts, process flow diagrams, compliance schedule (if applicable), and wastewater analytical data.

50. The wastewaters discharged from a papermaking process tend to have a low pH because the optimum pH range for paper forming is between 4.0 and 4.7. Some paper mills use hydrochloric acid for cleaning the machines; this tends to lower pH as well.

51. Sewer maps can be used to assist with emergency responses if they indicate the locations of high-risk facilities with the route of flow to the treatment plant clearly marked. Maps should reflect the locations of relief lines, flow control structures, stoplogs, cross connections or changes implemented by the discharger. The capabilities to divert flow throughout the collection system should be noted on the maps.

52. A single contact person should be designated to deal with the media so that misinformation is not released and response personnel will not be distracted.

53. Spill material can be removed from a sewer by using vacuum trucks that can extract the material from the sewer. Wet wells can be used to trap floating spills so they can be removed.

54. POTWs need a written spill notification procedure in order to respond quickly and efficiently to an emergency situation of a material spilled into a sanitary sewer.

True-False

1. True — Unannounced inspections are a deterrent to noncompliant dischargers.
2. True — EPA's program requires POTWs to notify and help industry.
3. False — Cooperation with industrial personnel is easier than enforcement actions.
4. False — Permit should include means for disposal of rainwater other than to sanitary sewer.
5. True — All samples must be properly documented to meet legal requirements.
6. False — Mass-based limits and concentration-based limits are *NOT* the same.
7. False — EPA gives authority to the state to implement and enforce the national pretreatment program.
8. True — All IUs must notify the POTW immediately of all problem discharges.
9. False — IUs are *NOT* allowed to use dilution to achieve compliance.
10. True — Criminal penalties may be imposed for any violation of Ordinance.
11. False — A minimum of two people should work in congested areas.
12. False — Entry into confined spaces is not recommended unless absolutely necessary.
13. False — Current flow through the body causes problems and injuries.
14. True — The major portion of an inspector's sampling work load involves sampling industrial discharges.
15. True — Analyze all samples as soon as possible after collection.
16. False — Composite samples may be appropriate for Categorical Standards.
17. True — To determine industry's impact, measure each industry's discharge.
18. False — Mass-based waste limits require flow measurements.

19. True   Flow is amount of water going past a point over time.

20. False   Depth of water must be measured at a particular location.

21. False   Chemical containment areas must *NOT* have drains.

22. True   The pretreatment inspector's job is to prevent slug discharges.

23. True   Disposal of POTW sludge is affected by industrial discharges.

24. True   Pretreatment standards require pretreatment of discharges to POTWs.

25. True   Pollution prevention focuses on prevention of the generation of wastes.

26. False   Material or chemical distribution should be by pumps and pipes.

27. False   Evaporation is a proven but expensive industrial process.

28. False   Anaerobic digestion occurs in the absence of free or dissolved oxygen.

29. True   Review spill containment facilities throughout inspection.

30. True   Excessive raw material usage may indicate leaks of process chemicals.

31. True   Storage must be provided for impounded rainwater from pretreatment areas.

32. True   Most common emergency sewer condition is gasoline in sewer.

33. False   Internal combustion engines may be a source of a spark.

34. True   Spills or slug loads should be documented for the industry involved.

Best Answer

1. 3   A representative sample is similar to the larger body of wastestream being sampled.

2. 2   The main focus of pollution prevention is minimizing the amount of waste generated.

3. 4   Permits allow discharge to wastewater collection system.

4. 2   Laboratory chemical analyses are best to determine industrial discharges.

5. 4   Tanks must be within berm that prevents leaks from reaching a sewer.

6. 1   Direct and indirect are two types of dischargers recognized by EPA.

7. 4   Oil and grease are conventional pollutants.

8. 4   Receiving water standards are based on tolerance of stream to pollutants.

9. 3   Inspectors work with sewer maintenance crews to detect and correct potential problems.

10. 1   The purpose of an unannounced industrial inspection is to check effluent quality.

11. 2   When a clarifier's capacity to hold solids or oil is exceeded, materials must be pumped out.

12. 2   Floor drains are *NOT* allowed in chemical storage areas due to risk of spills.

13. 4   Remove yourself from the area and warn others.

14. 3   The most common toxic gas encountered during sampling work is hydrogen sulfide.

15. 1   Composite samples measure average amount of pollutants.

16. 4   Storage is to maintain integrity of sample until analysis.

17. 3   Chain of custody is a record of every person who has access to a sample.

18. 1   Characteristics of a flume are distinct and easy to observe.

19. 3   Measure floor of flume to water surface at measuring point.

20. 3   A bubbler measures depth of water.

21. 4   Electromagnetic flowmeters measure flows by measuring voltage.

22. 4   Compatible pollutants are those normally removed by a POTW treatment system.

23. 1   Chromium is the most heavily used toxic pollutant in leather tanning.

24. 1   Acetic acid is an organic acid.

25. 2   Products spilled on the grounds are the sources of pollutants in storm water runoff.

26. 1   Prevent release of chemicals by installing some type of containment.

27. 3   Equalization basins increase the effectiveness of downstream treatment equipment.

28. 3   Lamella settlers require less space than traditional clarifier with same area.

29. 1   Residues remaining after incineration are disposed of in a land disposal facility.

30. 1   Collect wastewater samples before entering an industrial facility.

31. 1   Inspector should discuss violations with industry before leaving site.

32. 1   Collect samples first to obtain unbiased samples.

33. 2   Oily wastes are treated by centralized waste treatment facilities.

34. 3   Inspector should activate procedures by calling one person in home office.

35. 2   Notes will be useful for court if enforcement action is initiated.

36. 2   Explosion-proof pumps must be used when drawing liquid out of a sewer.

37. 3   Try to trace unknown material back to industrial source.

Multiple Choice

1. 1, 4, 5   Conventional pollutants include biochemical oxygen demand (BOD), oil and grease and suspended solids (SS).

2. 1, 2, 3, 4, 5   The objectives of the General Pretreatment Regulations are to improve opportunities to reclaim and recycle sludge, prevent interference with operation of POTW, prevent interference with use or disposal of municipal sludge, prevent introduction of pollutants which pass through treatment works, and reduce the health and environmental risks from pollution.

3. 1, 2, 4   Industrial wastes could cause clogging or corrosion of the POTW's sewers or explosions from flammable wastes.

4. 1, 5   Amount of industrial waste flow and size of agency determine staff.

5. 1, 2, 3, 4, 5   Basic elements of a pretreatment program include frequency of inspection, inspection of companies, monitoring of discharges, permission to discharge, and pollution prevention.

6. 1, 2, 3, 4, 5   Factors that should be considered when determining the level of enforcement response include consistency, duration of noncompliance, equity, fairness, and integrity of the program.

7. 1, 2, 3   Pollutants or conditions that contaminate sludge, interfere with treatment or pass through the treatment system are regulated by the General Pretreatment Regulations.

8. 1, 3, 4, 5   Basic elements of a toxic management plan include identification of toxic organics used, method of disposal, quantity of each toxic organic used, and the use of each toxic organic.

9. 2, 3, 4, 5   Factors that should be considered when assessing penalties include making penalties large enough to deter future noncompliance, calculating penalties on a logical basis, making penalties uniform, and ensuring recovery of the economic benefit for noncompliance.

10. 1, 2, 4, 5   Circumstances that may require an investigation of sewer lines include emergency response to explosive conditions, foaming in a sewer line, scaling or corrosion in a sewer line, and stoppages that cause flow onto streets.

11. 1, 2, 3, 4   If a request for entry to perform an inspection is denied, the operator could call a supervisor for direction, exercise authority to enter the facility without permission, issue a citation for denial of access, or leave and immediately collect samples for analysis.

12. 1, 2, 4, 5   A complete inspection includes a general tour of the facility and inspection of effluent treatment equipment, in-plant wastewater control equipment, and the outfall.

13. 3, 4, 5   When collecting industrial effluent samples, follow chain of custody procedures, preserve samples properly, and use proper sampling techniques.

14. 2, 3, 4   A fire extinguisher, first aid kit and spill control pillows must be in each inspector's vehicle.

15. 2, 3, 4, 5   Factors that influence traffic control strategies include congestion, local regulations, time of day, and traffic speed.

16. 1, 2, 4   Accidental discharges, leaking underground storage tanks and spills are possible sources of flammable or explosive gases, mists and vapors in sewers.

17. 1, 2, 3, 4, 5   An inspector entering a confined space should wear eye protection, foot protection, a full-body, parachute-type harness, gloves, and a hard hat.

18. 1, 2, 3, 4   Major categories of hazardous materials encountered during sampling work include corrosive materials, inorganic compounds, poisonous or toxic gases, and solvents and flammable materials.

19. 2, 3, 4, 5   A Material Safety Data Sheet (MSDS) for a particular chemical lists first aid measures, hazardous ingredients, personal protection needed, and spill, leak and disposal procedures.

20. 1, 2, 3, 4, 5   Samples are collected to check compliance with local standards, determine strength to accurately assess discharge/use fees, search for prohibited wastes, verify industry monitoring points, and verify self-monitoring data.

21. 1, 3, 4   Container material, shape and size must be considered when selecting a sample container.

22. 1, 3, 4   Acids, bases and refrigeration are used to preserve samples.

23. 1, 3, 4, 5   Field tests are the *ONLY* accurate method for measuring chlorine residual, pH, sulfite and temperature.

24. 2, 4, 5   Information that must be recorded for each sample collected includes field observations, time the sample was collected, and who collected the sample.

25. 1, 3, 4   Flow is measured in CFS (cubic feet per second), GPM (gallons per minute), and MGD (million gallons per day).

26. 1, 2, 5   Stilling well eliminates effects of foam on water surface, eliminates the effects of rags, and is not affected by wave action in the flume or weir.

27. 1, 2, 3, 4   Errors in ultrasonic flowmeter readings can be caused by a dirty face on the transducer, excessive wave action on water surface, foam layer on water surface, and improper generation and receipt of acoustic pulses.

28. 1   Samples should be collected at least once every hour.

29. 2, 3, 5   Inspectors need to understand the sources and quantities of industrial wastewaters to define, identify, and solve problems caused by industrial discharges.

30. 1, 2, 3, 4, 5   Characteristics that must be considered when determining the acceptability of an industrial waste discharge to a sewer include concentration, odor, pH, temperature, and toxicity.

31. 1, 5   Plating bath rinses and storm water runoff are dilute solutions that may be discharged to the pretreatment system or POTW.

32. 2, 3, 5   Gasoline, hexane, and methyl ethyl ketone may cause explosions in sewers.

33. 1, 5   BOD and suspended solids are found in food industry effluents.

34. 1, 3, 4   When establishing treatment and discharge objectives, average flow, peak flow and total daily flow are important.

35. 2, 3, 4   Chemical constituents in wastewater, potential wastewater sources, and techniques for wastewater reduction are important to inspectors.

36. 1, 3, 4, 5 Benefits of source control include cost savings, protection of the public, reduced chemical costs, and reduced liability.

37. 1, 2, 4, 5 Wastewater treatment technologies used by industry include biological, chemical, and physical processes and land application.

38. 2, 3, 4   Neutralization, oxidation/reduction, and precipitation are chemical treatment processes.

39. 1, 3, 5   The most common and most dangerous hazards that pretreatment inspectors will encounter are atmospheric hazards, physical hazards, and traffic hazards.

40. 1, 2, 3, 4 Collect information on the amount, chemical nature, destination, and frequency of an industrial discharge.

41. 1, 2, 3   During an on-site inspection, check the accidental spill prevention control plan of the industry, conduct a peripheral examination of the industrial user, and observe the physical characteristics of the wastestream in the sewer.

42. 2, 3, 5   Oil and grease may appear in wastewater as emulsified oil, free oil, and soluble oil.

43. 1, 2, 3, 5 BOD, COD, dissolved sulfide and pH can change if samples are not properly chilled and sealed.

44. 1, 2, 3, 5 Discharges of concentrated acids or bases, large volumes of toxic materials, and poisonous gases could create emergencies.

45. 2, 4, 5   Chemical plants, fuel transfer stations, and oil refineries are the most likely sources of catastrophic discharges.

46. 1, 2, 4   Possible ways to determine the source of a spill include identifying the material, phoning industries and asking if spill occurred, and tracing material upstream.

47. 2, 5   The high-risk industry's emergency response plan and spill prevention plan must be reviewed.

APPENDIX II

PRETREATMENT ARITHMETIC

by

Bill Garrett

TABLE OF CONTENTS

Appendix II.   PRETREATMENT ARITHMETIC

APPENDIX II.   PRETREATMENT ARITHMETIC

A.   HOW TO STUDY THIS APPENDIX

This appendix may be worked early in your training program to help you gain the greatest benefit from your learning efforts. Whether to start this appendix early or wait until later is your decision. The chapters in this manual were written in a manner requiring very little background in arithmetic. You may wish to concentrate your efforts on the chapters and refer to this appendix when you need help. Some operators prefer to complete this appendix early so they will not have to worry about how to do the arithmetic when they are studying the chapters. You may try to work this appendix early or refer to it while studying the other chapters.

After you have worked a problem involving your job, you should check your calculations, examine your answer to see if it appears reasonable, and, if possible, have another inspector check your work before making any decisions or changes.

B.   STEPS IN SOLVING PROBLEMS

B.0   Identification of Problem

To solve any problem, you have to identify the problem, determine what kind of answer is needed, and collect the information needed to solve the problem. A good approach to this type of problem is to examine the problem and make a list of KNOWN and UNKNOWN information.

EXAMPLE: Find the theoretical detention time in a rectangular sedimentation tank 8 feet deep, 30 feet wide, and 60 feet long when the flow is 1.4 MGD.

Known	**Unknown**
Depth, ft = 8 ft	Detention Time, hours
Width, ft = 30 ft	
Length, ft = 60 ft	
Flow, MGD = 1.4 MGD	

Sometimes a drawing or sketch will help to illustrate a problem and indicate the knowns, unknowns, and possible additional information needed.

B.1   Selection of Formula

Most problems involving mathematics in pretreatment facility inspection can be solved by selecting the proper formula, inserting the known information, and calculating the unknown. Here is an example of how to calculate the detention time for a tank or clarifier in hours.

$$\text{Detention Time, hr} = \frac{(\text{Tank Volume, cu ft})(7.48 \text{ gal/cu ft})(24 \text{ hr/day})}{\text{Flow, gal/day}}$$

Converting the known information to fit the terms in a formula sometimes requires extra calculations. The next step is to find the values of any terms in the formula that are not in the list of known values.

Flow, gal/day = 1.4 MGD

= 1,400,000 gal/day

Tank Volume, cu ft = (Length, ft)(Width, ft)(Depth, ft)

= 60 ft x 30 ft x 8 ft

= 14,400 cu ft

Solution of Problem:

$$\text{Detention Time, hr} = \frac{(\text{Tank Volume, cu ft})(7.48 \text{ gal/cu ft})(24 \text{ hr/day})}{\text{Flow, gal/day}}$$

$$= \frac{(14,400 \text{ cu ft})(7.48 \text{ gal/cu ft})(24 \text{ hr/day})}{1,400,000 \text{ gal/day}}$$

$$= 1.85 \text{ hr}$$

The remainder of this section discusses the details that must be considered in solving this problem.

B.2   Arrangement of Formula

Once the proper formula is selected, you may have to rearrange the terms to solve for the unknown term.

$$\text{Velocity, ft/sec} = \frac{\text{Flow Rate, cu ft/sec}}{\text{Cross-Sectional Area, sq ft}}$$

OR $V = \dfrac{Q}{A}$

In this equation if Q and A were given, the equation could be solved for V. If V and A were known, the equation would have to be rearranged to solve for Q. To move terms from one side of an equation to another, use the following rule:

When moving a term or number from one side of an equation to the other, move the numerator (top) of one side to the denominator (bottom) of the other; or from the denominator (bottom) of one side to the numerator (top) of the other.

$$V = \frac{Q}{A} \quad OR \quad Q = AV \quad OR \quad A = \frac{Q}{V}$$

If the volume of a sedimentation tank and the desired detention time were given, the detention time formula could be rearranged to calculate the design flow.

$$\text{Detention Time, hr} = \frac{(\text{Tank Vol, cu ft})(7.48 \text{ gal/cu ft})(24 \text{ hr/day})}{\text{Flow, gal/day}}$$

By rearranging the terms,

$$\text{Flow, gal/day} = \frac{(\text{Tank Vol, cu ft})(7.48 \text{ gal/cu ft})(24 \text{ hr/day})}{\text{Detention Time, hr}}$$

B.3   Unit Conversions

Each term in a formula or mathematical calculation must be stated in the correct units. The area of a rectangular clarifier (Area, sq ft = Length, ft x Width, ft) can't be calculated in square feet if the width is given as 246 inches or 20 feet 6 inches. The width must be converted to 20.5 feet. In the previous example, if the tank volume were given in gallons, then the 7.48 gal/cu ft would not be needed. *TO AVOID TIME-CONSUMING MISTAKES, ALWAYS CHECK THE UNITS IN A FORMULA BEFORE PERFORMING ANY CALCULATIONS.*

$$\text{Detention Time, hr} = \frac{(\text{Tank Volume, cu ft})(7.48 \text{ gal/cu ft})(24 \text{ hr/day})}{\text{Flow, gal/day}}$$

$$= \frac{\cancel{\text{cu ft}}}{} \times \frac{\text{gal}}{\cancel{\text{cu ft}}} \times \frac{\text{hr}}{\cancel{\text{day}}} \times \frac{\cancel{\text{day}}}{\cancel{\text{gal}}}$$

$$= \text{hr (all other units cancel)}$$

NOTE: We have hours = hr. Note that the hour unit on both sides of the equation can be cancelled out and nothing would remain. This is one more check that we have the correct units. By rearranging the detention time formula, other unknowns could be determined.

If the design detention time and design flow were known, the required capacity of the tank could be calculated.

$$\text{Tank Volume, cu ft} = \frac{(\text{Detention Time, hr})(\text{Flow, gal/day})}{(7.48 \text{ gal/cu ft})(24 \text{ hr/day})}$$

If the tank volume and design detention time were known, the design flow could be calculated.

$$\text{Flow, gal/day} = \frac{(\text{Tank Volume, cu ft})(7.48 \text{ gal/cu ft})(24 \text{ hr/day})}{\text{Detention Time, hr}}$$

Rearrangement of the detention time formula to find other unknowns illustrates the need to always use the correct units.

B.4   Calculations

In general, do the calculations inside parentheses ( ) first and brackets [ ] next. All of the calculations above and below the division line should be done (working from left to right) before dividing.

$$\text{Detention Time, hr} = \frac{[(\text{Tank Volume, cu ft})(7.48 \text{ gal/cu ft})(24 \text{ hr/day})]}{\text{Flow, gal/day}}$$

$$= \frac{[(14,400 \text{ cu ft})(7.48 \text{ gal/cu ft})(24 \text{ hr/day})]}{1,400,000 \text{ gal/day}}$$

$$= \frac{2,585,088 \text{ gal-hr/day}}{1,400,000 \text{ gal/day}}$$

$$= 1.85, \text{ or}$$

$$= 1.9 \text{ hr}$$

B.5   Significant Figures

In calculating the detention time in the previous section, the answer is given as 1.9 hr. The answer could have been calculated:

$$\text{Detention Time, hr} = \frac{2,585,088 \text{ gal-hr/day}}{1,400,000 \text{ gal/day}}$$

$$= 1.846491429 \ldots \text{ hours}$$

How does one know when to stop dividing? Common sense and significant figures both help.

First, consider the meaning of detention time and the measurements that were taken to determine the knowns in the formula. Detention time in a tank is a theoretical value and assumes that all particles of water throughout the tank move through the tank at the same velocity. This assumption is not correct; therefore, detention time can only be a representative time for some of the water particles.

Will the flow of 1.4 MGD be constant throughout the 1.9 hours, and is the flow exactly 1.4 MGD, or could it be 1.35 MGD or 1.428 MGD? A carefully calibrated flowmeter may give a reading within 2% of the actual flow rate. Flows into a tank fluctuate and flowmeters do not measure flows extremely accurately; so the detention time again appears to be a representative or typical detention time.

Tank dimensions are probably satisfactory within 0.1 ft. A flowmeter reading of 1.4 MGD is less precise and it could be 1.3 or 1.5 MGD. A 0.1 MGD flowmeter error when the flow is 1.4 MGD is (0.1/1.4) x 100% = 7% error. A detention time of 1.9 hours, based on a flowmeter reading error of plus or minus 7%, also could have the same error or more, even if the flow was constant. Therefore, the detention time error could be 1.9 hours x 0.07 = ±0.13 hour.

In most of the calculations in the inspection of pretreatment plants, the inspector uses measurements determined in the lab or read from charts, scales, or meters. The accuracy of every measurement depends on the sample being measured, the equipment doing the measuring, and the inspector reading or measuring the results. Your estimate is no better than the least precise measurement. Do not retain more than one doubtful number.

To determine how many figures or numbers mean anything in an answer, the approach called "significant figures" is used. In the example the flow was given in two significant figures (1.4 MGD), and the tank dimensions could be considered accurate to the nearest tenth of a foot (depth = 8.0 ft) or two significant figures. Since all measurements and the constants contained two significant figures, the results should be reported as two significant figures or 1.9 hours. The calculations are normally carried out to three significant figures (1.85 hours) and rounded off to two significant figures (1.9 hours).

Decimal points require special attention when determining the number of significant figures in a measurement.

Measurement	Significant Figures
0.00325	3
11.078	5
21,000.	2

EXAMPLE: The distance between two points was divided into three sections, and each section was measured by a different group. What is the distance between the two points if each group reported the distance it measured as follows:

Group	Distance, ft	Significant Figures
A	11,300.	3
B	2,438.9	5
C	87.62	4
Total Distance	13,826.52	

Group A reported the length of the section it measured to three significant figures; therefore, the distance between the two points should be reported as 13,800 feet (3 significant figures).

When adding, subtracting, multiplying, or dividing, the number of significant figures in the answer should not be more than the term in the calculations with the least number of significant figures.

B.6 Check Your Results

After completing your calculations, you should carefully examine your calculations and answer. Does the answer seem reasonable? If possible, have another inspector check your calculations before making any recommendations.

C. BASIC CONVERSION FACTORS

UNITS

1,000,000	= 1 Million	1,000,000/1 Million
	= 1 M	1,000,000/1 M

LENGTH

12 in	= 1 ft	12 in/ft
3 ft	= 1 yd	3 ft/yd
5,280 ft	= 1 mi	5,280 ft/mi

AREA

144 sq in	= 1 sq ft	144 sq in/sq ft
43,560 sq ft	= 1 acre	43,560 sq ft/ac

VOLUME

7.48 gal	= 1 cu ft	7.48 gal/cu ft
1,000 mL	= 1 liter	1,000 mL/L
3.785 L	= 1 gal	3.785 L/gal
231 cu in	= 1 gal	231 cu in/gal

WEIGHT

1,000 mg	= 1 gm	1,000 mg/gm
1,000 gm	= 1 kg	1,000 gm/kg
454 gm	= 1 lb	454 gm/lb
2.2 lbs	= 1 kg	2.2 lbs/kg

POWER

0.746 kW	= 1 HP	0.746 kW/HP

DENSITY

8.34 lbs	= 1 gal	8.34 lbs/gal
62.4 lbs	= 1 cu ft	62.4 lbs/cu ft

DOSAGE

17.1 mg/L	= 1 grain/gal	17.1 mg/L/gpg
64.7 mg	= 1 grain	64.7 mg/grain

PRESSURE

2.31 ft water	= 1 psi	2.31 ft water/psi
0.433 psi	= 1 ft water	0.433 psi/ft water
1.133 ft water	= 1 in Mercury	1.133 ft water/in Mercury

FLOW

694 GPM	= 1 MGD	694 GPM/MGD
1.55 CFS	= 1 MGD	1.55 CFS/MGD

TIME

60 sec	= 1 min	60 sec/min
60 min	= 1 hr	60 min/hr
24 hr	= 1 day	24 hr/day

NOTE: In our equations the values in the right-hand column may be written either as 24 hr/day or 1 day/24 hours depending on which units we wish to convert to obtain our desired results.

D. BASIC FORMULAS

Listed below are 57 formulas relating to pretreatment process operations and inspections. Beginning with Section E, each of these formulas is applied to a hypothetical situation to help you work through the calculations.

FLOW EQUALIZATION

1. Process Time, hours $= \dfrac{\text{Wastewater Volume, gal}}{(\text{Treatment Flow, GPM})(60 \text{ min/hr})}$

2. Waste Load, % $= \dfrac{(\text{Discharge Waste Load, lbs/day})(100\%)}{\text{Plant Capacity, lbs/day}}$

   Discharge Time, % $= \dfrac{(\text{Discharge Time, hr/day})(100\%)}{\text{Total Time, hr/day}}$

MASS EMISSION RATE

3. Mass Emission Rate, lbs/day $= (\text{Flow, MGD})(\text{Conc, mg/}L)(8.34 \text{ lbs/gal})$

   OR $= \dfrac{(\text{Flow, GPD})(\text{Conc, mg/}L)(3.785\ L\text{/gal})}{(1,000 \text{ mg/gm})(454 \text{ gm/lb})}$

COMPOSITE SAMPLING

4. No. of Samples $= \dfrac{(\text{Sampling Time, hr})(60 \text{ min/hr})}{\text{Sample Interval, min/sample}}$

   Sample Size, mL $= \dfrac{(\text{Total Sample Volume, }L)(1,000 \text{ m}L/L)}{\text{Number of Samples}}$

5. Sample Size, mL $= \dfrac{(\text{Portion})(\text{Sample Vol, }L)(1,000 \text{ m}L/L)}{\text{Number of Samples}}$

6. Sample Size, mL $= \dfrac{(\text{Sample Volume, }L)(1,000 \text{ m}L/L)}{\text{Number of Samples}}$

7. Sampling Interval, gal $= \dfrac{\text{Flow Volume, gal}}{\text{No. of Samples}}$

   Sampling Interval, min $= \dfrac{(\text{Sampling Time, hr})(60 \text{ min/hr})}{\text{No. of Samples}}$

8. Portion $= \dfrac{\text{Wastestream Flow, GPD}}{\text{Total Flow, GPD}}$

   Sample Size, mL $= (\text{Portion})(\text{Sample Volume, m}L)$

9. Sampling Time Interval, min $= \dfrac{(\text{Sample Interval, gal})(24 \text{ hr/day})(60 \text{ min/hr})}{\text{Flow, GPD}}$

   No. Samples $= \dfrac{(\text{Total Time, hr})(60 \text{ min/hr})}{\text{Sampling Time Interval, min}}$

   Sample Vol, mL $= (\text{Sample Size, m}L\text{/sample})(\text{No. Samples})$

SPILL CONTAINMENT

10. Total Containment Volume, cu ft $= (\text{Length, ft})(\text{Width, ft})(\text{Height, ft})$

IMPACT OF TOXIC WASTE ON SEWER SYSTEM

11. Load, lbs/day $= (\text{Flow, MGD})(\text{Conc, mg/}L)(8.34 \text{ lbs/gal})$

    OR $= \dfrac{(\text{Flow, GPD})(\text{Conc, mg/}L)(3.785\ L\text{/gal})}{(1,000 \text{ mg/gm})(454 \text{ gm/lb})}$

12. Effluent, mg/L $= \dfrac{\text{Effluent Load, lbs/day}}{(\text{Flow, MGD})(8.34 \text{ lbs/gal})}$

    OR $= \dfrac{(\text{Effl Load, lbs/day})(454 \text{ gm/lb})(1,000 \text{ mg/gm})}{(\text{Flow, GPD})(3.785\ L\text{/gal})}$

A
R
I
T
H
M
E
T
I
C

WASTE STRENGTH MONITORING

13. $\dfrac{\text{Load,}}{\text{lbs/day}}$ = (Flow, MGD)(Conc, mg/L)(8.34 lbs/gal)

COMPARE WITH

$\dfrac{\text{Downstream Load,}}{\text{lbs/day}}$ = Upstream Load, lbs/day + Company Load, lbs/day

14. $\dfrac{\text{Total Load,}}{\text{lbs BOD/day}}$ = $\dfrac{\text{Industry Load,}}{\text{lbs BOD/day}}$ + $\dfrac{\text{POTW Load,}}{\text{lbs BOD/day}}$

CHEMICAL LOADING ON A POTW

15. $\dfrac{\text{Load, \%}}{\text{(from Company)}}$ = $\dfrac{\text{(Company Load, lbs/day)(100\%)}}{\text{POTW Load, lbs/day}}$

16. $\dfrac{\text{BOD, \%}}{\text{(Manuf.)}}$ = $\dfrac{\text{(Manuf. BOD, lbs/day)(100\%)}}{\text{POTW BOD, lbs/day}}$

17. COD, mg/L = $\dfrac{\text{(Solution, gm/mole)(Isopropyl, eq/mole)}}{\text{(COD, mg/eq)(1,000 m}L/L\text{)}}{\text{Isopropyl, gm/mole}}$

$\dfrac{\text{COD Increase,}}{\text{mg}/L}$ = $\dfrac{\text{(COD, mg}/L\text{)(Dump, gal)(24 hr/day)}}{\text{(Flow, GPD)(Dump Time, hr)}}$

SEWER-USE FEES

18. $\dfrac{\text{Load Fee,}}{\text{\$/yr}}$ = (Load, lbs/day)(Load Fee, \$/lb)(Work, days/yr)

$\dfrac{\text{Flow Fee,}}{\text{\$/yr}}$ = (Flow, MGD)(Flow Fee, \$/MG)(Work, days/yr)

19. $\dfrac{\text{Excess,}}{\text{lbs/day}}$ = (Flow, MGD)(Excess, mg/L)(8.34 lbs/gal)

$\dfrac{\text{Penalty Fee,}}{\text{\$/day}}$ = (Excess, lbs/day)(Penalty, \$/day)

20. Flow, gal/yr = (Flow, GPD)(Work, days/yr)

$\dfrac{\text{Amount,}}{\text{lbs/yr}}$ = (Flow, MG/yr)(Waste, mg/L)(8.34 lbs/gal)

Fee, \$/yr = (Amount, lbs/yr)(Charge, \$/lb)

21. $\dfrac{\text{Sulfite,}}{\text{mg/day}}$ = (Flow, GPD)(Sulfite, mg/L)(3.785 L/gal)

$\dfrac{\text{COD,}}{\text{lbs/day}}$ = $\dfrac{\text{(Sulfite, mg/day)(Sulfite, eq/mole)(COD, mg/eq)}}{\text{(Sulfite, mg/mole)(1,000 mg/gm)(454 gm/lb)}}$

$\dfrac{\text{Sewer-Use Fee, \$/yr}}{}$ = (COD, lbs/day)(Work, days/yr)(Fee, \$/lb)

DETENTION TIME

22. $\dfrac{\text{Detention Time, min}}{}$ = $\dfrac{\text{Tank Volume, gal}}{\text{Flow Rate, GPM}}$

23. $\dfrac{\text{Flow Rate, GPM}}{}$ = $\dfrac{\text{Flow Volume, gal}}{\text{(Work Day, hr)(60 min/hr)}}$

24. Volume, gal = (Flow Rate, GPM)(Detention Time, min)

25. $\dfrac{\text{Detention Time, min}}{}$ = $\dfrac{\text{(Pipe Vol, gal)(24 hr/day)(60 min/hr)}}{\text{Flow Rate, GPD}}$

pH NEUTRALIZATION

26. $\dfrac{\text{NH}_3,}{\text{lbs/day}}$ = $\dfrac{\text{(H}^+\text{, moles/day)(NH}_3\text{, gm/mole)}}{\text{454 gm/lb}}$

27. $\dfrac{\text{H}^+\text{, moles/}}{\text{day}}$ = (Flow, GPD)(3.785 L/gal)(H$^+$, mole/L)

$\dfrac{\text{NaOH,}}{\text{lbs/day}}$ = $\dfrac{\text{(H}^+\text{, moles/day)(NaOH, gm/mole)}}{\text{454 gm/lb}}$

28. $\dfrac{\text{Soda Ash,}}{\text{lbs/day}}$ = $\dfrac{\text{(H}^+\text{, moles/day)(106 gm/mole Soda Ash)}}{\text{(2 moles H}^+\text{/mole Soda Ash)(454 gm/lb)}}$

29. $\dfrac{\text{NaOH,}}{\text{gal/day}}$ = $\dfrac{\text{(H}^+\text{, moles/day)(40 gm/mole NaOH)}}{\text{(250 gm NaOH/}L\text{)(3.785 }L\text{/gal)}}$

30. $\dfrac{\text{Ammonia,}}{\text{lbs/day}}$ = $\dfrac{\text{(H}^+\text{, moles/day)(17 gm/mole NH}_3\text{)}}{\text{454 gm/lb}}$

31. $\dfrac{\text{Lime,}}{\text{gal/day}}$ = $\dfrac{\text{(H}^+\text{, moles/day)(56 gm/mole Lime)}}{\text{(2 moles H}^+\text{/mole Lime)(Lime, gm/}L\text{)(3.785 }L\text{/gal)}}$

pH ADJUSTMENT

32. $\dfrac{\text{Sulfuric Acid}}{}$ = $\dfrac{\text{(Waste, gal)(Waste, Normality)}}{\text{Sulfuric Acid, Normality}}$

$\dfrac{\text{Waste Normality}}{}$ = $\dfrac{\text{(Acid, m}L\text{)(Acid, }N\text{)}}{\text{Waste Sample, m}L}$

33. $\dfrac{\text{Caustic, gal}}{\text{(To treat waste)}}$ = $\dfrac{\left(\dfrac{\text{Caustic to}}{\text{Sample, gal}}\right)\left(\dfrac{\text{Caustic to}}{\text{Sample, \%}}\right)\left(\dfrac{\text{Waste to}}{\text{Treatment, gal}}\right)}{\text{(Sample, gal)(Treatment Caustic, \%)}}$

34. $\dfrac{\text{Acid Feed, GPD}}{}$ = $\dfrac{\dfrac{\text{(Waste Flow, GPM)(Waste Normality)}}{\text{(60 min/hr)(24 hr/day)}}}{\text{Acid Feed, }N}$

HYDROXIDE PRECIPITATION

35. NaOH, gal = $\dfrac{\text{(Wastewater, gal)(m}L\text{ NaOH)}}{\text{(1,000 m}L/L\text{)(1 }L\text{ Wastewater)}}$

36. $\dfrac{\text{NaOH Feed, GPD}}{}$ = $\dfrac{\text{(Zinc Flow, GPM)(m}L\text{ NaOH)(60 min/hr)(24 hr/day)}}{\text{(1,000 m}L/L\text{)(1 }L\text{ Zinc)}}$

COMPLEXED METAL PRECIPITATION

37. NaOH, gal = $\dfrac{\text{(Wastewater, gal)(m}L\text{ NaOH)}}{\text{(1,000 m}L/L\text{)(1 }L\text{ Wastewater)}}$

38. $\dfrac{\text{NaOH Feed, GPD}}{}$ = $\dfrac{\text{(Copper Flow, GPM)(m}L\text{ NaOH)(60 min/hr)(24 hr/day)}}{\text{(1,000 m}L/L\text{)(1 }L\text{ Copper)}}$

REDUCTION OF HEXAVALENT CHROMIUM

39. $\dfrac{\text{Cr}^{6+}\text{ Treated, lbs}}{}$ = (Waste, MG)(Cr^{6+}, mg/L)(8.34 lbs/gal)

$\dfrac{\text{Dosage, lbs SO}_2}{}$ = (Cr^{6+} Treated, lbs)(Treatment, lbs SO$_2$/lb Cr^{6+})

40. $\dfrac{\text{Sulfonator Feed, lbs/min}}{}$ = $\dfrac{\text{(Flow, GPM)(Waste, mg/}L\text{)(8.34 lbs/gal)(Treat, lb/lb)}}{\text{1,000,000/M}}$

OR

$\dfrac{\text{Sulfonator Feed, lbs/day}}{}$ = (Flow, MGD)(Waste, mg/L)(8.34 lbs/gal)(Treat, lb/lb)

41. Time, min = $\dfrac{\text{(Waste, lbs Cr}^{6+}\text{)(Treat, lbs SO}_2\text{/lb Cr}^{6+}\text{)(1,440 min/day)}}{\text{Feed, lbs SO}_2\text{/day}}$

42. Ferrous Sulfate, lbs $= \text{(Chromium, lbs)(Ferrous Sulfate Dose, mg/}L\text{/mg/}L\text{)}$

Time, hr $= \dfrac{\text{(Ferrous Sulfate, lbs)(100\%)}}{\text{(Ferrous Sulfate, GPH)(8.34 lbs/gal)(Ferrous Sulfate, \%)}}$

43. Ferrous Sulfate Feed, GPD $= \dfrac{\text{(Ferrous Sulfate Feed, lbs/day)(100\%)}}{\text{(8.34 lbs/gal)(Ferrous Sulfate, \%)}}$

44. Conc, mg/L $= \dfrac{\text{(Conc, oz/gal)(454 gm/lb)(1,000 mg/gm)}}{\text{(16 oz/lb)(3.785 }L\text{/gal)}}$

45. Conc, oz/gal $= \dfrac{\text{(Conc, mg/}L\text{)(16 oz/lb)(3.785 }L\text{/gal)}}{\text{(1,000 mg/gm)(454 gm/lb)}}$

46. Cyanide, lbs $= \dfrac{\text{(CN Vol, gal)(CN Conc, mg/}L\text{)(8.34 lbs/gal)}}{\text{1,000,000/M}}$

Chlorine Req'd, lbs $= \text{(Cyanide, lbs)(Chlorine Dose, lb/lb)}$

Time, hr $= \dfrac{\text{(Chlorine Req'd, lbs)(100\%)(24 hr/day)}}{\text{(Flow, GPD)(8.34 lbs/gal)(Hypochorite, \%)}}$

47. Cyanide, lbs/day $= \dfrac{\text{(Flow, GPM)(Cyanide, mg/}L\text{)(8.34 lbs/gal)}}{\text{694 GPM/MGD}}$

Caustic Feed, lbs/day $= \text{(Cyanide, lbs/day)(Caustic Dose, lb/lb)}$

Caustic Pump, GPD $= \dfrac{\text{(Caustic Feed, lbs/day)(100\%)}}{\text{(8.34 lbs/gal)(Caustic, \%)}}$

48. Chlorine, lbs/hr $= \dfrac{\text{(Flow, GPD)(Hypochorite, \%)(8.34 lbs/gal)}}{\text{(24 hr/day)(100\%)}}$

COUNTERFLOW RINSING

49. $R^n = \dfrac{C_p}{C_n}$

COMBINED WASTESTREAM FORMULA

50.

$$C_m, \text{mg/}L = \dfrac{\left[\displaystyle\sum_{r=1}^{N} C_r F_r\right]\left[F_t - F_d\right]}{\left[\displaystyle\sum_{r=1}^{N} F_r\right]\left[F_t\right]}$$

where C_m = the alternative concentration limit for the combined wastestream

C_r = the categorical pretreatment standard concentration limit for a pollutant in the regulated stream r

F_r = the average daily flow (at least a 30-day average) of stream r to the extent that it is regulated for such pollutant

51 and 52. Same formula as for 50.

53. Final Conc, mg/L $= \dfrac{\begin{array}{c}\text{(Flow 1, GPD)(Conc 1, mg/}L\text{)} + \text{(Flow 2, GPD)}\\ \text{(Conc 2, mg/}L\text{)} + \text{(Flow 3, GPD)(Conc 3, mg/}L\text{)}\end{array}}{\text{Flow 1, GPD} + \text{Flow 2, GPD} + \text{Flow 3, GPD}}$

54.

$$M_t = \dfrac{\left[\displaystyle\sum_{i=1}^{N} M_i\right]\left[F_t - F_d\right]}{\left[\displaystyle\sum_{i=1}^{N} F_i\right]}$$

where M_t = the alternative mass limit for a pollutant in a combined wastestream

M_i = the categorical pretreatment standard mass limit for a pollutant in the regulated stream r (the categorical pretreatment mass limit multiplied by the appropriate measure of production)

F_i = the average flow (at least a 30-day average) of stream r to the extent that it is regulated for such pollutant

AVERAGE LIMITATIONS

55. Four-day Avg, mg/L $= \dfrac{\text{Sum of Four Concentrations, mg/}L}{\text{4 Measurements}}$

56. Monthly Avg, mg/L $= \dfrac{\text{Sum of Monthly Concentrations, mg/}L}{\text{Number of Monthly Measurements}}$

E.   FLOW EQUALIZATION

1. An industry works an eight-hour shift five days a week. During the eight-hour shift 100,000 gallons of wastewater are generated. The pretreatment system has a capacity of 150 gallons per minute. How long must the pretreatment facility operate each day to process the wastewater?

Known		Unknown
Wastewater Vol, gal	= 100,000 gal	Process Time, hr
Treatment Flow, GPM	= 150 GPM	

Calculate the treatment process time each day in hours.

Process Time, hr $= \dfrac{\text{Wastewater Vol, gal}}{\text{(Treatment Flow, GPM)(60 min/hr)}}$

$= \dfrac{\text{100,000 gallons}}{\text{(150 gal/min)(60 min/hr)}}$

$= 11.1 \text{ hours}$

2. A conventional activated sludge wastewater treatment system at an adhesives, resins and dye chemical plant can treat 8,000 pounds of BOD per day. A three percent isopropyl alcohol wastestream (0.8 mg BOD/mg alcohol) is discharged from a water-cooled condenser for one hour every 12 hours. Determine the percent of the plant BOD waste load discharged during the two hours each day and the percent of time the waste is discharged.

Known		Unknown
Waste Flow, GPM	= 100 GPM	1. Waste Load, lbs BOD/day
Discharge Time, min/day	= 1 hr/12 hr	
OR	= 2 hr/24 hr	2. Waste Load, %
	= 120 min/day	3. Discharge Time, %
Waste Conc, %	= 3%	
, mg/L	= 30,000 mg/L	
BOD, $\frac{\text{mg BOD}}{\text{mg alcohol}}$	= $\frac{0.8 \text{ mg BOD}}{\text{mg alcohol}}$	
Plant Capacity, lbs BOD/day	= 8,000 lbs BOD/day	

a. Calculate the BOD waste load contributed by the alcohol in pounds of BOD per day.

$$\begin{array}{l}\text{Waste} \\ \text{Load, lbs} \\ \text{BOD/day}\end{array} = \frac{\text{(Flow, gal/min)(Discharge Time, min/day)}}{\text{(Waste Conc, mg/L)(BOD, mg BOD/mg Waste)}} \\ \text{(3.785 L/gal)(1 lb/454,000 mg)}$$

$$= \text{(100 gal/min)(120 min/day)} \\ \text{(30,000 mg/L)(0.8 mg BOD/mg)} \left(\frac{3.785 \text{ L/gal}}{454,000 \text{ mg/lb}} \right)$$

$$= 2,401 \text{ lbs BOD/day}$$

Conversion factors

1% = 10,000 mg/L
1 gal = 3.785 Liters
1 lb = 454 grams
    = 454,000 milligrams

b. Determine the percent of waste load discharged during the two hours.

$$\text{Waste Load, \%} = \frac{\text{(Discharge Waste Load, lbs/day)(100\%)}}{\text{Plant Capacity, lbs/day}}$$

$$= \frac{\text{(2,401 lbs BOD/day)(100\%)}}{8,000 \text{ lbs BOD/day}}$$

$$= 30\%$$

c. Calculate the percent of time the waste is discharged.

$$\text{Discharge Time, \%} = \frac{\text{(Discharge Time, hr/day)(100\%)}}{\text{Total Time, hr/day}}$$

$$= \frac{\text{(2 hr/day)(100\%)}}{24 \text{ hr/day}}$$

$$= 8.3\%$$

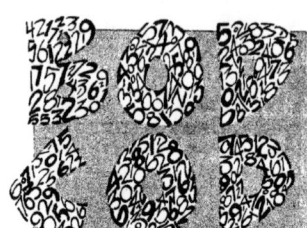

F. MASS EMISSION RATE

3. An industry has a wastewater flow of 10,000 gallons per day containing 4.5 milligrams per liter (mg/L) of copper. What is the mass emission rate of copper in pounds per day?

Known	Unknown
Effluent Flow, GPD = 10,000 gal/day	Mass Emission Rate, lbs/day
Waste Conc, mg/L = 4.5 mg/L	

Calculate the mass emission rate in pounds per day.

$$\begin{array}{l}\text{Mass Emission} \\ \text{Rate, lbs/day}\end{array} = \frac{\text{(Flow, gal/day)(Conc, mg/L)(3.785 L/gal)}}{454,000 \text{ mg/lb}}$$

$$= \frac{\text{(10,000 gal/day)(4.5 mg/L)(3.785 L/gal)}}{454,000 \text{ mg/lb}}$$

$$= 0.38 \text{ lbs/day}$$

$$OR \qquad = \text{(Flow, MGD)(Conc, mg/L)(8.34 lbs/gal)}$$

$$= \text{(0.010 MGD)(4.5 mg/L)(8.34 lbs/gal)}$$

$$= 0.38 \text{ lbs/day}$$

G. COMPOSITE SAMPLING

4. An industrial plant is required to collect a 24-hour composite sample for analysis and include the results in its monthly report. The composite sample volume must be at least two liters. A grab sample must be collected every 30 minutes. How many milliliters must be collected every 30 minutes? Assume the wastewater flow is constant during the 24-hour sampling period.

Known	Unknown
Sample Vol, L = 2 Liters	Sample Size, mL
Sampling Time, hr = 24 hours	
Sample Interval, min = 30 minutes	

a. Calculate the number of samples required.

$$\text{No. Samples} = \frac{\text{(Sampling Time, hr)(60 min/hr)}}{\text{Sample Interval, min/sample}}$$

$$= \frac{\text{(24 hr)(60 min/hr)}}{30 \text{ min/sample}}$$

$$= 48 \text{ Samples}$$

b. Calculate the size of sample.

$$\text{Sample Size, mL} = \frac{\text{(Sample Volume, L)(1,000 mL/L)}}{\text{Number of Samples}}$$

$$= \frac{\text{(2 L)(1,000 mL/L)}}{48 \text{ Samples}}$$

$$= 42 \text{ mL/sample}$$

NOTE: If you have trouble getting a representative sample because 42 mL is too small, it's OK to collect 100 mL sample sizes and mix them all together. Be sure that the samples are well mixed when analyzed for constituents in the wastewater.

5. A company has four (4) separate in-house wastestreams. Each stream discharges to the sewer through a separate outlet. The wastestream flow rate is steady from each outlet. Each wastestream must be sampled twelve (12) times. A five (5) liter composite must be prepared which represents the composite flow from all wastestreams. Determine the size of each grab sample from each wastestream to prepare the composite sample. The flows from the four streams are as follows:

Stream	A	B	C	D
Flow, GPM	25	180	50	245

Known	Unknown
Sample Vol, L = 5 L	Sample Size, mL
No. Wastestreams = 4	
No. Samples = 12	

a. Calculate the total wastestream flow in gallons per minute.

$$\text{Total Flow, GPM} = \text{Sum of Flow from Each Stream, GPM}$$
$$= 25 \text{ GPM} + 180 \text{ GPM} + 50 \text{ GPM} + 245 \text{ GPM}$$
$$= 500 \text{ GPM}$$

b. Find the portion of the five-liter sample needed from each wastestream.

$$\text{Portion} = \frac{\text{Wastestream Flow, GPM}}{\text{Total Flow, GPM}}$$

Stream A

$$\text{Portion} = \frac{25 \text{ GPM}}{500 \text{ GPM}} = 0.05$$

Stream B

$$\text{Portion} = \frac{180 \text{ GPM}}{500 \text{ GPM}} = 0.36$$

Stream C

$$\text{Portion} = \frac{50 \text{ GPM}}{500 \text{ GPM}} = 0.10$$

Stream D

$$\text{Portion} = \frac{245 \text{ GPM}}{500 \text{ GPM}} = 0.49$$

c. Calculate the size of grab sample for each sample from each wastestream.

$$\text{Sample Size, m}L = \frac{(\text{Portion})(\text{Sample Vol, } L)(1{,}000 \text{ m}L/L)}{\text{No. Samples}}$$
$$= \frac{(\text{Portion})(5 \ L)(1{,}000 \text{ m}L/L)}{\text{No. Samples}}$$
$$= \frac{(\text{Portion})(5{,}000 \text{ m}L)}{\text{No. Samples}}$$

Stream A

$$\text{Sample Size, m}L = \frac{(0.05)(5{,}000 \text{ m}L)}{12 \text{ Samples}} = 21 \text{ m}L/\text{Sample}$$

Stream B

$$\text{Sample Size, m}L = \frac{(0.36)(5{,}000 \text{ m}L)}{12 \text{ Samples}} = 150 \text{ m}L/\text{Sample}$$

Stream C

$$\text{Sample Size, m}L = \frac{(0.10)(5{,}000 \text{ m}L)}{12 \text{ Samples}} = 42 \text{ m}L/\text{Sample}$$

Stream D

$$\text{Sample Size, m}L = \frac{(0.49)(5{,}000 \text{ m}L)}{12 \text{ Samples}} = 204 \text{ m}L/\text{Sample}$$

6. A pretreatment facility has a wastewater flow of 50,000 gallons per day (GPD) and works an eight-hour day. The facility wishes to sample every 20 minutes and have a three-liter composite sample for analysis by the laboratory. How many milliliters should be collected every 20 minutes if the wastestream flow rate is constant?

Known	Unknown
Flow, GPD = 50,000 GPD	Sample Size, mL
Total Time, hr = 8 hr	
Sample Time, min = 20 min	
Sample Vol, L = 3 L	

a. Determine the number of samples to be collected.

$$\text{Number of Samples} = \frac{(\text{Total Time, hr})(60 \text{ min/hr})}{\text{Sample Time, min/sample}}$$
$$= \frac{(8 \text{ hr})(60 \text{ min/hr})}{20 \text{ min/sample}}$$
$$= 24 \text{ Samples}$$

b. Calculate the size of sample in milliliters.

$$\text{Sample Size, m}L = \frac{(\text{Sample Vol, } L)(1{,}000 \text{ m}L/L)}{\text{Number of Samples}}$$
$$= \frac{(3 \ L)(1{,}000 \text{ m}L/L)}{24 \text{ Samples}}$$
$$= 125 \text{ m}L/\text{sample}$$

7. A pretreatment facility discharges 60,000 gallons at a constant flow rate during a 16-hour work day. A composite sample of four liters is needed for analysis by a laboratory. The sampler is set to collect a 125 mL sample every time the sampling pump is activated. A sample should be collected after how many gallons of waste flow, and how many minutes should elapse between each sample?

Known	Unknown
Flow Vol, gal = 60,000 gal	1. Sampling Interval, gal
Time, hr = 16 hr	2. Sampling Interval, min
Sample Vol, L = 4 L	
Sample Size, mL = 125 mL	

a. Determine the number of samples that must be collected.

$$\text{No. of Samples} = \frac{(\text{Sample Vol, } L)(1{,}000 \text{ m}L/L)}{\text{Sample Size, m}L/\text{sample}}$$
$$= \frac{(4 \ L)(1{,}000 \text{ m}L/L)}{125 \text{ m}L/\text{sample}}$$
$$= 32 \text{ Samples}$$

b. Calculate the sampling interval in gallons.

$$\text{Interval, gal} = \frac{\text{Flow Volume, gal}}{\text{No. of Samples}}$$

$$= \frac{60,000 \text{ gal}}{32 \text{ Samples}}$$

$$= 1,875 \text{ gallon interval/sample}$$

c. Calculate the sampling interval in minutes.

$$\text{Sampling Interval, min} = \frac{(\text{Sampling Time, hr})(60 \text{ min/hr})}{\text{No. of Samples}}$$

$$= \frac{(16 \text{ hr})(60 \text{ min/hr})}{32 \text{ Samples}}$$

$$= 30 \text{ minutes/sample}$$

8. A company has five sewer connections with the following discharges:

Discharge	A	B	C	D	E
Flow, GPD	350	50	700	1,200	200

A flow-proportional composite sample of four liters is desired. How many milliliters should be collected from each connection for the composite sample?

Known	Unknown
Flow from each discharge, GPD	Sample Size, mL from each discharge
No. Connections = 5	
Sample Vol, L = 4 L	

a. Determine the total flow rate in gallons per day.

$$\text{Total Flow Rate, GPD} = \text{Sum of Each Discharge, GPD}$$

$$= 350 \text{ GPD} + 50 \text{ GPD} + 700 \text{ GPD} + 1,200 \text{ GPD} + 200 \text{ GPD}$$

$$= 2,500 \text{ GPD}$$

b. Convert sample volume from liters to milliliters.

$$\text{Sample Vol, mL} = (\text{Sample Vol, } L)(1,000 \text{ mL/}L)$$

$$= (4 \text{ } L)(1,000 \text{ mL/}L)$$

$$= 4,000 \text{ mL}$$

c. Find the portion of the total flow rate that is from each wastestream.

$$\text{Portion} = \frac{\text{Wastestream Flow, GPD}}{\text{Total Flow, GPD}}$$

Station A

$$\text{Portion} = \frac{350 \text{ GPD}}{2,500 \text{ GPD}} = 0.14$$

Station B

$$\text{Portion} = \frac{50 \text{ GPD}}{2,500 \text{ GPD}} = 0.02$$

Station C

$$\text{Portion} = \frac{700 \text{ GPD}}{2,500 \text{ GPD}} = 0.28$$

Station D

$$\text{Portion} = \frac{1,200 \text{ GPD}}{2,500 \text{ GPD}} = 0.48$$

Station E

$$\text{Portion} = \frac{200 \text{ GPD}}{2,500 \text{ GPD}} = 0.08$$

d. Calculate the sample size in milliliters from each connection.

$$\text{Sample Size, mL} = (\text{Portion})(\text{Sample Vol, mL})$$

Station A

$$\text{Sample Size, mL} = (0.14)(4,000 \text{ mL})$$

$$= 560 \text{ mL}$$

Station B

$$\text{Sample Size, mL} = (0.02)(4,000 \text{ mL})$$

$$= 80 \text{ mL}$$

Station C

$$\text{Sample Size, mL} = (0.28)(4,000 \text{ mL})$$

$$= 1,120 \text{ mL}$$

Station D

$$\text{Sample Size, mL} = (0.48)(4,000 \text{ mL})$$

$$= 1,920 \text{ mL}$$

Station E

$$\text{Sample Size, mL} = (0.08)(4,000 \text{ mL})$$

$$= 320 \text{ mL}$$

Summary

Station	Sample Size, mL
A	560 mL
B	80 mL
C	1,120 mL
D	1,920 mL
E	320 mL

9. A company produces a constant wastestream flow volume of 15,000 gallons during an eight-hour work day. A sampler collects a 125 milliliter sample for every 1,500 gallons of flow. What is the time interval between collection of samples and what is the final total volume of the composite sample?

Known	Unknown
Flow Volume, gal = 15,000	1. Time Interval, min
Work Time, hr = 8 hr	2. Sample Vol, mL
Sample Size, mL = 125 mL	
Sample Interval, gal = 1,500 gal	

a. Determine the sampling time interval in minutes.

$$\text{Sampling Time, min} = \frac{(\text{Sample Interval, gal})(\text{Work Time, hr})}{\text{Flow Volume, gal}}$$

$$= \frac{(1{,}500 \text{ gal})(8 \text{ hr})(60 \text{ min/hr})}{15{,}000 \text{ gal}}$$

$$= 48 \text{ min}$$

b. Calculate the number of samples to be collected.

$$\text{No. Samples} = \frac{(\text{Total Time, hr})(60 \text{ min/hr})}{\text{Sampling Time Interval, min}}$$

$$= \frac{(8 \text{ hr})(60 \text{ min/hr})}{48 \text{ min/Sample}}$$

$$= 10 \text{ Samples}$$

c. Find the total volume of the composite sample.

$$\text{Sample Vol, m}L = (\text{Sample Size, m}L/\text{sample})(\text{No. Samples})$$

$$= (125 \text{ m}L/\text{Sample})(10 \text{ Samples})$$

$$= 1{,}250 \text{ m}L$$

H. SPILL CONTAINMENT

10. A company is required to construct an outdoor spill containment system. This containment area must be capable of holding double the volume of the following tanks:

a. Square Tank 6 ft wide, 6 ft long and 6 ft high,

b. Rectangular Tank 4 ft wide, 20 ft long and 3 ft high,

c. Cylindrical Tank 8 ft in diameter and 9 ft high, and

d. A 12-inch rainfall during a 24-hour period.

The company is proposing to construct a 200-foot by 200-foot level containment pad surrounded by a 15-inch wall. Will the containment system be adequate?

Known	Unknown
Dimensions of 3 tanks	Is containment system adequate?
12-inch rainfall	
24-hour duration	
200' x 200' x 15" Pad	

a. Calculate the volume of each tank in cubic feet.

Square Tank
$$\text{Volume, cu ft} = (\text{Length, ft})(\text{Width, ft})(\text{Height, ft})$$
$$= (6 \text{ ft})(6 \text{ ft})(6 \text{ ft})$$
$$= 216 \text{ cu ft}$$

Rectangular Tank
$$\text{Volume, cu ft} = (\text{Length, ft})(\text{Width, ft})(\text{Height, ft})$$
$$= (20 \text{ ft})(4 \text{ ft})(3 \text{ ft})$$
$$= 240 \text{ cu ft}$$

Cylindrical Tank
$$\text{Volume, cu ft} = (0.785)(\text{Diameter, ft})^2(\text{Height, ft})$$
$$= (0.785)(8 \text{ ft})^2(9 \text{ ft})$$
$$= 452 \text{ cu ft}$$

b. Double the volume to determine required total tank storage volume.

$$\text{Tank Storage Volume, cu ft} = (\text{Double})(\text{Volume of All Tanks, cu ft})$$

$$= (2)(216 \text{ cu ft} + 240 \text{ cu ft} + 452 \text{ cu ft})$$

$$= (2)(908 \text{ cu ft})$$

$$= 1{,}816 \text{ cu ft}$$

c. Calculate the volume of 12 inches of rainfall on the 200 ft by 200 ft containment pad. Assume no evaporation, seepage or percolation.

$$\text{Volume Rain on Pad, cu ft} = \frac{(\text{Length, ft})(\text{Width, ft})(\text{Height, in})}{12 \text{ in/ft}}$$

$$= \frac{(200 \text{ ft})(200 \text{ ft})(12 \text{ in})}{12 \text{ in/ft}}$$

$$= 40{,}000 \text{ cu ft}$$

d. Determine the total required containment volume.

$$\text{Total Required Volume, cu ft} = \text{Tank Storage Volume, cu ft} + \text{Rain on Pad Volume, cu ft}$$

$$= 1{,}816 \text{ cu ft} + 40{,}000 \text{ cu ft}$$

$$= 41{,}816 \text{ cu ft}$$

e. Calculate the total volume of the containment area with a 15-inch high wall.

$$\text{Total Containment Volume, cu ft} = \frac{(\text{Length, ft})(\text{Width, ft})(\text{Height, in})}{12 \text{ in/ft}}$$

$$= \frac{(200 \text{ ft})(200 \text{ ft})(15 \text{ in})}{12 \text{ in/ft}}$$

$$= 50{,}000 \text{ cu ft}$$

The total available containment volume of 50,000 cu ft is greater than the total required containment volume of 41,816 cu ft; therefore, the proposed containment volume is adequate. The volume occupied by the tanks to the 15-inch depth is neglected.

I. IMPACT OF TOXIC WASTE ON SEWER SYSTEM

11. A plating company has an underground hard chrome waste recirculation tank located close to an underground sewer. The company's wastestream flow is 25,000 gallons per day with a chromium content of 10.5 mg/L. During routine sewer monitoring upstream and downstream of this company, the following information was obtained:

	Upstream	Downstream
Flow, GPD	150,000 GPD	175,000 GPD
Chromium Conc, mg/L	0.85 mg/L	3.2 mg/L

Could the inspector suspect a leak in the underground chromium tank migrating into the sewer system?

Known	Unknown
Flow, GPD = 25,000 GPD	Leak from tank to sewer?
Chromium Conc, mg/L = 10.5 mg/L	
Sewer Flow and Chromium Conc	

NOTE: There are at least two possible solutions or approaches to solving this problem. We will show you both approaches and you can use whichever one you like.

The chromium load discharged to the sewer or flowing in the sewer may be calculated in either of two ways. They will both give the same answer.

$$\text{Load, lbs/day} = (\text{Flow, MGD})(\text{Conc, mg/L})(8.34 \text{ lbs/gal})$$

To make the units check out, we use the following conversions.

One liter weighs 1,000,000 mg or 1 M mg.

Therefore, a concentration of 1 mg/L = 1 mg/M mg

$$= 1 \text{ lb/M lb}$$

Since milligrams are a weight measure, mg/M mg is the same as lbs/M lbs.

OR

$$\text{Load, lbs/day} = (\text{Flow, MGD})(\text{Conc, mg/L})(8.34 \text{ lbs/gal})$$

$$= \frac{(\text{M Gal})}{(\text{day})} \frac{(\text{lbs})}{(\text{M lbs})} \frac{(\text{lbs})}{(\text{gal})}$$

$$= \text{lbs/day}$$

The other approach is similar, but uses different conversion factors.

$$\text{Load, lbs/day} = \frac{(\text{Flow, GPD})(\text{Conc, mg/L})(3.785 \text{ L/gal})}{(1,000 \text{ mg/gm})(454 \text{ gm/lb})}$$

$$= \text{lbs/day}$$

a. Compare the upstream flow plus the industry's flow with the downstream flow.

$$\text{Total Flow, GPD} = \text{Upstream Flow, GPD} + \text{Industry Flow, GPD}$$

$$= 150,000 \text{ GPD} + 25,000 \text{ GPD}$$

$$= 175,000 \text{ GPD}$$

Since total calculated flow of 175,000 GPD is the same as the downstream flow, this indicates that the industry's flow reporting is accurate.

b. Calculate the company's chromium discharge or load to the sewer in pounds per day.

$$\text{Load, lbs/day} = (\text{Flow, MGD})(\text{Conc, mg/L})(8.34 \text{ lbs/gal})$$

$$= (0.025 \text{ MGD})(10.5 \text{ mg/L})(8.34 \text{ lbs/gal})$$

$$= 2.19 \text{ lbs chromium/day}$$

OR

$$\text{Load, lbs/day} = \frac{(\text{Flow, GPD})(\text{Conc, mg/L})(3.785 \text{ L/gal})}{(1,000 \text{ mg/gm})(454 \text{ gm/lb})}$$

$$= \frac{(25,000 \text{ GPD})(10.5 \text{ mg/L})(3.785 \text{ L/gal})}{(1,000 \text{ mg/gm})(454 \text{ gm/lb})}$$

$$= 2.19 \text{ lbs chromium/day}$$

c. Calculate the chromium load carried by the sewer upstream of the discharge in pounds per day.

$$\text{Load, lbs/day} = (\text{Flow, MGD})(\text{Conc, mg/L})(8.34 \text{ lbs/gal})$$

$$= (0.150 \text{ MGD})(0.85 \text{ mg/L})(8.34 \text{ lbs/gal})$$

$$= 1.06 \text{ lbs chromium/day}$$

OR

$$\text{Load, lbs/day} = \frac{(\text{Flow, GPD})(\text{Conc, mg/L})(3.785 \text{ L/gal})}{(1,000 \text{ mg/gm})(454 \text{ gm/lb})}$$

$$= \frac{(150,000 \text{ GPD})(0.85 \text{ mg/L})(3.785 \text{ L/gal})}{(1,000 \text{ mg/gm})(454 \text{ gm/lb})}$$

$$= 1.06 \text{ lbs chromium/day}$$

d. Calculate the chromium load carried by the sewer downstream from the discharge in pounds per day.

$$\text{Load, lbs/day} = (\text{Flow, MGD})(\text{Conc, mg/L})(8.34 \text{ lbs/gal})$$

$$= (0.175 \text{ MGD})(3.2 \text{ mg/L})(8.34 \text{ lbs/gal})$$

$$= 4.67 \text{ lbs chromium/day}$$

OR

$$\text{Load, lbs/day} = \frac{(\text{Flow, GPD})(\text{Conc, mg/L})(3.785 \text{ L/gal})}{(1,000 \text{ mg/gm})(454 \text{ gm/lb})}$$

$$= \frac{(175,000 \text{ GPD})(3.2 \text{ mg/L})(3.785 \text{ L/gal})}{(1,000 \text{ mg/gm})(454 \text{ gm/lb})}$$

$$= 4.67 \text{ lbs chromium/day}$$

e. Compare the chromium load upstream plus the company's discharge with the total chromium load downstream.

$$\text{Total Load, lbs/day} = \text{Upstream Load, lbs/day} + \text{Industry Load, lbs/day}$$

$$= 1.06 \text{ lbs/day} + 2.19 \text{ lbs/day}$$

$$= 3.25 \text{ lbs chromium/day}$$

Downstream Load, lbs/day = 4.67 lbs chromium/day

Since the downstream chromium load is greater than expected, there appears to be a small, concentrated leak of chromium into the sewer or an error in the chromium concentration lab results.

12. A plating company has a total chromium level of 25 mg/L in a discharge of 40,000 gallons per day to a sewer. The downstream POTW has a total flow of 45 MGD and 45 percent of the chromium is removed by the treatment processes. What would be the expected concentration of

chromium in the POTW's effluent in milligrams per liter? Assume no other sources of chromium flowing into the POTW.

Known		Unknown
Chromium Conc, mg/L	= 25 mg/L	POTW Effluent Chromium, mg/L
Industry Flow, MGD	= 0.04 MGD	
POTW Flow, MGD	= 45 MGD	
Chromium Removed, %	= 45%	

a. Calculate the chromium discharged to the sewer in pounds per day.

$$\text{Chromium Load, lbs/day} = \text{(Flow, MGD)(Conc, mg/L)(8.34 lbs/gal)}$$

$$= \text{(0.04 MGD)(25 mg/L)(8.34 lbs/gal)}$$

$$= 8.34 \text{ lbs/day}$$

b. Calculate the chromium load in the effluent from the POTW. If the plant removes 45% of the chromium, then 55% of the chromium remains in the effluent.

$$\text{Effluent Load, lbs/day} = \frac{\text{(Chromium Load, lbs/day)(Remain, %)}}{100\%}$$

$$= \frac{\text{(8.34 lbs/day)(55%)}}{100\%}$$

$$= 4.587 \text{ lbs/day remain}$$

c. Calculate the POTW chromium concentration in milligrams per liter by rearranging the loading term in the lbs per day formula.

$$\text{Effl Conc, mg/L} = \frac{\text{Effluent Load, lbs/day}}{\text{(Flow, MGD)(8.34 lbs/gal)}}$$

$$= \frac{4.587 \text{ lbs chromium/day}}{\text{(45 MGD)(8.34 lbs/gal)}}$$

$$= 0.0122 \text{ mg/L}$$

OR

$$\text{Effl Conc, mg/L} = \frac{\text{(Effl Load, lbs/day)(454 gm/lb)(1,000 mg/gm)}}{\text{(Flow, GPD)(3.785 L/gal)}}$$

$$= \frac{\text{(4.587 lbs/day)(454 gm/lb)(1,000 mg/gm)}}{\text{(45,000,000 gal/day)(3.785 L/gal)}}$$

$$= 0.0122 \text{ mg/L}$$

J. WASTE STRENGTH MONITORING

13. A company is suspected of having an illegal connection to a sewer to dispose of high BOD wastes. The company reports a flow of 100,000 GPD with a BOD of 500 mg/L. Monitoring equipment is installed upstream and downstream from where the company's discharge enters the sewer. The results are as follows:

	Upstream	Downstream
Flow, GPD	300,000 GPD	400,000 GPD
BOD, mg/L	350 mg/L	450 mg/L

Does the company appear to have an illegal connection?

Known		Unknown
Flows, GPD		Illegal Connection?
Upstream	= 300,000 GPD	
Company	= 100,000 GPD	
Downstream	= 400,000 GPD	
BOD, mg/L		
Upstream	= 350 mg/L	
Company	= 500 mg/L	
Downstream	= 450 mg/L	

NOTE: There are at least two possible solutions or approaches to solving this problem. We will show you both approaches and you can use whichever one you like.

The BOD load discharged to a sewer or conveyed by a sewer may be calculated in either of two ways. They will both give the same answer.

$$\text{Load, lbs/day} = \text{(Flow, MGD)(BOD, mg/L)(8.34 lbs/gal)}$$

To make the units check out, we use the following conversions:

One liter of water weighs 1,000,000 mg or 1 M mg.

Therefore, a BOD of 1 mg/L = 1 mg/1 M mg

$$= 1 \text{ lb/1 M lb}$$

Since milligrams are a weight measurement, mg/M mg is the same as lbs/M lbs.

OR

$$\text{Load, lbs/day} = \text{(Flow, MGD)(BOD, mg/L)(8.34 lbs/gal)}$$

$$= \frac{\text{(M Gal)}}{\text{(day)}}\frac{\text{(lbs BOD)}}{\text{(M lbs)}}\frac{\text{(lbs)}}{\text{(gal)}}$$

$$= \text{lbs BOD/day}$$

The other approach is similar, but uses different conversion factors.

$$\text{Load, lbs/day} = \frac{\text{(Flow, GPD)(BOD, mg/L)(3.785 L/gal)}}{\text{(1,000 mg/gm)(454 gm/lb)}}$$

$$= \text{lbs BOD/day}$$

a. Compare the flow upstream plus the company's flow with the downstream flow.

$$\text{Total Flow, GPD} = \text{Upstream Flow, GPD} + \text{Company Flow, GPD}$$

$$= 300,000 \text{ GPD} + 100,000 \text{ GPD}$$

$$= 400,000 \text{ GPD}$$

Since the total flow of 400,000 GPD is the same as the downstream flow, apparently the company's reported flow is accurate.

b. Calculate the company's reported BOD load to the sewer in pounds of BOD per day.

$$\text{Load, lbs/day} = \text{(Flow, MGD)(BOD, mg/L)(8.34 lbs/gal)}$$

$$= \text{(0.10 MGD)(500 mg/L)(8.34 lbs/gal)}$$

$$= 417 \text{ lbs BOD/day}$$

OR

$$\text{Load, lbs/day} = \frac{(\text{Flow, GPD})(\text{BOD, mg}/L)(3.785\ L/\text{gal})}{(1{,}000\ \text{mg/gm})(454\ \text{gm/lb})}$$

$$= \frac{(100{,}000\ \text{GPD})(500\ \text{mg}/L)(3.785\ L/\text{gal})}{(1{,}000\ \text{mg/gm})(454\ \text{gm/lb})}$$

$$= 417\ \text{lbs BOD/day}$$

c. Calculate the BOD load carried by the sewer upstream of the company's discharge in pounds of BOD per day.

$$\text{Load, lbs/day} = (\text{Flow, MGD})(\text{BOD, mg}/L)(8.34\ \text{lbs/gal})$$

$$= (0.30\ \text{MGD})(350\ \text{mg}/L)(8.34\ \text{lbs/gal})$$

$$= 876\ \text{lbs BOD/day}$$

OR

$$\text{Load, lbs/day} = \frac{(\text{Flow, GPD})(\text{BOD, mg}/L)(3.785\ L/\text{gal})}{(1{,}000\ \text{mg/gm})(454\ \text{gm/lb})}$$

$$= \frac{(300{,}000\ \text{GPD})(350\ \text{mg}/L)(3.785\ L/\text{gal})}{(1{,}000\ \text{mg/gm})(454\ \text{gm/lb})}$$

$$= 876\ \text{lbs BOD/day}$$

d. Calculate the BOD load carried by the sewer downstream of the company's discharge in pounds of BOD per day.

$$\text{Load, lbs/day} = (\text{Flow, MGD})(\text{BOD, mg}/L)(8.34\ \text{lbs/gal})$$

$$= (0.40\ \text{MGD})(450\ \text{mg}/L)(8.34\ \text{lbs/gal})$$

$$= 1{,}501\ \text{lbs BOD/day}$$

OR

$$\text{Load, lbs/day} = \frac{(\text{Flow, GPD})(\text{BOD, mg}/L)(3.785\ L/\text{gal})}{(1{,}000\ \text{mg/gm})(454\ \text{gm/lb})}$$

$$= \frac{(400{,}000\ \text{GPD})(450\ \text{mg}/L)(3.785\ L/\text{gal})}{(1{,}000\ \text{mg/gm})(454\ \text{gm/lb})}$$

$$= 1{,}501\ \text{lbs BOD/day}$$

e. Compare the BOD load upstream plus the company's reported BOD discharge with the total BOD load downstream.

$$\frac{\text{Total BOD Load,}}{\text{lbs/day}} = \frac{\text{Upstream Load,}}{\text{lbs/day}} + \frac{\text{Company Load,}}{\text{lbs/day}}$$

$$= 417\ \text{lbs/day} + 876\ \text{lbs/day}$$

$$= 1{,}293\ \text{lbs BOD/day}$$

$$\frac{\text{Downstream}}{\text{Load, lbs/day}} = 1{,}501\ \text{lbs BOD/day}$$

Since the downstream load is greater than expected, this implies there is a small additional flow of high-strength BOD being discharged by the company into the sewer.

14. A POTW has the capacity to treat 150,000 pounds of BOD per day. A food processing company is investigating locating within the service district. The company's planned industrial waste discharge to the sewer is a flow of 48,000 GPD with a BOD of 3,500 mg/L. If the current loading on the POTW is a flow of 30 MGD with a BOD of 450 mg/L, could this new industry come on line without causing a BOD overload?

Known		Unknown
POTW Capacity, lbs/day	= 150,000 lbs/day	POTW BOD overload?
Industry Flow, GPD	= 48,000 GPD	
Industry BOD, mg/L	= 3,500 mg/L	
POTW Flow, MGD	= 30 MGD	
POTW BOD, mg/L	= 450 mg/L	

a. Calculate the BOD loading from the industry in pounds of BOD per day.

$$\frac{\text{Industry Load,}}{\text{lbs BOD/day}} = (\text{Flow, MGD})(\text{BOD, mg}/L)(8.34\ \text{lbs/gal})$$

$$= (0.048\ \text{MGD})(3{,}500\ \text{mg}/L)(8.34\ \text{lbs/gal})$$

$$= 1{,}401\ \text{lbs BOD/day}$$

b. Calculate the current BOD loading on the POTW in pounds of BOD per day.

$$\frac{\text{POTW Load,}}{\text{lbs BOD/day}} = (\text{Flow, MGD})(\text{BOD, mg}/L)(8.34\ \text{lbs/gal})$$

$$= (30\ \text{MGD})(450\ \text{mg}/L)(8.34\ \text{lbs/gal})$$

$$= 112{,}590\ \text{lbs BOD/day}$$

c. Determine the total expected BOD loading on the POTW by summing the anticipated industrial food processing BOD load and the current POTW load.

$$\frac{\text{Total Expected}}{\text{Load, lbs BOD/day}} = \frac{\text{Industry Load,}}{\text{lbs BOD/day}} + \frac{\text{POTW Load,}}{\text{lbs BOD/day}}$$

$$= 1{,}401\ \text{lbs BOD/day} + 112{,}590\ \text{lbs BOD/day}$$

$$= 113{,}991\ \text{lbs BOD/day}$$

Since the POTW capacity of 150,000 lbs of BOD per day is greater than the total expected load of 113,991 lbs BOD/day, the food processing industry will not cause BOD overloading problems.

K. CHEMICAL LOADING ON A POTW

15. A pickle packaging plant discharges a sodium chloride brine waste amounting to 10,000 pounds of salt per seven-day work week. If the POTW has a 30 MGD flow containing 100 mg/L of sodium, what percentage of the sodium is from the pickle company? The molecular weight of sodium chloride is 58.44 and sodium is 22.99.

Known		Unknown
Discharge, lbs/wk	= 10,000 lbs salt/wk	Sodium, % from Company
POTW Flow, MGD	= 30 MGD	
POTW Conc, mg/L	= 100 mg Na/L	
Mol Wt NaCl	= 58.44	
Mol Wt Na	= 22.99	

a. Calculate the total sodium load entering the POTW in pounds of sodium per day.

$$\text{Sodium, lbs/day (POTW)} = (\text{Flow, MGD})(\text{Na, mg}/L)(8.34\ \text{lbs/gal})$$

$$= (30\ \text{MGD})(100\ \text{mg Na}/L)(8.34\ \text{lbs/gal})$$

$$= 25{,}020\ \text{lbs sodium/day}$$

b. Calculate the daily sodium discharge to the POTW from the pickle packaging plant in pounds of sodium per day.

$$\text{Pickle Sodium, lbs/day (POTW)} = \frac{(\text{Discharge, lbs NaCl/wk})(22.99\ \text{gm/mole Na})}{(7\ \text{days/wk})(58.44\ \text{gm/mole NaCl})}$$

$$= \frac{(10{,}000\ \text{lbs NaCl/wk})(22.99\ \text{gm/mole Na})}{(7\ \text{days/wk})(58.44\ \text{gm/mole NaCl})}$$

$$= 562\ \text{lbs/day}$$

c. Calculate the percent sodium at the POTW contributed by the pickle plant.

$$\text{Sodium, \% (POTW)} = \frac{(\text{Pickle Sodium, lbs/day})(100\%)}{\text{POTW Sodium, lbs/day}}$$

$$= \frac{(562\ \text{lbs/day})(100\%)}{25{,}020\ \text{lbs/day}}$$

$$= 2.25\%$$

16. A shampoo manufacturer has a flow rate of 20,000 gallons per day with a BOD of 2,000 mg/L. If the downstream POTW treats a flow of 30 MGD with a BOD of 250 mg/L, what percent of the influent BOD comes from the shampoo manufacturer?

Known	Unknown
Manuf. Flow, GPD = 20,000 GPD	BOD, % from Shampoo
Manuf. BOD, mg/L = 2,000 mg/L	
POTW Flow, MGD = 30 MGD	
POTW BOD, mg/L = 250 mg/L	

a. Calculate the BOD loading in pounds of BOD per day at the POTW.

$$\text{BOD, lbs/day (POTW)} = (\text{Flow, MGD})(\text{BOD, mg}/L)(8.34\ \text{lbs/gal})$$

$$= (30\ \text{MGD})(250\ \text{mg}/L)(8.34\ \text{lbs/gal})$$

$$= 62{,}550\ \text{lbs BOD/day}$$

b. Calculate the BOD loading from the shampoo manufacturer in pounds of BOD per day.

$$\text{BOD, lbs/day (Manuf.)} = (\text{Flow, MGD})(\text{BOD, mg}/L)(8.34\ \text{lbs/gal})$$

$$= (0.02\ \text{MGD})(2{,}000\ \text{mg}/L)(8.34\ \text{lbs/gal})$$

$$= 333.6\ \text{lbs BOD/day}$$

c. Determine the percentage of BOD entering the POTW from the shampoo manufacturer.

$$\text{BOD, \% (Manuf.)} = \frac{(\text{Manuf. BOD, lbs/day})(100\%)}{\text{POTW BOD, lbs/day}}$$

$$= \frac{(333.6\ \text{lbs/day})(100\%)}{62{,}550\ \text{lbs/day}}$$

$$= 0.53\%$$

17. A 70 percent solution of isopropyl alcohol, C_3H_8O, molecular weight of 60.1, would contain 70 grams per 100 milliliters of solution. There are 18 equivalents per mole of isopropyl alcohol and 8,000 milligrams of COD per equivalent. If 300 gallons of isopropyl alcohol are dumped into a sewer with a flow of 5 MGD, what would be the expected increase in COD in the sewer? The chemical reaction is as follows:

$$5H_2O + C_3H_8O \rightarrow 3CO_2 + 18H^+ + 18e^-$$

Known		Unknown
Mol Wt	= 60.1	COD Increase, mg/L
Isopropyl	= 70 gm/100 mL	
Isopropyl	= 18 equiv/mole	
COD	= 8,000 mg COD/equiv	
Dump, gal	= 300 gal	
Flow, MGD	= 5 MGD	

a. Calculate the COD of isopropyl alcohol in milligrams per liter.

$$\text{COD, mg}/L = \frac{(\text{Solution, gm/m}L)(\text{Isopropyl, eq/mole})(\text{COD, mg/eq})(1{,}000\ \text{m}L/L)}{(\text{Isopropyl, gm/mole})}$$

$$= \frac{(70\ \text{gm})(18\ \text{eq/mole})(8{,}000\ \text{mg COD/eq})(1{,}000\ \text{m}L/L)}{(100\ \text{m}L)(60.1\ \text{gm/mole})}$$

$$= 1{,}677{,}205\ \text{mg}/L$$

b. Calculate the COD increase in the sewer in milligrams per liter. Assume the 300-gallon dump occurred over 24 hours or one day.

$$\text{COD Increase, mg}/L = \frac{(\text{COD, mg}/L)(\text{Dump, gal/day})}{\text{Flow, gal/day}}$$

$$= \frac{(1{,}677{,}205\ \text{mg}/L)(300\ \text{gal/day})}{5{,}000{,}000\ \text{gal/day}}$$

$$= 101\ \text{mg}/L$$

NOTE: If the dump had occurred during 12 hours instead of 24 hours, the COD increase would be 202 mg/L. If the dump occurred during one hour instead of 24 hours, the increase would be 2,424 mg/L. As the flow moves down the sewer toward the POTW, the maximum concentrations tend to drop with time of travel due to dilution and dispersion.

L.   SEWER-USE FEES

18. A meat packing plant discharges a waste flow of 45,000 GPD with a BOD of 3,500 mg/L and suspended solids of 1,300 mg/L. If the local POTW agency assesses the following charges, what would be the company's annual sewer service fee if the company works 260 days per year?

BOD = $68.00/lb BOD

SS = $59.00/lb SS

Flow = $218.00/MG

Known		Unknown
Flow, GPD	= 45,000 GPD	Service Fee, $/yr
	= 0.045 MGD	
BOD, mg/L	= 3,500 mg/L	
SS, mg/L	= 1,300 mg/L	
Work, days/yr	= 260 days/yr	

Sewer Service Charges

a. Calculate the BOD discharged in pounds of BOD per day.

BOD, lbs/day = (Flow, MGD)(BOD, mg/L)(8.34 lbs/gal)

= (0.045 MGD)(3,500 mg/L)(8.34 lbs/gal)

= 1,313.55 lbs BOD/day

b. Calculate the annual BOD sewer service fee in dollars per year.

BOD Fee, $/yr = (BOD, lbs/day)(BOD, $/lb)(Work, days/yr)

= (1,313.55 lbs/day)($68/lb)(260 days/yr)

= $23,223,564/yr

c. Calculate the suspended solids discharged in pounds of SS per day.

SS, lbs/day = (Flow, MGD)(SS, mg/L)(8.34 lbs/gal)

= (0.045 MGD)(1,300 mg/L)(8.34 lbs/gal)

= 487.89 lbs SS/day

d. Calculate the annual SS sewer service fee in dollars per year.

SS Fee, $/yr = (SS, lbs/day)(SS Fee, $/lb)(Work, days/yr)

= (487.89 lbs SS/day)($59/lb)(260 days/yr)

= $7,484,233/yr

e. Calculate the annual flow sewer service fee in dollars per year.

Flow Fee, $/yr = (Flow, MGD)(Flow Fee, $/MG)(Work, days/yr)

= (0.045 MGD)($218/MG)(260 days/yr)

= $2,551/yr

f. Calculate the total sewer service fee for the meat packing plant in dollars per year by summing the charges for BOD, SS and flow.

BOD, $/yr	=	$23,223,564
SS, $/yr	=	7,484,233
Flow, $/yr	=	2,551
Total Fee, $/yr	=	$30,710,348

19. A metal finisher has the following discharge limitations: copper, 3.0 mg/L; lead, 0.7 mg/L; chromium, 2.7 mg/L; and nickel, 3.3 mg/L. The metal finisher had the following metal concentrations in a recent 24-hour discharge sample collected and analyzed for sewer-use fees: Cu, 15.0 mg/L; Pb, 3.2 mg/L; Cr, 18.3 mg/L; and Ni, 6.3 mg/L. The metal finisher discharges a flow of 30,000 gallons per day (0.03 MGD). The sewer-use penalties in dollars per pound discharged over the limitations are: Cu, $225/lb; Pb, $325/lb; Cr, $250/lb; and Ni, $375/lb. What would be the metal finisher's daily monetary penalty for exceeding the discharge limitations?

Known		Unknown
Discharge Limitations, mg/L		Penalty Fee, $/day
Discharge Values, mg/L		
Discharge Penalties, $/lb		
Flow, GPD	= 30,000 GPD	
	= 0.03 MGD	

a. Calculate the excess value in milligrams per liter for each metal by subtracting the limit from the discharge.

Metal	Discharge	–	Limit	=	Excess
Cu	15.0	–	3.0	=	12.0 mg/L
Pb	3.2	–	0.7	=	2.5 mg/L
Cr	18.3	–	2.7	=	15.6 mg/L
Ni	6.3	–	3.3	=	3.0 mg/L

b. Calculate the excess pounds per day of metal discharge to the sewer for each metal.

Excess, lbs/day = (Flow, MGD)(Excess, mg/L)(8.34 lbs/gal)

Metal	(Flow, MGD)(Excess, mg/L)(8.34 lbs/gal)	= Excess, lbs/day
Cu	(0.03 MGD)(12.0 mg/L)(8.34 lbs/gal)	= 3.002 lbs/day
Pb	(0.03 MGD)(2.5 mg/L)(8.34 lbs/gal)	= 0.626 lbs/day
Cr	(0.03 MGD)(15.6 mg/L)(8.34 lbs/gal)	= 3.903 lbs/day
Ni	(0.03 MGD)(3.0 mg/L)(8.34 lbs/gal)	= 0.751 lbs/day

c. Calculate the total penalty fees for the metal finisher in dollars per day by adding up the fee for each metal.

Fee, $/day = (Excess, lbs/day)(Penalty, $/lb)

Metal	(Excess, lbs/day)(Penalty, $/lb)	=	Fee, $/day
Cu	(3.002 lbs/day)($225/lb)	=	$ 675.45/day
Pb	(0.626 lbs/day)($325/lb)	=	$ 203.45/day
Cr	(3.903 lbs/day)($250/lb)	=	$ 975.75/day
Ni	(0.751 lbs/day)($375/lb)	=	$ 281.63/day
	Total Penalty Fee	=	$2,136.28/day

20. A company uses 20,000 gallons of water per day. The company claims the following water loss percentages:

Evaporative:	10%
Landscaping:	13%
Other:	5%

The company's wastestream BOD is 650 mg/L and suspended solids are 800 mg/L. The company is subject to the following sewer-use fees:

BOD = $65.00/lb BOD

SS = $72.00/lb SS

Flow = $225.00/MG

Determine the company's annual sewer-use fee if it works 312 days per year.

Known	Unknown
Water Loss Percentages	Sewer-Use Fee, $/yr
Sewer-Use Fees	
In Flow, GPD = 20,000 GPD	
= 0.02 MGD	
BOD, mg/L = 650 mg/L	
SS, mg/L = 800 mg/L	
Work, days/yr = 312 days/yr	

a. Calculate the total flow to the sewer in million gallons per year.

Sum of Losses, % = Evap, % + Landscape, % + Other, %

$$= 10\% + 13\% + 5\%$$

$$= 28\%$$

$$\text{Flow to Sewer, gal/yr} = \frac{(\text{Total} - \text{Loss})(\text{Flow, GPD})(\text{Work, days/yr})}{\text{Total}}$$

$$= \frac{(100\% - 28\%)}{100\%}(20{,}000 \text{ GPD})(312 \text{ days/yr})$$

$$= 4{,}492{,}800 \text{ gallons/yr}$$

$$= 4.4928 \text{ MGal/yr}$$

b. Calculate the amounts discharged to the sewer per year.

Amount, lbs/yr = (Flow, MG/yr)(Waste, mg/L)(8.34 lbs/gal)

Waste	(Flow, MG/yr)(Waste, mg/L)(8.34 lbs/gal) = Amount, lbs/yr
BOD	(4.4928 MG/yr)(650 mg/L)(8.34 lbs/gal) = 24,355 lbs/yr
SS	(4.4928 MG/yr)(800 mg/L)(8.34 lbs/gal) = 29,976 lbs/yr

c. Calculate the sewer-use fee in dollars per year.

Fee, $/yr = (Amount, lbs/yr)(Charge, $/lb)

Waste	(Amount, lbs/yr)(Charge, $/lb) = Fee, $/yr
BOD	(24,355 lbs/yr)($65/lb) = $1,583,075/yr
SS	(29,976 lbs/yr)($72/lb) = 2,158,272/yr
Flow	(4.4928 MG/yr)($225/MG) = 1,011/yr
	Total Sewer-Use Fee = $3,742,358/yr

21. A company that reclaims lead has an industrial waste discharge of 65,000 GPD with an average sulfite concentration of 7,500 mg/L (measured as sulfite sulfur). The company works 250 days per year. The local POTW has an industrial waste sewer-use fee that charges $23.50 per thousand pounds of COD. How much does this company pay each year for the sulfite discharge? The molecular weight of sulfur is 32. There are two equivalents per mole of sulfite and 8,000 milligrams of COD per equivalent.

Known	Unknown
Flow, GPD = 65,000 GPD	Sewer-Use Fee, $/yr
Sulfite, mg/L = 7,500 mg/L	
Work, days/yr = 250 days/yr	
Fee, $/COD = $23.50/1,000 lbs COD	
Mol Wt S = 32	
Sulfite = 2 equiv/mole	
COD = 8,000 mg COD/equiv	

a. Calculate the amount of sulfite discharged in milligrams per day.

Sulfite, mg/day = (Flow, GPD)(Sulfite, mg/L)(3.785 L/gal)

$$= (65{,}000 \text{ GPD})(7{,}500 \text{ mg/L})(3.785 \text{ L/gal})$$

$$= 1{,}845{,}187{,}500 \text{ mg/day}$$

b. Calculate the pounds of COD discharged per day.

$$\text{COD, lbs/day} = \frac{(\text{Sulfite, mg/day})(\text{Sulfite, eq/mole})(\text{COD, mg/eq})}{(\text{Sulfite, mg/mole})(1{,}000 \text{ mg/gm})(454 \text{ gm/lb})}$$

$$= \frac{(1{,}845{,}187{,}500 \text{ mg/day})(2 \text{ eq/mole})(8{,}000 \text{ mg/eq})}{(32{,}000 \text{ mg/mole})(1{,}000 \text{ mg/gm})(454 \text{ gm/lb})}$$

$$= 2{,}032 \text{ lbs COD/day}$$

c. Calculate the sewer-use fee in dollars per year for the COD contributed by the sulfite discharge.

$$\text{Sewer-Use Fee, } \$/yr = \frac{(\text{COD, lbs/day})(\text{Work, days/yr})(\text{Fee, }\$/1{,}000 \text{ lbs})}{1{,}000 \text{ lbs/thousand}}$$

$$= \frac{(2{,}032 \text{ lbs/day})(250 \text{ days/yr})(\$23.50/1{,}000 \text{ lbs COD})}{1{,}000 \text{ lbs/thousand}}$$

$$= \$11{,}938/yr$$

M. DETENTION TIME

22. A pretreatment facility has a cylindrical tank eight feet in diameter and ten feet tall. This tank is used as a detention tank for chromium treatment. The total flow during an eight-hour work day is 15,000 gallons. What is the theoretical detention time for this tank in minutes?

Known	Unknown
Diameter, ft = 8 ft	Detention Time, min
Height, ft = 10 ft	
Work Day, hr = 8 hr	
Flow Vol, gal = 15,000 gal	

a. Calculate the volume of the cylindrical tank in gallons.

Tank Vol, gal = (0.785)(Diameter, ft)²(Height, ft)(7.48 gal/cu ft)

$$= (0.785)(8 \text{ ft})^2(10 \text{ ft})(7.48 \text{ gal/cu ft})$$

$$= 3{,}758 \text{ gal}$$

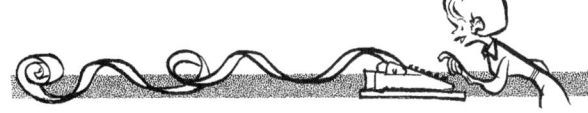

b. Calculate the flow rate in gallons per minute.

$$\text{Flow Rate, GPM} = \frac{\text{Flow Volume, gal}}{(\text{Work Day, hr})(60 \text{ min/hr})}$$

$$= \frac{15,000 \text{ gal}}{(8 \text{ hr})(60 \text{ min/hr})}$$

$$= 31.25 \text{ GPM}$$

c. Determine the theoretical detention time in minutes.

$$\text{Detention Time, minutes} = \frac{\text{Tank Volume, gal}}{\text{Flow Rate, gal/min}}$$

$$= \frac{3,758 \text{ gal}}{31.25 \text{ gal/min}}$$

$$= 120 \text{ minutes}$$

23. A pretreatment facility has a rectangular clarifier 10.5 feet long, 6.25 feet wide and a water depth of 4.75 feet. If the total flow during an eight-hour work day is 30,000 gallons, what is the detention time in minutes for the clarifier?

Known		Unknown
Length, ft	= 10.5 ft	Detention Time, min
Width, ft	= 6.25 ft	
Depth, ft	= 4.75 ft	
Total Flow, gal	= 30,000 gal	
Work Day, hr	= 8 hr	

a. Calculate the volume of the rectangular clarifier in gallons.

$$\text{Tank Vol, gal} = (\text{Length, ft})(\text{Width, ft})(\text{Depth, ft})(7.48 \text{ gal/cu ft})$$

$$= (10.5 \text{ ft})(6.25 \text{ ft})(4.75 \text{ ft})(7.48 \text{ gal/cu ft})$$

$$= 2,332 \text{ gal}$$

b. Calculate the flow rate in gallons per minute.

$$\text{Flow Rate, GPM} = \frac{\text{Flow Volume, gal}}{(\text{Work Day, hr})(60 \text{ min/hr})}$$

$$= \frac{30,000 \text{ gal}}{(8 \text{ hr})(60 \text{ min/hr})}$$

$$= 62.5 \text{ gal/min}$$

c. Determine the theoretical detention time in minutes.

$$\text{Detention Time, minutes} = \frac{\text{Tank Volume, gal}}{\text{Flow Rate, gal/minute}}$$

$$= \frac{2,332 \text{ gal}}{62.5 \text{ GPM}}$$

$$= 37.3 \text{ min}$$

24. If a pretreatment facility must provide a 30-minute detention time to treat a flow of 100,000 GPD during a 24-hour work day, what size clarifier is needed in gallons and in cubic feet?

Known		Unknown
Detention Time, min	= 30 min	Clarifier Size in:
Flow, GPD	= 100,000 GPD	1. Gallons
Work Day, hr	= 24 hr	2. Cubic feet

a. Calculate the flow rate in gallons per minute.

$$\text{Flow Rate, GPM} = \frac{\text{Flow, GPD}}{(24 \text{ hr/day})(60 \text{ min/hr})}$$

$$= \frac{100,000 \text{ GPD}}{(24 \text{ hr/day})(60 \text{ min/hr})}$$

$$= 69.4 \text{ GPM}$$

b. Calculate the clarifier size in gallons.

$$\text{Size, gallons} = (\text{Flow Rate, GPM})(\text{Detention Time, min})$$

$$= (69.4 \text{ GPM})(30 \text{ min})$$

$$= 2,082 \text{ gallons}$$

c. Convert the clarifier size in gallons to size in cubic feet.

$$\text{Size, cu ft} = \frac{\text{Size, gal}}{7.48 \text{ gal/cu ft}}$$

$$= \frac{2,082 \text{ gal}}{7.48 \text{ gal/cu ft}}$$

$$= 278.3 \text{ cu ft}$$

25. Determine the time in minutes for a flow of 40,000 GPD to pass through an interceptor with a volume of 1,500 gallons. Assume the interceptor is flowing full and flow moves as a slug or plug flow.

Known		Unknown
Flow Rate, GPD	= 40,000 GPD	Time, minutes
Volume, gal	= 1,500 gal	

Calculate the time in minutes in the interceptor.

$$\text{Detention Time, min} = \frac{(\text{Volume, gal})(24 \text{ hr/day})(60 \text{ min/hr})}{\text{Flow Rate, GPD}}$$

$$= \frac{(1,500 \text{ gal})(24 \text{ hr/day})(60 \text{ min/hr})}{40,000 \text{ gal/day}}$$

$$= 54 \text{ min}$$

N. pH NEUTRALIZATION

NOTE: Industrial wastes normally have their pH adjusted to meet discharge limits or to get a specific pH level for a chemical reaction to take place. The pH adjustment dose is determined in the lab by titrating with a chemical (acid or base) until the desired pH is reached. This procedure is essential and considers the "mix" of chemicals in the waste being treated. The procedures in this section are strictly theoretical. However, these calculations have been used by pretreatment inspectors to determine the approximate amount of neutralizing agent that should be on hand at an industrial pretreatment site or to determine the feasibility of neutralizing a slug dump or accidental spill of an acid or base when titration results are not available.

26. How many pounds of anhydrous ammonia are required to neutralize an industrial waste flow of 40,000 gallons per day having a pH of 2.0?

Known	Unknown
Flow, GPD = 40,000 GPD	Anhydrous Ammonia, lbs/day
pH = 2.0	

CHEMICAL REACTIONS

$NH_3 + H_2O \rightarrow NH_4^+ + OH^-$

and

NH_3 = 17 grams NH_3 per mole

A pH of 2.0 has a hydrogen ion concentration of 10^{-2} moles H^+ per liter or 0.01 moles H^+ per liter.

a. Calculate the moles of hydrogen ion to be neutralized.

H^+, moles/day = (Flow, GPD)(3.785 L/gal)(H^+, moles/L)

= (40,000 gal/day)(3.785 L/gal)(0.01 mole H^+/L)

= 1,514 moles H^+/day

b. Determine the anhydrous ammonia dose in pounds per day to neutralize the industrial waste flow.

$$NH_3, \text{lbs/day} = \frac{(H^+, \text{moles/day})(NH_3, \text{gm/mole})}{454 \text{ gm/lb}}$$

$$= \frac{(1,514 \text{ moles/day})(17 \text{ gm/mole})}{454 \text{ gm/lb}}$$

= 56.7 lbs NH_3/day

NOTES: 1. Many POTW ordinances regulate the discharge of ammonia and lime may be the preferred neutralizing agent.

2. In the metal finishing industry and electroplating industry, few would ever use anhydrous ammonia for acid neutralization because it would complex (bind) copper which might be present in the wastestream.

27. How many pounds of sodium hydroxide (NaOH) are needed to neutralize an industrial wastewater flow of 35,000 GPD with a pH of 1.0?

Known	Unknown
Flow, GPD = 35,000 GPD	Sodium Hydroxide, lbs/day
pH = 1.0	

CHEMICAL REACTION

In water $NaOH \rightarrow Na^+ + OH^-$

and $NaOH$ = 40 grams NaOH per mole

A pH of 1.0 has a hydrogen ion concentration of 10^{-1} moles H^+ per liter or 0.1 moles H^+ per liter.

a. Calculate the moles of hydrogen ion to be neutralized.

H^+, moles/day = (Flow, GPD)(3.785 L/gal)(H^+, moles/L)

= (35,000 GPD)(3.785 L/gal)(0.1 mole H^+/L)

= 13,248 moles H^+/day

b. Determine the sodium hydroxide dose in pounds per day to neutralize the industrial waste flow.

$$NaOH, \text{lbs/day} = \frac{(H^+, \text{moles/day})(NaOH, \text{gm/mole})}{454 \text{ gm/lb}}$$

$$= \frac{(13,248 \text{ moles } H^+/\text{day})(40 \text{ gm/mole})}{454 \text{ gm/lb}}$$

= 1,167 lbs NaOH/day

28. How many pounds per day of soda ash (Na_2CO_3) will be needed to neutralize an industrial waste flow of 30,000 gallons per day with a pH of 3.0? The molecular weight of soda ash is 106.

Known	Unknown
Flow, GPD = 30,000 GPD	Soda Ash, lbs/day
pH = 3.0	
Mol Wt Na_2CO_3 = 106	

CHEMICAL REACTION

$Na_2CO_3 + 2H^+\text{-}R \rightarrow 2Na^+\text{-}R + CO_2 + H_2O$

A pH of 3.0 has a hydrogen ion concentration of 10^{-3} moles H^+ per liter or 0.001 moles H^+ per liter.

a. Calculate the moles of hydrogen ion to be neutralized.

H^+, moles/day = (Flow, GPD)(3.785 L/gal)(H^+, moles/L)

= (30,000 GPD)(3.785 L/gal)(0.001 mole H^+/L)

= 114 moles H^+/day

b. Determine the pounds per day of soda ash needed. One mole of soda ash will neutralize two moles of hydrogen ion.

$$\text{Soda Ash, lbs/day} = \frac{(H^+, \text{moles/day})(106 \text{ gm/mole } Na_2CO_3)}{(2 \text{ moles } H^+/\text{mole } Na_2CO_3)(454 \text{ gm/lb})}$$

$$= \frac{(114 \text{ moles } H^+/\text{day})(106 \text{ gm/mole } Na_2CO_3)}{(2 \text{ moles } H^+/\text{mole } Na_2CO_3)(454 \text{ gm/lb})}$$

= 13.3 lbs Soda Ash/day

29. An industrial manufacturing process discharges 32,000 gallons per day. The wastewater pH is 2.0. How many gallons per day of a 25 percent solution of sodium hydroxide (NaOH) will be necessary to maintain a discharge pH of 7.0? The molecular weight of sodium hydroxide is 40.

Known	Unknown
Flow, GPD = 32,000 GPD	NaOH, GPD
Waste pH = 2.0	
NaOH, % = 25%	
Discharge pH = 7.0	
Mol Wt NaOH = 40	

CHEMICAL REACTION

$NaOH + H^+\text{-}R \rightarrow Na^+\text{-}R + H_2O$

A pH of 2.0 has a hydrogen ion concentration of 10^{-2} moles H^+ per liter or 0.01 moles H^+ per liter.

a. Calculate the moles of hydrogen ion to be neutralized.

H^+, moles/day = (Flow, GPD)(3.785 L/gal)(H^+, moles/L)

= (32,000 GPD)(3.785 L/gal)(0.01 mole H^+/L)

= 1,211 moles H^+/day

b. Determine the gallons per day of 25 percent sodium hydroxide solution needed. A 25 percent solution contains 250 grams of sodium hydroxide in one liter.

$$\frac{NaOH,}{gal/day} = \frac{(H^+, \text{ moles/day})(40 \text{ gm/mole NaOH})}{(250 \text{ gm NaOH}/L)(3.785 \text{ } L/gal)}$$

$$= \frac{(1,211 \text{ moles } H^+/day)(40 \text{ gm/mole})}{(250 \text{ gm NaOH}/L)(3.785 \text{ } L/gal)}$$

= 51 gal 25% NaOH/day

30. A plating company discharges a waste flow of 25,000 gallons per day with a pH of 5.0. The discharge pH limit is 6.5. How many pounds per day of anhydrous aqueous ammonia will be necessary to produce a pH of 7.0? The molecular weight of ammonia, NH_3, is 17.

Known		Unknown
Flow, GPD	= 25,000 GPD	Ammonia, lbs/day
pH Waste	= 5.0	
pH Limit	= 6.5	
pH Treat	= 7.0	
Mol Wt NH_3	= 17	

CHEMICAL REACTION

NH_3 (in water, H_2O) $\rightarrow NH_4^+ + OH^-$

A pH of 5.0 has a hydrogen ion concentration of 10^{-5} moles H^+ per liter or 0.00001 moles H^+ per liter.

a. Calculate the moles of hydrogen ion to be neutralized.

H^+, moles/day = (Flow, GPD)(3.785 L/gal)(H^+, moles/L)

= (25,000 GPD)(3.785 L/gal)(0.00001 mole H^+/L)

= 0.946 moles H^+/day

b. Determine the pounds per day of anhydrous aqueous ammonia needed.

$$\frac{Ammonia,}{lbs/day} = \frac{(H^+, \text{ moles/day})(17 \text{ gm/mole } NH_3)}{454 \text{ gm/lb}}$$

$$= \frac{(0.946 \text{ moles } H^+/day)(17 \text{ gm/mole})}{454 \text{ gm/lb}}$$

= 0.035 lbs/day

31. A battery manufacturer uses lime to neutralize sulfuric acid wastes. The wastestream flow is 40,000 gallons per day. The acid is neutralized using an eighty-five percent slurry of lime and the initial pH is 1.0. The molecular weight of lime, CaO, is 56. How many gallons per day of 85 percent lime solution will be needed?

Known		Unknown
Flow, GPD	= 40,000 GPD	Lime, gal/day
Lime, %	= 85%	
Waste pH	= 1.0	
Mol Wt Lime	= 56	

CHEMICAL REACTION

$H_2SO_4 + CaO \rightarrow CaSO_4 + H_2O$

A pH of 1.0 has a hydrogen ion concentration of 10^{-1} moles H^+ per liter or 0.1 moles H^+ per liter.

a. Calculate the moles of hydrogen ion to be neutralized.

H^+, moles/day = (Flow, GPD)(3.785 L/gal)(H^+, moles/L)

= (40,000 GPD)(3.785 L/gal)(0.1 mole H^+/L)

= 15,140 moles H^+/day

b. Determine the gallons of 85 percent lime required. One mole of lime is required for every two moles of acid. An 85 percent solution contains 850 grams of lime in one liter.

$$\frac{Lime,}{gal/day} = \frac{(H^+, \text{ moles/day})(56 \text{ gm/mole Lime})}{(2 \text{ moles } H^+/\text{mole Lime})(Lime, \text{ gm}/L)(3.785 \text{ } L/gal)}$$

$$= \frac{(15,140 \text{ moles } H^+/day)(56 \text{ gm/mole Lime})}{(2 \text{ moles } H^+/\text{mole Lime})(850 \text{ gm}/L)(3.785 \text{ } L/gal)}$$

= 132 gal 85% Lime/day

O. pH ADJUSTMENT

32. The effluent from a filter press treating an industrial sludge must be neutralized before additional treatment. If a 50 mL sample of effluent requires 6.2 mL of 0.5 N sulfuric acid to lower the pH to 7.0, how many gallons of 2 N sulfuric acid will be required to neutralize 3,200 gallons of filter press effluent to a pH of 7.0?

NOTE: Frequently a permit will specify a maximum pH limit of 8.0 or 8.5. Use the same procedure to determine the amount of acid to adjust the pH to the necessary level.

Known		Unknown
Sample, mL	= 50 mL	2 N Acid, gal
Acid, mL	= 6.2 mL	
Acid, N	= 0.5 N for sample	
Waste, gal	= 3,200 gal	
Acid, N	= 2 N for waste treatment	

a. Determine normality of waste.

Millequivalents = (mL)(N)

$$\text{Waste Normality} = \frac{(Acid, \text{ mL})(Acid, \text{ } N)}{Sample, \text{ mL}}$$

$$= \frac{(6.2 \text{ mL})(0.5 \text{ } N)}{50 \text{ mL}}$$

= 0.062 N

b. Calculate the gallons of 2 *N* sulfuric acid to neutralize the filter press effluent.

$$\frac{2\ N\ \text{Sulfuric}}{\text{Acid, gal}} = \frac{(\text{Waste, gal})(\text{Waste, } N)}{\text{Sulfuric Acid, } N}$$

$$= \frac{(3{,}200\ \text{gal})(0.062\ N)}{2\ N}$$

$$= 100\ \text{gal}$$

33. A metal wastestream has a pH of 7.2. A one-quarter gallon (0.25 gal) sample of the wastestream required 0.1 gallon of 1 percent caustic to increase the pH to 10 for hydroxide precipitation. How many gallons of 4 percent caustic are required for 100 gallons of the wastestream using a batch treatment process?

Known	Unknown
Sample, gal = 0.25 gal	4% Caustic, gal
Caustic, gal = 0.1 gal	
Caustic, % = 1% for sample	
Caustic, % = 4% for treatment	
Waste, gal = 100 gal	

Calculate the gallons of 4% caustic needed.

$$\frac{4\%\ \text{Caustic,}}{\text{gal}} = \frac{\left(\begin{smallmatrix}\text{Caustic to}\\\text{Sample, gal}\end{smallmatrix}\right)\left(\begin{smallmatrix}\text{Caustic to}\\\text{Sample, \%}\end{smallmatrix}\right)\left(\begin{smallmatrix}\text{Waste to}\\\text{Treatment, gal}\end{smallmatrix}\right)}{(\text{Sample, gal})(\text{Treatment Caustic, \%})}$$

$$= \frac{(0.1\ \text{gal})(1\%)(100\ \text{gal})}{(0.25\ \text{gal})(4\%)}$$

$$= 10\ \text{gal}$$

34. An industrial wastewater with a pH of 10.8 flows from an equalization tank to a neutralization mixing tank at a rate of 9 GPM. Lab tests indicate that a 100 m*L* sample of the waste requires 11.3 m*L* of 0.5 *N* sulfuric acid to lower the pH to 7.0. Determine the setting in gallons per day on a chemical feed pump which is pumping 2 *N* sulfuric acid to the neutralization mixing tank.

Known	Unknown
Waste Flow, GPM = 9 GPM	Acid Feed, GPD
Waste Sample, m*L* = 100 m*L*	
Acid Vol, m*L* = 11.3 m*L*	
Acid, *N* = 0.5 *N*	
Acid Feed, *N* = 2 *N*	

a. Determine normality of waste.

$$\text{Millequivalents} = (\text{m}L)(N)$$

$$\text{Waste Normality} = \frac{(\text{Acid Vol, m}L)(\text{Acid, } N)}{\text{Waste Sample, m}L}$$

$$= \frac{(11.3\ \text{m}L)(0.5\ N)}{100\ \text{m}L}$$

$$= 0.0565\ N$$

b. Calculate the flow to be delivered by the chemical feed pump.

$$\frac{\text{Acid Feed,}}{\text{GPD}} = \frac{(\text{Waste Flow, GPM})}{(\text{Waste Normality})(60\ \text{min/hr})(24\ \text{hr/day})}{\text{Acid Feed, } N}$$

$$= \frac{(9\ \text{GPM})(0.0565\ N)(60\ \text{min/hr})(24\ \text{hr/day})}{2\ N}$$

$$= 366\ \text{GPD}$$

P. HYDROXIDE PRECIPITATION

35. Zinc is being removed in a plating wastestream by hydroxide precipitation. Laboratory tests indicate that 1.8 milliliters of a four percent or one normal sodium hydroxide solution will increase the pH of one liter of wastewater to 10 and precipitate the zinc. How many gallons of the four percent sodium hydroxide are required to treat 500 gallons of zinc wastewater?

Known	Unknown
Lab Results	NaOH, gal
1.8 m*L* NaOH/*L* Wastewater	
Wastewater, gal = 500 gal	

Calculate the gallons of sodium hydroxide needed.

$$\text{NaOH, gal} = \frac{(\text{Wastewater, gal})(\text{m}L\ \text{NaOH})}{(1{,}000\ \text{m}L/L)(1\ L\ \text{Wastewater})}$$

$$= \frac{(500\ \text{gal})(1.8\ \text{m}L\ \text{NaOH})}{(1{,}000\ \text{m}L/L)(1\ L\ \text{Wastewater})}$$

$$= 0.9\ \text{gal NaOH}$$

36. Zinc is being removed in a plating wastestream by hydroxide precipitation. Laboratory tests indicate that 1.8 milliliters of a four percent or one normal sodium hydroxide solution will increase the pH of one liter of wastewater to 10 and precipitate the zinc. Determine the setting on the sodium hydroxide feed pump in gallons per day to treat a zinc wastewater flow of 10 GPM.

Known	Unknown
Lab Results	NaOH Feed, GPD
1.8 m*L* NaOH/*L* Wastewater	
Zinc Flow, GPM = 10 GPM	

Calculate the sodium hydroxide feed rate in gallons per day.

$$\frac{\text{NaOH}}{\text{Feed,}}{\text{GPD}} = \frac{(\text{Zinc Flow, gal/min})(\text{m}L\ \text{NaOH})(60\ \text{min/hr})(24\ \text{hr/day})}{(1{,}000\ \text{m}L/L)(1\ L\ \text{Zinc})}$$

$$= \frac{(10\ \text{GPM Zinc})(1.8\ \text{m}L\ \text{NaOH})(60\ \text{min/hr})(24\ \text{hr/day})}{(1{,}000\ \text{m}L/L)(1\ L\ \text{Zinc})}$$

$$= 26\ \text{GPD}$$

Q.   COMPLEXED METAL PRECIPITATION

37. Complexed copper is being removed from a plating wastestream by hydroxide precipitation. Laboratory results indicate that 10 milliliters of four percent or one normal sodium hydroxide will increase the pH of one liter of wastewater to 12 and precipitate the copper. How many gallons of four percent sodium hydroxide are required to treat 400 gallons of complexed copper wastewater?

Known	Unknown
Lab Results	NaOH, gal
10 mL NaOH/L Wastewater	
Wastewater, gal = 400 gal	

Calculate the gallons of sodium hydroxide needed.

$$\text{NaOH, gal} = \frac{(\text{Wastewater, gal})(\text{m}L\text{ NaOH})}{(1{,}000\text{ m}L/L)(1\ L\text{ Wastewater})}$$

$$= \frac{(400\text{ gal})(10\text{ m}L\text{ NaOH})}{(1{,}000\text{ m}L/L)(1\ L\text{ Wastewater})}$$

$$= 4\text{ gal NaOH}$$

38. Complexed copper is being removed from a plating wastestream by hydroxide precipitation. Laboratory results indicate that 10 milliliters of four percent or one normal sodium hydroxide will increase the pH of one liter of wastewater to 12 and precipitate the copper. Determine the setting on the sodium hydroxide feed pump in gallons per day to treat a complexed copper wastewater flow of 12 GPM.

Known	Unknown
Lab Results	NaOH Feed, GPD
10 mL NaOH/L Wastewater	
Copper Flow, GPM = 12 GPM	

Calculate the sodium hydroxide feed rate in gallons per day.

$$\text{NaOH Feed, GPD} = \frac{(\text{Copper Flow, gal/min})(\text{m}L\text{ NaOH})(60\text{ min/hr})(24\text{ hr/day})}{(1{,}000\text{ m}L/L)(1\ L\text{ Copper})}$$

$$= \frac{(12\text{ GPM Copper})(10\text{ m}L\text{ NaOH})(60\text{ min/hr})(24\text{ hr/day})}{(1{,}000\text{ m}L/L)(1\ L\text{ Copper})}$$

$$= 173\text{ GPD}$$

NOTE: If the copper was complexed with ammonia, it would not precipitate.

R.   REDUCTION OF HEXAVALENT CHROMIUM

39. How much sulfur dioxide is required to treat 1,100 gallons of chromic acid containing 1,400 mg/L of hexavalent chromium? Assume that one pound of hexavalent chromium is reduced to the trivalent state by the addition of three pounds of sulfur dioxide.

Known	Unknown
Waste, gal = 1,100 gal	Dosage, lbs SO$_2$
Conc, mg/L = 1,400 mg Cr^{6+}/L	
Treat, lb/lb = 3 lbs SO$_2$/lb Cr^{6+}	

a. Calculate the pounds of Cr^{6+} to be treated.

$$\text{Cr}^{6+}\text{ Treated, lbs} = (\text{Waste, MG})(\text{Cr}^{6+}, \text{mg}/L)(8.34\text{ lbs/gal})$$

$$= (0.0011\text{ MG})(1{,}400\text{ mg}/L)(8.34\text{ lbs/gal})$$

$$= 12.8\text{ lbs Cr}^{6+}$$

b. Calculate the dosage of sulfur dioxide.

$$\text{Dosage, lbs SO}_2 = (\text{Cr}^{6+}\text{ Treated, lbs})(\text{Treatment, lbs SO}_2/\text{lb Cr}^{6+})$$

$$= (12.8\text{ lbs Cr}^{6+})(3\text{ lbs SO}_2/\text{lb Cr}^{6+})$$

$$= 38.4\text{ lbs SO}_2$$

NOTE: Sulfur dioxide is used where the amount of hexavalent chromium to be reduced is high. In the case where the amount of hexavalent chromium is low, the more economical method is to use sodium bisulfite in acid media, or sodium hydrosulfite in neutral to alkaline media.

40. A chrome waste flowing at a rate of 50 GPM from a manufacturing process contains 240 mg/L of hexavalent chromium. Determine the sulfonator feed rate in pounds of sulfur dioxide per minute if one pound of hexavalent chromium is reduced to the trivalent state by the addition of three pounds of sulfur dioxide.

Known	Unknown
Flow, GPM = 50 GPM	Sulfonator Feed, lbs/min
Waste, mg/L = 240 mg Cr^{6+}/L	
Treat, lb/lb = 3 lbs SO$_2$/lb Cr^{6+}	

Calculate the sulfonator feed rate in pounds of sulfur dioxide per minute.

$$\text{Sulfonator Feed, lbs/min} = \frac{(\text{Flow, GPM})(\text{Waste, mg}/L)(8.34\text{ lbs/gal})(\text{Treat, lb/lb})}{1{,}000{,}000/\text{M}}$$

$$= \frac{(50\text{ GPM})(240\text{ mg Cr}^{6+}/L)(8.34\text{ lbs/gal})(3\text{ lbs SO}_2)}{(1{,}000{,}000/\text{M})(1\text{ lb Cr}^{6+})}$$

$$= \frac{(50)(240)(8.34)(3)}{1{,}000{,}000}$$

$$= 0.3\text{ lbs SO}_2/\text{min}$$

OR

$$\text{Sulfonator Feed, lbs/day} = (0.3\text{ lbs SO}_2/\text{min})(60\text{ min/hr})(24\text{ hr/day})$$

$$= 432\text{ lbs/day}$$

If the flow of 50 GPM is converted to MGD (50 GPM/694 GPM/MGD = 0.072 MGD), then the sulfonator feed rate of sulfur dioxide could be calculated as shown below.

$$\text{Sulfonator Feed, lbs/day} = (\text{Flow, MGD})(\text{Waste, mg}/L)(8.34\text{ lbs/gal})(\text{Treat, lb/lb})$$

$$= (0.072\text{ MGD})(240\text{ mg Cr}^{6+}/L)(8.34\text{ lbs/gal})(3\text{ lbs SO}_2/1\text{ lb Cr}^{6+})$$

$$= 432\text{ lbs/day}$$

41. How much time will be required to completely reduce 15,000 gallons of chrome waste with a concentration of 240 mg/L of hexavalent chromium if the sulfonator is set to feed 1,200 pounds of sulfur dioxide per day? Assume one pound of hexavalent chromium is reduced to the trivalent state by three pounds of sulfur dioxide.

Known	Unknown
Waste Vol, gal = 15,000 gal	Time, min
Conc, mg/L = 240 mg Cr^{6+}/L	
Feed, lbs/day = 1,200 lbs SO_2/day	
Treat, lb/lb = 3 lbs SO_2/lb Cr^{6+}	

a. Calculate the pounds of hexavalent chromium to be treated.

$$\text{Waste, lbs } Cr^{6+} = (\text{Waste Vol, MG})(\text{Conc, mg } Cr^{6+}/L)(8.34 \text{ lbs/gal})$$

$$= (0.015 \text{ MG})(240 \text{ mg } Cr^{6+}/L)(8.34 \text{ lbs/gal})$$

$$= 30 \text{ lbs } Cr^{6+}$$

b. Calculate the time required to add the sulfur dioxide.

$$\frac{\text{Time,}}{\text{min}} = \frac{(\text{Waste, lbs } Cr^{6+})(\text{Treat, lbs } SO_2/\text{lb } Cr^{6+})(1,440 \text{ min/day})}{\text{Feed, lbs } SO_2/\text{day}}$$

$$= \frac{(30 \text{ lbs } Cr^{6+})(3 \text{ lbs } SO_2/\text{lb } Cr^{6+})(1,440 \text{ min/day})}{1,200 \text{ lbs } SO_2/\text{day}}$$

$$= 108 \text{ min}$$

42. A batch treatment process is used by a plating shop to treat 800 gallons of a chrome plating waste by the reduction and precipitation process to reduce the hexavalent chromium in the waste to the trivalent stage. Assume one mg/L of hexavalent chromium will require 16 mg/L of ferrous sulfate ($FeSO_4 \cdot 7H_2O$), 6 mg/L of sulfuric acid, and 9.5 mg/L lime. The hexavalent chromium concentration in the wastes is 320 mg/L.

1. If the ferrous sulfate is fed as a four percent solution at a rate of 80 gallons per hour, how long should the ferrous sulfate feed pump operate?

2. If the sulfuric acid is fed as a 2.8 percent or 0.57 Normal sulfuric acid solution at a rate of 35 gallons per hour, how long should the sulfuric acid feed pump operate?

3. If the lime slurry feeder mixes 0.5 pounds of lime per gallon of water and delivers a lime slurry of 50 gallons per hour to the mixing tank, how long should the lime slurry feeder operate?

Known

Volume Cr Waste, gal	= 800 gal
Cr^{6+} Conc, mg/L	= 320 mg/L
Ferrous Sulfate Dose, mg/L/mg/L	= 16 mg/L/mg/L
Ferrous Sulfate, %	= 4%
Ferrous Sulfate Feed, GPH	= 80 GPH
Sulfuric Dose, mg/L/mg/L	= 6 mg/L/mg/L
Sulfuric Acid, %	= 2.8%
Sulfuric Feed, GPH	= 35 GPH
Lime Dose, mg/L/mg/L	= 9.5 mg/L/mg/L
Lime Conc, lbs/gal	= 0.5 lbs/gal
Lime Feed, GPH	= 50 GPH

Unknown

Time to Operate:	1. Ferrous Sulfate Feed, hr
	2. Sulfuric Feed, hr
	3. Lime Feed, hr

a. Calculate the pounds of hexavalent chromium to be treated.

$$\frac{\text{Chromium,}}{\text{lbs}} = \frac{(\text{Vol Cr Waste, gal})(Cr^{6+} \text{ Conc, mg/}L)(8.34 \text{ lbs/gal})}{1,000,000/M}$$

$$= \frac{(800 \text{ gal})(320 \text{ mg/}L)(8.34 \text{ lbs/gal})}{1,000,000/M}$$

$$= 2.14 \text{ lbs hexavalent chromium}$$

b. Calculate the pounds of ferrous sulfate needed to treat the chromium.

$$\frac{\text{Ferrous}}{\text{Sulfate, lbs}} = (\text{Chromium, lbs})(\text{Ferrous Sulfate Dose, mg/}L/\text{mg/}L)$$

$$= (2.14 \text{ lbs } Cr^{6+})(16 \text{ mg/}L/\text{mg/}L)$$

$$= 34.3 \text{ lbs ferrous sulfate}$$

c. Calculate the length of time to operate the ferrous sulfate feed pump.

$$\frac{\text{Time,}}{\text{hr}} = \frac{(\text{Ferrous Sulfate, lbs})(100\%)}{(\text{Ferrous Sulfate, GPH})(8.34 \text{ lbs/gal})(\text{Ferrous Sulfate, \%})}$$

$$= \frac{(34.3 \text{ lbs})(100\%)}{(80 \text{ GPH})(8.34 \text{ lbs/gal})(4\%)}$$

$$= 1.28 \text{ hr}$$

$$= 1 \text{ hr and } (0.28 \text{ hr})(60 \text{ min/hr})$$

$$= 1 \text{ hr and } 17 \text{ min}$$

d. Calculate the pounds of sulfuric acid needed to treat the chromium.

$$\text{Sulfuric Acid, lbs} = (\text{Chromium, lbs})(\text{Acid Dose, mg/}L/\text{mg/}L)$$

$$= (2.14 \text{ lbs } Cr^{6+})(6 \text{ mg/}L/\text{mg/}L)$$

$$= 12.8 \text{ lbs sulfuric acid}$$

e. Calculate the length of time to operate the sulfuric acid feed pump.

$$\text{Time, hr} = \frac{(\text{Sulfuric Acid, lbs})(100\%)}{(\text{Acid Feed, GPH})(8.34 \text{ lbs/gal})(\text{Acid, \%})}$$

$$= \frac{(12.8 \text{ lbs})(100\%)}{(35 \text{ GPH})(8.34 \text{ lbs/gal})(2.8\%)}$$

$$= 1.57 \text{ hr}$$

$$= 1 \text{ hr and } (0.57 \text{ hr})(60 \text{ min/hr})$$

$$= 1 \text{ hr and } 34 \text{ min}$$

f. Calculate the pounds of lime needed to treat the chromium.

$$\text{Lime, lbs} = (\text{Chromium, lbs})(\text{Lime Dose, mg/}L/\text{mg/}L)$$

$$= (2.14 \text{ lbs } Cr^{6+})(9.5 \text{ mg/}L/\text{mg/}L)$$

$$= 20.3 \text{ lbs lime}$$

ARITHMETIC

g. Calculate the length of time to operate the lime slurry feeder.

$$\text{Time, hr} = \frac{\text{Lime, lbs}}{(\text{Lime Feed, gal})(\text{Lime, lbs/gal})}$$

$$= \frac{20.3 \text{ lbs}}{(50 \text{ gal/hr})(0.5 \text{ lbs/gal})}$$

$$= 0.81 \text{ hr}$$

$$= (0.81 \text{ hr})(60 \text{ min/hr})$$

$$= 49 \text{ min}$$

43. A continuous flow treatment process is used by a plating shop to treat a chrome plating waste by the reduction and precipitation process to reduce the hexavalent chromium in the waste to the trivalent state. One mg/L of hexavalent chromium requires 16 mg/L of ferrous sulfate, 6 mg/L of sulfuric acid, and 9.5 mg/L of lime. The hexavalent chromium concentration is 320 mg/L and is fed to the mixing tank by a five-gallon-per-minute pump.

1. If the ferrous sulfate is fed as a four percent solution, what should be the feed setting on the ferrous sulfate pump?

2. If the sulfuric acid is fed as a 2.8 percent or 0.57 Normal sulfuric acid solution, what should be the feed setting on the sulfuric acid pump?

3. If the lime slurry feeder mixes 0.5 pound of lime per gallon of water, what should be the lime slurry feeder setting in pounds of lime per hour?

Known		Unknown
Chromium Pump, GPM	= 5 GPM	1. Ferrous Sulfate Feed, GPD
Chromium Conc, mg/L	= 320 mg/L	2. Acid Feed, GPD
Ferrous Sulfate Dose, mg/L/mg/L	= 16 mg/L/mg/L	3. Lime Feed, lbs/hr
Ferrous Sulfate, %	= 4%	
Acid Dose, mg/L/mg/L	= 6 mg/L/mg/L	
Acid, %	= 2.8%	
Lime Dose, mg/L/mg/L	= 9.5 mg/L/mg/L	
Lime Slurry, lbs/gal	= 0.5 lbs/gal	

a. Calculate the hexavalent chromium feed rate.

$$\text{Chromium Feed, lbs/day} = \frac{(\text{Cr}^{6+}\text{ Pump, GPM})(\text{Cr}^{6+}\text{ Conc, mg/L})(8.34 \text{ lbs/gal})}{694 \text{ GPM/MGD}}$$

$$= \frac{(5 \text{ GPM})(320 \text{ mg/L})(8.34 \text{ lbs/gal})}{694 \text{ GPM/MGD}}$$

$$= 19.2 \text{ lbs/day}$$

b. Calculate the ferrous sulfate feed in pounds per day.

$$\text{Ferrous Sulfate Feed, lbs/day} = \frac{(\text{Chromium Feed, lbs/day})}{(\text{Ferrous Sulfate Dose, mg/L/mg/L})}$$

$$= (19.2 \text{ lbs/day})(16 \text{ mg/L/mg/L})$$

$$= 307.2 \text{ lbs/day}$$

c. Calculate the setting on the ferrous sulfate feed pump in gallons per day.

$$\text{Ferrous Sulfate Feed, GPD} = \frac{(\text{Ferrous Sulfate Feed, lbs/day})(100\%)}{(8.34 \text{ lbs/gal})(\text{Ferrous Sulfate, \%})}$$

$$= \frac{(307.2 \text{ lbs/day})(100\%)}{(8.34 \text{ lbs/gal})(4\%)}$$

$$= 921 \text{ GPD}$$

d. Calculate the sulfuric acid feed in pounds per day.

$$\text{Acid Feed, lbs/day} = (\text{Chromium Feed, lbs/day})(\text{Acid Dose, mg/L/mg/L})$$

$$= (19.2 \text{ lbs/day})(6 \text{ mg/L/mg/L})$$

$$= 115.2 \text{ lbs/day}$$

e. Calculate the setting on the sulfuric acid feed pump in gallons per day.

$$\text{Acid Feed, GPD} = \frac{(\text{Acid Feed, lbs/day})(100\%)}{(8.34 \text{ lbs/gal})(\text{Acid, \%})}$$

$$= \frac{(115.2 \text{ lbs/day})(100\%)}{(8.34 \text{ lbs/gal})(2.8\%)}$$

$$= 493 \text{ GPD}$$

f. Calculate the lime feed in pounds per day.

$$\text{Lime, lbs/day} = (\text{Chromium Feed, lbs/day})(\text{Lime Dose, mg/L/mg/L})$$

$$= (19.2 \text{ lbs/day})(9.5 \text{ mg/L/mg/L})$$

$$= 182 \text{ lbs/day}$$

g. Calculate the lime slurry feeder setting in pounds of lime per hour.

$$\text{Lime Feed, lbs/hr} = \frac{\text{Lime, lbs/day}}{24 \text{ hr/day}}$$

$$= \frac{182 \text{ lbs day}}{24 \text{ hr/day}}$$

$$= 7.6 \text{ lbs/hr}$$

44. A plating solution contains 2 ounces per gallon of hexavalent chromium. What is the concentration in milligrams per liter?

Known		Unknown
Hexavalent Chromium, oz/gal	= 2 oz/gal	Hexavalent Chromium, mg/L

Convert the hexavalent chromium solution from ounces per gallon to milligrams per liter.

$$\text{Conc, mg/L} = \frac{(\text{Conc, oz/gal})(454 \text{ gm/lb})(1{,}000 \text{ mg/gm})}{(16 \text{ oz/lb})(3.785 \text{ L/gal})}$$

$$= \frac{(2 \text{ oz/gal})(454 \text{ gm/lb})(1{,}000 \text{ mg/gm})}{(16 \text{ oz/lb})(3.785 \text{ L/gal})}$$

$$= 14{,}993 \text{ mg/L}$$

45. The concentration of hexavalent chromium in the effluent from a reduction process is 2 mg/L. What is the concentration in ounces per gallon?

Known		Unknown
Hexavalent Chromium, mg/L	= 2 mg/L	Hexavalent Chromium, oz/gal

Convert the hexavalent chromium solution from milligrams per liter to ounces per gallon.

$$\text{Conc, oz/gal} = \frac{(\text{Conc, mg/}L)(16 \text{ oz/lb})(3.785 \text{ }L\text{/gal})}{(1{,}000 \text{ mg/gm})(454 \text{ gm/lb})}$$

$$= \frac{(2 \text{ mg/}L)(16 \text{ oz/lb})(3.785 \text{ }L\text{/gal})}{(1{,}000 \text{ mg/gm})(454 \text{ gm/lb})}$$

$$= 0.00027 \text{ oz/gal}$$

S. CYANIDE DESTRUCTION

46. A cyanide-bearing waste is to be treated by a batch process using the alkaline chlorination method. The cyanide holding tank contains 12,000 gallons with a cyanide concentration of 15 mg/L. Seven pounds of caustic soda and eight pounds of chlorine are required to oxidize one pound of cyanide to nitrogen gas.

1. How many pounds of chlorine are needed?

2. How long will a hypochlorinator have to operate if the hypochlorite solution is 12.0 percent chlorine and the hypochlorinator can deliver 250 GPD?

3. How long should the caustic soda feed pump operate if the pump delivers 120 gallons per day of a ten percent caustic soda solution?

Known		Unknown
Cyanide Vol, gal	= 12,000 gal	1. Chlorine Required, lbs
Cyanide Conc, mg/L	= 15 mg/L	2. Hypochlorinator Time, hr
Chlorine Dose, lb/lb	= 8 lbs Cl/lb CN	3. Caustic Pump Time, hr
Caustic Dose, lb/lb	= 7 lbs NaOH/lb CN	
Hypochlorite, % Cl	= 12.0% Cl	
Hypochlorinator, GPD	= 250 GPD	
Caustic Soda, %	= 10%	
Caustic Pump, GPD	= 120 GPD	

a. Calculate the pounds of cyanide to be treated.

$$\frac{\text{Cyanide,}}{\text{lbs}} = \frac{(\text{CN Vol, gal})(\text{CN Conc, mg/}L)(8.34 \text{ lbs/gal})}{1{,}000{,}000/M}$$

$$= \frac{(12{,}000 \text{ gal})(15 \text{ mg/}L)(8.34 \text{ lbs/gal})}{1{,}000{,}000/M}$$

$$= 1.50 \text{ lbs cyanide}$$

b. Calculate the pounds of chlorine and caustic needed.

$$\frac{\text{Chlorine}}{\text{Required, lbs}} = (\text{Cyanide, lbs})(\text{Chlorine Dose, lb/lb})$$

$$= (1.50 \text{ lbs CN})(8 \text{ lbs Cl/lb CN})$$

$$= 12.0 \text{ lbs chlorine}$$

$$\frac{\text{Caustic}}{\text{Required, lbs}} = (\text{Cyanide, lbs})(\text{Caustic Dose, lb/lb})$$

$$= (1.50 \text{ lbs CN})(7 \text{ lbs NaOH/lb CN})$$

$$= 10.50 \text{ lbs caustic}$$

c. Calculate the time of operation of the hypochlorinator in hours.

$$\text{Time, hr} = \frac{(\text{Chlorine Required, lbs})(100\%)(24 \text{ hr/day})}{(\text{Flow, GPD})(8.34 \text{ lbs/gal})(\text{Hypochlorite, }\%)}$$

$$= \frac{(12.0 \text{ lbs})(100\%)(24 \text{ hr/day})}{(250 \text{ GPD})(8.34 \text{ lbs/gal})(12.0\%)}$$

$$= 1.15 \text{ hours}$$

$$= 1 \text{ hr} + (0.15 \text{ hr})(60 \text{ min/hr})$$

$$= 1 \text{ hr } 9 \text{ min}$$

d. Calculate the time of operation of the caustic soda pump.

$$\text{Time, hr} = \frac{(\text{Caustic Required, lbs})(100\%)(24 \text{ hr/day})}{(\text{Flow, GPD})(8.34 \text{ lbs/gal})(\text{Caustic Soda, }\%)}$$

$$= \frac{(10.50 \text{ lbs})(100\%)(24 \text{ hr/day})}{(120 \text{ GPD})(8.34 \text{ lbs/gal})(10\%)}$$

$$= 2.52 \text{ hours}$$

$$= 2 \text{ hr} + (0.52 \text{ hr})(60 \text{ min/hr})$$

$$= 2 \text{ hr } 31 \text{ min}$$

47. A cyanide-bearing waste is to be treated by a continuous flow process using the alkaline chlorination method. The cyanide holding tank contains 6,000 gallons with a cyanide concentration of 15 mg/L. Seven pounds of caustic soda and eight pounds of chlorine are required to oxidize one pound of cyanide to nitrogen gas. The cyanide wastes are delivered to a mixing tank by a 25-GPM pump.

1. What should be the setting on the hypochlorinator in gallons per day if the hypochlorite solution is 12.0 percent chlorine?

2. What should be the setting on the caustic feed pump in gallons per day if the caustic soda is a 10 percent solution?

Known		Unknown
Cyanide Vol, gal	= 6,000 gal	1. Hypochlorinator, GPD
Cyanide Conc, mg/L	= 15 mg/L	2. Caustic Pump, GPD
Cyanide Flow, GPM	= 25 GPM	
Chlorine Dose, lb/lb	= 8 lbs Cl/lb CN	
Caustic Dose, lb/lb	= 7 lbs NaOH/lb CN	
Hypochlorite, % Cl	= 12.0%	
Caustic Soda, %	= 10%	

a. Calculate the cyanide feed rate in pounds per day.

$$\frac{\text{Cyanide,}}{\text{lbs/day}} = \frac{(\text{Flow, GPM})(\text{Cyanide Conc, mg/}L)(8.34 \text{ lbs/gal})}{694 \text{ GPM/MGD}}$$

$$= \frac{(25 \text{ GPM})(15 \text{ mg/}L)(8.34 \text{ lbs/gal})}{694 \text{ GPM/MGD}}$$

$$= 4.5 \text{ lbs/day}$$

b. Determine the chlorine feed rate in pounds per day.

$$\begin{aligned} \text{Chlorine Feed,} \atop \text{lbs/day} &= (\text{Cyanide, lbs/day})(\text{Chlorine Dose, lb/lb}) \\ &= (4.5 \text{ lbs CN/day})(8 \text{ lbs Cl/lb CN}) \\ &= 36 \text{ lbs Cl/day} \end{aligned}$$

$$\begin{aligned} \text{Caustic Feed,} \atop \text{lbs/day} &= (\text{Cyanide, lbs/day})(\text{Caustic Dose, lb/lb}) \\ &= (4.5 \text{ lbs CN/day})(7 \text{ lbs NaOH/lb CN}) \\ &= 31.5 \text{ lbs caustic/day} \end{aligned}$$

c. Determine the hypochlorinator setting in gallons per day.

$$\begin{aligned} \text{Hypochlorinator,} \atop \text{GPD} &= \frac{(\text{Chlorine Feed, lbs/day})(100\%)}{(8.34 \text{ lbs/gal})(\text{Hypochlorite, }\%)} \\ &= \frac{(36 \text{ lbs/day})(100\%)}{(8.34 \text{ lbs/gal})(12.0\%)} \\ &= 36 \text{ GPD} \end{aligned}$$

d. Determine the caustic pump setting in gallons per day.

$$\begin{aligned} \text{Caustic Pump,} \atop \text{GPD} &= \frac{(\text{Caustic Feed, lbs/day})(100\%)}{(8.34 \text{ lbs/gal})(\text{Caustic, }\%)} \\ &= \frac{(31.5 \text{ lbs/day})(100\%)}{(8.34 \text{ lbs/gal})(10\%)} \\ &= 38 \text{ GPD} \end{aligned}$$

48. A hypochlorinator is used in the cyanide destruction process. How many pounds of chlorine are delivered per hour if the hypochlorinator pumps a 10 percent chlorine solution of sodium hypochlorite (NaOCl) at a rate of 40 gallons per day (GPD)?

Known	Unknown
Hypochlorite, % Cl = 10%	Chlorine, lbs/hr
Pumping Rate, GPD = 40 GPD	

Calculate the chlorine delivered in pounds per hour.

$$\begin{aligned} \text{Chlorine,} \atop \text{lbs/hr} &= \frac{(\text{Flow, GPD})(\text{Hypochlorite, }\%)(8.34 \text{ lbs/gal})}{(24 \text{ hr/day})(100\%)} \\ &= \frac{(40 \text{ GPD})(10\%)(8.34 \text{ lbs/gal})}{(24 \text{ hr/day})(100\%)} \\ &= 1.39 \text{ lb Cl/hr} \end{aligned}$$

T. COUNTERFLOW RINSING

49. An electroplating facility has a plating bath with a metal concentration of 20 ounces per gallon. The maximum metal concentration allowed in the last rinse tank is 0.005 ounce of metal per gallon. The drag out flow rate is two gallons per hour.

1. If only one rinse tank is available, what is the rinse rate in gallons per minute?

2. How many rinse tanks must be used if the rinse water flow rate must be less than five gallons per minute?

For a multiple-tank counterflow rinse system, the following relationship exists:

$$R^n = \frac{C_p}{C_n}$$

where

C_p = Plating bath metal concentration

C_n = Metal concentration in the nth rinse tank

R = Rinse ratio (ratio of rinse water volumetric flow rate to the drag out volumetric flow rate)

n = Number of rinse tanks

Known		Unknown
C_p, oz/gal	= 20 oz/gal	1. Rinse flow, GPM with only one tank, and
C_n, oz/gal	= 0.005 oz/gal	
Drag Out Flow, gal/hr	= 2 gal/hr	2. Number of tanks if rinse flow <5 GPM

a. Calculate the rinse flow in gallons per hour with only one rinse tank.

$$\frac{C_p}{C_n} = \left(\frac{\text{Rinse Flow Rate, gal/hr}}{\text{Drag Out Flow Rate, gal/hr}} \right)^n$$

$$\frac{20 \text{ oz/gal}}{0.005 \text{ oz/gal}} = \left(\frac{\text{Rinse, gal/hr}}{2 \text{ gal/hr}} \right)^1$$

Rearranging the terms in the equation gives

$$\begin{aligned} \text{Rinse, gal/hr} &= \frac{(2 \text{ gal/hr})(20 \text{ oz/gal})}{0.005 \text{ oz/gal}} \\ &= 8,000 \text{ gal/hr} \end{aligned}$$

b. Convert the rinse flow from gallons per hour to gallons per minute.

$$\begin{aligned} \text{Rinse, GPM} &= \frac{\text{Rinse, gal/hr}}{60 \text{ min/hr}} \\ &= \frac{8,000 \text{ gal/hr}}{60 \text{ min/hr}} \\ &= 133 \text{ GPM} \end{aligned}$$

c. Convert the drag out flow from gallons per hour to gallons per minute.

$$\text{Drag Out Flow, gal/min} = \frac{\text{Drag Out Flow, gal/hr}}{60 \text{ min/hr}}$$

$$= \frac{2 \text{ gal/hr}}{60 \text{ min/hr}}$$

$$= 0.033 \text{ gal/min}$$

d. Determine the required number of rinse tanks if the rinse flow must be less than five gallons per minute.

$$\frac{C_p}{C_n} = \left(\frac{\text{Rinse Flow Rate, GPM}}{\text{Drag Out Flow Rate, GPM}}\right)^n$$

$$\frac{20 \text{ oz/gal}}{0.005 \text{ oz/gal}} = \left(\frac{5 \text{ GPM}}{0.0333 \text{ GPM}}\right)^n$$

$$4{,}000 = 150^n$$

This equation can be solved for n using logarithms, or by trial and error by assuming n values, or directly by using your electronic calculator.

$$\text{Log } 4{,}000 = n \text{ Log } 150$$

$$n = \frac{\text{Log } 4{,}000}{\text{Log } 150}$$

$$= \frac{3.602}{2.176}$$

$$= 1.66 \text{ tanks}$$

Therefore, 2 tanks will do the job.

NOTE: There are many types of rinsing rate calculations. Since the pretreatment inspector is not expected to be a finishing professional, application of this type of process control calculation should be used with great caution. Below find a listing of several papers with different approaches to this subject.

1. "Enviroscope," by Dr. Rebecca M. Spearot, *Plating and Surface Finishing,* November 1985, pages 22 and 23.

2. "Calculating Water Flow for Rinsing," by J. B. Mohler, *Plating and Surface Finishing,* November 1980, pages 50 and 51.

3. "Dragout Recovery with a Double-Use Rinse," by G. Vertes, *Plating and Surface Finishing,* April 1983, pages 53-55.

4. "Dual Purpose Rinsing," by J. B. Mohler, *Plating and Surface Finishing,* September 1979, pages 48 and 49.

5. "The Performance of a Rinsing Tank," by J. B. Mohler, *Metal Finishing,* August 1981, pages 21-28.

U. COMBINED WASTESTREAM FORMULA

50. Calculate the modified daily maximum concentration limit for cadmium for a job shop electroplating facility which discharges to the local POTW. This facility combines 7,000 gallons per day of electroplating wastewater with 1,000 gallons per day of noncontact cooling water prior to pretreatment. The categorical concentration limit for cadmium is 1.2 mg/L. This is the pretreatment standard for cadmium for the electroplating category. Job shops are not subject to pretreatment standards for the metal finishing category.

Known	**Unknown**
F_r, GPD = 7,000 GPD	C_m, mg/L
F_d, GPD = 1,000 GPD	
F_t, GPD = $F_r + F_d$	
= 8,000 GPD	
C_r, mg/L = 1.2 mg/L	

where C_m = the alternative concentration limit for the combined wastestream

C_r = the categorical pretreatment standard concentration limit for a pollutant in the regulated stream r

F_r = the average daily flow (at least a 30-day average) of stream r to the extent that it is regulated for such pollutant

Calculate the modified daily maximum concentration limit for cadmium.

$$C_m, \text{mg/L} = \frac{\left[\sum_{r=1}^{N} C_r F_r\right]\left[F_t - F_d\right]}{\left[\sum_{r=1}^{N} F_r\right]\left[F_t\right]}$$

$$= \frac{[(1.2 \text{ mg/L})(7{,}000 \text{ GPD})][8{,}000 \text{ GPD} - 1{,}000 \text{ GPD}]}{[7{,}000 \text{ GPD}][8{,}000 \text{ GPD}]}$$

$$= \frac{(1.2 \text{ mg/L})(7)}{(8)}$$

$$= 1.05 \text{ mg/L}$$

51. Calculate the modified concentration limit for nickel (Ni) for a captive electroplater which discharges to the local POTW. This facility combines 80,000 gallons per day of plating wastewater with 20,000 gallons per day of non-contact cooling water. The categorical pretreatment standard for nickel is 3.98 mg/L in the metal finishing category. Captive shops are subject to pretreatment standards for the metal finishing category.

Known	**Unknown**
F_r, GPD = 80,000 GPD	C_m, mg/L
F_d, GPD = 20,000 GPD	
F_t, GPD = $F_r + F_d$	
= 100,000 GPD	
C_r, mg/L = 3.98 mg/L	

Calculate the modified categorical concentration limit for nickel.

$$C_m, mg/L = \frac{\left[\begin{array}{c}N \\ \Sigma C_r F_r \\ r=1\end{array}\right]\left[F_t - F_d\right]}{\left[\begin{array}{c}N \\ \Sigma F_r \\ r=1\end{array}\right]\left[F_t\right]}$$

$$= \frac{[(3.98\ mg/L)(80,000\ GPD)][100,000\ GPD - 20,000\ GPD]}{[80,000\ GPD]\quad\quad[100,000\ GPD]}$$

$$= \frac{(3.98\ mg/L)(80)}{(100)}$$

$$= 3.18\ mg/L$$

NOTE: The total flow of 100,000 GPD is treated in the wastewater treatment system. The combined wastestream formula applies only to the combination of a regulated wastestream with any other wastestream, be it some other regulated wastestream, dilution stream or unregulated stream *PRIOR TO TREATMENT.*

52. Calculate the modified concentration limit for zinc for a captive integrated facility which combines 50,000 gallons per day of plating wastewater with 20,000 gallons per day of porcelain enameling wastewater and 30,000 gallons of sanitary wastewater prior to discharge to the local wastewater treatment plant. The concentration limit for zinc is 2.61 mg/L for the metal finishing category and 1.33 mg/L for the porcelain enameling category.

Known	**Unknown**
F_1, GPD = 50,000 GPD	C_m, mg/L
F_2, GPD = 20,000 GPD	
F_d, GPD = 30,000 GPD	
F_t, GPD = $F_1 + F_2 + F_d$	
= 100,000 GPD	
C_1, mg/L = 2.61 mg/L	
C_2, mg/L = 1.33 mg/L	

Calculate the modified categorical concentration limit for zinc.

$$C_m, mg/L = \frac{\left[\begin{array}{c}N = 2 \\ \Sigma C_r F_r \\ r=1\end{array}\right]\left[F_t - F_d\right]}{\left[\begin{array}{c}N = 2 \\ \Sigma F_r \\ r=1\end{array}\right]\left[F_t\right]}$$

$$= \left[\frac{(2.61\ mg/L)(50,000\ GPD) + (1.33\ mg/L)(20,000\ GPD)}{[50,000\ GPD + 20,000\ GPD]}\right]\frac{[100,000\ GPD - 30,000\ GPD]}{[100,000\ GPD]}$$

$$= [(2.61\ mg/L)\left(\frac{5}{7}\right) + (1.33\ mg/L)\left(\frac{2}{7}\right)]\frac{[7]}{[10]}$$

$$= [1.86\ mg/L + 0.38\ gm/L][0.7]$$

$$= (2.24\ mg/L)(0.7)$$

$$= 1.57\ mg/L$$

NOTE: All flows (100,000 GPD) are combined prior to treatment. The combined wastestream formula does not apply where wastestreams are combined after treatment. In these cases, sampling and flow measurements of the separate wastestreams would have to take place and a flow-weighted calculation performed to determine the final regulated pollutant concentration or mass.

53. Three wastestreams are combined AFTER treatment. Flow 1 is 25,000 GPD with a zinc concentration of 1.10 mg/L, Flow 2 is 18,000 GPD with a zinc concentration of 0.98 mg/L and Flow 3 is 9,000 GPD with a zinc concentration of 1.25 mg/L. Determine the final zinc concentration using a flow-weighted calculation.

Known	**Unknown**
Flow 1, GPD = 25,000 GPD	Final Zinc Conc, mg/L
Conc 1, mg/L = 1.10 mg/L	
Flow 2, GPD = 18,000 GPD	
Conc 2, mg/L = 0.98 mg/L	
Flow 3, GPD = 9,000 GPD	
Conc 3, mg/L = 1.25 mg/L	

Calculate the final zinc concentration using a flow-weighted calculation.

$$\text{Final Zinc Conc, mg/L} = \frac{\begin{array}{c}(Flow\ 1, GPD)(Conc\ 1, mg/L) + (Flow\ 2, GPD) \\ (Conc\ 2, mg/L) + (Flow\ 3, GPD)(Conc\ 3, mg/L)\end{array}}{Flow\ 1, GPD + Flow\ 2, GPD + Flow\ 3, GPD}$$

$$= \frac{\begin{array}{c}(25,000\ GPD)(1.10\ mg/L) + (18,000\ GPD) \\ (0.98\ mg/L) + (9,000\ GPD)(1.25\ mg/L)\end{array}}{25,000\ GPD + 18,000\ GPD + 9,000\ GPD}$$

$$= \frac{27,500 + 17,640 + 11,250}{52,000}$$

$$= \frac{56,390}{52,000}$$

$$= 1.08\ mg/L$$

V.   ALTERNATIVE MASS LIMIT FORMULA

54. A facility manufacturing aluminum water tanks forms the parts by forging metal into the appropriate shape. After forging, the formed parts are welded. To prepare the parts for welding, they are cleaned and rinsed. The wastewaters generated are treated in a combined treatment facility to comply with the standards of the Aluminum Forming Point Source Category (40 CFR Part 467). In addition, wastewaters from a parts washer on a paint line are treated in the combined treatment facility. Daily wastewater flow rates through the combined treatment facility are as follows:

Cleaning bath	900 gallons per day
Cleaning rinse	2,700 gallons per day
Parts washer	1,400 gallons per day

The applicable zinc limits from 40 CFR Part 467 are as follows:

	Maximum (for any one day)
Cleaning or etching bath	0.26 mg/off-kg (lb/million off-lbs)
Cleaning or etching rinse	2.03 mg/off-kg (lb/million off-lbs)

Calculate the alternative mass limit using the following equation:

$$M_t = \left[\sum_{i=1}^{N} M_i \right] \times \dfrac{\left[F_t - F_d \right]}{\left[\sum_{i=1}^{N} F_i \right]}$$

M_t = Alternative mass limit for the pollutant in the combined wastestream (mass per day)

M_i = Production-based categorical pretreatment standard for the pollutant in regulated stream i (or the standard multiplied by the appropriate measure of production if the standards being combined contain different units of measurement)

F_i = Average daily flow (at least 30-day average) of regulated stream i

F_d = Average daily flow (at least 30-day average) of dilute wastestream(s)

F_t = Average daily flow (at least 30-day average) through the combined treatment facility (including regulated, unregulated and dilute wastestreams)

N = Total number of regulated streams

Known		**Unknown**
M_1, mg/off-kg	= 0.26 mg/off-kg	Alternative Mass
M_2, mg/off-kg	= 2.03 mg/off-kg	Limit, mg/off-kg
F_1, GPD	= 900 GPD	
F_2, GPD	= 2,700 GPD	
F, GPD (unregulated)	= 1,400 GPD	
F_0, GPD (dilution)	= 0	
N	= 2 regulated streams	

Calculate the alternative mass limit for the zinc on a per-day basis.

$$M_t = \left[\sum_{i=1}^{N=2} M_i \right] \times \left[\dfrac{F_t - F_d}{\sum_{i=1}^{N=2} F_i} \right]$$

$$= (0.26 \text{ mg/off-kg} + 2.03 \text{ mg/off-kg}) \times \left[\dfrac{900 \text{ GPD} + 2{,}700 \text{ GPD} + 1{,}400 \text{ GPD*-0}}{900 \text{ GPD} + 2{,}700 \text{ GPD}} \right]$$

$$= \dfrac{(2.29 \text{ mg/off-kg})(5{,}000 \text{ GPD})}{3{,}800 \text{ GPD}}$$

$$= 3.18 \text{ mg Zn/off-kg (lb Zn/million off-pounds) of aluminum cleaned or etched}$$

---

* The parts washer flow is an unregulated wastestream, not a dilution wastestream.

W.   AVERAGE LIMITATIONS

Average Limitations

Categorical standards establish daily maximum limitations and, in most cases, also set maximum average limitations. The structure of these average limits varies among categories. For example, in the Electroplating category, there is a four-day average, while the Metal Finishing category establishes a monthly average. These two types of averages apply to numerous industrial users.

Four-Day Average

In developing the Electroplating four-day average, the POTW performs a statistical analysis that examines independent groups of four consecutive sampling days. Implementation of the Electroplating four-day average calls for comparison of the standard with independent results from four consecutive sampling days. For the sampling days to be independent, each calculated four-day average should not include sampling data used in another four-day average. For example, if there were eleven days of sampling, samples 1, 2, 3 and 4 constitute a four-day average; samples 5, 6, 7 and 8 produce the next four-day average; and samples 9, 10 and 11 will have to wait until an additional sample is taken so that the next four-day average can be calculated. These sampling days are not necessarily calendar days, but reflect the sampling frequency; namely, weekly sampling produces a four-day average every four weeks, and monthly sampling produces a four-day average every four months.

Monthly Average

A monthly average is used in the Metal Finishing category and many other categories, such as Porcelain Enameling, Coil Coating, Battery Manufacturing, Copper Forming, and Aluminum Forming. In developing these monthly averages, the POTW performs a statistical analysis based on a fixed number of samples being taken per month (10 for Metal Finishing). To implement these regulations, the average of the samples taken in a calendar month constitutes the monthly average and should be compared to the standard. This could mean a monthly average based on only one sample or as many as 31 sampling events. As stated in the preamble to the Metal Finishing rule, 48 FR 32478 (July 15, 1983):

> Although it is not anticipated that a monitoring frequency of 10 times per month will always be required, the cost of this frequency of monitoring is presented in the economic impact analysis to the

metal finishing regulation. That frequency was selected because if facilities sample 10 times per month, they can expect a compliance rate of approximately 99 percent, if they are operating at the expected mean and variability. Plant personnel, in agreement with the Control Authority, may choose to take fewer samples if their treatment system achieves better long-term concentrations or lower variability than the basis for the limits, or if plant personnel are willing to accept a statistical possibility of increased violations.

55. Calculate the four-day average cadmium concentration in the effluent from an electroplating facility if the concentrations from four samples were 0.38, 0.97, 0.66 and 0.51 mg/L.

Known	Unknown
Four Cadmium Concentrations, mg/L	Four-day Average, mg/L

Calculate the four-day average cadmium concentration in milligrams per liter.

$$\text{Four-day Cd Avg, mg/}L = \frac{\text{Sum of Four Cadmium Concentrations, mg/}L}{\text{Number of Measurements}}$$

$$= \frac{0.38 \text{ mg/}L + 0.97 \text{ mg/}L + 0.66 \text{ mg/}L + 0.51 \text{ mg/}L}{4 \text{ Measurements}}$$

$$= \frac{2.52 \text{ mg/}L}{4}$$

$$= 0.63 \text{ mg/}L$$

56. Calculate the monthly average copper concentration in the effluent from a metal finishing facility if the copper concentrations measured during the month were 0.28, 0.76, 0.38, 0.41, 0.56, 0.21, 0.61 and 0.55 mg/L.

Known	Unknown
Eight Copper Concentrations, mg/L	Monthly Average Copper Conc, mg/L

Calculate the monthly average copper concentration in milligrams per liter.

$$\text{Monthly Average Conc, mg/}L = \frac{\text{Sum of Concentrations, mg/}L}{\text{Number of Measurements}}$$

$$= \frac{0.28 + 0.76 + 0.38 + 0.41 + 0.56 + 0.21 + 0.61 + 0.55 \text{ mg/}L}{8 \text{ Measurements}}$$

$$= \frac{3.76 \text{ mg/}L}{8}$$

$$= 0.47 \text{ mg/}L$$

57. Using the following definition for significant noncompliance (SNC) and the data in Table A-1, the SIU (significant industrial user) is (circle all correct answers):

a. in SNC for chronic violations of applicable pretreatment standards.

b. not in SNC for chronic violations of applicable pretreatment standards.

c. in SNC for TRC violations of applicable pretreatment standards.

d. not in SNC for TRC violations of applicable pretreatment standards.

- Chronic violations of wastewater discharge limits, defined here as those in which sixty-six percent (66%) or more of the wastewater measurements taken during a six (6) month period exceed (by any magnitude) the daily maximum limit or the average limit for the same pollution parameter.

- Technical Review Criteria (TRC) violations, defined here as those in which thirty-three percent (33%) or more of the wastewater measurements taken for each pollutant parameter during a six (6) month period equal or exceed the product of the daily maximum limit or the average limit multiplied by the applicable TRC. (TRC = 1.4 for BOD, TSS, fats, oil and grease, and 1.2 for all other pollutants except pH.)

- Any other discharge violation that the Director believes has caused, alone or in combination with other discharges, interference or pass-through, including endangering the health of POTW personnel or the general public.

- Any discharge of a pollutant that has caused imminent endangerment to the public or to the environment, or has resulted in the Director's exercise of emergency authority to halt or prevent such a discharge.

- Failure to meet, within ninety (90) days of the scheduled date, a compliance schedule milestone contained in a consent order, compliance order, or wastewater discharge permit for starting construction, completing construction, or attaining final compliance, unless due to good and valid cause.

- Failure to provide, within thirty (30) days after the due date, any reports required by local ordinance or orders or permits issued under local ordinance, including baseline monitoring reports, reports on compliance with categorical pretreatment standard deadlines, periodic self-monitoring reports, and reports on compliance with compliance schedules.

- Failure to accurately report noncompliance.

- Any other violations that the Director determines will adversely affect the operation or implementation of the local pretreatment program.

The correct answers are b. and c. From Table A-1 you will find 11 daily maximum violations of the mass standard daily average and 4 violations of the mass standard monthly average (disregard the concentration data). These violations are boxed in the table. There are 31 daily data points and 6 monthly average data points for 37 total data points. Therefore the noncompliance rate with wastewater discharge limits is (11 + 4)/(31 + 6) = 38%. This is less than the 66% guideline for chronic violations. However, there are 9 violations of the mass standard daily maximum (1.2* x 2.66 kg = 3.19 kg) and 4 violations of the mass standard monthly average (1.2* x 1.37 kg = 1.6 + kg) (again, ignore the concentration data). There are

the same number of total data points (37). Therefore the noncompliance rate with wastewater discharge limits is (9 + 4)/(31 + 6) = 35%. This is greater than the 33% guideline for TRC violations. This is a difficult problem, but it illustrates the difference between chronic SNC (significant noncompliance) and SNC due to violations of technical review criteria (TRC).

---

* To calculate TRC values, multiply by 1.4 for BOD, TSS, fats, oil and grease, and by 1.2 for all other pollutants except pH.

TABLE A-1. SNC EVALUATION DATA

ACME ORGANIC CHEMICALS — BENZENE
(Evaluation Period 1 — October – March)

Date	Concentration (µg/L)	Flow (MGD)	Mass (kg)	IU or POTW Data	Daily Maximum Conc (µg/L)	Daily Maximum Mass (kg)	Monthly Average Conc (µg/L)	Monthly Average Mass (kg)
04-Oct-94	16	5.73	0.35	IU	117	2.66		
11-Oct-94	23	5.67	0.49	IU	117	2.66		
18-Oct-94	35	4.99	0.66	IU	117	2.66		
25-Oct-94	20	5.61	0.42	IU	117	2.66		
31-Oct-94	140	5.692	3.02	POTW	117	2.66		
Month Avg	**47**		**0.99**				50	1.37
01-Nov-94	514	6.87	13.37	IU	117	2.66		
08-Nov-94	91	5.46	1.88	IU	117	2.66		
15-Nov-94	210	5.82	4.63	IU	117	2.66		
22-Nov-94	111	5.9	2.48	IU	117	2.66		
29-Nov-94	89	6.01	2.03	IU	117	2.66		
28-Nov-94	318	5.794	6.98	POTW	117	2.66		
Month Avg	**222**		**5.23**				50	1.37
06-Dec-94	128	6.19	3.00	IU	117	2.66		
13-Dec-94	119	5.62	2.53	IU	117	2.66		
20-Dec-94	425	5.33	8.58	IU	117	2.66		
28-Dec-94	23	5.24	0.46	IU	117	2.66		
14-Dec-94	6	5.491	0.12	POTW	117	2.66		
Month Avg	**140**		**2.94**				50	1.37
03-Jan-95	13	5.6	0.28	IU	117	2.66		
10-Jan-95	44	5.06	0.84	IU	117	2.66		
17-Jan-95	21	4.63	0.37	IU	117	2.66		
24-Jan-95	27	5.13	0.52	IU	117	2.66		
Month Avg	**26**		**0.50**				50	1.37
02-Feb-95	118	5.81	2.60	IU	117	2.66		
07-Feb-95	253	5.3	5.08	IU	117	2.66		
14-Feb-95	478	5.39	9.76	IU	117	2.66		
20-Feb-95	93	5.05	1.78	POTW	117	2.66		
21-Feb-95	81	5.06	1.55	IU	117	2.66		
28-Feb-95	222	6.08	5.11	IU	117	2.66		
Month Avg	**208**		**4.31**				50	1.37
06-Mar-95	31	5.55	0.65	POTW	117	2.66		
07-Mar-95	17	5.57	0.36	IU	117	2.66		
16-Mar-95	150	5.67	3.22	IU	117	2.66		
21-Mar-95	655	5.24	13.00	IU	117	2.66		
28-Mar-95	144	3.8	2.07	IU	117	2.66		
Month Avg	**199**		**3.86**				50	1.37

SNC Calculation: Evaluation Period 1	Conc (µg/L)	Flow (MGD)	Mass (kg)
Average	149	5	3.17
Number of Data Points (n)	31		31

Number of mass values greater than the applicable pretreatment standard in this quarter (both daily maximum and monthly average): _____.
Percent of mass values greater than the applicable pretreatment standard in this quarter:

Number of mass values greater than the applicable pretreatment standard (1.2) in this quarter (both daily maximum and monthly average): _____.
Percent of mass values greater than the applicable pretreatment standard in this quarter:

APPENDIX III.   PRETREATMENT WORDS

40 CFR 403

EPA's General Pretreatment Regulations appear in the Code of Federal Regulations under 40 CFR 403. 40 refers to the numerical heading for the environmental regulations portion of the Code of Federal Regulations (CFR). 403 refers to the section which contains the General Pretreatment Regulations. Other sections include 413, Electroplating Categorical Regulations.

A

ABC

ABC

See Association of Boards of Certification.

ACEOPS

ACEOPS

See ALLIANCE OF CERTIFIED OPERATORS, LAB ANALYSTS, INSPECTORS, AND SPECIALISTS (ACEOPS).

ABSORPTION (ab-SORP-shun)

ABSORPTION

The taking in or soaking up of one substance into the body of another by molecular or chemical action (as tree roots absorb dissolved nutrients in the soil).

ACID

ACID

(1)   A substance that tends to lose a proton.

(2)   A substance that dissolves in water with the formation of hydrogen ions.

(3)   A substance containing hydrogen which may be replaced by metals to form salts.

(4)   A substance that is corrosive.

ACIDITY

ACIDITY

The capacity of water or wastewater to neutralize bases. Acidity is expressed in milligrams per liter of equivalent calcium carbonate. Acidity is not the same as pH because water does not have to be strongly acidic (low pH) to have a high acidity. Acidity is a measure of how much base must be added to a liquid to raise the pH to 8.2.

ACUTE HEALTH EFFECT

ACUTE HEALTH EFFECT

An adverse effect on a human or animal body, with symptoms developing rapidly.

ADSORPTION (add-SORP-shun)

ADSORPTION

The gathering of a gas, liquid, or dissolved substance on the surface or interface zone of another material.

AEROBIC (AIR-O-bick)

AEROBIC

A condition in which atmospheric or dissolved molecular oxygen is present in the aquatic (water) environment.

AIR STRIPPING

AIR STRIPPING

A physical treatment process used to remove volatile substances from wastestreams. The process transfers volatile pollutants from a high concentration in the wastestream into an air stream with a lower concentration of the pollutant. The process requires the wastestream containing the volatile pollutant to come in contact with large volumes of air.

ALIQUOT (AL-li-kwot)

ALIQUOT

Portion of a sample. Often an equally divided portion of a sample.

WORDS

ALKALINITY (AL-ka-LIN-it-tee)   ALKALINITY

The capacity of water or wastewater to neutralize acids. This capacity is caused by the water's content of carbonate, bicarbonate, hydroxide, and occasionally borate, silicate, and phosphate. Alkalinity is expressed in milligrams per liter of equivalent calcium carbonate. Alkalinity is not the same as pH because water does not have to be strongly basic (high pH) to have a high alkalinity. Alkalinity is a measure of how much acid must be added to a liquid to lower the pH to 4.5.

ALLIANCE OF CERTIFIED OPERATORS,   ALLIANCE OF CERTIFIED OPERATORS,
   LAB ANALYSTS, INSPECTORS,   LAB ANALYSTS, INSPECTORS,
   AND SPECIALISTS (ACEOPS)   AND SPECIALISTS (ACEOPS)

A professional organization for operators, lab analysts, inspectors, and specialists dedicated to improving professionalism; expanding training, certification, and job opportunities; increasing information exchange; and advocating the importance of certified operators, lab analysts, inspectors, and specialists. For information on membership, contact ACEOPS, 1810 Bel Air Drive, Ames, IA 50010, phone (515) 663-4128 or e-mail: ACEOPS@aol.com.

ANAEROBIC (AN-air-O-bick)   ANAEROBIC

A condition in which atmospheric or dissolved molecular oxygen is *NOT* present in the aquatic (water) environment.

ANALOG   ANALOG

The readout of an instrument by a pointer (or other indicating means) against a dial or scale. Also the continuously variable signal type sent to an analog instrument (for example, 4–20 mA).

ANALOG READOUT   ANALOG READOUT

The readout of an instrument by a pointer (or other indicating means) against a dial or scale.

ANODIZING   ANODIZING

An electrochemical process which deposits a coating of an insoluble oxide on a metal surface. Aluminum is the most frequently anodized material.

ANTAGONISTIC REACTION   ANTAGONISTIC REACTION

An interaction between two or more individual compounds that produces an injurious effect upon the body (or an organism) which is *LESS* than either of the substances alone would have produced.

APPURTENANCE (a-PURR-ten-nans)   APPURTENANCE

Machinery, appliances, structures and other parts of the main structure necessary to allow it to operate as intended, but not considered part of the main structure.

ASSOCIATION OF BOARDS OF CERTIFICATION (ABC)   ASSOCIATION OF BOARDS OF CERTIFICATION (ABC)

An international organization representing over 150 boards which certify the operators of waterworks and wastewater facilities. For information on ABC publications regarding the preparation of and how to study for operator certification examinations, contact ABC, 208 Fifth Street, Ames, IA 50010-6259. Phone (515) 232-3623.

AUTOIGNITION TEMPERATURE   AUTOIGNITION TEMPERATURE

The temperature at which a material will spontaneously ignite and sustain combustion.

B

BOD (pronounce as separate letters)   BOD

Biochemical **O**xygen **D**emand. The rate at which organisms use the oxygen in water or wastewater while stabilizing decomposable organic matter under aerobic conditions. In decomposition, organic matter serves as food for the bacteria and energy results from its oxidation. BOD measurements are used as a measure of the organic strength of wastes in water.

BASE   BASE

(1) A substance which takes up or accepts protons.

(2) A substance which dissociates (separates) in aqueous solution to yield hydroxyl ions (OH^-).

(3) A substance containing hydroxyl ions which reacts with an acid to form a salt or which may react with metals to form precipitates.

BASELINE MONITORING REPORT (BMR)

BASELINE MONITORING REPORT (BMR)

All industrial users subject to Categorical Pretreatment Standards must submit a baseline monitoring report (BMR) to the Control Authority (POTW, state or EPA). The purpose of the BMR is to provide information to the Control Authority to document the industrial user's current compliance status with a Categorical Pretreatment Standard. The BMR contains information on (1) name and address of facility, including names of operator(s) and owner(s), (2) list of all environmental control permits, (3) brief description of the nature, average production rate and SIC (or NAICS) code for each of the operations conducted, (4) flow measurement information for regulated process streams discharged to the municipal system, (5) identification of the pretreatment standards applicable to each regulated process and results of measurements of pollutant concentrations and/or mass, (6) statements of certification concerning compliance or noncompliance with the pretreatment standards, and (7) if not in compliance, a compliance schedule must be submitted with the BMR that describes the actions the industrial user will take and a timetable for completing these actions to achieve compliance with the standards.

BATCH PROCESS

BATCH PROCESS

A treatment process in which a tank or reactor is filled, the wastewater (or other solution) is treated or a chemical solution is prepared, and the tank is emptied. The tank may then be filled and the process repeated. Batch processes are also used to cleanse, stabilize or condition chemical solutions for use in industrial manufacturing and treatment processes.

BEST AVAILABLE TECHNOLOGY (BAT)

BEST AVAILABLE TECHNOLOGY (BAT)

A level of technology represented by a higher level of wastewater treatment technology than required by Best Practicable Technology (BPT). BAT is based on the very best (state of the art) control and treatment measures that have been developed, or are capable of being developed, and that are economically achievable within the appropriate industrial category.

BEST PRACTICABLE TECHNOLOGY (BPT)

BEST PRACTICABLE TECHNOLOGY (BPT)

A level of technology represented by the average of the best existing wastewater treatment performance levels within the industrial category.

BIOCHEMICAL OXYGEN DEMAND (BOD)

BIOCHEMICAL OXYGEN DEMAND (BOD)

The rate at which organisms use the oxygen in water or wastewater while stabilizing decomposable organic matter under aerobic conditions. In decomposition, organic matter serves as food for the bacteria and energy results from its oxidation. BOD measurements are used as a measure of the organic strength of wastes in water.

BIODEGRADABLE (BUY-o-dee-GRADE-able)

BIODEGRADABLE

Organic matter that can be broken down by bacteria to more stable forms which will not create a nuisance or give off foul odors is considered biodegradable.

BIOLOGICAL PROCESS

BIOLOGICAL PROCESS

A waste treatment process by which bacteria and other microorganisms break down complex organic materials into simple, non-toxic, more stable substances.

BIOSOLIDS

BIOSOLIDS

A primarily organic solid product, produced by wastewater treatment processes, that can be beneficially recycled. The word biosolids is replacing the word sludge.

BLOWDOWN

BLOWDOWN

The removal of accumulated solids in boilers to prevent plugging of boiler tubes and steam lines. In cooling towers, blowdown is used to reduce the amount of dissolved salts in the recirculated cooling water.

BUCHNER FUNNEL

BUCHNER FUNNEL

A special funnel used to separate solids (sludge) from a mixture. A perforated plate or a filter paper is placed in the lower portion of the funnel to hold the mixture (wet sludge). The funnel is placed in a filter flask and a vacuum is applied to remove the liquid (dewater the sludge).

C

CERCLA (SIRK-la)

CERCLA

Comprehensive Environmental Response, Compensation, and Liability Act of 1980. This Act was enacted primarily to correct past mistakes in industrial waste management. The focus of the Act is to locate hazardous waste disposal sites which are creating problems through pollution of the environment and, by proper funding and implementation of study and corrective activities, eliminate the problem from these sites. Current users of CERCLA-identified substances must report releases of these substances to the environment when they take place (not just historic ones). This Act is also called the Superfund Act. Also see SARA.

WORDS

COD (pronounce as separate letters)   COD

Chemical **O**xygen **D**emand. A measure of the oxygen-consuming capacity of organic matter present in wastewater. COD is expressed as the amount of oxygen consumed from a chemical oxidant in mg/L during a specific test. Results are not necessarily related to the biochemical oxygen demand (BOD) because the chemical oxidant may react with substances that bacteria do not stabilize.

CAPTIVE SHOP   CAPTIVE SHOP

Those electroplating and metal finishing facilities which in a calendar year own more than 50 percent (based on area) of material undergoing metal finishing.

CARCINOGEN (CAR-sin-o-JEN)   CARCINOGEN

Any substance which tends to produce cancer in an organism.

CATEGORICAL LIMITS   CATEGORICAL LIMITS

Industrial wastewater discharge pollutant effluent limits developed by EPA that are applied to the effluent from any industry in any category anywhere in the United States that discharges to a POTW. These are pollutant effluent limits based on the technology available to treat the wastestreams from the processes of the specific industrial category and normally are measured at the point of discharge from the regulated process. The pollutant effluent limits are listed in the Code of Federal Regulations.

CATEGORICAL STANDARDS   CATEGORICAL STANDARDS

Industrial waste discharge standards developed by EPA that are applied to the effluent from any industry in any category anywhere in the United States that discharges to a Publicly Owned Treatment Works (POTW). These are standards based on the technology available to treat the wastestreams from the processes of the specific industrial category and normally are measured at the point of discharge from the regulated process. The standards are listed in the Code of Federal Regulations.

CAUTION   CAUTION

This word warns against potential hazards or cautions against unsafe practices. Also see DANGER, NOTICE, and WARNING.

CEMENTATION   CEMENTATION

(1) A spontaneous electrochemical process that involves the reduction of a more electropositive (noble) species, for example, copper, silver, mercury, or cadmium, by electronegative (sacrificial) metals such as iron, zinc, or aluminum. This process is used to purify spent electrolytic solutions and for the treatment of wastewaters, leachates, and sludges bearing heavy metals. Also called ELECTROLYTIC RECOVERY.

(2) The process of heating two substances that are placed in contact with each other for the purpose of bringing about some change in one of them such as changing iron to steel by surrounding it with charcoal and then heating it.

CEMENTATION TANK   CEMENTATION TANK

A tank in which metal ions are precipitated onto scrap aluminum, steel or other metals. The collected metal can be sent to a smelter for recovery. This process does not require electric current.

CENTRALIZED WASTE TREATMENT (CWT) FACILITY   CENTRALIZED WASTE TREATMENT (CWT) FACILITY

A facility designed to properly handle treatment of specific hazardous wastes from industries with similar wastestreams. The wastewaters containing the hazardous substances are transported to the facility for proper storage, treatment and disposal. Different facilities treat different types of hazardous wastes.

CERTIFICATION EXAMINATION   CERTIFICATION EXAMINATION

An examination administered by a state agency or professional association that pretreatment facility inspectors take to indicate a level of professional competence. Today pretreatment facility inspector certification exams may be voluntary and administered by professional associations. In some states inspectors are recognized as environmental compliance inspectors. Current trends indicate that more employers will encourage inspector certification in the future.

CHAIN OF CUSTODY   CHAIN OF CUSTODY

A record of each person involved in the handling and possession of a sample from the person who collected the sample to the person who analyzed the sample in the laboratory and to the person who witnessed disposal of the sample.

CHELATING (key-LAY-ting) AGENT   CHELATING AGENT

A chemical used to prevent the precipitation of metals (such as copper).

CHEMICAL OXYGEN DEMAND (COD)   CHEMICAL OXYGEN DEMAND (COD)

A measure of the oxygen-consuming capacity of organic matter present in wastewater. COD is expressed as the amount of oxygen consumed from a chemical oxidant in mg/L during a specific test. Results are not necessarily related to the biochemical oxygen demand (BOD) because the chemical oxidant may react with substances that bacteria do not stabilize.

CHEMICAL PROCESS

A treatment process involving the addition of chemicals to achieve a desired level of treatment. Any given process solution in metal finishing is also called a "chemical process."

CHRONIC HEALTH EFFECT

An adverse effect on a human or animal body with symptoms that develop slowly over a long period of time or that recur frequently.

CLARIFIER

Settling Tank, Sedimentation Basin. A tank or basin in which wastewater is held for a period of time during which the heavier solids settle to the bottom and the lighter materials float to the water surface.

CLEAN WATER ACT

An act passed by the U.S. Congress to control water pollution. The Federal Water Pollution Control Act passed in 1972 (Public Law [PL] 92-500). It was amended in 1977 (the Clean Water Act, PL 95-217) and again in 1987 (the Water Quality Act, PL 100-4).

CODE OF FEDERAL REGULATIONS (CFR)

A publication of the United States Government which contains all of the proposed and finalized federal regulations, including environmental regulations.

COMBUSTIBLE LIQUID

A liquid whose flash point is at or above 100°F (38°C). Flammable liquids present a greater fire or explosion hazard than combustible liquids. Also see FLAMMABLE LIQUID.

COMMUNITY RIGHT-TO-KNOW

The **S**uperfund **A**mendments and **R**eauthorization **A**ct (SARA) of 1986 provides statutory authority for communities to develop "right-to-know" laws. The Act establishes a state and local emergency planning structure, emergency notification procedures, and reporting requirements for facilities. Also see SARA and RIGHT-TO-KNOW LAWS.

COMPATIBLE POLLUTANTS

Those pollutants that are normally removed by the POTW treatment system. Biochemical oxygen demand (BOD), suspended solids (SS), and ammonia are considered compatible pollutants.

COMPETENT PERSON

A competent person is defined by OSHA as a person capable of identifying existing and predictable hazards in the surroundings, or working conditions which are unsanitary, hazardous or dangerous to employees, and who has authorization to take prompt corrective measures to eliminate the hazards.

COMPLIANCE

The act of meeting specified conditions or requirements.

COMPOSITE (PROPORTIONAL) SAMPLE

A composite sample is a collection of individual samples obtained at regular intervals, usually every one or two hours during a 24-hour time span. Each individual sample is combined with the others in proportion to the rate of flow when the sample was collected. Equal volume individual samples also may be collected at intervals after a specific volume of flow passes the sampling point or after equal time intervals and still be referred to as a composite sample. The resulting mixture (composite sample) forms a representative sample and is analyzed to determine the average conditions during the sampling period.

CONFINED SPACE

Confined space means a space that:

A. Is large enough and so configured that an employee can bodily enter and perform assigned work; and

B. Has limited or restricted means for entry or exit (for example, tanks, vessels, silos, storage bins, hoppers, vaults, and pits are spaces that may have limited means of entry); and

C. Is not designed for continuous employee occupancy.

Also see definitions of DANGEROUS AIR CONTAMINATION and OXYGEN DEFICIENCY.

(Definition from the Code of Federal Regulations (CFR) Title 29 Part 1910.146.)

CONFINED SPACE, NON-PERMIT

A non-permit confined space is a confined space that does not contain or, with respect to atmospheric hazards, have the potential to contain any hazard capable of causing death or serious physical harm.

CONFINED SPACE, PERMIT-REQUIRED (PERMIT SPACE)

A confined space that has one or more of the following characteristics:

- Contains or has a potential to contain a hazardous atmosphere,
- Contains a material that has the potential for engulfing an entrant,
- Has an internal configuration such that an entrant could be trapped or asphyxiated by inwardly converging walls or by a floor which slopes downward and tapers to a smaller cross section, or
- Contains any other recognized serious safety or health hazard.

(Definition from the Code of Federal Regulations (CFR) Title 29 Part 1910.146.)

CONSERVATIVE POLLUTANT

A pollutant found in wastewater that is not changed while passing through the treatment processes in a conventional wastewater treatment plant. This type of pollutant may be removed by the treatment processes and retained in the plant's sludges or it may leave in the plant effluent. Heavy metals such as cadmium and lead are conservative pollutants.

CONTAMINATION

The introduction into water of microorganisms, chemicals, toxic substances, wastes, or wastewater in a concentration that makes the water unfit for its next intended use.

CONTINUOUS PROCESS

A treatment process in which wastewater is treated continuously in a tank or reactor. The wastewater being treated continuously flows into the tank, is treated as it flows through the tank, and flows out as treated wastewater.

CONVENTIONAL POLLUTANTS

Those pollutants which are usually found in domestic, commercial or industrial wastes such as suspended solids, biochemical oxygen demand, pathogenic (disease-causing) organisms, adverse pH levels, and oil and grease.

CONVENTIONAL TREATMENT

(1) The common treatment processes such as preliminary treatment, sedimentation, flotation, trickling filter, rotating biological contactor, activated sludge and chlorination wastewater treatment processes used by POTWs.

(2) The hydroxide precipitation of metals processes used by pretreatment facilities.

CORROSIVE MATERIAL

A material which through its chemical action is destructively injurious to body tissues or other materials.

CRADLE TO GRAVE

A term used to describe a hazardous waste manifest system used by regulatory agencies to track a hazardous waste from the point of generation to the hauler and then to the ultimate disposal site.

D

DANGER

The word *DANGER* is used where an immediate hazard presents a threat of death or serious injury to employees. Also see CAUTION, NOTICE, and WARNING.

DECIBEL (DES-uh-bull)

A unit for expressing the relative intensity of sounds on a scale from zero for the average least perceptible sound to about 130 for the average level at which sound causes pain to humans. Abbreviated dB.

DELETERIOUS (DELL-eh-TEAR-ee-us)

Refers to something that can be or is hurtful, harmful or injurious to health or the environment.

DIALYSIS (die-AL-uh-sis)

The selective separation of dissolved or colloidal solids on the basis of molecular size by diffusion through a semipermeable membrane.

DIELECTRIC (DIE-ee-LECK-trick)   DIELECTRIC

Does not conduct an electric current. An insulator or nonconducting substance.

DIGITAL READOUT   DIGITAL READOUT

The use of numbers to indicate the value or measurement of a variable. The readout of an instrument by a direct, numerical reading of the measured value. The signal sent to such readouts is usually an ANALOG signal.

DIRECT DISCHARGER   DIRECT DISCHARGER

A point source that discharges a pollutant(s) to waters of the United States, such as streams, lakes or oceans. These sources are subject to the National Pollutant Discharge Elimination System (NPDES) program regulations.

DRAG OUT   DRAG OUT

The liquid film (plating solution) that adheres to the workpieces and their fixtures as they are removed from any given process solution or their rinses. Drag out volume from a tank depends on the viscosity of the solution, the surface tension, the withdrawal time, the draining time and the shape and texture of the workpieces. The drag out liquid may drip onto the floor and cause wastestream treatment problems. Regulated substances contained in this liquid must be removed from wastestreams or neutralized prior to discharge to POTW sewers.

E

EPA   EPA

United States **E**nvironmental **P**rotection **A**gency. A regulatory agency established by the U.S. Congress to administer the nation's environmental laws. Also called the U.S. EPA.

EFFLUENT   EFFLUENT

Wastewater or other liquid—raw (untreated), partially or completely treated—flowing *FROM* a reservoir, basin, treatment process, or treatment plant.

EFFLUENT LIMITS   EFFLUENT LIMITS

Pollutant limitations developed by a POTW for industrial plants discharging to the POTW system. At a minimum, all industrial facilities are required to comply with federal prohibited discharge standards. The industries covered by federal categorical standards must also comply with the appropriate discharge limitations. The POTW may also establish local limits more stringent than or in addition to the federal standards for some or all of its industrial users.

ELECTRODIALYSIS   ELECTRODIALYSIS

The selective separation of dissolved solids on the basis of electrical charge, by diffusion through a semipermeable membrane across which an electrical potential is imposed.

ELECTROLYTIC RECOVERY   ELECTROLYTIC RECOVERY

A spontaneous electrochemical process that involves the reduction of a more electropositive (noble) species, for example, copper, silver, mercury, or cadmium, by electronegative (sacrificial) metals such as iron, zinc, or aluminum. This process is used for purifying spent electrolytic solutions and for the treatment of wastewaters, leachates, and sludges bearing heavy metals. Also called CEMENTATION.

ELUTRIATION (e-LOO-tree-A-shun)   ELUTRIATION

The washing of digested sludge with either fresh water, plant effluent or other wastewater. The objective is to remove (wash out) fine particulates and/or the alkalinity in sludge. This process reduces the demand for conditioning chemicals and improves settling or filtering characteristics of the solids.

EMPTY BED CONTACT TIME   EMPTY BED CONTACT TIME

The time required for the liquid in a carbon adsorption bed to pass through the carbon column assuming that all liquid passes through at the same velocity. It is equal to the volume of the empty bed divided by the flow rate.

ENTRAIN   ENTRAIN

To trap bubbles in water either mechanically through turbulence or chemically through a reaction.

EXISTING SOURCE   EXISTING SOURCE

An industrial discharger that was in operation at the time of promulgation of the proposed pretreatment standard for the industrial category.

F

FIRE POINT FIRE POINT

The lowest temperature of a liquid at which a mixture of air and vapor from the liquid will continue to burn.

FLAMMABLE LIQUID FLAMMABLE LIQUID

A liquid which by itself, or any component of it present in greater than one percent concentration, has a flash point below 100°F (38°C). Also see COMBUSTIBLE LIQUID.

FLASH POINT FLASH POINT

The minimum temperature of a liquid at which the liquid gives off a vapor in sufficient concentration to ignite when tested under specific conditions.

FRICTION LOSSES FRICTION LOSSES

The head, pressure or energy (they are the same) lost by water flowing in a pipe or channel as a result of turbulence caused by the velocity of the flowing water and the roughness of the pipe, channel walls, and restrictions caused by fittings. Water flowing in a pipe loses pressure or energy as a result of friction losses. Also see HEAD LOSS.

G

GIS GIS

Geographic Information System. A computer program that combines mapping with detailed information about the physical locations of structures such as pipes, valves, and manholes within geographic areas. The system is used to help operators and maintenance personnel locate utility system features or structures and to assist with the scheduling and performance of maintenance activities.

GRAB SAMPLE GRAB SAMPLE

A single sample of water collected at a particular time and place which represents the composition of the water only at that time and place.

H

HARMFUL PHYSICAL AGENT HARMFUL PHYSICAL AGENT

Any chemical substance, biological agent (bacteria, virus or fungus) or physical stress (noise, heat, cold, vibration, repetitive motion, ionizing and non-ionizing radiation, hypo- or hyperbaric pressure) which:

(A) Is regulated by any state or federal law or rule due to a hazard to health;

(B) Is listed in the latest printed edition of the National Institute for Occupational Safety and Health (NIOSH) Registry of Toxic Effects of Chemical Substances (RTECS);

(C) Has yielded positive evidence of an acute or chronic health hazard in human, animal or other biological testing conducted by, or known to, the employer; or

(D) Is described by a Material Safety Data Sheet (MSDS) available to the employer which indicates that the material may pose a hazard to human health.

Also called a TOXIC SUBSTANCE. Also see ACUTE HEALTH EFFECT and CHRONIC HEALTH EFFECT. (Definition from California General Industry Safety Orders, Section 3204.)

HAZARD COMMUNICATION HAZARD COMMUNICATION

Employee "Right-to-Know" legislation requires employers to inform employees (pretreatment inspectors) of the possible health effects resulting from contact with hazardous substances. At locations where this legislation is in force, employers must provide employees with information regarding any hazardous substances which they might be exposed to under normal work conditions or reasonably foreseeable emergency conditions resulting from workplace conditions. OSHA's Hazard Communication Standard (HCS) (Title 29 CFR Part 1910.1200) is the federal regulation and state statutes are called Worker RIGHT-TO-KNOW LAWS. Also see COMMUNITY RIGHT-TO-KNOW and SARA.

HAZARDOUS MATERIAL MANAGEMENT PLAN (HMMP) HAZARDOUS MATERIAL MANAGEMENT PLAN (HMMP)

A document prepared by an industry which contains copies of MSDSs (Material Safety Data Sheets) as well as additional information regarding the storage, handling and disposal of all chemicals used on site by the industry.

HAZARDOUS WASTE HAZARDOUS WASTE

A waste, or combination of wastes, which because of its quantity, concentration, or physical, chemical, or infectious characteristics may:

1. Cause, or significantly contribute to, an increase in mortality or an increase in serious, irreversible, or incapacitating reversible illness; or

2. Pose a substantial present or potential hazard to human health or the environment when improperly treated, stored, transported, or disposed of or otherwise managed; and

3. Normally not be discharged into a sanitary sewer; subject to regulated disposal.

(Resource Conservation and Recovery Act (RCRA) definition.)

HEAD HEAD

The vertical distance (in feet) equal to the pressure (in psi) at a specific point. The pressure head is equal to the pressure in psi times 2.31 ft/psi.

HEAD LOSS HEAD LOSS

The head, pressure or energy (they are the same) lost by water flowing in a pipe or channel as a result of turbulence caused by the velocity of the flowing water and the roughness of the pipe, channel walls, or restrictions caused by fittings. Water flowing In a pipe loses head, pressure or energy as a result of friction losses. Also see FRICTION LOSSES.

HYDRAULIC JUMP HYDRAULIC JUMP

The sudden and usually turbulent abrupt rise in water surface in an open channel when water flowing at high velocity is suddenly retarded to a slow velocity.

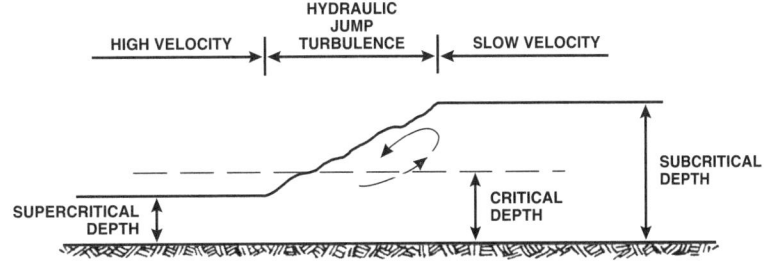

I

IDLH IDLH

Immediately Dangerous to Life or Health. The atmospheric concentration of any toxic, corrosive, or asphyxiant substance that poses an immediate threat to life or would cause irreversible or delayed adverse health effects or would interfere with an individual's ability to escape from a dangerous atmosphere.

IMMISCIBLE (im-MISS-uh-bull) IMMISCIBLE

Not capable of being mixed.

INDIRECT DISCHARGER INDIRECT DISCHARGER

A nondomestic discharger introducing pollutants to a POTW. These facilities are subject to the EPA pretreatment regulations.

INDUSTRIAL PRETREATMENT (WASTE) INSPECTOR INDUSTRIAL PRETREATMENT (WASTE) INSPECTOR

A person who conducts inspections of industrial pretreatment facilities to ensure protection of the environment and compliance with general and categorical pretreatment regulations. Also called an INSPECTOR and a PRETREATMENT INSPECTOR.

INDUSTRIAL WASTE SURVEY INDUSTRIAL WASTE SURVEY

A survey of all companies that discharge to a POTW. The survey identifies the magnitude of the wastewater flows and pollutants in the discharge.

INFLUENT INFLUENT

Wastewater or other liquid—raw (untreated) or partially treated—flowing *INTO* a reservoir, basin, treatment process, or treatment plant.

INHIBITORY SUBSTANCES INHIBITORY SUBSTANCES

Materials that kill or restrict the ability of organisms to treat wastes.

INORGANIC WASTE

INORGANIC WASTE

Waste material such as sand, salt, iron, calcium, and other mineral materials which are only slightly affected by the action of organisms. Inorganic wastes are chemical substances of mineral origin; whereas organic wastes are chemical substances usually of animal or plant origin. Also see NONVOLATILE MATTER.

INSPECTOR

INSPECTOR

An Industrial Pretreatment Inspector is a person who conducts inspections of industrial pretreatment facilities to ensure protection of the environment and compliance with regulations adopted by the POTW in addition to the General and Categorical Pretreatment Regulations. Also called an INDUSTRIAL PRETREATMENT (WASTE) INSPECTOR and a PRETREATMENT INSPECTOR.

INTEGRATED FACILITY

INTEGRATED FACILITY

Plants which, prior to treatment or discharge, combine the metal finishing wastewaters with significant quantities (more than 10 percent of total volume) of wastewaters not covered by the metal finishing category. Also refers to any facilities covered by any of the industrial pretreatment categories and various combinations of categories.

INTEGRATOR

INTEGRATOR

A device or meter that continuously measures and calculates (adds) a process rate variable in cumulative fashion; for example, total flows displayed in gallons, million gallons, cubic feet, or some other unit of volume measurement. Also called a TOTALIZER.

INTERFACE

INTERFACE

The common boundary layer between two substances such as water and a solid (metal); or between two fluids such as water and a gas (air); or between a liquid (water) and another liquid (oil).

INTERFERENCE

INTERFERENCE

Interference refers to the harmful effects industrial compounds can have on POTW operations, such as killing or inhibiting beneficial microorganisms or causing treatment process upsets or sludge contamination.

INTERSTICES (in-TOUR-stee-sees)

INTERSTICES

Small crevices on parts where concentrated plating solutions gather during the plating process. This liquid solution becomes the drag out when the part is removed from the plating bath.

J

JSA

JSA

Job **S**afety **A**nalysis. A supervisor selects the job to be analyzed and then, with the help of the inspectors, the job is subdivided into individual steps and each step of the job is critically examined. During the examination process any potential hazards or problems associated with that step are identified by reviewing how each inspector performs the tasks of this step. The intent of this review is to see what can happen and how any potential accidents and injuries can be prevented.

JOB SHOP

JOB SHOP

Those electroplating and metal finishing facilities which in a calendar year do not own more than 50 percent (based on area) of material undergoing metal finishing.

K

(NO LISTINGS)

L

LAMINAR

LAMINAR

Laminar flow is smooth or viscous flow; not turbulent flow.

LITMUS PAPER

LITMUS PAPER

A strip of paper that is chemically treated with a dye named litmus. This strip is placed in the liquid whose pH is being determined. The use of litmus paper, which is red at any pH less than 5 and blue at any pH greater than 8, allows an inspector to determine only whether a solution is acidic or basic. Also see pH TEST PAPER.

LOWER EXPLOSIVE LIMIT (LEL)

LOWER EXPLOSIVE LIMIT (LEL)

The lowest concentration of gas or vapor (percent by volume in air) that explodes if an ignition source is present at ambient temperature. At temperatures above 250°F the LEL decreases because explosibility increases with higher temperature.

LOWER FLAMMABLE LIMIT (LFL)

LOWER FLAMMABLE LIMIT (LFL)

The lowest concentration of a gas or vapor (percent by volume in air) that burns if an ignition source is present.

M

MASS EMISSION RATE

MASS EMISSION RATE

The rate of discharge of a pollutant expressed as a weight per unit time, usually as pounds or kilograms per day.

MATERIAL SAFETY DATA SHEET (MSDS)

MATERIAL SAFETY DATA SHEET (MSDS)

A document which provides pertinent information and a profile of a particular hazardous substance or mixture. An MSDS is normally developed by the manufacturer or formulator of the hazardous substance or mixture. The MSDS is required to be made available to employees and inspectors whenever there is the likelihood of the hazardous substance or mixture being introduced into the workplace. Some manufacturers are preparing MSDSs for products that are not considered to be hazardous to show that the product or substance is *NOT* hazardous.

MERCAPTANS (mer-CAP-tans)

MERCAPTANS

Compounds containing sulfur which have an extremely offensive skunk-like odor; also sometimes described as smelling like garlic or onions.

MISCIBLE (MISS-uh-bull)

MISCIBLE

Capable of being mixed. A liquid, solid, or gas that can be completely dissolved in water.

MUTAGENIC (MUE-ta-JEN-ick)

MUTAGENIC

Any substance which tends to cause mutations or gene changes prior to conception.

WORDS

N

NAICS

NAICS

North **A**merican **I**ndustry **C**lassification **S**ystem. A code number system used to identify various types of industries. This code system replaces the SIC (Standard Industrial Classification) code system used prior to 1997. Use of these code numbers is often mandatory. Some companies have several processes which will cause them to fit into two or more classifications. The code numbers are published by the Superintendent of Documents, U.S. Government Printing Office, PO Box 371954, Pittsburgh, PA 15250-7954. Stock No. 041-001-00509-9; price, $31.50. There is no charge for shipping and handling.

NIOSH (NYE-osh)

NIOSH

The **N**ational **I**nstitute of **O**ccupational **S**afety and **H**ealth is an organization that tests and approves safety equipment for particular applications. NIOSH is the primary federal agency engaged in research in the national effort to eliminate on-the-job hazards to the health and safety of working people. The NIOSH Publications Catalog, Sixth Edition, NIOSH Pub. No. 84-118, lists the NIOSH publications concerning industrial hygiene and occupational health. To obtain a copy of the catalog, write to National Technical Information Service (NTIS), 5285 Port Royal Road, Springfield, VA 22161. NTIS Stock No. PB-86-116-787, price, $103.50, plus $5.00 shipping and handling per order.

NPDES PERMIT

NPDES PERMIT

National **P**ollutant **D**ischarge **E**limination **S**ystem permit is the regulatory agency document issued by either a federal or state agency which is designed to control all discharges of pollutants from point sources and storm water runoff into U.S. waterways. NPDES permits regulate discharges into navigable waters from all point sources of pollution, including industries, municipal wastewater treatment plants, sanitary landfills, large agricultural feedlots and return irrigation flows.

NAPPE (NAP)

NAPPE

The sheet or curtain of water flowing over a weir or dam. When the water freely flows over any structure, it has a well-defined upper and lower water surface.

NEW SOURCE

NEW SOURCE

Any building, structure, facility or installation from which there is or may be a discharge of pollutants. Construction of the facility must have begun after publication of the applicable Pretreatment Standards. The building, structure, facility or installation must also be constructed at a site at which no other source is located; or, must totally replace the existing process or production equipment producing the discharge at the site; or, must be substantially independent of an existing source of discharge at the same site.

NONBIODEGRADABLE (non-BUY-o-dee-GRADE-able)

NONBIODEGRADABLE

Substances that cannot readily be broken down by bacteria to simpler forms.

NONCOMPATIBLE POLLUTANTS

NONCOMPATIBLE POLLUTANTS

Those pollutants which are normally *NOT* removed by the POTW treatment system. These pollutants may be a toxic waste and may pass through the POTW untreated or interfere with the treatment system. Examples of noncompatible pollutants include heavy metals such as copper, nickel, lead, and zinc; organics such as methylene chloride, 1,1,1-trichloroethylene, methyl ethyl ketone, acetone, and gasoline; or sludges containing toxic organics or metals.

NONINTEGRATED FACILITY

NONINTEGRATED FACILITY

Plants that produce wastewaters from different pretreatment categorical processes but do not combine the wastestreams prior to pretreatment or discharge to sewers.

NONVOLATILE MATTER

NONVOLATILE MATTER

Material such as sand, salt, iron, calcium, and other mineral materials which are only slightly affected by the actions of organisms and are not lost on ignition of the dry solids at 550°C. Volatile materials are chemical substances usually of animal or plant origin. Also see INORGANIC WASTE and VOLATILE MATTER or VOLATILE SOLIDS.

NOTICE

NOTICE

This word calls attention to information that is especially significant in understanding and operating equipment or processes safely. Also see CAUTION, DANGER, and WARNING.

O

O & M MANUAL

O & M MANUAL

Operation and **M**aintenance Manual. A manual that describes detailed procedures for operators to follow to operate and maintain a specific wastewater treatment or pretreatment plant and the equipment of that plant.

ORP (pronounce as separate letters)

ORP

Oxidation-**R**eduction **P**otential. The electrical potential required to transfer electrons from one compound or element (the oxidant) to another compound or element (the reductant); used as a qualitative measure of the state of oxidation in wastewater treatment systems. ORP is measured in millivolts, with negative values indicating a tendency to reduce compounds or elements and positive values indicating a tendency to oxidize compounds or elements.

OSHA (O-shuh)

OSHA

The Williams-Steiger **O**ccupational **S**afety and **H**ealth **A**ct of 1970 (OSHA) is a federal law designed to protect the health and safety of industrial workers and also the operators and inspectors of pretreatment facilities. OSHA regulations require employers to obtain and make available to workers the Material Safety Data Sheets (MSDSs) for chemicals used at industrial facilities and treatment plants. OSHA also refers to the federal and state agencies which administer the OSHA regulations.

OFF POUNDS

OFF POUNDS

The amount of material which is off loaded from a machine with one or more regulated process flows of water. It is *NOT* the number of pounds shipped. It is also *NOT* the number of pounds loaded into the process.

ORGANIC WASTE

ORGANIC WASTE

Waste material which may come from animal or plant sources. Natural organic wastes generally can be consumed by bacteria and other small organisms. Manufactured or synthetic organic wastes from metal finishing, chemical manufacturing, and petroleum industries may not normally be consumed by bacteria and other organisms. Also see INORGANIC WASTE and VOLATILE SOLIDS.

OSMOSIS (oz-MOE-sis)

OSMOSIS

The passage of a liquid from a weak solution to a more concentrated solution across a semipermeable membrane. The membrane allows the passage of the water (solvent) but not the dissolved solids (solutes). This process tends to equalize the conditions on either side of the membrane.

OUTFALL

(1)   The point, location or structure where wastewater or drainage discharges from a sewer, drain, or other conduit.

(2)   The conduit leading to the final disposal point or area.

OXIDATION STATE/OXIDATION NUMBER

In a chemical formula, a number accompanied by a polarity indication (+ or −) that together indicate the charge of a substance as well as the extent to which the substance has been oxidized or reduced in a REDOX REACTION.

Due to the loss of electrons, the charge of a substance that has been oxidized would go from negative toward or to neutral, from neutral to positive, or from positive to more positive. As an example, an oxidation number of 2+ would indicate that a substance has lost two electrons and that its charge has become positive (that it now has an excess of two protons).

Due to the gain of electrons, the charge of the substance that has been reduced would go from positive toward or to neutral, from neutral to negative, or from negative to more negative. As an example, an oxidation number of 2− would indicate that a substance has gained two electrons and that its charge has become negative (that it now has an excess of two electrons). As a substance gains electrons, its oxidation state (or the extent to which it is oxidized) lowers; that is, its oxidation state is reduced.

OXIDATION-REDUCTION POTENTIAL (ORP)

The electrical potential required to transfer electrons from one compound or element (the oxidant) to another compound or element (the reductant); used as a qualitative measure of the state of oxidation in wastewater treatment systems. ORP is measured in millivolts, with negative values indicating a tendency to reduce compounds or elements and positive values indicating a tendency to oxidize compounds or elements.

OXIDATION-REDUCTION (REDOX) REACTION

A two-part reaction between two substances involving a transfer of electrons from one substance to the other. Oxidation is the loss of electrons by one substance, and reduction is the acceptance of electrons by the other substance. Reduction refers to the lowering of the OXIDATION STATE or OXIDATION NUMBER of the substance accepting the electrons.

In a redox reaction, the substance that gives up the electrons (that is oxidized) is called the reductant because it causes a reduction in the oxidation state or number of the substance that accepts the transferred electrons. The substance that receives the electrons (that is reduced) is called the oxidant because it causes oxidation of the other substance. Oxidation and reduction always occur simultaneously.

P

POTW

Publicly **O**wned **T**reatment **W**orks. A treatment works which is owned by a state, municipality, city, town, special sewer district or other publicly owned and financed entity as opposed to a privately (industrial) owned treatment facility. This definition includes any devices and systems used in the storage, treatment, recycling and reclamation of municipal sewage or industrial wastes of a liquid nature. It also includes sewers, pipes and other conveyances only if they carry wastewater to a POTW treatment plant. The term also means the municipality (public entity) which has jurisdiction over the indirect discharges to and the discharges from such a treatment works. (Definition adapted from EPA's General Pretreatment Regulations, 40 CFR 403.3.)

PASSIVATING

A metal plating process which forms a protective film on metals by immersion in an acid solution, usually nitric acid or nitric acid with sodium dichromate.

PATHOGENIC (PATH-o-JEN-ick) ORGANISMS

Organisms, including bacteria, viruses or cysts, capable of causing diseases (giardiasis, cryptosporidiosis, typhoid, cholera, dysentery) in a host (such as a person). There are many types of organisms which do *NOT* cause disease. These organisms are considered non-pathogenic. Many beneficial bacteria are found in wastewater treatment processes actively cleaning up organic wastes.

PERMEATE (PURR-me-ate)

(1)   To penetrate and pass through, as water penetrates and passes through soil and other porous materials.

(2)   The liquid (demineralized water) produced from the reverse osmosis process that contains a *LOW* concentration of dissolved solids.

pH (pronounce as separate letters)

pH is an expression of the intensity of the basic or acidic condition of a liquid. Mathematically, pH is the logarithm (base 10) of the reciprocal of the hydrogen ion activity.

$$pH = Log \frac{1}{[H^+]} \quad\quad or = -Log\ [H^+]$$

The pH may range from 0 to 14, where 0 is most acidic, 14 most basic, and 7 neutral.

pH TEST PAPER

A strip of paper which is treated with several dyes that change color at narrow and different pH ranges. To determine the pH of a solution, place a drop of the solution on the test paper. Then compare the color that develops with the colors on a chart which relates color to pH. Also see LITMUS PAPER.

PHYSICAL WASTE TREATMENT PROCESS

Physical wastewater treatment processes include use of racks, screens, comminutors, clarifiers (sedimentation and flotation) and filtration. Chemical or biological reactions are important treatment processes, but *NOT* part of a physical treatment process.

PICKLE

An acid or chemical solution in which metal objects or workpieces are dipped to remove oxide scale or other adhering substances.

PITOT (pea-TOE) TUBE

An instrument used to measure fluid (liquid or air) velocity by means of the differential pressure between the tip (dynamic) and side (static) openings.

POLLUTANT

Any substance which causes an impairment (reduction) of water quality to a degree that has an adverse effect on any beneficial use of the water.

POLLUTANT

Dredged spoil, solid waste, incinerator residue, sewage, garbage, sewage sludge, munitions, chemical wastes, biological materials, radioactive materials, heat, wrecked or discarded equipment, rock, sand, cellar dirt and industrial, municipal and agricultural waste discharged into water and onto the ground where subsurface waters can become contaminated by leaching. (Definition from Section 502(6) of the Clean Water Act.)

POLLUTION

The impairment (reduction) of water quality by agricultural, domestic or industrial wastes (including thermal and radioactive wastes) to a degree that the natural water quality is changed to hinder any beneficial use of the water or render it offensive to the senses of sight, taste, or smell or when sufficient amounts of wastes create or pose a potential threat to human health or the environment.

POLYELECTROLYTE (POLY-ee-LECK-tro-lite)

A high-molecular-weight substance that is formed by either a natural or synthetic process. Natural polyelectrolytes may be of biological origin or derived from starch products, cellulose derivatives, and alignates. Synthetic polyelectrolytes consist of simple substances that have been made into complex, high-molecular-weight substances. Often called a POLYMER.

POLYMER (POLY-mer)

A long chain molecule formed by the union of many monomers (molecules of lower molecular weight). Polymers are used with other chemical coagulants to aid in binding small suspended particles to larger chemical flocs for their removal from water.

PONT-A-MOUSSON (MOO-san) PRESSURE CELL

A device that works like a filter press and is used to perform filter tests. A cell applies pressurized air (7 to 225 psig) with a piston (instead of using a vacuum as with a Buchner funnel) to separate solids from a mixture.

POTTING COMPOUNDS

Sealing and holding compounds (such as epoxy) used in electrode probes.

POZZOLANIC (POE-zoe-LAN-ick) PROCESS

The pozzolanic process mixes lime and other fine-grained siliceous materials such as fly ash, cement kiln dust, or blast furnace slag with a liquid hazardous waste to form a cement-like solid.

PRECIPITATE (pre-SIP-uh-TATE)

(1) An insoluble, finely divided substance which is a product of a chemical reaction within a liquid.

(2) The separation from solution of an insoluble substance.

PRETREATMENT FACILITY

Industrial wastewater treatment plant consisting of one or more treatment devices designed to remove sufficient pollutants from wastewaters to allow an industry to comply with effluent limits established by the US EPA General and Categorical Pretreatment Regulations or locally derived prohibited discharge requirements and local effluent limits. Compliance with effluent limits allows for a legal discharge to a POTW.

PRETREATMENT INSPECTOR PRETREATMENT INSPECTOR

A person who conducts inspections of industrial pretreatment facilities to ensure protection of the environment and compliance with general and categorical pretreatment regulations. Also called an INDUSTRIAL PRETREATMENT (WASTE) INSPECTOR and an INSPECTOR.

PRIMARY ELEMENT PRIMARY ELEMENT

The hydraulic structure used to measure flows. In open channels, weirs and flumes are primary elements or devices. Venturi meters and orifice plates are the primary elements in pipes or pressure conduits.

PRIORITY POLLUTANTS PRIORITY POLLUTANTS

The EPA has proposed a list of 126 priority toxic pollutants. These substances are an environmental hazard and may be present in water. Because of the known or suspected hazards of these pollutants, industrial users of the substances are subject to regulation. The toxicity to humans may be substantiated by human epidemiological studies or based on effects on laboratory animals related to carcinogenicity, mutagenicity, teratogenicity, or reproduction. Toxicity to fish and wildlife may be related to either acute or chronic effects on the organisms themselves or to humans by bioaccumulation in food fish. Persistence (including mobility and degradability) and treatability are also important factors.

Q

(NO LISTINGS)

R

RCRA (RICK-ruh) RCRA

The Federal **R**esource **C**onservation and **R**ecovery **A**ct (10/21/76), Public Law (PL) 94-580, provides technical and financial assistance for the development of plans and facilities for recovery of energy and resources from discarded materials and for the safe disposal of discarded materials and hazardous wastes. This Act introduces the philosophy of the "cradle to grave" control of hazardous wastes. RCRA regulations can be found in Title 40 of the Code of Federal Regulations (40 CFR) Parts 260-268, 270 and 271.

RAW WASTEWATER RAW WASTEWATER

Treatment facility or plant influent or wastestream *BEFORE* any treatment.

REAGENT (re-A-gent) REAGENT

A pure chemical substance that is used to make new products or is used in chemical tests to measure, detect, or examine other substances.

RECEIVER RECEIVER

A device that indicates the result of a measurement. Most receivers use either a fixed scale and movable indicator (pointer) such as a pressure gage, or a moving chart with a movable pen like those used on a circular flow-recording chart. Also called an indicator.

RECEIVING WATER RECEIVING WATER

A stream, river, lake, ocean, or other surface or groundwaters into which treated or untreated wastewater is discharged.

RECORDER RECORDER

A device that creates a permanent record, on a paper chart or magnetic tape, of the changes in a measured variable.

RECYCLE RECYCLE

The use of water or wastewater within (internally) a facility before it is discharged to a treatment system. Also see REUSE.

REDOX (ree-DOCKS) REDOX

Oxidation-reduction reactions in which the oxidation state of at least one reactant is raised while that of another is lowered.

REDOX REACTION REDOX REACTION

(See OXIDATION-REDUCTION (REDOX) REACTION)

REJECT REJECT

The liquid produced from the reverse osmosis process that contains a *HIGH* concentration of dissolved solids.

WORDS

REPRESENTATIVE SAMPLE

A sample portion of material or wastestream that is as nearly identical in content and consistency as possible to that in the larger body of material or wastestream being sampled.

RETURN SLUDGE

The recycled sludge in a POTW that is pumped from a secondary clarifier sludge hopper to the aeration tank.

REUSE

The use of water or wastewater after it has been discharged and then withdrawn by another user. Also see RECYCLE.

REVERSE OSMOSIS (oz-MOE-sis)

The application of pressure to a concentrated solution which causes the passage of a liquid from the concentrated solution to a weaker solution across a semipermeable membrane. The membrane allows the passage of the water (solvent) but not the dissolved solids (solutes). In the reverse osmosis process two liquids are produced: (1) the reject (containing high concentrations of dissolved solids), and (2) the permeate (containing low concentrations). The "clean" water is not always considered to be demineralized. Also see OSMOSIS.

RIGHT-TO-KNOW LAWS

Employee "Right-to-Know" legislation requires employers to inform employees (pretreatment inspectors) of the possible health effects resulting from contact with hazardous substances. At locations where this legislation is in force, employers must provide employees with information regarding any hazardous substances which they might be exposed to under normal work conditions or reasonably foreseeable emergency conditions resulting from workplace conditions. OSHA's Hazard Communication Standard (HCS) (Title 29 CFR Part 1910.1200) is the federal regulation and state statutes are called Worker Right-to-Know Laws. Also see COMMUNITY RIGHT-TO-KNOW and SARA.

S

SARA

Superfund **A**mendments and **R**eauthorization **A**ct of 1986. The Comprehensive Environmental Response, Compensation, and Liability Act (CERCLA), commonly known as Superfund, was enacted in 1980. The Superfund Amendments increase Superfund revenues to $8.5 billion and strengthen the EPA's authority to conduct short-term (removal), long-term (remedial) and enforcement actions. The Amendments also strengthen state involvements in the cleanup process and the Agency's commitments to research and development, training, health assessments, and public participation. A number of new statutory authorities, such as Community Right-to-Know, are also established. Also see CERCLA.

SIC CODE

Standard **I**ndustrial **C**lassification Code. A code number system used to identify various types of industries. In 1997, the United States and Canada replaced the SIC code system with the North American Industry Classification System (NAICS); Mexico adopted the NAICS in 1998. See NAICS.

SANITARY SEWER

A pipe or conduit (sewer) intended to carry wastewater or waterborne wastes from homes, businesses, and industries to the treatment works. Storm water runoff or unpolluted water should be collected and transported in a separate system of pipes or conduits (storm sewers) to natural watercourses.

SCUM

The material which rises to the surface of a liquid and forms a layer or film on the surface.

SECONDARY ELEMENT

The secondary measuring device or flowmeter used with a primary measuring device (element) to measure the rate of liquid flow. In open channels bubblers and floats are secondary elements. Differential pressure measuring devices are the secondary elements in pipes or pressure conduits. The purpose of the secondary measuring device is to (1) measure the liquid level in open channels or the differential pressure in pipes, and (2) convert this measurement into an appropriate flow rate according to the known liquid level or differential pressure and flow rate relationship of the primary measuring device. This flow rate may be integrated (added up) to obtain a totalized volume, transmitted to a recording device, and/or used to pace an automatic sampler.

SEDIMENTATION BASIN

Clarifier, Settling Tank. A tank or basin in which wastewater is held for a period of time during which the heavier solids settle to the bottom and the lighter materials float to the water surface.

SENSOR SENSOR

A device that measures (senses) a physical condition or variable of interest. Floats and thermocouples are examples of sensors.

SEWAGE SEWAGE

The used household water and water-carried solids that flow in sewers to a wastewater treatment plant. The preferred term is WASTEWATER.

SIGNIFICANT INDUSTRIAL USER (SIU) SIGNIFICANT INDUSTRIAL USER (SIU)

A Significant Industrial User (SIU) includes:

1. All categorical industrial users, and

2. Any noncategorical industrial user that

   a. Discharges 25,000 gallons per day or more of process wastewater ("process wastewater" excludes sanitary, noncontact cooling and boiler blowdown wastewaters), or

   b. Contributes a process wastestream which makes up five percent or more of the average dry weather hydraulic or organic (BOD, TSS) capacity of a treatment plant, or

   c. Has a reasonable potential, in the opinion of the Control or Approval Authority, to adversely affect the POTW treatment plant (inhibition, pass-through of pollutants, sludge contamination, or endangerment of POTW workers).

SIGNIFICANT NONCOMPLIANCE SIGNIFICANT NONCOMPLIANCE

An industrial user is in significant noncompliance if its violation meets one or more of the following criteria:

(A) Chronic violation of wastewater discharge limits, defined here as those in which sixty-six percent or more of all of the measurements taken during a six-month period exceed (by any magnitude) the daily maximum limit or the average limit for the same pollutant parameter;

(B) Technical Review Criteria (TRC) violations, defined here as those in which thirty-three percent or more of all of the measurements for each pollutant parameter taken during a six-month period equal or exceed the product of the daily maximum limit or the average limit multiplied by the applicable TRC (TRC = 1.4 for BOD, TSS, fats, oil and grease, and 1.2 for all other pollutants except pH);

(C) Any other violation of a pretreatment effluent limit (daily maximum or longer-term average) that the Control Authority determines has caused, alone or in combination with other discharges, interference or pass through (including endangering the health of POTW personnel or the general public);

(D) Any discharge of a pollutant that has caused imminent endangerment to human health, welfare or to the environment or has resulted in the POTW's exercise of its emergency authority under paragraph (f)(1)(vi)(b) of this section of the regulations to halt or prevent such a discharge;

(E) Failure to meet, within 90 days after the schedule date, a compliance schedule milestone contained in a local control mechanism or enforcement order for starting construction, completing construction, or attaining final compliance;

(F) Failure to provide, within 30 days after the due date, required reports such as baseline monitoring reports, 90-day compliance reports, periodic self-monitoring reports, and reports on compliance with compliance schedules;

(G) Failure to accurately report noncompliance; or

(H) Any other violation which the Control Authority determines will adversely affect the operation or implementation of the local pretreatment program.

SLUDGE (sluj) SLUDGE

(1) The settleable solids separated from liquids during processing.

(2) The deposits of foreign materials on the bottoms of streams or other bodies of water.

SOLVENT SOLVENT

Any substance that is used to dissolve another substance in it.

SOLVENT EXTRACTION SOLVENT EXTRACTION

The process of dissolving and separating out particular constituents of a liquid by treatment with solvents specific for those constituents. Extraction may be liquid-solid or liquid-liquid.

SOLVENT MANAGEMENT PLAN SOLVENT MANAGEMENT PLAN

A strategy for keeping track of all solvents delivered to a site, their storage, use and disposal. This includes keeping spent solvents segregated from other process wastewaters to maximize the value of the recoverable solvents, to avoid contamination of other segregated wastes, and to prevent the discharge of toxic organics to any wastewater collection system or the environment. The plan should describe measures to control spills and leaks and to ensure that there is no deliberate dumping of solvents. Also known as a TOXIC ORGANIC MANAGEMENT PLAN (TOMP).

STANDARD METHODS   STANDARD METHODS

STANDARD METHODS FOR THE EXAMINATION OF WATER AND WASTEWATER, 20th Edition. A joint publication of the American Public Health Association (APHA), American Water Works Association (AWWA), and the Water Environment Federation (WEF) which outlines the accepted laboratory procedures used to analyze the impurities in water and wastewater. Available from Water Environment Federation, Publications Order Department, 601 Wythe Street, Alexandria, VA 22314-1994. Order No. S82010. Price to members, $164.25; nonmembers, $209.25; price includes cost of shipping and handling.

STILLING WELL   STILLING WELL

A well or chamber which is connected to the main flow channel by a small inlet. Waves and surges in the main flow stream will not appear in the well due to the small-diameter inlet. The liquid surface in the well will be quiet, but will follow all of the steady fluctuations of the open channel. The liquid level in the well is measured to determine the flow in the main channel.

STORM SEWER   STORM SEWER

A separate pipe, conduit or open channel (sewer) that carries runoff from storms, surface drainage, and street wash, but does not include domestic and industrial wastes. Storm sewers are often the recipients of hazardous or toxic substances due to the illegal dumping of hazardous wastes or spills created by accidents involving vehicles and trains transporting these substances. Also see SANITARY SEWER.

SUMP   SUMP

The term "sump" refers to a facility which connects an industrial discharger to a public sewer. The facility (sump) could be a sample box, a clarifier or an intercepting sewer.

SUMP CARD   SUMP CARD

A 3x5 reference card which identifies the location of a sump, lists the monitoring devices located in the sump (pH for example), and indicates which treatment processes are connected to the sump.

SUPERFUND   SUPERFUND

The **C**omprehensive **E**nvironmental **R**esponse, **C**ompensation, and **L**iability **A**ct of 1980 (CERCLA). This Act was enacted primarily to correct past mistakes in industrial waste management. The focus of the Act is to locate hazardous waste disposal sites which are creating problems through pollution of the environment and, by proper funding and implementation of study and corrective activities, eliminate the problem from these sites. Current users of CERCLA-identified substances must report releases to the environment when they take place (not just historic ones).

SUPERNATANT (sue-per-NAY-tent)   SUPERNATANT

Liquid removed from settled sludge. Supernatant commonly refers to the liquid between the sludge on the bottom and the scum on the surface of an anaerobic digester. This liquid is usually returned to the influent wet well or to the primary clarifier.

SYNERGISTIC (SIN-er-GIST-ick) REACTION   SYNERGISTIC REACTION

An interaction between two or more individual compounds which produces an injurious effect upon the body (or an organism) which is *GREATER* than either of the substances alone would have produced.

T

TOC (pronounce as separate letters)   TOC

Total **O**rganic **C**arbon. TOC measures the amount of organic carbon in water.

TECHNOLOGY-BASED STANDARDS   TECHNOLOGY-BASED STANDARDS

Refers to discharge limits that a specific industry could economically achieve if they were to use the best pretreatment technology currently available, such as neutralization, precipitation, clarification, and filtration.

TERATOGENIC (TEAR-a-toe-JEN-ick)   TERATOGENIC

Any substance which tends to cause birth defects after conception.

THRESHOLD LIMIT VALUE (TLV)   THRESHOLD LIMIT VALUE (TLV)

The average concentration of toxic gas or any other substance to which a normal person can be exposed without injury during an average work week.

TIME WEIGHTED AVERAGE (TWA)
TIME WEIGHTED AVERAGE (TWA)

The average concentration of a pollutant based on the times and levels of concentrations of the pollutant. The time weighted average is equal to the sum of the portion of each time period (as a decimal, such as 0.25 hour) multiplied by the pollutant concentration during the time period divided by the hours in the workday (usually 8 hours). 8TWA PEL is the Time Weighted Average permissible exposure limit, in parts per million, for a normal 8-hour workday and a 40-hour workweek to which nearly all workers may be repeatedly exposed, day after day, without adverse effect.

TIMER
TIMER

A device for automatically starting or stopping a machine or other device at a given time.

TOTALIZER
TOTALIZER

A device or meter that continuously measures and calculates (adds) a process rate variable in cumulative fashion; for example, total flows displayed in gallons, million gallons, cubic feet, or some other unit of volume measurement. Also called an INTEGRATOR.

TOXIC
TOXIC

A substance which is poisonous to a living organism. Toxic substances may be classified in terms of their physiological action, such as irritants, asphyxiants, systemic poisons, and anesthetics and narcotics. Irritants are corrosive substances which attack the mucous membrane surfaces of the body. Asphyxiants interfere with breathing. Systemic poisons are hazardous substances which injure or destroy internal organs of the body. The anesthetics and narcotics are hazardous substances which depress the central nervous system and lead to unconsciousness.

TOXIC ORGANIC MANAGEMENT PLAN (TOMP)
TOXIC ORGANIC MANAGEMENT PLAN (TOMP)

A strategy for keeping track of all solvents delivered to a site, their storage, use and disposal. This includes keeping spent solvents segregated from other process wastewaters to maximize the value of the recoverable solvents, to avoid contamination of other segregated wastes, and to prevent the discharge of toxic organics to any wastewater collection system or the environment. The plan should describe measures to control spills and leaks and to ensure that there is no deliberate dumping of solvents. Also known as a SOLVENT MANAGEMENT PLAN.

TOXIC POLLUTANT
TOXIC POLLUTANT

Those pollutants or combinations of pollutants, including disease-causing agents, which after discharge and upon exposure, ingestion, inhalation or assimilation into any organism, either directly from the environment or indirectly by ingestion through food chains, will, on the basis of information available to the Administrator of EPA, cause death, disease, behavioral abnormalities, cancer, genetic mutations, physiological malfunctions (including malfunctions in reproduction) or physical deformations, in such organisms or their offspring. (Definition from Section 502(13) of the Clean Water Act.)

TOXIC SUBSTANCE
TOXIC SUBSTANCE

Any chemical substance, biological agent (bacteria, virus or fungus) or physical stress (noise, heat, cold, vibration, repetitive motion, ionizing and non-ionizing radiation, hypo- or hyperbaric pressure) which:

(A) Is regulated by any state or federal law or rule due to a hazard to health;

(B) Is listed in the latest printed edition of the National Institute for Occupational Safety and Health (NIOSH) Registry of Toxic Effects of Chemical Substances (RTECS);

(C) Has yielded positive evidence of an acute or chronic health hazard in human, animal or other biological testing conducted by, or known to, the employer; or

(D) Is described by a Material Safety Data Sheet (MSDS) available to the employer which indicates that the material may pose a hazard to human health.

Also called a HARMFUL PHYSICAL AGENT. Also see ACUTE HEALTH EFFECT and CHRONIC HEALTH EFFECT. (Definition from California General Industry Safety Orders, Section 3204.)

TOXIC SUBSTANCES CONTROL ACT (TSCA)
TOXIC SUBSTANCES CONTROL ACT (TSCA)

The **T**oxic **S**ubstances **C**ontrol **A**ct of 1976 gave EPA the responsibility of controlling the entry of toxic, carcinogenic or otherwise biologically active compounds into the environment. The Act placed a heavy reporting burden on industry and contained provisions allowing EPA to demand premarket testing of some chemicals.

TRANSDUCER (trans-DUE-sir)
TRANSDUCER

A device that senses some varying condition measured by a primary sensor and converts it to an electrical or other signal for transmission to some other device (a receiver) for processing or decision making.

U

U.S. EPA

U.S. EPA

United **S**tates **E**nvironmental **P**rotection **A**gency.

ULTRAFILTRATION

ULTRAFILTRATION

A membrane filter process used for the removal of some organic compounds in an aqueous (watery) solution.

UPPER EXPLOSIVE LIMIT (UEL)

UPPER EXPLOSIVE LIMIT (UEL)

The point at which the concentration of a gas in air becomes too great to allow an explosion upon ignition due to insufficient oxygen present.

UPPER FLAMMABLE LIMIT (UFL)

UPPER FLAMMABLE LIMIT (UFL)

The point at which the concentration of a gas in air becomes too great to sustain a flame upon ignition due to insufficient oxygen present.

V

VOLATILE (VOL-uh-tull)

VOLATILE

(1) A volatile substance is one that is capable of being evaporated or changed to a vapor at relatively low temperatures. Volatile substances also can be partially removed by air stripping.

(2) In terms of solids analysis, volatile refers to materials lost (including most organic matter) upon ignition in a muffle furnace for 60 minutes at 550°C. Natural volatile materials are chemical substances usually of animal or plant origin. Manufactured or synthetic volatile materials such as ether, acetone, and carbon tetrachloride are highly volatile and not of plant or animal origin. Also see NONVOLATILE MATTER.

VOLATILE ACIDS

VOLATILE ACIDS

Fatty acids produced during digestion which are soluble in water and can be steam-distilled at atmospheric pressure. Also called organic acids. Volatile acids are commonly reported as equivalent to acetic acid.

VOLATILE LIQUIDS

VOLATILE LIQUIDS

Liquids which easily vaporize or evaporate at room temperature.

VOLATILE SOLIDS

VOLATILE SOLIDS

Those solids in water, wastewater, or other liquids that are lost on ignition of the dry solids at 550°C. Also called organic solids.

W

WARNING

WARNING

The word *WARNING* is used to indicate a hazard level between *CAUTION* and *DANGER*. Also see CAUTION, DANGER, and NOTICE.

WASTEWATER

WASTEWATER

A community's used water and water-carried solids that flow to a treatment plant. Storm water, surface water, and groundwater infiltration also may be included in the wastewater that enters a wastewater treatment plant. The term "sewage" usually refers to household wastes, but this word is being replaced by the term "wastewater."

WASTEWATER FACILITIES

WASTEWATER FACILITIES

The pipes, conduits, structures, equipment, and processes required to collect, convey, and treat domestic and industrial wastes, and dispose of the effluent and sludge.

WASTEWATER ORDINANCE

WASTEWATER ORDINANCE

The basic document granting authority to administer a pretreatment inspection program. This ordinance must contain certain basic elements to provide a legal framework for effective enforcement.

WET OXIDATION

A method of treating or conditioning sludge before the water is removed. Compressed air is blown into the liquid sludge. The air and sludge mixture is fed into a pressure vessel where the organic material is stabilized. The stabilized organic material and inert (inorganic) solids are then separated from the pressure vessel effluent by dewatering in lagoons or by mechanical means.

X

(NO LISTINGS)

Y

(NO LISTINGS)

Z

ZETA POTENTIAL

In coagulation and flocculation procedures, the difference in the electrical charge between the dense layer of ions surrounding the particle and the charge of the bulk of the suspended fluid surrounding this particle. The zeta potential is usually measured in millivolts.

APPENDIX IV.   SUBJECT INDEX

INDEX

I
N
D
E
X

INDEX

INDEX

INDEX

NOTES

NOTES

NOTES

NOTES